Wireless Internet of Things: Principles and Practice

Other World Scientific Titles by the Author

Telecommunications Engineering: Principles and Practice
ISBN: 978-981-120-025-0
ISBN: 978-981-120-147-9 (pbk)

Wireless Internet of Things: Principles and Practice

Amoakoh Gyasi-Agyei
School of Science, Engineering & Information Technology,
Federation University Australia

World Scientific

NEW JERSEY • LONDON • SINGAPORE • BEIJING • SHANGHAI • HONG KONG • TAIPEI • CHENNAI • TOKYO

Published by

World Scientific Publishing Co. Pte. Ltd.
5 Toh Tuck Link, Singapore 596224
USA office: 27 Warren Street, Suite 401-402, Hackensack, NJ 07601
UK office: 57 Shelton Street, Covent Garden, London WC2H 9HE

Library of Congress Cataloging-in-Publication Data
Names: Gyasi-Agyei, Amoakoh, author.
Title: Wireless internet of things : principles and practice / Amoakoh Gyasi-Agyei, Melbourne Institute of Technology, Australia.
Description: Singapore : World Scientific Publishing Co. Pte. Ltd., 2019. | Includes bibliographical references and index.
Identifiers: LCCN 2019027167 | ISBN 9789811201325 (hardcover) | ISBN 9789811202056 (paperback) | ISBN 9789811201332 (ebook)
Subjects: LCSH: Internet of things.
Classification: LCC TK5105.8857 .G93 2019 | DDC 004.67/8--dc23
LC record available at https://lccn.loc.gov/2019027167

British Library Cataloguing-in-Publication Data
A catalogue record for this book is available from the British Library.

For any available supplementary material, please visit
https://www.worldscientific.com/worldscibooks/10.1142/11308#t=suppl

Desk Editor: Tay Yu Shan

Dedication

To Michaela, Priscilla, Benedict and my students (past, current and future)

About the Author

Amoakoh Gyasi-Agyei has been lecturing at universities in New Zealand and Australia. He obtained his PhD from the University of Adelaide in Australia, Masters from Chalmers University of Technology in Sweden, Bachelors from TUHH Germany and High School at Prempeh College. He is a Chartered Engineer with the United Kingdom Engineering Council, a senior member of the IEEE (USA), and a member of the Australian Computer Society (ACS). He was the 2011 Chair of IEEE Activities in Africa, and founded the IEEE Ghana Section. He served as Dean of the Faculty of Engineering, Science and Computing at Pentecost University from 1999–2014. Gyasi-Agyei is a curriculum design consultant in electrical/electronic engineering, computing and data science, as well as the director of Gnosis Community Foundation.

Preface

The Internet of Things (IoT) is one of the disruptive technologies of the 21st century. It is also a killer application of the Internet which is poised to be the impetus for the full-scale realisation of the connected world and the Digital Revolution. In the connected world, people and things communicate with each other. Things in IoT include smart mobile phones, home consumer devices, autonomous cars, utility meters, car park meters, robots, satellites and radars. Even animals, birds and trees can be tagged with sensors with radio transceivers to transfer data on, for example, their geographical locations and health status, to a monitoring station.

This book treats IoT from an engineering perspective. **It teaches the principles and technologies implemented in the physical infrastructure to enable IoT**. Some people may think of only miniature devices and sensors as citizens of IoT. However, as the word 'Things' in IoT does not have a strict limit, we should not be surprised to see radar systems, satellites and wireless positioning systems being IoT citizens. Positioning devices, satellites and radars are used to gather data and are thus sensors.[1]

Motivation for Writing this Book

I have taught several courses at both undergraduate and postgraduate levels in information systems, enterprise systems, information technology, computer networking, telecommunications engineering, electrical/electronic engineering and computer science since 2003 at four universities in Australia and New Zealand. One thing stands out: many of the students struggle to understand the textbooks in the libraries. The main reason is that some authors probably inadvertently write books for their peers, rather than for students. Therefore, many students repeatedly asked me in class "Is there any book that you can recommend to us?" Another motivation is I have hardly found a book which treats all the topics in a course that I have been teaching. For these reasons, I started to write full lecture notes for my students, which have over time become this book. Consequently, this book has been written with the purpose of making it accessible to even the busiest student, who, unfortunately, for various reasons has little time to study.

Intended Audience of Book

The curriculum differs across institutions. What one may teach at undergraduate level, another may teach at the postgraduate level. For example, the following are advanced topics which are suited for postgraduate-level courses: MIMO systems, part of the chapter on wireless channels, mobile cellular system design, the later part of the multiple access schemes (e.g. non-orthogonal MAC and OFDM/OFDMA), radar systems design, satellite communications and wireless positioning systems.

Book's Features

This book has been written as a textbook right from the onset to be adopted for courses in Internet of Things, electrical/electronic engineering,

[1]There can be glory in failure and despair in success. Abraham Lincoln, USA President 4 March 1861–15 April 1865.

information systems, information technology, computer networking, computer communications, computer science, and related disciplines. Each chapter is divided into sections, and each section ends with review questions put under a section referred to as *Check Your Understanding* to quiz the comprehension of the topic in a progressive manner. I advise non-experts in the field, especially students, not to skip these sections. Almost all sections contain at least one solved example. We have made an effort to use case studies and parameters of realistic technologies to make the text practically relevant. Wherever necessary, we have reduced the theory to make the book accessible to an average reader. A good background of High School mathematics is all that is needed to access the contents of this book.

Furthermore, the book has been written in simple English to improve the accessibility of its content. The chapters discuss to a good level communications systems, principles, standards and protocols. Also, there are several system design case studies in the book, as well as in-class group research for collaborative learning. Institutions which cover all the topics in this book will provide a solid background to their students, upon which they can build using more advanced books. Hopefully, both students and practicing professionals will benefit from this book.

The website accompanying this book is hosted at

https://jessieacademy.com.au/textbooks/

The main purposes of this site are the following:

- **To keep errata:** Although the book has been written with great care, there is no guarantee of being free of typos. Readers are encouraged to email any typos or errors found in the text to the author via **info@jessieacademy.com.au**.
- **Study resources:** Any study resources, including power point slides of all chapters and PDF files of figures and tables in the book will be kept at this website. Lecturers around the world who have adopted the book as textbooks are encouraged to post to me their slides so we add them to the list. Such lecturers will be recognised for their inputs.
- **Interactions between readers and author:** Readers are encouraged to join the online forum at this page to make it a kind of global center of community learning from each other regardless of our geographical locations and/or universities. The author will stay active to explain topics if necessary.
- A list of institutions adopting or have previously adopted the book as textbook.

Overview of the Book

Chapter 1 motivates us into the IoT industry. Topics treated include history of the Internet and IoT, history of wireless communications and IoT enabling technologies. Chapter 2, Wireless IoT Fundamentals provides a brief survey of basic topics on Internet and IoT. Topics here include nascent technologies relating to IoT (e.g. cloud computing, fog computing, big data from IoT devices and virtualisation), Internet Protocol, subnetting of IoT networks, and the basic electrical circuits supporting IoT at the physical layer. Chapter 3 discusses analog modulation to set up the tone for digital modulation in Chapter 4.

We discuss radio wave propagation in Chapter 5. This is followed by the discussion of wireless channels in Chapter 6. Antennas, including multiple-input, multiple-output (MIMO) are mission critical for wireless IoT. We discuss them in Chapter 7. Chapter 8 deals with multiplexing and multiple access. The sharing of communications resources is the norm for public networks, IoT not being an exception. This cannot be achieved without multiple access protocols. Chapter 9 is dedicated to the design of mobile wireless networks.

Chapters 10 to 13 deal with microwave wireless technologies. Chapter 10 handles point-to-point microwave wireless systems design. We then discuss radar systems in Chapter 11. It may not be clear to some people how the sensed data from radar systems travel over the IoT. This is proved to readers. Satellites are major IoT sensors. We thus discuss them in Chapter 12. The concluding chapter in the book discusses wireless positioning technologies.

Reviewers

I am grateful to the following professors who, in spite of their busy schedules, reviewed this book:

- Professor Jinho Choi, School of Electrical Engineering and Computer Science (EECS), Gwangju Institute of Science and Technology (GIST), Republic of Korea. I met Jinho during my PhD candidature at the University of Adelaide where he was a lecturer then. Jinho has been very kind and supportive to me since then.
- Professor Seong-Lyun Kim, Radio Resource Management & Optimization Laboratory (RAMO) and the Center for Flexible Radio (CFR+), School of Electrical & Electronic Engineering, Yonsei University, Seoul, Korea. Interestingly, I met Seong-Lyun in 2002 through the recommendation of Jinho. Since then he has been very supportive.

Acknowledgments

I cannot have enough words to express my gratitude to Professor Jinho Choi and Professor Seong-Lyun Kim for reviewing the text. I have been privileged to be working with Professor Johnson Inyeh Agbinya who has always been available to discuss technical issues when contacted. With patience Professor Emeritus Reginal Coutts mentored me into the academic work. Professor Lang White hosted me in his research laboratory for over 12 months during my candidature the University of Adelaide. My students, both past and present, deserve my appreciation for motivating me to write this book. I am grateful to the team at World Scientific Publishing, including Ng Yan Hong, Yu Shan Tay, Rajesh Babu and Janice Sim, for believing in my ability to complete this project, and surprisingly, on time. I would be ungrateful not to thank my family for freeing me (may be involuntarily :-)) to spend sleepless nights on this project. In this regard, my open apologies to Master Benedict, Priscilla and Michaela.

Contents

Chapter 1

Introduction to Internet of Things

Chapter's Learning Outcomes

Presumably, those who invented the United States of America (USA) Defense Department's Advanced Research Projects Agency Network (ARPANET) in 1969 could not imagine that they were creating the Internet of Things (IoT) of the 21st century, an enabler of the Digital Transformation. Also, while coining the term IoT in 1999, Kevin Ashton might have not imagined the eventual state of the IoT. The reason being that at its conception the IoT was viewed as the Internet's offshoot enabling data sharing among things. The 'things' were probably thought to be only radio frequency identification (RFID) devices and miniature sensors. Today, the things in IoT is any sensor, small or large. It includes accelerometers, thermostats, wearable devices to track fitness (e.g., Fitbit, Pebble, Jawbone Up), smart utility meters, radar, satellite, mobile phones, industrial asset monitors (e.g., GE, AGT Intl.), and devices for home automation (e.g., Nest, 4Control, Lifx). Other terms related to IoT include the Internet of Everything (IoE), Industry 4.0, machine-to-machine communications (M2MC), Industrial Internet[1] and Web of Things. This introductory chapter motivates readers into the vibrant nascent wireless IoT industry. In particular, we study:[2]

(1) Abbreviated history of wireless communications which provide connectivity between IoT devices.
(2) Abbreviated history of the Internet, which is the foundation for the IoT.
(3) A review of the enabling technologies of the IoT.

1.1 Evolution in Communications Technologies

The desire to exchange information, namely communications, is as old as Adam and Eve. Consider Fig. 1.1. Communications technologies have evolved from birds, through smoke, drums, errand boys, Morse code, analogue technology to the current digital technology. It is known that pigeons were moved from their rearing locations. Messages were then attached to them. Upon released, they would fly to their home locations with the messages. Another story also claims that baits were attached to the messages to attract the birds to pick them up, as illustrated in Fig. 1.1. It was hoped that the birds would fly to the desired locations prior to releasing the messages. The next evolution in communications technology was the smoke signal. Seeing a smoke indicated the presence of humans. My grand mum used to leave a little fire with smoke at the exit to her farm to indicate her presence in the farm to neighboring farmers who might pass by.

The next evolution was the talking drums (or tom-tom drums). You might have heard about the town crier, who moves around cities beating drums or ringing bells to communicate information. Then

[1] Visit https://www.iiconsortium.org/ for more information.
[2] It is more honourable to fail than to cheat — Abraham Lincoln, USA President 4 March 1861–15 April 1865.

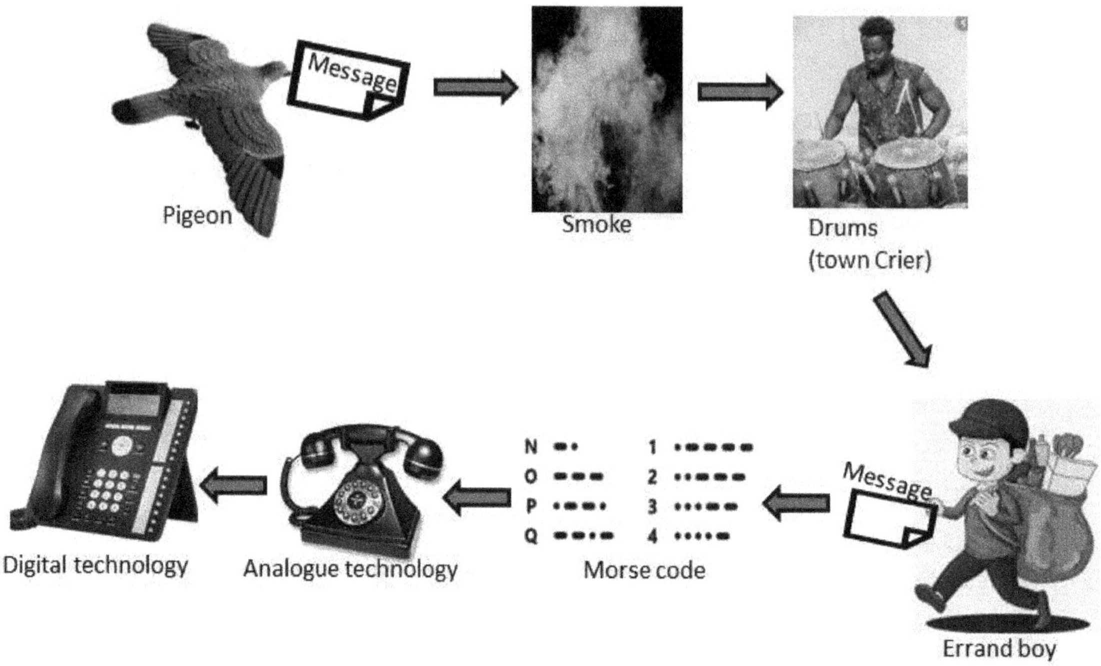

Fig. 1.1 Evolution in communications technologies.

came the errand boys. Of course, these archaic technologies had their weaknesses. For example, a bird could leave the message to unauthorized recipients, and an errand boy could read the messages en route.

The next technology was the invention of the Morse code. Then followed analogue communications technologies. Lastly, we are in the era of digital communications technologies. We note that some of these archaic communications technologies are still alive. For example, in occasions like festivals and enstoolment of kings and queens in Africa, the drums are beating to communicate messages. Also, we still have errand boys, except that they do not walk, the use mot or bikes. With the growing interest in IoT, it would not be surprising to use animals and trees as radio repeaters, signal switches or routers. If we implant a wireless repeater to birds, then we can monitor their movements regardless of their location. Signal from one bird would be intercepted by another bird at a distance, and so forth.

1.2 Introduction

The **Internet-of-Things** (IoT) has been the destructive technology of the 21st century due to its revolutionary power in both the social and business lives of humanity. The bottom line of the evolution of today's Internet is connected things or IoT, and the bottom line of IoT is sensors. There are a plethora of sensors, which are differentiated in size, shape and application area (see Fig. 1.19).

This book treats the principles, systems, architectures and standards supporting the wireless IoT from the engineering perspective. Most of the use cases of IoT are operated in environments without mains power supply and/or wired connectivity. Therefore, wireless IoT will be much more popular than wired IoT, the reason why this book focuses on the former. IoT is also referred to variously, including Internet of Everything (IoE), Internet of Systems (IoS), and the Web of Things (WoT). As IoT has a wide range of use cases, different terms have been coined for its specific application areas. For example, *Industrial IoT* (IIoT or I^2oT) refers to the IoT use cases in the industry, such as:

(1) Precision agriculture: low-power sensors monitor the condition of soil, etc.
(2) Smart metering: low-power smart sensors record the consumption of utilities (e.g. elec-

tricity, water and gas) and then communicate the information to the utility supplier for billing, troubleshooting of faults and monitoring purposes.

(3) Smart car parks: smart low-power sensors record empty bays and communicate to drivers.

(4) Environmental protection in smart cities: smart rubbish bins communicate their condition to a monitoring station.

(5) Smart advertising: digital signage, digital promo boards, video walls and virtual displays that display information dynamically from remote locations.

(6) Healthcare and manufacturing.

(7) Internet of Vehicles (IoV): a moving network comprising IoT-enabled automobiles. IoV provides several advantages, including the management of fleet and traffic flow, and for the avoidance or minimisation of road accidents.

(8) Internet of Robotic Things (IoRT): a network of smart devices to serve several purposes, including:

   (a) Monitor events.
   (b) Fuse sensor data from a variety of sources.
   (c) Use local and distributed intelligence to determine a best course of action.

Another example is *Consumer IoT* (CIoT), which is the IoT in the context of the consumer use cases, applications and devices. A popular example of CIoT is smart home (see Fig. 1.11). Figure 1.2 shows a few more IoT use cases. These use cases alone verify the versatility of IoT. Examples of IoT vendors are Canary, 3D Robotics, Athos, Sigfox, Ring, Jawbone, Razer, Greenwave, Fitbit, FreedomPop, Google Alphabet and Amazon.

The term **IoT**, thought to have been coined by Kevin Ashton in 1999, is an Internet in which data generation, processing and forwarding or sharing can occur without human involvement. Under IoT, machines can also use the data they gather to trigger an action, without human intervention. In order to generate data, an entity needs data gatherer which are sensors. Many people think of sensors as only miniature devices. However, sensors range from magnetic sensors of size $1.4 \times 1.4 \times 0.45$ $\mathtt{mm}^3$ (e.g., RedRock's TMR RR121) and humidity and temperature sensors of size $2 \times 2 \times 0.8$ $\mathtt{mm}^3$ (e.g., SHTC1) through accelerometers to huge satellite and radar systems.

Wireless Internet is a converged network of the Internet (computer communications) and wireless communications. These are the most important technologies for information collection or generation, processing, storage and transfer supporting

Industry sector	Sample use cases of IoT
Agriculture	Soil analysis, crop management
Automotive	Tracking (e.g., anti-theft tracking), traffic flow, fleet management, parking
Consumer	Home automation (smart home), energy use, maintenance
Environmental	Weather prediction, species tracking, animal migration, waste management, resource management
Industrial	Smart meters, wear-out monitoring, fleet management, manufacturing
Medical	Wearable health monitoring devices, telehealth services, implant devices
Military	Troop management, resource allocation, threat detection and analysis
Retail	Inventory control, focused marketing, product tracking

Fig. 1.2 Sample IoT use cases.

businesses and social life in our modern times. Almost all communications devices of today are Internet Protocol (IP) enabled, and are thus IoT citizens or **internetizens**. IoT is a powerful enabler of the global Digital Transformation (DX) or Digital Revolution (DR). What is DX? Let us examine some of the definitions.

> Salesforce defines *Digital Transformation* as the process of using digital technologies to create new — or modify existing — business processes, culture and customer experiences to meet changing business and market requirements. This re-imagining of business in the digital age is digital transformation.

> i-SCOOP defines *Digital Transformation* as the profound transformation of business and organisational activities, processes, competencies and models to fully leverage the changes and opportunities of a mix of digital technologies and their accelerating impact across society in a strategic and prioritised way, with present and future shifts in mind.

> Bounfour [Bounfour (2015)] and Wikipedia define *Digital Transformation* as the novel use of digital technology to solve traditional problems. These digital solutions enable inherently new types of innovation and creativity, rather than simply enhance and support traditional methods. DX includes going paperless.

DX is changing traditional business models, consumption patterns, socio-economic structures, legal and policy measures, organisational patterns, cultural barriers, for example [Cochoy (2017)]. Transducers and artificial intelligence (AI), including machine learning and business intelligence or business analytics, are among the core technological advances fueling the current DX. Three of the important components of DX are: IoT with AI, digital reality (augmented reality and virtual reality), and Symbiotic Autonomous Systems (SAS).

An interconnection of two or more networks is called *internet* (written in lower case), and the largest internet is the global network called the *Internet* (written capitalised and with a definite article as there are many internets but only one Internet). There are many rivers, all of which flow into one sea or ocean, just as there are many internets which interconnect to form the Internet. An internet results from the interconnection of two or more local area networks (LANs) and/or metropolitan area networks (MANs). An interconnection of LANs or MANs forms a wide area network, of which the Internet is the largest.

Not too long ago, we had separate networks for video, data and voice communications. These were single-media or one-play networks. A network which can be used to transmit video, data and voice is called **triple play**, multimedia network or multi-application network or multi-service network. The Internet is one of such networks. Multimedia is categorised into:

- Streaming media, which play as they are being downloaded. Examples are online radio and Internet TV (IPTV).
- Non-streaming media: these must be fully downloaded before they can start playing. An example is Internet music.

We use media players such as VLC multimedia player, PotPlayer and Windows media player to play back these media. Some websites are embedded with media players to allow playback from web browsers. Such websites have programs, called *plug-in*, which enable them to play the media. Examples of plug-ins are flash media and quicktime player. Multimedia communications has been enabled by the convergence of communications technologies, an advantage of which is the use of a single device, called smartphone, to access a myriad of services, as illustrated in Fig. 1.3. This figure also illustrates the power of wireless connectivity, as well as the smart phone as a powerful sensor. Smartphones of today perform the functions of dozens of electronic devices. Clearly, the functions shown in Fig. 1.3 are inexhaustive.

Networks' convergence is one of the four paradigm shifts that the ICT industry underwent between the 1990s and mid-2000. The remaining

Fig. 1.3 Convergence of technologies and electronic devices.

three are:

(1) Media convergence.
(2) End-to-end adoption of the Internet Protocol (IP) to so-called all-IP networking which is now referred to by the 3GPP as the evolved packet core (EPC) networks.
(3) 'Anything' seeking to be connected to a communication network.

1.2.1 *Brief History of Wireless Communications*

Clearly, wireless IoT cannot be realised without wireless communications. Exchanging information in the form of signals between two points or entities using the ether instead of wires as the transmission medium is called *wireless communications*. If all the entities enabling wireless communications are located on the Earth's surface then we have *terrestrial wireless communications*. However, if any of the components in the wireless system hangs in the sky or space then we have *satellite communications*.

How did wireless communications begin? Hans Christian Oersted and Andre-Marie Ampere discovered electromagnetism in the early 1820s to pave the way for the development of the wired telegraphy by Samuel Morse in 1832. Following the postulation of wireless transmission of signals by James Clerk Maxwell in 1864, Heinrich Hertz demonstrated the concept in 1880 and 1887. Guglielmo Marconi transmitted signals wirelessly to cover a distance of about 1.8 km in Italy in 1895, and about 29 km over the English channel in 1897. In those days, *wireless* was used in British English as synonymous to *radio* used then in the United States. Wireless communications experienced slow developments until Nippon Telegraph and Telephone launched the world's first mobile cellular network in Japan in 1979. Currently, developments in wireless technologies occur so fast that we have a new generation of mobile wireless systems every 10 years on the average. Wireless communications has many pros and cons. Table 1.1 lists some of them.

Every wireless technology is developed to operate over a certain band of frequencies. Devices must also operate in the same frequency bands to be able to reliably access such a wireless technology. Globally, the International Telecommunication Union (ITU) has divided the frequency spectrum into bands (see Fig. 1.4). Each band has been

Table 1.1 Advantages and disadvantages in wireless communications compared with wired communications.

	Advantages	Disadvantages
1.	Quicker to deploy than wired networks	Generally has lower bandwidth and data rate than wired links
2.	Initial network launch costs is lower	Capacity is limited by the finite RF spectrum. Capacity of wired networks is limited by only budget
3.	Enables user mobility during communications	Wireless links are more liable to disturbances (e.g., interference) than wired counterparts
4.	Some buildings and terrain (e.g., historic buildings and those containing asbestos) practically allow only wireless transmissions	Wireless links are more vulnerable to security attacks

Band Number	Name	Symbol	Frequency Range, f	Wavelength Range, λ
4	Very Low Frequency	VLF	3 to 30 kHz	10 to 100 km
5	Low Frequency	LF	30 to 300 kHz	1 to 10 km
6	Medium Frequency	MF	300 to 3000 kHz	100 to 1000 m
7	High Frequency	HF	3 to 30 MHz	10 to 100 m
8	Very High Frequency	VHF	30 to 300 MHz	1 to 10 m
9	Ultra High Frequency	UHF	300 to 3000 MHz	10 to 100 cm
10	Super High Frequency	SHF	3 to 30 GHz	1 to 10 cm
11	Extremely High Frequency	EHF	30 to 300 GHz	1 to 10 mm
12	Tremendously High Frequency	THF	300 to 3000 GHz	0.1 to 1 mm

Fig. 1.4 ITU frequency bands.

earmarked for specific wireless services. Frequency bands are usually referred to by names assigned to them as listed in Fig. 1.4. We shall therefore follow convention and refer to them as such. For example, we shall say High Frequency (HF) band antenna in the chapter on antennas in this book.

1.2.2 *The Internet and Its Genesis*

How did the Internet come into being? The Advanced Research Project Agency (ARPA), an arm of the USA army, created a network called ARPA Network (aka ARPANET) which in 1969 had only four switching nodes. The USA Army established the MILNET (Military Network) as part of the ARPANET, but separated the two in 1983 about the same time that IPv4 was released. The MILNET later became Defense Data Network (DDN). The nodes in ARPANET were located on the campuses of the University of California Los Angeles (UCLA), the University of California San Barbara (UCSB), Stanford Research Institute Augmentation Research Center (SRI ARC) and the University of Utah, all in the USA. This is illustrated in Fig. 1.5. The switching nodes were called Interface Message Processors (IMP) which were manufactured by a company called Bolt Beranek and Newman (BBN). A story is told that BBN located the IMPs at the campuses of the four universities named in Fig. 1.5 because those were the organisations that presumably had the expertise to manage them at that time. The trunks that interconnected the IMPs had the maximum data rate of only 50 kbps, something which is too low to imagine today. A few people contributed to the invention and/or development of the Internet in different ways. For example, Paul Baran and Donald Davies invented packet switching in the 1960s. Packet switching is the basic principle of all data networks. Leonard Kleinrock researched into packet switching to support the development of the ARPANET. In the 1970s, Vinton Cerf and Robert Kahn invented the TCP/IP protocol suite. Ray Tomlinson invented the network email in 1972.

Owing to its origin, the Internet was then called the ARPA Network. The objective of ARPA was to create a network which was robust against nuclear attack. Indeed, the Internet has been so architectured that it has no bottleneck. Therefore, it cannot

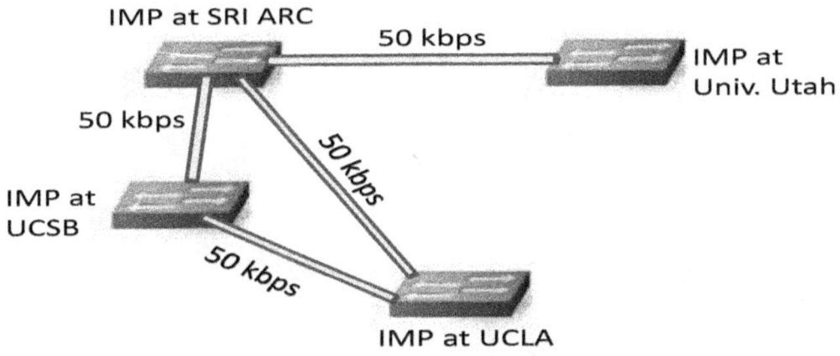

Fig. 1.5 The Internet of 1969.

be destroyed. The worst cyberattack cannot bring down the entire Internet.

Around 1969 when the ARPANET started there was no formal authority to coordinate the standardisation of communication protocols. Therefore, proponents of protocols used to run the IMPs established an informal discussion group which they called Network Working Group (NWG). The NWG has evolved to the present-day IETF (Internet Engineering Task Force). Furthermore, as the NWG had not been formally empowered to create standards, they could not call their protocols standards but drafts called **Request for Comments** (RFCs). This is the reason why most protocols used in the modern Internet are still called RFCs rather than standards. Also, the IETF is still somewhat an informal organisation.

> We can safely say that the standard protocols implemented on the Internet are called Request for Comments (RFCs), or Internet standards are called RFCs.

The ARPANET motivated the development of the BITNET (meaning *Because It's Time Network*) in the early 1980s. The National Science Foundation (NSF) in the United States developed the NSF Network in 1986. The development of the World Wide Web (WWW) in 1989 and Hypertext Markup Language (HTML) in 1990 by Sir Tim Berners-Lee expedited the realisation of the global Internet. There are two ways in which internetizens connect to the Internet, either always-on or dial-up with the latter becoming obsolete.

> Sir Tim Berners-Lee called the world's first web browser which he developed World-WideWeb (WWW). The WWW or Web is a virtual network of websites hosted on multiple servers which are attached to the physical Internet infrastructure. Therefore, the WWW is not the same as the Internet, but a virtual network of websites interconnected by hyperlinks running over the physical Internet.

The Internet had as few as only 213 hosts in 1979, and it became available commercially in 1988. Prior to 1988 its usage was restricted to only a few research and governmental institutions. Following the invention of the Web, the USA NSF sponsored the installation of access points to the T3 links of the Internet in 1992. The first graphic-enabled web browser, called Mosaic, was introduced in 1993 by Netscape Communications (then called Mosaic Communications). Later on Netscape Navigator, also designed by Netscape Communications in December 1994, replaced Mosaic. Microsoft joined the Web browser competition when it launched the Internet Explorer Version 1.0 on 16 August 1995, which successfully competed fiercely with Netscape Navigator for market share. The graphic browsers made the Web attractive to use by many. This, coupled with fallen prices of computers, boosted the Internet and Web uptake. The number of hosts attached to the Internet then grew exponentially from 213 in 1979 to 60 million in 2000, 100 million in early 2001 and over 1 billion in 3rd Quarter 2018. These are only the hosts formally registered in the domain name system (DNS). Figure 1.6 shows the growth of the global Internet hosts between 1993 and 2018. Counting all internetizens (incl. mobile phones, personal digital assistants and utility meters) the actual number of Internet hosts exceeds the data in Fig. 1.6. Gartner Inc.'s forecast was 6.3818 billion in 2016 and about 8.4 billion in 2017. The cornerstone of the Internet is the Internet Protocol suite, especially IP, which is discussed in Chapter 2 of this book.

The Internet is not an isolated single network. Rather, it is a conglomeration of a number of diverse self-autonomous networks, including personal

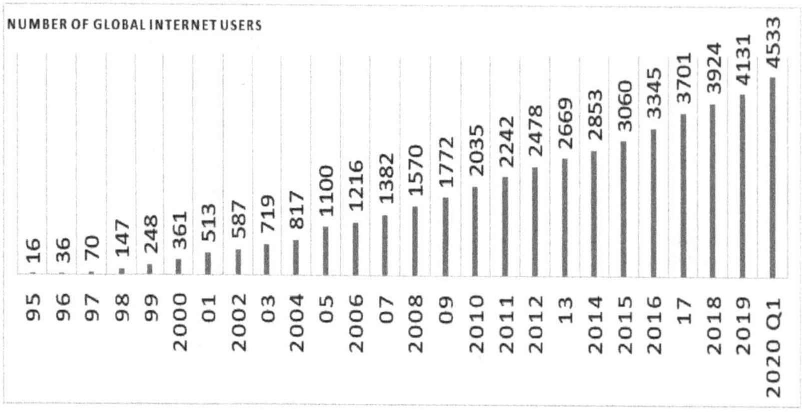

Fig. 1.6 The growth of the number of Internet users in millions worldwide [sources: https://www.statista.com/statistics/273018/number-of-internet-users-worldwide/ and https://www.internetworldstats.com/emarketing.htm].

area networks (PANs), local area networks (LANs), metropolitan area networks (MANs) and wide area networks (WANs). Therefore, the Internet cannot transform from being the largest computer network to Internet of Things (IoT) or the Internet of Everything (IoE) without its component networks undergoing a similar transformation. For this reason, modern networking books such as the one that you are reading now should have a title like *Network of Things: The New Paradigm in Computer Networking*. We can equally use *Network of Everything* instead of *Network of Things*.

A few years ago (i.e., before year 2000) devices networked were mainly printers and desktop computers. We referred to such a network as a 'computer network.' Now, networked devices are of diverse nature and include TV sets, CCTVs, car park meters, utility readers, monitoring sensors, vehicles (airborne, space borne and seaborne), iPads, smartphones and tablet computers. A question to ask is: can all these devices attached to contemporary networks be classified as computers? If not, then the term 'computer network' is obsolete and limiting, else we would need to redefine what a computer is. As it may be easier to update the name of modern networks, it appears logical to call modern networks 'network of things' (NoT), just as we have IoT. It is clear that NoT is a subset of IoT and the latter is the largest realisation of the former. When the IoT or the NoT becomes fully fledged in such a manner that all consumer devices (e.g., toasters, refrigerators, rice cookers, ovens, blenders and cooking pots) are sold as IP-enabled devices then the Internet would become a 'legal utility' just as water, electricity and gas are in many societies. We can therefore refer to what has been so far called 'computer networking' as 'network of things' (NoT) or 'Things Network' (TNet).

1.2.3 *Internet Number Resources*

There are three main Internet Number Resources (INR) and associated registries as shown in Fig. 1.7. The Internet Protocol (IP) addresses are discussed in many other sections of this book. A collection of IP-based networks which uses a single and clearly defined external routing policy is called an Autonomous System (AS). AS is also defined as a collection of networks which is administratively controlled by a single entity. A routing policy is used to control which routes on the network or Internet are stored in, and retrieved from, routing tables of routers. For example, we can configure routers in a network not to allow routes associated with a particular organisation to be placed in the routing tables of routers. Such a routing policy such routes are not used to forward data packets to any destinations and the routes are not advertised by the routing protocol to neighbours.

The networks making an AS can be operated by a single or multiple network operator(s). A pair of ASs exchange routing information using an exterior routing protocol such as BGP. Each AS is identified by an Autonomous System Number (ASN).

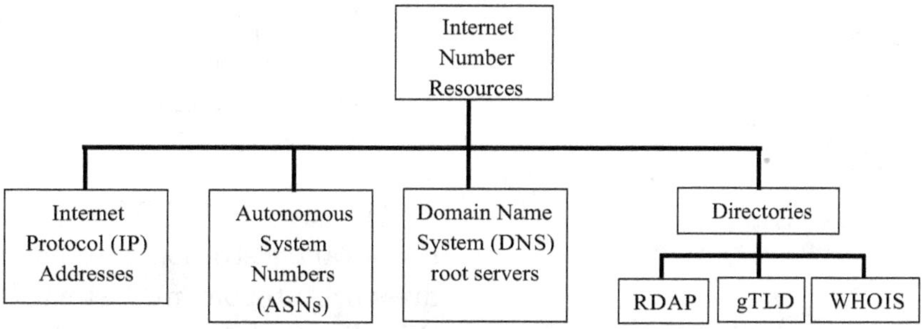

Fig. 1.7 Types of Internet number resources.

ASs also exchange routing information using the ASNs. ASNs can be either 16-bit or 32-bit long. An ASN can also be public or private. A public ASN uniquely identifies an AS globally. Routing using a private ASN is not visible on the Internet. Such routing can occur, for example, between an AS attached to only one Local Internet Registry (referred to as provider). The AS64512 through to AS65535 have been reserved as private AS Numbers by Internet Assigned Numbers Authority (IANA). ASNs obtained from a Regional Internet Registry (RIR) or National Internet Registry (NIR) are portable, but not those obtained from Local Internet Registry (LIRs). The list of ASNs can be found at the IANA.

The **Domain Name System** (DNS) provides a systematic way of naming computers and network services. Computers (including servers) are identified on internetworks using IP addresses. However, it is not very easy for humans to deal with IP addresses. Therefore, the DNS has been designed to map these machine-readable IP addresses to human-readable website addresses called domain names. For example, the domain name apcollege.net has the IP address 173.254.28.35. We can easily find the IP address of a website if we know the domain name using a command such as ping apcollege.net on Windows command prompt, for example. As another example, by executing the command ping yahoo.com we obtain the IP address 206.190.36.45. Therefore, whenever I want to visit the website *apcollege.net*, for example, I just type in a browser apcollege.com and a DNS server automatically translates the domain name into the IP address 173.254.28.35.

No website can operate without the DNS. Every website, such as gnosiscf.org, is linked to the DNS through certain DNS settings which are assigned to the domain name by its registrars through a web-hosting company. As illustrated in Fig. 1.8, the DNS is organised in a hierarchical fashion, just as the Internet Protocol address. The topmost DNS is called root DNS server or DNS root zone. The next level is called top-level DNS (TLD) servers. There are two types of TLD servers: the country codes (e.g. au for Australia, gh for Ghana, de for Germany and uk for the United Kingdom), and the generic TLDs (e.g. mi, edu, com, net, org, church). Next follows the second-level DNS (SLD) servers also called domain names. Then follows the subdomains and the computer names.

The Internet Routing Registry (IRR) is a distributed routing database. Data from the IRR are used to validate the contents of the border gateway protocol (BGP) announcement messages and/or mapping an origin autonomous system number to a list of networks. Anyone can access an IRR database to source data to assist in debugging, configuration and engineering of Internet routing and addressing. All the RIRs have mirrored databases of the IRR. As of July 2016 the largest public IRR is Merit RADb.

WHOIS is a protocol specified in RFC 3912 in September 2004 by L. Daigle for the accessing of Internet registry data. It is basically a query and response protocol which can be used to retrieve from databases the information about the registered users/owners or assignees of an Internet number resource. When queried it provides data including the name and email address of the one responsi-

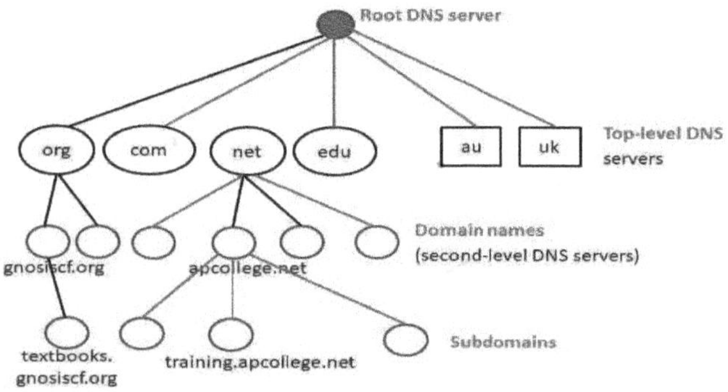

Fig. 1.8 The hierarchy of DNS servers.

ble for the Internet resource number. A number of weaknesses have been found in the WHOIS protocol. The below are listed in RFC 7482 [Hollenbeck (2016)]:

(1) Lack of standardised command structures.
(2) Lack of standardised output and error structures.
(3) Lack of support for internationalisation and localisation.
(4) Lack of support for user identification, authentication and access control.

A number of initiatives have been made to seek a replacement for the WHOIS protocol. These initiatives include the Registration Data Access Protocol (RDAP) proposed by a working group of the IETF on 25 March 2015. Details of RDAP can be found in several Request for Comments (RFCs) including RFC 7480 through RFC 7485. Another potential replacement of WHOIS is the Top-Level Domain (gTLD) Directory Services which is a recommendation from the Expert Working Group (EWG) constituted by ICANN in 2013. The American Registry for Internet Numbers (ARIN) website states the following as the differences between WHOIS and RDAP:

(1) RDAP is an HTTP-based REST-style protocol with standardised responses specified in JSON (JavaScript Object Notation), whereas Whois is a text based protocol, utilising a specialised protocol and port.
(2) RDAP response data objects are easily translated into languages other than English, whereas Whois response data objects may not
(3) RDAP responses offer direct referrals to other RIRs, whereas Whois defines no queries or responses, and interaction with DNRs and RIRs can vary significantly

The entry in the WhoWas registry contains the historical registration information for a given IP address or Autonomous System Number. Simply put, WhoWas contains data similar to those of WHOIS database, except that the former is obsolete. Only authorised users have access to this registry.

1.2.3.1 *Management of Internet Number Resources*

This section reviews the organisations that manage the numbers without which internetworking and the Internet would not be workable. The main global organisation behind it is the Internet Corporation for Assigned Names and Numbers (ICANN). Being a not-for-profit public-benefit corporation with globally-located volunteers, the main objectives of ICANN is to maintain security, stability and interoperability of the Internet. The ICANN was established on 18 September 1998 and it is headquartered in the USA city of Los Angeles. Actually, Internet Assigned Numbers Authority (IANA), an arm of ICANN, manages Internet Number Resources. IANA has five Regional Internet Registries (RIRs) as shown in Fig. 1.9.

AFRINIC (African Network Information Center) serves the African continent. APNIC (Asia Pacific Network Information Center) serves the entire Asia Pacific region (i.e. Asia and Oceania including Australia and NZ). ARIN (American Registry for Internet Numbers) serves Canada, some Caribbean and North Atlantic islands, and the United States. LACNIC (Latin American and Caribbean Network Information Center) is the regional Internet registry for Latin America and the Caribbean. RIPE NCC (Reseaux IP Europeens Network Coordination Centre) serves Europe, the Middle East and Russia.

Each RIR has a number of National Internet Registries (NIRs), which on the other hand, has a number of Local Internet Registries (LIRs) or network operators. The LIRs are usually Internet Service Providers (ISPs). We note that an RIR can

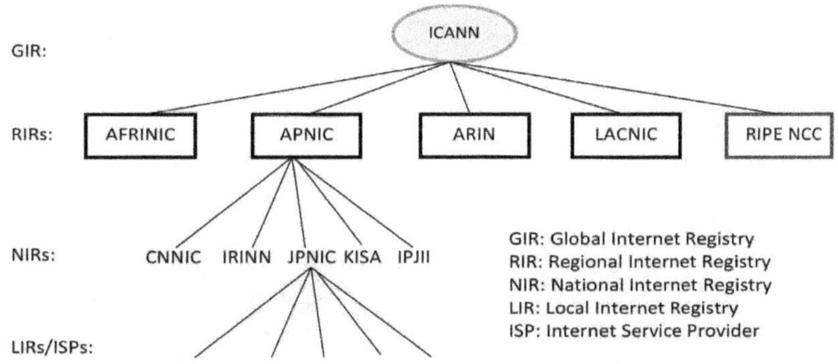

Fig. 1.9 Internet number resources management hierarchy.

choose to allocate addresses directly to LIRs/ISPs. For example, APNIC has the RIRs in China (CNNIC), India (IRINN), Japan (JPNIC), South Korea (KISA), Indonesia (IPJII) and Australia where it is headquartered. Small organisations usually acquire Internet Number Resources from Internet Service Providers while big organisations can acquire them directly from registration authorities.

1.2.4 *The Internet Governance*

The Internet is an amorphous network without any defined form or topology. Strictly speaking, the Internet started with no universal governance as portions of it are owned and managed by different organisations. However, USA established the Internet Society (ISOC) in 1992 as a non-for-profit organisation to provide leadership for the standardisation, education, accessibility, and policies related to the Internet. ISOC activities are divided into working groups. Figure 1.10 shows the components of the ISOC. Internet Architecture Board (IAB) supervises the technical development of the Internet. It is the advisory body for ISOC, which advises the IETF on best practices. The IAB is further divided into the Internet Research Task Force (IRTF) and the Internet Engineering Task Force (IETF). The technical work on the protocols supporting the Internet occur within the IETF. Currently, the Internet is governed by the so-called **I* Partners**, comprising:

- The Internet Architecture Board (IAB).
- The Internet Engineering Task Force (IETF).
- The World Wide Web Consortium (W3C).
- The Internet Society.
- The Internet Corporation for Assigned Names and Numbers (ICANN).
- The Regional Internet Registries (RIRs).

The IRTF is divided into 14 research groups as at May 2019. Three of the groups are shown in Fig. 1.10, namely the Thing-to-Thing Research Group (T2TRG), the Quantum Internet Proposed Research Group (QIPRG) and the Global Access to the Internet for All research group (GAIA). The activities of each research group is supervised by a committee with a chair. The chairs of each research group forms another overseeing committee called the Internet Research Steering Group (IRSG). According to the ISOC documents the IRTF encourages research which is relevant to the further development of the Internet by creating focused, long-term Research Groups working on topics related to Internet protocols, applications, architecture and technology.

The IETF is divided into eight areas, each of which is overseen by a committee with a director as a chair. The committee of the area directors forms the steering committee called the Internet Engineering Steering group (IESG). According to the ISOC documents the functions of the IESG are "the technical management of IETF activities, the Internet standards process, and for the actions associated with entry into and movement along the Internet standards track, including final approval of specifications as Internet Standards and publication as an RFC."

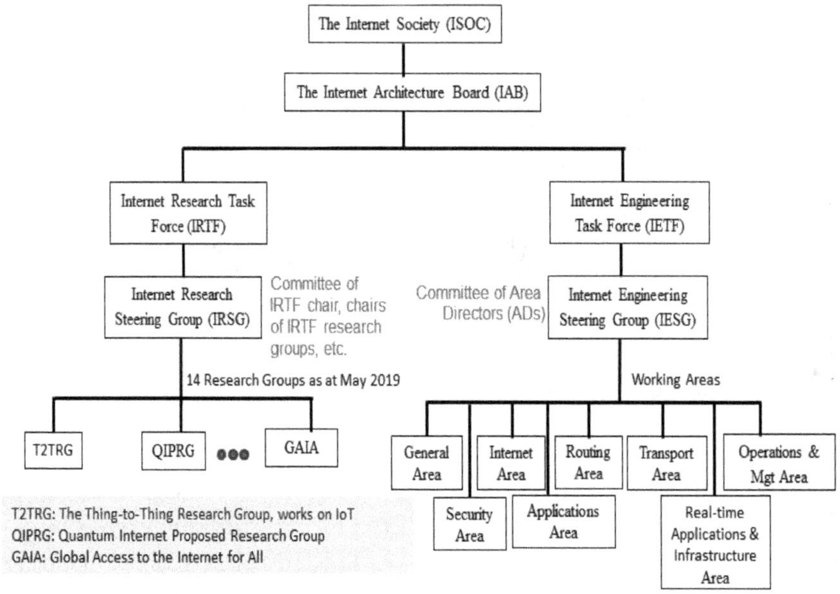

Fig. 1.10 Internet governance architecture.

1.2.5 *What is the Internet of Things?*

In the beginnings of the Internet the internetizens were gigantic computers, one of which filled a large room. Also, one needed a computer science degree to operate them. Mini-computers, microcomputers and mobile devices became internetizens in that order. Now, almost every electronic device is seeking to be an internetizen. All that a device needs to completely qualify to be an internetizen are the four abilities:

- Computing or processing engine which utilises artificial intelligence (AI).
- Communication or connectivity.
- Sensing or data acquisition.
- Understand the Internet Protocol (IP), the common language for internetizens.

For easiness of reference, let us refer to these four abilities as C2SI. The IoT enables three types of data exchange: people-to-people communications (P2PC), machine-to-people communications (M2PC) and machine-to-machine communications (M2M) or even massive M2M communications (μMMC).

Some people may think of only miniature transducers (i.e., sensors and actuators) as IoT-enabled devices. However, every electrical device can be an IoT citizen, regardless of its size or shape. That is the real meaning of the word *things* in IoT. This book explores satellites, wireless positioning systems, radars and cellular mobile handsets as IoT devices. Today, the precise readings of utility meters (e.g., for water, gas and electricity) still requires a technician to physically visit homes. IoT-enabled smart metering is changing the game altogether. These meters are equipped with sensors which collect the readings and send to the billing offices of utility companies.

IoT has an uncountable number of use cases. We consider only two of them here: one for home and one for the public. Let us consider the application of IoT scenario in Fig. 1.11. Here, a person uses smart consumer devices, which can be programmed to make life easier. For example, the refrigerator can send text messages to the mobile phone of its owner outside home to inform him of the lack of sufficient milk in it (i.e., the fridge) so that some is acquired on the way home. The autonomous vehicle (aka self-driving vehicle or driverless vehicle) like Tesla Model 3 receives directions from the Internet and satellites to move around without a driver. In another example shown in the figure, a person momentarily outside home can interrogate a microwave oven or pressure cooker to start cooking so that food is ready by the time he reaches home.

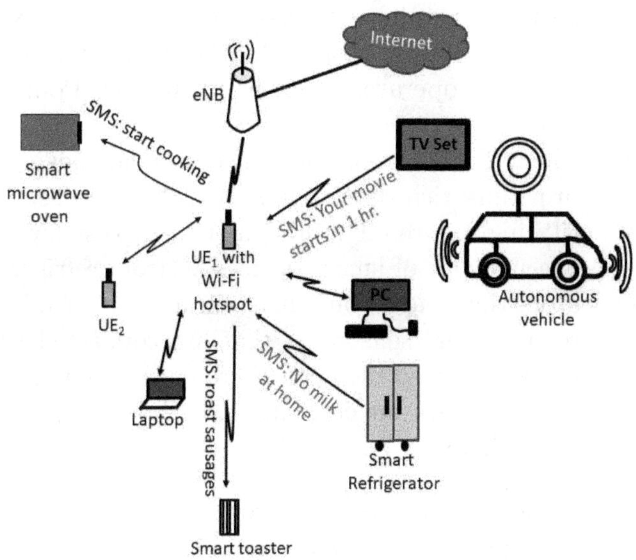

Fig. 1.11 A connected world of objects through IoT.

We all know that one can forget switching off an air-conditioning system while departing home or office. Traditionally, one would need to travel a distance to switch it off or else waste resources. An IoT application like the one in Fig. 1.11 can be used to switch off the air-conditioning system over distance using the smartphone, for example.

Another application of the IoT is illustrated in Fig. 1.12. This is a network of solar-powered smart rubbish bins (SRBs). Such a network has several purposes. Rubbish bins are traditionally used to keep our environment tidy. However, traditionally, the bins are dumb as they are not equipped with any intelligence. For this reason, a bin can overflow without the knowledge of the City Council. Such a situation makes the environment untidy, stinky and can cause diseases. Second, currently we unfortunately witness the spate of criminals hiding detonations in bins in the attempt to destroy both people and property. For this reason, many City Councils have depleted streets, train stations and other public places of rubbish bins. Therefore, people have only two choices: litter the environment or keep their rubbish and dispose of them at home. None of these two options is attractive. Public places are increasingly becoming littered consequently. Therefore, crime is causing other problems, one of which is environmental filth which can lead to diseases.

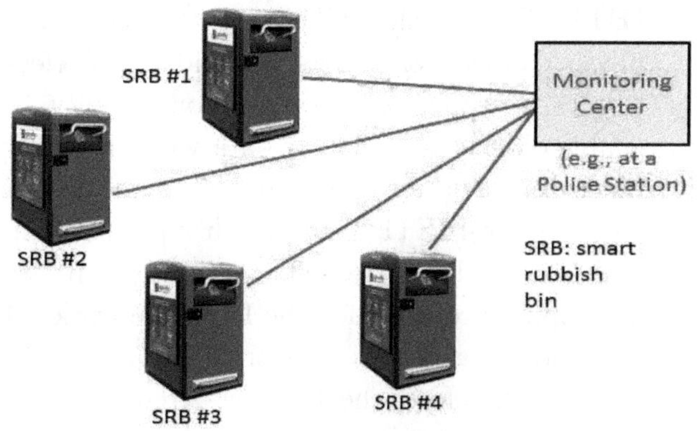

Fig. 1.12 An application of IoT in public places: a network of solar-powered smart rubbish bins (SRBs).

We need to find practical ways of keeping crime out of our environments.

We need strategies to keep the rubbish bins for their needed purpose while keeping away the criminals. An approach to this effect is to equip the rubbish bins with sensing, communications and computing capabilities, as well as empowering them with the Internet Protocol so they become internetizens. That way, we can network the bins. Once the networked smart bins are equipped with voice recognition and video recorders, they can record activities around them, such as who dumps what into them and at what time. With this, we can easily track down criminals, and eventually deter people away from crime and restore public freedom. The smart bins can be networked so that a monitoring station, possibly at local police stations, can observe what is happening around each of them 24/7. If we power the SRBs with solar panels then electricity would not be much of a cost. Solar-powered bins are already common. Each SRB can be numbered and linked with its exact geographical location to make identification of events easier.

We can return to the subject of this section after familiarising ourselves with some of the applications of the IoT. *Now, what is IoT?* The IoT is the offshoot of the Internet. The traditional Internet, referred to as Internet Phase 1, enabled communications between only humans. IoT, however, enables communications between things or entities. Let us **quote** here some of the definitions of IoT in the trusted open literature:

(1) **IETF:** The Internet of Things is the network of physical objects or 'things' embedded with electronics, software, sensors and connectivity to enable objects to exchange data with the manufacturer, operator and/or other connected devices. Several IETF Working Groups spanning multiple Working Areas shown in Fig. 1.10 are developing protocols for the IoT. The works of these working groups are somewhat uncoordinated. Therefore, the Internet Area directors and the Directorate Chairs have chosen experts to form the IoT Directorate to coordinate the work of the various IETF arms relating to IoT. The second function of the IoT Directorate is to promote the the visibility and adoption of IoT standards made by the IETF to organisations and external standards development organisations (SDOs) working on IoT.
(2) **Gartner:** The IoT is the network of physical objects that contain embedded technology to communicate and sense or interact with their internal states or the external environment.
(3) **Cisco:** Billions of devices are connected to the Internet. Each of these devices includes sensors that collect data, interact with the environment and communicate over a network. The IoT is the network of these connected devices. These smart, connected devices generate data that IoT applications use to aggregate, analyse and deliver insight, which helps drive more informed decisions and actions.
(4) **IBM:** The IoT is the concept of connecting any device (so long as it has an on/off switch) to the Internet and to other connected devices. The IoT is a giant network of connected things and people — all of which collect and share data about the way they are used and about the environment around them [Clark (2016)].
(5) **Wikipedia:** The IoT refers to the concept of extending Internet connectivity beyond conventional computing platforms such as personal computers and mobile devices, and into any range of traditionally dumb or non-Internet-enabled physical devices and everyday objects.
(6) The IoT is a system of physical things embedded with sensors, software, electronics and connectivity to allow it to perform better by exchanging information with other connected devices, the operator or the manufacturer [Poulos (2015)].
(7) The IoT is a network of connected smart devices providing rich data, but it can also be a security nightmare. The IoT, at its simplest level, is a network of smart devices — from refrigerators that warn you when you are out of milk to industrial sensors — that are connected to the Internet so they can share data, but IoT is far from a simple challenge for IT departments [Gold (2017)].

A newly-introduced technology which gives birth to a completely new industry or eventually supplants an established technology and change the way things are done is referred to as *disruptive technology* or *disruptive innovation*. These terms were coined by Professor Clayton M. Christensen of Harvard Business School in 1997 [Christensen (1997)]. The book [Christensen (1997)] explains why some organisations fold up when a new technology is introduced into the market while others survive the competition. Examples of disruptive technologies are: (a) the personal computer (PC) which supplanted the typewriter, (b) the electronic mail (Email) which has almost supplanted manual letter writing and has made the post offices less relevant, and (c) bitcoin/blockchain which is disrupting the banking and finance industry.

Figure 1.13 shows the rate at which the number of devices attached to the IoT is increasing. Data for this figure was obtained from the website [Statista (2019)]. Within less than seven years there is over 132% growth. Considering this growth, some people think that IoT will eventually become a disruptive technology, especially if prices of IoT-enabled devices fall and/or increase in utility. A fully developed IoT will make some established infrastructure redundant. For example, self-driving cars (e.g., trains and buses) will relieve human drivers of their jobs. IoT-enabled smart meters will make physical visits of homes by utility readers unnecessary. IoT-enabled smart buildings will automatically switch off and on lights, air-conditioning, etc. IoT-enabled smart grid will optimally dis-

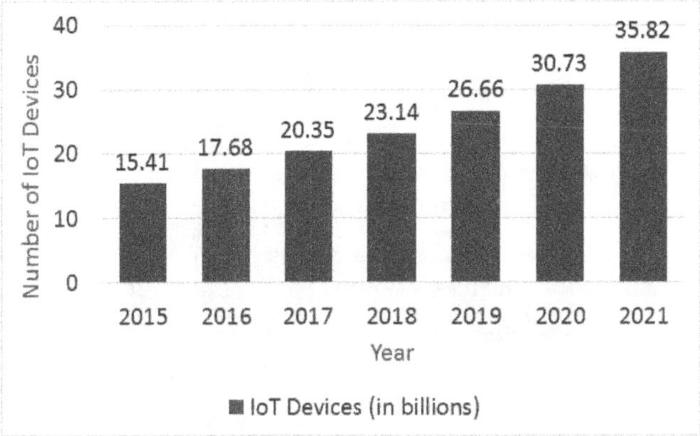

Fig. 1.13 The growth of the number of devices connected to the IoT [Statista (2019)].

tribute power in a smart city. We end this section with another application of IoT in a smart city, i.e., the smart parcel locker supporting 24/7 postal services.

IoT Case Study: AusPost Smart Parcel Locker

Most post offices in Australia are open 9am to 5pm on weekdays and 9am to 12pm on Saturdays. They are closed on Sundays and public holidays. This is a restriction for many people, especially those of us who cannot stay in post office queues during core working hours to engage in private transactions. An innovative solution to this restrictive physical post office opening hours is a smart box called parcel locker, which is a use case of IoT. Parcel lockers, as virtual post offices, can be used to both post and receive parcels. Parcel lockers are usually open 24/7, i.e., all the time. This is the beauty of it. They enable access to postal services 24/7.

What is a parcel locker? A parcel locker is referred to by various names, such as intelligent express cabinet, intelligent self-reporting cabinet and robot locker. Australian Post (aka AusPost) has mounted them around cities and suburbs in the country for the exchange of parcels between customers 24/7. They are located at strategic places like inside shopping malls and at filling stations. Usually, multiple lockers are kept in the same place just as postal letterboxes are. How long can a parcel be stored in the smart parcel locker? Usually, a parcel can be kept in the smart parcel locker for up to 5 days (for Priority Mail Express^{TM}) and 15 days for standard parcels. The sender can, however, request the storage to be extended to up to 30 days.

How can a person use a parcel locker? One must register with the AusPost at www.auspost.com.au/parcellockers to receive a 6-digit access code (or parcel locker number) and a customer number. Then, select a parcel locker at a preferred physical location. The access code is received over the Internet into an email account or over a text message. To post a parcel, one opens the locker with the 6-digit access code, place the parcel in it and then lock it up. Post office staff collects the parcel from the locker and forwards it to its intended recipient just as done with parcels posted from physical post offices. A recipient of a parcel kept in the locker receives a text message or an email providing information about the parcel. Part of the data in the message is a unique 6-digit access code and the address of the locker's location. One has to go to the locker, press a button and enter the 6-digit access code to open the locker in which the parcel is kept. The parcel recipient just picks the package if the locker opens and then closes the latter. The smart parcel locker is a smart logistics technology which is part of the smart city or smart community initiative. The components of a parcel locker are:

(1) Storage facility (or warehouse).
(2) Platform management system with information release, intelligent storage, intelligent pickup and information management.

Table 1.2 Comparison between Web 1.0 and Web 2.0.

Feature	Web 1.0	Web 2.0
Content	Static and unchanging, and provided by developer of the website.	Real-time content is created by web users. A user can create a content (e.g. blog) for other users to add to it.
Nature	Commercial and technical.	Socially oriented (i.e. strong presence of social media platforms).
Dominant users	Mainly large corporations, institutions and governmental agencies.	Individual people.
Interactivity	Limited to none. Users could not post any content to the Web.	Enables interactions between websites and users through social media applications like Facebook, Twitter, Youtube, Wikipedia, Whatsapp, blogs, podcasts and wikis.

1.2.6 *Web 1.0, Web 2.0 and Web 3.0*

Since its launch in 1969 the Internet has experienced a dramatic evolution. World Wide Web (WWW), usually abbreviated just Web, is a virtual network of websites overlaid on the Internet. The early phases of the Internet is referred to as Web 1.0 or Read-only Web (ROW). The main users of Web 1.0 were big organisations and government agencies. Web 2.0 is also referred to as Read-Write Web (RWW). Table 1.2 compares some features of Web 1.0 and Web 2.0. For example, Web 1.0 enabled only one-way communications (viz. from websites to users), while Web 2.0 enabled two-way communications (viz. users can talk to websites).

The successor of *Web 2.0* is called Web 3.0 or *Semantic Web* or *Read-Write-Execute Web*. Web 3.0 does everything capable of Web 2.0 plus advanced intelligence. This is the Internet of the future which has just began. The intelligence in Web 3.0 is due to its extensive use of artificial intelligence (AI). Computers and other machines are designed to imitate humans as close as possible. How does Web 2.0 differ from Web 3.0? Web 3.0 provides content through curation, while Web 2.0 does so through sharing. We search for products on the Internet using context or relevance in Web 3.0, but keywords or tags in Web 2.0. Web 2.0 enables interactive communications, while Web 3.0 supports engaged or invested communications. Web 2.0 enables dynamic information transfer, while Web 3.0 supports personal and portable information exchange. While Web 3.0 focuses on individuals, Web 2.0 is community oriented. While Web 3.0 uses lifestreams, Web 2.0 uses social media, blogs and wikis to serve personal interests. Users interact with smart applications in Web 3.0, but Web applications in Web 2.0. Technologies supporting Web 3.0 are OWL (Ontology Web Language), RDF (Resource Description Framework) and RDFS (RDF Schema), while Flash, Java and XML support Web 2.0.

1.2.7 *Internet Architecture: NDN/ICN/CCN versus TCP/IP*

At the inception of the Internet the only global network that was in existence was the Public Switched Telephone Network (PSTN), also referred to as Plain Old Telephony Service (POTS). PSTN/POTS was developed around 1876 by Alexander Graham Bell for only a single service, that is to transport human voice point to point (P2P). The Internet, which started with the ARPANET (Advanced Research Projects Agency Network) invented by Robert E. Kahn and Vint Cerf in 1960s, borrowed the P2P nature of the POST/PSTN. Therefore, the Transmission Control Protocol/Internet Protocol (TCP/IP) architecture was developed to support P2P transfer of datagrams using the Internet Protocol (IP). Traditionally, the TCP/IP is the Internet architecture. In modern times, however, most of the traffic carried by the Internet is multicast in nature, rather than P2P. Multicasting can be point-to-multipoint (P2M), aka one-to-many, or multipoint-to-multipoint (M2M), aka many-to-many. Examples of multicast communications are teleconferencing, sending an e-mail message to a mailing list, social media chats (e.g., Whatsapp and Youtube) and videoconferencing. This has aroused the need to seek an architecture suited to the modern use of the Internet.

The current Internet architecture is based on the TCP/IP protocol model. This model is hierarchically organised with the IP being on Layer 3. Each of the layers above and below the Layer 3 have multiple protocols. However, the Layer has more or less only one protocol, which is the IP. Such an architecture is referred to as **hour-glass model** or **thing-waist model**. Named Data Networking (NDN) is meant to maintain this hour-glass and the end-to-end philosophy of the Internet. However, an IP address in a packet referencing a communication endpoint will be replaced by a name. The name can refer to anything, such as an endpoint (as with IP), data in a written document or a command to execute an action. Routing in NDN uses a packet called *interest packet.* A person seeking access to a content, such as a movie, puts the name of the desired content in this interest packet and sends it into the network. Routers then use the name of the desired content to request it from its vendor for the consumer.

The proposed future architecture for the Internet, which is still being researched, is referred to by many names, including Named Data Networking (NDN), Information-Centric Networking (ICN), Content-Centric Networking (CCN), Data-Oriented Networking (DON), Publish/Subscribe Networking (P/SN) and Network of Information (NetInfo). Proposed by Van Jacobson in 2006, NDN is poised to transition the Internet from its current host-centric network architecture based on IP to a data-centric network architecture. Technologies related to NDN include IP-over-ICN and Information-Centric Access Network (I-CAN). Information-Centric Networking Research Group, one of the 14 IRTF research groups, is coordinating the work relating to NDN within the Internet Society. What are the implications of data-centric networking? Cisco Systems divides NDN into three main components:

(1) *Named content*: break content into discrete pieces that can be given names independent of location, storage, application and transfer method, to enable multicasting and protection of the content rather than the container of the content. NDN embeds security inside a packet, rather than in the network or packets' endpoints.
(2) *Named-based routing*: to enable a name to refer to any number of entities, names but not IP addresses are used for routing for an efficient multicasting and dynamic content-based routing. Packets have no addresses in named-based routing. This requires the design of named-based routing protocols. The IP address used currently will be replaced by an object identifier.
(3) *Transport enhancements*: NDN addresses problems associated with mobility support by TCP. Also, in-network caching/storage and replication are allowed to enable error recovery using local cache. In-network caching reduces network congestion as well as improving data delivery speed.

The NDN architecture is being investigated for the IoT under the name such as **NDN-IoT Framework**.

1.2.8 *IoT Protocol Stack and Architecture*

Figure 1.14 compares the TCP/IP and IETF's IoT protocol stacks. The IPv6 over Low power Wireless Personal Area Networks (6LoWPAN) is a layer which adapts IPv6 over wireless links based on IEEE802.15.4. The 6LoWPAN operates only in the 2.4 GHz frequency range the maximum bit rate of 250 kbps. Most IoT devices are resource-constrained. For this reason, the Constrained Application Protocol (CoAP) has been developed as their application layer protocol. CoAP is simple, has very low overheads and is able to translate to HTTP for simplified integration with the web, while

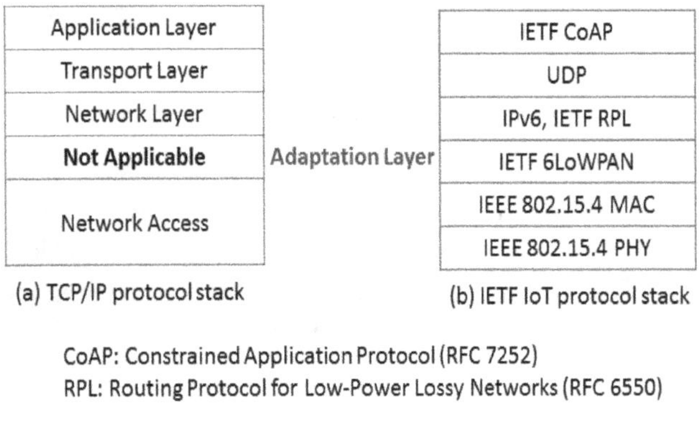

Fig. 1.14 Comparing the TCP/IP and IETF's IoT protocol stacks.

also meeting specialised requirements such as multicast support. A new routing protocol, called routing protocol for low-power lossy networks (RPL), has been specified in RFC 6550 to support IoT.

The IoT architecture comprises unaccountable number entities, including protocols, cloud services, and layers and transducers (e.g. sensors and actuators). A sensor converts a non-electrical input signal into an electrical signal. An actuator, however, converts an electrical input signal into a non-electrical output signal in the form of action, motion or heat. For example, an actuator can control lights and temperature of a place. An IoT architecture must possess these features: security, functionality, scalability, availability and maintainability. Many different architectures have been proposed for the IoT. For brevity reasons, we discuss only the one proposed in [Wu *et al.* (2010)]. This architecture comprises three layers:

- The perception layer (incl. sensors and actuators).
- The Network layer (incl. data centers, cloud).
- The application layer.

The **perception layer** contains the data sources, which are the transducers (sensors and actuators). These data collectors include satellites, radars, RFID, GPS, smart utility meters, wearable health trackers and mobile phones. The data collectors are equipped with data processing techniques to be able to, for example, convert the sensed data into digital form. The network layer can be a cloud network. It processes the data received from the perception layer and then sends it to the application layer. The signal transfer occurs via many networking technologies, both wired and wireless. The application layer concerns itself with the practical applications of the IoT, such as for smart metering, smart home and smart city.

1.2.9 *Check Your Understanding*

Students are encouraged to check their understanding of the content of this section with the following questions prior to proceeding to the next section.

(1) Where did the global Internet originate from?
(2) Discuss the main differences between Web 1.0, Web 2.0 and Web 3.0.
(3) State the alternative names used to refer to Web 1.0, Web 2.0 and Web 3.0.
(4) Describe the purpose of Autonomous System Numbers (ASN).
(5) Visit the website of your employer or institution.
   (a) What is the name of the top-level DNS server used?
   (b) Write two subdomains being used at the website.
   (c) Write the domain name used by the website.
(6) We seek to find the IP address, IP address version and type, as well as the ASN number of a given website.
   (a) Open Windows command prompt and execute the command ping *www.gnosiscf.org* to find the IP address for the website.
   (b) What version of IP address is used?
   (c) Is the IP address private or public address?
   (d) Find the ASN corresponding to the found IP address.

(e) Determine the country where the ISP which allocated the IP address is located.

(7) Which were the four institutions with connections to the Internet when it first came into being?
(8) Which were the original devices that connected to the Internet?
(9) Who invented the Internet?
(10) Which are the four main Internet number resources?
(11) How are the Internet Number Resources managed?
(12) What is the name and address location of the organisation managing Internet Number Resources in your country or continent?
(13) **In-class group work:** In groups of 3 to 4 students investigate into the reasons why a new architecture, called Named Data Networking (NDN), has been proposed for the future Internet. What other synonymous names are used to refer to the NDN? Which Internet Society group is coordinating the NDN work? Who and when was the NDN initiated?
(14) **In-class research:** In groups of 3 to 4 students, research into the meaning of *Digital Transformation* and its impact on economies.
(15) Discuss the role of IoT in Digital Transformation.
(16) **In-class research:** In groups of 3 to 4 students, research to find the differences, relationship and similarities between IoT and the Internet.
(17) What is the name of the world's first graphic web browser? When was it released commercially? Which company developed it?
(18) Describe your understanding of IoT. Include in your description two advantages and two disadvantages in IoT.
(19) Study Fig. 1.11. Reflect, list and describe two applications of this IoT use case that are not mentioned in this book.
(20) Reflect on the IoT use case in Fig. 1.12. List and describe two applications of this IoT use case that are not mentioned in this book.
(21) Analyse the various IoT definitions quoted in this section. Compare and contrast them.
(22) Explain the meaning of the phrase *disruptive technology*. Discuss whether or not IoT is a disruptive technology. Substantiate your claims.
(23) Reflect on how the AusPost parcel lockers use IoT.
(24) List the names of the research groups and working groups/areas of the Internet Society (ISOC) whose activities relate to IoT.
(25) Find the name of the ICANN Regional Internet Registry which allocates Internet Number Resources (INR) to ISPs in your country of residence.

1.3 IoT Enablers and Standards

We discuss some enabling technologies of IoT in this section. The basic components in IoT are processing engine and artificial intelligence (AI), network connectivity and sensors.

1.3.1 *AI and Machine Learning in IoT*

Machine Learning (ML), a branch in AI, is increasingly becoming a game changer in many industries, including information and communications technologies (ICT). They have become an impetus for innovation. For example, ML and AI are used to extract business intelligence from organisational data kept in data warehouses, data marts or databases to build new products and maintain customers. The data sources are mostly IoT devices in modern times. ML/AI enables organisations to extract insights from data on how to identify profitable opportunities, avoid unknown risks, reduce operational costs and optimise customer experience. A main reason is that ML and AI enable a fast processing of huge amounts of data, referred to as big data.

ML and AI can be exploited to optimise mobile tower operations, for example. For example, we can mount IoT-enabled sensors at base station tower locations to monitor them. Then, interconnect the sensors using the principles of IoT. That way the sensors can process sensed data using AI/ML algorithms to extract intelligence from the collected data. The sensors then generate the appropriate alert and forward to a monitoring station. Events that the sensor can monitor and alert authorities include smoke, intrusion and fire. Examples of analytics-enabled IoT platforms are General Electric Predix, which has been developed for the Industrial IoT.

One of the most popular applications of AI is business intelligence (BI), which applies analytical tools on organisational data to extract actionable insights to make data-driven decisions in real time. Modern trends in BI include data storytelling, automated BI (ABI), mobile BI (MBI), analytics everywhere and data querying in natural language which has enabled analytics for everyone.

1.3.2 *Wireless Connectivity Between IoT Devices*

The connectivity between IoT devices can be either wired or wireless. However, wireless connectivity is much more versatile and important. The reason being that in many IoT use cases the devices must be interconnected using only wireless technologies. Wireless connectivity is indeed very important for the IoT, at least for its full-scale realisation. There are two basic types of wireless connectivity: short-range and long-range technologies (Fig. 1.15). Some of the technologies are licensed, while others are not. The long-range standards include Long-Term Evolution for machine-type communications (LTE-M), Narrowband IoT (NB-IoT) and Extended Coverage GSM IoT (EC-GSM-IoT). All these are 3GP standards.

Wireless communications technologies developed for large-area IoT deployments are collectively referred to as Low-Power Wide Area (LPWA), low-power network (LPN) or low-power wide-area network (LPWAN). LPWA is a wide area network (WAN) of IoT devices interconnected using LPWA-compatible wireless technology. LPWA has no definite defintion. However, in order to be classified as LPWA a technology must fulfill the five basic requirements shown in Fig. 1.16, namely ultra-high energy or power efficiency, low data rate links or low bandwidth, large area coverage (i.e. the radius of a cell), support massive number of IoT-enabled devices and low-cost technologies delivering devices at about \$10 apiece. Note, in particular, the 10 appearing in each of the five requirements in Fig. 1.16. They are collectively referred to as *the five-tens in LPWA*.

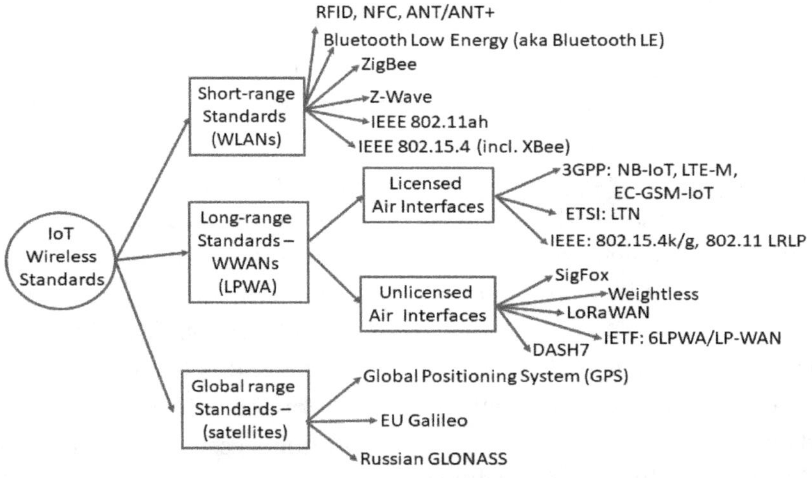

Fig. 1.15 A sample of wireless technologies for IoT.

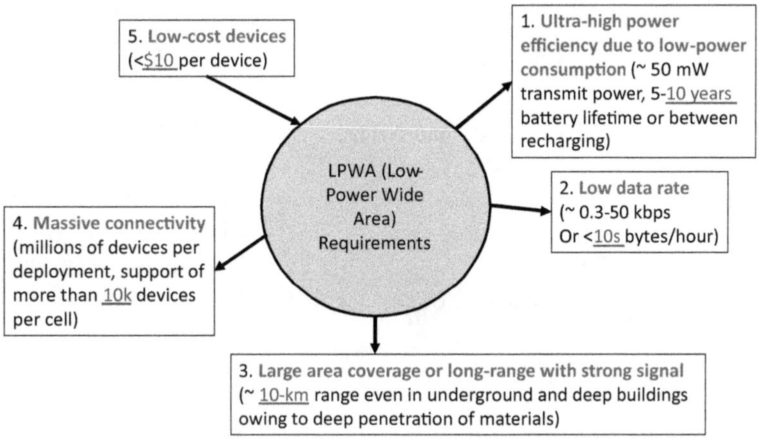

Fig. 1.16 Requirements of Low-Power Wide Area (LPWA).

There are many competing wireless standards fulfilling the LPWA requirements in Fig. 1.16. Some of the popular wireless standards are low-energy Bluetooth (LE-BT), RFID, Ingenu, ZigBee, DASH 7 from DASH7 Alliance, Low Throughput Networks (LTN) from ETSI, 6LPWA/LP-WAN from IETF, IEEE 802.11ah/802.15.4k/802.15.4g/802.11-LRLP developed by the IEEE, Z-Wave, Long Range Wide Area Network (LoRaWAN) supported by LoRa Alliance, Extended Coverage GSM IoT (EC-GSM-IoT), enhanced MTC (eMTC) and Narrowband IoT (NB-IoT) developed by 3GPP, and Sigfox. See Fig. 1.15. Future wireless IoT devices will use eSIM instead of the traditional SIM cards. See Chapter 9 for further discussions on eSIM.

1.3.2.1 *IEEE Standards Related to IoT*

The Institute of Electrical and Electronics Engineers or IEEE (www.ieee.org) has several working groups, each of which engages on a specific set of technologies. For example, the Working Group (WG) 21451 (formerly 1451) has released a set of standards to serve as interfaces between smart transducers and between smart transducers and networks. One of the set of specifications of the WG is the IEEE 21451.5, which is the wireless interface between networks and smart sensors and actuators. The IEEE 802.16 WG has released the IEEE 802.16s, nicknamed mission-critical IoT (MC-IoT), to support IoT in vertical industries such as utilities (electric/energy, gas and water) and intelligent transportation. Further, 802.16p supports machine-to-machine communications.

Another IEEE WG developing standards for IoT is 802.15. By convention, its standards are coded IEEE 802.15.x. Of specific interest to IoT is the IEEE 802.15.4 standard to support low-rate wireless personal area networks (LR-WPAN). The IEEE 802.15.4f has been designed for active RFID systems, IEEE 802.15.4g for smart metering utility networks (SMUN), and IEEE 802.15.4j for medical body area networks (MBAN), for example. Note that other non-IEEE wireless standards are built upon IEEE standards. For example, ISO/OSI Layers 1 and 2 of ZigBee, 6LoWPAN and WirelessHART are based on 802.15.4.

1.3.2.2 *LTE-M and Narrowband IoT*

Long-Term Evolution for Machine-Type Communications (LTE-M or LTE-MTC) is one of the three wireless standards developed by the 3GPP to support wide-area deployment of IoT. LTE-M enables battery-powered IoT devices to connect directly to a 4G network without a gateway. Another LPWA standard developed by the 3GPP is the Narrowband IoT (NB-IoT). Both LTE-M and NB-IoT are open standards based on the LTE standard. NB-IoT can be deployed on an existing LTE network or co-exist with non-LTE networks. These two standards are compared in Table 1.3. Both LTE-M and NB-IoT are 5G standards.

Table 1.3 Comparison between LTE-M and NB-IoT.

Feature	NB-IoT	LTE-M
Categories	LTE Cats NB1 and NB2	LTE Cat 0 (Rel. 12), M1 (Rel. 13) and M2 (Rel. 14)
Air interface	New from release 13	Regular E-UTRAN, reuses existing LTE eNBs
Duplexing	Half duplex	Full duplex
Cell type	Standalone and LTE-integrated cells	Integrated with LTE cells
Latency	1.6-10 seconds	10-15 ms (Cat M1)
Peak data rate	250 kbps	1 Mbps
Mobility	No	Yes
Voice communications	No	Yes
Number of antennas	One	One
Power consumption	Best at low data rates	Best at medium daata rates
Initial rollout	North America	Europe

Live trials by Vodafone has proved that NB-IoT signals can penetrate two to three double-brick walls, making it suitable to interconnect smart rubbish bins mounted in underground train stations and shopping malls. This live trial also found the relationship

$$R_{NB-ioT} \approx 100^{\frac{1}{\gamma}} R_{GSM} \tag{1.1}$$

where R_{NB-IoT} and R_{GSM} are the ranges of NB-IoT and GSM signals, respectively, and γ is the pathloss exponent. Table 1.4 compares more wireless technologies which are suited for IoT. As at May 2019 NB-IoT, LoRaWAN ans SigFox are the popular wireless technologies in IoT. Advantages in NB-IoT which have been promoting its popularity are shown in Fig. 1.17.

1.3.2.3 *LoRaWAN*

LoRaWAN is being developed for low-power wide area networking by the LoRa Alliance whose website is at https://lora-alliance.org/. The development of IoT globally is expected to receive a boost if this standard receives the expected acceptance. It can be deployed for the interconnection of battery-operated IoT devices. It will enable machine-to-machine communications (M2MC), support smart cities and industrial applications. Desirable features in LoRaWAN include:

(1) *Massive connectivity*: millions of devices can be interconnected, something which is a desire in IoT.
(2) *Long distance connectivity*: devices can be separated by up to 2 to 5 km in urban areas and 15 km in suburban areas.
(3) *Highly energy efficient*: a battery can last 5 to 10 years. The maximum transmit powers consumption is 50 mW for Class C devices, Classes A and B consume less power. Typically, 14 dBm (Europe) and 20-30 dBm (North America).
(4) *Data rates*: 0.3 kbps to 50 kbps.
(5) *Modulation scheme*: chirp spread spectrum (CSS), which is a modified frequency-shift keying (FSK).
(6) *Specification*: only ISO/OSI Layer 1 or physical layer.
(7) *Recommended network architecture:* a star-of-stars topology.
(8) *Security*: provided by the Advanced Encryption Standard (AES) algorithm using 128-bit keys.
(9) Frequency bands:

    (a) Europe: 867-869 MHz divided into 10 channels.
    (b) USA: 902-928 MHz divided into 8 uplink channels each of 500 kHz bandwidth, and 64 200-kHz plus 8 500-kHz channels for downlink.

Table 1.4 Basic properties of popular IoT connectivity.

Standard	Range	Operating frequency (MHz)	Power consumption	Data rate (kbps)
NB-IoT	1 km (urban) to 10 km (rural)	LTE bands	0.2 W	0.3-180 (UL)/ 0.5-200 (DL)
LTE-M	Global	LTE bands	Varies	Varies
SigFox	10 km (urban) to 40 km (rural)	868 (EU), 915 (North America), 433 (Asia)		
LoRaWAN	2-5 km (urban), 15-20 km (suburban/rural)	169 & 433, 868 (EU), 915 (USA), 433 (Asia)	Low	0.3-50
IEEE 802.11af (White-Fi)	≤ 100 m	470-710	Low	$\leq 35,600$
IEEE 802.11ah (HaLow)	≤ 20 km	700 (China), 850 (EU), 900 (USA)	Medium	150
ZigBee (IEEE 802.15.4)	10-100 m (LOS)	868-868.6 (EU), 902-928 (USA)	Low	20, 40, 250
IEEE 802.11ax	70-240?	2400, 5000		$\leq 4.8 \times 10^6$
NFC	10 cm	13.56	Very low	100-420
BLE	50-150 m	2400-2483.5	Low	1000
Z-Wave (ITU-T G.9959)	30 m	900 (ISM band)	Low	9.6, 40,100

BLE: Bluetooth Low Energy; NFC: Near-Field Communications.

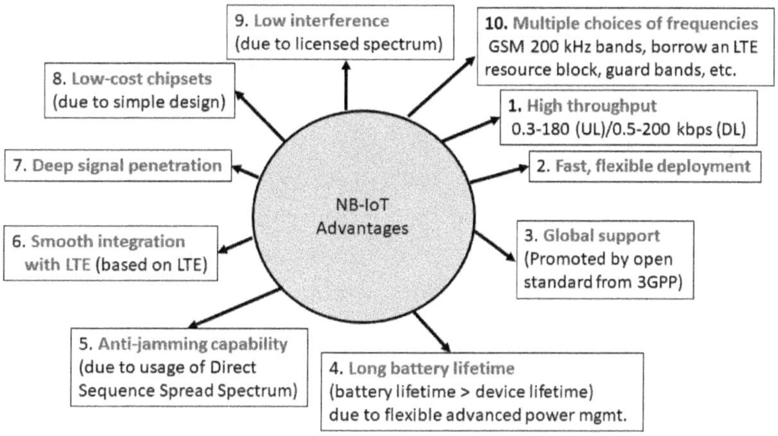

Fig. 1.17 Some promotional advantages in NB-IoT.

1.3.2.4 *Near-Field Communications*

Near-Field Communications (NFC) is a family of wireless technologies used for short-range communications within about 10 cm range. It operates in the Industrial Scientific Medical (ISM) band at $13,560 \pm 7$ kHz. NFC has the three data transmission data rates 106 kbps, 212 kbps and 424 kbps using amplitude shift keying (ASK) modulation.

NFC uses the ISO/IEC 18000-3 air interface which is also used by RFID technology. Therefore, NFC operates like RFID. Its link comprises the two device types: an active device called the initiator or interrogator and a passive device called the target. The target is powered by the electromagnetic field generated by the wireless signal emitted by the interrogator. Examples of RFC targets are stickers, cards, key fogs and tags.

1.3.2.5 *ANT and ANT+*

These standards are described in [Dynastream (2014)]. They are short-range air interfaces developed for low data rate wireless sensor networks (WSN) by ANT Wireless, a subsidiary of Garmin Ltd. This company develops technologies based on the Global Positioning System (GPS), and it is known by its GPS receivers. ANT+ is used in heart rate monitors, speed and distance monitors, sport bike speed and cadence sensors, sport bike power sensors, weight scales, fitness equipment data sensors and temperature sensors [Dynastream (2014)]. Many modern smartphones support ANT+, which enables people to monitor on the phone in real time data on health, fitness, etc. For example, a heart rate monitor strapped around the chest measures signals generated by the heart during constrictions and sends these over ANT+ to a mobile phone or a computer. ANT-based WSN is network topology agile. They allow star, connected star, peer-to-peer, tree and fixed mesh topologies and more [Dynastream (2014)].

ANT/ANT+ are are broadcast technologies which transmit signals with ultra-low power. ANT+ operates at 2.457 GHz, while ANT operates at 2.4-2.524 GHz (excluding 2.457 GHz) in the ISM band. Each channel with bandwidth of 1 MHz can be divided into up 2^{16} time slots to enable TDMA-based multiuser communications over Gaussian Frequency Shift Keying (GFSK) modulation. A time slot in ANT/ANT+ contains a 64-bit payload and a 16-bit cyclic redundancy check (CRC) for error detection. Although ANT and ANT+ are proprietary standards, they are supported by a special interest group (SIG) of companies, called ANT+ Alliance, which have adopted them in their products.

1.3.2.6 *Weightless for IoT*

This is a wireless technology developed to interconnect smart machines to the Internet to realise machine-to-machine (M2M) communications. Weightless is an open standard for wide area communications with a maximum transmission range of about 10 km. The details of the Weightless standard is available at www.try.weightless.org. Weightless can operate at any frequency band. However, it is currently defined for operation in license-exempt ISM frequency bands including 138 MHz, 433 MHz, 470 MHz, 780 MHz, 868 MHz, 915 MHz and 923 MHz.

1.3.3 *LTE-Advanced and 5G Use Cases in IoT*

The first generation (1G) of mobile cellular technologies were launched to provide the low-rate single service, voice. Second generation (2G) technologies were developed to support two services: voice and text messaging. 3G systems came to support voice, text and data services. 4G systems provided the services of 3G but with higher data rates. 5G systems poised for commercial release in the year 2020 will support faster 4G services with massive connectivity, ultra-low latency to realise virtual reality, ultra-reliability, ubiquitous coverage and high-mobility connections. With these features, especially the massive machine-to-machine connectivity (mMMC), 5G will be an enabling platform for wireless IoT. 5G is the first mobile cellular technology that will integrate seamlessly with IEEE 802.11 (aka Wi-Fi) technologies. Cellular mobile technologies are traditionally developed by telecommunications pundits, while Wi-Fi are developed by computer networking professionals. These two industries are becoming increasingly united through 5G and Wi-Fi 6.

The generation of mobile cellular network used in 2018 is called 4G (i.e., fourth generation), of which LTE-Advanced is the main technology. The next generation of mobile cellular network poised for release into commercial operation around year 2020 is referred to as IMT 2020 or 5G. There are many initiatives working on 5G, such as:

- Mobile and wireless communications Enablers for Twenty-twenty Information Society (METIS)
- MIllimetre-Wave Evolution for backhaul and Access (MiWEBA)

- ITU-R and ITU-T
- COST2100
- IEEE 802.11
- NYU WIRELESS: interdisciplinary academic research center
- Fraunhofer HHI

The following are a sample of the basic features of the 5G standard:

(1) mmWaves: 5G systems are poised to use frequencies in the GHz to THz range, much higher than those of 4G systems. Signals at such high frequencies have wavelengths in the millimeter range, and thus referred to as *millimeter waves* (mmWaves). The reason for using mmWaves is to achieve high data rates. mmWaves provides high capacity in two ways:

  (a) The higher the frequency, the higher the available bandwidth.
  (b) The low wavelengths of mmWaves causes the absorption over short distances, allowing small frequency reuse factors. The smaller the reuse factor, the higher the achievable capacity if all other parameters remain the same.

(2) Spectrum sharing: in order to increase the utilisation efficiency of the spectrum, 5G systems allow spectrum sharing.
(3) Optimised OFDM: 4G systems used OFDM, but 5G systems are to use optimised OFDM waveforms.
(4) Massive MIMO: 4G systems use MIMO. 5G systems build upon 4G MIMO by using a larger number of antennas.
(5) Femto cells: 5G will use small cells to achieve network densification that is required to provide the required data capability. Small cells are achieved by low-power transmitting NR and mmWave.

OFDM techniques insert GIs (guard intervals) between symbols to eliminate the inter-symbol interference (ISI). An example of GI implementation is zero padding (ZP). A zero-paded OFDM traansmission is referred to as ZP-OFDM. ZP-OFDM can guarantee symbol recovery regardless of the presence of channel spectral nulls. However, ZP-OFDM suffers from a complexity problem.

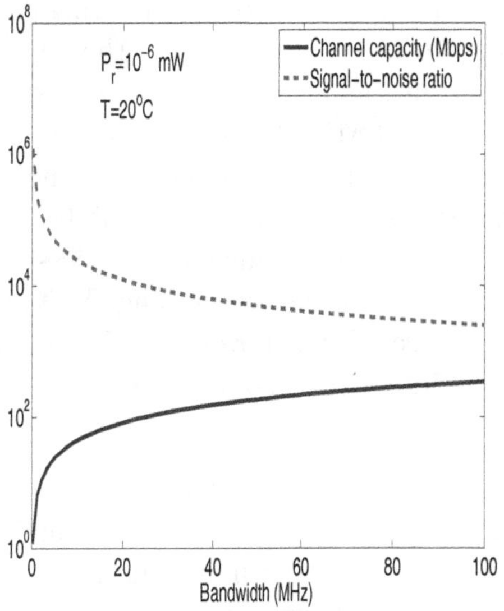

Fig. 1.18 Channel capacity vs SNR at fixed received power.

Figure 1.18, based on the classical Shannon-Hartley channel capacity formula, shows an interesting observation. The increase in capacity by increasing the bandwidth starts to decrease as the bandwidth increases. This means that for a given receive power, the background noise or signal-to-noise ratio limits the capacity gains achievable by increasing the channel bandwidth. Consequently, achieving a higher bandwidth by exploiting signal transmissions over mmWaves is not a panacea for increasing data rates.

Every communications network provider needs an interface to monitor, control, analyse and manage the communications network. Such a set of tools is called an operations support system (OSS). Some of the tasks of the OSS are tracking and ordering network components (including IP addresses), usage and traffic patterns, billing and reporting.

The applications of 5G systems in a specific industry is referred to as *5G verticals*. 5G industry verticals include energy (e.g., smart metering), smart agriculture, automotive (e.g., self-driving vehicles and automated violet parking), smart city management, government, manufacturing, public transportation and healthcare. A main application of IoT in the automative industry is the self-autonomous vehicles (SAV). An SAV is expected to be driverless and be able to book parking place, pay for it and park. It then drives to the owner when needed following an instruction over public land mobile network (PLMN) or satellite. 5G is expected to provide the PLMN services to connect SAVs, carparks, roads, etc. An SAV will be delivered from factory with embedded IoT devices. 5G will strengthen the vehicle-to-everything (V2X) communication service provided over the LTE (Long-Term Evolution) interface. Foprecasted advantages in SAV include:

- Reduction in road congestion.
- Reduction in road accidents by reducing accidents caused by drunkenness and human error.
- Reduction in fuel consumption to reduce the carbon footprint.

Another important 5G vertical is smart agriculture, which is meant to improve food security, efficiency in agricultural processes, as well as increasing crop yield. According to the Alliance for IoT Innovation (AIOTI) smart agriculture embodies many things, including:

(1) Precise farming: acquisition of real-time data on conditions of soil, air and crops.
(2) Precision livestock farming: real-time monitoring of health, welfare and productions of livestock in order to achieve optimum yield.
(3) Smart irrigation: keeps an eye on air humidity, soil moisture, light intensity and temperature in order to maintain the precisely-required water.

1.3.4 *Transducers, Sensors, Actuators and Microcontrollers*

A **transducer** is a device which changes one physical quantity in one energy form at its input into a physical quantity in another form of energy at its output. The energy can be any of the following: acoustic energy, electrical energy, mechanical energy, electromagnetic energy, chemical energy, thermal energy and light energy. Transducers can be passive or active. Unlike passive transducers (e.g., photocell and potentiometer), active transducers do not need an external power source. Examples of active transducers are piezoelectric transducers, moving coil generator and thermocouple. Transducers can also be digital or analogue. The output of an analogue transducer is analogue, regardless of the nature of its input. Examples are thermistors. A digital transducer, however, produces digital outputs. Based on the nature of their input, transducers can be categorised into heat, stress, strain, inductive, resistive and capacitive transducers. Examples are digital encodes. The output of transducers is in one of the forms: electrical (i.e., current and voltage), displacement, sound, force, pressure or magnetic flux. Examples of transducers are:

(1) *Solar panels*: convert light energy into electrical energy.
(2) *Microphones*: convert sound energy into electrical energy.
(3) *Speakers*: convert electrical energy into sound energy.
(4) *Thermocouple*: converts thermal/heat energy into electrical energy.
(5) *Generator*: converts kinetic/motion energy into electrical energy.
(6) *Piezoelectric transducer*: converts mechanical energy into electrical energy.
(7) *Pressure transducer*: e.g., piezometers, pressure indicators, manometers and piezometers convert pressure into electrical energy.

1.3.4.1 *Sensors*

A **sensor** is a transducer whose output is an electrical signal or energy. In general, a sensor converts a physical parameter or signal into an electrical parameter or signal. There are many types of sensors, depending on the nature of their input. Some examples of sensors are shown in Fig. 1.19. As stated above, solar panels, microphones, generators, piezoelectric transducers and thermocouples are sensors. The basic components in a sensor include radio module, power management module, sensing or data collection module and energy management module. Sensors are the interface between the IoT and the physical world of physical devices.

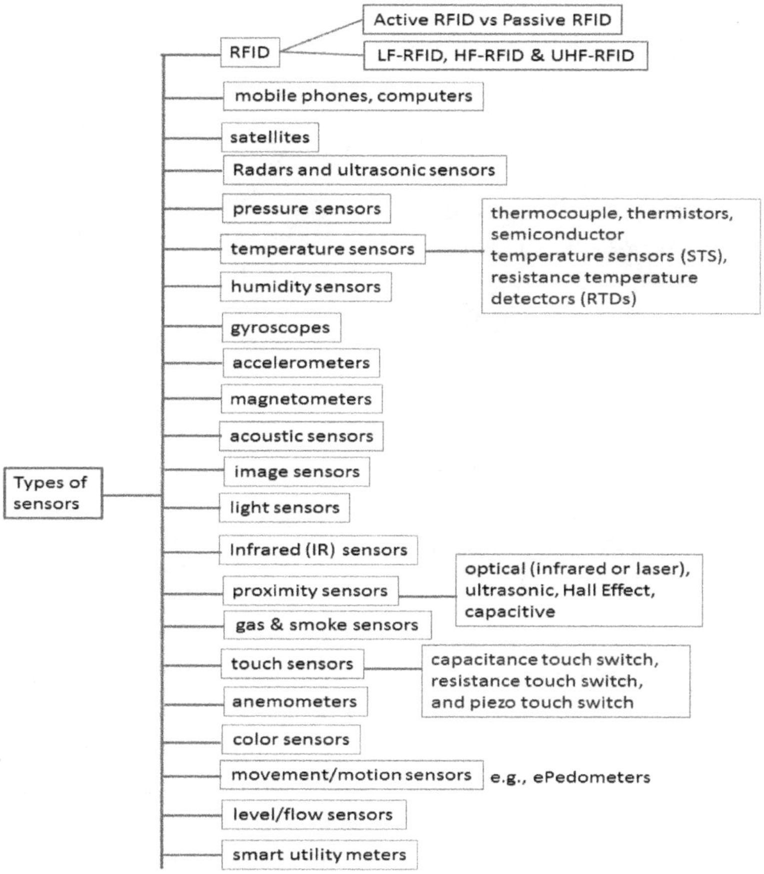

Fig. 1.19 Examples of sensors used for data collecting and data exchange in IoT.

1.3.4.2 *Actuators*

An **actuator** is a transducer which converts an input signal or energy into a physical output or action. Examples of the actions performed by actuators are dimming of lights, closing of heating valves and display of information. There are many types of actuators. For example, motion actuators convert an energy fed into them into motion. There are also many types of motion-generating actuators. Hydraulic actuators generate motion from input liquid. Thermal actuators generate motion from heat. Electrical actuators generate motion from an input electrical power, either from mains power supply or from a battery.

1.3.4.3 *Microcontrollers*

A **microcontroller** is a device containing a microprocessor, memory and programmable input/output peripherals and implemented on a single integrated circuit, called chip. The inclusion of the microprocessor makes a microcontroller a low-cost, small computer on a chip. There are several types of microcontrollers, depending on their bus width (e.g., 16-bit and 32-bit bus), type of memory structure (i.e., use of external or embedded memory) and type of instruction set (i.e., RISC-Reduced Instruction Set Computer or CISC-Complex Instruction Set Computer) . A cable interconnecting the internal components of a microcontroller or a computer is called bus. Most popular vendors of microcontrollers include Altera, Atmel, Xilinx and Intel. Transducers (both sensors and actuators), microcontrollers and single-board computers (SBC) are important enablers of the IoT. Two of the most popular SBCs in IoT are:

- Arduino Uno (www.arduino.cc) is a microcontroller motherboard. It thus allows running of just one program at a time. It is easy and simple to use, making it a preferential choice for beginners in IoT. However, it is not suited for complicated projects.

- Raspberry Pi (www.raspberrypi.org) is a general-purpose computer (GPC) and so it allows the programming in many languages, including Python, C++, HTML and Java. As a GPC, it can run multiple programs simultaneously. However, it has a learning curve compared with Arduino.

1.3.4.4 *RFID Systems*

Three types of sensors are used for automatic tracking and identification of objects. These are radio frequency identification (RFID) system, barcodes and magnetic strips. As illustrated in Fig. 1.20(a), an RFID system comprises:

- RFID tag: a smart electronic chip with a 96-bit number called serialised global trade identification numbers (SGTIN). The SGTIN can also be encoded in 24 hexadecimal characters. The composition of the SGTIN is illustrated in 1.20(c). It is the part of the RFID system which is attached to the object to be identified, monitored and/or tracked. The SGTIN is used to identify the objects.
- RFID reader (aka interrogator): a chip with inbuilt antennas and a radio transceiver. It interrogates the tag, captures data in the tag and sends it to a computer in the RFID system.

Both the tag and the reader contains antennas to enable wireless communications. Objects with these tags are tracked and identified using electromagnetic waves through the reader. The objects being tracked using RFID technology can be books (in library), animals (in livestock tracking), products in a supermarket, etc.

Products must be uniquely identified using barcodes (also bar code) to enable retailers to track sales of products from different vendors within their inventory system. The barcode enables the unique identification of a specific product type, in a specific packaging configuration and from a specific manufacturer. Barcodes are also referred to as article number. The two most popular barcode systems are:

- Universal Product Code (UPC): usually 12 decimal digits in length. It is mostly used in the USA.
- EAN-13: the most popular European Article Number (EAN) standard. It was developed as a superset of UPC by the addition of an extra decimal digit to the beginning to increase its scope of numbering. The EAN-13 has become a global standard.

An average barcode scanner can read both the UPC and EAN-13 codes. Both UPC and EAN-13 are used in Australia. The barcodes are obtained by product manufacturers who attach them to their products before delivering them to retailers. A retailer has to associate the barcode to the product (by product name, barcode number, price and reordering info) in its retailing software application. The standard size of an EAN-13 barcode image is 37.3 mm wide $\times$ 25.9 mm high. Although a barcode is unique globally, it does not indicate the origin of a product. It reveals only the origin of the barcode itself. EAN and UPC barcodes are used to represent the serialised global trade identification numbers (SGTIN) on trade items so they can be scanned at Electronic Funds Transfer at Point-of-Sale (EFTPOS) and distribution centers globally. There is a one-to-one conversion between UPC/EAN-13 and SGTIN.

RFIDs systems can be classified into passive RFID and active RFID. The tag used in an active RFID system contains battery which supplies it with energy. Active tags operate at 433 MHz and 915 MHz. Active tags have much longer range (up to about 100 meters) than passive tags. A popular use of active RFID systems is pay toll booths which enable a car with an active tag to drive through without the need to stop. Passive RFID systems use tags with no inherent power supply. They receive power from the EM wave beamed unto them from the reader through *electromagnetic coupling*. Passive RFID systems transfer electrical energy from the reader to the tag through either *inductive coupling* or *capacitive coupling*. Passive tags operate in the three frequency bands: 125-134 kHz (low

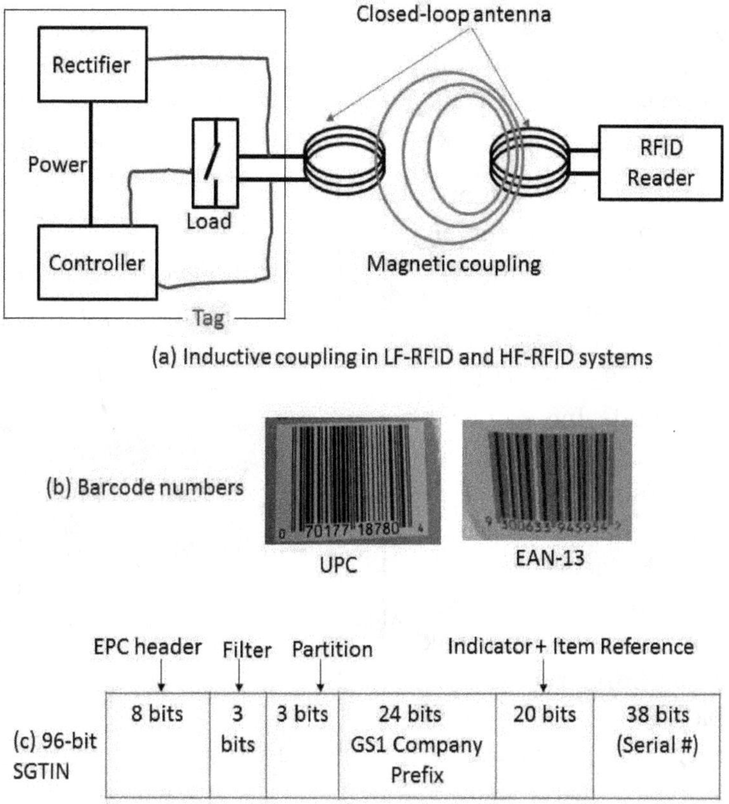

Fig. 1.20 Illustration of the operation of passive LF and HF RFID systems.

frequency or LF), 13.56 MHz (high frequency or HF) and 865-960 MHz (ultra-high frequency or UHF). Passive LF-RFID and HF-RFID tags obtain power supply through *inductive coupling* (aka near-field coupling). Passive UHF-RFID tags, however, obtain power supply through electromagnetic coupling (aka far-field coupling). The modulation scheme used by UHF-RFID is called *backscatter modulation*, which is ASK or PSK in the reverse link and DSB-ASK, SSB-ASK or phase-reversal ASK (PR-ASK) with pulse interval encoding (PIE) in the forward link. The modulation scheme used by LF-RFID and HF-RFID is called *load modulation*, which is a modified ASK modulation. Thus, the EM wave transmitted from the reader to a passive tag serves the trio purposes:

- Acts as the carrier of the data returned by the passive tag when queried.
- Induces enough electric power into the tag for it to be able to respond to the interrogation.
- Synchronises the clock of the tag to that of the reader.

RFID systems can also be classified into low-frequency RFIDs (LF-RFID), high-frequency RFIDs (HF-RFID) and ultra-high-frequency RFIDs (UHF-RFID), as indicated in Fig. 1.19. LF-RFID systems operate in the frequency range 125-134 kHz with a transmission range of 1-10 cm. A popular application of LF-RFID system is livestock tracking owing to its long-frequency signal which easily penetrates objects. HF-RFID systems are popularly used for passport control at airports. They operate at 13.56 MHz with the transmission range of about 1 cm to 1 meter. Lastly, the UHF-RFID systems operate in the frequency band 865-960 MHz with average transmission range of about 5 to 15 meters. In good propagation conditions, a signal generated by a large UHF-RFID system can cover more than 30 meters. Sample application areas of UHF-RFID systems are asset tracking and race timing.

1.3.5 *Case Study: Coca Cola's IoT-Enabled Beverage-Vending Machine Freestyle*

The beverage-vending giant Coca Cola has developed a machine called *Freestyle* to provide hundreds of drink combinations just at the touch of a soft button. The development of Freestyle started in 2005, and was launched in 2009. The Freestyle machine exploits radio frequency identification (RFID) technology, built by Impinj, Inc., a USA company headquartered in Seattle, Washington. Impinj manufactures RAIN RFID systems. RAIN (meaning RAdio frequency IdentificatioN) is a global alliance which is promoting the penetration of UHF-RFID systems. The aim of RAIN is to develop a link between cloud computing infrastructure and UHF-RFID, to enable data collected from RFID tags to be managed, stored and shared through the Internet. RAIN RFID technology enables the identification, locationing, authentication and engagement of objects attached to the Internet of Things.

RAIN RFID systems are based on the GS1 UHF Gen2 protocol, which has been standardised by IEC/ISO as 18000-63. The RFID built inside each Freestyle vending machine uses Impinj Monza tag and Impinj Indy SiP readers. Freestyle has a touchscreen display which enables consumers to effortlessly customise the drink of their choice. Advantages in Freestyle include:

(1) Simplified operating procedures for employees who service the vending machines.
(2) Enhanced customer experience, as consumers have more choices of beverages.
(3) Automated monitoring and tracking of dispenser consumables. This is performed remotely because the Freestyle machines are connected to Coca Cola manufacturing plant over both wired and wireless media.
(4) Real-time business analytics about product. For example, Coca Cola is able to locate cartridges getting empty and replace them in time by analysing the data collected and transferred by the RFIDs over the Internet to the manufacturing plant.

Pepsi developed its beverage-vending machine called *Spire* to compete with Coca Cola's Freestyle.

1.3.6 *Check Your Understanding*

Students are encouraged to check their understanding of the content of this section with the following questions prior to proceeding to the next section.

(1) **In-class group study**: In groups of 3 to 5 students research to find the relevance of artificial intelligence and machine learning in IoT.
(2) Research to find the meaning of *5G verticals.*
(3) Assume that the pathloss exponent in the coverage area of a mobile wireless network is about 2.4. Using the formula in Eq. (1.1), find how much that NB-IoT coverage exceeds that of the 2G wireless technology GSM.
(4) **In-class group study**: In groups of 3 to 5 students research to find the differences and similarities between sensors, actuators and transducers. Give at least one practical example of each.
(5) Explain the differences in the operation of LF-RFID/HF-RFID system and UHF-RFID system.
(6) Discuss the three main functions provided by the electromagnetic waves used by the reader to interrogate a passive RFID tag.
(7) Discuss the importance of sensors, actuators and transducers in Internet of Things.

(8) Analyse transducers, sensors and actuators. State the differences between these three devices.
(9) **In-class group study**: In groups of 3 to 5 students research to find the role of artificial intelligence in IoT.
(10) Research to find two single-board computers other than Arduino Uno and Raspberry Pi which are popularly used by IoT devices.
(11) **In-class collaborative learning**: In groups of 3 to 5 students research to find the differences between LoRaWAN, LTE-M and NB-IoT. Include in your comparison their prices, operating frequencies, bandwidth or data rate, repeater-less maximum link transmission range, popularity, major vendors of their products and organisations developing each of them.
(12) Discuss the meaning of the phrase *the five-tens in LPWA* technology.
(13) Study the sensors listed in Fig. 1.19. List those that you are familiar with. Describe how each of the sensors you have listed operate.
(14) State the name of the machine that Pepsi developed to compete with Coca Cola's Freestyle.
(15) The 12-decimal-digit number 0 70177 18780 5 is the universal product code (UPC) found in the barcode of a product. Compute the equivalent serialised global trade ID number (SGTIN).

1.4 Software Packages for Wireless IoT Systems Design

There are a plethora of software packages used for the design, modeling, analysis and prototyping of wireless IoT systems. We survey briefly in this section only three of them, namely CelPlanner™, MATLAB® and R®. CelPlanner™ and MATLAB® are used in some parts of the book.

1.4.1 *CelPlanner™*

CelPlanner™ is a proprietary suite of software developed and distributed by CelPlan Wireless Global Solutions, Inc. with headquarters in Reston Virginia, USA. This company has been in operation since 1992. CelPlanner™ is a software used to plan, design, model, analyse and optimise wireless communication systems. The main features of CelPlanner™ are:

- Integrated Geographic Information System (GIS) platform.
- Simultaneous multi-resolution 3D models for topography and morphology.
- Morphology height defined in pixels.
- Multiple image layers.
- Regions and sub-regions with attributes.
- Multi-layered traffic information.

We characterise radio wave propagation mathematically using radio propagation models. Some of the radio propagation models implemented in CelPlanner™ are Lee, Okumura-Hata, European COST 231, five-intercept microcell model, the Stanford University Interim model, line-of-sight (LOS), TCA and ITU-R P.1812/1546. CelPlanner™ has support for many wireless technologies, including 4G/LTE, LTE-A and 5G. The CelPlanner™ includes:

(1) **CellTrace™**: An indoor prediction tool based on ray tracing is a set of software tools used for Radio Frequency (RF) planning, design, modeling, analysis and optimisation of wireless communication systems in indoor environments, such as inside buildings, stadiums, tunnels and in campus environments.
(2) **CellDesigner™**: Software uses to plan and design Wireless Communication Systems. It can be used to design networks based on many wireless standards, including LTE, LTE-A and 5G.
(3) **CellSpectrum™**: A software platform for the analysis of radio frequency spectrum and channel with respect to interference.
(4) **Geographic Information System (GIS) Databases**: The design of wireless systems require data on morphology and topographic

of the coverage area. CelPlanner™ contains a complete portfolio of databases containing these geographic data. The GIS databases contain data on morphology, topography and maps. The scale used for the first two are 1 ArcSec ($\approx$ 30 m), 3 ArcSec ($\approx$ 90 m)) and 6 ArcSec ($\approx$ 180 m). The resolution of the images are 1:24000, 1:100,000 and 1:250,000.

More information about CelPlanner™ can be found at http://www.celplan.com/. Sample CelPlanner™ laboratories are provided at the Appendix to be used as a jump-start.

1.4.2 *MATLAB®*

MATLAB® is an abbreviation for MATrix LABoratory. Therefore, all parameters used in MATLAB® can be formulated as matrices and vectors, as done in linear algebra. Written in Java, C and C++, MATLAB® is a proprietary software developed for numerical computing. It is maintained and distributed by MathWorks, Inc., which was founded by John N. Little, Cleve Moler, Steven Bangert and headquartered in Natick, Massachusetts in USA. The design of MATLAB® is attributed to Cleve Moler. The software comprises the MATLAB® backbone and accompanied software package called Simulink®. There are packages, referred to as toolboxes, for each application, besides the main backbone. One needs to acquire only the base and the packages needed for one work. All MATLAB® toolboxes are divided into the three product families:

- MATLAB® Product Family.
- Simulink® Product Family.
- Polyspace® Product Family.

More information about these products can be found at www.mathworks.com.

MATLAB® combines a desktop environment tuned for iterative analysis and design processes with a programming language that expresses matrix and array mathematics directly. It includes the Live Editor for creating scripts that combine code, output and formatted text in an executable notebook.

It is pretty easy to learn how to program in MATLAB®. This is one of the reasons why almost every curriculum in Electrical and Electronic Engineering uses it for laboratories, especially courses on signal processing and digital communications. The only problem that I personally find with MATLAB® is its costs. While it is cheap for students, it is relatively expensive for professionals especially if one needs many of its packages. Fortunately, many universities acquire them for academic use by both students and lecturers. Another interesting feature of MATLAB® is that many people build packages for it and make them freely available on the Internet. The third advantage in MATLAB® is that it has a wide or even a global industry support. Therefore, knowing it enhances one's employability.

Simulink® is used for model building and simulation. Simulink is a graphical programming environment which is used to model, simulate and analyse multi-domain dynamical systems. It has a library of graphical block diagramming tools which its users can customise to suit the system under design.

Polyspace® Product Family combines checkers for coding rules, security vulnerabilities, code metrics and hundreds of additional classes of bugs to prevent critical run-time errors under all possible control flows and data flows.

1.4.3 *R Package*

R is a programming language and environment used for statistical computing and graphics. The package was designed by Ross Ihaka and Robert Gentleman. The designers of R were influenced by a high-level general-purpose dynamic programming language called Julia. R is available as Free Software under the terms of the Free Software Foundation's GNU General Public License in source code form. Therefore, R is wholly free to use for all purposes. This is a very important feature compared with commercial software such as MATLAB. Also, R can be used to do pretty much everything that MATLAB does. R is considered as a different implementation of a programming language called S, which was developed at Bell Laboratories (formerly AT&T, now Lucent Technologies) by a team of people led by John Chambers. More details of R can be found at https://www.r-project.org.

R was initially used mainly by statisticians. However, it is an environment enabling statistical computing as well as other purposes. Like MATLAB, users of R can build new functions to extend the functionality of the base R. For this reason, many people have built customised functions and made them freely available on the Internet for public use. This is the reason why the applications of R has been growing exponentially. The popularity of R has received a boost in recent years because of its suitability for machine learning, artificial intelligence and data analytics. We continue with the discussion of IoT fundamental topics in the next chapter.

1.5 Review of Important Terms and Acronyms

Acronyms, abbreviations and important terms		
Web x.0 (x=1, 2, 3)	Semantic Web	Read-Write-Execute Web
podcast	Wikis	blog
AI	LPWA	Arduino Uno
ARPANET	artificial intelligence	BITNET
disruptive innovation	disruptive technology	LTE-M
HTML	Internet	internet
Internet of Things	internetizen	IoT
IoT enablers	IoT standards	Smart City
machine learning	eUICC	MATLAB
microcontroller	eSIM	NB-IoT
Netscape	M2M	Raspberry Pi
Consumer IoT	sensor	SNR/SINAD
wireless communications	wireless connectivity	WWW
Industrial IoI	the five-tens of LPWA	smart bins
RFID	NFC	BLE
IETF	ETSI	IEEE
3GPP	SigFox	LP-WAN/6LPWA
transducer	actuator	DASH7

1.6 Supplementary Reading

(1) David Hanes, *IoT Fundamentals: Networking Technologies, Protocols, and Use Cases for the Internet of Things*, Cisco Press, 2017.

(2) Jan Holler and Vlasios Tsiatsis, *Internet of Things*, Academic Press, 2014.

(3) Frank Gustrau, *RF and Microwave Engineering: Fundamentals of Wireless Communications*, John Wiley & Sons Inc., 2012.

(4) Clarence W. de Silva, *Sensors and Actuators: Engineering System Instrumentation*, CRC Press, 2015.

(5) Charles R. Severance and Sue Blumenberg, *Introduction to Networking: How the Internet Works*, CreateSpace Independent Publishing Platform, 2015.

(6) Harvey M. Deitel and Paul J. Deitel, *Wireless Internet and Mobile Business How to Program*, Standards Information Network, 2018.

(7) Simone Cirani and Gianluigi Ferrari, *Internet of Things: Architectures, Protocols and Standards*, Wiley, 2018.

(8) Harvey M. Deitel and Paul J. Deitel, *The Complete Wireless Internet and Mobile Business Programming Training Course*, Prentice Hall PTR, 2002.

(9) Devi Kamal, *Mobile Computing*, Oxford University Press, 2012.

Chapter 2

Wireless IoT Fundamentals

Chapter's Learning Outcomes

IoT is the disruptive technology of the 21st century. As wireless IoT impacts almost every aspect of social and business lives, it relates to many other technologies. This chapter studies some fundamentals of the Internet, as well as technologies relating to the IoT. In particular, we study:[1]

(1) Internet fundamental principles, including packet switching technique and the Internet Protocols (viz. IPv4 and IPv6).
(2) Nascent technologies relating to IoT, including cloudification of network resources and services, network function virtualisation (NFV), software-defined networking (SDN), orchestration, slicing, big data capability of IoT, applications of machine learning in IoT, network tactility, virtual reality and augmented reality.
(3) IP subnetting in network design.
(4) Wide area networking (WAN) technologies competing with the Internet-based WAN, including MPLS and software-defined WAN (SD-WAN).
(5) Basic electrical networks which support the wireless connectivity required of wireless IoT.

2.1 Nascent Technologies Relating to IoT

This chapter discusses the desired features of next-generation networks (NGNs), such as IoT and 5G. Some of the modern trends in communications networks are cloud computing and network softwarisation in which network functions are increasingly implemented in software instead of hardware. Aspects of network softwarisation are software-defined networking (SDN) and network function virtualisation (NFV). We review some desired features of next-generation networks (NGNs), such as 5G and 6G. Many of the features are interrelated or even similar. As in current 4G networks, these networks comprise mobile telecommunication networks and fixed networks. The term WWWW or W^4 has been coined to describe 5G networks because of the fact that they are expected to possess features such as coverage, capacity and latency that are much better than those of their predecessor networks.

Wireless communications, including IP-enabled wireless networking, has gone through several developments, including both revolutions (paradigm shifts) and evolutions, since its beginnings in the 18th century. One of the modern trends is the desire to implement as much as possible the system's functionalities in software to make it easily reconfigurable or programmable. This paradigm shift is referred to generically as software XX (i.e. SXX), where S means software and XX can mean defined networking as in software-defined networking or defined radio as in software-defined radio communications. Another trend related to SXX is network function virtualisation (NFV), which separates the control plane from the data plane and installs only the latter on individual telecommunications devices while the former function is centralised.

Over time developments in telecommunications proceed at a faster pace. The *Internet* is now the backbone or core network of all communications

[1]The discerning heart seeks knowledge, but the mouth of a fool feeds on folly.

networks. One of the current trends is the *Internet of Things* (IoT), also called Internet of Everything (IoE), which seeks to enable communications between everything electronic. When this paradigm is fully realised then all consumer electronic devices, such as TV sets, refrigerators, toasters and microwave ovens shall have Internet Protocol (IP) addresses and thus attach to the Internet. A development in IoT is the Internet of Underwater Things (iOUT), which seeks to enable communication between underwater bodies, including submarines, scuba divers, aquatic robots and undersea sensors. NATO (North Atlantic Treaty Organization) has standardised the first international protocol, referred to as Janus, to speed up the realisation of iOUT [Nordrum (2017)]. Janus achieves a maximum of 10 km transmission rate at 11.5 kHz using FH-BFSK modulation.

All the networks mentioned above together with the legacy PSTN are being replaced by the multi-service next-generation network (NGN), as illustrated in Fig. 2.1. **What is NGN?** The ITU-T broadly defines NGNs as the migration from circuit-switched networks to packet-switched networks using the Internet Protocol as the data transport technology. In the NGNs rich communications services (RCS) replaces SMS (simple messaging service), for example. RCS has more features than just text messaging. We discuss some of the important current trends in the succeeding sections.

2.1.1 *QoS and QoE*

Modern and future networks collect, process, store and transfer all types of traffic, such as audio or music, voice-over-IP (VoIP), video (both still and moving), text, images, virtual reality, augmented reality and machine-to-machine (M2M) or vehicle-to-vehicle (V2V) communication. These traffic differ in nature and thus have varying demands on network performance, such as data rates, throughput, packet loss rate, delay (or latency) and jitter (i.e. delay variance). QoS (quality-of-service) or QoE (quality-of-experience) depends on network reliability and network performance. Appropriate network slicing must be used to provide the infrastructure required to meet the required QoS of specific applications or traffic.

2.1.2 *Sustainability Through Energy Conservation*

The objective is to reduce the energy consumed by the network infrastructure while guaranteeing the expected QoS for each VNO (virtual network operator) or end user through energy budget for the virtual infrastructure. To this effect we need to monitor the energy parameters, such as CPU loads and power consumption, of radio access networks (RANs), backhaul elements, fronthaul elements and virtual network functions that are used to support internal processes. Strategies used to

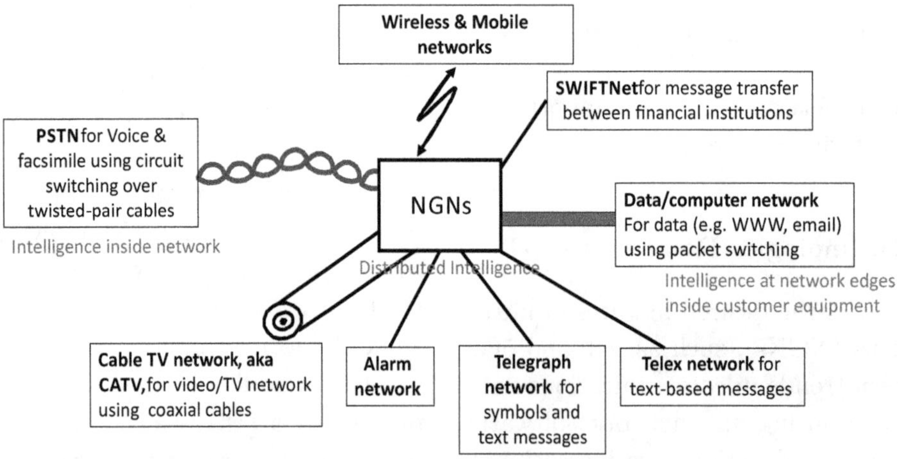

Fig. 2.1 Convergence of all networks into NGN.

conserve energy include implementing virtual network functions on less-loaded servers or less power-hungry servers.

Efficient energy-conserving method should consider all the three issues:

(1) minimise operating costs of networks
(2) maximise network performance, and
(3) reduce harmful emissions from network infrastructure.

Traditional energy-conserving methods have been hardware based and thus focused on only the third issue. Future networks (e.g. 5G) will use software-based solutions to consider all the three issues above. Further, software-based solutions align with the modern trend of softwarisation of networks. As in SDN the data plane of devices are decoupled from the control plane, software-based energy conservation is managed at the control plane.

2.1.3 *Network Scalability and Network Agility*

Traditionally, a communication network is implemented to meet anticipated peak loads through overprovisioning, in which the network infrastructure installed is more than what is needed to support normal traffic load. This practice is wasteful as resource utilisation is not optimised. Further, regardless of being overprovisioned, a network may not be able to support all types of peak loads. With the large size and complexity of future networks overprovisioning becomes even more expensive to support. A better choice is an elastic network which scales to meet peak demands. The result is the birth of scalable networks.

Scalability can be easily understood from an everyday life example. A pair of trousers (aka pants) with elastic waistband adjusts to suit the waist size of the individual wearing it. Similarly, a computer system or facility which is scalable can process data with increasing rate as its CPU power is boosted. It can also store and process a larger amount of data as its memory and hard disk drive are increased. Scalability is one of the most important features of computer systems supporting Big Data processing and storage.

Scalability can be horizontal or vertical. Horizontal scalability increases or decreases the capability of a network by adding or removing nodes/computers or servers to the network. Adding more computers to the system is called scale out or out-scaling, which is the basis of cluster computing. Removing computers from the network is called scale in. Vertical scalability can be upscaling or downscaling. The former adds resources to the same number of computers to boost the overall network capability. The resources added are memory, processors (CPU) or hard disk pace. In downscaling we remove resources.

Almost everything is liable to change. The amount of data passing through a computer network grows by the second. Furthermore, the number and types of applications that the network has to support change even faster than the rate of change of the user population. Thus, one important feature of today's, and especially future, communications networks is the ability to expand or scale up to accommodate growth or scale down to absorb reduced demand in any dimension and function as expected. A network possessing such a virtue is referred to as a *scalable network*. The ability of a communications network to cope with a growing demand on it to accommodate growth is referred to as network scalability.

An *agile network* is a responsive network in that it is able to change and adapt to the current needs of businesses in a quick manner. Thus, agility is a measure of how quick a network scales. Agility requires quick deployment or decommission of infrastructure, applications and services. Such a network remains usable and resourceful. The proliferation of software-defined networking (SDN), cloud computing and associated technologies such as network function virtualisation (NFV) have improved and will continue to improve the agility of networks. This means that an agile network is a scalable network. However, the opposite is not necessarily true.

2.1.4 *Network Tactility*

Latency is the time it takes to send a single packet of data between two points in a network. Fourth-generation mobile cellular communication networks (aka 4G) have a latency of 25 milliseconds (ms). The targeted latency in next-generation networks, referred to as fifth generation (5G) systems, is at most 1 ms. In order to achieve this latency target, next-generation networks must deliver data transmission rates hundred times those of 4G networks. Low latency networks are crucial for certain application areas, such as robot-aided surgeries and unmanned automated vehicles.

A professor of the Dresden Technical University in Germany by name Gerhard Fettweis coined the term **tactile Internet** in 2012 to describe a network that enables cordless control of both virtual and real objects and processes in a perceived *real time* [Pretz (2017)]. The real-time or virtual reality aspect requires very low-latency data transmission. The 5G use case called Ultra-Low Latency Reliable Communications (uLLRC) is poised to support this low-latency service. The average visual response time and the average auditory response time of humans are 10 milliseconds and 100 milliseconds, respectively. Thus, the average ear does not detect any delay if a sound takes less than 100 ms from its source to reach the ear. The same is true of images taken less than 10 ms to reach the eyes. Thus, a network with 1 ms latency can create a virtual reality that enables surgeons to perform using robots or assist other doctors to perform operations over distance. Virtual reality also enables online gaming and self-driving automobiles.

VM_1 (OS_1)	VM_2 (OS_2)	...	VM_n (OS_n)
Hypervisor (Virtualization)			
Pool of RAM	Pool of Storage	Pool of Servers	Pool of CPU

Fig. 2.2 Virtualisation of computing resources.

2.1.5 *Virtualisation and Hypervisor*

Something which is not actual but a simulated or computer-generated version of the actual is referred to as *virtual*. The act of creating a virtual instead of the actual or physical version of a computing resource is called virtualisation. Through virtualisation we can run multiple operating systems on a single hardware as illustrated in Fig. 2.2. The original operating system is called host OS, while those on the virtual machines are called guest OS. Resources that can be virtualised in computing include storage devices, server and computer network resources. In virtualisation users interact with an abstract instead of the real computing resource. Virtualisation allows us to run the unix operating system on a Windows computer, for example. An emulation (or virtualisation) of a computer system is called virtual machine. Virtual machines are created and managed by software called *hypervisor*, also called *virtual machine monitor* (VMM). There are two types of hypervisors. Type 1 is installed on a real hardware and thus runs as a real operating system. Examples of these are Hyper-V and Oracle OVM for SPARC. Type 2 hypervisors are also called hosted hypervisor. It is installed just as an ordinary application software on a host existing operating system. Examples of this type are Solaris Zones, Oracle Virtual Machine Virtualbox, Oracle VM for x86, VMware Workstation and Parallels.

Virtualisation is used to reduce costs and increase the utilisation of network infrastructure. For example, we can use a virtualisation software to divide a single physical server into multiple smaller servers called virtual servers. Each virtual server runs its own operating system and appears to users as a physical server. Popular virtualisation software include RedHat SPICE, Microsoft Hyper-V, Amazon Elastic Compute Cloud (EC2), Oracle (Solaris Zones, LDoms and xVM), VMware (vSphere and VMware Server), Huawei's FusionSphere, and Citrix (Xen and XenServer).

There are three types of virtualisation technologies: abstraction, partitioning and pooling. In each case a set of physical components are combined to form a single logical or virtual entity. Figure 2.4(c) illustrates the virtualisation of storage devices, servers, central processing units (CPUs) and random-access memories (RAMs). In partitioning technology the physical infrastructure is partitioned into independent logical entities which maintain the features of the hosting physical device to improve resource utilisation. In the abstraction technology, the logical infrastructure does not inherit the features of its hosting physical infrastructure but maintains the features of the emulated devices in order to simplify operation efficiency. The pooling technology is the logical entity which maintains the characteristics of the original physical computing devices. The hypervisor (or virtualisation) in Fig. 2.4(c) is the heart of the virtualisation. It hides from the virtual machines (VMs) the fact that they are sharing a single physical resource. Thus, each VM can be thought of as an independent physical entity. An interesting feature in virtualisation is scalability. The VMs can be deleted or more added on-demand. With the virtualisation we can run multiple operating systems on a single hard drive without the need to partition it.

Virtualisation, coupled with cloudification and softwarisation (VCS), are exploited in every modern and future communications network, and IoT is not an exception. The demand for services is dynamically changing. Service providers must find strategies to quickly deploy services, as well as modifying or even decommissioning them without much costs and/or complexity. VCS enable on-demand provisioning of network services and functions. Also, VCS enables automation of networks which reduces OPEX owing to the reduced manpower.

2.1.6 *Software-Defined Networking*

Software-defined networking (SDN) is the realisation of softwarisation in networking, besides NFV. It combines automation and programmability to create agile networks which can be quickly reconfigured, scaled and/or customised on-demand. SDN allows the creation of network abstractions. NFV and SDN do not strictly need each other, but SDN is able to enhance the performance of NFV, ease the compatibility of NFV with legacy systems, and also facilitate the operation and maintenance procedures [Mullins and Barros (2017)].

SDN is a networking paradigm which focuses on traffic shaping real time by separating the network traffic into the three components: the purpose of the data, raw data, how the data is sent through the network. Some people and organisations (e.g. Cisco) refer to SDN as network programmability, although there is no consensus at this stage on one definition. In general, SDN separates the network into control plane or layer, data plane (or infrastructure layer) and services plane (or applications layer). The data plane contains the internetworking devices (e.g. hubs, switches and routers), the control plane contain the servers which manage the devices in the data plane, while the services plane contain functions like firewalls.

The heart of SDN is network programmability and operational agility. A popular protocol used to achieve network programmability or create SDNs is the OpenFlow which is supported by Cisco. SDN applications created by software developers can extract real-time intelligence from a network using data science techniques to develop appropriate policies. We then use tools, such as OpenFlow, to run the policies on the network elements (i.e. switches and routers) to control their routing tables remotely from a central location.

How does SDN differ from traditional networking paradigm? The functions of network devices occur in two planes: the control plane and the data plane, as said earlier. In the control plane device management occurs. Devices also learn how to handle incoming data by creating the network topology. They use static or dynamic routing protocols to collect information on paths to remote networks to build their packet forwarding tables. In the data plane devices forward (i.e. route or switch) PDUs (frames or packets) using the information they collect in the control plane. Other data plane functions include rule-based content filtering, and contention to access transmission links. In traditional networking paradigm both planes are implemented directly on the networking device. In SDN, however, the control plane functions are taken away from the physical devices and installed centrally on the SDN controller. What is the advantage in the SDN then? Benefits of SDN over traditional networking cited include:

(1) SDN creates a flexible network architecture which can change dynamically as the business requirements or an organisation change.
(2) We can allocate network resources in a better manner.

(3) By using an API, the central controller can implement network commands on devices from multiple vendors. This saves network administrators from the need to learn the command line syntax of multiple vendor products. This is the reason why some people have nicknamed SDN Cisco killer.
(4) Improved operational flexibility and enhanced transparency.
(5) Reduced long-term costs.
(6) Packet prioritisation for traffic shaping.
(7) Enables applications to maintain information about the capabilities of network elements.
(8) Enables networks to respond to application status and resource requirements.

2.1.7 *Network Function Virtualisation*

Network functions include routing, switching, firewalling, caching, encryption, Domain name Service (DNS), network address translation, packet filtering and (deep) packet inspection. Network resources include storage, networking and processing or computation. Under network virtual functions (NVFs) or network function virtualisation (NFV) network entities become network of virtual functions, referred to as *virtual machines* (VMs). In NVFs we decouple or isolate network functions from the:

- infrastructure upon which they run and the
- resources they need to run.

Each function is traditionally implemented in autonomous hardware. The problems in such an approach include difficulty, cost and time required to install new or upgrade existing networked functions. Also, a technician would need to physically visit the site where the network function must be implemented. NVF implements on-the-fly and on-demand desired network functions using a set of elementary (building) functions or blocks over the cloud. A single network infrastructure in the cloud can be shared by multiple organisations (called tenants or clients or customers) on demand by instantiating the multiple logical versions of the physical resources. This process is called *virtualisation.* Two of the advantages in NFV are reduced costs and quick deployment of new network functions. The main advantages in NFV are:

(1) **Reduced network infrastructure costs**: NFV replaces legacy hardware with software functions running over commoditized hardware to reduce costs on acquiring network infrastructure.
(2) **Network scalability**: network functions can be scaled up or down easily using cloud resources based on demand.
(3) **Network performance**: as demand on a network increases, its performance reduces unless the supporting infrastructure is scaled up. This scaling of infrastructure to support the needed performance is inherent in the NFV model.

2.1.8 *Network Slicing, Virtual EPC and NaaS*

In the network-as-a-service (NaaS) paradigm network products are replaced by network services. This is achieved through network slicing, which is a logical instantiation of a network on-demand to meet a needed service. A set of specific radio access technologies (RAT) and network functions are used to implement a specific business model or use case under network slicing.

Network slicing is a new technology which enables the creation of multiple virtual networks over a single physical network infrastructure. The virtual networks (aka network slices or logical networks) use separate user planes, control planes and management planes. Also, they can be configured according to desired quality of service parameters such as data transmission rates, latency, throughput and capacity.

A technology that can be used to achieve sliced networks is the virtual Evolved Packet Core (EPC) designed by Ericsson. Currently, most functions of a network are performed by dedicated hardware such as switches, routers, firewalls and load balancers. **Network function virtualisation (NFV)** enable us to take the network functions from these dedicated hardware devices and performed them on virtual machines under the control of a virtual machine monitor called **hypervisor**. The Ericsson virtual EPC enables the virtualisation of all the components of the Evolved Packet Core of the 4G LTE technology, including vMME, vSGSN, vPGW, vSGW (virtual serving gateway), vGGSN, vPCRF, vDPI, vProbe, vePDG and vTWAG. This enables the operation of a telecommunications service company without owning any physical infrastructure. Such organisation trading over the Ericsson virtual EPC are called Mobile Virtual Network Operator (MVNO).

2.1.9 *Big Data Analytics in IoT and AaaS*

It has been estimated that about 90% of today's data has been generated within just the last two years. The main driver of this trend is the exponential increase of the number of data generating sources, and the trend is not relenting anytime soon. Devices inconceivable a few months ago to be data sources are now so. Sensors are penetrating every industry, each of which generate enormous amounts of data. Examples of unconventional data sources are smart watches, fitness trackers and smart eyeglasses. This data deluge is referred to as Big Data.

The characteristics of Big Data is traditionally described by words beginning with the letter V. We describe eight of them here.

(1) **Volume:** the amount of data classified as big data is voluminous or gigantic. Big data handles datasets with sizes in the range of petabyte (PB). As a comparison, $1PB = 10^{15}$ `bytes` = 1000 terabytes.
(2) **Variety:** the datasets have a wide variety in many dimensions, such as the different sources of data (e.g. video surveillance, mobile phones, smart grid meters, space exploration, sensors, medical imaging, GPS, RFID, genomic sequencing), different structures and formats (e.g. structured, semi-structured, quasi-structured and unstructured), and different data types (e.g. voice, video, images, photos). Organisations collect and keep data about their prospects, competitors, business practices, business threats, strengths, opportunities, weaknesses, customers, suppliers and partners. Semi-structured and quasi-structured data is a mixture of both structured and unstructured data. Most data in the world is not structured.
(3) **Velocity:** new data is generated at high rates and so the amount of data generated grows fast too. Velocity also refers the rate at which data must be processed. If voluminous data is generated at a high rate, then it makes no sense storing it for later processing as storage capacity may not keep up. One way of solving this problem is stream processing which analyses data as it enters the system. Stream processing is therefore a realtime data processing.
(4) **Veracity:** much of the data generated from many sources are messy having questionable reliability, accuracy and authenticity. Anyone can generate any data and send it across the Internet and there is none to check the veracity of the data.
(5) **Value or viability:** there is wealth hidden inside the massive datasets just as minerals are hidden under the oceans and under the earth awaiting exploration. The task of data scientists is to create business value out of the massive datasets. Most owners of datasets think there may be some useful information hiding in it which may create business value.
(6) **Visualisation:** We seek to represent data in a manner that is easily understood by non-data scientists. Therefore, data visualisation has become a needed feature of data.
(7) **Variability:** Another feature of data which is very important for sentiment analysis is variability. This feature refers to the fact that the meaning of data depends on context.
(8) **Versatility:** Big Data is versatile because we analyse it to extract many things, including patterns, trends, and associations. It is also versatile because almost every industry can benefit from it.

Clearly, the number of Vs that can be used to describe Big Data seems inexhaustible. There is an argument that not all the Vs are peculiar to Big Data.

The big data generated in the IoT is now the most precious asset for organisations. The data contains actionable knowledge that we seek in order to improve business outcomes. The processes, methodologies and tools used to extract actionable knowledge from big data sets is called *big data analytics*. Thus, IoT is increasing the utility in analytics. Many organisations now have an enormous amount of data, and seek to find actionable knowledge from it. However, maintaining in-house analytic infrastructure and manpower may not be affordable. For this reason **Analytics-as-a-Service (AaaS)** has arisen. AaaS integrates two technologies, cloud computing and big data analytics, to extract actionable insights from enterprise data at reduced costs. One of the analytics areas of much current interest is *immersive analytics* (for more details see e.g. https://www.immersiveanalytics.com/).

ITU has identified artificial intelligence (including machine learning or ML) and 5G networks as fields of innovation required to realise smarter societies. It has therefore established the Focus Group on Machine Learning for Future Networks, including 5G (FG ML5G). The terms of reference of this group includes:

(1) Studying use cases.
(2) Developing protocols and algorithms.
(3) Creating ML-aware network architecture.
(4) Developing data formats.

2.1.10 *Basic Virtual Network Infrastructure*

We can create logical divisions of a single physical network into multiple logical divisions each of which has its independent policies such as on security, routing policies and quality-of-service. Each of the logical divisions is called a virtual network. A virtual network can be used to provision the most suitable and efficient networking topology for the applications it hosts. We can change its topology on-demand using software rather than requiring hardware rewiring. A virtual network simulates the features of a physical network. However, its components are not physical but virtual. For example, the virtual machine (VM) takes the place of the physical computer and is thus interconnected by virtual switches (vSwitch) rather than physical switches.

The infrastructure in cloud computing is made of the six basic components: servers, storage devices, the network interconnecting the cloud services using the Internet as backbone, hypervisor to manage the virtual machines, cloud deployment software for applications deployment and integration, and cloud management software which is used to configure and maintain the infrastructure.

Scalability: Organisations subscribing to cloud computing scan increase (i.e scale up) or reduce (i.e. scale down) their computing resources quickly according to their business demands. The speedy commissioning or decommissioning of computing resources based on demand is referred to as *agility*.

2.1.11 *Cloud Computing*

Traditionally, we use cloud to represent a network in diagrams. This terminology has been inherited by cloud computing. Thus, the *cloud* in the name refers to a network including the Internet which is a network of networks. What then is *cloud computing*? Cloud computing is a form of *utility computing* in which computational resources are offered to clients as a metered service just like electricity, gas and water. It has been enabled by the Internet. The art or ability to create, access, manipulate, manage or configure software applications over the Internet from any location is what has been christened cloud computing. Popular cloud computing platforms include Amazon Web Services and Microsoft Azure. Cloud computing is the electronic form of employment agencies which keep a pool of prospective employees for organisations to hire and dump on demand. They do so for similar reasons behind cloud computing.

There are nine components in cloud computing, as illustrated in Fig. 2.4(c). These components can be organised into a seven-layer stack as done in the ISO/OSI layering. An interconnected pool of servers storing, processing and distributing a large amount of data is called *data center*. Data centers are a centerpiece in cloud computing. The length of time that a program takes to run is referred to as *runtime*. A special software which serves as a bridge between applications and the operating system or database is referred to as a *middleware*. Middlewares provide services unavailable in operating systems to applications. Examples of middlewares are IBM WebSphere, JBoss Enterprise Application Platform (EAP) of Red Hat and Oracle WebLogic.

Topics covered in this section include the CCNA v3 Topic 1.3 *Describe the effects of cloud resources on enterprise network architecture.* The world of computing is moving away from the traditional hosting to cloud hosting. In this model companies focus on services and leave computing resource acquisition and management to a third-party company. In cloud computing organisations and individuals buy or lease on-demand software applications, computer processing capacity and data storage from a third-party company over the Internet.

2.1.11.1 *Deployment Models of Cloud Computing*

There are different types of cloud computing deployment models (see Fig. 2.3). The basic deployment models are public cloud, private cloud, hosted private cloud (HPC) and virtual private cloud (VPC). We briefly describe each below.

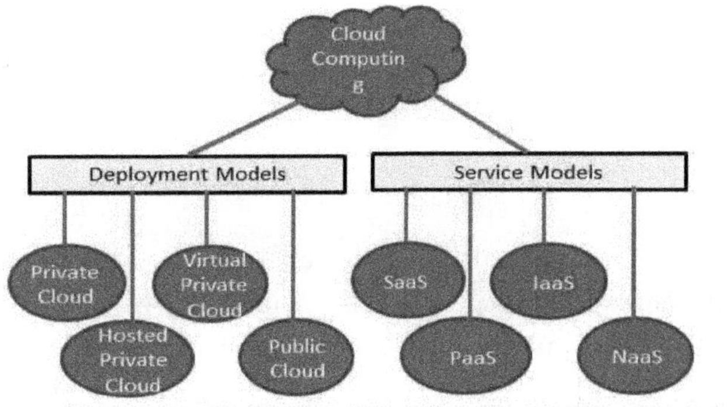

Fig. 2.3 Cloud computing models.

(1) *Public cloud* provides services that are accessible and shared by the general public over the public Internet at a fee. Computing services in a public cloud can be rented on demand. Organisations which subscribe to public cloud are not responsible for managing and maintaining the data centers. The results is cost saving. However, the main problem in public cloud is the security of data which cannot be ensured as multiple organisations share a data center. Major providers of public cloud services include Microsoft Azure, Amazon Web Services (AWS) and Google Cloud. As multiple customers share the resources of a public cloud, it is also referred to as *multi-tenant cloud environment.*

(2) *Private cloud*, also referred to as internal cloud or enterprise cloud, is where an enterprise hosts and controls all the components of cloud computing in its own dedicated hosted data center or intranet. It is also referred to as *single tenant cloud environment* as the computing resources or cloud infrastructure are used by a single organisation. End-users of private cloud are connected to the resource either via the public Internet or private line. The main advantage in this model is security to organisational data as it may not share resources with any other organisation. The main disadvantages in this deployment model is infrastructural costs. Another disadvantage is the maintenance and management of the data centers. Examples of private cloud providers are Microsoft Azure Private Cloud, Hewlett Packard Enterprise (HPE), VMware and IBM Cloud Private for Data.

(3) *Hosted Private Cloud* (HPC) providers include Rackspace, HP and IBM. As in public cloud the data centers in HPC are owned by cloud providers. For this reason, HPC is also called managed cloud. HPC allows customers to choose between a number of criteria, including support for development (Dev) and information technology operations (Ops) or DevOps, prior IT relationships, levels of infrastructure and system support, and geographic location.

(4) *Virtual Private Cloud* (VPC) is more of a private cloud running on a publicly-shared computing infrastructure. VPC uses encrypted tunnels, IP subnets and virtual private networks (VPNs). Customers of such a service access the computing resource through VPNs to ensure data security. VPC combines the resource sharing feature in public cloud and the security feature in private cloud. It is thus also referred to as *hybrid cloud.* Providers of VPC include

Microsoft Azure, Microsoft Windows Server, Microsoft System Center, Rackspace and Hewlett Packard Enterprise.

2.1.11.2 *Basic Service Models of Cloud Computing*

In the traditional Information Technology model, an enterprise controls and usually owns all the nine components of cloud computing shown in Fig. 2.4(c). This is different in cloud computing. We discuss the popular basic service models of cloud computing in this section, including software-as-a-service (SaaS), platform-as-a-service (PaaS) and infrastructure-as-a-service (IaaS). These service models are shown in Fig. 2.4. Each service model has two two types of infrastructure: back-end infrastructure and front-end infrastructure. The **back-end infrastructure** is owned and managed by the cloud services provider, while the **front-end infrastructure** is managed by the **cloud client** or cloud customer. The cloud services are together referred to as XaaS, meaning X-as-a-Service. The X can mean software, infrastructure, platform, desktop, etc. We now have in addition to these desktop-as-a-service (DaaS), mobile backend as a service (MBaaS), managed software as a service (MSaaS), Database-as-a-Service (DBaaS) and information technology management as a service (ITMaaS). The generic term for all cloud computing models is XaaS. Everything begins with IaaS, as we need infrastructure to build or run any applications.

2.1.11.3 *Software-as-a-Service (SaaS)*

Traditionally, vendors of software would sell them outright to users who would install and manage them on their own computers or servers. The current trend is that software vendors store their application software on their own servers attached to the global Internet. Users then use the software on-demand usually, at a subscription fee, without installing them on their local computers. Thus, the software is not sold as a product but as a service. The subscription fee can be paid monthly or access time. SaaS is the easiest form of cloud computing. As shown in Fig. 2.4, customers do not own any of the nine components of cloud computing. Other terms used to refer to SaaS are on-demand software, software plus services, Web-based software, on-demand software, or hosted software. Salesforce, a USA-originated company, started SaaS with its Customer Relationship Management (CRM) software. A sub-category of SaaS is database-as-a-service (DaaS). Centralised software started in the 1960s and developed through Application Service Providers (ASP) in 1990 and has now evolved to SaaS. ASPs usually did not own the software, but

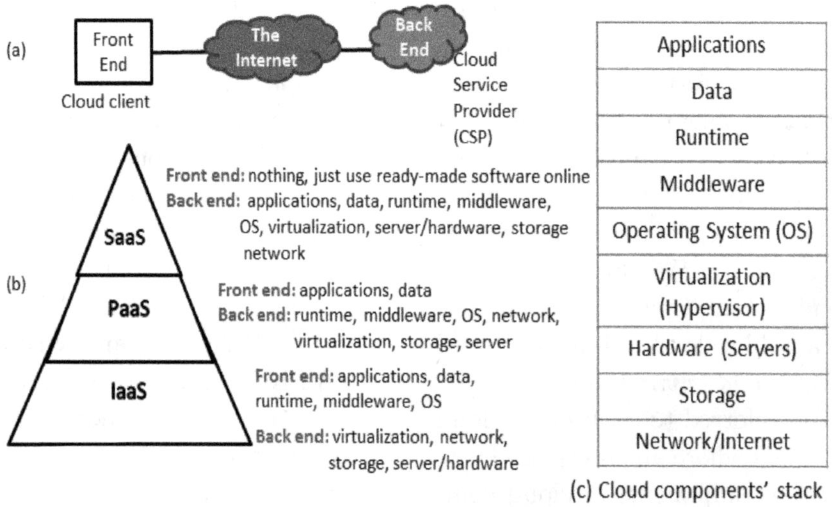

Fig. 2.4 The basic service models of cloud computing.

made available through their data centres software developed by other organisations. This means that ASPs are liable to software updating issues. Another SaaS vendor is Workday which offers human capital management, financial management and cloud integration applications from America. A main disadvantage of SaaS is the requirement of always-on Internet connectivity to use the service. Thus, SaaS may increase digital divide by cutting off locations without good Internet access.

There are a plethora of advantages to SaaS, and examples are [Salesforce (2017)]:

(1) **Smooth software upgrades and updates:** Usually, vendors of SaaS are the developers or license holders of the software being used in the cloud. This means that the most current version of a given software is kept in the cloud and thus used by its users. This means that users do not need to worry about software upgrading. Also, people collaborating on a file use the same version of the software. Further, a document looks different when opened with a different version of the software. Attempt to open Word 2010 file using Word 2007, for example, to feel this.
(2) **Lower initial capital investment:** The initial costs customers incur to use SaaS-based software is lower in both monetary and in-house expertise than buying the software outright. This is because customers do not have to buy software or hardware which also frees them from software installation, maintenance and updating. However, over long terms, customers pay more to increase profits of vendors. As the software is owned and managed by its vendor, customers do not have to pay for upgrades.
(3) **Global availability of same software:** Subscribers to SaaS have access to the same resource irrespective of where they are geographically so long as they can connect to the cloud. Problem is, of course, places without Internet access.
(4) **Multitenant architecture:** A single instance of software running on a centralised server serves multiple tenants, i.e. a group of users sharing a common access with specific privileges to the software instance. Multitenancy reduces costs of applications usage.

2.1.11.4 *Platform-as-a-Service (PaaS)*

As shown in Fig. 2.4, the Platform-as-a-Service model is used by software developers, but not end-users, to exploit its main advantage of rapid application development. The PaaS provider makes available to developers the cloud's infrastructure, including servers, operating system, storage, runtime environment, software development framework, virtualisation, middleware, networking and service management mechanisms (e.g. workflow management and monitoring). PaaS customers own only the data and the applications they develop. PaaS vendors include Oracle, Microsoft Windows Azure which is a total cloud computing platform, Google's App Engine and Salesforce.

There are two implementations of PaaS. First, a cloud vendor provides the infrastructure to host application developers' activities. In the alternative model, software is installed in private data centres which is managed by internal IT personnel. What are the advantages in this model? PaaS makes application development easier and cheaper for app developers as the operations and all infrastructure they need for their applications' life cycle (create, test, deploy) are provided by a third-party company at a fee. Thus, app developers focus on their actual business solutions by managing only data and apps.

2.1.11.5 *Infrastructure-as-a-Service (IaaS)*

Telecom equipment manufacturers, such as Ericsson and Huawei, used to build equipment and sell them outright to service providers who managed them in-house. Currently, the business model changed dramatically. The manufacturers install the infrastructure and manage them for the service providers to just run their applications and services. This is a kind of IaaS. So, what is IaaS? As shown in Fig. 2.4, IaaS offerings such as Ericsson, IBM SmartCloud Enterprise, Amazon Web Services, Google Compute Engine, Rackspace Open Cloud and Windows Azure provide all the infrastructure needed by organisations in the cloud for access over the Internet.

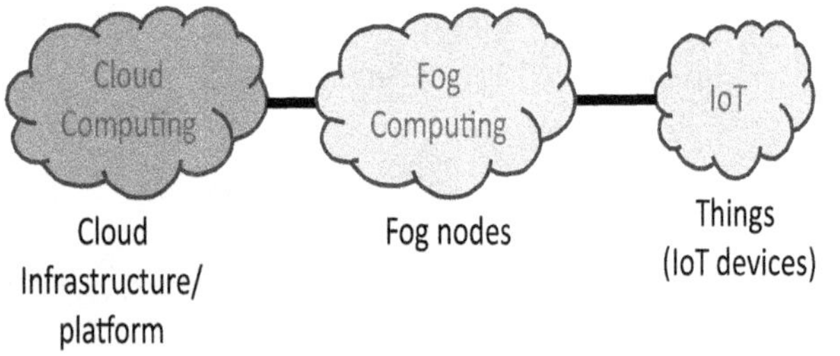

Fig. 2.5 The relationship between cloud and fog computing.

The benefits in IaaS include high resource availability and reliability, faster disaster recovery, and scalability as we can scale up or down by just clicking a button.

2.1.12 *Fog Computing*

Also referred to as fogging or edge computing, *fog computing* enables computing functions (i.e., computation, storage and networking) between data centers in cloud computing architectures and IoT devices. In IoT, fog computing serves as a bridge between the cloud infrastructure and IoT devices. This is illustrated in Fig. 2.5. Fog computing extends cloud computing services to the edge of an enterprise's network. The term fog computing was coined by Cisco Systems in January 2014. Both fog nodes and IoT things have storage, computing and connectivity capabilities. The main advantage in fog computing is that it enables a more efficient processing of data as data does not need to be transferred from the end points of the network to the cloud before processing it. The analysis of the data generated by the *things* or IoT devices is performed by the fog devices. Therefore, data analytics occurs closer to the IoT devices. This way bandwidth is conserved and processing also occurs quicker.

2.1.13 *Infrastructure Sharing and Cloud Telecom*

Outsourcing is the business model in which one organisation contracts another organisation to perform a business operation that can be done inhouse. Traditionally, a service that is commonly outsourced by many companies is customer service through contact centers or call centers.[2] The opposite to outsourcing is inshoring or insourcing, i.e. organisations perform business operations that were previously done overseas. For example, a company can bring its call center from overseas to its country of operation. Also, a car company that has been outsourcing parts overseas can do that in its country. Two main types of outsourcing can be identified. In *nearshoring* an organisation performs a business operation in a nearby country. The practice becomes *offshoring* if the country in which the business operation is transferred to is far away. For example, if a Singaporean company transfers a business operation to Malaysia it is nearshoring because the two countries share a border. However, the practice becomes offshoring if the operation is done in the UK, for example.

Cloud computing and its associated technologies have enabled the outsourcing of networking infrastructure in the form of a service rather than a product. This feature is exploited in the outsourcing model of Bahrti Airtel from India in Fig. 2.7. Why do companies engage in outsourcing? There are many reasons. However, the main reason is usually that of cutting costs. Other reasons are to exploit better talents available overseas, exploit tax benefits as some countries have lower taxes, reduce the time to market a product by cutting production time achieved through workload sharing, focus on core business, and to exploit operational best practices that would be too time consuming or/and expensive to develop in-house.

In the past, the value chain in telecommunications was made of only telecommunications equipment manufacturers and telecommunications services providers. Examples of the world's largest telecommunication equipment vendors are Huawei from China, Ericsson from Sweden, Cisco Systems from USA, Nokia Networks from Finland and ZTE Corporation from China. Examples of the world's largest service providers are China Mobile, Verizon and AT&T from USA, Vodafone from UK, NTT

[2]This trend is also changing, especially in the era of virtual service providers (VSP). For example, some mobile communications service providers, referred to as Virtual Telecommunications Operators (VTO), own only the customer service center as they outsource everything else from third-party companies.

from Japan, Deutsche Telekom, Bharti Airtel from India, Telstra from Australia and Sonera from Finland. In the past, especially when they were called PTTs, most of the telecom service providers were owned by the government of the country where they conducted business and usually operated in only one country. The service providers owned the telecom infrastructure and operated it using their in-house engineers and technologists. Therefore, the telecom service providers were also telecom infrastructure operators.

How do current trends differ from the past trend? Currently, many of the service providers have exploited *globalisation* to operate in many countries. This is the reason why we have used '*from* country' instead of '*in* country' above. Second, the service providers have began to exploit *outsourcing* by giving out some of their business operations to third-party companies. We shall use the cloud computing and outsourcing business model of Bharti Airtel illustrated in Fig. 2.7 as our case study in this chapter. Technically-speaking, we call it cloud telecommunications service provider (CTSP) and client, respectively. The consequence of the outsourcing is that the service providers do not want or need to employ engineers and technologists any more. Thus, under the cloud telecom business model those brainy people are employed by the equipment vendors who provide the infrastructure and its management to the service providers as a service (Fig. 2.6). This means that the telecom equipment vendors no longer sell products but services. As illustrated in Fig. 2.6(b), modern telecom service providers

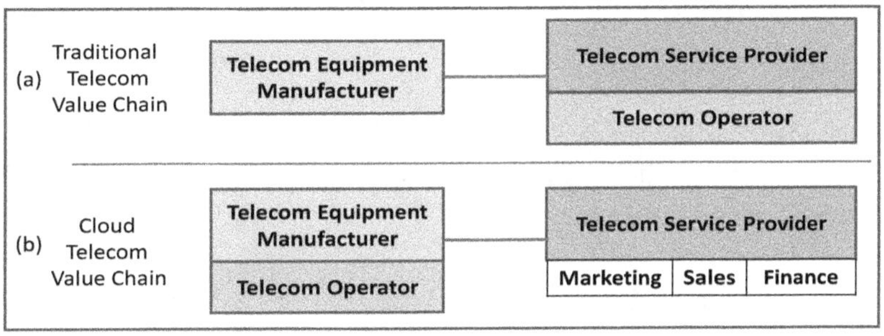

Fig. 2.6 Traditional versus cloud communications value chains.

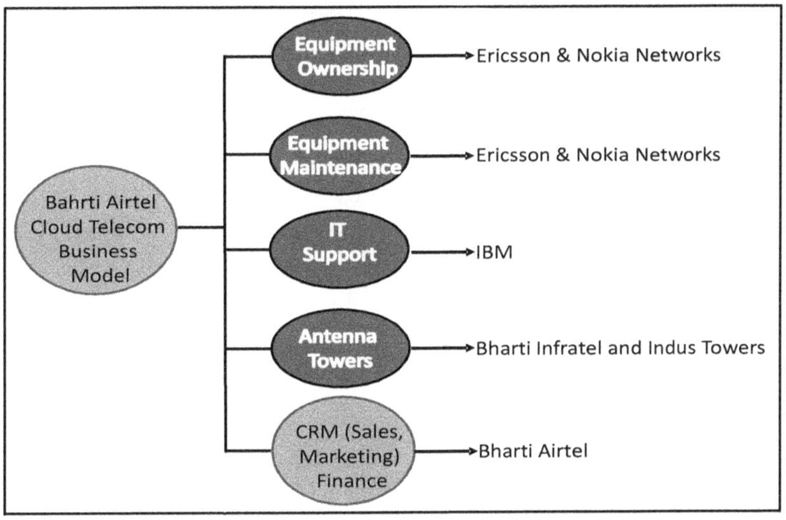

Fig. 2.7 Bahrti Airtel's outsourcing business model.

are focusing only on the customer interface portion of the business, such as sales, marketing and finance, while the equipment manufacturers take over the commissioning and maintenance of the network infrastructure.

How do CTSPs charge their clients? There are several options, including:

(1) Fixed charge based on use.
(2) Fixed periodic charge (e.g. minute, daily, weekly, fortnightly, monthly or yearly)
(3) Fixed periodic charge plus charge based on use.

For example, it is believed that Ericsson charges Bharti Airtel by the minute to commission and maintain its telecommunications infrastructure for the latter to use. Airtel then concentrates on the customer interface, comprising marketing, finance and sales.

2.1.14 *Self-Organising Network (SON)*

Companies offering SON solutions include Ericsson, Intucell and Celcite. As its name connotes, SON (Self-Organising Network or Self-Optimizing Network), is a communication network which configures and reconfigures itself automatically according to network dynamics. SON is poised to supplant traditional network optimisation by automating functions. SON is more tailored to wireless networks which are key components in tactile Internet. SON reduces the lifecycle cost of communication networks through:

(1) Non-manual configuration of network equipment at deployment.
(2) Automated dynamic optimisation of network performance during operation based on demand, e.g. automated shutdown and activation of radio access network (RAN) or mobile wireless networks.
(3) Automated troubleshooting of networks during operation.
(4) Adaptation to unpredicted dynamics automatically.

There are many advantages in SON, including:

(1) Reduction in roll-out time of networks.
(2) Reduction in the ratio of operational expenditure (OpEx) to revenue, thus increasing net revenue or profitability. Tier 1 wireless network operators applying SON techniques in the radio access network, mobile core and backhaul can reduce their OpEx by 32%.
(3) The dynamic and automated activation and shutdown of radio access networks (RANs) alone enabled by SON is expected to reduce electrical power consumption cost by up to 35% compared to the costs of running a similar network which is not using SON.
(4) Reduction in call dropping in SON-enabled wireless access networks by up to 40% compared to similar non-SON-enabled networks.
(5) Increase in data transmission rate by over 20% in SON-enabled network compared to an equivalent conventional network.

2.1.15 *Cognitive Radio: AI-Enabled Networks*

Communications networks are increasingly becoming intelligent. This is attributed to the increasing adoption of AI in networks. We briefly survey cognitive radio networks (CRN), one of today's AI-enabled networks. Such networks are able to interact with their environments and adjust their operations. Owing to the scarcity of radio spectrum, some frequency bands (especially the unlicensed bands) are shared by multiple technologies. Finding the best available frequency at any given time is important. The *best* can be measured in terms of minimum interference, largest data rate and/or maximum overall spectrum efficiency. A CRN senses its environment to find spectrum holes (i.e. currently idle radio frequencies), selects the best in the list and use it. This feature of *dynamic spectrum access* increases spectral efficiency.

A radio transceiver which is equipped with AI and can be configured and programmed dynamically is called cognitive radio network (CRN). A CRN is capable of dynamic radio frequency spectrum management. It selects the best wireless channel or frequency from the list of the available to minimise interference and optimise data transfer.

Transceivers in a cognitive radio network can be instructed to configure its parameter, such as waveform and operating frequency. Below we quote how CRN is defined by several organisations:

(1) **Wireless World Research Forum (WWRF)**: CRN employs a dynamic time-frequency-power based radio measurement and analysis of the RF environment, to make an optimum choice of carrier frequency and channel bandwidth to guide the transceiver in its end-to-end communications, with quality of service being an important design requirement.
(2) **International Telecommunications Union (ITU)**: A radio or system which senses, and is aware of, its operational environment and can dynamically and autonomously adjust its radio operating parameters accordingly.
(3) **National Telecommunication and Information Administration's (NTIA) of USA**: A radio or system which senses its operational electromagnetic environment and can dynamically and autonomously adjust its radio operating parameters to modify system operation, such as maximise throughput, mitigate interference, facilitate interoperability and access secondary markets.

Another AI-enabled network is *autonomic networks*, which are self-managing networks. They can configure or/and optimise their operations automatically.

2.1.16 *Virtual Reality and Augmented Reality*

Virtual reality (VR) has been in movies for years. It has now become a real technology. Anything which seeks to replace physical things is VR, especially if done digitally. VR is a computer simulated environment which tricks the human brain to feel that it is in that environment rather than the actual one. The requirement of VR is to trick the brain to create an *immersive experience* (IE). IE is defined by the Immersive Analytics Community as "the realistic feeling of being there or being present within a virtual space with others."

Popular tools used to experience VR are headsets or goggles fitted with lenses like Oculus Rift S developed by Oculus VR (now acquired by Google) and Google Cardboard. Looking into both lenses in the goggles at the same time creates a sense of three dimensions. The goggles also contain gyroscopes which track movements to create the feeling of being in that setting. VR is now used in many fields, including:

(1) Military to engage in virtual combat training.
(2) Driving schools to train learners in a simulated environment prior to taking them to the roads.
(3) Medical schools to teach students anatomy.

Augmented Reality (AR), also referred to as **mixed reality** (MR), is the technology which adds digital elements (e.g. graphics or information) into the real environment to enhance it to appear as a newly-created artificial environment. Like VR, AR is also best experienced using smart glasses, such as Microsoft Hololens, Sony SmartEyeglasses and Epson Moeverio BT-300. AR can also be experienced using apps on smart phones. For example, the Gatwick airport passenger app enables passengers to use AR maps on a mobile phone to navigate the airport.

2.1.17 *Check Your Understanding*

Students are encouraged to check their understanding of the content of this section with the following questions prior to proceeding to the next section.

(1) Discuss the meaning of *tactile Internet.*
(2) Describe the importance of over-the-air (OTA) as used in 5G mobile networks.
(3) **In-class group workshop**: In groups of 2 to 4 students, discuss the differences and any relationship between network agility and network scalability.
(4) **In-class group workshop**: In groups of 2 to 4 students, analyse how network function virtualisation (NFV) and

software-defined networking (SDN) enhance or even enable network agility and scalability, especially in relation to cloud computing.

(5) Discuss in groups the difference between a product and a service.

(6) What is your position on cloud telecom business model? Consider its pros and cons from the perspectives of both technical employees and business owners (i.e. the equipment vendors and service providers). Also, consider the effects of the cloudification business model on capital expenditure (CapEx) and operational expenditure (OpEx) in providing electronic communications services, as well as the agility of service provisioning.

(7) Discuss in groups of 2 to 4 students the differences and similarities between infrastructure sharing and cloud computing.

(8) How does SON differ from SDN? How is SON similar to SDN? Pair up the class and research on these, then discuss to see if any consensus can be reached.

(9) Discuss the differences and any relationship between fog computing and cloud computing.

(10) Justify the relationship between IoT and fog computing.

(11) List at least two advantages in fog computing for IoT.

(12) Briefly describe each of the following, and state an example of each:

   (a) Network agility.
   (b) Network scalability.
   (c) Network function virtualisation (NFV).
   (d) Network slicing.

(13) Considering network function virtualisation (NFV), list four (4) network functions which can be virtualised.

(14) **In-class collaborative exercise**: In groups of 2 to 4 students, research to find the purpose of cognitive radio, and how it can be used in wireless IoT.

(15) List two key factors in autonomic networks.

(16) Analyse the operation of fog computing technology, and show how it is important in IoT.

(17) The computing infrastructure provided by cloud computing providers (CCP) as a service can be any combination of software, information technology (IT) support and hardware. Consider each of the following cloud computing service models:

   (a) SaaS.
   (b) PaaS.
   (c) IaaS.
   (d) AaaS (i.e. analytic-as-a-service).

   List and describe which of the three infrastructure is provided by the service provider and which by the service user in each of the four service models listed above.

(18) Assume that someone seeks to build and test a business intelligence application using SAP Lumira cloud at minimal capital expenditure. State the name and discuss the cloud computing service model which best meets the developer's needs.

(19) Discuss the functions of the ITU focus group FG ML5G.

2.2 Core Principles of the Internet

We discuss the fundamental principles of the Internet in this section. As IoT can be viewed as child of the Internet, these topics apply also to the IoT. Topics discussed include packet switching and the Internet Protocol (IP) as a means for addressing devices and networks attached to the Internet.

2.2.1 *Packet Switching*

The Internet is a network of networks. There are two possible ways by which data can be transferred between two devices attached to the Internet: via switched links or non-switched links. Non-switched networks interconnect devices using

Table 2.1 Types of interconnecting devices and their basic properties.

ISO/OSI Layer	Interconnecting Device	PDU Type/ Data Format	Connection Type	Addressing Type
Physical layer	Repeater, hub, multiplexer	Bit, symbol or signal	Physical transmission medium	None
Data link layer	Bridge, switch	Frame	Hop-to-hop	Hardware/physical/ MAC/NIC address
Network layer	Router	Packet	Network-to-network	Internet Protocol (IP)
Transport layer	Transport gateway	Segment (TCP)/ datagram (UDP)	Hosts-to-host	Socket (i.e. IP address and port number)
Session	–	Data	–	None
Presentation	–	Data	–	None
Application layer	Application gateway	Data	–	Uniform Resource Identifier (URI)

point-to-point links. Switched networks, however, interconnect communicating devices over switching nodes. Switching nodes are also referred to as *internetworking devices* or intermediate nodes. Switching nodes are usually switches (running up to ISO/OSI Layer 2 protocols) and routers (running up to ISO/OSI Layer 3 protocols). Table 2.1 lists the popular networking devices and some of their basic features. Note that the TCP/IP protocol model implemented in the Internet performs the functions of the session and presentation layers at the application layer.

How is data transferred between devices attached to a switched network? There are three types of switching techniques: circuit switching, message switching and packet switching. As illustrated in Fig. 2.8, each of these three switching techniques has variants. Circuit switching originated from the Public Switched Telephone Network (PSTN) which provided Plain Old Telephone Service (POTS). Both circuit switching and PSTN are obsolete as telecommunications service providers are migrating from PSTN to Internet-based Next Generation Network (NGN). As we replace the PSTN/POTS with NGNs, all Private Automatic Branch Exchanges (PABX), Private Branch Exchanges (PBX) and Centrex are being replaced by Internet Protocol-based PBX or IP-PBX. See Fig. 2.9 for a possible IP-PBX architecture. For this reason, we do not discuss the details of circuit switching. Readers may note, however, that some features of circuit switching have been carried over to modern networks. This is the reason why PSTN and circuit switching are referred to as *legacy technologies*. Different countries refer to their NGN by different names. For example, Australia's NGN is called National Broadband Network (NBN). Message switching routes an entire message through the Internet, from source to destination. The main problem with this switching technique is that even a single erroneous bit requires the entire message to be retransmitted.

In packet switching the entire message to be transferred between a source and the destination is divided into segments, each with a maximum length. There are two types of packet switching techniques: datagram packet switching (DPS) and virtual-circuit packet switching (VPS). Routing using either VPS or DPS adds routing information to each message segment. The addition of routing information and other signaling data to a message segment forms a *packet*. Clearly, message switching is a specialised form of packet switching in which the entire message is transmitted in a single packet. Internet, IoT and local area networks use DPS, while VPS is used in Multi-Protocol Label Switching (MPLS), Asynchronous Transfer Mode (ATM), Frame Relay and X.25. Be reminded that ATM, X.25 and Frame Relay are currently obsolete wide area networking (WAN) technologies.

Like circuit switching, VPS is connection oriented. The DPS, however, is connectionless. Thus,

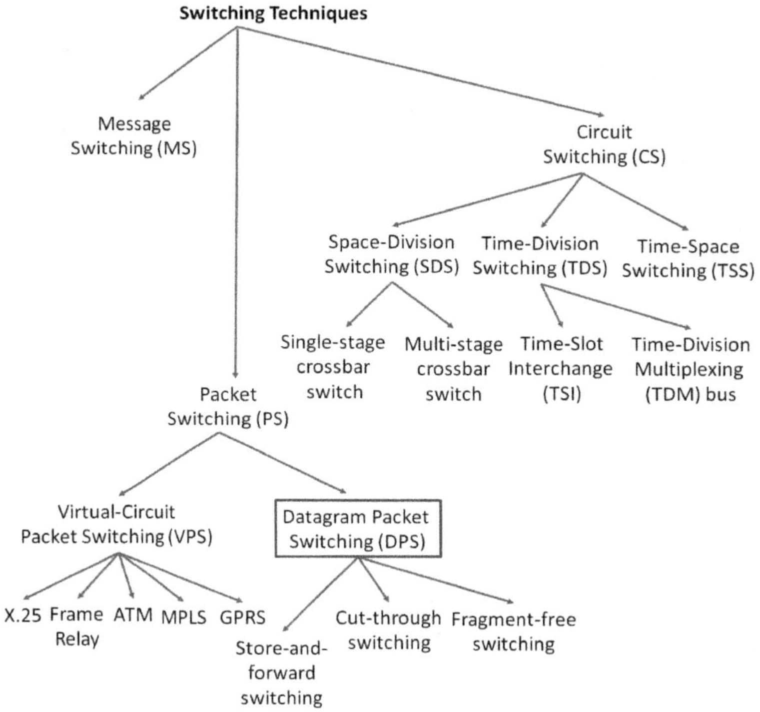

Fig. 2.8 Types of switching techniques.

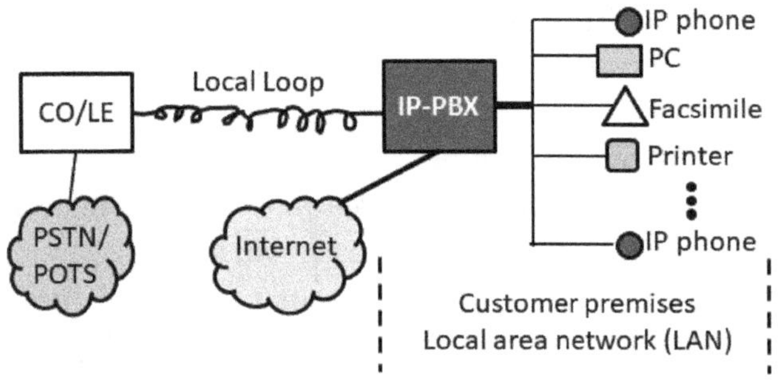

Fig. 2.9 A topology of an IP-based PBX.

no signalling traffic is needed to set up, maintain and tear down DPS links. Thus, switches in DPS-based networks do not keep information about the states of connections. This is a feature which promoted its adoption in the Internet. Circuit switching, VPS and DPS are implemented at ISO/OSI Layer 1 (physical layer), ISO/OSI Layer 2 (data link layer) and ISO/OSI Layer 3 (network layer), respectively. Thus, all the technologies based on VPS such as those in Fig. 2.8 are Layer 2 technologies.

Both VPS and DPS require routing information inside each packet. The routing information in virtual-circuit packet-switched networks is called virtual-circuit (VC) numbers, while routing information in datagram packet-switched networks is the Internet Protocol (IP) address. Datagram networks use just one IP address to route a packet from source network to destination network. However, each link in the source-destination path of a VC packet-switched network uses a different VC number for two reasons. It reduces setup time of VCs as it does not require routers to exchange information to ensure that unique VCs are used in the network. Secondly, it enables the use of a few bits in the packet to encode the VC number. For example, frame relay uses 10-bit data link connection identifier (DLCI) to identify its virtual circuits. ATM uses 12-bit to 16-bit virtual circuit identifier (VCID) to identify its virtual circuits and 8-bit to 16-bit virtual path identifier (VPI) to identify channels or paths over different ATM switches. IPv4 and IPv6 use 32-bit and 128-bit IP addresses, respectively, to identify packets.

Datagram packet switching usually occurs at ISO/OSI Layer 3 where the intermediate nodes are called routers. DPS treats the packets belonging to the same message as independent entities within the network. Packets belonging to the same message can use different routes between the source and the destination. For that reason, the packets are liable to different treatments. As illustrated in Fig. 2.8, there are three types of DPS: cut-through switching, fragment-free switching and store-and-forward switching. In **store-and-forward switching**, the switch copies the entire incoming protocol data unit (PDU) into its memory, checks for any errors by computing the Cyclic Redundancy Check (CRC). If there are errors then the PDU is discarded. In the absence of errors the switch forwards the PDU over its appropriate interface towards the intended destination.

Cut-through switching is also called cut-through forwarding. In **cut-through switching**, the switch needs to read only up to the destination address of an incoming PDU. For example, a switch in an Ethernet-based network reads the first 14 bytes of the Ethernet frame. Therefore, it needs memory space for only the destination address in the header of the PDU. The switch begins to forward the PDU as soon as it reads the destination address. This means that a cut-through switch starts to forward PDU before it receives the entire PDU to expedite forwarding. A problem in cut-through switching is the possibility of forwarding garbled frames.

In computer networking a PDU with a very small size is called *runt*. Local area networking (LAN) based on Ethernet (IEEE 802.3) requires the Layer 2 PDU (referred to as a frame) to have a minimum size of 64 bytes. Therefore, any frame less than 64 bytes is called a *runt frame*. **Fragment-free switching** requires the switches to read at least the first 64 bytes of a PDU before making a forwarding decision. This is made to avoid forwarding a *runt PDU*. Owing to its operation, fragment-free switching is also referred to as *runtless switching*.

Unlike circuit switching, packet switching allows multiple users to share communications links. The sizes of packets transmitted over a packet-switched network is not fixed. However, networks usually impose a maximum limit on the packet size, referred to as **maximum transfer unit (MTU)**. MTU of Ethernet-based networks is 1500 bytes. A packet exceeding 1500 bytes in size is called a **jumbo packet**. IPv4-based routers could be configured to divide a jumbo packet. IPv6-based routers, however, are to throw them away.

Certain features in DPS have contributed to its success, and adoption by the Internet. First, DPS-based networks are *fault tolerant* or resilient in that a failure of a switch does not cause the loss of an entire message, unlike networks based on VPS and circuit switching (CS). The main reason is that packets of the same message can use different routes, enabling routing around a failed route. Second, DPS-based networks make more efficient use of network resources as multiple traffic streams share resources. Circuit switched networks dedicate resources end-to-end for a connection, which is wasted when the communicating party do not use it. Third, CS-based networks bill users based on the distance between the communications endpoints and duration of connection. Packet-switched networks do not consider distance in their billing.

2.2.2 *Internet Protocols (IP)*

Packets are routed through internets (i.e. interconnection of networks with different administrative domains) using network layer address. Examples of network layer addresses are AppleTalk, the Internet Protocol (IP) and the Internet Packet Exchange (IPX). By their functionality, network layer addresses are also referred to as **routed protocols**. On the Internet (i.e. the largest internet), the dominating routed protocol is IP, so we focus on it. The IP has gone through a series of transitions, with its history summarised in Table 2.2. Prior to August 1977 the functions of IP and TCP were combined and referred to as *Internet Transmission Control Program* (ITCP). We treat IPv4 in this section. IPv6 is treated in a later section. We do not treat IPv1 to IPv3 any further.

Table 2.2 Evolution of the Internet Protocol.

Date	Development
December 1974	IPv1 = TCPv1 (RFC 675)
March 1977	IPv2 = TCPv2
August 1977	functions of IP and TCP are separated
1978	IPv3 = IP and TCP (IP and TCP are separate protocols)
September 1981	IPv4 (RFC[3] 791)
October 1990	IPv5 Internet Stream Protocol (RFC 1190)
December 1998	IPv6 (RFC 2460)

RFC = Request for Comment.

The boundary between a network node (e.g. hosts and intermediate systems[4]) and a physical transmission media is called interface. A host usually has one interface, while, in order to perform its packet forwarding and routing functions, a packet switch needs at least two interfaces. Usually, each interface of a network nodes requires a globally-unique routed protocol address, except, of course, nodes in private networks using private network addresses such as those behind NAT (network address translator) or proxy server. Thus, for example, a router with five interfaces attached to the Internet requires five globally unique IP addresses.

IP addresses are assigned to network nodes either manually or automatically. Nowadays hosts are usually configured automatically using protocols such as dynamic host configuration protocol (DHCP). During host configuration, DHCP assigns to the host (a) temporary IP address, (b) subnet mask, (c) IP address of host's default gateway (i.e. first-hop router) and (d) IP address of host's local domain name system (DNS) server. DHCP is said to be a client-server protocol with the host requiring network configuration serving as the client. DHCP is a plug-and-play (PnP) protocol because it allows networked devices to seamlessly discover the presence of a network and seek attachment to it.

IPv4 was designed in 1981 for end-to-end packet routing (i.e. from packet source to destination). It is a software protocol by which data units called packets are forwarded by routers from one point (source) to another (destination) through the Internet. The forwarding of packets is connectionless, i.e., packets belonging to the same source/destination pair are forwarded independently, and can traverse different routes. The connectionless trait of IP makes the Internet robust, as one link cannot be a point of failure. The network does not guarantee the time taken for a forwarded packet to reach its intended destination, and whether or not it really gets there at all. This art of unguaranteed packet forwarding service is widely referred to as **best effort service**. The best-effort IP forwarding, although robust and efficient (due to statistical multiplexing), raises the issue of unavailability of service guarantees (QoS), especially now that the Internet handles traffic of various applications, some of which are strictly QoS-sensitive. Better QoS-enabled forwarding mechanisms were then introduced into the IPv4-based Internet lately using technologies such as multiprotocol label switching (MPLS), Integrated Services (IntServ) and Differentiated Services (DiffServ). The IP featured in this chapter operates at Layer 3 of the ISO/OSI reference model. Together with the other Layer 3 functions illustrated

[3]There are only a few Internet standards, and most of the protocols used on the Internet are called Request for Comments (RFC). Therefore, some people define RFCs as Internet standards.

[4]As a reminder, devices attached to the Internet are classified into intermediate systems (e.g. routers and switches) and end systems or hosts (e.g. personal computers).

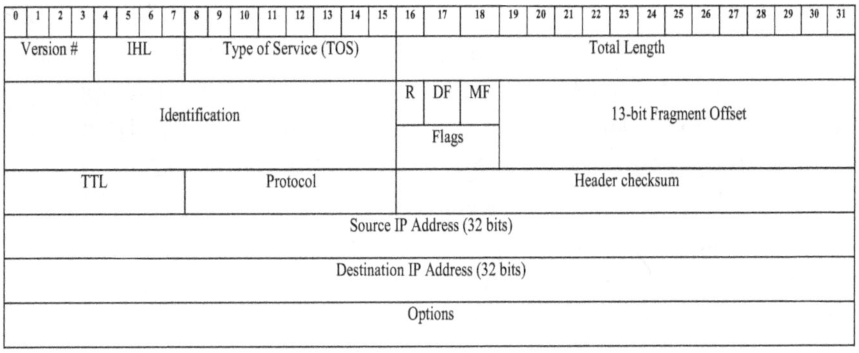
0 1 2 3 4 5 6 7 8 9 10 11 12 13 14 15 16 17 18 19 20 21 22 23 24 25 26 27 28 29 30 31
Version #
IHL
Type of Service (TOS)
Total Length
Identification
R
DF
MF
Flags
13-bit Fragment Offset
TTL
Protocol
Header checksum
Source IP Address (32 bits)
Destination IP Address (32 bits)
Options

Fig. 2.10 IPv4 header.

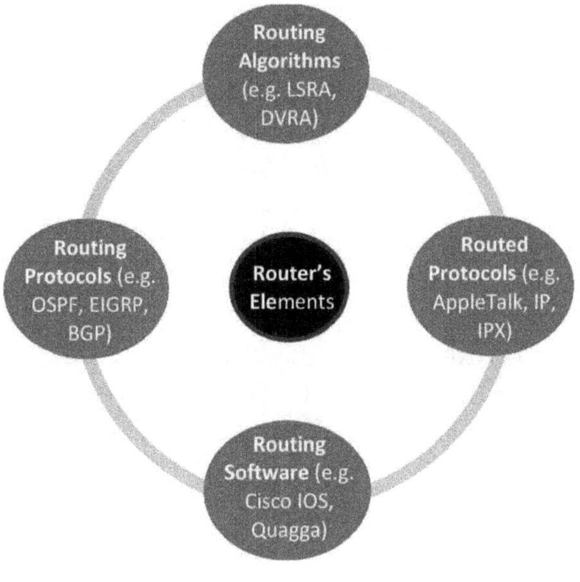

Fig. 2.11 Functions of ISO/OSI Layer 3.

in Fig. 2.11 enables the Internet to function as it is.

2.2.2.1 *IPv4*

The IPv4 was specified in September 1981 as RFC 781, and first deployed in 1983 in the then Internet, the ARPANET. The IPv4 header and its fields are shown in Fig. 2.10 with the number of bits for each field indicated. We discuss these fields briefly sequentially. Routers use the 4-bit version field to learn the version of the IP protocol and thus how to interpret the remaining fields in the header. The binary value in this field is 0100 which is equivalent to the decimal number 4. Owing to its optional field, the IPv4 header has a variable length. In order to assist routers to operate faster a header length field is provided. The 4-bit internet header length (IHL) field indicates the length of the IPv4 header in 32-bit words. Thus, the maximum IPv4 header is $2^4 (= 16)$ 32-bit words or 64 bytes. Thus, the maximum options should be 44 bytes. The type-of-service TOS field is meant to provide quality-of-service (QoS) or different service levels to different packets.

Total length field: The length of IPv4 datagram (header plus variable-sized options field plus transport layer PDU) is variable. Therefore, a 16-bit total length field is defined to indicate the total datagram size in bytes. Clearly, the theoretical maximum datagram size is 2^{16} bytes or 65,536 bytes and the minimum is 20 bytes. However, the minimum datagram size that all hosts in the Internet must be prepared to accept is 576 bytes for IPv4 and 1280 bytes for IPv6. Noting that Ethernet (a popular link-layer technology) has the MTU of 1500 bytes, it is important to keep IP datagrams a few bytes smaller than that, at least, to cater for headers. Other notable MTUs in bytes are:

- 68-64,000 IPv4 path in the Internet.
- 1,289-64,000 IPv6 path in the Internet.
- 150-9198 for Ethernet Jumbo Frames.
- 7981 for WLANs based on IEEE 802.11.
- 1492for PPPoE over Ethernet v2.

The three *fragmentation fields* are the 16-bit identifier field, the 3-bit flag field and the 13-bit fragmentation offset field. IPv4 routers can break a datagram exceeding the MTU of the outgoing link into multiple packets/datagrams, each of which is called a *fragment.* All fragments of the same datagram carry the same 16-bit identifier number. The *flag field* has 3 bits. The first or most significant bit is reserved and always set to binary 0. The middle bits is called don't fragment bit (DF). Routers cannot fragment a packet whose DF is set to 1 (i.e. DF=1). A router discards a packet with DF=1 and length in excess of the MTU of the interface where it should be forwarded to. The third or least significant bit in the flag is called more fragment(s) bit (MF). If it is set to zero in a fragment then it tells the host reconstructing the original datagram that that fragment is the last in the fragmented datagram. If the more bit is set to 1 then it means more fragments of the datagram are in transit.

The 13-bit *fragmentation offset field* specifies the position of a fragment in the original datagram. The fragmentation offset value is measured in 8 bytes. For example, an offset of 62 means 62×8 `bytes` = 496 bytes. Consider a packet switch with four interfaces of different link-layer technologies with differing MTUs. Consider a datagram of size 1544 bytes arriving at one interface to be forwarded to another interface with MTU of 496 bytes (excluding headers). This datagram is divided into four fragments by the router. The last fragment may be padded with zeros if it is smaller than the smallest-size datagram permissible on a link.

TTL (time-to-live): communication networks usually use redundant links. Such links create loops, which in turn can cause a packet to loop around in a network forever if nothing is done. The 8-bit TTL field is used to avoid packet looping by limiting its lifetime. It is specified in seconds, but time intervals less than 1 second are rounded up to 1. In practice, this field is interpreted as a hop count. Any router through which a datagram passes decrements the TTL field by one. When the TTL field becomes zero, the router discards the packet and may send an ICMP Time Exceeded message to the source of the packet.

Protocol: This indicates upper layer whose PDU is carried in the payload or data field of the datagram. It can be a transport layer segment or signaling data (e.g. ICMP messages). The protocol whose data is carried in the payload portion of IPv4 datagram can be identified in the 8-bit 'upper-layer' protocol field in IPv4 header. Sampled protocols and their network layer protocol numbers in decimal are:

- 1 for ICMP (Internet Control Message).
- 4 for IPv4 encapsulation.
- 6 for TCP (Transmission Control Protocol).
- 17 for UDP (User Datagram Protocol).
- 41 for IPv6 encapsulation.
- 58 for IPv6-ICMP (ICMP for IPv6).

The *16-bit header checksum*: this field is defined in RFC 791 and it is used to check but not correct any errors in only the header field of a datagram. When a packet arrives at a router, the router calculates the checksum of the header and compares it to the checksum field. The full name for this checksum is **Internet Checksum**. If the values do not match, the router discards the packet. As each router must decrease the TTL of any datagram it processes, the checksum of each packet must be recomputed at each router. This increases the workload on the routers. The checksum algorithm initially sets the checksum field to zero. It then computes the 16-bit one's complement of the one's complement sum of all 16-bit words in the header. The source and destination IP addresses are used to route a datagram from its source to its intended destination. See Section 5.2.2 in [Gyasi-Agyei (2019a)] for examples of how the Internet Checksum is computed.

IPv4 Address Classes

IPv4 addressing can be classful or classless. Classful addressing is the original design, while classless addressing is the current trend. Why? Read on. Classless IPv4 addressing is also called *classless inter-domain routing* (CIDR). The IPv4 classes in

Class	1st high-order bits	Private IPv4 address ranges			Public IPv4 address ranges		Class broadcast address
		Address range	CIDR	Network mask	Address range	Default network mask	
A	0	10.0.0.0 - 10.255.255.255	/8	255.0.0.0	1.0.0.0 - 9.255.255.255; 11.0.0.0 - 126.255.255.255	/8	x.255.255.255
B	10	172.16.0.0 - 172.31.255.255	/12	255.240.0.0	128.0.0.0 - 172.15.255.255; 172.32.0.0 - 191.255.255.255	16	172.16.255.255
C	110	192.168.0.0 - 192.168.255.255	/16	255.255.0.0	192.0.0.0 - 192.167.255.255; 192.169.0.0 - 223.255.255.255	/24	x.x.x.255
D	1110	Reserved for multicasting					
E	1111	Reserved for experiments and research					

Fig. 2.12 The five IPv4 classful addresses.

the classful addressing are summarised in Fig. 2.12. We can observe many things from the data in this figure. First, each class can support different number of networks and number of hosts in each network. Second, a Class A address supports more hosts than all the other classes, but supports the least number of networks. Therefore, organisations with big number of employees but fewer networks favours that class.

It is also clear that the classful addressing wastes a lot of IPv4 addresses. To understand that let us consider a company with 50 employees altogether. Class C supports the least number of hosts, which is 254. Even allocating a Class C address to such a company would waste 204 IP addresses! This is one of the reasons why IPv4 addresses are said to be in exhaustion. It is also a motivation behind classless (i.e. CIDR) addressing, which is the default currently. Several other important observations can be made on Fig. 2.12. For each class the number of addressable hosts reduces by 2 because the first address is the **network number** while the last address is the **directed broadcast number**. Therefore, we have the following formula:

$$\begin{aligned} \texttt{Number of networks} &= 2^N \\ \texttt{Number of hosts per network} &= 2^H \end{aligned} \tag{2.1}$$

where N and H mean the number of bits in the IPv4 address used to number networks and hosts, respectively.

Class D block is used for multicasting. For example, RIPv2 requires its routers by default to multicast their entire routing tables every 30 seconds to RIP-enabled routers using the multicast address 224.0.0.9. Class E address with first byte having a decimal value of 255 is reserved for local broadcast to devices in a broadcast domain like network segment or virtual LAN. Class A addresses 0.0.0.0 to 0.255.255.255 are reserved as default gateway address. This address range is also referred to as zero network address or all-IP address. Class A addresses 127.0.0.0 to 127.255.255.255 are reserved for loopback and local diagnostic functions and thus cannot be used for packet forwarding. For example, I can check the functionality of the TCP/IP stack of my computer by executing the command below on windows command prompt: *ping 127.5.6.2.*

An echo reply to this command confirms that the TCP/IP suit of protocols is installed well on the native computer. The ping protocol uses the ICMP to echo back packets from routers in the end-to-end path.

Most of the IPv4 drawbacks are solved and the address exhaustion issue has been tackled aggressively through mechanisms such as:

(1) Reclaiming of allocated but unassigned IP addresses.
(2) Usage of network address translation (NAT).
(3) Usage of private IPv4 addresses. For example, usually IP addresses used for networking laboratories are private addresses.
(4) Deployment of classless interdomain routing (CIDR).
(5) Introduction of the real specific IP (RSIP).

For example, Stanford University relinquished its Class A network IP addresses in 2000. NAT assigns IP addresses dynamically to hosts of a network for internal usage. NAT's weaknesses include: (a) violation of IP's fundamental design (i.e. end-to-end routing), requiring an application layer gateway by some applications, (b) failure of some security protocols to operate with modified TCP socket, and (c) possible problems with address translation servers between the packet source and destination hosts. These NAT-related issues promoted the design of RSIP by 3Com in 1998. CIDR unfortunately increases the size of core routers' routing table, and complicates packet routing. Therefore, route summarisation was born. The IPv4 enhancement mechanisms, however, can only delay the full transition to IPng or IPv6. As at Q2 2019, many organisations are already using IPv6.

2.2.2.2 *IPv6*

As said earlier, the address space available from IPv4 is not enough to support the Internet of Things. Moreover, due to inefficient (and perhaps unfair) allocation, only about 10 million addresses are available today. This practical address limit was supposedly reached in January 2001. The IP address exhaustion scalability issue, coupled with the lack of security and autoconfiguration, inefficient use of addresses, and the lack of service quality in IPv4, spawned the introduction of the next-generation IP (IPng) or IP version 6 (IPv6) in 1994. In June 2002 some organisations deployed IPv6 to work in parallel with IPv4.

IPv6 has a 128-bit address space and a variable packet size in the range 576 bytes to 65 575 bytes, including a 40-byte header. An IPv6 node must be able to receive a packet of size 1500 bytes, but it must not send fragments which when assembled exceeds 1500 bytes, unless it knows that the receiving node is capable of processing packets of such lengths. The minimum IPv6 header has the size of 40 bytes as optional extension headers may be in use. The IPv6 header includes a 4-bit priority field and a 24-bit flow label field. The priority bits can be used by a source to indicate the relative delivery priorities of packets it injects into the network, while the source can use the flow label fields to label packets for differentiated QoS. Also, mechanisms such as packet encapsulation increase the packet size. Unlike IPv4, IPv6 restricts packet fragmentation to only packet sources and routers are not allowed to do so.

The IPv6 header has two portions: compulsory basic header field of size 40 bytes, and optional extension headers. See Fig. 2.13. Both the source and destination addresses have the fixed length of 128 bits or 16 bytes. The 4-bit version field is binary coded 0110 in IPv6 packets. The *Payload Length* is a 16-bit unsigned integer, measured in bytes, to indicate the size of the packet without the IPv6 header. Thus, the theoretical maximum payload is $2^{16} = 65,536$ bytes. The *Next Header* is an 8-bit selector with two purposes: it indicates either the type of extension header that immediately follows the IPv6 header or the type of upper layer PDU. The *hop limit* field contains an 8-bit unsigned integer number used as a counter. Any switch which forwards the packet reduces its value by one. The packet is discarded if the hop limit reaches zero.

The 20-bit *Flow Label* field is used to identify packets belonging to a given communication. The 8-bit *Traffic Class* field has two main purposes:

- The six most significant bits are used to encode the type of service that routers should provide to the packet.

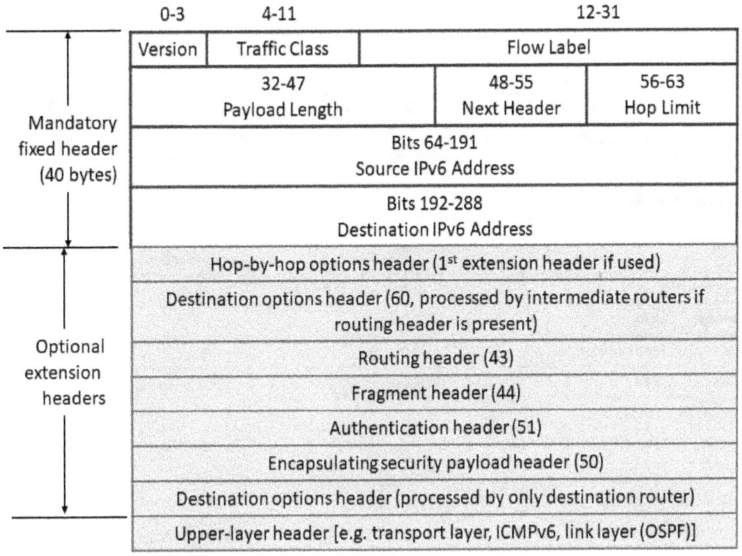

Fig. 2.13 IPv6 header with the optional extension headers mandated by RFC 2460.

- The two least significant bits are used for Explicit Congestion Notification (ECN) signals.

The 40-byte fixed header contains all the information needed to route a packet. Any extension header is inserted between the fixed header and the header of the upper layer PDU. The size of each extension header is a multiple of 8 bytes. The Next Header field indicates to routers whether a packet contains extension headers or not. Each extension header is identified by a distinct value. These values are indicated in Fig. 2.13. For example the routing extension header, used for source routing, has the next header value of 43.

The 128-bit IPv6 addresses are written in 32 hexadecimal (i.e. base 16) characters, but not in dotted decimal notation as done in IPv4. These 32 hex characters are written in four groups separated by colon (:), such as 80AF:03BF:D79C:ABC3. IPv6 addressing uses certain shorthand notation. For example 80AF:0000:D79C:ABC3 and 80AF:0000:0000:CFE6 are written as 80AF:0:D79C:ABC3 and 80AF::CFE6, respectively. We compare some basic features of IPv6 with those of IPv4 in Fig. 2.14.

As illustrated in Fig. 2.15, the 128-bit IPv6 address is divided into two equal parts. The 64 most significant bits make up the network ID, while the latter 64 bits make up the host or interface ID. Unlike IPv4, we do not borrow bits from the host ID for subnetting. Rather, the 16 least significant bits in the network ID are used for subnetting. The 64-bit host ID is obtained from the physical address (aka Layer 2 address of MAC address). The MAC address is made of 48 bit, while the host address is 64 bits. How do we convert between the two? We must obtain the equivalent IEEE's Extended Unique Identifier (EUI-64). The process is:

(1) Divide the 48-bit MAC address into two equal parts.
(2) Insert the 16-bit hexadecimal number FFFE in the middle to obtain a 64-bit number.
(3) Invert (i.e. convert bit 0 to bit 1 and vice versa) the 7th bit (called universal unique bit) in the host ID. The 7th bit is found in the first two hexadecimal characters in the host ID. We need to convert these two characters into their equivalent binary, invert the 7th bit and convert the results back to 2-digit hexadecimal number. The result is EUI-64.

2.2.3 *FLSM Subnetting in IPv4 Networks*

Subnetting is the process of dividing an IP-based network into two or more logical portions referred

Feature	IPv4	IPv6
Release date	1974	1998
Address size	4 bytes	16 bytes
Compulsory Header size	20 bytes	40 bytes
Number of fields in header	13 fields	8 fields
Classes	Yes, 5 classes	No
Means of identifying subnet ID	Subnet mask	Prefix length
Security	Optional external IPsec	Built-in (encryption and authentication)
Fragmentation of packets	Both sending host and the router	Only sending host
Inherent mobility	No, needs Mobile IP	yes
Representation of addresses	Dotted decimal	Hexadecimal with colon
Addressing types	Private, public and multicast	Global, unicast and anycast
Checksum in header	yes	no
Mapping between IP addr and MAC address	Through ARP	Through Neighbor Discovery Protocol (NDP)
Host IP address configuration	Manual or via DHCP	Automatic configuration available
Packet flow identification	None	Via the Flow Label
Broadcast addresses	Yes, directly	No, via link local multicast and anycast addresses

Fig. 2.14 Comparison between IPv6 and IPv4.

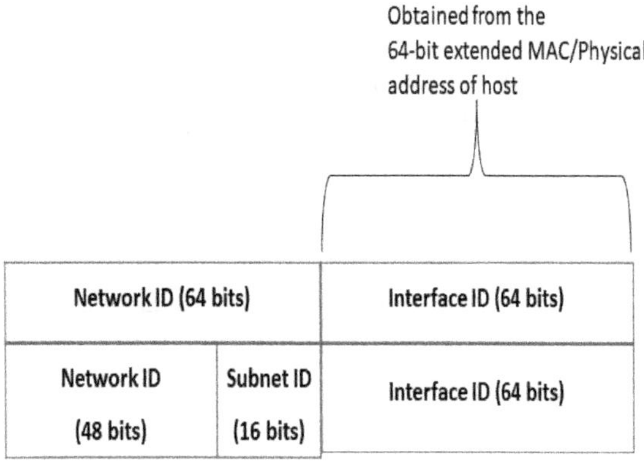

Fig. 2.15 The division of IPv6 address into network, subnet and host parts.

to as **subnets**. There are multiple reasons for subnetting a network. Let us note three of them here. First, it can be used to confine certain traffic to certain users. Second, an organisation can use subnetting to provide logically different networks to its departments or divisions. Third, an organisation with multiple offices can also create a subnet at each site. Subnetting can be done in networks based on both IPv4 or IPv6, although we focus on IPv4 subnetting here.

The first and the last subnets in a subnetted network are referred to jointly as **subnet 0**. All the subnets of a network appear as a single network on the outside such as on the Internet. Actually, every network (e.g. LANs and MANs) which is part of the Internet can be viewed as a 'subnet' or 'segment' of the Internet. Each of these 'subnets' can be further partitioned into smaller networks. **Broadcasting** is the process in which a device attached to an IP network sends a packet to all devices attached to the network. IPv4 networks have the following two types of broadcasting in the network layer:

- **Directed broadcast**: by default, routers may forward this broadcast packet to other network segments. We note, however, that a system's administrator can configure routers not to forward directed broadcast packets. Directed broadcast IP addresses have all the bits in the host ID portion of the address set to 1s. This makes the directed broadcast address the very last IP address in the network segment.
- **Limited broadcast** (aka *flooded broadcast* or *local broadcast*): this broadcast packet does not go beyond its subnet or network segment. Thus, it never passes any router it encounters. All 32 bits of the IPv4 address is set to 1, and thus has the dotted-decimal equivalent of 255.255.255.255. A local broadcast packet has

> this as its destination address. This local broadcast is the broadcast address of the zero network or 0.0.0.0, which in Internet Protocol standards stands for *this network*, i.e. the local network.

IPv6 does not implement the broadcasting found in IPv4-based networks. Broadcasting is done via the link-local all nodes address.

Subnetting requires us to know that each IP address is divided into two parts (Fig. 2.16). The first part counting from left identifies a network, and it is called **network ID** (or netID). The latter part identifies a specific host within that network, and it is thus called **hostID**. Given an IP address we can use a **network mask** (or netmask) to separate the netID from the hostID. Sometimes subnet mask is used incorrectly as the synonym for netmask. Either is often referred to simply as mask. The netmask for IPv4 is also 32-bit long. All the bits in the netmask identifying the netID are 1s, while the remaining rightmost bits identifying the hostID are all 0s. The default netmasks for the IPv4 classes A, B and C are /8, /16 and /24, respectively. We observe that, for example, the netmask for Class B is 255.255.0.0. This is written in binary form as 11111111.11111111.00000000.00000000. Assume that a host has the Class B address 173.45.80.12. To determine the netID of the address we need to apply a bitwise AND operation between the given IPv4 address and the Class B netmask. A bitwise AND operation saves the netID portion of the address while discarding the host portion. The results of the bitwise AND operation is netID = 173.45.0.0 and hostID = 80.12.

There are two types of subnet masks: fixed-length subnet mask (FLSM) and variable-length subnet mask (VLSM), as illustrated in Fig. 2.17. FLSM is also referred to as *classful subnetting*. We discuss in this section only FLSM.

2.2.3.1 *Subnetting Example*

We use an example to explain the subnetting process. Assume that a company has three sites each with up to five workers altogether. Assume that such an organisation has acquired from a local ISP the IPv4 Class C network number 201.89.23.0.

Net ID	Host ID	
Net ID	Subnet ID	Host ID
The same for all subnets of a network	Different for each subnet of a network	Different for each host

Fig. 2.16 The three IP address components of a subnet.

Net ID	Host ID	
Net ID	Subnet ID	Host ID
The same for all subnets of a network	Different for each subnet of a network	Different for each host

Mask Type / Feature	FLSM (Classful Subnetting)	VLSM (Classless Subnetting)
Subnet size	Same for all subnets	Different for each subnet
Available # of host addresses	Same for all subnets	Different for each subnet
Efficiency	Wastes IP addresses	Economizes IP address usage

Fig. 2.17 Types of subnet masks.

Assume that altogether each branch of the organisation needs to interconnect six additional network devices (e.g. routers, gateway, printer, scanner, facsimile). How do we build the network?

Step 1: We know that we need $N_{\mathrm{S}} > 3$ subnets or segments (one subnet for each site) and $N_{\mathrm{H}} > 5 + 6 = 11$ hosts for the largest site/subnet.

Step 2: Number of subnet bits (S) and number of host bits (H) needed? $S \geq log_2(N_{\mathrm{S}}) = log_2(3) \approx 1.6$ and $H \geq log_2(N_{\mathrm{H}}) = log_2(11) \approx 3.6$. Therefore, we need at least 2 bits for subnets and 4 bits for hosts in each subnet. An IPv4 Class C address uses 8 bits for hostID. As $2 + 4 = 6 < 8$, this is possible. We have room in the network design as to how to use the remaining 2 bits. Let us choose $S = 3$ and $H = 5$ to give room for the organisation to grow in both branches and employees. Thus, we have $2^H = 32$ host addresses (including the subnet address and directed broadcast address) for each subnet.

Step 3: Determine the subnet mask. Now, the total bits for network address and the subnet address is $24 + 3 = 27$. Therefore, the network mask is 201.89.23.0/27. This can also be written as

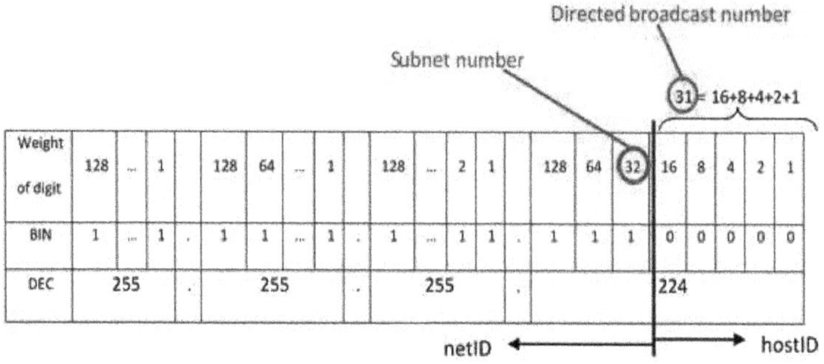

Fig. 2.18 Subnetting in networks using IPv4 Class C.

255.255.224.0 or in its equivalent binary and hexadecimal notations.

Step 4: Determine the subnet addresses (aka network addresses or network numbers). We note that the network (or subnetwork) address has all the host bits (here the last 5 bits) set to zeros (0s). We are going to use binary arithmetic to compute the *directed broadcast number* and the *subnetwork number* of the first subnet, and then build upon it. Note that subnetting plays around only the host part of the given IPv4 address. Here, we have a Class C address so the playing field is the last byte or octet. This octet is partitioned into host part and subnet part as shown in Fig. 2.16 and Fig. 2.18.

The sum of the weights of all the bits in the host part is called the direct broadcast number, which is here $31 = 16 + 8 + 4 + 2 + 1$. This number (i.e. 31) forms the last byte of the first direct broadcast address, which is also the highest address in the first subnet, aka subnet 0 (see Fig. 2.19). Noting that the given address being subnetted is 201.89.23.0/27, we obtain the first direct broadcast address as 201.89.23.31. Given 201.89.23.0/27, we know that the first subnet has the address 201.89.23.0. These two addresses are entered in Fig. 2.19. Recalling from above, the subnet 201.89.23.0 and 201.89.23.224 in Fig. 2.19 are the subnet 0 which cannot be used unless subnet 0 is enabled on the router.

The weight of the last bit in the subnet part is called **network number**, which is here $2^5 = 32$. Now, increasing the first subnet 0's address in steps of 32 (i.e. the network number) gives the subnetwork numbers of the remaining subnets. These are entered in Fig. 2.19. Note from Fig. 2.19 that the subnetwork number or address occurs in the last byte of the address. We also know that each subnet has up to $2^H = 32$ hosts. Therefore, increasing the first directed broadcast address by 32 produces the directed broadcast addresses of the remaining seven subnets. So, these values are entered in Fig. 2.19. Now that we know the beginning address (i.e. subnet address) and ending address (i.e. directed broadcast address) of each subnet, it is straightforward to know that what lie between the two numbers are the range of host addresses. This completes the design as seen in Fig. 2.19. Figure 2.20 shows the subnets (i.e. logical divisions) of the network 201.89.23.0.

2.2.4 *VLSM and Classless Subnetting in IPv4*

The variable-length subnet masking (VLSM) is said to be the advanced type of subnetting as it allows subnet masks of varying lengths for subnets of a given IP network. Simply put, we use VLSM to divide a given IP network into subnets which support varying number of hosts. This is impossible if we required all subnets to use the same subnet mask. There are two main reasons why VLSM is used: (a) reduction in the wastage of IPv4 addresses and its lifetime extension, and (b) reduce the size of routing tables in routers by using path aggregation (aka path summarisation). We shall explain these terms in this section. Addresses are wasted in subnetting using fixed-length subnet masks. This is because all subnets under a network use the same number

Subnet #	Subnet address/number	Hosts address range		Directed broadcast address
		From	To	
0.	201.89.23.0	201.89.23.1	201.89.23.30	201.89.23.31
1.	201.89.23.32	201.89.23.33	201.89.23.62	201.89.23.63
2.	201.89.23.64	201.89.23.65	201.89.23.94	201.89.23.95
3.	201.89.23.96	201.89.23.97	201.89.23.126	201.89.23.127
4.	201.89.23.128	201.89.23.129	201.89.23.158	201.89.23.159
5.	201.89.23.160	201.89.23.161	201.89.23.190	201.89.23.191
6.	201.89.23.192	201.89.23.193	201.89.23.222	201.89.23.223
7.	201.89.23.224	201.89.23.223	201.89.23.254	201.89.23.255

Fig. 2.19 Creating subnets for an IPv4 address.

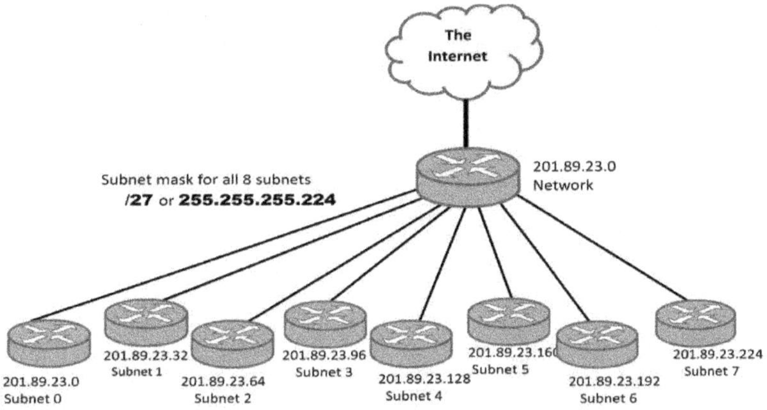

Fig. 2.20 Subnetted network layout for 201.89.23.0.

of bits to number their hosts, although the number of hosts attached to each subnet may vary widely. The VLSM was designed to solve this problem in order to extend the IPv4 address exhaustion date. VLSM is also referred to as **classless subnetting**.

Again, we use a case study to explain classless subnetting. Consider a university with the five departments: Engineering (ENG) with 20 staff, Business (BUS) with 31 staff, Administration (ADMIN) with 5 staff, Information Technology helpdesk (IT) with 3 staff and Library (LIB) requiring 121 hosts. What IPv4 address is the best both technically and economically? There are several alternatives in this IPv4 planning task. Assume that we have the network with the physical topology shown in Fig. 2.21. There are five point-to-point wide-area network (WAN) links interconnecting the five departments or segments. The departments can be located in different cities or even different countries. Each WAN connection requires two IP addresses. Also, each WAN link is a subnet or segment. Therefore, the network in Fig. 2.21 has ten subnets, which altogether requires $2 \times 5+5+20+31+3+121 = 190$ host addresses.

A network with 10 segments and 190 hosts can be planned in different ways. We can use FLSM or VLSM. First let us use FLSM in order to compare the two in IPv4 address usage efficiency. For FLSM we need 4 bits for the 10 subnets and 7 bits for the hosts of the largest segment. This makes 4+7=11 bits. Clearly, one Class C address would not be enough. The IP planning choices can be:

(1) Two Class C addresses /25 and /27, such as 221.80.45.0/25 and 221.80.46.0/27. The /25 address serves $2^1 = 2$ subnets each with up to $2^7 - 2 = 126$ hosts. This is enough to serve the

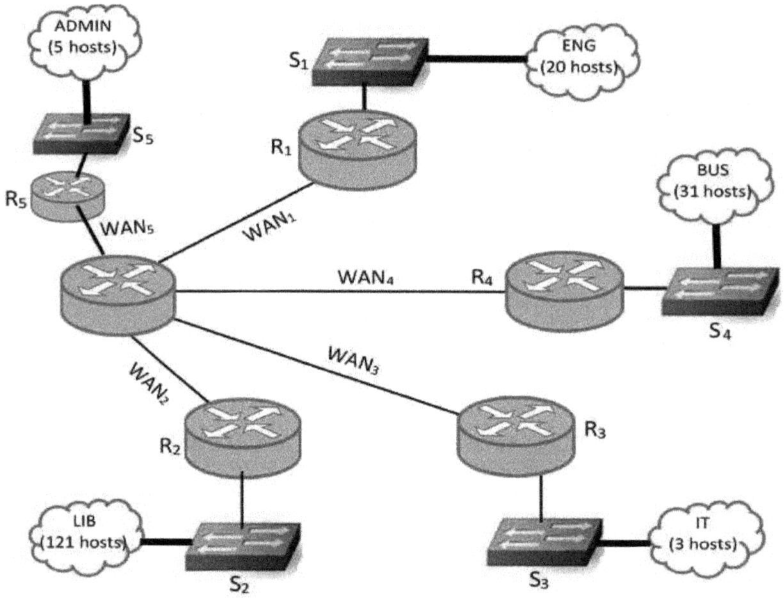

Fig. 2.21 An IP network for a company with five departments separated geographically.

two largest segments LIB with 121 hosts and BUS with 31 hosts. Again, the 2 IP addresses deducted are for the directed broadcast and subnet numbers. The /27 address serves the remaining $2^3 = 8$ segments, each with $2^5 - 2 = 30$ hosts. These serve the five WAN links (each requiring 2 addresses), ENG (20 hosts), ADMIN (5 hosts) and IT (3 hosts). The number of IP addresses wasted in this IP planning is $(126 - 121) + (126 - 31) + 5 * (30 - 2) + (30 - 20) + (30 - 5) + (30 - 3) = 302$. Note that if subnet zero is not enabled on the network then we will need more than two Class C addresses.

(2) One Class B address /20, such as 184.47.0.0/20. Here, there are 4 bits for subnets and 12 bits for hosts. This gives $2^4 = 16$ subnets, each with up to $2^{12} - 2 = 4,094$ host addresses. Altogether, we need only 10 subnets and 190 IPv4 addresses. However, we have to pay for $2^{16} = 65,536$ addresses. Assuming each address costs \$11.25 we would waste $(65,536 - 190) \times \$11.25 = \$735,142.50$! The next step would be your boss eyeing to make the network architect redundant.

We have observed that it would be cheaper buying two Class C addresses than buying a single Class B address, although both waste some money. This alone gives enough motivation to use VLSM.

VLSM is about subnetting of networks in a manner which saves IPv4 addresses and then money. The algorithm used to subnet a network using VLSM is:

(1) Start with the largest network segment. This is the subnet requiring the largest number of host addresses.
(2) Find the appropriate subnet mask for the largest subnet found in step 1.
(3) List all subnets for the subnet mask found above and their appropriate host and directed broadcast addresses.
(4) Assign a subnet to the largest segment. Keep the remaining subnets to be further subnetted.
(5) Continue with the next largest subnet/segment.
(6) Make the second largest subnet a subnet of the largest subnet above. Find the appropriate subnet mask for it.
(7) Repeat from Step 5 until the appropriate subnet mask has been found for all the segments in the network.

We assume we have the single Class C address 221.80.45.0/24 and apply the VLSM algorithm to our case study in Fig. 2.21. We begin with the

Subnet #	Subnet/network ID	Valid host IP address range	Directed broadcast address	Purpose
0	221.80.45.0/25	221.80.45.1 - 221.80.45.126	221.80.45.127	LIB (Library)
1	221.80.45.128/25	221.80.45.129 - 221.80.45.254	221.80.45.255	To be subnetted

(a) Finding the largest subnet

Subnet #	Subnet/network ID	Valid host IP address range	Directed broadcast address	Purpose
2	221.80.45.128/26	221.80.45.129 - 221.80.45.190	221.80.45.191	BUS
3	221.80.45.192/26	221.80.45.193 - 221.80.45.254	221.80.45.255	To be subnetted

(b) Finding the second largest subnet

Subnet #	Subnet/network ID	Valid host IP address range	Directed broadcast address	Purpose
4	221.80.45.192/27	221.80.45.193 - 221.80.45.222	221.80.45.223	ENG
5	221.80.45.224/27	221.80.45.225 - 221.80.45.254	221.80.45.255	To be subnetted

(c) Finding the third largest subnet

Fig. 2.22 Illustrating classless subnetting process.

largest subnet requiring the largest number of hosts. In our example above, the largest subnet is Library which has 121 hosts. Then, we find the appropriate subnet mask for this largest subnet. In our case we need 7 bits to number 121 hosts. Thus, we need the subnet mask /25 or 255.255.255.128. As we have robbed only one bit from the host ID part, we have two subnets with the details in Fig. 2.22(a).

We are going to further subnet the subnet 1 with the subnet ID 221.80.45.128/25 to accommodate the remaining network segments beginning with the second largest network segment. The second largest network segment in our case study is BUS with 31 hosts, requiring 6 bits. Thus the appropriate subnet mask is /26 or 255.255.255.192, as illustrated in Fig. 2.22(b). Now, our next largest network segment is ENG with 20 hosts. As $2^4 = 16$ and $2^5 = 32$, we need 5 bits to number the 20 hosts. Thus, the appropriate subnet mask is /27 or 255.255.255.224. Noting that we are subnetting 221.80.45.192/26, we obtain the results in Fig. 2.22(c).

We are going to subnet the subnet 221.80.45.224/27 to accommodate the remaining subnets beginning with the largest segment among the remaining segment. In our case study the largest segment among the remaining segments is ADMIN with 5 hosts. We need 3 bits to number 5 hosts. Thus, the required subnet mask is /29 or 255.255.255.248. We note that the next largest subnet is IT with 3 hosts which can also be served with the mask /29. If we subnet 221.80.45.224/27 to 221.80.45.224/29 we obtain 4 subnets, each of which supports 6 hosts. We allocate the first two of the 4 subnets to ADMIN and IT as shown in Fig. 2.23. The remaining two subnets are to be used for the five point-to-point WAN links.

The WAN links can be allocated IP addresses using the mask /30 or 255.255.255.252 as follows. We subnet 221.80.45.240/29 into four subnets with the subnet mask /30 as shown in Fig. 2.23. This yields the subnets in Fig. 2.23. We are left with the WAN link 5. We can purchase a Class C address such as 221.80.46.0/30 for it. In recap, we obtain the address allocations in Fig. 2.23 for the 10 segments of the network. The resulting IP-planned network with 5 segments and 5 WAN links is shown in Figs. 2.23 and 2.24.

As can be observed in Fig. 2.23, there is much less wastage in using VLSM compared to the FLSM design. The penalty in conserving IP addresses is the lack of much room for network growth. For example, if the number of hosts in Library, Business and Engineering departments grow by more than 5, 31 and 10, respectively, the computers have to be renumbered.

Segment	Subnet ID	Host IP address range	Directed broadcast address	Wasted IP addresses
LIB	221.80.45.0/25	221.80.45.1 - 221.80.45.126	221.80.45.127	5
BUS	221.80.45.128/26	221.80.45.129 - 221.80.45.190	221.80.45.191	31
ENG	221.80.45.192/27	221.80.45.193 - 221.80.45.222	221.80.45.223	10
ADMIN	221.80.45.224/29	221.80.45.225 - 221.80.45.230	221.80.45.231	1
IT	221.80.45.232/29	221.80.45.233 - 221.80.45.238	221.80.45.239	3
WAN_1	221.80.45.240/30	221.80.45.241 - 221.80.45.242	221.80.45.243	0
WAN_2	221.80.45.244/30	221.80.45.245 - 221.80.45.246	221.80.45.247	0
WAN_3	221.80.45.248/30	221.80.45.249 - 221.80.45.250	221.80.45.251	0
WAN_4	221.80.45.252/30	221.80.45.253 - 221.80.45.254	221.80.45.255	0
WAN_5	221.80.46.0/30	221.80.46.1 - 221.80.46.2	221.80.46.3	0

Fig. 2.23 Addresses for each of the 10 subnets under VLSM subnetting.

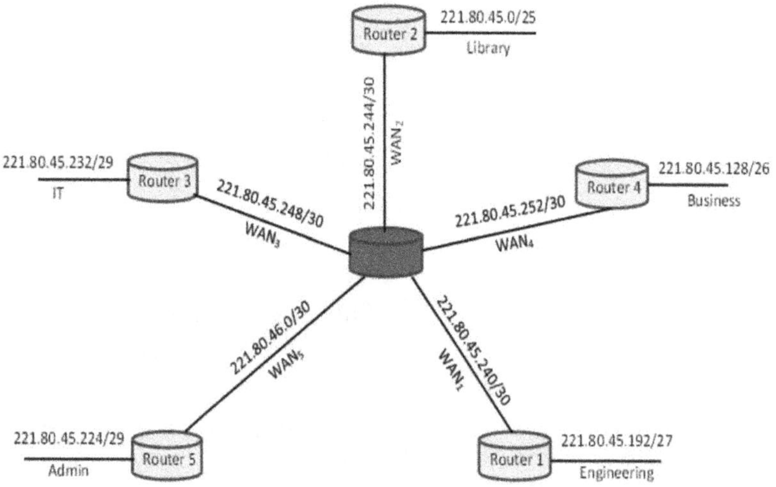

Fig. 2.24 Ten subnets (5 WAN links and 5 segments) with VLSMs.

2.2.5 *Classless Interdomain Routing (CIDR)*

As we observed earlier, one disadvantage in classful IPv4 addressing is the wastage of already near-exhaustion IPv4 addresses. Therefore, CIDR (classless inter-domain routing), also referred to as *supernetting*, extends the traditional subnetting to make elastic the number of bits in the IPv4 addresses used as netID or hostID. This way the IPv4 can be configured based on the size of subnets and hosts per subnet. It is worthy of note that the default masks are no more valid in a CIDRized network. CIDR is a generalisation of subnetting addressing using the notation $w.x.y.z/n$ (or just/n) where $w, x, y, z < 255$ in decimal and

- the first $n < 32$ bits of the 32-bit IPv4 address is the network number (aka netid, prefix or network prefix).
- The remaining rightmost $32 - n$ bits identify a specific host within the network and thus referred to as hostID.

In the CIDR notation $w.x.y.z/n$, n is called **prefix length**. We note that the CIDR notation

Direct broadcast number
63 = 32+16+8+4+2+1

Subnet number

Weight of digit	128	64	...	1		128	64	...	2	1		128	...	2	1		128	64		32	16	8	4	2	1
BIN	1	1	...	1	.	1	1	...	1	1	.	1	...	1	1	.	1	1		0	0	0	0	0	0
DEC	255				.	255					.	255				.									

netID ← → hostID

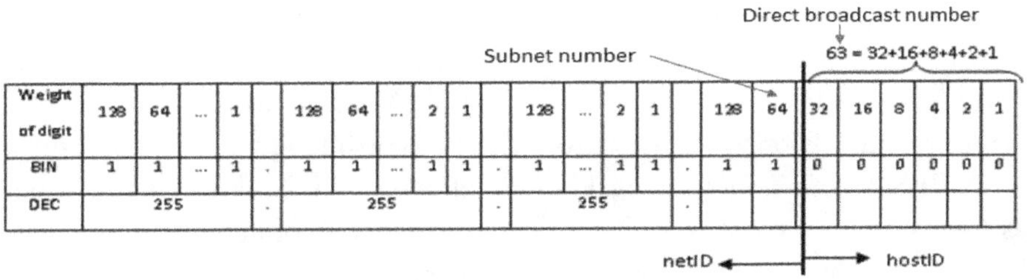

(a) Finding direct broadcast and subnet numbers

Subnet ID/address	Address range for hosts in subnet	Subnet directed broadcast IP address	Number of hosts per subnet
x.x.x.0	x.x.x.1 – x.x.x.62	x.x.x.63	$62=2^6-2$
x.x.x.64	x.x.x.65 – x.x.x.126	x.x.x.127	62
x.x.x.128	x.x.x.129 – x.x.x.190	x.x.x.191	62
x.x.x.192	x.x.x.193 – x.x.x.254	x.x.x.255	62

(b) Computing subnets' identities

Fig. 2.25 Illustration of classless interdomain routing (CIDR).

is shorthand which merges an IP address and its subnet mask. For example, 181.54.167.0/21 means 181.54.167.0 255.255.248.0. In another example, assume that we are given the CIDR /26 for a network with netmask 255.255.255.0/24. We need to (a) find the number of subnets supportable, and for each subnet find (b) the subnet ID, (c) the broadcast address and the (d) address range for hosts. We solve the problem using the two illustrations in Fig. 2.25.

From Fig. 2.25(a) we find the direct broadcast number and the subnet number as 63 and 64, respectively. Therefore, /26 has the three uniquely-identifiable subnets shown in Fig. 2.25(b), each of which can host up to 62 hosts plus reserved broadcast and subnet addresses. The last octet of the first subnet ID equals the weight of the last bit in the netID. The last bytes of the remaining subnet IDs are obtained by adding 64 to the subsequent one. The broadcast address of the first subnet is just the sum of the hostID weight (i.e. 63) and the last byte of the first subnet ID (i.e. 64). The remaining broadcast addresses are obtained by progressively adding 64 to the subsequent address. Note that the subnet addresses and the broadcast addresses are reserved and cannot be assigned to hosts.

2.2.6 *WAN Connectivity Technologies*

The most popular wide area network (WAN) technologies are:

(1) Internet-based connectivity.

(2) Software-Defined WAN (SD-WAN).

(3) MPLS (multi-protocol label switching).

(4) Carrier Ethernet: including Virtual Private LAN Services (VPLS), Gigabit Ethernet (GE) and Metro Ethernet (ME).

The most popular WAN technology is Internet and MPLS. However, there is a growing trend of companies migrating from MPLS services to Internet-based WAN connectivity. The following four reasons have been found to be the causes:

- *Network flexibility*: the ability to change the configuration of a LAN connection without the involvement of its provider is essential to organisations. Internet-based WAN provide this feature, while MPLS does not.

- *Connectivity to the cloud*: the Internet provides a public WAN connectivity, and it is the simplest and cheapest connectivity to the cloud infrastructure. MPLS connections are private. Although MPLS can provide direct connections

to the cloud, it is a pricy and complicated option.

- *Fast installation of connection*: Internet broadband services can be commissioned within three business days. MPLS connection installation, however, takes weeks or even months.
- *Bandwidth/data rate*: the Internet offers a cheaper bandwidth than MPLS. Also, high-rate MPLS connections are not available everywhere, especially in remote locations.

One of the main concerns in Internet-based WAN connectivity has been with security as the Internet is a public infrastructure. A solution to this problem is the creation of site-to-site virtual private network (VPN) technologies. Another drawback in Internet-based WAN arises from the business model in which multiple Internet Service Providers (ISPs) may provide WAN connections in the geographical location that an enterprise operates. Such a situation would require dealing with multiple ISPs. MPLS service providers can partner to provide single contract for a global WAN connection. Third, ISPs do not provide service level agreements (SLAs), but MPLS providers do. The reason being that Internet WAN is connectionless while MPLS WAN is connection oriented, making it easier to guarantee service levels. This means that network managers must be responsible for the monitoring of Internet-based WAN.

2.2.6.1 *Software-Defined WAN*

A WAN is a network with a large coverage area. Its applications include interconnecting branches of a company widely separated geographically or connecting a corporate network to a data center located at a distance away. What is SD-WAN? It is an application of software-defined networking (SDN) in WAN connections based on MPLS, broadband Internet, cellular networks, etc. It decouples control mechanism, and traffic management and monitoring functions from the hardware to simplify the management and operation of a WAN. Traditional WAN connections like MPLS uses proprietary hardware, making reconfiguration of connections complicated and costly. The trend now in WAN is geared toward open, flexible and cloud-based WAN technologies which are secure. Table 2.3 compares SD-WAN with MPLS.

According to Cisco Systems, SD-WAN is a software-defined approach to managing the WAN in order to exploit several advantages, including the following (Cisco Systems):

(1) Reducing costs with transport independence across MPLS, 3G/4G LTE, etc.
(2) Improving business application performance and increasing agility.
(3) Optimising the user experience and efficiency for SaaS (Software-as-a-Service) and public cloud applications.
(4) Simplifying operations with automation and cloud-based management.

2.2.6.2 *MPLS*

Multiprotocol Label Switching (MPLS) combines the good features of ISO/OSI Layer 2 switching and Layer 3 routing. Layer 3 routing is flexible and scalable while Layer 2 switching has good performance and traffic-management capabilities. MPLS comprises a set of protocols providing many functions, such as:

- Interconnecting multiple sites of an organisation to create a virtual private network (VPN).
- Providing connection-oriented traffic engineering in packet-switched networks that is faster and more efficient than what the native IP protocol provides.
- Providing varying quality-of-service to different traffic types.

Figure 2.26 shows a snapshot of the development of MPLS.

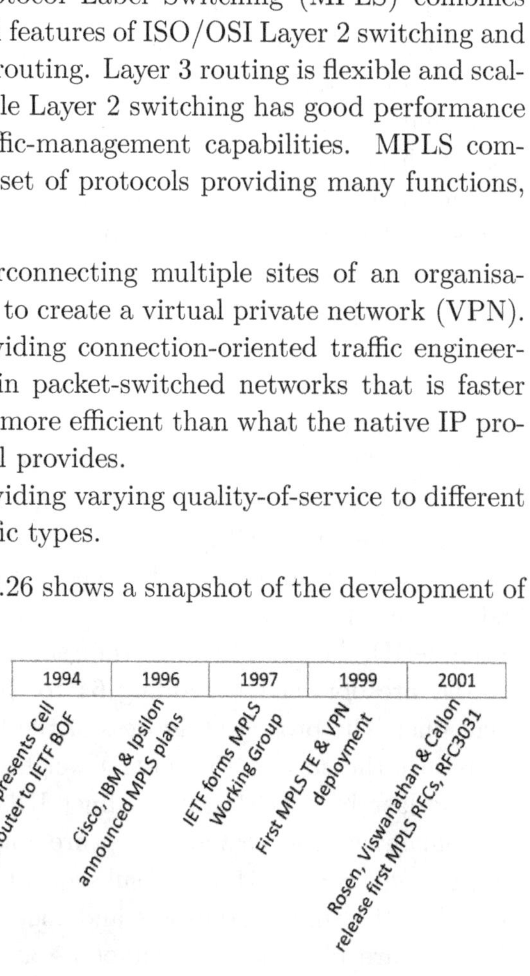

Fig. 2.26 An abbreviated history of MPLS.

Table 2.3 Comparison between MPLS and SD-WAN.

Feature	SD-WAN	MPLS
Cloud access	Direct	Via backhaul to data center
QoS	Defined by business policies	High QOS due to dedicated connections
Security	Use 3rd party security providers or embed	Good because of dedicated connection and traffic passes through datacenter for inspection
Connectivity choices	Hybrid connections from LTE, MPLS and broadband	Only MPLS
Scalability	Can be done on demand	Takes months to scale capacity

The word *multiprotocol* in MPLS indicates its ability to provide a Layer 2 method to forward protocol data units (PDUs) of various protocols over a packet-switched networks. As an ISO/OSI Layer 2.5 protocol, the process by which MPLS forwards PDUs between a pair of MPLS-enabled routers, called label switched routers (LSRs), is called switching but not routing, the reason for *switching* in its name. Switching is faster than routing, as the latter processes more information than the former. The *label* in MPLS means that a label instead of a Layer 3 IP address or a Layer 2 MAC address is used to forward a PDU. Also, MPLS labels have a fixed length of 32 bits, contrasting the variable-length 32-bit IPv4 address and 128-bit IPv6 address. Fixed-length address is faster to process than variable-length address just as a shorter address is processed quicker than a longer address. MPLS can forward the PDUs of ISO/OSI Layer 2 protocols, including DSL, T1/E1, PDH, SONET/SDH, Ethernet, PPP (Point-to-Point Protocol), ATM and Frame Relay.

2.2.7 *Check Your Understanding*

Students are encouraged to check their understanding of the content of this section with the following questions prior to proceeding to the next section.

(1) Describe how circuit switching differs from packet switching.

(2) Analyse the main differences between virtual-circuit packet switching and datagram packet switching.

(3) Compare and contrast cut-through switching and store-and-forward switching.

(4) What type of IP address is 10.23.50.89?

(5) In IPv4-based networks, discuss the purposes of, and the differences between:

  (a) Network number/address.
  (b) Directed broadcast number/address.
  (c) Subnet number/address.
  (d) Limited broadcast number/address.
  (e) Default gateway number/address.
  (f) Loopback number/address.
  (g) Local number/address.

(6) List three mechanisms used to extend the lifetime of IPv4.

(7) Discuss three differences between IPv4 and IPv6.

(8) Analyse how broadcasting in IPv6-based networks differ from that of IPv4-based networks.

(9) Cite three reasons to defend the process of subnetting a network. Briefly discuss each advantage given.

(10) List and briefly discuss three reasons while MPLS was developed.

(11) You are required to create subnets for the Class-B network address 172.16.0.0/16. It is required that you subnet this network using CIDR /19. Answer the following

questions:

(a) What is the first subnet address?

(b) What is the address of the third host on the first subnet?

(c) What is the valid host range in the fourth subnet?

(12) If 192.168.1.8 needs to subnetworked into 11 subnets what will be the subnet mask, subnet addresses, address range for each subnet and the directed broadcast address for each subnet?

(13) Assume that you are given the Class A network 21.0.0.0. You are expected to use it to design a network to serve a company with 12,000 branches each requiring 5 to 923 hosts.

(a) What is the suitable subnet mask?

(b) For each subnetwork, what is the direct broadcast address and the address range for hosts?

(14) Assume that a bank with 1158 branches each with 5 to 24 workers altogether, has been issued the IPv4 Class B network number 161.50.0.0. (Note that Class C address would not be enough for 1158 network segments.) Assume that altogether each branch of the organisation needs to interconnect additional 5 network devices (e.g. routers, gateways, firewalls, printers, scanners, facsimile).

(a) What subnet mask should we use?

(b) For each subnet, compute the:

i. Subnetwork number/address.

ii. Direct broadcast address.

iii. Address range for hosts.

(15) Assume that the physical address printed in the network interface card of a computer is D7-4D-A3-F1-B2-5F. We seek to assign an IPv6 address to this device and use as IoT device. Compute the 64-bit (i.e. EUI-64) host ID for this device.

2.3 Basic Electrical Networks in Wireless Networks

An interconnection of one or more electrical components or elements such as resistors, capacitors, inductors, antennas, filters, amplifiers, duplexers, oscillators, mixers, current sources and voltage sources is called an **electrical network**. An electrical network is said to be **linear** if the relationship between its input and output is a linear function. There are many ways of categorising electrical networks. An interesting categorisation in microwave communications is the number of ports that the network has. As a *port* is a pair of terminals, an electrical network with $2n$ terminals has n ports. A LEN is connected to other electrical components through its ports. Most electrical components have two ports. Table 2.4 lists common examples of one-port through four-port networks. Some of these electrical devices, components or elements are used in all wireless communications systems. We briefly review them in a collective manner in this section.

2.3.1 *Parameters Describing Linear Electrical Networks*

A linear electrical network (LEN) produces an output if at least one of its ports is excited. The excitations or incident waves are called independent variables. The coefficients of the independent variables in the function describing the input/output relationship of a LEN are called **parameters**. These are the common six sets of parameters used to describe LENs:

(1) S-parameters
(2) Y-parameters
(3) Z-parameters
(4) h-parameters
(5) g-parameters
(6) T-parameters

Each of the above set of parameters can be used to uniquely characterise the terminals or ports of a LEN without considering its internal composition. In other words, we view the LENs as black boxes and study only their ports using these parameters. We shall use the block diagrams of the electrical

Table 2.4 Examples of electrical networks.

Network Type	Number of Terminals	Examples
One-port networks	2	Resistors, inductors, capacitors
Two-port networks	4	Antennas, filters, oscillators, impedance matching networks (e.g., $\frac{\lambda}{4}$ transformers, lumped elements (i.e., resistors, inductors, capacitors) and matching stubs), transmission lines (e.g., coaxial cables), transformers, amplifiers, transistors (small signal analysis)
Three-port networks	6	Duplexers (circulators), power dividers
Four-port networks	8	(Directional) couplers

networks shown in Fig. 2.27 for our discussions here. Note that some authors use instead small letters s, z and y.

2.3.1.1 *S-Parameters*

Scattering parameters (aka S-parameters) is one of the ways of describing the input/output characteristics on a LEN. A matrix whose elements are the S-parameters of a LEN is called S-matrix. The S-matrix for an n-port LEN is a square matrix of size n with the elements S_{ij}, where i is the output port and j is the input port. The S-parameters can be measured using a spectrum analyser. Using the variables in Fig. 2.27(d), we can write the general form of the input/output relationship of the LEN as

$$\mathbf{J} = \begin{bmatrix} J_1 \\ J_2 \\ \vdots \\ J_n \end{bmatrix} = \mathbf{S}_n\mathbf{I} = \begin{bmatrix} S_{11} & S_{12} & \dots & S_{1n} \\ S_{21} & S_{22} & \dots & S_{2n} \\ \vdots & \vdots & \ddots & \vdots \\ S_{n1} & S_{n2} & \dots & S_{nn} \end{bmatrix} \cdot \begin{bmatrix} I_1 \\ I_2 \\ \vdots \\ I_n \end{bmatrix}. \tag{2.2}$$

There are special cases for the S-parameters. For example, the reflection coefficient at the kth port is

$$S_{kk} = \Gamma_k = \frac{output}{input} = J_k/I_k\,. \tag{2.3}$$

Therefore, if $S_{kk} = 0$, then there is a perfect impedance match at the port. The **return loss** is defined as

$$L_{r,k} = -20\log_{10}|\Gamma_k| = -20\log_{10}|S_{kk}|\,. \tag{2.4}$$

Also, for *two-port LENs* the following definitions are valid:

`Passive network:`

$$|S_{ij}| \leq 1, \forall i \neq j \tag{2.5}$$

`LEN's phase shift:`

$$\phi = atan\left(\frac{Im\{S_{21}\}}{Re\{S_{21}\}}\right) \tag{2.6}$$

`Insertion losses (IL):`

$$\alpha_j = -20\log_{10}|S_{ij}|, j \neq i \tag{2.7}$$

`Transmission coefficients:`

$$T_{21} = S_{21} \texttt{ and } T_{12} = S_{12} \tag{2.8}$$

`Symmetric/reciprocal network:`

$$S_{ij} = S_{ji}, \forall i \neq j \tag{2.9}$$

`kth port of symmetric lossless network:`

$$\sum_{i=1}^{n} |S_{ik}|^2 = 1 \tag{2.10}$$

The insertion loss is also called attenuation. A lossless circuit has no power dissipation inside it. Therefore, ideally a lossless LEN does not heat up. This also means that the power entering one of its ports equals the sum of the powers exiting all its other ports plus any power reflected back from the input port.

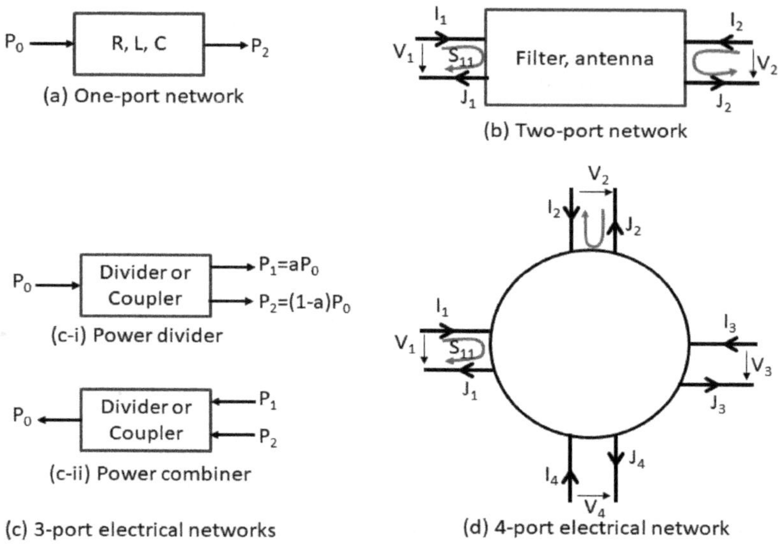

Fig. 2.27 Examples of electrical networks.

2.3.1.2 *Y-Parameters*

The variable Y is used to refer to the admittance, measured in Siemens (S), of an electrical element. From Fig. 2.27(d), we can define the input/output functions using the Y-parameters as

$$\mathbf{I} = \begin{bmatrix} I_1 \\ I_2 \\ \vdots \\ I_n \end{bmatrix} = \mathbf{Y}_n\mathbf{V} = \begin{bmatrix} Y_{11} & Y_{12} & \dots & Y_{1n} \\ Y_{21} & Y_{22} & \dots & Y_{2n} \\ \vdots & \vdots & \ddots & \vdots \\ Y_{n1} & Y_{n2} & \dots & Y_{nn} \end{bmatrix} \cdot \begin{bmatrix} V_1 \\ V_2 \\ \vdots \\ V_n \end{bmatrix}. \tag{2.11}$$

The Y-parameters for a two-port LEN can be obtained by short-circuiting one of the two ports, which are

$$\begin{aligned} V_1 = 0: \; Y_{12} = \frac{I_1}{V_2} \text{ and } Y_{22} = \frac{I_2}{V_2} \\ V_2 = 0: \; Y_{21} = \frac{I_2}{V_1} \text{ and } Y_{11} = \frac{I_1}{V_1}. \end{aligned} \tag{2.12}$$

As all the Y-parameters are obtained from short circuits, they are also referred to as *short-circuit admittances*.

2.3.1.3 *Z-Parameters*

The variable Z denotes the impedance of an electrical element with the units ohm (Ω). Using the Ω as the circuits parameter, we obtain the input/output functions

$$\mathbf{V} = \begin{bmatrix} V_1 \\ V_2 \\ \vdots \\ V_n \end{bmatrix} = \mathbf{Z}_n\mathbf{I} = \begin{bmatrix} Z_{11} & Z_{12} & \dots & Z_{1n} \\ Z_{21} & Z_{22} & \dots & Z_{2n} \\ \vdots & \vdots & \ddots & \vdots \\ Z_{n1} & Z_{n2} & \dots & Z_{nn} \end{bmatrix} \cdot \begin{bmatrix} I_1 \\ I_2 \\ \vdots \\ I_n \end{bmatrix}. \tag{2.13}$$

The Z-parameters for a two-port LEN can be obtained by open-circuiting one of the two ports, which are

$$\begin{aligned} I_1 = 0 \texttt{ (open circuit)}: \; Z_{12} = \frac{V_1}{I_2} \text{ and } Z_{22} = \frac{V_2}{I_2} \\ I_2 = 0 \texttt{ (open circuit)}: \; Z_{21} = \frac{V_2}{I_1} \text{ and } Z_{11} = \frac{V_1}{I_1}. \end{aligned} \tag{2.14}$$

As all the Z-parameters are obtained from open circuits, they are also referred to as *open-circuit impedances*.

2.3.1.4 *ABCD Parameters*

Also referred to as transmission parameters (T-parameters) or chain parameters, the ABCD parameters are used to describe two-port LENs. Using the two-port LEN in Fig. 2.27(b) and (V_1, I_1) and $(V_2, -I_2)$ as independent and dependent vectors, respectively, we obtain the input/output functions

$$\begin{bmatrix} V_1 \\ I_1 \end{bmatrix} = \begin{bmatrix} A & B \\ C & D \end{bmatrix} \cdot \begin{bmatrix} V_2 \\ -I_2 \end{bmatrix}. \tag{2.15}$$

We obtain the ABCD parameters from Eq. (2.15) as follows

$$I_2 = 0 \text{ (open circuit)}: \ A = \frac{V_1}{V_2} \text{ and } C = \frac{I_1}{V_2}$$

$$V_2 = 0 \text{ (short circuit)}: \ B = -\frac{V_1}{I_2} \text{ and } D = -\frac{I_1}{I_2}. \tag{2.16}$$

The negative sign on I_2 means that the direction of the current I_2 is opposite to what is shown in Fig. 2.27(b). Therefore, it enters the load instead of the electrical network as desired. The conversion between the ABCD-parameters and S-parameters are

$$S_{11} = \frac{C}{A}, \qquad S_{12} = \frac{AD - BC}{A},$$
$$S_{21} = \frac{1}{A}, \qquad S_{22} = -\frac{B}{A}. \tag{2.17}$$

2.3.1.5 *h-Parameters*

Hybrid parameters (or h-parameters) are used in the modelling of transistors. The excitations or independent variables in the circuits used to compute the h-parameters are I_1 and V_2. For a two-port LEN, such as illustrated in Fig. 2.27(b), the input/output characteristics using the h-parameters are

$$\begin{bmatrix} V_1 \\ I_2 \end{bmatrix} = \begin{bmatrix} h_{11} & h_{12} \\ h_{21} & h_{22} \end{bmatrix} \cdot \begin{bmatrix} I_1 \\ V_2 \end{bmatrix}. \tag{2.18}$$

We obtain the h-parameters from Eq. (2.18) as follows

$$I_1 = 0 \text{ (open circuit)}: \ h_{12} = \frac{V_1}{V_2} \text{ and } h_{22} = \frac{I_2}{V_2}$$

$$V_2 = 0 \text{ (short circuit)}: \ h_{21} = \frac{I_2}{I_1} \text{ and } h_{11} = \frac{V_1}{I_1}. \tag{2.19}$$

We can observe that the h-parameters are obtained from the *mixture* of open and short circuits. This is the reason for the word *hybrid*, meaning mixture.

2.3.1.6 *g-Parameters*

The g-parameters are referred to as *inverse hybrid parameters*. The excitations here are I_2 and V_1. Figure 2.27(b), the input/output characteristics using the g-parameters are

$$\begin{bmatrix} I_1 \\ V_2 \end{bmatrix} = \begin{bmatrix} g_{11} & g_{12} \\ g_{21} & g_{22} \end{bmatrix} \cdot \begin{bmatrix} V_1 \\ I_2 \end{bmatrix}. \tag{2.20}$$

We obtain the h-parameters from Eq. (2.20) as follows

$$I_2 = 0 \text{ (open circuit)}: \ g_{11} = \frac{I_1}{V_1} \text{ and } g_{21} = \frac{V_2}{V_1}$$

$$V_1 = 0 \text{ (short circuit)}: \ g_{22} = \frac{V_2}{I_2} \text{ and } g_{12} = \frac{I_1}{I_2}. \tag{2.21}$$

Example

We use a transmission line (e.g., coaxial cables) to connect a voltage source (e.g., mains power supply) to a load (electrical device). Every transmission line has a characteristic impedance, Z_0. $Z_0 = 50\,\Omega$ and $Z_0 = 75\,\Omega$ are common for coaxial cables. Let the load's impedance be Z_L. Part of the power reflects back into the transmission line if the load and the line are not matched, i.e., $Z_0 \neq Z_L$. In such a case a lossless impedance matching network such as in Fig. 2.28 is placed between the load and the line to match the system. For a perfect match the combined impedance of the load and that of the matching network, referred to as *input impedance* $Z_{\mathtt{in}}$, must be equal to Z_0. All the ports of a multiport network must be matched to reduce signal reflection at junctions.

We use the electrical network in Fig. 2.28 to illustrate how each of the parameters discussed above are computed from other parameters.

(1) Find the Z-parameters of the matching network.
(2) Evaluate the Y-parameters of the matching network.
(3) Compute the h-parameters of the matching network.
(4) Compute the g-parameters of the matching network.

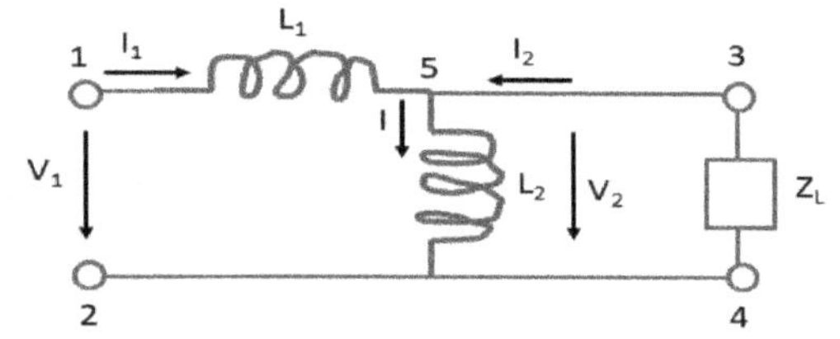

Fig. 2.28 An impedance matching network with lumped elements.

Solution

We apply three laws — Kirchhoff's current law (KCL), Kirchhoff's voltage law (KVL) and Ohm's law — to solve electrical circuit problems. The KCL states that the current flowing into a node (or a junction) of an electrical circuit must be equal to current flowing out of it. The KVL states that the sum of all voltages around any closed loop in a circuit equals zero. Ohm's law states that the voltage across an ideal conductor is proportional to the current through it. The proportionality constant is the impedance of the conductor.

(1) Z-parameters

$$\texttt{KVL:}\quad V_1 = Z_1 I_1 + V_2 = Z_1 I_1 (I_1 + I_2) Z_2 = (Z_1 Z_2) I_1 + Z_2 I_2$$

`Ohm's law:`

$$V_2 = Z_2 I = Z_2(I_1 + I_2) = Z_2 I_1 + Z_2 I_2$$

In matrix form these equations become

$$\begin{bmatrix} V_1 \\ V_2 \end{bmatrix} = \begin{bmatrix} Z_1 + Z_2 & Z_2 \\ Z_2 & Z_2 \end{bmatrix} \cdot \begin{bmatrix} I_1 \\ I_2 \end{bmatrix}. \qquad (2.22)$$

Thus, $Z_{11} = Z_1 + Z_2$, $Z_{22} = Z_{12} = Z_{21} = Z_2$.

(2) Y-parameters

$$V_1 = 0: \; I_1 = \frac{V_2}{Z_1} \Leftrightarrow Y_{12} = \frac{I_1}{V_2} = \frac{1}{Z_1}$$

$V_1 = 0$ `and KCL:`

$$I_2 = I - I_1 = \frac{V_2}{Z_2} - \frac{V_2}{Z_1} \Leftrightarrow Y_{22} = \frac{I_2}{V_2} = \frac{1}{Z_2} - \frac{1}{Z_1}$$

$V_2 = 0$ `and KVL:`

$$V_1 = I_1 Z_1 \Leftrightarrow Y_{11} = \frac{I_1}{V_1} = \frac{1}{Z_1}$$

and finally,

$V_2 = 0$ `and KCL:`

$$I_2 = I_1 = \frac{V_1}{Z_1} \Leftrightarrow Y_{21} = \frac{I_2}{V_1} = \frac{1}{Z_1}$$

(3) h-parameters

$V_2 = 0$ `and KCL:`

$$I_2 = I_1 \Rightarrow h_{21} = \frac{I_2}{I_1} = 1$$

$V_2 = 0$ `and KVL:`

$$V_1 = I_1 Z_1 \Rightarrow h_{11} = \frac{V_1}{I_1} = Z_1$$

$I_1 = 0$ `and KVL:`

$$V_1 = V_2 \Rightarrow h_{12} = \frac{V_1}{V_2} = 1$$

and lastly

$I_1 = 0$ `and KCL:`

$$I_2 = I = \frac{V_2}{Z_2} \Rightarrow h_{22} = \frac{I_2}{V_2} = \frac{1}{Z_2}$$

(4) g-parameters

$$I_2 = 0 \texttt{ and KCL: } I_1 = I = \frac{V_2}{Z_2} \qquad (2.23)$$

$I_2 = 0$ `and KVL:`

$$V_1 = I_1 Z_1 + V_2 \overset{(2.23)}{=} (Z_1 + Z_2) I_1 \Rightarrow g_{11} = \frac{I_1}{V_1} = \frac{1}{Z_1 + Z_2}$$

$I_2 = 0$ `and KVL:`

$$V_1 = I_1 Z_1 + V_2 \overset{(2.23)}{=} \frac{V_2}{Z_2} Z_1 + V_2 \Rightarrow g_{21} = \frac{V_2}{V_1} = \frac{Z_2}{Z_1 + Z_2}$$

$$V_1 = 0 \texttt{ and KVL: } I_1 Z_1 + V_2 = 0 \Rightarrow V_2 = -Z_1 I_1 \qquad (2.24)$$

$V_1 = 0$ `and KCL:`

$$I_2 = I - I_1 = \frac{V_2}{Z_2} - I_1 \overset{(2.24)}{=} -I_1(1 + Z_1/Z_2) \Rightarrow g_{12} = -\frac{Z_2}{Z_1 + Z_2}$$

and $V_1 = 0$ `and KCL:`

$$I_2 = I - I_1 \overset{(2.24)}{=} \frac{V_2}{Z_2} + \frac{V_2}{Z_1} \Rightarrow g_{22} = \frac{Z_1 Z_2}{Z_1 + Z_2}$$

2.3.2 *Filters*

A filter is a signal processing operation which removes frequency components from a signal which passes through it. A filter has four terminals and thus an example of two-port electrical networks. Filters are classified depending on the frequencies they suppress into low-pass filters (LPFs), band-pass filters (BPFs), band-reject filters (BRFs) and highpass filters (HPFs). BRFs are also called band-stop filters (BSF). A highly-selective BRF (i.e., with narrow stopband) is called notch filters. Notch filters are used in, for example, xDSL technologies to block frequencies for amateur radio (aka ham) transmissions.

As illustrated in Fig. 2.29(a), an ideal LPF passes frequencies from 0 Hz to a certain maximum frequency W, referred to as cut-off frequency. For example, if $W = 60$ Hz, then all frequencies in a signal passing through it which exceed 60 Hz are blocked. A BPF passes frequencies between two cut-off frequencies (Fig. 2.29(c) and (d)). Figure 2.29(g) and (h)) illustrate the BRF, which blocks only a given frequency band. Such a filter is used to block amateur radio signals from disturbing ADSL service, for example. Finally, a HPF passes frequencies from a given cut-off frequency.

Sharp transitions in any aspect of a time-domain signal, be it phase, amplitude or frequency, causes the signal to consume a large bandwidth, usually larger than the allocated bandwidth. The use of the smallest bandwidth to transfer a signal is important owing to the cost and limited nature of the frequency spectrum. Also, a signal covering a large bandwidth has the potential of interfering with other signals operating in neighbouring frequency bands. Filtering is used to avert abrupt changes in waveforms to reduce bandwidth consumption and its associated ill-effects. Another advantage in filtering is the reduction in the amount of thermal noise in the receiver to increase signal-to-noise ratio (SNR).

The SNR measures the quality of the signal, and it is an important figure-of-merit in communications systems. Another signal quality measure is the signal-to-noise and distortion ratio (SINAD), which is defined as

$$\begin{aligned} SINAD &= \frac{\texttt{total device output power}}{\texttt{noise} + \texttt{distortion}} \\ &= \frac{\texttt{signal + noise + distortion}}{\texttt{noise} + \texttt{distortion}} \\ &\approx \frac{\texttt{signal + noise}}{\texttt{noise}} \approx \frac{\texttt{signal}}{\texttt{noise}} = SNR. \end{aligned} \tag{2.25}$$

As the adage "there's no such thing as a free lunch" goes, filtering has its negative effects. First, as they limit bandwidth, filters in a circuit can cause inter-symbol interference (ISI). For example, the ISI introduced by the pre-modulation filtering using Gaussian-shaped LPF to obtain GMSK in GSM technology degrades the bit error rate by about 0.7 dB compared to its non-filtered version MSK [Kuchi and Prabhu (1999)]. Second, filters make circuits more complex and, possibly, larger and more power consuming. Third, filtering can cause the transition points between the states of a signal to overshoot. The problem in overshoot is that it requires the electronic circuit to transmit more power in the carrier signal than what is usually needed to transmit a symbol. For example, if we use a cosine filter with the roll-off factor of $\alpha = 0.2$, we reduce the bandwidth consumption from $2R_s$ to $1.2R_s$. The penalty in this is to increase the transmitted peak power in the carrier signal by 5 dB [Agilent Technologies (2018)]. We review some basic features of two classes of filters which are commonly used in communications systems. These are Nyquist filters and Gaussian filters. Examples of Nyquist filters are raised cosine (RC) filter and square-root-raised cosine (RRC) filter. Digital data to be transmitted usually pass through one of these filters to smoothen their sharp transitions prior to modulating the carrier signal in digital modulation systems (Fig. 4.3).

2.3.2.1 *Examples of Passive Filters and Their Bode Diagrams*

A filter can be passive or active. A passive filter contains no active elements, such as transistors. It contains only passive electrical elements: resistors, capacitors and inductors. For this reason, a passive filter does not need an external power

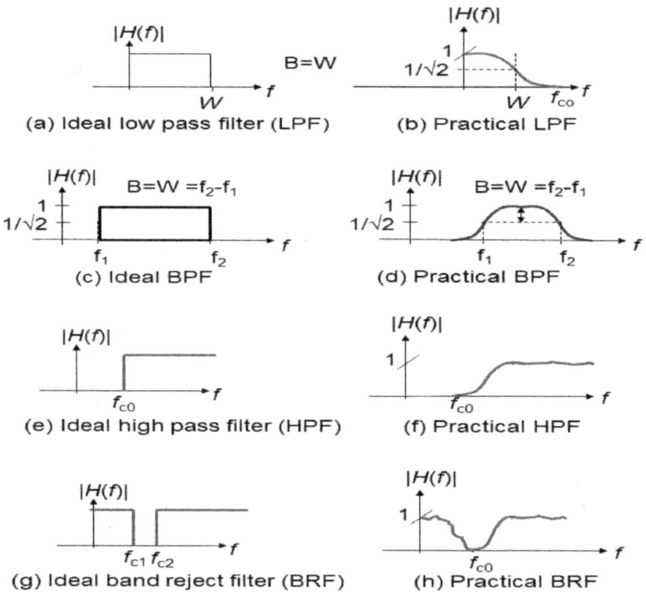

Fig. 2.29 Types of filters.

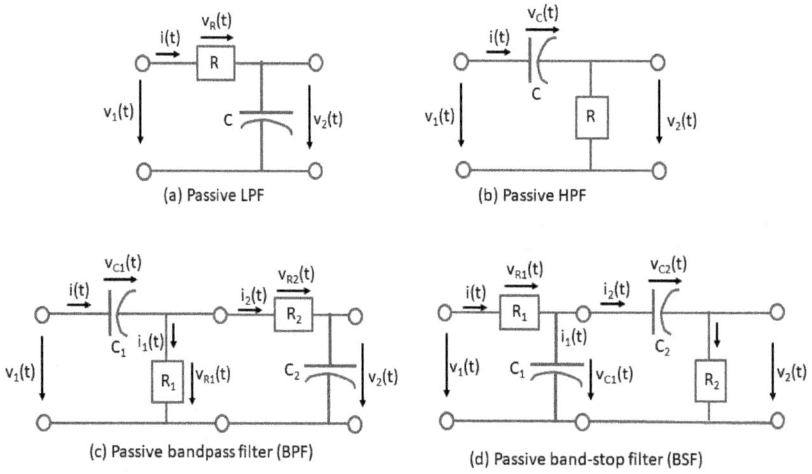

Fig. 2.30 Block diagrams of passive RC filters.

supply. An active filter contains at least one active element, which enables it to amplify input signals. Figure 2.30 shows circuits for simple passive filters. We use them as case studies for practical passive filters.

2.3.2.2 *Low-Pass Filter*

Let us analyse the simple passive LPF in Fig. 2.30(a). The voltage across the capacitor is $v_2(t)$ and the current is $i(t) = C\frac{dv_2(t)}{dt}$. Applying the KVL on Fig. 2.30(a) we obtain

$$\begin{aligned} v_1(t) &= v_R(t) + v_2(t) = i(t)R + v_2(t) \\ &= RC\frac{d}{dt}v_2(t) + v_2(t)\,. \end{aligned} \tag{2.26}$$

By noting the Fourier identity

$$\mathcal{FT}\{x(t)\} = X(f) \rightarrow \mathcal{FT}\left\{\frac{dx(t)}{dt}\right\} = j2\pi f X(f)\,,$$

we can transform Eq. (2.26) into the frequency domain as

$$\begin{aligned} V_1(f) &= j2\pi RCfV_2(f) + V_2(f) \\ &= (1 + j2\pi RCf)V_2(f)\,. \end{aligned} \tag{2.27}$$

The **transfer function** of an electrical circuit is defined as

$$H(f) = \frac{\texttt{output or response signal}}{\texttt{input or excitation signal}}$$
$$= \frac{V_2(f)}{V_1(f)}. \tag{2.28}$$

Applying Eq. (2.28) on (2.27), we obtain the transfer function of the passive LPF as

$$H(f) = \frac{V_2(f)}{V_1(f)} = \frac{1}{1 + j2\pi RCf}$$
$$= \frac{1}{1 + j2\pi f\tau_c} \tag{2.29}$$

where $\tau_c = RC$ is the time constant of the filter. The magnitude response of the LPF is then

$$|H(f)| = \sqrt{H(f)\overline{H(f)}} = \frac{1}{\sqrt{1 + (2\pi f\tau_c)^2}} \tag{2.30}$$

where $\overline{H(f)}$ is the complex conjugate of $H(f)$, while the phase response is

$$\phi(f) = \texttt{arg}\{H(f)\} = \texttt{tan}^{-1}\left(\frac{\mathcal{I}\{H(f)\}}{\mathcal{R}\{H(f)\}}\right)$$
$$= -\texttt{tan}^{-1}(2\pi f\tau_c). \tag{2.31}$$

The maximum magnitude of the LPF is $\texttt{maxH} = 1$ and it occurs at $f = 0$ Hz. Of particular importance is the 3-dB frequency, also called cut-off frequency, of the LPF. This is the frequency at which the power falls by half (or 3 dB) compared with its maximum value. That is

$$|H(f_{3dB})|^2 = \frac{1}{2}\texttt{maxH} \Leftrightarrow f_{3dB} = \frac{1}{2\pi RC}\ \texttt{[Hz]}. \tag{2.32}$$

By noting that $s = j2\pi f$ the Laplace transform version of the LPF is

$$H(s) = \frac{1}{1 + s\tau_c}. \tag{2.33}$$

The magnitude response in decibels and the phase response in degrees or radians as a function of frequency of an electrical circuit is referred to as **bode diagram** or **Bode plot**. Figure 2.31 shows the Bode plot of the LPF using its Laplace form and the MATLAB code:

```
clear all;
R=2; C=26.5*1e-6;
tc=R*C; % time constant
H=tf([1], [tc 1]); % call to MATLAB native function tf
bode(H, {1000, 25000}); % plots both magnitude and phase responses
```

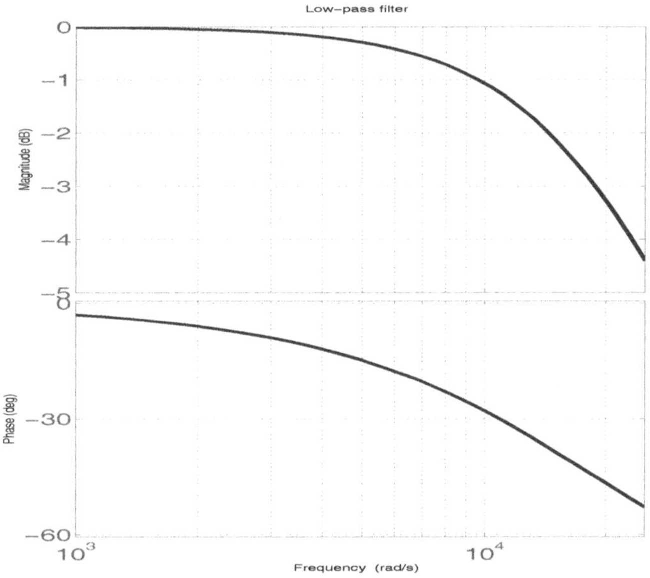

Fig. 2.31 The responses of the LPF filter in Fig. 2.30(a): $R = 2\,\Omega$, $C = 26.5\,\mu$F.

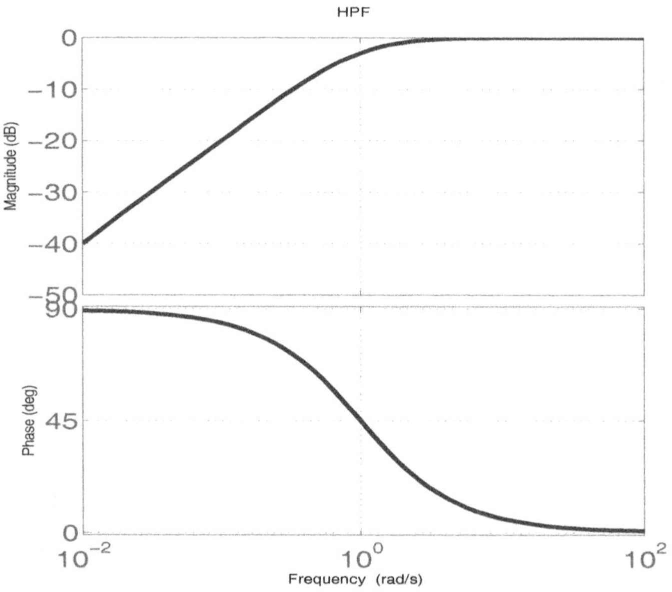

Fig. 2.32 The responses of the HPF filter in Fig. 2.30(b): $R = 1$ k$\Omega\,\Omega$, $C = 1$ mF.

Note that the first argument of the MATLAB function *tf* are the coefficients of numerator of the filter, while the second arguments are the coefficients of denominator of the filter.

2.3.2.3 *High-Pass Filter*

Let us analyse the simple passive HPF in Fig. 2.30(b). Note the voltage division

$$v_C(t) = \frac{Z_C}{Z_C + R} v_1(t)\,.$$

Then, the KVL yields

$$v_1(t) = v_C(t) + v_2(t) = \frac{Z_C}{Z_C + R} v_1(t) + v_2(t)\,. \tag{2.34}$$

We can apply Fourier transform on Eq. (2.34) to obtain

$$V_1(f)\left[1 - \frac{Z_C}{Z_C + R}\right] = V_2(f) \Leftrightarrow H(f) = \frac{V_2(f)}{V_1(f)}$$

$$= \frac{j2\pi f \tau_C}{1 + j2\pi f \tau_C}\,. \tag{2.35}$$

The magnitude and the phase responses are respectively

$$|H(f)| = \frac{2\pi f \tau_C}{\sqrt{1 + (2\pi f \tau_C)^2}} \tag{2.36}$$

$$\phi(f) = \texttt{arg}\{H(f)\} = \texttt{tan}^{-1}\left(\frac{\mathcal{I}\{H(f)\}}{\mathcal{R}\{H(f)\}}\right)$$

$$= \texttt{tan}^{-1}\left(\frac{1}{2\pi f \tau_C}\right). \tag{2.37}$$

Just as with the LPF, the gain of the HPF is low for frequencies below the cut-off frequency

$$f_{\texttt{3dB}} = \frac{1}{2\pi RC}\ \texttt{Hz}\,. \tag{2.38}$$

Figure 2.32 shows the responses of the filter. The plots were obtained using the MATLAB code in Table 2.5.

2.3.2.4 *Bandpass Filter*

A BPF circuit is shown in Fig. 2.30(c). We can obtain the following relationships:

$$\texttt{Ohm's law:}\quad i_2(t) = \frac{v_2(t)}{Z_{C2}} \tag{2.39}$$

Table 2.5 MATLAB code for Fig. 2.32.

```
clear all;
R=1000; C=1e-3;
tc=R*C; % time constant
H=tf([tc 0], [tc 1]); % call to MATLAB native function tf
bode(H); % plots both magnitude and phase responses
```

$$\texttt{KVL}: \ v_{\texttt{R1}}(t) = i_2(t)R_2 + v_2(t) \overset{(2.39)}{=} \frac{R_2 + Z_{\texttt{C2}}}{Z_{\texttt{C2}}} v_2(t) \,. \quad (2.40)$$

From the KVL, we obtain

$$\begin{aligned} v_1(t) &= v_{\texttt{C1}}(t) + v_{\texttt{R1}}(t) = \frac{Z_{\texttt{C1}}}{R_1 + Z_{\texttt{C1}}} v_1(t) + v_{\texttt{R1}}(t) \\ &\overset{(2.40)}{=} \frac{Z_{\texttt{C1}}}{R_1 + Z_{\texttt{C1}}} v_1(t) + \frac{R_2 + Z_{\texttt{C2}}}{Z_{\texttt{C2}}} v_2(t) \\ &\Leftrightarrow \ v_1(t)\left[\frac{R_1}{R_1 + Z_{\texttt{C1}}}\right] v_1(t) = \frac{R_2 + Z_{\texttt{C2}}}{Z_{\texttt{C2}}} v_2(t) \,. \end{aligned} \quad (2.41)$$

Transforming Eq. (2.41) into the frequency domain using Fourier transform and rearranging terms, we obtain

$$\begin{aligned} H(f) = \frac{V_2(f)}{V_1(f)} &= \frac{R_1 Z_{\texttt{C2}}}{(R_1 + Z_{\texttt{C1}})(R_2 + Z_{\texttt{C2}})} \\ &= \frac{j2\pi f \tau_1}{(1 + j2\pi f \tau_1)(1 + j2\pi f \tau_2)} \\ &= \frac{j2\pi f \tau_1}{1 - (2\pi f)^2 \tau_1 \tau_2 + j2\pi f(\tau_1 + \tau_2)} \end{aligned} \quad (2.42)$$

where we have substituted into Eq. (2.42) $Z_{\texttt{C1}} = \frac{1}{j2\pi f C_1}$, $Z_{\texttt{C2}} = \frac{1}{j2\pi f C_2}$, and $\tau_k = RC_k$, $k = 1, 2$.

The magnitude response of the BPF is

$$|H(f)| = \frac{2\pi \tau_1 f}{\sqrt{(1 - 4\pi^2 \tau_1 \tau_2 f^2)^2 + 4\pi^2(\tau_1 + \tau_2)^2 f^2}} \quad (2.43)$$

while the phase response is

$$\begin{aligned} \phi(f) &= \texttt{arg}\{H(f)\} \\ &= \texttt{tan}^{-1}\left(\frac{1 - 4\pi^2 f^2 \tau_1 \tau_2}{2\pi f(\tau_1 + \tau_2)}\right). \end{aligned} \quad (2.44)$$

Using the Laplace transform, the BPF's frequency response is

$$H(s) = \frac{\tau_1 s}{\tau_1 \tau_2 s^2 + (\tau_1 + \tau_2)s + 1} \,. \quad (2.45)$$

Figure 2.33 shows the responses of the BPF. The figure was obtained using the MATLAB code in Table 2.6.

2.3.2.5 *Raised Cosine and RRC Filters*

Nyquist filters are ISI-free filters. Popular Nyquist filters are the raised cosine (rc) and root-raised cosine (rrC) pulses. Both are LPFs. The impulse response of the raised-cosine filter is

$$\begin{aligned} h_{\texttt{rc}}(t) &= \frac{\cos(\pi\alpha\frac{t}{T_{\texttt{s}}})}{1 - 4\alpha^2(t/T_{\texttt{s}})^2} \frac{\sin(\pi\frac{t}{T_{\texttt{s}}})}{\pi\frac{t}{T_{\texttt{s}}}} \\ &= \frac{\cos(\pi\alpha\frac{t}{T_{\texttt{s}}})}{1 - 4\alpha^2(t/T_{\texttt{s}})^2} \texttt{sinc}(t/T_{\texttt{s}}) \end{aligned} \quad (2.46)$$

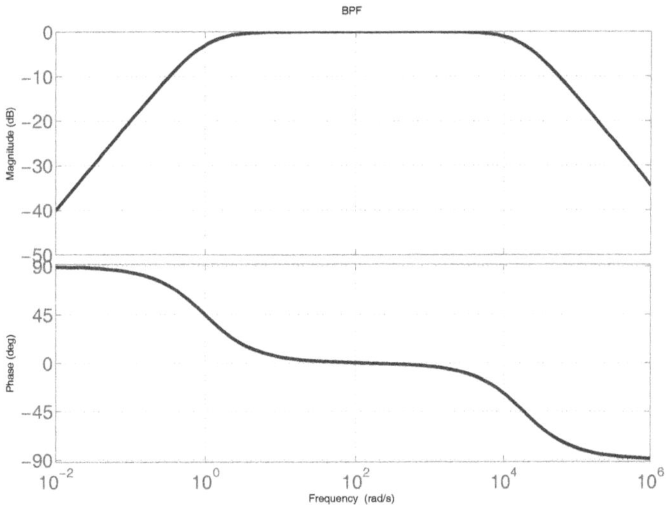

Fig. 2.33 The responses of the BPF in Fig. 2.30(c): $R = 1$ kΩ, $R_2 = 2\,\Omega$, $C = 1$ mF, $C_2 = 26.5\,\mu$F.

Table 2.6 MATLAB code for Fig. 2.33.

```
clear all;
R1=1000; R2=2; % resistances in Ohms
C1=1e-3; C2=26.5*1e-6; % capacitances in Farad
t1=R1*C1; t2=R2*C2; % time constants
H=tf([t1 0], [t1*t2 t1+t2 1]); % call to MATLAB native function tf
bode(H); % plots both magnitude and phase responses
```

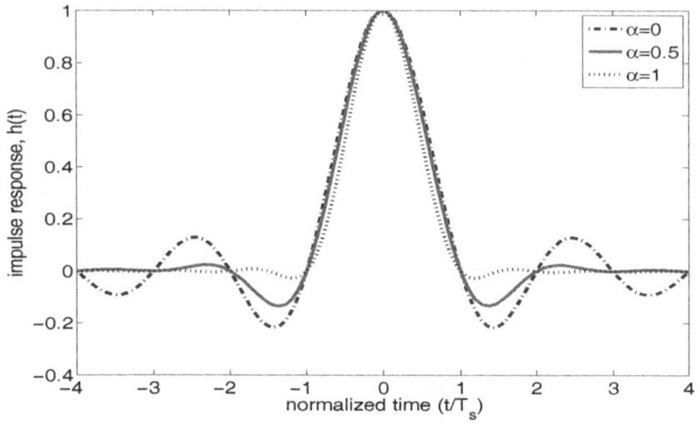

Fig. 2.34 Impulse response of raised cosine pulses.

where T_s is the symbol duration and $\alpha \in [0, 1]$ is called roll-off factor or excess bandwidth factor, a parameter which determines the sharpness and the bandwidth consumption of the filter. The impulse responses of raised cosine pulses for different roll-off factors are shown in Fig. 2.34.

The Fourier transform of $h_{rc}(t)$ results in the transfer function

$$H_{rc}(f) = \begin{cases} T_s, & 0 \le |f| \le \frac{1-\alpha}{2T_s} \\ \frac{T_s}{2}\left[1+\cos\left((|f|-\frac{1-\alpha}{2T_s})\frac{\pi T_s}{\alpha}\right)\right], & \frac{1-\alpha}{2T_s} \le |f| \le \frac{1+\alpha}{2T_s} \\ 0, & |f| > \frac{1+\alpha}{2T_s} \end{cases}.$$

Transceivers in communication systems seek that the overall transfer function, comprising the transmitting pulse-shaping filter $H_T(f)$, the transfer function of the transmission medium $H_C(f)$ and the filter in the receiving circuit $H_R(f)$ should be a Nyquist filter in order to obtain ISI-free signal detection. The ideal Nyquist ISI-free condition is

$$H(f) = H_T(f)H_C(f)H_R(f) = 1 \tag{2.47}$$

which in the time-domain becomes

$$h(t) = \mathcal{F}^{-1}\{H(f)\} = h_T(t) * h_C(t) * h_R(t) = \delta(t). \tag{2.48}$$

The root-raised cosine filter fulfills the Nyquist ISI-free condition. Its transfer function is

$$H_{rrc}(f) = \begin{cases} \sqrt{T_s}, & 0 \le |f| \le \frac{1-\alpha}{2T_s} \\ \sqrt{\frac{T_s}{2}\left[1+\cos\left((|f|-\frac{1-\alpha}{2T_s})\frac{\pi T_s}{\alpha}\right)\right]}, & \frac{1-\alpha}{2T_s} \le |f| \le \frac{1+\alpha}{2T_s} \\ 0, & |f| > \frac{1+\alpha}{2T_s} \end{cases}.$$

Note that the square root is taken in the frequency domain, i.e., $H_{rrc}(f) = \sqrt{H_{rc}(f)}$. This results in the impulse response

$$h_{rrc}(t) = \mathcal{F}^{-1}\{H_{rrc}(f)\} = \frac{2\alpha}{\pi\sqrt{T_s}} \frac{\cos[(1+\alpha)\pi t/T_s] + \frac{\sin[(1-\alpha)\pi t/T_s]}{4\alpha t/T_s}}{1-(4\alpha t/T_s)^2}. \tag{2.49}$$

Figures 2.35 and 2.36 show, respectively, the transfer functions of the RC and RRC with four different roll-off factors. The factor $\frac{1}{2T_s} = \frac{1}{2}R_s$ is called **Nyquist bandwidth**, while the bandwidth of the RC filters is

$$B = \frac{1+\alpha}{T_s} = (1+\alpha)R_s. \tag{2.50}$$

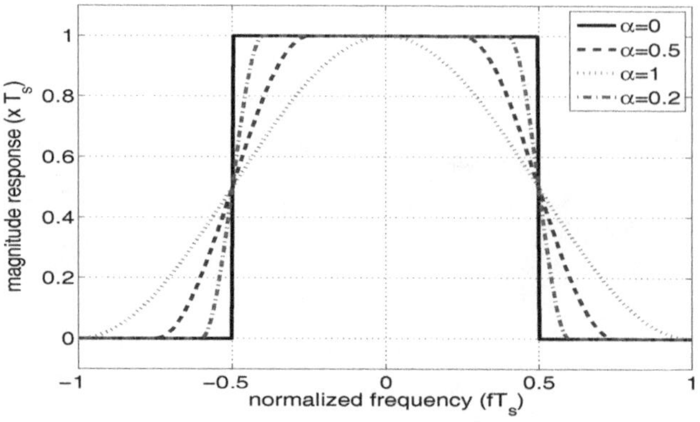

Fig. 2.35 Transfer functions of Nyquist ISI-free pulses.

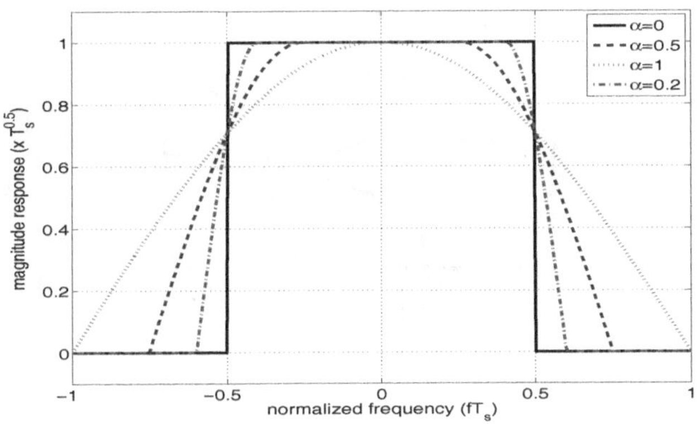

Fig. 2.36 Transfer functions of root-raised cosine pulses.

2.3.2.6 *Gaussian Filter*

The Gaussian filter achieved its name from its bell-shaped impulse response and transfer function. This filter is used, for example, to shape the digital data modulating the carrier in the binary frequency-shift keying (FSK) modulator used in GSM mobile technology. The result of the BFSK modulator using the modulating data shaped a priori by Gaussian filter is called GMSK (Gaussian Minimum Shift Keying) modulation. Gaussian filters are also used in image processing.

The impulse response of the one-dimensional Gaussian filter is

$$h_g(t) = \frac{1}{\sigma\sqrt{2\pi}} e^{-\frac{t^2}{2\sigma^2}} \quad (2.51)$$

where σ is the standard deviation, which is the square root of the variance. The standard deviation is inversely proportional to the amount of filtering. The smaller the variance, the more the frequencies suppressed and vice versa. Thus, the design of a Gaussian filter requires the manipulation of just a single variable, which is the standard deviation. Fourier transform of Eq. (2.51) yields the transfer function

$$H_g(f) = e^{-\frac{f^2}{2\sigma^2}} . \quad (2.52)$$

The half-power bandwidth is $B_{3dB} = \sigma\sqrt{8\log_e 2}$ Hz. Another form of the Gaussian filter is

$$h_g(t) = \frac{\alpha}{\sqrt{\pi}} e^{-\alpha^2 t^2} \leftrightarrow H_g(f) = e^{-(\pi f/\alpha)^2} \quad (2.53)$$

where $\alpha = \pi B_{3dB}\sqrt{\frac{2}{\log_e(2)}}$ and B_{3dB} is the filter's single-sided half-power bandwidth. This means

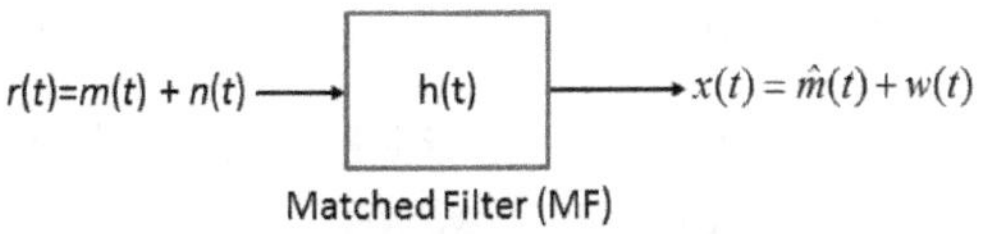

Fig. 2.37 Illustration of matched filtering.

that the total 3-dB bandwidth of the filter is $2B_{3dB}$. A close observation of the analytic formulas for the three filters discussed earlier show that they are all lowpass filters.

2.3.2.7 *Matched Filter*

Assume that we transmit the signal $m(t)$ through an AWGN channel and the signal is $r(t) = m(t) + n(t)$ reached the input of the receiver, where $n(t) \sim \mathcal{N}(0, \frac{N_0}{2})$ is the zero-mean additive white Gaussian noise. As illustrated in Fig. 2.37, let the signal at the output of the detector be $x(t) = \hat{m(t)} + w(t)$. Then, $w(t) \sim \mathcal{N}(0, \sigma_w^2)$ where

$$\sigma_w^2 = \frac{N_0}{2} \int_{-\infty}^{\infty} |h(t)|^2 dt$$

is the AWGN at the output of the detector.

The signal $x(t)$ at the output of the circuit in Fig. 2.37 is

$$\begin{aligned} x(\tau) &= h(t) * [m(t) + n(t)] \\ &= \int_{-\infty}^{\infty} m(\tau - t)h(t)dt + h(t) * n(t) . \end{aligned} \quad (2.54)$$

We seek a filter $h(t)$ which yields the global maximum of $x(t)$ at a given time instant, say T_0, subject to the constraint $E_h = \int_{-\infty}^{\infty} |h(t)|^2 dt < \infty$. A filter fulfilling the above condition is called matched filter. Its impulse response is

$$h(t) = m(T_0 - t), \;\; T_0 \geq T_m \quad (2.55)$$

where T_m is the duration of the signal $m(t)$. The condition $T_0 \geq T_m$ is needed to ensure that the matched filter is causal, else it cannot be practically implemented. The impulse response of the optimal filter is matched to the input information-bearing waveform. If $M(f)$ is the transfer function of $m(t)$, then the transfer function of $h(t)$ is $H(f) = M^*(f)$. The matched filter $h(t) = m(T_0 - t)$ maximises the SNR at time instant T_0. That means that the maximum SNR of the circuit in Fig. 2.37 is $SNR_{max} = \frac{|x(T_0)|^2}{\sigma_w^2}$.

2.3.3 *Antennas*

Antennas are transducers which convert electric signals into electromagnetic waves and vice versa. Antennas are designed to operate in a certain frequency band, and for certain systems. For example, Cassegrain and Gregorian shaped parabolic reflectors are used in satellite communications. Both the transmitter and receiver in a wireless link must have an antenna, although it may not be visible. As antennas are so important, we dedicate Chapter 6 for it. An antenna can be modelled as a two-port electrical network.

2.3.4 *Oscillators for Carrier Generation*

An electronic device which is common in modulation and other electronic circuits is the oscillator, especially crystal oscillator. An oscillator can be modelled as a two-port electrical network. An oscillator is an electronic circuit which, when excited with a direct current, produces an alternating current signal. For this reason, oscillators are also referred to as inverters. Usually, the output signal of an oscillator is a sinusoidal wave or a square wave. Carrier signals in modulation circuits are generated with oscillators. Based on how fast its output signal changes (i.e., the number of oscillations per second or frequency), oscillators are classified into:

- **Radio-frequency oscillators** (RFOs): these produce output signals which oscillate at a rate of over about 100 kHz.
- **Audio-frequency oscillators** (AFOs): these produce output signals with audio frequencies, about 30 Hz to 20 kHz.
- **Low-frequency oscillators** (LFOs): these produce output signals which oscillate at a rate below 30 Hz.

A special class of oscillators which are very important in electronic circuits used in communications systems, such as modulators, are the voltage-controlled oscillators (VCOs). The input or excitation signals of a VCO is an electrical signal, either current or voltage signal. These input signals determine the range of frequencies that the output signals of VCO can oscillate.

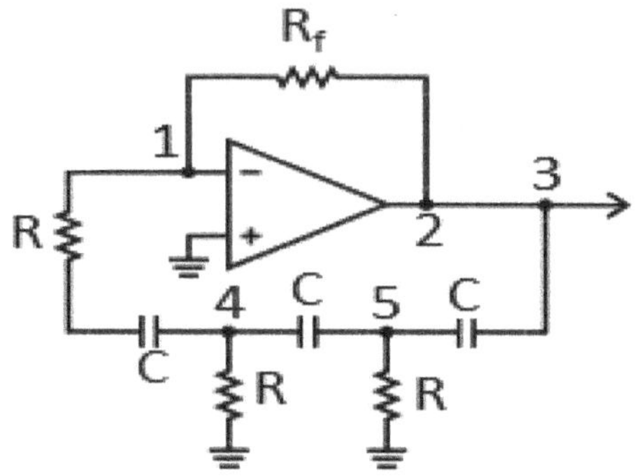

Fig. 2.38 An RC feedback oscillator.

A simple resistor-capacitor (RC) feedback oscillator is shown in Fig. 2.38. This circuit is used as a phase shifter. In order to provide a 360°-phase shift, the RC network provides 180° phase shift while the operational amplifier provides the remaining 180° phase shift. Thus, each of the three identical and cascaded RC sections in the circuit delivers 60-degree phase shift at the resonance/oscillation frequency, which is

$$f_0 = \frac{1}{\sqrt{24}\pi RC}. \tag{2.56}$$

The OpAMP in the circuit must have a gain of 29 to offset the feedback attenuation of 1/29 in order to achieve the required unit loop gain. Thus, $R_f = 29R$.

2.3.5 *Mixers*

A mixer is a frequency converter. It receives signals at two different frequencies and produce another two different frequencies. All modulators have mixers. Usually, one of the input frequencies of a mixer is a carrier frequency generated by an oscillator.

Let us use Fig. 2.39 to illustrate the operation of the mixer. Assume that the input signal to

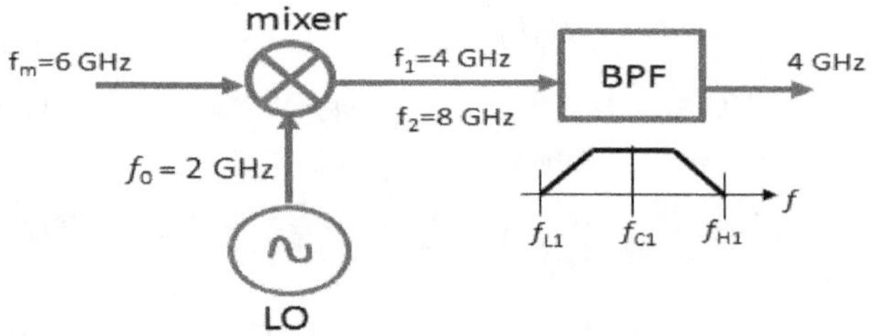

Fig. 2.39 An illustration of a mixer.

the mixer is a single-tone signal with the frequency $f_m = 6$ GHz. The second input frequency is generated by a local oscillator (LO) which is $f_0 = 2$ GHz. The mixer 'mixes' these two frequencies to obtain

$$f_1 = f_m - f_0 = (6 - 2)\ \texttt{GHz} = 4\ \texttt{GHz} \quad (2.57)$$

$$f_2 = f_m + f_0 = (6 + 2)\ \texttt{GHz} = 8\ \texttt{GHz}\,. \quad (2.58)$$

As illustrated in Fig. 2.39, usually we need only one of the frequencies at the output of the mixer in practical situations. Therefore, we need to do post-filtering using a bandpass filter to get rid of the spurious frequency. In our example in Fig. 2.39 we filter out the higher frequency. In such a situation we have reduced the frequency so it is referred to as downconversion. If we have instead filtered out the 4 GHz using the appropriately-dimensioned BPF then we would have performed upconversion.

2.3.6 *Amplifiers: LNA and TWTA*

Signals lose part of their energy or power in transmission. Also, signals exiting transmitters usually do not have enough power to reach their intended destination for good detection. For these two reasons, there is a device called amplifier which is used to increase the strength of signals. There are many types of amplifiers. Those commonly used in microwave communications are low noise amplify (LNA) and traveling wave tube amplifier (TWTA). Usually, the first stage in a receiver and the last stage of a transmitter are LNA and TWTA amplifiers, respectively. TWTA is a wideband amplifier which uses non-resonant wave circuits. TWTA are heavily used in satellite communications.

There are two classes of amplifiers: solid state devices based on solid-state physics and tubes based on thermionic emission. Devices which generate and amplify signals in the microwave range of the frequency spectrum are called microwave vacuum tubes. There are two types of microwave tubes: microwave linear-beam tubes (aka O-type tubes) and microwave crossed-field tubes (aka M-type tubes). An example of M-type tubes is magnetron (portmanteau word from magnetic and electron) which finds application in home microwave ovens.

Linear-beam tubes are further classified into cavity types and slow beam structure. TWTA belongs to the class of slow beam structure tubes, while klysterons belong to the cavity tubes. TWTA has two types: couple cavity TWT and helix TWT. A main advantage in TWTA is its wideband feature compared with klysterons, enabling it to amplify signals in a wide range of frequencies. TWTA can achieve gains in the range 40–70 dB in about 0.3–50 GHz frequency range with power output from a few watts to megawatts.

Signals reaching receivers, especially those that have travelled a distance as it is in satellite communications, have very low power levels. We need to boost such signals prior to detection. LNA amplifiers are placed at the front-end of a receiver to do the job. LNA can amplify weak signals while keeping the SNR fairly constant.

2.3.7 *Duplexer*

The *duplexer* is a three-port microwave multiplexer or switch which enables the transmitter and the receiver sides of a wireless device to share a single antenna while operating at similar frequencies. The duplexer is used in VSAT satellite systems and monostatic radar systems, for example. A duplexer can be realised using a PIN diode switch or a circulator. It is therefore also referred to as a 3-port RF circulator. The TR (transmit/receive) switch connects in turn the transmitter or receiver circuitry to the antenna. It ensures that power from the transmitter does not enter the sensitive receiver circuitry to destroy it. The operation of the duplexer is illustrated in Fig. 2.40.

The duplexer shuts down the receiver during transmission in order to prevent it from being destroyed by stray power from the transmitter. For example, a monostatic radar uses a duplexer to switch its single antenna between the transmit and receive circuitry.

There are two basic types of duplexers: orthomode transducers (OMT) and circulators. Figure 2.40 illustrates the two types of duplexers. The OMT is better than the circulator for two reasons. First, the OMT has a wider bandwidth of operation than the circulator. Second, the isolation between the transmit and the receive ports in OMT is achieved using orthogonal polarisations. Therefore,

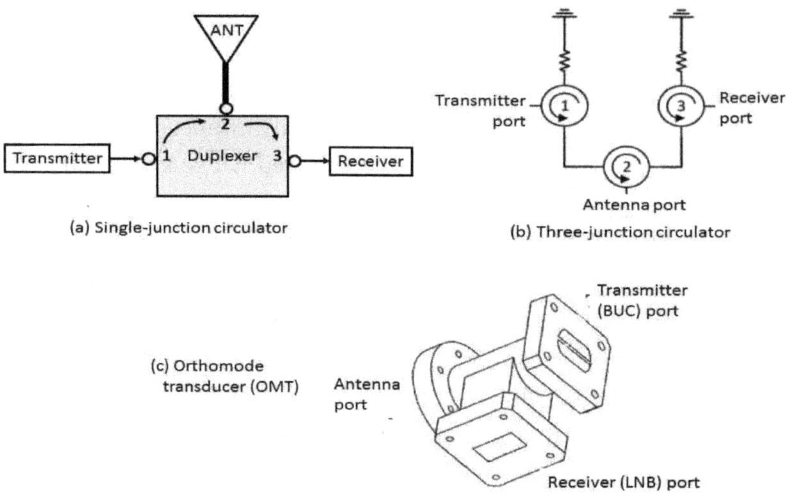

Fig. 2.40 An illustration of the two types of duplexers.

the OMT duplexer achieves a better isolation between the two ports than the circulator. A single-junction circulator achieves only 20-dB ports isolation, a three-junction circulator achieves up to 35-dB isolation, while an OMT achieves 40 dB port isolation.

The duplexer is illustrated in Fig. 2.27(c). It is a three-port electrical network with all its ports matched. Its S-matrix is non-reciprocal, but it is lossless. Here, $a = 1$, resulting in its S-matrix

$$S_{c1} = \begin{bmatrix} 0 & 0 & 1 \\ 1 & 0 & 0 \\ 0 & 1 & 0 \end{bmatrix} \tag{2.59}$$

or

$$S_{c2} = \begin{bmatrix} 0 & 1 & 0 \\ 0 & 0 & 1 \\ 1 & 0 & 0 \end{bmatrix}. \tag{2.60}$$

2.3.8 *Diplexer and Splitter*

Note that there is only one letter differentiating the diplexer and the duplexer. This is part of the reason why the two devices are easily confused. A diplexer is a three-port frequency-domain signal multiplexer, as illustrated in Fig. 2.41(a). There are two variants of diplexers: passive diplexer and active diplexer. A passive diplexer is an RF combiner as it does only multiplexing. An active diplexer, however, amplifies its input signals and can translate them into a different frequency band. This means that an active diplexer contains a modulator.

A diplexer is a reciprocal device. It can also split an input signal into two output signals, as illustrated in Fig. 2.41(b). In such an application, it is referred to as RF splitter. A diplexer with four ports (three inputs and one output) is called triplexer. Table 2.7 presents the important parameters of diplexers. Figure 2.41 illustrates the operation of the diplexer and splitter.

Diplexer, triplexer duplexer and multiplexer can designed by appropriately combining low-pass, high-pass and bandpass filters. A popular application of diplexers is in multi-band mobile phone handsets. For example, a dual-band 3G/4G or 4G/5G handset needs diplexer to discriminate between the two technologies.

2.3.9 *Coupler*

A coupler is a four-port passive electrical network or device which splits the power in an input electromagnetic signal over a transmission line into two components. Each power component is fed into another circuit through transmission lines. A special kind of couplers is the directional coupler. Figure 2.42 illustrates the block diagram of a coupler. Note that the numbering of the four ports is arbitrary: it can be done either anti-clockwise or clockwise.

Couplers are evaluated using the three metrics: directivity, isolation and coupling. Using the pa-

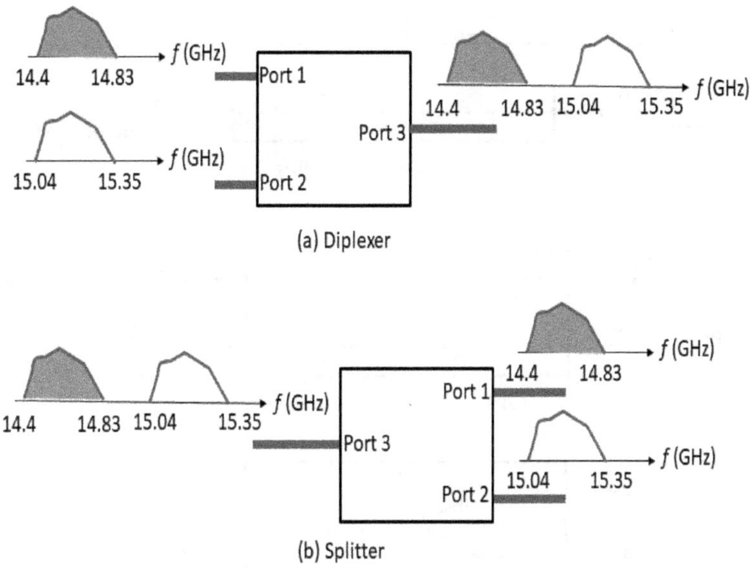

Fig. 2.41 Illustration of the principle of diplexers and splitters.

Table 2.7 Diplexer case study: Diplexer D0114G91.

Feature	Value
Manufacturer	Microwave Circuits
Passbands	Channel 1: 14.4–14.83 GHz; Channel 2: 15.04–15.35 GHz
Bandwidth	Channel 1: 430 MHz; Channel 2: 310 MHz
Center frequencies	Channel 1: 14.615 GHz; Channel 2: 15.195 GHz
Crossover frequency	14.87–14.98 GHz
Insertion loss	≤ 1.8 dB
Impedance	50 Ω
VSWR	≤ 1.5
Peak RF power	20 W
Operating temperature	−40° C to +85° C
Storage temperature	−55° C to +90° C

rameters in Fig. 2.42, these metrics are defined as

$$\begin{aligned}
\texttt{Directivity:}\quad & D = 10\log_{10}(P_3/P_4) \quad \texttt{(a)} \\
\texttt{Isolation:}\quad & I = 10\log_{10}(P_1/P_4) \\
& \quad = -20\log_{10}(|S_{14}|) \quad \texttt{(b)} \\
\texttt{Coupling:}\quad & C = 10\log_{10}(P_1/P_3) \quad \texttt{(c)}
\end{aligned} \tag{2.61}$$

where P_k is power out of port k and S_{ij} are the S-parameters. For example, CBR16-0003 high-directivity bridge coupler manufactured by Marki Microwave and operating in the frequency range 200 kHz to 3 GHz has 30–40 dB directivity, 16 dB mean coupling, 1.1 VSWR and up to 2.2 dB insertion loss.

We have studied some of the fundamentals in wireless IoT. The next chapter builds upon this foundation. All topics studied in Chapter 6 are also very basic topics in Internet Protocol (IP) based wireless communications, which is the most interesting connectivity between IoT devices. The reason is that a large quantity of IoT devices are deployed in environments which do not allow the use of wires. Consider, for example, IoT sensors commissioned under the ocean to monitor ocean pollution. Another example is the consumer application of IoT. Consider an IoT-enabled home which uses an alarm watch to wake up its owner and then automates the coffee machine to prepare coffee. The refrigerator in the same home alerts the owner at work of insufficient foodstuff in it. Such an automation can hardly be achieved using wires.

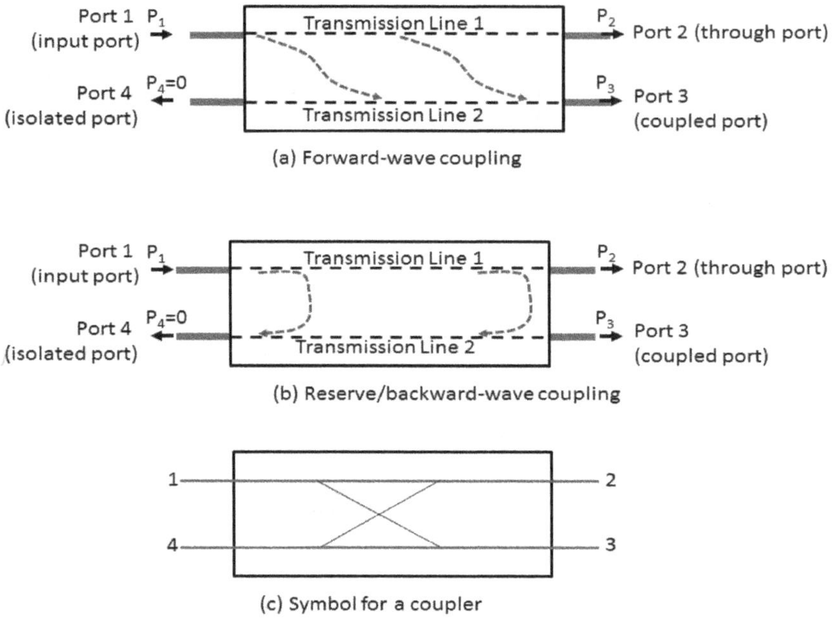

Fig. 2.42 An illustration of a differential coupler.

2.3.10 *Check Your Understanding*

Students are encouraged to check their understanding of the content of this section with the following questions prior to proceeding to the next section.

(1) Discuss the importance of oscillators in modulation systems.

(2) From first principles, derive Eq. (2.56).

(3) **In-class group study**: in groups of 2 to 4 students, discuss.

  (a) The purpose of duplexers.

  (b) The differences between the two types of duplexing: OMT and circulator.

(4) Assume that Fig. 2.43(a) is the the equivalent circuit for the small signal model of a transistor.

  (a) Find the Z-parameters of the transistor.

  (b) Evaluate the Y-parameters of the transistor.

  (c) Find the ABDC parameters of the transistor.

  (d) Compute the S-parameters of the transistor.

  (e) Compute the h-parameters of the transistor.

  (f) Compute the g-parameters of the transistor.

(5) Repeat Problem (4) using the electrical matching network in Fig. 2.43(b).

(6) Repeat Problem (4) using the electrical matching network in Fig. 2.43(c).

(7) The block diagram of a passive band-stop filter (BSF) is shown in Fig. 2.30(d).

  (a) Compute the transfer function of the BSF.

  (b) Using MATLAB or any other software package, plot the magnitude response and the phase response of this filter. Select the filter's parameters R_1, R_2, C_1 and C_2 so as to block the 60 Hz frequency. The filter's stopband should be as low as possible, but should not exceed 20 kHz.

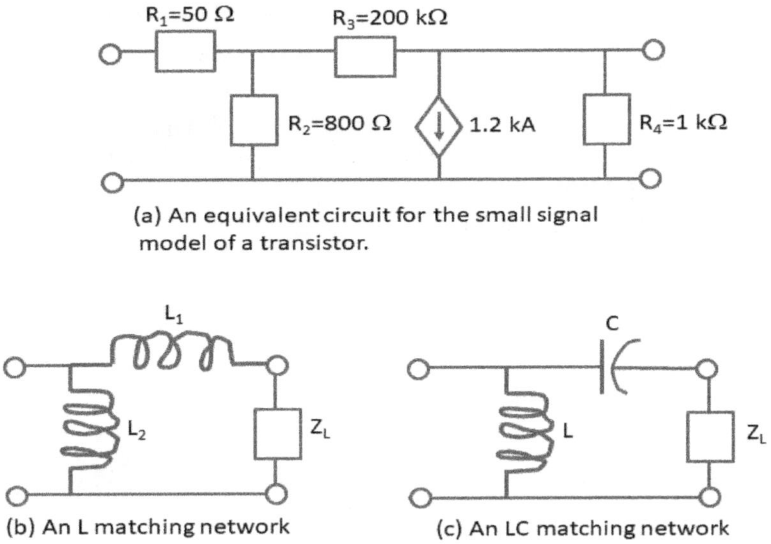

Fig. 2.43 Three electrical networks for Problems 2.3.9, 2.3.9 and 2.3.9.

2.4 Review of Important Terms and Acronyms

Acronyms, abbreviations and important terms		
ABCD parameters	actuator	amplifier
antenna	augmented reality	big data analytics
BPF	BRF	CIDR
classful subnetting	classless subnetting	cloud computing
cognitive network	coupler	data analytics
datagram	directed broadcast	duplexer
electrical network	filter	FLSM
fog computing	Gaussian filter	g-parameter
hosted private cloud	h-parameter	HPF
hypervisor	IaaS, PaaS, SaaS	Internet Protocol (IP)
IoT (Internet-of-Things)	IPv4	IPv6
limited broadcast	middleware	mixed reality
MPLS	network agility	network slicing
NFV	NGN	oscillator
packet switching	private cloud	public cloud
QoS, QoE	rrc filter	SDN
SD-WAN	SON	S-parameter
subnet	subnet mask	subnetting
tactile Internet	transducer	VLSM
virtual private cloud	virtual reality	virtualisation
WAN	Y-parameter	Z-parameter

2.5 Supplementary Reading

(1) Qusay F. Hassan (editor), *Internet of Things A to Z: Technologies and Applications*, Wiley-IEEE Press, 2018.
(2) James Kurose and Keith Ross, *Computer Networking: A Top-Down Approach*, Pearson, 2016.
(3) Keith B. Darrell, *Issues In Internet Law: Society, Technology, and the Law*, Amber Book Company LLC, 2017.
(4) Agus Kurniawan, *Smart Internet of Things Projects*, Packt Publishing, 2016.

Chapter 3

Analog Modulation and Demodulation Methods

Chapter's Learning Outcomes

Modulation is the process of piggybacking an information-bearing signal over another signal to achieve effective transfer between two locations. Every transmission medium or channel has some undesirable filtering effects on signals, such as blocking of parts of signal, interference, noise, attenuation and distortion. Modulation is used to condition a signal to suit the characteristics of the channel. The reverse process of modulation is called demodulation. Both modulation and demodulation (aka modem) are ISO/OSI Layer 1 (i.e., physical layer) techniques. Information transfer over communications networks cannot work without modem. For this reason, the understanding of the basic principles and practice of modem schemes is fundamental in the communications disciplines. This chapter presents the fundmentals of analog modulation and demodulation methods. Topics treated include:[1]

(1) Basic principles and practical applications of analog modulation methods.
(2) Amplitude modulation and its variants.
(3) Phase modulation.
(4) Frequency modulation.
(5) Baseband modulation methods versus broadband modulation methods.
(6) Performance metrics of modulation schemes, e.g., bandwidth efficiency and power efficiency.

3.1 Introduction

In modulation we use a signal, referred to as *modulating signal* or message signal or intelligence, to change some aspects of another signal called *carrier signal* or bearer signal or carrier wave or just a carrier. The resulting modified carrier signal is referred to as the *modulated signal*. Modulation is also often somewhat loosely referred to as **waveform**. In this chapter we use $c(t)$ to refer to the carrier signal, $m(t)$ for the modulating signal, and $y(t)$ for the modulated carrier signal. The modulation principle is illustrated in Fig. 3.1. The process by which the modulating signal modifies the carrier signal to obtain the modulated signal is referred to as the *modulation method* or *modulation scheme* or *modulation technique*. There are several types of modulation schemes, each with certain desirable features for specific applications. We shall study in the following sections both conventional and futuristic modulation schemes.

The spectral space occupied by the message signal prior to modulation is called *baseband frequency*. Hence, the original message signal is referred to as *baseband signal*. The reverse process to modulation is called *demodulation* in which the original signal is recovered from the wave produced by modulation. Modulation can be *carrierless modulation* or *carrier modulation*.

[1]Though wood may be covered with gold and silver, there is no life in it.

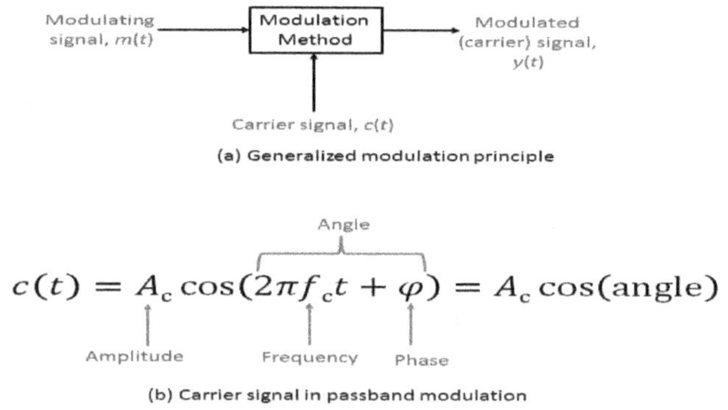

Fig. 3.1 The generalised principle of modulation.

The design of modulation schemes strives to optimise the seven competing objectives:

- **High power efficiency**: aka energy efficiency. We seek modulation schemes which consume as low power as possible. The simpler the modulation scheme, the lower its power consumption.
- **High bandwidth efficiency**: aka spectral efficiency. This is the number of bits that we can transfer using 1 Hz of bandwidth at a given bit error rate. We seek as high as possible spectral efficiency. Bandwidth efficiency is increased by increasing the number of bits used to encode each symbol or signalling element. By increasing the unique number of symbols, we increase the required transmit power, reducing power efficiency.
- **Least cost**: the cost of the transceiver circuitry increases with the complexity of the modulation scheme used. Therefore, simple modulation schemes are preferable to reduce costs.
- **Low bit-error rate**: at a given bit energy, we seek the modulation scheme with the lowest bit error rate.
- **Low sensitivity to multipath fading**: most communications channels, especially wireless channels, suffer from fading. For this modulations schemes which are robust against multipath fading are desirable.
- **Constant envelope**: modulation schemes with constant amplitude, such as PSK is desirable. Such modulation schemes allow the use of nonlinear amplifiers which have higher gains than linear amplifiers. The high gain is needed so that electronic transmitters can transmit signals with minimum power and still achieve the required effective radiated power. Using nonlinear amplifiers in non-constant envelope modulators creates spectral sidelobes.
- **Low out-of-band radiation**: requires modulation schemes with low sidelobes.

Improving upon one these above objectives degrades the other. This is the reason why an optimised trade-off between them is always necessary. For example, mobile cellular systems use hand-held devices with limited battery-supplied energy. Therefore, power efficiency is important in such systems. As the radio spectrum is an inelastic and tightly-regulated commodity, bandwidth efficiency is also important in mobile networks. Costs of the hand-held devices and mobile services must also be low enough to attract uptake or subscriptions. Communications systems which do not have to be mobile (e.g., line-of-sight microwave systems) can sacrifice power to the advantage of bandwidth or costs. Modulation methods are designed with the above three efficiencies in mind. Simple methods enabling simple transceiver implementations are good in cost savings and power consumption. However, simple transceivers are in general inefficient in bandwidth consumption.

3.1.1 *Baseband versus Broadband Channels*

Every channel or transmission medium has a filtering characteristic. Every channel has its pros and cons, none is the best for all applications, and each of them is suitable for some applications. Some behave as a low-pass filter (Fig. 3.2(a) and (b)) while others behave as bandpass filter (Fig. 3.2(c) and (d)). Baseband transmission media are also referred to as **lowpass channels** as they pass frequencies from dc (i.e., 0 Hz) to a maximum frequency, referred to as cut-off frequency, which is indicated in Fig. 3.2(a) and (b) as f_{co}. Examples of lowpass channels are the metallic cables called coaxial cables (aka coax) and twisted-pair cables. These cables are commonly used for short-distance communications as in local area networks. Table 3.1 contains a

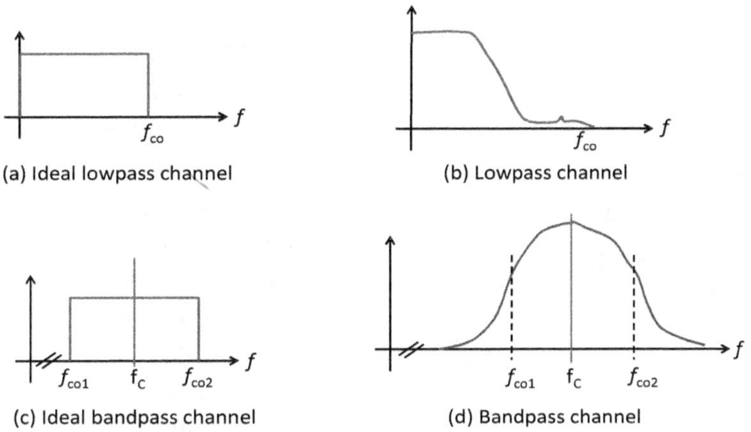

Fig. 3.2 Lowpass and bandpass channels.

Table 3.1 Frequency bands of some baseband and broadband transmission media.

Transmission Medium Category	Transmission Medium Type	Frequency Band/Bandwidth (GHz) or wavelength
Twisted pair	CAT-3	0–0.016
Twisted pair	CAT-5e	0–0.1
Twisted pair	CAT-6a	0–0.25
Coaxial cable	LMR-1700	0–2.5
Coaxial cable	RG-214	0–1.1
Coaxial cable	TTL-26	0–26.5
Waveguide	WR-187	3.95–5.85
Waveguide	WR-137	5.85–8.2
Waveguide	WR-90	8.2–12.4
Waveguide	WR-62	12.4–18
Waveguide	WR-42	18.26.5
Planar lines	Stripline	
Planar lines	Single-mode microstrip	ca. 0–30
Fiber-optic	Step-index single-mode	1.25–1.35 (1.31) μm; 1.5–1.6 (1.55) μm
Fiber-optic	Step-index multi-mode	800–900 (850) nm
Fiber-optic	Graded-index	800–1600 (850, 1310, 1550) nm
Wireless channels	EM spectrum	Several bandpass bands

UTP: unshielded twisted pair, BASE/Base: baseband, WR: rectangular waveguide, CAT: category, 'e' for *enhanced* performance means lower crosstalk between the four pairs of wires, 'a' for augmented performance means lower crosstalk than 'e', RG: radio guide, TTL: transistor-transitor logic, LMR: Lockheed Martin Radar, IC/PC: integrated/printed circuit.

sample of both baseband and broadband channels [Keysight Technologies (2017)]. Baseband channels can transfer only baseband signals using baseband modulation methods.

Broadband media are also referred to as **bandpass channels** as they do not pass dc signals and have a maximum frequency that they pass without excessive attenuation, which is indicated in Fig. 3.2 as f_{co2}. Typical examples are wireless channels, waveguides and optical fiber cables. Broadband channels can transfer only broadband signals using broadband modulation methods or a combination with baseband methods, as illustrated in Fig. 3.3. For example, we can use a baseband signal as the output of a pulse amplitude modulator to modulate a high-frequency wave to enable the transfer

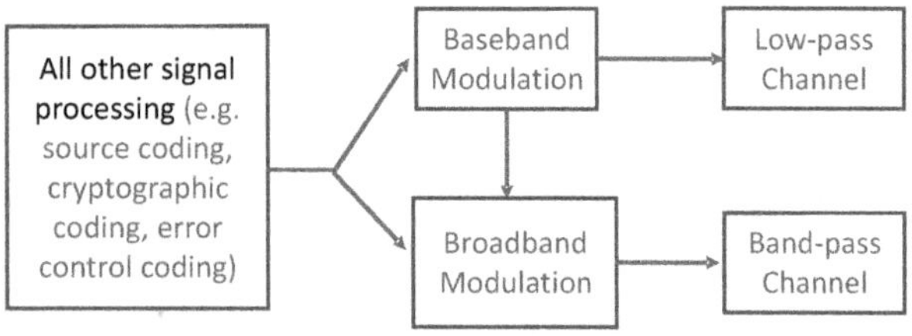

Fig. 3.3 The appropriate modulation types for baseband and broadband channels.

over bandpass channels. Note, however, that the term broadband is loosely used to mean data rate larger than V.90/V.92 data rate of 56 kbps.

Table 3.1 shows the categories of transmission media and their frequency spectra. These are waveguides, twisted-pair, coaxial cables, microstrip, stripline and wireless at various frequencies. Waveguides are used in low-loss microwave communications. Striplines are used to interconnect components in integrated circuits and printed boards working at microwave frequencies. Microstrips are used to form microwave components such as patch antennas, filters, couplers and power dividers. Wireless channels are used in various types of wireless communications.

Fiber-optic cables are of two types: step-index and graded-index. The step-index can be single-mode or multi-mode fiber. Step-index single-mode fiber-optic cables are used in low-loss long-distance high-bandwidth links, e.g., backbone of high-rate LANs and WAN, PSTN trunks, cable TV head-end. Graded-index fiber-optic cables are used for links requiring less than 1 Gbps data rate over short distance (< 3 km). CAT-3 UTP cables are used in local area networking (LAN), such as 10Base-T and token ring. CAT-5e are used in LAN technologies such as 10BASE-T, 100BASE-TX and 1000BASE-T, and so is CAT-6e used in for example 10BaseT, 100Base-TX, 1000Base-T and 10GBase-T. The coax cable LMR-1700 has applications in antenna feeder, GPS, WLAN, WiMaX, SCADA and mobile antennas.

3.1.2 *Types of Modulation Methods*

Figure 3.4 categorises modulation methods that are used to transfer signals into four basic types: analog modulation, digital modulation, waveform coding (aka pulse modulation) and line coding. Analog modulation is used to transfer analog data over analog transmission systems. An example is the transfer of human voice over the analog legacy PSTN network. Digital modulation methods transfer digital data over analog transmission systems. An example is the connection of voiceband modem to a computer to send data over the analog PSTN network. Line coding is used to transfer digital data over digital transmission systems. Lastly, pulse modulation methods are used to transfer analog data over digital transmission systems. An example is the pulse code modulation (PCM) used to transfer analog human voice over digital networks. Both analog and digital modulation schemes produce analog output. They are therefore referred to as **continuous wave modulation techniques**. Transmission of data over long distances requires continuous wave (CW) modulation as pulses cannot travel long distances.

3.1.2.1 *Baseband Modulation Methods*

Pulse modulation and line coding produce digital output signals. Digital-output modulation schemes do not move the modulating signal from its original frequency spectrum into another frequency spectrum. They are therefore referred to as **baseband modulation** or **baseband communication** or **baseband transmission**. The original frequency spectrum occupied by an information-bearing modulating signal is referred to as **baseband**. Baseband includes low frequencies and often includes (or is near to) 0 Hz. For this reason, signals that have not been relocated to another frequency band are referred to as **baseband signals**. For example, baseband for human speech is about 3.0 Hz to 3.2 kHz, audio/music about 0–20 kHz, and about 0–4.3 MHz for video. Most of the energy in human speech falls between about 200–300 Hz and about 2700–2800 Hz.

As with any technology, baseband transmission has some drawbacks, some of which are:

- It cannot be used over wireless communication media. The reason is that wireless communication media are in general bandpass in na-

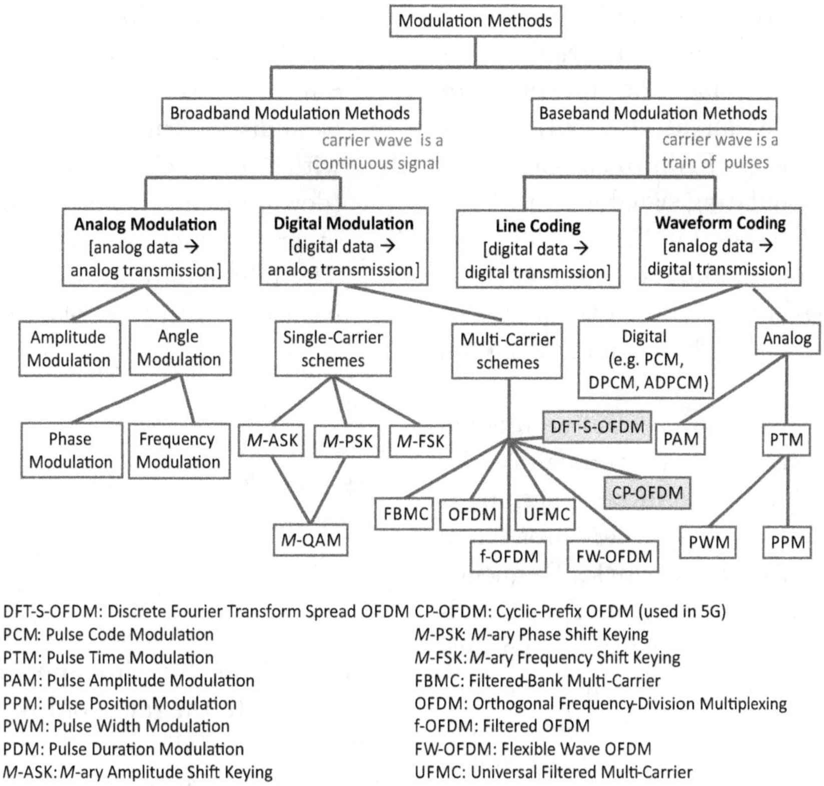

Fig. 3.4 Types of modulation methods.

ture and do not efficiently pass low-frequency signals.

- It makes inefficient use of transmission medium. The bandwidth of the baseband signal is usually smaller than that of the transmission medium. As different baseband signals overlap in frequency, more than one of them cannot be transferred over the same transmission medium at the same time, resulting in bandwidth wastage.

Baseband signal transmission uses pulses (i.e., signals with short duration) as carrier signals, while broadband modulation uses sinusoidal waves as carrier signals. The word *modulation* means changing an aspect of a signal. As we use the modulating signal to change an aspect of a train of pulses in baseband signal transmission, we call the process modulation. As shown in Fig. 3.4, there are two types of baseband modulation methods: line coding and pulse modulation (aka waveform coding). We treat line coding in the chapter dedicated to data transmission and waveform coding in the chapter on digitisation. The reason being that pulse modulation schemes convert analog data or signals into digital signals. They are therefore part of the analog signal digitisation process.

We use waveform coding to transfer analog data over a digital transmission system. A waveform coder uses a signal of finite duration, referred to as a *pulse*, as the carrier signal. For this reason, waveform coders are also referred to as **pulse modulations**. Popular examples of waveform coding are Pulse Code Modulation (PCM), Adaptive Differential PCM (ADPCM) and Delta Modulator (DM). Waveform coders convert analog signals into digital data by producing a coded form of the original analog waveform. This is the reason for the *waveform* in their name.

3.1.2.2 *Broadband Modulation Methods*

Both analog and digital modulation methods produce analog outputs. Analog-output modulation schemes move the modulating signal from its original frequency spectrum into another frequency spectrum. These are therefore referred to as **carrier modulation** or **passband modulation** methods. Note that we can use the output of a baseband modulator such as PAM (pulse amplitude modulation) as the modulating signal in broadband modulation. This is illustrated in Fig. 3.3. The carrier signal in broadband modulation methods is at a frequency much higher than that of the modulating signal. This is not so with baseband modulation methods.

A signal that has been moved from its original frequency band into another band through carrier modulation is referred to as **broadband signal**. A communication system which transfers broadband signals is called **broadband network**. Frequency translation via carrier modulation used in broadband networks enable them to multiplex multiple signals and transfer them over the same transmission medium at the same time. For this reason, broadband network is also used to refer to a communications network which uses multiplexing. Broadband is also used to refer to high-data-rate networks, just as baseband is used to refer to low-data-rate networks.

In recap, broadband and baseband are defined along the three related lines:

(1) *Frequency translation*: a signal relocated from its original frequency band into another band is called broadband signal, and the communication systems transferring broadband signals are referred to as broadband networks. Baseband signals are not shifted in frequency, and networks which transfer them are baseband or narrowband communications networks.
(2) *Multiplexed links*: communications networks which allow multiple signals to share a medium at the same time are called broadband networks. Otherwise, they are baseband systems. If the multiplexing method is frequency-division multiplexing then frequency translation is required.
(3) *Data rate*: communications networks using fat pipes (i.e., high-data-rate transmission media) are called broadband networks, while those using narrow pipes (i.e., low-data-rate transmission media) are called baseband networks. Since 2015 the FCC (Federal Communications Commission) in the USA classifies connections with the minimum data rates of 25 Mbps (downlink) and 3 Mbps (uplink) as broadband, otherwise they are baseband or narrowband. The minimum data rate for this classification moves up with technological developments.

Some texts use *carrier* modulation methods instead of *broadband* modulation methods. However, as illustrated in Fig. 3.1, every modulation method uses a carrier signal. The difference lies in the nature of the carrier signal used. In order to avoid confusion, we refer to modulation schemes which translate the modulating signal into another frequency band as a **broadband modulation method** or **passband modulation method**. The word passband refers to the range of frequencies in which the transmission medium do not remarkably attenuate signals transferred over it. Every channel has a filtering characteristic.

3.1.3 *Importance of Modulation Methods*

Why do we need modulation? Every transmission medium has some undersirable filtering effects on signals, such as distortion, blocking of signal components, noise, attenuation and distortion. Modulation is used to condition a signal to suit the characteristics of the transmission medium. A simple analogy will clarify the answer to this question.

Consider the desire to get a light-weight object like cotton wool to a person a few meters away from you by throwing it when it is windy. The cotton wool will surely be blown away by the wind towards another direction. However, if we tie the cotton wool on a stone and throw it, then we may be able to achieve the intended results. In this analogy, the cotton wool represents the modulating signal, the stone represents the carrier signal, and the cotton wool tied to the stone represents the modulated (carrier) signal.

An Experiment

Before we delve into modulation, let us consider a simple motivating experiment. *Take two pieces of a light-weight object, such as a cotton wool or a feather of a bird. Tie one of the objects to a heavy-weight object, such as a stone and call this object 1. Leave the other light-weight object unchanged, and call it object 2. With equal pressure or force, throw the two objects in a non-windy environment.*

What do you observe? Certainly, object 1 will travel a longer distance than object 2 and object 1 can be easily steered in the intended direction than object 2, unless it (object 1) is too heavy to throw. If it is very windy object 2 would most probably travel a much longer distance than object 1, but would be easily steered in a very wrong direction by the wind. Object 1, however, would be able to resist the disturbance from the wind to some extent. This is all what carrier or radio modulation is about. But, please wait, there is something more to it. Assume for now the following equivalence:

- The light-weight object is the information-bearing signal or modulating signal of interest to be sent over a distance.
- The heavy-weight object is the carrier signal which helps the information-bearing signal to get to its intended destination efficiently.
- The wind is a disturbance in the way of the information-bearing signal of which the 'natural' wind itself is a contributing factor.

The above simple experiment has at least motivated us to embrace the importance of modulation methods. Thus, practical communications systems do not just transmit message signals as they are but move them from their original frequency bands to another via carrier modulation? Some of the common reasons are:

(1) We use carrier modulation to shift a signal from its original frequency band into another desired frequency band. This technique is needed to allow multiple signals to share a frequency band, as done in frequency-division multiplexing (FDM). FDM cannot work without modulation.
(2) Many transmission media, such as wireless channels, used for signal transfer cannot pass signals near zero frequency. Such transmission media are bandpass in nature. Some message signals, referred to as baseband signals, have non-zero components near 0 Hz. Baseband signals cannot be transferred over passband channels without carrier modulation.
(3) Signals at zero frequency have theoretically infinite wavelength.
(4) There are many natural and man-made sources which generate signals in the low-frequency bands message signals usually contain. Hence, there would be too much interference if message signals were transmitted unmodified.
(5) For effective transmission and reception of electromagnetic signals, antennas should have dimensions at least in the order of the wavelengths of the message signals. For example, an antenna with dimension 100 km would be required to receive a 3-kHz signal assuming a free-space radio propagation, an impractical antenna size. Modulation allows the design of realistic antenna to transmit and receive signals occupying a wide range of spectrum.
(6) Many of the message signals themselves may be generated with overlapping frequency bands. Hence, they would interfere with each other if transmitted unmodulated over the same channel. Thus modulation allows multiplexing of many signals with overlapping baseband onto the same channel simultaneously.
(7) Modulation allows the same receiver that is tuned to a given frequency to be used to detect/transmit signals with different baseband spectra.

Carrier modulation always shifts or translates the message signal from its original spectral band to another band. Usually, both signals (at least the carrier) are identified by three parameters: amplitude A, phase θ and frequency f. The modulating signal modifies one or two of these three parameters of the carrier signal to generate a modulated signal. The carrier signal is a carrier modulation usually at much higher frequency than the modulating signal[2], i.e. $f_c \gg f_m$. Usually, the carrier signal used in passband modulation is sinusoidal in nature and hence has the general form

$$c(t) = A_c \cos(2\pi f_c t + \theta_c) = A_c \cos(angle(t)) \quad (3.1)$$

where f_c is the **carrier frequency**, A_c is the amplitude and θ_c is the phase. The carrier frequency is also referred to as **rest frequency**.

The original modulating signal can be sound or electromagnetic radiation,[3] including optical waves. The carrier wave can be a direct current an alternating current or a pulse chain. As the carrier wave is usually at much higher frequency than the intelligence it carries, it is also referred to as *radio-frequency (RF) signal*. Note that modulation (up conversion) and demodulation (down conversion), i.e. change of center frequencies of signals, of communication signals are forms of frequency translation.

Some of the most interesting questions in carrier modulation are: (1) How much of the total power in the modulated signal is consumed by the carrier signal alone? and (2) How much bandwidth is consumed by the modulated signal? Of course, we want to use as minimum bandwidth as possible, and waste as minimum power as possible in the carrier.

As said earlier, the parameter of the carrier signal that we use the message signal to modify is one or a combination of:

(1) Amplitude, as in e.g., amplitude modulation and amplitude-shift keying.
(2) Frequency, as in e.g., frequency modulation and frequency-shift keying.
(3) Phase, as in e.g., phase modulation and phase-shift keying.
(4) Phase and amplitude, as in quadrature amplitude modulation (QAM).
(5) Position, as in e.g., pulse position modulation.
(6) Width, as in e.g., pulse width/duration modulation.

The aspect of the carrier signal that we modify thus contains or 'carries' the message signal that we transfer from one point or user to another.

We use *analog modulation* method if we have analog data which has to be transferred using analog transmission system. Popular applications of analog modulation techniques have been in radio (e.g. AM radio and FM radio) broadcasting, television broadcasting and telecommunications. Currently, most transmission systems used in telecommunications are digital. Also, digital audio and video broadcasting (DAVB) are globally phasing out analog broadcasting. An interesting question then is: is it still necessary to study analog modulation schemes? My answer is yes for at least the two reasons:

- First, although the popularity of technologies decrease over time, they still find applications in some niche areas. For example, frequency modulation (FM) is used in continuous-wave frequency-modulated (CWFM) radar systems. Such systems are used to measure the velocity and range of targets (such as airplanes landing at airports). Also, instrument landing system (ILS) used to guide the approach and landing of aircrafts at airports uses 150-Hz and 90-Hz tones to amplitude modulate carriers in the bands 108.1-111.95 MHz and 329.15-335.0 MHz.
- Second, new technologies do not fall from vacuum, they are usually evolutions of old technologies. Also, having expertise in old technologies enable us to think about their advancements which lead to new technologies. A technology may expire, but not the basic principles underpinning it.

[2]A simple analogy would help here. Consider a person P1 carrying another person P2 on his/her shoulders or head. P1 should be at least as tall as P2 in order that the latter's feet do not touch the ground and get hurt. Similarly, the carrier frequency is usually much higher than the maximum frequency content of the modulating signal in order to shift it to the spectrum band most suitable for message transmission.

[3]An electromagnetic radiation is an energy which is radiated in the form of a wave as a result of the motion of electric charges.

For the reasons above, students are encouraged to enjoy studying old technologies as well as modern ones. At least some people spend their lives studying history, and knowing the history of technological developments is an expertise in itself.

Digital modulation is the method used to transfer digital data over analog transmission system. This is how the MODEM that we connect to a computer to send data over the phone line operates. An example of technologies using this technique is xDSL over PSTN/POTS infrastructure.

Modulation schemes can be divided into linear modulation and nonlinear modulation. Modulation schemes which use the message signal to vary a parameter of the carrier wave in a linear manner are referred to as **linear modulation schemes**. Modulating the amplitude, whether analog or digital, results in a linear modulation. Note that phase modulation, either analog or digital, and QAM are also classified as linear modulation. Modulating the frequency of a carrier wave, however, is a **nonlinear modulation** as we obtain the phase of the carrier wave by integrating the message signal.

We will sample the applications of digital modulation schemes prior to proceeding. Some of the applications of phase-shift keying (PSK) are:

(1) Binary PSK (BPSK):
   (a) $\frac{\pi}{2}$-BPSK has been standardised in 5G mobile cellular technology.
   (b) Cable modems and RFID's.
   (c) WLAN IEEE 802.11b: 1 Mega bits/second.
   (d) Deep space satellite telemetry.

(2) QPSK (Quadrature Phase Shift Keying):
   (a) 4G and 5G mobile cellular systems.
   (b) Digital Video Broadcasting: DVB-Sx, DVB-T with OFDM.
   (c) Terrestrial Trunk Radio (TETRA) for emergency services and public safety.
   (d) Satellite and aircraft communications.
   (e) WLAN IEEE 802.11b: 2 M bps, 5.5 M bps, 11 M bps.

Some applications of QAM are:

(1) Mobile cellular systems, e.g., 4G and 5G.
(2) Terrestrial microwave links, e.g., DVB-Terrestrial.
(3) Downstream of cable modem: 64QAM, 256QAM.
(4) WLAN IEEE 802.11a/g: 16-QAM for 24 Mbps and 36 Mbps, 64-QAM for 38 Mbps bps and 54 Mbps.
(5) Telephone modems, e.g., 16-QAM.
(6) High Speed Packet Access: HSDPA & HSUPA for LTE Cellular Systems.

Low-rate (e.g., 1.2 kbps) voiceband modems and storage of data on audio cassettes use frequency-shift keying (FSK). First generation of cellular mobile technologies transmitted control signals using FSK. FSK is also used for communicating alerts in emergency situations. Applications of amplitude-shift keying (ASK) include fibre-optic data transmission in which the carriers are modulated by light intensities. ASK (precisely, on-off keying) was also used in Morse code for telegraphy. Clearly, PSK and QAM are much more popular than ASK and FSK.

3.1.4 *Check Your Understanding*

Students are encouraged to check their understanding of the content of this section with the following questions prior to proceeding to the next section.

(1) Describe in details the differences between baseband and broadband digital communications networks. Include in your description the differences between the types of signals transferred over each of the two types of networks. Also, cite examples of each case.

(2) Consider a signal with the baseband bandwidth $B = 200$ kHz transferred using an 16-QAM digital modulation scheme and rate $\frac{2}{3}$ forward-error control (FEC). The physical layer data rate is 2 Mbps. Compare the maximum modulation efficiency

with the bandwidth efficiency if this signal is transmitted using:

(a) Baseband transmission.
(b) Single-sideband (SSB) passband transmission.
(c) Double-sideband (DSB) passband transmission.

Hint. Use the Nyquist rate, which is $R_s \leq 2B$ for baseband and SSB passband transmissions, and $R_s \leq B$ for DSB passband transmissions.

(3) List and discuss the three efficiencies that the design of communications systems seek to optimise?
(4) Assume that someone who has no idea about electronic communications wants you to explain to her the meaning of the word *modulation*. Write two statements to that effect.
(5) **In-class research:** In groups of two to three students, brainstorm about the *importance* of modulation methods in communications systems. Include in your discussions the types of modulation methods and their relative differences.
(6) In a tabular form, differentiate between broadband and baseband communications systems, as well as the difference between broadband and baseband signals.
(7) State the minimum data rates for both uplink (or reverse link) and downlink (or forward link) that is required of a communications network to be classified as a broadband network. Refer in particular to the definitions of USA Federal Communications Commission (FCC).
(8) Why do we also refer to broadband modulation methods as passband modulation?
(9) Describe how carrier signals used in baseband modulation methods differ from those used in broadband modulation methods.
(10) Give at least examples each of baseband and broadband transmission media. Research to find some of the practical applications of each type.
(11) How do we call the signal which changes another signal in modulation?
(12) How do we call the signal which is changed by another signal in modulation?
(13) How do we call the signal at the output of a modulation method?
(14) What are the possible characteristics of the carrier signal which can be varied to transfer an information between two points?
(15) State atleast two reasons why we cannot stop teaching technologies which are getting obsolete?
(16) **In-class research:** In groups of two to three students, research to find at least one example of a practical application of each of the following modulation methods:

(a) Line coding.
(b) Digital modulation.
(c) Waveform coding or pulse modulation.
(d) Analog modulation.

(17) Discuss the reasons why power efficiency and bandwidth efficiency are desired of modulation methods.
(18) State and explain the reason(s) why amplitude modulation is a linear but frequency modulation is nonlinear.
(19) Compute and plot the power spectrum of the signal $s(t) = 2\cos(100\pi t)$.

3.2 Amplitude Modulation

Amplitude Modulation (AM) is among the oldest (analog) modulation schemes, and the most simplest of them. AM is still widely used in radio broadcasting, for example. AM is also referred to as **linear modulation**, and it has many variants. The common characteristic among the AM variants is that all of them *carry* the *message signal* in their amplitude. Hence, it is the amplitude parameter of the carrier that is changed, i.e. modulated. For that reason, AM is also referred to as **envelope modulation**. By reason of its nature, the phase of the modulated signal is not relevant in AM, and hence

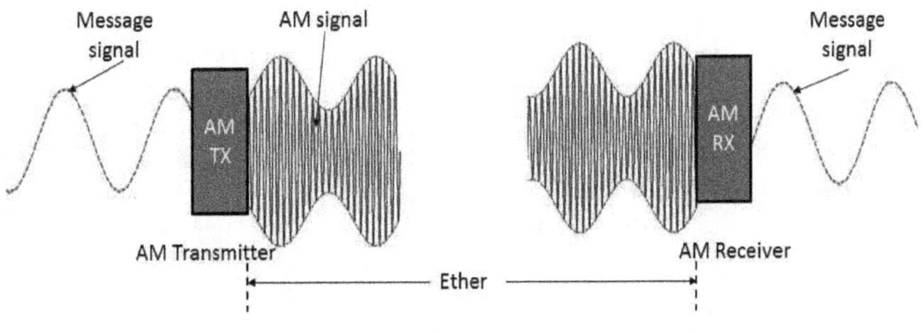

Fig. 3.5 Pictorial view of amplitude modulation.

omitted in our discussions of AM here. Figure 3.5 illustrates AM transmission. This figure provides a pictorial understanding of why the modulated signal is called carrier. The modulating signal is the envelope being carried.

The inventors and invention date of AM are not quite clear as many people contributed to it. Lee de Forest, Landell de Moura and Reginald Fesseden are known to have invented AM in early 1900s. On 1st Dec 1915 John Renshaw Carson filed a US patent for single-sideband[4] AM which was first used in practical systems on 7 January 1927 in a radiotelephone between London and New York. Prior to these dates, several people proposed AM. Examples are Mayer in 1875, Lablanc in 1886 and Rayleigh in 1894. The most popular application of AM is in radio broadcasting. AM radio signals are transmitted in the frequency band around 535 kHz to 1605 kHz. Each AM radio station is assigned a bandwidth of 10 kHz, which is far less than FM radio bandwidth of 200 kHz. Unfortunately, FM radio is being replaced worldwide by digital audio broadcasting (DAB). FM was also used in first-generation of mobile technologies. It is also used in two-way radio communications systems.

If you speak into a microphone (i.e. transducer) of an AM transmitter, the microphone converts the voice into a varying voltage. This voltage is amplified and then used to vary the strength of the transmitter's output. Amplitude modulation adds power to the carrier, with the amount added depending on the strength of the modulating voltage. Amplitude modulation results in the transmission of up to three separate frequencies: the original carrier frequency, a *lower sideband* (LSB) below the carrier frequency, and an *upper sideband* (USB) above the carrier frequency. The sidebands are 'mirror images' of each other about the carrier frequency and contain the same intelligence. When an AM signal is received, these frequencies are combined to produce the sounds we hear. Each sideband occupies as much frequency space as the highest audio frequency being transmitted. If the highest audio frequency being transmitted is B Hz, then the total frequency space occupied by an AM signal will be $2B$ Hz. Exception to this bandwidth requirements is single sideband AM, which suppresses either the LSB or the USB. Note that the carrier, if transmitted, consumes a negligible frequency spectrum, but a larger part of the signal's power.

AM has the advantages of being easy to produce in a transmitter and AM receivers are simple in design. Its main disadvantage is its power inefficiency. About two-thirds of an AM signal's power is concentrated in the carrier, which contains no intelligence. One-third of the power is in the sidebands, which contain the signal's intelligence. Since the sidebands contain the same intelligence, however, one is essentially 'wasted.' Of the total power output of an AM transmitter, only about one-sixth is

[4]Frequency domain representation of modulated signals have sidebands. These are the frequency bands below and above the carrier frequency. Sidebands below the carrier frequency are called **lower sidebands** (LSBs) or **inverted sidebands**, while those above the carrier frequency are referred to as **upper sidebands** (USBs) or **erect sidebands**. The LSB is also called inverted sideband as its highest frequency corresponds to the lowest frequency in the baseband signal. It is the sidebands which contain the information of interest.

actually productive, useful output! AM signals require transceivers with linear amplifiers, preventing amplifiers to operate in their power-efficient nonlinear modes. Other disadvantages of AM include the relatively wide amount of frequency space an AM signal occupies and its susceptibility to static and other forms of electrical noise. Despite this, AM is simple to tune on ordinary receivers, and that is why it is used for almost all shortwave broadcasting. As more AM receivers are needed than AM transmitters (i.e. radio stations), it is reasonable to design simpler receivers than transmitters.

It is interesting to note a priori that the single sideband suppressed carrier AM consumes both the least power in carrier (as the carrier is not transmitted at all) and the least bandwidth (equals that of the modulating signal). This is achieved at the cost of higher system complexity.

3.2.1 *Continuous Wave Modulation*

The simplest form of analog modulation is the continuous wave (CW) modulation. The transceiver of a CW modulation toggles its transmitter between on and off modes, typically to form the characters of the Morse code. CW transmitters are simple and inexpensive, and the bandwidth requirement of CW signals is usually less than half a megahertz. However, the CW signals are difficult to detect on a normal receiver; only the faint quieting of the background noise can be heard as the CW signals are transmitted. Some transceivers, e.g. shortwave and ham radio receivers, are designed with embedded beat frequency oscillator (BFO) circuit to mitigate this effect. The BFO circuit produces an internally-generated second carrier that 'beats' against the received CW signal, producing a tone that turns on and off in step with the received CW signal. This is how Morse code signals are received on shortwave.

3.2.2 *Full-Carrier AM*

A Full-Carrier Amplitude Modulation (FC-AM) radio transceiver transmits every part of the modulated signal, i.e. the carrier wave and both sidebands. For this reason, it is also referred to as *double sideband transmitted carrier AM* (DSBTC-AM). As AM is commonly used in commercial AM radios we find everywhere, it is also dubbed *conventional AM* (C-AM). FC-AM signal can be represented mathematically as

$$\begin{aligned} y(t) &= [A_c + m(t)]\cos(2\pi f_c t) \\ &= A_c[1 + \eta_{AM} m_n(t)]\cos(2\pi f_c t), \end{aligned} \tag{3.2}$$

where $m_n(t) = \frac{m(t)}{\max\{|m(t)|\}}$, $|m_n(t)| \leq 1$ is the normalised version of $m(t)$, $\max\{|m(t)|\}$ is the maximum value of $m(t)$, and

$$\eta_{AM} = \frac{\max\{|m(t)|\}}{A_c} \tag{3.3}$$

is referred to as the AM **modulation index**. We can also derive the modulation index from the AM waveform displayed on an oscilloscope as shown in Fig. 3.6 as

$$\eta_{AM} = \frac{A_2 - A_1}{A_2 + A_1}. \tag{3.4}$$

The modulation index is also referred to as the *percentage modulation* or *modulation factor*. The process of multiplying a message signal by an auxiliary carrier wave as shown in Eq. (3.2) translates the message signal into another frequency band. For that reason, the process is called *frequency translation, mixing* or *heterodyning*. Hence, the carrier is sometimes referred to as a *mixer, heterodyning signal* or *local oscillator (LO) signal*. By observing the following Fourier transform pairs

$$\begin{aligned} y_1(t) &= A_c[1 + \eta_{AM} m_n(t)] \\ &\leftrightarrow A_c[\delta(f) + \eta_{AM} M_n(f)] = Y_1(f) \end{aligned}$$

and

$$\cos(2\pi f_c t) \leftrightarrow \frac{1}{2}[\delta(f - f_c) + \delta(f + f_c)] = Y_2(f),$$

we can write the Fourier transform of the FC-AM wave in Eq. (3.2) as

$$\begin{aligned} Y(f) = Y_1(f) \star Y_2(f) &= \int_{-\infty}^{\infty} Y_1(\eta) Y_2(f - \eta) d\eta \\ &= \frac{1}{2} Y_1(f - f_c) + \frac{1}{2} Y_1(f + f_c), \end{aligned}$$

which results

$$\begin{aligned} Y(f) &= \frac{A_c}{2}[\delta(f - f_c) + \delta(f + f_c)] \\ &\quad + \frac{A_c}{2}\eta_{AM}[M_n(f - f_c) + M_n(f + f_c)]. \end{aligned} \tag{3.5}$$

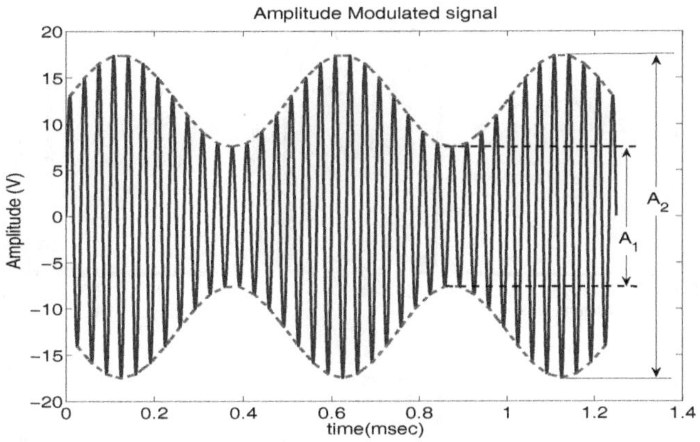

Fig. 3.6 Deriving amplitude modulation index from the AM waveform.

Note that the first term on the right of Eq. (3.5) contains no information but only the carrier. Hence, by observing Eq. (3.5), we see that the modulating signal carrying the information we are interested in is just shifted from its original frequency band to another frequency band which depends only on the carrier frequency. Such an operation can be achieved by simply multiplying the modulating signal by a sinusoidal carrier wave. Figure 3.7(b) illustrates Eq. (3.5), assuming a triangular modulating signal and showing only the positive frequency components. Figure 3.8 also illustrates Eq. (3.5) using a 120-Hz sinusoidal modulating signal and a 900-Hz carrier. The frequency band of the AM spectra $f_c + f_1 \leq |f| \leq f_c + f_2$ is referred to as the **upper sideband** (USB) while the band $f_c - f_2 \leq |f| < f_c - f_1$ is called **lower sideband** (LSB). For obvious reasons, the USB and the LSB are also called sum and difference frequencies, respectively. Note that if $f_c < f_2$ then $M_n(f - f_c)$ would overlap $M_n(f + f_c)$ to cause distortion. Hence, prevention of distortions in AM spectra requires $f_c \gg f_2$. It is also obvious from Fig. 3.7 that the *transmission bandwidth* B_T of the AM signal is

$$B_T = 2 \times (f_2 - f_1) = 2B_m\,, \tag{3.6}$$

which is twice the bandwidth B_m of the message signal $m(t)$.

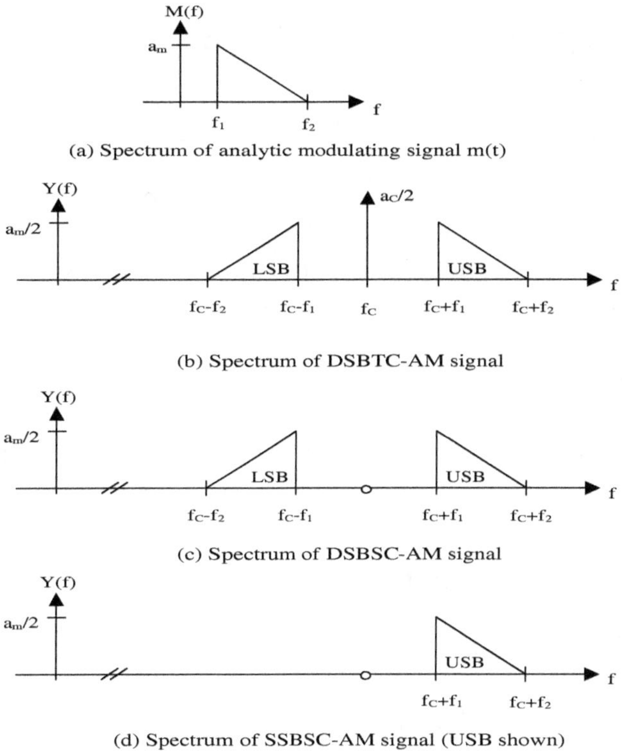

Fig. 3.7 Magnitude spectrum of a modulating signal $m(t)$ and spectra of AM modulated signals using $m(t)$ as modulating signal. Note that only the amplitudes of the spectra are shown.

3.2.2.1 *Effective Modulation Index*

As many signals can be converted to sums of sinusoids, we will use a sinusoidal modulating signal for further illustration of analog modulation principles. Hence, in the following, assume that

$$m(t) = A_m \sin(2\pi f_m t)\,. \tag{3.7}$$

Hence, from Eq. (3.3) the modulation index is $\eta_{AM} = \frac{A_m}{A_c}$, where both A_m and A_c are peak voltages. Their corresponding root-mean-square

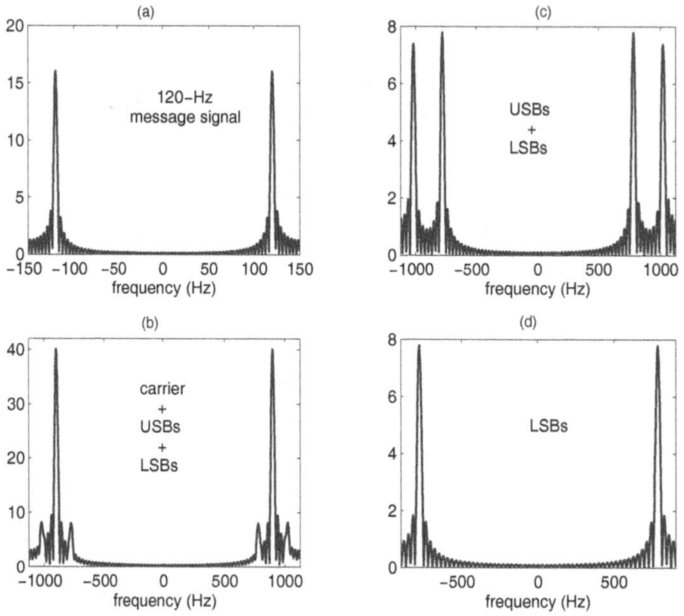

Fig. 3.8 Magnitude spectra of (a) a 120-Hz sinusoidal modulating signal $m(t)$, (b) DSBTC-AM, (c) DSBSC-AM and (d) LSBs of SSBSC-AM. Modulated carrier is at 900 Hz, the LSBs are at frequencies $\pm(f_c - f_m)$ and sampling rate $f_s = 1.25 f_N$.

(RMS) values are $\frac{A_m}{\sqrt{2}}$ and $\frac{A_c}{\sqrt{2}}$. However, we are interested in the **effective modulation index** η_{eff} in case a single RF signal is modulated by more than one, say N, sinusoidal waves. This modulation factor is defined as

$$\eta_{eff} = \sqrt{\sum_{k=1}^{N} \eta_k^2} \leq 1\,. \tag{3.8}$$

Example: Effective Modulation Index

In the following, assume that m sinusoidal waves jointly modulate an RF signal. Assume that the effective modulation index of the AM system is α if $m-1$ of the same set of sinusoidal waves modulate the carrier. What should be the range of the modulation index of the AM system if the carrier is modulated by only the remaining sine wave? Assume that splatter distortion due to overmodulation is forbidden in the locality where the AM radio station operates.

Solution

From Eq. (3.8) we can straightaway write $\eta_{eff}^2 = \sum_{k=1}^{m-1} \eta_k^2 + \eta_m^2$. Hence, by rearranging terms, we obtain $\eta_{eff}^2 - \sum_{k=1}^{m-1} \eta_k^2 = \eta_m^2 \Leftrightarrow \eta_m = \sqrt{\eta_{eff}^2 - \alpha^2}$. As the condition to prevent the occurrence of overmodulation and its associated splatter noise is $0 \leq \eta_{eff} \leq 1$, we obtain $0 \leq \eta_m \leq \sqrt{1-\alpha^2}$, as $\eta_{AM} \in \Re_+$.

Example: Measurement of AM parameters

This example illustrates some ways of estimating the key AM parameters in the time domain, such as the modulation index, the modulating signal and the carrier wave from the AM modulation envelope displayed on oscilloscope, such as in Fig. 3.9. It is obvious from this figure that the four AM transmissions have different levels of modulation, but the same peak carrier amplitude A_c. The modulation indices from this figure are

$$\begin{aligned} \eta_1 &= \frac{A_{max} - A_{min}}{A_{max} + A_{min}} = \frac{1.4 - 0.6}{1.4 + 0.6} = 0.4\,, \\ \eta_2 &= \frac{A_{max} - A_{min}}{2A_0} = \frac{1.9 - 0.1}{2 \times 1} = 0.9\,, \end{aligned} \tag{3.9}$$

Similarly, we obtain $\eta_3 = (2-0)/(2+0) = 1$ and $\eta_4 = (2.5 - -0.5)/(2.5 - 0.5) = 3/2 = 1.5$.

It can also be observed from Fig. 3.9 that the carrier wave is a cosine wave as $y(t=0) \neq 0$. The modulating wave in each case is a sine (not cosine) wave

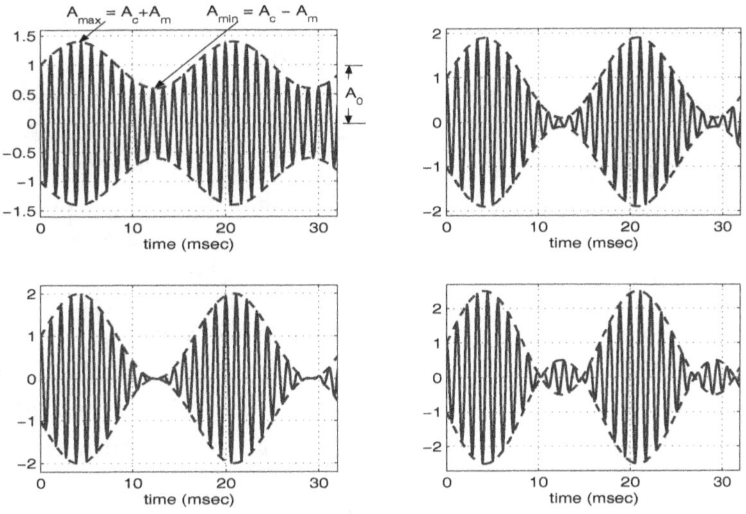

Fig. 3.9 FC-AM waves $y(t) = A_c[1 + \eta_{AM}\sin(2\pi f_m t)]\cos(2\pi f_c t)$ with the same carrier and the same modulating signal but different modulation factors. Note that the figures are normalised by A_c.

as each of the envelopes has zero at $t = 0$. Hence, we can write $y(t) = A_c[1+\eta_{AM}\sin(2\pi f_m t)]\cos(2\pi f_c t)$. As $y(t = 0) = 1$ in each case, we have $A_c = 1$. Hence, the modulating and carrier signals in each of the four cases are $x_c(t) = \cos(2\pi t \times 900 \text{ Hz })$, $m_k(t) = \eta_k \sin(2\pi t \times 60 \text{ Hz }), k \in \{1, 4\}$. As nine periods of the carrier wave last 10 ms, the carrier frequency is $f_c = (\frac{10}{9} \text{ ms})^{-1} = 900$ Hz. It can also be observed that two periods of the modulating signal lasts about $T_m = 33.5$ ms, yielding the frequency $f_m = 1/T_m \approx 60$ Hz.

3.2.2.2 *Power Requirements of FC-AM*

As said ealier, one of the most interesting efficiency metrics in AM is the proportion of signal power consumed by the carrier wave $x_c(t) = A_c\cos(2\pi f_c t+\theta)$ itself. Recall that we usually use voltages in electrical engineering as any other physical quantity can be converted to voltages by a transducer. Plugging the modulating signal Eq. (3.7) in Eq. (3.5) yields the spectra in Fig. 3.8.

The mean power in the carrier signal if it appears across a resistor of resistance R ohms is

$$P_c = \frac{1}{RT_c}\int_{t_0}^{T_c+t_0} x_c^2(t)dt = \frac{A_c^2}{2R}. \tag{3.10}$$

From Fig. 3.8(b) we can calculate the power in each sideband of the FC-AM wave as

$$P_{usb} = P_{lsb} = 2 \times \frac{1}{R}(A_m/4)^2 = \frac{A_m^2}{8R}.$$

By recalling that $A_m = \eta_{AM} A_c$, we can rewrite P_{usb} as $P_{usb} = P_{lsb} = \frac{A_c^2}{8R}\eta_{AM}^2 = \frac{1}{4}\eta_{AM}^2 P_c$. Hence, the total sideband power is

$$P_{sb} = P_{usb} + P_{lsb} = \frac{1}{2}\eta_{AM}^2 P_c. \tag{3.11}$$

This brings the total FC-AM signal power to

$$\begin{aligned} P_T &= P_c + P_{sb} = P_c + \frac{1}{2}\eta_{AM}^2 P_c \\ &= \left(1 + \frac{1}{2}\eta_{AM}^2\right)P_c. \end{aligned} \tag{3.12}$$

As electric power P measured in watts is related to electric current i measured in amperes by $P \propto i^2$, we obtain the current version of Eq. (3.12) as

$$i_T = \sqrt{(1 + \frac{1}{2}\eta_{AM}^2)i_c}, \tag{3.13}$$

where i_T is the total current transmitted from AM transmitter to the antenna and i_c is the carrier current. Note that a similar form of Eq. (3.13) holds for voltages. From the above argument we obtain the fraction of total power embedded in the sidebands as

$$\rho_{sb} = \frac{P_{sb}}{P_T} = \frac{\eta_{AM}^2}{2 + \eta_{AM}^2}. \tag{3.14}$$

ρ_{sb} is called the AM *power efficiency*. Figure 3.10 shows ρ_{sb} and $\rho_c = P_c/P_T$ versus η_{AM}. It is clear from the figure that the carrier signal consumes at least $\frac{2}{3}$ of the total AM power.

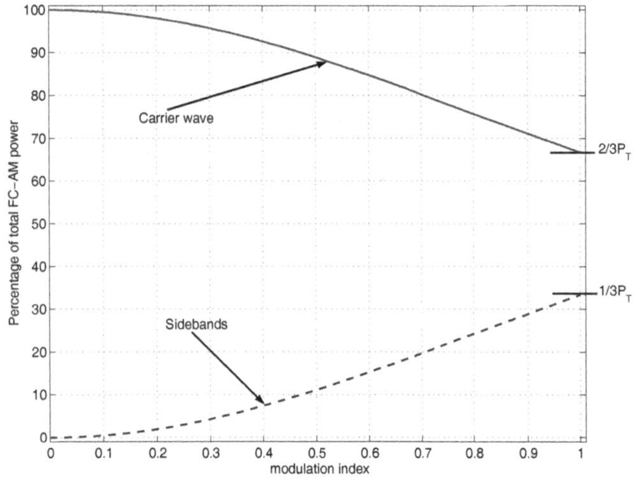

Fig. 3.10 FC-AM power efficiency: ρ_{sb} and ρ_c versus η_{AM}.

Example: Frequency-domain estimation of η_{AM}

We can express the modulation index as a function of the power parameters from Eq. (3.12), i.e. $\eta_{AM} = \sqrt{\frac{2(P_T - P_c)}{P_c}} = \sqrt{\frac{2P_{sb}}{P_c}}$. Hence, a plot of η_{AM} versus $\frac{P_{sb}}{P_c}$ as shown in Fig. 3.11 should enable the estimation of η_{AM}. For example, if we want to evaluate η_{AM} at $r_1 = \frac{P_{sb}}{P_c} = 10^{-2}$, we need just to read off the value of η_{AM} given $\frac{P_{sb}}{P_c} = 10^{-2}$ from Fig. 3.11, which is $\eta_{AM} \approx 0.14$.

3.2.2.3 *Overmodulation and Its Ill-Effects*

As can be observed in Fig. 3.12, overmodulation distorts the AM envelope so that it does no longer resembles the shape of the message signal. As discussed earlier, this also prevents the use of simple envelope detection as the envelope falls below zero. For these reasons we usually expect $0 \leq \eta_{AM} \leq 1$ in practical AM systems. If $\eta_{AM} > 1$ then the system is said to be *overmodulated*, and $\eta_{AM} = 1$ is referred to as 100% modulation. The distortion caused by overmodulation is referred to as **splatter** or **spurious sidebands**. Spurious sidebands in AM transmitters can usually be heard as severely distorted or even unintelligible sounds when tuning between stations. Overmodulation is forbidden by law in many countries, as it causes

- Distortion and interference to neighbouring AM stations and/or receivers, as overmodulation increases bandwidth consumption of the AM signal.
- Voltage and current surges in the AM transmitter which can overheat the transmitter, destroy transformer windings and capacitor plates. Hence, AM transmitters usually use circuit breakers to prevent the effects of overmodulation.

Usually, we want η_{AM} to be as close as possible to unity in order to waste the least power in the carrier. Figure 3.9 illustrates FC-AM with different levels of modulation indices.

Example: AM power efficiency

Consider an AM broadcast station which operates at its maximum allowable power of 52 kW. Assume that the antenna voltage of this AM transmitter is 6 volts when unmodulated but increases to 7.2 volts after modulation. Calculate the modulation factor of this transmitter. What percentage of the AM power is used to transmit the carrier? How much of the total AM transmitted power is used to transmit intelligence?

Solution: As $P \propto v^2$ we obtain a relation similar to Eq. (3.13) between the antenna voltages before modulation v_c and after modulation v_T, i.e. $v_T = \sqrt{1 + \frac{1}{2}\eta_{AM}^2} v_c$. Hence, the AM modulation index is $\eta_{AM} = \sqrt{2[(\frac{v_T}{v_c})^2 - 1]} = \sqrt{2[(\frac{7.2}{6})^2 - 1]} \approx 0.94$. The fraction of total AM power used to transmit the carrier alone can be obtained after rearranging Eq. (3.12) as $\rho_c = P_c/P_T = (1 + \frac{1}{2}\eta_{AM}^2)^{-1} \approx (1 + \frac{1}{2}0.94^2)^{-1} \approx 69.4\%$. The part of the AM signal that contains the intelligence are the sidebands. As 69.4% of P_T is wasted in the carrier, the total sidebands' power is $P_{sb} = (1 - \rho_c)P_T = 0.306 P_T \approx 15.9$ kW.

3.2.2.4 *Generation and Detection of FC-AM Waves*

Figure 3.13 illustrates the generation and detection of DSBTC-AM waves. It is clear that $y(t)$ in Fig. 3.13(a) is equivalent to the DSBTC-AM wave $y(t) = [A_c + m(t)] \sin(2\pi f_c t)$. Neglecting any possible distortion in the transmission system, including

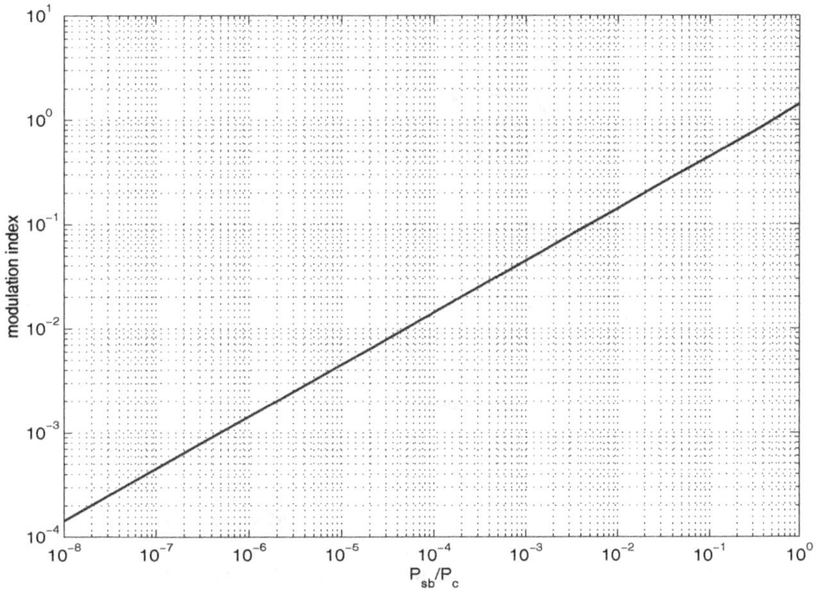

Fig. 3.11 AM modulation index η_{AM} versus fraction of sidebands' power $\frac{P_{sb}}{P_c}$.

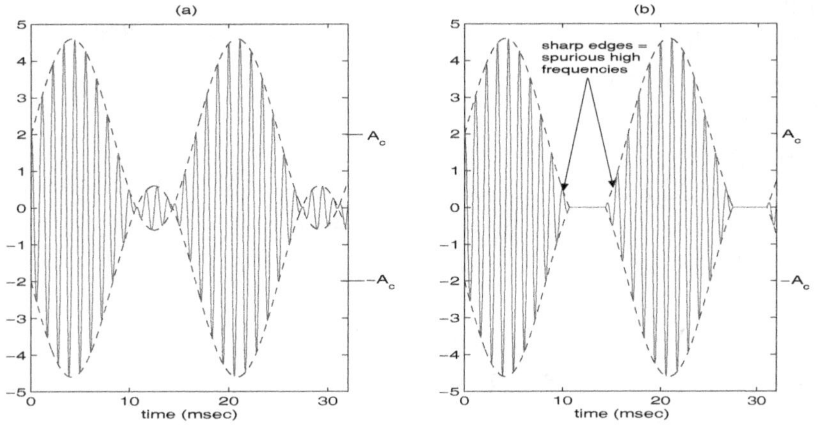

Fig. 3.12 Overmodulated FC-AM wave $A_c[1+2.2\sin(2\pi f_m t)]\cos(2\pi f_c t)$: (a) theoretical output of transmitter and (b) output of practical FC-AM transmitter.

the transmission channel, $p(t)$ in Fig. 3.13(b) is of the form

$$p(t) = y(t)\sin(2\pi f_c t) = [A_c + m(t)]\sin^2(2\pi f_c t)$$
$$= \frac{A_c + m(t)}{2}[1 - \cos(4\pi f_c t)]\,.$$

Hence, any low pass filter (LPF) with cut-off frequency f_{co} in the range $f_m < f_{co} < 2f_c - f_m$ will be able to block the frequency component at $2f_c$ and deliver at its output

$$z(t) = \frac{1}{2}m(t) + \frac{1}{2}A_c\,, \qquad (3.15)$$

as shown in Fig. 3.13(b). Hence, if we are able to block the dc component and scale the output by 2, we will be in a position to recover the message signal $m(t)$. As can be discerned from Eq. (3.2) so long as overmodulation does not occur the positive envelope in $y(t)$ always remains positive, as shown in Fig. 3.9. This allows the usage of simple demodulators such as the envelope detector shown in Fig. 3.13. Hence, although the RF carrier consumes too much of the signal power, it sustains the positive FC-AM envelope above zero voltage.

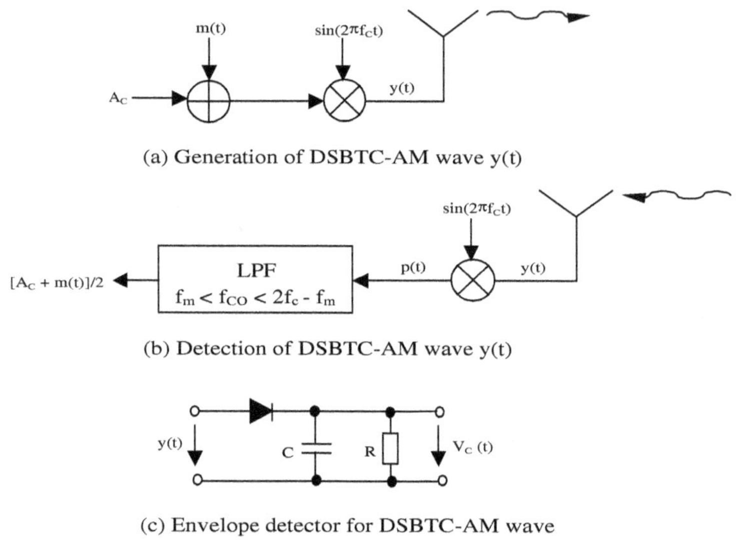

Fig. 3.13 Simplified illustrations of the generation and detection of DSBTC-AM waves.

The envelope detection in Fig. 3.13 operates as follows. In each positive cycle of the AM wave, the diode (assumed to be ideal, i.e. has infinite reverse biased resistance and a zero forward biased resistance) is forward biased so that the capacity is charged to the peak value of the wave. In the negative cycle of the AM wave, however, the capacitor discharges over the resistor. In the ideal case, the output of the envelope detector is $V_c(t) = |1 + \eta_{AM} m_n(t)| = 1 + \eta_{AM} m_n(t)$, as $1 + m_n(t) \geq 0$, allowing easy detection of $m(t)$. For efficient operation of the envelope detection, an appropriate time constant $\tau_c = RC$ must be chosen for the RC circuit. That is the RC must ensure that the change in $V_c(t)$ between diode cycles is at least equal to the change in carrier amplitude between cycles.

3.2.3 *Suppressed-Carrier AM*

The DSBSC-AM was designed to mitigate the issue of wasting at least 66% of the total AM power in transmitting the carrier. DSBSC-AM is also referred to simply as DSB-AM. Owing to the lack of the RF carrier, DSBSC-AM waves cannot be detected using simple envelope demodulators. Hence, they require coherent detectors, which require strict synchronisation between the phases and the carrier frequencies at both the AM receiver and transmitter. The DSBSC-AM wave can be written as

$$\begin{aligned} A_c m(t)\cos(2\pi f_c t) &= y_{dsb}(t) \leftrightarrow Y_{dsb}(f) \\ &= \frac{A_c}{2} M(f - f_c) + \frac{A_c}{2} M(f + f_c). \end{aligned} \tag{3.16}$$

A DSBSC-AM wave and the corresponding damped sinusoidal message signal are shown in Fig. 3.14.

The power contained in the DSB-AM is

$$\begin{aligned} P_{dsb} &= \lim_{T\longrightarrow\infty} \frac{1}{T} \int_{t_0}^{t_0+T} y_{dsb}^2(t)dt \\ &= \frac{A_c{}^2}{2} \lim_{T\longrightarrow\infty} \frac{1}{T} \int_{t_0}^{t_0+T} m^2(t) \\ &\quad + \frac{A_c{}^2}{2} \lim_{T\longrightarrow\infty} \frac{1}{T} \int_{t_0}^{t_0+T} m^2(t)\cos(4\pi f_c t) \\ &\approx \frac{A_c^2}{2} P_m , \end{aligned}$$

where it has been assumed that the last term tends to zeros as T increases towards infinity. Hence, the power in an DSBSC-AM signal is

$$P_{dsbsc} \approx \frac{1}{2} A_c P_m . \tag{3.17}$$

3.2.3.1 *Generation and Detection of DSB-AM*

Figure 3.15(a) shows a product modulator which can be used to generate Eq. (3.16). A simple

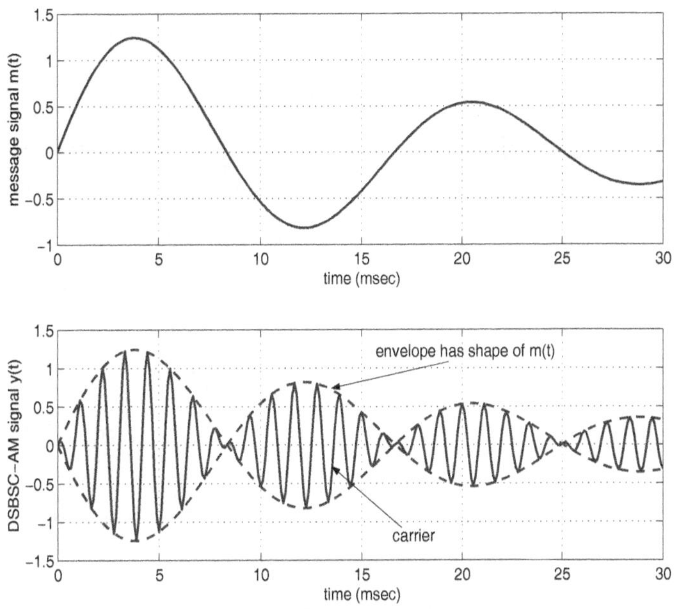

Fig. 3.14 Double sideband suppressed carrier AM: modulating signal (top figure) and modulated signal (bottom figure) [$f_m = 60$ Hz, $T_m = 1/f_m$, $f_c = 900$ Hz, $a = 50$, $A_c = 1$, $m(t) = e^{-a(t-T_m/2)} \sin 2\pi f_m t$].

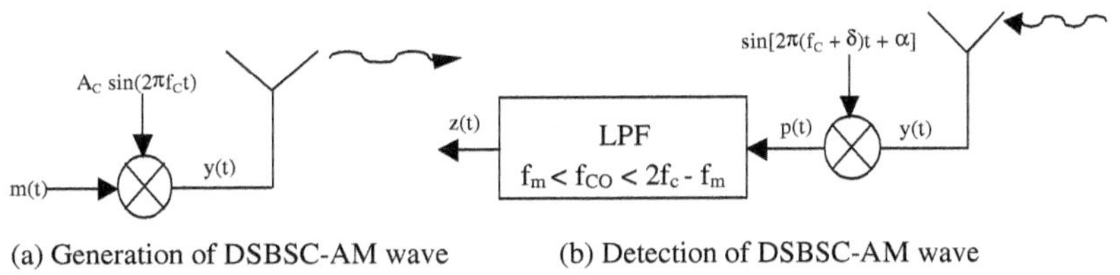

Fig. 3.15 Double sideband suppressed carrier AM (DSBSC-AM): (a) Generation and (b) product demodulator.

detector for DSBSC-AM wave is the product demodulator shown in Fig. 3.15(b). However, as will be clear soon, such a detector requires strict synchronisation of phase and frequency of the carrier. Any slight error in these carrier parameters can cause unpleasant effects.

Assuming an ideal transmission between the AM transmitter and receiver stations, the signal $p(t)$ in Fig. 3.15(b) can be written as $p(t) = y(t)\sin[2\pi(f_c + \delta)t + \alpha]$, where $y(t) = A_c m(t)\sin(2\pi f_c t)$ and δ and α are frequency and phase errors or deviations in the generated carrier at the receiver due to synchronisation errors. Hence, we obtain $p(t) = A_c m(t)\sin(2\pi f_c t)\sin[2\pi(f_c + \delta)t + \alpha] = \frac{1}{2}A_c m(t)\cos(2\pi\delta t + \alpha) - \frac{1}{2}A_c m(t)\cos[(2\pi(2f_c + \delta)t + \alpha)]$. After passing $p(t)$ through the LPF with cut-off frequency $f_m \le f_{co} \le 2f_c - f_m$, we obtain the demodulated signal

$$z(t) \approx \frac{1}{2}A_c m(t)\cos(2\pi\delta t + \alpha)\,. \tag{3.18}$$

Note that, as usually $f_c >> f_m$, a slight deviation in producing f_c can create a frequency distortion in the order of f_m. If $\delta = 0, \alpha = \pi/2$ then $z(t) = 0$, i.e. no signal is detectable at all. Different levels of carrier errors and their effect on $z(t)$ is depicted in Fig. 3.16.

Figure 3.17 illustrates two circuits used to demodulate and detect double sideband AM signal. They are the coherent detector and the Costas loop, respectively. The output of both circuits are the same. A coherent demodulator uses ths same carrier wave used at the modulator to demodulate the modulated signal. Assume that the phase difference between the carrier wave at the transmitter

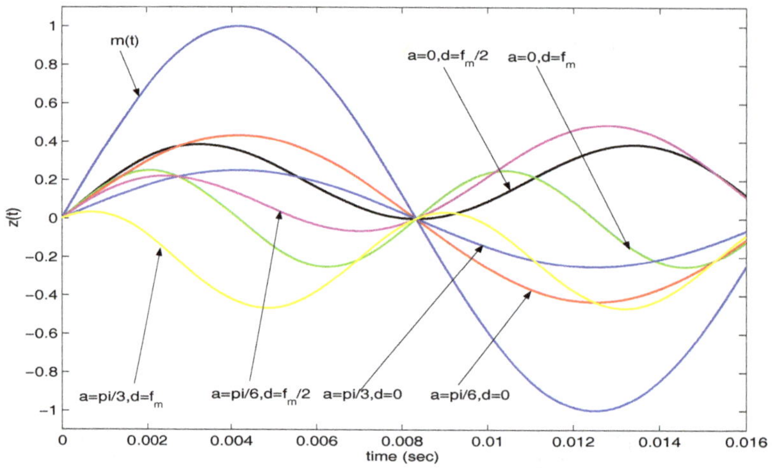

Fig. 3.16 $z(t)$ in Eq. (3.18) for different levels of phase and frequency errors: $a \equiv \alpha$, $d \equiv \delta$.

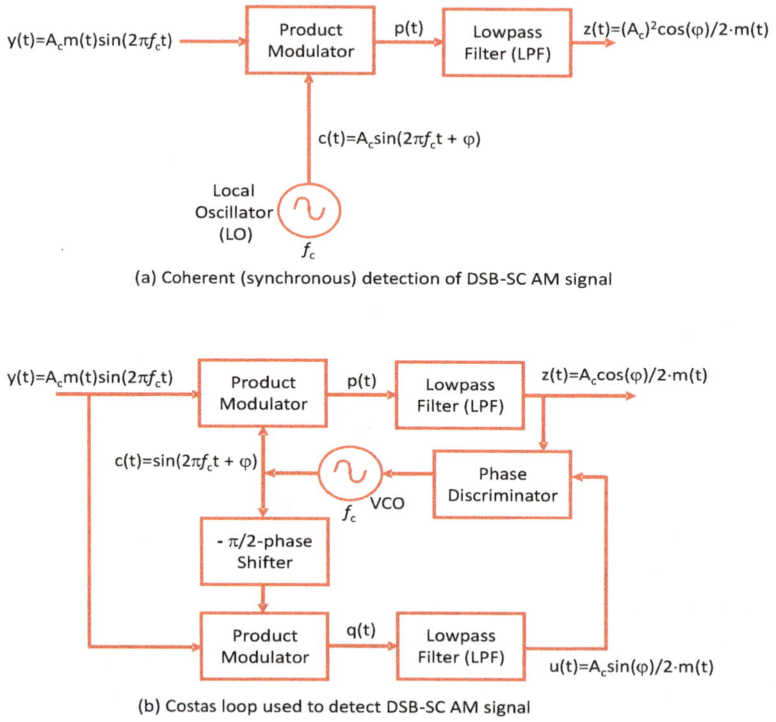

Fig. 3.17 Two circuits for DSB-SC AM demodulation.

and receive is φ, then the carrier at the detector is $A_c \sin(2\pi f_c t + \varphi)$. If the DSB signal is $y(t) = A_c m(t) \sin(2\pi f_c t)$, then the output of the product modulator in Fig. 3.17(a) has the analytic form

$$\begin{aligned} p(t) &= A_c^2 m(t) \sin(2\pi f_c t) \sin(2\pi f_c t + \varphi) \\ &= \frac{A_c^2}{2} m(t) [\cos(\varphi) - \cos(4\pi f_c t)]. \end{aligned} \tag{3.19}$$

Passing $p(t)$ through the LPF in Fig. 3.17(a) yields the output

$$z(t) = \frac{A_c^2}{2} \cos(\varphi) \cdot m(t) \tag{3.20}$$

which is a scaled version of the baseband signal sent from the transmitter. Therefore, we have demodulated the DSB signal. If the detector is coherent then we should have $\varphi \approx 0$, resulting in

Table 3.2 Comparison of AM formats (B : maximum frequency component in message signal, P_T: total AM signal power).

Parameter	DSBTC	DSBSC	SSBTC	SSBSC	VSB
f_c synchronization	No	Yes			
Phase synchronization	No	Yes			
Transmission bandwidth	2B	2B	B	B	$\in (B, 2B)$
Carrier power consumption	$\geq \frac{2}{3}P_T$	0		0	
Detector complexity	Coherent envelope detection	Coherent		Coherent	
Generation complexity	Simple	Simple		Average	

$z(t) = \frac{A_c^2}{2} \cdot m(t)$. In the worse case, if $\varphi = \pm\frac{\pi}{2}$, then $z(t) = 0$. Such an undesirable situation is referred to as `quadrature null effect`.

Costas loop, invented by John P. Costas in the 1950s, is a type of **phase-locked loop** (PLL) circuit. A PLL circuit is used to recover the carrier frequency at the receiver in modulation schemes which do not send the carrier to the receiving circuit. The Costas loop comprises two product modulators, two lowpass filters (LPFs), a phase shifter and a phase discriminator. If the input DSB-SC signal is $y(t) = A_c m(t)\sin(2\pi f_c t)$ then the input of the upper LPF in Fig. 3.17(b) has the analytic form

$$\begin{aligned} p(t) &= A_c m(t)\sin(2\pi f_c t)\sin(2\pi f_c t + \varphi) \\ &= \frac{A_c}{2} m(t)[\cos(\varphi) - \cos(4\pi f_c t)]\,. \end{aligned} \quad (3.21)$$

Passing $p(t)$ through the top LPF in Fig. 3.17(b) yields the output

$$z(t) = \frac{A_c}{2}\cos(\varphi)\cdot m(t) \quad (3.22)$$

which is just a scaled version of the baseband signal sent from the transmitter. Therefore, we have demodulated the DSB-SC signal.

Noting the the input to the bottom product modulator in Fig. 3.17(b) are $y(t) = A_c m(t)\sin(2\pi f_c t)$ and $\sin(2\pi f_c t + \varphi - \frac{\pi}{2}) = -\cos(2\pi f_c t + \varphi)$, we obtain the input to the bottom LPF as

$$\begin{aligned} q(t) &= -A_c m(t)\sin(2\pi f_c t)\cos(2\pi f_c t + \varphi) \\ &= \frac{A_c}{2} m(t)[\sin(\varphi) - \sin(4\pi f_c t + \varphi)]\,. \end{aligned} \quad (3.23)$$

Passing $q(t)$ through the bottom LPF in Fig. 3.17(b) yields the output

$$u(t) = \frac{A_c}{2}\sin(\varphi)\cdot m(t) \quad (3.24)$$

which, again, is the scaled version of the transmitted message signal. The signals $u(t)$ and $z(t)$ differ just by their phase difference of 90°. These signals excite the phase discriminator circuit to produce a direct-current signal which is used to correct the phase error in the voltage-controlled oscillator (VCO) used to synthesise the carrier wave. The result is that the carrier signal synthesised locally using the VCO is phase synchronised with the carrier signal used to produce the DSB-SC signal at the transmitter. We conclude our discussions on AM by comparing important features of its basic formats or types. This is summarised in Table 3.2.

3.2.4 *Check Your Understanding*

Students are encouraged to check their understanding of the content of this section with the following questions prior to proceeding to the next section.

(1) What are the main disadvantages and advantages in each form of AM transmissions? List them in a tabular form.

(2) Discuss why overmodulation in AM transmitters need to be avoided.

(3) For a conventional AM system with 60% modulation calculate the power in the carrier and sidebands if the total output power of the AM transmitter to the antenna is 730 watts.

(4) In full-carrier AM, what is the difference between the content of the LSB and that of the USB? Do we lose any information if we transmit only one of the sidebands?

(5) Does the carrier in AM carry any useful information? Discuss.

(6) What feature of the carrier wave is used to carry the information-bearing signal in amplitude modulation scheme?

(7) Prove the validity of Eq. (3.18) using Fig. 3.16.

(8) Consider the baseband signal $x(t) = 2\sin(2\pi f_0 t)$, $0 \leq t \leq \frac{1}{2f_0}$, $f_0 = 50$ Hz which is modulating the carrier signal $c(t) = \sin(2\pi f_c t)$, $f_c = 50$ kHz.

(a) Plot $x(t)$ in the time domain.
(b) Compute the Fourier transform $X(f)$ of $x(t)$ and plot it.
(c) Sketch the spectrum of the carrier signal $c(t) = \sin(2\pi f_c t)$ for an arbitrary f_c.
(d) Sketch the spectrum of the DSB-SC signal $x(t)c(t)$.
(e) Indicate in the spectrum of the DSB-SC signal $x(t)c(t)$ the upper sideband and the lower sideband spectra.
(f) Identify and write down the frequency range of the USB, LSB and the baseband signal in the spectrum of the DSB-SC signal $x(t)c(t)$.
(g) Compute the percentage of the overall transmitted power that is in the USB.

(9) The modulating signal $X(f)=\cos(2\pi f\tau)$, $-\frac{\pi}{4\tau} \leq f \leq \frac{\pi}{4\tau}$ modulates the carrier wave $c(t) = \cos(2\pi f_c t)$, $f_c = 50\pi$ Hz.

(a) Plot $X(f)$) in the frequency domain.
(b) Compute the inverse Fourier transform $x(t)$ of $X(f)$ and plot it.
(c) Sketch the spectrum of the carrier signal $c(t) = \cos(2\pi f_c t)$, $f_c = 50\pi$ Hz.
(d) Sketch the spectrum of the DSB-SC signal $x(t)c(t)$.
(e) Indicate in the spectrum of the DSB-SC signal $x(t)c(t)$ the upper sideband and the lower sideband spectra.
(f) Identify and write down the frequency range of the USB, LSB and the baseband signal in the spectrum of the DSB-SC signal $x(t)c(t)$.
(g) Compute the percentage of the overall transmitted power that is in the USB.

(10) Prove that the architecture in Fig. 3.13 can be used to demodulate DSBTC-AM signal.

(11) Consider the message signal with the spectrum shown in Fig. 3.18, which is used to modulate the carrier signal $c(t) = A_c\cos(2\pi f_c t)$. Assume that $a = 5$V, $A_c = 4$V, $b = \frac{1}{2}c = 30$ Hz, $f_c = 30$ Hz.

(a) Sketch the full-carrier AM spectrum.
(b) Compute the power efficiency of the FC-AM signal.

(12) The output of an AM transmitter is given by

$$y(t) = 5[1+0.6\sin(5\pi t\times 10^3)]\sin(4\pi t\times 10^3)\,.$$

(a) Evaluate the modulation index of the AM signal.
(b) Is the AM signal undermodulated and overmodulated?
(c) Compute the power used to transmit the carrier if it falls across a resistor of resistance $R = 1\,\Omega$.
(d) Compute the power required to transmit the upper sideband (USB).
(e) Compute the total power used to transmit the AM signal.
(f) Find the bandwidth consumed by the AM signal.
(g) What is the achieved power efficiency in the AM system?

(13) An AM transmitter has to send the message signal $m(t) = 2\cos(1000\pi t)$ using the carrier wave $c(t) = 20\cos(100000\pi t)$.

(a) Compute the modulation index.
(b) Is the system undermodulated? Justify your answer.
(c) How much power is used to transmit only the carrier signal assuming it falls across a resistor of resistance $R = 1\,\Omega$?
(d) Compute the power required for transmitting each sideband.
(e) Compute the power required to transmit the entire AM signal.
(f) What is the achieved power efficiency in the AM system?

(14) Consider a full-carrier AM transmission system using the modulating signal $m(t) = A_{\mathtt{m}}\cos(2\pi f_{\mathtt{m}}t)$ to modulate the carrier wave $c(t) = A_{\mathtt{c}}\sin(2\pi f_{\mathtt{c}}t)$.
(a) Write the analytic formula for the full-carrier AM signal.
(b) Write the expression for the modulation index.
(c) Under what condition(s) is the system overmodulated, i.e., modulation index is larger than unity?
(d) Derive the relationship between $A_{\mathtt{m}}$ and $A_{\mathtt{c}}$ for 100% modulation.
(e) Assume that $f_{\mathtt{m}} = 60$ Hz and $f_{\mathtt{c}} = 500$ Hz and plot the full-carrier AM waveform for the three cases: (i) undermodulation, (ii) 100% modulation, and (iii) overmodulation.
(f) Plot the spectrum of the FC-AM wave.

(15) Consider Fig. 3.19. Figure 3.19(a) is the spectrum of the message signal modulating a carrier with the frequency $f_{\mathtt{c}} = 1$ kHz. Assume that $a = 5$ V, $b = 4$ V, $c = 60$ Hz.
(a) State the type of amplitude modulation that each of the spectra shown in (b) to (f) in Fig. 3.19 represents.
(b) Fully calibrate each of the five spectra. That is, compute and indicate on each figure the important amplitudes and frequencies.

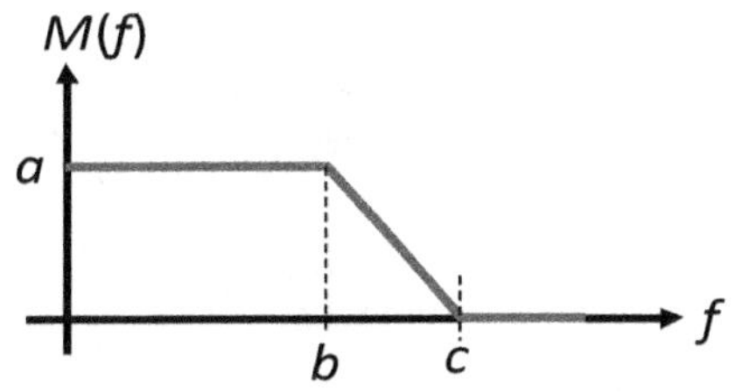

Fig. 3.18 Spectrum of an information-bearing signal.

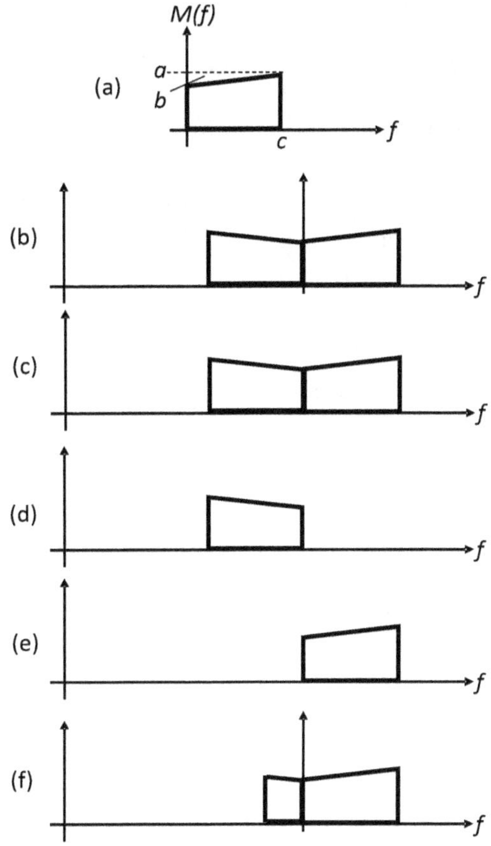

Fig. 3.19 Types of amplitude modulation.

3.3 Angle Modulation: PM and FM

Amplitude modulation is simple. However, it has the weakness that every interference affects its envelope which contains the information-bearing signal. Angle modulation was invented to solve that problem. Frequency and phase are together referred to as angle (Fig. 3.1(b)). Modulation techniques which vary the phase and/or frequency to convey information is referred to as **angle modulation**. As said earlier, amplitude modulation schemes are linear while angle modulation schemes are nonlinear schemes.

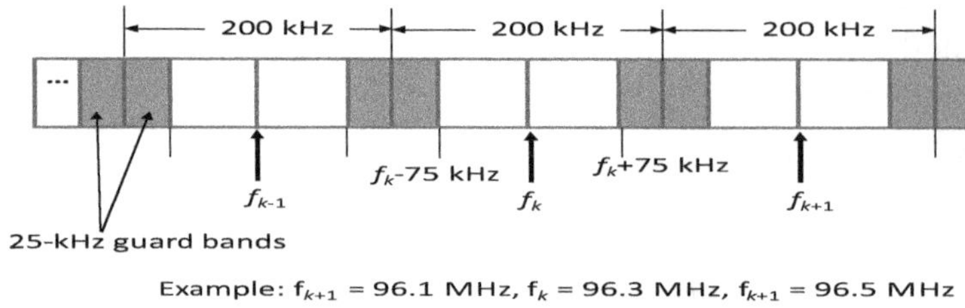

Fig. 3.20 Example of FM radio frequency bands and carriers.

3.3.1 *Frequency Modulation*

In frequency modulation (FM) the message signal is carried in the frequency of the carrier wave. First generation of cellular mobile technologies transmitted audio signals using FM. The most popular application of FM is in radio broadcasting in which the band 87.5 MHz to 108 MHz has been reserved for that purpose. Each FM radio station is allocated 200 kHz[5] (Fig. 3.20) in this band, which is far larger than the 10 kHz bandwidth of AM signals. The date of invention and the invention of FM, like most technologies, is not cast in stones as many people contributed to it either directly or indirectly. Edwin H. Armstrong is known to have invented the superheterodyne radio receiver in 1918 and then the FM in 1933 [Armstrong (1918)].

FM has the following advantages:

- **No need for linear amplifiers:** as FM is a constant-envelope scheme its transceiver do not need a linear amplifier. This allows the operation of the amplifiers in the system in their most-efficient Class C mode. Using high-gain amplifiers is specially important for devices which transmit low power and/or must conserve their power. Examples are battery-powered handheld devices for mobile communications.
- **Robust against signal variations**: the message signal in FM is contained in the frequency, but not the amplitude. This means that FM signals are not disturbed by amplitude variations. This is particularly important if FM signals are transferred over wireless links which varies the signal's amplitude. Also, FM signals can be passed through amplitude limiter circuit to reduce noise.

The weaknesses in FM include the need for detectors that are more complicated than those used to detect AM signals. Also, an FM signal consumes a larger bandwidth than an equivalent AM signal. This is especially so if the modulation index is large, as used in wideband FM systems.

The **instantaneous frequency** of the carrier wave in FM is

$$f(t) = f_c + k_f m(t) = f_c + \delta(t) \tag{3.25}$$

where $m(t)$ is the signal to be carried from the transmitter to the receiver, $\delta(t)$ is the **instantaneous frequency deviation**, f_c is the rest frequency, and k_f, measured in Hz/volt, is the frequency deviation constant (aka **frequency sensitivity**) of the FM modulator. Each commercial FM radio station is allocated a 200-kHz band in 87.5-108 MHz, messages signals have a maximum baseband frequency of 15 kHz, and must have a maximum frequency deviation of 75 kHz. This results in a minimum guard band of 50 kHz at either side of each FM radio carrier to minimise adjacent channel interference (ACI). This results in the **instantaneous phase** of the FM signal

$$\varphi(t) = 2\pi \int_0^t f(x)dx = 2\pi f_c t + 2\pi k_f \int_0^t m(x)dx + \varphi_0 \tag{3.26}$$

where φ_0 is the initial phase. The FM waveform is

$$y_{FM}(t) = A_c \cos\left[2\pi f_c t + 2\pi k_f \int_0^t m(x)dx\right]. \tag{3.27}$$

It can be observed from Eq. (3.27) that increasing the frequency of the modulating signal only

[5]Note that GSM, a 2G mobile wireless technology, also uses 200-kHz bandwidth.

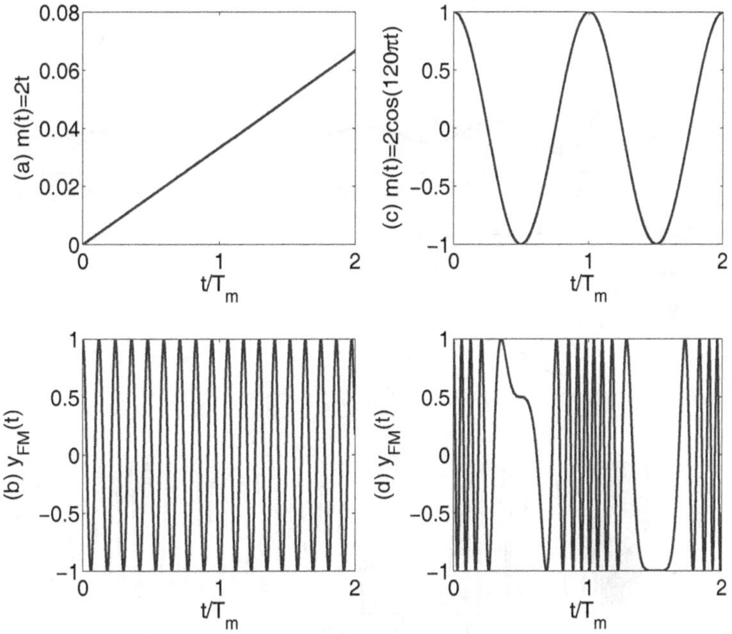

Fig. 3.21 Two baseband signals and their corresponding FM signals [$f_c = 500\texttt{Hz}$, $F_s = 1000 f_c$, $A_c = \sqrt{2}$ V, $k_f = 500$ Hz/V].

changes the speed at which the FM signal's frequency changes between its minimum and maximum values. However, changing the amplitude of the modulating signal changes the frequency of the FM signal. The amplitude of the modulated carrier wave in angle modulation schemes does not change even if the message signal changes. It is only the frequency or phase of the carrier which is modulated. For this reason, such modulation schemes are referred to as **constant-envelope modulation**. We can also observe from Eq. (3.27) that the FM signal $y_{\texttt{FM}}(t)$ depends on the message signal $m(t)$ in a nonlinear manner as the latter must be integrated to obtain the phase of the former. For this reason, FM is classified as a **nonlinear modulation scheme**. The nonlinearity in FM makes its analysis more challenging than that of AM waveform.

Figure 3.21 shows two baseband signals and their corresponding FM signals. Figure 3.21(a) is the baseband signal $m(t) = 2t$ while Fig. 3.21(b) is its corresponding FM signal. Figure 3.21(c) is the baseband signal $m(t) = 2\ \texttt{V}\cos(30\pi t \times 10^3)$ while Fig. 3.21(d) is its corresponding FM signal. Clearly, it is impossible to differentiate one baseband signal from other baseband signals from their FM equivalents.

Example

The message signal in a frequency-modulated radio transmitter is $m(t) = 2.5\texttt{Vsinc}(2f_m t)$, $|t| \le \frac{5}{f_m}$, where $f_m = 50$ Hz. Assume that the frequency deviation constant is $k_f = 1000$ Hz/V and the carrier wave is $c(t) = \cos(2\pi t \times 350\ \texttt{Hz})$.

(1) Find the FM modulation index.
(2) Plot the modulating signal $m(t)$, the instantaneous frequency $f(t)$ and the FM signal in the time domain.
(3) Plot the power spectrum of the modulating signal $m(t)$ and the FM signal $y_{\texttt{FM}}(t)$ in the frequency domain.

Solution

The FM modulation index is $\beta = \frac{k_f A_m}{f_m} \approx 41.7$. Using the sampling frequency of $F_s = 100 f_c$, we obtain the waveforms shown in Fig. 3.22. The corresponding power spectra, measured in watts, are shown in

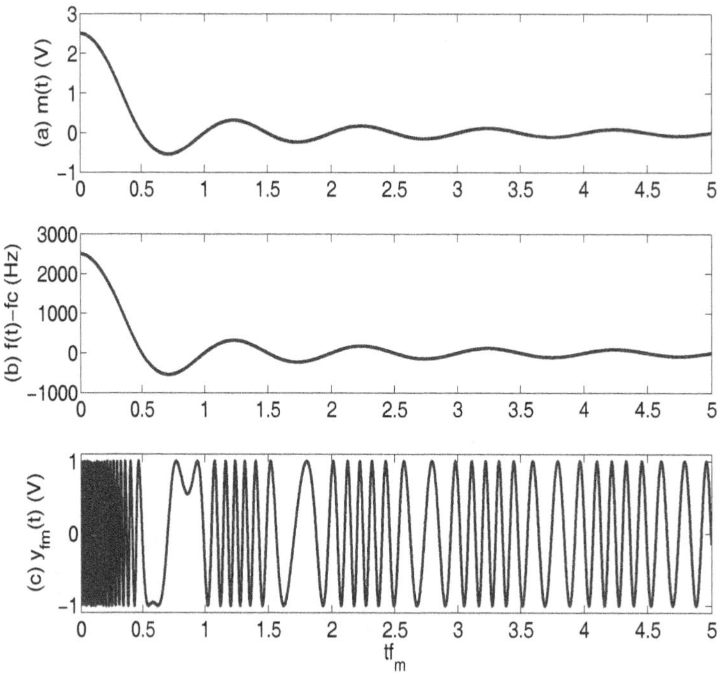

Fig. 3.22 Waveforms of (a) modulating signal, (b) instantaneous frequency and (b) corresponding FM signal: $f_m = 50$ Hz, $c(t) = \cos(2\pi t \times 350 \text{ Hz})$ and $m(t) = 2.5\text{V}\,\text{sinc}(2f_m t)$, $|t| \le \frac{5}{f_m}$.

Fig. 3.23. We can observe the sidebands of the FM signal in Fig. 3.23.

3.3.1.1 *Carson's Rule for FM Bandwidth*

According to Fourier theory we know that an information-bearing signal usually contains multiple tones or frequencies. Assume that the maximum frequency in the message signal to be sent by frequency modulation is f_m. The easiest way to estimate the bandwidth of FM signals is to compute the bandwidth consumed by a single-frequency signal in the frequency spectrum. We then generalise the results by assumption that the tone is the maximum frequency in multi-tone signals. For example, Fig. 3.24 shows the spectrum of the message signal used in practical FM broadcasting. The signal $s_{L+R}(t) = \frac{L(t)+R(t)}{2}$, occupying about 40 Hz to 15 kHz spectrum, is the sum of the left and right channels which is the mono radio signal. A signal generated from the difference between the left and right channels, $s_{L-R}(t) = \frac{L(t)-R(t)}{2}$, occupies the band 23-53 kHz and it is transmitted at 38 kHz using double-sideband suppressed-carrier (DSB-SC) AM. The signal $s_{L-R}(t) = L(t) - R(t)$ is used for the transmission of the stereo-audio signal. A pilot single-frequency signal at 19 kHz is transmitted to lock the receiver onto the 38-kHz subcarrier. The baseband signal used in FM stereo modulation is then of the form

$$\begin{aligned} m_{FM}(t) = {} & \alpha_0 m(t) + \alpha_1 \cos(2\pi t \times 19 \text{ kHz}) \\ & + \alpha_0 s(t) \cos(2\pi t \times 38 \text{ kHz}) \\ & + \alpha_2 SDR(t) \cos(2\pi t \times 57 \text{ kHz}) . \end{aligned} \tag{3.28}$$

Note also that the $SDR(t)$ signal is also a DSB-SC AM waveform. This means that practical FM systems contain amplitude modulation as well. A typical modulation level for each of the components of $m_{FM}(t)$ are shown in Fig. 3.24. The coefficients α_i, $i = 0, 1, 2$ are used to adjust the amplitudes of the corresponding signals to achieve the desired modulation levels.

RDS in Fig. 3.24 means radio data system. It is an ancillary signal transmitted on a 57-kHz subcarrier to convey static information (such as track titles and station identification) which appears on receiver displays.

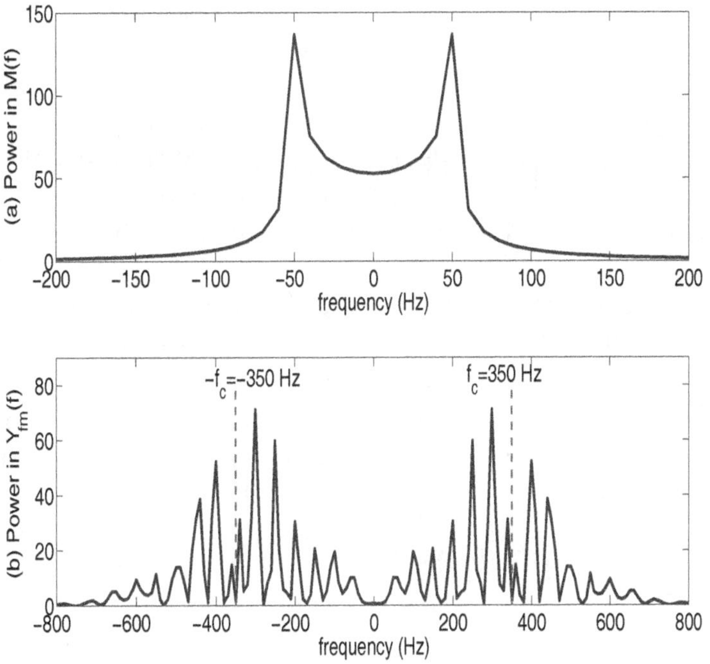

Fig. 3.23 Power spectrum of a modulating signal and corresponding FM signal.

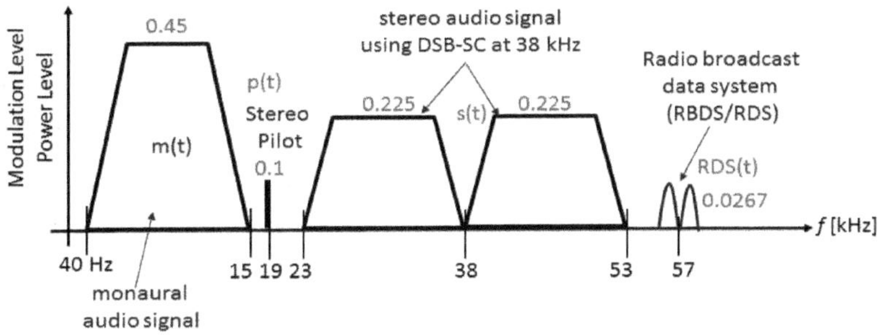

Fig. 3.24 Spectrum of the baseband message signal used in practical FM broadcasting.

For the sake of simplicity of illustration, we use a single-frequency signal in the following. Let the single-tone message signal be

$$m(t) = A_{\mathtt{m}} \cos(2\pi f_{\mathtt{m}} t) \,. \tag{3.29}$$

Plugging Eq. (3.29) into Eq. (3.27) yield

$$\begin{aligned} y_{\mathtt{FM}}(t) &= A_{\mathtt{c}} \cos[2\pi f_{\mathtt{c}} t + \frac{k_{\mathtt{f}} A_{\mathtt{m}}}{f_{\mathtt{m}}} \sin(2\pi f_{\mathtt{m}} t)] \\ &= A_{\mathtt{c}} \cos[2\pi f_{\mathtt{c}} t + \frac{\Delta f}{f_{\mathtt{m}}} \sin(2\pi f_{\mathtt{m}} t)] \\ &= A_{\mathtt{c}} \cos[2\pi f_{\mathtt{c}} t + \beta \sin(2\pi f_{\mathtt{m}} t)] \\ &= A_{\mathtt{c}} \cos(\omega_{\mathtt{c}} t) \cos[\beta \sin(\omega_{\mathtt{m}} t)] \\ &\quad - A_{\mathtt{c}} \sin(\omega_{\mathtt{c}} t) \sin[\beta \sin(\omega_{\mathtt{m}} t)] \end{aligned} \tag{3.30}$$

where

$$\Delta f = k_{\mathtt{f}} \cdot \max\{|m(t)|\} = k_{\mathtt{f}} A_{\mathtt{m}} \tag{3.31}$$

is the peak frequency deviation, and

$$\beta = \frac{\texttt{peak frequency deviation}}{\texttt{modulating frequency}} = \frac{\Delta f}{f_{\mathtt{m}}} = \frac{k_{\mathtt{f}} A_{\mathtt{m}}}{f_{\mathtt{m}}} \tag{3.32}$$

is the **frequency modulation index**. A related term to the modulation index is the **deviation ratio** (DR), which is

$$DR = \frac{\max\{\Delta f\}}{\max\{f_{\mathtt{m}}\}} \,. \tag{3.33}$$

For single-frequency signal $DR = \beta$. FM systems can be classified into **narrowband FM** (NBFM) and **wideband FM** (WBFM) as follows

$$\texttt{NBFM:} \ = DR < 1 \ \texttt{or} \ \left| k_f \int_{-\infty}^{t} m(x)dx \right| \ll 1 \,. \quad (3.34)$$

$$\texttt{WBFM:} \ = DR \geq 1 \ \texttt{or} \ \left| k_f \int_{-\infty}^{t} m(x)dx \right| > 1 \,. \quad (3.35)$$

A typical frequency deviation used in WBFM systems is ±75 kHz, while that of NBFM systems is ±3 kHz. Given the message signal $m(t)$, k_f must be appropriately chosen to achieve the desired FM signal. WBFM is usually used for radio broadcasting, while NBFM is good enough for radio communication.

Mathematicians have found that

$$\cos[\beta \sin(\omega_{\mathtt{m}} t)] = J_0(\beta) + 2 \sum_{k>0,\ \mathtt{even}}^{\infty} J_k(\beta) \cos(k\omega_{\mathtt{m}} t)$$

and

$$\sin[\beta \sin(\omega_{\mathtt{m}} t)] = 2 \sum_{k>0,\ \mathtt{odd}}^{\infty} J_k(\beta) \sin[k\omega_{\mathtt{m}} t]$$

where $J_k(\beta)$ is the Bessel functions of the first kind and order k with argument β. Using these Bessel functions, we can express Eq. (3.30) as

$$\begin{aligned} y_{\mathtt{FM}}(t) &= A_{\mathtt{c}} \sum_{k=-\infty}^{\infty} J_k(\beta) \cos[(\omega_{\mathtt{c}} + k\omega_{\mathtt{m}})t] \\ &= A_{\mathtt{c}} J_0(\beta) \cos(\omega_{\mathtt{c}} t) \\ &\quad + A_{\mathtt{c}} J_1(\beta)[\cos((\omega_{\mathtt{c}} + \omega_{\mathtt{m}})t) - \cos((\omega_{\mathtt{c}} - \omega_{\mathtt{m}})t)] \\ &\quad + A_{\mathtt{c}} J_2(\beta)[\cos((\omega_{\mathtt{c}} + 2\omega_{\mathtt{m}})t) - \cos((\omega_{\mathtt{c}} - 2\omega_{\mathtt{m}})t)] \\ &\quad + A_{\mathtt{c}} J_3(\beta)[\cos((\omega_{\mathtt{c}} + 3\omega_{\mathtt{m}})t) - \cos((\omega_{\mathtt{c}} - 3\omega_{\mathtt{m}})t)] \\ &\quad + A_{\mathtt{c}} J_4(\beta)[\cos((\omega_{\mathtt{c}} + 4\omega_{\mathtt{m}})t) - \cos((\omega_{\mathtt{c}} - 4\omega_{\mathtt{m}})t)] \cdots \end{aligned} \quad (3.36)$$

The Fourier transformed version of Eq. (3.36) is

$$\begin{aligned} Y_{\mathtt{FM}}(f) &= A_{\mathtt{c}} \sum_{k=-\infty}^{\infty} J_k(\beta)[\delta(f - f_{\mathtt{c}} - kf_{\mathtt{m}}) + \delta(f - f_{\mathtt{c}} + kf_{\mathtt{m}})] \\ &= A_{\mathtt{c}} J_0(\beta)[\delta(f - f_{\mathtt{c}}) + \delta(f - f_{\mathtt{c}})] \\ &= A_{\mathtt{c}} J_1(\beta)[\delta(f - f_{\mathtt{c}} - f_{\mathtt{m}}) + \delta(f - f_{\mathtt{c}} + f_{\mathtt{m}})] \\ &= A_{\mathtt{c}} J_2(\beta)[\delta(f - f_{\mathtt{c}} - 2f_{\mathtt{m}}) + \delta(f - f_{\mathtt{c}} + 2f_{\mathtt{m}})] \cdots \end{aligned} \quad (3.37)$$

The term $k = 0$ in Eq. (3.37) is the carrier which contains no information. The remaining terms are the sidebands containing the message signal. Clearly, an FM signal has an infinite number of sidebands. In order to reduce the bandwidth consumption of the FM signal, only some of the sidebands are transmitted.

Figure 3.25 shows the first ten orders of the Bessel function of the first kind for modulation indices up to 10 rounded to four decimal places. For example, only the carrier (i.e., J_0) exists if $\beta = 0$, which means no modulating signal. However, the number of sidebands increases with the modulation index. The FM spectrum for $\beta = 0.5$, $f_{\mathtt{c}} = 96.3$ MHz, $A_{\mathtt{c}} = \sqrt{2}$ is shown in Fig. 3.27. As can be observed, only the first four sidebands contain a significant part of the FM signal's power. As in AM, the sidebands to the right of the carrier frequency are the upper sidebands, while those to the left of the carrier frequency make up the lower sidebands. Other values which are not in Fig. 3.25 can be estimated from Fig. 3.26.

The **Carson's rule** is a rule-of-thumb used to estimate the bandwidth containing 98% of the FM signal's power. Assume that the maximum frequency in the message signal to be sent by frequency modulation is $f_{\mathtt{m}}$ and the frequency modulation index is $\beta = \frac{\Delta f}{f_{\mathtt{m}}}$. The FM bandwidth according to the **Carson's rule** is

$$B_{\mathtt{FM}} \approx 2(1+\beta)f_{\mathtt{m}} = 2(\Delta f + f_{\mathtt{m}}) \,. \quad (3.38)$$

Example

Assume that the maximum frequency deviation of a FM transmitter is $\Delta f = 75$ kHz and the maximum frequency in the audio signal to be broadcast is 15 kHz. (a) Is the FM system NBFM or WBFM? (b) Calculate the effective bandwidth consumed by the FM radio signal.

Solution

Given are: $\Delta f = 75$, $f_{\mathtt{m}} = 15$ kHz. (a) Therefore, the frequency modulation index is $\beta = \frac{\Delta f}{f_{\mathtt{m}}} = \frac{75\ \mathtt{kHz}}{15\ \mathtt{kHz}} = 5$. Thus, the system is a WBFM. (b) From the Carson's rule, the FM bandwidth is

$$B_{\mathtt{FM}} \approx 2(1+\beta)f_{\mathtt{m}} = 2(1+5)15\mathtt{kHz} = 180\ \mathtt{kHz} \,.$$

Practical FM radio broadcasting channels are allocated a bandwidth of 200 kHz. The difference

β	$J_0(\beta)$	$J_1(\beta)$	$J_2(\beta)$	$J_3(\beta)$	$J_4(\beta)$	$J_5(\beta)$	$J_6(\beta)$	$J_7(\beta)$	$J_8(\beta)$	$J_9(\beta)$	$J_{10}(\beta)$
0	1.0000	0	0	0	0	0	0	0	0	0	0
0.25	0.9844	0.1240	0.0078	0.0003	0.0000	0.0000	0.0000	0.0000	0.0000	0.0000	0.0000
0.5	0.9385	0.2423	0.0306	0.0026	0.0002	0.0000	0.0000	0.0000	0.0000	0.0000	0.0000
0.75	0.8642	0.3492	0.0671	0.0085	0.0008	0.0001	0.0000	0.0000	0.0000	0.0000	0.0000
1.0	0.7652	0.4401	0.1149	0.0196	0.0025	0.0002	0.0000	0.0000	0.0000	0.0000	0.0000
1.25	0.6459	0.5106	0.1711	0.0369	0.0059	0.0007	0.0001	0.0000	0.0000	0.0000	0.0000
1.5	0.5118	0.5579	0.2321	0.0610	0.0118	0.0018	0.0002	0.0000	0.0000	0.0000	0.0000
1.75	0.3690	0.5802	0.2940	0.0919	0.0209	0.0038	0.0006	0.0001	0.0000	0.0000	0.0000
2.0	0.2239	0.5767	0.3528	0.1289	0.0340	0.0070	0.0012	0.0002	0.0000	0.0000	0.0000
2.25	0.0827	0.5484	0.4047	0.1711	0.0515	0.0121	0.0023	0.0004	0.0001	0.0000	0.0000
2.5	-0.0484	0.4971	0.4461	0.2166	0.0738	0.0195	0.0042	0.0008	0.0001	0.0000	0.0000
2.75	-0.1641	0.4260	0.4739	0.2634	0.1007	0.0297	0.0071	0.0015	0.0003	0.0000	0.0000
3.0	-0.2601	0.3391	0.4861	0.3091	0.1320	0.0430	0.0114	0.0025	0.0005	0.0001	0.0000
3.25	-0.3328	0.2411	0.4811	0.3510	0.1669	0.0599	0.0174	0.0042	0.0009	0.0002	0.0000
3.5	-0.3801	0.1374	0.4586	0.3868	0.2044	0.0804	0.0254	0.0067	0.0015	0.0003	0.0001
3.75	-0.4014	0.0332	0.4191	0.4138	0.2430	0.1046	0.0359	0.0103	0.0025	0.0006	0.0001
4.0	-0.3971	-0.0660	0.3641	0.4302	0.2811	0.1321	0.0491	0.0152	0.0040	0.0009	0.0002
4.25	-0.3692	-0.1556	0.2960	0.4341	0.3169	0.1624	0.0652	0.0217	0.0062	0.0015	0.0003
4.5	-0.3205	-0.2311	0.2178	0.4247	0.3484	0.1947	0.0843	0.0300	0.0091	0.0024	0.0006
4.75	-0.2551	-0.2892	0.1334	0.4015	0.3738	0.2280	0.1063	0.0405	0.0131	0.0037	0.0009
5.0	-0.1776	-0.3276	0.0466	0.3648	0.3912	0.2611	0.1310	0.0534	0.0184	0.0055	0.0015
5.25	-0.0931	-0.3450	-0.0384	0.3158	0.3993	0.2926	0.1581	0.0687	0.0252	0.0080	0.0022
5.5	-0.0068	-0.3414	-0.1173	0.2561	0.3967	0.3209	0.1868	0.0866	0.0337	0.0113	0.0034
5.75	0.0760	-0.3179	-0.1866	0.1882	0.3829	0.3446	0.2164	0.1070	0.0441	0.0156	0.0049
6.0	0.1506	-0.2767	-0.2429	0.1148	0.3576	0.3621	0.2458	0.1296	0.0565	0.0212	0.0070
6.25	0.2131	-0.2207	-0.2837	0.0391	0.3213	0.3721	0.2741	0.1541	0.0712	0.0281	0.0097
6.5	0.2601	-0.1538	-0.3074	-0.0353	0.2748	0.3736	0.2999	0.1801	0.0880	0.0366	0.0133
6.75	0.2895	-0.0803	-0.3133	-0.1053	0.2196	0.3656	0.3220	0.2069	0.1070	0.0468	0.0178
7.0	0.3001	-0.0047	-0.3014	-0.1676	0.1578	0.3479	0.3392	0.2336	0.1280	0.0589	0.0235
7.25	0.2920	0.0686	-0.2731	-0.2192	0.0916	0.3204	0.3502	0.2593	0.1506	0.0729	0.0305
7.5	0.2663	0.1352	-0.2303	-0.2581	0.0238	0.2835	0.3541	0.2832	0.1744	0.0889	0.0390
7.75	0.2252	0.1916	-0.1758	-0.2823	-0.0428	0.2382	0.3501	0.3039	0.1989	0.1068	0.0490
8.0	0.1717	0.2346	-0.1130	-0.2911	-0.1054	0.1858	0.3376	0.3206	0.2235	0.1263	0.0608
8.25	0.1092	0.2622	-0.0456	-0.2843	-0.1611	0.1281	0.3164	0.3321	0.2472	0.1473	0.0742
8.5	0.0419	0.2731	0.0223	-0.2626	-0.2077	0.0671	0.2867	0.3376	0.2694	0.1694	0.0894
8.75	-0.0259	0.2672	0.0870	-0.2274	-0.2430	0.0053	0.2490	0.3362	0.2889	0.1921	0.1063
9.0	-0.0903	0.2453	0.1448	-0.1809	-0.2655	-0.0550	0.2043	0.3275	0.3051	0.2149	0.1247
9.25	-0.1474	0.2091	0.1926	-0.1258	-0.2743	-0.1114	0.1539	0.3110	0.3168	0.2370	0.1444
9.5	-0.1939	0.1613	0.2279	-0.0653	-0.2691	-0.1613	0.0993	0.2868	0.3233	0.2577	0.1650
9.75	-0.2273	0.1048	0.2488	-0.0028	-0.2505	-0.2028	0.0425	0.2551	0.3238	0.2763	0.1862
10	-0.2459	0.0435	0.2546	0.0584	-0.2196	-0.2341	-0.0145	0.2167	0.3179	0.2919	0.2075

Fig. 3.25 Normalised amplitudes of FM carrier and sidebands.

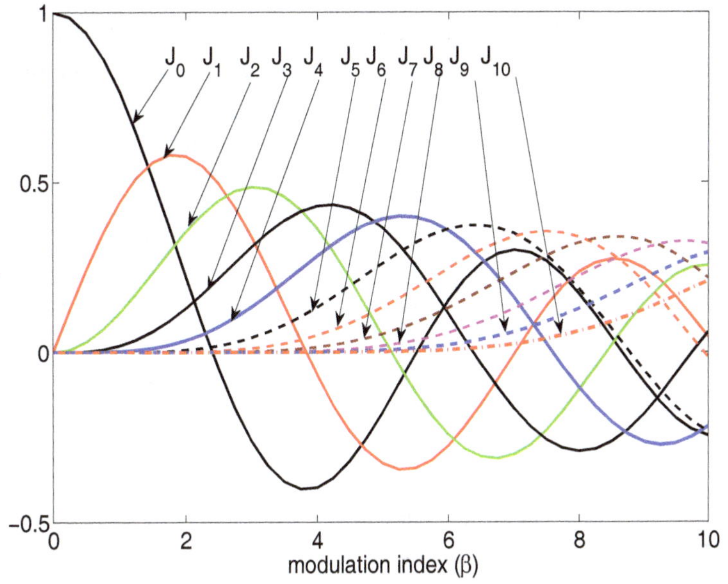

Fig. 3.26 Bessel function of the first kind of order ν, $J_\nu(\beta)$.

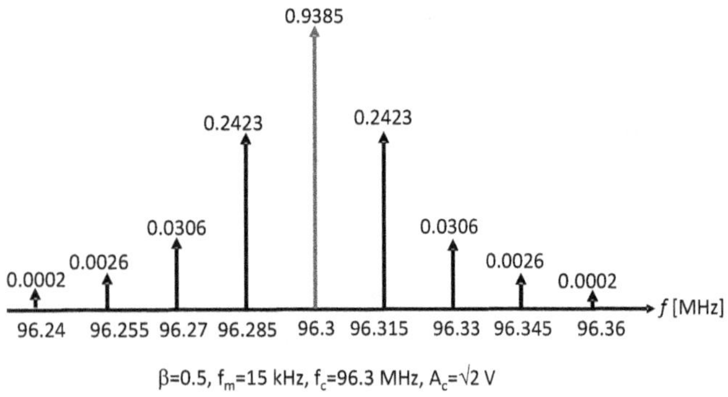

Fig. 3.27 FM voltage spectrum.

20 kHz is used as guardband to reduce inter-channel interference (ICI) or adjacent channel interference (ACI). From Fig. 3.25 there are at least 10 sidebands in either side of the carrier if the modulation index is $\beta = 5$. Therefore, the bandwidth consumption is $(2 \times 10 + 1)15\ \mathtt{kHz} = 315\ \mathtt{kHz}$. Transmitting all the sidebands would make the FM signal consume more than the 200 kHz available to it. This is the reason why only a few sidebands are transmitted.

3.3.1.2 *FM Transmitters and Receivers*

Figure 3.28 shows two FM transmitters. These two circuits differ only in the pre-emphasis filtering preceding the modulator. Figure 3.29(a) shows a simple FM detector, comprising a differentiator and an envelope detector. The differentiator converts the FM signal to an AM signal with the general form

$$\begin{aligned} x(t) &= \frac{dy_{\mathtt{FM}}(t)}{dt} \\ &= -2\pi A_{\mathtt{c}}[f_{\mathtt{c}} + k_{\mathtt{f}} m(t)] \\ &\quad \times \sin\left(2\pi f_{\mathtt{c}} t + 2\pi k_{\mathtt{f}} \int_{-\infty}^{t} m(x)dx\right) \end{aligned} \quad (3.39)$$

which is an amplitude-modulated signal. Therefore, AM detection techniques can be used to retrieve the message signal $m(t)$. Passing $x(t)$ through the envelope detector yields as output

$$z_1(t) = -2\pi A_{\mathtt{c}}[f_{\mathtt{c}} + k_{\mathtt{f}} m(t)] + w_1(t) \quad (3.40)$$

where $w_1(t)$ is the channel's noise at the detector.

The transmission medium (aka channel) introduces noise to the useful signal. The channel noise is a wideband signal, with most of its energy concentrated at high frequencies. The FM detector shown in Fig. 3.29(a) is simple, except that the differentiator, being a high-pass filter, enhances the channel noise to reduce the overall signal-to-noise ratio (SNR). A solution to this problem is the use of a filter called pre-emphasis, which is a high-pass filter such as a differentiator, at the FM transmitter to amplify the amplitude of the high-frequency sounds in the message signal, while reducing the low-frequency components. This **pre-emphasis** must be reversed at the FM detector using the reciprocal filter called **de-emphasis filter**, which is a low-pass filter such as an integrator. This results in the circuit shown in Fig. 3.29(b). The combined pre-emphasis and de-emphasis filters reduces the

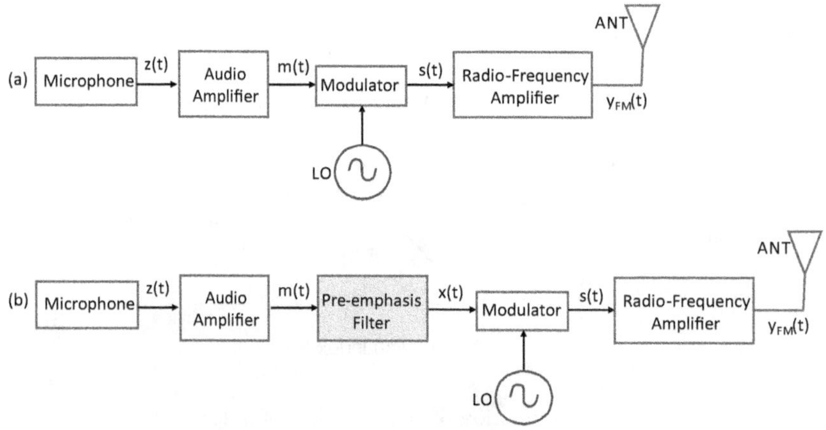

Fig. 3.28 Block diagrams of two FM transmitters.

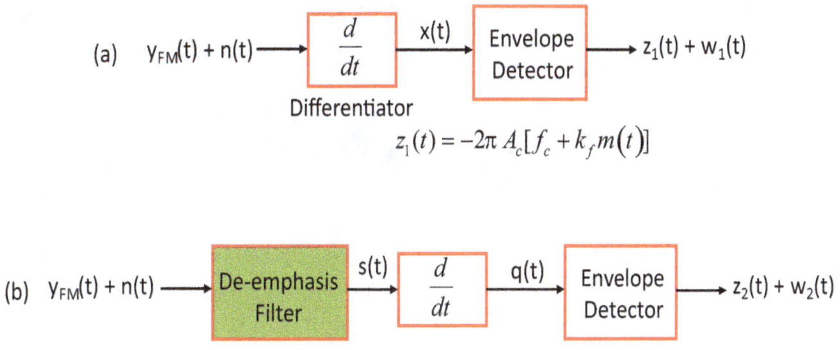

Fig. 3.29 Block diagrams of two FM detectors.

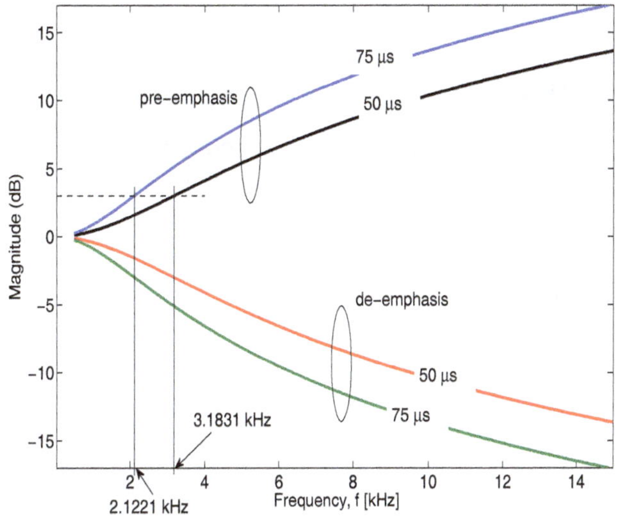

Fig. 3.30 Frequency response of the pre-emphasis and de-emphasis filters.

channel noise to increase the overall SNR. Why? This reason is simple. The noise occurs after the FM leaves the transmitter and therefore it is not boosted by the pre-emphasis filter. However, the noise is attenuated by the de-emphasis filter at the receiver. The message signal is left unchanged as the de-emphasis reverses the pre-processing by the pre-emphasis filter.

The frequency response of the pre-emphasis filter (which is a high-pass filter) used at the FM transmitter is

$$H_{\mathtt{pef}} = 1 + j2\pi f \tau_{\mathsf{c}} \tag{3.41}$$

while that of the de-emphasis filter used at the FM detector is

$$H_{\mathtt{def}} = \frac{1}{H_{\mathtt{pef}}} = \frac{1}{1 + j2\pi f \tau_{\mathsf{c}}} \tag{3.42}$$

where τ_{c} is the time constant of the filter, which is usually an RC circuit. The relation in Eq. (3.42) is a first-order low-pass filter. Values used in the world are $\tau_{\mathsf{c}} = 50\,\mu\mathbf{s}$ for European and Australian systems and $\tau_{\mathsf{c}} = 75\,\mu\mathbf{s}$ for North American systems. The frequency responses are plotted in Fig. 3.30 showing the 3-dB corner frequencies for both time constants.

The impulse response of the pre-emphasis filter is

$$h_{\mathtt{pef}}(t) = \mathcal{F}^{-1}\{1 + j2\pi f \tau_{\mathsf{c}}\} = \delta(t) + \tau_{\mathsf{c}} \frac{d}{dt}\delta(t) \tag{3.43}$$

where $\delta(t)$ is the Dirac delta impulse function, while impulse response of the de-emphasis filter is

$$h_{\mathtt{def}}(t) = \mathcal{F}^{-1}\{H_{\mathtt{def}}(f)\} = \frac{1}{\tau_{\mathsf{c}}} e^{-\frac{t}{\tau_{\mathsf{c}}}} u(t) \tag{3.44}$$

where $u(t)$ is the unit step function.

3.3.1.3 *FM Signal Power and SNR*

As FM is a constant-envelope modulation scheme, the amplitude of the carrier does not change but remains constant. Therefore, the power that an FM signal dissipates into a resistor of resistance R Ohm is

$$P_{\mathtt{FM}} = \frac{A_{\mathtt{c}}^2}{2R} \tag{3.45}$$

where $A_{\mathtt{c}}$ is the amplitude of the carrier wave.

Example

The FM signal

$$y_{\mathtt{FM}}(t) = \sqrt{2}\mathtt{V}\cos[192.6\pi t \times 10^6 + 0.5\sin(30\pi t \times 10^3)]$$

falls across a $1 - \Omega$ resistor. The modulation constant is $k_{\mathtt{f}} = 3750$ $\mathtt{Hz/V}$. (a) What is the modulating signal $m(t)$? (b) Calculate the power that the FM signal dissipates into the resistor.

Solution

Comparing the given AM signal with Eq. (3.27), we can find that $A_{\mathtt{c}} = \sqrt{2}$ $\mathtt{V}$ and $f_{\mathtt{c}} = 96.3 \times 10^6$ Hz, $A_{\mathtt{m}} = 2$ $\mathtt{V}$ and $f_{\mathtt{m}} = 15$ $\mathtt{kHz}$. (a) The modulating signal is obtained from

$$\begin{aligned} 2\pi k_{\mathtt{f}} \int_0^t m(x)dx &= 0.5\sin(30\pi t \times 10^3) \\ &\rightarrow m(t) = 2\cos(30\pi t \times 10^3) \end{aligned}$$

(b) Using Eq. (3.45) where $R = 1\Omega$ we obtain

$$P_{\mathtt{FM}} = \frac{A_{\mathtt{c}}^2}{2R} = \frac{(\sqrt{2})^2}{2 \times 1}\ \mathtt{W} = 1\ \mathtt{W}.$$

Summing the powers in each sideband and the carrier in Fig. 3.27 yields

$$\begin{aligned} P_{\mathtt{FM}} &= 2(0.2423^2 + 0.0306^2 + 0.0026^2 + 0.0002^2) \\ &\quad + 0.9385^2\ \mathtt{W} = 1.00008715 \approx 1\ \mathtt{W} \end{aligned}$$

which confirms the earlier results.

The signal-to-noise ratio (SNR) of an FM signal relates to that of the AM signal as follows:

$$SNR_{\mathtt{FM}} = 3\beta^2 \times SNR_{\mathtt{AM}}. \tag{3.46}$$

FM mono without pre-/de-emphasis filtering achieves the SNR

$$\begin{aligned} SNR_{\mathtt{FM}} &= 3\beta^2(1+\beta)CNR \\ &= \frac{3A_{\mathtt{c}}^2}{4kTB_{\mathtt{FM}}B_{\mathtt{m}}}\left[\frac{B_{\mathtt{FM}}}{2B_{\mathtt{m}}} - 1\right] \end{aligned} \tag{3.47}$$

where $B_{\mathtt{m}}$ is the message signal's bandwidth which is equal to $f_{\mathtt{m}}$ for a single-tone message signal. The thermal noise power is $N_0 = kTB_{\mathtt{FM}}$. Using pre-emphasis at the transmitter and de-emphasis at the receiver further increases the FM signal's SNR by the factor

$$\Delta SNR_{\mathtt{FM}} = \frac{B_{\mathtt{m}}/f_{\mathtt{3dB}}}{3[B_{\mathtt{m}}/f_{\mathtt{3dB}} - \mathtt{tan}^{-1}(B_{\mathtt{m}}/f_{\mathtt{3dB}})]} \tag{3.48}$$

where $B_{\mathtt{m}}/f_{\mathtt{3dB}}$ is the 3-dB break frequency (aka corner frequency) of the pre-emphasis/de-emphasis filters. The corner frequency is obtained from

$$f_{\mathtt{3dB}} = \frac{1}{2\pi\tau_{\mathtt{c}}} \tag{3.49}$$

where $\tau_{\mathtt{c}}$ is the time constant of the pre-emphasis/de-emphasis filters. The two values used in the world are $\tau_{\mathtt{c}} = 50\,\mu\mathtt{s}$ for European systems and $\tau_{\mathtt{c}} = 75\,\mu\mathtt{s}$ for North American systems. This yields the approximate respective corner frequencies 2.1 kHz (for $\tau_{\mathtt{c}} = 75\,\mu\mathtt{s}$) and 3.2 kHz (for $\tau_{\mathtt{c}} = 50\,\mu\mathtt{s}$).

3.3.2 *Phase Modulation*

The **instantaneous phase** in phase modulated signal is

$$\varphi_{\mathtt{PM}}(t) = 2\pi f_{\mathtt{c}}t + k_{\mathtt{p}}m(t) \tag{3.50}$$

where $k_{\mathtt{p}}$, measured in rad/V, is the phase deviation constant of the PM circuit. We obtain the instantaneous frequency from Eq. (3.50) as

$$f(t) = \frac{1}{2\pi}\frac{d\varphi_{\mathtt{PM}}(t)}{dt} = f_{\mathtt{c}} + \frac{k_{\mathtt{p}}}{2\pi}\frac{dm(t)}{dt}. \tag{3.51}$$

The bandwidth of the PM signal is

$$B_{\mathtt{PM}} = 2(1 + k_{\mathtt{p}}A_{\mathtt{m}})B_{\mathtt{m}}. \tag{3.52}$$

Comparing the phase FM signal in Eq. (3.26) and that of PM signal in Eq. (3.50), we can observe the reciprocal relationship between them. We

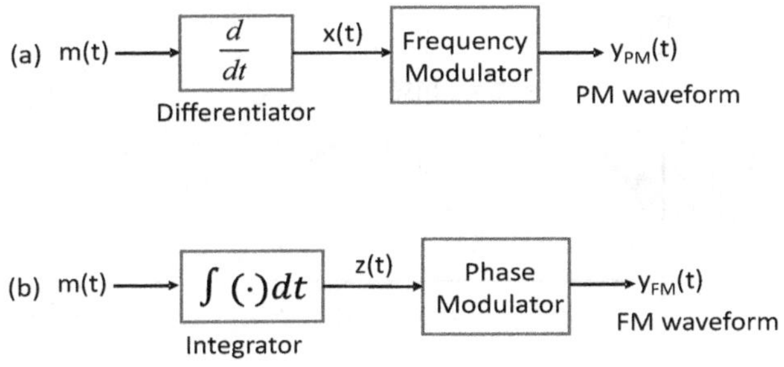

Fig. 3.31 The reciprocity of FM and PM modulators.

can generate an FM signal using a phase modulator by first passing the message through a differentiator before using it to modulate the carrier wave. Similarly, we can produce a PM signal using an FM modulator by first passing the message through an integrator before using it to modulate the carrier wave. These are illustrated in Fig. 3.31, respectively.

3.3.3 *Check Your Understanding*

Students are encouraged to check their understanding of the content of this section with the following questions prior to proceeding to the next section.

(1) A frequency-modulated signal is given by

$$y_{\mathrm{FM}}(t) = 50\sin\left(192.2\pi t \times 10^6 + 80\pi \int_{-\infty}^{t} m(x)dx\right)$$

where $m(t)$ is shown in Fig. 3.32.

(a) Compute the frequency deviation constant k_{f} in Hz/V.
(b) Determine the frequency deviation Δf in Hz.
(c) Sketch the instantaneous frequency as a function of time using any software, such as MATLAB.
(d) Sketch the FM signal as a function of time using any software, such as MATLAB.
(e) How many sidebands fall within the band of frequencies plotted in Part (d)?
(f) Compute the power contained in the band of frequencies plotted in Part (d). Hint. Use the Bessel function coefficients.

(2) The message signal $m(t) = 5\sin(3000\pi t)$ modulates the carrier wave $c(t) = 50\cos(177.4\pi t \times 10^6)$. Assume that the frequency deviation constant is $k_{\mathrm{f}} = 3000$ Hz/V.

(a) Compute the FM modulation index β.
(b) Evaluate the FM signal's bandwidth B_{FM} using Carson's rule.
(c) Using any software, plot the waveform of the FM signal.
(d) Using any software, plot the frequency spectrum of the FM signal in the frequency band $0 \le f \le B_{\mathrm{FM}}$.
(e) How many sidebands fall within the band of frequencies plotted in Part (d)?
(f) Compute the power contained in the band of frequencies plotted in Part (d). Hint. Use the Bessel function coefficients.

(3) Assume that the signal shown in Fig. 3.33 is used to modulate the carrier with amplitude 5 V and frequency 81.1 MHz in an FM system.

(a) Write down the analytic formula for the modulating signal.
(b) Compute the FM modulation index β.
(c) Evaluate the FM signal's bandwidth B_{FM} using Carson's rule.
(d) Using any software, plot the waveform of the FM signal.
(e) Using any software, plot the frequency spectrum of the FM signal in the frequency band $0 \le f \le B_{\mathrm{FM}}$.
(f) How many sidebands fall within the band of frequencies plotted in Part (d)?
(g) Compute the power contained in the band of frequencies plotted in Part (d). Hint. Use the Bessel function coefficients.

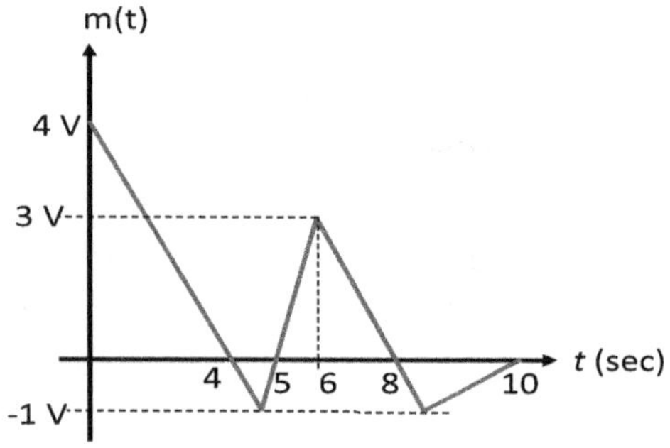

Fig. 3.32 Modulating signal in an FM system.

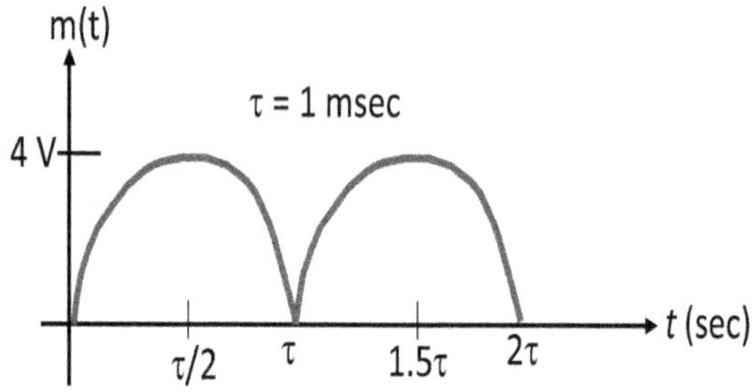

Fig. 3.33 A rectified sinusoidal wave as modulating signal in an FM system.

3.4 Review of Important Terms and Acronyms

Acronyms, abbreviations and important terms		
amplitude modulation	analog demodulation	analog modulation
angle modulation	bandwidth	bandwidth efficiency
baseband frequency	baseband modulation	baseband signal
broadband channel	broadband modulation	carrier frequency
carrier modulation	carrierless modulation	Carson bandwidth
Carson's rule	constant envelope	Costas loop
de-emphasis filter	demodulation	double sideband AM
effective modulation index	emphasis filter	ether
frequency band	frequency deviation	frequency modulation
full-carrier AM	linear modulation	lower sideband
lowpass channels	message signal	modulated signal
modulating signal	modulation	modulation index
modulation method	modulation scheme	nonlinear modulation
out-of-band radiation	overmodulation	phase modulation
phase-locked loop	power efficiency	quadrature null effect
rest frequency	sidebands	transmission medium
upper sideband		

3.5 Supplementary Reading

(1) Alister Burr, *Modulation and Coding for Wireless Communications*, Prentice Hall PTR, 2001.
(2) Harry L. Van Trees, *Detection, Estimation, and Modulation Theory. Part I: Detectio n, Estimation, and Linear Modulation Theory (Part 1)*, John Wiley, 1968.
(3) G. M. Miller, *Modern Electronic Communication*, 3rd ed., Prentice-Hall, 1988.
(4) J. D. Gibson, *Principles of Digital and Analog Communications*, Macmillan Publishing, 1990.

Chapter 4

Digital Modulation and Demodulation Methods

Chapter's Learning Outcomes

Modulation is the process of piggybacking an information-bearing signal over another signal to achieve effective transfer between two locations. Every transmission medium or channel has some undesirable filtering effects on signals, such as blocking of parts of signal, interference, noise, attenuation and distortion. Modulation is used to condition a signal to suit the characteristics of the channel. The reverse process of modulation is called demodulation. Both modulation and demodulation (aka modem) are ISO/OSI Layer 1 (i.e., physical layer) techniques. Information transfer over communications networks cannot work without modem. For this reason, the understanding of the basic principles and practice of modem schemes is fundamental in the communications disciplines. This chapter presents a wide-scope treatment of both *single-carrier* and *multicarrier modulation* techniques, and topics treated include:[1]

(1) Basic principles and practical applications of digital modulation methods.
(2) Basic digital modulation methods:
   (a) Amplitude shift keying (ASK).
   (b) Frequency shift keying (FSK), including minimum shift keying (MSK) and Gaussian MSK.
   (c) Phase shift keying (PSK), including BPSK, QPSK and M-ary PSK.
(3) Quadrature amplitude modulation (QAM).
(4) Baseband modulation methods and broadband modulation methods.
(5) Trellis-coded modulation (TCM), integrating error control and modulation.
(6) Performance metrics of modulation schemes, e.g., bit error rate, bandwidth efficiency and power efficiency.
(7) Multi-carrier modulation (MCM) methods (e.g., FBMC, UFMC, f-OFDM) for LTE-A/4G and 5G mobile systems.

4.1 Introduction

In modulation we use a signal, referred to as *modulating signal* or message signal or intelligence, to change some aspects of another signal called *carrier signal* or bearer signal or carrier wave or just a carrier. The resulting modified carrier signal is referred to as the *modulated signal.* Modulation is also often somewhat loosely referred to as **waveform**. In this chapter we use $c(t)$ to refer to the carrier signal, $m(t)$ for the modulating signal and $y(t)$ for the modulated carrier signal. The modulation principle is illustrated in Fig. 4.1. The process by which the modulating signal modifies the carrier signal to obtain the modulated signal is referred to as the *modulation method* or *modulation scheme* or *modulation technique*. There are several types of modulation schemes, each with certain desirable features for specific applications. We shall study in the following sections both conventional and futuristic modulation schemes.

The spectral space occupied by the message signal prior to modulation is called *baseband frequency.* Hence, the original message signal is referred to as

[1]Since they rejected knowledge and instruction they perished.

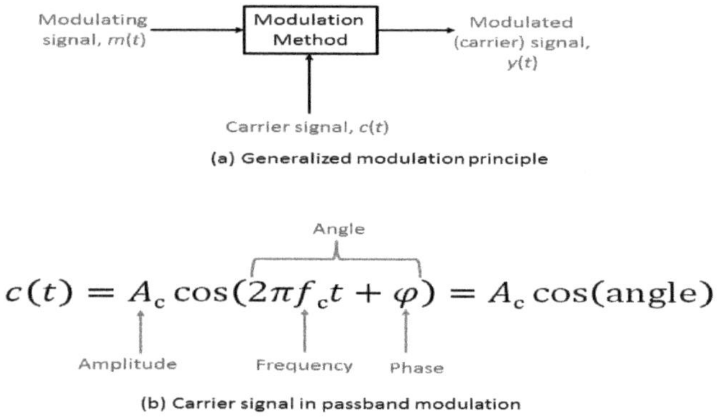

Fig. 4.1 The generalised principle of modulation.

baseband signal. The reverse process to modulation is called *demodulation* in which the original signal is recovered from the wave produced by modulation. Modulation can be *carrierless modulation* or *carrier modulation*.

The design of modulation schemes strives to optimise the seven competing objectives:

- **High power efficiency**: aka energy efficiency. We seek modulation schemes which consume as low power as possible. The simpler the modulation scheme, the lower its power consumption.
- **High bandwidth efficiency**: aka spectral efficiency. This is the number of bits that we can transfer using 1 Hz of bandwidth at a given bit error rate. We seek as high as possible spectral efficiency. Bandwidth efficiency is increased by increasing the number of bits used to encode each symbol or signalling element. By increasing the unique number of symbols, we increase the required transmit power, reducing power efficiency.
- **Least cost**: the cost of the transceiver circuitry increases with the complexity of the modulation scheme used. Therefore, simple modulation schemes are preferable to reduce costs.
- **Low bit-error rate**: at a given bit energy, we seek the modulation scheme with the lowest bit error rate.
- **Low sensitivity to multipath fading**: most communications channels, especially wireless channels, suffer from fading. For this modulations schemes which are robust against multipath fading are desirable.
- **Constant envelope**: modulation schemes with constant amplitude, such as PSK is desirable. Such modulation schemes allow the use of nonlinear amplifiers which have higher gains than linear amplifiers. The high gain is needed so that electronic transmitters can transmit signals with minimum power and still achieve the required effective radiated power. Using nonlinear amplifiers in non-constant envelope modulators creates spectral sidelobes.
- **Low out-of-band radiation**: requires modulation schemes with low sidelobes.

Improving upon one of these above objectives degrades the other. This is the reason why an optimised trade-off between them is always necessary. For example, mobile cellular systems use hand-held devices with limited battery-supplied energy. Therefore, power efficiency is important in such systems. As the radio spectrum is an inelastic and tightly-regulated commodity, bandwidth efficiency is also important in mobile networks. Costs of the hand-held devices and mobile services must also be low enough to attract uptake or subscriptions. Communications systems which do not have to be mobile (e.g., line-of-sight microwave systems) can sacrifice power to the advantage of bandwidth or costs. Modulation methods are designed with the above three efficiencies in mind. Simple methods enabling simple transceiver implementations are good in cost savings and power consumption. However, simple transceivers are in general inefficient in bandwidth consumption.

4.1.1 *Signal Constellation Diagram*

BER is one of the interesting performance metrics of modulation schemes. A technique used to compute the BER is to represent the $M = 2^m$ waveforms of a modulator as vectors in the two-dimensional XY plane. Such a representation is referred to as **signal constellation diagram** or signal space. Signal constellation diagram is a way of conceptualising the modulation process geometrically. Vectors in geometry are expressed as functions of the three *orthonormal* **basis functions** or vectors i, j, k. Similarly, basis functions are needed in signal constellation diagram. We usually denote such basis functions by $\phi_1(t)$, $\phi_2(t)$, etc. The basis functions must be mutually orthonormal, meaning each has unit energy and are mutually orthogonal, i.e.,

$$r_{ij} = \int^{T_b} \phi_i(t)\phi_j(t)dt = \begin{cases} 1, & \texttt{if } i = j \\ 0, & \texttt{if } i \neq j \end{cases}.$$

Waveforms of all symbols of a digital modulation scheme are expressed as the linear combinations of the set of basis functions of the modulator. A popular method used to find the basis functions of a given signal set is Gram Schmidt orthogonalisation. The dimensionality of a modulation scheme equals the number of basis functions that it uses.

Figure 4.2 shows two signal vectors and the three important components of one of them. Each signal vector has in-phase component s_I, the quadrature component s_Q, and the phase angle α that the signal vector makes with the positive horizontal axis.

Figure 4.3 shows QPSK signal constellation diagram with pulse shaping (Fig. 4.3(b)) and without pulse shaping (Fig. 4.3 (a)). We can observe that the constellation diagram with pulse shaping has smoother transitions because the digital data was passed through either a RC or RRC filter prior to modulation. This means that the bandwidth consumption has been reduced, albeit at the cost of increased transmit power.

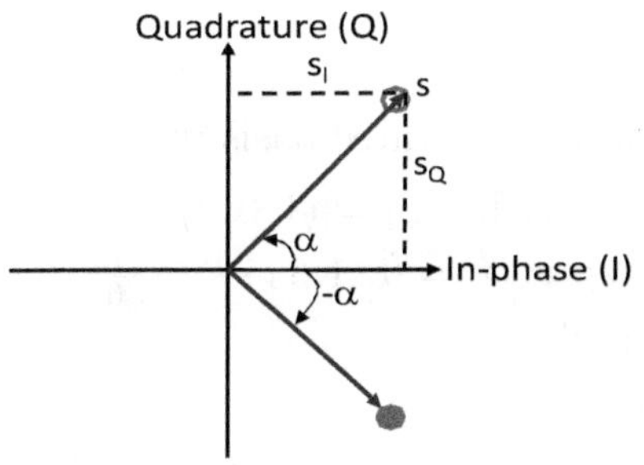

Fig. 4.2 Elements of the signal constellation diagram.

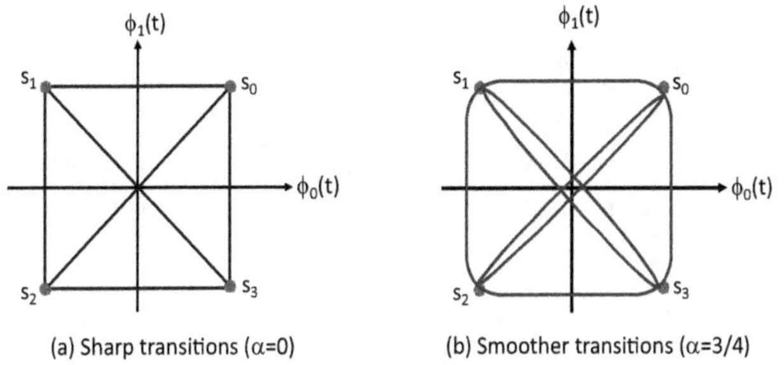

Fig. 4.3 QPSK signal constellation diagram: (a) with and (b) without pulse shaping.

4.1.1.1 *Gram-Schmidt Orthogonalisation Algorithm*

As discussed in the following, we need the set of basis functions for a modulation scheme in order to draw its signal space. The basis functions must be orthogonal. A popular algorithm used to compute orthogonal basis functions for a given set of signals is the Gram-Schmidt orthogonalisation algorithm. If $\mathcal{S} = \{\phi_1, \phi_2, \ldots, \phi_n\}$ is an orthogonal basis for an inner space Φ, then every vector $u \in \Phi$ can be expressed as the linear combination of the basis vectors in the set $\mathcal{S}$, i.e., $u = \sum_{i=1}^{n} \beta_i \phi_i$, where $\{\beta_i\}$ are scalars. Note that the basis functions are orthogonal, and thus linearly independent.

Assume that $\mathcal{A} = \{\mathbf{a}_1, \mathbf{a}_2, \mathbf{a}_3, \ldots, \mathbf{a}_k\}$ is a given set of vectors. We can find the set of basis functions $\{\phi_1, \phi_2, \ldots, \phi_n\}$ for $\mathcal{A}$, such that $n \leq k$. Assume that the space spanned by the vector ϕ_i is $\mathcal{S}_i$. The

Gram-Schmidt orthogonalisation algorithm has the following steps:

- Step 1: Let $\phi_1 = a_1$.
- Step 2: Let $\phi_2 = \mathbf{a}_2 - \mathtt{proj}_{\mathcal{S}_1}\mathbf{a}_2 = \mathbf{a}_2 - \frac{\langle \mathbf{a}_2, \phi_1 \rangle}{||\phi_1||^2}\phi_1$. Note that $proj_{\mathcal{S}_1}\mathbf{a}_2$ is the orthogonal projection of $\mathbf{a}_2$ onto the space $\mathcal{S}_1$ spanned by the vector ϕ_1.
- Step 3: Let $\phi_3 = \mathbf{a}_3 - \frac{\langle \mathbf{a}_3, \phi_1 \rangle}{||\phi_1||^2}\phi_1 - \frac{\langle \mathbf{a}_3, \phi_2 \rangle}{||\phi_2||^2}\phi_2$.
- Step 4: Let $\phi_4 = \mathbf{a}_4 - \frac{\langle \mathbf{a}_4, \phi_1 \rangle}{||\phi_1||^2}\phi_1 - \frac{\langle \mathbf{a}_4, \phi_2 \rangle}{||\phi_2||^2}\phi_2 - \frac{\langle \mathbf{a}_4, \phi_3 \rangle}{||\phi_3||^2}\phi_3$.
- Continue to project the remaining vectors onto the space already found.

Example

Consider the four vectors $\{\mathbf{a}_1 = (1, 0, 2), \mathbf{a}_2 = (0, 1, 1), \mathbf{a}_3 = (-4, 2, 2), \mathbf{a}_4 = (2, 0, 4)\}$.

(1) Find the basis functions for given vectors.
(2) Normalise the basis functions so that each has a unit energy.
(3) Express the given vectors as functions of the basis functions found.

Solution

(1) Computing the three basis functions:

- Step 1: Let $\phi_1 = (1,0,2)$.
- Step 2: $\phi_2 = (0,1,1) - \frac{\langle(0,1,1),(1,0,2)\rangle}{||(1,0,2)||^2}(1,0,2) = \frac{1}{5}(-2,5,1)$.
- Step 3:

$$\phi_3 = (-4,2,2) - \frac{\langle(-4,2,2),(1,0,2)\rangle}{||(1,0,2)||^2}(1,0,2) - \frac{\left\langle(-4,2,2),\frac{1}{5}(-2,5,1)\right\rangle}{||\frac{1}{5}(-2,5,1)||^2}\frac{1}{5}(-2,5,1) = \frac{4}{3}(-2,-1,1).$$

The given vectors are three-dimensional. Therefore, we need only three basis functions.

(2) The normalised basis functions are: $\phi_1 = \frac{1}{\sqrt{5}}(1,0,2)$, $\phi_2 = \frac{1}{\sqrt{30}}(-2,5,1)$ and $\phi_3 = \frac{1}{\sqrt{6}}(-2,-1,1)$. These are orthonormal vectors as they are orthogonal and each has a unit energy.

(3) It is clear that $\mathbf{a}_1 = \sqrt{5}\phi_1$, $\mathbf{a}_2 = \frac{2}{\sqrt{5}}\phi_1 + \frac{6}{\sqrt{30}}\phi_2$, $\mathbf{a}_3 = \frac{20}{\sqrt{30}}\phi_2 + \frac{8}{\sqrt{6}}\phi_3$, $\mathbf{a}_4 = 2\sqrt{5}\phi_1$.

Example

Express the five signals shown in Fig. 4.4 as functions of the two orthonormal functions $X(t)$ and $Y(t)$. We can express the signals as $S_1(t) = AX(t)$, $S_2(t) = -AX(t)$, $S_3(t) = AY(t)$, $S_4(t) = -AY(t)$ and $S_5(t) = aX(t) - bY(t)$, $a, b < 1$. We will use the knowledge acquired in this section in the analysis of digital modulation schemes.

4.1.2 *Bandwidth, Symbol Rate, Energy, Power and Power Spectrum*

Power spectrum, or power spectral density (PSD), of a signal is a plot of its power density over frequency. It is used to estimate at least the following:

- Bandwidth consumption of the signal.
- How the power in the signal is distributed over frequency.
- Adjacent channel interference (ACI) and co-channel interference (CCI).

For this reason, power spectral density (PSD) of communication signals is a very important topic. An important power-related metric in modulated signals is **power efficiency**. It measures the percentage of the total transmitted power that is used to transfer the actual information-bearing signal. The carrier signal consumes part of the energy, although it is not the intelligence of interest. Power efficiency also relates to the amount of power needed to achieve a given error performance, usually below 10^{-5}, i.e., at most one in every 100,000 bits transmitted can be received in error. If $\sigma_n^2 = \frac{N_0}{2}$ is the two-sided power spectral density in W/Hz of the AWGN on channel, and E_b is the energy used to transmit a bit, then the power efficiency is

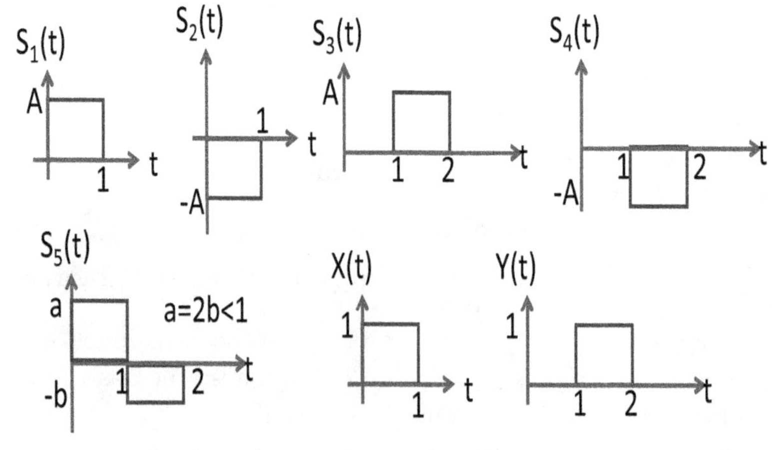

Fig. 4.4 Computing orthonormal basis vectors for a set of signals.

measured by

$$\eta_{\mathrm{p}} = \frac{E_{\mathrm{b}}}{N_0}\,. \tag{4.1}$$

We use waveforms with reduced pea-to-average power ratio such as OQPSK and $\frac{\pi}{4}$-QPSK or constant-envelope modulations techniques (such as GMSK) whenever power efficiency is a priority. This is usually the situation with handsets used in mobile wireless communications.

A signal can be classified into one of the following: energy signal, power signal or none of the two. There are two ways by which we can determine the power spectrum of a signal. These are:

- Compute the Fourier transform of the signal and then derive the power spectral density from it using Eq. (4.8). Unfortunately, in general, we cannot compute the Fourier transform of random signals and power signals. Thus, this method is benign to only deterministic non-power signals.
- Compute the autocorrelation function (ACF) of the signal and then compute the power spectral density as the Fourier transform of the ACF. This is alright for statistical signals and power signals. A random signal cannot be correctly described as a function of time. It is thus usually described by its average parameters, such as its ACF.

The energy in a continuous-time signal $y(t)$ is defined as

$$E_y = \lim_{\tau\to\infty}\int_{-\tau}^{\tau} |y(t)|^2 dt\,. \tag{4.2}$$

The energy definition for the corresponding digital signal $y[k]$ is

$$E_y = \lim_{K\to\infty}\sum_{k=-K}^{K} |y[k]|^2 dt\,. \tag{4.3}$$

Similarly, the power in the continuous-time signal $y(t)$ is defined as

$$P_y = \lim_{\tau\to\infty}\frac{1}{2\tau}\int_{-\tau}^{\tau} |y(t)|^2 dt\,. \tag{4.4}$$

The power definition for the corresponding digital signal $y[k]$ is

$$P_y = \lim_{K\to\infty}\frac{1}{2K+1}\sum_{k=-K}^{K} |y[k]|^2 dt\,. \tag{4.5}$$

The signal $y(t)$ is an energy signal if E_y is finite, i.e., $E_y < \infty$. It is a power signal if P_y is finite, i.e., $0 < P_y < \infty$. The signal is neither an energy nor power signal if both E_y and P_y infinite. A given signal cannot be both an energy and power signal. As power is the time average of energy, an energy signal has zero power, while a power signal has an infinite energy. Let us define a snapshot of $y(t)$ as

$$y_\tau(t) = \begin{cases} y(t), & -\tau \le t \le \tau \\ 0, & \texttt{otherwise} \end{cases}.$$

If $Y_\tau(f)$ is the Fourier transform of $y_\tau(t)$ then the Perseval's theorem states that

$$\int_{-\infty}^{\infty} |y_\tau(t)|^2 dt = \int_{-\infty}^{\infty} |Y_\tau(f)|^2 df\,. \tag{4.6}$$

This means that we can rewrite Eq. (4.4) as

$$\begin{aligned} P_y &= \lim_{\tau\to\infty}\frac{1}{2\tau}\int_{-\tau}^{\tau} |y_\tau(t)|^2 dt \\ &= \lim_{\tau\to\infty}\frac{1}{2\tau}\int_{-\tau}^{\tau} |Y_\tau(f)|^2 df \\ &= \int_{-\infty}^{\infty}\lim_{\tau\to\infty}\frac{|Y_\tau(f)|^2}{2\tau} df \\ &= \int_{-\infty}^{\infty} S_y(f) df \end{aligned} \tag{4.7}$$

where we have defined the **power spectral density** (PSD) or power spectrum of $y(t)$ as

$$S_y(f) = \lim_{\tau\to\infty}\frac{|Y_\tau(f)|^2}{2\tau}\,. \tag{4.8}$$

This results in the energy spectral density (ESD) of $y(t)$ being

$$E_y(f) = |Y_\tau(f)|^2\,. \tag{4.9}$$

Assume that $y(t)$ is a value or realisation of a wide-sense stationary (WSS) process $\mathcal{Y}$ at time t. Then $y(t)$ is a random variable. Let μ_y be the average value and σ_y^2 be the variance of the process $\mathcal{Y}$. For a WSS process the mean and the variance do not change with time, and the autocorrelation function of two realisations depend on only the time difference (say, τ) between them. Therefore, the autocorrelation function measuring the similarity between the two realisations of $\mathcal{Y}$ at two time instants separated by τ is defined as

$$r_{yy}(\tau) = E[y(t)y(t+\tau)]\,. \tag{4.10}$$

Another definition for the ACF of a WSS process is

$$r_{yy}(\tau) = \frac{E[(y(t) - \mu_y)(y(t+\tau) - \mu_y)]}{\sigma_y^2}. \quad (4.11)$$

Let the infinite sequence of M-PAM symbols A_k, $k = 1, 2, \ldots$ be a WSS process. Map these M-PAM symbols onto another infinite sequence of K-dimensional signal-space vectors $\mathbf{v}_l$, $l = 1, 2, \ldots, K$ each of duration T, which is the signalling or symbol period. Choose the K-dimensional basis functions $\theta_k(t)$, $k = 1, 2, \ldots, K$, each of duration T. Then, the carrier-modulated M-PAM signal has the form

$$y(t) = \sum_{l} \sum_{k=1}^{K} \mathbf{v}_{l,k} \theta_k(t - lT - \alpha) \quad (4.12)$$

where α is a random variable uniformly distributed in the interval $[0, T)$ which is chosen to keep $y(t)$ WSS.

4.1.2.1 *Bandwidth, Bandwidth Efficiency, Symbol Rate*

Let us review the following important definitions on digital modulation schemes prior to proceeding:

- B: measured in Hertz (Hz), is the bandwidth of the signal or transmission medium.
- $R_{\mathtt{b}}$: measured in bits per second (bps or b/s), is the data rate.
- $T_{\mathtt{b}} = \frac{1}{R_{\mathtt{b}}}$: measured in seconds (s), is the bit duration.
- M: the number of *different* symbols (aka signalling levels, baseband signals, signalling elements, conditions or unique combinations) of a digital modulation scheme.
- m: the number of bits used to encode each symbol.
- $R_{\mathtt{s}} = \frac{R_{\mathtt{b}}}{\log_2 M}$: measured in symbols per second (or baud), is referred to as symbol rate, modulation rate, signalling rate or baud rate.
- $T_{\mathtt{s}} = \frac{1}{R_{\mathtt{s}}} = T_{\mathtt{b}} \log_2 M$: measured in seconds (s), is the symbol duration.

We have the following further relationships:

$$m = \log_2(M) \text{ or } M = 2^m. \quad (4.13)$$

These three equations on maximum data rates are important in digital communications, including modulation:

`Nyquist rate:`

$$C_{\mathtt{N}} \leq 2B \log_2 M \Leftrightarrow R_{\mathtt{s}} = \frac{C_N}{\log_2 M} \leq 2B \quad (4.14)$$

`Nyquist sampling theorem:`

$$F_{\mathtt{s}} \geq 2 f_{\mathtt{max}} = 2B \quad (4.15)$$

`Shannon-Hartley theorem:`

$$C_{\mathtt{S}} \leq B \log_2(1 + SNR) \quad (4.16)$$

where $f_{\mathtt{max}}$ is the maximum frequency component in the analog signal being sampled, and $F_{\mathtt{s}}$ is the sampling frequency. We will make use of the Nyquist rate to compute the minimum bandwidth required to support a given symbol rate, or the maximum supportable symbol rate for a given bandwidth. Digital modulation schemes are categorised as follows:

- $M = 2$ or $m = 1$: Binary modulation
- $M > 2$ or $m > 1$: Multi-level modulation or M-ary modulation

Bandwidth efficiency, also referred to as link **spectral efficiency** or **spectral bit rate**, $\eta_{\mathtt{b}}$, is one of the important metrics of modulation schemes. It is defined as

$$\eta_{\mathtt{b}} = \frac{\texttt{maximum achievable net bit rate}}{\texttt{bandwidth used}} = \frac{R_{\mathtt{b,net}}}{W} = \frac{k}{n} \cdot \frac{R_{\mathtt{s}} \log_2 M}{W} \quad \texttt{[bps/Hz]} \quad (4.17)$$

where $r = \frac{k}{n}$ is the rate of the error-control encoder which encodes every k bits into n bits. We can observe from Eq. (4.17) that the higher the value of M, the higher the bandwidth efficiency. This is the motivation behind multi-level modulation schemes. The price paid for higher bandwidth efficiency is the complexity in the detection circuits. Methods besides multi-level modulation used to increase the bandwidth efficiency include:

- Frequency reuse, small cells and small cluster size in cellular mobile networks.
- Spectrum sharing using, e.g., multiplexing and multiple access schemes.

Another term metric of importance is the **modulation efficiency**, which is defined as

$$\eta_m = \frac{\texttt{gross data rate}}{\texttt{bandwidth used}} = \frac{\texttt{data rate after error control coding}}{\texttt{bandwidth used}} = \frac{R_s \log_2 M}{W} \ \texttt{[bps/Hz]}\,. \quad (4.18)$$

The relationship between the modulation efficiency η_m and the bandwidth efficiency is

$$\eta_b \leq \frac{k}{n}\eta_m\,. \quad (4.19)$$

Usually, the gross bit rate is just quoted as the bit rate, and denoted R_b, without mentioning the word **gross**. It is the data rate at the physical layer after processing by all the layers above ISO/OSI Layer 1. As modulation efficiency accounts for the redundant bits used to error protect the data, it is expected to be larger than the bandwidth efficiency. In modulation schemes which transmit the carrier frequency, such as amplitude modulation (AM) schemes, the modulation efficiency is defined along transmit power as

$$\eta_m = \frac{\texttt{power in sidebands}}{\texttt{total power in AM signal}} \quad (4.20)$$

as only the sidebands contain the information-bearing signal.

4.1.3 *Check Your Understanding*

Students are encouraged to check their understanding of the content of this section with the following questions prior to proceeding to the next section.

(1) Describe in details the differences between baseband and broadband digital communications networks. Include in your description the differences between the types of signals transferred over each of the two types of networks. Also, cite examples of each case.

(2) Consider a signal with the baseband bandwidth $B = 200$ kHz transferred using a 16-QAM digital modulation scheme and rate $\frac{2}{3}$ forward-error control (FEC). The physical layer data rate is 2 Mbps. Compare the maximum modulation efficiency with the bandwidth efficiency if this signal is transmitted using:

   (a) Baseband transmission.
   (b) Single-sideband (SSB) passband transmission.
   (c) Double-sideband (DSB) passband transmission.

   Hint. Use the Nyquist rate, which is $R_s \leq 2B$ for baseband and SSB passband transmissions, and $R_s \leq B$ for DSB passband transmissions.

(3) A digital modulation scheme uses 16 unique symbols. How many bits does it use to encode each symbol?

(4) A digital modulation method uses symbols with duration 1 mus.

   (a) How many symbols must the modulator use to deliver a maximum data rate of 32 Mbps?
   (b) If the half-power bandwidth of the modulation scheme is twice the symbol rate, compute the bandwidth efficiency of the modulation scheme.

(5) List and discuss the three efficiencies that the design of communications systems seek to optimise?

(6) Assume that someone who has no idea about electronic communications wants you to explain to her the meaning of the word *modulation.* Write two statements to that effect.

(7) **In-class research:** In groups of two to three students, brainstorm about the *importance* of modulation methods in communications systems. Include in your discussions the types of modulation methods and their relative differences.

(8) In a tabular form, differentiate between broadband and baseband communications systems, as well as the difference between broadband and baseband signals.

(9) State the minimum data rates for both uplink (or reverse link) and downlink (or forward link) that is required of a communications network to be classified as a broadband network. Refer in particular to the definitions of US Federal Communications Commission (FCC).

(10) Why do we also refer to broadband modulation methods as passband modulation?

(11) Describe how carrier signals used in baseband modulation methods differ from those used in broadband modulation methods.

(12) Give at least two examples each of baseband and broadband transmission media. Research to find some of the practical applications of each type.

(13) How do we call the signal which changes another signal in modulation?

(14) How do we call the signal which is changed by another signal in modulation?

(15) How to we call the signal at the output of a modulation method?

(16) What are the possible characteristics of the carrier signal which can be varied to transfer an information between two points?

(17) State at least two reasons why we cannot stop teaching technologies which are getting obsolete?

(18) **In-class research:** In groups of two to three students, research to find at least one example of a practical application of each of the following modulation methods:

(a) Line coding.
(b) Digital modulation.
(c) Waveform coding or pulse modulation.
(d) Analog modulation.

(19) Discuss the reasons why power efficiency and bandwidth efficiency are desired of modulation methods.

(20) State and explain the reason(s) why amplitude modulation is linear but frequency modulation is nonlinear.

(21) Discuss the importance of oscillators in modulation systems.

(22) From first principles, derive Eq. (2.56).

(23) Compute and plot the power spectrum of the signal $s(t) = 2\cos(100\pi t)$.

4.2 Amplitude-Shift Keying

In digital modulation schemes we use digital data to change one or two of the three parameters of an analog carrier signal to produce an analog waveform. Contrary to analog modulation, the names for digital modulation schemes use the word *keying* instead of *modulation*. Frequency shift keying and phase shift keying, like their analog counterparts, have a better noise immunity than amplitude shift keying (ASK). However, ASK is simpler to demodulate. ASK is the most simplest form of digital modulation. In ASK the message signal is contained in the carrier's amplitude, as done in analog amplitude modulation. ASK is a digital modulation format in which we use data bits to vary or modulate the amplitude of an analog or continuous carrier signal. As noise can change the amplitude of the modulated signal easily, ASK is highly susceptible to noise and interference and nonlinearities in the end-to-end transmission system. Color-shift keying (CSK) is one of the modulation schemes adopted in visible light communication (VLC) systems based on the IEEE 802.15.7 standard, referred to as light fidelity (Li-Fi). CSK is a form of ASK.

Figure 4.5 shows a block diagram of an ASK modulator, comprising an oscillator to generate the carrier signal and a multiplier. The entire circuit is called product modulator. We obtain from this figure the general form of the ASK signal as

$$s_{\mathtt{ASK},k}(t) = m_k \sin(2\pi f_c t)\,,$$
$$kT_{\mathtt{s}} \le t \le (k+1)T_{\mathtt{s}},\ k = 0, 1, 2, \ldots, M-1 \tag{4.21}$$

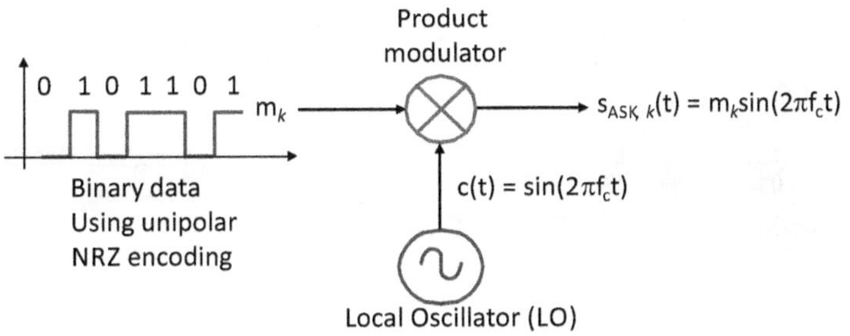

Fig. 4.5 Block diagram of an ASK waveform generator.

where is $m_k = m(t = kT_b) \in \{0, 1\}$ are the digital data, representing the sampled amplitudes of the analog message signal $m(t)$, to be communicated. Note that we could use a cosine instead of a sine wave in Eq. (4.21). By defining the orthonormal basis function $\phi(t) = \sqrt{\frac{2}{T_b}}\sin(2\pi f_c t)\prod(\frac{t-T_b/2}{T_b})$, we can express Eq. (4.21) as

$$s_{ASK,k}(t) = m_k\sqrt{\frac{T_b}{2}}\phi(t)\,,$$
$$kT_s \le t \le (k+1)T_s,\ k = 0, 1, 2, \ldots, M-1\,. \tag{4.22}$$

Hence, ASK is a one-dimensional modulation scheme as all baseband signals are linearly dependent, and each can be expressed as an amplitude-scaled version of the unit-energy basis function $\phi(t)$. The energy in the signal $s_k(t)$ is

$$E_s = \int_{kT_b}^{(k+1)T_b} |s_k(t)|^2 dt = |m_k|^2 \frac{T_b}{2}\,.$$

ASK has the following pros and cons:

- **Advantages**: simpler generation and detection than PSK and FSK. The byproduct of ASK's simplicity is its lower cost than PSK and FSK.
- **Disadvantages**: information-bearing data is in the amplitude of the carrier, which is the aspect of the three parameters of the carrier most affected by noise. Thus, ASK has the poorest noise immunity of all modulation techniques. Second, ASK is affected severely by any nonlinearity in the communication system. Any nonlinearity in the transceiver produces harmonic or spurious frequencies. For example, amplifiers in the transceiver cannot operate in their most efficient mode (which is nonlinear), requiring high transmit powers to meet a given effective radiated power. This is a problem for hand-held devices which must transmit low powers to conserve their battery. BER and signal-to-noise ratio of ASK are also poor. ASK is also bandwidth inefficient.

ASK is used in only low-data-rate applications, usually in binary form. For this reason, we analyse in the following only binary ASK, specifically OOK.

4.2.1 *On-Off Keying*

The on-off keying (OOK) is the simplest form of the ASK. Manchester-encoded OOK is a popular scheme in Li-Fi communications. In binary OOK, we sent nothing for bit 0 (also called *space*) and send a sinusoidal carrier wave for bit 1 (also called *mark*) for the duration of a bit period, as and shown in Fig. 4.6(b). Some features of OOK are listed in Table 4.1. An example of digital data and its corresponding on/off keying waveforms are shown in Fig. 4.6. The data in Fig. 4.6(a) is encoded using unipolar NRZ line coding. The figure also shows the waveforms of BFSK (binary frequency shift keying) and BPSK (binary phase shift keying) as a way of comparing them with the BASK.

Assume that we transmit $s_k(t)$ over an AWGN channel, then we shall receive

$$r_k(t) = s_k(t) + w(t), \quad k = 0, 1 \text{ and } 0 \le t \le T \tag{4.23}$$

where $w(t) = \mathbb{R}e\{n(t) = w(t) + jv(t)\}$ is the real part of the AWGN with mean zero and variance $\sigma_w^2 = \frac{N_0}{2}$, i.e., $w(t) \sim \mathcal{N}(0, \frac{N_0}{2})$. Thus, the

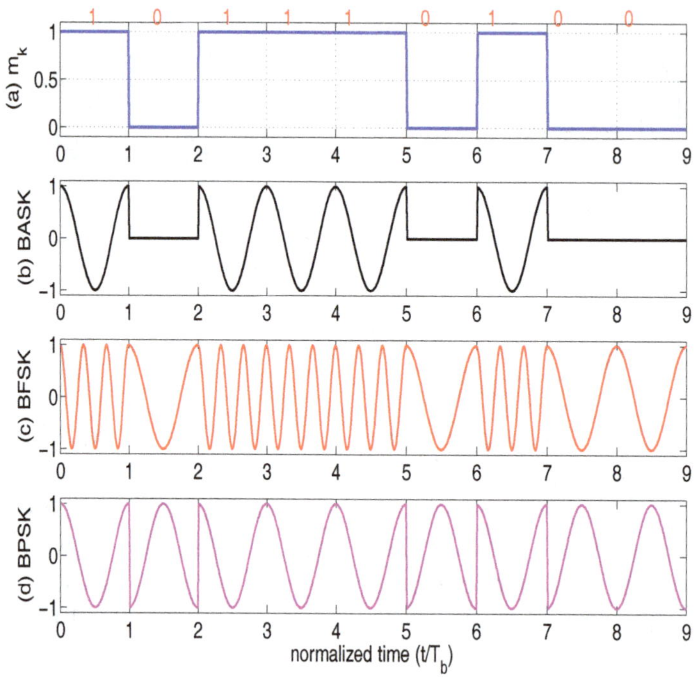

Fig. 4.6 Comparing the waveforms of binary digital modulators for a given unipolar NRZ-encoded binary data.

Table 4.1 On/off keying features.

Bit	Amplitude a_k	$s_{ook}(t),\ kT_b \le t \le (k+1)T_b$
0	1 V	$\sin(2\pi f_c t)$
1	0 V	0

probability density function (pdf) of the AWGN is

$$f_w(x) = \frac{1}{\sqrt{\pi N_0}} e^{-\frac{x^2}{N_0}}. \qquad (4.24)$$

Assume that the sampled version of $r_k(t)$ is

$$r_k = s_k + w, \quad k = 0, 1. \qquad (4.25)$$

On-Off Keying: Detection and BER on AWGN Channel

An integrated circuit which can be used to achieve ASK (both modulation and detection) is MC1496 Balanced Modulator and Demodulator, which is manufactured by ON Semiconductor. One of the performance measures of digital modulation schemes is the bit error rate (BER). BER is the ratio of the number of bits received in error to the total number of bits transmitted. We can receive bit 1 if we transmit bit 0 and vice versa because of the non-ideal nature of the transmission media. There are two main methods of detecting the transmitted bits, namely:

- **Coherent detection**: also referred to as synchronous detection, requires the receiver to know both the phase and the frequency of the carrier signal used at the modulator. It is important in angle-modulation schemes, such as PSK, FSK, PM and FM.
- **Non-coherent detection**: is simpler than coherent detection, except that it performs poorer at a given signal-to-noise ratio. A popular non-coherent detection scheme is the envelope detector used in amplitude modulation schemes, such as ASK and AM. It is also implemented in, for example, differential PSK (DPSK).

Figure 4.7 shows the block diagram of three circuits, each of which can be used to detect ASK waveform. Coherent detectors are used if the phase of the received modulated wave is random (i.e., unknown). If the phase $\phi(t)$ of the received signal is randomly but uniformly distributed in $[0, 2\pi]$ then Fig. 4.7(c) is an optimum detecting circuit.

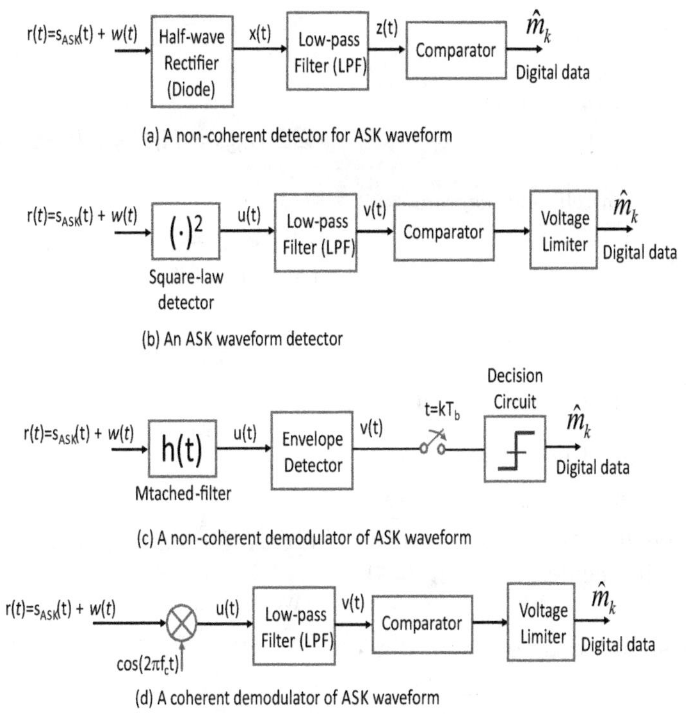

Fig. 4.7 Detectors for ASK waveform.

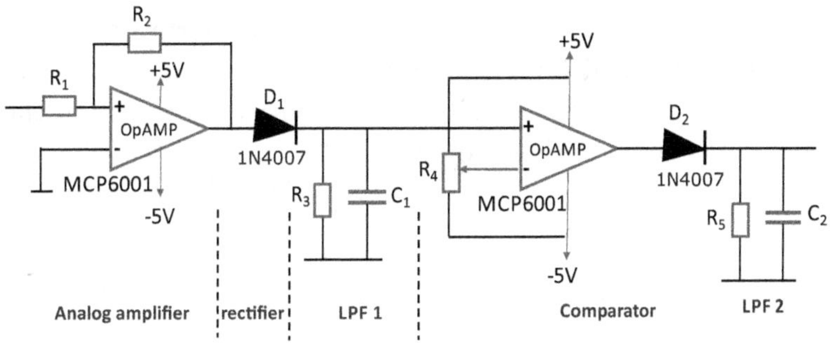

Fig. 4.8 A circuit for the ASK waveform detector in Fig. 4.7(a) $[R_2 = 22R_1]$.

In Fig. 4.7(a), the ASK waveforms pass through a half-wave rectifier which can be realised using a single diode. The signal $r(t)$ at the input of the rectifier is an alternating voltage, which is converted to the pulsed positive direct voltage signal $x(t)$. The signal $x(t)$ passes through a low-pass filter, made of resistors and capacitors, to be smoothed. The results are fed into the comparator circuit, which comprises an operational amplifier (OpAMP), capacitors and resistors. The comparator is a decision device, which compares its input signal with a reference signal and takes a decision. The decision of the comparator is based on the ASK signal constellation diagram in Fig. 4.9. The comparator decides that $m_k = 1$ bit if its input fulfills $z(t) \geq 0.5\sqrt{E_s}$, else it decides that $m_k = 0$ bit. A circuit which can implement Fig. 4.7(a) is shown in Fig. 4.8. MCP6001 is a general-purpose OpAMP, while 1N4007 is a general-purpose diode which can be used as the rectifier. Note that R_4 is a variable resistor.

Figure 4.7(b) can also demodulate an ASK signal. For now, let us forget the channel noise $w(t)$ in Eq. (4.23). Therefore, the output of the square-law

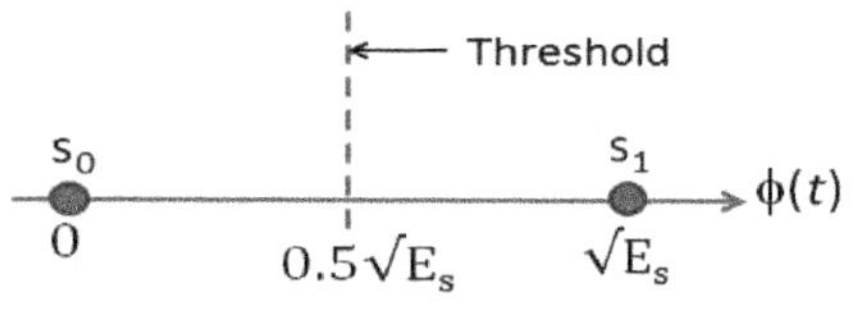

Fig. 4.9 On-off keying signal constellation diagram.

detector is

$$\begin{aligned} u(t) &= s^2_{\mathtt{ASK}}(t) = m_k^2 \sin^2(2\pi f_c t) \\ &= \frac{m_k^2}{2} - \frac{m_k^2}{2}\cos(4\pi f_c t)\,. \end{aligned} \tag{4.26}$$

The LPF then blocks the high-frequency component in $u(t)$ to produce at its output $v(t) = \frac{m_k^2}{2}$. The comparator decides that $m_k = 1$ bit if its input fulfills $v(t) \geq 0.5\sqrt{E_s}$, else it decides that $m_k = 0$ bit.

Figure 4.9 shows the signal constellation diagram of OOK. We use it to derive the formula for the bit error rate of OOK. Note that for OOK over AWGN channel we obtained the signal sample

$$r_k = \begin{cases} s_1 + w = \sqrt{E_s} + w, & \texttt{if bit 1 is sent} \\ s_0 + w = w, & \texttt{if bit 0 is sent} \end{cases}.$$

Let P_{e1} be the probability that we send bit 1 but receive bit 0. Also, let P_{e0} be the probability that we send bit 0 but receive bit 1. These are the error probabilities. They are defined as

$$\begin{aligned} P_{e1} &= Pr\left[r_k < \frac{\sqrt{E_s}}{2}\middle| s_1 \texttt{ was sent}\right] \\ &= Pr\left[\sqrt{E_s} + w < \frac{\sqrt{E_s}}{2}\right] \\ &= Pr\left[w < -\frac{\sqrt{E_s}}{2}\right] \\ &= \int_{-\infty}^{-\frac{\sqrt{E_s}}{2}} f_w(x)dx = Q(\sqrt{E/N_0}). \end{aligned} \tag{4.27}$$

Similarly

$$\begin{aligned} P_{e0} &= Pr\left[r_k \geq \frac{\sqrt{E_s}}{2}\middle| s_0\right] \\ &= Pr\left[w \geq \frac{\sqrt{E_s}}{2}\right] \\ &= \int_{\frac{\sqrt{E_s}}{2}}^{\infty} f_w(x)dx = Q(\sqrt{E/N_0})\,. \end{aligned} \tag{4.28}$$

Let P_0 and P_1 be the probabilities of transmitting a bit 0 and a bit 1, respectively. We assume that $P_1 = P_0 = \frac{1}{2}$. Thus, the bit error probability of OOK is

$$BER_{ook} = P_0 P_{e0} + P_1 P_{e1} = Q(\sqrt{\gamma}) = \frac{1}{2}\mathtt{erfc}\left(\frac{\sqrt{\gamma}}{2}\right) \tag{4.29}$$

where $\gamma = E_s/N_0$ is the signal-to-noise ratio and the error functions are defined as

$$\begin{aligned} Q(A) &= \frac{1}{\sqrt{2\pi}}\int_A^{\infty} e^{-\frac{x^2}{2}}dx \texttt{ and} \\ \mathtt{erfc}(A) &= \frac{2}{\sqrt{\pi}}\int_A^{\infty} e^{-x^2}dx\,. \end{aligned} \tag{4.30}$$

OOK Power Spectrum

Two of the interesting features of modulated signals are the distribution of the transmitted average power over frequency, as well as their bandwidth consumption. Both metrics can be estimated using the power spectral density (PSD). The PSD can be computed from either the magnitude-squared of the Fourier transform of the waveform or from the Fourier transform of the autocorrelation function (ACF) of the waveform. We use the first method. The Fourier transform of the OOK waveform is

$$\begin{aligned} S_{\mathtt{ook}}(f) &= \int_{-\infty}^{\infty} m_k \cos(2\pi f_c t)e^{-j2\pi f t}dt \\ &= \frac{1}{2}\int_{-\infty}^{\infty} m_k[e^{j2\pi f_c t} + e^{-j2\pi f_c t}]e^{-j2\pi f t}dt \\ &= \frac{1}{2}M(f - f_c) + \frac{1}{2}M(f + f_c) \end{aligned} \tag{4.31}$$

where $M(f)$ is the Fourier transform of the modulating digital data m_k. The corresponding power spectrum is

$$\begin{aligned} S_{\mathtt{ask}}(f) &= |S_{\mathtt{ook}}(f)|^2 \\ &= \frac{1}{4}|M(f - f_c)|^2 + \frac{1}{4}|M(f + f_c)|^2 \\ &\quad + \frac{1}{2}|M(f - f_c) \cdot M(f + f_c)| \\ &= \frac{1}{4}|M(f - f_c)|^2 + \frac{1}{4}|M(f + f_c)|^2 \\ &= \frac{1}{4}S_m(f - f_c) + \frac{1}{4}S_m(f + f_c) \end{aligned} \tag{4.32}$$

where $S_m(f)$ is the baseband PSD of the unipolar NRZ-encoded bitstream, which is

$$\begin{aligned} S_m(f) &= \frac{A^2}{2}\mathtt{sinc}^2(\pi f T_\mathsf{b})\sum_{k=0}^{\infty}\cos(2\pi f T_\mathsf{b} k) \\ &= \frac{A^2}{2}\delta(f) + \frac{A^2 T_\mathsf{b}}{2}\mathtt{sinc}^2(\pi f T_\mathsf{b})\,. \end{aligned} \tag{4.33}$$

The last line in Eq. (4.32) has exploited the fact that $M(f - f_\mathsf{c})$ and $M(f + f_\mathsf{c})$ are orthogonal, resulting in their product being zero. Plugging Eq. (4.33) into (4.32) yields the PSD of the OOK waveform as

$$\begin{aligned} S_\mathsf{ook}(f) = &\frac{A^2}{8}\mathtt{sinc}^2[\pi(f - f_\mathsf{c})T_\mathsf{b}] \\ &\times \sum_{k=0}^{\infty}\cos[2\pi(f - f_\mathsf{c})T_\mathsf{b} k] \\ &+ \frac{A^2}{8}\mathtt{sinc}^2[\pi(f + f_\mathsf{c})T_\mathsf{b}] \\ &\times \sum_{k=0}^{\infty}\cos[2\pi(f + f_\mathsf{c})T_\mathsf{b} k]\,. \end{aligned} \tag{4.34}$$

We can use another method to arrive at (4.34). The PSD of the cosine wave $c(t) = A_\mathsf{c}\cos(2\pi f_\mathsf{c} t)$ is

$$S_\mathsf{c}(f) = \frac{A_\mathsf{c}^2}{4}\delta(f - f_\mathsf{c}) + \frac{A_\mathsf{c}^2}{4}\delta(f + f_\mathsf{c})\,. \tag{4.35}$$

We make use of the convolution theorem of Fourier transform

$$\begin{aligned} \mathcal{F}\{x(t) * y(t)\} &= \mathcal{F}\{x(t)\}\cdot\mathcal{F}\{y(t)\} = X(f)Y(f) \\ \mathcal{F}\{x(t)\cdot y(t)\} &= \mathcal{F}\{x(t)\} * \mathcal{F}\{y(t)\} = X(f) * Y(f) \end{aligned} \tag{4.36}$$

to obtain the PSD of the OOK signal as

$$\begin{aligned} S_\mathsf{ook}(f) &= S_\mathsf{c}(f) * S_m(f) = \frac{A_\mathsf{c}^2}{4}\delta(f - f_\mathsf{c}) * S_m(f) \\ &\quad + \frac{A_\mathsf{c}^2}{4}\delta(f + f_\mathsf{c}) * S_m(f) \\ &= \frac{A_\mathsf{c}^2}{4}S_m(f - f_\mathsf{c}) + \frac{A_\mathsf{c}^2}{4}S_m(f + f_\mathsf{c}) \\ &\overset{(4.33)}{=} \frac{A^2}{8}\mathtt{sinc}^2[\pi(f - f_\mathsf{c})T_\mathsf{b}] \\ &\quad \times \sum_{k=0}^{\infty}\cos[2\pi(f - f_\mathsf{c})T_\mathsf{b} k] \\ &\quad + \frac{A^2}{8}\mathtt{sinc}^2[\pi(f + f_\mathsf{c})T_\mathsf{b}] \\ &\quad \times \sum_{k=0}^{\infty}\cos[2\pi(f + f_\mathsf{c})T_\mathsf{b} k] \end{aligned} \tag{4.37}$$

where we have set $Ac = 1$. The power spectrum of OOK according to Eq. (4.37) is shown in Fig. 4.10. We can observe the carrier at $f_\mathsf{c}T_\mathsf{b} = 10$. Clearly, from the power spectrum we observe that the absolute bandwidth of OOK is infinite. However, we consider only the portion of the spectrum containing about 90% of the energy or the half-power bandwidth. The null-to-null bandwidth is

$$B_\mathsf{n2n} = (1 + r)R_\mathsf{b},\ 0 \le r \le 1 \tag{4.38}$$

where r is the excess bandwidth of the pulse shaping filtering prior to modulation. Figure 4.11 shows the periodogram of OOK corresponding to the waveform in Fig. 4.6(c) using MATLAB simulations. The periodogram is a nonparametric estimate of the power spectral density (PSD) of a wide-sense stationary random process. We can observe the peak

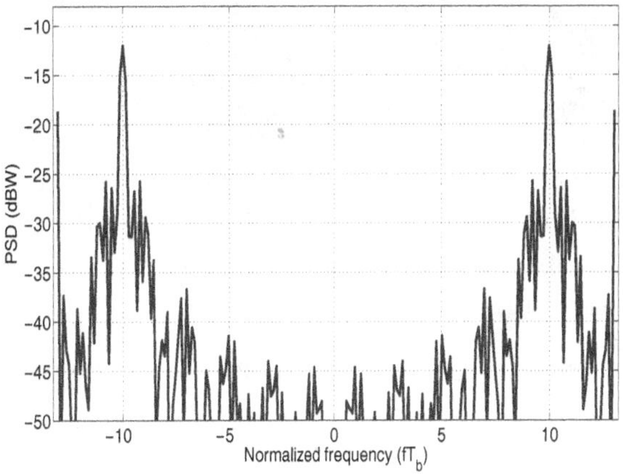

Fig. 4.10 On-off keying power spectrum according to Eq. (4.34) $[f_\mathsf{c}T_\mathsf{b} = 10]$.

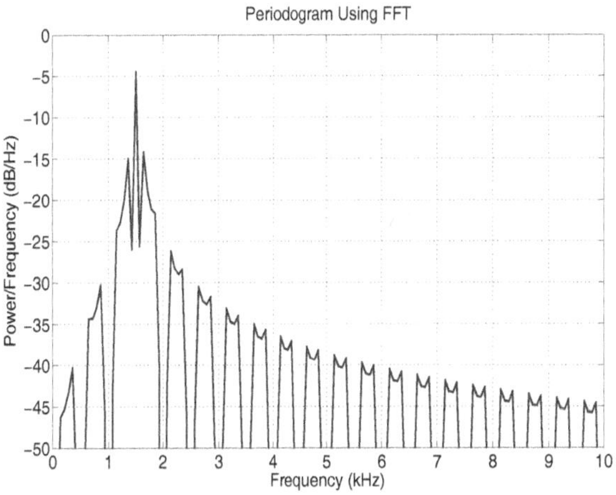

Fig. 4.11 On-off keying power spectrum.

at 1.5 kHz, which corresponds to the frequency of the carrier wave used at the ASK modulator.

8-VSB Modulation

8-VSB is the modulation method adopted by the Advanced Television Systems Committee (ATSC) in USA for the transmission of digital television signals over both satellite and terrestrial networks. Vestigial Sideband (VSB) is a type of amplitude shift keying modulation method. In this scheme every symbol contains 3 bits, resulting in $2^3 = 8$ different symbols. This is the reason for the '8' in the name. 8-VSB uses each set of three bits to vary the amplitude of a sinusoidal carrier to one of eight levels. For this reason, 8-VSB is said to be based on the three-level pulse amplitude modulation (3-PAM).

4.2.2 *Check Your Understanding*

Students are encouraged to check their understanding of the content of this section with the following questions prior to proceeding to the next section.

(1) Which aspect of the carrier signal contains the intelligence in amplitude shift keyed signal?

(2) The binary data 100010101 is used to modulate the amplitude of carrier wave $c(t) = 5\cos(4.2\pi t \times 10^9)$. Assume that the bit rate of the binary modulation system is 10 Mbps.

   (a) Write down the analytical formula for the output of the modulator.
   (b) Draw the signal constellation diagram of ASK.
   (c) How do we call the output of the modulator?
   (d) Sketch the carrier wave, digital data and the output of the modulator.
   (e) Assume that the modulator operates at 27°C, compute the bit error rate of the system.

(3) Consider the circuit in Fig. 4.8. Find the expression for the transfer function of the first low-pass filter (LPF 1) in the circuit. If $R_3 = 6800\,\Omega$ and $C_1 = 10$ nF, compute the 3-dB cut-off frequency of the LPF.

(4) Consider the circuit in Fig. 4.8. Find the expression for the transfer function of the second low-pass filter (LPF 2) in the circuit. If $R_5 = 1000\,\Omega$ and $C_1 = 100$ nF, compute the 3-dB cut-off frequency of the LPF.

(5) Which of the two LPFs (i.e., LPF 1 and LPF 2) in Fig. 4.8 blocks much of the additive noise in an ASK transceiver?

(6) If the upper part of the variable resistor in Fig. 4.8 is 2 kΩ and the entire resistor is 8 kΩ, what is the power level at any point between the two portions of the variable resistor?

(7) Using Eq, (4.37), plot the power spectrum of OOK. Use the figure to estimate the half-power (i.e., 3 dB) bandwidth of OOK as a function of the symbol duration (T_{s}) or data rate (R_{b}). Estimate the

   (a) half-power (i.e., 3-dB) bandwidth
   (b) null-to-null bandwidth

   of OOK waveform from Fig. 4.10 or Fig. 4.11.

(8) Using the estimated bandwidths from previous question, compute the bandwidth efficiency of OOK.

4.3 Frequency Shift Keying

A digital modulation scheme which transfers the intelligence in the form of digital data by the variations in the frequency of a continuous signal is called frequency shift keying (FSK). The general form of an M-ary FSK waveform is

$$s_{\mathtt{fsk}}(t) = A_{\mathsf{c}}\cos(2\pi f_k t)\,, \quad 0 \le t \le T_{\mathsf{s}},\ k = 0, 1, \ldots, M-1 \tag{4.39}$$

or

$$s_{\mathtt{fsk}}(t) = A_{\mathsf{c}}\cos[2\pi(f_{\mathsf{c}} + a_k\Delta f)t]\,, \quad k = 0, 1, 2, \ldots, M-1 \tag{4.40}$$

where Δf is the frequency deviation, and M is the number of unique symbols used in the modulator. The energy in the FSK waveform is

$$E_s = \int_0^T |s_{\texttt{fsk}}(t)|^2 dt \approx \frac{A_c^2 T}{2} \tag{4.41}$$

where T is the symbol duration. Note that we obtain the best bit error rate if the frequencies f_k are orthogonal to each other. M-ary FSK consumes more bandwidth than M-PSK. This is one of the reasons why M-ary FSK for $M > 2$ is unpopular.

4.3.1 *BFSK*

The simplest FSK is the binary FSK (BFSK) in which a single bit is transmitted in each symbol, resulting in $M = 2$ distinct symbols. In this case

`for bit` 0:

`send` $s_0(t) = A_c \cos(2\pi f_0 t),\ f_0 = f_c - \Delta f$

`for bit` 1:

`send` $s_1(t) = A_c \cos(2\pi f_1 t),\ f_1 = f_c + \Delta f\,.$ (4.42)

Therefore

$$\begin{aligned} s_0(t) &= A_c[\cos(2\pi f_c t)\cos(2\pi\Delta f t) \\ &\quad + \sin(2\pi f_c t)\sin(2\pi\Delta f t)] \\ s_1(t) &= A_c[\cos(2\pi f_c t)\cos(2\pi\Delta f t) \\ &\quad - \sin(2\pi f_c t)\sin(2\pi\Delta f t)] \end{aligned} \tag{4.43}$$

The BFSK waveform is compared with that of the ASK in Fig. 4.12.

For further analysis of the FSK, let us define the two orthonormal basis functions

$$\begin{aligned} \phi_1(t) &= \sqrt{\frac{2}{T}}\cos(2\pi f_c t) \;\; \texttt{and} \\ \phi_2(t) &= \sqrt{\frac{2}{T}}\sin(2\pi f_c t) \end{aligned} \tag{4.44}$$

to obtain

$$\begin{aligned} s_0(t) &= \sqrt{E_s}[\cos(2\pi\Delta f t)\phi_1(t) + \sin(2\pi\Delta f t)\phi_2(t)] \\ s_1(t) &= \sqrt{E_s}[\cos(2\pi\Delta f t)\phi_1(t) - \sin(2\pi\Delta f t)\phi_2(t)]\,. \end{aligned} \tag{4.45}$$

Thus, FSK is a two-dimensional modulation scheme. An alternative to Eq. (4.44) is to define

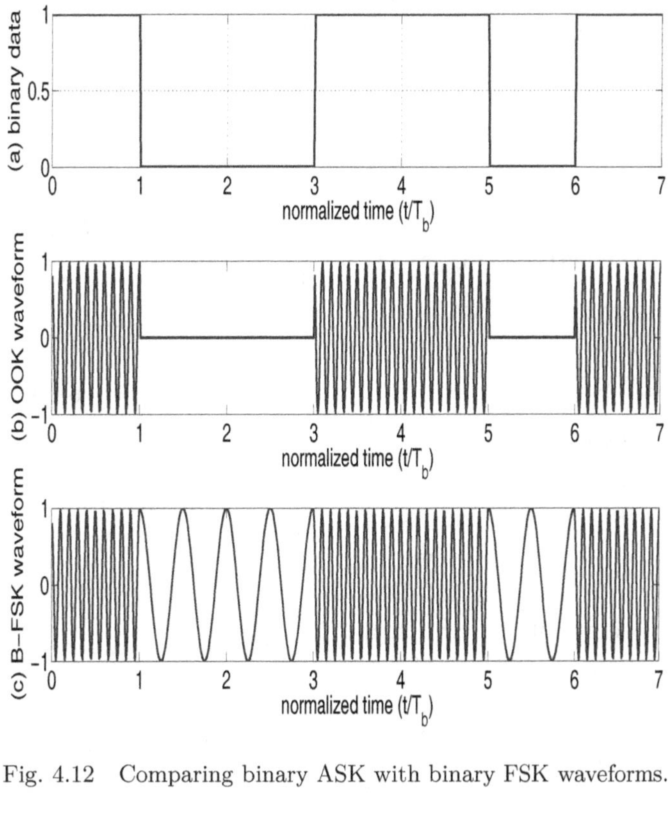

Fig. 4.12 Comparing binary ASK with binary FSK waveforms.

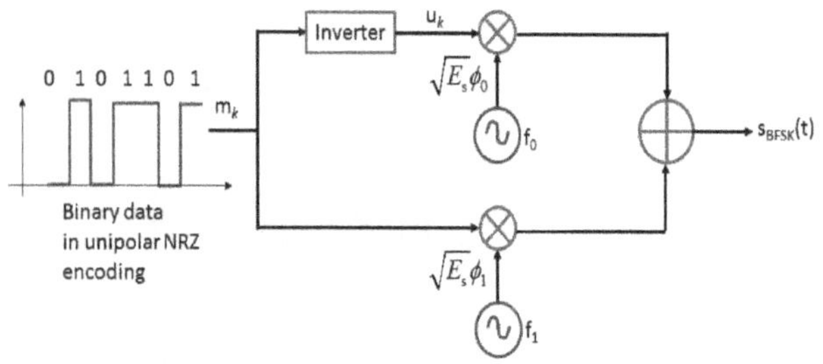

Fig. 4.13 A modulator for BFSK.

the basis functions as

$$\phi_0(t)=\sqrt{\frac{2}{T}}\cos(2\pi f_0 t) \text{ and } \phi_1(t)=\sqrt{\frac{2}{T}}\cos(2\pi f_1 t) \quad (4.46)$$

so that the two BFSK waveforms in Eq. (4.42) become

$$s_0(t) = \sqrt{E_s}\phi_0(t) \text{ and } s_1(t) = \sqrt{E_s}\phi_1(t)\,. \quad (4.47)$$

Figure 4.13 shows a block diagram of a circuit which can be used to achieve Eq. (4.47). As the phase of the modulated signal is discontinuous at the switching times, such a modulator produces a discontinuous phase FSK signal. This BFSK modulator is a switch, which is implemented using voltage-controlled oscillators (e.g., HMC1166 and CVC055), product modulators, an adder and an inverter. Assume that an input bit is '1' then the inverter converts it to bit '0', resulting in

$$s_{BFSK}(t)=0\times\sqrt{E_s}\phi_0(t)+1\times\sqrt{E_s}\phi_1(t)=\sqrt{E_s}\phi_1(t)\,.$$

On the other hand, if the input bit is '0' the inverter converts it to bit '1', resulting in

$$s_{BFSK}(t)=1\times\sqrt{E_s}\phi_0(t)+0\times\sqrt{E_s}\phi_1(t)=\sqrt{E_s}\phi_0(t)\,.$$

This verifies the effectiveness of the simple circuit.

Detection and BER of BFSK

FSK waveform can be detected either coherently (synchronously) or non-coherently (asynchronously), as illustrated in Fig. 4.14. Figure 4.14 shows the block diagram of two circuits which can be used to demodulate BFSK waveform. Figure 4.14(a) is the coherent demodulator. Such a circuit requires the estimation of the phases of the waveforms and the bit intervals. A popular circuit used to achieve that is PLL (phased-locked loop) such as the Costas loop. For now, let us neglect the additive noise term $w(t)$ in the receiver. Note that $\phi_1(t)$ and $\phi_0(t)$ are usually orthogonal. This means that $\int^{T_b}\phi_0(t)\phi_1(t)dt \approx 0$. The outputs of the two integrators in the coherent demodulator of Fig. 4.14(a) are respectively

$$I_k = \int^{T_b} s_k(t)\phi_1(t)dt = \begin{cases}\sqrt{E_s}, & \texttt{if } s_k = s_1\\ 0 & \texttt{if } s_k = s_0\end{cases}$$

and

$$Q_k = \int^{T_b} s_k(t)\phi_0(t)dt = \begin{cases}\sqrt{E_s}, & \texttt{if } s_k = s_0\\ 0 & \texttt{if } s_k = s_1\end{cases}.$$

In each bit period either I_k or Q_k is zero, depending on whether bit 1 or bit 0 is received.

Figure 4.14(b) shows an alternative demodulator, a non-coherent one, based on bandpass filters (BPFs) and envelope detectors. The BPFs have the respective center frequencies f_0 and f_1 and bandwidth equal to the symbol rate $\frac{1}{T_s}$. Note that the BPFs in Fig. 4.14(b) can be replaced by matched filters defined as $h_0(t) = \phi_0(T_b - t)$ and $h_1(t) = \phi_1(T_b - t)$. If the two frequencies used are mutually orthogonal then during each symbol period the output of one of the matched filters will be E_s while that of the other will be zero. The outputs are compared to a threshold value by the comparator. Neglecting the additive noise, the output of the matched filters would be respectively

$$x_1(t) = h_1(t)s_k(t) = \phi_1(T_b - t)s_k(t)$$

and

$$x_0(t) = h_0(t)s_k(t) = \phi_0(T_b - t)s_k(t)\,.$$

If ϕ_0 and ϕ_1 are mutually orthogonal, then either $x_1(t)$ or $x_0(t)$ is zero and the other is E_s in each symbol duration.

The signal constellation diagrams of BFSK according to Eqs. (4.45) and (4.47), respectively, showing the two signal vectors s_0 and s_1 are shown in Fig. 4.15. We use this figure to compute the BER of BFSK, which is

$$P_{b,\texttt{BFSK}} = \begin{cases}\frac{1}{2}e^{-\gamma/2}, & \texttt{non-coherent demodulation}\\ \frac{1}{2}\texttt{erfc}(\sqrt{\frac{\gamma}{2}}) = Q(\sqrt{\gamma}), & \texttt{coherent demodulation}\end{cases}.$$

For equiprobable symbols the symbol error probability of M-FSK over AWGN channels has the

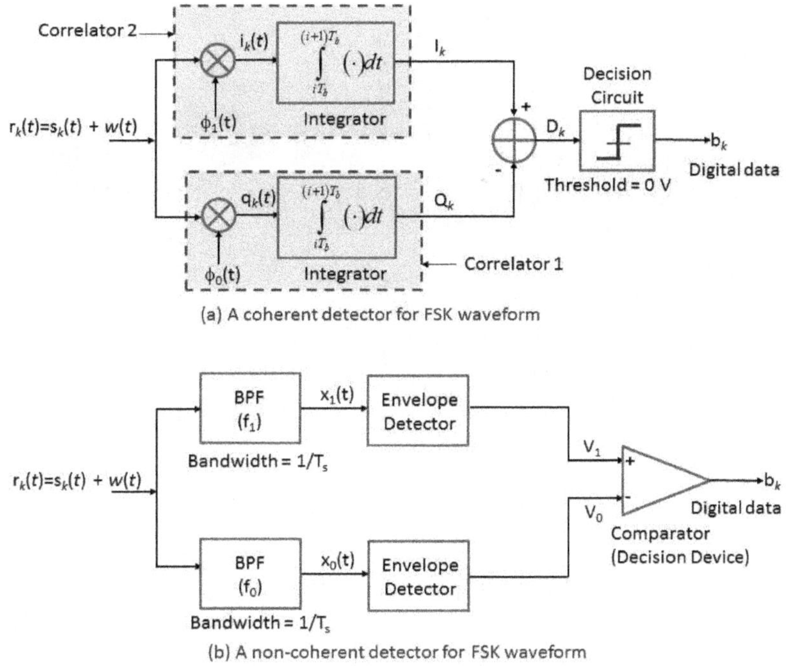

Fig. 4.14 Coherent and non-coherent demodulators for FSK waveform.

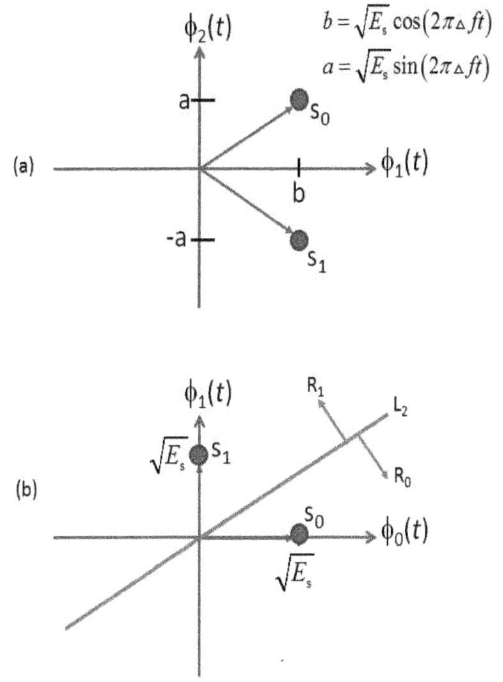

Fig. 4.15 Signal constellation diagrams of orthogonal BFSK according to (a) Eq. (4.45) and (b) Eq. (4.47).

bound

$$P_{\mathtt{e,\ MFSK}} \leq (M-1)\mathtt{Q}\left(\sqrt{\frac{E_{\mathrm{b}}\log_2 M}{N_0}}\right). \qquad (4.48)$$

Power Spectrum of BFSK

We can rewrite the BFSK waveform as the sum of two BASK waveforms as

$$s_{\mathtt{BFSK}}(t) = \sum_{-\infty}^{\infty}(1-b_k)g(t-kT_{\mathrm{b}})\cos(2\pi f_0 t) + \sum_{-\infty}^{\infty} b_k g(t-kT_{\mathrm{b}})\cos(2\pi f_1 t) \qquad (4.49)$$

where $b_k \in \{0,1\}$ are the digital data to be transmitted, and $g(t)$ is a pulse shaping filter, which is here NRZ line code using rectangular-shaped pulses. As done with the BASK PSD, the PSD of the BFSK is

$$S_{\mathtt{BFSK}}(f) = \frac{1}{4}[S_m(f-f_0)+S_m(f+f_0)] + \frac{1}{4}[S_m(f-f_1)+S_m(f+f_1)] \quad (4.50)$$

where $S_m(f)$ is the baseband PSD of the NRZ-encoded bitstream which is given in Eq. (4.33).

Example: BFSK

Assume that the digital data fed into a B-FSK modulator is baseband encoded using polar non-return-

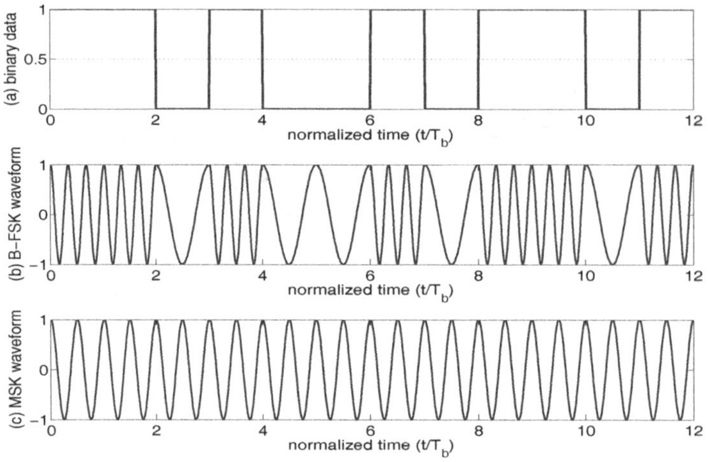

Fig. 4.16 Comparing the waveforms of MSK and BFSK for a given data.

to-zero (NRZ) line code at the rate of 50 kbps. Assume further that the two frequencies used are separated by 500 Hz. Compute that bandwidth consumption of the FSK signal. *Hint.* Use the first null-to-null bandwidth of the baseband signal.

Solution: B-FSK

The bit duration is $T_b = \frac{1}{R_b} = 20\,\mu s$. The null-to-null bandwidth of polar NRZ line code is $B = \frac{1}{2T_b} = 25$ kHz. We also know from the question that $2\Delta f = 500$ kHz. Therefore, the bandwidth of the BFSK signal is

$$B_{\mathtt{bfsk}} = 2(\Delta f + B) = 2(250 + 25)\ \mathtt{kHz} = 550\ \mathtt{kHz}\,. \tag{4.51}$$

4.3.2 *Minimum Shift Keying and GMSK*

The minimum shift keying (MSK) is a special form of BFSK. BFSK system in which we intentionally require the fulfillment of the condition

$$|f_1 - f_0| = \frac{R_b}{2} = \frac{1}{2T_b} = 2\Delta f \tag{4.52}$$

results in MSK. We refer to the parameter

$$h = (f_1 - f_0)T_b = 2\Delta f T_s \tag{4.53}$$

as the **h-factor** which is the equivalence of FM modulation index and Δf as the **peak frequency deviation**. Note that $h = \frac{1}{2}$ for MSK. The frequency separation $\frac{1}{2T_b}$ is the smallest required to achieve orthogonality between the two waveforms used in BFSK. If the condition Eq. (4.52) is fulfilled then the frequency changes in MSK occur at the zero crossings of the carrier wave, resulting in no phase discontinuities (Fig. 4.16). For this reason, MSK is also referred to as **continuous phase FSK** (CPFSK). As phase modulation and frequency modulation are reciprocal, CPFSK is also a special form of PSK. Unlike pure FSK schemes, MSK has no discrete components.

Figure 4.16 compares the waveforms of MSK and conventional BFSK. It is clear that the phase transitions in MSK are smoother than those in BFSK. For this reason, MSK can cope with nonlinearities in transceivers than BFSK. The sidelobes in MSK are futher reduced by passing the baseband data through a lowpass filter prior to using it to modulate the carrier.

Like FSK, MSK uses phase changes in the carrier to transform information. Furthermore, the pulse sent in MSK depends also on the bit sent in the previous bit interval. The waveform of MSK is

$$s_{\mathtt{MSK}}(t) = \sqrt{\frac{2E_b}{T_b}}\cos(2\pi f_c t + \phi(t)) \tag{4.54}$$

where

$$\phi(t) = \begin{cases} \phi(0) + \pi(|f_1 - f_0|)t \\ \quad = \phi(0) + \frac{\pi t}{2T_b}, & \texttt{if bit 1 was sent} \\ \phi(0) - \pi(|f_1 - f_0|)t \\ \quad = \phi(0) - \frac{\pi t}{2T_b} & \texttt{if bit 0 was sent} \end{cases}$$

in the previous bit interval. The carrier frequency is $f_c = (f_0 + f_1)/2$.

Example

The initial phase of an MSK signal can be either 0 or 180 degrees. Assume that the initial phase is 0 degrees. (a) Determine the terminal phase state after transmitting the 6-bit data 011101. (b) Compare the phase transitions of MSK with that of GMSK. Figure 4.17(a) shows the results for MSK. We can observe that the terminal state of the phase is π. The bold curve in Fig. 4.17(b) shows roughly the phase developments in GMSK. We can observe that the Gaussian pulse shaping filtering has removed the corners or sharp edges. These positions translate to high frequencies. Thus, the pulse shaping filtering reduces the bandwidth consumption of GMSK compared with MSK. It is clear that both MSK and GMSK have continuous phases. This is especially obvious if we compare them to the BPSK shown in Fig. 4.17(c), which has abrupt phase transitions between 0 and 180 degrees.

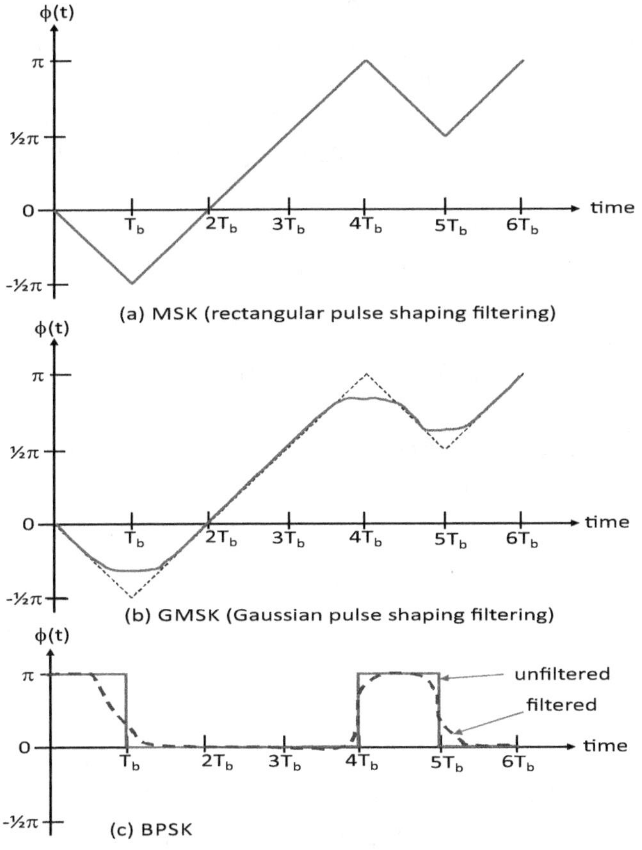

Fig. 4.17 Comparing the phase states of BPSK, MSK and GMSK for the data stream 011101.

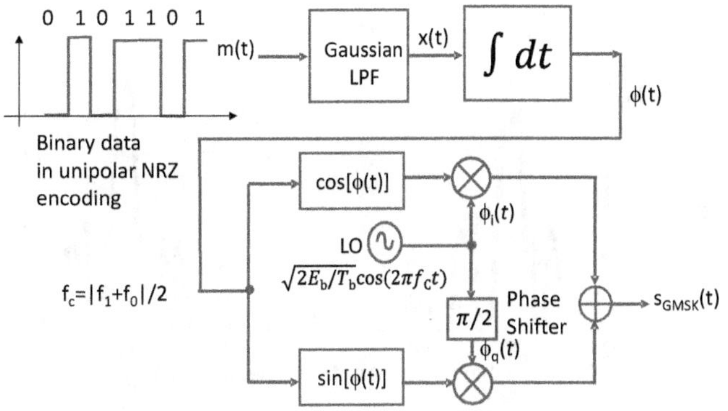

Fig. 4.18 A Gaussian MSK modulator.

The power spectral density of MSK is [Gronemeyer and McBride (1976)]

$$S_{\mathrm{MSK}}(f) = \frac{16P_{\mathrm{s}}T_{\mathrm{s}}}{\pi^2}\left(\frac{\cos(2\pi fT_{\mathrm{s}})}{16f^2T_{\mathrm{s}}^2 - 1}\right)^2 \tag{4.55}$$

where T_{s} is the symbol duration and P_{s} is the average power in the modulated signal.

As discussed earlier in this chapter, passing the digital data through a Gaussian-shaped filter prior to using it to modulate the analog carrier in MSK transceiver results in GMSK signal. GMSK with the time-bandwidth product (aka bandwidth-delay product) of $B_{3\ \mathrm{dB}}T_{\mathrm{b}} = 0.3$ is used in GSM mobile technology, while that in DECT cordless technology is $B_{3\ \mathrm{dB}}T_{\mathrm{b}} = 0.5$, where $B_{3\ \mathrm{dB}}$ is the half-power bandwidth of the GMSK. The GMSK is also used in the IEEE 802.15.6, which is the standard for wireless body area networking (WBAN). Figure 4.18 illustrates a self-explanatory circuit producing GMSK waveform.

Example

GMSK with the time-bandwidth product (aka bandwidth-delay product) of $B_{3\ \mathrm{dB}}T_{\mathrm{b}} = 0.3$ is used in GSM mobile technology. If channels spacing is 200 kHz and data rate is 270.833 kbps, compute the:

(1) Half-power bandwidth of the GMSK.
(2) Spectral efficiency of GSM.

Solution

The bit duration is $T_{\mathrm{b}} = 1/R_{\mathrm{b}}$, while the time-bandwidth product is given as $B_{3\ \mathrm{dB}}T_{\mathrm{b}} = 0.3$.

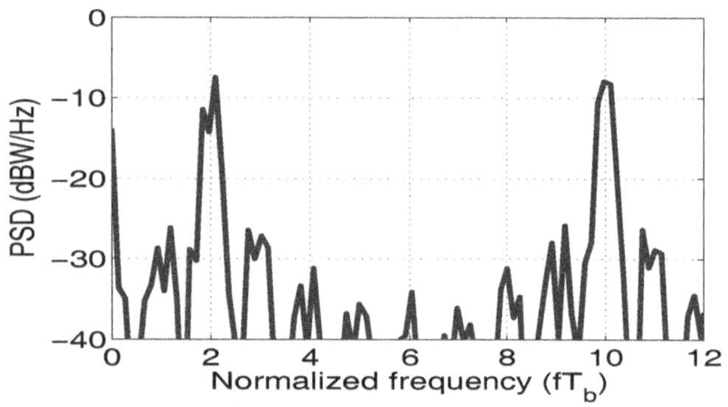

Fig. 4.19 The spectrum of a B-FSK signal.

(1) Thus, half-power bandwidth $B_{3\text{ dB}} = 0.3/T_b = 0.3R_b \approx 81.25$ kHz.
(2) The bandwidth efficiency of GSM is then $\eta_b = \frac{R_b}{W} = \frac{270.833}{200} \approx 1.35$ bps/Hz.

4.3.3 *FH-BFSK*

This simple modulation scheme has been adopted in the Janus protocol which has been standardised by NATO for undersea communications. Janus is the first international standard for communications between underwater entities, such as scuba divers, submarines and aquatic robots, using acoustic signals. Janus achieves transmission range up to 10 km at 11.5 kHz [Nordrum (2017)].

Example: B-FSK

Figure 4.19 shows the spectrum of a B-FSK signal. Assume that the bit duration is $T_b = 1.5\,\mu\text{s}$, Also, assume that f_0 and f_1, with $f_1 > f_0$, are the frequencies for bit 0 and bit 1, respectively.

(1) Describe the given power spectral density (PSD).
(2) Compute the two frequencies used, i.e., f_0 and f_1.
(3) Compute the peak frequency deviation Δf.
(4) Show the peak frequency deviation Δf in the given PSD figure.
(5) Find the carrier frequency f_c.
(6) Compute the h-factor (i.e., the FSK modulation index).
(7) Compute the time-bandwidth product of the modulation scheme.
(8) Compute the half-power bandwidth $B_{3\text{dB}}$ from the given figure.

Solution: B-FSK

(1) Power spectral density (PSD), measured in W/Hz or dBW/Hz, presents the distribution of the power in the signal over frequency. We can use it to find the power contained in each unit of frequency. The PSD has two types of sidebands: the main lobe and the sidelobes. The sidelobes are the undesirable out-of-band radiations (OOBR) which can cause adjacent channel interference (ACI) to disturb the transmissions on neighbouring frequency bands. Regulators use spectral masks (or channel masks) to limit the intensity of sidelobes that a modulation scheme can generate into a given out-of-band frequency. Sidelobes also consume bandwidth. For this reason, many techniques (including filtering) are available to reduce the sidelobes.
(2) The two carrier frequencies are located in the middle of the two main lobes of the PSD, as shown in Fig. 4.20. We have $f_0 T_b = 2 \Rightarrow f_0 = 2/T_b = 2$ MHz. Also, $f_1 T_b = 10 \Rightarrow f_1 = 10/T_b = 10$ MHz.
(3) As illustrated in Fig. 4.20, the peak frequency deviation can be obtained as
$$2\Delta f = f_1 - f_0 \Leftrightarrow \Delta f = \frac{f_1 - f_0}{2} = 4\ \texttt{MHz}\,.$$
(4) The peak frequency deviation Δf is shown in Fig. 4.20.
(5) The carrier frequency f_c can be found from the relationships
$$f_c = f_0 + \Delta f = f_1 - \Delta f = 6\ \texttt{MHz}\,.$$
This can also be found from Fig. 4.20.
(6) The h-factor (i.e., the FSK modulation index) is
$$h = (f_1 - f_0)T_b = 8\,.$$
(7) As illustrated in Fig. 4.20, the time-bandwidth product of the modulation scheme is
$$B_{3\text{dB}} T_b = (f_H - f_L) \approx 10.15 - 9.85 = 0.3\,.$$
This is the time-bandwidth product of GSM, a 2G mobile technology.

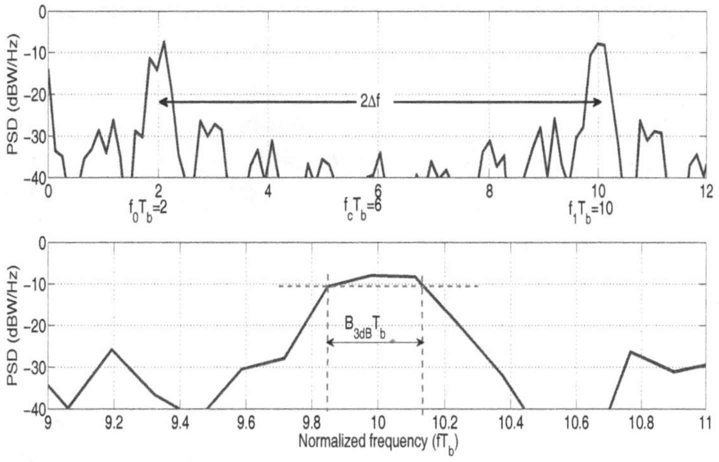

Fig. 4.20 Solution to BFSK problem.

(8) The half-power bandwidth is obtained from

$$B_{3dB}T_b = 0.3 \Leftrightarrow B_{3dB} = \frac{0.3}{T_b} = \frac{0.3}{1.5\,\mu s} = 200 \text{ kHz}.$$

The bandwidth of the BFSK signal can be found from Fig. 4.20 as

$$B_{BFSK} = 2\Delta f + 2B = 2(\Delta f + B) \tag{4.56}$$

where B is the bandwidth of the baseband modulating signal.

4.3.4 *Check Your Understanding*

Students are encouraged to check their understanding of the content of this section with the following questions prior to proceeding to the next section.

(1) Which aspect of the carrier signal is used to convey the intelligence in FSK? What about in MSK?

(2) Discuss the main differences between B-FSK and MSK.

(3) Discuss the main differences between MSK and GMSK.

(4) In order to avoid interference the Bluetooth radio standard uses frequency-hopping spread spectrum technology. GFSK is one of the modulation schemes used to modulate the 79 Bluetooth carriers in the 2.4 GHz ISM band (i.e., 2.4 to 2.4835 GHz). For the sake of simplicity, we assume that only two of the 79 carriers are used and that the BFSK modulator in Fig. 4.13 is adopted. Assume that $f_0 = 2.402$ GHz, $f_1 - f_0 = 1$ MHz and $T_b = 5\,\mu s$. Assume that the digital data to be transmitted is 11001.

(a) Compute and draw the output of the inverter. Assume that the rightmost bit is transmitted first, and unipolar signalling is used.

(b) Sketch the output of the modulator.

(5) Assume that the BFSK used in a GSM mobile system uses the frequencies 876.2 MHz for bit 1 and 876.0 MHz for bit 0. The data rate is 270.833 kbps. Compute:

(a) Peak frequency deviation.

(b) h-factor (i.e., BFSK modulation index).

(c) The baud rate (aka symbol rate).

(d) Minimum bandwidth consumption of the BFSK signal.

(e) Bandwidth efficiency of the system.

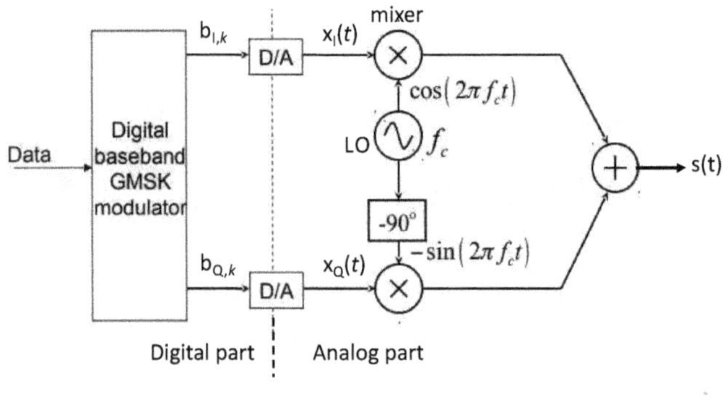

Fig. 4.21 Block diagram of the GMSK moderator in GSM.

(6) The initial phase of an MSK signal can be either 0 or π radians. Assume that the initial phase is π rad. Determine the terminal phase state after transmitting the 8-bit data 10011011. *Hint.* Use a diagram.

(7) Study the circuit in Fig. 4.18. Write out the analytic formula for the GMSK signal $s_{\texttt{GMSK}}(t)$ as a function of the input signal $m(t)$.

(8) Assume that the BFSK used in a DECT cordless system uses the carrier frequency at 1.8 GHz and the time-bandwidth product of 0.5. The data rate is 1 Mbps and the channel spacing is 1.728 MHz. Compute:
   (a) Peak frequency deviation.
   (b) h-factor (i.e., BFSK modulation index).
   (c) The baud rate (aka symbol rate).
   (d) Minimum bandwidth consumption of the BFSK signal.
   (e) Bandwidth efficiency of the system.

(9) Figure 4.21 shows the block diagram of the GMSK moderator used in 2G mobile cellular technology called GSM. Write the analytical formulas for the waveforms $x_I(t), x_Q(t)$ and $s(t)$ shown in the figure. Assume that the binary data fed into the circuit is 10011010.

4.4 *M*-ary Phase Shift Keying: *M*-PSK

In Phase Shift Keying (PSK) the information-bearing data is contained in the phase of the carrier signal. The M-PSK is a multi-level modulation scheme in which $m = \log_2 M$ bits are used to encode each of the $M = 2^m$ signalling levels. One signalling element equals one symbol, which is also equal to one sample of the signal. The general form of a PSK scheme with M symbols is

$$\begin{aligned}
s_k(t) &= A\cos\left(2\pi f_c t + \frac{2\pi k}{M}\right), \quad k = 0, 1, \ldots, M-1. \\
&= A\cos\left(\frac{2\pi k}{M}\right)\cos(2\pi f_c t) \\
&\quad - A\sin\left(\frac{2\pi k}{M}\right)\sin(2\pi f_c t) \\
&= A\sqrt{\frac{T}{2}}\cos\left(\frac{2\pi k}{M}\right)\phi_1(t) \\
&\quad - A\sqrt{\frac{T}{2}}\sin\left(\frac{2\pi k}{M}\right)\phi_2(t) \\
&= i_k\phi_1(t) + q_k\phi_2(t)
\end{aligned}$$

where we have defined the two basis functions (i.e., orthornormal signals)

$$\begin{aligned}
\phi_1(t) &= \sqrt{\frac{2}{T}}\cos(2\pi f_c t)\Pi\left(\frac{t-T/2}{T}\right) \quad \textbf{and} \\
\phi_2(t) &= \sqrt{\frac{2}{T}}\sin(2\pi f_c t)\Pi\left(\frac{t-T/2}{T}\right) \qquad (4.57)
\end{aligned}$$

$\Pi(\frac{t-T/2}{T}) = 1$ for $0 \le t \le T$ and zero otherwise. We have also defined the in-phase and quadrature components

$$i_k = A\sqrt{\frac{T}{2}}\cos\left(\frac{2\pi k}{M}\right) \quad \textbf{and} \quad q_k = -A\sqrt{\frac{T}{2}}\sin\left(\frac{2\pi k}{M}\right). \qquad (4.58)$$

It is clear that all the M symbols have the same amplitude A and frequency f_c, but different phases. The phase variations contain the data of interest. For this reason, the M symbols are distributed uniformly over the signal constellation diagram which is a circle (Fig. 4.22).

The signal constellation diagrams for some M-PSK waveforms are shown in Fig. 4.22. Note that the constellation of QPSK can also be rotated by $\frac{\pi}{4}$. We use Gray coding to map bit patterns to symbols so that a symbol error most probably causes

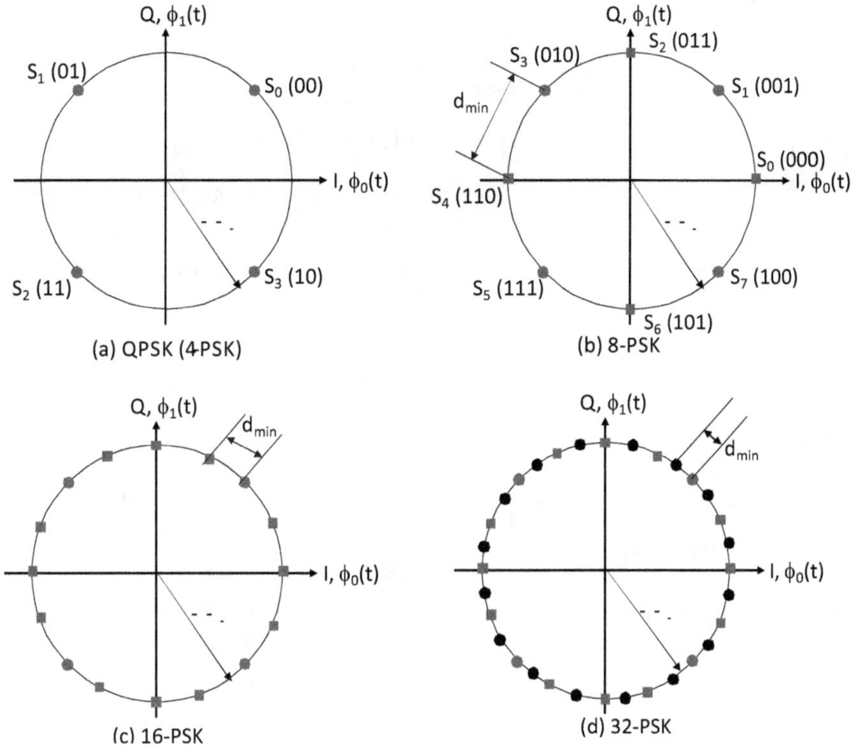

Fig. 4.22 Signal constellation diagrams of M-PSK.

just a bit error. The reason is that the bit patterns assigned to neighbouring symbols differ in only one bit position in Gray coding, and the most probable error is for noise to blow a symbol to its neighbour. It is clear that as the number of bits per symbol increase, we increase the number of distinct symbols in the system. The consequence of increasing the number of distinct symbols is the decrease in the Euclidean distance (i.e., straight line distance between two points in a vector space) between the closest symbols. This means that a little noise can cause the detector to confuse between neighbouring symbols, leading to a symbol error. The distance between the closest pair of symbols in the constellation diagram can be obtained using the Cosine Rule as

$$d_{\mathtt{min}}(M) = \sqrt{2E_{\mathtt{s}}}[1-\cos(2\pi/M)]^{\frac{1}{2}} \\ = 2\sqrt{2E_{\mathtt{s}}}\sin\left(\frac{\pi}{M}\right) \tag{4.59}$$

where $E_{\mathtt{s}}$ is the energy in a symbol. Table 4.2 summarises some M-ary PSK and their minimum distances.

Table 4.2 Some properties of some M-ary PSK.

Modulator	m (No. bits/ symbol)	$M = 2^m$ (No. of distinct symbols)	Minimum inter-symbol distance $d_{\mathtt{min}}$
BPSK	1	2	$2\sqrt{E_{\mathtt{s}}}$
QPSK	2	4	$\sqrt{2E_{\mathtt{s}}}$
8-PSK	3	8	$\sqrt{0.59E_{\mathtt{s}}}$
16-PSK	4	16	$\sqrt{0.15E_{\mathtt{s}}}$
32-PSK	5	32	$\sqrt{0.04E_{\mathtt{s}}}$

As discussed earlier, we use s_k to refer to the signal vector representing the waveform $s_k(t)$ in the signal constellation diagram. The energy in the signal $s_k(t)$, $\forall k$ is

$$E_s = \int_0^T |s_k(t)|^2 dt = \frac{A^2T}{2}\,. \tag{4.60}$$

The power spectrum of the baseband M-PSK signal has the analytic form [Gronemeyer and McBride (1976)]

$$S_{\mathtt{M\text{-}PSK}}(f) = 2E_{\mathtt{b}}\log_2(M)\mathtt{sinc}^2(fT_b\log_2 M)\,. \tag{4.61}$$

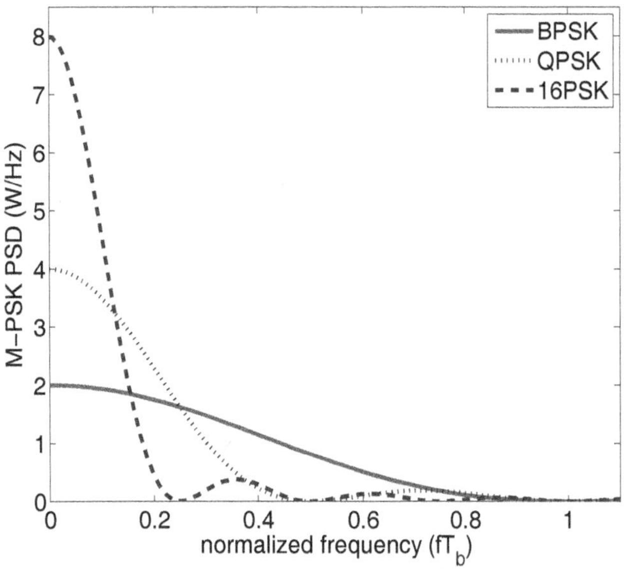

Fig. 4.23 Baseband power spectrum of M-PSK.

Example

The baseband PSD of M-PSK for $M = 2, 4, 16$ is shown in Fig. 4.23. Using this figure, compute the null-to-null bandwidth consumption of BPSK, 4-PSK (i.e., QPSK) and 16-PSK.

Solution

The first null of BPSK PSD occurs at $fT_{\text{b}} = 1$. This means that the BPSK null-to-null bandwidth is $f_{\text{BPSK}} = \frac{1}{T_{\text{b}}} = R_{\text{b}}$. The first null of QPSK PSD occurs at $fT_{\text{b}} = 0.5$, yielding the bandwidth of $f_{\text{QPSK}} = \frac{0.5}{T_{\text{b}}} = \frac{1}{2}R_{\text{b}}$. Similarly, the null-to-null bandwidth of 16-PSK is $f_{\text{16-PSK}} = \frac{0.25}{T_{\text{b}}} = \frac{1}{4}R_{\text{b}}$.

If we transmit the symbol $s_k(t)$ over an additive white Gaussian noise (AWGN) channel then the signal which reaches the receiver is

$$r_k(t) = s_k(t) + w(t), \quad k = 0, 1 \text{ and } 0 \leq t \leq T \tag{4.62}$$

where $w(t) = \mathbb{R}e\{n(t) = w(t) + jv(t)\}$ is the real part of the AWGN with mean zero and variance $\sigma_w^2 = \frac{N_0}{2}$, i.e. $w(t) \sim \mathcal{N}(0, \frac{N_0}{2})$. Thus, the probability density function (pdf) of the AWGN is

$$f_w(x) = \frac{1}{\sqrt{\pi N_0}} e^{-\frac{x^2}{N_0}}. \tag{4.63}$$

Assume that the sampled version of $r_k(t)$ is

$$r_k = s_k + w, \quad k = 0, 1, 2, \ldots, M-1. \tag{4.64}$$

4.4.1 *BPSK*

The simplest M-PSK is binary PSK (BPSK) in which $M = 2$. Also, referred to as 2-PSK or **bi-phase modulation**, BPSK is used in IEEE802.11, for example. As $\log_2(M)|_{M=2} = 1$, BPSK transmits just a single bit in each symbol or signalling period T_{s}. For that reason, we have $T_{\text{s}} = T_{\text{b}}$. A simple circuit which can generate BPSK waveform is illustrated in Fig. 4.24. Digital data made of 0s and 1s (i.e., unipolar signalling) are first converted to a polar signal taken the values ± 1. The polar signal and a carrier wave produced by the reference local oscillator are fed into a product modulator to obtain the BPSK waveform $s_{\text{BPSK}}(t)$.

Two circuits which can be used to detect BPSK waveform are shown in Fig. 4.25. Given the the received signal $r)t) = s_{\text{BPSK}}(t) + w(t)$, the output of the product modulator in Fig. 4.25(a) is

$$\begin{aligned} x(t) &= s_{\text{BPSK}}(t)\cos(2\pi f_{\text{c}}t) + w(t)\cos(2\pi f_{\text{c}}t) \\ &= \pm A\cos(2\pi f_{\text{c}}t)\cos(2\pi f_{\text{c}}t + \varphi) \\ &\quad + w(t)\cos(2\pi f_{\text{c}}t) \\ &= \pm\frac{1}{2}A\cos(\varphi) \pm \frac{1}{2}\cos(4\pi f_{\text{c}}t + \varphi) \\ &\quad + w(t)\cos(2\pi f_{\text{c}}t). \end{aligned} \tag{4.65}$$

The output of the bandpass filter[2] (BPF) is then $\pm\frac{1}{2}A\cos(\varphi)$, where φ is a phase error. The detector decides for bit 1 if its input signal envelope is larger than zero, else it decides in favour of bit 0. The sampled version of the received signal is

$$r_k = s_k + w, \quad k = 0, 1. \tag{4.66}$$

For BPSK we have $M = 2$. This simplifies Eq. (4.58) to

$$\begin{aligned} i_k &= A\sqrt{\frac{T}{2}}(-1)^k \\ &= \sqrt{E_s}(-1)^k \text{ and } q_k = 0, \quad k = 0, 1. \end{aligned} \tag{4.67}$$

This results in the signal constellation diagram and the conditional probability density functions shown in Fig. 4.26.

[2]Actually, even a lowpass filter would do the job.

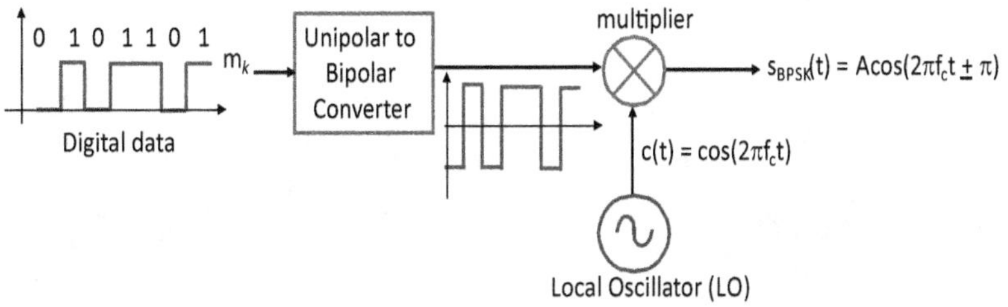

Fig. 4.24 A BPSK modulator.

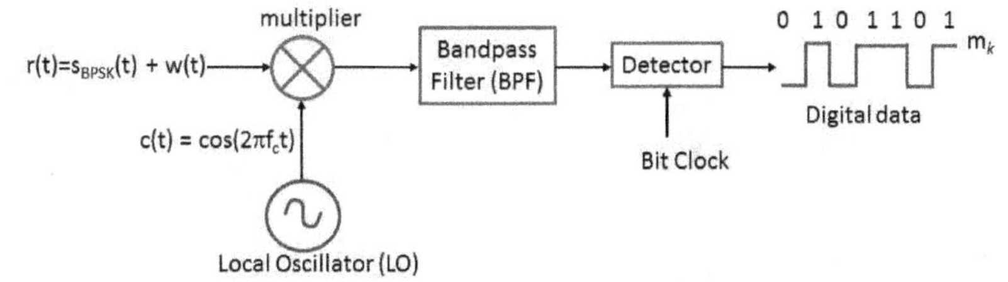

(a) BPSK demodulator

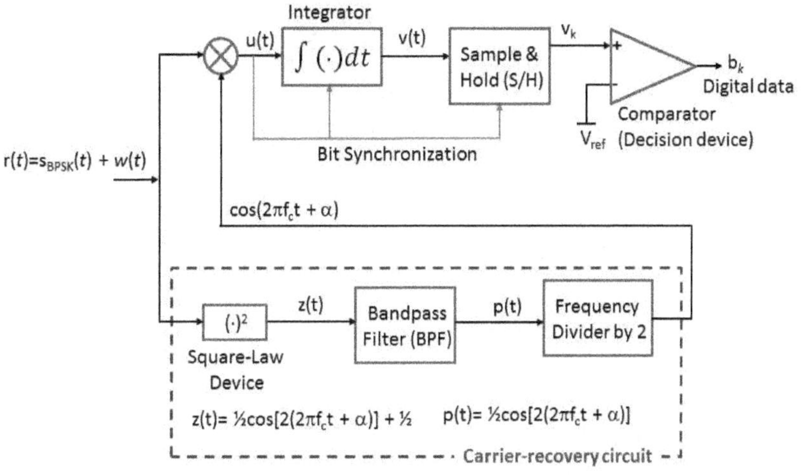

(b) A coherent BPSK demodulator

Fig. 4.25 Two BPSK detectors.

4.4.1.1 *BER of BPSK over AWGN Channel*

Using the signal constellation diagram in Fig. 4.26, a hard-decision demodulation (aka maximum likelihood demodulation) decides as follows:

$$r_k \geq 0 \Rightarrow s_0 \quad \texttt{(i.e. bit 0) was sent}$$

$$r_k < 0 \Rightarrow s_1 \quad \texttt{(i.e. bit 1) was sent}$$

Thus, we have detection error if bit 0 was sent but $r_k < 0$. We denote such a bit error as $p_{e0} = Pr[r_k < 0|s_0]$. Similarly, detection error results if bit 1 was sent but $r_k \geq 0$. We denote such a bit error as $p_{e1} = Pr[r_k \geq 0|s_1]$. Noting that $E_s = \frac{A\sqrt{T}}{2}$ we obtain from Eq. (4.66)

$$r_k = \sqrt{E_s} + w \quad \texttt{if bit 0 was sent}$$

$$r_k = -\sqrt{E_s} + w \quad \texttt{if bit 1 was sent}$$

Assuming that w is the sample of an AWGN, the pdfs of r_k are shown in Fig. 4.26. Using these pdfs

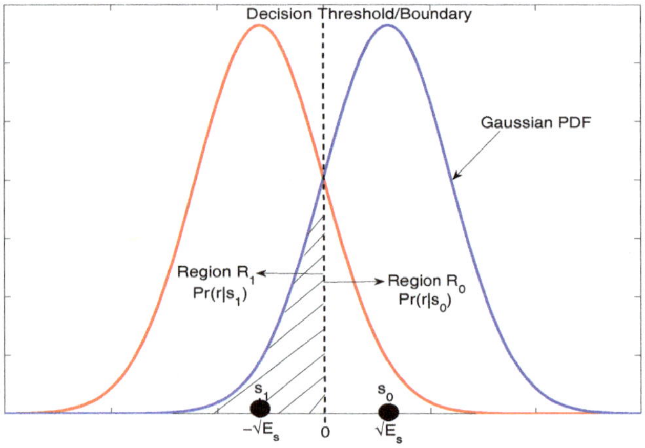

Fig. 4.26 Conditional probability density functions for BPSK superimposed on its signal constellation diagram.

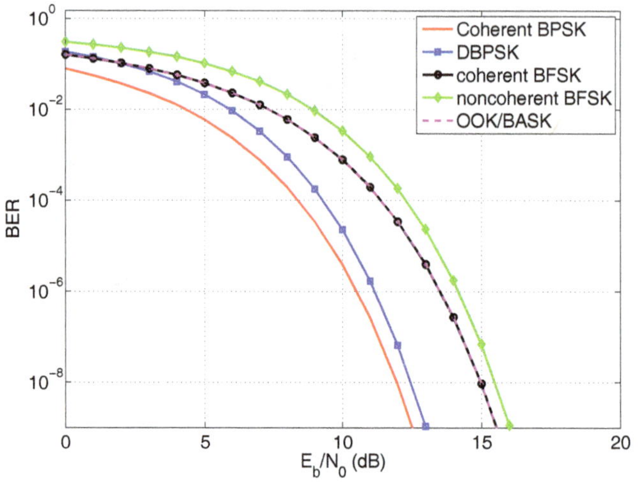

Fig. 4.27 Comparison of bit error performance of digital binary modulators.

we obtain the bit error probabilities

$$\begin{aligned} p_{e0} = p_{e1} &= Pr[r_k \geq 0|s_1] = Pr[-\sqrt{E_s} + w \geq 0] \\ &= Pr[w \geq \sqrt{E_s}] = \int_{\sqrt{E_s}}^{\infty} f_w(x)dx \\ &= \frac{1}{\sqrt{\pi N_0}} \int_{\sqrt{E_s}}^{\infty} e^{-\frac{x^2}{N_0}} dx \\ &= \frac{1}{\sqrt{2\pi}} \int_{\sqrt{2\frac{E_s}{N_0}}}^{\infty} e^{-\frac{y^2}{2}} dy \\ &= Q\left(\sqrt{2\frac{E_s}{N_0}}\right) = Q(\sqrt{2\gamma}) \end{aligned} \tag{4.68}$$

where $\gamma = \frac{E_s}{N_0}$ is symbol energy to noise power spectral density ratio. It is usually referred to as signal-to-noise ratio. Finally, the BER of BPSK is

$$\begin{aligned} BER_{\text{BPSK}} &= p_0 \cdot p_{e0} + p_1 \cdot p_{e1} \\ &= \frac{1}{2}Q(\sqrt{2\gamma}) + \frac{1}{2}Q(\sqrt{2\gamma}) \\ &= Q(\sqrt{2\gamma}) \end{aligned} \tag{4.69}$$

where we have assumed that, in a given data stream, the probability of the occurrence of bit 1 (i.e., $p_1 = \frac{1}{2}$) equals that of bit 0 (i.e., $p_0 = \frac{1}{2}$). Figure 4.27 compares the BER of binary modulators. It is observed that coherent BPSK outperforms all modulators. Next to it is the differentially-encoded coherent BPSK (DBPSK). The worst performance is delivered by the noncoherent BFSK. The generalised bit error rate for coherently demodulated M-PSK (CPSK) at large signal-to-noise ratio using equally like symbols is [Korn (1985)]

$$\begin{aligned} P_b(M) &\approx \frac{2}{\log_2 M} Q\left(\sqrt{\frac{2E_b \log_2(M)}{N_0}} \sin\frac{\pi}{M}\right) \\ &= \frac{1}{\log_2 M} \texttt{erfc}\left(\sqrt{\frac{E_b \log_2(M)}{N_0}} \sin\frac{\pi}{M}\right), \quad M \geq 4 \end{aligned} \tag{4.70}$$

while the symbol error propabaility for differentially coherent demodulated M-PSK (DPSK) is [Korn (1985)]

$$P_e(M) \approx 2Q\left(\sqrt{\frac{2E_b \log_2(M)}{N_0}} \sin\frac{\pi}{\sqrt{2}M}\right). \tag{4.71}$$

Figure 4.28 compares the error performance of M-ary PSK with M-ary FSK modulation schemes. We can observe that FSK has a higher error rate than both coherent and differentially coherent PSK.

4.4.1.2 *Power Spectrum and Bandwidth of BPSK*

The power spectrum of the bandpass BPSK signal has the analytic form

$$\begin{aligned} S_{\text{BPSK}}(f) &= \frac{E_b}{2}(\texttt{sinc}^2[(f - f_c)T_b)] \\ &\quad + \frac{E_b}{2}(\texttt{sinc}^2[(f + f_c)T_b)]. \end{aligned} \tag{4.72}$$

The BPSK spectrum is shown in Fig. 4.29. We can observe that the null-to-null bandwidth of BPSK

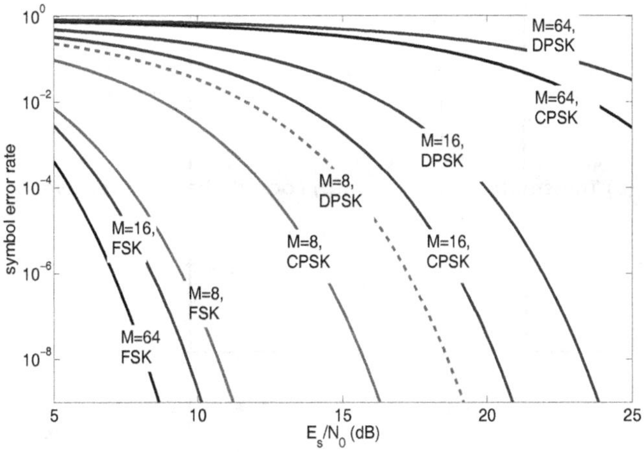

Fig. 4.28 Comparison of symbol error probabilities of M-ary digital modulation methods.

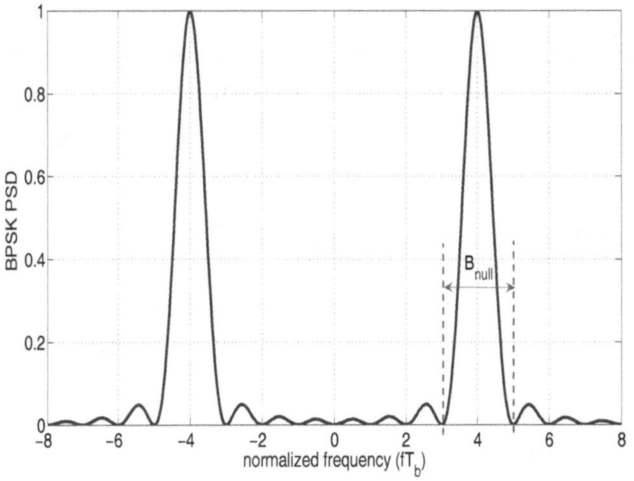

Fig. 4.29 Power spectrum of BPSK ($f_c T_b = 4$).

is $B_{\texttt{null}} = \frac{2}{T_b} = 2R_b$. Thus, the bandwidth efficiency of BPSK is 0.5 bps/Hz. The bandwidth containing 99% of the energy in the BPSK signal is $20R_b$. We can also observe in Fig. 4.29 that the output of a (digital) modulator is a double-sideband suppressed-carrier signal. The reason is that the bandwidth of the modulated signal is twice that of the baseband modulating signal, and no carrier is transmitted.

Example

Consider a coherent BPSK receiver operating at 15°C and receiving data at the rate 1 Mbps. The system uses the waveforms $s_0(t) = -A_c \cos(2\pi f_c t)$ and $s_1(t) = A_c \cos(2\pi f_c t)$ to transmit bit 0 and bit 1, respectively, where $A_c = 0.25\,\mu\text{V}$. Assume that the energy per bit is normalised relative to a $1-\Omega$ resistive load. Also, assume that the transmission medium causes only additive white Gaussian noise (AWGN).

(1) Compute the signal energy per bit, E_b.
(2) Compute the single-sided power spectral density of the AWGN, N_0.
(3) Compute the signal-to-noise ratio (i.e., bit energy to noise power spectral density), E_b/N_0.
(4) Compute the bit error rate (BER) of the coherent BPSK detector.
(5) Assuming that the detector operates continuously over a period of two days, find the expected number of bits received in error.

Note: The Boltzmann constant is about 1.38065×10^{-23} J/K.

Solution

(1) The energy per bit over $1-\Omega$ resistive load is

$$E_b = \int_0^{T_b} |s_k(t)|^2 dt = \frac{A_c^2 T_b}{2} = \frac{A_c^2}{2R_b} = 3.125 \times 10^{-20} \text{ J}.$$

(2) The operating temperature is given as $T = 15°\text{C}$. Note that rooms hosting RF receivers are normally cooled to reduce additive noise. This temperature is equivalent to $T = 15 + 273.15 = 288.15$ K. Thus, the single-sided power spectral density of the AWGN is $N_0 = kT \approx 3.9783 \times 10^{-21}$ W/Hz.

(3) The bit energy to noise power spectral density is

$$\gamma = E_b/N_0 = \frac{3.125 \times 10^{-20} \text{ J}}{3.9783 \times 10^{-21} \text{ W/Hz}} \approx 7.855.$$

We have noted the equivalence of the units joule (J) and watt per Hertz (W/Hz).

(4) The bit error rate (BER) of coherent BPSK is given in Eq. (4.69) as

$$BER = Q(\sqrt{2\gamma}) = Q(\sqrt{2 \times 7.855}) \approx 3.6916 \times 10^{-5}.$$

This result can also be read off Fig. 4.27.

(5) Continuous two-day data transmission lasts $t_0 = 2 \times 24 \times 60^2$ s $= 172800$ s. With the bit rate of $R_b = 1$ Mbps, this results in the total number of transmitted bits

$$N_b = t_0 \times R_b = 172800 \text{ s} \times 10^6 \text{ bps} = 1728 \times 10^8 \text{ bits}.$$

The expected (or average) number of bits received in error is thus

$$N_e = N_b \times BER = 1728 \times 10^8 \text{ bits} \times 3.6916 \times 10^{-5} \approx 6,379,084.8 \text{ bits}.$$

4.4.2 $\frac{\pi}{2}$-BPSK

This modulation scheme has been adopted in the upcoming 5G mobile cellular systems. It is also used in IEEE 802.15.6 standard for wireless body area networks (WBANs). The constellation of M-PSK is $\frac{\pi}{M}$ rotation invariant. This means that the detector cannot know if a transmitted PSK signal undergoes a phase change of π in transit, resulting in phase ambiguity. This is illustrated in Fig. 4.30. **Phase ambiguity** is a main problem in M-PSK, and its known causes are:

- **Phase slips**: which is caused by the time-varying nature of the transmission medium. A timing error produces a phase error or slip, something which causes the detector to be out of synchronisation with the transmitter.
- **Erroneous phase lock at receiver**: the detector at the receiver may lock onto a phase orientation which is different from what the transmitter used. This is illustrated in Fig. 4.30.

The variations in the amplitude or envelope of $\frac{\pi}{2}$-BPSK are smaller than those in conventional BPSK. The smaller amplitude variations in $\frac{\pi}{2}$-BPSK are achieved by multiplying the symbol sequence of BPSK with a rotating phasor which has phase increment per symbol duration of $\frac{\pi}{2}$. The linear feature in $\frac{\pi}{2}$-BPSK, compared with BPSK, renders it more benign for circuits using nonlinear elements like Class C amplifiers. The most important thing is not the actual power but the effective power

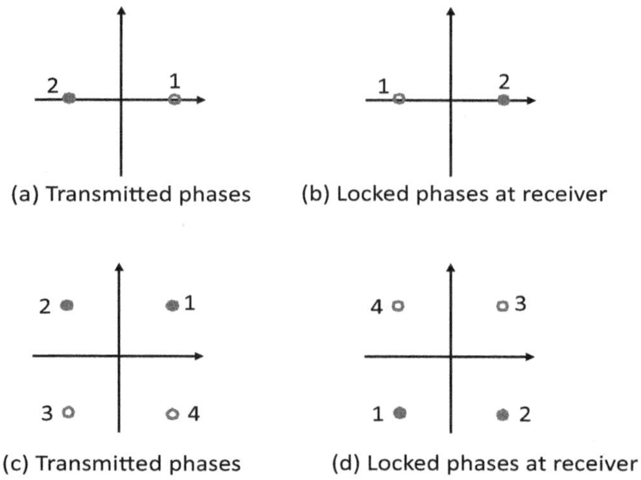

Fig. 4.30 PSK and QAM phase ambiguity: the phase constellations used at transmitter and detector are 180° out of phase with each other.

radiated from the antenna of a wireless device. The effective radiated power (ERP) in decibels (dB) is

$$ERP = \texttt{transmit power} + \texttt{antenna gain} = P_T + G_T. \tag{4.73}$$

Therefore, for a given ERP, the higher the gain of the antenna, the smaller the power output required of the electronic device. This means that using a high-gain antenna to achieve a given ERP conserves the battery of the device. Amplifiers with high gain (e.g., travelling wave tube and Class C amplifiers) have nonlinear features. If the envelope variations are large then the sidelobes in the modulated signal's power spectrum produce significant **spectral regrowth** when they pass through nonlinear device. Spectral regrowth occurs in any modulation which has non-constant amplitude (or envelope) when they pass through a nonlinear device. This is the reason why modulation schemes benign to nonlinear amplifications are of special interest in wireless communications. Every feature in $\frac{\pi}{2}$-BPSK is the same as BPSK, except the reduced envelope variations.

The generalised waveform of $\frac{\pi}{2}$-BPSK is [Petenaude and Moher (1992)]

$$s(t) = \sum_{n=-\infty}^{\infty} m_k e^{-jn\frac{\pi}{M}} \tag{4.74}$$

where $m_k \in \{\pm 1\}$ for $\frac{\pi}{2}$-BPSK and $m_k \in \{\pm 1, \pm j\}$ for $\frac{\pi}{4}$-QPSK. This equation is valid for both

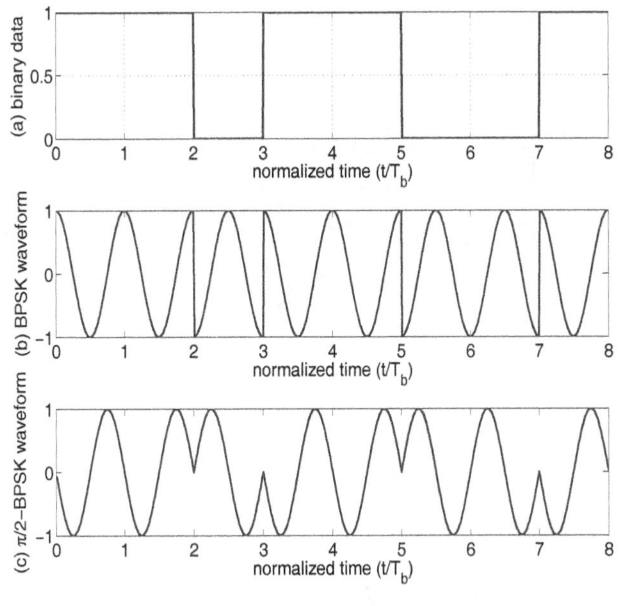

Fig. 4.31 Comparing waveforms of BPSK and $\frac{\pi}{2}$-BPSK.

$\frac{\pi}{2}$-BPSK and $\frac{\pi}{4}$-QPSK. Thus, in BPSK we transmit $s_1(t) = A_c \cos(2\pi f_c t + \frac{\pi}{2})$ for bit 1 and $s_0(t) = A_c \cos(2\pi f_c t - \frac{\pi}{2})$ for bit 0, or vice versa. Figure 4.31 compares the waveforms of BPSK and $\frac{\pi}{2}$-BPSK. We can observe that, unlike BPSK, $\frac{\pi}{2}$-BPSK avoids phase transitions through the origin. Figure 4.32 shows a circuit which can generate $\frac{\pi}{2}$-BPSK waveform [Ningdalli and Sujatha (2015)].

4.4.3 *QPSK and Its Variants*

Quaternary phase shift keying (QPSK) is a very important modulation scheme as it is used in modern mobile technologies, such as 4G/LTE-A. In communications engineering parlance, the word *quaternary* means four, while *quadrature* means at right angles. Thus, QPSK is a special form of M-PSK modulation with four symbols or states which are at right angles to each other (i.e., separated by 90°). Also, the quadratures relate to the fact that QPSK has a two-dimensional constellation diagram formed by two basis functions which are perpendicular to each other. Having four symbols also means that the QPSK carrier signal must have four distinct phase changes, each representing two binary bits (aka dibit) of data. QPSK can thus achieve the data rate of BPSK using half of the bandwidth of BPSK, or use the same bandwidth to achieve twice the bit rate. The data bits to be transmitted are first divided into **dibits**. In the QPSK modulator, each dibit causes the carrier signal to assume one of the four possible phases. Two carrier signals are used: the in-phase (I) and the quadrature (Q). The I carrier has the same phase as the reference oscillator, while the Q carrier is 90° out of phase with the reference oscillator. The four symbols in QPSK can be encoded with the four phases

$$\theta_k = \theta_0 + \frac{\pi}{2}k, \; k = 0, 1, 2, 3\,. \tag{4.75}$$

For example, if $\theta_0 = 0$, we have the phases $\{0, 90°, 180°, 270°\}$. If $\theta_0 = 45°$, we have the phases $\{45°, 135°, 225°, 315°\}$. Thus, the mapping of phases to dibit or symbols in QPSK is not unique. The signal constellation diagram of QPSK with phases $\{45°, 135°, 225°, 315°\}$ is shown in Fig. 4.33.

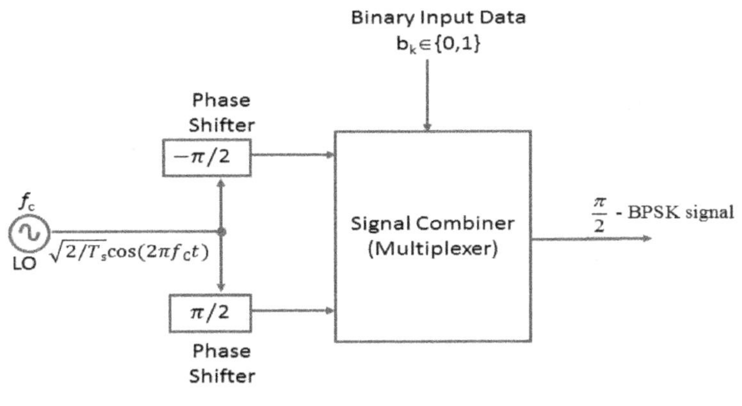

Fig. 4.32 A modulator for $\frac{\pi}{2}$-BPSK.

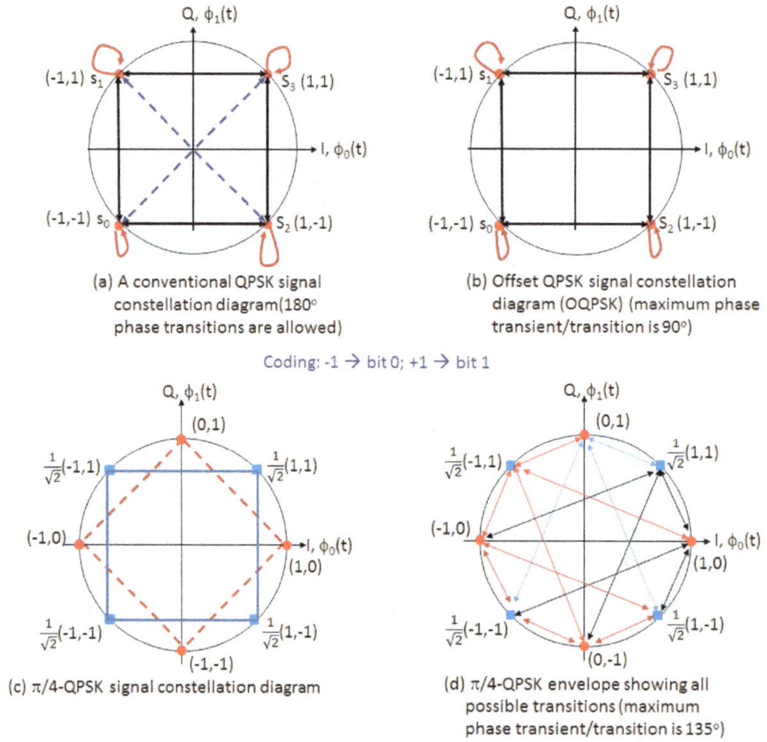

Fig. 4.33 Constellation diagrams of QPSK and its variants showing possible phase transitions.

Table 4.3 Symbols for bit patterns in LTE's QPSK modulator [3GPP (2009)].

	Modulated Symbol		
Input Dibit	In-phase Component (I)	Quadrature Component (Q)	Output Phase
00	$\frac{1}{\sqrt{2}}$	$\frac{1}{\sqrt{2}}$	45°
01	$\frac{1}{\sqrt{2}}$	$-\frac{1}{\sqrt{2}}$	315°
10	$-\frac{1}{\sqrt{2}}$	$\frac{1}{\sqrt{2}}$	135°
11	$-\frac{1}{\sqrt{2}}$	$-\frac{1}{\sqrt{2}}$	225°

This is what is adopted in 4G LTE mobile standard (Table 4.3). Table 4.4 also shows the 16-PSK adopted in LTE. Any of the four states in QPSK can transition to any of the three states. Thus, 180° transitions between states causes the amplitude (or envelope) of the carrier signal to be zero as the transition passes through the origin of the signal constellation diagram. In Fig. 4.33(a) each of the two state transitions $s_1 \leftrightarrow s_2$ and $s_0 \leftrightarrow s_3$ passes through the origin. Amplitude change to zero is an abrupt change in time domain and thus causes large spectral sidelobes. The trellis diagrams of QPSK

Table 4.4 Symbols for bit patterns in LTE's 16-PSK modulator [3GPP (2009)].

	Modulated Symbol	
Input Bit Pattern	In-phase Component (I)	Quadrature Component (Q)
0000	$\frac{1}{\sqrt{10}}$	$\frac{1}{\sqrt{10}}$
0001	$\frac{1}{\sqrt{10}}$	$\frac{3}{\sqrt{10}}$
0010	$\frac{3}{\sqrt{10}}$	$\frac{1}{\sqrt{10}}$
0011	$\frac{3}{\sqrt{10}}$	$\frac{3}{\sqrt{10}}$
0100	$\frac{1}{\sqrt{10}}$	$-\frac{1}{\sqrt{10}}$
0101	$\frac{1}{\sqrt{10}}$	$-\frac{3}{\sqrt{10}}$
0110	$\frac{3}{\sqrt{10}}$	$-\frac{1}{\sqrt{10}}$
0111	$\frac{3}{\sqrt{10}}$	$-\frac{3}{\sqrt{10}}$
1000	$-\frac{1}{\sqrt{10}}$	$\frac{1}{\sqrt{10}}$
1001	$-\frac{1}{\sqrt{10}}$	$\frac{3}{\sqrt{10}}$
1010	$-\frac{3}{\sqrt{10}}$	$-\frac{1}{\sqrt{10}}$
1011	$-\frac{3}{\sqrt{10}}$	$\frac{3}{\sqrt{10}}$
1100	$-\frac{1}{\sqrt{10}}$	$-\frac{1}{\sqrt{10}}$
1101	$-\frac{1}{\sqrt{10}}$	$-\frac{3}{\sqrt{10}}$
1110	$-\frac{3}{\sqrt{10}}$	$-\frac{1}{\sqrt{10}}$
1111	$-\frac{3}{\sqrt{10}}$	$-\frac{3}{\sqrt{10}}$

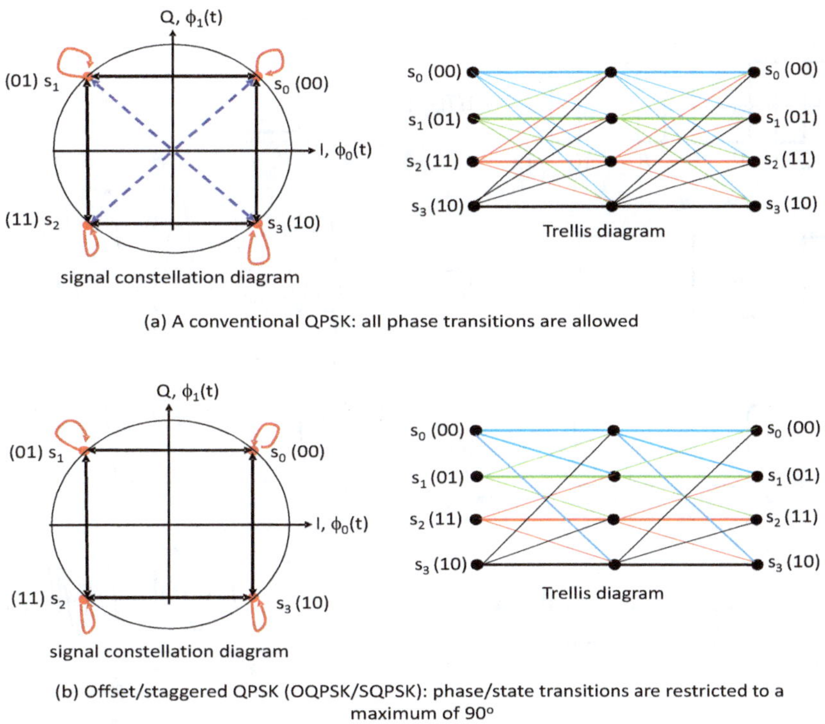

Fig. 4.34 Trellis and signal constellation diagrams of QPSK and OQPSK.

and OQPSK are shown in Fig. 4.34 to confirm their state changes.

The QPSK waveform has the general form

$$s_k(t) = \underbrace{A}_{\texttt{constant}} \cos(\underbrace{2\pi f_c t}_{\texttt{changes with time}} + \underbrace{\theta_0 + \frac{\pi}{2}k}_{\texttt{changes with dibit}}), \quad k = 0, 1, 2, 3\,. \tag{4.76}$$

OQPSK (offset QPSK), also known as staggered QPSK (SQPSK), was developed to solve the problem of transitioning through the origin in conventional QPSK. QPSK has large amplitude (or envelope) fluctuations because of its unrestricted phase transitions. The in-phase and the quadrature components in QPSK are intentionally misaligned by T_b or half the symbol duration to obtain OQPSK in order to avoid 180° phase transitions. Stated differently, as observable in Fig. 4.33(b), no transition in the constellation diagram of OQPSK passes through the origin because only one bit of the dibit making a symbol changes at a time. By allowing only one of the bits in a dibit to change at a time, the magnitude of the amplitude fluctuations in OQPSK is much lower than non-offset QPSK. This can be observed in Fig. 4.33(b), which also shows the only permissible state transitions which is a maximum phase transient transition of 90°. Although all amplitude modulators need linear amplification, QPSK transceivers must use highly linear devices including amplifiers to avoid spectral regrowth.

$\frac{\pi}{4}$-QPSK is an alternative to OQPSK to avoid state transitions passing through the origin in order to avoid abrupt changes in the modulated waveform. By rotating every other symbol by $\frac{\pi}{4}$ to restrict the phase transitions to a maximum of 135°, $\frac{\pi}{4}$-QPSK avoids transitioning through the origin. We can observe the signal constellation diagram in Fig. 4.33(c) and (d) that no state transitions pass through the origin.

4.4.3.1 *QPSK Modulators and Demodulators*

QPSK uses two carrier waves referred to as in-phase (I) and quadrature (Q). The bitstream used to modulate the carrier waves are divided into two parts: even-numbered bits and odd-numbered bits

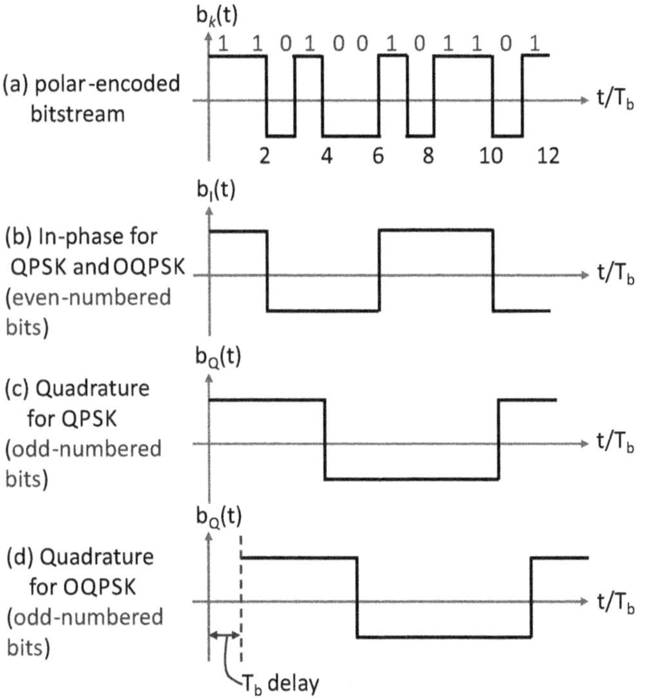

Fig. 4.35 Quadrature and in-phase components of QPSK and OQPSK for a given data.

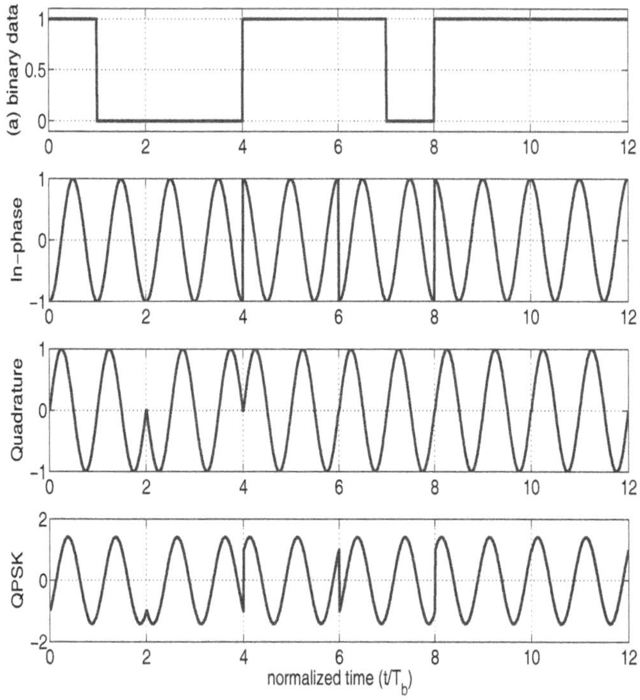

Fig. 4.36 QPSK waveform for a given data stream.

(Fig. 4.35). The former is used to modulate the I carrier which is in-phase with the reference oscillator, while the latter modulates the quadrature (or Q) carrier. Figure 4.37 shows a block diagram of a modulator which can be used to produce QPSK and OQPSK waveforms. The serial to parallel converter divides the incoming bitstream into dibits. It then sends one bit in each dibit to the in-phase (I) channel and the other to the quadrature (Q) channel of the modulator. The Q and I channels independently process their inputs. Each bit is encoded into baseband bipolar NRZ pulse and then passed through the lowpass filter (LPF). The LPF smoothens the sharp transitions in the bipolar pulses to reduce the bandwidth consumption of the resulting QPSK signal. Each lowpass-filtered bipolar pulse is then fed into a mixer which multiplies it with the carrier signal from the oscillator to achieve the modulation. The mixing translates the baseband signal to a desired frequency of the carrier signal, resulting in a bandpass signal. The output of the I or Q channel is a BPSK signal. As the two carriers used by the I and Q channels are orthogonal, the two BPSK signals can be summed and transmitted over the same infrastructure without mutual interference. Thus, QPSK comprises two orthogonal BPSK waves. QPSK allows the maximum possible phase transitions, which is 180°. Such a phase transition causes 100% change in the carrier signal.

The offset in OQPSK means that there is a delay of $T_b = T_s/2$ between the I (even-numbered bits) and Q (odd-numbered bits) of a bitstream which is used to modulate the two quadrature carriers. This allows only two possible phase transitions (i.e., $\pm 90°$) in each state. Compared to QPSK, OQPSK has a maximum spectral spreading of 33%.

Figure 4.38 illustrates the block diagram of a modulator for $\frac{\pi}{4}$-QPSK. As observed in the figure, $\frac{\pi}{4}$-QPSK uses two sets of QPSK signal constellations: the conventional one and its $\frac{\pi}{4}$ offset version. Even-numbered symbols, each containing two bits, is used to modulate one of these two constellations, while the odd-numbered symbols modulate the other constellation. The result is that $\frac{\pi}{4}$-QPSK allows up to 135° phase transition at a go.

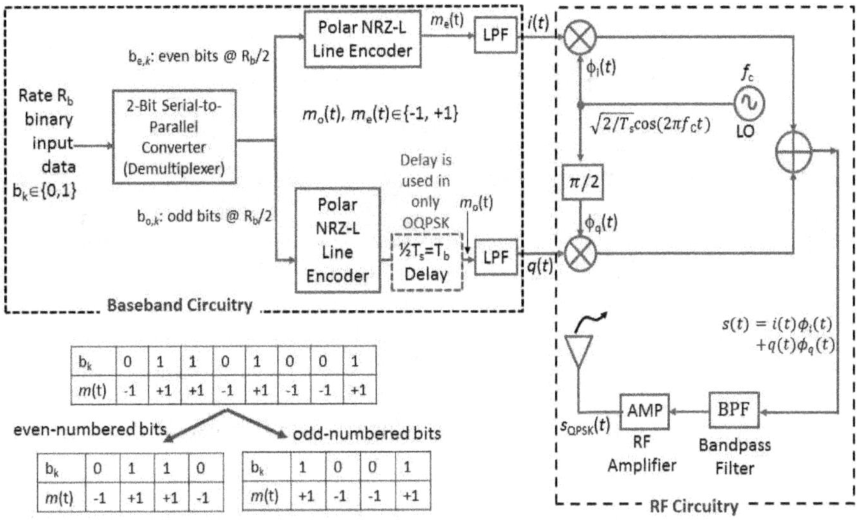

Fig. 4.37 Block diagram of a modulator for QPSK and OQPSK.

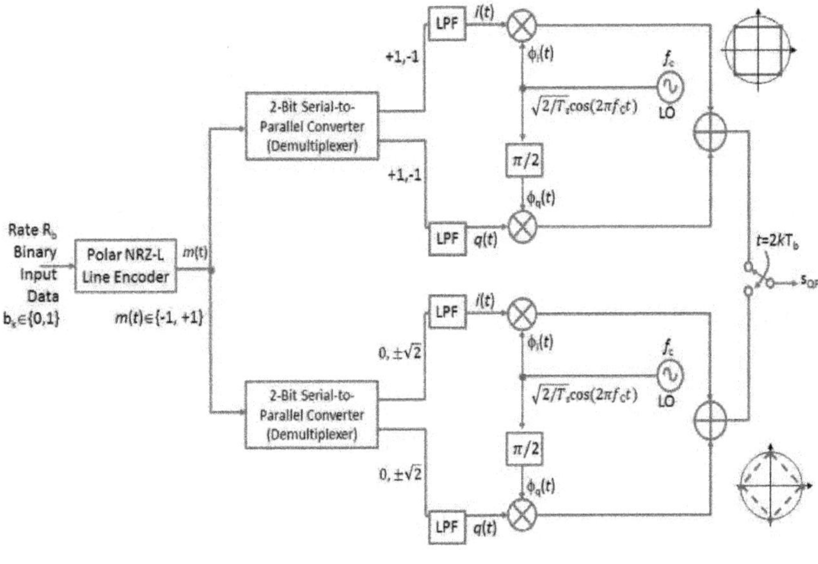

Fig. 4.38 Block diagram of a modulator for $\frac{\pi}{4}$-QPSK.

A coherent demodulator of QPSK is shown in Fig. 4.39. The integrator integrates the input signal over the time interval $[0, T_s]$. Neglecting noise effects, the output of the integrators are

$$r_i(t) = \int_0^{T_s} r(t)\phi_1(t)dt$$
$$= \begin{cases} \sqrt{E_s}\sin(\theta_k), & \texttt{if } r(t) \sim \cos(\cdot)) \\ 0, & \texttt{if } r(t) \sim \sin(\cdot) \end{cases}$$

where $\theta_k = \theta_0 + \frac{\pi}{2}k$.

Example

Consider the block diagram of QPSK moderator shown in Fig. 4.37.

(1) Write the analytic formula of the waveform for each of the four dibits of the QSK.
(2) Construct the truth table.
(3) Draw the signal constellation diagram.
(4) Draw the phasor diagram of the system.

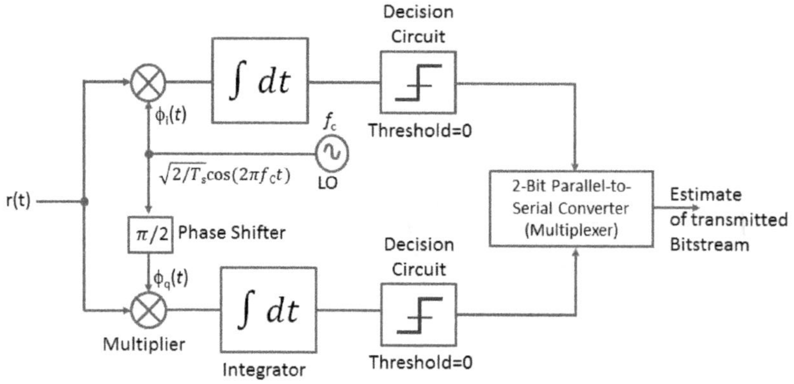

Fig. 4.39 Block diagram of a coherent QPSK demodulator.

Solution

The carrier waves used for the in-phase (I) and the quadrature (Q) channels of the moderator are, respectively

$$\phi_i(t) = \sqrt{\frac{2}{T_s}}\cos(\omega_c t) \text{ and}$$

$$\phi_q(t) = \sqrt{\frac{2}{T_s}}\sin(\omega_c t)$$

where $\sqrt{\frac{2}{T_s}}$ is a normalisation constant. The output of the summer in the modulator is

$$s(t) = i(t)\cos(\omega_c t) + q(t)\sin(\omega_c t)\,.$$

This is the QPSK waveform. Noting that bit 1 is bipolar NRZ line-encoded as +1 while bit 0 is encoded as −1, we obtain the waveforms for the four dibits as follows:

- Dibit 00: $\rightarrow i(t) = -1, q(t) = -1$, resulting in the waveform

$$s(t) = \sqrt{\frac{2}{T_s}}[-\cos(\omega_c t) - \sin(\omega_c t)]$$
$$= \frac{1}{\sqrt{T_s}}\sin(\omega_c t - 135^\circ)$$

- Dibit 01: $\rightarrow i(t) = 1, q(t) = -1$, resulting in the waveform

$$s(t) = \sqrt{\frac{2}{T_s}}[\cos(\omega_c t) - \sin(\omega_c t)]$$
$$= \frac{1}{\sqrt{T_s}}\sin(\omega_c t - 45^\circ)$$

- Dibit 10: $\rightarrow i(t) = -1, q(t) = 1$, resulting in the waveform

$$s(t) = \sqrt{\frac{2}{T_s}}[-\cos(\omega_c t) + \sin(\omega_c t)]$$
$$= \frac{1}{\sqrt{T_s}}\sin(\omega_c t + 135^\circ)$$

- Dibit 11: $\rightarrow i(t) = 1, q(t) = 1$, resulting in the waveform

$$s(t) = \sqrt{\frac{2}{T_s}}[\cos(\omega_c t) + \sin(\omega_c t)]$$
$$= \frac{1}{\sqrt{T_s}}\sin(\omega_c t + 45^\circ)$$

With the above data, we can compile the truth table, phasor diagram and constellation diagram in Fig. 4.40.

Example

In a QPSK system a dibit is encoded into the waveform $s(t) = \frac{1}{\sqrt{T_s}}\sin(w_c t - 135^\circ)$. Find the complex envelope of the waveform.

Solution

We can express the given waveform as

$$s(t) = \frac{1}{\sqrt{T_s}}\sin(w_c t - 135^\circ) = Im\left\{\frac{1}{\sqrt{T_s}}e^{j(w_c t - 135^\circ)}\right\}$$
$$= Im\left\{\frac{1}{\sqrt{T_s}e^{-j135^\circ}\,e^{jw_c t} = Im\{\tilde{s}(t)e^{jw_c t}\}}\right\} \quad (4.77)$$

Dibit		QPSK output phase
Quadrature (Q)	In-phase (I)	
0	0	-135°
0	1	-45°
1	0	+135°
1	1	+45°

(a) QPSK truth table

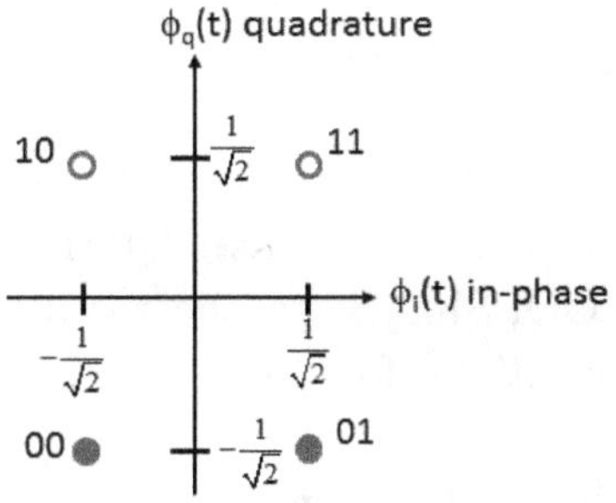

(b) QPSK signal constellation diagram

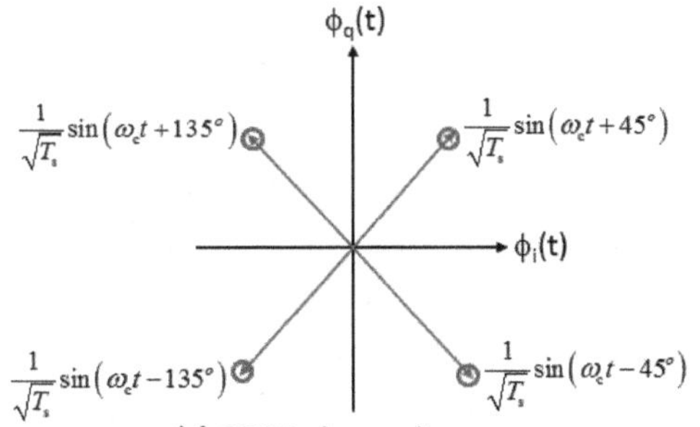

(c) QPSK phasor diagram

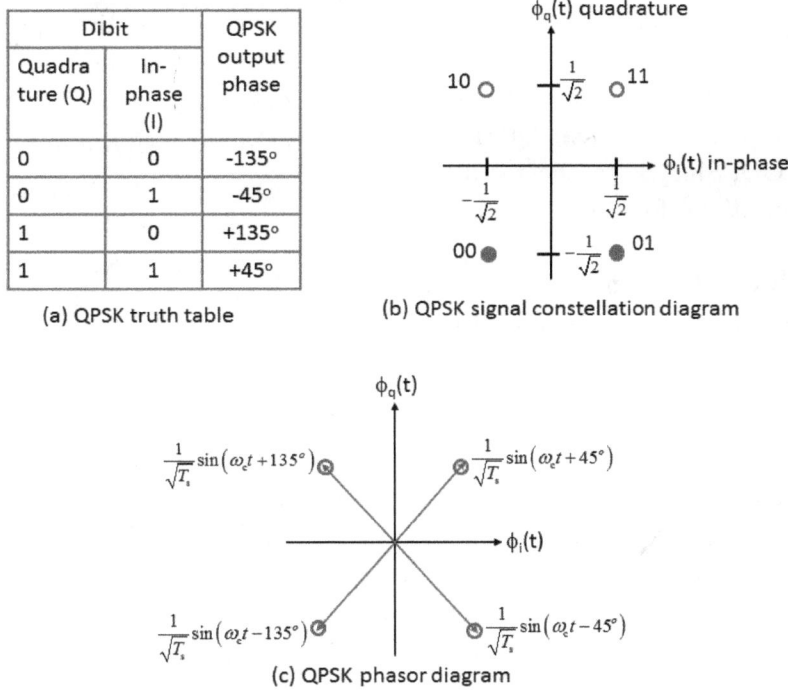

Fig. 4.40 QPSK features: truth table, phasor diagram and constellation diagram.

resulting in the complex envelope $\tilde{s}(t) = \frac{1}{\sqrt{T_s}}e^{-j135°} = -\frac{1}{\sqrt{2T_s}}(1+j)$. The complex envelope is the baseband modulating message expressed in complex form, while e^{jw_ct} is the complex form of the carrier signal.

4.4.3.2 *Differential QPSK*

As discussed earlier, the phase of the carrier signal contains the intelligence to be transmitted in all phase modulators. Therefore, any ambiguity in the phase creates a detection challenge. Phase modulation can use either changes in phase or absolute phases. The former is robust against phase ambiguities. Conventional PSK modulator uses a fixed phase for a set of data. For example, in QPSK, we can map the four dibits into the four phases: $00 \rightarrow 0°$, $01 \rightarrow 270°$, $10 \rightarrow 90°$ and $11 \rightarrow 180°$. However, the carrier's phase can change in the transmission systems between the transmitter and the receiver to cause phase error or ambiguity. Differential modulation, such as DQPSK, is used to solve this problem by changing the carrier's phase by a fixed amount depending on the data to be transmitted. The differential modulator/demodulator depends on the difference between successive phases, rather than using fixed phases. Unlike pure phase modulation, differential phase modulation can be detected non-coherently.

4.4.4 *Check Your Understanding*

Students are encouraged to check their understanding of the content of this section with the following questions prior to proceeding to the next section.

(1) Describe the meaning of the word *quaternary* as used in QPSK.

(2) How many distinct symbols are used in BPSK?

(3) How many bits does each symbol/carrier in QPSK carry?

(4) In a QPSK system a dibit is encoded into the waveform $s(t) = \frac{1}{\sqrt{T_s}}\sin(w_ct + 45°)$.

Find the complex envelope of the waveform.

(5) Which of these digital modulation schemes allow the largest phase transition: QPSK, $\frac{\pi}{4}$-QPSK and OQPSK? State the maximum phase transitions for each modulation schemes.

(6) How does BPSK differ from QPSK, in terms of bit error rate and bandwidth efficiency?

(7) How many bits are contained in one QPSK symbol?

(8) A communication system uses coherent BPSK modulation to achieve a BER of 10^{-6}. If we replace the modulator with coherent BFSK, how much more signal-to-noise ratio (i.e., E_b/N_0) is needed to maintain the error performance? *Hint.* Use Fig. 4.27.

(9) Explain the reason why the output of a digital modulator is effectively a double-sideband suppressed-carrier signal.

(10) Consider a coherent BPSK receiver operating at 10°C and receiving data at the rate 500 kbps. The system uses the waveforms $s_0(t) = -A_c \cos(2\pi f_c t)$ and $s_1(t) = A_c \cos(2\pi f_c t)$ to transmit bit 0 and bit 1, respectively, where $A_c = 2\,\text{mV}$. Assume that the energy per bit is normalised relative to a $1-\Omega$ resistive load. Also, assume that the BPSK signal is disturbed by only additive white Gaussian noise (AWGN) as it traverses the transmission medium.

(a) Compute the signal energy per bit, E_b.
(b) Compute the single-sided power spectral density of the AWGN, N_0.
(c) Compute the signal-to-noise ratio (i.e., bit energy to noise power spectral density), E_b/N_0.
(d) Compute the bit error rate (BER) of the coherent BPSK detector.
(e) Assuming that the detector operates continuously over a period of two days, find the expected number of bits received in error.

(11) Consider the block diagram of QPSK moderator shown in Fig. 4.38.

(a) Write the analytic formula of the waveform for each of the four dibits of the QSK.
(b) Construct the truth table.
(c) Draw the signal constellation diagram.
(d) Draw the phasor diagram of the system.

(12) Compare the bit error performance of BPSK, BFSK and BASK at the same E_b/N_0.

(13) Assume that a coherent 16-PSK system operates with an $E_b/N_0 = 15$ dB over an AWGN channel.

(a) If the symbols are equally likely, compute the bit error probability.
(b) If the above system operates continuously over 2 hours at the data rate of 20 kbps, how many bits will be expected to be detected erroneously?

(14) A given QPSK system operates at the bit error probability of 10^{-5} over an AWGN channel at the rate of 2 Mbps. The bandwidth of the system is 15 kHz.

(a)
(b) Compute the E_s/N_0 in decibels needed to achieve the given error performance.
(c) If Gray coding is used to map bits onto symbols, compute the E_b/N_0 in decibels needed to achieve the given error performance.

(15) Consider a 64-PSK system.

(a) Draw the signal constellation diagram.
(b) Make a table to map symbols to bit patterns and angles using Gray coding.

4.5 M-QAM

Quadrature amplitude modulation (QAM) is a two-dimensional digital modulation method comprising two orthogonal carriers which are independently amplitude modulated by the data to be transmitted. QAM can be viewed from different but equivalent angles, including:

- An extension to QPSK in the sense that in each two orthogonal carrier waves are independently amplitude modulated by bipolar NRZ line encoded bits. For this reason, M-QAM signal can be generated using a circuit similar to that of QPSK.
- A combination of amplitude and phase modulations. For this reason, QAM is also referred to as amplitude phase keying (APK).
- Two-dimensional amplitude shift keying using two orthogonal carrier signals. This is the reason why QAM is also referred to as quadrature amplitude shift keying (QASK), or quadrature multiplexing.

As said earlier about modulation schemes, M-QAM is a double-sideband suppressed-carrier (DSB-SC) signal. However, by modulating two dimensions of the carrier signal (i.e., amplitude and phase), M-QAM is much more bandwidth efficient than other modulation schemes. The bandwidth efficiency of QAM has made it popular in practical systems. For example, QAM is used in DOCSIS (Data Over Cable Service Interface Specification), 4G and 5G mobile cellular networks, and DVB-C (Digital Video Broadcasting-Cable).

The signal constellation diagram (aka signal space) of M-QAM is either a square or rectangular grid with the origin being the center. In the square grid the symbols are placed

$$\pm 1, \pm 3, \pm 5, \ldots, \pm(\log_2 M - 1) \qquad (4.78)$$

in each of the two dimensions of the signal space. Thus, each of the two coordinates in M-QAM rectangular signal space is a one-dimensional pulse amplitude modulation (PAM). We note, however, that 8-QAM is unpopular as 16-QAM is much more bandwidth efficient and it can achieve the BER of 8-QAM using just about 0.5 dB more E_b/N_0. The reason being the fact that 8-QAM has a rectangular (Fig. 4.41(a)) but not a square signal constellation, resulting in smaller distance between closest symbols.

The multi-level quadrature amplitude modulation is the most popular passband modulation scheme in mobile communications. The error performance of M-QAM is better than that of M-PSK for $M \geq 16$. The reason being that the distance between closest symbols in M-PSK is smaller than that in M-QAM. Therefore, M-QAM can use a smaller signal-to-noise ratio to achieve the data rate of M-PSK. For this reason, M-QAM is preferred to M-PSK. We have discussed 4-QAM (which is equivalent to QPSK) in previous sections. Other popular QAM schemes are 16-QAM, 64-QAM and 256-QAM. These are the M-QAM versions most popularly used in modern mobile networks. All of them are used in 4G (i.e., LTE-Advanced) systems, as well in the upcoming 5G New Radio. LTE uses adaptive modulation and coding (AMC) in order to match modulation scheme to the instantaneous condition of the wireless channel. Good wireless channels have high signal-to-noise ratio (SNR), while noisy channels have low SNR. The higher the SNR, the larger the value of M used to achieve a higher data rate. The Amplitude Phase Shift Keying (APSK) was investigated for adoption in 5G cellular networks. It has a desirable feature of low peak-to-average power ratio (PAPR), which is a main limitation in conventional OFDM.

The M-QAM symbols are of the form

$$s_m(t) = \mathtt{Re}\{v_m p(t) e^{j\omega_c t}\}, \; m = 1, 2, \ldots, M \qquad (4.79)$$

where $p(t)$ is the amplitude-shaping pulse (such as Nyquist pulses), and $v_m = v_{\mathtt{I},m} + jv_{\mathtt{Q},m}$ is the set of data symbols to be transmitted. The M-QAM waveform Eq. (4.79) can also be expressed in the form

$$\begin{aligned} s_m(t) &= v_{\mathtt{I},m} p(t)\cos(\omega_c t) - v_{\mathtt{Q},m} p(t)\sin(\omega_c t) \\ &= a_m p(t)\cos(\omega_c t + \varphi_m), \; m = 1, 2, \ldots, M \end{aligned} \qquad (4.80)$$

where $\varphi_m = \mathtt{atan}(v_{\mathtt{Q},m}/v_{\mathtt{I},m})$ and $a_m = \sqrt{v_{\mathtt{I},m}^2 + v_{\mathtt{Q},m}^2}$. By defining the orthogonal basis functions $\phi_{\mathtt{I}}(t) = p(t)\cos(\omega_c t)$ and $\phi_{\mathtt{Q}}(t) =$

$p(t)\sin(\omega_c t)$, we obtain the signal vectors of M-QAM as

$$\begin{aligned} s_m &= (a_m \cos\varphi_m, a_m \sin\varphi_m) \\ &= (\alpha_m, \beta_m), \ m = 1, 2, \ldots, M\,. \end{aligned} \tag{4.81}$$

This results in the transmitted M-QAM waveform

$$\begin{aligned} s(t) &= \sum_k \alpha_k p(t - kT_s)\cos(\omega_c t) \\ &\quad + \sum_k \beta_k p(t - kT_s)\sin(\omega_c t) \\ &= i(t)\cos(\omega_c t) + q(t)\sin(\omega_c t) \end{aligned} \tag{4.82}$$

where $(\alpha_k, \beta_k) \in \{s_m\}_{m=1,2,\ldots,M}$. The signal constellation of M-QAM is a square grid if $M = 2^{\texttt{even integer}}$ and a rectangular grid if $M = 2^{\texttt{odd integer}}$. The rectangular grid is a subset of a square grid, as illustrated in Fig. 4.41. The signal constellation diagrams of standard 4-QAM (aka QPSK), 8-QAM, 16-QAM, 32-QAM and 64-QAM are shown in Fig. 4.41. We can observe the differences.

The Euclidean distance between the closest symbols in the rectangular M-QAM signal space is

$$d_{\texttt{min}} = \sqrt{2E_p} \tag{4.83}$$

where E_p is the energy in the amplitude-shaping pulse $p(t)$.

Does QAM have any disadvantages? Sure. First, by modulating the amplitude of the carrier it cannot use nonlinear devices, such as high-power Class C amplifiers. Second, QAM is more susceptible to noise than constant-envelope modulation schemes. Owing to its non-constant amplitude we cannot use limiting amplifiers to limit noise to enhance signal-to-noise ratio.

Figure 4.42 shows a simplified demodulator for QAM. This modulator works under certain conditions. First, the two waves multiplexed to form QAM must remain in quadrature when they reach the demodulator. Second, the local oscillator circuit should generate a carrier which has exactly the frequency and phase of the carrier used by the

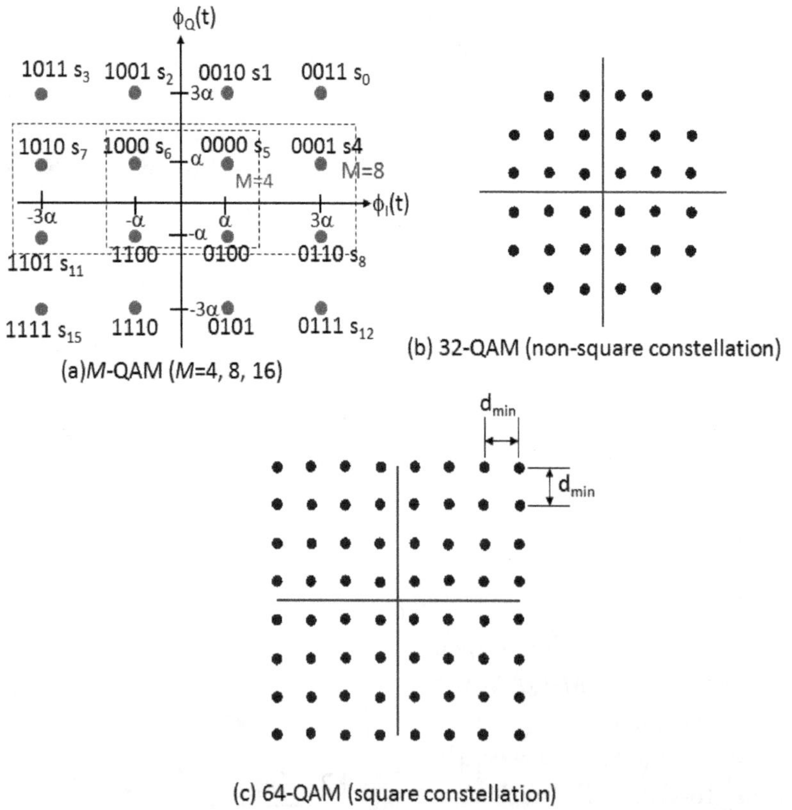

Fig. 4.41 M-QAM signal constellation diagrams with Gray coding.

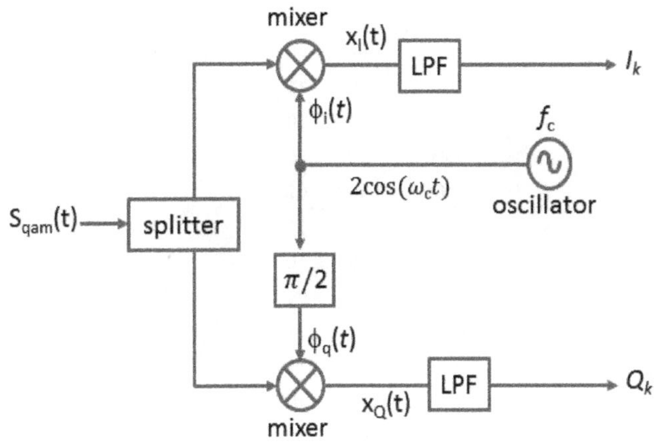

Fig. 4.42 A simplified QAM demodulator.

transmitter. A popular carrier recovery circuit is the phase locked loop (PLL). Neglecting noise, the QAM signal at the input to the demodulator has the form $s_{\mathtt{qam}}(t) = I_k \cos(\omega_c t) + Q_k \sin(\omega_c t)$. This produces the waveforms at the outputs of the two product modulators

$$\begin{aligned} x_I(t) &= 2s_{\mathtt{qam}}(t)\cos(\omega_c t) = 2I_k \cos^2(\omega_c t) \\ &\quad + 2Q_k \sin(\omega_c t)\cos(\omega_c t) \\ &= I_k + I_k \cos(2\omega_c t)] + Q_k \sin(2\omega_c t) \qquad (4.84) \end{aligned}$$

and

$$\begin{aligned} x_Q(t) &= 2s_{\mathtt{qam}}(t)\sin(\omega_c t) = 2I_k \sin(\omega_c t)\cos(\omega_c t) \\ &\quad + 2Q_k \sin^2(\omega_c t) \\ &= I_k \sin(2\omega_c t) - Q_k \cos(2\omega_c t) + Q_k. \qquad (4.85) \end{aligned}$$

Exciting the lowpass filters (LPFs) with $x_I(t)$ and $x_Q(t)$ produces the baseband data I_k and Q_k, respectively. This is because the LPF blocks its input components with frequency $2f_c$.

M-QAM Example

A message has been digitised into the bit stream 11110000101001010011 which we want to send to another destination using 16-QAM modulation. As usual, the leftmost bit is transmitted first. Assume that the 16-QAM modem is designed to use the bits to waveform mapping shown in Fig. 4.43.

(1) How many bits are transmitted in each symbol?
(2) How many symbols must be transmitted?
(3) Assuming a bit duration of $T_b = 1\,\mu\mathtt{s}$, compute the duration of the data transmission.
(4) Derive the mathematical formula for each of the symbols transmitted if no redundant bits are used.
(5) Plot a calibrated graph for the 16-QAM waveform for the entire message. Use the carrier frequency $f_c = \frac{2}{T_s}$ and the amplitude-shaping pulse $p(t) = 1$, $0 \le t \le T_s$ and zero otherwise, where T_s is the symbol duration.

Solution

We can observe from Fig. 4.43 that QAM is not a constant-envelope modulation scheme as the amplitudes of the symbols used do not have the same size. Note that each symbol in 16-QAM contains $m = \log_2(16) = 4$ bits, which are equally divided between the in-phase (I) and the quadrature (Q) components. Note that transmission and reception occur in symbols, but not bits. Thus, we can partition the given binary message into the five nibbles: $u_0 = 1111$, $u_1 = 0000$, $u_2 = 1010$, $u_3 = 0101$, $u_4 = 0011$. Each of these nibbles is used to moderate the two QAM carriers independently and sequentially. We can now answer the given questions.

(1) Each 16-QAM symbol transmits 4 bits.
(2) The number of symbols which must be transmitted is 5 as the message is made of 20 bits and 16-QAM transmits 4 bits at a time.
(3) The time taken to transmit the given message is

$$\begin{aligned} T_d &= (\mathtt{number\ of\ symbols}) \\ &\quad \times (\mathtt{symbol\ duration}) \\ &= 5 \times 4T_b = 20\,\mu\mathtt{s}\,. \end{aligned}$$

(4) The generalised formula for the waveforms of M-QAM symbols is given in Eq. (4.86) as

$$\begin{aligned} s_m(t) &= v_{\mathtt{I},m} p(t)\cos(\omega_c t) - v_{\mathtt{Q},m} p(t)\sin(\omega_c t) \\ &= a_m p(t)\cos(\omega_c t + \varphi_m),\ m = 1, 2, \ldots, M\,. \end{aligned}$$

We need to find the phase and the amplitude of the five 16-QAM symbols which are transmitted. The first step is to identify from Fig. 4.43 the symbol transmitted for each of the five nibbles. We observe the following mapping $u_0 \to s_{12}$, $u_1 \to s_6$, $u_2 \to s_1$, $u_3 \to s_{11}$ and $u_4 \to s_4$. The amplitude and phase for the first

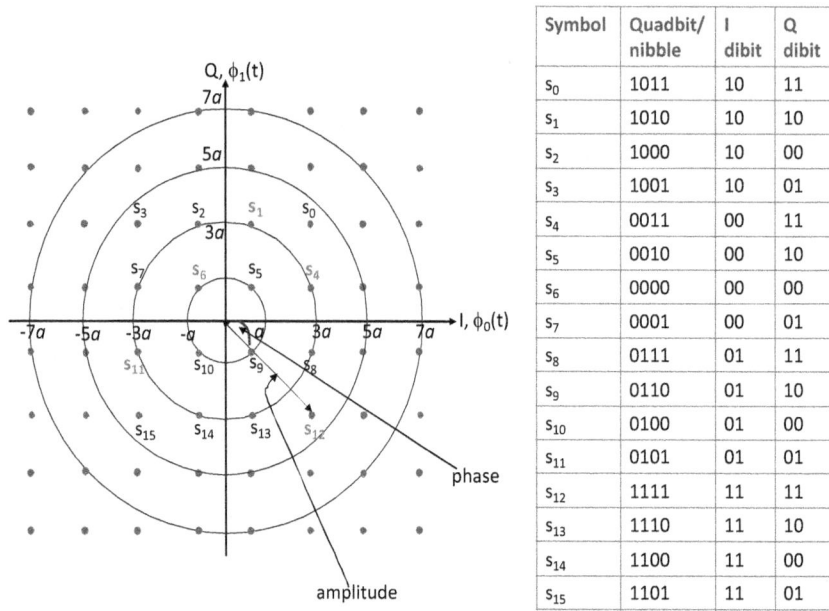

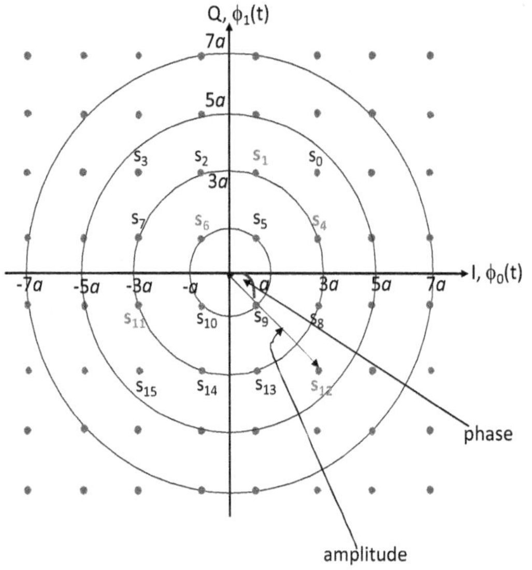

Symbol	Quadbit/ nibble	I dibit	Q dibit
s_0	1011	10	11
s_1	1010	10	10
s_2	1000	10	00
s_3	1001	10	01
s_4	0011	00	11
s_5	0010	00	10
s_6	0000	00	00
s_7	0001	00	01
s_8	0111	01	11
s_9	0110	01	10
s_{10}	0100	01	00
s_{11}	0101	01	01
s_{12}	1111	11	11
s_{13}	1110	11	10
s_{14}	1100	11	00
s_{15}	1101	11	01

Fig. 4.43 QAM signal constellation diagram and bit mapping to symbols.

Nibble	Quadbit/ nibble	I dibit	Q dibit	16-QAM Symbol	Amplitude	Phase
u_0	1111	11	11	s_{12}	$3\sqrt{2}a$	-¼π
u_1	0000	00	00	s_6	$\sqrt{2}a$	¾π
u_2	1010	10	10	s_1	$\sqrt{10}a$	atan(3)
u_3	0101	01	01	s_{11}	$\sqrt{10}a$	π + atan(1/3)
u_4	0011	00	11	s_4	$\sqrt{10}a$	atan(1/3)

Fig. 4.44 QAM example.

nibble $u_0 = 1111$ is $a_0 = \sqrt{(3a)^2 + (-3a)^2} = 3\sqrt{2}a$ and $\varphi_0 = \mathtt{atan}(-1) = -45°$, respectively. Thus, the waveform for the symbol transferred for the nibble $u_0 = 1111$ is $s_{12}(t) = 3\sqrt{2}a\cos(\omega_c t - 45°)$. Proceeding similarly, from Fig. 4.43, the amplitude and phase for the second nibble $u_1 = 0000$ is $a_1 = \sqrt{(-a)^2 + (a)^2} = \sqrt{2}a$ and $\varphi_1 = \mathtt{atan}(-1) = 135°$, respectively. Figure 4.44 gives the phases and amplitudes of all the five symbols. Plugging these into the formula Eq. (4.86) yields the waveform transmitted for each symbol.

(5) The 16-QAM waveform transmitted for the entire message is shown in Fig. 4.45. We can observe the varying amplitudes and phases which carry the relevant information in the waveform.

M-QAM PSD, Bandwidth and Bandwidth Efficiency

The PSD of M-QAM is the same as that of M-PSK. The passband power spectral density of M-QAM is thus

$$S_{\mathtt{QAM}} = \frac{E_{\mathtt{s}}}{2}\mathtt{sinc}^2[(f - f_{\mathtt{c}})T_{\mathtt{s}}] + \frac{E_{\mathtt{s}}}{2}\mathtt{sinc}^2[(f + f_{\mathtt{c}})T_{\mathtt{s}}]. \quad (4.86)$$

The bandwidth consumption of every modulation scheme depends on the type of amplitude-shaping

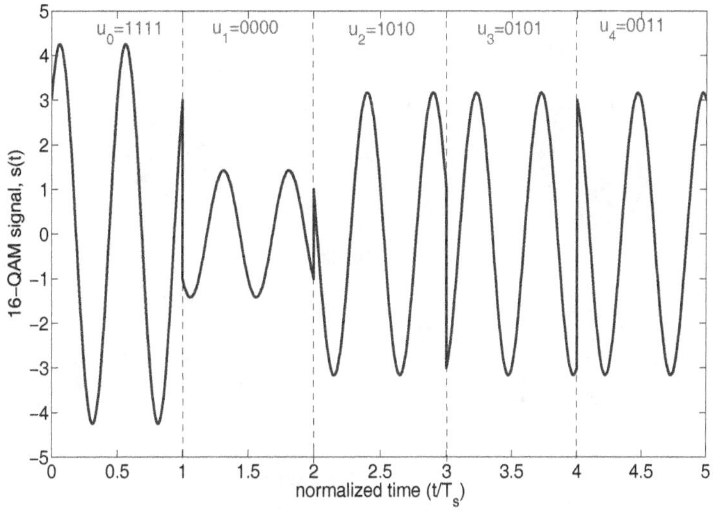

Fig. 4.45 Waveform for QAM example ($a = 1$).

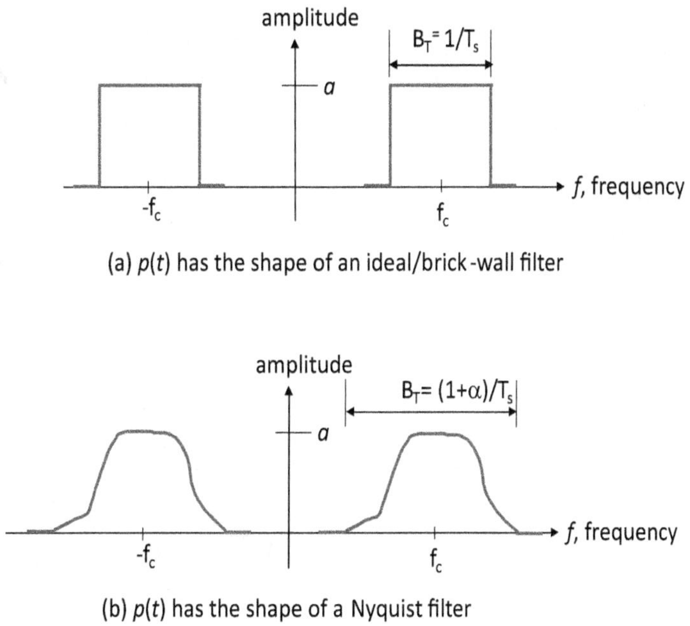

Fig. 4.46 Some amplitude-shaping pulses.

pulse $p(t)$ used. Figure 4.46 shows a rough shape of two of them. The bandwidth consumption is

$$\text{Ideal pulse in Fig. 4.46}(a): B_{\mathrm{T}} = \frac{1}{T_{\mathrm{s}}} = R_{\mathrm{s}}$$

$$\text{Nyquist pulse in Fig. 4.46}(b): B_{\mathrm{T}} = \frac{1+\alpha}{T_{\mathrm{s}}} = (1+\alpha)R_{\mathrm{s}} \tag{4.87}$$

where R_{s} is the symbol rate and α is the roll-off factor of the Nyquist pulse.

Example

Assume that we are transmitting data over a band-pass channel with center frequency at $f_{\mathrm{c}} = 1.8$ GHz and bandwidth 200 kHz. Compare the maximum data rate that the data transmission can occur for

64-QAM and 256-QAM for the two cases in which the amplitude-shaping pulse $p(t)$ used is:

(1) A root-raised cosine pulse with the excess bandwidth (aka roll-off factor) 0.4.
(2) An ideal rectangular pulse.

Solution

The bandwidth efficiency is $\eta = \frac{R_b}{B_T} = m$ bps/Hz if the amplitude-shaping filter $p(t)$ is an ideal rectangular pulse, but $\eta = \frac{R_b}{B_T} = \frac{m}{1+\alpha}$ bps/Hz for Nyquist pulses. This results in the respective maximum data rates $R_b = mB_T$ and $R_b = \frac{m}{1+\alpha}B_T$.

(1) For 64-QAM: $m = \log_2(64) = 6$. Therefore, the maximum data rates are $R_b = 1.6$ Mbps for ideal pulses and $R_b = 857.1$ kbps for Nyquist pulses.
(2) For 256-QAM: $m = \log_2(256) = 8$. Therefore, the maximum data rates are $R_b = 1.2$ Mbps for ideal pulses and $R_b = 1.1429$ Mbps for Nyquist pulses.

It is clear that the larger the number of symbols used, the higher the achievable rate. The high data rates are achievable only at high signal-to-noise ratios.

QAM Bit Error Performance

As discussed earlier, error performance is one of the most important features of modulation schemes. The bit error rates of QAM over AWGN channels using matched-filter detector are [Korn (1985)]

$$\begin{aligned} \texttt{4-QAM:} \quad & P_b = 0.5\texttt{erfc}(\sqrt{\gamma_b}) \\ \texttt{16-QAM:} \quad & P_b \approx \frac{3}{8}\texttt{erfc}(\sqrt{0.4\gamma_b}) \qquad (4.88) \\ \texttt{64-QAM:} \quad & P_b \approx \frac{7}{24}\texttt{erfc}(\sqrt{\gamma_b/7}) \end{aligned}$$

$$M\texttt{-QAM:} \quad P_b \approx \frac{2(M^{0.5}-1)}{M^{0.5}\log_2 M} \times \texttt{erfc}\left(\sqrt{\frac{3\gamma_b \log_2 M}{2(M-1)}}\right), \quad M > 64 \qquad (4.89)$$

where $\gamma_b = E_b/N_0$. Figure 4.47 compares the BER of M-QAM and M-PSK. It can be observed that

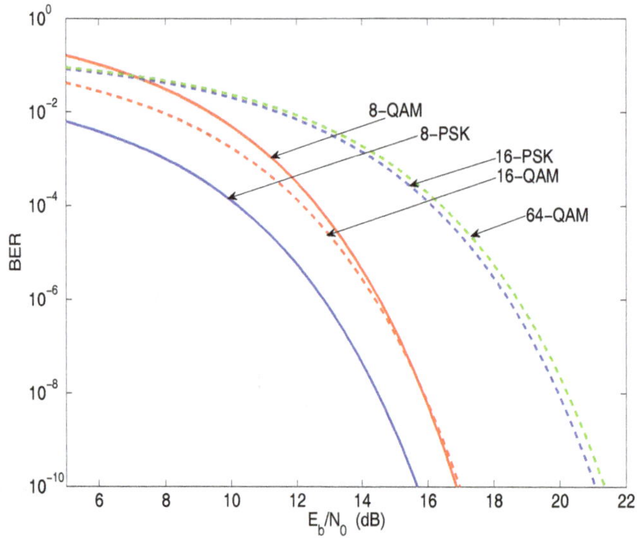

Fig. 4.47 Bit error probabilities of M-QAM and M-PSK.

M-QAM needs a smaller signal-to-noise ratio to achieve the same error rate of M-PSK. This is the reason why M-QAM is preferable to M-PSK at high data rates in which $M \geq 16$. In general, the relationship between the E_b/N_0 needed to achieve the same bit error performance is approximately

$$\gamma_b|_{M\texttt{-PSK}} = \frac{3\gamma_b|_{M\texttt{-QAM}}}{2(M-1)\sin^2(\pi/M)}, \quad M > 16\,. \qquad (4.90)$$

The reason for this is the larger minimum distance in QAM compared with PSK and APSK (amplitude phase shift keying), as illustrated in Fig. 4.48.

4.5.1 *Check Your Understanding*

Students are encouraged to check their understanding of the content of this section with the following questions prior to proceeding to the next section.

(1) Which feature(s) of the carrier wave carries/carry the information to be transferred in QAM?
(2) The detection of M-QAM signals require strict synchronisation between the transmitter and the receiver. Assume that the

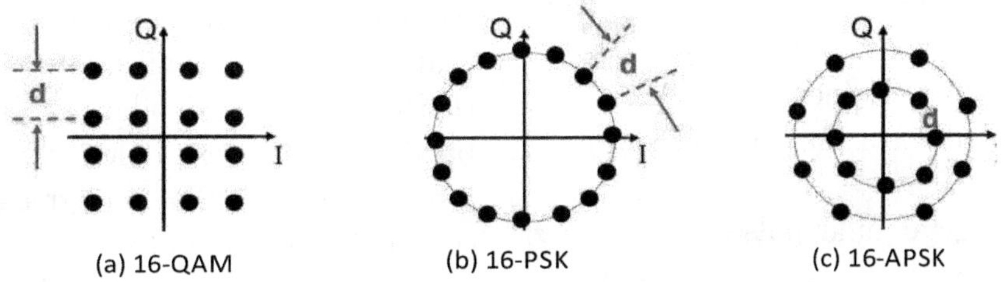

Fig. 4.48 Comparing signal spaces of 16-QAM, 16-PSK and 16-APSK.

M-QAM demodulator in Fig. 4.42 uses the carrier $\cos(2\omega_c t + \varphi)$ with the phase offset of φ from that of the transmitter. Show analytically that the in-phase component interferes with the quadrature component and vice versa. *Hint.* Derive the demodulated signals.

(3) Consider the M-QAM signal space shown in Fig. 4.41. Let us assume that the distance between the closest pair of points in the signal space is α.

 (a) Compile in a tabular form the Cartesian coordinates of the symbols in (i) 4-QAM, (ii) 8-QAM and (iii) 16-QAM.
 (b) Compute the average energy in each symbol of 16-QAM, assuming that they are equiprobable.

(4) Discuss the advantages of M-QAM over M-PSK.

(5) State two disadvantages of M-QAM compared with M-PSK.

(6) Figure 4.41(b) shows the signal constellation diagram of a modulation method.

 (a) Discuss concisely but precisely the type of modulation method that the constellation diagram represents.
 (b) If the data transfer rate of the system using this modulation method is 10 Mbps, compute the duration of each symbol emitted by the modulator.
 (c) There are two basic ways of assigning bit patterns to the symbols of a digital modulation method. (i) List them. (ii) State which of them is used in the constellation diagram of Fig. 4.41(b) and (iii) explain why that choice is the common practice.

(7) A message has been digitised into the bit stream 111000110011001 which we want to send to another destination using 8-QAM modulation. As usual, the leftmost bit is transmitted first. Assume that the 8-QAM modem is designed to use the bits to waveform mapping shown in Fig. 4.43.

 (a) How many bits are transmitted in each symbol?
 (b) How many symbols must be transmitted?
 (c) Design the 8-QAM modem by appropriately mapping between bits and symbols using Gray encoding. Then draw the signal constellation diagram.
 (d) Assuming a bit duration of $T_b = 1\,ns$, compute the duration of the data transmission.
 (e) Derive the mathematical formula for each of the symbols transmitted if no redundant bits are used.
 (f) Plot a calibrated graph for the 16-QAM waveform for the entire message. Use the carrier frequency $f_c = \frac{4}{T_s}$ and the amplitude-shaping pulse $p(t) = 1,\ 0 \leq t \leq T_s$ and zero otherwise, where T_s is the symbol duration.

(8) Assume that we are transmitting data over a bandpass channel with center frequency at $f_c = 2.1$ GHz and bandwidth 15 kHz. Compare the maximum data rate that the data transmission can occur for 64-QAM and 256-QAM for the two cases in which

the amplitude-shaping pulse $p(t)$ used is:

(a) A root-raised cosine pulse with the excess bandwidth (aka roll-off factor) 0.4.

(b) An ideal rectangular pulse.

(9) Compare the bit error rate of 16-QAM with that of 16-PSK.

(10) Study Fig. 4.47. Compare the E_b/N_0 in decibels needed by 8-PSK, 8-QAM, 16-PSK and 16-QAM to achieve the bit error probability of 10^{-6}. Comment on your results.

(11) Study Fig. 4.47.

(a) Compare the BER of 64-QAM with that of 16-PSK.

(b) From your study of the figure, give reason(s) why QAM is often preferable to PSK.

(c) State one advantage of M-PSK over M-QAM. That is, under what condition(s) is/are M-PSK a better choice than M-QAM?

4.6 TCM: Combined Coding and Modulation

Almost all (if not all) processing in communications systems, such as error control coding, modulation, equalisation and interleaving are used to achieve the same purpose. Which is to overcome the possible ill-effects of the communications channel on the information-bearing signals passing through it.

Traditionally, error control coding and modulation are performed and optimised independent of each other, and thus separately. Error control coding (abbreviated coding) is performed on bits in the digital domain, while digital modulation produces analog waveforms. In TCM, both error control and modulation are performed together and at the level of symbols or waveforms. An error control coding with rate $\frac{k}{n}$ prior to modulation requires the symbol transmission rate (aka modulation rate) to increase from R_s to $\frac{n}{k}R_s$ in order to maintain the same data transfer rate as an equivalent system without error control. The increased symbol rate increases the required signal transmission bandwidth. Worse of it all is in systems which use a fixed bandwidth and thus cannot accommodate the demand for a higher bandwidth. For example, telephony systems using a fixed circa 3.1 kHz for signals thus could not accommodate error control.

An alternative to symbol rate increase is to increase the number of symbols used by the modulation scheme on bandlimited systems. The best solution found in 1976 by G. Ungerboeck is the integration of baseband modulation with error control coding, resulting in the technique now referred to as **trellis-coded modulation** (TCM) [Ungerboeck (1982)], [Ungerboeck (1987a)], [Ungerboeck (1987b)]. TCM performs coding and modulation together, and it comprises a rate $r = \frac{k}{k+1}$ **trellis encoder** and an M-ary digital modulation scheme as a constellation mapper. Trellis encoder is also referred to as convolutional encoder. Another method similar to TCM is the block-coded modulation (BCM). We discuss only TCM. Another name for convolutional code is trellis code. Therefore, TCM is actually a combination of rate $\frac{n}{n+1}$ convolutional code and digital modulation.

4.6.1 *Some TCM Terminology*

Let us pause to refresh ourselves of some terminologies used in TCM.

(1) Two types of distances are popularly used in coding and modulation. These are:

(a) **Hamming distance**: the number of bit positions that a pair of sequences differ. For example, the Hamming distance between the two bit sequences 10110 and 11010 is two. Note that the sequences being compared must have the same length. The Hamming distance is used in hard-decision decoding/detection (HDD).

(b) **Euclidean distance** d: the straight line distance between any two points in the Euclidean space, where d is the amplitude of the signal. If point p_0 at (x_0, y_0) and p_1 at (x_1, y_1) are two points in the two-dimensional Euclidean space, then the Euclidean distance between them is $d =$

$\sqrt{(x_0 - x_1)^2 + (y_0 - y_1)^2}$. The Euclidean distance (ED) is used in soft-decision decoding/detection (SDD). **TCM is based on the Euclidean distance between signal sequences.**

(2) Minimum squared Euclidean distance (MSED): designated $d^2_{\texttt{min}}$, is the minimum of all SEDs of a given constellation.

(3) Free distance ($d^2_{\texttt{free}}$): the sum of the SEDs from a reference sequence, which is usually the all-zero sequence.

(4) Squared Euclidean distance (SED): d^2. The coding gain of a TCM is computed in reference to an uncoded digital modulation scheme which has the same spectral efficiency as the TCM. For example, the spectral efficiency of uncoded BPSK is unity which is the same as TCM based on QPSK and rate $\frac{1}{2}$ convolutional code. The coding gain measures the efficiency of a TCM, and it is the main motivation behind TCM.

(5) **Coding gain**: the difference in signal-to-noise ratio (SNR) between an uncoded system and a coded system having the same bit error rate and the same data rate. A higher SNR is needed for uncoded system than coded system in order to achieve the same bit error rate at a given data rate. The coding gain is expressed as

$$\eta = SNR_{\texttt{uncoded}} - SNR_{\texttt{coded}} \,. \qquad (4.91)$$

We refer to the coding gain at a high SNR as **asymptotic coding gain**, which is

$$\eta_\infty = 10\log_{10}\frac{(d^2_{\texttt{free}}/E_{\texttt{s}})_{\texttt{coded}}}{(d^2_{\texttt{free}}/E_{\texttt{s}})_{\texttt{uncoded}}}$$

$$= 10\log_{10}\left(\frac{d^2_{\texttt{free}}|_{\texttt{TCM}}}{d^2_{\texttt{free}}|_{\texttt{uncoded}}}\right) \qquad (4.92)$$

where E_s is the average signal energy (Fig. 4.49). For high signal-to-noise ratio, Eq. (4.92) is equivalent to Eq. (4.91). TCM can achieve a coding gain as much as 6 dB without:

(a) Reducing the effective data transmission rate.

(b) An increase in symbol rate and hence the transmission bandwidth.

(c) Changing the power spectrum of the signal, hence keeping bandwidth consumption the same.

How does TCM achieve the coding gain? It uses a signal constellation diagram with more points than used if there were no coding.

Example

For each of the three signal constellations shown in Fig. 4.49, as a function of the signal energy that $E_{\texttt{s}}$, compute:

(1) All the SEDs.
(2) The minimum SED (MSED) or $d^2_{\texttt{min}}$.
(3) The sum of the SEDs (SSED) or $d^2_{\texttt{free}}$.

Solution

Note that all the three Euclidean distances needed are invariant of the reference sequence used for a given signal constellation. Without loss of generality, and by convention, we use the all-zero sequence as the reference sequence. In the following, let d_{ij} be the Euclidean distance between two points i and j in the Euclidean space.

(1) Squared Euclidean distances (SEDs):
   (a) BPSK has only two points in its constellation diagram. Therefore, there is only one SED, which is $d^2_{01} = 4E_{\texttt{s}}$.
   (b) QPSK has 4 points in its constellation. Therefore, there are three SEDs in reference to the all-zero sequence (or symbol zero). These are $d^2_{01} = d^2_{03} = 2E_{\texttt{s}}$ and $d^2_{02} = 4E_{\texttt{s}}$.
   (c) 8-PSK has 8 symbols and thus 7 SEDs from the all-zero sequence. These are $d^2_{01} = d^2_{07} = (2-\sqrt{2})E_{\texttt{s}} \approx 0.586E_{\texttt{s}}$, $d^2_{02} = d^2_{06} = 2E_{\texttt{s}}$, $d^2_{05} = d^2_{03} = (2+\sqrt{2})E_{\texttt{s}} \approx 3.414E_{\texttt{s}}$ and $d^2_{04} = 4E_{\texttt{s}}$.

(2) The minimum SEDs (MSEDs) or $d^2_{\texttt{min}}$ are:
   (a) BPSK: $d^2_{\texttt{min}} = 4E_{\texttt{s}}$.
   (b) QPSK: $d^2_{\texttt{min}} = 2E_{\texttt{s}}$.
   (c) 8-PSK: $d^2_{\texttt{min}} = (2-\sqrt{2})E_{\texttt{s}} \approx 0.586E_{\texttt{s}}$.

(3) The sum of the SEDs (SSED) in reference to the all-zero sequence or $d^2_{\texttt{free}}$ are:
   (a) BPSK: $d^2_{\texttt{free}} = d^2_{\texttt{min}} = 4E_{\texttt{s}}$.
   (b) QPSK: $d^2_{\texttt{free}} = 2 \times 2E_{\texttt{s}} + 4E_{\texttt{s}} = 8E_{\texttt{s}}$.

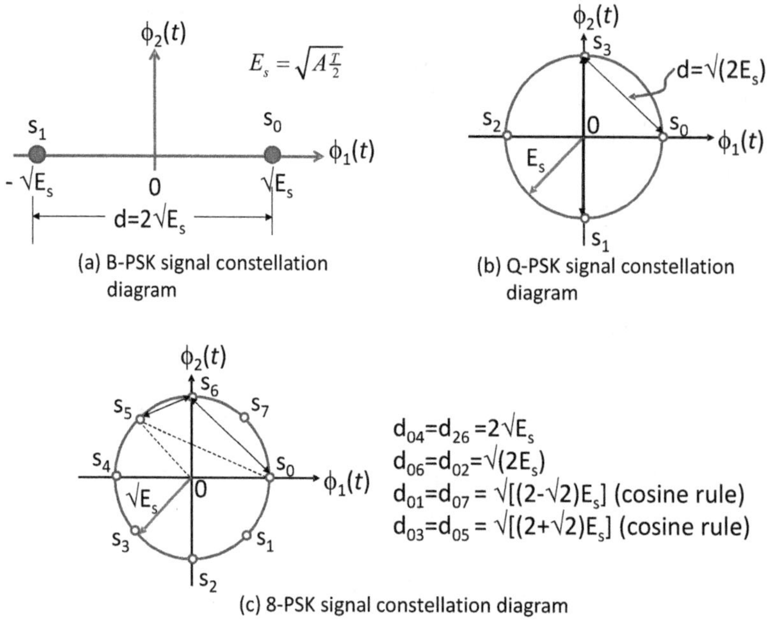

Fig. 4.49 Euclidean distances in BPSK, QPSK and 8-PSK signal constellation.

(c) 8-PSK: $d^2_{\texttt{free}} = 2\times(2-\sqrt{2})E_{\text{s}} + 2\times(2+\sqrt{2})E_{\text{s}} + 2\times 2E_{\text{s}} + 4E_{\text{s}} = 16E_{\text{s}}$.

4.6.2 *Bandwidth Efficiency*

A fundamental basis of TCM is bandwidth efficiency. The minimum bandwidth required to support a symbol rate of R_{s} symbols/sec is $B = R_{\text{s}}$ Hz. This minimum bandwidth does not depend on how many bits that we transfer within a symbol duration. Can we therefore transmit any number of bits in a symbol period? The restriction is the Euclidean distance between the symbols. The higher the number of bits transferred per symbol, the lower the smallest Euclidean distance. The spectral efficiency for non-error coded M-ary digital modulation scheme is

$$\eta_{\texttt{uncoded}} = \frac{\texttt{data rate } R_{\text{b}}}{\texttt{bandwidth}B} = \frac{R_{\text{b}}}{R_{\text{b}}/m} = m \texttt{ bps/Hz} \tag{4.93}$$

whereas the spectral efficiency for an M-ary digital modulation scheme preceded by a traditional rate $r = \frac{k}{n}$ error control performed digitally over bits is

$$\eta_{\texttt{coded}} = \frac{\texttt{data rate } R_{\text{b}}}{\texttt{bandwidth}B} = \frac{R_{\text{b}}}{R_{\text{b}}/(rm)} = r\cdot m \texttt{ bps/Hz} = \frac{k}{n}m \texttt{ bps/Hz}. \tag{4.94}$$

This means that the bandwidth efficient of the uncoded system is higher than that of the error-coded system by the factor $\frac{n}{k}$. The only problem is that the uncoded system may not be able to transmit any bits successfully if the transmission medium is very noisy.

Example

We seek to compare the bandwidth efficiency of coded and uncoded digital modulation schemes. Let us start with data transmission with BPSK at 64 kbps. An uncoded BPSK transmits in each symbol. Therefore, the symbol rate is $R_{\text{s}} = R_{\text{b}}/m = R_{\text{b}}$, resulting in the bandwidth efficiency $\eta_{\texttt{coded}} = 1$ bps/Hz. Now, if we are using BPSK with rate $\frac{1}{2}$ convolutional encoder, then we have the three choices:

(1) Transmit data at $\frac{1}{2}$ of the encoded data rate, i.e., $R_{\text{b}} = \frac{1}{2}64 = 32$ bps. The modulation rate $R_{\text{s}} = R_{\text{b}}/(mr) = 2.0R_{\text{b}}$, resulting in $\eta_{\texttt{coded}} = 0.5$ bps/Hz.
(2) Maintain the original data rate. This requires us to transmit at the symbol rate 128 BPSK kilosymbols per second, doubling the bandwidth needed and reducing the bandwidth efficient to 0.5 bps/Hz.

(3) Maintain the original data rate of $R_\mathtt{b} = 64$ bps, but use a higher-level modulation scheme (e.g. QPSK) to transmit 2 bits in each symbol. Thus, the effective bit rate is $R_\mathtt{b} = 128$ bps. Therefore, the symbol rate is 64 symbols per second for QPSK, but 128 symbols per second for BPSK, resulting in $\eta_{\mathtt{coded}} = \frac{R_\mathtt{b}}{R_\mathtt{b}/(m\cdot r)}) = 1$ bps/Hz. The third option is the rationale behind TCM.

4.6.3 *Set Partitioning*

The number of distinct symbols used by a digital modulation scheme make up a set. The symbols are represented by vectors or points in the signal constellation diagram of the digital modulation scheme. For example, the set of symbols in

- BPSK is $\mathcal{S}_2 = \{s_0, s_1\}$.
- QPSK is $\mathcal{S}_4 = \{s_0, s_1, s_2, s_3\}$.
- 8-PSK is $\mathcal{S}_8 = \{s_0, s_1, s_2, s_3, s_3, s_5, s_6, s_7\}$.

Each of these sets can be divided (or partitioned) into disjointed subsets of equal size, in such a manner than the SED between symbols in a subset is maximised. It should be remembered that every symbol appears in a subset and only in one subset as the subsets are disjointed. Also, neighbouring symbols never appear in the same subset as we seek to maximise SED.

Set partitioning as proposed by Ungerboeck for TCM is illustrated in Fig. 4.50 for 8-PSK and in Fig. 4.51 for 16-QAM. The computation of the Euclidean distances in Fig. 4.51 assumes that the average value of the squared amplitudes of the symbols is unity. We start from the original set of symbols. We partition the set successively into two disjointed subsets at each stage until only one symbol remains in each subset. We can observe from Fig. 4.50 that, the SED increases from top to bottom.

4.6.4 *TCM*

TCM uses a digital modulation scheme with 2^{m+1} symbols so as to avoid bandwidth increase by the redundant bit used for error control coding. As we learned from the digital modulation schemes, the larger the symbol set used, the smaller the Euclidean distance (ED) between the symbols become. TCM compensates for this via the free ED between valid code sequences by mapping symbols to state transitions and avoiding some state transitions. As can be understood from the above discussions, in order to specify TCM we need only a set of waveforms from a suitably-chosen digital modulation method and a trellis diagram whose state transitions are encoded to the waveforms. For this reason, TCM is a waveform encoding scheme. Figure 4.52(a) and (b) compares traditional coding and modulation to TCM. The trellis diagram of TCM must fulfill the following conditions:

(1) All symbols or waveforms occur with equal frequency.
(2) There must be 2^m possible transitions from each state to the next state in the trellis diagram, where $m = \log_2 M$ is the number bits carried in each symbol of the M-ary digital modulator.
(3) The assignment of waveforms to state transitions of the trellis must be done in such a manner that:
    (a) Waveforms from the same subset are assigned to the state transitions merging into, or originating from, the same state. With respect to the 8-PSK example of Fig. 4.50 waveforms are selected from either the subset G_{11} or G_{12}, but never a mixture of them.
    (b) Multiple transitions can occur between a pair of states. Any parallel transitions (i.e., multiple transitions originating from the same state and joining the same state) must be assigned waveforms from the same subset. With respect to the 8-PSK example of Fig. 4.50 waveforms are selected from either the subset G_{21} or G_{22} or G_{23} or G_{24}, but never a mixture of them.

As mentioned above, a TCM comprises rate $r = \frac{m}{m+1}$ convolutional encoder and a digital modulation scheme, which can be M-ary PSK, M-ary QAM or M-ary PAM. The convolutional encoder has memory and therefore it is a finite-state

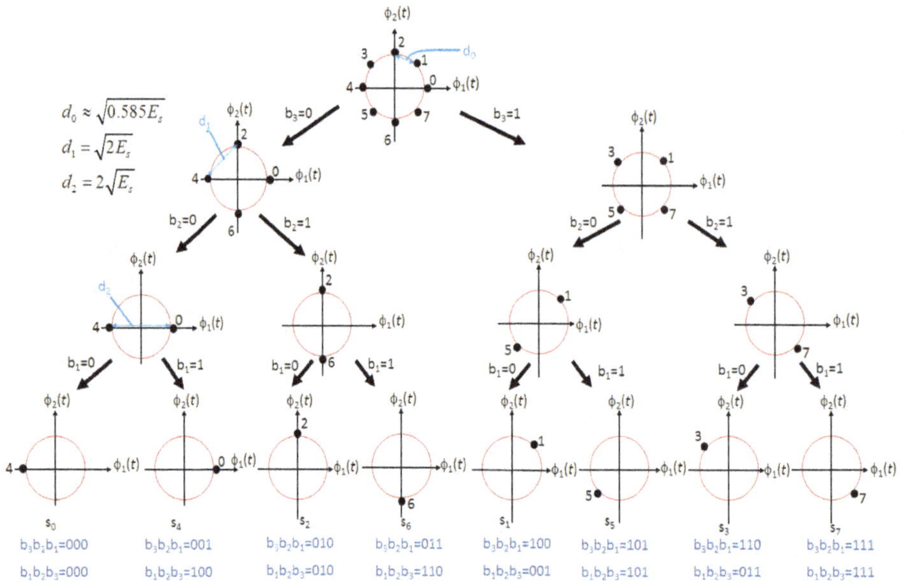

Fig. 4.50 Set partitioning of 8-PSK with bit labeling by Ungerboeck.

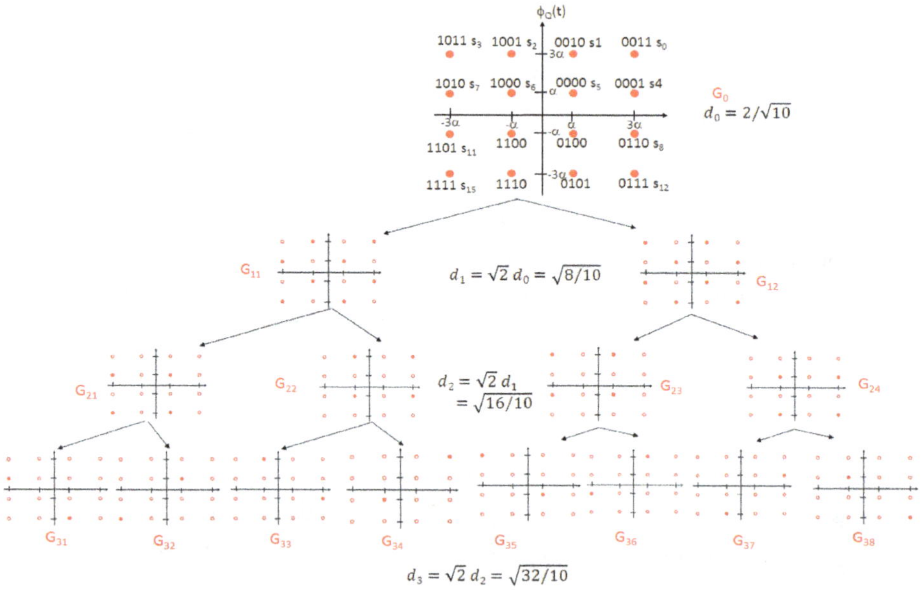

Fig. 4.51 Set partitioning of 16-QAM symbols by Ungerboeck [d_{ij} is the Euclidean distance between symbols in each subset].

machine (FSM). Its behaviour can be described by the trellis diagram. The convolutional encoder always adds only a single redundant bit to its input. It therefore converts an 2^m-ary modulation scheme to an equivalent modulation scheme with twice the number of symbols. For example, BPSK is mapped into QPSK using rate 1/2 trellis encoder, QPSK is mapped into 8-PSK using rate 2/3 trellis encoder, 8-PSK is mapped into 16-PSK using rate 3/4 trellis encoder, and 16-QAM is mapped into 32-QAM using rate 4/5 trellis encoder. The modulation scheme maps bit blocks into modulated symbols. The mapping is implemented using set partitioning (see Section 4.6.3) in order to maximise the SED.

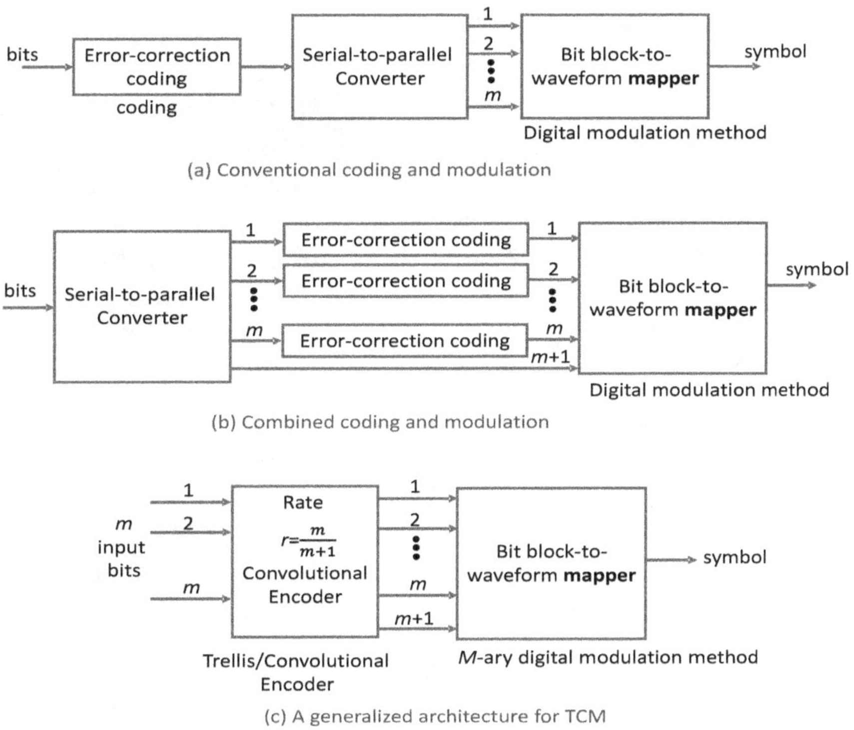

Fig. 4.52 Illustrating the principle of trellis-coded modulation.

For the example shown in Fig. 4.52, b_3 and b_2 are the error-coded bits, while bit b_1 is uncoded. The Euclidean distance between symbols increase as we move down the tree. Therefore, it is much more probable for the TCM decoded to confuse a pair of symbols in the top two branches than the bottom branch. For this reason, the two encoded bits are assigned to the top two branches, while the uncoded bit is assigned to the bottom branch. This means that we use the two coded bits to select one subset of the four subsets. We then use the uncoded bit to pick one of the two symbols in the chosen subset. Note that two symbols making up a subset differ in only one bit position, and they are separated by 180°. This separation between them results in a small chance of confusing one for the other.

TCM assigns symbols or waveforms from the extended set 2^{m+1} to transitions in the trellis diagram in such a manner that the free Euclidean distance between waveforms which can most probably be confused is maximised. This is the reason why the state transitions in the trellis diagram of TCM are labelled with waveforms, contrasting it from the trellis used in conventional convolutional encoder which is labelled with bits.

4.6.5 *Check Your Understanding*

Students are encouraged to check their understanding of the content of this section with the following questions prior to proceeding to the next section.

(1) What other name is used to describe a convolutional encoder?
(2) State the reason why TCM is considered a waveform encoding scheme.
(3) Describe how the trellis diagrams used in conventional convolutional code differs from that used in TCM.
(4) Compute the free distance for 16-QAM.
(5) Describe motivation behind TCM.
(6) Construct a TCM code based on the convolutional encoder shown in Fig. 4.53.

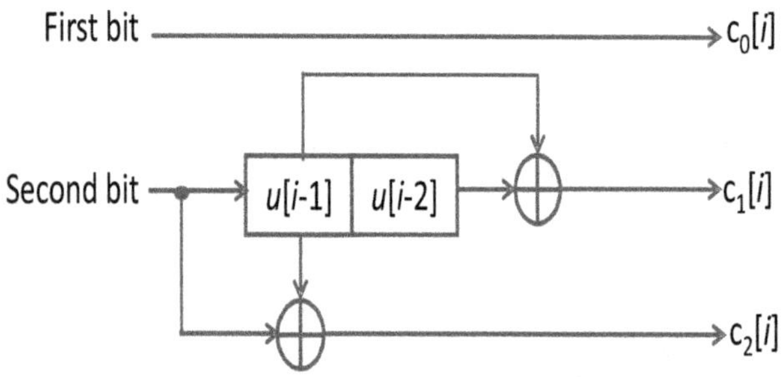

Fig. 4.53 Convolutional encoder for TCM.

(a) Which digital modulation scheme should we use?
(b) Construct the state table of the encoder.
(c) Using the state table, construct the trellis diagram of the encoder.
(d) Map symbols or waveforms to the state transitions in the trellis so as to maximise the SED.

4.7 Review of Important Terms and Acronyms

AMC	analog distance	analog modulation
asymptotic coding gain	asynchronous detection	AWGN
bandwidth efficiency	baseband channel	baseband modulation
basis functions	Bessel functions	BFSK
bit error probability	bit error rate (BER)	BPF
broadband channel	broadband modulation	broadband network
carrier	carrier signal	carrier wave
demodulator	coding gain	coherent detection
constant-envelope modulation	constellation diagram	constellation mapping
continuous wave modulation	Costas loop	data rate
digital distance	digital modulation	dibit
Euclidean distance	excess bandwidth	filter
free distance (d_{free})	frequency	frequency sensitivity
GMSK	Gram-Schmidt algorithm	Hamming distance
in-phase	linear modulation scheme	frequency translation
lowpass filter (LPF)	M-ASK	matched filter
M-FSK	mixer	modulation
modulation index	modulator	M-PSK
M-QAM	MSK	nibble
non-coherent detection	nonlinear modulation scheme	Nyquist pulse
on-off keying	orthonormal basis	oscillator
phase transition	passband channel	passband modulation
phase-locked loop (PLL)	power efficiency	power spectral density
spectral regrowth	pulse modulation	Q-function
QPSK/OQPSK/$\frac{p}{4}$-QPSK	quadbit	quadrature
quaternary	rest frequency	roll-off factor
root-raised cosine pulse	signal power	sequence coding
set partitioning	sideband	sidelobe
signal constellation diagram	signal energy	signal space
synchronous detection	trellis diagram	trellis-coded modulation (TCM)

4.8 Supplementary Reading

(1) Alister Burr, *Modulation and Coding for Wireless Communications*, Prentice Hall PTR, 2001.
(2) Harry L. Van Trees, *Detection, Estimation, and Modulation Theory. Part I: Detection, Estimation, and Linear Modulation Theory (Part 1)*, John Wiley, 1968.
(3) Fuqin Xiong, *Digital Modulation Techniques*, Artech House Publishers, 2000.
(4) G. M. Miller, *Modern Electronic Communication*, 3rd ed., Prentice-Hall, 1988.
(5) J. D. Gibson, *Principles of Digital and Analog Communications*, Macmillan Publishing, 1990.
(6) H. Zarrinkoub, *Understanding LTE with MATLAB: From Mathematical Modeling to Simulation and Prototyping*, Chichester: Wiley, 2014.
(7) Travis F Collins and Pu Di, *Software-Defined Radio for Engineers*, Artech House Publishers, 2018.

Chapter 5

Electromagnetic Wave Propagation

Chapter's Learning Outcomes[1]

The interest in the transmission of information over wireless media has been on the increase, and the trend will continue. Wireless signal transmission cannot occur without electromagnetic (EM) waves. An antenna is used to convert an electrical signal into an EM wave. EM waves are the type of waves with three components (electric field, magnetic field and direction of wave travel) and which can travel through both media and in vacuum.

This chapter treats the physical principles behind wireless signal propagation. In particular, we study:

(1) The importance of EM waves in wireless communications.
(2) The frequency spectrum.
(3) The components of an EM wave and polarisation.
(4) Maxwell's equations and electromagnetism.
(5) Types of EM wave propagation environments, including atmospheric divisions.
(6) EM wave propagation modes (viz., ground waves and sky waves).
(7) EM wave propagation mechanisms (e.g., reflection, refraction, diffraction, scattering, absorption, interference and transmission).
(8) Attenuation of EM waves, e.g., by rain.
(9) EM wave disturbances and losses, e.g., multipath propagation and free-space loss.
(10) Probabality distributions used to model EM wave propagation.

5.1 Introduction

Wireless communications occur through a special form of waves, called electromagnetic (EM) waves, which travel through the Earth's atmosphere. This means that the proper understanding of EM waves and its propagation behaviour are very vital for communications specialists. A *wave* is a disturbance in a medium which causes the transfer of energy from one point to another without a permanent displacement of any of the particles in the medium. Examples of waves are EM radiation, sound waves, sea waves and gravitational radiation.

What is an EM wave? A wave which comprises both magnetic and electric fields which travel through the atmosphere at right angles to each other and to the direction of travel of the wave is called **EM wave**. Examples of EM waves are radio waves, infrared light waves and visible light waves. Radio communications usually occur in the frequency range 3 kHz to 300 GHz. For this reason, this band is referred to as *radio frequency* (RF), and EM waves generated at such frequencies are called **radio waves**. History records that there was an unhealthy debate about the identity of EM waves over a period of time. One cohort identified it as a magnetic entity, while the other cohort thought it was an electric entity. This argument supposedly continued until it was established that it is the same entity which possesses both magnetic and electric behaviours.

A moving EM wave is produced if an alternating current (a.c.) flows through a conductor. The

[1]Humans seek the unknowns through the five scientific principles: time, force, energy, space, and matter — Hubert Spencer, 1820.

production of the wave is much more efficient if a specialised conductor called antenna is used. The antenna works as a transducer as it converts signals between two energy forms, namely, electrical energy and EM radiation energy. A wireless communications' link requires two electronic devices, each equipped with an antenna. The space between the transmitting antenna and the receiving antenna is called **channel** or wireless channel. Inside the an electronic device, such as a smart phone, the signal is an electrical signal, but the signal travels along the channel as EM waves.

A (radio) wave which is sufficiently far away from its generating antenna appears to be flat. Such a wave is referred to as **plane wave**. Our discussions on EM waves focus on plane waves.

5.1.1 *The Frequency Spectrum*

The ordering of waves with respect to their wavelengths or frequencies from 0 Hz to infinity is referred to as the **frequency spectrum** or sometimes **electromagnetic spectrum**. Each country has an agency for radio frequency management. There are also international or world-wide organisations for radio management. These international and national radio frequency management agencies have divided the frequency spectrum into ranges called *frequency bands*. Each band has been allocated for specific purposes and tightly controlled for adherence. Figure 5.1 presents a snapshot of the EM spectrum and their sample applications. Electromagnetic waves with frequencies in the range 0.3 GHz to 300 GHz (equivalently 1 m to 1 mm) are classified as **microwaves**, but the range currently used for practical communications is 450 MHz to 40 GHz. The higher the operating frequency, the higher the potential bandwidth, but the higher the attenuation of the waves. Furthermore, the higher the required data rates or capacity, the higher should be the bandwidth used. The entire frequency spectrum comprises EM waves, light waves and sound waves.

Owing to the depleting, expensive and congested radio frequency, as well as the demand for very high data rates, future mobile cellular systems seek to transmit signals at frequencies beyond 4 GHz and up to as high as 100 GHz. By noting that

Wavelength Band	Frequency Band	Band Name	Wave Propagation Mode	Sample Services Supported
100 – 10 km	3-30 kHz	Very Low Frequency (VLF)	Ground wave	Maritime navigational instruments
10 – 1 km	30 – 300 kHz	Low Frequency (LF)	Ground wave	Radio navigation aids
1 km – 100 m	300 kHz – 3 MHz	Medium Frequency (MF)	Ground wave/ sky wave	AM radio
100 – 10 m	3 – 30 MHz	High Frequency (HF)	Sky wave	Radio telephony, short-wave radio
10 – 1 m	30 – 300 MHz	Very High Frequency (VHF)	Space wave	FM radio, radio navigation aids, VHF TV
1 m – 10 cm	0.3 – 3 GHz	Ultra High Frequency (UHF)	Space wave	Mobile communications, GPS, UHF TV
10 – 1 cm	3 – 30 GHz	Super High Frequency (SHF)	Space wave	Microwave communications, satellite
1 cm – 1 mm	30 – 300 GHz	Extremely High Frequency (EHF)	Space wave	Radar systems, radio astronomy, 5G wireless systems

Fig. 5.1 A snapshot of the electromagnetic spectrum.

$c = f\lambda$ and the speed of waves remaining constant at about the speed of light in vacuum $c = 299.7925 \times 10^6$ m/s $\approx 3 \cdot 10^8$ m/s, the wavelengths of the signals fall in the range 5 mm to 75 mm. This is the reason why we refer to such signals as **millimeter waves**. For example, the 5G mobile cellular systems poised to be operational around the year 2020 is exploring millimeter waves (mmWaves) at frequencies around 24–86 GHz. Unfortunately, high-frequency waves such as microwaves cannot easily pass through obstacles. Also, EM waves spread out between the transmitting and receiving antennas. Part of the stray signals hit obstacles (if there are any) and get reflected to the receiver, and can be out-of-phase with the direct path to cancel each other.

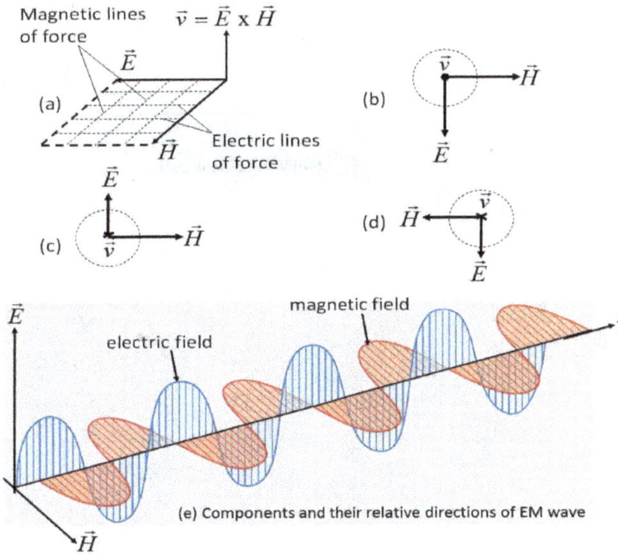

Fig. 5.2 Relative directions of the EM wave components.

5.1.2 *EM Wave Components and Polarisation*

EM waves are special form of waves. They are made of electrical and magnetic fields, and arise if an electrical current flows through a metallic conductor. The behaviour of EM waves is analytically described by Maxwell's equations for electromagnetism. The movement of an EM wave is made of three components, all of which are vectors: the direction of travel of the wave fronts ($\vec{v}$), the direction of the magnetic field component ($\vec{H}$) and the direction of the electric field component ($\vec{E}$). Waves can be categorised into transverse waves and longitudinal waves. An example of transverse waves is all EM waves (e.g. light waves and radio waves), while sound wave is an example of longitudinal waves. The particles in the EM wave propagation medium oscillate in the same direction as the direction of the wave in **longitudinal waves**, while in the perpendicular direction in **transverse waves**.

These three components (i.e. $\vec{E}$, $\vec{H}$ and $\vec{v}$) in an EM wave are usually mutually perpendicular to each other (Fig. 5.2). However, this rule that electric and magnetic field components are always perpendicular to the direction of propagation does not hold in waveguides. We can find their relative directions via **the right-hand rule** in which the thumb is the direction of the wave, the index finger is the direction of the electric-field vector and the remaining three fingers point towards the direction of the magnetic-field vector. In Fig. 5.2(b), the wave is moving out of the paper towards the reader which is indicated by the dot symbol. In Fig. 5.2(c) and (d), however, the wave is moving into the paper away from the reader which is indicated by the cross symbol. As illustrated in Fig. 5.2(e), the directions of both the electric field vector $\vec{E}$ and the magnetic field vector $\vec{H}$, change by 180 degrees in each half-cycle of the EM wave without changing the propagation direction of the wave.

There is another form of right-hand rule in which the thumb points to the direction of the electric field component but all other fingers being at right angle with the thumb point to the direction of the magnetic field component. The right hand must rotate in clockwise manner to maintain the correct relative directions of the fields.

The plane which is perpendicular to the direction of travel of the wave is referred to as the **wavefront** (also wave front). The wavefront is the surface which contains all points that are affected in the same manner by the EM wave at a given time instant. The plane stretched by the electric and magnetic field vectors in Fig. 5.2(a) is a wavefront, for example.

The direction of the electric-field vector $\vec{E}$ of a transverse EM wave is called its **polarisation**.

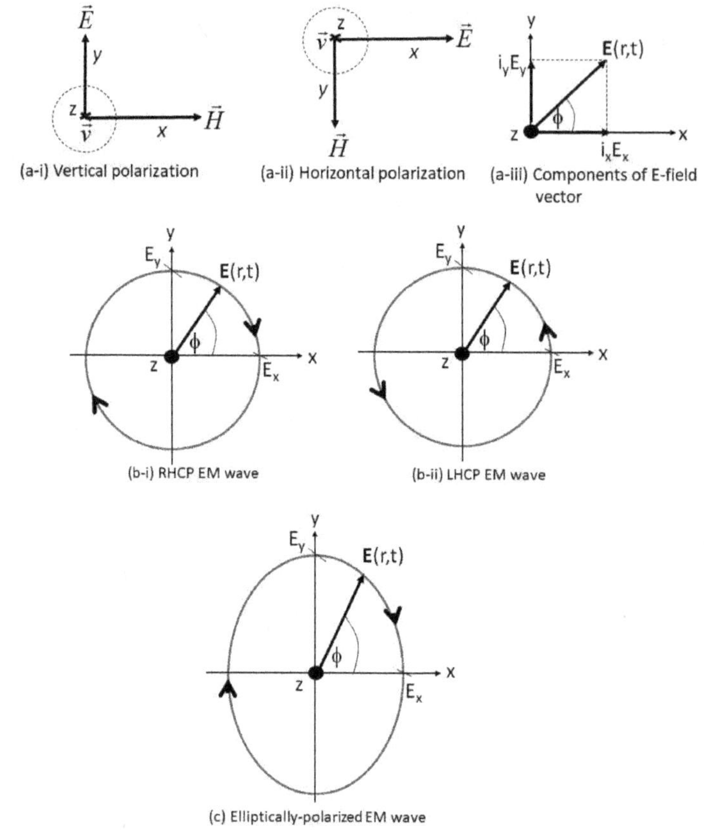

Fig. 5.3 The basic types of polarisations of EM waves.

Note that, unlike transverse waves, longitudinal waves cannot be polarised. There are three types of polarisations: linear, circular and elliptical polarisations. These polarisations are differentiated in Fig. 5.3. The E-field vector of a linearly-polarised wave moves in the plane lying in the direction that the wave is traveling. There are two types of linear polarisation: vertical polarisation (VP) and horizontal polarisation (HP). If the orientation of the E-field vector $\vec{E}$ is perpendicular to the Earth's surface then we have VP. HP, however, results if the E-field vector $\vec{E}$ is parallel to the Earth's surface. Linear polarisations are illustrated in Fig. 5.3(a).

The locus of the E-field vector of a linearly-polarised EM wave is a fixed straight line perpendicular to the wave's direction of movement. Let i_x, i_y, i_z be unit vectors in the x, y and z directions, respectively. We can view a linearly-polarised EM wave moving in the z-direction as illustrated in Fig. 5.3(a-iii) as a superposition of the two perpendicular plane waves which are in phase

$$\mathbf{E}_x = i_x E_x e^{j(\omega t - \kappa x)} \text{ and } \mathbf{E}_y = i_y E_y e^{j(\omega t - \kappa x)} \tag{5.1}$$

to yield the plane wave

$$\mathbf{E}(r,t) = \mathbf{E}_x + \mathbf{E}_y = (i_x E_x + i_y E_y) e^{j(\omega t - \kappa x)} \tag{5.2}$$

where $\kappa = 2\pi/\lambda$ is the wave number when the wavelength is λ. Note that the requirement for Eq. (5.2) to be a plane-polarised wave is that E_x and E_y should have the same phase.

The electric-field vector of a **circularly polarised** EM wave has a constant magnitude but rotates at a constant rate in a plane which is at right angles with the direction of travel of the wave. There can be either right-hand circular polarisation (RHCP) if the E-field rotates clockwise or left-hand circular polarisation (LHCP) if the E-field rotates anti-clockwise looking at the direction of the waves movement.

A circularly polarised wave radiates energy in both horizontal and vertical planes as well as every plane between them. Circular polarisations are illustrated in Fig. 5.3(b). Note that the magnitude of the EM wave is real. Therefore, we take the real part of Eq. (5.2) to obtain

$$\mathbf{E}_x(r,t) = i_x E_x \cos(\omega t - \kappa z) \text{ and}$$
$$\mathbf{E}_y(r,t) = i_y E_y \sin(\omega t - \kappa z)\,. \tag{5.3}$$

The locus of the circularly-polarised EM wave is a circle. Therefore, $E_x = E_y = E_0$. Thus,

$$|\mathbf{E}_x(r,t)|^2 + |\mathbf{E}_y(r,t)|^2$$
$$= E_0^2[\cos^2(\omega t - \kappa z) + \sin^2(\omega t - \kappa z)] = E_0^2 \tag{5.4}$$

which is the equation of a circle. *Thus, the superposition of two linearly-polarised EM waves which are perpendicular to each other and have the same magnitude produces a circularly-polarised EM wave.*

We have *elliptical polarisation* if the E-field vector moves along an elliptically-shaped path. The elliptical polarisation is illustrated in Fig. 5.3(c). Just as in linearly- and circularly-polarised EM waves, elliptically-polarised waves results from two linearly-polarised waves which are at right angles with each other. The difference is that the magnitudes of the two EM waves are different in elliptically-polarised waves, i.e., $E_x \neq E_y$. Consequently, it follows from Eq. (5.3) as

$$\frac{|\mathbf{E}_x(r,t)|^2}{E_x^2} + \frac{|\mathbf{E}_y(r,t)|^2}{E_y^2}$$
$$= \cos^2(\omega t - \kappa z)\sin(\omega t - \kappa z) = 1 \tag{5.5}$$

which is the equation of an ellipse, as illustrated in Fig. 5.3(c).

Signal reception is best if both the receive antenna and the incoming EM wave have the same polarisation, so it is made so practically. Unfortunately, the polarisation of a wave can change as it travels along the channel. Such an occurrence poses challenges in wireless communications in the form of cross-polarisation distortion. The EM wave possessing the desired polarisation is referred to as co-polarised wave, while the undesired polarisation is called cross-polarisation. These polarisations are at right angles to each other. Note that the two linear polarisations, vertical and horizontal polarisations, are perpendicular to each other, and so are LHCP and RHCP.

Cross-polarisation can also occur from the non-ideal nature of the source of EM waves, the antenna. No practical antenna generates waves at a single polarisation. Cross-polarisation distortion (XPD) measures the ratio of the energy in the cross-polarised wave to that of the co-polarised wave at the transmitting antenna. At the receiving antenna, however, it measures how the antenna is able to discriminate the co-polarised wave from the cross-polarised wave.

5.1.3 *Maxwell's Equations for Electromagnetism*

We cannot complete discussing EM wave propagation without mentioning Maxwell's equations as they are the basis for wireless communications. Here, we do not discuss their solutions, but just present the basic four equations. These equations exist in many forms, but we discuss just one of them. Around 1861–1862 James Clerk Maxwell found that the same phenomenon exhibits itself as magnetism, electricity and light by establishing that electric and magnetic fields propagate at the velocity of light. This finding paved the way for the discovery of EM waves. An electrically-charged conductor produces an electric field. Also, a magnetic field induces current through a conductor lying in it. Thus, the two fields are intertwined in a manner akin to the relationship between chicken and egg.

In the following, let $\vec{E}$ be the electric field vector, $\vec{B}$ or $\vec{H}$ be the magnetic field vector, μ_0 and ϵ_0 be the permittivity and permeability of free space, q be the electric charge, I be electric current, l and A be the length and area of a conductor. Maxwell's equations are made of four equations, with only the last one given their name. These are the Gauss' law for electric fields

$$\int \vec{E} \cdot d\vec{A} = \frac{q}{\epsilon_0} \tag{5.6}$$

the Gauss' law for magnetic fields

$$\int \vec{B} \cdot d\vec{A} = 0 \tag{5.7}$$

the Faraday's law of magnetic induction

$$\oint \vec{E} \cdot d\vec{l} = -\frac{d}{dt}\left(\int \vec{B} \cdot d\vec{A}\right) \tag{5.8}$$

and the Ampere's law combined with the Maxwell's law for current displacement

$$\oint \vec{B} \cdot d\vec{l} = \mu_0\Big(I + \frac{d}{dt}(\epsilon_0 \int \vec{E} \cdot d\vec{A})\Big). \qquad (5.9)$$

5.1.4 *Check Your Understanding*

Students are encouraged to check their understanding of the content of this section with the following questions prior to proceeding to the next section.

(1) Describe what electromagnetic waves are.

(2) What does the polarisation of a wave mean?

(3) How does a radio wave differ from other types of electromagnetic waves?

(4) Explain the term microwave. How does microwaves differ from radio waves?

(5) How do longitudinal waves differ from transverse waves? Which of the two do radio waves belong to?

(6) The three main types of polarisation, namely linear (horizontal and vertical), circular (LHCP and RHCP) and elliptical, have been discussed. Differentiate between them.

(7) Study Fig. 5.2 to answer this question. Assume that the directions of the three vectors of an EM wave $\vec{E}$ (electric field vector), $\vec{H}$ (the magnetic field vector) and $\vec{v}$ (wave's velocity) are the positive vertical axis, positive horizontal axis and into the paper, respectively. Assume further that the EM wave's wavelength is λ_0. Draw a figure similar to Fig. 5.2(a) to indicate the directions of the three EM wave's vectors after the wave has travelled the distance:
(a) $\frac{\lambda_0}{2}$.
(b) $\frac{\lambda_0}{4}$

(8) What do we call the arrangement of the EM waves according to their wavelengths?

(9) State the term used to refer to the time interval occupied by one complete cycle of an oscillating radio signal.

(10) The speed at which a radio wave moves is measured in cycles per second. What other term is commonly used instead of cycles per second?

(11) What is the distance between a pair of troughs or crests of an EM wave called?

(12) Explain the meaning of wavefronts of an EM wave.

(13) What do we call the set of four equations underpinning the theory and behaviour of electromagnetic waves?

5.2 EM Wave Propagation

The way wireless signals move in space between the transmitter and the receiver and the effects on them of the impediment in the channel is referred to as radio wave propagation. The radio path impediments or obstacles are called **clutter**. Microwave propagation is well-suited for *point-to-point communications*, and usually uses antennas mounted on high towers of ca. 30–60 m high to achieve **direct-wave propagation** or **line-of-sight propagation** (LOS) (LOS). Wave propagation in the microwave frequency bands is mainly by direct wave propagation. However, ground wave propagation can affect direct wave propagation by diffraction and reflection. The movement of the EM wave used to carry information occurs in many ways and depends on many factors. We briefly discuss them in this section.

5.2.1 *Types of Wave Propagation Environments*

The propagation mechanisms of radio waves depend on the specific propagation environment. These environments are classified based on both natural terrain and human developments. Figure 5.4 shows the most popular classifications of propagation environments. With respect to natural terrain we have open area, flat terrain, hilly terrain and mountainous terrain. These are self-explanatory.

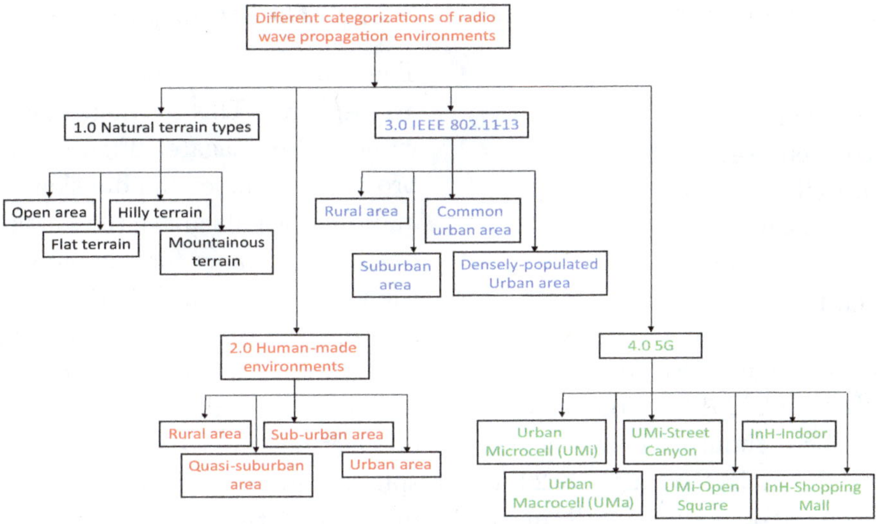

Fig. 5.4 Types of radio wave propagation environments.

With respect to human developments, we have rural area, quasi-suburban area, suburban area and urban area. An open area is space without obstructions between the transmitter and the receiver, such as factory floors or deserts. *Suburban area* refers to non-busy highway along which trees and houses are scattered. There can also be some obstacles near the mobile but not very congested. *Urban area* refers to heavily built-up city centers with large buildings. It can also model a village with closely-packed houses and tall trees. The IEEE 802.11-13 categorisation is described in Table 5.1 [Liu *et al.* (2013)].

The terrain causes fluctuations and long-term fading (aka local-mean attenuation) of the signal, while the artificial obstructions cause local-mean attenuation and short-term fading. [Lee (1993), p. 8]. Radio propagation in the residential environment is liable to penetration losses from wall and floor. In enterprise environment we consider cubic wall penetration, while outdoor propagation looks at wall and building penetration losses [Liu *et al.* (2013)].

Table 5.1 Classification of radio propagation scenarios/environments [Liu *et al.* (2013)].

Radio Propagation Environment Type	Description	Typical Penetration Loss
Densely-populated urban area	Densely-populated areas with average building height exceeding 30 m. Buildings are closely distributed with the average inter-building spacing varying in the range circa 10 to 20 meters.	18–25 dB
Common urban area	There are some open spaces and foliage. Average building height and average inter-building spacing are about the same, which is about 20 m.	15–18 dB
Suburban area	Buildings are scattered with the average inter-building spacing of 30–50 m, wide streets and circa 10-m average building height.	10–12 dB
Rural area	Average building height of about 5 m, and buildings are scattered over a large area containing large open spaces, roads and foliage.	6–8 dB

APL: aggregate penetration loss.

Typical building penetration losses at 1.8–2.6 GHz are [Liu *et al.* (2013)]

(1) 20–30 dB for lift door.
(2) 15–30 dB for concrete wall.
(3) 10 dB for brick wall.
(4) 1–3 dB for thin glass wall.
(5) 3–5 dB for thick glass door.
(6) 5 dB for wooden floor.

Some aggregate penetration losses for homes are given in Table 5.2 [Liu *et al.* (2013)]. Note, however, that some LTE-Advanced simulation environments use a fixed penetration loss in the range 18–20 dB, while ITU recommends 20 dB for macrocells in suburban environments.

5.2.2 *The Atmospheric Divisions*

The space above the Earth's surface is called the atmosphere. This is where EM waves transferred wirelessly propagate. Figure 5.5 illustrates in approximate manner the division of the atmosphere, and their heights above the Earth's surface. Of particular importance for microwave communication is the ionosphere, which is the ionised portion of the atmosphere. It is so called as it contains electrically-charged ions. The ions are formed from ionisation, which occurs if ultraviolet light from the sun breaks off electrons from the atoms of gases in the atmosphere and cause them to be charged ions. The intensity of the ionisation depends on two factors: the amount of atoms in the atmosphere and the

Table 5.2 Aggregate penetration loss for homes at 5.85 GHz using 5.5-m transmitter height [Liu *et al.* (2013)].

Home	Exterior	Insulation Type	TX/RX Distance	APL
Tranter	Brick	Foiled-backed	48 m	21.1 dB
			160 m	15.5 dB
Rappaport	Brick	Paper-backed	30 m	13.3 dB
			150 m	16.4 dB
Woerner	Wood sliding	Paper-backed	30 m	13.1
			210 m	7.2 dB

APL: aggregate penetration loss.

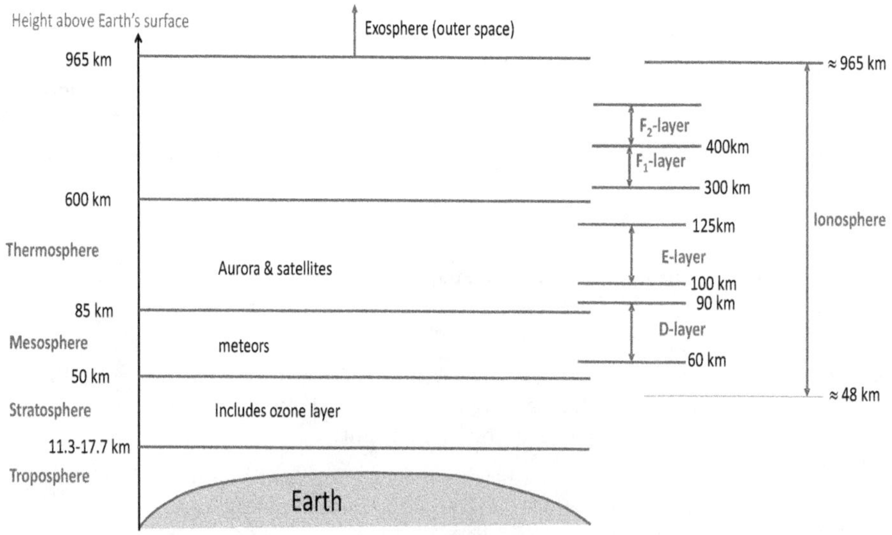

Fig. 5.5 Atmospheric layers.

strength of the UV light, both of which depend on time of day and seasons. Ionisation is high from sunrise to sunset, but decreases from late afternoon to its trough just prior to sunrise.

The ionosphere (**ion**isation + atm**osphere**) is divided into three regions: D, E and F. Region D is the lowest ionised part of the ionosphere and spans roughly 60 km to 92 km above the Earth's surface. This region appears at sunrise to achieve its maximum intensity at local noon and disappears at sundown until the next sunrise. Thus it can affect radio signal propagation only during the day when there are plenty of ions in the ionosphere caused by the sun's radiation. The D layer has low ionisation and refracts low-frequency signals, but allows high-frequency signals to pass right through it to be attenuated. The D region's attenuation of radio signals is more pronounced on signals operating in the band 0.03 to 3 MHz. However, its attenuation A_D of signals decrease with frequency according to the formula

$$A_D \propto \frac{1}{f^2}. \tag{5.10}$$

The ionosphere has some irregular variations, one of which is called sudden ionospheric disturbance (SID). SID is caused by bright solar eruption and results in abnormal ionisation of the D layer. SID causes total absorption of all frequencies above about 1 MHz.

The region above D is called E, and spans roughly 100 km to 115 km above the Earth's surface. This region barely exists around midnight, but can support wireless communication at HF band (i.e. 3–30 MHz) at day time to cover distances as far as to circa 2,400 km. It serves as a refractor of signals passing through it while slightly attenuating them.

The F region of the atmosphere lies above the E region, and it is used for HF band (i.e. 3–30 MHz) radio communications. It is the most important region for long-distant radio communications. It spans roughly 136 km to about 500 km above the Earth's surface, as it depends on seasons, time of day, latitudes, etc. In the day time when ionisation is high the sun's radiation split the F region into two regions: lower F_1 between 136 km and 248 km region and upper F_2 region lying between circa above 248 km to 400 km. Note that there is no consensus about the exact altitudes of the atmospheric regions. The F region is a good reflector of signals transmitted with frequencies 3 to 30 MHz.

The variations in the features of the ionosphere are attributed to the following four factors:

- Cyclic ionospheric disturbance: sudden ionospheric disturbances, sunspot cycle, etc. have effects on the ionosphere and the maximum of sunspot repeats every 11 years.
- Seasonal changes: during the summer season the heat energises the atmosphere to heat up the layers and create static which pushes the layers to higher altitudes.
- Time of day: as the temperature on the earth's surface changes over the day, so do the ionospheric layers. Some vanish during the night.
- Geographical location: the differing geography of locations (e.g. land mass and the height above sea level) effect radio signal propagation in both negative and positive ways.

5.2.3 *EM Wave Propagation Modes*

EM waves used to transfer information over wireless links travel the distance between the transmitting and receiving antennas by one of three ways, referred to as **wave propagation modes**. These are:

- Surface-wave propagation.
- Space-wave propagation (aka tropospheric scatter).
- Sky-wave propagation or ionospheric propagation.

These three wave propagation modes are compared in Fig. 5.6. The propagation mode of an EM wave depends on its frequency or wavelength. As Fig. 5.1 shows, the EM waves at low frequencies (i.e. below 300 kHz) usually propagate in ground wave mode. An example is AM radio operating in the band 535–1605 kHz. The sky wave propagation mode is used for 3–30 MHz EM waves. An example is FM radio operating in the band 87.8–108 MHz. EM waves at the extreme end of frequencies (i.e. above 30 MHz) usually propagate as

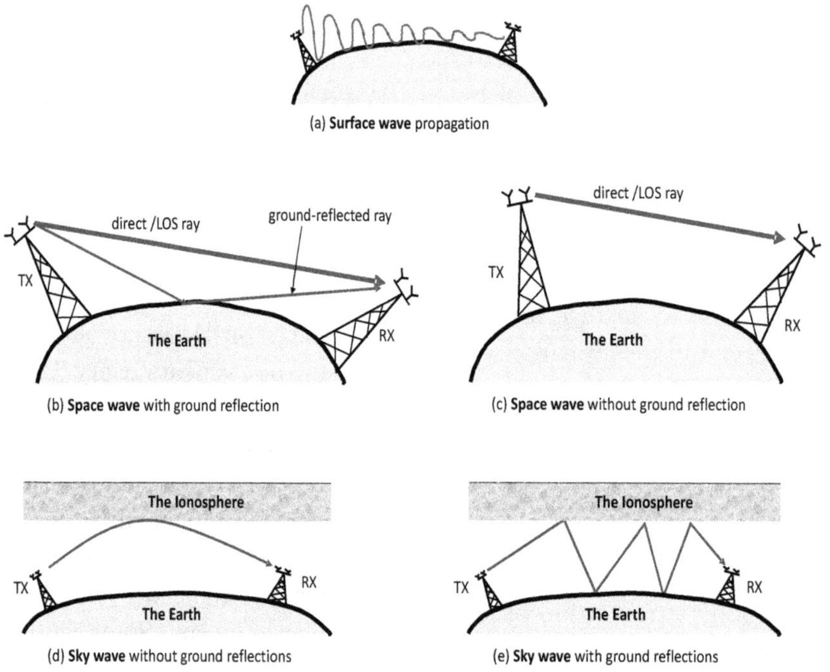

Fig. 5.6 EM wave propagation modes.

space waves. Signals transferred in terrestrial wireless communications systems propagate as only ground waves. Sky-wave propagation is used for satellite communications.

5.2.3.1 *Ground Wave Propagation Mode*

Waves which propagate near the surface of the Earth are referred to as ground waves. Note that ground waves have two types: surface wave (Fig. 5.6(a)) and space wave (Fig. 5.6(b) and (c)). The surface wave depends on diffraction off the Earth's surface, while space wave depends on LOS and reflection off the ground. It is most suitable for low-frequency signals operating in the frequency band below 3 MHz and 0.3 to 3000 MHz. Ground waves travel longer distances:

(1) For flat than hilly terrain.
(2) At lower than higher frequencies.
(3) At higher than lower transmit powers.
(4) If the conductivity of the ground/Earth is higher.

Surface waves (Fig. 5.6(a)) move along the Earth's curvature by diffraction off the ground to travel beyond the optical horizon. Surface wave propagation does not require line-of-sight propagation. It travels longer distances if its polarisation is vertical rather than parallel with respect to the Earth's surface. The ground absorbs the energy in the wave intensely to shorten its propagation distance if its electric field vector travels parallel to the Earth's surface. For this reason, vertical polarisation is preferable for space waves. A good conductive surface also prolongs the distance travelled by surface waves.

As illustrated in Fig. 5.7, the distance travelled by a surface wave also depends very much on its frequency. Low frequency waves have long wavelengths which are better diffracted by the Earth's surface than shorter waves. This is the reason why surface waves used in AM radio operate at low frequencies. Surface waves at AM radio frequencies, for example, can travel from one continent to the other, except that they are very sensitive to bad weather. For example, AM radio is spontaneously affected by lightening. Surface waves use the space between the Earth's surface and the ionosphere as its conduit. As illustrated in Fig. 5.7 and exploited in AM radio, for example, at low frequencies ground waves can cover thousands of kilometers.

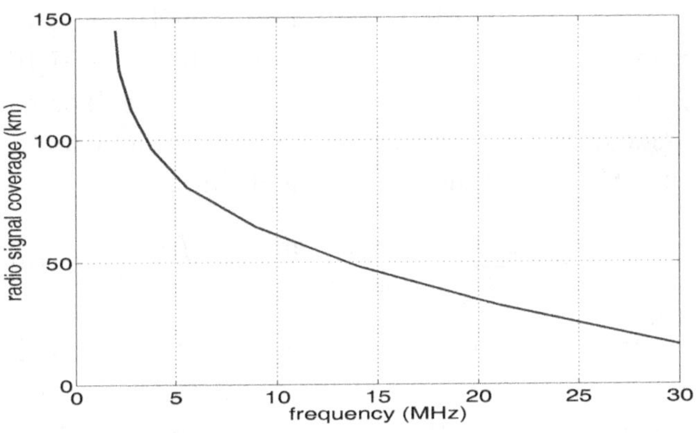

Fig. 5.7 Ground wave coverage versus operating frequency.

Space-wave propagation mode (Fig. 5.6(b) and (c)) occurs as the operating frequencies increase to make the wavelengths small enough to pass between the ions in the ionosphere. Signals which are reflected by the ionosphere propagate as sky waves, while those which penetrate straight through the ionosphere propagate as space waves. Space wave propagation relies on line-of-sight (LOS) propagation, and signal reflection off the ground. Unfortunately, destructive interference can result from the two rays at the receiver if their phase difference is an odd multiple of 180 degrees.

Only the LOS propagation mode is feasible when the wavelengths of the EM wave are below about 10 m (i.e. frequencies above 30 MHz). Thus, microwaves must always travel in LOS mode. LOS means that there is a clear link between the transmitting and receiving antennas. In other words, the transmitting and the receiving antennas must be in sight of each other. This requires that the link contains no object which are able to block the EM wave. LOS path is the straight direct path between the transmitting and receiving antennas. The range of a LOS path depends on the heights of the transmitting h_T and receiving h_R antennas as

$$d_{LOS}\ \texttt{[km]} = \sqrt{13h_T\ \texttt{[m]}} + \sqrt{13h_R\ \texttt{[m]}}\,. \quad (5.11)$$

As illustrated in Fig. 5.6(b) and (c), space wave propagation can be one of two types. Just a single direct ray can reach the receiver, or the direct ray and ground-reflected ray. The troposphere can be divided into layers. A layer containing air at a higher temperature than those below and above it forms a duct like a cable for radio signal transmissions over large distances. Such a situation is referred to as **tropospheric ducting**.

5.2.3.2 *Sky-Wave Propagation Mode*

Sky wave propagation (Fig. 5.6(d) and (e)) relies on radio signal reflection off the atmospheric layer called the ionospheric layer (Fig. 5.5), and thus also referred to as **ionospheric propagation**. As discussed earlier, the D region vanishes during the night time, rendering sky-wave propagation much more effective during the night than during the day. Sky-wave propagation is used for long-distant wireless communications, and can cover thousands of kilometers at 3–30 MHz by reflection by the F region. Thus, sky waves travel farther than ground waves. As illustrated in Fig. 5.6(d) and (e), sky waves can have a single hop or multiple hops. The multihop propagation relies on ground reflections of the wave.

Sky wave propagation allows transmitted signals to be reflected (bounced) off a portion of the Earth's ionosphere to reach a radio receiver located hundreds or even thousands of kilometers away. However, ionospheric reflection of waves is more probable to occur at low frequencies, such as 30–300 kHz.

Sky-wave propagation is based on the signal reflection by the ions in the ionosphere, the reason why it is also referred to as ionospheric propagation. This mode of wave propagation is possible only during the night when the ionosphere has just enough ions in it to reflect the EM waves, rather than heavily attenuating them. Signals operating at frequencies in the band 3 MHz to 30 MHz are exclusively transferred using sky waves. Frequencies in the range 3 kHz to 3 MHz can also be transferred by ground waves, but better with sky waves. The distance covered by a sky wave depends on many parameters, such as the operating frequency, transmit power, ionospheric condition and the radiation angle between the sky and the antenna.

The sky-wave propagation of EM waves relies on ionospheric reflection of the waves. However, the ionosphere is unable to effectively reflect short EM waves having wavelengths below circa 10 m. Ground-wave propagation also uses low- to medium-frequency signals to propagate around the surface of the earth. Such propagation mechanisms do not work with high frequencies, leaving only line-of-sight wave propagation as the only choice.

Each of the ionospheric sublayers has a maximum frequency called *critical frequency* $f_c(t)$ at a given time t at which signals refract back to Earth if transmitted vertically. A radio signal at a frequency higher than $f_c(t)$ passes through the layer. Signals at frequencies lower than $f_c(t)$, however, are refracted back to the Earth. The sharpness of the reflection is higher at lower signal frequencies. Further, the intensity of the refraction experienced by a radio signal at a given frequency depends on the angle at which it touches the layer. If the angle is about 90 degrees then it penetrates the layer. Refraction to the Earth occurs at lower angles.

The refractive index of the ionosphere decreases with increasing frequency, and thus bounds the maximum frequency that can be used for sky wave propagation. Also, the refractive index depends on the density of ions in the ionosphere, which also depends on solar activity, season and time of day. The highest possible frequency that is supported by ionospheric reflection for communication between two points on the Earth's surface is called **maximum usable frequency** (MUF). Signals transmitted at frequencies higher than the MUF penetrate the ionosphere into outer space. For an angle of wave's incidence θ at an ionospheric layer, the MUF at a time instant t is defined as

$$MUF = \frac{\texttt{critical frequency}}{\cos\theta} = \frac{f_c(t)}{\cos\theta}. \quad (5.12)$$

The maximum angle of incidence at a layer with a given ionisation density that a radio signal at a given frequency returns to the Earth is called *critical angle* θ_c (Fig. 5.8). An EM wave which leaves an antenna at an angle greater than θ_c penetrates the layer, otherwise it is refracted off the ionosphere. However, a wave refracted off the ionosphere does not return to the Earth if its angle of incidence is too small. Figure 5.8 illustrates how wave W1 at 3 MHz and wave W2 at 5 MHz penetrate the ionosphere into outer space until their respective angle of incidence was reduced to the required critical angle. We can also observe that the higher the frequency, the lower the critical angle and that higher frequencies have longer range than lower frequencies if all else is held the same.

The distance from the transmitting antenna to the point where the first sky wave returns to the

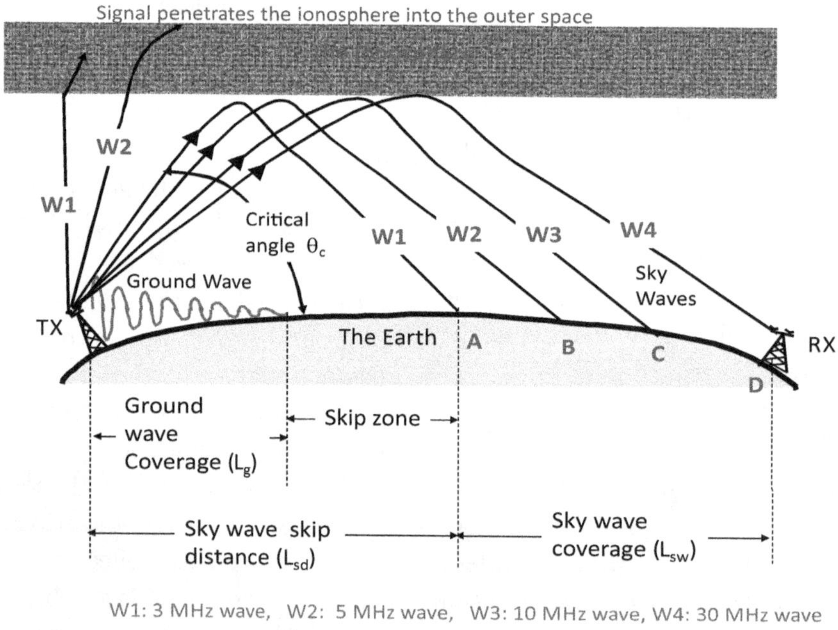

Fig. 5.8 Skip zone, skip distance and critical angles.

ground is referred to as the *skip distance* L_{sd}, as illustrated in Fig. 5.8. This distance depends on the incidence angle and the wave's frequency, as well as the strength of the ionospheric ionisation. Ground wave has a maximum coverage. The space between the ground wave's coverage and the skip distance is called **skip zone**. There is no or only poor ionospherically-propagated radio coverage within the skip zone. If the coverage of the ground wave is about the same as the skip distance of the sky wave then there is no skip zone. However, zero skip zone also has a problem. There can be destructive interference between the ground wave and the sky wave if their phase difference is an odd multiple of 180 degrees.

Considering Fig. 5.8, sending a radio signal at the given frequency and ionisation density at the critical angle can enable ionospheric radio communication between the transmitting antenna and point A. An incidence angle lower than the critical angle is required to communicate with any other point. This is where antenna's tilts is important.

The region D of the ionosphere can refract only low-frequency signals, allowing signals at high frequencies to penetrate through it. The region E of the ionosphere supports wireless communications at 3–30 MHz during the day over distances as far as 2,400 km. The region F of the ionosphere reflects radio signals to support long-distance wireless communications.

In recap, the degree at which a layer in the ionosphere bends an incident EM wave increases with increasing ionisation density at a given wavelength. Also, wave bending increases as the wavelength is increased at a given ionisation density. The two extreme situations are then:

- Wavelength is low or the ionisation density falls below a minimum limit: regardless of the angle of incidence the ionosphere does not reflect enough energy in the wave back to the Earth.
- Wavelength is sufficiently high and the ionisation density is above a minimum limit: the ionosphere reflects back to Earth waves which touch it even at right angles.

5.2.4 *Free-Space EM Wave Propagation*

The phrase *free-space propagation* means that there is no matter in the path. This requires the path to be a vacuum. In communications engineering, however, free space means we are using wireless transmission media in contrast to wired or guided medium for the signal transmission. Line-of-sight (LOS) radio propagation, of course, can occur only in free space as we use the wireless medium.

In LOS propagation the sending and the receiving antennas of the signal are expected to 'see' each other in order to achieve effective communications. Hence, the name line of sight propagation. Ideal LOS propagation is an idealisation, as it is impossible to clear the signal propagation path of all obstacles, both natural and man-made. Notwithstanding, ideal LOS propagation is a good starting point for radio propagation studies. Ideal LOS signal propagation assumes a perfectly clear transmission path between signal originating antenna and the receiving antenna. The path length, path loss, etc. of an ideal LOS propagation is calculated using the **FRIIS formula**

$$P_r = A_e \frac{P_t G_t}{4\pi d^2} = P_t G_t G_r \left(\frac{c}{4\pi d f}\right)^2 = P_t G_t G_r \left(\frac{\lambda}{4\pi d}\right)^2 \tag{5.13}$$

where

- P_r is the received power
- P_t is the radiated power from the transmitting antenna
- G_t is the gain (directivity) of the transmitting antenna
- G_r is the gain (directivity) of the receiving antenna
- $A_e = \frac{\lambda^2 G_R}{4\pi}$ is the effective area of the receiving antenna
- d is the distance between the transmitting and receiving antennas
- c, λ and f are the velocity of light, operating wavelength and frequency, respectively.

The antenna gain depends on its *directivity* (D) according to

$$G = \eta \times D \tag{5.14}$$

where $\eta \leq 1$ is the antenna's *radiation efficiency*. It is often more convenient to handle relations such as Eq. (5.13) in the logarithmic domain. In decibel domain, Eq. (5.13) is equivalent to

$$\begin{aligned} P_r(\texttt{dBm}) &= P_\texttt{t}(\texttt{dBm}) + G_\texttt{t}(\texttt{dB}) + G_\texttt{r}(\texttt{dB}) - 32.44 \\ &\quad - 20\log_{10} d\ (\texttt{km}) - 20\log_{10} f \\ &\approx P_\texttt{t}(\texttt{dBm}) + G_\texttt{t}(\texttt{dB}) + G_\texttt{r}(\texttt{dB}) - 22 \\ &\quad - 20\log_{10} d\ (\texttt{km}) + 20\log_{10} \lambda\ (\texttt{km}) \end{aligned} \tag{5.15}$$

where d is in kilometers (km), f is in Megahertz (MHz) and $c \approx 3 \cdot 10^8$ m/s is assumed. In the special case of ideal omnidirectional antennas the gains become $G_\texttt{t} = G_\texttt{r} = 0$ dB. The expression

$$\begin{aligned} L(\texttt{dB}) &= P_t(\texttt{dBm}) - P_r(\texttt{dBm}) \\ &= 32.44 + 20\log_{10} d\ (\texttt{km}) + 20\log_{10} f\ (\texttt{MHz}) \\ &= 22 + 20\log_{10} d\ (\texttt{km}) - 20\log_{10} \lambda\ (\texttt{km}) \end{aligned} \tag{5.16}$$

is referred to as *free-space path loss*. This is the attenuation experienced by a single ray emitted from gainless antenna passing through an unobstructed LOS channel. Its reciprocal in ratio (not dB) is the *path gain*. If the path loss has a negative sign (i.e. less than unity) then it is a path gain instead, and vice versa.

Example

Terrestrial wireless systems using FDD (frequency-division duplexing) usually use lower frequency band for uplink or reverse link than the frequency band for downlink or forward link. For example, the FDD Band 3 at 1800 MHz that is used in LTE and LTE-Advanced uses 1710–1785 MHz for the uplink and 1805–1880 MHz for the downlink. In satellite communications, however, the opposite is usually the case. The uplink frequency band is higher than the downlink. For example, C band (4–8 GHz) frequencies are 5.850–6.425 GHz for uplink and 3.625–4.2 GHz for downlink. Why is it so?

Solution

There are two reasons for the choice of frequency bands for the uplink and downlink of the FDD mode. These are acceptable path losses and antenna dimensions. We can observe from Eq. (5.16) that the path loss is proportional to frequency, and precisely $L \propto 20\log_{10} f\ (\texttt{MHz})$. This means the path loss increases as the frequency increases if all other parameters are the same. The distance between the base station and the mobile station is almost the same. This means that the difference in path loss for the duplex directions depends very much on the carrier frequency used.

Transmit power and antenna gain: The argument is, which of the two ends of a duplex link is practically best to transmit at higher power and higher antenna gain? In terrestrial wireless communications it is the base station. Thus, the base station can transmit higher powers using higher-gain antennas to offset the higher path loss caused by transmitting at a higher frequency. Mobile station uses batteries for power and must transmit at lower power to conserve the battery. Transmitting at low power is also good for human health. In satellite communications, however, the earth station can be powered by the mains power supply, making transmit power no issue. In remote areas battery banks can be used as well. The space segment hanging in the air is powered by either batteries and/or solar cells. The lifespan of satellite space segment is usually limited by how long their power supply lasts, otherwise astronauts would need to climb high to replenish them. Refurnishing these power supply sources is expensive.

Antenna gain and weight: earth stations of satellite systems can be equipped with large antennas with high gains so that a high EIRP (effective isotropic radiated power) is emitted to offset the high path loss on the uplink owing to the use of higher frequencies. The same thing can be down at the base station of terrestrial wireless systems. Mobile phone sizes should be kept small with light weight to make it easy to carry around.

5.2.5 *Non-Free Space Wave Propagation Mechanisms*

In the following analysis we consider an EM wave to contain a single frequency for ease of analysis. The Friis equation can be used to evaluate the power

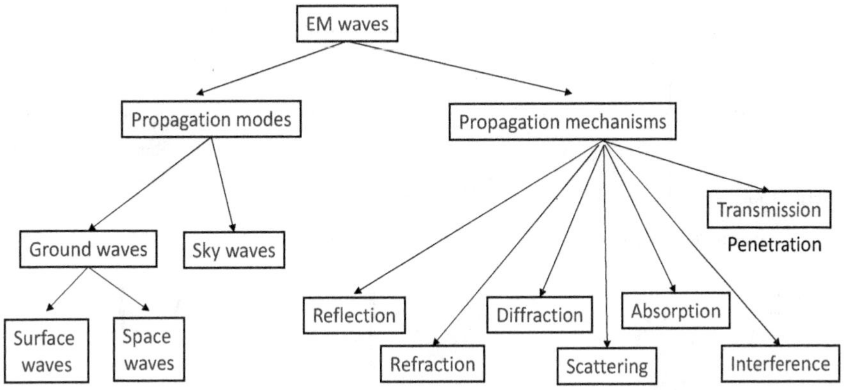

Fig. 5.9 EM wave propagation properties.

budget of idealised free-space propagation. However, terrestrial microwave propagation practically deviates far from the free-space idealisation due to the following potential obstacles in the propagation path between transmitter and receiver:

- Proximity or presence of obstructions (e.g. trees, hills, wind, mounds) to signal propagation path. These cause shadowing, reflection, scattering, diffraction, refraction and absorption of wireless signals.
- Interference from other neighbouring wireless systems.
- Earth curvature (radio waves 'see' the Earth to be 4/3 of its real radius).
- Environmental noise, natural or man-made.
- Absorption and scattering effects of the atmosphere due to oxygen and water vapour.
- Effects of the atmospheric refractivity of signals.
- EM wave absorption and scattering by precipitation (i.e., rainfall, snow and graupel).

We classify the propagation modes and propagation mechanism of EM waves in non-free space in the following sections (Fig. 5.9). Some of the wave propagation mechanisms in non-free space are illustrated in Fig. 5.10.

5.2.5.1 *Wave Reflection*

The phenomena by which an EM wave hits an object and returns back to the original medium is called **reflection**. If I throw an object X against another object Y it may bounce back. Technically, we say that object X has reflected off object Y. Assume that X is an EM wave pulse and Y is an obstruction in the channel. For a reflection to occur, the dimensions of the surface of Y at which X hits must be larger than the wavelength (λ) of X. The amount of reflection also depends on many other factors, such as the angle of incidence (i.e. the angle at which X touches the surface of Y), the hardness of the surface of Y, the roughness or smoothness of the surface of Y, and the geometry and material composition of the object Y. If object Y is hard (like a stone) then X can make an 180-degree phase change, i.e. reflects back in opposite direction, but maintains its strength and travelling speed. If object Y is soft, however, then the phase change of the reflected wave can be near zero degrees but maintains its amplitude and velocity.

EM wave reflection is guided by the **laws of reflection** of light. This is illustrated in Fig. 5.10(a). The laws of reflection state that: (a) the incident wave, the reflected wave and the normal to the surface of the reflecting object lie in the same plane. (b) The angle of incidence (θ_{i}) equals the angle of reflection (θ_{r}), and (c) the incident wave and the reflected wave lie on the opposite sides of the normal. The **normal** is the line which is at right angles with the boundary of the obstruction where the incident wave touches.

Let us use the impedance Z of an object, which is neither electrically soft nor hard, to measure its electrical roughness. If the obstruction Y is neither hard nor soft then part of the energy in the

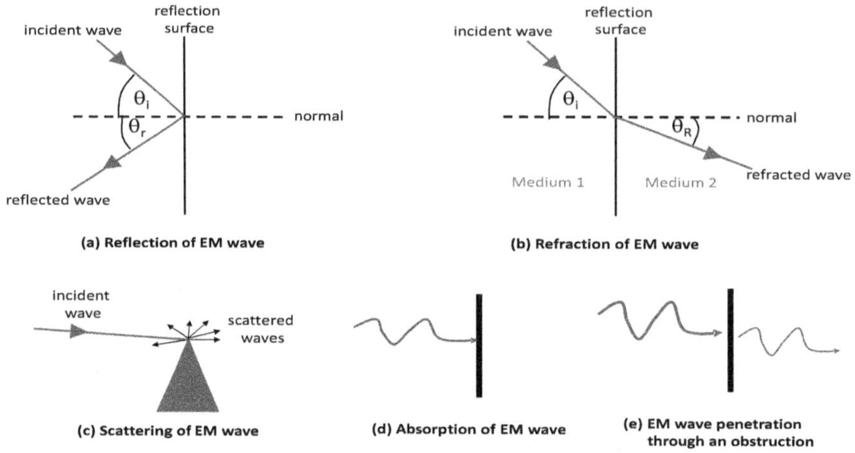

Fig. 5.10 Wave propagation mechanisms.

incident wave pulse is reflected back, while the remaining passes through the object Y. EM waves usually travel through the air. So, let $Z_0 \approx 120\pi$ ohms be the characteristic impedance of air, and Z_1 be that of the obstructing object Y. Let A_0, A_r and A_t be the amplitudes of the incident, reflected and transmitted wave pulses, respectively, then we have

$$A_r = \frac{Z_0 - Z_1}{Z_0 + Z_1} A_0 \text{ and } A_t = \frac{2Z_1}{Z_0 + Z_1} A_0. \quad (5.17)$$

The speed of an EM wave depends on the density of the propagating medium. An EM wave travels faster in a lighter medium than a denser medium. The refractive index or coefficient indicating the reflective ability of a medium depends on its density. EM wave reflection off objects is not always bad. It has many positive applications in communications. For example, it is used in radar systems for weather forecasting. EM wave reflection occurs if the edges of the obstruction is more than 5–10 wavelengths away from the point at which the wave touches it. A wave incident on an object is reflected if:

- the natural oscillation frequencies of the electrons in the atoms of the obstruction are different from the frequencies in the EM wave, and
- the obstructive object is optically opaque to the EM waves.

If the object obstructing the EM waves is transparent and its electrons' vibrating frequencies differ from the frequencies in the wave, then the waves penetrate it. This is the reason why EM waves at a given frequency penetrate some objects, but not others. *EM wave reflection or bouncing is usually exploited in ground-wave communications.*

5.2.5.2 *Wave Scattering*

Wave scattering occurs if, upon hitting an obstruction, the wave breaks up into multiple waves each with a weaker strength than the incident wave. Wave scattering occurs if the edges of the obstructing object are closer to each other than the wavelength of the wave. Thus, the wave is larger than the obstructing object, and so it cannot be reflected. In such a case the effect of the obstruction on the incident wave depends on the latter's frequency. A general rule between the width of the obstructing object (w_o) and the wavelength of the incident EM wave (λ) is:

$$\begin{array}{rl} \texttt{Wave reflection:} & w_o > 10\lambda \\ \texttt{Complicated model of wave behaviour:} & 0.1\lambda < w_o < 10\lambda \\ \texttt{Wave scattering:} & w_o < 0.1\lambda \end{array} \quad (5.18)$$

5.2.5.3 *Wave Refraction*

Wave refraction occurs if a wave moves from one medium to another medium having a different optical density, which is referred to as refractive index. The bending of the wave is caused by the change in its velocity of propagation, and hence its wavelength. Figure 5.10(b) shows a situation for refraction of a wave pulse. The bending of the wave is larger the greater the density of the medium that it enters relative to the density of the medium from which it came. The refracted ray bends further away from the surface of the medium the denser the medium the wave enters. Vacuum has the lowest optical density (aka refractive index) which is 1. The refractive index of air is about 1.0003, and that of water is about 1.333. The law guiding the refraction of EM waves is called **Snell's law**, which is

$$\texttt{Snell's law:} \quad n_1 \sin(\theta_{\mathtt{i}}) = n_2 \sin(\theta_{\mathtt{R}}) \qquad (5.19)$$

where

- n_1 is the refractive index of the incident medium (i.e. medium 1 in Fig. 5.10(b)).
- n_2 is the refractive index of the refractive medium (i.e. medium 2 in Fig. 5.10(b)).
- $\theta_{\mathtt{i}}$ is the angle of incidence of the EM wave.
- $\theta_{\mathtt{R}}$ is the angle of refraction of the EM wave.

We can reformulate Eq. (5.19) as $\frac{n_1}{n_2} = \frac{\sin\theta_{\mathtt{R}}}{\sin\theta_{\mathtt{i}}}$. This means that $\sin\theta_{\mathtt{R}} < \sin\theta_{\mathtt{i}}$ if $n_1 > n_2$, which results in $\theta_{\mathtt{R}} < \theta_{\mathtt{i}}$. Thus, the wave moves away from the refractive medium towards the normal as it is optically lighter than the incident medium. The reverse is also true. *EM wave refraction or bending is usually exploited in skywave communications.*

5.2.5.4 *Wave Absorption*

An obstructive object can absorb the energy in an EM wave that incidents on it. The absorption of the energy can be partial or full. The energy in the incident wave that the obstruction absorbs is converted into heat inside the obstruction or dissipated into the ground (Fig. 5.10(d)). If the EM wave comprises multiple frequencies then the effect of the obstruction on it differs for each of its frequency components. Every object is made of atoms, which, in turn, contain electrons. The electrons in a given object vibrate at some natural frequencies. Assume that an EM wave containing the frequency f_k incidents on an object whose electrons vibrate with the natural frequencies f_0. If $f_0 = f_k$, then the object absorbs the EM wave and converts all its energy into heat.

Foliage, such as the leaves of trees, absorb energy in radio waves. Thus, in summer when trees have a lot of leaves radio propagation in place with trees can be poorer than in winter when the trees have shed their leaves. For this reason, radio propagation quality also depends on seasons.

Depending on the operating frequency (and hence wavelength) of an EM wave and the material composition of an obstruction, an EM wave can penetrate the obstruction. This is illustrated in Fig. 5.10(e). The energy in the penetrated wave is, of course, expected to be less than that in the incident wave. The difference in the two energy levels is referred to as **penetration loss**, whose level depends on the type of obstruction. Another term for wave penetration through an object is wave transmission.

5.2.5.5 *Wave Diffraction*

An EM wave changes its direction of propagation if it passes around a corner, a barrier or through a slit in its path. Such a bending is called diffraction. Diffraction aids in getting radio waves to non-line-of-sight locations, such as under hills or a building. Diffraction requires that the wavelength of the EM wave is larger than the dimensions of the obstruction. The amount of diffraction (i.e. the sharpness of the bending of the EM wave) decreases with decreasing wavelength, and increases with increasing wavelength. Figure 5.11 illustrates the diffraction of radio signals. It is obvious from the figure how through diffraction we are able to receive signals, although the receiver is blocked by an obstruction.

5.2.5.6 *Wave Interference*

If two or more EM waves travel in the same channel at the same time then they end up intruding into each other's freedom. The waves interfering are expected to have the same wavelength, about the same amplitude, and travel in the same direction. Such an occurrence is referred to as wave interference. The interference can be constructive or destructive. **Constructive interference** occurs if the waves build upon each other to produce a single wave which is stronger in intensity than any of the

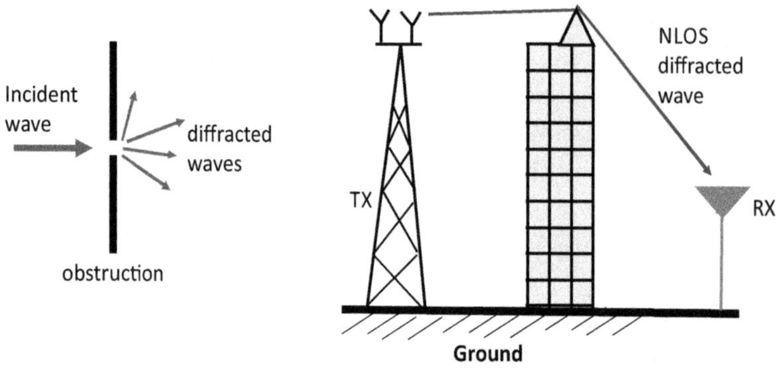

Fig. 5.11 Diffraction of radio signals.

constituent waves. This is what is desired in communications systems. Unfortunately, that is the exception rather than the norm. Constructive interference requires the constituent waves to possess the same phases or displacement at the same time. If their phases differ then their sum becomes weaker than the strongest component in them. Such a situation is called **destructive interference** of the waves. We shall see later in this chapter that a main cause of destructive interference in wireless communications is multipath signal propagation, which is one of the most troublesome disturbances. It causes inter-symbol interference (ISI) or time dispersion in the time domain and frequency-selective fading or frequency dispersion in the frequency domain.

Finding the effect of interference of multiple waves is obtained through the **superposition principle** of waves. This principle states that if two waves interfere with each other, then the strength of the resulting EM wave at a given location and time is the algebraic sum of the strengths of the individual waves at that same location and time. Figure 5.12 illustrates the superposition principle through the interference between two single-frequency waves. The two waves $s_1(t) = \sin(2\pi f_1 t)$ and $s_2(t) = \sin(2\pi f_2 t + \alpha)$ are shown in the top part of Fig. 5.12. For simplicity, we assume that $f_1 = f_2 = 50$ Hz, while both waves have unit amplitudes. The phase difference α determines whether there is constructive or destructive interference. This is illustrated in the bottom part of Fig. 5.12. Destructive interference occurs so long as $\alpha \neq 0$ and increases to attain its maximum value at $\alpha = \pi$, as shown in Fig. 5.12 (b). For constructive interference between two waves their phase difference $\Delta\alpha$ and path difference Δd should be

$$\Delta\alpha = 2\pi n \texttt{ and } \Delta d = n\lambda,\ n = 0, 1, 2, \ldots \quad (5.20)$$

Destructive interference occurs between two waves if their phase difference $\Delta\alpha$ and path difference Δd fulfill

$$\Delta\alpha = (2n+1)\pi, \texttt{ and}$$
$$\Delta d = (2n+1)\frac{\lambda}{2},\ n = 0, 1, 2, \ldots \quad (5.21)$$

As the angle covered by a wave is 360 degrees for each of its cycle, we can express the relationship between phase difference and the path length difference of two waves as

$$\Delta\alpha = 2\pi\frac{\Delta d}{\lambda} \texttt{ radians.} \quad (5.22)$$

5.2.6 *Rain Attenuation of Radio Signals*

Precipitation can have a remarkable effect on wireless signals depending on its intensity and the signals' wavelength. Precipitation includes snow, graupel, sleet, hail, rain and drizzle. In this section, we review rain effects on radio signal propagation. Rain attenuation of wireless signals depends on:

- The rate or intensity of the rainfall.
- Size of rain drops.
- Frequency or wavelength of the wireless signal.

Rain attenuation of wireless signals occur by scattering and absorption propagation mechanisms. The higher the frequency of signal, the higher the effect of rain on it. Owing to the demand for

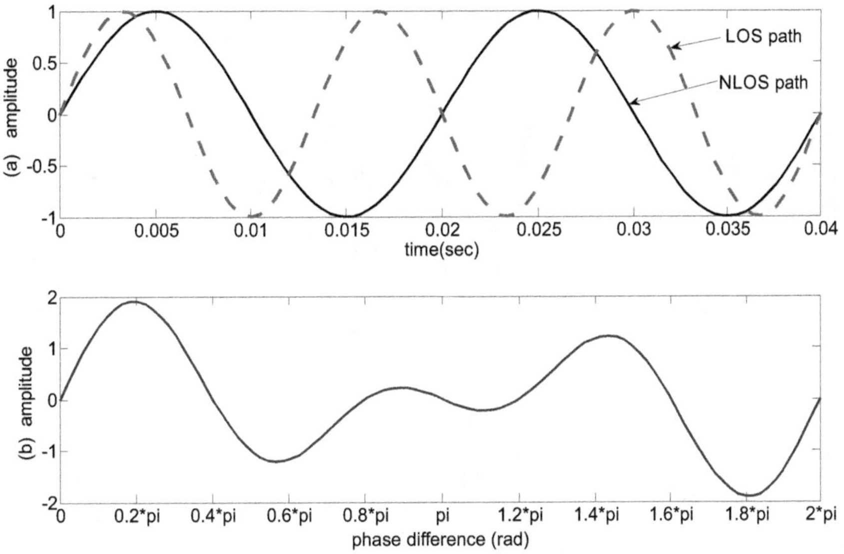

Fig. 5.12 Interference of two single-frequency waves.

high data rates and hence large bandwidths, future wireless systems will operate at gigahertz frequencies. Electromagnetic waves at such high frequencies have short wavelengths and thus the absorption and scattering effects of rain are more pronounced than low-frequency signals. This requires accurate models of rain attenuation of signals in the design wireless link for mmWave systems.

There are several models proposed to quantify the relationship between the amount of rainfall R (measured in millimeter per hour (mm/hr)) and the attenuation (measured in decibels or dB) that it causes to wireless signals. We will focus on the model proposed by the radio committee of the International Telecommunications Union (ITU-R). The ITU-R model in [ITU-R (2005)] states

$$\gamma_{\mathtt{R}}(f) = k(f) \cdot R^{\alpha(f)} \; \texttt{[dB/km]} \tag{5.23}$$

where f is the operating frequency in gigahertz (GHz) and takes values in the range 1–1000 GHz, R is the rainfall rate measured in millimeter per hour (mm/hr), and $k(f)$ and $\alpha(f)$ are frequency-dependent coefficients. These coefficients are given in [ITU-R (2005)] as

$$\log_{10} k = = \sum_{i=1}^{4} a_i \exp\left[\left(\frac{\log_{10} f - b_i}{c_i}\right)^2\right] + m_k \log_{10} f + c_k, \quad 1 \leq f \leq 1000 \; \texttt{GHz} \tag{5.24}$$

and

$$\alpha = \sum_{i=1}^{5} a_i \exp\left[\left(\frac{\log_{10} f - b_i}{c_i}\right)^2\right] + m_\alpha \log_{10} f + c_\alpha, \quad 1 \leq f \leq 1000 \; \texttt{GHz} \tag{5.25}$$

where $k \in \{k_H, k_V\}$ and $\alpha \in \{\alpha_H, \alpha_V\}$ for horizontal (H) and vertical (V) polarisations, respectively. The values of these four coefficients for frequencies between 1 GHz and 1000 GHz are shown in Figs. 5.13, 5.14, 5.15 and 5.16. The values of the parameters a_i, b_i, c_i, m_α, m_k, c_k and c_α are found in [ITU-R (2005)].

Plugging the coefficients Eq. (5.24) and Eq. (5.25) into (5.23) results in Figs. 5.17 and 5.18. Some important wireless system design information can be extracted from the figures. First, the rain attenuation on wireless signals increase with the operating frequency. Second, the rain attenuation increases with the signal's propagation distance through the rain.

Two different professions have specific interest in a model such as Eq. (5.23). These are communication professionals and meteorologists. Meteorologists seek to extract rainfall rate from the attenuation that the rainfall causes to radio signals. Such signal attenuation is usually recorded by telecommunications service providers. Commun-

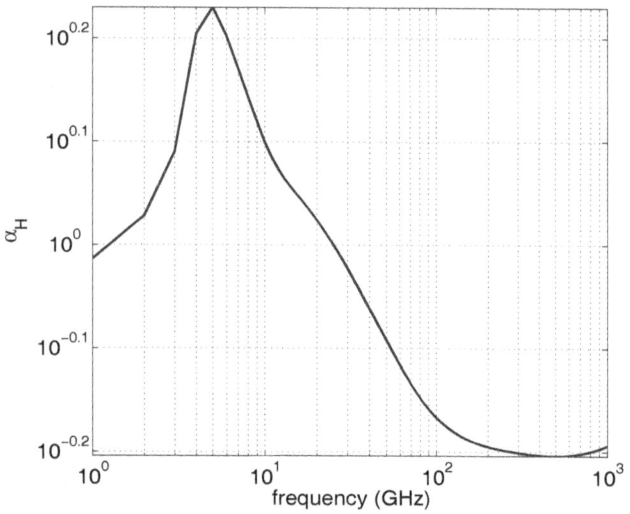

Fig. 5.13 Horizontal polarisation coefficient α_H versus frequency f.

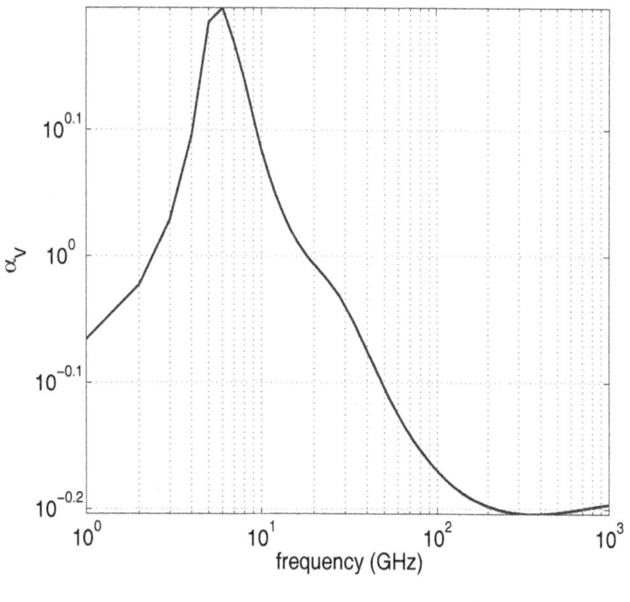

Fig. 5.14 Vertical polarisation coefficient α_V versus frequency f.

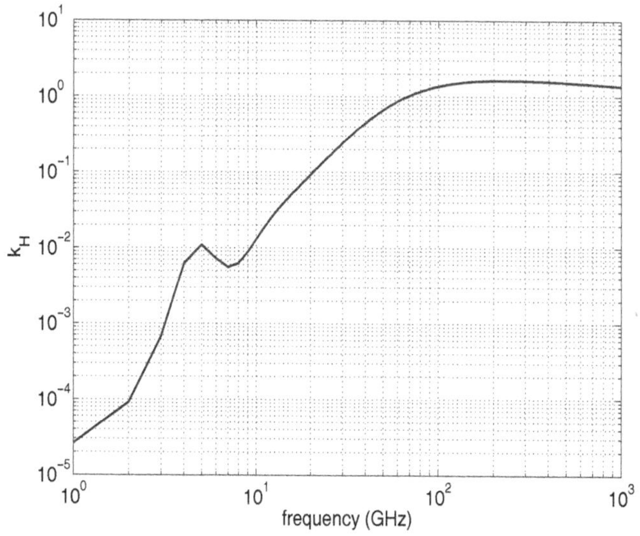

Fig. 5.15 Horizontal polarisation coefficient k_H versus frequency f.

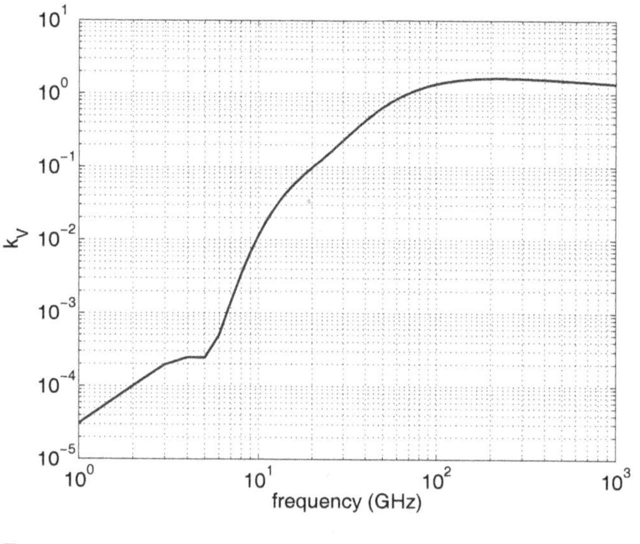

Fig. 5.16 Vertical polarisation coefficient k_V versus frequency f.

ications professionals, however, seek rainfall rates in specific operating regions in order to dimension the communications system in a manner that the heaviest rainfall does not easily cause outages and disrupt communications. The rainfall data is kept by meteorologists. The idea of extracting rainfall rate from the attenuation data of wireless communications signals by meteorologists is newer than extracting attenuation from rainfall data. The trend has been motivated by the fact that it is easier and cheaper than estimating rainfall data via other technologies requiring data from satellite or radar.

The worldwide heaviest rainfalls so far recorded, besides the rain that caused Noah's flood around 2348 BC,[2] are 38 mm within a minute in Barot

[2]For an interested reader, it is believed that the universe was created around 4004 BC and the worldwide flood occurred some 1656 years later, which is 1656 AM or 2348 BC. *AM* here means *Anno Mundi* which is the Latin words for *Year After Creation*, while BC means Before Christ. The rain fell for 40 days and covered the highest mountain on Earth to a depth of almost 7 m.

Guadeloupe on 26 November 1970, 401 mm/hr in Shangdi Nei Monggol China on 3 July 1975, 2,493 mm for 48 hours (i.e., 51.9375 mm/hr) in Cherrapunyi India on 15–16 June 1995 and 43 inches in 24 hours (i.e., ca. 45.51 mm/hr) in Alvin Texas USA on 24–25 July 1979. Australia recorded its heaviest rainfall has been around 240 mm/hr.

Example: Cellular System Design

A telecommunications operator seeks to launch a cellular wireless network to serve a certain city. Assume that the heaviest rainfall rate in the city is 305 mm/hr. The operator has choice of the following two mobile technologies:

(1) 4G LTE-Advanced: operating at any of the frequencies 800 MHz, 1.8 GHz, 1.9 GHz and 2.2 GHz.
(2) 5G mobile cellular technology: operating at 3.75 GHz and 38.5 GHz.

The operator also has a choice between the four types of cells:

(1) Picocell with cell radius $r \leq 500$ m.
(2) Microcell with cell radius $0.5 < r \leq 1$ km.
(3) Small macrocell with cell radius $1 < r \leq 3$ km.
(4) Large macrocell with cell radius $3 < r \leq 30$ km.

Assume that the wireless system's link budget has a maximum of 26 dB margin for rain attenuation. What are the choices of cell sizes, radio signal linear polarisation (i.e., vertical or horizontal) and technology combination that are available to the telecommunications operator for an optimum network operation? Justify your answer quantitatively. *Hint.* Use the data in Figs. 5.17 and 5.18.

Solution

We use the given operating frequencies, linear polarisations, cell sizes and the rainfall rate of 305 mm/hr to find the corresponding attenuation from Figs. 5.17 and 5.18. The values obtained are shown in Table 5.3.

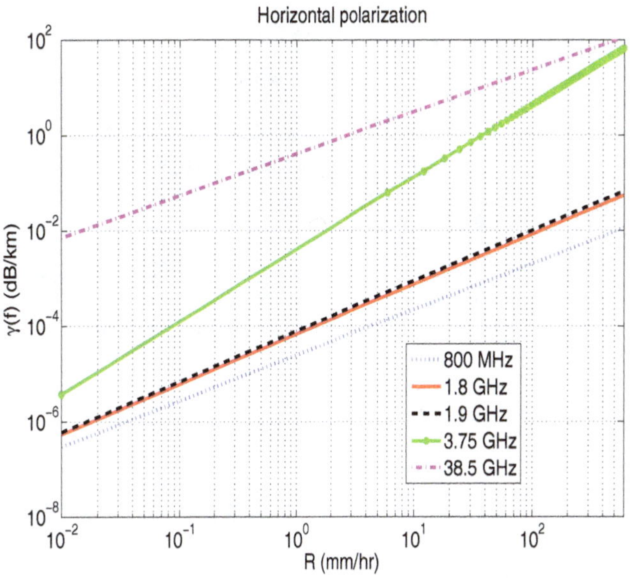

Fig. 5.17 Rain attenuation versus frequency f for horizontal polarisation.

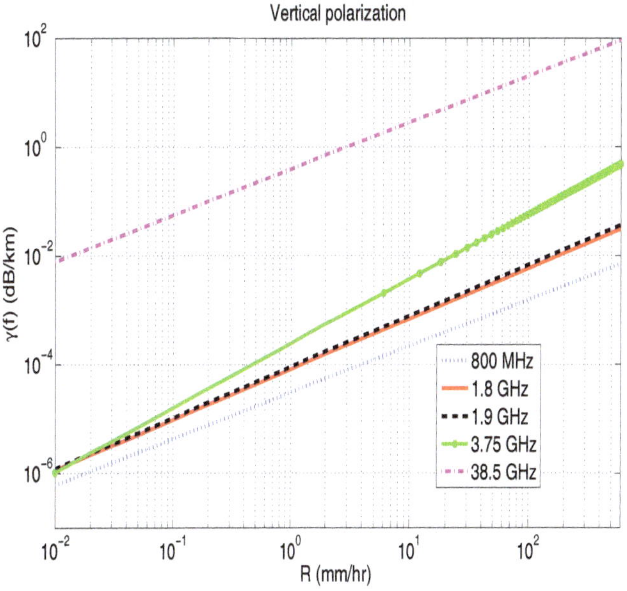

Fig. 5.18 Rain attenuation versus frequency f for vertical polarisation.

We can observe a few important things from the data in Table 5.3. First, the attenuation is higher for horizontal polarisation than vertical polarisation. This is always so for ground wave propagation. Second, the attenuation increases with increasing frequency and cell radius. The cell size indicates the maximum possible distance between the eNodeB and user equipment for effective communications.

Table 5.3 Rain attenuation of radio signals at different frequencies at 305 mm/hr rainfall rate.

f (GHz)	γ (dB/km)	$r = 0.5$ km V/H (dB)	$r = 1$ km V/H (dB)	$r = 3$ km V/H (dB)	$r = 30$ km V/H (dB)
0.8	0.0040/0.0059	0.002/0.003	0.004/0.0059	0.012/0.0177	0.12/0.168
1.8	0.0164/0.0269	0.0082/0.0135	0.0164/0.0269	0.0492/0.0808	0.492/0.807
1.9	0.0194/0.0329	0.0097/0.0164	0.0194/0.0329	0.0581/0.0987	0.582/0.987
3.75	0.217/23.6395	0.1085/11.8198	0.217/23.6395	0.651/70.9186	6.51/709.185
38.5	51.634/62.3154	25.817/31.1577	51.634/62.3154	154.9019/186.9463	1,549.02/1,869.462

V: vertical polarisation, H: horizontal polarisation.

Now to answer the specific question posed:

(1) The operator can use both cellular technologies at all the five frequencies if it chooses picocells with radius not exceeding 500 m and vertical wave polarisation. Horizontal polarisation does not work at frequencies 38.5 GHz and above as the rain attenuation exceeds 26 dB.
(2) Microcell can be used at 800 MHz, 1.8 GHz, 1.9 GHz and 3.75 GHz for both linear polarisations. However, it does not work at 38.5 GHz and above as the rain loss exceeds 26 dB.
(3) Both small and large macrocells can be used at 800 MHz, 1.8 GHz and 1.9 GHz for both linear polarisations, but at 3.75 GHz for only vertical polarisation. However, none works at 38.5 GHz and above.

5.2.7 *Multipath Signal Propagation*

Consider the situation in communications in which the signal $s(t)$ leaves the antenna of the transmitter, but multiple time-delayed and amplitude-scaled versions of $s(t)$ reach the receiving antenna. Such a signal propagation is referred to as **multipath signal propagation**. Causes of multipath propagation are the reflections of the transmitted signal off obstacles (both natural and man-made objects) and the Earth's surface (i.e. ground). The different delays and attenuation experienced by the multipath components are attributed to the fact that they travel along different paths from the transmitter to the receiver. A signal transmission delay causes a phase change in the signal.

Figure 5.19 illustrates three multipath channels. In Fig. 5.19(a) we have a two-ray situation in which a ground-reflected signal and the direct line-of-sight (LOS) signal reach the receiving antenna. If the two antennas are close to each other then ground reflections are negligible, resulting in the situation in Fig. 5.19(b). However, there is a reflection either off a man-made object or a natural object such as a hill, tree or the ions in the atmosphere. The third situation in Fig. 5.19(c) is a three-ray model comprising the LOS path, the ground reflected path and the obstruction-reflected path. Ground reflections usually occur under two different conditions: the antennas are farther apart from each other, or they are too close to the Earth's surface. This is a

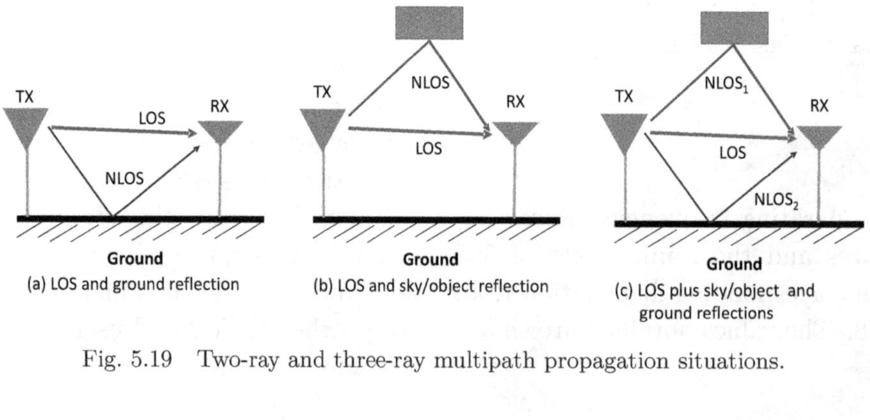

Fig. 5.19 Two-ray and three-ray multipath propagation situations.

reason for mounting antennas on tall masts or towers to achieve the Earth's clearance necessary to avoid ground reflection of signals.

In all multipath cases, the reflective objects must have dimensions larger than the wavelength of the incident wave, as discussed earlier. As illustrated in Fig. 5.12, the multipath signal can result in either constructive interference or destructive interference at the receiving antenna. One of the most popular analytical model for multipath propagation is the tapped-delay line model in Eq. (6.45). Multipath signal propagation is the norm rather than the exception in mobile cellular communications, and it results in constructive and destructive interference called fading.

5.2.8 *Check Your Understanding*

Students are encouraged to check their understanding of the content of this section with the following questions prior to proceeding to the next section.

(1) What is radio frequency? How did the term come about?

(2) What is the cause of the ionisation in the ionosphere?

(3) Two people separated by 100 km have to communicate using a radio system operating at 25 MHz. Can they use ground wave propagation? Justify your answer. If your answer is no, which mode of wave propagation is most suited for the system and why?

(4) Discuss the causes of path loss or large-scale fading of wireless signals.

(5) Discuss the conditions that cause a radio signal to undergo each of the following propagation mechanisms upon encountering an obstruction:

   (a) Scattering
   (b) Absorption (both partial or full)
   (c) Refraction
   (d) Reflection
   (e) Diffraction
   (f) Penetration

(6) Differentiate between the two types of ground-wave propagation modes.

(7) Surface waves are one of the two types of ground wave propagation modes. Explain the reason why surface waves must be vertically polarised for better transmission.

(8) Discuss the two main causes of (or conditions for) radio wave reflections off the ground in LOS communications.

(9) Which of the EM wave propagation mechanisms is most suited for each of the following radio signal propagation modes:

   (a) Surface-wave propagation.
   (b) Space-wave propagation.
   (c) Sky-wave propagation.

(10) Using illustrations, explain the meaning of skip distance and skip zone.

(11) A radio signal at 2 GHz has been emitted from an antenna inclined at 45° to the ground. If the radio signal penetrated the ionosphere into outer space when the antenna was at right angle with the ground, calculate the MUF of the system.

(12) Which frequencies or wavelengths are most suitable for each of the radio signal propagation modes:

   (a) Surface-wave propagation.
   (b) Space-wave propagation.
   (c) Sky-wave propagation.

(13) Using illustration(s), reflect on the meaning of multipath signal propagation. State its causes, effects and remedies.

(14) Explain the difference between constructive and destructive interference.

(15) How do we call the maximum frequency at which radio waves can be transmitted vertically towards the ionosphere and still be refracted back to the Earth?

(16) Explain the meaning of maximum usable frequency as used in wireless communications.

(17) Describe the main cause(s) of multipath signal propagation.

(18) Two people, separated by 5 km, are communicating over a wireless system operating at 1800 MHz. Assume that both user equipments have perfect omnidirectional antennas. Compute the path loss experienced by the signal from one user to the other. Assume that the two antennas are in radio line-of-sight to each other.

(19) A telecommunications operator wants to establish a line-of-sight wireless link between two places. Assume that the wireless system is coverage limited.

(a) What is the maximum possible distance that the two positions can be separated if the antennas are mounted on 30-meter and 45-meter towers?

(b) If one antenna measures 50 meters above ground, what should be the height of the other antenna to achieve a communications' coverage of 2 km?

(20) Using the data in Figs. 5.17 and 5.18, compute the maximum possible cell size to be used by a cellular wireless system if the maximum tolerable rain attenuation is 50 dB. Assume that the maximum rainfall rate in the coverage area is 60 mm/hr.

5.3 Probability Distributions for Radio Wave Modelling

We discuss probability density functions (pdfs) used for the analysis and modelling of the propagation of radio waves. The discussion here is based mostly on the ITU-R document [ITU-R (2017)]. We quote the models for the sake of making the text self-contained. Let X be a random variable (r.v.) describing the magnitude of a channel's impulse response (i.e., $X = |h(t)|$) and x be a single value or realisation of it. The magnitude of a channel's impulse response $|h(t)|$ is referred to as the channel's envelope. Let $f_X(x)$ be the pdf of X and $F_X(x)$ be its cumulative distribution function (CDF). The following definitions are then valid

$$F_X(x) = Pr[X \leq x] = \int_{-\infty}^{x} f_X(y)dy \quad \texttt{and}$$

$$f_X(x) = \frac{d}{dx}F_X(x) \tag{5.26}$$

The following functions are also used in the analysis of communications systems

`(Gauss) error function:`

$$\frac{2}{\sqrt{\pi}}\int_0^x e^{-u^2}du\,. \tag{5.27}$$

`Complementary error function:`

$$erfc(x) = \frac{2}{\sqrt{\pi}}\int_x^\infty e^{-u^2}du\,. \tag{5.28}$$

`Q-function:`

$$Q(x) = \frac{1}{\sqrt{2\pi}}\int_x^\infty e^{-u^2/2}du = \frac{1}{2}erfc\left(\frac{x}{\sqrt{2}}\right). \tag{5.29}$$

Note that $erfc(x) = 1 - erf(x)$ and $Q(x) = \frac{1}{2}[1 - erf(\frac{x}{\sqrt{2}})]$. MATLAB implements these functions as erf, erfc and qfunc, respectively. The functions used by MATLAB to compute their inverses are respectively erfinv, erfcinv and qfuncinv. There is also a scaled complementary error function, which is defined as $erfcx(u) = e^{u^2}erfc(u)$, and implemented in MATLAB as erfcx with the inverse erfcxinv.

5.3.1 *Uniform Probability Distribution*

A random variable X is continuous and uniformly distributed in the interval $[x_1, x_2]$, where $x_2 > x_1$, if its pdf and CDF are respectively

$$f_X(x) = \begin{cases} \frac{1}{x_2 - x_1}, & x_1 \leq x \leq x_2 \\ 0, & \texttt{otherwise} \end{cases}.$$

$$F_X(x) = \int_{-\infty}^{x} f_X(y)dy = \begin{cases} 0, & x < x_1 \\ \frac{x - x_1}{x_2 - x_1}, & x_1 \leq x \leq x_2 \\ 1, & x \geq x_2 \end{cases}.$$

A short analytical expression for the above is $X \sim U[x_1, x_2]$.

5.3.2 *Normal Distribution*

This distribution is used, for example, to model thermal noise effects in communications systems. Every electronic system is liable to thermal noise, making this distribution widely useful. According to the central limit theorem, the sum of a large number of random variables results in another random variable whose distribution approaches the normal distribution, which is also referred to as the Gaussian distribution. A random variable X is normal distributed if its pdf and CDF are respectively

$$f_X(x) = \frac{1}{\sqrt{2\pi}\sigma_X} \exp\left(-\frac{(x-m_X)^2}{2\sigma_X^2}\right). \quad (5.30)$$

$$F_X(x) = \int_{-\infty}^{x} f_X(y)dy = \frac{1}{2}\left[1 + \texttt{erf}\left(\frac{x-m_X}{\sqrt{2}\sigma_X}\right)\right]$$
$$= 1 - \frac{1}{2}\texttt{erfc}\left(\frac{x-m_X}{\sqrt{2}\sigma_X}\right). \quad (5.31)$$

Note that σ_X and m_X are the standard deviation and mean of the random variable X. We use the symbolism $X \sim \mathcal{N}(m_X, \sigma_X^2)$ to indicate that X is a random variable with mean m_X, standard deviation σ_X and is normal distributed. If $\sigma_X^2 = 1$ and $m_X = 0$, then $X \sim \mathcal{N}(0,1)$. We say that X has the standard (or normalised) normal distribution.

5.3.3 *Log-Normal Distribution*

We use the log-normal distribution to model continuous random quantities when the distribution is skewed. For example, it is used to model the effects of shadowing in cellular communications. Shadowing causes slow fading of wireless signals. It is also used in modelling the duration of signal fading. A random variable X, $X > 0$ is log-normal distributed if its pdf and CDF are respectively

$$f_X(x) = \frac{1}{x\sigma_X\sqrt{2\pi}} \exp\left(-\frac{(\log_e x - m_X)^2}{2\sigma_X^2}\right). \quad (5.32)$$

Similarly, a random variable X, $X > 0$ is log-normal distributed with mean μ and standard deviation σ if $X = e^Z$ and Z is normally distributed with mean μ and standard deviation σ, i.e., $Z \sim \mathcal{N}(\mu, \sigma)$.

$$F_X(x) = \int_{0}^{x} f_X(y)dy = \frac{1}{2}\left[1 + \texttt{erf}\left(\frac{\log_e x-m_X}{\sqrt{2}\sigma_X}\right)\right]$$
$$= 1 - \frac{1}{2}\texttt{erfc}\left(\frac{\log_e x-m_X}{\sqrt{2}\sigma_X}\right). \quad (5.33)$$

Note that σ_X and m_X are the standard deviation and mean of the natural logarithm of the random variable X. That is, σ_X and m_X are the standard deviation and mean of the random variable $\log_e(X)$. We use the symbolism $X \sim \mathcal{LN}(m_X, \sigma_X^2)$ or $\log_e(X) \sim \mathcal{N}(m_X, \sigma_X^2)$ to indicate that X is a random variable with mean m_X, standard deviation σ_X and is log-normal distributed.

5.3.4 *Rayleigh Probability Distribution*

This is one of the distributions used to model small-scale fading caused by multipath signal propagation. Given the two independent but identically-distributed (i.i.d.) r.vs. $X_1 \sim \mathcal{N}(0,\sigma^2)$ and $X_2 \sim \mathcal{N}(0,\sigma^2)$, the r.v.

$$X = \sqrt{X_1^2 + X_2^2} \quad (5.34)$$

is Rayleigh distributed. Note that X equals the magnitude of the complex number

$$Z = X_1 + jX_2\,. \quad (5.35)$$

Also, assume that we have a large number of signals X_i, $i = 1, 2, \ldots$, with the amplitudes a_i and phases ϕ_i. If

$$a_i \approx a_j,\ i \neq j \ \texttt{and}\ \phi_i \sim U(0, 2\pi),\ \forall i \quad (5.36)$$

then the envelope of the composite signal $X = \sum_i X_i$ is Rayleigh distributed. This means that none of the signals summed dominates the others in strength as they have about the same amplitudes. Such a situation occurs, for example, if a single ray from a transmitting antenna results in multiple rays reaching the receiving antenna, but none of the rays followed the LOS path. Sometimes, the amplitudes are required to be normal distributed.

A random variable X is said to be Rayleigh distributed if its pdf and CDF are respectively

$$f_X(x) = \frac{x}{\sigma^2} \exp\left(\frac{-x^2}{2\sigma^2}\right). \quad (5.37)$$

$$F_X(x) = \int_0^x f_X(y)dy = 1 - \exp\left(\frac{-x^2}{2\sigma^2}\right). \quad (5.38)$$

Note that the values of the Rayleigh-distributed random variable X can only be positive. The Rayleigh model is used if the receiving antenna is far away from the transmitting antenna, such that signals reaching the receiver contain no LOS path. The mean level $E[X]$, the average power P_X and the variance $Var(X)$ in the fading signal, respectively, are

$$E[X] = \int_0^\infty x f_X(x)dx = \sqrt{\frac{\pi}{2}}\sigma \tag{5.39}$$

$$P_X = E[X^2] = \int_0^\infty x^2 f_X(x)dx = 2\sigma^2 \tag{5.40}$$

$$\begin{aligned} Var(X) &= E[X^2] - (E[X])^2 \\ &= \frac{1}{2}(4-\pi)\sigma^2 \end{aligned} \tag{5.41}$$

We shall refer to a r.v. X which is Rayleigh distributed as $X \sim R(\sigma^2)$. There are two distributions which generalise the Rayleigh distribution. These are the Nakagami-m and Nakagami-Rice distributions.

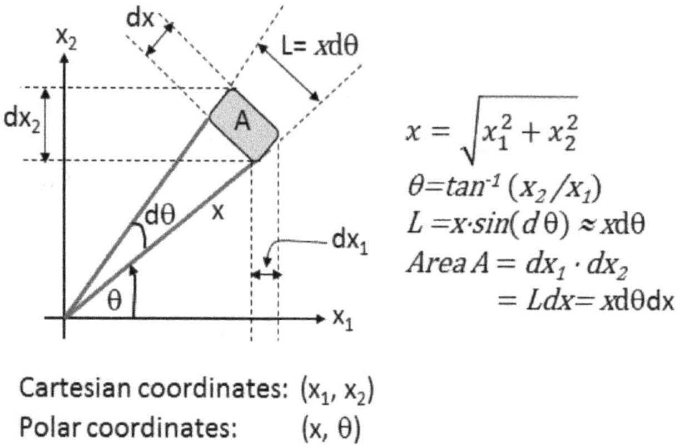

Fig. 5.20 Deriving the Rayleigh distribution.

Deriving the Rayleigh Distribution

If $X_k \sim \mathcal{N}(0, \sigma^2)$, $k = 1, 2$ and X_k are i.i.d. random variables, then the joint probability density function (pdf) is

$$\begin{aligned} f_{X_1,X_2}(x_1, x_2) &= f_{X_1}(x_1) \cdot f_{X_2}(x_2) \\ &= \frac{1}{2\pi\sigma^2} e^{-\frac{x_1^2+x_2^2}{2\sigma^2}} \end{aligned} \tag{5.42}$$

We convert Eq. (5.42) from the Cartesian coordinates system (x_1, x_2) into the polar coordinates system (x, Θ). This is illustrated in Fig. 5.20.

The conversion is

$$(x_1, x_2) \leftrightarrow (x, \Theta)$$

and $x = \sqrt{x_1^2 + x_2^2}$ and $\Theta = \mathtt{tan}^{-1}(x_2/x_1)$. The area $dx_1 dx_2$ in the Cartesian coordinate system equals the area $xdxd\theta$ in the polar coordinates system. This is illustrated in Fig. 5.20. We therefore have the equivalent relationship

$$\begin{aligned} &Pr[x_1 \le X_1 \le x_1 + dx_1;\ x_2 \le X_2 \le x_2 + dx_2] \\ &= \frac{1}{2\pi\sigma^2} e^{-\frac{x_1^2+x_2^2}{2\sigma^2}} dx_1 dx_2 \\ &= \frac{1}{2\pi\sigma^2} e^{-\frac{x^2}{2\sigma^2}} xdxd\theta \\ &= \frac{x}{\sigma^2} e^{-\frac{x^2}{2\sigma^2}} dx \cdot \frac{1}{2\pi} d\theta \\ &= f_{X,\Theta}(x,\theta)dxd\theta = f_X(x)dx \cdot f_\Theta(\theta)d\theta \end{aligned} \tag{5.43}$$

The last expression in Eq. (5.43) results from the fact that Θ and X are independent r.vs. Therefore, the phase of $X_1 + jX_2$ is uniformly distributed as

$$f_\Theta(\theta) = \frac{1}{2\pi}, \quad -\pi \le \theta \le \pi \tag{5.44}$$

while the amplitude $X = |X_1 + jX_2| = \sqrt{X_1^2 + X_2^2}$ is Rayleigh distributed with the pdf

$$f_X(x) = \frac{x}{\sigma^2} e^{-\frac{x^2}{2\sigma^2}}, \quad x > 0. \tag{5.45}$$

We apply a simple trick on the Rayleigh pdf in Eq. (5.45) to obtain the CDF. Note that

$$\frac{d}{dx} F_X(x) = f_X(x)$$

and

$$\frac{d}{dx}[1 - e^{-\frac{x^2}{2\sigma^2}}] = \frac{x}{\sigma^2} e^{-\frac{x^2}{2\sigma^2}} \overset{(5.45)}{=} f_X(x).$$

Thus,

$$F_X(x) = 1 - e^{-\frac{x^2}{2\sigma^2}}. \tag{5.46}$$

5.3.5 *Combined Rayleigh and Log-Normal Distribution*

In certain radio wave propagation environments, such as tropospheric scatter, the propagation medium is inhomogeneous, causing remarkable long-term variations. Modelling such situations require a hybrid distribution combining Rayleigh distribution to capture the fast variations and log-normal distribution to capture slow variations in the signal's intensity. We then model the random variable at a given time as Rayleigh distributed, but with variance or mean value which is also a random variable having a log-normal distribution. A random variable X is both Rayleigh and log-normal distributed if its pdf and CDF are respectively

$$f_X(x) = \sqrt{\frac{2}{\pi}}kx\int_{-\infty}^{\infty}\exp\left(-kx^2\exp(-2(\sigma t+m)) - 2(\sigma t+m) - \frac{t^2}{2}\right)dt \tag{5.47}$$

$$F_X(x) = \int_0^x f_X(y)dy = 1 - \frac{1}{\sqrt{2\pi}}\int_{-\infty}^{\infty} \times\exp\left(-kx^2\exp(-2(\sigma t+m)) - \frac{t^2}{2}\right)dt \tag{5.48}$$

where m and σ, expressed in nepers, are the mean and the standard deviation of the normal probability distribution associated with the log-normal distribution. The value of k depends on the interpretation of m and σ [ITU-R (2017)]:

(1) If m and σ are the standard deviation and mean of the natural logarithm of the most probable value of the Rayleigh probability distribution, then $k = 1/2$;
(2) If m and σ are the standard deviation and mean of the natural logarithm of the median value of the Rayleigh probability distribution, then $k = \log_e(2)$;
(3) If m and σ are the standard deviation and mean of the natural logarithm of the mean value of the Rayleigh probability distribution, then $k = \pi/4$; and
(4) If m and σ are the standard deviation and mean of the natural logarithm of the root mean square value of the Rayleigh probability distribution, then $k = 1$.

We shall refer to a r.v. X which is both Rayleigh and normal distributed with the parameters σ and m as $X \sim RN(\sigma, m)$.

5.3.6 *Nakagami-m Probability Distribution*

This is a generalised model for fast fading which is proposed in [Nakagam (1958)]. The amplitude of the sum of multiple independent but identically distributed Rayleigh-fading signals has the Nakagami-m distribution. The pdf is

$$f_X(x) = \frac{2m^m}{\Gamma(m)\rho^m}x^{2m-1}e^{-\frac{mx^2}{\rho}}, \quad x \geq 0 \tag{5.49}$$

where the scale parameter $\rho = E[X^2]$ is the mean power in the fading signal, and the fading factor is $m = \frac{\rho^2}{E[(X-\rho)^2]}$. Note the interesting properties of the Nakagami-m distribution:

- If the random variable X is Nakagami-m distributed, then the r.v. X^2 is gamma distributed.
- If $m = 1$ (i.e. Nakagami-1) equals the Rayleigh distribution.
- If $m = \frac{1}{2}$ (i.e. Nakagami-$\frac{1}{2}$) equals the one-sided normal distribution.

We can use the Nakagami-m model to approximate the pdf of the power in a Rician fading signal using the substitution

$$m = \frac{(1+K)^2}{2k+1} \Leftrightarrow k = (m-1) + \sqrt{m^2 - m}, \; m > 1 \tag{5.50}$$

where k is the Rician factor. This approximation is very useful for analysis in situations where the multipath signal contains a *specular path* (aka LOS path), but the presence of the Bessel functions in the Rician model makes it too difficult to use. We shall refer to a r.v. X which is Nakagami-m distributed with the parameters ρ and m as $X \sim Nm(\rho, m)$. Note that in the Nakagami-Rice distribution X is viewed as signal-to-noise ratio (SNR).

5.3.7 *Nakagami-Rice Probability Distribution*

This distribution is also referred to as *Rician distribution*. This model is more appropriate for radio wave propagation in a microcell where there is a LOS path (aka specular wave component) in the multipath signal. Sometimes one of the multipath signals which reach the receiving antenna in a wireless communication system is the LOS component. The strength of such a component is higher than that of any of the other multipath components. In such radio propagation environments the time-varying envelope of the received composite signal can be modelled by the Nakagami-Rice probability distribution. Also referred to as Nakagami-n distribution or Rician distribution, the Nakagami-Rice distribution is a generalisation of the Rayleigh probability distribution. This distribution is used to model the following situations:

- Given two vectors X_1 and X_2, where X_1 is a fixed deterministic vector while the length of X_2 is Rayleigh distributed, the composite vector $X = X_1 + X_2$ has a length which is Nakagami-Rice distributed.
- The received signal X is the sum of multiple signals, each with uniformly distributed phase and normal distributed amplitude, and one of them is LOS path, then X is Nakagami-Rice distributed.

The pdf of a random variable X which is Nakagami-Rice distributed is

$$f_X(x) = \frac{x}{\sigma^2} \exp\left(- \frac{x^2+\alpha^2}{2\sigma^2} \right) I_0\left(\frac{\alpha x}{\sigma^2} \right) \tag{5.51}$$

where $I_0(z)$ is the modified Bessel function of the first kind and of zero order, which is defined as

$$I_0(z) = \frac{1}{\pi} \int_0^{\pi} e^{iz\cos\theta} d\theta = \sum_{n=0}^{\infty} (-1)^n \frac{(0.25z^2)^n}{(n!)^2} .$$

Note that X is the envelope of the multipath or fading signal. The total mean power in all paths, referred to as scale parameter, is $\Omega = \alpha^2 + 2\sigma^2$, while the power in only the specular or LOS path is $P_{\mathtt{LOS}} = \alpha^2$. The parameter

$$k = \frac{\textit{power in LOS path}}{\texttt{power in all non-LOS paths}} = \frac{\alpha^2}{2\sigma^2} \tag{5.52}$$

is referred to as the **shape parameter** or **Rician factor** or k-factor of the distribution. If the amplitude of the LOS path is strong compared with all other paths (i.e. $\alpha \approx \sigma$) then the Nakagami-Rice distribution becomes normal distribution. On the other hand, if there is no LOS path (i.e. $\alpha \approx 0$), then the Nakagami-Rice distribution becomes Rayleigh distribution. Thus, larger the value of k is, the stronger the LOS path relative to the non-LOS path. In-door signal propagation usually has $k > 1$. We shall refer to a r.v. X which is Nakagami-Rice distributed with the parameters σ and α as $X \sim NR(\sigma, \alpha)$. Figures 5.21 and 5.22 illustrate graphically the CDFs and pdfs of Gaussian, log-normal, Rayleigh, Nakagami-m with $m = 1/2$ and Nakagami-Rice with $\alpha = 2$ random variables.

An estimate of the k-factor is ([Celplan (2017)], p. 24)

$$k = \frac{10(h_{\mathtt{a}}/3)^{0.46}(\theta_{\mathtt{a}}/17)^{0.62} F_{\mathtt{s}}}{\sqrt{d}} \tag{5.53}$$

where $F_{\mathtt{s}}$ is the terrain-dependent morphology factor, $h_{\mathtt{a}}$ is antenna height, $\theta_{\mathtt{a}}$ is antenna beamwidth and d is distance. The morphology factors for some terrains are: 2 for water and suburban area, 3 for low vegetation, 2.5 for medium and dense vegetation, 1.5 for urban area and 1 for dense urban area.

5.3.8 *Gamma Probability Distribution*

This is a generalisation of the exponential probability distribution. The pdf of a random variable X which is Gamma distributed is

$$f_X(x) = \frac{\alpha^{\nu}}{\Gamma(\nu)} x^{\nu-1} e^{-\alpha x} \tag{5.54}$$

where α is a scale parameter for the variable x and Γ is the Euler function of second order. The corresponding CDF is difficult to evaluate in closed form. An approximation used for practical system design is

$$F_X(x) \approx 1 - \frac{\nu e^{-\alpha x}}{0.68 + \alpha x + 0.28 \log(\alpha) x}, \quad \nu < 0.1,\ \alpha x > 0.03 . \tag{5.55}$$

We shall refer to a r.v. X which is gamma distributed with the parameters ν and α as $X \sim G(\nu, \alpha)$.

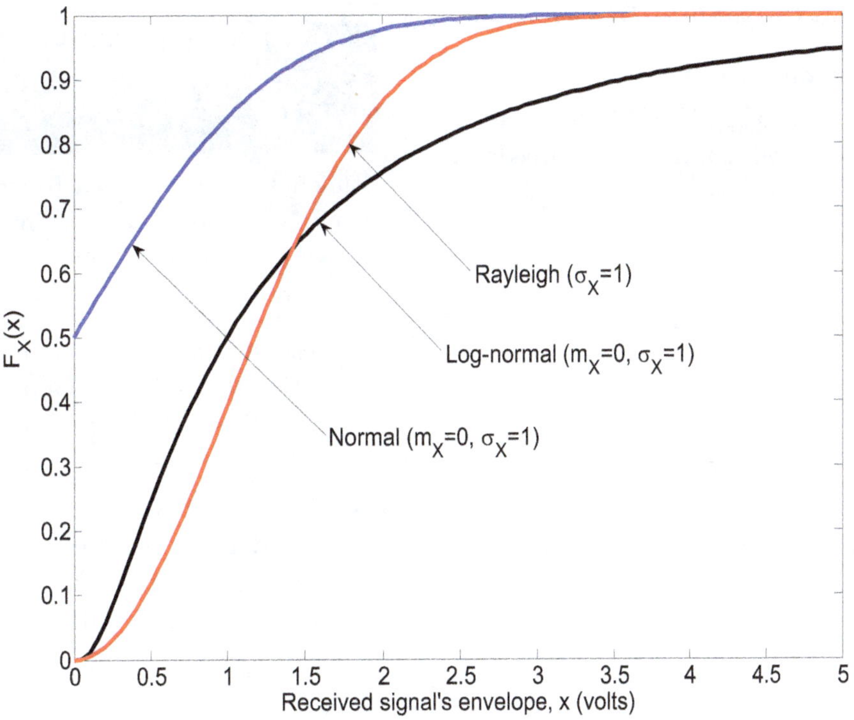

Fig. 5.21 Cumulative distribution functions ($\sigma_X = 1, m_X = 0, \alpha = 2, \rho = 1$).

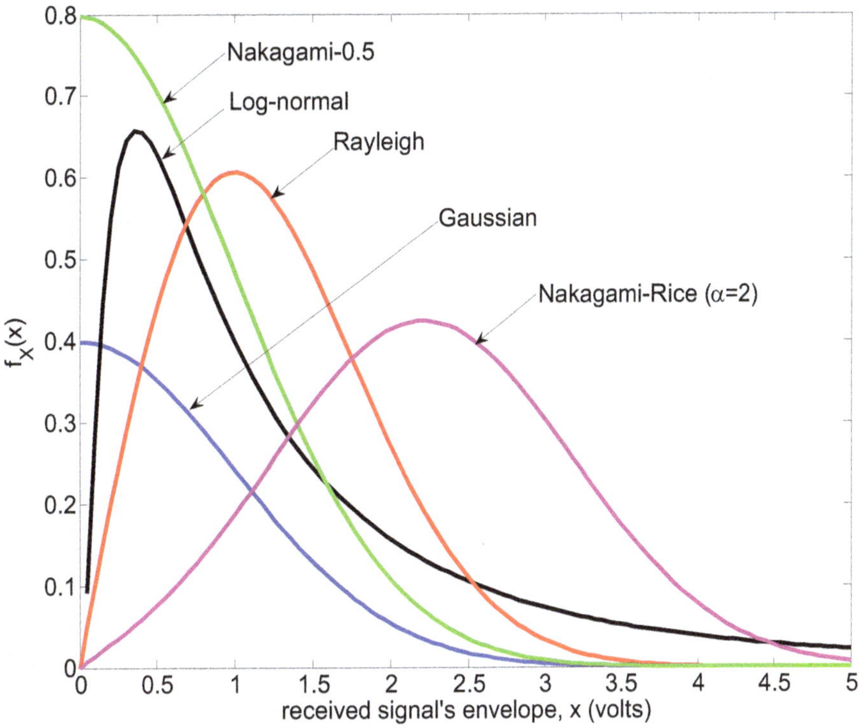

Fig. 5.22 Probability density functions ($\sigma_X = 1, m_X = 0, \alpha = 2, \rho = 1$).

5.3.9 *Pearson χ^2 Probability Distribution*

Given the n random variables $X_i \sim \mathcal{N}(m_i, \sigma_i)$, $i = 1, 2, \ldots, n$ then the r.v. $\sum_{i=1}^{n} \frac{(X_i - m_i)^2}{\sigma_i}$ has the χ^2 probability distribution with n degrees of freedom. The pdf and CDF of a random variable X which is χ^2 distributed are respectively

$$f_X(\chi^2) = \frac{1}{2^{\frac{\nu}{2}}\Gamma(\frac{\nu}{2})} \exp\left(-\frac{\chi^2}{2}\right)(\chi^2)^{\frac{\nu}{2}-1} \quad (5.56)$$

$$F_X(\chi^2) = \frac{1}{2^{\frac{\nu}{2}}\Gamma(\frac{\nu}{2})} \int_0^{\chi^2} \exp\left(-\frac{t^2}{2}\right) t^{(\frac{\nu}{2}-1)} dt \quad (5.57)$$

where $\Gamma(\frac{\nu}{2})$ is the Euler functions with the definition

$$\Gamma\left(\frac{\nu}{2}\right) = \begin{cases} (\frac{\nu}{2} - 1)!, & \nu \texttt{ is even} \\ (\frac{\nu}{2} - 1)(\frac{\nu}{2} - 2)\cdots\frac{1}{2}\sqrt{\pi}, & \nu \texttt{ is odd} \end{cases}.$$

The Pearson χ^2 distribution and the Nakagami-m convert into each other if

$$\nu = 2m \texttt{ and } \chi^2 = \frac{2m}{\rho}x^2. \quad (5.58)$$

We shall refer to a r.v. X which is Pearson χ^2 distributed with the parameters χ^2 and ν as $X \sim P(\chi^2, \nu)$.

5.3.10 *Weibull Probability Distribution*

This is another generalised distribution for small-scale fading. We previously learned that given the two independent but identically-distributed r.vs $X_1 \sim \mathcal{N}(0, \sigma^2)$ and $X_2 \sim \mathcal{N}(0, \sigma^2)$, the r.v. $Y = \sqrt{X_1^2 + X_2^2}$ is a Rayleigh distributed r.v. However, the r.v. $X = (X_1^2 + X_2^2)^{\frac{1}{n}}$ is Weibull distributed with the pdf

$$f_X(x) = \frac{nx^{n-1}}{2\sigma^2} e^{-\frac{x^n}{2\sigma^2}} \quad (5.59)$$

where $E[x^2] = 2\sigma^2$. Note that Y and X are envelopes of radio signals.

5.3.11 *Check Your Understanding*

Students are encouraged to check their understanding of the content of this section with the following questions prior to proceeding to the next section.

(1) Assume that X is a random variable indicating the duration of a phone call. Assume that $X \sim U(2, 4)$. Find the:
  (a) pdf of X.
  (b) CDF of X.
  (c) Mean duration of a phone call, i.e., the mean of X.
  (d) Standard deviation of X.

(2) Assume that X is a random variable indicating the path loss experienced by a radio signal, and that $X \sim \mathcal{N}(2, 8)$. Find the probability that $X < 5$. All numbers are in ratio form, but not in decibels.

(3) Describe how the Rayleigh distribution differs from the Nakagami-Rice probability distribution.

(4) Under what conditions does the Nakagami-m distribution degenerate to the Rayleigh distribution?

(5) Let us study the probability distributions in Fig. 5.21. Assume that the random variable X is the envelope of the impulse response of a wireless channel. Assume further that the standard deviation of X is unity and its mean is zero. Compute the probability that X is 2.5 volts if X is
  (a) Gaussian distributed.
  (b) Rayleigh distributed.
  (c) Log-normal distributed.

  Which of the situations is most probable? What is the probability that X is a Rayleigh-distributed r.v. and lies in the range 1.5 to 2.5?

5.4 Review of Important Terms and Acronyms

atmospheric divisions	circular polarisation	constructive interference
critical frequency	destructive interference	diffraction
direct wave	distribution	E-field
electric field	electromagnetic spectrum	EM wave
frequency spectrum	FRIIS formula	Gamma distribution
ground wave	horizontal polarisation	interference
line-of-sight (LOS)	log-normal distribution	longitudinal wave
magnetic field	Maxwell's equations	M-field
microwaves	mmWaves	MUF
multipath propagation	Nakagami-m distribution	Nakagami-Rice distribution
normal distribution	Pearson-Chi distribution	plane wave
polarisation of EM wave	probability distribution	radio waves
rain attenuation	Rayleigh distribution	reflection
refraction	Rice distribution	Rician factor
right-hand rule	shape factor	sky wave
Snell's laws	space wave	superposition principle
surface wave	transverse wave	uniform distribution
vertical polarisation	wave	wave scattering
wavelet	wavefront	Weibull distribution

5.5 Supplementary Reading

(1) Akira Ishimaru, *Electromagnetic Wave Propagation, Radiation, and Scattering: From Fundamentals to Applications*, Wiley-IEEE Press, 2017.

(2) Tapan K. Sarkar and Magdalena Salazar Palma, *The Physics and Mathematics of Electromagnetic Wave Propagation in Cellular Wireless Communication*, Wiley-IEEE Press, 2018.

(3) Donald Dearholt and William McSpadden, *Electromagnetic Wave Propagation*, McGraw-Hill Inc., 1973.

(4) Eugene I. Nefyodov and Sergey M. Smolskiy, *Electromagnetic Fields and Waves: Microwave and mmWave Engineering with Generalized Macroscopic Electrodynamics*, Springer, 2018.

(5) Gokhan Apaydin and Levent Sevgi, *Radio Wave Propagation and Parabolic Equation Modeling*, Wiley-IEEE Press, 2017.

(6) John A. Richards, *Radio Wave Propagation: An Introduction for the Non-Specialist*, Springer, 2008.

Chapter 6

Wireless Channels

Chapter's Learning Outcomes[1]

The interest in the transmission of information over wireless media has been on the increase, and the trend will continue. For this reason, a solid knowledge of wireless channels and how they treat signals propagating through them is very fundamental in communications engineering. The wireless channel, in general, fades in both time and frequency. **Time fading** is caused by Doppler spread due to relative motion between the two antennas communicating. It causes a signal's quality to change over time. **Frequency fading** is caused by multipath signal propagation, in which the wireless channel varies over frequency. The fading property of wireless media renders mobile system design more challenging than wireline systems.

This chapter treats the principles behind wireless signal propagation, wireless channel behaviour and wireless channel models used for system design and analysis. In particular, we study:

(1) Physical principles affecting radio signals as they move through the atmosphere.
(2) The three types of fading affecting radio signals, namely, large-scale fading (aka path loss), medium-scale fading (aka lognormal fading) caused by shadowing and small-scale fading caused by multipath signal propagation and Doppler shifts. Path loss results from transmitter to receiver travel distance, signal absorption by oxygen and water vapour, signal spreading, rain and other types of precipitation such as snow.
(3) Extensive radio propagation models relating to operating frequency, transmission distance and other environmental factors for all the three types of signal fading for both below 6 GHz and above 6 GHz radio signals. Propagation models are used to predict the effects of radio channels on signals under different constraints, which is useful for system design and analysis. For example, path loss prediction determines the effective coverage area of a eNodeB in LTE-based cellular network.
(4) Wireless channel models proposed for next-generation mobile wireless system referred to as 5G. This is especially essential for system designers and research students.

6.1 Introduction

Wireless communications is a branch of communications in which the transmission medium is not a cable, but the ether. Wireless communications occur through a special form of waves, called electromagnetic (EM) waves, which travel through the Earth's atmosphere. This means that the proper understanding of EM waves and its propagation behaviour are very vital for communications specialists. For this reason, readers without background in wave propagation are advised to first read the preceding chapter on EM wave propagation. A wireless communications' link requires two electronic devices, each equipped with an antenna. The space between the transmitting antenna and the receiving antenna is called **channel** or wireless channel.

[1]Eliminating the distinction between information rich and information poor countries is critical to eliminating other inequalities between the North and the South — President Nelson Mandela, 1996.

Inside an electronic device, such as a smart phone, the signal is an electrical signal, but the signal travels along the channel as EM waves. A (radio) wave which is sufficiently far away from its generating antenna appears to be flat. Such a wave is referred to as **plane wave**. Our discussions on EM waves focus on plane waves. Every wireless signal occupies a certain range of frequency.

The ether or air through which information is transferred from one point to another is called a wireless channel. Wireless channels can be **broadcast channels** or **non-broadcast channels**. Common examples of the former are channels for AM and FM radio broadcasting, wireless LAN links, TV broadcasting, and downlink signalling and control channels of cellular communications networks. Traffic channels in cellular communications networks are among the non-broadcast channels. Communications networks using broadcast channels are called **broadcast networks**. In broadcast networks, many receivers are tuned to the transmissions of a single radio transmitter. Mobile stations tune to the downlink transmissions of base stations over broadcast signalling channels.

The characteristics, and hence the effects, of a wireless channel depends on many factors, including time, spatial location or position, operating carrier frequency, topography or terrain in the channel (e.g. hills, valleys, water), heights of the transmitting and receiving antennas, and obstructions. The profile of the obstructions are very important. *What is a channel model?* A channel model attempts to capture in analytical form all the physical processes which affect a signal as it moves along the channel. We seek analytical models that we can use to quantify the interdependence between the instantaneous channel's behaviour and factors such as frequency, moving velocity of the user, bandwidth of the channel, Doppler spread and multipath delay spread. A channel model represents the cumulative impairments of the transmission medium (or channel) on signals transferred through it. Channel modeling is very crucial for the proper design of communications systems. Its importance can be summarised in two statements. It is used for system design and analysis and for the design of algorithms.

The ill-effects of a channel of signals is usually quantified using the channel's impulse response (in time domain) or frequency response (in the frequency domain). However, it may be very difficult or practically impossible to find the impulse response of a fast fading channel. Therefore, another approach used to quantify the characteristics of a channel is called **channel sounding**. This is achieved by transmitting a known signal, referred to as training sequence or a single-tone wave, through the channel and studying the statistics of the output of the channel. This section presents the channel models commonly used for the analysis and design of wireless communications systems.

6.1.1 *Classification of Wireless Fading Channels*

The reduction in the strength or power in a signal as it travels through the channel is referred to as **fading**. Fading is usually modelled analytically as a signal with random phase and amplitude which depends on time and location, among other factors. Wireless communication signals are affected by three types of fading, as illustrated in Fig. 6.1. These are large-scale fading (aka path loss or long-term fading), medium-scale fading (aka slow fading or shadowing), and small-scale fading (aka fast fading or short-term fading). Note that some texts classify channel fadings into two: large-scale fading and small-scale fading; with large-scale fading referring to both path loss and shadowing. Still other texts classify channels into path loss, shadowing as large-scale fading and small-scale fading. Path loss is deterministic while both slow fading and fast fading are stochastic in nature.

All three types of fading are measured at the wireless receiver. Large-scale fading is the attenuation endured by the signal after covering a large distance (e.g. hundreds of wavelengths) from transmitting to receiving antennas, or over a large time scale. Small-scale fading, however, is the random attenuation experienced by a signal over short time period or over small distances (in the order of a few wavelengths or even a fraction of the wavelength). Medium-scale fading lies between the two extremes.

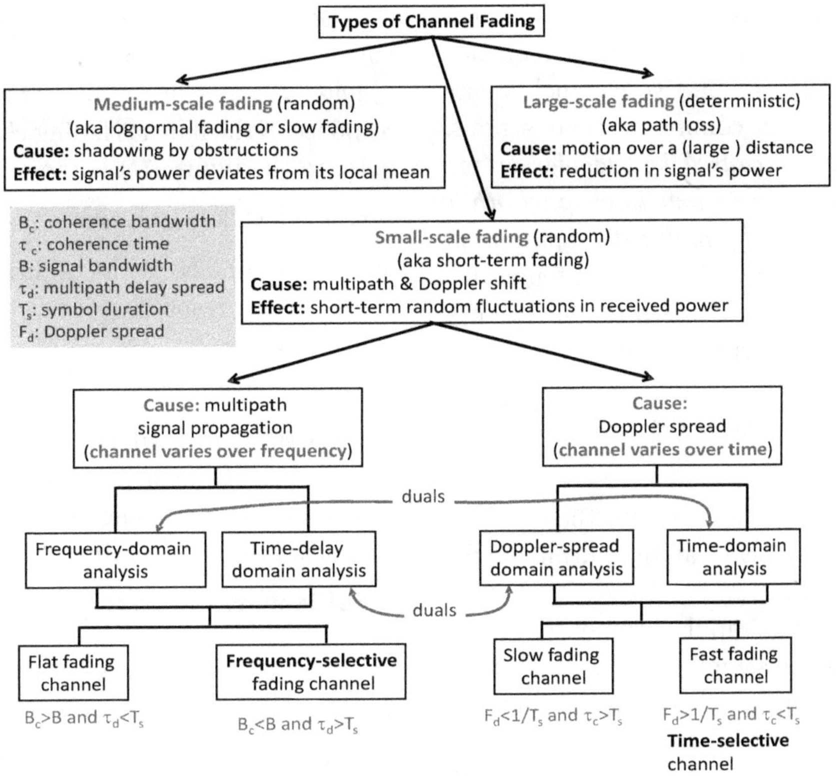

Fig. 6.1 The three basic types of wireless fading and their causes and effects.

Small-scale fading is also called **microscopic fading**, while medium-scale fading is also referred to as **macroscopic fading**.

Assume that a signal leaves the transmitting antenna with the effective isotropic radiated power $EIRP = P_t G_t$, then the received power P_r has the general form

$$P_r = \alpha_T P_t G_t G_r, \texttt{ where } \alpha_T = L_{\text{PL}} L_{\text{MF}} L_{\text{SF}} < 1 \tag{6.1}$$

where α_T is the total attenuation experienced by the signal through the wireless channel, L_{PL} is attenuation due to path loss, L_{MF} is attenuation due to medium-scale fading and L_{SF} is attenuation due to small-scale fading. Each of the three fading types contributes to α_T. We shall explore them in turn later in this chapter.

Let us note the following definitions prior to proceeding:

- T_{s}: symbol period or duration, measured in seconds
- $R_{\text{s}} = \frac{1}{T_{\text{s}}}$: symbol/signalling rate, measured in baud
- τ_{d}: root-mean-square multipath delay spread, measured in seconds
- f_{d}: Doppler shift (aka Doppler-shifted frequency), measured in Hertz
- f_{m}: maximum Doppler shift, measured in Hertz
- $F_{\text{d}} = 2f_{\text{m}}$: Doppler spread (aka bandwidth of Doppler spectrum), measured in Hertz
- B: bandwidth of the signal, measured in Hertz
- B_{c}: wireless channel's coherence bandwidth, measured in Hertz
- τ_{c}: wireless channel's coherence time, measured in seconds

These parameters are used to classify or characterise wireless channels, with respect to the information-bearing signals passing along them, into the four types: non-dispersive channels, time-dispersive channels, frequency-dispersive channels, and time- and frequency-dispersive channels.

A wireless channel is a **time-dispersive channel** if $\tau_{\text{d}} \gg \frac{1}{B}$ or $\tau_{\text{d}} \gg T_{\text{s}}$, otherwise it is non-time dispersive. A time-dispersive channel causes frequency-selective fading to communications signals. A wireless channel is frequency-dispersive if $F_d \gg \frac{1}{T_s}$ or $F_d \gg B$, otherwise it is non-frequency dispersive. A frequency-dispersive channel causes time-dependent fading to communications signals. *Frequency dispersion is caused by Doppler spread due to motion and time-variations in channel, while time dispersion is caused by multipath signal propagation.* A small Doppler shift due to antenna displacements can change the carrier frequency of a signal and cause inter-carrier interference (ICI) or fast fading.

Let us note the following definitions prior to proceeding:

- $x(t)$, $x[n]$: signal transmitted into the channel.
- $r(t)$, $r[n]$: signal received at the output of the channel if $x(t)$ is sent.
- $h(t;\tau)$, $h[n;k]$: the impulse response of the channel. This is the signal at the output of the channel at time t in response to a unit impulse $\delta(t)$ sent through it at an earlier time $t-\tau$.
- $n(t)$: additive noise which is not correlated with the signal of interest.

Note that $x[n]$ is the discrete time version of $x(t)$ and so on, which can be obtained through sampling. The impulse response, or its frequency-domain equivalent called frequency response, is usually used to analytically characterise the channel.

6.1.1.1 *Non-dispersive Wireless Channels*

These are also referred to as **flat-flat fading channels**. Their impulse response, in the ideal case, does not vary over time, neither does their frequency response depend on frequency. Channels of this type are also referred to as non-time-selective, non-frequency-selective channels. A wireless channel is classified a flat-flat fading channel if

`Non-dispersive wireless channel:`

$$\tau_d \ll \frac{1}{B} \text{ or } F_d \ll \frac{1}{T_s}. \tag{6.2}$$

The impulse response of a channel is non-frequency-selective if it fulfills $h(t;\tau) = h(t)\delta(\tau)$. If the channel is also non-time-selective then its impulse response does not depend on time. Hence, a non-frequency-selective, non-time-selective channel has the impulse response $h(t;\tau) = h(t)\delta(\tau) = h\delta(\tau)$ in the continuous-time domain and $h[n;k] = h[n]\delta[k] = h\delta[k]$ in the discrete-time domain. The response of the flat-flat fading channel with additive noise to the input $x(t)$ is then

$$r(t) = hx(t) + n(t) \text{ and } r[i] = hx[i] + n[i] \tag{6.3}$$

where $n(t)$ or $n[i]$ is the additive channel noise at times t and i respectively.

6.1.1.2 *Time-dispersive Wireless Channels*

A wireless channel which is time-dispersive but frequency non-dispersive is also referred to as non-time-selective, frequency-selective channel. Such channels fulfill the condition

`Time-dispersive wireless channel:`

$$\tau_d \gg \frac{1}{B} \text{ or } F_d \ll \frac{1}{T_s}. \tag{6.4}$$

Another equivalent expression is frequency-selective but time-flat fading channel. The main cause of time dispersion in a channel is **multipath signal propagation**. Time dispersion in a channel decides whether the channel causes frequency-selective or frequency-flat fading to communications signals. A frequency-selective channel causes a distortion called **inter-symbol interference** (ISI) at the receiver. ISI is the spreading of symbols in time to disturb neighbouring symbols.

The level of time dispersion in a channel is quantified by the channel's coherence bandwidth (B_c) and the multipath delay spread (τ_d) within a local area. Small-scale fading based on time dispersion has the following types

`Flat fading:`

$$B_c > B \text{ and } \tau_d < T_s \tag{6.5}$$

`Frequency-selective fading:`

$$B_c < B \text{ and } \tau_d > T_s \tag{6.6}$$

Flat fading channel is modelled with a single-tapped-delay model, while frequency-selective fading is modelled with multi-tapped-delay model. A frequency-selective channel affects the frequency components in a signal in a varying manner. A frequency-selective channel causes the symbols or signals passing over it to spread in a manner that, depending on the level of dispersion, can disturb each other in the form of ISI.

Non-time-selective channel has an impulse response which does not depend on time as it is time invariant. Hence, $h(t;\tau) = h(0;\tau) \equiv h(\tau)$. Therefore, the response of a non-time-selective, frequency-selective channel with additive noise to the input $x(t)$ is

$$\begin{aligned} r(t) &= \int_{\mathcal{T}} h(\tau)x(t-\tau)d\tau + n(t) \\ &= h(t) * x(t) + n(t)\,. \end{aligned} \tag{6.7}$$

$$\begin{aligned} r[i] &= \sum_j h[j]x[i-j] + n[i] \\ &= h[i] * x[i]n[i]\,. \end{aligned} \tag{6.8}$$

6.1.1.3 *Frequency-dispersive Wireless Channels*

The third type of channel is the non-time-dispersive, but frequency-dispersive channel. Such channels are also referred to as time-selective, non-frequency-selective channels. Frequency spreading or frequency dispersion is caused by Doppler spread. Named after Christian Doppler, the Austrian physicist who first documented the phenomenon, Doppler spread describes the variation of the wavelength or frequency of an EM wave due to relative movement between the wave and the observer. Doppler shift has two causes: (a) the relative motion between the transmitting and the receiving antennas, and (b) relative movements of objects in the channel between the transmitter and receiver. Doppler shift depends on velocity and angle of arrival of multipath signals. Doppler spread can cause ICI, in which the Doppler shifts change the carrier frequencies of different signals to cause them to overlap. If the maximum Doppler shift is $f_{\mathtt{m}}$ and the transmit frequency is $f_{\mathtt{t}}$, then the received frequency $f_{\mathtt{r}}$ lies in the range $f_{\mathtt{r}} = f_{\mathtt{T}} \pm f_{\mathtt{m}}$. If $f_{\mathtt{m}}$ exceeds the band gaps between carriers then ICI results. The condition for that is

`Frequency-dispersive channel:`

$$F_{\mathtt{d}} \gg \frac{1}{T_{\mathtt{s}}} \ \texttt{or}\ \tau_{\mathtt{d}} \ll \frac{1}{B}\,. \tag{6.9}$$

The level of frequency dispersion is quantified by the Doppler spread ($F_{\mathtt{d}}$) and the channel's coherence time ($\tau_{\mathtt{c}}$) within a small-scale geographical area. Frequency dispersion decides whether the channel causes slow fading or fast fading to communications signals. Small-scale fading based on frequency dispersion has the following types

$$\texttt{Slow fading:}\ \tau_{\mathtt{c}} > T_{\mathtt{s}} \ \texttt{and}\ F_{\mathtt{d}} < R_{\mathtt{s}}\,. \tag{6.10}$$

$$\texttt{Fast fading:}\ \tau_{\mathtt{c}} < T_{\mathtt{s}} \ \texttt{and}\ F_{\mathtt{d}} > R_{\mathtt{s}}\,. \tag{6.11}$$

Time selectivity means that the frequency response of the channel varies with time. Thus, $h(t;\tau) = h(t)\delta(\tau)$. The discrete-time equivalent is $h(n;k) = h[n]\delta[k]$. Therefore, the response of a time-selective, non-frequency-selective channel with additive noise to the input $x(t)$ is

$$\begin{aligned} r(t) &= \int_{\mathcal{T}} h(t)\delta(\tau)x(t-\tau)d\tau + n(t) \\ &= h(t)x(t) + n(t)\,. \end{aligned}$$

Its discrete-time equivalent is

$$\begin{aligned} r[i] &= \sum_j h[i]\delta[j]x[i-j] + n[i] \\ &= h[i]x[i] + n[i]\,. \end{aligned} \tag{6.12}$$

6.1.1.4 *Time-dispersive, Frequency-dispersive Wireless Channels*

The fourth type of wireless channel is the extreme case of being both frequency-dispersive and time-dispersive. Such a channel is also referred to as time-selective, frequency-selective channel. Such a channel causes both fast fading and frequency-selective fading. The required conditions for such a wireless channel are

$$\begin{aligned} &\texttt{Frequency dispersion:} \\ &\qquad F_{\mathtt{d}} \gg \frac{1}{T_{\mathtt{s}}} \ \texttt{or}\ F_{\mathtt{d}} \gg B \\ &\texttt{Time dispersion:} \\ &\qquad \tau_{\mathtt{d}} \gg \frac{1}{B} \ \texttt{or}\ F_{\mathtt{d}} \gg \frac{1}{T_{\mathtt{s}}}\,. \end{aligned} \tag{6.13}$$

The input-output characteristics of a time-selective, frequency-selective channel can be described by the generalised channel model

$$\begin{aligned} r(t) &= \int_{\mathcal{T}} h(t;\tau)x(t-\tau)d\tau + n(t); \\ r[i] &= \sum_j h[i;j]x[i-j] + n[i]\,. \end{aligned} \tag{6.14}$$

The fading caused by a wireless channel can be categorized into slow fading or fast fading depending on the magnitude of its Doppler spread $F_{\mathtt{d}}$. We have **slow fading** if $T_{\mathtt{s}} < \tau_{\mathtt{c}}$, but **fast fading** otherwise. A wireless channel is classified into flat fading or frequency-dependent fading depending on its magnitude of the multipath delay spread $\tau_{\mathtt{d}}$ relative to the signalling rate or symbol period $T_{\mathtt{s}}$. We have flat-fading channel if $B < B_{\mathtt{c}}$ or $\tau_{\mathtt{d}} < T_{\mathtt{s}}$, where $B_{\mathtt{c}}$ is the channel's coherence bandwidth, and $\tau_{\mathtt{c}}$ is its coherence time. Two frequencies, say f_1 and f_2, in a signal fulfilling $|f_1 - f_2| \leq B_{\mathtt{c}}$ are treated by the channel similarly. This is the reason why a signal with bandwidth B, where $B \leq B_{\mathtt{c}}$ undergoes flat fading.

Two probability density functions commonly used to model the delay τ experienced by a single transmitted EM wave in a multipath signal are

$$f_\tau(t) = U(0, 2\tau_{\mathtt{m}}) \text{ and } f_\tau(t) = \frac{1}{\tau_{\mathtt{m}}} e^{-\frac{t}{\tau_{\mathtt{m}}}} \tag{6.15}$$

where $\tau_{\mathtt{m}}$ is the **average delay spread**, which is defined as

$$\tau_{\mathtt{m}} = \int_0^\infty t f_\tau(t) dt\,. \tag{6.16}$$

The instantaneous multipath delay spread is rather

$$T_{\mathtt{d}} = \max_{k,l} |\tau_k(t) - \tau_l(t)| \tag{6.17}$$

which is the time difference between the earliest and the latest arriving paths to the receiver. The **root-mean-square multipath delay spread** is then

$$\tau_{\mathtt{d}} = \sqrt{\int_0^\infty (t - \tau_{\mathtt{m}})^2 f_\tau(t) dt}\,. \tag{6.18}$$

Power delay profile (PDP) measures the power in a received multipath signal as a function of the time taken to propagate from the transmitter to the receiver. If $P_{\mathtt{h}}(\tau)$ is the PDF of a discretised multipath channel, then we can define the average delay as

$$\tau_{\mathtt{m}} = \frac{\sum \tau P_{\mathtt{h}}(\tau)}{\sum \mathtt{h}(\tau)} \tag{6.19}$$

and the root-mean-square multipath delay spread as

$$\tau_{\mathtt{d}} = \sqrt{\bar{\tau^2} - \tau_{\mathtt{m}}^2} \tag{6.20}$$

where $\frac{\sum \tau^2 P_{\mathtt{h}}(\tau)}{\sum \mathtt{h}(\tau)}$. *The multipath delay spread is used to determine the length of the cyclic prefix (CP) used in OFDM (orthogonal frequency-division multiplexing) technologies. The CP should not be shorter than the delay spread of the wireless channel.*

The **coherence bandwidth** $B_{\mathtt{c}}$ of a wireless channel is the maximum deviation between two frequency components to experience the same disturbance from the channel. In other words, it is the range of frequency over which the frequency response of the channel remains fairly constant. Coherence bandwidth indicates how fast a channel's behaviour varies with frequency. **Coherence time** is the time duration over which the channel's impulse response remains approximately constant. It indicates how fast the channel varies with time. Assume that r_{12} is the cross-correlation coefficient between the amplitudes (i.e. fading envelopes) of two frequencies f_1 and f_2 in a signal passing through a wireless channel at given time. Then, approximate relationships between the coherence bandwidth and the root-mean-square multipath delay spread are

$$B_{\mathtt{c}} \approx \frac{1}{2\pi\tau_{\mathtt{d}}}, \quad \text{if } r_{12} \approx 0.5 \text{ or } |f_1 - f_2| < B_{\mathtt{c}}\,. \tag{6.21}$$

$$E[B_{\mathtt{c}}] \approx \frac{1}{4\pi\tau_{\mathtt{d}}}, \quad \text{if } r_{12} \approx 0 \text{ or } |f_1 - f_2| > B_{\mathtt{c}}\,. \tag{6.22}$$

In some situations the approximate relation used is

$$B_{\mathtt{c}} \approx \frac{1}{\tau_{\mathtt{d}}}\,. \tag{6.23}$$

A similar approximation between the coherence time of the channel and the Doppler spread is

$$F_{\mathtt{d}} \approx \frac{1}{2\tau_{\mathtt{c}}} \Leftrightarrow \tau_{\mathtt{c}} \approx \frac{1}{4 f_c \frac{v}{c}} \tag{6.24}$$

where $c \approx 3 \cdot 10^8$ m/s is assumed to be the approximate speed of radio signals in the atmosphere and v is the speed at which the mobile user is moving. The approximation in Eq. (6.24) is used, for example, in CelPlanner®software used to design and manage mobile networks. The relation

$$S_{\mathtt{f}} = \tau_{\mathtt{c}} B_{\mathtt{c}} = \frac{1}{F_{\mathtt{d}} \tau_{\mathtt{d}}} \tag{6.25}$$

is referred to as the *spread factor* of the wireless channel. The channel is overspread if $S_{\mathtt{f}} > 1$, otherwise it is said to be underspread.

The **coherence distance** D_c of a wireless channel is the maximum separation between a pair of antennas for which the channel's quality remains the same. If the spatial separation D between a pair of antennas is such that $D > D_c$ then we say that the channel is space selective. A communications signal experiences an ergodic channel if the symbol duration is much higher than the channel's coherence time to experience all its behaviour.

Multipath signal propagation is one of the main causes of ISI in wireless communications systems. Inter-symbol Interference increases the bit-error rate (BER) performance of the system, and thus must be mitigated using, for example, equalisation. Some rules-of-thumb for an acceptably low ISI are

$$\texttt{Acceptably low ISI: } T_s > 10\tau_d \texttt{ and } R_b < \frac{1}{\tau_d} \tag{6.26}$$

where R_b is the bit rate. The maximum delay spread recorded is about $\tau_d = 25$ microseconds. Some typical delay spread values in microseconds are

$$\tau_d = \begin{cases} < 0.2, & \texttt{open areas} \\ \approx 0.5, & \texttt{suburban areas}\,. \\ \approx 3, & \texttt{urban areas} \end{cases}$$

The range of delay spreads and coherence bandwidths are compiled in Table 6.1 for different environments. These values are used, for example, in the CelPlanner® software used to design and manage mobile networks.

Example: Fading Channels

The three major telecommunications operators in Australia as at Q1 2018 are Telstra, Optus and Vodafone. The 4G (i.e. LTE-Advanced) network of all three companies uses 1.8 GHz (referred to as band B3) as one of its frequency bands. Assume that you are travelling on a train at the speed of 110 km/hr while talking to someone over the LTE network. Assume that the radio signal reaches the smartphone's antenna at an angle of 0° with respect to the direction at which the train is travelling.

(1) How far do you travel before you expect the quality of your radio link to change remarkably?

(2) Wireless systems do channel sounding by periodically transmitting a training sequence to estimate the channel's quality during information transfer. The channel information is used to revert its undesirable effects on signals at the receiver, sometimes also at the transmitter. How frequently do we need to transmit the training sequence for optimum signal detection over the wireless channel under discussion?

Solution: The given data are $f_c = 1.8$ GHz, $v = 110\ \texttt{km/hr} \approx 30.6\ \texttt{m/s}$, and the speed of radio wave in the atmosphere (assumed clear) is $v \approx 3 \cdot 10^8$ m/s. We need to compute the coherence time τ_c of the wireless channel using the given data. This parameter depends on the Doppler spread F_d experienced by signals transferred over the wireless link. The maximum Doppler shift is $f_m = (v/c) * f_c \approx 183.3$ Hz. Thus, the Doppler spread is $F_d = 2f_m = 366.7$ Hz.

(1) Therefore, the channel's coherence time is

$$\tau_c \approx \frac{1}{2F_d} \approx 1.4\ \texttt{msec}\,.$$

The coherence time is the period of time in which the channel's quality remains fairly con-

Table 6.1 Common multipath delay spread and coherence bandwidth for different signal propagation environments.

Environment Type	Multipath Delay Spread Range τ_d (μs)	RMS Multipath Distance Spread $c \cdot \tau_d$ (m)	Coherence Bandwidth $B_c \approx \frac{1}{\tau_d}$ (kHz)
Indoors	0.01–0.27	3–81	100–3703.7
Urban	0.2–2.1	60–630	5000–476.2
Suburban	0.6–3.5	180–1050	1666.7–285.7
Rural	2–10	600–3000	500–100

stant. This means that the wireless link can change every 1.4 msec. The coherence time reduces as the speed of motion increases. It also reduces as the carrier frequency increases.

(2) The channel must be estimated periodically with the period not longer than the coherence time. This means that the wireless system must transmit its training sequence at intervals less than the coherence time.

6.1.2 *Radio Signal Fading Statistics: LCR and AFD*

There are two parameters that are of importance on fading signals. These are the level crossing rate (LCR) and the average fade duration (AFD). These are second-order statistical parameters on fading wireless signals, which depend on the speed at which the mobile user is moving as well as the scattering environment. These two parameters are used in the appropriate selection of error control mechanisms for the system. The longer the AFD, the more the number of bits affected by the channel, and the poorer the performance of the wireless system. Long AFD requires a strong error control and/or diversity schemes. We can estimate from the AFD the *outage probability*, i.e., the percentage of time that the received signal's level is lower than the minimum level required for signal detection with acceptable bit error rate.

A signal which is transferred over a fading wireless channel is referred to as a *fading signal*. As discussed in previous sections, the envelope of the received fading signal is a random process. Let us again denote the envelope as X. The **level crossing rate** (LCR) of X at the threshold value, say X_0, is defined as the rate (i.e., number of times per second) at which X crosses the level X_0 either in the downward (i.e., negative) or upward (i.e., positive) direction. Let us denote the LCR at level X_0 by $N_X(X_0)$. The LCR at the threshold X_0 is defined analytically as

$$N_X(X_0) = \int_0^\infty \dot{x} f_{X,\dot{X}}(X_0, \dot{x}) d\dot{x} \tag{6.27}$$

where X is the signal's envelope and $\dot{X} = \frac{dX}{dt}$ is envelope's slope or gradient. Equation (6.27) is the general definition of the LCR. Hence, it is applicable to any probability distribution of the envelope X.

Assume that X is a Rician fading signal, and thus X is a Rician distributed random variable. Let $S_{X_0X_0}(f)$ be the continuous portion of the Doppler power spectrum of $X = X_0$, $f_\mathrm{q} = f_\mathrm{m} \cos\theta_\mathrm{q}$ be the Doppler frequency of the specular or LOS multipath component, and $f_\theta(x)$ be the pdf of the azimuth scatter component. We assume that $S_{X_0X_0}(f)$ is symmetrical about f_q. This is the case, for example, if $\theta_\mathrm{q} = 90°$ and $f_\theta(x) = U(-\pi, \pi)$. Under the above assumptions Eq. (6.27) simplifies to

$$N_X(X_0) = \sqrt{2(1+k)\pi} f_\mathrm{m} \rho e^{-k-(1+k)\rho^2} \times I_0(2\rho\sqrt{(1+k)k}) \tag{6.28}$$

where $I_0(\cdot)$ is the modified Bessel function of the first kind of order zero, k is the Rician factor, $\rho = \frac{X_0}{X_\mathrm{rms}} = \frac{X_0}{\sqrt{E[X^2]}}$, where $\sqrt{E[X^2]}$ is the root-mean-square (rms) envelope level. The special case where $k = 0$ (i.e. absence of a specular or LOS path) results in the LCR for a Rayleigh fading signal, which is

$$N_X(X_0) = \sqrt{2\pi} f_\mathrm{m} \rho e^{-\rho^2} = \sqrt{2\pi} \frac{v f_\mathrm{c} X_0}{c X_\mathrm{rms}} e^{-(X_0/X_\mathrm{rms})^2} \tag{6.29}$$

where we have noted that $I_0(0) = 1$. Clearly, the LCR increases with the carrier frequency and the mobile user's speed of motion. It is also higher for a higher threshold value X_0. Knowing the level crossing rate we can compute the Doppler spread and the speed of user's movement. Some LCRs at different mobile user's speed of motion and fading channels are shown in Fig. 6.2.

The **average fade duration** (AFD) of the envelope X at the threshold value, say X_0, is defined as the average period of time that X stays below the threshold X_0 after crossing it in a downward direction. AFD is usually used instead of the instantaneous fade duration as there is no known analytical expression for the latter as at the time of compiling this book. Let us observe the fading of a signal over a sufficiently long time interval of duration, say T. During this time frame the observed

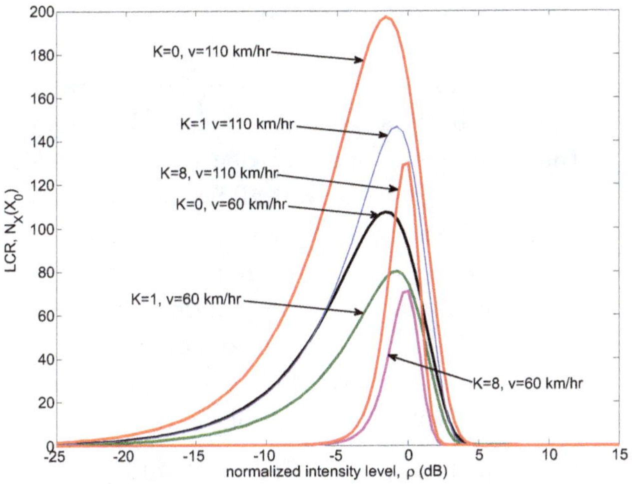

Fig. 6.2 Level crossing rates versus normalised signal's envelope level ρ at $f_c = 1.8$ GHz.

signal's level will be changing about a given threshold, say X_0. Let t_k denote the time duration that the kth fade stays below the threshold value X_0. Therefore, the probability that the envelope X of the received signal does not exceed X_0 is

$$Pr[X \leq X_0] = \frac{\sum_k t_k}{T}. \tag{6.30}$$

Analytically, the AFD is defined as

$$\begin{aligned} T_X(X_0) &= \frac{Pr[X \leq X_0]}{N_X(X_0)} = \frac{F_X(X_0)}{N_X(X_0)} \\ &= \frac{\int_0^{X_0} f_X(x)dx}{N_X(X_0)} \end{aligned} \tag{6.31}$$

where $F_X(x)$ and $f_X(x)$ are the cumulative distribution function and the probability density functions of X, respectively. In the following, we will make use of the Marcum-Q-function $Q_M(\alpha, \beta)$, which is defined as

$$\begin{aligned} Q_M(\alpha, \beta) = &\int_\beta^\infty y\left(\frac{y}{\alpha}\right)^{M-1} \\ &\times \exp\left(-\frac{y^2+\alpha^2}{2}\right) I_{M-1}(\alpha y)dy \end{aligned} \tag{6.32}$$

where α and β are nonnegative real numbers, M is a positive integer and $I_{(M-1)}(\cdot)$ is the modified Bessel function of the first kind of order $M-1$. An alternative definition of the Marcum-Q-function is

$$Q_M(\alpha, \beta) = \exp\left(-\frac{\beta^2+\alpha^2}{2}\right) \sum_{k=1-M}^{\infty} \left(\frac{\alpha}{\beta}\right)^k I_k(\alpha\beta). \tag{6.33}$$

MATLAB implements the Marcum-Q-function using the function name *marcumq*. If the envelope X is a Rician fading signal then

$$\begin{aligned} Pr[X \leq X_0] &= \int_0^{X_0} f_X(x)dx \\ &= 1 - Q_M(\sqrt{2k}, \sqrt{2(1+k)\rho^2}) \end{aligned} \tag{6.34}$$

where $f_X(x)$ is the Rician pdf given in Eq. (5.51). Using the same assumptions on the Doppler shift and the pdf of the azimuth scatter component made above, we can plug Eqs. (6.34) and (6.28) into Eq. (6.31) to obtain the AFD

$$\begin{aligned} T_X(X_0) &= \frac{Pr[X \leq X_0]}{N_X(X_0)} \\ &= \frac{1 - Q_M(\sqrt{2k}, \sqrt{2(1+k)\rho^2})}{\sqrt{2(1+k)\pi} f_{\mathtt{m}} \rho e^{-k-(1+k)\rho^2} I_0(2\rho\sqrt{(1+k)k})}. \end{aligned} \tag{6.35}$$

Again, in the special case of $k = 0$ we obtain the AFD for Rayleigh fading signals as

$$\begin{aligned} T_X(X_0) &= \frac{Pr[X \leq X_0]}{N_X(X_0)} = \frac{\exp(\rho^2) - 1}{\rho f_{\mathtt{m}} \sqrt{2\pi}} \\ &= \frac{\exp((X_0/X_{\mathtt{rms}})^2) - 1}{\sqrt{2\pi}\frac{X_0}{X_{\mathtt{rms}}} \cdot \frac{v}{c} f_c}. \end{aligned} \tag{6.36}$$

Example

A mobile communications system has the following parameters: carrier frequency $f_c = 1.8$ GHz, mobile user's speed of motion $v = 100$ km/h and the signal reaching the user is affected by a Rayleigh fading channel. The bit rate of the system is just $R_b = 80$ bits per second.

(1) Compute the level crossing rate (LCR) for $\rho = \frac{X_0}{E[X^2]} = 0.1, 0.2, 1, 2$.

(2) Compute the average fade duration (AFD) for $\rho = \frac{X_0}{E[X^2]} = 0.1, 0.2, 1, 2$.

(3) How many bits N_b are disturbed by each of the four fade situations?

(4) In which of the four fading scenarios does a fading signal undergo fast fading and in which does it experience slow fading?

Table 6.2 Statistics of fading channels: LCR and AFD.

$\rho = \frac{X_0}{X_{\text{rms}}}$	LCR, $N_X(X_0)$	AFD (msec)	N_b	Fading type
0.1	41.3614	0.2406	0	fast fading
0.2	80.2781	0.4884	0	fast fading
1.0	153.6895	4.1130	0	fast fading
2.0	15.3035	64.1477	5	slow fading

Solution

The given data are $v = 100$ km/hr ≈ 27.8 m/s, $c \approx 3 \cdot 10^8$ m/s, $\rho = \frac{X_0}{E[X^2]} = 0.1, 0.2, 1, 2$ and $f_c = 1.8$ GHz. Also, the bit duration is $T_b = 1/R_b = 12.5$ msec. The maximum Doppler frequency is $f_m = vf_c/c = 500/3$ Hz ≈ 166.7 Hz. We apply Eq. (6.36) for the AFD and Eq. (6.29) for the LCR and the results are shown in the second and third columns in Table 6.2.

The number of bits affected by each fade situation is obtained by dividing the corresponding AFD by the bit duration, i.e., $N_b = \frac{T_X(X_0)}{T_b}$. The results are shown in the fourth column in Table 6.2. The fading signal experiences fast fading if its bit duration T_b is longer than the AFD, otherwise it experiences slow fading. Fast fading means the channel's quality or state changes during the transfer of a single bit. The bit duration is 12.5 msec. It is thus longer than the AFD in the first three fading situations and so fast fading affects the signal. The fourth channel scenario causes slow fading as T_b is smaller than the AFD.

6.1.3 *AWGN Wireless Channel Model*

This is the most basic wireless channel model. It results from the flat-flat channel model in Eq. (6.3) in which the channel gain h is a deterministic constant. For simplicity, we set $|h| = 1$. Additive white Gaussian noise (AWGN) is the widely accepted model for the background noise affecting communications systems. Such a thermal noise model assumes the validity of the following:

- It is called noise because it is an undesired signal which disturbs the useful signals.
- The noise is white. That is, like white light which emits all colours (or frequencies) in the visible spectrum in equal intensity, it has the same power or intensity at all frequencies used in the communications system. For this reason, the AWGN is a wideband noise. This means that the power spectral density of the noise has a constant value over all frequencies, meaning that its autocorrelation function is nonzero only at zero lag. As communications systems use limited bandwidth, in practical systems the noise is filtered, making it no more white noise.
- The noise is additive. This means that it is not correlated with the signal of interest.

AWGN is the cumulative effect of noise from several sources, including thermal noise, black body noise, shot (aka Poisson) noise and from celestial bodies like the sun. All these noise sources are random in nature. The central limit theorem used in statistics states that the sum of a large number of random variables results in another random variable which is normal distributed. This is the reason for Gaussian in the name AWGN. If the signal of interest is $s(t)$ and the AWGN is $n(t)$, then the received signal on an AWGN-only channel is

$$r(t) = s(t) + n(t)\,. \tag{6.37}$$

The AWGN has the one-sided power spectral density of N_0 and two-sided power spectral density of $N_0/2$. The pdf of $n(t)$ is given in Eq. (5.30) with $\sigma_X^2 = \sigma_n^2 = \frac{N_0}{2}$ and $m_X = m_n = 0$. Thus, $n(t) \sim \mathcal{N}(0, \frac{N_0}{2})$. As we shall see in this book, communications systems are affected by other types of noise. Some of them are so high that they make AWGN irrelevant.

6.1.4 *LTI Wireless Channel Model*

A word that is commonly used when discussing the fading of wireless signals is a **signal's envelope**. A letter is usually inserted in an envelope made of paper before sending it through the postal system. So, the envelope is the outermost, or the only publicly exposed, part of the letter. The amplitude of a communication signal varies as it passes through the channel. Like a paper-based envelope, the **envelope of the signal** is the smooth curve which traces its extreme levels or amplitudes as it oscillates along the channel. A **baseband signal**

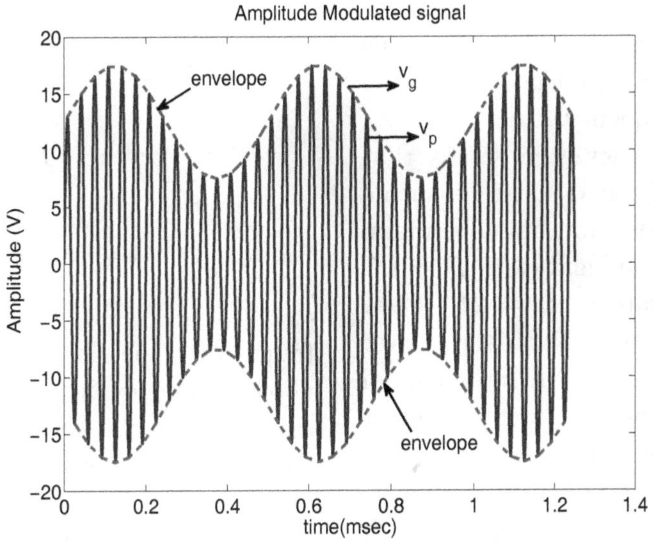

Fig. 6.3 Illustration of a wave and its envelope ($f_0 = 2$ kHz, $f_c = 16$ kHz).

$s(t)$ has nonzero spectral magnitude at frequencies around 0 Hz, and zero elsewhere. A **passband signal** $y(t)$, however, has zero spectral magnitude around 0 Hz, but nonzero around the frequency f_c, referred to as center or carrier frequency. A band-pass signal $y(t)$ can be represented analytically as

$$y(t) = Re\{s(t)e^{j2\pi f_c t}\} = r(t)\cos(2\pi f_c t + \alpha(t)) \tag{6.38}$$

where $Re\{x\}$ means the real part of x, and $s(t)$ is the complex envelope of $y(t)$. The complex envelope can be expressed as

$$s(t) = s_i(t) + js_q(t) = |s(t)|e^{\angle s(t)} = r(t)e^{j\alpha(t)}\,.$$

If we set $r(t) = \sin(2\pi f_0 t)$, $\alpha(t) = 0$, $f_c = 1.8$ GHz, we obtain Fig. 6.3.

The velocity, measured in meters per second, at which a signal's envelope propagate through the channel is referred to as **group velocity**, and denoted v_g. A communication's signal is usually made of multiple frequencies. Each frequency component of the signal has a phase and an amplitude. The rate at which the phase of a single frequency component or wave in the signal moves is referred to as **phase velocity**, and denoted v_p. If T is the wave's period, λ is its wavelength, $\omega = 2\pi f$ its circular frequency, then the phase and group velocities are defined as

$$v_p = \frac{\omega}{k} = \frac{\lambda}{T} \text{ and } v_g = \frac{\Delta\omega}{\Delta k} = \frac{d\omega}{dk} \tag{6.39}$$

where $k = 2\pi/\lambda$ is the angular wave number. Figure 6.3 illustrates the relationship between the envelope and individual waves in a multi-frequency signal. The high amplitudes can be seen as representing constructive interference of multipath signals, while the low amplitudes represent destructive interference. This is the reason why a **fading signal** can be defined as the envelope of a signal passing through a fading channel.

The behaviour of a wireless channel can be modelled as a linear filter with the frequency response

$$\begin{aligned} H(f) &= a(f)e^{j\phi(f)} \\ &= |H(f)|e^{j\angle H(f)},\ f_0 \le f \le f_0 + W \end{aligned} \tag{6.40}$$

over a certain bandwidth W, where f_0 is a fixed frequency and $\phi(f)$ is the phase response and $a(f)$ is the amplitude response over the bandwidth W. In general, both parameters of the channel depend on frequency f. The group delay (aka envelope delay) that channel introduces to multi-frequency signals passing through it is defined as

$$\tau_g(f) = -\frac{d}{df}\phi(f)\,. \tag{6.41}$$

The channel is classified as a **distortionless channel** or **linear channel** if both its amplitude and phase responses are constant and thus do not change with frequency. That is the desired ideal case. The amplitude of a signal passing through the channel experiences distortion if the $a(f)$ changes with frequency. Furthermore, the phase of signals passing through the channel experiences distortion if the $\phi(f)$ changes with frequency.

If $\phi(f)$ depends on frequency then we say that the channel has nonlinear phase response. In such a case the group delay is not constant, but depends on frequency. If a signal has the bandwidth $B = W$, then the Nyquist symbol rate is $R_s = 2B = \frac{1}{T_s}$. Transmitting data at or above the Nyquist rate over a nonlinear-phase channel causes ISI. Inter-symbol interference is a main contributing factor to error rates in communications systems.

A channel is classified as **LTI channel** (linear-time-invariant channel) if its phase response $\phi(f)$ as in Eq. (6.40) is linear and its amplitude response $a(f)$ is constant over the frequency band of interest. These two conditions on the LTI channel

$$a(f) = C_1 \text{ and } \phi(f) = -C_2 \cdot 2\pi\tau f \tag{6.42}$$

where C_1 and C_2 are constants and τ is the channel's constant delay. LTI channel delays are frequency components in a multi-frequency signal by the same amount and scales their amplitudes by the same amount. This means that signal experience no distortion after propagating along such a channel. This is the desired ideal case. It is obvious from Eq. (6.41) the group delay experienced by signals on an LTI channel is $\tau_g \propto 2\pi\tau$, which is independent on frequency. Figure 6.4 illustrates the frequency response of an LTI channel.

Assuming that we send the signal $s(t)$ through the above LTI system which delays all signal components by τ seconds and scale their amplitudes C_1, then the output of the channel is

$$r_1(t) = C_1 s(t-\tau)\,. \tag{6.43}$$

A common simplified model used for communications system design and analysis is LTI channel plus AWGN. The output of such a channel model if $s(t)$ were sent through it would be

$$r_2(t) = C_1 s(t-\tau) + n(t) \tag{6.44}$$

where $n(t)$ is the additive noise.

A channel which is not time-invariant is a **time-varying channel**. Such a channel has time-dependent impulse response which is caused by the

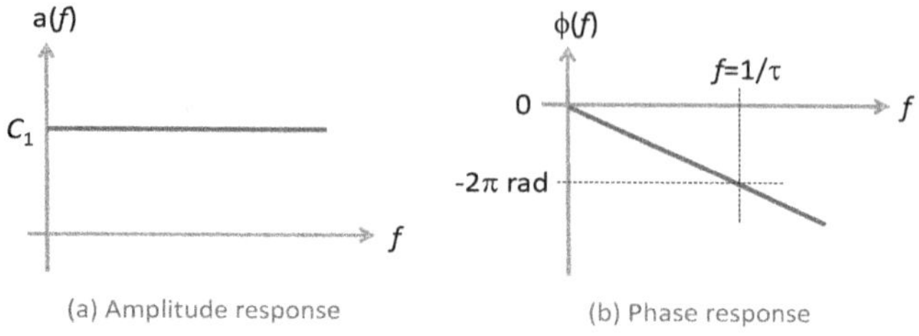

Fig. 6.4 The idealised response of an LTI channel with $C_2 = 1$.

frequency dispersion in the channel. As discussed above, frequency dispersion is rooted in relative motion between the communicating antennas or objects in the channel.

6.1.5 *Tapped-Delay Line Wireless Channel Model*

This is a discrete-time model used to model **time-varying** *wideband wireless channels* causing multipath signal propagation. The channel is usually characterised in time domain by its bandpass impulse response $h_B(t,\tau)$. The equivalent baseband complex-envelope channel impulse response is $h(t;\tau)$, and in the frequency domain by its transfer function $H(f;\tau)$. We can use Fourier analysis to convert one into the other. The impulse response of a discretised multipath channel in which K impulses reach the receiver if only one impulse is transmitted has the general form

$$\begin{aligned} h(t;\tau) &= \sum_{k=1}^{K} h_k(t;\tau)\delta(\tau-\tau_k) \\ &= \sum_{k=1}^{K} a_k(t;\tau)e^{j\alpha_k(t;\tau)}\delta(\tau-\tau_k(t)) \end{aligned} \tag{6.45}$$

where $\alpha_k(t;\tau)$, $\tau_k(t)$ and $a_k(t;\tau)$ are the phase shift, excess delay and amplitude of the kth multipath signal. Note the time-dependence of the channel's complex coefficients $h_k(t;\tau)$. Each of these coefficients, referred to as **channel gain**, are usually modelled with a fast-fading model such as the Rayleigh distribution. The taps of the channel can also be modelled differently. The model in Eq. (6.45) is referred to as K-tap delay line model.

Another form of the tapped-delay line model is

$$h(t;\tau) = \sum_{k=1}^{K} h_k e^{-j2\pi f\tau_k} e^{j2\pi f_d t} \tag{6.46}$$

where h_k is the gain of the kth multipath component (which can also be time varying). The time-varying nature of the channel in Eq. (6.46) is captured by Doppler shift f_d in $e^{j2\pi f_d t}$, while the multipath delay is captured in $e^{-j2\pi f \tau_k}$.

The problem caused by multipath propagation to information signals is attributed to the $\tau_k(t)$ in Eq. (6.45) being different for each tap or impulse received. The different delays result in differing phases of the impulses at the receiver, which can result in destructive interference. The *autocorrelation function* of the time-varying channel's impulse response is

$$r_{hh}(\tau) = E\{h_k(\tau)h_k^*(t+\tau)\}. \qquad (6.47)$$

Correlation is used to measure the similarity between two path components. A low value of $r_{hh}(\tau)$ means that the kth multipath signal has experienced a remarkable change within the time period of duration τ. A high value of $r_{hh}(\tau)$, on the other hand, indicates that the channel changes slowly.

Let us analyse the autocorrelation function using a single ray (or path) of the channel's impulse response definition in Eq. (6.46). We have

$$\begin{aligned} r_{hh}(\tau) &= E\{h_k(\tau)h_k^*(t+\tau)\} \\ &= E\{h_k e^{-j2\pi f \tau_k} e^{j2\pi f_d t} \cdot h_k^* e^{j2\pi f \tau_k} e^{-j2\pi f_d (t+\tau)}\} \\ &= E\{|h_k|^2 e^{j2\pi f_d \tau}\} = |h_k|^2 E\{e^{j2\pi \frac{v}{c} \cos\phi_k \tau}\} \\ &= |h_k|^2 e^{j2\pi \frac{v}{c}\tau E\{\cos\phi_k\}} \qquad (6.48) \end{aligned}$$

where ϕ_k is the angle at which the kth path reaches the receiving antenna. This angle is uniformly distributed in the interval 0 to 2π, i.e., $\phi_k \sim U(0, 2\pi)$. Thus,

$$\begin{aligned} r_{hh}(\tau) &= |h_k|^2 e^{j2\pi \frac{v}{c}\tau} E\{\cos\phi_k\} \\ &= J_0(2\pi f_m \tau) \cdot |h_k|^2 \\ &= J_0\left(\frac{\pi\tau}{2\tau_c}\right) \cdot |h_k|^2 \qquad (6.49) \end{aligned}$$

where τ_c is the channel's coherence time. We used the approximation between the coherence time and Doppler spread, i.e. $\tau_c \approx 1/(2F_d) = 1/(4f_m) = 1/(4vf/c)$, while deriving Eq. (6.49). In Eq. (6.49), $J_0(\cdot)$ is the Bessel function of the first kind of zeroth order. It is implemented in MATLAB as *besselj (order, parameter)*. The autocorrelation function is illustrated in Fig. 6.5. We can observe that the correlation decreases with the delay, meaning that the change in the channel's characteristics is higher over longer time scales.

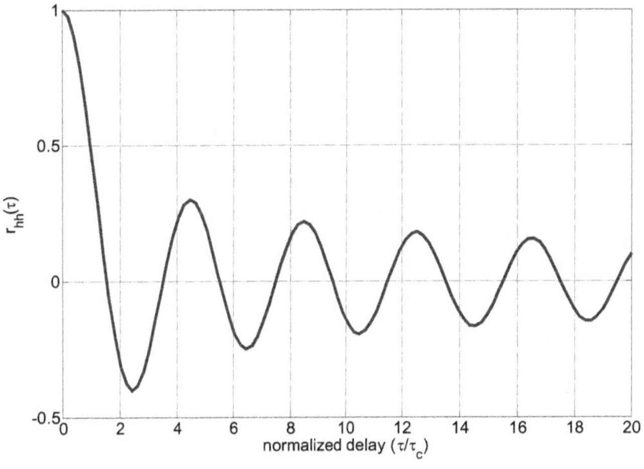

Fig. 6.5 Autocorrelation function of a wireless channel ($|h_k|^2 = 1$, τ_c is channel's coherence time).

Doppler Spectrum

The Doppler spectrum can be used to estimate the Doppler spread, and hence how fast the wireless channel varies. The range of frequencies in which the Doppler spectrum is significantly nonzero measures the Doppler spread. We can estimate the coherence time of the channel from its Doppler spread using, for example, Eq. (6.24). The Doppler power spectrum can be obtained from the Fourier transform of the autocorrelation function of $h(t;\tau)$, i.e.,

$$\begin{aligned} S_{hh}(f, f_d) &= \mathcal{F}\{r_{hh}(\tau)\} = \int_{-\infty}^{\infty} r_{hh}(\tau) e^{-j2\pi f\tau} d\tau \\ &= |h_k|^2 \int_{-\infty}^{\infty} J_0\left(\frac{\pi\tau}{2\tau_c}\right) e^{-j2\pi f\tau} d\tau \\ &= \frac{rect(0.5f/f_d)}{\pi f_m \sqrt{1-(f/f_d)^2}} \qquad (6.50) \end{aligned}$$

where $rect(t) = 1, -1/2 \leq t \leq 1/2$ and zero otherwise.

Another Way of Computing the Doppler Spectrum

Assume that the angle of incidence of a multipath signal on the receiving antenna is ϕ. A common assumption is that ϕ is uniformly distributed in the interval 0 to 2π, i.e., $\phi \sim U(0, 2\pi)$. If the receiver is moving at the speed of v m/s, then the received signal experiences the Doppler shift $f_d = \frac{v}{c} f_c \cos\phi = f_m \cos\phi$, where $f_m = \frac{v}{c} f_c$ is the maximum Doppler shift and f_c is the carrier frequency. Thus, $\phi = \cos^{-1}(f_d/f_m)$.

Note that if $z = \cos(y)$ then $\frac{dy}{dz} = -\frac{1}{\sqrt{1-\cos^2 y}} = -\frac{1}{\sqrt{1-z^2}}$ for $|z| \le 1$. The received power P_r in angle $d\phi$ about ϕ is proportional to $\frac{d|\phi|}{2\pi}$. The Doppler power spectrum is then defined as

$$S_{hh}(f, f_d) \propto \frac{1}{2\pi}\left|\frac{d\phi}{df_d}\right| = \frac{1}{2\pi}\left|\frac{d}{df_d}\cos^{-1}(f_d/f_m)\right| = \frac{1}{2\pi\sqrt{1-(f_d/f_m)^2}}. \quad (6.51)$$

Thus, using a proportional constant, say A, we obtain the Doppler spectrum

$$S_{hh}(f, f_d) = \frac{A}{2\pi\sqrt{1-(f_d/f_m)^2}}. \quad (6.52)$$

A frequency-dispersive channel broadens the spectrum of a single tone transmitted at the carrier frequency f_c to cover the spectrum $f_c - f_m \le f \le f_c + f_m$. In this case of a single tone, the Doppler spread is $F_d = (f_c + f_m) - (f_c - f_m) = 2f_m$.

We define the multipath intensity profile or **power delay profile** (PDP) as

$$P_h(\tau) = r_{\mathbf{hh}}(0, \tau) = E\{|h_k(\tau)|^2\} = \int_{-\infty}^{\infty} S_{\mathbf{hh}}(f;\tau)df \quad (6.53)$$

where f is the Doppler frequency, τ is the multipath propagation delay, and $S_{\mathbf{hh}}(f;\tau)$ is the scattering function. The PDP measures the strength in a multipath signal as a function of time delay between the multipath components. The power delay profile is the channel's power spectral density as a function of delay. It is used to estimate the maximum excess delay, a parameter which is used to classify channels into flat-fading channel and frequency-selective channel. A wireless channel is frequency selective if its maximum delay spread exceeds the symbol duration. Frequency selectivity in a channel results in inter-symbol interference. The power delay profile around a local spatial position can be estimated as

$$P_h(\tau) = \begin{cases} \frac{1}{L}\sum_{t=x}^{x+L} |h(t;\tau)|^2, & \texttt{for discrete-time} \\ \frac{1}{L}\int_{x}^{x+L} |h(t;\tau)|^2 dt, & \texttt{for continuous-time} \end{cases}.$$

6.1.6 *COST 207 Wireless Channel Models*

COST 207 defined the following power delay profiles which are normalised so that $\int_0^\infty P_h(\tau)d\tau = 1$:

(1) For rural (non-hilly) areas (RA)

$$P_h(\tau) = \begin{cases} \frac{9.2}{1-\exp(-6.44)} e^{-9.2\tau}, & 0 \le \tau \le 0.7 \\ 0, & \texttt{otherwise} \end{cases} \quad (6.54)$$

(2) For typical urban (TU) (non-hilly) areas

$$P_h(\tau) = \begin{cases} \frac{1}{1-\exp(-7)} e^{-\tau}, & 0 \le \tau \le 7 \\ 0, & \texttt{otherwise} \end{cases} \quad (6.55)$$

(3) For bad urban (BU) (non-hilly) areas

$$P_h(\tau) = \begin{cases} \frac{2}{3(1-e^{-5})} e^{-\tau}, & 0 \le \tau \le 5 \\ \frac{1}{3(1-e^{-5})} e^{5-\tau}, & 5 \le \tau \le 10 \\ 0, & \texttt{otherwise} \end{cases} \quad (6.56)$$

(4) For hilly terrain (HT) areas

$$P_h(\tau) = \begin{cases} \frac{1}{(1-e^{-7})/3.5+0.1(1-e^{-5})} e^{-3.5\tau}, & 0 \le \tau \le 2 \\ \frac{0.1}{(1-e^{-7})/3.5+0.1(1-e^{-5})} e^{15-\tau}, & 15 \le \tau \le 20 \\ 0, & \texttt{otherwise} \end{cases} \quad (6.57)$$

Figure 6.6 illustrates these four power delay profiles, where τ is measured in microseconds.

COST 207 also defined the following Doppler spectra $S_{hh}(f)$ for GSM

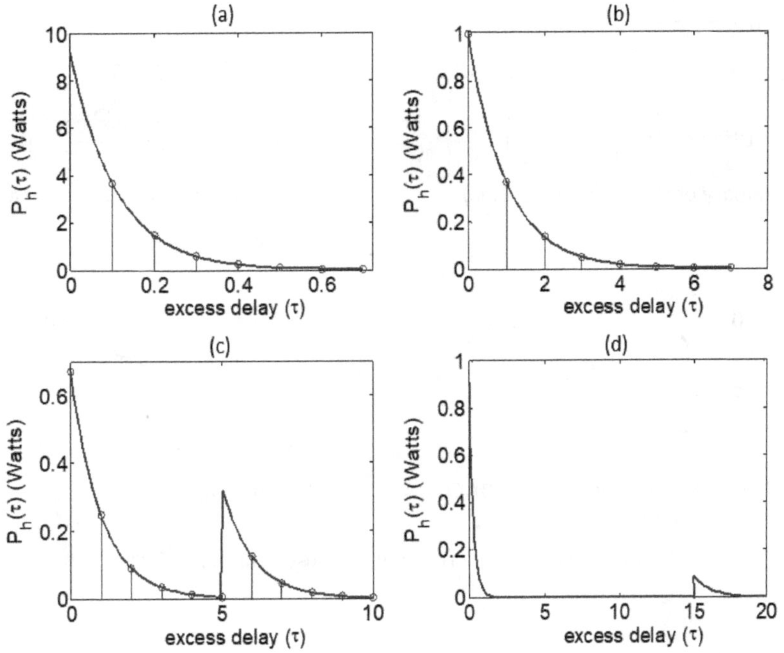

Fig. 6.6 COST 207 power delay profiles for (a) rural (non-hilly) areas, (b) typical urban (non-hilly) areas, (c) bad urban (non-hilly) areas and (d) hilly terrain.

$$\begin{aligned}
\texttt{CLASS:}\quad & S_{hh}(f) = \frac{A}{\sqrt{1-(f/f_m)^2}}, \\
& |f| \leq f_m,\ \tau_k \leq 500\ \texttt{ns} \\
\texttt{GAUS1:}\quad & S_{hh}(f) = G(A, -0.8f_m, 0.05f_m) \\
& \quad + G(A_1, 0.4f_m, 0.1f_m), \\
& 0.5 \leq \tau_k \leq 2\ \mu\text{s} \\
\texttt{GAUS2:}\quad & S_{hh}(f) = G(B, 0.7f_m, 0.1f_m) \\
& \quad + G(B_1, -0.4f_m, 0.15f_m), \\
& \tau_k > 2\ \mu\text{s} \\
\texttt{RICE:}\quad & S_{hh}(f) = \frac{0.41}{2\pi f_m\sqrt{1-(f/f_m)^2}},\ |f| \leq f_m
\end{aligned} \tag{6.58}$$

where $A_1 = A - 10$ dB, $B_1 = B - 15$ dB, and $G(A, f_1, f_2) = A\exp(-\frac{(f-f_1)^2}{2f_2^2})$.

6.1.7 *Binary Discrete Memoryless Noisy Channel Models*

We discuss common simple models used to characterise noisy communications channels, namely BEC (binary erasure channel), BSC (binary symmetric channel), BAC (binary assymetric channel) and ZC (Z-channel). As these models are simple and can be used to build more realistic but complex models, they are often used in information and coding theory. Every channel has an input alphabet, say $\mathcal{X}$, and an output alphabet, say $\mathcal{Y}$. A channel is said to be a **discrete channel** if its input and output alphabets are finite distinct symbols, or digital signals. A discrete channel is called a **binary channel** if its input alphabet $\mathcal{X}$ contains only two possible symbols, which are usually denoted by $\mathcal{X} = \{0, 1\}$. Note that the output alphabet does not need to be binary. A channel is said to be a **memoryless channel** if its current output depends on only its current input, but not earlier inputs. A channel which is discrete and memoryless is referred to as **discrete memoryless channel** (DMC).

In the following, let us assume that we transmit the n-bit codeword $\mathbf{c} = [c_1, c_2, \ldots, c_n]$ through any of the channels discussed in this section and that we receive the n-bit word $\mathbf{r} = [r_1, r_2, \ldots, r_n]$.

Binary erasure channel (BEC) model, developed by Peter Elias at MIT in 1955, is a discrete-

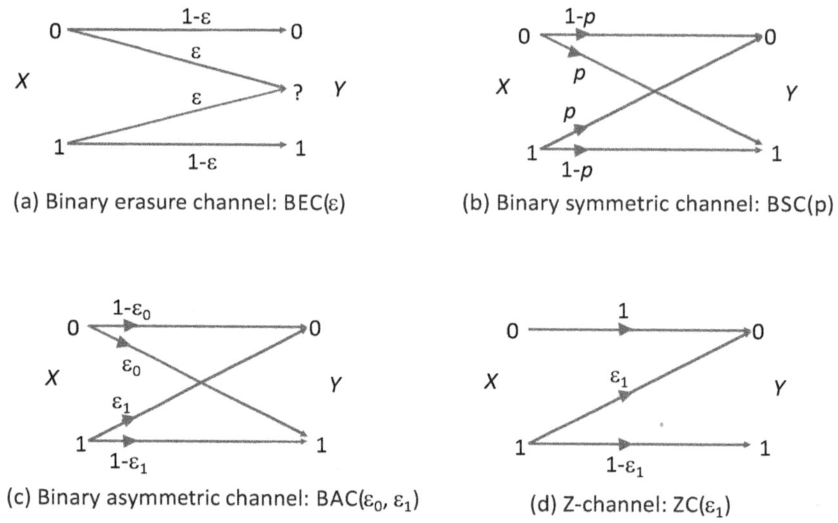

(a) Binary erasure channel: BEC(ε)
(b) Binary symmetric channel: BSC(p)
(c) Binary asymmetric channel: BAC(ε0, ε1)
(d) Z-channel: ZC(ε1)

Fig. 6.7 Binary discrete memoryless channel models.

time model used to model the effects of the noise in communications channels on transmitted signals in the form of packet losses. In this case we are interested in only two outcomes: the data gets to the intended destination or gets so garbled in the channel that the receiver cannot make sense of this. The latter case is referred to as *erasure*. Symbolically, the data bit gets lost within the channel. A model used to illustrate the behaviour of the BEC is shown in Fig. 6.7(a). As noted from the figure, the output alphabet of BEC is $\mathcal{Y} = \{0, 1, ?\}$, where the question mark indicates the difficulty the receiver faces to guess which symbol left the transmitter. In that case only retransmission can correct the erased bit.

The performance of the BEC is quantified by the probability of a bit being unrecognisable at the receiver, which is referred to as erasure probability ϵ. Then, the properties of the BEC channel are described by the conditional probabilities

$$\begin{aligned} Pr[r_k = 0|c_k = 0] &= Pr[r_k = 1|c_k = 1] = 1 - \epsilon \text{ and} \\ Pr[r_k = ?|c_k = 1] &= Pr[r_k = ?|c_k = 0] = \epsilon\,. \end{aligned} \tag{6.59}$$

The achievable maximum data transfer rate (aka channel capacity) over the BEC is

$$C_{\mathtt{BEC}(\epsilon)} = 1 - \epsilon\,. \tag{6.60}$$

The **binary symmetric channel** (BSC) is a discrete-time model used to model the effects of the noise in communications channels on transmitted signals in the form of bit flips or bit errors. In this BSC model, the channel surely delivers the data to intended destinations, except that the chance of errors is non-negligible. A model used to illustrate the behaviour of the BSC is shown in Fig. 6.7(b). A crossover probability p is defined to quantify the chance of the BSC delivering a data bit in error. The properties of the BSC channel are described by the conditional probabilities

$$\begin{aligned} Pr[r_k = 0|c_k = 0] &= Pr[r_k = 1|c_k = 1] = 1 - p \text{ and} \\ Pr[r_k = 0|c_k = 1] &= Pr[r_k = 1|c_k = 0] = p\,. \end{aligned} \tag{6.61}$$

The achievable maximum data transfer rate (aka channel capacity) over the BSC is

$$C_{\mathtt{BSC}(p)} = 1 + p\log_2 p + (1-p)\log_2(1-p)\,. \tag{6.62}$$

The **binary asymmetric channel** (BAC) model is shown in Fig. 6.7(c). It is the most generalised DMC model. Its properties are described by the conditional probabilities

$$\begin{aligned} Pr[Y = 0|X = 0] &= 1 - \epsilon_0 \quad Pr[Y = 1|X = 0] = \epsilon_0 \\ Pr[Y = 1|X = 1] &= 1 - \epsilon_1 \quad Pr[Y = 0|X = 1] = \epsilon_1\,. \end{aligned} \tag{6.63}$$

The BAC reduces to BSC model if $\epsilon_1 = \epsilon_0$ and to the Z-channel if $\epsilon_0 = 0$. If we assume that $\epsilon_1 \geq \epsilon_0$, then the capacity of BAC is

$$C_{BAC} = \log_2 \left[1 + 2^{\frac{H_2(\epsilon_0)-H_2(\epsilon_1)}{1-\epsilon_1-\epsilon_0}}\right] - \frac{1-\epsilon_1}{1-\epsilon_1-\epsilon_0} H_2(\epsilon_0) + \frac{\epsilon_0}{1-\epsilon_1-\epsilon_0} H_2(\epsilon_1) \tag{6.64}$$

where $H_2(\epsilon_k) = -\epsilon_k \log_2(\epsilon_k) - (1-\epsilon_k)\log_2(\epsilon_k)$, $k = 0, 1$.

The **Z-channel** (ZC) flips binary 1 to binary 0 with a finite probability, but behaves transparently to binary 0. Its behaviour can thus be described by the conditional probabilities

$$Pr[Y=0|X=0] = 1 \quad Pr[Y=1|X=0] = 0$$
$$Pr[Y=1|X=1] = 1-\epsilon_1 \quad Pr[Y=0|X=1] = \epsilon_1 \,. \tag{6.65}$$

By setting $\epsilon_0 = 0$ in Eq. (6.64) we obtain the capacity of the Z-channel as

$$C_{ZC} = \log_2 \left(1 + 2^{-\frac{H_2(\epsilon_1)}{1-\epsilon_1}}\right). \tag{6.66}$$

Note that C_{ZC} tends to unity as ϵ_1 tends to zero, but goes to zero as ϵ_1 increases towards unity. The four DMC models discussed in this section are based on bits. They can, however, be used for packet-level transmissions. For example, using packets instead of bits, we can convert BEC to packet-erasure channel (PEC) model. In this case the transition probabilities will be computed at the packet level instead of bit level.

Claude Shannon established in 1948 that if the coding rate $r = \frac{k}{n}$ does not exceed the channel's capacity C, then we can find a channel coding scheme that achieves near zero probability of decoding error. Here, k is the number of data bits converted into n bits. An AWGN channel, for example, the channel capacity is $C = B\log_2(1+SNR)$, where B is the channel's bandwidth and SNR is the signal-to-noise ratio. We shall spend more effort in these in the two chapters dedicated to channel coding and error control. The four DMCs that have been discussed are random-error channel models, as they are memoryless. We discuss next a popular model model for burst-error channels.

6.1.8 *Gilbert-Elliot Burst-Noise Channel Model*

Communications channels can be classified as random channels and bursty channels. Random channels cause only uncorrelated random errors while bursty channels multiply correlated bit errors in a transmitted codeword. Wireless channels, especially when in deep fade, are bursty in nature. However, error control of a random-error channel is simpler than that on burst-error channel. For this reason, there are strategies, such as time diversity (e.g. interleaving), that are used to convert a burst-error channel to a random-error channel.

The Gilbert-Elliot model is a discrete channel model based on the hidden Markov model with two states. The two states of the channel are bad state S_{b} and good state S_{g}. Let $\mathcal{S}_t$ denote the state of the channel at time t. Then, $\mathcal{S}_t = \{S_{\mathsf{b}}, S_{\mathsf{g}}\}$. Assume that the probability of the channel being in good state or bad state during signal transmission at time t are respectively, then

$$P_{\mathsf{g}} = Pr[\mathcal{S}_t = S_{\mathsf{g}}] \text{ and}$$
$$P_{\mathsf{b}} = Pr[\mathcal{S}_t = S_{\mathsf{b}}] \,. \tag{6.67}$$

Let BER_{b} be the error probability when the channel is in its bad state, and BER_{g} be the equivalent in the good state. The channel disturbs signals higher when it is in its bad state than the good state. Hence, $BER_{\mathsf{g}} < BER_{\mathsf{b}}$. In the good state signals are affected by only additive noise. The transition or crossover probabilities measuring how likely the channel moves from one state into the other are

- P_{gb}: probability of channel moving from good state to its bad state.
- P_{bg}: probability of channel moving from bad state to its good state.
- P_{gg}: probability of channel remaining in its good state.
- P_{bb}: probability of channel remaining in its bad state.

Figure 6.8 shows the two-state burst-error channel model with its transition probabilities.

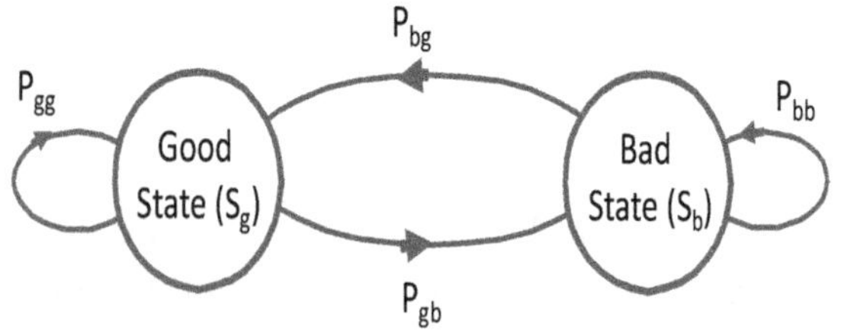

Fig. 6.8 Gilbert-Elliot two-state hidden Markov model for burst-error channels.

6.1.9 *Check Your Understanding*

Students are encouraged to check their understanding of the content of this section with the following questions prior to proceeding to the next section.

(1) State two reasons why channel modelling is important in communications.
(2) Describe the meaning of the term *fading* of a radio signal.
(3) Briefly state the reason why the average fade duration (AFD) of the envelope of a fading signal is usually quantified instead of the instantaneous fade duration.
(4) What does the envelope of a radio signal mean?
(5) Describe the relationship between the multipath delay spread and the length of the cyclic prefix used in OFDM technologies.
(6) Briefly differentiate between macroscopic fading and microscopic fading.
(7) What are the three basic types of fading discussed in this section? Differentiate between them with respect to their causes and how they affect radio signals.
(8) Differentiate between time dispersion and frequency dispersion with respect to wireless channels.
(9) Assume that the delay experienced by a multipath signal is modelled by the random variable τ whose pdf are given in Eq. (6.15). Assume that the average delay spread is $\tau_{\mathrm{m}} = 4\mu$s.
  (a) What is the likelihood that a signal experiences a delay larger than $1.2\tau_{\mathrm{m}}$ for each of the two distributions?
  (b) Compute the root-mean-square multipath delay spread τ_{d} for each of the two cases.
  (c) What should be the maximum bandwidth of a radio signal which propagates on the wireless channel described above to experience a flat fading?
(10) What conditions should a communications channel fulfill to be classified as AWGN channel?
(11) Analyse the term an LTI communications channel.
(12) Compare the order (or magnitude) of *coherence time* with that of `multipath delay spread` of wireless channels used for commercial mobile cellular communications networks.
(13) Let us consider a wireless channel whose behaviour is described by the impulse response

$$h(t) = a\delta(t) + b\delta(t - \tau_1) + c\delta(t - \tau_2)\,.$$

  (a) Compute the frequency response $H(f; \tau_1, \tau_2)$ of the wireless channel.
  (b) Compute the magnitude response $a(f; \tau_1, \tau_2)$ and the phase response $\phi(f; \tau_1, \tau_2)$ of the channel.
  (c) Find the group delay and the phase delay of the channel.
  (d) Is the wireless channel an LTI channel? Justify your opinion quantitatively.

  Assume that $a = 1, b = 0.6, c = 0.4, \tau_1 = 2\mu$s, and $\tau_2 = 5\mu$s.
(14) Let us consider the four discrete memoryless channels studied in this section.
  (a) Compute the capacity of a BSC with 0.2 cross-over likelihood.
  (b) Compute the capacity of a BEC with the parameter $\epsilon = 0.2$.
  (c) Compute the capacity of a Z-channel with the parameter $\epsilon_1 = 0.2$.

(d) Compute the capacity of a BAC channel with the parameters $\epsilon_1 = 0.2$ and $\epsilon_0 = 0.02$.

(e) Which of the four channels discussed is best suited for communications, and why?

(15) We seek to reflect on burst-error and random-error channels.

(a) Discuss in groups of two to three students the differences between burst-error channel and random-error channel.

(b) How can we convert a burst-error channel into a random-error channel?

(c) What it the motivation behind randomising a burst-error channel?

6.2 Slow Fading or Shadowing Models

Shadowing occurs when an obstruction blocks the radio path between the transmitting and the receiving antennas. The effect of shadowing on the received signal is that the path loss that it experiences deviates from the average value, which is the median or mean path loss computed using path loss models such as the Okumura-Hata model. Just as a light source casts the shadow of an object in front it, so does the radio wave cast a shadow of its obstructions. The right side of Fig. 5.11 illustrates an example of shadowing. In this particular case, the receiver relies on only diffraction to receive radio signals from the eNodeB or base station. The effect of shadow fading is quantified by how the path loss experienced by a signal (or the received signal strength) at a given location changes in its average value. The attenuation experienced by a radio signal over a transmission distance d is

$$Y(d) = L_{50}(d) + X_\sigma \ \texttt{[dB]} \qquad (6.68)$$

where $L_{50}(d)$ is the median local path loss computed using any path loss model such as those of Okumura and Okumura-Hata, and X_σ in decibels is a lognormally-distributed random variable, representing the attenuation caused by shadowing, with the standard deviation σ in decibels. Note that in some parts of the book we use just L to mean L_{50}, especially in the section discussing path loss models. Typical range of values for σ is 4 dB $\leq \sigma \leq$ 10 dB.

It is reported in [Segata *et al.* (2013)] that slow fading due to shadowing and fast fading are correlated. They also reported that shadowing depends on the distance between the transmitting and the receiving antennas, as well as the size of the shadowing-causing obstruction. Shadowing reduces the power in the received signal by about 10 dB at 80-meter distance between the transmitter and the receiver, but decreases to about 5 dB at 120-meter separation [Segata *et al.* (2013)]. Shadowing is usually modelled as a random variable having a log-normal probability distribution in Eqs. (5.32) and (5.33).

Shadowing causes the instantaneous path loss Y experienced by the received signal to fluctuate in a random manner. The deviation of the path loss is such that approximately 50% of the time the path loss Y exceeds the mean path loss L_{50}, while falling below it 50% of the time. The random variable X_σ in decibels is normal or Gaussian distributed with zero mean. That is X_σ [dB] $\sim \mathcal{N}(0, \sigma^2)$. This means that X where $X_\sigma = \log_{10} X$ is normal distributed. Therefore, Y in Eq. (6.68) is normal distributed with mean L_{50} and standard deviation σ, i.e. $Y \sim \mathcal{N}(L_{50}, \sigma^2)$. The pdf of Y with $L_{50} = 40$ dB and $\sigma = 8$ dB is shown in Fig. 6.9. Figure 6.10 shows the probability of the total path loss taken on a given value.

Mobile wireless link design must contain a margin to cater for slow fading. Assume that σ in dB is the standard deviation of the log-normal distributed random variable X modelling slow fading. Also, let us assume that we design the mobile network to have the *cell edge coverage probability* of P_c, and L_{SFM} be the slow fading margin in dB. Then, we have

$$L_{SFM} = \sigma \cdot Q^{-1}(1 - P_c) \qquad (6.69)$$

where Q^{-1} is the inverse of the Q-function. For example, if $P_c = 90\%$, then $L_{SFM} \approx 1.2816\sigma$ dB, which results in 10.3 dB for $\sigma = 8$ dB. Figure 6.11 illustrates two slow fade margins for different cell edge coverages.

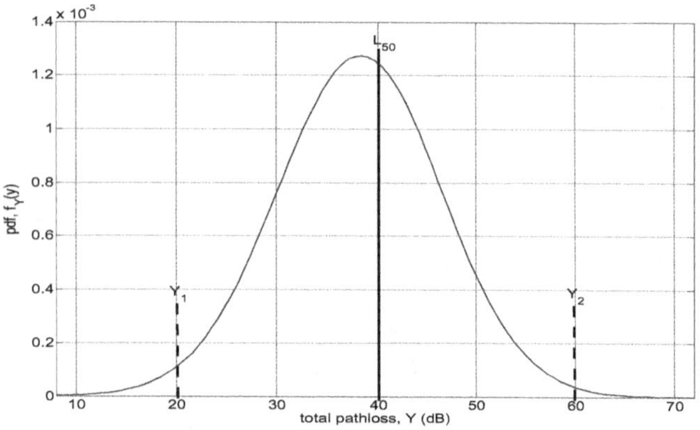

Fig. 6.9 Estimating path loss due to shadowing and TX-RX distance ($L_{50} = 40$ dB, $\sigma = 8$ dB).

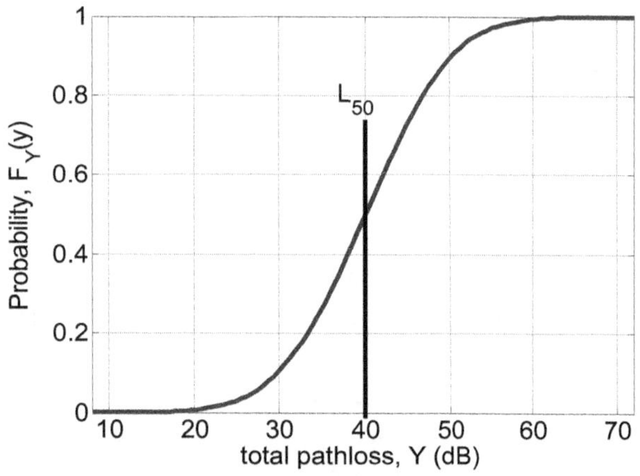

Fig. 6.10 Estimating probability of path loss due to shadowing and TX-RX distance ($L_{50} = 40$ dB, $\sigma = 8$ dB).

Example

Compute the likelihood that the path loss Y exceeds $Y_2 = 60$ dB, if $Y \sim \mathcal{N}(40\texttt{dB}, \sigma^2)$ with $\sigma = 8$ dB.

Solution

The probability that the path loss exceeds a given value, say $Y_2 = 60$ dB is given by

$$\begin{aligned}
P_1 &= Pr[Y > Y_2] = Pr[L_{50} + X_\sigma > Y_2] \\
&= Pr[X_\sigma > Y_2 - L_{50}] \\
&= \int_{x=Y_2-L_{50}}^{\infty} f_Y(x)dx \\
&= Q\left(\frac{Y_2 - L_{50}}{\sigma}\right) \\
&= Q\left(\frac{60-40}{8}\right) \\
&= Q(2.5 \texttt{ dB}) = 0.5\texttt{erfc}(2.5/\sqrt{2} \texttt{ dB}) \\
&= 0.5\texttt{erfc}(1.7678 \texttt{ dB}) \approx 0.0062. \quad (6.70)
\end{aligned}$$

Note that all the values used are in decibels. Note that our results agree with the CDF in Fig. 6.10. We can read off Fig. 6.10 that $F_Y(y)|_{y=60 \texttt{ dB}} \approx 0.9938$. This means that $Pr[Y > 60 \texttt{ dB}] = 1 - F_Y(y)|_{y=60 \texttt{ dB}} = 1 - 0.9938 = 0.0062$.

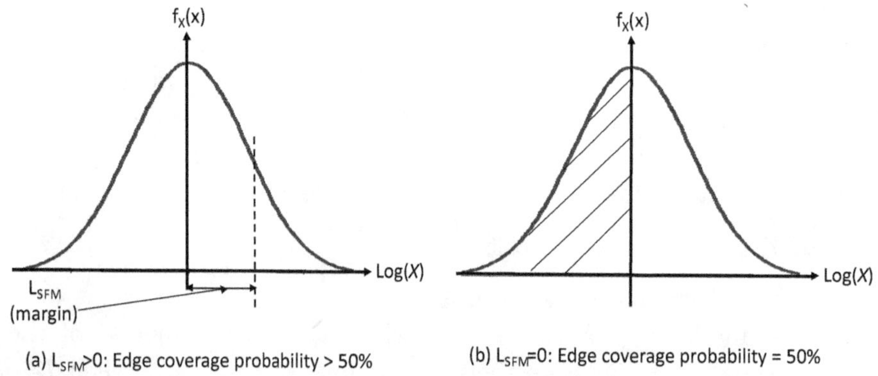

Fig. 6.11 Slow fade margins for different cell edge coverages.

6.2.1 *Check Your Understanding*

Students are encouraged to check their understanding of the content of this section with the following questions prior to proceeding to the next section.

(1) Consider Fig. 6.9. Compute the probability that the instantaneous path loss exceeds $Y = 50$ dB, if $L_{50} = 40$ dB and $\sigma = 8$ dB.
(2) Consider Fig. 6.9. Compute the probability that the maximum instantaneous path loss is $Y = 20$ dB, if $L_{50} = 40$ dB and $\sigma = 8$ dB.
(3) Consider Fig. 6.10. Compute the probability that the instantaneous path loss is $Y = 40$ dB, if $L_{50} = 40$ dB and $\sigma = 8$ dB. Discuss the implications of your answer.
(4) Consider Fig. 6.10. Compute the fraction of time that the path loss exceeds $Y = 40$ dB. Discuss the implications of your answer.
(5) Consider Fig. 6.10. Compute the fraction of time that the path loss falls below $Y = 40$ dB. Discuss the implications of your answer.

6.3 Path Loss Models for Frequencies Below 6 GHz

As a radio signal travels a distance from the transmitter towards the receiver its power intensity reduces. Such a distance-caused attenuation is referred to as **path loss.** Path loss is also referred to as path attenuation. Path loss is caused mainly by the propagation distance travelled by the wireless signal through the channel. Path loss causes deterministic **large-scale fading** or long-term fading, the reason why it is also popularly called so. Path loss models are used, for example, to dimension cells in a cellular system, as well as predicting received signal strengths. We are studying path loss models for non-free space propagation environments in which the line-of-sight path between the transmitting and the receiving antennas are obstructed by man-made and/or natural objects. Path loss models are categorised into three types: deterministic models, empirical models and semi-empirical models. The main causes of path loss are:

(1) Free-space loss owing to the spreading in space of the waves as it moves, as well as wave's energy absorption by ions in the atmosphere.
(2) Diffraction, scattering, reflection and refraction of signals.
(3) Obstacle penetration or absorption loss.

The applicable path loss model depends on the size of the cell. There are models for large macro-cells with circa 1 km to 30 km radius, small macro-cells with circa 0.5 km to 3 km radius, micro-cells with circa less than 1 km radius, and pico-cells with circa up to 0.5 km radius. Path loss models are used to estimate signal power loss for a given system and propagation environment and they are used in system design such as the placement of radio base stations. In the following, we use path loss to refer to the 50th percentile (i.e. median) value of the propagation path loss, which is usually denoted by L_{50}. However, we shall be using L instead of L_{50} just for the sake of brevity.

Deterministic models are analytical models derived from ray tracing or by solving Maxwell's electromagnetic equations with boundary conditions. Exact solution methods used to solve the Maxwell's equations include the finite element method, the finite-difference time domain and the method of moments. Ray tracing is usually based on high frequency approximations to solution of Maxwell's equations, and they are of two types: ray launching and ray imaging. Empirical models are obtained from field measurements of radio propagation. As their name says, semi-empirical models are hybrid models combining some aspects of the other two types of models. A number of path loss models have been proposed in the literature. Figure 6.12 attempts to classify them. We study a sample of each type of path loss model in this section.

Statistical channel models are very versatile in communications systems design, optimisation and performance analysis as many things cannot be

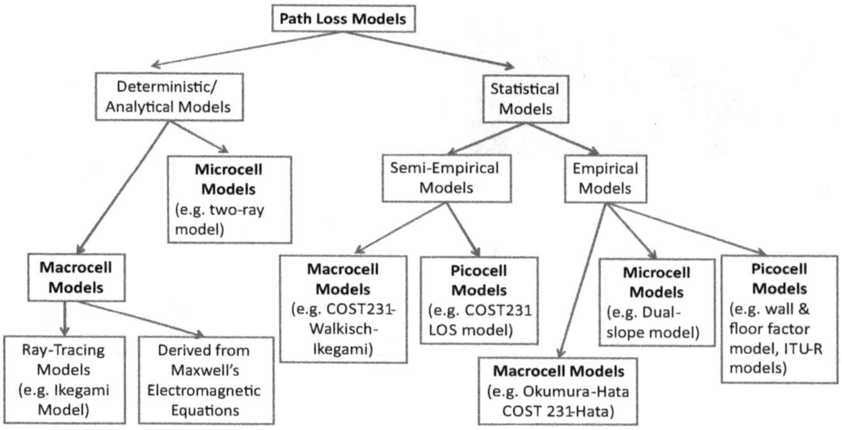

Fig. 6.12 Classification of path loss models.

defined concretely. A popular statistical model is called Monte Carlo simulations.

6.3.1 *Deterministic Path Loss Models*

Deterministic path loss models are derived from Maxwell's electromagnetic equations, and ray tracing for macrocells and microcells. A common example of micro-cell deterministic model is the two-ray model, and Ikegami model for macrocells. A simple approximation used for path loss is

$$L(d) = \bar{L}(d_0) + 10\gamma \log_{10}(d/d_0) \tag{6.71}$$

where $\bar{L}(d_0)$ is the average path loss at a reference distance d_0 from the transmitting antenna, and γ is called path loss exponent. Typical values for the path loss exponent are $2 \leq \gamma \leq 8$. The following sections discuss more precise and detailed path loss models.

6.3.1.1 *Motley-Keenan Indoor Model*

This model has been proposed for indoor radio wave propagation. The total path loss is defined as

$$L = L_{\mathtt{wall}} + L_{\mathtt{floor}} + L(d_0) + 10n \log_{10}(d/d_0) \tag{6.72}$$

where $L(d_0)$ is the path loss at a distance d_0 from the transmitting antenna, $L_{\mathtt{wall}}$ is the cumulative signal attenuation by the walls (about 1–20 dB per wall) and $L_{\mathtt{floor}}$ is the total attenuation due to the floor. As walls and floors of different buildings use different materials, this model is site specific.

6.3.1.2 *Walfisch-Bertoni Outdoor Model*

Proposed in [Walfisch and Bertoni (1988)], the Walfisch-Bertoni Outdoor Model considers diffraction and absorption of radio signals by buildings which are viewed as cylinders lying on the earth. The total path loss experienced by a signal after covering the distance d in this model is

$$L = a_{\mathtt{fsl}} a_{\mathtt{rts}} Q^2 \tag{6.73}$$

where Q^2 quantifies the attenuation experienced by the roof-top signal due to shadowing by a row of buildings at street level, $a_{\mathtt{rts}}$ refers to the signal attenuation caused by roof-top to street diffraction and scattering, and $a_{\mathtt{fsl}}$ is the free-space attenuation. The other details of the model can be found in, for example, [Walfisch and Bertoni (1988)].

6.3.2 *Empirical Path Loss Models*

Empirical models are based on measurements collected from the field. We describe the Okumura model and Okumura-Hata model for macro-cells. These are based on data collected from in and around Tokyo city in Japan by Okumura. They are among the most popular models for systems operating with RF carriers below 3 GHz.

6.3.2.1 *3GPP Path Loss Model*

Let FSL be the free-space path loss, $\mathcal{X}_0 \sim \mathcal{LN}(0, 8\mathtt{dB})$ be a lognormal-distributed random variable with zero mean and 8 dB standard deviation to model shadowing and α is the signal propagation path loss exponent. Also, let $L_{\texttt{pen}}$ be the loss resulting from signal passing through obstructions, and d be the distance between the UE and the base station. Then, 3GPP defines the path loss [3GPP (2010)]

$$PL\,[\texttt{dB}] = FSL + \mathcal{X}_0 + L_{\texttt{pen}} + \alpha \log_{10}(d\,[\texttt{km}]) \tag{6.74}$$

with some typical values $\alpha \approx 3.76$ and $L_{\texttt{pen}} \approx 18 - 20$ dB for urban propagation environment.

6.3.2.2 *Erceg Path Loss Model*

Proposed by in [Erceg *et al.* (1999)] in 1999, Erceg model is an empirical model based on data collected by AT&T (now Lucent) Wireless Services from 95 operational macrocells operating at 1.9 GHz in the United States. The model divides the terrain into the three categories:

(1) Type A: hilly terrain with moderate to heavy tree density, causing a high path loss.
(2) Type B: hilly terrain with light tree density or flat terrain with moderate to heavy tree density.
(3) Type C: mostly flat terrain with light tree density, causing a low path loss.

Under this model, the median path loss is

$$\begin{aligned} L_m\ [\texttt{dB}] &= 10\gamma \log_{10}\left(\frac{d}{d_0}\right) \\ &\quad + 20\log_{10}\left(\frac{4\pi d_0}{\lambda}\right) \\ &\quad + 6\log_{10}\frac{f}{2000} + \alpha_{ms} + X_\sigma \end{aligned} \tag{6.75}$$

where $d > d_0$, $f \leq 2$ GHz, $h_{ms} \leq 2$ m, λ is the operating wavelength, X_σ is a random variable modelling shadowing which has lognormal distribution with a standard deviation in the interval 8.2 dB to 10.6 dB, reference distance $d_0 = 100$ m, and γ is the path loss exponent, which is given by

$$\gamma = a - bh_{bs} + \frac{c}{h_{bs}} \tag{6.76}$$

where h_{bs} is base station height in meters, and a, b, c are constants given in Table 6.3 for each of the three terrain types defined in the model. The variable

Table 6.3 Erceg radio propagation model's constants.

Model Parameter	Terrain Type A	Terrain Type B	Terrain Type C
a	4.6	4	3.6
$b[\texttt{m}^{-1}]$	0.0075	0.0065	0.005
c [m]	12.6	17.1	20

α_{ms} in Eq. (6.75) is the correction factor for the height of the mobile handset. It is given by

$$\alpha_{ms} = \begin{cases} -10.8\log_{10}\frac{h_{ms}}{2000}, & \texttt{Types A and B models} \\ -20\log_{10}\frac{h_{ms}}{2000}, & \texttt{Types C model} \end{cases}$$

where h_{ms} is the mobile station's height.

6.3.2.3 *Okumura Macrocell Path Loss Model*

This model has versions for open areas, quasi-open area and suburban areas. Let us define the following parameters:

- L(dB): Median value of propagation path loss (statistical value)
- $L_{\texttt{FS}}$: Free-space path loss in decibels
- $A_{mu}(f, d)$: Median attenuation relative to free-space in an urban area in dB
- $G(h_{bs})$: Base antenna height gain correction factor
- $G(h_{ms})$: Mobile station antenna height gain correction factor
- G_{AREA}: The gain in dB due to the type of signal propagation environment, e.g. over water, hilly area

The Okumura path loss model is defined as

$$L = L_{\texttt{FS}} + A_{mu}(f, d) - G(h_{bs}) - G(h_{ms}) - G_{AREA} \tag{6.77}$$

where

$$\begin{aligned} G(h_{ms}) &= 20\log_{10}(h_{ms}\,[\texttt{m}]/3)\ \texttt{dB}, \\ &\qquad 1\ \texttt{m} < h_{ms} < 3\ \texttt{m} \qquad (6.78) \\ G(h_{bs}) &= 20\log_{10}(h_{bs}\,[\texttt{m}]/200)\ \texttt{dB}, \\ &\qquad 30\ \texttt{m} < h_{bs} < 1000\ \texttt{m}. \qquad (6.79) \end{aligned}$$

This model is valid for the distance 1 km to 100 km and frequency range of 150 to 1920 MHz but

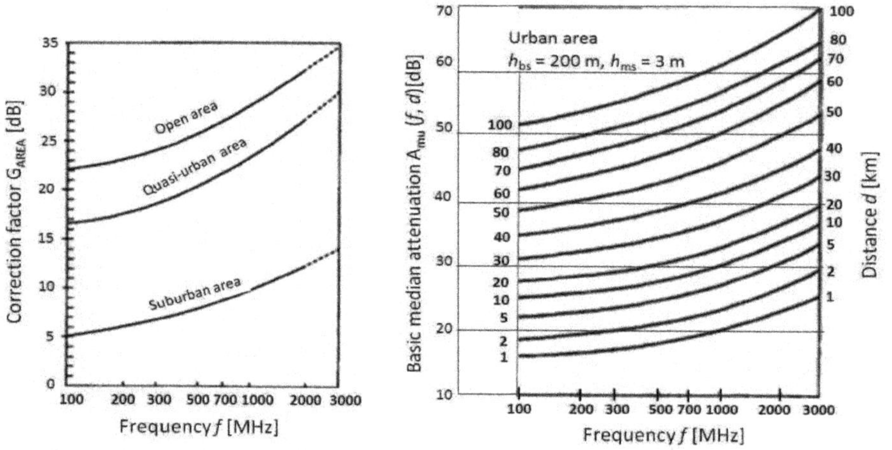

(a) Gain for specific propagation environments (b) Median attenuation

Fig. 6.13 Correction factors for Okumura path loss model [Okumura *et al.* (1968)].

often extrapolated to 3 GHz. Figure 6.13 shows the correction factors G_{AREA} and $A_{mu}(f, d)$ for the Okumura model. Note that the world's tallest radio tower is the 628.8-meter KVLY-TV mast located in North Dakota, USA.

Example: Path Loss Estimation Using Okumura Model

Assume that a mobile cellular system operates in a suburban area at the carrier frequency 1 GHz. Assume further that the user equipment (UE) and the eNodeB have the respective heights of 1.8 m and 60 m above ground. If the cell radius is 2 km and UEs at cell fringes must receive sufficient signal strength, how much median path loss must the system consider in the design to ensure optimum operation? Assume that the system designers have access to only the Okumura path loss model.

Solution

The correction factor for the mobile equipment is

$$G(h_{\mathtt{ms}}) = 20 \log 10(h_{\mathtt{ms}}\,[\mathtt{m}]\,/3)$$
$$= 20 \log 10(1.8/3) \approx -4.437 \;\mathtt{dB}$$

while that for the base station is

$$G(h_{\mathtt{bs}}) = 20 \log 10(h_{\mathtt{bs}}\,[\mathtt{m}]\,/200)$$
$$= 20 \log 10(260/200) \approx 2.2789 \;\mathtt{dB}\,.$$

We estimate the value of G_{AREA} at 1 GHz off Fig. 6.13(a) for a suburban area as 9.8 dB. Similarly, we estimate the median attenuation relative to free-space $A_{\mathtt{mu}}(f, d)$ in dB at 1 GHz and 2-km distance off Fig. 6.13(b) as $A_{\mathtt{mu}}(f, d) \approx 7$ dB. We assume ideal isotropic antennas for both transmitter and receiver (i.e., $G_{\mathrm{T}} = G_{\mathrm{R}} = 0$ $\mathtt{dB}$) and apply Eq. (5.16) to compute the free-space path loss as

$$L_{\mathtt{FS}} = 20 \log_{10} d(\mathtt{km}) + 20 \log_{10} f(\mathtt{MHz}) + 32.44$$
$$= 20 \log_{10} 2 + 20 \log_{10} 1000 + 32.44 \approx 98.5 \;\mathtt{dB}\,.$$

Thus, the median path loss of the 2-km link is

$$L = L_{\mathtt{FS}} + A_{mu}(f, d) - G(h_{bs}) - G(h_{ms}) - G_{AREA}$$
$$= 98.5 + 7 - 2.3 - 4.437 - 9.8 = 88.96 \;\mathtt{dB}\,.$$

6.3.2.4 *Okumura-Hata Macrocell Path Loss Model*

This model is often referred to as just Hata model. Hata used the data collected by Okumura from Tokyo to build these models. Thus, these models are a set of analytical models approximately fitted onto Okumura's data. This model differs from the Okumura model in the following ways:

(1) It extends the graphical information from the Okumura model to cover the effects of diffraction, reflection and scattering caused by city structures.
(2) Uses the urban area as reference and then applies correction factors for applications in

suburban and rural environments while the Okumura model is used for only urban environments.

(3) Valid for only 150–1500 MHz, while Okumura covers 150–1920 MHz.

(4) Applicable to both point-to-point and broadcast communications.

Let us first define the parameters used in this model.

- L: Path loss in urban areas in decibels (dB)
- h_{bs}: Height of base station antenna in meters (m)
- h_{ms}: Height of mobile station antenna in meters (m)
- f: Frequency of transmission in Megahertz (MHz)
- A_{C}: Antenna height correction factor
- d_{TR}: Distance between the base and mobile stations in kilometers (km)
- L_{SU}: Path loss in suburban areas in decibels (dB)
- LU: Path loss in urban areas, measured in decibels (dB)
- LO: Path loss in open areas, measured in decibels (dB)

The path loss in urban areas is

$$\begin{aligned} L_{\text{U}} &= 69.55 + 26.16\log_{10} f \\ &\quad + (44.9 - 6.55\log_{10} h_{\text{bs}})\log_{10} d_{\text{TR}} \\ &\quad - 13.82\log_{10} h_{\text{bs}} - A_{\text{c}} \end{aligned} \tag{6.80}$$

where for small- to medium-sized cities

$$A_{\text{c}} = 0.8 + (1.1\log_{10} f - 0.7)h_{\text{ms}} - 1.56\log_{10} f$$

and for large cities

$$A_{\text{c}} = \begin{cases} 9.29(\log_{10}(1.54h_{\text{ms}}))^2 \\ \quad -1.1, & 150 \le f \le 200 \text{ MHz} \\ 3.2(\log_{10}(11.7554h_{\text{ms}}))^2 \\ \quad -4.97, & 200 < f \le 1500 \text{ MHz} \end{cases}.$$

Path loss for suburban areas is

$$\begin{aligned} L_{\text{SU}} &= 69.55 + 26.16\log_{10} f \\ &\quad + (44.9 - 6.55\log_{10} h_{\text{bs}})\log_{10} d_{\text{TR}} \\ &\quad - 13.82\log_{10} h_{\text{bs}} - 2(\log_{10}(f/28))^2 + 5.4\,. \end{aligned} \tag{6.81}$$

Path loss for open areas is

$$\begin{aligned} L_{\text{O}} &= 69.55 + 26.16\log_{10} f \\ &\quad + (44.9 - 6.55\log_{10} h_{\text{bs}})\log_{10} d_{\text{TR}} \\ &\quad - 4.78(\log_{10} f)^2 + 18.33\log_{10} f + 40.94\,. \end{aligned} \tag{6.82}$$

6.3.2.5 *Lee's Path Loss Model*

Developed at Bell Labs by William C. Y. Lee, a pioneer in mobile communications, from data collected in North-Eastern United States in the 1980s, Lee's model is a macroscopic propagation model. This model was originally developed for 900 MHz transmissions, but has been extended to 2 GHz. It is a slope-intercept propagation prediction model assuming that the factors contributing to radio propagation path loss can be divided into:

- Clutter and man-made structures.
- Natural terrain.

Other assumptions underlying Lee's model are:

(1) Base station: height of 30.48 m, transmit power of 10 W and antenna gain of 6 dB.

(2) Mobile station: height of 3 m, and antenna gain of 0 dB (i.e., omni-directional antenna).

(3) Transmission distance or cell coverage: 1.6 km (i.e., microcell).

The mathematical formulation of Lee's model is

$$P_r = 10\log_{10}\left[P_0\left(\frac{d_0}{d}\right)^{\gamma}\left(\frac{f_C}{f}\right)^{n}\alpha_0\right] \tag{6.83}$$

where γ is the path loss exponent, P_r is the received power when the power at a reference distance $d_0 = 1$ mile ≈ 1.6 km away from the transmitter is P_0, d is the distance of receiver from the transmitter, $f_C = 900$ MHz and f are frequencies,

$$n = \begin{cases} 2, & f_c < 450 \text{ MHz, suburban area} \\ 3, & f_c > 450 \text{ MHz, urban area} \end{cases}$$

and the correction factor which can be used to tailor the model to different environments is

$$\alpha_0 = \alpha_1\alpha_2\alpha_3\alpha_4\alpha_5\,. \tag{6.84}$$

The components in the correction factor Eq. (6.84) are

- Base station antenna height correction factor:

$$\alpha_1 = \left(\frac{\texttt{height of new base station [m]}}{30.48\ \texttt{m}}\right)^2. \tag{6.85}$$

- Mobile station antenna height correction factor:

$$\alpha_2 = \left(\frac{\texttt{height of new mobile station [m]}}{3\ \texttt{m}}\right)^m \tag{6.86}$$

where

$$m = \begin{cases} 1, & h_{ms} > 3\ \texttt{m} \\ 2, & h_{ms} \leq 3\ \texttt{m} \end{cases}.$$

- Transmit power correction factor:

$$\alpha_3 = \left(\frac{\texttt{new transmit power [W]}}{10\ \texttt{W}}\right)^2. \tag{6.87}$$

- Base station gain correction factor:

$$\alpha_4 = \left(\frac{\texttt{new base station gain}}{4}\right)^2. \tag{6.88}$$

- α_5: correction factor for using different antennas, which equals the mobile station's antenna gain in dBd.

Some sample data for Lee's model are given in Table 6.4. Using the sampled data in Table 6.4, Lee computed the median path loss for the six propagation scenarios above from Eq. (6.83) as

$$L_{50} = \begin{cases} 85 + 20\log_{10}\left(\frac{d}{1.6\ \texttt{km}}\right) \\ \quad +10n\log_{10}\left(\frac{f}{900\ \texttt{MHz}}\right) - \alpha_0, & \texttt{Terrain 1} \\ 89 + 43.5\log_{10}\left(\frac{d}{1.6\ \texttt{km}}\right) \\ \quad +10n\log_{10}\left(\frac{f}{900\ \texttt{MHz}}\right) - \alpha_0, & \texttt{Terrain 2} \\ 101.7 + 38.4\log_{10}\left(\frac{d}{1.6\ \texttt{km}}\right) \\ \quad +10n\log_{10}\left(\frac{f}{900\ \texttt{MHz}}\right) - \alpha_0, & \texttt{Terrain 3} \\ 110 + 36.8\log_{10}\left(\frac{d}{1.6\ \texttt{km}}\right) \\ \quad +10n\log_{10}\left(\frac{f}{900\ \texttt{MHz}}\right) - \alpha_0, & \texttt{Terrain 4} \\ 104 + 43.11\log_{10}\left(\frac{d}{1.6\ \texttt{km}}\right) \\ \quad +10n\log_{10}\left(\frac{f}{900\ \texttt{MHz}}\right) - \alpha_0, & \texttt{Terrain 5} \\ 124 + 30.5\log_{10}\left(\frac{d}{1.6\ \texttt{km}}\right) \\ \quad +10n\log_{10}\left(\frac{f}{900\ \texttt{MHz}}\right) - \alpha_0, & \texttt{Terrain 6} \end{cases} \tag{6.89}$$

6.3.2.6 *COST 231-Hata Macrocell Path Loss Model*

COST 231 was a working group of an EU Forum for cooperative scientific research called COST in the time period 1986 to 1996. Also known as COST 231-Hata model, it is an extension of the Hata model for urban areas for up to 2 GHz. The model is based on extensive measurement in Lithuania at the frequencies 160, 450, 900 and 1800 MHz and measurements at 800 and 900 MHz in Brazil. The conditions for the validity of this model are:

- Height of base station is higher than all rooftops in the channel.
- Frequency: 1500 MHz to 2000 MHz.
- Mobile station antenna height: 1 m to 10 m.
- Base station antenna height: 30 to 200 m.
- Link distance between transmitter and receiver: 1 km to 20 km.

The path loss is $L = 46.3 + 33.9\log_{10} f - 13.82\log_{10} h_{\texttt{bs}} + [44.9 - 6.55\log_{10} h_{\texttt{bs}}]\log_{10} d_{\texttt{TR}} - A_{\texttt{c}} + g$ where

$$g = \begin{cases} 0\ \texttt{dB, suburban and medium-sized cities} \\ 3\ \texttt{dB, metropolitan areas} \end{cases}$$

and for small cities (aka rural) to medium-sized cities (aka suburban areas)

$$A_{\texttt{c}} = 0.8 + (1.1\log_{10} f - 0.7)h_{\texttt{ms}} - 1.56\log_{10} f$$

and for large cities

$$A_{\texttt{c}} = \begin{cases} 9.29(\log_{10}(1.54h_{\texttt{ms}}))^2 \\ \quad -1.1, & 150 \leq f \leq 200\ \texttt{MHz} \\ 3.2(\log_{10}(11.7554h_{\texttt{ms}}))^2 \\ \quad -4.97, & 200 < f \leq 1500\ \texttt{MHz} \end{cases}.$$

The standard deviation of the error in this model is about 5–7 dB in urban and suburban areas but as much as 15 dB in rural areas.

6.3.3 *Semi-Empirical Path Loss Models*

We discuss only the COST-231 Walfisch/Ikegami Model as a case study.

Table 6.4 Some sample data for Lee's model.

Terrain Type	Terrain	Path Loss Exponent γ	Reference Power P_0 (dBm)
1	Free space	2	−45
2	Open area	4.35	−49
3	Suburban (America)	3.84	−61.7
4	Urban (Philadelphia)	3.68	−70
5	Urban (Newark)	4.31	−64
6	Urban (Tokyo)	3.05	−84

6.3.3.1 *COST-231 Walfisch/Ikegami Model*

This is a semi-empirical path loss model, which considers both the terrain profile and buildings' profile. Diffraction loss of buildings is modelled using multiple signal-absorbing half-screens spaced uniformly by a distance which is much larger than the wavelength. This model is valid under the following conditions: base station height of 4 to 50 meters, mobile station height of 1 to 3 meters, 20 meters to 5 km transmission range and the frequency range of 800–2000 MHz. The path loss is defined for LOS and NLOS paths separately. They are for LOS path

$$L_{\texttt{los}} = 42.6 + 26\log_{10} d\,\texttt{[km]} + 20\log_{10} f,\ d > 20\ \texttt{m} \tag{6.90}$$

and for the NLOS path

$$L_{\texttt{nlos}} = \begin{cases} L_{\texttt{fs}}, & \texttt{if } L_{\texttt{rts}} + L_{\texttt{msd}} < 0 \\ L_{\texttt{fs}} + L_{\texttt{rts}} + L_{\texttt{msd}}, & \texttt{otherwise} \end{cases}$$

where the free-space path loss is $L_{\texttt{fs}} = 32.4 + 20\log_{10} d\,\texttt{[km]} + 20\log_{10} f\,\texttt{[MHz]}$ and

- LRTS: rooftop-to-street diffraction and scatter loss (aka multiscreen loss), represents the coupling of wave propagating along the multiscreen path into the street mobile station is located.
- $L_{\texttt{msd}}$ is multi-diffraction loss.
- $h_{\texttt{r}}$: height of rooftop.
- $h_{\texttt{ms}}$: height of mobile handset.
- k_a denotes the increase of the path loss for base station antennas below the rooftops of adjacent buildings.
- The terms k_d and k_f control the dependence of the multi-screen diffraction loss versus distance and radio frequency.
- b is the spacing between buildings.

These parameters are defined as

$$L_{\texttt{rts}} = \begin{cases} 0, & \texttt{if } L_{\texttt{rts}} < 0 \\ -16.9 - 10\log_{10} w \\ \quad +10\log_{10} f \\ \quad +20\log_{10}\Delta h_{\texttt{m}} + L_{\texttt{ori}}, & \texttt{if } h_{\texttt{r}} > h_{\texttt{ms}} \end{cases}$$

where

$$L_{\texttt{rts}} = \begin{cases} -10 + 0.354\varphi, & 0^\circ \le \varphi \le 35^\circ \\ 2.5 + 0.075(\varphi - 35), & 35^\circ \le \varphi \le 55^\circ \\ 4 - 0.114(\varphi - 55), & 55^\circ \le \varphi \le 90^\circ \end{cases}$$

$$L_{\texttt{msd}} = L_{\texttt{bsh}} + k_a + k_d \log_{10} d\,\texttt{[km]} + k_f \log_{10} f\,\texttt{[MHz]} - 9\log_{10} b\,\texttt{[m]} \tag{6.91}$$

$$L_{\texttt{bsh}} = \begin{cases} -18\log_{10}(1 + \Delta h_{\texttt{bs}}), & h_{\texttt{bs}} > h_{\texttt{r}} \\ 0, & \texttt{otherwise} \end{cases}$$

$$k_d = \begin{cases} 18, & h_{\texttt{bs}} > h_{\texttt{r}} \\ 18 - 15\frac{\Delta h_{\texttt{bs}}}{h_{\texttt{r}}}, & \texttt{otherwise} \end{cases}$$

$$k_a = \begin{cases} 54, & h_{\texttt{bs}} > h_{\texttt{r}} \\ 54 - 0.8\Delta h_{\texttt{bs}}, & d \ge 0.5\ \texttt{km and } h_{\texttt{bs}} \le h_{\texttt{r}} \\ 54 - 1.6\Delta h_{\texttt{bs}}, & d < 0.5\ \texttt{km and } h_{\texttt{bs}} \le h_{\texttt{r}} \end{cases}$$

and

$$k_f = -4 + \begin{cases} 0.7(f/925 - 1), & \texttt{for medium-sized city} \\ 1.5(f/925 - 1), & \texttt{for metropolitan city} \end{cases}.$$

Table 6.5 WINNER II LOS probability models.

Scenario	LOS probability as a function of TX-RX distance d [m]
A2,B2,B4,C2,C3	$P_{LOS}(d) \approx 0$
A1	$P_{LOS}(d) = \begin{cases} 1, & d \leq 2.6 \\ 1 - 0.9[1 - (1.16 - 0.4\log_{10} d)^3]^{1/3}, & d > 2.6 \end{cases}$
B1	$P_{LOS}(d) = \min(18/d, 1) \cdot (1 - e^{-d/36}) + e{-}d/36$
B3	$P_{LOS}(d) = \begin{cases} 1, & d \leq 1 \\ \exp(-(d-1)/9.4), & d > 1 \end{cases}$
C1	$P_{LOS}(d) = e^{-d/200}$
C2	$P_{LOS}(d) = \min(18/d, 1) \cdot (1 - e^{-d/63}) + e{-}d/63$
D1	$P_{LOS}(d) = e^{-d/1000}$

6.3.4 *WINNER II Channel Models*

WINNER is the acronym for Wireless World Initiative New Radio. The WINNER II models are valid for operating frequencies in the range 2–6 GHz with up to 100 MHz RF bandwidth. The WINNER II channel models can be used for both link-level and system-level simulations of local area, metropolitan area and wide area wireless communication networks.

6.3.4.1 *WINNER II LOS Probability Models*

A plethora of LOS probability models have been developed by the WINNER II initiative and reported in [Kyösti *et al.* (2008)]. The LOS probability models are compiled in Table 6.5 ([Kyösti *et al.* (2008)], p. 48). According to [Kyösti *et al.* (2008)], the WINNER model divides radio propagation environments into thirteen (13) types, which are:

(1) A1: indoor office.
(2) A2: indoor-to-outdoor.
(3) B1: urban micro-cell.
(4) B2: bad urban micro-cell.
(5) B3: indoor hotspot.
(6) B4: outdoor-to-indoor.
(7) B5: stationary feeder, both terminals communicating are immobile.
(8) C1: suburban macro-cell.
(9) C2: urban macro-cell.
(10) C3: bad urban macro-cell.
(11) C4: urban macro outdoor-to-indoor.
(12) D1: rural macro-cell.
(13) D2: rural moving networks, both end-points communicating are moving, and possibly at high speeds.

The indoor hotspot model B3, for example, is suitable for big factory halls, airports and train stations.

6.3.4.2 *WINNER II Path Loss Models*

The general WINNER II path loss model is ([Kyösti *et al.* (2008)], p. 43)

$$PL = A\log_{10} d \;\texttt{[m]} + B + C\log_{10}(f_c\,\texttt{[GHz]}/5) + X \qquad (6.92)$$

and uses the following form of the free-space path loss

$$PL_{\texttt{free}} = 46.4 + 20\log_{10} d + 20\log_{10}(f_c/5)\,. \qquad (6.93)$$

The parameters used in Eq. (6.92) have the meanings:

- X: an optional, environment-specific term (e.g., wall attenuation in the A1 NLOS scenario).
- A: A variable which depends on the path-loss exponent.
- B: the intercept.
- C: describes the frequency dependence of the path loss.
- For all LOS deployment scenarios, and for all distances smaller than or equal to the

breakpoint distance, $d'_{\mathtt{BP}}$: $C = 20$. Beyond the breakpoint distance, the frequency dependence is defined by the formulas in Table 6.6.
- For rural NLOS environments: $C = 20$.
- For urban and suburban NLOS macro-cells: $C = 23$.
- For indoor-to-outdoor and outdoor-to-indoor environments: C is the same as in the corresponding outdoor scenario.
- For fixed NLOS feeder scenarios: in urban and suburban scenarios $C = 23$, otherwise $C = 20$.

Table 6.6 is a snapshot of the model parameters in ([Kyösti *et al.* (2008)], p. 44).

Example

Figure 6.14 shows the parameters of a link in a mobile wireless network operating at 5.0 GHz. The radio propagation environment is modelled by the WINNER II C1 LOS 2 path loss model (see Table 6.6). The receive antenna height is $h_{\mathtt{r}} = 1.5$ m, while the transmit antenna height is $h_{\mathtt{t}} = 25$ m.

(1) If the receiver's sensitivity is -101 dBm, compute the maximum cell radius of the mobile network.
(2) Compute the probability that the wireless link described above is a line-of-sight link.

Solution

First, we need to do a link budget using the parameters and values given in Fig. 6.14 to find the maximum acceptable path loss PL_{max}. We must note that we obtain the maximum acceptable path loss if the power in the received signal equals the receiver's sensitivity P_r. Plugging PL_{max} into the WINNER II path loss model C1 LOS 2, we can find the receiver to transmitter distance d, which is the maximum cell radius. In the link budget, we add any gains and subtract any losses as shown in Table 6.7. Doing so yields

$$P_r = 28.5 - PL_{max} \ \mathtt{[dBm]}\,. \qquad (6.94)$$

With the given receiver sensitivity $P_r = 101$ dBm, we obtain the maximum path loss $PL_{max} = 129.5$ dB.

The WINNER II C1 LOS 2 path loss model can be read off Table 6.6 as

$$PL = 11.65 + 40\log_{10} d - 16.2\log_{10}(h_{\mathtt{MS}} h_{\mathtt{BS}}) + 3.8\log_{10}(f_c/5)$$

where $d_{\mathtt{BP}} < d < 5$ km and $h_{\mathtt{BS}} = 25m, h_{\mathtt{MS}} = 1.5\ \mathtt{m}$. With the given parameters, we obtain $d_{\mathtt{BP}} = 4h_r h_t f_c/c = 2.5\ \mathtt{km}$. At $f_c = 5$ GHz, the path loss reduces to

$$PL = 40\log_{10} d - 13.85\ \mathtt{dB} \overset{!}{\leq} PL_{\mathtt{max}} = 129.5\ \mathtt{dB}\,. \qquad (6.95)$$

Hence, the maximum cell radius is

$$d = 10^{\frac{129.5+13.85}{40}}\ \mathtt{m} \approx 10^{3.584}\ \mathtt{m} \approx 3.84\ \mathtt{km}\,. \qquad (6.96)$$

Thus, the largest radius of the cell is about 3.84 km, which is a macro-cell.

The probability of the link being in LOS for the WINNER II model C1 is

$$P_{\mathtt{LOS}}(d) = \exp -\frac{d}{200} = \exp -\frac{3840}{200} \approx \exp -19.19 \approx 4.6 \times 10^{-9}\,.$$

6.3.5 *Check Your Understanding*

Students are encouraged to check their understanding of the content of this section with the following questions prior to proceeding to the next section.

(1) What is the reason why path loss is referred to as large-scale fading?
(2) Assume that the path loss for an indoor radio propagation is given by

$$L = 30 + 20\log_{10} R + \sum_i L_i,\ \ R \geq 2\ \mathtt{m}\ [dB]$$

Table 6.6 WINNER II path loss models.

Scenario	Path loss [dB]	Shadow fading std dev. (σ) [dB]	Applicability range, antenna height default values
A1 LOS	$A = 18.7, B = 46.8, C = 20$	3	$3\text{m} < d < 100$ m, $1\text{m} \leq h_{BS} = h_{MS} \leq 2.5$ m
A1 NLOS[a]	$A = 36.8, B = 43.8, C = 20$ and $X = 5(n_w - 1)$ (light walls) or $X = 12(n_w - 1)$ (heavy walls)	4	Same as A1 LOS, n_w is the number of walls between the BS and the MS ($n_w > 0$ for NLOS)
A1 NLOS[b] (light walls)	$A = 20$, $B = 46.4$, NLOS 2 $C = 20$, $X = 5n_w$	6	Same as A1 LOS
A1 NLOS (heavy walls)	$A = 20$, $B = 46.4$, $C = 20$, $X = 12n_w$	8	Same as A1 LOS
A1 FL	For any of the cases above, add the floor loss (FL), if the BS and MS are in different floors: $FL = 17 + 4(n_f - 1)$, $n_f > 0$		n_f is the number of floors between the BS and the MS ($n_f > 0$)
A2 LOS	$A = 22.7$, $B = 41.0$, $C = 20$	3	10 m $< d_1 < d'_{BP}$
A2 LOS	$PL = 9.45 + 40\log_{10} d_1 - 17.3\log_{10}(h'_{BS}h'_{MS}) + 2.7\log_{10}(f_c/5)$	3	$d'_{BP} < d_1 < 5$, $h_{BS} = 10$ m and $h_{MS} = 1.5$ m
B3 LOS	$A = 13.9$, $B = 64.4$, $C = 20$	3	$5\text{m} < d < 100$ m, $h_{BS} = 6$ m, $h_{MS} = 1.5$ m
B3 NLOS	$A = 37.8$, $B = 36.5$, $C = 23$	4	Same as B3 LOS
C1 LOS 1	$A = 23.8$, $B = 41.2$, $C = 20$	4	30 m $< d < d_{BP}$
C1 LOS 2	$PL = 11.65 + 40\log_{10} d - 16.2\log_{10}(h_{MS}h_{BS}) + 3.8\log_{10}(f_c/5)$	6	$d_{BP} < d < 5$ km, $h_{BS} = 25m, h_{MS} = 1.5$ m
C1 NLOS	$PL = 31.46 + [44.9 - 6.55\log_{10}(h_{BS})]\log_{10} d + 5.83\log_{10}(h_{BS}) + 23\log_{10}(f_c/5)$	8	50 m $< d < 5$ km, $h_{BS} = 25m, h_{MS} = 1.5$ m

$d'_{BP} = 4h'_{BS}h'_{MS}f_c/c$, $d_{BP} = 4h_{BS}h_{MS}f_c/c$, where f_c is the center frequency in Hz, c is the propagation velocity in free space, and h'_{BS} and h'_{MS} are the effective antenna heights at the BS and the MS, respectively. The effective antenna heights h'_{BS} and h'_{MS} are computed as follows: $h'_{BS} = h_{BS} - 1.0\text{m}$, $h'_{MS} = h_{MS} - 1.0\text{m}$, where h_{BS} and h_{MS} are the actual antenna heights, and the effective environment height in urban environments is assumed to be equal to 1.0 m.

[a] Actual A1 NLOS scenario (Corridor-to-Room)
[b] Optional A1 NLOS scenario (Room-to-Room through wall)

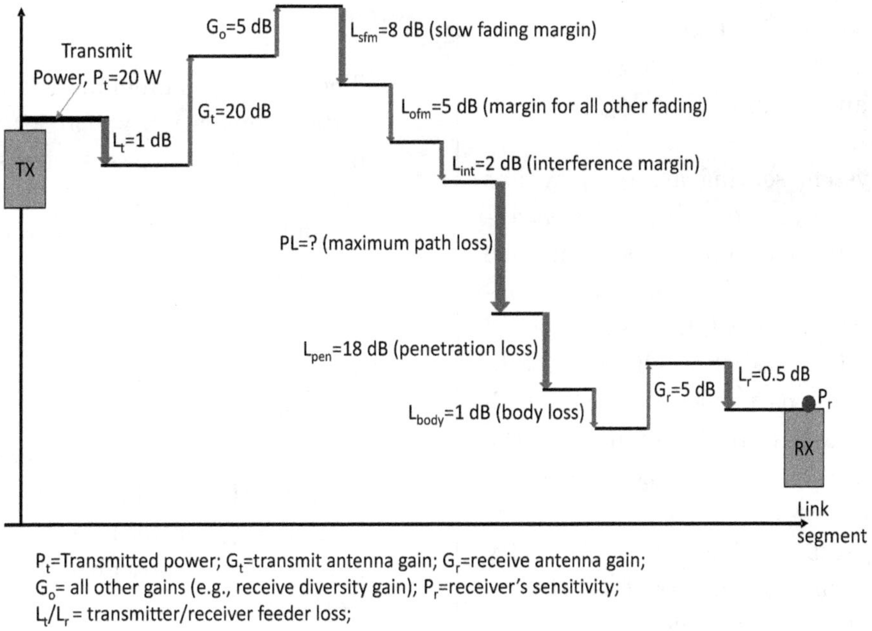

Fig. 6.14 Finding cell radius using link budget and path loss model.

Table 6.7 Link budget to find maximum path loss.

Parameter	P_t	L_t	G_t	G_O	L_{sfm}	L_{ofm}	L_{int}	PL	L_{pen}	L_{body}	G_r	L_5
Value	+43 dB m	−1 dB	+20 dB	+5 dB	−8 dB	−5 dB	−2 dB	−PL dB	−18 dB	−1 dB	+5 dB	−0.5 dB

where $L_i \approx 12$ dB is the attenuation per floor, which is assumed constant for all floors. Assume further that the transmit power of a radio signal is $P_T = 15$ dBm, and the receive antenna has gain of 3 dB. Compute the power in the signal reaching your smartphone if you are located four floors away from the Wi-Fi hotspot. The total radio distance R between you and the hotspot is 200 m. Neglect any other noise besides path loss, and assume that the transmit antenna is an ideal omnidirectional antenna.

(3) Consider the simplified path loss model in Eq. (6.71). A mobile user is at a standstill 2 km away from the serving eNodeB (or base station). If the power in the transmitted signal reduces by 20 dB at a distance 800 m away from the eNodeB, what should be the maximum value for the path loss exponent to keep the total path loss within 32 dB?

(4) Find the area correction factor needed to dimension the path loss of a radio link serving an open area using the Okumura path loss model. Assume that the radio link operates at 850 MHz.

(5) A cellular mobile system serving a suburban area has the following parameters: base station antenna height above ground level is 250 m, mobile station antenna height above ground level is 2 m, and the operating frequency is 1.8 GHz. What should be the largest radius of a cell in the cellular mobile network in order to keep the total link path loss within 80 dB? Assume that the path loss for the coverage area is

best described by the Okumura path loss model. Neglect any other losses besides path loss, and assume that $A_{\mathtt{mu}}(f, d) \approx A_{\mathtt{mu}}(f, 1\ \mathtt{km})$.

(6) A mobile system serving a large city has the following parameters: base station antenna height above ground is 200 m, mobile station antenna height is 2.2 m, and the operating frequency is 0.8 GHz. What should be the largest radius of a cell in order to avoid a total link path loss exceeding 70 dB if the Okumura-Hata path loss model is the only path loss model available to the cell planners?

(7) A cellular system serving a small-sized city has the following parameters: base station antenna height above ground is 180 m, mobile station antenna height is 5.2 m, and the operating frequency is 1.9 GHz. How far away can a mobile user be from the serving eNodeB to avoid keeping his total link path loss to a maximum of 60 dB if the COST-231-Hata path loss model is used?

6.4 Channel Models Above 6 GHz for 5G Networks

How high the frequency used to transmit radio signals depends on the speed of the oscillators used by conventional electronic devices to generate radio waves. In 2006, United Microelectronics Corp in Taiwan developed a voltage-controlled silicon CMOS oscillator operating at 192 GHz. The theoretical limit on the speed of oscillators is the reciprocal of the Planck's time, which is about 1.86×10^{43} Hz. As of Q3 2018, the fastest oscilloscope, LabMaster 10-100Zi, has the sampling rate of 240 GS/s, allowing carrier frequencies of up to 100 GHz. It is important to note that research on developing models for carrier frequencies above 6 GHz is ongoing. Therefore, the models discussed here are just for information of what is happening in the research area about the topic, as these models can be modified in future. Existing channel models used for current wireless systems, such as 3G and 4G, are unsuitable for next-generation wireless networks for several reasons, including:

- The existing channel models were developed for radio propagation at carrier frequencies below 6 GHz. 5G wireless systems are expected to use carrier frequencies beyond 6 GHz and even as high as 100 GHz. Currently, AT&T, Verizon and T-mobile are exploring the use of 37–40 GHz band, while several telecommunications service providers (including SK Telecom and NTT Docomo) are exploring the lower range 3.3–4.2 GHz.
- 5G systems aim to support large bandwidths (e.g. 2 GHz individual carrier bandwidth plus carrier aggregation), while existing systems support low bandwidths (e.g. 4G LTE-A supports up to 20 MHz).
- Poised to operate at gigahertz frequencies, 5G systems will use millimeter waves (mmWaves) which have very short wavelengths. EM waves with such short wavelengths can support larger antenna arrays (e.g. massive MIMO) and narrower beamforming techniques than earlier wireless systems using lower-frequency carriers. These are good for system performance. 5G systems seek to support any type of antenna arrays, including linear, cylindrical spherical and planar, with an arbitrary polarisation.
- 5G wireless systems will support device-to-device (D2D) or machine-to-machine (M2M) communications in which both communications endpoints are moving. This contrasts with earlier systems in which one point of the communications link is usually a base station and thus fixed.
- 5G wireless systems will support wireless communications in areas with a crowd of people. Using mmWaves, human bodies, buildings and vehicles can temporarily block or even absorb radio waves to disturb communications.
- The channel model must support vehicular speeds up to 350 km/hr.

The above features of 5G require accurate channel models with high-resolution radio wave propagation directions than using existing models designed for systems with different characteristics.

[Haneda *et al.* (2016)] proposes a model to estimate attenuation endured by a radio signal as it encounters building (i.e. *building penetration loss*) as

$$BPL[\mathtt{dB}] = 10\log_{10}(A + Bf^2) \tag{6.97}$$

where f is frequency in gigahertz, $A = 5, B = 0.03$ for low-loss buildings and $A = 10, B = 5$ for high-loss buildings.

The radio propagation environments in 5G are divided into:

(1) Urban: both urban microcell (UMi) or urban macrocell (UMa).
(2) Urban microcell street canyon (UMi-Street Canyon): in a street canyon (aka urban canyon) the street is flanked by buildings (usually skyscrapers) on either side as it is in most city centers.
(3) Urban microcell open square (UMi-Open Square).
(4) Indoor-hotspot-indoor (InH-Indoor).
(5) Indoor-hotspot-shopping mall (InH-Shopping Malls).

Each of the five environments listed above is further divided into line-of-sight (LOS) and non-line-of-sight (NLOS) scenarios. Figure 6.16 lists the parameters for the path loss models discussed in the following for the five propagation environments.

6.4.1 *Probability of Wireless Link Being LOS*

Wireless channel modelling for use in 5G networks consider the likelihood of the channel between the two radio terminals being a line-of-sight (LOS) link. LOS probability models are needed for system-level simulation of wireless networks. Several LOS probability models for indoor radio signal propagation have been proposed in the literature. We sample in this section some of the most popular models. In the following, $p(d)$ means the probability that the link over the distance d meters is a line-of-sight (LOS) link. We have the $\frac{d_1}{d_2}$ LOS probability model [3GPP (2015a)], [ITU-R (2015)], [Haneda *et al.* (2016)]

$$p(d) = \min\left\{\frac{d_1}{d}, 1\right\}(1 - e^{-d/d_2}) + e^{-d/d_2} \tag{6.98}$$

where d is the distance in meters between the two radio nodes communicating, and d_1 and d_2 are also distances which can be fitted onto a given data. The second LOS probability model is the NYU (squared) LOS probability model [Samimi *et al.* (2015)], [Haneda *et al.* (2016)]

$$p(d) = \left(\min\left\{\frac{d_1}{d}, 1\right\}(1 - e^{-d/d_2}) + e^{-d/d_2}\right)^2. \tag{6.99}$$

Another LOS probability model is the 3GPP model for urban macrocell environment given by [3GPP (2015a)]

$$p(d) = \left[\min\left\{\frac{18}{d}, 1\right\}(1 - e^{-d/63}) + e^{-d/63}\right][1 + C(d, h_{\mathtt{UE}})] \tag{6.100}$$

where $h_{\mathtt{UE}}$ is the height of the user equipment with the values $C(d, h_{\mathtt{UE}}) = 0$ for $h_{\mathtt{UE}} < 13$ m and $C(d, h_{\mathtt{UE}}) = g(d)(\frac{h_{\mathtt{UE}}-13}{10})^{1.5}$ for 13 m $\leq h_{\mathtt{UE}} \leq$ 23 m where

$$g(d) = \begin{cases} 1.25d^2 e^{-(6+d/150)}, & d > 18 \text{ m} \\ 0, & \mathtt{otherwise} \end{cases}.$$

One of the indoor LOS probability models proposed by the ITU is

$$P_{\mathtt{ITU}}(d) = \begin{cases} 1, & d \leq 1.1 \\ \exp(-(d-1)/4.9), & 1.1 \leq d < 9.8 \\ 0.17, & d \geq 9.8 \end{cases}.$$

A new model useful for an arbitrary indoor (aka indoor-mixed office) radio signal propagation environment is ([ETSI-3GPP (2017)], p. 29)

$$P_{\mathtt{new}}(d) = \begin{cases} 1, & d \leq 1.2 \\ \exp(-(d-1.2)/4.7), & 1.2 \leq d < 6.5 \\ 0.32\exp(-(d-6.5)/32.6), & d \geq 6.5 \end{cases}.$$

Figure 6.15 compares four of the LOS probability models discussed earlier.

Example: Cellular Mobile Network Design Case Study

Wireless communications systems perform better when the link between the transmitter and the receiver is line-of-sight (LOS). This is especially so

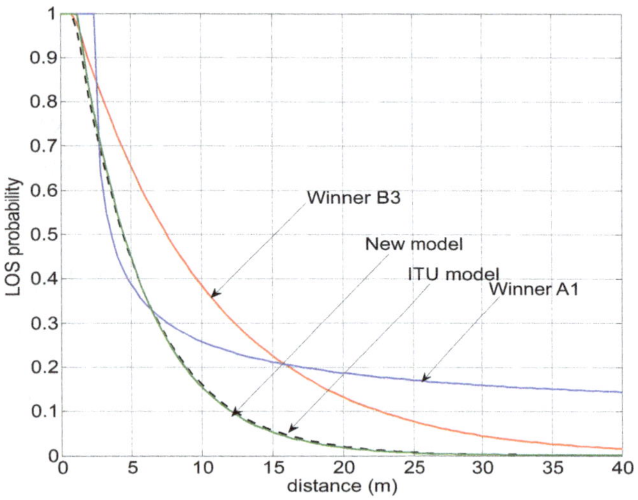

Fig. 6.15 Comparison of line-of-sight probability models.

for mmWave communications expected in 5G cellular mobile networks. For this reason, there are models to compute the probability of a link being LOS. Four of such LOS probability models are superimposed in Fig. 6.15.

(1) The four models shown in Fig. 6.15 describe different geographical locations. With justification, which of the locations is most favourable for LOS wireless communications?
(2) If the probability of the link being LOS should not be less than 0.15 (or 15%), compute the maximum distance between a user equipment (UE) and its serving eNodeB in a cellular mobile network operating in the environment found in Part (1).
(3) Assume that the cellular mobile network is operating at 4.0 GHz. Compute the maximum free-space path loss experienced by a connection between a UE and its serving eNBs in the system described in Parts (1) and (2) in both decibels and ratio form.
(4) If all other link losses besides the free-space path loss is 40 dB and the receiver sensitivity of a UE is −104 dBm, what should be the minimum transmit power of the eNodeB for maintaining an acceptably good link? Compute the power in both dBm and mW.
(5) Assume that the network with features described earlier should cover an area with size 1.5 km^2. Compute the number of cells or eNBs that the PLMN operator needs for 100% coverage. Assume that shapes of cells are hexagonal, and the network uses uniform cells.
(6) If each eNB serves on the average 200 users and each user generates 0.4 Erlang of traffic on the average, how many physical resource elements (or channels) should each eNB be equipped with? Each eNB blocks a connection request at a time that it has no idle resource elements. The operator must keep the grade of service (GOS) of the entire network within 20%.
(7) Assume that each physical resource element occupies $B = 15$ kHz bandwidth, and the mobile network uses a cluster size of $C = 19$, compute how much bandwidth that a PLMN operator must acquire from a local regulator to run the cellular mobile network described earlier. Assume that two control channels are used in each cell.

Solution

(1) The environment described by the Winner A1 model in Fig. 6.15 is most favourable for LOS wireless communications. The reason being that it has the highest probability of having a LOS link with the largest coverage distance.
(2) At LOS probability of 15%, the LOS distance in the environment described by the Winner A1 model is about $d_{\text{T-R}} = 40$ m. All the other environments achieve much less distance at 15% LOS probability.
(3) At 4.0 GHz, the operating wavelength is $\lambda = 75$ mm. The maximum free-space path loss (FSL) occurs at the cell fringe, which is

$$\begin{aligned} FSL &\approx 32.44 + 20\log_{10} d\,(\texttt{km}) \\ &\quad + 20\log_{10} f\,(\texttt{MHz})\ \texttt{dB} \\ &\approx 76.52\ \texttt{dB}\,. \end{aligned} \tag{6.101}$$

(4) Let S_{r} be the receiver sensitivity, P_{t} be the transmit power and P_{r} be the received power. We need to perform a link budget here, which is

$$\begin{aligned} P_{\text{r}} &= P_{\text{t}} - FSL - \textit{other losses} \geq S_{\text{r}} \\ &\Leftrightarrow P_{\text{t}} \geq -104\ \texttt{dBm} + (76.52 + 40)\ \texttt{dB} \\ &= 12.52\,. \end{aligned} \tag{6.102}$$

(5) The maximum cell radius $R_{\texttt{cell}}$ equals the maximum distance between a UE and its serving eNodeB, which is $d_{\texttt{T-R}} = 40$ m. The area of a cell with hexagonal shape is $A_{\texttt{cell}} = \frac{\sqrt{27}}{2} R^2_{\texttt{cell}} \approx 0.00416\ \texttt{km}^2$. Under the given assumptions of uniform cells for the network and hexagonal-shaped cells, we obtain the number of eNBs for 100% coverage as $N_{\texttt{cells}} = \frac{1.5\texttt{km}^2}{0.00416\ \texttt{km}^2} \approx 361$.

(6) Given are $Nu = 200$ users, each with traffic intensity of $Au = 0.4$ E. Thus, the traffic intensity to be handled by each eNB is $A = Nu \times Au = 80$ E. If each eNB blocks a connection request at a time that it has no idle resource elements, then the Erlang B formula can be used to dimension each eNB. At GOS of 20% we read off Erlang B table 80.83 E using 68 channels.

(7) Given are $B = 15$ kHz bandwidth for each physical resource element, $C = 19$ cluster size and $N = 68$ number of physical resource elements needed at each cell. The minimum bandwidth required for a PLMN operator to launch the network dimensioned earlier is

$$W \geq (N \times C + 2) \times B = (68 \times 19 + 2) \times 15 \text{ kHz}$$
$$= 19.41 \text{ MHz}.$$

Many PLMN operators are issued 20 MHz bandwidth by local radio spectrum management organisations.

6.4.2 *Path Loss Models for Frequencies Above 6 GHz*

Signals transferred over carrier frequencies above 6 GHz are referred to as millimeter waves (mmWaves). One of the current research on mmWaves is whether cellular mobile networks based on mmWaves will be thermal noise limited or interference limited. Alternatively, under which conditions will such networks be noise-limited networks of interference-limited networks. Owing to their tiny wavelengths, mmWaves are liable to high isotropic path loss. Therefore, they need to be transmitted over high-gain directional antennas to offset the high path loss. The good news in this is that directional transmissions reduce co-channel interference. Their high absorption properties means that mmWaves cannot be used over long transmission ranges.

We sample a few path loss models which have been proposed for systems using carrier frequencies above 6 GHz. We first look at the close-in (CI) free-space reference distance path loss model [Rappaport *et al.* (2015)], [MacCartney *et al.* (2015)]

$$L^{\texttt{CI}}(f, d) = L_{\texttt{FSPL}}(f, 1\ \texttt{m}) + 10n \log_{10} d\ \ [\texttt{m}]$$
$$+ X_\sigma^{\texttt{CI}}\ \ [\texttt{dB}] \qquad (6.103)$$

where f is frequency in hertz, n is the path loss exponent, d is the link distance in meters, $X_\sigma^{\texttt{CI}}$ is path loss due to shadowing with the standard deviation σ in decibels, and $L_{\texttt{FSPL}}(f, 1\ \texttt{m})$ is the free-space path loss at the distance 1 m from the transmitter, which is given as

$$L_{\texttt{FSPL}}(f, 1\ \texttt{m}) = 20 \log_{10}(4\pi f / c) \qquad (6.104)$$

with c being the speed at which the radio signal propagates through the atmosphere.

There is the alpha-beta-gamma (ABG) path loss model [MacCartney *et al.* (2015)], [Haneda *et al.* (2016)]

$$L^{\texttt{ABG}}(f, d) = 10\alpha \log_{10} d + \beta + 10\gamma \log_{10} f$$
$$+ X_\sigma^{\texttt{ABG}}\ \ [\texttt{dB}] \qquad (6.105)$$

where $X_\sigma^{\texttt{ABG}}$ is the path loss due to shadowing with the standard deviation σ in decibels, α is the path loss exponent, d is the transmit-receive distance in meters, f is frequency in GHz and β is a floating offset in decibels.

We also have the close-in free-space reference distance path loss model with frequency-dependent path loss exponent (CIF), which is an extension of the CI model Eq. (6.103) in [Rappaport *et al.* (2015)], [MacCartney *et al.* (2015)]. It is given by

$$L^{\texttt{CIF}}(f, d) = L_{\texttt{FSPL}}(f, 1\ \texttt{m})$$
$$+ 10n\left[1 + b\left(\frac{f - f_0}{f_0}\right)\right] \log_{10} d[\texttt{m}]$$
$$+ X_\sigma^{\texttt{CIF}}\ \ [\texttt{dB}] \qquad (6.106)$$

where n is the path loss exponent, and b is an optimization parameter used to capture the frequency dependence of the path loss exponent. Finally, the fixed frequency reference is given by $f_0 = \sum_{k=1}^{K} f_k N_k / \sum_{k=1}^{K} N_k$, where K is the number of unique frequencies, N_k is the number of path

Scenario	CI/CIF Model Parameters	ABG Model Parameters
UMa-LOS	n=2.0, SF=4.1 dB	Not applicable
UMa-NLOS	n=3.0, SF=6.8 dB	φ=3.4, β=19.2, γ=2.3, SF=6.5 dB
UMi-Street Canyon-LOS	n=1.98, SF=3.1 dB	Not applicable
UMi-Street Canyon-NLOS	n=3.19, SF=8.2 dB	φ=3.48, β=21.02, γ=2.34, SF=7.8 dB
UMi-Open Square-LOS	n=1.85, SF=4.2 dB	Not applicable
UMi-Open Square-NLOS	n=2.89, SF=7.1 dB	φ=4.14, β=3.66, γ=2.43, SF=7.0 dB
InH-Indoor-LOS	n=1.73, SF=3.02 dB	Not applicable
InH-Indoor-NLOS single slope (FFS)	n=3.19, b=0.06, f_0=24.2 GHz, SF=8.29 dB	φ=3.83, β=17.3, γ=2.49, SF=8.03 dB
InH-Indoor-NLOS dual slope	n_1=2.51, n_2=4.25, b_1=0.12, b_2=0.04, f_0=24.1 GHz, SF=7.65 dB, d_{BP}=7.8 m	$φ_1$=1.7, $φ_2$=4.17, $β_1$=33.0, γ=2.49, SF=7.78 dB, d_{BP}=7.78 m
InH-Shopping Malls-LOS	n=1.73, SF=2.01 dB	Not applicable
InH-Shopping Malls-NLOS single slope (FFS)	n=2.59, b=0.01, f_0=39.5 GHz, SF=7.40 dB	φ=3.21, β=18.09, γ=2.24, SF=6.97 dB
InH-Shopping Malls-NLOS dual slope	n_1=2.43, n_2=8.36, b_1=-0.01, b_2=0.39, f_0=39.5 GHz, SF=6.26 dB, d_{BP}=110 m	$φ_1$=2.9, $φ_2$=11.47, $β_1$=22.17, γ=2.24, SF=6.36 dB, d_{BP}=147.0 m

Fig. 6.16 Environment-dependent parameters for the closed-in (CI), alpha-beta-gamma and CI with frequency-dependent path loss exponent path loss models.

loss data points corresponding to the k-th frequency f_k. The parameter f_0 is the weighted frequencies of all measurements (or ray-tracing) data used for the model.

Dual-slope large-scale fading models have also been proposed to serve indoor hotspot environments. Two types have been proposed by the 3GPP. We have the *dual-slope ABG model*

$$L_{\mathtt{DS}}^{\mathtt{ABC}} = \begin{cases} 10\alpha_1 \log_{10} d + \beta_1 + 10\gamma \log_{10} f, & 1 < d \leq d_{\mathtt{BP}} \\ 10\alpha_1 \log_{10} d_{\mathtt{BP}} + 10\alpha_2 \log_{10}(d/d_{\mathtt{BP}}) & \\ \quad + \beta_1 + 10\gamma \log_{10} f, & d > d_{\mathtt{BP}} \end{cases}$$

and the *dual-slope CIF model*

$$L_{\mathtt{DS}}^{\mathtt{CIF}} = \begin{cases} L_{\mathtt{FSPL}}(f, 1\ \mathtt{m}) & \\ \quad + 10n_1[1 + b_1(\frac{f - f_0}{f_0})] \log_{10}(d[\mathtt{m}]), & 1 < d \leq d_{\mathtt{BP}} \\ L_{\mathtt{FSPL}}(f, 1\ \mathtt{m}) & \\ \quad + 10n_1\left[1 + b_1\left(\frac{f - f_0}{f_0}\right)\right] \log_{10}(d_{\mathtt{BP}}) & \\ \quad + 10n_2\left[1 + b_2\left(\frac{f - f_0}{f_0}\right)\right] \log_{10}(\frac{d}{d_{\mathtt{BP}}}), & d > d_{\mathtt{BP}} \end{cases}$$

Figure 6.16 lists the parameters for the 5G path loss models discussed earlier in different radio signal propagation environments.

6.4.3 *Check Your Understanding*

Students are encouraged to check their understanding of the content of this section with the following questions prior to proceeding to the next section.

(1) Consider the building penetration loss experienced by radio signals in Eq. (6.97). Compare the losses for low-loss and high-loss buildings at 800 MHz and 6 GHz.

(2) How do radio propagation at frequencies below 6 GHz differ from those above 6 GHz as proposed for use in 5G wireless systems?

(3) Consider Fig. 6.15. Find the probability that a mobile user who is 25 m away from the serving eNodeB has a line-of-sight connection if the:

(a) WINNER B3 model is used.

(b) WINNER A1 model is used.

(c) ITU model is used.
(d) The new model is used.
How do the above values compare with a user who is 35 m away from the serving eNodeB?

(4) Assume that a mobile cellular system operates at 3.75 GHz in a cluttered environment in which the path loss exponent is 3.5. The link between the antennas of user equipment and the eNodeB is blocked by a tall building to cause a shadow fading. Assume that the shadow fading is a random variable X_σ which is lognormal distributed with the standard deviation $\sigma = 8$ dB. If the mobile user is 200 m away from the serving eNodeB, compute the total path loss due to distance and shadowing that his signal experiences if the:
(a) CI path loss model applies.
(b) Alpha-beta-gamma path loss model applies.
(c) CIF path loss model applies.
(d) Dual-slope alpha-beta-gamma path loss model applies.
(e) Dual-slope CIF path loss model applies.
The model parameters are compiled in Fig. 6.16.

6.5 Small-Scale Fading Models

Also referred to as short-term fading and sometimes fast fading, small-scale fading has two main causes. One cause is **multipath signal propagation** which results from multiple versions of a ray from the transmitting antenna reaching the receiving antenna. The second cause of small-scale fading is Doppler shift, which is caused by any relative movement of any of the communicating antennas or their environment. Wireless signal fading due to multipath propagation is modelled using ray tracing or statistical distributions, of which classical ones are:

- **Ricean probability density function:** used if one of the multipath signals is the line-of-sight path. Such a path is called specular path.
- **Rayleigh probability density function:** used when all paths in the received multipath signals have about the same strengths.
- **Nakagami-m probability density function:** is the generalised statistical model. It has as special cases the Ricean and Rayleigh distributions.

6.5.1 *Two-Ray Multipath Fading Model*

Let us analyse a simple case of small-scale fading in which two rays reach the receiver as illustrated in Fig. 5.19(a) and (b). If both the transmitter and the receiver are standstill then we have the static two-path model

$$r(t) = s(t) + as(t - \tau)\,. \tag{6.107}$$

In Eq. (6.107), the signal $s(t)$ was sent from the transmitting antenna but $r(t)$ reached the receiving antenna, where $s(t)$ is the direct LOS path and $as(t - \tau)$ is the NLOS path (which is an amplitude scaled and time delayed version of $s(t)$) in Fig. 5.19(a) and (b). If $s(t)$ is a single-frequency signal then we have the situation illustrated in Fig. 5.12. We can transform Eq. (6.107) into the frequency domain using Fourier transform to obtain

$$\begin{aligned} R(f) &= S(f)[1 + ae^{-j2\pi f\tau}] \\ &= S(f)H(f;\tau) \end{aligned} \tag{6.108}$$

where $S(f) = \mathcal{FT}\{s(t)\}$ and the channel's transfer function is

$$H(f;\tau) = 1 + ae^{-j2\pi f\tau}. \tag{6.109}$$

Through inverse Fourier transform we obtain the impulse response of the channel from Eq. (6.109) as

$$\begin{aligned} h(t;\tau) &= \mathcal{F}^{-1}\{H(f;\tau)\} \\ &= \int_{-\infty}^{\infty} H(f;\tau)e^{j2\pi ft}df \\ &= \delta(t) + a\delta(t-\tau)\,. \end{aligned} \tag{6.110}$$

We use the symbology

$$\begin{aligned} &h(t;\tau) \xrightarrow{\mathcal{F}} H(f;\tau) \ \texttt{or} \\ &h(t;\tau) \supset H(f;\tau) \ \texttt{or} \\ &h(t;\tau) \mapsto H(f;\tau)\,. \end{aligned} \tag{6.111}$$

Example: Lowpass channel characteristics

Assume that a wireless channel is characterised by the transfer function $H(f;\tau) = 1 + ae^{-j2\pi f\tau}$ which is modelling the two-ray multipath radio propagation shown in Fig. 5.19(a) and (b), where $a = 0.4$. Assume that the length of the path travelled by the non-line-of-sight (NLOS) signal to reach the receiver is 60 m longer than the direct LOS path. Find the:

(1) Path delay τ experienced by the NLOS path relative to the direct path.
(2) Phase difference $\Delta\phi$ between the direct path and the reflected NLOS path if the signal is a 60 MHz single-tone signal. What kind of interference results?
(3) Half-power (or 3 dB) bandwidth of the wireless channel.
(4) Type of wireless channel that it is and the maximum frequency in signals that it passes without a significant attenuation.
(5) Which of the statistical small-scale fading models, e.g. Rayleigh, Rician and Nakagami-m, best describes the multipath situation discussed in this case study?
(6) Which radio propagation environment (i.e. urban, suburban and rural) could have most probably produced the radio signal propagation situation discussed in this case study? Justify your opinion.

Solution

(1) Assuming that the radio signals travel through the atmosphere with the speed of light $c \approx 3{\cdot}10^8$ m/s, and the given path difference is $\Delta d = 20$ m, we obtain the relative path delay

$$\tau = \frac{\Delta d}{c} = \frac{60\ \mathtt{m}}{3\cdot 10^8\ \mathtt{m/s}} = 0.2\ \mu\mathtt{s}\,.$$

(2) The wavelength of the radio signals assuming negligible Doppler shift is

$$\lambda = \frac{c}{f} = \frac{3\cdot 10^8\mathtt{m/s}}{60\ \mathtt{MHz}} = 5\ \mathtt{m}\,.$$

Therefore, the phase difference is

$$\Delta\phi = 2\pi \times \frac{\Delta d}{\lambda} = 2\pi \times \frac{60\ \mathtt{m}}{5\ \mathtt{m}} = 24\pi\,.$$

Thus, there is constructive interference according to Eq. (5.20).

(3) The magnitude of the channel's transfer function is

$$|H(f;\tau)|^2 = |1{+}ae^{-2\pi f\tau}|^2 = 1{+}a^2{+}2a\cos(2\pi f\tau)$$

with its maximum being $\max\{|H(f;\tau)|\} = |H(f;\tau)|_{f=0} = (1+a)^2$. We obtain the half-power bandwidth $f_{3\mathrm{dB}}$ from the relation

$$|H(f;\tau)|^2_{f=f_{3\mathrm{dB}}} \stackrel{!}{=} \frac{1}{2}|H(f;\tau)|^2_{f=0} = \frac{(1+a)^2}{2}\,. \qquad (6.112)$$

This results in the equation

$$1 + a^2 + 2a\cos(2\pi f_{3\mathrm{dB}}\tau) = \frac{(1+a)^2}{2}$$

which yields the solution

$$f_{3\mathrm{dB}} = \frac{1}{2\pi\tau}\cos^{-1}\left(\frac{2a - a^2 - 1}{4a}\right). \qquad (6.113)$$

Plugging $a = 0.4$ and $\tau = 0.2$ μs into Eq. (6.113) yields $f_{3\mathrm{dB}} \approx 143.1$ MHz. The channel's magnitude response is shown in Fig. 6.17. We can observe that the half-power bandwidth is about 143 MHz, which agrees with our result from manual calculations.

(4) The channel is a low pass wireless channel with half-power bandwidth of circa $f_{3\ \mathrm{dB}} = 143$ MHz. Thus, it attenuates significantly signals containing frequencies above 143 MHz, while behaving relatively transparent to lower-frequency signals.

(5) One of the received multipath signals is a direct path, also called specular path. Also, the reflected path gain is less than unity, i.e. $a = 0.4$, indicating that it has a lower gain than the other path. Thus, Rician distribution best matches the radio propagation situation in question.

(6) We hardly get the direct path to reach the receiver in the urban environment which is heavily cluttered with obstructions. Thus, if a specular path is received then the most probable propagation environment is rural. The Rician model is not suitable for urban environment.

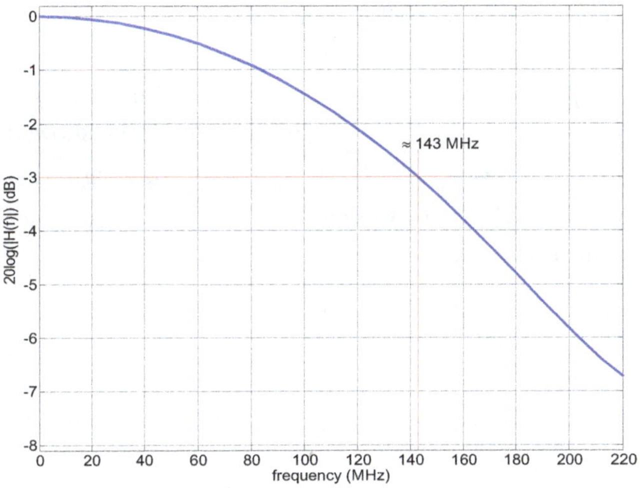

Fig. 6.17 Two-ray multipath channel's magnitude response.

Now, assume that $s(t) = b\cos(2\pi f_0 t)$ and that the receiver is now moving at the velocity $v(t)$ in the direction which makes the angle ϕ with respect to the direction of the incident radio signal. This causes the **Doppler shift** (aka Doppler frequency)

$$f_d(t) = \frac{v(t)f_0}{c}\cos\phi \tag{6.114}$$

where $c \approx 3 \cdot 10^8$ m/s is the speed of EM waves in space. Note that in general the velocity changes with time and so can the angle of EM wave incidence ϕ. This means that the Doppler frequency is in general time varying. For simplicity of analysis, however, we drop the time dependence for now and write $f_d = \frac{vf_0}{c}\cos\phi$. Now Eq. (6.107) becomes

$$r(t) = \cos\left[2\pi f_0\left(1 - \frac{v}{c}\cos\phi\right)t\right] + a\cos\left[2\pi f_0\left(1 - \frac{v}{c}\cos\phi\right)(t-\tau)\right] \tag{6.115}$$

We can observe in (6.115) that the Doppler shift due to the relative movement has spread the signal in frequency, while the multipath propagation due to signal reflection has spread the signal in time.

Assume now that instead of just two multipath signals we have L where $L \to \infty$ and that the Doppler shift for path l is $f_l = \frac{vf_0}{c}\cos\phi_l$, $l = 1, 2, \ldots, L$, where v is the speed of travel of the receiving antenna. Further, let a_l, ϕ_l, β_l be the amplitude, the angle of incidence and phase of path l. Assume now that the transmitted signal is an unmodulated carrier wave, i.e., $s(t) = e^{j2\pi f_0 t}$. Then, the received multipath signal is

$$r(t) = \sum_{l=1}^{L} a_l(t)e^{j[2\pi(f_0 - f_l)t + \beta_l]}. \tag{6.116}$$

The complex envelope of Eq. (6.116) is

$$X(t)e^{j\Theta(t)} = \sum_{l=1}^{L} a_l(t)e^{-j(2\pi f_l t - \beta_l)}. \tag{6.117}$$

For a wireless channel the instantaneous phases $2\pi f_l t - \beta_l$ and amplitudes $a_l(t)$ are random variables, which can be assumed to be independent but identically distributed. Thus, if L is large then we can assume, following the central limit theorem, that Eq. (6.116) is a Gaussian random variable. This means that Eq. (6.117) is a lowpass Gaussian process with the imaginary and real components being independent and possessing zero average values. It then follows that $\Theta(t) \sim U(0, 2\pi)$ and $X(t) \sim R(\sigma^2)$ with $\sigma^2 = E[X^2]$. That is, $X(t)$ has the Rayleigh distributions in Eqs. (5.37) and (5.38).

If there is also a LOS path in the multipath signals then the resulting complex envelope can be modelled as

$$X(t)e^{j\Theta(t)} = p(t)e^{j\gamma(t)} + \alpha(t)e^{j\theta(t)} \tag{6.118}$$

where $\gamma(t) \sim U(0, 2\pi)$, $p(t)$ is Rayleigh distributed while $\alpha(t)$ and $\theta(t)$ are non-random signals representing the specular path. In this case, $X(t)$ is Rician distributed with the pdf given in Eq. (5.51).

6.5.2 *Clarke's Model for Rayleigh Fading Channel*

This is a multipath model based on scattering without a specular path that is proposed in [Clarke (1968)]. Its basic assumptions are:

- The transmitter is fixed at a position, something which simulates a conventional eNodeB or radio base station.
- A single-frequency vertically-polarised wave leaves the transmitter but K copies reach the receiver. This model is thus one of the models called sum-of-sinusoid models.
- The angle at which all K multipath components reach the receiver are independent of each other, and can assume an arbitrary distribution. Often the angle of arrivals are assumed to be uniformly distributed in $(0, 2\pi)$.

- All K multipath components have the same constant amplitude at the receiver. Thus, Clarke's model models generates a Rayleigh channel model.
- The EM waves incident on the receiver are vertically polarised.
- Assumes an isotropic scattering in which there is a linear dependence of the received signal on the transmitted signal. The former is a sum of multiple scaled and delayed versions of the latter.
- Doppler shifts are small compared to the carrier frequency.
- The relative delays between the multipath components is assumed negligible to result in flat fading situation.
- The number of EM waves reaching the receiver L is large to validate the application of the central limit theorem.

Let E_z be the electric field component of the multipath signal reaching the receiver, H_x and H_y be the magnetic field components in the X and Y directions, respectively. The receiver is travelling in the X-Y plane at constant velocity v km/hr. Further, for the lth multipath component arriving at the receiver, let B_l be its real amplitude, ϕ_l be its random phase, and α_l be its angle of incidence relative to the travel direction. Then, we have the field components

$$\begin{aligned} H_x &= -\frac{A_0}{377}\sum_{l=1}^{L} B_l \sin\alpha_l \cos(2\pi f_c t + \phi_l) \\ H_y &= -\frac{A_0}{377}\sum_{l=1}^{L} B_l \cos\alpha_l \cos(2\pi f_c t + \phi_l) \qquad (6.119) \\ E_z &= A_0 \sum_{l=1}^{L} B_l \cos(2\pi f_c t + \phi_l) \end{aligned}$$

$$\sum_{l=1}^{L} \bar{B}_l^2 = 1, \quad \phi_l = 2\pi f_l t + \theta_l \ \text{ and } \ f_l = \frac{v}{\lambda}\cos\alpha_l$$

where $\sum_{l=1}^{L} \bar{B}_l^2$ is called ensemble average, A_0 is a real constant denoting the amplitude of the local mean value of the E-field and f_c is the carrier frequency. For a large L the central limit theorem allows us to approximate the H_x, H_y and E_z as Gaussian random variables, while the phase angles are uniformly distributed over 0 to 2π. If $A_{in}(t)$ and $A_q(t)$ are the amplitudes of the in-phase and the quadrature components of E_z fulfilling $A_{in}(t) \sim \mathcal{N}(0,\sigma^2)$ and $A_q(t) \sim \mathcal{N}(0,\sigma^2)$, then Clarke showed that $R(t) = \sqrt{A_{in}(t)^2 + A_q(t)^2}$ is Rayleigh distributed with the parameter $\sigma^2 = A_0^2/2$. Another form of the Clarke's model is

$$\begin{aligned} H_x &= -\frac{A_0}{377}\sum_{l=1}^{L} \sin\alpha_l e^{j\phi_l}, \\ H_y &= -\frac{A_0}{377}\sum_{l=1}^{L} \cos\alpha_l e^{j\phi_l}, \qquad (6.120) \\ E_z &= A_0 \sum_{l=1}^{L} e^{j\phi_l}. \end{aligned}$$

6.5.3 *Jake's Simulation Model for Rayleigh Fading*

This model is based on Clarke's model. Specifically, it provides a model for the simulation of the Clarke's model of Rayleigh fading. There are two popular alternative techniques used to simulate fading radio channels. These are:

- *Complex Gaussian noise process method:* The envelope of a complex-valued noise process which is Gaussian distributed has a Rayleigh distribution. If $n_I(t) \sim \mathcal{N}(m,\sigma^2)$ and $n_Q(t) \sim \mathcal{N}(m,\sigma^2)$ are the in-phase and quadrature-phase random processes which are Gaussian distributed and low-pass filtered, then the envelope of $n(t) = n_I(t) + jn_Q(t)$ is Rayleigh distributed.
- *Sum-of-sinusoid (SOS) method*: Generate a large number of complex sinusoidal waves and sum them. This is the approach used by Jakes to build his model of fading channels. It has been proved that as low as eight sinusoids provide a good enough approximation.

Jakes' model differs from that of Clarke in that it considers path attenuation, includes Doppler shift and explicit time dependence of the channel [Jakes (1994)], [Dent *et al.* (1993)]. Assume that we transmit over a multipath fading channel the single-tone signal

$$m(t) = A_0 \cos(2\pi f_c t), \ A_0 = \sqrt{2} \qquad (6.121)$$

and receive the

$$r(t) = A_0 \sum_{k=1}^{K} a_k \cos(2\pi f_c t + 2\pi f_m t \cos \phi_k + \gamma_k) \tag{6.122}$$

where for the kth path f_c is the carrier frequency, γ_k is the phase shift, ϕ_k is the angle of arrival, a_k is the channel gain, and the Doppler shift experienced is

$$f_k = 2\pi f_m t \cos \phi_k, \; k = 1, 2, \ldots, K \tag{6.123}$$

$$K = 4M + 2, \; M \texttt{ is a positive integer} \tag{6.124}$$

$$\phi_k = \frac{2\pi k}{K}, \; k = 1, 2, \ldots, K \tag{6.125}$$

where $f_m = vf_c/c$ is the maximum Doppler shift at the receiver's travelling speed of v m/s. The restriction on K reduces the number of unique Doppler shifts from K to $M + 1$ and thus reduces accordingly the number of oscillators used to generate the Doppler shifts.

Now, we can rewrite the multipath signal in Eq. (6.122) as

$$\tilde{r}(t) = \tilde{X}_i(t) \cos(2\pi f_c t) + \tilde{X}_q(t) \sin(2\pi f_c t) \tag{6.126}$$

where

$$\tilde{X}_i(t) = \frac{2}{\sqrt{K}} \left(\sqrt{2} \cos \beta_{M+1} \cos(\omega_m t) + 2 \sum_{m=1}^{M} \cos \beta_m \cos(\omega_m t) \right)$$

$$\tilde{X}_q(t) = \frac{2}{\sqrt{K}} \left(\sqrt{2} \sin \beta_{M+1} \cos(\omega_m t) + 2 \sum_{m=1}^{M} \sin \beta_m \cos(\omega_m t) \right)$$

and the gains of the oscillators are

$$\beta_{M+1} = 0, \; \beta_m = \frac{\pi m}{M+1}, \; m = 1, 2, \ldots, M.$$

The values of β_k are chosen to ensure that the phase of $\tilde{r}(t)$ is uniformly distributed in $(0, 2\pi)$.

Another form of Jakes model found in the literature is

$$r(t) = \frac{1}{\sqrt{K}} \times \sum_{k=1}^{K} [\cos(2\pi F_d t \cos \alpha_k + a_k) + j \sin(2\pi F_d t \cos \alpha_k + b_k)] \tag{6.127}$$

where α_k is the angle of arrival, and a_k and b_k are random phases, all uniformly distributed in the interval $[0, 2\pi]$. In the ideal case, the sinusoids in the sum-of-sinusoids methods, such as that of Clarke and Jakes models, must be uncorrelated. In practice, however, it is not so. The correlation is higher in the Clarke's model than in the Jakes model. Jakes proved that for as low as $K = 34$ or $M = 8$ Eq. (6.126) approximate fairly well a Rayleigh distribution [Jakes (1994)], [Dent *et al.* (1993)].

There are many other modern channel models that we did not discuss. These include:

- The QuaDRiGa (QUAsi Deterministic RadIo channel GenerAtor) developed at the Fraunhofer Heinrich Hertz Institute in Germany. This model was developed to enable the modeling of MIMO radio channels.
- METIS channel models are developed for 5G networks developed by the European Union in partnership with industries and academic institutions. METIS means Mobile and wireless communications Enablers for Twenty-twenty (2020) Information Society.
- 3GPP-3D channel models were developed by the 3GPP for 4G mobile networks.
- MiWEBA (Millimetre-Wave Evolution for Backhaul and Access) is a collaborative research project developing models for mmWave communications.

6.5.4 *Check Your Understanding*

Students are encouraged to check their understanding of the content of this section with the following questions prior to proceeding to the next section.

(1) Consider the channel with the frequency response shown in Fig. 6.17. If a transmission link cannot endure more than 5 dB attenuation by the channel, what should be the maximum usable frequency? Justify your answer.
(2) Consider the two multipath cases shown in Fig. 6.18.

(a) Which of them includes the direct path and which does not?
(b) What are the two basic causes of the ground reflection of radio signals as illustrated in Fig. 6.18(a)?
(c) Discuss with justification the appropriate statistical model for small-scale fading which best describes each of them.
(d) Which of the two scenarios illustrated in Fig. 6.18 is most appropriate for rural propagation environment and which for urban environment? Justify your answer.

(3) **Students' project:** Simulation of wireless channels is very essential for system design and analysis. Students can be asked to simulate the Clarke's flat fading model given analytically in Eq. (6.119). Software tools, such as MATLAB and Simulink, can be used. The simulation can also be done in hardware using common electrical engineering laboratory equipment, such as mixers, Doppler filters (both RF and baseband types), Gaussian noise generators and accumulators or adders.

We have observed by now that wireless channels are liable to many forms and kinds of noise and their associated potential impairments. The channel disturbances include thermal noise (which is always present and much more important in satellite communications), impulse noise, crosstalk, intermodulation distortion, multipath signal propagation and atmospheric effects. Many communications engineers would be jobless if the wireless channel were nice to communications signals. Much of the work in communications systems is about how to prepare the signal to make it robust enough to withstand the channel's undesirable behaviour.

Figure 6.19 shows the three broad categories of techniques used to mitigate the potential ill-effects

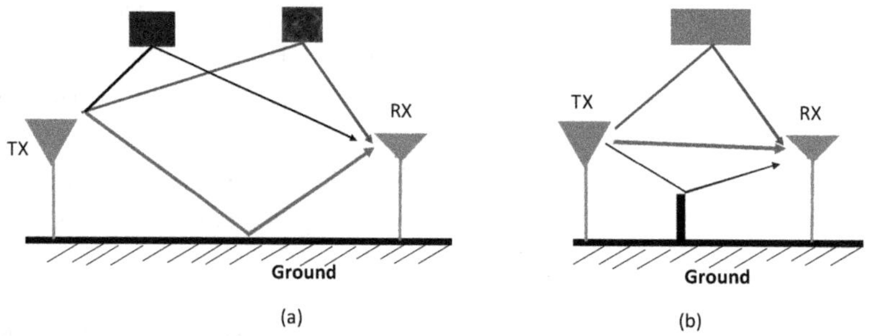

Fig. 6.18 Two multipath propagation scenarios.

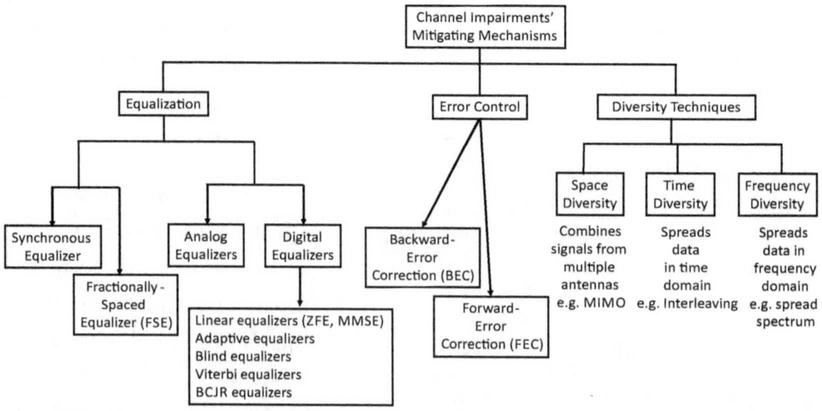

Fig. 6.19 Techniques used to combat channel impairments and noise.

of the channel. They are error control, equalisation and diversity techniques. These topics are treated in depth in many books, such as [Gyasi-Agyei (2019b)]. Note that there is a form of space diversity called **macrodiversity** which enables a user to receive the same information from different cells or base stations. We also have multipath diversity in which multiple versions of the transmitted signals are detected and combined. A popular receiver used to this effect is called **rake receiver**.

6.6 Review of Important Terms and Acronyms

5G path loss model	macrodiversity	rake receiver
autocorrelation function (ACF)	Z-channel model	wireless channel
AWGN channel	BAC channel	BEC
BSC	CDF	Clarke channel model
coherence bandwidth	coherence time	COST 207 model
critical angle	critical frequency	delay spread
diffraction	discrete memoryless channel (DMC)	Doppler spectrum
Doppler spread	EM wave	envelope of signal
fading	fast fading	flat fading
free-space	frequency-selective fading	Gamma distribution
Gaussian distribution	Gilbert-Elliot model	ground wave
interference	intersymbol interference (ISI)	ionospheric propagation
Jakes channel model	large-scale fading	lognormal distribution
longitudinal waves	LOS/NLOS path	LTI channel
LTI channel model	Maxwell's equations	medium-scale fading
microwaves	multipath propagation	path loss
polarisation	power delay profile radio waves	Rayleigh fading
Rician fading	scattering	shadowing
skip distance	skip zone	sky wave
slow fading	small-scale fading	space wave
surface wave	tapped-delay model	transverse waves
uniform distribution	wave propagation	wave propagation modes
wavefront	Weibull distribution	WINNER

6.7 Review Questions and Problems

(1) What is a fading channel?

(2) Explain with illustrations the meaning of a fading signal.

(3) Describe the meaning of the envelope of a signal.

(4) In a given cellular system operating at 2 GHz the height of the eNodeB (i.e. radio base station) is 300 m and mobile station height is 2 m. The mobile use is located 25 km away from the serving eNodeB. Assume that the system is operating in a city such as Melbourne. Use the Okumura model to estimate the following:

(a) The free-space path loss.

(b) The median attenuation relative to free-space $A_{mu}(f, d)$.

(c) The area gain G_{AREA}.

(d) The correction factors for the eNodeB and the user equipment.

(e) The total path loss in decibels experienced by the user's signal.

(5) At any given time instant, each ionospheric layer has a maximum frequency at which radio signals can hit it at right angles and be

refracted back to Earth. How do we call such a frequency?

(6) Consider Fig. 6.6. Find the:
   (a) Percentage of power in the second multipath signal in Fig. 6.6(a).
   (b) Power in the third multipath signal in Fig. 6.6(b).
   (c) How much time taken by the second multipath signal to reach the receiver in Fig. 6.6(c).
   (d) The multipath delay spread of the channel with the power delay profile in Fig. 6.6(a).

(7) Consider Fig. 6.6. Which of the four channels causes inter-symbol interference if the symbol duration is 2.5 μsec?

(8) Assume that a mobile cellular system operates in an urban area at the carrier frequency 1.8 GHz. Assume further that the user equipment (UE) and the eNodeB have the respective heights of 2.2 m and 80 m above ground. Mobile users require good signal strength even at cell fringes. Assume that the total median path loss due to the distance between the transmitter and receiver is limited to 120 dB. How large can each cell in the cellular system be to ensure optimum operation even at cell edges? Assume that the cellular system designers have access to only the *Okumura path loss model.*

(9) Assume that a mobile cellular system operates in an urban area at the carrier frequency 1.8 GHz. Assume further that the UE and the eNodeB have the respective heights of 2.2 m and 80 m above ground. Mobile users require good signal strength even at cell fringes. Assume that the total median path loss due to the distance between the transmitter and receiver is limited to 120 dB. How large can each cell in the cellular system be to ensure optimum operation even at cell edges? Assume that the cellular system designers have access to only the *Okumura-Hata path loss model.* Compare your result with that of the previous question using the Okumura model.

6.8 Supplementary Reading

(1) Theodore S. Rappaport, Robert W. Heath Jr., Robert C. Daniels and James N. Murdock, *Millimeter Wave Wireless Communications*, Prentice Hall, 2014.

(2) Andreas F. Molisch, *Wireless Communications*, Wiley-IEEE Press, 2010.

(3) Afif Osseiran, Jose F. Monserrat and Patrick Marsch (Editors), *5G Mobile and Wireless Communications*, Cambridge University Press, 2016.

(4) Theodore S. Rappaport, *Wireless Communications: Principles and Practice*, Prentice Hall, 2008.

(5) David Tse and Pramod Viswanath, *Fundamentals of Wireless Communication*, Cambridge University Press, 2005.

(6) Andrea Goldsmith, *Wireless Communications*, Cambridge University Press, 2005.

Chapter 7

Antennas and Spatial Diversities

Chapter's Learning Outcomes

An antenna is a type of transducer which converts signals between the electrical domain and electromagnetic (EM) waves. The relationship between antenna and EM waves is akin to that between chicken and egg. At the transmitter, electrical signals are converted to EM waves so they can be transferred over the wireless medium. At the receiver, EM waves are converted into electrical signals so that they can be processed by electronic circuits. This chapter discusses the basic principles, types and applications of antennas. In particular, we study:[1]

(1) General introduction to antennas, including their basic operation.
(2) The link between EM waves and antennas, near-field and far-field of antennas, and their implications in signal transmission and reception.
(3) Antenna types: e.g., aperture antennas (e.g., horn and parabolic antennas), monopole, dipole, panel antennas, microstrip patch antennas, Yagi-Uda antennas, array antennas, dumb antennas and smart antennas (aka adaptive array antennas or digital antenna arrays).
(4) Channel models for spatial diversity systems, power allocation (i.e., water-filling principle).
(5) Array antenna and adaptive array antennas and their applications in radio communications.
(6) Antenna diversity types and their benefits: SIMO (receive diversity), MISO (transmit diversity), MIMO (transmit and receive diversities), single-user MIMO (SU-MIMO), multi-user MIMO (MU-MIMO) and massive MIMO systems.
(7) General introduction to space-time communications, including space-time coding (STC).

7.1 Introduction

An antenna, also referred to as *aerial*, is a conductive material (a wire or a rod) which functions as a transducer by converting between the two energy forms: electrical energy and electromagnetic (EM) waves. The plural form of antenna is antennae or antennas. A transmitting antenna converts a guided electrical signal fed to it from the attached electronic circuitry through a cable into unguided EM waves. This is illustrated in Fig. 7.1. The EM waves then propagate through the wireless medium towards the receiver.

The feeder cable is lossy and so reduces the power in the signal. This loss is called *feeder loss*. There is also a loss, called *insertion loss*, at the intersection between the cable and the antenna. This loss results if the characteristics impedance of the cable is not exactly the same as the input impedance of the antenna. Antennas are manufactured using electrically-conductive material, usually copper wire. Antennas can be one dimensional or two dimensional.

At the receiver the antenna intercepts the EM waves which induces electrical current in it (i.e., the antenna). The receiver attached to the antenna then processes the electrical signal to retrieve the

[1]Gold and rubies might be in abundance, but the lips that speak knowledge are a rare jewel.

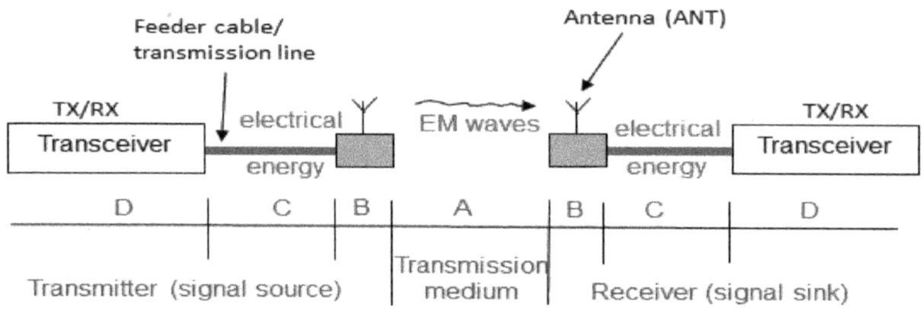

Fig. 7.1 Architecture of a typical wireless communications link with antennas.

data it is carrying. Every antenna is associated with two types of fields: radiation field and induction field. The latter is linked with the energy stored in the antenna which creates a magnetic field around it. The radiation field emits electromagnetic radiation from the antenna, whose intensity decreases with distance from the antenna. As usually the same antenna is used for both signal transmission and reception, antennas are referred to as **reciprocal devices**.

> The relationship between antenna and EM waves is akin to that between chicken and egg. From the perspective of the wireless transmitter, the antenna produces EM waves. From the perspective of the wireless receiver, however, EM waves produce antenna. The reason is that the receiving antenna would not be needed if there were no EM waves. Thus, there is no consensus whether to study antenna before EM waves or the other way round. This is the choice of lecturers and/or curriculum developers.

7.1.1 *Generation of EM Waves*

The linkage between electricity and magnetism was discovered by the Danish physicist Hans Christian Oersted in 1820. He realised that when a current flows through a conductor it deflects the needle of a compass. The current produces a magnetic field (or H-field). Similarly, an electric field is produced in the presence of a magnetic field. As an antenna is a conductor, likewise it produces an H-field when current passes through it. The intensity or magnitude of the H-field is measured in Amperes per meter (A/m). As a vector, the H-field has also a direction.

An antenna also produces an electric field (aka E-field), whose intensity or magnitude is measured in volts per meter (V/m). An E-field is produced whenever there is a change in potential or voltage between two points of a conductor or between two conductors. Different parts of an antenna or between an antenna and the earth have different voltages. This potential difference produces an E-field.

If the E-field and H-field are produced by a direct current (dc) or direct voltage (dv), then the field produced is called **electrostatic field**. The current flowing through an antenna, however, is an alternating current (ac) with frequency f. Therefore, the fields produced by an antenna are oscillatory fields which change in the same manner as the current producing them, and with the same oscillation frequency f. The coupling or combination of the E-field and H-field produces electromagnetic (EM) waves. The two fields composed in the EM wave are perpendicular to each other and to the direction of travel of the wave. EM waves are produced by the movement of electrons in a conductive material. The electrons, on the other hand, are produced by an ac in, or voltage drop across, the conductor.

EM waves are travelling waves with the velocity of light $c \approx 3 \cdot 10^8$ m/s. The relationship between the velocity c in m/s, the wavelength λ in meters, and the frequency f in Hertz (Hz) of the waves is given by

$$c = f\lambda. \tag{7.1}$$

The relationship between the intensity of the E-field E, measured in V/m, and the power density of an

EM wave is

$$P = \frac{E^2}{Z_0} = \frac{E^2\ (\mathtt{V/m})^2}{\mu_0 c_0\, \Omega} = \frac{E^2}{\sqrt{\frac{\mu_0}{\epsilon_0}}} = \frac{E^2}{377}\ \mathtt{W/m}^2 \quad (7.2)$$

where Z_0 is the impedance of free space.

Example

Assume that a radio transceiver operating at 3 GHz emits EM waves with the intensity $E = 5$ V/m. Compute the:

(1) Wavelength λ of the waves.
(2) Distance the waves have moved away from the transmitting antenna by the time they complete three periods of their oscillations.
(3) Maximum power density in the EM waves.

Solution

Given are $f = 3 \cdot 10^9$ Hz, $c \approx 3 \cdot 10^8$ m/s and $E = 5$ V/m.

(1) $\lambda = c/f = 0.1$ m.
(2) The distance covered by the wave for each complete cycle is its wavelength. Therefore, the wave moves away from the transmitting antenna three wavelengths in three cycles, i.e., $d = 3\lambda = 300$ mm.
(3) $P = E^2/Z_0 = 25/377 \approx 66.31\ \mathtt{mW/m}^2$.

7.1.2 *Isotropic, Omnidirectional and Directional Antennas*

With respect to the shape and type of the radiated beam, antennas are classified into four types: isotropic radiator, omnidirectional antenna, one directional and array antennas. Ideally, an omnidirectional antenna (aka omni) radiates the power in the signal it transmits in all directions or 360° with equal intensities. Therefore, the gain of the ideal omni is 0 dB (i.e., no gain as there is no directivity). The ideal omni is a point-source referred to as an **isotropic radiator**. The isotropic radiator or ideal omni cannot be realised in practice. It is rather used as the benchmark to measure the gain of practically-realisable antennas. The unit dBi, meaning decibels in reference to an isotropic radiator, is used.

Although the prefix 'omni' means 'all directions,' a practical omnidirectional antenna has some directivity as it radiates or receives signals in two dimensions. The omni-directional antenna receives or transmits signals in all directions except in the azimuthal plane. It thus has a non-zero gain in all directions except in the azimuthal plane. For example, the gain of a half-wave dipole can be up to 2.15 dBi, which occurs perpendicular to its axis.

Also referred to as beam antenna, a directional antenna (D-ANT) radiates EM waves in a specific direction, resulting in a higher gain than omni. The gain of a D-ANT increases as its angle of radiation or beamwidth decreases. A common example of D-ANT is parabolic dish antennas used for satellite TV reception. Their gain increases with their diameter. Another popular D-ANT is Yagi-Uda antenna. A directional antenna focuses the transmitted energy in one direction just as a glowing light bulb which is enclosed in a container with a small slit does. An advanced form of directional antennas is the array antenna which has multiple beams, each of which can be used to serve a different user.

7.1.3 *Passive versus Active Antennas*

An antenna can be a **passive** or **active** device. A passive device contains no energy source nor requires an external power. Therefore it cannot amplify a signal traversing it. For this reason, the power at the input port $P_{\mathtt{in}}$ of a passive device cannot be less than the power at its output port $P_{\mathtt{out}}$. The opposite holds for an active antenna. From the law of energy conservation (i.e., the first law of thermodynamics), a passive device cannot oscillate. Let $L_{\mathtt{f}}$ be the loss in the feeder cable connecting the transmitter to the antenna, and $L_{\mathtt{c}}$ be the connector loss, then

`Passive ANT:`

$$P_{\mathtt{in}}\ [\mathtt{dBm}] = P_{\mathtt{out}}\ [\mathtt{dBm}] - L_{\mathtt{f}}\ [\mathtt{dB}] - L_{\mathtt{c}}\ [\mathtt{dB}]\,. \quad (7.3)$$

`Active ANT:`

$$P_{\mathtt{in}}\ [\mathtt{dBm}] < P_{\mathtt{out}}\ [\mathtt{dBm}] - L_{\mathtt{f}}\ [\mathtt{dB}] - L_{\mathtt{c}}\ [\mathtt{dB}]\,. \quad (7.4)$$

$$\mathtt{Active\ ANT} = \mathtt{Passive\ ANT} + \mathtt{Built\text{-}in\ Amplifier}\,. \quad (7.5)$$

Passive antennas are more common than active antennas. Depending on their shape, some passive antennas have non-zero gain because they exhibit some directionality. Such a gain is referred to as **passive gain** or **directional gain** in contrast to **active gain** or **power gain** of an active antenna.

7.1.4 *Smart versus Dumb Antennas*

Antennas can be classified into **dumb antennas** [Viswanath *et al.* (2002)] and **smart antennas**. A smart antenna is a directional antenna which tracks the direction of a mobile user and focuses the transmitted energy only in the right direction. It is smart because it learns the direction of a mobile user from signals received from it and track it. Antennas which are not smart are called dumb antennas. The smart antenna technology is able to deliver a higher data rate and longer transmission range without requiring a higher transmit power or channel bandwidth.

Smart antennas are of two types: phased array antennas (aka adaptive antennas or adaptive array antennas) and switched beam antennas. Note that multiple input multiple output (MIMO) is a smart antenna architecture. Smart antennas are commonly used at base stations of cellular mobile systems. One by-product of focusing signals only in the right direction is the avoidance of unnecessary egress and ingress EM interference.

7.1.5 *Operating Frequency, Bandwidth and Size of Antennas*

We use the characteristics of the HP 30-044 S antenna in Table 7.1 as a case study to explain the basic (electrical) parameters of antennas, starting from the operating frequency band. The P 30-044 S is a dish antenna manufactured by Faini Telecommunication Systems, Italy. The key electrical parameters of antennas include:

Every antenna is tuned to operate in a certain band of frequencies. The operating frequency band and how the antenna radiates or receives electromagnetic waves depend on its size and shape. While some antennas can be identified by their physical shapes, it is not always possible. In summary, the determining factors of an antenna's size and shape are:

Table 7.1 Case study antenna: HP 30-044 S dish antenna.

Parameter	Value
Frequency range	4.4-5.0 GHz
Gain, low band	40.5 dBi
Gain, mid band	41.0 dBi
Gain, top band	41.5 dBi
Return Loss	30.7 dB
VSWR	1.06
HPBW	1.6°
Front to back ratio	68 dB
Isolation	35 dB
XPD	30 dB
Polarisation	Single
Electrical Compliance	Class 3
Diameter	3.0 m
Radome	Complan fabric
Pole (mounting pipe) diameter	115 mm
Elevation-fine adjustment	±5°
Azimuth-fine adjustment	±5°
Wind velocity Operational	30 m/s
Wind velocity Survival	55 m/s
Wind deflection	$< 0.3 \times \texttt{HPBW} = 0.48°$

(1) *Antenna's sensitivity*: this is a measure of how low the power in the signal incident on an antenna can be and still be detected by the antenna.
(2) *Effective radiated power (ERP)*: this is the product of the antenna's gain and the power fed into it. The ERP thus depends on the transmission range of the wireless link.
(3) *Operating frequency*: the higher the operating frequency, the smaller the antenna, all other things being the same.
(4) *Directionality of the antenna*: for example parabolic dish antennas are directional and the higher its directionality, the larger its physical dimensions should be.

The physical size of an antenna depends on its operating frequency and hence wavelength λ of signals it handles effectively. For optimum reception and transmission, the size of the antenna should be equal to the wavelength of signals it is designed to receive or transmit. However, this is not practical as otherwise we would need very long antennas, especially for low-frequency transmissions. A sub-optimum operation requires that the length of the antenna l is a perfect fraction of the wavelength, i.e.,

$$l = \frac{n}{2^k}\lambda,\ n, k \in \mathcal{Z}^+ . \tag{7.6}$$

For example, we have half-wave ($\frac{\lambda}{2}$) antennas, quarter-wave ($\frac{\lambda}{4}$) antennas and eighth-wave ($\frac{\lambda}{8}$) antennas. The longer the antenna compared with the signal's wavelength, the higher the gain. A diminishing effect occurs if the antenna is larger than the wavelength, if practically possible at all.

The frequency range over which the interesting parameters of an antenna change within the prescribed limits is called *antenna bandwidth.* The bandwidth of an antenna can be defined using any of the following:

(1) Half-power bandwidth (HPBW), aka 3-dB bandwidth.
(2) Beamwidth.
(3) Impedance bandwidth (aka return loss bandwidth): this is the range of frequencies over which the antenna achieves impedance matching with a reference impedance. The VSWR should be less than 2:1 or $S_{11} < -10$ dB.
(4) Axial ratio (AR). Expressed in decibels, the AR is used to quantify the circularity of antenna's polarisation. Its value of $AR = 0$ dB means perfectly circular polarisation, while $AR = \infty$ means perfectly linear polarisation.

7.1.6 *Antenna Impedance, Radiation Resistance and VSWR*

Every transmission medium has a characteristic impedance Z_0. This impedance is defined as

$$Z_0 = \sqrt{\frac{\texttt{line inductance}}{\texttt{line capacitance}}} = \sqrt{\frac{L}{C}}. \tag{7.7}$$

If the inductance is measured in Henry (H) and the capacitance is measured in farad (F), then Z_0 is purely resistive with the unit Ohm (Ω). Why then do we refer to it as impedance rather than resistance? It is due to industry terminology. There are many terms used in technological disciplines that are not correct strictly speaking.

Also, every electronic device, including antennas, has an impedance, $Z_{\mathtt{A}}$. Let us use the circuit in Fig. 7.2 for our analysis. Here, an antenna with impedance $Z_{\mathtt{A}}$ is connected to a signal source with impedance $Z_{\mathtt{S}}$ using a transmission line with length L and characteristic impedance Z_0. The impedance that a source connected to an antenna sees when looking into the antenna is called **antenna input impedance** or *feedpoint impedance*, $Z_{\mathtt{in}}$. For a lossless transmission line it is defined as

$$\begin{aligned} Z_{\mathtt{in}} &= \frac{\texttt{antenna input voltage}}{\texttt{antenna input current}} \\ &= Z_0 \frac{Z_{\mathtt{A}} + jZ_0 \mathtt{tan}(2\pi L/\lambda)}{Z_0 + jZ_{\mathtt{A}} \mathtt{tan}(2\pi L/\lambda)} \end{aligned} \tag{7.8}$$

where λ, measured in meters (m), is the wavelength of the EM wave. Note that the input impedance is a function of four parameters:

- Frequency or wavelength.
- Length of the feeder cable, L.

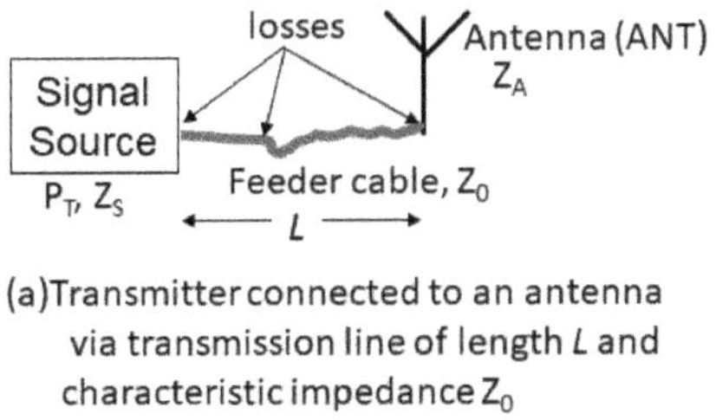

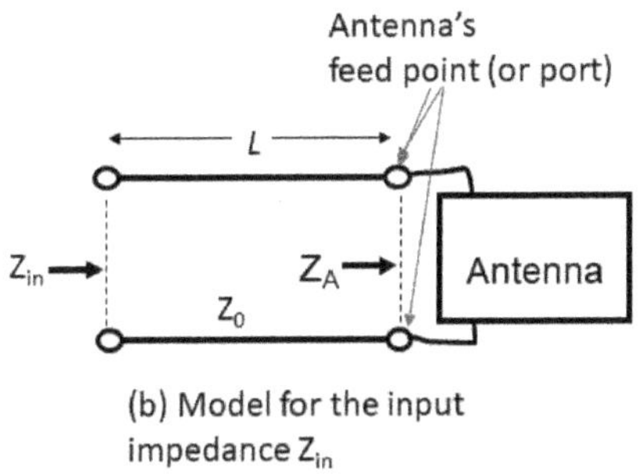

Fig. 7.2 High-frequency characteristics of an antenna connection.

- Antenna impedance, $Z_{\mathtt{A}}$.
- Characteristic impedance of the feeder cable, Z_0.

There are alternative tools available to measure $Z_{\mathtt{in}}$. An example is a vector network analyser (VNA).

If $Z_{\mathtt{A}} \neq Z_0$ and no matching network is used, then there will be signal reflection at the interface between the antenna's port and the transmission line. If $Z_{\mathtt{S}} \neq Z_{\mathtt{in}}$ then there is signal reflection at the interface between the signal source and the transmission line. Also, we observe from Eq. (7.8) that, if the feeder cable (or transmission line) is perfectly matched to the antenna, i.e., $Z_0 = Z_{\mathtt{A}}$, then $Z_{\mathtt{in}} = Z_0$. In such a matching situation no part of the input signal fed into the antenna is reflected back into the cable. Unfortunately, the ideal case is hardly achieved. The amount of reflected power is measured by the wave reflection reflection coefficient (aka load reflection coefficient), $\Gamma_{\mathtt{L}}$, which is defined as

$$\Gamma_{\mathtt{L}} = \sqrt{\frac{\begin{array}{c}\texttt{part of the incident power}\\ \texttt{reflected off the antenna}\end{array}}{\begin{array}{c}\texttt{total power incident}\\ \texttt{on the antenna}\end{array}}}\,.$$

$$= \sqrt{\frac{P_{\mathtt{r}}}{P_{\mathtt{in}}}} = \frac{Z_{\mathtt{A}} - Z_0}{Z_{\mathtt{A}} + Z_0}\,. \tag{7.9}$$

The desired ideal case is $\Gamma_{\mathtt{L}} = 0$. With the reflection coefficient, we define the signal *return loss* as

$$L_{\mathtt{r}} = -20\log_{10}|\Gamma_{\mathtt{L}}| = -20\log_{10}|S_{11}|\,. \tag{7.10}$$

where S_{11} is an S-parameter of the antenna's input port. The desired ideal case is $L_{\mathtt{r}} = 0$ dB. Any signal reflection at the interface between the antenna and the cable causes oscillations on the cable. We measure the level of signal reflection or oscillations using *voltage standing wave ratio* (VSWR). It is defined as

$$VSWR = \frac{1 + |\Gamma_{\mathtt{L}}|}{1 - |\Gamma_{\mathtt{L}}|} = \frac{Z_{\mathtt{A}}}{Z_0}\,. \tag{7.11}$$

A matched interface has no signal reflection, resulting in unity VSWR. The closer the VSWR to unity is, the better the performance of the antenna. The frequency range over which an antenna has a specified VSWR is called **VSWR bandwidth**. If $P_{\mathtt{T}}$ is the power in the signal from the transmitter, then the power coupled to the antenna and radiated as EM wave is

$$P_{\mathtt{EM}} = (1 - |\Gamma_{\mathtt{L}}|^2)P_{\mathtt{T}}\,. \tag{7.12}$$

The remaining power $|\Gamma_{\mathtt{L}}|^2\, P_{\mathtt{T}}$ is reflected back into the feeder line.

Example

The case study dish antenna in Table 7.1 is rated with VSWR of 1.06.

(1) Compute the range of possible values for the voltage reflection coefficient at the boundary between the feeder cable and the antenna.
(2) What percentage of the electrical power fed to the antenna is radiated as EM waves?
(3) Compute the signal return loss between the antenna and the feeder cable. How does your result match with the value in Table 7.1?

Solution

(1) Upon re-arranging Eq. (7.11), we obtain the voltage reflection coefficient

$$|\Gamma_a| = \frac{VSWR - 1}{VSWR + 1} = \frac{1.06 - 1}{1.06 + 1} \approx 0.03\,. \tag{7.13}$$

Thus, the range of values of the reflection coefficient is $-0.03 \leq \Gamma_a \leq 0.03$.

(2) The percentage of the electrical power fed to the antenna which is radiated as EM waves is obtained from Eq. (7.12) as

$$\frac{P_{\mathtt{EM}}}{P_{\mathtt{T}}} = 1 - |\Gamma_{\mathtt{L}}|^2 = 1 - \left(\frac{0.06}{2.06}\right)^2 \approx 99.92\%\,.$$

(3) The return loss is obtained from Eq. (7.10) as

$$L_{\mathtt{r}} = -20\log_{10}|\Gamma_{\mathtt{L}}| = L_{\mathtt{r}}$$
$$= -20\log_{10}(6/206) \approx 30.71\ \mathtt{dB}\,.$$

The return loss in Table 7.1 is 30.7 dB, which agrees with the results of the above computations.

Let us return to the impedance seen at the terminals of the antenna, Z_A. It has the components

$$Z_{\mathtt{A}} = R_{\mathtt{A}} + jX_{\mathtt{A}} = R_r + R_l + jX_{\mathtt{A}} \tag{7.14}$$

where $R_{\mathtt{A}}$ is the resistance at the terminals of the antenna, $X_{\mathtt{A}}$ is the antenna's input reactance, R_r

is the radiation resistance and R_l is the net ohmic resistance (aka loss resistance) of the wires used to construct the antenna. R_l is the source of Ohmic losses in the antenna owing to the imperfect material used for its construction. The **radiation resistance** is a fictitious or 'virtual' resistance which radiates part of the power fed into the antenna as EM waves. The actual power radiated in the EM wave is

$$P_{\mathtt{EM}} = I^2 R_r \tag{7.15}$$

where I is the current flowing through the antenna. Thus, the total power supplied to the antenna is

$$P_{\mathtt{T}} = P_{\mathtt{EM}} + P_{\mathtt{loss}} = I^2(R_r + R_l)\,. \tag{7.16}$$

7.1.6.1 *Self Impedance and Mutual Impedance*

The terminal impedance $Z_{\mathtt{A}}$ of an antenna operating in a practical environment comprises two components:

- Self impedance or feed point impedance, $Z_{\mathtt{F}}$.
- Mutual impedance, $Z_{\mathtt{M}}$.

Therefore, we can write

$$Z_{\mathtt{A}} = Z_{\mathtt{M}} + Z_{\mathtt{F}}\,. \tag{7.17}$$

The **self impedance** is the impedance of an antenna operating in an environment which is free of conductive objects. Such an environment is practically impossible as it requires a free space. Self impedance is the impedance at the antenna's feed point. It is defined as

$$Z_{\mathtt{F}} = \frac{\texttt{voltage at antenna's feed point}}{\texttt{current flowing into antenna feed point}}\,. \tag{7.18}$$

If ϕ_v, ϕ_i are the phases of the voltage and current, respectively, then

$$Z_{\mathtt{F}} = \begin{cases} R_{\mathtt{F}} \ \texttt{(purely resistive)}, & \texttt{if } \phi_v = \phi_i \\ R_{\mathtt{F}} + jX_{\mathtt{F}} \ \texttt{(complex)}, & \texttt{if } \phi_v \neq \phi_i \end{cases}.$$

Mutual impedance is produced by the parasitic effects of conductive materials lying within the reactive near-field region on the antenna's radiation pattern. Note that even the ground is an electrically-conducting object. This is the reason why the ground surrounding the antenna must be covered with a non-conductive material especially for antennas operating very close to the ground. Current flowing through the antenna produces coupled voltage in near-by conductive materials. Current through any near-by conductive materials also induce voltage in the antenna. The mutual impedance is defined as

$$Z_{\mathtt{M}} = \frac{\texttt{voltage induced in one conductor}}{\texttt{current in other coupled conductor}}\,. \tag{7.19}$$

The mutual impedance is either capacity or inductive. What are the effects of the mutual impedance? Mutually-coupled conductors cause two main effects on the desired antenna:

- It changes the impedance of the antenna, according to Eq. (7.17). We desire only the self impedance.
- It alters the radiation pattern of the antenna, especially that of directive antennas. This feature is exploited in array antennas, in which the mutual coupling between the elements on an array antenna is used to shape the radiation pattern to a desired shape and direction.

7.1.7 *Antenna Radiation Pattern*

The generation or reception of a wavefront by an antenna is called radiation. An important feature of an antenna is its *radiation pattern* (aka radiation diagram or antenna pattern), which is a plot showing the dependence of the antennas gain on direction or angle. The American Radio Relay League (ARRL) defines the radiation pattern as the graphical representation of the directivity of an antenna in three-dimensional polar coordinates. The radiation pattern depends on operating frequency and antenna size, and it is characterised by the beam width of the main lobe, sidelobes, backlobes and the gain. It is a measure of the selectivity of the antenna around ±180° of its main axis. The radiation pattern of an antenna is the shape of the EM wave intercepted or transmitted by the antenna. It is measured using EM compatibility (EMC) chamber called **anechoic chamber**. The radiation pattern can also be measured using radiation measurement systems (RMS), such as RMS-0640 of MegiQ BV in the Netherlands.

The radiation pattern (or radiation diagram) of an antenna can be divided in three main regions:

- Reactive near-field region (RENF) or induction field.
- Radiative near-field region (RANF) or Fresnel field.
- Radiative far-field region (RFF) or Fraunhöfer field or Fraunhofer field.

These regions are illustrated in Fig. 7.3. Note that all the three fields co-exist in the region with radius R_1, although the reactive near-field dominates. Also, both the radiative near-field and the far-field co-exist in the region with radius R_2, but radiative near-field dominates the far-field.

The boundary between these three fields is blur, and moreso for the induction and the Fresnel fields. For simple wire antennas, such as a dipole, the induction field region is approximately

$$R_1 \leq \frac{\lambda}{2}. \tag{7.20}$$

The boundary between the two radiating fields (Fresnel and Fraunhöfer fields) is approximately

$$R_2 = \max\left\{\frac{2D^2}{\lambda_0}, 3\lambda_0\right\} \tag{7.21}$$

where $\lambda_0 = \frac{3\cdot 10^8}{f(\mathtt{Hz})}$ m and D is the largest of the physical dimensions of the antenna. Note that the far-field region $R_3 - R_2$ is limited by the transmission range of the wireless transmitter. The distance

$$d = \frac{2D^2}{\lambda_0} \tag{7.22}$$

is referred to as the **Fraunhofer distance**.

7.1.7.1 *Reactive Near-Field Region*

As illustrated in Fig. 7.3, this is the region closest to the antenna. Any conductive materials in this region causes mutual impedance. The word reactive in the name of this region refers to the fact that the mutual impedance produced by nearby conductors is reactive, either capacitive or inductive. As the intensity of the H-field in this region is higher than that of the E-field, this region is also referred to as **induction field** (INF) region. The antenna acts as the storage of the energy fed into it, as an inductor or capacitor does, rather than radiating it into EM waves in this region.

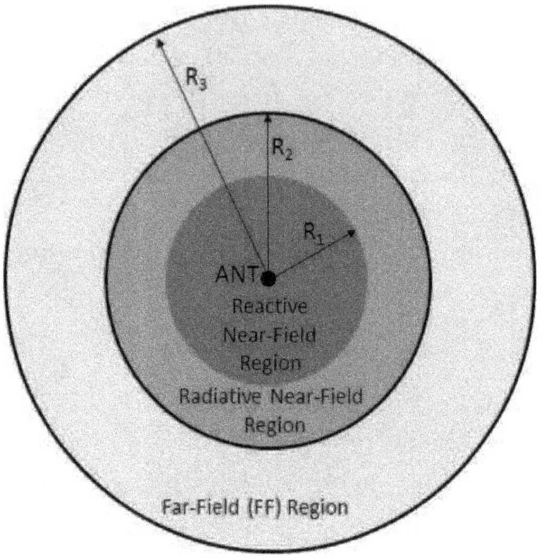

Fig. 7.3 The three regions of an antenna's radiation diagram.

7.1.7.2 *Radiating Far-Field Region*

Wireless communications is usually interesting only in the far field region of an antenna's radiation diagram. This is obvious from the example of the far-field region of the monopole antenna that we calculated above as 672.7 mm $\approx 5.83\lambda$. Even Wi-Fi hotpot mounted at the roof of a residential unit achieves this distance for far-field radio signal propagation. The radiating far-field region is usually simply referred to as far-field. The far-field region is about a few wavelengths away from the antenna. In this region:

- The intensity of the field decreases with the square of the distance away from the antenna.
- The field is called EM field, comprising E-field and H-field. These fields are at right angles with each other and none dominates the other. They have equal strengths.

7.1.7.3 *Planes and Coordinates Systems for Antenna Pattern*

The variations of the E-field and H-field of the EM waves generated by an antenna are vectors which are represented as functions of the azimuthal angle

θ and elevation angle ϕ in the spherical polar coordinates system. As the spherical polar coordinates system is three dimensional, the radiation diagram of any antenna has three dimensions. This is because to some extent every antenna radiates waves in all directions.

The 3D radiation patterns are also shown in Figs. 7.6 and 7.7. We can use multiple 2D planar cuts instead of the 3D to represent the antenna radiation pattern. As a 3D figure cannot be represented in a single plane, we usually describe the antenna pattern using two planar patterns, referred to as *principal plane patterns*. The principal plane patterns are obtained by direct measurements or as cross-sections of the 3D pattern through the antenna's axis.

Two sets of principal plane patterns are commonly used to describe the radiation pattern of antennas over ground but not in free space. These are:

- The azimuth plane (or horizontal plane) and elevation plane (or vertical plane) patterns. See Figs. 7.6 and 7.7.
- The E-plane and H-plane patterns. See Fig. 7.5. The E-plane is the plane in which the electric field dominates the magnetic field. Conversely, the H-plane is the plane in which the magnetic field dominates the electric field.

Antenna which can be fairly well described by the principal plane patterns without the loss of much information are referred to as **well-behaved antennas**. For a linear vertically-polarised antenna, the E-plane radiation pattern coincides with the elevation plane, while the H-plane radiation pattern equals the azimuthal radiation pattern. For a linear horizontally-polarised antenna, the E-plane coincides with the azimuthal plane. The H-plane is always perpendicular to the E-plane.

The appearance of an antenna's radiation pattern depends on the coordinates system used. Unfortunately, there is no fixed coordinate system used to plot the antenna pattern. The common systems include:

(1) *Linear polar coordinate system*: concentric circles are equally or linearly spaced.
(2) *Logarithmic polar coordinate system*: concentric circles are spaced according to the logarithm of the voltage of the EM wave.
(3) *Rectangular or Cartesian coordinates*: this plot has gain in dB on the vertical axis and angle (azimuth or elevation) on the horizontal axis.

The radiation pattern of an antenna is commonly described using the spherical coordinates system, which is a three-dimensional space. Why? The EM waves radiated from an antenna are practically spherical waves in the far field region. Figures 7.4(c) and (d) compare spherical with plane waves. As illustrated in Fig. 7.4(c), the wavefronts of a spherical wave propagate in all directions away from the radiator. Figure 7.4(d) shows that the wavefronts of a plane wave propagate in only one direction away from the radiator. Pure plane waves cannot be realised in practice.

The position of a point P in this coordinates system is specified by the vector (r, θ, ϕ), where

- r is the radial distance from a reference point (usually the origin of the coordinates system) to the point.
- θ is the azimuthal angle of the projection P' of the point P onto the x-y plane (i.e., horizontal plane).
- ϕ is the polar angle measured from the z axis.

The spherical coordinates system in relation to the Cartesian coordinates system is illustrated in Fig. 7.4(a). The relationship between the spherical coordinates system and the Cartesian coordinates system are

$$\begin{aligned} x &= r\sin(\phi)\cos(\theta), \\ y &= r\sin(\phi)\sin(\theta), \\ z &= r\cos(\phi) \end{aligned} \tag{7.23}$$

$$r = \sqrt{x^2+y^2+z^2}, \quad \theta = \mathtt{atan}\left(\frac{y}{x}\right), \quad \phi = \mathtt{atan}\left(\frac{\sqrt{x^2+y^2}}{z}\right). \tag{7.24}$$

Example

A flexible whip monopole antenna which can be used by mobile handsets for the 2G/3G/4G

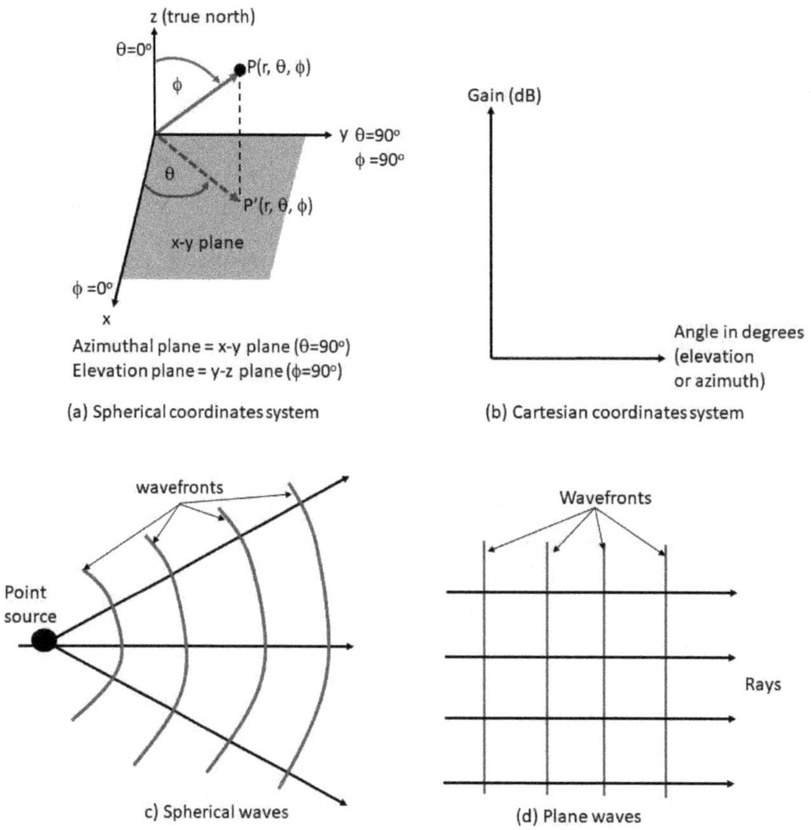

Fig. 7.4 Waves types and coordinates systems for antenna radiation pattern.

LTE/CDMA cellular mobile technologies has the dimensions:

- Model: AC-Q6027-L20D.
- Operating frequency range: 698-960/1710-2170/2500-2700 MHz.
- Dimensions: 13-mm diameter, 197-mm length.
- Weight: 35 g.
- Impedance: 50 Ω; polarisation: vertical.
- Maximum input power: 50 W.
- Gain: 3 dBi; VSWR≤ 2.0.
- Operating temperature: −40°C to +70°C.

(1) Compute the induction field region of the antenna at 2.6 GHz.
(2) The Fresnel field region of the antenna at 2.6 GHz.
(3) The Fraunhöfer field region of the antenna at 2.6 GHz.
(4) If a voltage source feeds the antenna with a 2.6 GHz signal by a 100-m cable with 75 Ω characteristic impedance, calculate the input impedance observed by the voltage source. Assume that the source is matched to the feeder cable.
(5) Compute VSWR on the feeder cable.

Solution

The maximum dimension of the monopole antenna is its length, which is $D = 197$ mm. Other given parameters are: $f = 2.6$ GHz or $\lambda = 3/26$ m, $Z_0 = 75\,\Omega$, $Z_{\mathtt{A}} = 50\,\Omega$ and $L = 100$ m.

(1) Induction field region at 2.6 GHz: $R_1 \le \frac{\lambda}{2} = \frac{c}{2f} \approx 57.7$ mm.
(2) Fresnel field region at 2.6 GHz: $R_2 = \frac{2D^2}{\lambda} = \frac{2D^2 f}{c} \approx 672.7$ mm. The Fresnel field region is $57.7\ \mathtt{mm} = R_1 \ge R \le R_2 = 672.7$ mm.
(3) The Fraunhöfer field region at 2.6 GHz: all spherical regions located over $672.7\ \mathtt{mm} \approx 5.83\lambda$ away from the antenna.

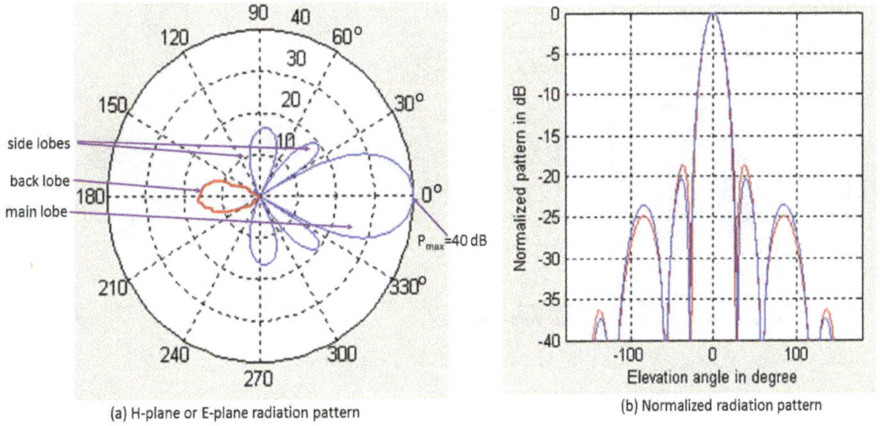

(a) H-plane or E-plane radiation pattern (b) Normalized radiation pattern

Fig. 7.5 The far-field radiation pattern of a parabolic antenna.

(4) We use Eq. (7.8) to compute the input impedance

$$\begin{aligned} Z_{\text{in}} &= Z_0 \frac{Z_{\text{A}} + jZ_0 \texttt{tan}(2\pi L/\lambda)}{Z_0 + jZ_{\text{A}} \texttt{tan}(2\pi L/\lambda)} \\ &= 50 \frac{50 + j75\texttt{tan}(2\pi \times 100/(3/26))}{75 + j50\texttt{tan}(2\pi \times 100/(3/26))} \\ &\approx 92.9 + j37.1 \approx 100 \exp(j21.8^\circ) . \end{aligned} \tag{7.25}$$

(5) The voltage reflection coefficient is $\Gamma_a = \frac{Z_{\text{A}} - Z_0}{Z_{\text{A}} + Z_0} = -0.2$, resulting in the VSWR $1.2/0.8 = 1.5$.

Figure 7.5 illustrates the radiation pattern of a parabolic antenna. The local maxima of the radiation pattern are referred to as lobes or beams. Three main types of lobes are identifiable: the *main lobe* or *primary beam*, the *side lobes* (or sidelobes) and the back lobe (aka backlobe). The back lobes and the side lobes are jointly referred to as *minor lobes*. We desire that there are no minor lobes as they are the source of both ingress and egress electromagnetic interference (EMI) which are undesirable. If P_{T} is the total power radiated or received by the antenna, $P_{\texttt{minor}}$ is the power radiated or received via the minor lobes and $P_{\texttt{main}}$ is the power radiated or received via the main lobe, then we define the antenna's *beam efficiency* as

$$\eta_B = \frac{P_{\texttt{main}}}{P_{\text{T}}} = 1 - \frac{P_{\texttt{minor}}}{P_{\text{T}}} . \tag{7.26}$$

We desire the ideal case $\eta_B \approx 1$.

Three more radiation patterns are shown in Figs. 7.6 and 7.7 for comparison. We can observe in Fig. 7.6(a) that the omni-directional antenna has no gain in the azimuthal plane, but has some gains at 0° and 180° in the elevation plane. It is therefore clear that every practical antenna has some gain in one of the three-dimensional planes. All these radiation patterns were obtained from Celplanner™ software. We also produce them using MATLAB®.

The backlobe is caused by the radiation from the feeder which misses the main reflector. The backlobe of a *directional antenna* is determined by the **front-to-back ratio (F/B)** of the antenna, which indicates how much of the radiation is emitted at the back of a *directional antenna* opposite its main beam's direction at 180° ± 40° degrees across the band. It is defined as

$$F/B = 10 \log_{10} \left(\frac{P_{front}}{P_{back}} \right) \tag{7.27}$$

where P_{front} is the power in the main lobe or the power in the signal propagating in the forward direction and P_{back} is the power in the backlobe or signal propagating in the backward direction. The residual power of the radiations from the main lobe causes the sidelobes.

The desire is to avoid the sidelobes and the back lobes altogether, as they are the source of ingress and egress radio frequency interference as well as power losses. Thus, antenna with smaller sidelobes and back lobe performs better than another antenna whose are larger. ETSI has categorised antennas

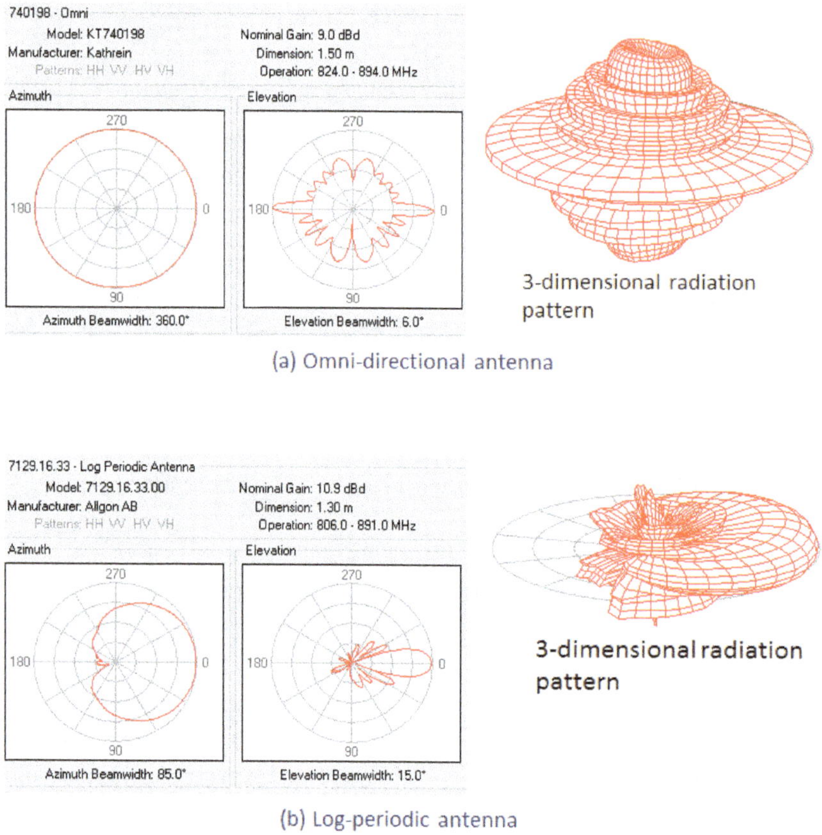

Fig. 7.6 Radiation patterns of omni and log-periodic antennas.

into classes based partly on their radiation pattern envelopes (RPEs). For example, Class 4 antennas have tighter RPEs than Class 3 antennas and thus have better selectivity and better interference immunity. The main lobe has the highest gain and it is the desirable lobe. The main lobe usually lies in the *boresight* of the antenna, which is the axis or direction of the antenna where its gain is maximum.

Example

Consider the case study parabolic antenna with features in Table 7.1. If the power radiated via the main lobe is 3 W, compute the power radiated through the back lobe in both dBm and watts.

Solution

We read off Table 7.1 the F/B of 68 dB. From Eq. (7.27), we obtain

$$\begin{aligned} P_{back} &= (10\log_{10} P_{front} - F/B \approx 34.8 - 68)\ \texttt{dBm} \\ &\approx -33.2\ \texttt{dBm} \end{aligned}$$

which is equivalent to about 0.48 nW.

7.1.7.4 *Finding the Radiation Pattern*

One thing to note is that we have unintentional antennas and intentional antennas. Every electronic device (such as light bulbs, fluorescent tubes and microwave oven) radiate EM waves to some extent. Such EM wave radiators are the *unintentional antennas* or *virtual antennas*. There are industry standards, called electromagnetic compatibility (EMC) guidelines, which must be followed in order to keep unintentional radiations below a certain limit as we cannot eliminate them altogether. This book focuses on intentional EM wave radiators or physical antennas.

Which methods are used to find the EM fields of an antenna? We need to find the distribution of electric current over the surface of the antenna.

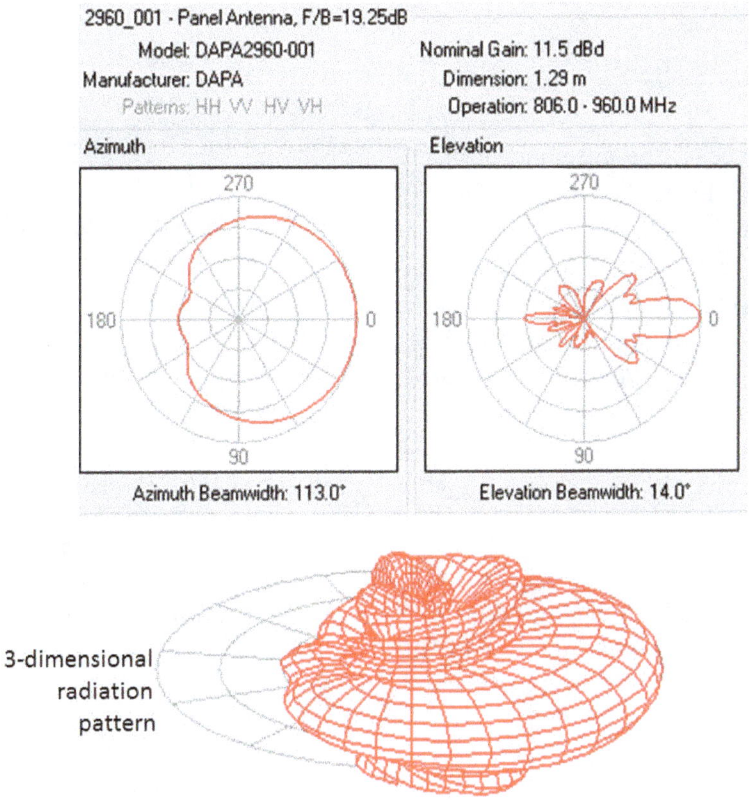

Fig. 7.7 Radiation pattern of a panel antenna.

Then, we solve a system of four independent equations, referred to as **Maxwell's equations**:

`Faraday's law of induction:`

$$\nabla \times \mathbf{E}(\mathbf{r}, t) = -\frac{\partial}{\partial t}\mathbf{B}(\mathbf{r}, t) \tag{7.28}$$

`Ampere's law:`

$$\nabla \times \mathbf{H}(\mathbf{r}, t) = \frac{\partial}{\partial t}\mathbf{D}(\mathbf{r}, t) + \mathbf{J}(\mathbf{r}, t) \tag{7.29}$$

`Gauss's law:`

$$\nabla \cdot \mathbf{D}(\mathbf{r}, t) = \rho_{\mathtt{free}}(\mathbf{r}, t) \tag{7.30}$$

`Gauss's law for magnetism:`

$$\nabla \cdot \mathbf{B}(\mathbf{r}, t) = 0 \tag{7.31}$$

where $\mathbf{E}$ is the electric field intensity, $\mathbf{H}$ is the magnetic field intensity, $\mathbf{D}$ is the electric flux intensity (aka electric displacement), $\mathbf{B}$ is the magnetic flux intensity (aka magnetic induction), $\rho_{\mathtt{free}}(\mathbf{r}, t)$ is the total free electric charge density, $\mathbf{J}(\mathbf{r}, t)$ is the total electric current density (i.e., total current per unit volume) and $I = \iint \mathbf{J} d\mathbf{S}$ is the total electric current with $d\mathbf{S}$ being a vector element of the surface area S. We observe that the above computations are quite involved, the reason why approximate solutions are often used. The solution to the Maxwell's equations are studied in courses in electromagnetism.

7.1.8 *Beamwidth, Directivity and Gain of Antennas*

The directivity of the main lobe determines the gain of the antenna. The higher the gain or directivity of the antenna, the narrower the main lobe is. The main lobe is the largest among the lobes of an antenna. We desire to have only the main lobe. The beamwidth, θ, measures the angle spanned by the main lobe. Popular beamwidths are the half-power beamwidth (HPBW), aka 3-dB beamwidth, and null-to-null beamwidth (NNBW). As illustrated in Fig. 7.5, the HPBW of an antenna is the angle between the two points on the main lobe where the power level is half that of the peak effective radiated power of the main lobe. The HPBW in degrees of a parabolic antenna is

$$\theta_{3dB} = q\frac{\lambda}{D} = q\frac{c}{Df} \tag{7.32}$$

where D is the diameter of the antenna, and λ is the wavelength of the EM signal, f is operating frequency and q (measured in degrees) is a constant which depends on the pattern that the feeder illuminates the main reflector as well as the shape of the reflector. Its value for dish antennas falls typically in the range $57.3° \leq q \leq 70°$, with the lowest value assumed in the ideal case of feeder uniformly illuminating the reflector.

Example

Consider the case study parabolic antenna with data in Table 7.1. Compute the parameter q of this antenna.

Solution

Re-arranging Eq. (7.32) gives $q = \frac{Df}{c}\theta_{3dB}$. We read off Table 7.1 $D = 3$ m, $f = 4.4$ GHz, $\theta_{3dB} = 1.6°$. Plugging these values into q, we obtain $q = 44 \times 1.6° = 70.4°$.

We need to define the Poynting power density $S(\theta, \phi)$ in order to define the directivity and gain of an antenna. It is defined as

$$S(\theta, \phi) = \frac{1}{2}Re\{\vec{E} \times \vec{H}^*\} \tag{7.33}$$

where $\vec{E}$ and $\vec{H}^*$ are the electric field and conjugate magnetic field vectors, respectively. The directivity of an antenna is defined as

$$D(\theta, \phi) = \frac{4\pi R^2 S(\theta, \phi)}{P_{\mathtt{T}}} = \frac{2\pi R^2 Re\{\vec{E} \times \vec{H}^*\}}{P_{\mathtt{T}}}. \tag{7.34}$$

The maximum of the directivity $D_{\mathtt{max}}$ for an isotropic antenna is unity. The gain, G, measures the directivity of the antenna. It is defined as

$$\mathtt{Raw\ Gain}\ (G_{\mathtt{r}}) = \frac{\mathtt{output\ power}}{\mathtt{input\ power}} = \frac{P_{\mathtt{out}}}{P_{\mathtt{in}}} = \eta D_{\mathtt{max}} \tag{7.35}$$

where η is the *radiation/illumination efficiency*, which is referred to as *aperture efficiency* for a parabolic antenna. It is defined as

$$\eta = \frac{\mathtt{radiated\ power\ by\ antenna}}{\mathtt{power\ coupled\ into\ antenna}} = \frac{P_{\mathtt{EM}}}{P_{\mathtt{EM}} + P_{\mathtt{loss}}}. \tag{7.36}$$

$P_{\mathtt{EM}}$ is the power in the radiated EM waves. $P_{\mathtt{loss}}$ is the total power dissipated in the antenna as heat and/or reflected away owing to impedance mismatch, i.e., $Z_{\mathtt{A}} \neq Z_0$. Relative to itself, the gain of a passive antenna, measured in decibels (dB), is less than unity (i.e., $G_{\mathtt{r}} < 1$), which is a loss. However, the gain relative to that of an isotropic radiator, measured in decibels isotropic (dBi), is

$$G = \frac{G_{\mathtt{r}}}{G_i} \tag{7.37}$$

which is greater than unity. The gain of a parabolic antenna, which is a directional antenna is

$$G = \frac{4\pi\eta A}{\lambda^2} \tag{7.38}$$

where A is the physical surface area of the antenna, and λ is the operating wavelength. For aperture antennas the efficiency factor η is typically in the range 55–70%. The radiation efficiency is determined by the conduction losses in the antenna's material and/or dielectric losses of conductive materials near the antenna.

Example

Consider the case study parabolic antenna with features in Table 7.1. Compute the aperture efficiency η of the antenna.

Solution

From Table 7.1 we extract the parameters:

- Operating frequency range: 4.4-5.0 GHz.
- Low-band gain of $G_L = 40.5$ dBi, mid-band gain of $G_M = 41.0$ dBi and top-band gain of $G_H = 41.5$ dBi.
- Antenna diameter (D): 3 m.
- $c_0 \approx 3 \cdot 10^8$ m/s: speed of EM waves in vacuum.

The physical surface area of the antenna is $A = \frac{\pi D^2}{4} = \frac{9\pi}{4}$ $\mathtt{m}^2$. From Eq. (7.38) we obtain the aperture efficiency

$$\eta = \frac{\lambda^2 G}{4\pi A} = \left(\frac{c_0}{\pi D f}\right)^2 G. \tag{7.39}$$

Plugging the appropriate values into Eq. (7.39), we obtain the aperture efficiencies 0.587, 0.577 and 0.579 for the low-band, mid-band and top-band frequencies, respectively. The average aperture efficiency for the case study antenna is then approximately 0.6.

7.1.9 *Antenna Polarisation and XPD*

Another parameter of an antenna is its *polarisation*, i.e., which is the direction of the electric field component of the EM waves it transmits or receive. As discussed in Chapter 5, EM waves transmitted or received by an antenna comprises three vectors: electric field (E-field) vector, magnetic field (H-field) vector and the direction of the wave travel. The direction of the E-field vector of an EM wave exiting or arriving at an antenna is called **polarisation**.

As illustrated in Fig. 7.8, there are three basic types of EM wave polarisation:

- Linear polarisation: can be horizontal polarisation (HP) or vertical polarisation (VP).
- Circular polarisation: can be left-hand (aka anti-clockwise) or right-hand (aka clockwise) polarisation.
- *Elliptical polarisation*: the polarisation rotates continually, and can be random. Linear and circular polarisations are special cases of the elliptical polarisation.

An antenna can be designed with multiple polarisation. An example is a dual-polarised antenna capable of working with vertical and horizontal polarisations. In *linear polarisation* the direction of the E-field vector is parallel (for horizontal polarisation) and perpendicular (for vertical polarisation) to the Earth's surface. For *circular polarisation* the locus of the E-field vector of the EM wave is circular. The E-field vector rotates through 360° as the wave propagates a distance equal to its wavelength. If the rotation of the E-field vector about the transmitting antenna is anti-clockwise then we have left-hand circular polarisation (LHCP). If the E-field vector

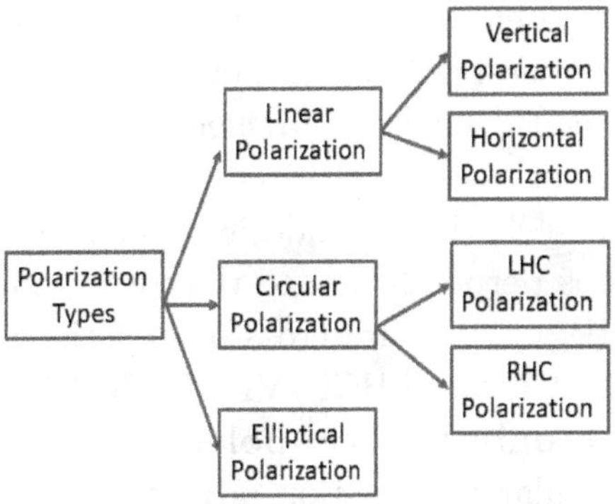

Fig. 7.8 Types of polarisations.

rotates in the clockwise direction relative to the transmitting antenna, then we have right-hand circular polarisation (RHCP).

Linear polarisation is more common than circular polarisation. For best reception both the transmit and receive antennas must have the same polarisation. Also, the polarisation of the receive antenna must be the same as that of the incoming EM wave for best operation. A circular polarisation can work with both horizontal and vertical polarisations. However, this is not the case for LHCP and RHCP. Both receive and transmit antennas must have the same polarisation, either both are RHCP or both are LHCP. An EM wave which is LHCP becomes RHCP if it bounces off a reflecting surface, and vice versa.

The polarisation of a wave can be rotated by a magnetic field. This occurrence is referred to as *Faraday's effect* or *Faraday's rotation*. Owing to Faraday's effect and the fact that the orientation of receive antennas relative to transmit cannot be known, circular polarisation is preferable in such situations. Why is circular polarisation insensitive to receive antenna's orientation relative to the wave? The E-field in circular polarisation changes in two orthogonal planes with the same magnitude. Such an E-field excites two H-field modes TM_{10} (in x direction) and TM_{01} (in y direction) simultaneously.

The polarisation of an EM wave or signal can be changed by the transmission medium. Thus, tuning both transmit and receive antennas with the same polarisation of signal leaving the transmitter does not guarantee that the signal reaching the receiver shall have the same polarisation as the receiving antenna. An EM wave which has a different polarisation than that of the antenna receiving it is called **cross-polarised** (XP) wave. The desired polarisation is referred to as *co-polarisation* (CP), while XP is the undesired polarisation. The XP wave is perpendicular to the corresponding CP wave. HP and VP are cross-polarisations, just as LHCP/RHCP and LHCP/VP are. **XP is a problem with only linearly-polarised waves.** This makes circular polarisation advantageous over linear polarisation. *Cross-polarisation discrimination* (XPD), measured in dB, is the figure used to check how much of the EM signal's power is received with the desired polarisation. It is defined as

$$XPD = 10\log_{10}\left(\frac{P_{acp}}{P_{axp}}\right) = 10\log_{10}\left(\frac{P_{cp}}{P_{xp}}\right) \quad (7.40)$$

where P_{cp} is the peak power in the signal co-polarised with the main beam of the antenna, P_{xp} is the maximum power in the cross-polarised signal, P_{axp} is the power received in the antenna's cross-polar port, and P_{acp} power received in the antenna's co-polar port.

Inter-port isolation (IPI), measured in dB, is the EM separation or isolation between the input ports of a dual-polarised antenna. An *antenna isolation* is a measure of how tightly coupled nearby antennas are. It is determined by the transmission coefficient S_{kl} which is the power coupled to, or received by, antenna k relative to the power input to antenna l. The larger the antenna isolation, the more efficient the antenna is. Our case study parabolic antenna with features in Table 7.1 has an isolation of 35 dB. For example, 35 dB antenna isolation means that the power coupled to an antenna k from the radiations of the nearby antenna l. Any of the following four measures can be used to increase the antenna isolation:

- Use different polarisations for the antennas interfering with each other.
- Increase the physical separation between the antennas.
- Reduce the correlation coefficient between the antenna's radiation patterns. This can be achieved by arranging antennas so their main beams are in orthogonal directions.
- For antennas operating at different frequencies, appropriate filtering can reduce the EMI.

Example

A microwave transceiver operating at 2.5 GHz uses a 4.5-meter parabolic antenna with the radiation efficiency of $\eta = 0.6$. The antenna has the input impedance of 60 Ω and connected to the radio transceiver by a coax cable with the characteristic impedance of 50 Ω.

(1) What is the 3-dB beam width of the antenna if $q = 60$?
(2) Compute the gain of the antenna.
(3) Find the fraction of the power incident on the antenna that it absorbs.
(4) What is the return loss (RL) of the antenna?
(5) Compute the VSWR on the cable connecting the antenna to the radio transceiver.
(6) Evaluate the maximum tolerable angular movement of the antenna caused by cumulative environmental effects.
(7) Assume that the antenna is mounted on a tower of height 30 m above ground level. If the terrain factor is $T = 0.25$ and the mean wind speed 10 meters above the ground is 40 m/s, what is the wind speed experienced by the antenna?

Solution

Given in the problem are:

- Operating frequency: $f = 2.5$ GHz and so $\lambda = c/f = \frac{3}{25}$ m.
- Characteristic impedance of transmission line or coax: $Z_0 = 50\ \Omega$.
- Impedance of load (input impedance of antenna): $Z_L = 60\ \Omega$.
- Diameter of dish antenna: $D = 4.5$ m.

(1) $\theta_{3dB} = q\lambda/D = 60 \times (3/25)/4.5° = 1.6°$.
(2) Gain of antenna: $G = \frac{4\pi A\eta}{\lambda^2} = \frac{(\pi D)^2\eta}{\lambda^2} \approx 99.3 \approx$ 30 dB.
(3) Reflection coefficient $\Gamma_a = \frac{Z_L - Z_0}{Z_L + Z_0} = \frac{1}{11}$. Thus, the fraction of the incident power absorbed by the antenna is $1 - \Gamma_a = \frac{10}{11}$.

(4) The return loss at the antenna is $RL = -20\log_{10}\Gamma_a \approx 20.8$ dB.
(5) The VSWR on the coax cable is $VSWR = \frac{1+\Gamma_a}{1-\Gamma_a} = 1.2$.
(6) The maximum angular movement of the antenna is $\theta_a \approx 0.3 \times \theta_{3dB} = 0.3 \times 1.6° = 0.48°$.
(7) Wind speed at antenna fixed at the height $h = 30$ m if $T = 0.25, v_{10} = 40$ m/s is $v(h) = 40(0.1 \times 20)^{0.25} \approx 47.6$ m/s.

7.1.10 *Weather Effects on Antennas*

Antennas are sensitive devices and thus must be protected against weather effects, such as wind, dust, precipitation (e.g., snow/ice and water) and solar radiation. A protective material used to cover aperture antennas is called **radar dome**, but usually referred to by its portmanteau word *radome*. A radome also conceals the electronics in the antenna from public view. The third purpose of a radome is to protect people from accidental hurts by the antennas, especially for antennas with rotating components. The material used to manufacture a radome is important as it should absorb minimal wave's energy. However, without a radome, weather effects can corrode the antenna to reduce its transmission efficiency. Note that radomes can be used to cover antennas for all applications rather than only radar antennas.

Wind can have a dramatic effect on antennas mounted on masts. Wind load is the quantity of force that wind exerts on an exposed structure, such as antenna. Wind load depends on wind speed, height of the antenna, terrain, and climate and thus depends on the specific site where the antenna is mounted. The mean wind speed at the antenna height h meters above the ground surface is

$$v(h) = v_{10}(0.1h)^T \ \ \texttt{[m/s]}, \quad 0.1 \le T \le 0.4 \quad (7.41)$$

where T is the terrain factor and v_{10} in m/s is the mean wind speed at 10 m above ground level. This formula accounts for the differing wind speeds at the bottom and top of the mast holding the antenna in the desired position. The natural frequency of wind is circa 8 Hz. Thus, wind blowing around antenna structures with similar natural frequency can cause vibrations which can lead to resonance and structural fatigue or mechanical damage.

Example

Assume that the average wind speed at the height 10 m above the foot of a mobile antenna tower is 408 km/h. The terrain factor of the environment where the antenna mast is mounted is $T = 0.2$. The antenna tower has been designed to withstand the maximum wind speed of 743 km/h without experiencing a remarkable damage. Compute what should be the maximum height of the antenna tower.

Solution

The parameters in the question are: $T = 0.2$, $v_{10} = 408$ km/h and $v_h = 743$ km/h. The equation relating these parameters is given in Eq. (7.41) as $v(h) = v_{10}(0.1h)^T$ `[m/s]`. Thus,

$$v(h) = v_{10}(0.1h)^T \Leftrightarrow h = 10\left(\frac{v_h}{v_{10}}\right)^{1/T}$$
$$= 10\left(\frac{743}{408}\right)^{1/0.2} \texttt{ m} \approx 200.3 \texttt{ m}. \quad (7.42)$$

Comment: The fastest wind speed which is not linked with tornadoes recorded in Australia so far occurred in 1996 on Barrow Island. Its speed was 408 km/h. The tallest radio mast in Australia is the Naval Communication Station Harold E. Holt located in Exmouth in Western Australia. Its height is 389 m. One of the tallest antenna mast is the Warsaw radio mast in Poland. Its height is about 646.4 m. The average height of mobile antenna towers lies at 200–250 m. A special authorisation is required to mount tall towers in many countries. For example, in the USA, towers above 61 m fall in this category. As a way of case study, the Cambium Network's point-to-point radio, PTP50650, can withstand the maximum wind speed of 322 kph.

Wind can change the direction of an antenna's main beam. The amount of wind tolerable by an antenna without going off course its target is referred to as *operational wind load*. The case study dish antenna in Table 7.1 is rated with the operational wind speed of 30 m/s (equivalent to 1,080 km/h). Wind can exert twisting and shifting effects on an antenna to cause angular movements. The maximum allowable angular movement or deflection of an antenna with the half-power (or 3-dB) beamwidth θ_{3dB} is

$$\theta_a = 0.3\theta_{3dB} \texttt{ [degrees]}. \quad (7.43)$$

The maximum wind speed tolerable by an antenna without experiencing permanent damage is referred to as *survival wind load* or survival wind velocity. The case study dish antenna in Table 7.1 is rated with the survival wind speed of 55 m/s (equivalent to 1,980 km/h). Thus, one of the ratings of antennas is their survival wind load. Designers of antenna masts use struts to support the antenna to endure wind speeds up to its survival wind load rating.

As antennas for point-to-point microwave links are usually mounted outdoors, they are exposed to corrosion. Corrosion can destroy the structural integrity of antennas. Corrosion of outdoor units can be caused by salty air (if antenna is mounted near the ocean) and air pollutants (such as dust and exhausts of cars). A type of corrosion which occurs when different metals come in contact with each other is called galvanic corrosion. Also, the ultraviolet radiation from the sun can destroy non-UV-resistant materials in the antenna over time.

7.1.11 *Antennas for IoT Devices*

There is no limit on what devices can be attached to the IoT. Therefore, all forms of antennas, whether discussed here or not, is important for wireless IoT. For example, industrial IoT (IIoT) uses Yagi antennas in Supervisory Control and Data Acquisition (SCADA) systems. Owing to their wide beamwidth, multiple panel antennas can be mounted at base stations of WLAN, SigFox and LoRa systems to achieve 360° coverage. Embedded antennas, such as chip antennas and printed circuit boards (incl. folded dipole, inverted-F and inverted-L antennas) also find applications in IoT. Embedded antennas for portable IoT devices must be compact, lightweight and consume low power.

7.1.12 *Check Your Understanding*

Students are encouraged to check their understanding of the content of this section with the following questions prior to proceeding to the next section.

(1) Describe antenna polarisation? State the reason why circular polarisation is better than linear polarisation.
(2) Describe the importance of antennas in wireless Internet of Things.
(3) Describe your understanding of the radiation pattern of an antenna.
(4) List two equipments which can be used to measure the radiation pattern of an antenna.
(5) Explain the meaning of 3-dB beamwidth of an antenna.
(6) Discuss the implications of cross-polarisation discrimination (XPD) of an antenna.
(7) How does the main lobe of an antenna differ from its sidelobes and back lobes?
(8) Discuss the meaning of F/B ratio.
(9) What does the boresight of an antenna mean?
(10) Describe how to improve the directivity of a dish antenna.
(11) A 2-meter RF-9913 coaxial cable rated with 4.5 dB/30.5 m loss connects a transmitter to an antenna. The transmitter generates a signal with power 18 dBm. The loss at each of the two connectors at both ends of the cable is 1.2 dB.
    (a) Find the power at the input of the antenna in both mW and dBm.
    (b) If the gain of the antenna is 15 dBi, compute the EIRP in the EM wave exiting the antenna.
(12) Describe the meaning of passive gain or directional gain in regard to antennas.

(13) Differentiate between passive gain and active gain (aka power gain) of antennas.
(14) Explain the reason why an antenna is a transducer.
(15) Justify the phrase *an antenna is a reciprocal device.*
(16) Assume that a flexible whip monopole antenna which can be used by mobile handsets of cellular mobile technologies has the dimensions:
   - Operating frequency: 3.5 GHz.
   - Dimensions: 13-mm diameter, 197-mm length.
   - Weight: 35 g.
   - Impedance: $75\,\Omega$; polarisation: vertical.
   - Maximum input power: 50 W.
   - Gain: 3 dBi; VSWR$\leq$ 2.0

   (a) Compute the induction field region of the antenna.
   (b) The Fresnel field region of the antenna.
   (c) The Fraunhöfer field region of the antenna.
   (d) If a voltage source feeds the antenna with a 3.5 GHz signal by a 200-m cable with $50\,\Omega$ characteristic impedance, calculate the input impedance observed by the voltage source as a function of frequency. Plot the input impedance over the frequency range 1.8-8 GHz. Assume that the source is matched to the feeder cable.
   (e) Compute VSWR on the feeder cable.

7.2 Case Studies of Antennas

We review a sample of antennas in this section using practical examples as case studies. There are many types of antennas and the type of antenna used depends on the application area and very much on the frequency of operation. For example, parabolic antennas would be too large to deploy in mobile devices so it is not done. Antennas of mobile handsets are generally omnidirectional as they cannot be pointed to the direction of incoming signals. In microwave communications systems, however, parabolic (aka dish) antennas, which are highly directional antennas, are used in order to tightly focus the transmitted energy, and also to receive signals mainly from one specific direction. Thus, high-gain parabolic dish antennas enable a small transmit power to cover long-distance transmissions.

7.2.1 *Classification of Antennas*

Antennas can be classified along many several lines. A sample classification is found in Fig. 7.9. We discuss just a few of them here. For example, we have low-gain and high-gain antennas. With respect to their radiation pattern, antennas are subdivided into isotropic radiator, omnidirectional and directional. Based on their shape and size, antennas can be categorised into aperture antennas (e.g., horn and parabolic), wire antennas (e.g., dipole, loop and helix), rod antennas (e.g., monopole) and printed antennas (e.g., patch, spiral and printed circuit boards).

Based on the nature of the beam used, antennas can use electrical scanning, mechanical scanning or no scanning at all. In regard to the mounting location, there are indoor antennas (e.g., rabbit ear and loop antennas) which are mounted inside buildings and outdoor antennas (e.g., Yagi and log-periodic) which are mounted outside buildings.

Antennas can be subdivided into *resonant antennas* and *non-resonant antennas.* A magnetic field (or an EM field) around an antenna generates an electric current in the antenna. At the resonant frequency of the antenna the electrical current through the antenna achieves its maximum value, and it is 90° out of phase with the voltage in the antenna. The resonance of a dipole antenna occurs if its length L fulfills

$$L = \frac{2k-1}{2}\lambda, \quad k = 1, 2, \ldots \tag{7.44}$$

i.e., an odd multiple of half of the operating wavelength. This means that we can change the length of an antenna to make it resonate or otherwise. This is the science behind the reason why an antenna can be tuned to a given frequency or wavelength.

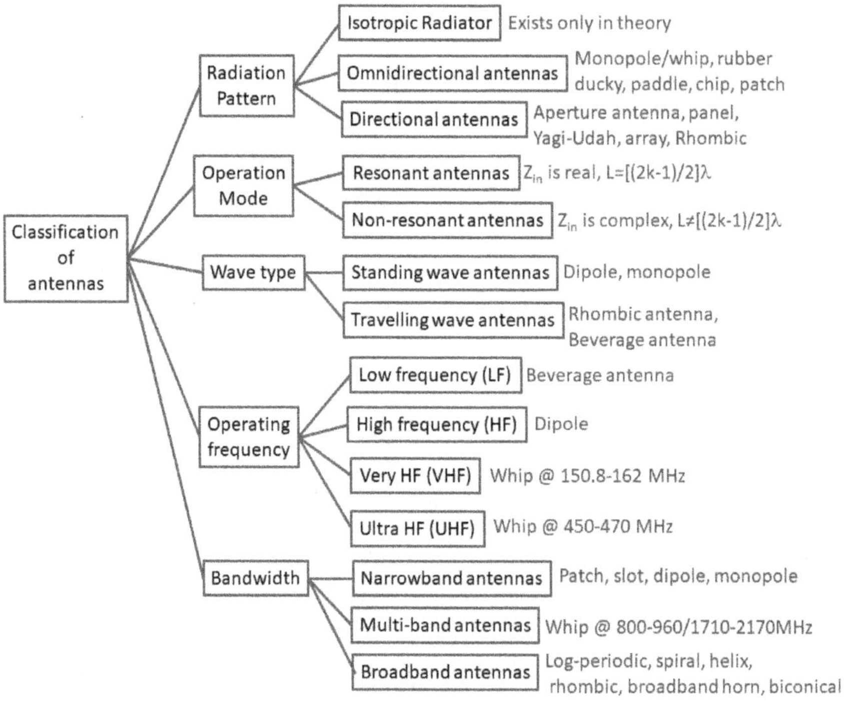

Fig. 7.9 Classification of antennas.

The input impedance of an antenna at resonance is purely resistive, i.e., $Z_{\text{in}} = R_{\text{in}}$. The S_{11} parameter of an antenna decreases remarkably at its resonant frequency (or frequencies).

Depending on the frequencies of operation, antennas can be categorised into narrowband, multi-frequency and broadband antennas. A broadband antenna is also referred to as a *wideband antenna*, while a multi-frequency antenna is also called *multi-band antenna*. A *narrowband antenna* has a single resonant frequency, while *multi-frequency antenna* has multiple resonant frequencies. Note that we can convert a narrowband antenna to a broadband antenna by increasing its volume. For example, the operating frequency range or bandwidth of a dipole antenna can be increased by using elements with a larger diameter. We can also increase the bandwidth of a dipole by using conical instead of cylindrical elements.

Also referred to as multi-band antenna, a multi-frequency antenna operates on several bands. The trade-off for this is reduced gain or larger size. Also, depending on operating frequencies, antennas are classified into HF antenna, VHF antenna, etc. See Fig. 1.4 in Chapter 1 for the global naming convention of the frequency bands.

Antenna can be standing wave antenna or travelling wave antenna. Unlike standing wave antennas, travelling wave antennas are non-resonant antennas. The advantage in this is that they possess wider bandwidths than their equivalent resonant antennas. The electric current used to generate the EM waves in travelling wave antennas travel in just one direction through the antenna. The standing wave-generating current travels in both directions between the two ends of the antenna.

7.2.2 *Dipole and Monopole Antennas*

Dipole and monopole antennas are the most basic and most common type of antennas. They are also used to realise more complex antennas. Common dipole antennas include the *half-wave dipole* which is also called *Hertz antenna*, and *quarter-wave dipole* which is also referred to as *Marconi antenna*. These were named after their inventors. Heinrich Hertz invented the half-wave dipole around 1886. The half-wave dipole is the simplest realisable

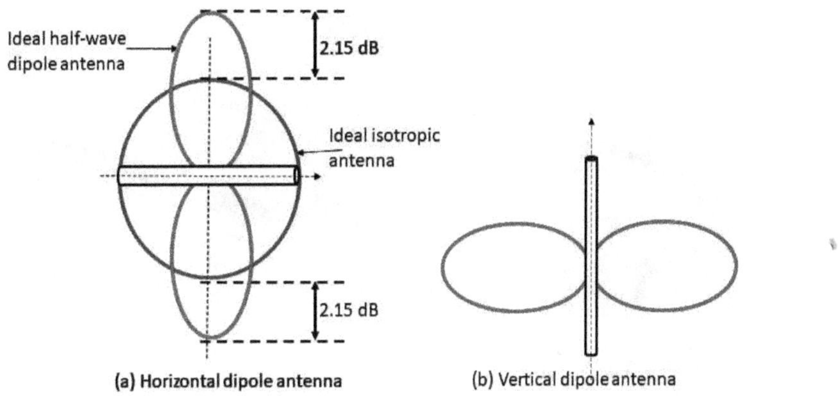

Fig. 7.10 Radiation patterns of ideal omnidirectional and dipole/monopole antennas.

antenna. It is a directional antenna. The relationship between the gains of a half-wave dipole and that of an isotropic antenna is

$$G_d = 1.64G_i \Leftrightarrow G_d \text{ [dBd]} = G_i + 2.15 \text{ [dBi]} . \quad (7.45)$$

We can increase the gain of a dipole by installing reflectors at a distance d away from it, where

$$\frac{\lambda}{16} \le d \le \frac{\lambda}{4} . \quad (7.46)$$

The gain of a monopole is about 3 dB less than that of the equivalent dipole. This difference is attributed to ground plane signal reflection used by a monopole. Figure 7.10(a) compares the radiation patterns of ideal isotropic antenna with that of a half-wave dipole antenna.

A dipole antenna has two rods of equal length, as shown in Fig. 7.11(a). One rod serves as the radiator while the other serves as the ground to the former. As a dipole is a balanced antenna and coax is unbalanced, using a coax as the feedline requires a balun to match the two. Balun, meaning balanced to unbalanced, is a transformer used to match an unbalanced line to a balanced circuit, and vice versa. Applications of dipole antennas include:

- The elements of array antennas such as Yagi-Uda.
- Rabbit ear antenna for television reception. It is called rabbit ear because of its shape.
- Antennas for FM radio receivers. The length of the antenna is usually fitted onto the center frequency of the FM band.

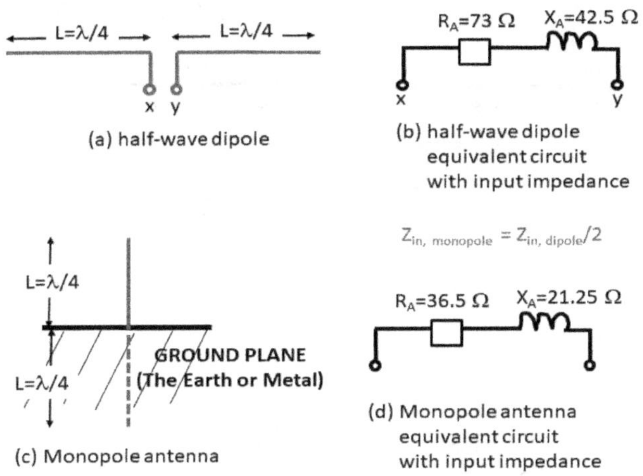

Fig. 7.11 Monopole and dipole antennas.

- Antennas for two-way radio such as walkie-talkie.

Two practical examples of dipole antennas are shown in Fig. 7.12.

Figure 7.13 shows the distribution of the electric current and voltage along a half-wave dipole antenna. Whenever the current is maximum the voltage is zero, and vice versa. A center-fed dipole is illustrated. We can also feed the power through any point, except that doing so would increase the feedpoint impedance remarkably. For example, using the Ohm's law

$$Z = \frac{voltage}{current} = \frac{V}{I} , \quad (7.47)$$

the input impedance becomes infinite. This is because the current is zero at the end points of the

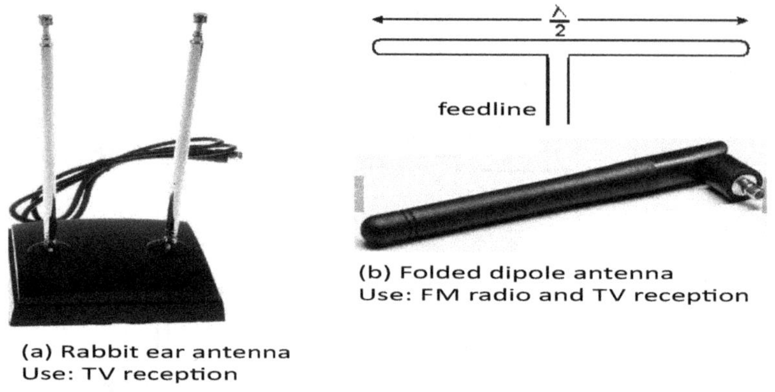

Fig. 7.12 Examples of dipole antennas.

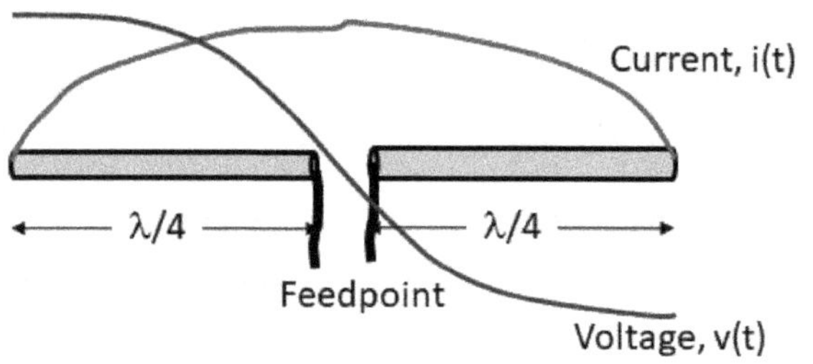

Fig. 7.13 Current and voltage distribution along a half-wave dipole.

dipole. A dipole antenna is usually fed at its points where the input impedance is low.

The situation shown in Fig. 7.13 in which the voltage leads the current results in the impedance being an inductive load. The impedance becomes a capacitive load if the current leads the voltage. We have a pure resistive load if the voltage and current are in phase with each other. This is the resonance condition. If the current and phase are completely out-of-phase with each other (i.e., at right angles) then the antenna would radiate no EM waves regardless of the level of the current flowing through it.

Figure 7.11(b) shows the equivalent circuit of the half-wave dipole. Its input impedance is

$$Z_{\mathtt{A}} = R_{\mathtt{A}} + jX_{\mathtt{A}} = 73 + j42.5\,. \tag{7.48}$$

The total power radiated by a half-wave dipole is

$$P_{\mathtt{rad}} = |I_{\mathtt{in,\,rms}}|^2 R_{\mathtt{A}} = 73|I_{\mathtt{in,\,rms}}|^2 \tag{7.49}$$

where $|I_{\mathtt{in,\,rms}}|^2$ is the root-mean squared current at the terminal of the antenna. The power density at a distance r away from the broadside[2] of the half-wave dipole antenna is

$$S = 9.54\frac{|I_{\mathtt{in,\,rms}}|^2}{r^2}\,. \tag{7.50}$$

The general expression for the electric field of half-wave dipole antennas and monopole antennas mounted on the ground plane at a distance r in the far-field region of the antenna is

$$\begin{aligned} E_\theta(r,\theta) &= j\eta_0\frac{\sqrt{2}I_{\mathtt{in,\,rms}}}{2\pi r}e^{-j2\pi r/\lambda}F(\theta) \\ &= \hat{F}(\theta)I_{\mathtt{in,\,rms}}\frac{e^{-j\beta r}}{r} \end{aligned} \tag{7.51}$$

where $\beta = \frac{2\pi}{\lambda}$, $\eta_0 = 377\,\Omega$ is the impedance of free space, $\hat{F}(\theta) = j\frac{\eta_0}{\sqrt{2\pi}}F(\theta)$, and the *antenna pattern factor* is defined as

$$F(\theta) = \frac{\cos\left(\frac{\pi l}{\lambda}\cos\theta\right) - \cos\left(\frac{\pi l}{\lambda}\right)}{\sin\theta} \tag{7.52}$$

where l is the length of the antenna, θ is the azimuthal angle with $\theta = 90°$ at the broadside of the antenna and ϕ is the elevation angle. The antenna pattern factor determines the shape of the antenna's radiation pattern. It is therefore specific to antennas. The magnetic field is perpendicular to the electric field. It is therefore

$$H_\phi(r,\theta) = \frac{E_\theta(r,\theta)}{\eta_0} = j\frac{\sqrt{2}I_{\mathtt{in,\,rms}}}{2\pi r}e^{-j2\pi r/\lambda}F(\theta). \tag{7.53}$$

[2]The broadside of an antenna is the plane perpendicular to its axis. For example, the broadside of a vertical dipole antenna is the horizontal plane.

For a half-wave dipole $l = \lambda/2$. Thus,

$$E_\theta(r,\theta) = j\eta_0 \frac{\sqrt{2}I_{\mathtt{in,\,rms}} \cos(\frac{\pi}{2}\cos\theta)}{2\pi r \sin\theta} e^{-j2\pi r/\lambda} \tag{7.54}$$

and

$$\begin{aligned} H_\phi(r,\theta) &= \frac{E_\theta(r,\theta)}{\eta_0} \\ &= j\frac{\sqrt{2}I_{\mathtt{in,\,rms}} \cos(\frac{\pi}{2}\cos\theta)}{2\pi r \sin\theta} e^{-j2\pi r/\lambda}. \end{aligned} \tag{7.55}$$

Example

A half-wave dipole is used to tune to FM radio stations. An alternating current oscillating at 88.5 MHz and with the root-mean-squared magnitude 80 mA is fed into the antenna through its feed point. Compute the

(1) Average power density at a distance 400 m away from the broadside of the antenna.
(2) Total power radiated by the dipole antenna.
(3) Electric field intensity at the broadside of the antenna.
(4) Magnetic field intensity at the broadside of the antenna.

Solution

(1) The average power density at a distance 400 m away from the broadside of the antenna is

$$\begin{aligned} S &= 9.54\frac{|I_{\mathtt{in,\,rms}}|^2}{R^2} \\ &= \frac{9.54 * 0.08^2}{400^2} \approx 38.2 \ \mathtt{nW}. \end{aligned}$$

(2) The total power radiated by the dipole antenna is

$$\begin{aligned} P_{\mathtt{rad}} &= |I_{\mathtt{in,\,rms}}|^2 R_{\mathtt{A}} = 73|I_{\mathtt{in,\,rms}}|^2 \\ &= 73(0.08)^2 \approx 0.467 \ \mathtt{W}. \end{aligned}$$

(3) At the broadside we have $\theta = 90°$. Plugging this angle into Eq. (7.54) gives

$$E_\theta\left(r,\theta=\frac{\pi}{2}\right) \approx j\frac{84.86 I_{\mathtt{in,\,rms}}}{r} e^{-j2\pi r/\lambda}. \tag{7.56}$$

(4) The magnetic field follows from Eq. (7.55) as

$$H_\phi\left(r,\theta=\frac{\pi}{2}\right) \approx j\frac{0.225 I_{\mathtt{in,\,rms}}}{r} e^{-j2\pi r/\lambda}. \tag{7.57}$$

Figure 7.11(c) shows a *monopole antenna*, comprising a single straight metal rode whose length is a fraction of the operating wavelength and half the length of an equivalent dipole. As a monopole antenna is half the size of its equivalent dipole, it is preferred where space is an issue. A monopole antenna is an unbalanced antenna, because one of its terminals is connected to the ground plane. For this reason, no balun is needed to interface it with the coax feedline. The prefix 'mono' in the name indicates 'single' as it contains a single conductor. For the monopole to function as expected, the ground plane should be a perfect electrical conductor (PEC) and of infinite size. The commonest monopole is the quarter-wave monopole whose length is $L \approx \frac{\lambda}{4}$. The quarter-wave monopole is popularly referred to as *whip antenna*. It has a wide application as vehicles' radio antennas. Quadrifilar helix (aka quad helix) is a type of monopole antenna popularly used in GPS receivers.

One end of the feedline connects to the rod while the other end connects to a conductor called ground plane, which is often the earth or a metal. The ground plane acts as a virtual quarter-wave dipole. Consequently, the combined effect of the ground plane and the quarter-wave monopole is equivalent to that of the half-wave dipole. The quarter-wave monopole was independently invented by Guglielmo Marconi and Alexander Popov about the same time around 1895/6.

Figure 7.11(d) shows the equivalent circuit of the quarter-wave monopole. Its input impedance is

$$Z_{\mathtt{A}} = R_{\mathtt{A}} + jX_{\mathtt{A}} = 36.5 + j21.25. \tag{7.58}$$

A popular example of monopole antennas is *rubber ducky antenna* (aka rubber duck aerial), which is popularly used in portable hand-held radio devices. There are many other types and applications of monopole antennas. Figure 7.14 shows some of them. The radiation resistance of the quarter-wave monopole is half that of an equivalent half-wave dipole. The gain of the monopole is about 3 dB higher than that of the equivalent half-wave dipole.

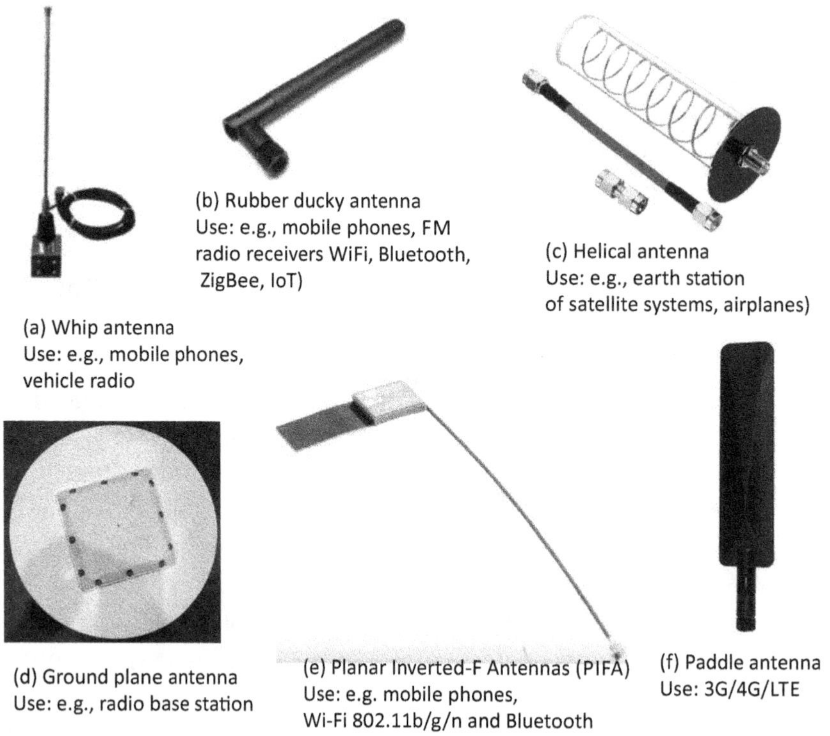

Fig. 7.14 Examples of monopole antennas and their applications.

Example

A half-wave dipole antenna was constructed using a pure copper wire with resistivity $\rho = 1.7 \times 10^{-8}\,\Omega\mathtt{m}$ and $A = 1.0\,\mathtt{mm}^2$ cross-sectional area. An alternating voltage source oscillating at 1.8 GHz with the internal resistance $Z_s = 20\,\Omega$ and peak voltage of 50 V drives the antenna. The configuration is illustrated in Fig. 7.15(a). Compute the total average power in the EM wave radiated by the antenna.

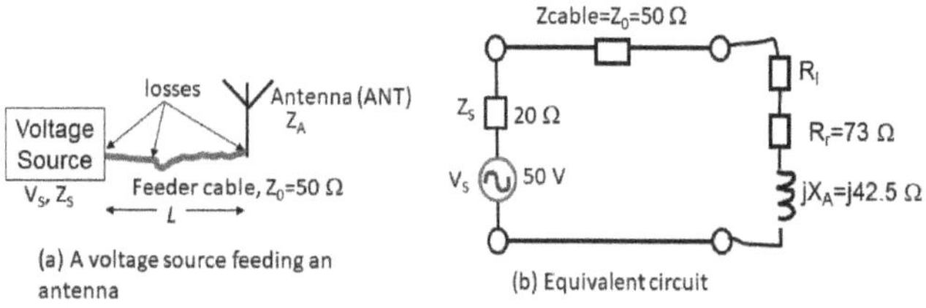

Fig. 7.15 Figure for illustrative example.

Solution

The equivalent circuit for Fig. 7.15(a) is shown in Fig. 7.15(b). Given are: $f = 1.8$ GHz and $A = 1.0\ \mathtt{mm}^2 = 10^{-6}\ \mathtt{m}^2$. The wavelength of the radiated EM wave is

$$\lambda = \frac{c}{f} = \frac{3 \cdot 10^8}{1.8 \cdot 10^9}\ \mathtt{m} = \frac{1}{6}\ \mathtt{m}.$$

Thus, the length of the wire used to construct the half-wave dipole is $l = \frac{1}{12}$ m. The ohmic loss resistance of the wire is

$$R_l = \rho\frac{l}{A} = 1.7 \times 10^{-8} \times \frac{\frac{1}{2} \cdot \frac{1}{12}}{10^{-6}} = 0.708\ \mathtt{m}\Omega. \quad (7.59)$$

The current flowing through the input terminals of the dipole antenna is

$$\begin{aligned} I &= \frac{V_s}{Z_s + Z_0 + R_l + R_r + jX_A} \\ &= \frac{50\angle 0^\circ}{20 + 50 + 708 \cdot 10^{-6} + 73 + j42.5} \\ &\approx \frac{50\angle 0^\circ}{143 + j42.5} \approx 0.3353\angle - 16.6^\circ\ \mathtt{A}. \quad (7.60) \end{aligned}$$

The total average radiated power is

$$\begin{aligned} P_{\mathtt{rad}} &= \frac{1}{2}|I|^2 R_r \\ &= \frac{1}{2}0.3353^2 \cdot 73\ \mathtt{W} \approx 4.1\ \mathtt{W}. \quad (7.61) \end{aligned}$$

7.2.3 *Log-Periodic Antenna*

Log-periodic antenna is a directive antenna with a large bandwidth, enabling it to operate at many different frequencies. This is its advantage over other directional antennas. The bandwidth is approximately the difference between the resonant frequencies of its longest and shortest element. Antennas can be classified into frequency-dependent and frequency-independent antennas. Owing to its wideband nature, log-periodic antenna belongs to the class of frequency-independent antennas. Frequency-independent antennas are broadband antennas.

The log-periodic antenna was invented by Dwight E. Isbell, Raymond DuHamel and Paul Mayes around 1957. There are many types of log-periodic antennas. Figure 7.16 illustrates one of the common types, called dipole array antenna, comprising a bunch of dipole antennas. Let L_n be the length of the nth dipole counting from the origin, and R_n be its distance from the origin. The lengths of the dipoles increase in proportional to their distance from the origin of the antenna. This causes the radiation pattern and input impedance to repeat periodically as the logarithm of the frequency. The relationship between the lengths and distances is

$$\beta = \frac{L_1}{L_2} = \frac{L_2}{L_3} = \frac{L_3}{L_4} = \cdots = \frac{R_n}{R_{n+1}}. \tag{7.62}$$

By noting that $L_n \propto \lambda_n = \frac{c}{f_n}$, we can rewrite Eq. (7.62) as

$$f_{n+1} = \beta f_n, \quad R_{n+1} = \frac{f_n}{f_{n+1}} R_n, \quad L_{n+1} = \frac{f_n}{f_{n+1}} L_n\,. \tag{7.63}$$

This equation is the source of the term log-periodic. Which means that if the antenna works well at a frequency f_n, then it should work at frequency $f_{n+1} = \beta f_n$, where β is some constant.

If the operating frequency range of the log-periodic antenna is $f_L \le f \le f_H$, then its bandwidth is

$$B = f_H - f_L \ \texttt{[Hz]}\,. \tag{7.64}$$

The lengths of the shortest element L_S and longest element L_L of the antenna are then

$$L_\mathrm{L} \approx \frac{3\cdot 10^8}{2f_L} \ \texttt{[m] and } L_\mathrm{S} \approx \frac{3\cdot 10^8}{2f_H} \ \texttt{[m]} \tag{7.65}$$

Example

Assume that you need to design a log-periodic antenna with 5-6 dB gain to operate in the frequency

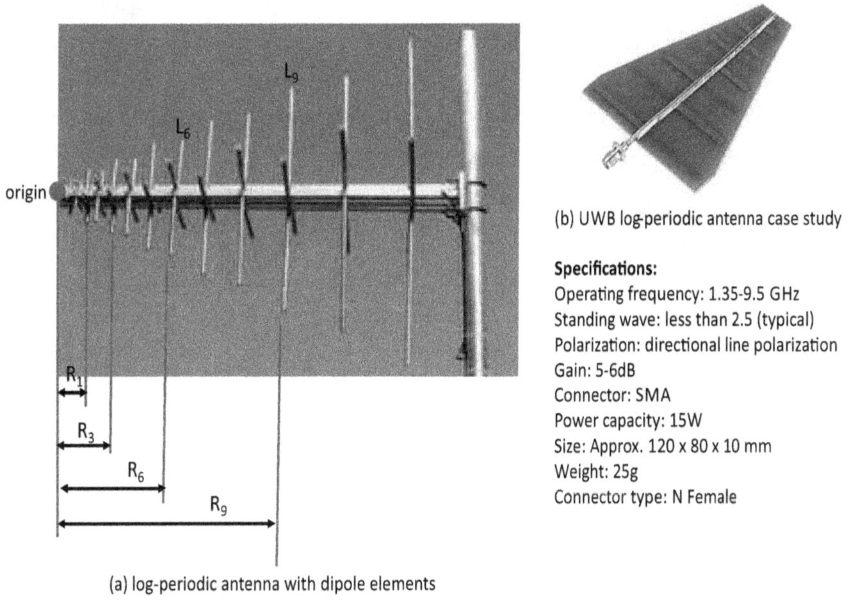
origin
L9
L6
R1
R3
R6
R9
(a) log-periodic antenna with dipole elements
(b) UWB log-periodic antenna case study
Specifications:
Operating frequency: 1.35-9.5 GHz
Standing wave: less than 2.5 (typical)
Polarization: directional line polarization
Gain: 5-6dB
Connector: SMA
Power capacity: 15W
Size: Approx. 120 x 80 x 10 mm
Weight: 25g
Connector type: N Female

Fig. 7.16 Log-periodic antennas.

band 1.35 GHz–9.5 GHz. Find the lengths of the longest and the shortest element of the antenna.

Solution

We are given $f_L = 1.35$ GHz $\leq f \leq f_H = 9.5$ GHz. We plug these frequencies into Eq. (7.65) to obtain $L_\mathrm{L} \approx \frac{3 \cdot 10^8}{2 \times 1.35 \cdot 10^9} \approx 11.1$ cm and $L_\mathrm{S} \approx \frac{3 \cdot 10^8}{2 \times 9.5 \cdot 10^9} \approx 1.6$ cm. The parameters of this antenna are shown in Fig. 7.16(b). The approximate dimensions are $120 \times 80 \times 10$ mm, which far exceeds what we have just calculated.

7.2.4 *Yagi-Uda Antenna*

Yagi-Uda is a basic form of array antennas, comprising of metallic rods or elements. Each element is a dipole antenna. The Yagi is a directional antenna, achieving a maximum gain of about 20 dB. This is a limitation, and requires more directors which increase its length. One of the popular applications of Yagis is for television reception. It is also used in array antennas. The Yagi is very easy to construct.

The invention of the Yagi antenna is attributed to the Japanese scientist Shintaro Uda in 1926. However, Prof. Uda published his invention in the Japanese language. Later, Hidetsugu Yagi, who is said to be Uda's student, publicised the invention in English language. This is the reason why the antenna is now called Yagi-Uda antenna. Both inventors were colleagues at Tohoku University in Japan.

As Fig. 7.17 illustrates, the Yagi-Uda (referred to simply as Yagi) has three types of elements: one element as the driven element (aka active element), two or more elements as directors and one element as the reflector. Power is supplied to the Yagi through the *driven element* over the feeder cable. The driven element is either a half-wave dipole or a folded dipole. The reflector and the directors draw energy from the driven element and re-radiate. They are therefore referred to as *parasitic elements*.

A Yagi antenna needs a single reflector, which is about 105% the length of the driven element. As illustrated in Fig. 7.17, if the length of the driven element is 0.5λ, then the length of the reflector of a practical Yagi-Uda is about 0.55λ. The reflector is positioned at a distance of about 0.23λ behind the driven element at the side away from the direction of the wave. Multiple reflectors can be used. However, they do not significantly increase the gain of the Yagi. The Yagis shown in Fig. 7.17 have vertical polarisation, as the elements are vertically mounted.

Each Yagi has two or more directors, which are shorter in length than the driven element. The length of the director closest to the driven element is about 0.45λ, and it is positioned about 0.15λ away from the driven element. The directors are positioned in front of the driven element at the side in the direction of the wave. Each director increases the gain by at most 1 dB. The contribution of a director to the gain of the Yagi decreases with its distance from the driven element. Each parasitic element is either a half-wave dipole or a full-wave dipole. The directivity and the gain of Yagi increases with the number of directors. For example, the 20-element Yagi with 600 mm length achieves a gain of up to 9 dB. Rules-of-thumb for the lengths of the elements of the Yagi are

$$\texttt{Length of driven element:} \quad L_\mathtt{e} = \frac{\lambda}{2} \tag{7.66}$$

`Length of nth reflector:`

$$L_{\mathtt{r},n} = (1 + 0.05n)L_\mathtt{e}, \ n = 1, 2\,, \tag{7.67}$$

`Length of nth director:`

$$L_{\mathtt{d},n} = (1 - 0.05n)L_\mathtt{e}, \ n = 1, 2\,, \tag{7.68}$$

Also, the distances between the elements are not arbitrary values. The wider the spacing between the elements are, the higher the achievable gain. Of particular importance are the positions of the first reflector and the first director relative to the driven element. Rules-of-thumb used are

`reflector-driven element spacing:`

$$ca.0.15\lambda - 0.25\lambda \tag{7.69}$$

`driven element-first director spacing:`

$$0.15\lambda - 0.125\lambda\,. \tag{7.70}$$

A parasitic dipole (reflectors and directors) is inductive if its length is equal to or slightly longer than $\lambda/2$. The phase of the voltage induced by the current in such a parasitic element leads the current. The opposite is true of shorter parasitic elements. A parasitic element whose length is shorter than $\lambda/2$ is capacitive. Therefore, the current flowing through

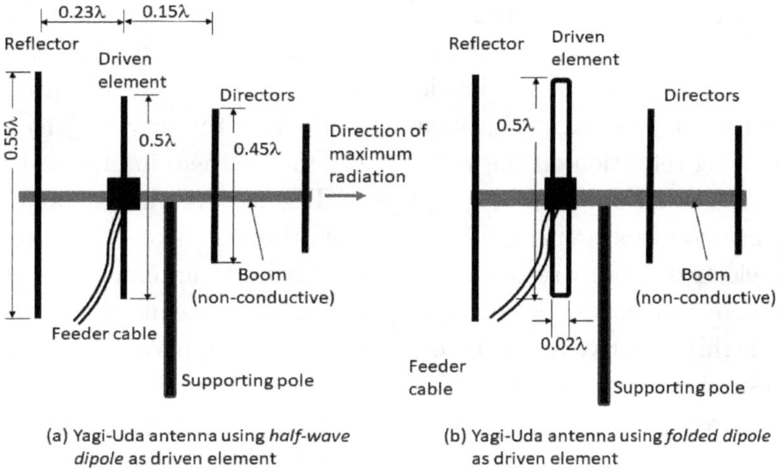

Fig. 7.17 Yagi-Uda antenna.

it leads the induced voltage. In general, reflectors behave electrically as inductors while directors behave as capacitors. The reflectors are made longer than the driven element so that they can re-enforce the signal generated by the driven element towards the latter. Also, the directors are so dimensioned that they re-enforce the intent to drive away the signal generated by the driven element away from the latter. This is the reason why the direction of maximum EM wave radiation is shown towards the right in Fig. 7.17.

7.2.5 *Aperture Antennas: Parabolic and Horn*

Aperture antennas are used by only communications systems operating at microwave frequencies. For example, point-to-point microwave links are used:

- To interconnect a pair of radio base stations in cellular mobile networks.
- For communications between an earth station and a space-borne satellite.
- In radar systems for target monitoring.

The free-space path loss increases with frequency and transmission distance. Systems operating at high microwave frequencies take some steps to reduce the impact of the path loss on signal detection quality. These include:

- Using high-gain antennas, which require the antennas to have very high directivity. High-directivity, on the other hand makes the width of the antenna's beam narrow. The narrower the beamwidth, the larger the cross-sectional area of the antenna. This is the reason why microwave communications systems use large antennas, usually of the aperture type.
- The transmission line connecting antennas to the radio transceiver should have minimum possible losses. The length of the feedline cable must be as short as possible to reduce attenuation. Among guided transmission media, waveguides have the least attenuation. This is the reason why waveguides are preferred over other types of transmission media such as coax cable at high frequencies. However, a waveguide can only be used for frequencies above its cut-off frequency. For example, a rectangular waveguide with the largest dimension x_0 has the cut-off frequency

$$f_C = \frac{0.5c_0}{x_0} \tag{7.71}$$

while a circular waveguide with internal radius r_0 has the cut-off frequency

$$f_C = \frac{0.9206c_0}{\pi r_0} \tag{7.72}$$

where $c_0 = 3 \cdot 10^8$ m/s.

Coax cable transfer signals in the form of electrical energy. Although current flows through the inside walls of the hollowconductor, waveguides transfer signals in the form of EM waves. Skin effect keeps the waves inside the hollow conductor so they propagate through the hollow portion using reflection off the walls of the waveguide.

- Receivers used in microwave systems must be very sensitive and selective. Highly sensitive receivers are able to detect signals with very low power levels, while highly selective receivers are able to pick the desired weak signal out of the clutter (i.e. noisy environment).

Antennas used for terrestrial microwave communications are usually aperture antennas, of which there are two basic types:

- Parabolic antenna (aka dish antenna) with circular shape (see Fig. 7.18(b)).
- Horn antenna with rectangular shape (see Fig. 7.18(c)).

Three types of aperture antennas are illustrated in Fig. 7.18. The shape of the reflectors of a parabolic antenna has the shape of a parabola. Because of their parabolic shape, parabolic antennas are among antennas with the highest directivity and thus highest gains. A high gain or directivity requires a narrow or small beamwidth. A smaller beamwidth also requires that the dimensions of the reflector are much higher than the wavelengths of the signals the antenna handles. This is the reason why parabolic antennas are used at high frequencies where the wavelengths are small as the huge size makes them impractical to use.

The feed of the antenna is the part that supplies the signal to the reflector. A common type of parabolic antenna is the Cassegrain antenna. Cassegrain antenna is a parabolic antenna in which the heavy complicated feed horn (aka feed antenna) is mounted on the main parabolic reflector of the dish rather than suspending it over the dish. This is seen as an advantage as the feed horn is heavy. Thus, Cassegrain antenna has two signal or energy reflectors, namely the secondary and primary reflectors, as illustrated in Fig. 7.18(b). The secondary reflector of Cassegrain antenna has a convex shape. A parabolic antenna is similar to a Cassegrain except that its secondary reflector has the shape of a concave and is called Gregorian antenna.

We discussed the gain and radiation pattern of parabolic antennas in previous sections. For dish antenna with diameter D, Eq. (7.38) becomes

$$G = \frac{4\pi\eta A}{\lambda^2} = \eta\left(\frac{\pi D}{\lambda}\right)^2. \tag{7.73}$$

A **horn antenna** has no resonant element, enabling it to operate over a wide range of frequencies. There are different types of horn antennas.

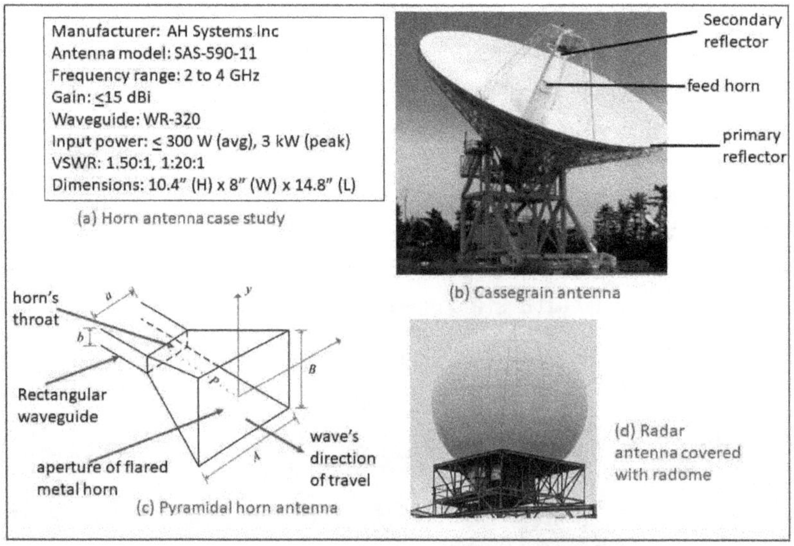

Fig. 7.18 A sample of aperture antennas.

We use as case study the pyramidal horn antenna in Fig. 7.18(c). It comprises two components: rectangular waveguide and flared metal horn. The rectangular waveguide is used as the feederline. Its impedance is about $Z_{\mathtt{rw}} \approx 50 \pm 10\Omega$. However, the impedance of free-space is about 377 Ω. Therefore, the unterminated rectangular waveguide would result in the reflection coefficient of about 0.75. Only a little portion of the energy fed through the waveguide would be radiated as EM waves. The flared waveguide is designed to be an impedance-matching network. The second function of the flared waveguide is to increase the directivity or gain by increasing the cross-sectional area. The gain of the rectangular horn antenna illustrated in Fig. 7.18(c) is

$$G = 8.1 + 10\log_{10}\left(\frac{AB}{\lambda^2}\right) \text{ [dBi]} \tag{7.74}$$

where A and B are the width and height of the flared flange respectively, as shown in Fig. 7.18(c). The half-power elevation (vertical) and azimuth (horizontal) *beamwidths* in degrees of a horn antenna are respectively

$$\phi_v = \frac{51\lambda}{b} \tag{7.75}$$

$$\theta_h = \frac{70\lambda}{a} \tag{7.76}$$

The shape of horn antennas is either rectangular or conical, while its polarisation is either circular or linear. A popular application of horn antennas is the feed for parabolic antennas.

Example

The rectangular horn antenna model SAS-590-11 operating in the band 2 GHz to 4 GHz has the dimensions: 10.4-inch height, 8-inch width and 14.8-inch length. Other parameters of the antenna are: 300 W average input power, 3 kW peak input power, 1.5 : 1 VSWR. Compute the gain of the antenna.

Solution

We use the wavelength at the mid-point of the band of frequency in which the antenna operates, which is 3 GHz. Hence, $\lambda = 0.1$ m. From Fig. 7.18(c), we can see that $A = 8$ inches and $B = 10.4$ inches. As the wavelength is in meters, we must convert A and B to meters as well. Thus, $A = 8'' = 0.2032$ m and $B = 10.4'' = 0.26416$ m. Plugging these values into Eq. (7.74) yields

$$G = 8.1 + 10\log_{10}\left(\frac{0.2032 \times 0.26416}{0.1^2}\right) \approx 15.4 \text{ dB}.$$

This antenna is rated with the maximum gain of 15 dBi.

7.2.6 *Array Antennas*

We have already been introduced to array antennas through the Yagi-Uda and log-periodic antennas. Every antenna has a specific radiation pattern. The radiation pattern of the basic antennas monopoles and dipoles are omnidirectional in their broadside, i.e., any plane perpendicular to their axes. Therefore, such antennas are unsuitable if either of the following conditions applies:

- We seek to minimise or even avoid interference, including both egress and ingress.
- We seek a higher gain antennas.

These conditions are satisfied by only directional antennas.

The original radiation pattern of an antenna is lost if multiples of them are combined. We can therefore combine multiple omnidirectional antennas in a manner in which their radiation patterns achieve nulls (i.e., destructive interference) and maxima (i.e., constructive interference) at desired directions. Beam shaping, as described above, is one of the purposes of *array antennas*.

Figure 7.19 lists seven more advantages in array antennas. *An array antenna is a set of antennas which synergise to realise these advantages.* Each antenna in an array is referred to as an *element*. A special form of array antennas is exploited in spatial diversity, and referred to as multiple input multiple output (MIMO). We discuss MIMO later in this chapter. An important application of array antennas is in the area of radio positioning systems. For example, antenna arrays are used to estimate the angle of arrival and direction of arrival of radio signals when tracking radio transmitters.

We can arrange the elements in an array antenna in one-dimensional or two-dimensional form.

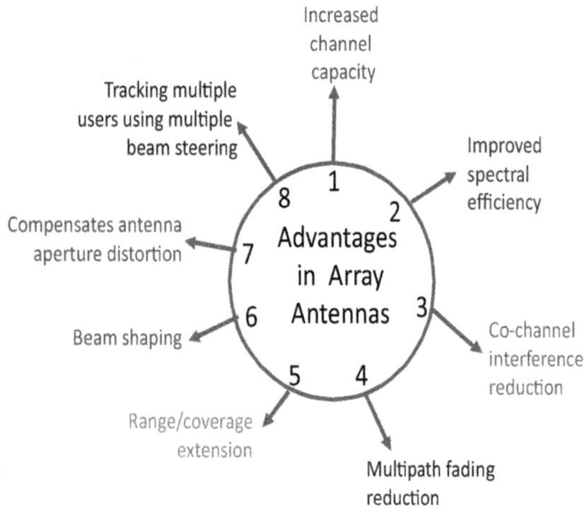

Fig. 7.19 Advantages in array antennas.

Figure 7.22(a) shows an array antenna constructed from four Yagi-Uda antennas. As discussed earlier, each Yagi-Uda antenna is also made of multiple dipole antennas.

There are several types of array antennas, which can be classified along different lines. A classification attempt is shown in Fig. 7.20. Based on their radiation pattern, array antennas can be subdivided into *broadside arrays* and *end-fire arrays*. How do they differ?

- *End-fire array* (EFA): As illustrated in Fig. 7.21(a), the direction maximum radiation generated by an end-fire array is in the same direction or parallel to the orientation of the array. Also, the array spacing is uniform, and the elements are fed by currents of the same magnitude but different phases.
- *Broadside array* (BSA): The main lobe or the direction of maximum radiation by the array is perpendicular to the orientation of the array (see Fig. 7.21(b)). Also, the array elements are excited by currents of the same magnitude and phase.

Based on their geometry, array antennas can be subdivided into conformal arrays, linear arrays and planar arrays. In a *conformal array*, the array elements are located in such a manner that they conform to a fixed but nonplanar geometry. As illustrated in Fig. 7.22, the centers of the elements of a *linear array* lie on a straight line. The centers of the elements in the array lie in a single plane in *planar*

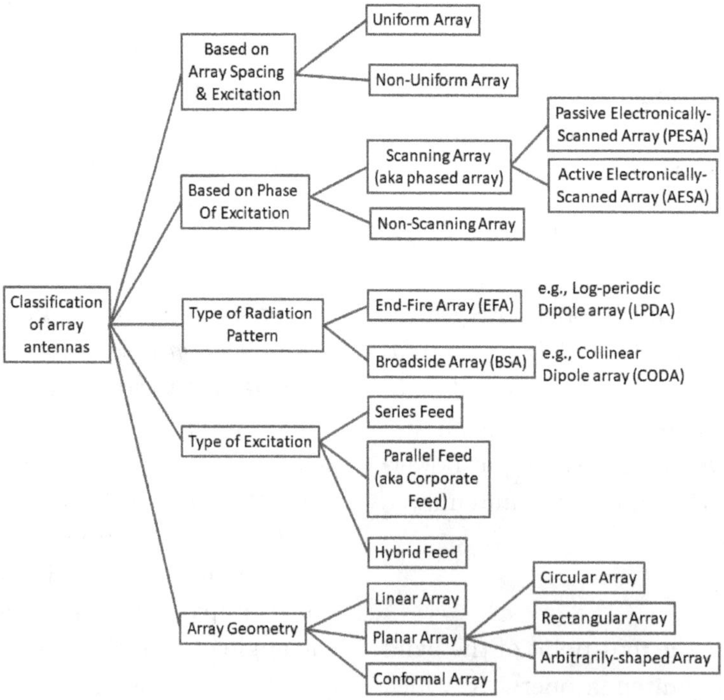

Fig. 7.20 A classification of array antennas.

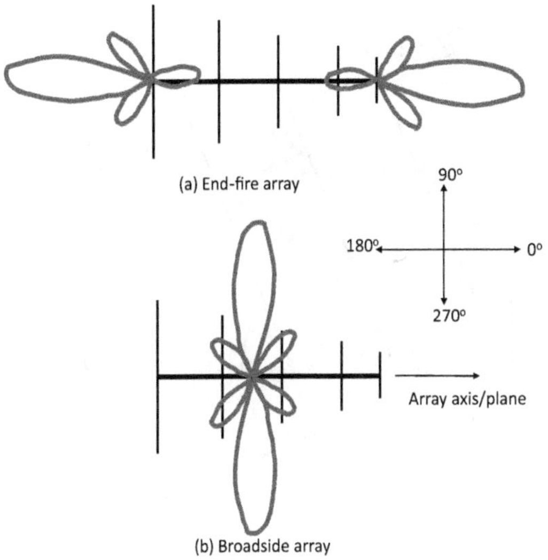

Fig. 7.21 Comparing broadside and end-fire arrays using Yagi-Uda antenna.

arrays. The elements in a planar array antenna can be arranged in a circular form to yield a *circular array*, rectangular form or an arbitrary form, such as a linear form to yield a *linear array*. Figure 7.22 shows a linear array.

The radiation pattern of an element in an array antenna is called *element factor*. The radiation pattern of an antenna array can be found by multiplying the element factor by a figure-of-merit called *array factor*, i.e.,

$$\texttt{array antenna radiation pattern} \\ = \texttt{element factor} \times \texttt{array factor} \quad (7.77)$$

Note that the *element factor* means the radiation pattern of an element in the array. If all the elements in the array are isotropic radiators, then the element factor is irrelevant as the radiation pattern is equivalent to the array factor. The array factor depends on many parameters, such as:

(1) The array geometry.
   (a) The number of elements in the array.
   (b) The distance between the array elements (i.e., relative positions of array elements), referred to as *array spacing*, and the orientation of the array elements.
(2) The phase and amplitude of the signal (i.e., current) fed to each element.

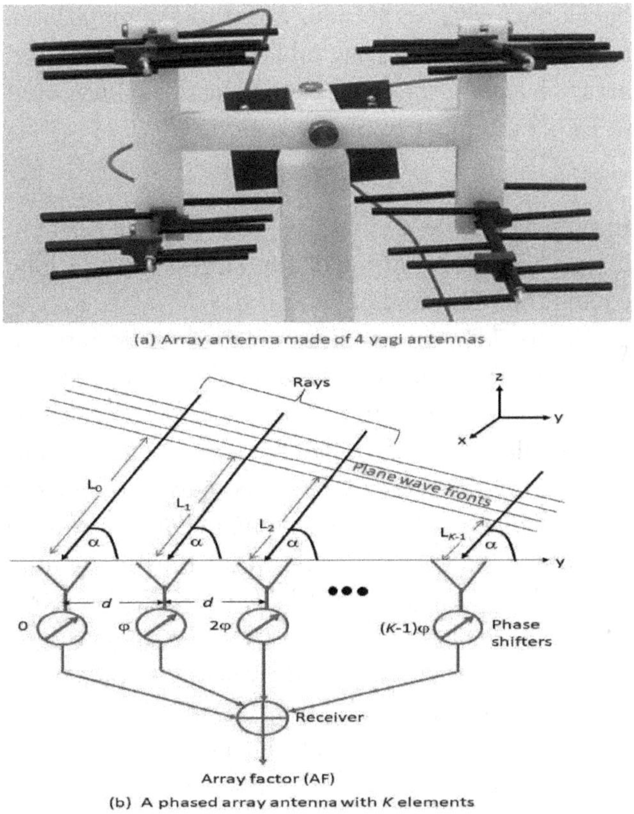

Fig. 7.22 A uniform linear phased-array antenna with K elements.

The array spacing can be made uniform or varying. The array antenna in Fig. 7.22 uses uniform array spacing. Also, the elements can be fed by the same signal source or fed independently. Based on array spacing and excitation of the elements, array antennas are classified into uniform arrays and non-uniform arrays. Besides the uniform array spacing, a *uniform array* uses identical elements which are excited with currents of the same magnitude and phase shift. As the elements are identical, they should all be dipoles or all be monopoles, for example.

The surface area of an array antenna is referred to as *aperture*. It is found from the element spacing and the number of elements. Increasing the aperture increases the directivity and thus gain of the array. This means that, in order to increase the gain, we must increase the element spacing, the number of elements or both. The limiting factor on the element spacing is the avoidance of grating lobes. *Grating lobes* are minor lobes (i.e., sidelobes and backlobes) with amplitudes equal to the main lobe. In order to avoid grating lobes in uniformly-spaced array elements, the element spacing d, as illustrated in Fig. 7.22, must fulfill the condition

$$d \leq \lambda/2. \tag{7.78}$$

In no way should $d \geq n\lambda$, where $n \geq 1$ is a positive integer. The magnitude of the largest grating lobe equals that of the main lobe if $d = \lambda$.

Figure 7.22 shows a linear phased array antenna with uniform array spacing and K elements receiving a plane wave which is oriented at the angle α to the plane of the antenna array. Such an antenna array is referred to as equally-space linear array (ESLA). All the K antenna elements receive the same plane wave. Therefore, the array antenna is called uniformly-excited ESLA (U-ESLA). However, the elements are liable to different phases as the wave travels different distances to reach each antenna element. Let φ_k be the phase of the EM wave reaching the kth array element. Under these assumptions, the geometry of the array in Fig. 7.22 becomes Fig. 7.23.

With the uniform array spacing of d and K array elements, the dimension of the array is

$$D = (K-1)d\,. \tag{7.79}$$

The maximum achievable directivity of a K-element array is K. This occurs at the broadside of the array. Therefore, the maximum array gain is

$$Gain \leq \eta D = \eta K \tag{7.80}$$

where η is the radiation efficiency. From the radiation in Fig. 7.23, we observe that

$$\Delta d = L_k - L_{k+1} = d\cos\alpha,\ k = 0, 1, 2, \ldots, K-1 \tag{7.81}$$

where L_k is the distance traveled by the wave exciting the kth array element. This path difference translates to the phase difference

$$\Delta\varphi = \frac{2\pi\Delta d}{\lambda} = \frac{2\pi d\cos\alpha}{\lambda} = \beta d\cos\alpha\,. \tag{7.82}$$

Therefore,

$$\varphi_k = \Delta\varphi + \varphi_{k-1} = k\beta d\cos\alpha,\ k = 0, 1, \ldots, K-1 \tag{7.83}$$

where we have assumed that the phases are measured with reference to the phase of the wave at the first array element where $k = 0$.

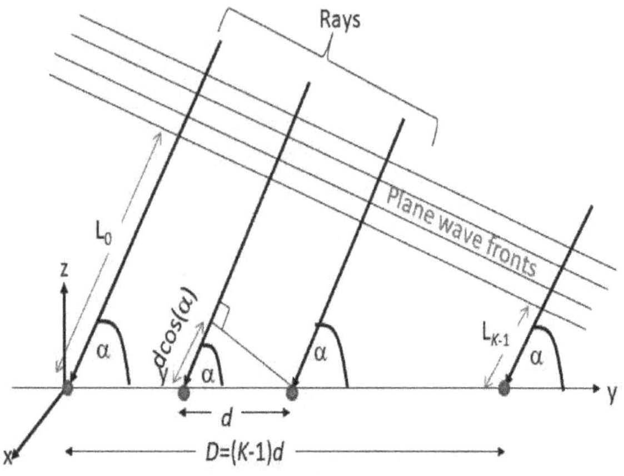

Fig. 7.23 Geometry of uniform linear phased-array in Fig. 7.22.

The array factor for the U-ESLA in Fig. 7.23 is defined as

$$AF = \sum_{k=0}^{K-1} e^{j\varphi_k} = \sum_{k=0}^{K-1} e^{jk\beta d\cos\alpha}\,. \tag{7.84}$$

Let us use a mathematical trick to find a closed form expression for AF. Let us multiply AF by $e^{j\beta d\cos\alpha}$ to obtain

$$\begin{aligned} e^{j\beta d\cos\alpha}\cdot AF &\overset{7.84}{=} \sum_{k=1}^{K} e^{j\varphi_k} \\ &= \sum_{k=0}^{K-1} e^{jk\beta d\cos\alpha}\,. \end{aligned} \tag{7.85}$$

We observe that the right side expressions of Eqs. (7.84) and (7.85) differ only in the first and last terms. Let us then subtract Eq. (7.85) from Eq. (7.84) to yield

$$(1 - e^{j\beta d\cos\alpha})AF = 1 - e^{jK\beta d\cos\alpha} \Leftrightarrow AF = \frac{1 - e^{jK\beta d\cos\alpha}}{1 - e^{j\beta d\cos\alpha}} \tag{7.86}$$

which simplifies to

$$\begin{aligned} AF &= \frac{\sin[K\beta d\cos(\alpha)/2]}{\sin[\beta d\cos(\alpha)/2]} e^{j(K-1)\beta d\cos\alpha/2} \\ &= \frac{\sin(K\phi/2)}{\sin(\phi/2)} e^{j(K-1)\phi/2} \end{aligned} \tag{7.87}$$

where

$$\phi = \beta d\cos(\alpha)\,. \tag{7.88}$$

The *normalised* magnitude of Eq. (7.87) is then

$$g(\phi) = |AF| = \frac{1}{K} \times \left|\frac{\sin(K\phi/2)}{\sin(\phi/2)}\right|. \tag{7.89}$$

The antenna radiation pattern is plotted using Eqs. (7.88) and (7.89). The E-field at a distance r from the array in the far-field region is

$$E(\theta = 90^\circ - \alpha, \phi) \propto AF \cdot \frac{e^{-j\beta r}}{r}. \tag{7.90}$$

The guidelines used to find the radiation pattern of the array are:

(1) The width of the minor lobes is $\theta_{ml} = 2\pi/K$, while the mainlobe's width is $2\theta_{ml} = 4\pi/K$.
(2) Compute the visible region, which is the region where

$$-1 \le \cos\alpha \le 1 \Leftrightarrow -\pi < \alpha < \pi$$
$$\Leftrightarrow -\beta d < \phi = \beta d \cos\alpha < -\beta d\,.$$

(3) Plot $g(\phi)$ as in Eq. (7.89) for a given d and K in the visible region where $-\beta d \le \phi = \beta d \cos\alpha \le -\beta d$.
(4) Plot a circle of radius βd right below $g(\phi)$.
(5) For each point in the visible region trace out the projected array diagram inside the circle.
   (a) For a given angle $-\pi < \alpha < \pi$, find the point where the radial line from the origin intercepts the circle with radius βd.
   (b) Then draw a vertical line from this point to cut the $g(\phi)$ curve. Read the value of $g(\phi)$ at this point, say $g_1 = g(\phi_1)$. This point is at the distance $g_1 = g(\phi_1)$ from the origin of the circle.
(6) Repeat the above steps for all values of K to obtain the array's radiation pattern.

Note that the array factor is maximum at the broadside of the array where $\alpha = 90^\circ$ or $\alpha = 270^\circ$. The width of the main beam decreases with increasing number of array elements. As K increases, the number of sidelobes increase, the level of sidelobes decrease, and there are $K - 2$ sidelobes in one period of $g(\phi)$. The location of the nulls of the antenna pattern can be found by setting the array factor to zero and solving for ϕ or α, i.e.,

$$g(\phi) = |AF| = \frac{1}{K}\left|\frac{\sin(K\phi/2)}{\sin(\phi/2)}\right| \overset{!}{=} 0$$
$$\Leftrightarrow \phi = \pm\frac{2\pi n}{K} \texttt{ or } \alpha = \cos^{-1}\left(\frac{n\lambda}{Kd}\right),\ n \in \mathcal{Z}\,. \tag{7.91}$$

The maximum of the main lobe occurs at $\phi = 0^\circ$. We can find the maxima of the array factor by differentiating Eq. (7.89) with respect to ϕ and finding the zeros of the differential, i.e.,

$$\frac{d}{d\phi}g(\phi) \overset{!}{=} 0 \Leftrightarrow \phi* = \pm\frac{2n-1}{K}\pi,\ n = 1, 2, \ldots\,. \tag{7.92}$$

These are the angles where the maxima of the sidelobes occur.

Example

Let us use a simple example to illustrate our discussions. Assume that $K = 4$, (i.e., an array with only four elements or dipole antennas), and the array spacing $d = \lambda/2$. Therefore, $\phi = \beta d \cos\alpha = \pi\cos\alpha$ and

$$g(\alpha) = |AF| = \frac{1}{4}\left|\frac{\sin(2\pi\cos\alpha)}{\sin\left(\frac{\pi}{2}\cos\alpha\right)}\right|,\ -\pi \le \alpha \le \pi\,.$$

Note that $g(\alpha = 90^\circ) = 4$. There are $K - 2 = 2$ sidelobes. The zeros of the pattern occur at

$$\phi = \pi\cos\alpha = \pm\frac{n\pi}{2},$$
$$n \in \mathcal{Z} \texttt{ or } \alpha = \cos^{-1}(n/2),\ n = 1, 2\,.$$

The maxima follows from Eq. (7.92) as

$$\phi^* = \pm 45^\circ,\quad \pm 135^\circ\,. \tag{7.93}$$

The radiation pattern is constructed in Fig. 7.24. We can observe the two sidelobes in relation to the main lobe.

A common type of antenna array is the collinear array, which is a type of broadside array. It is constructed from dipoles arranged in parallel and collinear. In order to avoid destructive interference of the radiations of the dipoles, there is 180° phase shift between them (i.e., the dipoles).

7.2.6.1 *Hansen-Woodyard End-Fire Array*

The phase of the ordinary EFA with the AF in Eq. (7.89) assumed that each element is excited with current of unit magnitude and zero phase. Now, let us assume that the phases are progressive as illustrated in Fig. 7.22. We can therefore write the current exciting the kth element as

$$i_k = e^{jk\varphi}\,. \tag{7.94}$$

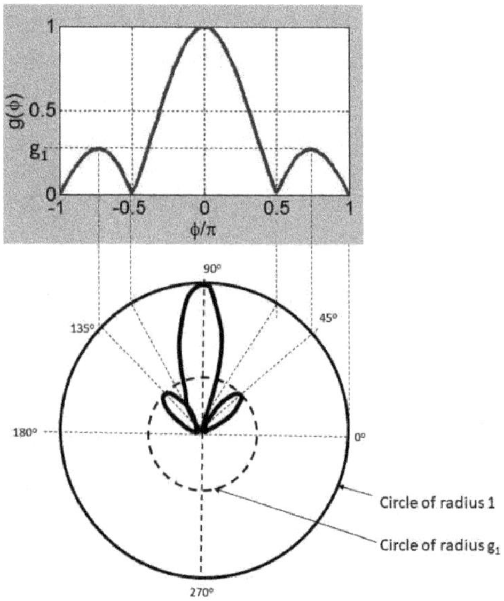

Fig. 7.24 Constructing the radiation pattern of a uniform linear array with $K = 4$, $d = \frac{\lambda}{2}$.

This changes the AF in Eq. (7.84) to the form

$$AF = \sum_{k=0}^{K-1} e^{j(\varphi_k + k\varphi)} = \sum_{k=0}^{K-1} e^{jk(\beta d \cos\alpha + \varphi)}. \quad (7.95)$$

Upon simplifying as before, we obtain

$$\begin{aligned} g(\alpha,\varphi) = |AF| &= \frac{\sin\left[\frac{K}{2}(\beta d\cos\alpha+\varphi)\right]}{K\sin\left[\frac{1}{2}(\beta d\cos\alpha+\varphi)\right]} \\ &\approx \frac{\sin\left[\frac{K}{2}(\beta d\cos\alpha+\varphi)\right]}{\frac{K}{2}(\beta d\cos\alpha+\varphi)} \end{aligned} \quad (7.96)$$

where we have used the approximation $\lim_{x\to 0}\sin(x) \approx x$ which holds here if $d \ll \lambda$. The maxima of $g\alpha)$ occurs at

$$\begin{aligned} \alpha = 0^\circ : \ \frac{K}{2}(\beta d\cos 0 + \varphi) &= -\frac{\pi}{2} \Leftrightarrow \varphi \\ &= -(\beta d + \pi/K) \\ \alpha = 180^\circ : \ \frac{K}{2}(\beta d\cos\pi + \varphi) &= \frac{\pi}{2} \Leftrightarrow \varphi \\ &= (\beta d + \pi/K)\,. \end{aligned} \quad (7.97)$$

Hansen and Woodyard found the exact maxima of Eq. (7.96) as

$$\begin{aligned} &\texttt{Maxima at } \alpha = 0^\circ : \varphi = -\left(\beta d + \frac{2.94}{K}\right) \\ &\texttt{Maxima at } \alpha = 180^\circ : \varphi = \left(\beta d + \frac{2.94}{K}\right). \end{aligned} \quad (7.98)$$

Equations (7.98) and their approximation (7.97) are referred to as Hansen-Woodyard conditions for improved gain or directivity over the ordinary EFA. Thus, in recap:

$$\varphi = \begin{cases} \pm\beta d, & \texttt{ordinary EFA} \\ \pm\left(\beta d + \frac{\pi}{K}\right) & \texttt{Hansen-Woodyard EFA} \end{cases}.$$

The *Hansen-Woodyard EFA* (HW-EFA) has a higher directivity than the ordinary EFA. The price for the increased directivity is the elongated array with closer-spaced elements, and higher sidelobes level.

7.2.7 *Panel Antennas*

These are compact antennas integrated into, or embedded in, wireless devices. With their high gain ($\approx 12 - 19dBi$) and focused beam, flat panel antennas can serve long range links. Unfortunately, they can be used only at high frequencies as their lengths are directly proportional to the operating wavelength. We can place a dipole in front of a reflector which is flat plate to realise a panel antenna. A popular application of panel antennas is at base stations of mobile cellular systems, where multiple dipoles can be mounted in parallel, one above the other, and connected to the same reflector to increase the gain.

7.2.8 *Microstrip Patch Antennas*

These are antennas built into devices using microstrip techniques and printed directly on electronic circuit boards. Owing to their implementation, these antennas are also referred to as *printed antennas*. Patch antennas are usually flat, the reason why they are also referred to as *planar antennas*. They are attractive because of their light weight, low cost and ability to host other electronic functions, such as amplifiers.

The first formal discussion on microstrip antennas is attributed to G. A. Deschamp in his presentation [Deschamp (1953)], but the first patent was awarded to H. Gutton and G. Baissinot in France in 1955. However, commercial practical application started not until 1970s after the development of printed circuit boards (PCBs) with controllable dielectric constant. The development of patch antennas to commercial use is attributed to Robert E. Munson [Munson (1972)], [Munson (1974)] and J. Q. Howell [Howell (1975)].

Printed microstrip patch antennas have a wide range of applications, including: Mobile Communication Systems, military aircraft, missiles, rockets and satellites, Direct Broadcast Satellite Service, radar, telemetry, radio altimetry, remote sensing and environmental instrumentation, satellite navigation receivers, biomedical radiators and intruder alarm systems.

The microstrip patch antenna usually has a planar geometry and consists of a ground, a substrate, a patch and a feed. A patch antenna can be fed in many different ways, such as microstrip line feed, coaxial probe feed, aperture coupling, electromagnetic coupling and coplanar waveguide (CPW) [Hasanujjaman *et al.* (2013)]. We study only microstrip-fed patch antennas, and specifically those with rectangular shape, just for brevity of discussions. Microstrip patch antenna arrays are among the most popular antennas used in GPS receivers.

Figure 7.25 illustrates a rectangular microstrip patch antenna. Let f_{r} be the resonant frequency of the antenna, ϵ_{r} be the dielectric constant (aka relative permittivity) of the substrate, and $c \approx 3 \cdot 10^8$ m/s be the speed of light, then the rectangular patch antenna has the width [Hasanujjaman *et al.* (2013)]

$$W = \frac{c}{\sqrt{2}(1+\epsilon_{\mathrm{r}})f_{\mathrm{r}}} \tag{7.99}$$

and the length

$$L = \frac{c}{2f_{\mathrm{r}}\sqrt{\epsilon_{\mathtt{eff}}}} - \frac{0.824h(0.3+\epsilon_{\mathtt{eff}})(0.264+W/h)}{(\epsilon_{\mathtt{eff}}-0.258)(0.8+W/h)} \tag{7.100}$$

where the effective dielectric constant is defined as

$$\epsilon_{\mathtt{eff}} = \frac{1+\epsilon_{\mathrm{r}}}{2} - \frac{1-\epsilon_{\mathrm{r}}}{2\sqrt{1+12h/W}}, \quad W > h. \tag{7.101}$$

If the dimensions of the radiating patch fulfills $\frac{L}{W} < 1$, then the dominant mode, which is the mode with the lowest resonant frequency, is the same as the mode in cavity and transmission line models. The resonant frequency of the dominant mode, i.e., TM_{01}, is

$$f_{\mathrm{r}} \approx 0.49\lambda_{\mathrm{d}} = \frac{0.49c}{L\sqrt{\epsilon_{\mathrm{r}}}} \tag{7.102}$$

where λ_{d} is the wavelength in the substrate.

An important step in the design of patch antennas is the selection of a substrate with the appropriate depth or thickness h and dielectric constant ϵ_{r}. Dielectric constants of substrates like ceramics can be as low as 4.0 and as high as 7000–9000.

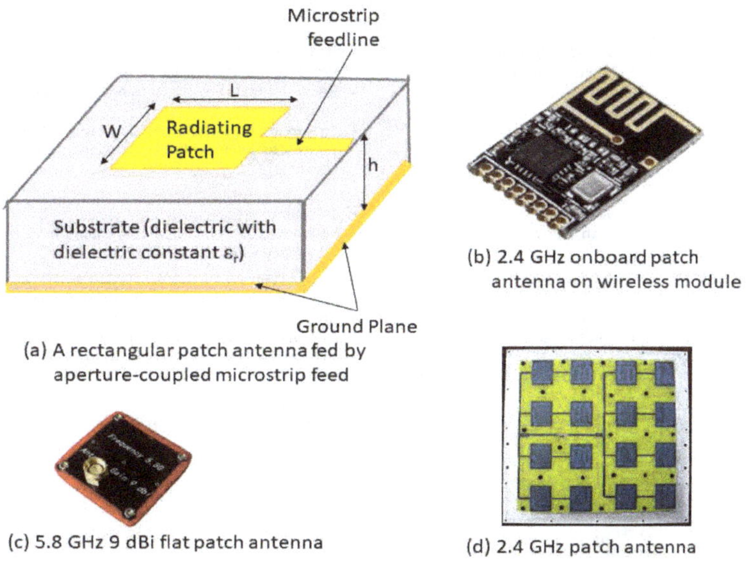

Fig. 7.25 Examples of patch antennas.

The bandwidth of the antenna is determined by h. The thickness is selected by an appropriate trade-off between:

- *Thicker substrate* is good for mechanically strong patch antenna, higher gain or higher radiated power, reduced conductor losses, and higher impedance bandwidth.
- *Thinner substrate* is good for lighter weight, lower dielectric losses, lower surface wave excitation and lower extraneous radiations from the probe field. Surface waves affect the antenna's radiation pattern and decreases the radiation efficiency.

A suitable substrate is glass epoxy with $4.4 \leq \epsilon \leq 4.7$.

Example

The dielectric constant of the substrate used to manufacture a rectangular microstrip patch antenna is $\epsilon_r = 4.4$. The dimensions of the radiating patch are 114 mm (width) and 50 mm (length). Compute the minimum frequency that the antenna can operate.

Solution

Equation (7.99) yields

$$f_r = \frac{c}{\sqrt{2}(1+\epsilon_r)W} \approx 344.6 \text{ MHz}\,.$$

7.2.9 *Antenna Manufacturing: Materials and Methods*

Antennas are built with highly-efficient electrically-conductive materials which are affordable. Copper is next to silver in conductivity. However, because of their higher price, aluminium and steel are being used in increasing manner. Therefore, the choice of antenna material is between copper, alumimium and steel (i.e., an alloy of iron, carbon and other metals). Wire antennas are usually made using forms of copper metal (e.g., FlexWeave, copperclad and hard drawn copper) and polystealth (i.e., bare copper-clad steel wire insulated with polyethylene jacket). Popular metals used to manufacture rod antennas include aluminium, copper and steel.

Figure 7.26 compares the common technologies used to manufacture antennas. Fillers used for the manufacture of three-dimensional antennas based on injection molding and selective metallization technologies include carbon fibers, glass fibers,

Techniques	Pros	Cons
Machined aluminium chassis (stamped metal design)	• Well-established technology • Good thermal heat dissipation	• Long machining time • Weight
Hand-built thermoset radomes (stamped metal design)	• Well-established technology • Large structures relatively easy to fabricate	• High touch-labor costs • Process repeatability due to touch labor • Prices are relatively constant regardless of production volumes • Shapes are typically smooth, and small features are difficult to create without secondary machining after molding
Selective plating using hand-trimmed masks (stamped metal design)	• Well-established technology • High conductivity or traces	• Difficult to achieve consistent parts. • Tolerances often do not meet requirements for high frequency applications
Antennas on printed circuit boards	• Well-established technology • Multi-layer boards relatively easy to achieve • Easy component integration onto RF traces	• Embedded antennas not the ideal shape. • Designs limited to two-dimensional structures • Conformal antennas are difficult to build
Injection molding and selective metallization technologies	• 3D antenna to obtain optimal gain and bandwidth. • Compact, lower weight and smaller aerodynamic drag • Possibility of better mechanical robustness to withstand harsh environments	• Relatively immature technologies

Fig. 7.26 Comparison between antenna manufacturing methods [Curran (2015)].

hollow microspheres, graphene, carbon nanotubes and foaming agents [Curran (2015)].

7.2.10 *Check Your Understanding*

Students are encouraged to check their understanding of the content of this section with the following questions prior to proceeding to the next section.

(1) How do the two units, dBi and dB, differ from each other? In which ways are they similar?
(2) Classify the following types of antennas into directional and omnidirectional antennas: dipole, sectored antennae, Yagi-Uda, monopole, phased array, horn antenna, parabolic dish antenna, log-periodic antenna and patch antenna.
(3) What is the difference between dBi and dBd? Write an analytical expression relating the two.
(4) A radio broadcasting system operates at 88.7 MHz and transmits signals with power 8 W.
   (a) Find the length of the half-wave dipole antenna which can be used to tune to this station.
   (b) Compute the length of the monopole antenna equivalent to the dipole found in Part (a) which can tune to this station.
   (c) Find the power density at a distance 2 km away from the broadcasting station.
   (d) Compute the root-mean-squared electric field intensity at a distance 2 km away from the broadcasting station.
(5) What is aperture antenna?
(6) How many types of aperture antennas are discussed in this section?
(7) Compute in decibels the range of gains achievable by a 2-m dish antenna used in a microwave system operating at 8 GHz.
(8) The rectangular horn antenna operating from 50 GHz to 75 GHz has the dimensions: 29.46-mm height, 45.47-mm width and 48.26-mm length. Compute the nominal gain of the antenna.
(9) Design an outdoor Ku-band dish antenna for the reception of satellite TV signals. The nominal gain of the antenna should be at least 36.7 dB and operates at 12.5 GHz.
   (a) Find the beamwidth and diameter of the antenna. Assume a radiation efficiency of 0.60.
   (b) Compute the distance from the antenna that you have designed where its far-field region begins.
(10) Assume that the antenna designed in Problem 9 is fed by a transmitter with 150-kW output power. Compute the minimum distance that someone has to be away from the antenna to be safe from electromagnetic radiation. *Hint.* The power density should be less than 10 mW/cm^2.
(11) A 100-m feeder cable connecting a source to an antenna with the impedance 50 ohm has the characteristics: 2.5 pF/m capacitance and 14.06 nH/m inductance. Compute the:
   (a) Characteristic impedance of the feeder cable.
   (b) Input impedance of the antenna.
(12) **Design laboratory**: design a 5-element Yagi antenna to receive free-to-air television signals at the frequency 226.5 MHz. Assume that the antenna has only one reflector.
(13) Consider a lossy quarter-wave monopole antenna with the ohmic resistance 12 Ω mounted on a perfect conductor and fed by 60 Hz 120 V ac source with an internal impedance of 20 Ω.
   (a) Compute both the magnitude and the direction of the electric field at the broadside of the antenna.

(b) Find both the magnitude and the direction of the magnetic field at the broadside of the antenna.
(c) Compute the average power dissipated by the antenna.

(14) The current flowing through the port of a half-wave dipole antenna is 80 mA and oscillates at 108 MHz. Assume that you are at a distance of 500 m away from the antenna and in the broadside plane of the antenna.
(a) Calculate the total power radiated by the antenna.
(b) Compute the average power density that you can measure.
(c) Find the magnitude and the direction of the electric field.
(d) Find the magnitude and the direction of the magnetic field.

(15) Assume that you need to design a log-periodic antenna with 5-6 dB gain to operate in the frequency band 87.5 MHz to 108 MHz. Find the lengths of the longest and the shortest elements of the antenna.

(16) The current with the root-mean-squared magnitude of 50 mA flows into an antenna through its terminals. Calculate the power density at a distance of 1 km broadside to the antenna.

(17) Consider the phased-array antenna in Fig. 7.22(b). Compute the scan angle α_0 if $K = 4$ and $\phi = 45°$, $d = \lambda/4$.

(18) Two identical dipole antennas are used to construct an array antenna with its plane parallel to the earth surface. The array spacing is d. Both antennas are fed with currents of equal amplitude. If $d = \lambda/2$ and $\alpha = 45°$,
(a) Find the nulls of the array radiation pattern.
(b) Find the maxima of the array radiation pattern.
(c) Sketch the radiation pattern of the array in a plane perpendicular to the earth.

(19) Assume that the array factor of a 2-element array is $g(\phi) = \sin(\pi \cos\phi/2)$. Plot or sketch the pattern of the array as a function of ϕ.

(20) An active patch antenna has the features: 1575.42 ± 1.02 MHz center frequency, 1.5 VSWR at center frequency, dimensions 15 × 15 × 6.8 mm($L \times W \times h$), 1.5 dBic gain, 5 MHz bandwidth, RHCP, 18-dB hosted low-noise amplifier with 1.5 noise figure. This antenna has a wide application areas including mobile phones, laptops, PDAs, automotive navigation, marine buoys, personal tracking, surveying equipment, healthcare and medical monitoring devices. The substrate is a ceramic dielectric material. Compute the dielectric constant of the substrate.

7.3 Introduction to Spatial Diversity Techniques

In general, diversity in communications is a technique of spreading a signal in some dimension to make it robust against the impairments of the transmission medium. There are three basic types of diversity in wireless communications:

- *Temporal diversity* (TD): aka time diversity spreads signals in time domain. Examples of TD techniques are channel coding and interleaving (or scrambling). These are used to make the signal robust against channel impairments.

- *Frequency diversity* (FD): spreads signals in the frequency domain over wider bandwidths. This can be achieved by transmitting the same signal over multiple carrier frequencies at the same time or at different times as in frequency-hopping spread spectrum or spread signal over a wider bandwidth as in CDMA.

- *Spatial diversity* (SD): transmit the same signal over multiple antennas. For this reason, SD is also referred to as antenna diversity.

This section explores spatial diversity, while the other diversities are treated in other books, such as [Gyasi-Agyei (2019a)].

A conventional wireless system which uses a single antenna at both the receiver and transmitter. This is referred to as single-input single-output (SISO) system, as only one signal is sent from the transmitter into the wireless channel and only one antenna intercepts the transmitted signal. Spatial diversity is an advancement over SISO in which multiple antennas are used at the transmitter, at the receiver or at both ends of the wireless system.

The use of multiple antennas synthesises multiple wireless propagation paths. Each of the signal propagation paths has a different behaviour, some may be in fade at a time that others are good. The technique is referred to variously as **spatial diversity**, space diversity, antenna diversity or just diversity. Note that we can mount multiple antennas on a single radio transmitter and/or receiver to realise spatial diversity. Thus, MIMO is not the same as antenna diversity. However, one of the applications of MIMO is antenna diversity.

The use of multiple antennas at the transmitter, receiver or both is called antenna diversity, space diversity or spatial diversity. As shown in Fig. 7.27, antenna diversity can be only transmit diversity (aka MISO), receive diversity (aka SIMO) or both transmit and receive diversities (aka MIMO). In *transmit diversity* we use multiple antennas at the transmitter, which is equivalent to the MISO discussed above. This can be implemented without the transmitter knowing the channel states. The two common types of transmit diversity techniques are transmit beamforming and space-time coding (STC). LTE applies transmit diversity on its common control channels and broadcast channels.

The origins of spatial multiplexing is widely attributed to the work of A. Roger Kaye and D.A. George in 1970s on multi-channel digital transmission systems [Kaye and George (1970)]. Branderburg and Wyner reported many research results relating MIMO in 1974 [Brandenburg and Wyner (1974)]. Van Etten researched into maximum likelihood reception of multi-channel transmission systems [Etten (1976)].

Multiple-input multiple-output (MIMO) is complementary technology to the OFDM technology that we studied in Chapter 4 on digital modulation. These two topics are among the foremost important and current technologies in wireless communications. Therefore, students and practitioners in the field must have a thorough understanding of both technologies. A conventional wireless communication system uses a single radio with a single antenna at both ends of a communication link, i.e., transmitter (TX) and receiver (RX). A MIMO wireless system, however, uses multiple radios each equipped with an antenna more for transmission and/or reception of signals, in the general case.

In the general MIMO system, multiple antennas at the transmitter transmit (or input) signals into the wireless transmission medium and thus the term *multiple input*. The multiple signals injected into

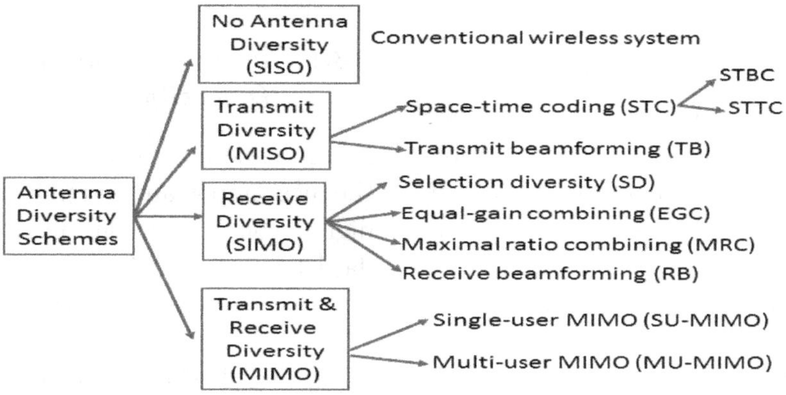

Fig. 7.27 Types of antenna diversity schemes.

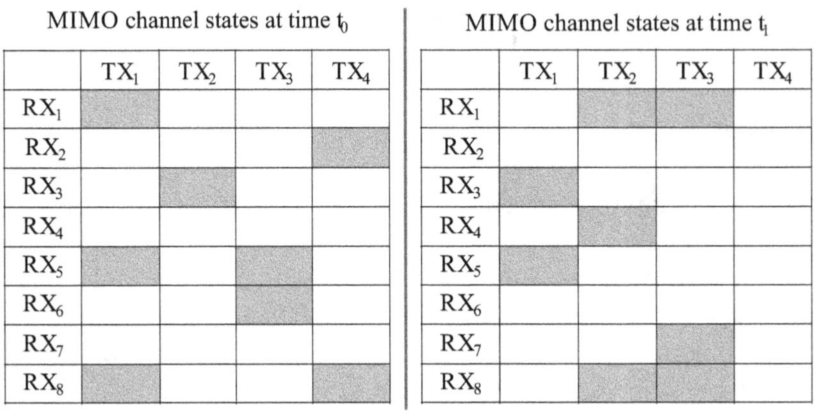

MIMO channel states at time t_0
MIMO channel states at time t_1
TX1 TX2 TX3 TX4
RX1 RX2 RX3 RX4 RX5 RX6 RX7 RX8

Fig. 7.28 States of a MIMO's channels at two different times t_0 and t_1.

the channel from the multiple antennas can be the same or different. A generalised MIMO system also uses multiple antennas to receive the same signal or multiple signals from the transmission medium, the reason for the term **multiple output**. Thus, we input multiple signals into the channel and receive multiple signals from it. The reason for the name is that we expect a different fading wireless channel between each pair of transmit and receive antennas: some can be in deep fade while others are in their most conducive states for signal transfer. This is illustrated in Fig. 7.28 [Gyasi-Agyei (2005)]. However, the fading situation of the various MIMO channels may be correlated or independent if there is a line-of-sight (LOS) signal transmission. It must be emphasised that MIMO works effectively only in non-LOS situations with multipath signal propagation causing frequency-selective fading on wideband channels.

The scenario illustrated in Fig. 7.28 is for a MIMO system with eight receive and four transmit antennas. Thus, we can have $8 \times 4 = 32$ channels as there are 32 transmit and receive antenna pairs. Of course, depending on the fading situation, not all these 32 channels may be uncorrelated. Let us denote the index of the channel between receive antenna l and transmit antenna k as lk, and its corresponding gain as h_{lk}. Each box in Fig. 7.28 indicates one channel, and the shaded boxes are those channels that are in fading and thus not in conducive states for signal transfer. We can observe that, at time t_0, the channels 11, 24, 32, 51, 52, 63, 81 and 84 are in deep fading and thus it would be better not to use the corresponding pairs of antennas for signal transfer. Interestingly, each of the receive antennas 4 and 7 has a good channel state with all the 4 transmit antennas at time t_0 but not at time t_1. As the wireless channel is in general spatio-temporally varying, the conditions of the MIMO channels at time t_1 are different from those at time t_0, as can be observed.

Knowledge of the channel state information at the transmitter (CSIT) enables advanced signal processing at the transmitter to improve the system throughput. Thus, one of the tasks in the maximum exploitation of MIMO techniques is the estimation of the channel state information at the transmitter. The only obvious *disadvantages of MIMO* are the increased transceiver complexity and associated cost. This is attributed to the multiple antennas and their associated increased signal processing.

7.3.1 *Practical Applications of Antenna Diversity*

As of September 2016 MIMO is counted as one of the most important and popular technologies in the digital communications industry. It is the standard transmission in modern systems. For example, the IEEE 802.11ax wireless local area network (WLAN) standard which is still in research laboratories and whose products are not expected to appear in the market until 2019 uses both MIMO and multiuser MIMO (MU-MIMO). In MU-MIMO a group of users with mobile terminals having multiple antennas communicate with another group of users using mobile terminals having multiple antennas. The MU-MIMO in IEEE 802.11ax is based on OFDMA. In single-user MIMO (SU-MIMO), however, an individual entity (person or robot) communicates with a single other entity, although both entities have multi-antenna radio terminals. With the combination of OFDM, wideband channels of width 20, 40, 80 or 160 MHz and 1024-QAM modulation, the IEEE 802.11ax is expected to deliver a whooping 10 Gbps data rate in the 2.4 and 5 GHz frequency bands when the channel quality is most conducive for information transfer.

MIMO is also used in both the 3G standard LTE (i.e., GSM Long-Term Evolution) and 4G standard LTE-Advanced for mobile communications. Other applications of MIMO include Wi-Fi standard (IEEE 802.11n), WiMAX (Worldwide Interoperability for Microwave Access) or IEEE 802.16 (a 4G proposal), power-line communication for 3-wire installations as part of ITU G.hn standard and HomePlug AV2 specification. The wide popularity of MIMO means that communications professionals and students need expert knowledge of it. In IEEE802.11ac, for example, 4×4 and 8×8 MIMO have been anticipated. Such a MIMO architecture allows up to four pair of users to communicate over the same radio frequency. The signals achieve the required orthogonality through the spatial dimension.

7.3.2 *Advantages in Antenna Diversity*

Figure 7.31 illustrates a MIMO system with K transmit antennas and L receive antennas. Such a system is referred to as $K \times L$ **MIMO channel**. The space between an antenna element in the transmitter and an antenna element at the receiver is called a channel. Thus, there are $K \times L$ channels. Each channel is made of the natural ether, natural obstructions (e.g. mountains, animals, trees and human beings) and artificial obstructions (e.g. buildings, cars and airplanes). **What is the purpose for the multiple antennas?** There are at least five reasons or advantages for this, as illustrated in Fig. 7.29. First, MIMO can be used to increase the signal transmission range or coverage at a fixed data rate [Friedlander (2005)].

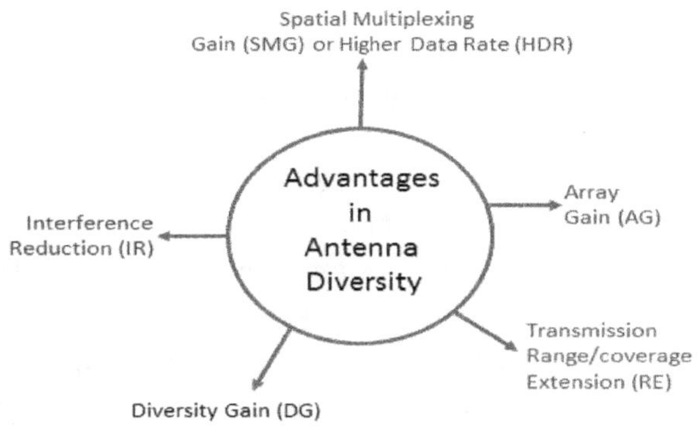

Fig. 7.29 Advantages in antenna diversity or smart antennas.

7.3.2.1 *Spatial Multiplexing Gain and Higher Data Rate*

Assume that we want to transmit the sequence of K symbols $s_0, s_1, \ldots, s_K$. We can choose to transmit just one symbol in a timeslot as done in conventional systems. That would take a time period of KT_s, assuming that the symbol period is T_s. Another way is to divide the K symbols into $n = K/N_T$ groups of N_T symbols and transmit each group simultaneously using N_T transmit antennas. That would require the total transmission time of $\frac{K}{N_T}T_s$, a reduction by a factor N_T compared to the conventional system as the data transmission rate increases from R_b to $N_T R_b$. The increase in data transmission rate or the reduction in the transmission time is referred to as *spatial multiplexing gain.* We call it so because we are using multiplexing in space instead of wavelength (as done in WDM) or time (as done in TDM) or frequency (as done in FDM). The spatial multiplexing gain is proportional to the number of antennas. For example, if $K = L = N_a$ then the gain is directly proportional to N_a. This means we can transmit N_a symbols in parallel over a maximum of $N_a \times N_a$ multipath fading wireless channels.

Spatial multiplexing gain is a better alternative to traditional methods of increasing channel capacity. Traditionally, we seek to increase the channel capacity by transmitting higher power, using larger bandwidth, reducing noise or any combinations thereof.

7.3.2.2 *Array Gain*

This is also called **antenna gain**. We can increase the average signal-to-noise ratio (SNR) at the receiver or decrease the transmit power at the transmitter without degrading the detection performance if we use multiple antennas to transmit the same signal. This is because: (a) we concentrate the antenna's gain on a smaller area, and/or (b) the radio link between a given pair of antennas have different characteristics: some can be in fade while others will be at their best quality at a given time. The array gain increases linearly with the number of antennas used and thus it is larger than the diversity gain for massive MIMO systems. Massive MIMO systems employ tens or hundreds of antennas, and thus are more amenable for implementation on fixed radio nodes (e.g. radio base stations and evolved node Bs) than on mobile handsets.

The problem in this technique is that the receiver must be privy of the states of the channels between each pair of antennas in order to at least fully exploit the array gain. For known channel states the receive can combine the multiple signals received coherently to maximise the received SNR for better detection. For example, assuming that $K = 1$ (i.e. SIMO) in the MIMO system illustrated in Fig. 7.31, we can achieve the output SNR γ_{out}

$$\gamma_{out} = \gamma_1 + \gamma_2 + \cdots + \gamma_L \tag{7.103}$$

if we are able to coherently receive the signal at all antenna elements, where $\gamma_l, l = 1, 2, \ldots, L$ is the signal-to-noise ratio at receive antenna l. The achievable array gain is a function of the singular value of the dominant channel matrix, as well as the number of the receive and transmit antennas.

7.3.2.3 *Interference Reduction*

In an attempt to increase the network capacity and spectral efficiency of a given wireless bandwidth, frequency reuse was born. This requires a geographical area to be covered by a wireless service provider to be subdivided into regions called *cells*. The set of neighboring cells is called cluster with the size of a cluster being the *cluster size*. The cluster size (C) is an important design parameter in cellular communications and should be chosen to fulfill

$$C = i^2 + j^2 + i \cdot j \tag{7.104}$$

where i and j are positive integers. For example, Fig. 7.30 shows a cellular system with the cluster size of seven. The *frequency reuse factor* is the reciprocal of the number of cells which cannot use the same set of frequencies for signal transmission. To appreciate the importance of frequency reuse, consider an operator which has been allocated a bandwidth of B Hz and the system requires W Hz for signal transmission. For example only $B = 79$ MHz is available in the 2.4 MHz ISM band in some countries. The Wi-Fi standard IEEE802.11n signal requires $W = 20$ MHz. Thus, we can have a maximum of only three non-overlapping frequencies which can never be sufficient for the users of the systems. A frequency reuse remains the only choice to support a capacity larger than three users.

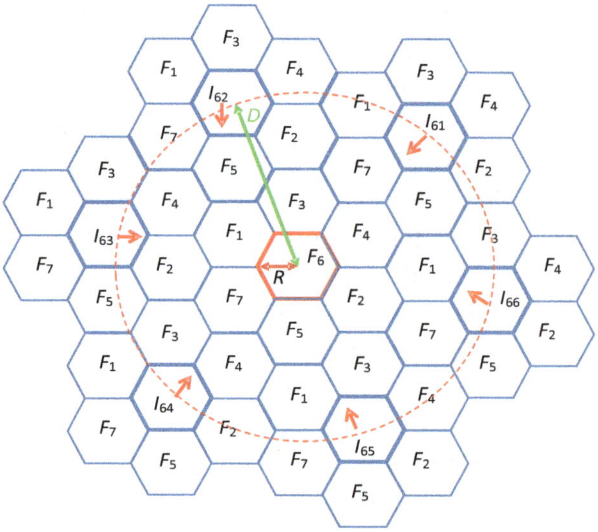

Fig. 7.30 Seven-cell frequency reuse and CCI in cellular networks.

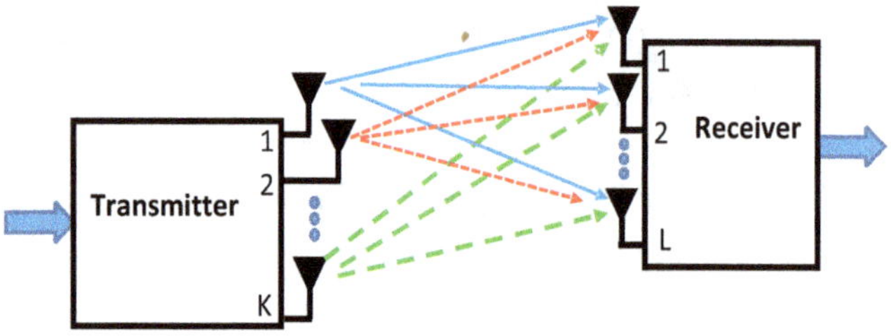

Fig. 7.31 K x L MIMO system model.

The available bandwidth is divided into frequencies which are also divided into sets. For example, we have the seven sets of frequencies $F_1, F_2, \ldots, F_7$ in Fig. 7.30. Each cell is served by a set of frequencies. We note that the same set of frequencies is used in different cells which are not direct neighbours. Cells using the same set of frequencies are called co-channel cells. The geographical distance between the centers of a pair of co-channel cells is called frequency **reuse distance** (D), which is a cellular systems' design parameter. It relates to the cluster size via the formula

$$D = \sqrt{3C}R = \sqrt{3(i^2 + j^2 + i \cdot j)}R \qquad (7.105)$$

where C is the cluster size and R is the radius of a cell which is assumed to be hexagonal in form. The disturbance caused between a pair of cells using the same set of frequencies is called *co-channel interference* (CCI). The main penalty in frequency reuse is the creation of CCI. For example, with respect to the cell 6 shaded red in Fig. 7.30 there are six first-tier co-channel interferers. These are the six cells within the dotted circle each of which uses the set of frequencies marked F_6. The first-tier CCI is the strongest as it is the closest to the cell of interest. CCI is a form of crosstalk that arises from two different radio transmitters using the same radio frequency. Just in passing, the disturbance between the set of neighbouring channels within the same cell is called *adjacent channel interference* (ACI). In cellular communications systems CCI is more of a problem than ACI as the latter can be mitigated using appropriate filtering.

Co-channel interference impairs signal detection just like any other type of noise. Therefore, great efforts are made to mitigate it. The larger the separation between co-channel cells the lower the CCI. However, increasing the distance between co-channel cells also reduces the efficient usage of the expensive wireless bandwidth thus reducing system capacity. Wireless designers have a competing design metric to reduce CCI and still achieve maximum system capacity. Thus, CCI mitigation is a major task in cellular systems in an attempt to gain maximum system capacity with a given bandwidth. CCI mitigation techniques have the central goal of making the power in the signal of interest to be relatively larger than the total power in the co-channel interferers at the receiver. The metric used is the maximum signal-to-interference ratio (SIR). Considering only the first tier of co-channel interferers in Fig. 7.30, the SIR is

$$SIR = \frac{C_6}{\sum_{k=1}^{6} I_{6k}} \qquad (7.106)$$

where I_{6k} is the interference power level of first-tier co-channel cell k and C_6 is the power level of the signal in the reference cell 6. CDMA-based systems do not experience CCI per se, but code channel interference which is mitigated using power control. Traditional CCI mitigation strategies include (a) the use of directional antennas and (b) increasing the reuse distance at the expense of bandwidth efficiency. Reducing CCI through MIMO exploits the differences in the *spatial signatures* of the signal of interest and those of the co-channel interferers. The challenge is that the receiver must know these spatial signatures to mitigate the CCI. The term **spatial signature** or **signal direction** is used to refer to the vector $[h_{1k}, h_{2k}, \ldots, h_{Lk}]^T$ of channel gains from the kth transmit antenna to multiple receive antennas.

7.3.2.4 *Diversity Gain*

Diversity gain can be achieved in both traditional wireless systems using a single antenna at both transmit and receive ends (aka SISO-based systems) and MIMO-based systems. However, unlike MIMO-based systems, SISO-based systems can exploit only *time diversity* at the cost of reduced transmission time or data rate or *frequency diversity* at the cost of increased transmission bandwidth. MIMO technology enables the improvement of system (i.e. reduction in bit error rate or BER) at a given signal-to-noise ratio (SNR) or achieve the same BER at a lower SNR. We achieve diversity gain by reducing the *fading margin* required to achieve a given BER by using multiple antennas. For example, a wireless system using binary phase-shift keying (BPSK) modulator system can achieve 1% BER with the fading margin of 9.5 dB using a single antenna as in traditional systems. However, we can employ two or four antennas to reduce the fading margin to 4.3 dB and 1.9 dB, respectively. Thus, the respective diversity gains achieved are 5.2 dB and 7.6 dB. The diversity gain can dominate the array gain if the number of antennas is small, it saturates at the fading margin. There are two types of diversity gains: transmit diversity gain and receive diversity gain. These are discussed later.

7.3.3 *Multi-User MIMO (MU-MIMO)*

All the spatial streams in a MIMO can be used to serve only a single user, leading to single-user MIMO (SU-MIMO). In order to increase system capacity, the different spatial streams or beams can be used to serve different users simultaneously. This advanced form of MIMO is called multi-user MIMO (MU-MIMO). Compared with SU-MIMO, MU-MIMO trades off data rate for system capacity.

7.3.4 *Massive MIMO*

This is MIMO with about tens to thousands of antenna elements at the transmitter (usually at the radio base station) to exploit spatial multiplexing to deliver gigabits of data rate in future mobile communication systems, such as 5G. A conventional MIMO uses about two to four antennas at the transmitter or receiver. There is no straight cut definition of what massive MIMO is. However, massive MIMO systems employing 96 to 128 antennas have been demonstrated by ZTE and Huawei, for example. We studied early in this chapter that the size of the antenna should match the wavelength of the signal that it handles. This raises a restriction on the minimum frequencies that can be used with MIMO. Too low frequencies (i.e., too long wavelengths) would not provide enough space to host many antennas. 5G and other systems poised to implement massive MIMO therefore have to use frequencies higher than SISO or standard MIMO based technologies. A large-scale massive MIMO implementation is poised for 5G. 4G/LTE-A systems and Wi-Fi (e.g., IEEE 802.11n) implement only the standard MIMO, with a few massive MIMO implementation in 4G.

7.3.5 *Check Your Understanding*

Students are encouraged to check their understanding of the content of this section with the following questions prior to proceeding to the next section.

(1) Discuss the general rationale behind diversity in information transfer systems.
(2) Describe the similarities and differences between the four types of antenna diversities.
(3) State a synonymous name for antenna diversity.
(4) List two practical applications of antenna diversity.
(5) State three advantages in antenna diversity.
(6) Compare and contrast spatial multiplexing gain with array gain.
(7) Antenna diversity can be used to reduce co-channel interference. Describe how this is achieved.
(8) Describe the differences between multi-user MIMO and massive MIMO.

7.4 MIMO System Model and Channel Capacity

Figure 7.31 depicts the generalised MIMO system whose transmit antenna has K elements while receive antenna has L elements. Let the signal emitted by the K antennas into the wireless channel be represented by the vector $\mathbf{s}(t) = [s_1(t), s_2(t), \ldots, s_K(t)]^T$, and the signal vector received by the L receiving antennas be represented by $\mathbf{r}(t) = [r_1(t), r_2(t), \ldots, r_L(t)]^T$. We represent the AWGN signal vector by $\mathbf{w}(t) = [w_1(t), w_2(t), \ldots, w_L(t)]^T$. The dimension of noise vector $\mathbf{w}(t)$ is the same as that of the number of receiving antennas. We can formulate the relationship between the input and the output signal vectors of the MIMO system in matrix-vector form as

$$\mathbf{r}(t) = \mathbf{H}(\tau, t) \otimes \mathbf{s}(t) + \mathbf{w}(t) \tag{7.107}$$

where $\otimes$ is the convolution operator and the $L \times K$ *channel matrix* (aka *channel transfer function*) is

$$\mathbf{H}(\tau,t) = \begin{bmatrix} h_{11}(\tau,t) & h_{12}(\tau,t) & \cdots & h_{1K}(\tau,t) \\ h_{21}(\tau,t) & h_{22}(\tau,t) & \cdots & h_{2K}(\tau,t) \\ \vdots & \vdots & \cdots & \vdots \\ h_{L1}(\tau,t) & h_{L2}(\tau,t) & \cdots & h_{LK}(\tau,t) \end{bmatrix}. \tag{7.108}$$

The kth column of the channel matrix is also referred to as the *signature* induced by the kth transmit antenna across the L receive antennas. The system described by Eq. (7.109) is referred to as $K \times L$ MIMO system. The channel element $h_{lk}(\tau,t)$ is a complex random variable which describes the effects on the signal transmission of the space between the lth receive antenna and the kth transmit antenna. $h_{lk}(\tau,t)$ is made of the gain or attenuation and the phase shift that the channel introduces to the signal which takes τ time units to move between transmitter's kth antenna element and the receiver's lth antenna element. The channel gains or fades $h_{lk}(\tau,t)$ are independent Gaussian random variables if the physical separation between the antennas is at least $\lambda/2$ (i.e., half the operating wavelength) [Lee (1985)]. The time variable indicates the time dependence of the wireless channel.

7.4.1 *The Additive Noise MIMO Channel*

Let us consider a MIMO system disturbed by only additive white Gaussian noise (AWGN) as modelled in Fig. 7.32. Thus, we neglect interference and multipath propagation and other types of noise. This means that we drop the channel's parameter τ to obtain

$$\mathbf{r}(t) = \mathbf{H}(t)\mathbf{s}(t) + \mathbf{w}(t) \tag{7.109}$$

where the $L \times K$ matrix of the channel coefficients is given by

$$\mathbf{H} = \begin{bmatrix} h_{11} & h_{12} & \cdots & h_{1K} \\ h_{21} & h_{22} & \cdots & h_{2K} \\ \vdots & \vdots & \cdots & \vdots \\ h_{L1} & h_{L2} & \cdots & h_{LK} \end{bmatrix}. \tag{7.110}$$

The channel gain or coefficient between the lth receive antenna and the kth transmit antenna is

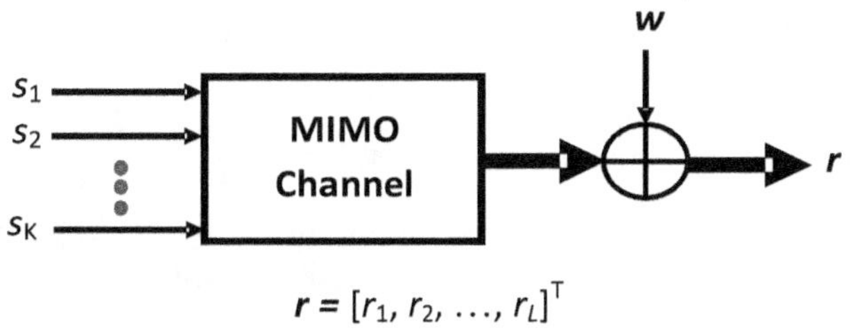

Fig. 7.32 $K \times L$ MIMO channel without multipath propagation.

h_{lk}. In the above we have omitted the time dependence of the channel gains. However, in practice the channel is time varying.

The signal at the lth receive antenna is

$$r_l(t) = \sum_{k=1}^{K} h_{lk}(t)s_k(t) + w_l(t), \; l = 1, 2, \ldots, L. \tag{7.111}$$

By the AWGN assumption we have $w_l \sim N(0, \sigma^2)$, i.e. has zero mean and the variance of σ^2. Also the noise signals are independent but identically distributed (iid) random variables, which means that the correlation between a pair of them is $E\{w_i w_j^*\} = \sigma^2 \delta_{ij}$. The Kronecker delta has the usual meaning of $\delta_{ij} = 0, \; i \neq j$ and $\delta_{ij} = 1, \; i = j$. This results in the $L \times L$ covariance matrix $R_w = E[\mathbf{w}\mathbf{w}^H] = \sigma^2 \mathbf{I}$, where $\mathbf{I}$ is the $L \times L$ identity matrix. As an example, let us consider the case where $K = 4, L = 3$ and $\sigma^2 = -8$dB. This results in $R_w = E[\mathbf{w}\mathbf{w}^H] \approx 0.16\mathbf{I}_{3\times3}$. Three types of multi-antenna systems are identified. These are: SISO, MISO, SIMO and MIMO.

7.4.2 *Flat Fading MIMO Channel with AWGN*

We assume that the fading experienced by signals over the wireless channel is Rayleigh fading, which is a rich scattering multipath propagation without a specular or LOS path between the transmitter and the receiver. If the symbol duration of the transmitted signal is larger than the channel's delay spread, or if the signal bandwidth is less than the channel's coherence bandwidth, then we have flat fading channel. In this case Eq. (7.109) becomes

$$\mathbf{r}(\tau) = a\mathbf{H}(\tau)\mathbf{s} + \mathbf{w}. \tag{7.112}$$

where $\mathbf{w} = \mathbf{w}_{\mathrm{R}} + j\mathbf{w}_I$, $\mathbf{w}_{\mathrm{R}}, \mathbf{w}_I \sim \mathcal{CN}(0, \sigma^2)$ and

$$a^2 = \alpha g = (\texttt{propagation path loss}) \times (\texttt{antenna gain}). \quad (7.113)$$

It is assumed that all antennas have the same gain. It has been shown in [Telatar (1999)] and [Shiu *et al.* (2000)] that the capacity of the flat fading MIMO channel is

$$C_{flatf} = \max_{tr(\mathbf{C}_{ss}) \le \rho_t} \log_2[\det(\mathbf{I}_{L\times L} + \mathbf{H}\mathbf{C}_{ss}\mathbf{H}^H)]. \quad (7.114)$$

where ρ_t is the maximum normalised transmit power and $\mathbf{C}_{ss} = cov(\mathbf{s}, \mathbf{s}^T) = E[(\mathbf{s} - E_s)(\mathbf{s} - E_s)^T]$ is the covariance matrix of the transmitted signal vector $\mathbf{s}(t)$. If the wireless channel is linear then we can perform linear transformations at both receiver and transmitter to convert the $L \times K$ MIMO channel into $k_p = \min\{L, K\}$ SISO subchannels.

Under the rich scattering assumption made above, the channel gains h_{ij} are complex Gaussian processes. Further assuming slow fading or quasi-static channel with large coherence time relative to the symbol duration, h_{ij} become Gaussian random variables.

7.4.3 *Capacity with Unknown Channel at Transmitter*

We assume that the total power available to the transmitter during each transmission time is P_{T}. Without knowing the channel's state, the transmitter allocates its power equally to the K transmit antennas. If the total power transmitted over each antenna is $\frac{P_{\mathrm{T}}}{K}$, the received power becomes $P_{\mathrm{R}} = \frac{a^2 P_{\mathrm{T}}}{K} = \frac{\alpha g P_{\mathrm{T}}}{K}$. With the noise power being σ^2, the received SNR is

$$\gamma = SNR = \frac{\alpha g P_{\mathrm{T}}}{K\sigma^2} = \frac{\gamma_{\mathrm{R}}}{K} \quad (7.115)$$

where γ_{R} is the total received SNR.

We shall exploit the singular value decomposition of the channel matrix

$$\mathtt{H} = \mathbf{U}_{\mathrm{H}}\mathbf{D}_{\mathrm{H}}\mathbf{V}_{\mathrm{H}} \quad (7.116)$$

where $\mathbf{U} \in \mathcal{C}^{L\times L}$ and $\mathbf{V} \in \mathcal{C}^{K\times K}$ are unitary matrices,[3] and $\mathbf{D} = \mathtt{diag}(\lambda_{\mathrm{H},n})$ is a diagonal matrix with the singular values of $\mathbf{H}$ as its elements. With $N = \min\{K, L\}$ and assuming that the signals exciting the K transmit antennas are statistically independent, we obtain the *MIMO channel capacity limit*

$$\begin{aligned} C &= B\sum_{n=1}^{N} \log_2(1 + \gamma\lambda_{\mathrm{H},n}^2)\ \mathtt{bps} \\ &= B\log_2 \prod_{n=1}^{N}(1 + \gamma\lambda_{\mathrm{H},n}^2)\ \mathtt{bps} \\ &= B\log_2\ \mathtt{det}[\mathbf{I}_{L\times L} + \gamma\mathbf{D}_{\mathrm{H}}\mathbf{D}_{\mathrm{H}}^{\dagger}] \\ &= B\log_2\ \mathtt{det}\left[\mathbf{I}_{L\times L} + \frac{\gamma_{\mathrm{R}}}{K}\mathbf{H}\mathbf{H}^{\dagger}\right]\ \mathtt{bps} \end{aligned} \quad (7.117)$$

where B is the channel bandwidth in Hertz and $\lambda_{\mathrm{H},n}$ are the singular values of the channel matrix $\mathbf{H} = (h_{ij})$. Note that $\mathbf{A}^{\dagger} = (\mathbf{A}T)^*$ means conjugate transpose of the matrix $\mathbf{A}$. The objective of information theory research is to approach the limit in Eq. (7.117). Architectures used to seek the MIMO capacity limit include:

- BLAST (Bell Labs Layered Space-Time architecture) with many versions, such as diagonal-BLAST (D-BLAST) and vertical-BLAST (V-BLAST) [Foschini (1996)].
- Space-time coding (STC), e.g., Alamouti STC with two transmit antennas [Alamouti (1998)].

The capacity at high SNR is referred to as asymptotic capacity. On a flat-fading (i.e., ISI-free) wireless channel, the asymptotic MIMO capacity with $N = \min\{K, L\}$ is derived from Eq. (7.117) as [Raleigh and Cioffi (1998)]

$$\lim_{\gamma_{\mathrm{R}}\to\infty} C \to N\log_2\gamma_{\mathrm{R}} - N\log_2 N + \sum_{n=1}^{N}\log_2(|\lambda_{\mathrm{H},n}|^2). \quad (7.118)$$

7.4.4 *Capacity with Known Channel at Transmitter*

Now, we assume that the channel state information is available at the transmitter. Then, the transmitter distributes the available total power P_{T} to the channels in a way to maximise overall capacity. Such a power or energy allocation is called optimal allocation. The process of this optimal power allocation is called *water-filling principle* (WFP) or

[3]A unitary matrix $\mathbf{A}$ is a square matrix whose conjugate transpose equals its inverse, i.e., $\mathbf{A}^{\dagger} = \mathbf{A}^{-1} \Leftrightarrow \mathbf{A}^{\dagger}\mathbf{A} = \mathbf{I}$.

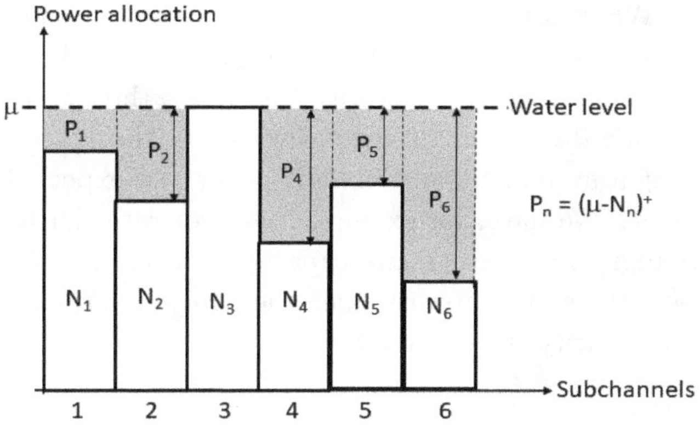

Fig. 7.33 Illustrating the water-filling principle for 6 parallel channels.

water-pouring principle (WPP). Let us note the following definitions prior to proceeding:

- $\mathbf{R}_s$: autocorrelation matrix of the transmitted symbols.
- $\lambda_{s,n}$: the eigenvalues of $\mathbf{R}_s$.
- $\lambda_{H,n}$: singular values of the channel matrix $\mathbf{H}$.
- Define $N_n = \frac{\sigma^2}{a^2\lambda^2_{H,n}}$ and $P_n = \lambda_{s,n}$.

The optimum power allocation using the water-filling principle is illustrated in Fig. 7.33. Analytically, it is defined as

$$P_{\mathrm{T}} = \sum_{n=1}^{N} P_n = \sum_{n=1}^{N} \lambda_{s,n} \qquad (7.119)$$

$$P_n = \lambda_{s,n} = (\mu - N_n)^+ = \left(\mu - \frac{\sigma^2}{a^2\lambda^2_{H,n}}\right)^+ \qquad (7.120)$$

where

$$(x)^+ = \begin{cases} x, & x > 0 \\ 0, & x \leq 0 \end{cases}. \qquad (7.121)$$

The achievable total channel capacity is [Khalighi and Raoof (2002)]

$$C = B \sum_{n=1}^{N=\min\{K,L\}} \log_2\left(1 + \frac{P_n}{N_n}\right) \text{ bps} \qquad (7.122)$$

subject to

$$\mathbf{R}_s = \mathtt{V}_H \mathbf{D}_s \mathtt{V}_H^{\dagger}. \qquad (7.123)$$

The water level μ in Fig. 7.33 is chosen to meet the total power constraint. It is found from the Lagrange optimisation problem

$$L(\{P_n\};\lambda) = B\sum_{n=1}^{N} \log_2\left(1+\frac{P_n}{N_n}\right) + \lambda\left(\sum_{n=1}^{N} P_n - P_{\mathrm{T}}\right) \qquad (7.124)$$

where λ is the Lagrange multiplier. From $\frac{\partial L(\{P_n\};\lambda)}{\partial P_n} \stackrel{!}{=} 0$, we obtain

$$N_n + P_n = constant(=\mu) \Leftrightarrow P_n = (\mu - N_n)^+. \qquad (7.125)$$

The optimal allocation of energy or power to channels using the waterfilling principle maximises the overall system capacity by allocating power to channels based on their effective noise level. A channel with less affective noise level is allocated more power than another channel with higher noise level. Waterfilling principle differs from opportunistic beamforming or scheduling in the sense that even poor channels are allocated some power. As illustrated in Fig. 7.33 and Eq. (7.125), the sum of the effective noise power and the allocated power for each channel is a constant, which is equal to the water level μ.

7.4.5 *No Spatial Diversity: SISO*

Here, we have $K = 1, L = 1$, resulting in a single-input single-output (SISO) system illustrated in Fig. 7.34. This is the conventional wireless system which uses only one antenna at either end of the two-way communication link as illustrated in Fig. 7.34. In this case Eq. (7.109) becomes

$$r(t) = h(t)s(t) + w(t). \qquad (7.126)$$

AWGN Channel

Let us assume that the channel is purely additive noise channel. In this case the channel gain is $|h(t)| = 1$ in Eq. (7.126). Let the energy in the input signal be $E_s = \mathbb{E}\{|s|^2\}$, its power be $P_s = \int_{t_0}^{t_0+T_s} |s(t)|^2\, dt$ where T_s is the symbol duration (aka symbol time), $h(X)$ denote the entropy of X, and $f_S(s)$ be the pdf of s. If $w(t) \sim N(0, \sigma^2)$, then $h(W) = \log_2(\pi e \sigma^2)$. Thus, $h(R) =$

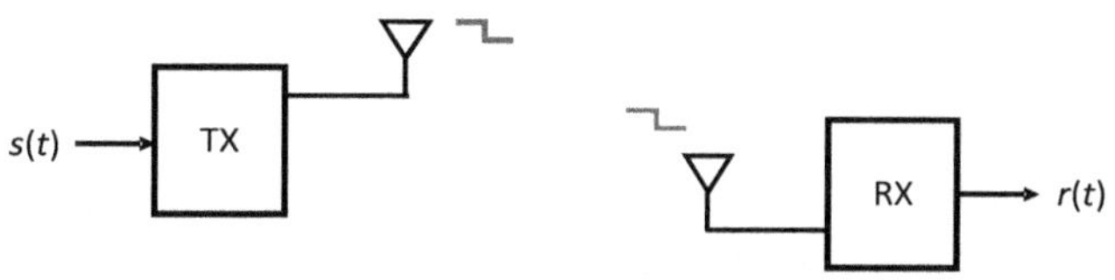

Fig. 7.34 SISO system model in pure AWGN channel without multipath

$\log_2[\pi e(E_s+\sigma^2)]$. The channel capacity is the mutual information

$$C/B = \max_{f_S(s)} I(S;R) = \max_{f_S(s)}[h(R) - h(R/S)] \tag{7.127}$$

where $h(R/S) = \log_2(\pi e\sigma^2)$ is the residual entropy in the additive noise w. Hence, Eq. (7.127) becomes

$$\begin{aligned} C &= B \max_{f_S(s)} I(S;R) \\ &= B\log_2[\pi e(E_s+\sigma^2)] - B\log_2(\pi e\sigma^2) \\ &= B\log_2\left(\frac{E_s+\sigma^2}{\sigma^2}\right) = B\log_2(1+\gamma) \end{aligned} \tag{7.128}$$

where $\gamma = E_s/\sigma^2$ is the signal-to-noise ratio without fading.

Fixed Random Fading with AWGN Channel

Now, let us assume that the wireless channel is fixed (i.e., non-time varying) but random with the gain h. We therefore amend the capacity Eq. (7.128) to

$$C = B\log_2(1+\gamma|h|^2) \tag{7.129}$$

which is the maximum channel capacity for error-free signal transmission. Here, B is the channel bandwidth in Hz.

Claude Shannon found that transmitting signals over the wireless channel with bandwidth B at a bit rate less than Eq. (7.129) we can use coding techniques to achieve a transmission with zero bit-error rate. Unfortunately, usually the wireless channel behaves randomly, rendering its impulse response $h(t)$ a random variable. For this reason the capacity in Eq. (7.129) is also a random variable, which can be zero. In order to find the non-random channel capacity the **ergodic channel capacity** was defined, which is the statistical mean of Eq. (7.129), i.e.

$$\begin{aligned} C_{erg} &= \mathcal{E}_h\left\{B\log_2\left(1+|h(t)|^2\frac{P_s}{\sigma^2}\right)\right\} \\ &\le B\log_2\left(1+\frac{P_s}{\sigma^2}E\{|h(t)|^2\}\right) \end{aligned} \tag{7.130}$$

where $\mathcal{E}_h(\cdot)$ is the expectation taken over the channel gains h. We have applied the Jensen's inequality which states that $\mathcal{E}\{g(y)\} \le g(y)$ when $g(y)$ is a concave function of the random variable y.

What is the probability that the required gain, signal-to-noise ratio or channel capacity is available when needed. We quantify this using the *outage probability*, which in general sense, is the chance that a system parameter falls below the expected value. We have, for example, bit-error rate (BER) outage, SNR outage and capacity outage. Let R_{th} be a threshold capacity, then the *outage probability* for capacity is defined as

$$\begin{aligned} P_{\mathrm{out}} &= Pr[C < R_{\mathrm{th}}] \\ &\overset{(7.129)}{=} Pr[B\log_2(1+\gamma|h|^2) < R_{\mathrm{th}}] \\ &= Pr\left[|h| < \sqrt{\frac{2^{R_{\mathrm{th}}/B}-1}{\gamma}}\right] \\ &= 1 - Pr\left[|h| > \sqrt{\frac{2^{R_{\mathrm{th}}/B}-1}{\gamma}}\right]. \end{aligned} \tag{7.131}$$

For a Rayleigh fading channel, the pdf of the channel gain is $f_{|h|}(x) = \frac{2x}{\sigma_h^2}e^{-x^2/\sigma_h^2}$. Therefore, Eq. (7.131) becomes

$$\begin{aligned} P_{\mathrm{out}} &= 1 - Pr\left[|h| > \sqrt{\frac{2^{R_{\mathrm{th}}/B}-1}{\gamma}}\right] \\ &= 1 - \int_{\sqrt{\frac{2^{R_{\mathrm{th}}/B}-1}{\gamma}}}^{\infty} \frac{2x}{\sigma_h^2}e^{-x^2/\sigma_h^2}dx\,. \end{aligned} \tag{7.132}$$

Let us make the substitution $y = \frac{x^2}{\sigma_h^2}$ to reduce Eq. (7.132) to

$$\begin{aligned} P_{\mathrm{out}} &= 1 - \int_{\frac{2^{R_{\mathrm{th}}/B}-1}{\gamma\sigma_h^2}}^{\infty} e^{-y}dy \\ &= 1 - \exp\left(-\frac{2^{R_{\mathrm{th}}/B}-1}{\gamma\sigma_h^2}\right). \end{aligned} \tag{7.133}$$

Now, let us exploit the Taylor's series expansion $e^{-x} = \sum_{i=0}^{\infty}\frac{(-x)^i}{i!}$ and

$$\lim_{x\to 0} e^{-x} \approx 1 - x\,. \tag{7.134}$$

Noting that $\lim_{\gamma\to\infty}\frac{2^{R_{\mathrm{th}}/B}-1}{\gamma\sigma_h^2} \to 0$, we can invoke Eq. (7.134) on Eq. (7.133) to obtain the *asymptotic outage probability*

$$P_{\infty,\mathrm{out}} = \lim_{\gamma\to\infty} P_{\mathrm{out}} \approx \frac{2^{R_{\mathrm{th}}/B}-1}{\gamma\sigma_h^2} \propto \frac{1}{\gamma}. \tag{7.135}$$

Example

Assume that a link in a wireless IoT system has the features: $\sigma_h^2 = 1$, $B = 15$ kHz and $R_{\text{th}} = 300$ kbps. Compute the minimum signal-to-noise ratio required to keep the outage probability within 5%.

Solution

The given system has the threshold bandwidth efficiency $\rho_{\text{th}} = R_{\text{th}}/B = 20$. We first need to rearrange Eq. (7.133) as

$$\gamma_{\text{out}} = -\frac{2^{R_{\text{th}}/B} - 1}{\sigma_h^2 \log_e(1 - P_{\text{out}})}. \tag{7.136}$$

We can refer to Eq. (7.136) as *outage SNR*. Plugging the given values into Eq. (7.136) yields $\gamma_{\text{out}} \approx 20.443 \times 10^6$ or 73.1 dB. Figure 7.35 shows the outage SNR as a function of the outage probability and bandwidth efficiency $\rho_{\text{th}} = \frac{R_{\text{th}}}{B}$. We can observe that we need a higher SNR to achieve a given outage probability as the bandwidth efficiency increases. For example, we need SNR of about 73.1 dB, 43 dB and 12.9 dB to achieve $P_{\text{out}} = 5\%$ at bandwidth efficiencies of 20, 10 and 1, respectively.

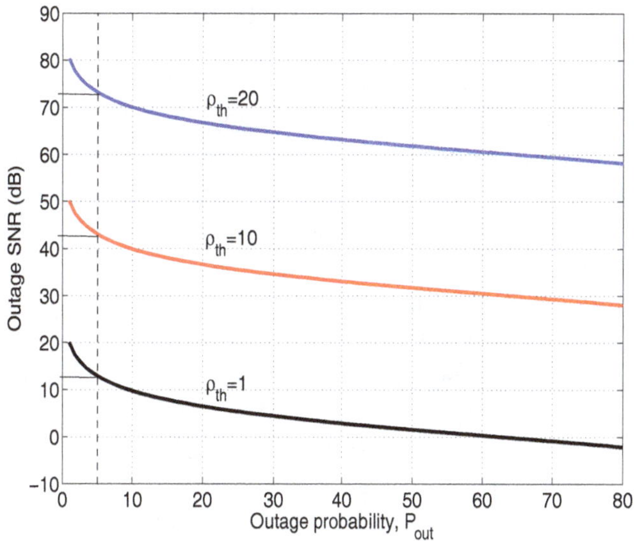

Fig. 7.35 Outage SNR versus outage probability at different bandwidth efficiencies: $\sigma_h^2 = 1, \rho_{\text{th}} = \frac{R_{\text{th}}}{B}$.

7.4.6 *Check Your Understanding*

Students are encouraged to check their understanding of the content of this section with the following questions prior to proceeding to the next section.

(1) Explain the meaning of ergodic channel capacity.

(2) Describe the usefulness of the outage probability.

(3) Consider the two wireless links: Link 1 with $P_{\text{out},\,1}(\gamma = 5\text{dB})$ and Link 2 with $P_{\text{out},\,2}(\gamma = 5\text{dB})$. If Link 1 is better than Link 2, state the analytical relationship between the outage probabilities of the two links.

(4) Derive Eq. (7.128) as the capacity of AWGN channel.

(5) A given link in a wireless IoT system has the features: $\sigma_h^2 = 2$, $B = 200$ kHz, and $R_{\text{th}} = 270$ kbps. Compute the minimum signal-to-noise ratio required to keep the outage probability within 2%.

(6) A given link in a wireless IoT system operates well with the parameters: $\sigma_h^2 = 20$, $B = 15$ kHz and $R_{\text{th}} = 1$ Mbps. The link requires the minimum signal-to-noise ratio of 8 dB to operate as expected. Compute the minimum achievable outage probability.

(7) Describe the water-filling principle.

(8) Justify the statement or otherwise: allocating power or energy to subchannels in a MIMO system using the water-filling principle is optimal.

(9) Discuss the main challenge in energy allocation to channels in a MIMO system using the water-pouring principle.

(10) Discuss the similarities and differences between opportunistic beamforming or scheduling and the water-filling principle.

7.5 Transmit Diversity: MISO

In cellular mobile communications systems transmit diversity, or multiple-input single-output (MISO) system, usually uses multiple antennas at the radio base station (aka eNodeB or gNB), but single antenna at the mobile handset. Therefore, transmit diversity is also referred to as spatial division multiple access (SDMA). The spectacular advantage in SDMA over MIMO is simpler and cheaper wireless handsets as they are designed with just one antenna element. In MISO we have $K \geq 2, L = 1$ as illustrated in Fig. 7.36. Such a system uses multiple antennas at the transmitter but only one antenna at the receiver. Therefore, we sent multiple signals into the wireless channel, but receive all using only one antenna. In this case Eq. (7.109) becomes

$$r(t) = \mathbf{h}(t)^H \mathbf{s}(t) + w(t) = \sum_{k=1}^{K} h_k^*(t) s_k(t) + w(t). \tag{7.137}$$

Equation (7.137) reflects the system model shown in Fig. 7.36(a). We can obtain the maximum array gain if the transmitter can know the channel state information (CSI) apriori. In the presence of CSI at the transmitter we can design it to match the channel's impulse response by selecting the weights in Fig. 7.36(b) as

$$a_k = \frac{h_k^*}{\|h_k\|} \Rightarrow \mathbf{a} = \frac{\mathbf{h}^*}{\|\mathbf{h}\|}. \tag{7.138}$$

Equation (7.138) is the matched filter relation. Thus, Eq. (7.137) now becomes

$$\begin{aligned} r(t) &= \sum_{k=1}^{K} h_k^*(t) a_k s(t) + w(t) \\ &= \mathbf{h}(t)^H \mathbf{a} s(t) + w(t) \\ &= \|\mathbf{h}\| s(t) + w(t) \end{aligned} \tag{7.139}$$

yielding the signal-to-noise ratio

$$\gamma_{miso} = \frac{\|\mathbf{h}\|^2 P_s}{\sigma^2} = \frac{P_s \sum_{k=1}^{K} |h_{1k}|^2}{\sigma^2} \propto K \tag{7.140}$$

where, again, P_s is the transmitted signal's power and $\sigma^2 = E[w(t)w^*(t)]$, i.e. noise variance. It is clear from Eq. (7.140) that increasing the number of antenna elements at the transmitter increases the signal-to-noise ratio.

With $K > 1$, $L = 1$, the channel matrix becomes a row vector of length K, i.e., $\mathbf{H} = (h_1, h_2, \ldots, h_K)$. On flat-fading wireless channels with bandwidth B Hz, the achievable capacity with MISO is [Foschini and Gans (1998)]

$$C = B \log_2 \left(1 + \frac{\gamma_R}{K} \sum_{k=1}^{K} |h_k|^2 \right) \, bps\,. \tag{7.141}$$

An assumption underlying Eq. (7.141) is that the transmitter is not privy of the state of the wireless channel. Knowing channel state by the transmitter, the achievable capacity of MISO is equivalent to that of SIMO.

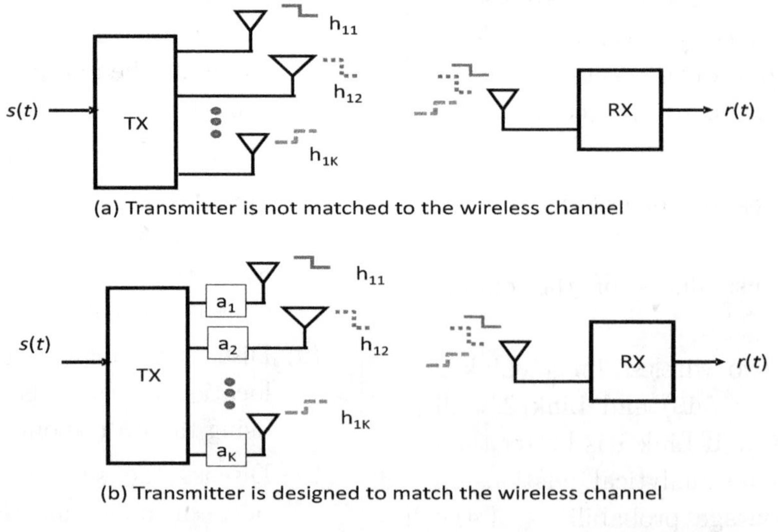

Fig. 7.36 MISO system model in pure AWGN channel without multipath.

7.5.1 *Transmit Beamforming*

This is a signal processing technique which uses an array of antennas to achieve directional signal transmission. By exploiting antenna reciprocity, beamforming also achieves directional signal reception. Beamforming, aka beamshaping, points an antenna array to specific directions by combining the phased array elements so as to achieve constructive and destructive interference at desired angles or directions. This is called *spatial selectivity*.

By limiting the beamwidth of the antenna, beamforming also reduces interference. The gain associated with beamforming can improve coverage and/or data rate. Beamforming can be achieved with dumb antennas or smart antennas. Beamforming with smart antennas uses beam steering to adaptively track a user.

Transmit beamforming provides power gain but requires knowledge of the channel. Transmit diversity can as well be used to create or enhance channel variations that are useful for, for example, opportunistic scheduling or opportunistic beamforming to gain power. Like space-time coding, the same dataset is sent over multiple antenna elements. However, instead of precoding, transmit beamforming uses phase/amplitude adjustment for each antenna array element. Transmit beamforming is capable of exploiting both array gain and diversity gain. For this reason, it is preferable to space-time coding. The main problems with transmit beamforming are its impairment degradation and the difficulty associated with the determination of the right phase/amplitude adjustment needed for each antenna.

7.5.2 *Space-Time Coding: Space-Time Block Code*

Space-time coding (STC) combines temporal diversity (i.e., error control coding) and spatial diversity (i.e., MISO). The STC was first introduced by V. Tarokh and team in 1998 [Tarokh *et al.* (1998)]. STC takes multiple copies of the same dataset, precode each differently from all others and then transmit each on a different antenna in order to improve data transmission reliability. The coding is done in two dimensions of time and space (the reason for its name) to achieve diversity gains in the presence of multipath fading. Array gain, however, eludes space-time coding as the location of the receiver is not known to the transmitter prior to injecting the signals into the wireless channel. A popular assumption in STC is that the transmitter does not have channel state information (CSI), but the receiver has perfect knowledge of CSI.

By combining channel coding and MIMO, STC exploits two diversities: spatial diversity through the multiple antennas and time diversity through the channel coding combined with interleaving. There are two types of STC:

- Space-time block codes (STBCs).
- Space-time trellis codes (STTCs).

Similar to block codes, a STBC combines the processing of a block of data at a time with spatial diversity. A STTC, on the other hand, distributes a trellis-coded data over multiple antennas and multiple time slots. Another difference between STBC and STTC is that, the former achieves only spatial diversity gain, while STTC achieves coding gain in addition. With respect to knowledge of the wireless channel by the receiver, STC can be categorised into three types:

(1) *Coherent STC*: the receiver knows the near instantaneous estimate of the channel's impairments. This is achieved using channel sounding or pilot tones.
(2) *Incoherent STC*: the receiver knows only the statistics of the wireless channel, but not its impairments.
(3) *Differential STC*: the receiver knows nothing about the wireless channel, not its statistics nor its impairments. Thus, the receiver is a blind detector of the transmitted signal.

We discuss further only STBC as it is simpler to encode and decode than the STTC.

7.5.2.1 *Space-Time Block Coding (STBC)*

As its name says, STBC encodes a block of data at a time. Remember that the Alamouti's scheme [Alamouti (1998)] is the simplest form of STBC in which two data symbols are encoded in each symbol period. We focus on orthogonal STBC (OSTBC), in which the transmitted symbols are mutually orthogonal [Tarokh *et al.* (1999)]. That way the receiver can detect the symbols independent of each other using the maximum-likelihood algorithm. OSTBC is based on the theory of *orthogonal designs* for both real-valued and complex-valued symbols.

An orthogonal matrix is a square matrix whose row vectors are mutually orthogonal vectors, just as its column vectors. For example, the matrix $\mathbf{A}$ is orthogonal if $\mathbf{A}\mathbf{A}^T = \mathbf{I}$, an identity matrix. An orthogonal design of size N is an $N \times N$ matrix whose rows are permutations of the numbers $\pm x_1, \pm x_2, \ldots, \pm x_N$. For example, $N = 2$ and $N = 4$ orthogonal designs are

$$\mathtt{X}_2 = \begin{bmatrix} x_1 & -x_2 \\ x_2 & x_1 \end{bmatrix}, \quad \mathtt{X}_4 = \begin{bmatrix} x_1 & x_2 & x_3 & x_4 \\ -x_2 & x_1 & -x_4 & x_3 \\ -x_3 & x_4 & x_1 & -x_2 \\ -x_4 & -x_3 & x_2 & x_1 \end{bmatrix}.$$

Specific examples of real orthogonal designs are

$$\mathtt{X}_2 = \begin{bmatrix} 1 & 2 \\ -2 & 1 \end{bmatrix}, \quad \mathtt{X}_4 = \begin{bmatrix} 1 & 2 & 3 & 4 \\ -2 & 1 & -4 & 3 \\ -3 & 4 & 1 & -2 \\ -4 & -3 & 2 & 1 \end{bmatrix}.$$

How do we apply this in OSTBC? The indeterminates in the orthogonal design are replaced by the information-bearing symbols so that we transmit N symbols over N antenna elements in each symbol period. For example, in the case of the 2×2 matrix, we transmit a block of $N = 2$ symbols in each symbol period as follows:

- First symbol period: transmit symbol s_1 from antenna element 1, and s_2 from antenna element 2.
- Second symbol period: transmit symbol s_2 from antenna element 1, and s_1 from antenna element 2.

The sequential procedure for building an OSTBC using the orthogonal design of size N is:

(1) Encoder operates on a block of size Nm bits.
(2) For $1 \leq n \leq N$, encoder selects a symbol s_n from a signal constellation $\mathbb{S}$ of size 2^m. This means that each symbol in the constellation contains m bits.
(3) Encoder uses the N symbols $s_1, s_2, \ldots, s_N$ to construct the orthogonal matrix $\mathbf{X}_N = \mathbb{O}(s_1, s_2, \ldots, s_N)$ based on the orthogonal design of size N.
(4) At time epoch t, we transmit the N symbols in the t-th row of orthogonal matrix $\mathbf{X}_N$ over the N antennas.

The code constructed with the above algorithm has the features:

- Code rate is m bps/Hz.
- Diversity order is NL, where L is the number of receive antenna elements.

By noting that

$$\mathbf{X}\mathbf{X}^T = \sum_{n=1}^{N} x_n^2 \cdot \mathbf{I}_N \tag{7.142}$$

where $\mathbf{I}_N$ is the $N \times N$ identity matrix. Thus, constructing orthogonal design is equivalent to constructing the *Hurwitz-Radon* family of matrices $\mathbf{B}_l$. A set of J $N \times N$ real-valued matrices $\{B_1, B_2, \ldots, B_J\}$ is referred to as *Hurwitz-Radon* family of matrices if

$$\begin{aligned} \mathbf{B}_j^{\dagger}\mathbf{B}_j = \mathbf{I}, \quad \mathbf{B}_j^{\dagger} = -\mathbf{B}_j, \quad 1 \leq j \leq J \\ \mathbf{B}_j\mathbf{B}_m = -\mathbf{B}_m\mathbf{B}_j, \quad 1 \leq j,\ m \leq J \end{aligned} \tag{7.143}$$

where $\mathbf{D}^{\dagger}$ denotes conjugate transpose of $\mathbf{D}$.

7.5.2.2 *Linear OSTBC Using Modulation Matrices*

An alternative method to the above used to construct linear STBC is proposed in [Sandhu *et al.* (2019)]. The method is based on the construction of *modulation matrices*. For MISO with K transmit antennas, construct the $K \times T$ modulation matrices $\mathbf{A}\}_{j=1}^{J}$. Then, we use these modulation matrices to spread the input data symbols s_j over KT spatio-temporal dimensions [Sandhu *et al.* (2019)]. Let $x_j = \mathbb{R}s_j$ and $x_{j+\frac{J}{2}} = \mathbb{I}s_j$ be the real and imaginary parts of the symbol s_j, respectively. We modulate $\mathbb{R}s_j$ with the matrix $\mathbf{A}_j$, but $\mathbb{I}s_j$ with the matrix $\mathbf{A}_{j+\frac{J}{2}}$. We then sum the modulated matrices to obtain the codeword $\mathbf{X}$ of dimension $K \times T$ as follows

$$\mathbf{X} = \sum_{j=0}^{\frac{J}{2}-1} [\mathbf{A}_j \mathbb{R}s_j + \mathbf{A}_{j+\frac{J}{2}} \mathbb{I}s_j]$$
$$= \sum_{j=0}^{J-1} \mathbf{A}_j \mathbb{R}s_j,\ J \leq 2KT\,. \qquad (7.144)$$

Below, we shall use the Alamouti's code to explain how this method is used to generate a simple linear OSTBC. Note that if $J < 2KT$ then we can design orthogonal modulation matrices. On the other hand, the maximum capacity is achieved if $J = 2KT$.

7.5.2.3 *Alamouti Space-Time Coding*

Alamouti proposed in [Alamouti (1998)] a simple MISO scheme employing two antennas at the transmitter and one antenna as the receiver. A transmission over two symbol periods is considered. Alamouti code is an example of a linear orthogonal space-time block code (OSTBC). Two symbols, say s_0 and s_1, are transmitted in a symbol period of duration T_s through the transmit antenna elements 0 and 1, respectively. In the next symbol period, we transmit $-s_1^*$ from antenna element 0 and s_0^* from antenna element 1, where x^* is the complex conjugate of x. This architecture is illustrated in Fig. 7.37. Assume that the total energy available at the transmitter in each symbol period is E_s, which must be equally shared among the two transmit antennas. The sharing of the transmitter's energy is the main drawback in the Alamouti's architecture. It requires 3 dB more SNR to achieve the same bit error performance as 2×2 MIMO system.

The received signal during two symbol periods are

$$r_0 = \sqrt{\frac{E_\mathrm{s}}{2}}[h_0 s_0 + h_1 s_1 + w_0] \qquad (7.145)$$

$$r_1 = \sqrt{\frac{E_\mathrm{s}}{2}}[-h_0 s_1^* + h_1 s_0^* + w_1] \qquad (7.146)$$

which can be written in matrix-vector form as

$$\mathbf{r} = \mathbf{Hs} + \mathtt{w} \qquad (7.147)$$

where $\mathbf{r} = [r_0\, r_1^*]^T$, $\mathbf{s} = [s_0\, s_1]^T$, $\mathbf{w} = [w_0\, w_1^*]^T$ and

$$\mathbf{H} = \begin{bmatrix} h_0 & h_1 \\ h_1^* & -h_0^* \end{bmatrix}.$$

Multiplying Eq. (7.147) on the left by $\mathbf{H}^\dagger$ yields

$$\mathbf{y} = \mathbf{H}^\dagger \mathbf{r} = \mathbf{H}^\dagger \mathbf{Hs} + \mathbf{H}^\dagger \mathtt{w} = \sqrt{\frac{E_\mathrm{s}}{2}} \mathbf{H}^\dagger \mathbf{Hs} + \mathbf{H}^\dagger \mathtt{w} \qquad (7.148)$$

or

$$y_0 = \sqrt{\frac{E_\mathrm{s}}{2}}(|h_0|^2 + |h_1|^2)s_0 + h_0^* w_0 + h_1 w_1^* \qquad (7.149)$$

$$y_1 = \sqrt{\frac{E_\mathrm{s}}{2}}(|h_0|^2 + |h_1|^2)s_1 - h_0^* w_1 + h_1^* w_0\,. \qquad (7.150)$$

We say that the Alamouti's architecture achieves order-2 diversity because of the factor $|h_0|^2 + |h_1|^2$ in Eq. (7.149). The Alamouti's scheme transmits two symbols over two symbol periods using

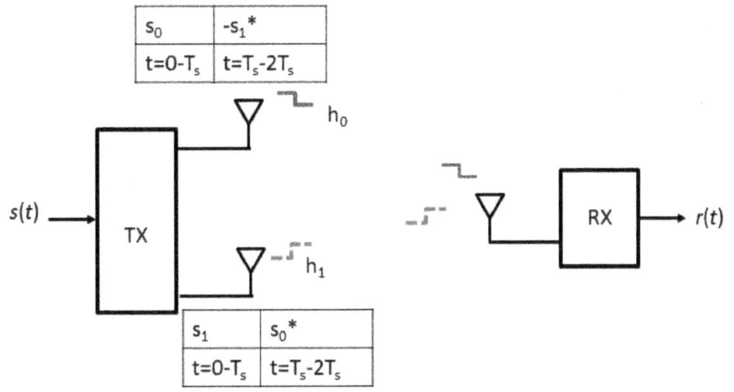

Fig. 7.37 Architecture of Alamouti STC.

transmit diversity, maintaining the system's bandwidth efficiency.

With $K = 2$, $L = 1$, $\mathbf{H} = [h_0, h_1]$ and thus $\mathbf{HH}^\dagger = (|h_0|^2 + |h_1|^2)\mathbf{I}_2$ in Eq. (7.117), the capacity of the Alamouti's scheme is

$$C_A = \frac{1}{2}\mathbb{E}\left\{ \log_2 \det\left(\mathbf{I}_2 + \frac{\gamma_{\mathrm{R}}}{2}\mathbf{HH}^\dagger\right)\right\}$$
$$= \frac{1}{2}\mathbb{E}\left\{ \log_2 \left[1 + \frac{\gamma_{\mathrm{R}}}{2}(|h_0|^2 + |h_1|^2)\right]\right\}. \quad (7.151)$$

The codeword generated by the Alamouti code is

$$\mathbf{X} = \begin{bmatrix} s_0 & -s_1^* \\ s_1 & s_0^* \end{bmatrix}$$
$$= \sum_{k=0}^{\frac{K}{2}-1} [\mathbf{A}_k \mathbb{R}s_k + \mathbf{A}_{k+\frac{K}{2}} \mathbb{I}s_k]$$
$$= \sum_{k=0}^{K-1} \mathbf{A}_k \mathbb{R}s_k$$
$$= \begin{bmatrix} 1 & 0 \\ 0 & 1 \end{bmatrix} \mathbb{R}s_0 + \begin{bmatrix} j & 0 \\ 1 & -j \end{bmatrix} \mathbb{I}s_0$$
$$+ \begin{bmatrix} 0 & -1 \\ 1 & 0 \end{bmatrix} \mathbb{R}s_1 + \begin{bmatrix} 0 & j \\ j & 0 \end{bmatrix} \mathbb{I}s_1 \quad (7.152)$$

where $\mathbb{R}s_k$ and $\mathbb{I}s_k$ are the real and imaginary parts of the symbol s_k, respectively. The matrices $\mathbf{A}_k$, $k = 0, 1, \ldots, K-1$ are referred to as *modulation matrices* [Sandhu *et al.* (2019)]. Note that, in general, the codeword $\mathbf{X}$ produced by an OSTBC is a unitary matrix.

7.5.3 *Check Your Understanding*

Students are encouraged to check their understanding of the content of this section with the following questions prior to proceeding to the next section.

(1) Discuss the meaning of spatial selectivity in wireless systems. Include in your discussions how it is achieved.

(2) Describe the advantage(s) and disadvantage(s) in beamforming in wireless communications.

(3) State the rationale behind space-time coding (STC). Who invented STC?

(4) List two differences between space-time trellis code and space-time block code.

(5) State the advantages of differential STC over coherent STC.

(6) How does coherent STC differ from incoherent STC?

(7) A wireless link, operating at signal-to-noise ratio of 8 dB, uses two antennas at the transmitter, but only one antenna at the receiver. The link bandwidth is 15 kHz. If the channel gains are $h_1 = 0.25 - j0.5$ and $h_2 = 0.5 + j0.2$, find how fast that data can be transferred over the link.

(8) We seek to transfer a file of size 50 bytes over the wireless link described in Problem 7. How much time is needed to achieve our aim?

(9) Explain the concept of orthogonal design as applied in space-time coding.

(10) State one orthogonal matrix of size 2 and another orthogonal matrix of size 4.

(11) Describe the Alamouti space-time code.

(12) Describe the reason why spatial diversity may be easier to implement with mmWaves than long waves transmissions.

(13) Using spatial diversity to increase data transmission rate exploits multipath propagation rather than combating it. Justify this statement or otherwise.

(14) The Alamouti STC achieves the same error performance as the MIMO system with two transmit and two receive antennas, albeit with 3 dB more signal-to-noise ratio. Describe the reason behind this penalty.

7.6 Receive Diversity: SIMO

In receive diversity we use multiple antennas at the receiver separated appropriately to receive the same signal transmitted by the transmitter but presumably having undergone different impairments in transit. This is equivalent to the SIMO system discussed earlier. We seek to process the multiple replicas of the same signal in such a manner as to eradicate or at least reduce the variability in the channel. Here, we have $K = 1, L \geq 2$, resulting in single-input multiple-output (SIMO) system. Such a system sends the signal through only one antenna into the wireless channel which is received using multiple antennas at the receiver. In this case Eq. (7.109) becomes

$$\begin{aligned} \mathbf{r}(t) &= \mathbf{h}(t)s(t) + \mathbf{w}(t), \text{ where} \\ r_l(t) &= s(t)h_l(t) + w_l(t),\ l = 1, 2, \ldots, L. \end{aligned} \tag{7.153}$$

7.6.1 *Maximal-Ratio Combining Receivers*

We explain the principles of the maximal-ratio combining (MRC) through its application in a receiver, called RAKE receiver, which was originally developed for CDMA-based systems. MRC is a diversity combining method in which the signals from all the spatial paths are unequally scaled before combining them. The scaling for each path or channel depends on the quality of the channel. It is inversely proportional to the mean square noise level in that channel, but proportional to the root-mean square signal level.

The RAKE receiver is a special form of correlation receiver in the sense that it comprises a bunch of correlators, say L, to exploit multipath fading which is otherwise a nuisance. Each of the L correlators processes one of the L strongest multipath signals arriving at the receiver. The RAKE receiver exploits multipath diversity to maximise the signal-to-noise ratio (SNR), and hence reduces the bit error rate at the output of the detector. The RAKE receiver was invented and patented by Robert Price and Paul Green in 1956 for wireless systems based on the spread-spectrum multiple access scheme. However, the technique became popular not until its application in commercial CDMA-based mobile wireless systems from the early 1990s. It was used predominantly in 3G systems.

The basic assumption underlying the operation of the RAKE receiver is that the relative time delays of the L strongest multipath signals arriving at the receiver exceeds the chip period, T_{c}. Under this assumption the multipath signals are uncorrelated. This means that we can use MRC to combine the outputs of all the L correlators to maximise the SNR.

Figure 7.38 illustrates the principle of the RAKE receiver. Assume that the signal $y(t)$ is transmitted, but at least L copies of it reach the receiver. Assume that the RAKE receiver has L correlators or branches, referred to as the **fingers** of the RAKE, as in the rake tool for gardening. Let $x_l(t)$ be the multipath component, which is processed by the lth correlator in the RAKE receiver. Let α_l be the weighting coefficient of the lth finger of the RAKE. Using the tapped-delay line (TDL) multipath channel model $h_{\mathsf{C}}(t) = \sum_{l=0}^{L-1} h_l \delta(t - \tau_l$, the gain of the lth path is

$$h_l = |h_l| e^{j\mathtt{arg}\{h_l\}} = g_l e^{j\theta_l}\,. \tag{7.154}$$

From Fig. 7.38 the received signals at the branches or fingers of the receiver are

$$x_l = z_l + w_l,\ l = 0, 1, \ldots, L-1 \tag{7.155}$$

where $w_l \sim \mathcal{N}(0, \sigma_{w,l}^2)$ is the AWGN at the lth finger, and z_l is the channel processed information-bearing signal. Therefore, the output of the maximal ratio combiner (MRC) is

$$Z = z_L + w_L = \underbrace{\sum_{l=0}^{L-1} \alpha_l z_l}_{\mathtt{signal}} + \underbrace{\sum_{l=0}^{L-1} \alpha_l w_l}_{\mathtt{noise}}\,. \tag{7.156}$$

The total AWGN power at the output of the MRC is

$$\sigma_L^2 = E\{[w_L - E\{w_l\}]^2\} = \sum_{l=0}^{L-1} |\alpha_l|^2 \sigma_{w,l}^2\,. \tag{7.157}$$

Furthermore, the total signal power at the output

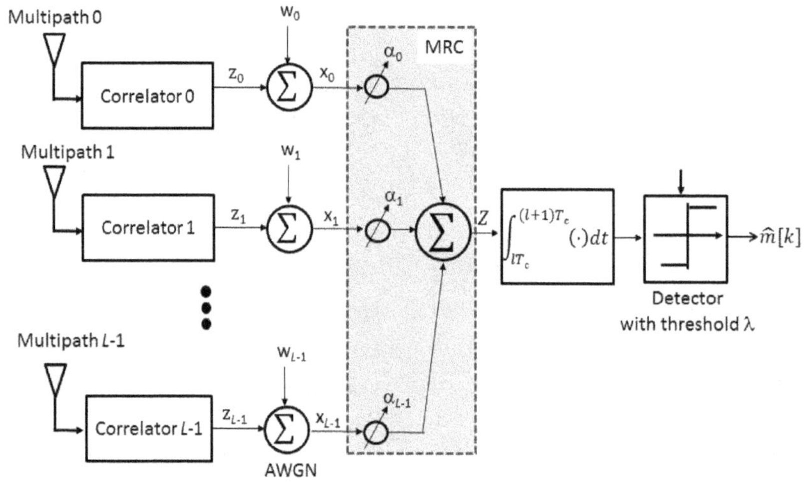

Fig. 7.38 Block diagram of the RAKE receiver with L correlators.

of the MRC is

$$P_s = E\{|z_L|^2\} = \overline{\left|\sum_{l=0}^{L-1} \alpha_l z_l\right|^2} = \left|\sum_{l=0}^{L-1} \alpha_l \sigma_{w,l} \frac{z_l}{\sigma_{w,l}}\right|^2$$

$$\leq \sum_{l=0}^{L-1} |\alpha_l \sigma_{w,l}|^2 \cdot \sum_{l=0}^{L-1} \left|\frac{z_l}{\sigma_{w,l}}\right|^2$$

$$= \sum_{l=0}^{L-1} |\alpha_l|^2 \sigma_{w,l}^2 \cdot \sum_{l=0}^{L-1} \left|\frac{z_l}{\sigma_{w,l}}\right|^2 \quad (7.158)$$

where the last-but-one expression exploits the Chebychev's inequality $|\sum_k xy|^2 \leq \sum_k |x|^2 \cdot \sum_k |y|^2$. We can now compute the SNR at the output of the MRC as

$$SNR_{mrc} = \frac{P_s}{\sigma_L^2} \leq \frac{\sum_{l=0}^{L-1} |\alpha_l|^2 \sigma_{w,l}^2 \cdot \sum_{l=0}^{L-1} |\frac{z_l}{\sigma_{w,l}}|^2}{\sum_{l=0}^{L-1} |\alpha_l|^2 \sigma_{w,l}^2}$$

$$= \sum_{l=0}^{L-1} |\frac{z_l}{\sigma_{w,l}}|^2 = \sum_{l=0}^{L-1} SNR_l \quad (7.159)$$

where SNR_l is the signal-to-noise ratio at the output of the lth finger. By noting that equality in Eqs. (7.158) and (7.159) holds if, and only if $\alpha_l \sigma_{w,l} = const \cdot \frac{z_l^*}{\sigma_{w,l}}$, we obtain

$$\alpha_l = const \cdot \frac{z_l^*}{\sigma_{w,l}^2}. \quad (7.160)$$

These are the weights for the optimum MRC, which is the combiner achieving the best error performance. By using the weights in Eq. (7.160), we multiply each output of the correlator by the corresponding complex-valued conjugate channel gain to achieve two purposes. First, we scale each correlator's output proportionally to its strength to maximise the combined SNR. Second, we compensate for any phase shifts in the channel to achieve *coherent combining*.

Example

We use this example to verify Eq. (7.160) and compute the maximum SNR.

(1) Prove that the equality in Eq. (7.158) or Eq. (7.159) is fulfilled if, and only if, Eq. (7.160) is fulfilled.
(2) Compute the maximum signal-to-noise ratio at the output of MRC.

Solution

(1) In the following, we assume $const = 1$ in Eq. (7.160). Plugging $\alpha_l = const \cdot \frac{z_l^*}{\sigma_{w,l}^2}$ into Eq. (7.158), we obtain

$$P_s = \left|\sum_{l=0}^{L-1} \alpha_l \sigma_{w,l} \frac{z_l}{\sigma_{w,l}}\right|^2 = \left|\sum_{l=0}^{L-1} \frac{z_l^*}{\sigma_{w,l}} \sigma_{w,l} \cdot \frac{z_l}{\sigma_{w,l}}\right|^2$$

$$= \left|\sum_{l=0}^{L-1} \frac{z_l}{\sigma_{w,l}}\right|^2 = \left|\sum_{l=0}^{L-1} SNR_l\right|^2. \quad (7.161)$$

Also, substituting Eq. (7.160) into Eq. (7.157) yields the noise power

$$\sigma_L^2 = \sum_{l=0}^{L-1} |\alpha_l|^2 \sigma_{w,l}^2 = \sum_{l=0}^{L-1} \frac{|z_l|^2}{\sigma_{w,l}^2} = \sum_{l=0}^{L-1} SNR_l\,. \tag{7.162}$$

(2) Now, the optimum SNR at the output of the MRC follows from Eq. (7.161) and Eq. (7.162) as

$$SNR_{mrc,opt} = \frac{P_s}{\sigma_L^2} = \frac{|\sum_{l=0}^{L-1} SNR_l|^2}{\sum_{l=0}^{L-1} SNR_l} = \sum_{l=0}^{L-1} SNR_l\,. \tag{7.163}$$

Capacity of MRC Receivers

Noting that $K = 1, L > 1$ in SIMO, the channel matrix becomes a vector of length L and with $rank(\mathbf{H}) = 1$, i.e., $\mathbf{H} = (h_1, h_2, \ldots, h_L)^T$. Thus, on a non-frequency-selective fading channel, the capacity achievable with MRC is [Foschini and Gans (1998)]

$$C = B\log_2\left(1 + \gamma_R \sum_{l=1}^{L} |h_l|^2\right). \tag{7.164}$$

The MRC is a linear and optimal combining method because it effectively captures as much of the transmitted power to stabilise the non-static wireless channel. As the number of receive antennas increases beyond 4 in single-user SIMO, the increase in the capacity is minimal considering the system complexity. Therefore, many antennas in SIMO is recommendable only in multi-user SIMO in which the spatial diversity is exploited to reduce the co-channel interference. MRC on flat fading channels performs equally with the minimum mean-sqaured error (MMSE) receiver [Winters (1987)], [Foschini and Gans (1998)].

7.6.2 *Equal-Gain Combining and Selection Diversity Receivers*

The MRC receiver is optimum. However, its requirement of the knowledge of the channel by the receiver makes it more challenging to use. For this reason, sub-optimal methods, such as the equal-gain combining (EGC) and selection diversity (SD), were developed.

The SD receiver picks the strongest signal among the L receive antennas and discards the remaining $L-1$ received signals. The selection can be based on the received signal-to-interference plus noise ratio. The capacity achieved by a maximum diversity selection is [Brennan (1959)].

$$C = B \max_{l\in\{1,L\}} \log_2(1 + \gamma_R |h_l|^2) = B\log_2\left(1 + \gamma_R \max_{l\in\{1,L\}} |h_l|^2\right). \tag{7.165}$$

The EGC receiver simply sums the signals impinging all the L receive antenna. Note, however, that co-phasing is necessary prior to combining if applying EGC on systems using non-constant phase modulation techniques. The limiting capacity of the EGC receiver is [Brennan (1959)].

$$C \le B\log_2\left(1 + \frac{\gamma_R}{L}\left(\sum_{l=1}^{L} |h_l|^2\right)^2\right). \tag{7.166}$$

7.6.3 *The Zero-Forcing MIMO Receiver*

In the following discussions we assume that the wireless channel has a single order. This means that the channel's impulse response has just a single tap delay. The MIMO receiver is expected to answer in optimum manner the question how do we recover the transmitted signal vector $\mathbf{s}(t)$ from the received signal vector $\mathbf{r}(t)$? In solving for the transmitted signal two cases must be differentiated. The first case is when $L = K$. This results in a square channel matrix. However, not every square matrix has an inverse. Thus, the channel gain matrix $\mathbf{H}(t)$ can be a square matrix but not invertible. If $\mathbf{H}(t)$ is an invertible square matrix then we can estimate the transmitted signal as:

$$\hat{\mathbf{s}} = \mathbf{H}^{-1}(\mathbf{r} - \mathbf{w})\,. \tag{7.167}$$

What about the detection of the transmitted message signal when $L > K$? As the channel gain matrix $\mathbf{H}(t)$ is not a square matrix we have to compute its left pseudoinverse using the least-squares method (LSM). In the LSM we seek the signal $\hat{\mathbf{s}}$ which minimises the detection error

$$||\mathbf{e}||^2 = ||\mathbf{r} - \mathbf{H}\mathbf{s}||^2\,. \tag{7.168}$$

Prior to proceeding let us note the following vector differentiation identities. If $f(\mathbf{x}) = \mathbf{c}^T\mathbf{x} = \sum_{n=1}^{N} c_n x_n$ and $g(\mathbf{y}) = \mathbf{y}^T\mathbf{H}\mathbf{y}$ then $\frac{\partial f(\mathbf{x})}{\partial \mathbf{x}} = [c_1, c_2, \ldots, c_N]^T$ and $\frac{\partial g(\mathbf{y})}{\partial \mathbf{x}} = 2\mathbf{H}\mathbf{y}$. We also note the following matrix-vector expansion

$$||\mathbf{e}||^2 = ||\mathbf{r} - \mathbf{H}\mathbf{s}||^2 = (\mathbf{r} - \mathbf{H}\mathbf{s})^T(\mathbf{r} - \mathbf{H}\mathbf{s})$$
$$= \mathbf{r}^T\mathbf{r} - 2\mathbf{s}^T\mathbf{H}^T\mathbf{r} + 2\mathbf{s}^T\mathbf{H}^T\mathbf{H}\mathbf{s}\,. \qquad (7.169)$$

The above identities assist us to obtain the partial differential of Eq. (7.168) as

$$\frac{\partial||\mathbf{e}||^2}{\partial \mathbf{s}} = 0 - 2\mathbf{H}^T\mathbf{r} + 2\mathbf{H}^T\mathbf{H}\mathbf{s}\,. \qquad (7.170)$$

Setting Eq. (7.170) to zero, i.e. $\frac{\partial||\mathbf{e}||^2}{\partial \mathbf{s}} \stackrel{!}{=} 0$, and rearranging terms yields the detection equation

$$\hat{\mathbf{s}} = (\mathbf{H}^T\mathbf{H})^{-1}\mathbf{H}^T\mathbf{r}. \qquad (7.171)$$

What about if the channel gain matrix $\mathbf{H}$ has complex values? In such a case we can replace the transposition with the Hermitian transposition to obtain the generalised formula

$$\hat{\mathbf{s}} = (\mathbf{H}^H\mathbf{H})^{-1}\mathbf{H}^H\mathbf{r}\,. \qquad (7.172)$$

The matrix $(\mathbf{H}^H\mathbf{H})^{-1}$ is referred to as the *left pseudoinverse* of the channel gain matrix $\mathbf{H}$. It is so called because $(\mathbf{H}^H\mathbf{H})^{-1}\mathbf{H}^H\mathbf{H} = \mathbf{H}^{-1}\mathbf{H}^{-H}\mathbf{H}^H\mathbf{H} = \mathbf{H}^{-1}\mathbf{I}\mathbf{H} = \mathbf{I}$.

In the above MIMO detection we know that the receive vector $\mathbf{r}(t)$ is known to the receiver. How does the receiver know the channel matrix or gains? Training sequence or channel sounding techniques should be used to estimate $\mathbf{H}(t)$.

7.6.4 *Check Your Understanding*

Students are encouraged to check their understanding of the content of this section with the following questions prior to proceeding to the next section.

(1) Consider the 2×2 MIMO system described by the of input-output equations system $\begin{bmatrix} y_1 \\ y_2 \end{bmatrix} = \begin{bmatrix} g_{11} & g_{12} \\ g_{21} & g_{22} \end{bmatrix} \begin{bmatrix} x_1 \\ x_2 \end{bmatrix} + \begin{bmatrix} w_1 \\ w_2 \end{bmatrix}$ where g_{kl} are the channel gains from transmit antenna l to receive antenna k, x_k is the binary symbol from transmit antenna k, w_k is the AWGN at receive antenna k, and y_k is the input to the receive antenna k. Assume a Rayleigh fading wireless channel with AWGN, and that the channel gains are i.i.d.

(a) Use MATLAB simulation to compute the 2×2 matrix of channel gains $\mathbf{G}$.

(b) Use MATLAB simulation to compute the vector $\mathbf{y}$ fed into the MIMO detector.

Hint. If $g_I \sim \mathcal{N}(0, \sigma^2)$ and $g_R \sim \mathcal{N}(0, \sigma^2)$. Then, $g = g_R + jg_I \sim \mathcal{CN}(0, \sigma^2)$ and $G = |g| = \sqrt{g_R^2 + g_I^2}$ is Rayleigh distributed with the pdf

$$f_G(g) = \frac{g}{\sigma^2} e^{-\frac{g^2}{2\sigma^2}}\,.$$

(2) Superimpose the curves of the MRC, EGC and MSD receivers according to Eqs. (7.164), (7.166) and (7.165). Order the achieveable capacities for a given signal-to-noise ratio, all other conditions being the same.

7.7 Student Projects

(1) Creation of spatial wireless channel models

(2) Modelling of MIMO channel other than AWGN with Rayleigh fading

(3) Estimation algorithms for OFDM-MIMO channels

(4) Determination of MIMO capacity in various transmission environments

(5) Software implementation of MIMO

(6) Hardware implementation of MIMO

(7) Co-channel interference (CCI) mitigation in MIMO systems

(8) Design of massive MIMO at eNodeB (aka base station) to minimise interference and save energy using narrow beams

7.8 Review of Important Terms and Acronyms

Acronyms, Abbreviations and Important Terms		
active antenna	adaptive antenna	aerial
Franhofer distance	antenna bandwidth	antenna diversity
aperture antenna	array antenna	array factor
backlobe	beamforming	beamwidth
boresight	broadside	broadside array antenna
circular polarisation	cross-polarisation	dipole antenna
directional antenna	directivity	dumb antenna
element factor	elliptical polarisation	EM waves
end fire	end-fire array (EFA)	equal-gain combining
ESLA	far-field region	feedpoint impedance
gain	grating lobes	Hansen-Woodyard EFA
horizontal polarisation	horn antenna	input impedance
IoT antennas	isotropic radiator	LHCP
linear polarisation	log-periodic antenna	MRC
MIMO	minor lobes	monopole antenna
near-field region	omnidirectional antenna	panel antenna
parabolic antenna	passive antenna	patched antenna
pattern factor	polarisation	primary lobe
printed antenna	radiation diagram	radiation resistance
radiative near-field	radiaton pattern	reactive near-field
receive diversity	reflection coefficient	RHCP
selection diversity	sidelobes	smart antenna
space-time coding	spatial diversity	transmit diversity
vertical polarisation	VSWR	water-filling principle
outage probability	whip antenna	wire antenna
Faraday's rotation	XPD	Yagi-Uda antenna

7.9 Supplementary Reading

(1) Constantine A. Balanis, *Antenna Theory: Analysis and Design*, John Wiley & Sons, Inc., 2016.

(2) Clayton R. Paul, *Introduction to Electromagnetic Compatibility*, John Wiley & Sons, Inc., 2006.

(3) ARRL, `The ARRL Antenna Handbook for Radio Communications`, ARRL, 2018.

(4) Qi Luo and Steven Shichang Gao, *Low-cost Smart Antennas*, John Wiley & Sons, Inc., 2019.

(5) Rakhesh Singh Kshetrimayum, *Fundamentals of MIMO Wireless Communications*, Cambridge University Press, 2017.

Chapter 8

Multiplexing and Multiple Access Methods

Chapter's Learning Outcomes

Both multiplexing and multiple access methods have three-fold purposes: allowing multiple users to share finite communications resources equitably and efficiently with acceptable mutual disturbance. The main resource shared is the transmission medium, which has a few degrees of freedom, such as space, frequency, spreading code, time and power. This chapter explores multiplexing and multiple access principles and practice, including:[1]

(1) The need for multiplexing and multiple access schemes, and how these two techniques differ from each other.
(2) Classification of multi-access schemes into orthogonal multiple access (OMA) and non-orthogonal multiple access (NOMA), and case studies of each type.
(3) Treatment of all fundamental multiplexing schemes, including FDM, TDM, CDM, WDM and OFDM.
(4) Treatment of all fundamental orthogonal multiple access schemes, including FDMA, TDMA, CDMA, WDMA and OFDMA.
(5) Principles of NOMA and study of some popular NOMA schemes, e.g., LDS, LDS-OFDM, PDMA, MUSA and SCMA.

The largest scale of multiplexing, specifically, frequency division multiplexing (FDM), is the division of the radio frequency spectrum and the consequential allocation of each portion to a specific set of wireless communications technologies, such as 3–30 kHz band for maritime communications, 3–30 GHz for satellite and Wi-Fi and 87.5–108 MHz for FM radio.

8.1 Introduction

Multiplexing/demultiplexing (MUX/DEMUX) and multiple access schemes (MAS) have common goals. Both schemes belong to the specialization called radio resource management (RRM) in wireless communications. First, to allow multiple users to share a finite resource in a *fair* manner. Second, to maximize communications resource utilisation. The communications resource is mainly the transmission facility. Given a wireless bandwidth B, signal power P, interference power I, noise power N and signal-to-noise plus interference ratio (SNIR) $\frac{P}{N+I}$, Claude Shannon found the maximum supportable data rate as

$$C = B\log_2\left(\frac{P}{N+I}+1\right) = B\log_2(SNIR+1). \tag{8.1}$$

The error rate on the system can never be reduced to zero if we transmit data at a rate higher than this limit. However, using appropriate signal conditioning techniques, we can transmit data over the channel with negligible errors if the transmission rate does not exceed Eq. (8.1).

There is a strong need to transmit data over the wireless channel at the maximum possible rate. How can we achieve this? We can glean this from Eq. (8.1), namely

(1) use a large bandwidth, but low SNIR. This is where millimeter wave (mmWave) communica-

[1] "There is good in everything, if only we look for it." — Laura Ingalls Wilder.

tions reigns supreme. We can obtain a huge bandwidth from the mmWave frequency bands to achieve high data rates.

(2) use a high SNIR, but small bandwidth. This can be achieved by either increasing the signal power and/or reducing the noise power. Improving the data rate through these two methods comes with penalties.

(3) another way of maximising the channel's use is to share it efficiently using algorithms. This is the main goal of multiple access schemes (MASs) and this chapter.

MUX/DEMUX and MAS are used in both wireless and wired communications systems. In wireless communications they are among the topics studied in radio resource management [Zander and Kim (2001)]. They are used to manage resources, such as power (as in power control), channel allocation, channel measurements and radio frequency interference (RFI). The direction from the user terminal to the network (or base station) on a duplex link is called uplink, reverse link or upstream, while the opposite direction is called downlink, forward link or downstream. For example, human voice and data are multiplexed onto a telephone line in xDSL technologies.

8.1.1 *Multiplexing versus Multiple Access*

A single-user system in which the communications resources are used by a single user at a time does not need multiplexing nor multiple access schemes. Communication resources, however, are usually finite and have to be shared by multiple users. This is where multiplexing and multiple access schemes come into play. Both multiplexing and multiple access schemes seek to achieve a similar purpose, but traditionally in different directions of a duplex communications link.

There are subtle differences between multiple access control (MAC) and multiplexing (MUX) schemes. Table 8.1 attempts to summarise these differences. Traditionally, the word *multiplexing* means the sharing of a communications resource in the downlink, while *multiple access* means the sharing of a communications resource in the uplink. These are the traditional definitions of the two types of resource-sharing schemes, as illustrated in Fig. 8.1(a). The downlink channel is referred to as **broadcast channel** as one device sends information to multiple devices simultaneously. The uplink channel is referred to as **multiple access channel** or **multi-access channel** as more than one device independently sends information to one receiver. Both multiplexing and MAC techniques are used to increase resource utilisation efficiency. In recent years, however, the distinction between multiplexing and MA is becoming blur and, somehow, confusing.

As illustrated in Fig. 8.1(b), LTE/LTE-A (a 4G wireless technology) uses orthogonal frequency division multiple access (OFDMA) to share its resources in the downlink and single-carrier frequency division multiple access (SC-FDMA) for its uplink. The blurring use of multiplexing and multiple access has created the need for more terminology. We now have **downlink multiplexing** which is OFDMA and **uplink multiplexing** which is SC-FDMA in 4G mobile technologies. Also, modern communications standards allow direct communications between wireless handsets. The direct link between user devices is called **sidelink**. The **backhaul** is used to refer to the portion of the network comprising the intermediate links between the core or backbone network and the small subnetworks at the edge of the entire network. For example, the link interconnecting eNodeBs are backhaul links.

Figure 8.1 mentions LTE and LTE-Advanced (aka LTE-A). What is the difference between them? We have summarised their basic differences in Fig. 8.2. The interested reader may consult the standard documents for further details.

Multiplexing simply refers to simultaneous sharing of network infrastructure, usually the transmission systems, by many different information sources. Transmission of signals from a single source over a dedicated transmission link whether the link type is radio, wire or cable, is inefficient.[2] Moreover, some telecommunications systems are rarely designed to monitor only one source (transducer).[3]

[2]Exception is a point-to-point system, in which each data source has a direct link to the other.

[3]A sensor is a transducer which translates physical data such as temperature into electrical signals.

Table 8.1 Differences between multiplexing and multiple access schemes.

	Multiplexing	Multiple Access (MA)
1)	Usually implemented in the physical layer (i.e., ISO/OSI Layer 1). However, TCP at Layer 4 multiplexes multiple sockets/processes to the network layer.	Implemented in both the physical and data link layers (i.e., ISO/OSI Layers 1 and 2).
2)	Traditionally used to refer to resource sharing in the downlink/forward link.	Traditionally used to refer to resource sharing in the uplink/reverse link.
3)	Resource needs of users sharing the resource are fairly fixed and resources are allocated to users a priori.	Resource needs of users sharing the resource are varying and resources are allocated dynamically.
4)	Concerns with *how* multiple signals can share a single communications channel.	Concerns with *which* signal among multiple signals is allowed to use/occupy which particular communications resource. Stated differently, MA prescribes how the resource sharing is coordinated or regulated.
5)	The resource sharing takes place at a local site. A device (e.g., base station) sends information (possibly different) to multiple devices (or users) simultaneously over the broadcast channel.	The resource sharing takes place remotely. Multiple devices send signals independently to one receiver.
6)	Signals multiplexed come from the same node. For example, in the downlink of a cellular mobile system, the base station multiplexes signals and sends them to multiple mobile. devices	Signals multiplexed come from different sources/transmitters. For example, in the uplink of a cellular mobile system, signals from multiple sources located at different positions share the transmission medium.

The classification of multiple access control (MAC) schemes into those operating at the physical layer and those at the data link layer is not straightforward. Some authors classify the channelisation schemes as Layer 1 schemes and the dynamic schemes as Layer 2 schemes. Others differentiate between a MAC protocol and a MAC technique/scheme, and claim that MAC protocols operate at Layer 2 while MAC techniques/schemes operate at Layer 1. With the popularity of channel-aware MAC and rate/link adaptation MAC, these categorizations may not be very relevant, even if the strict classification were valid.

In general, the number of data sources is large and using a separate transmission link for each one is not practical. Therefore, most communications systems combine the signals from a number of sources into a single composite signal for transmission over one transmission link (Fig. 8.3). The process of combining these signals is called **multiplexing**, and each individual signal occupies a **channel** in the multiplex system. The channel in this case can be a fraction of a period of transmission time, a fraction of a frequency spectrum, etc. The reverse process of multiplexing is called **demultiplexing** and it occurs at the receive end of the transmission system.

In recap, multiplexing

(1) Allows the *efficient* utilisation of an expensive transmission facility (e.g., bandwidth), as the latter usually has much wider bandwidth than required to transmit a single message signal. Hence, part of the channel is wasted without multiplexing.
(2) Allows simultaneous sharing of a single finite-capacity transmission channel by different users.
(3) Signals multiplexed to share a communications physical resource are traditionally *orthogonal* in

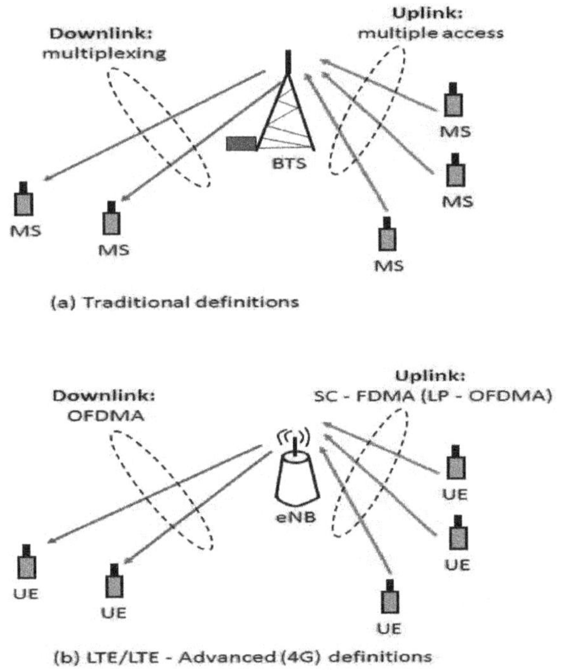

Fig. 8.1 The difference between multiplexing and multiple access schemes.

a remarkable parameter to enable their separation (i.e., demultiplexing) at the receiver. Orthogonality is achieved in frequency, time, wavelength, code sequence or space. The orthogonality is needed to eliminate multi-user interference (MUI) or multiple access interference (MAI), which is the mutual interference between signals sharing a physical radio resource.

For obvious reasons, multiplexors and demultiplexors are also referred to as *concentrator* and *deconcentrator*, respectively (Fig. 8.3). By their functionality, a demultiplexor is also called signal *splitter* or inverse multiplexor. The channel concentrator shown in Fig. 8.3 is referred to as $N : 1$ multiplexer, while its corresponding channel deconcentrator is called $1 : N$ demultiplexer.

Figure 8.4 shows the classification of the main multiplexing schemes. FDM, WDM, TDM and CDM achieves orthogonalization between signals in frequency, wavelength, time and code sequence, respectively. An antenna is used to convert signals between the electrical electromagnetic (EM) wave in wireless communications systems. We shall study each of these schemes sequentially in the following sections. As intimated earlier, traditionally, the uplink counterpart of the downlink is called multiple access scheme. Therefore, each of the multiplexing schemes shown in Fig. 8.4 ideally has its multiple access counterpart. For example, the multiple access version of FDM is frequency division multiple access (FDMA). Multiple access schemes being investigated for modern and future networks include both non-orthogonal multiple access (NOMA) and the traditional orthogonal multiple access (OMA). NOMA schemes include multi-user shared access (MUSA), sparse code multiple access (SCMA), resource spread multiple access (RSMA), pattern

Feature	LTE	LTE-Advanced
Specification	3GPP Releases 8/9	3GPP Release 10
Device categories (LTE Cat.)	1 through 5	6 or higher
Carrier aggregation (CA)	not allowed	allowed (can combine bandwidths in different bands)
System bandwidths (MHz)	1.4, 3, 5, 10, 15, 20 (symmetrical)	70 (DL), 40 (UL)
Subcarrier allocation	contiguous	Contiguous and clustered (uplink)
Peak data rate (UL/DL)	75/300 Mbps	500/1000 Mbps
Simultaneous PUCCH and PUSCH transmission	not allowed	allowed
Multiple access (UL/DL)	DFTS-OFDM/OFDMA	SC-FDMA/Hybrid OFDMA
MIMO (UL/DL)	1x1 (SISO) / 4x4	4x4 / 8x8
Peak spectral efficiency (bps/Hz) UL/DL	4.32 with 64-QAM SISO / 16.3 with 4x4 MIMO	15 with 4x4 MIMO / 30 with 8x8 MIMO
Coordinated Multi-Point transmission /reception (CoMP)	not supported	supported
Machine Type Communications (MTC)	not supported	supported
Multimedia Broadcast and Multicast Services (MBMS)	not supported	supported

Fig. 8.2 Some differences between LTE and LTE-Advanced.

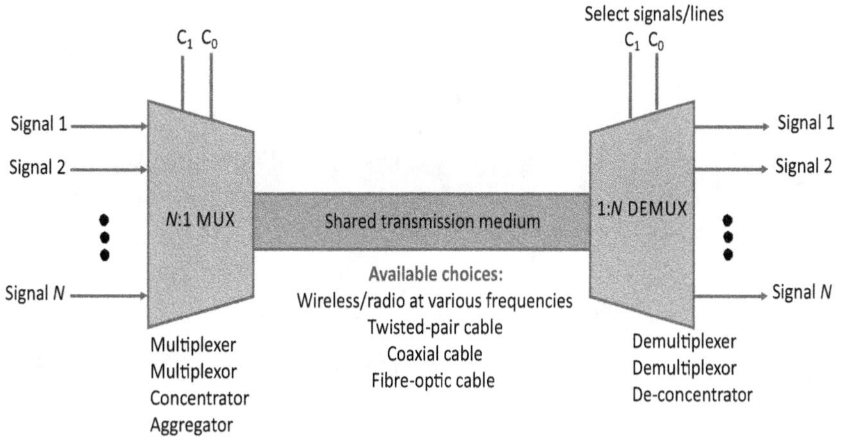

Fig. 8.3 Illustrating the multiplexing principle.

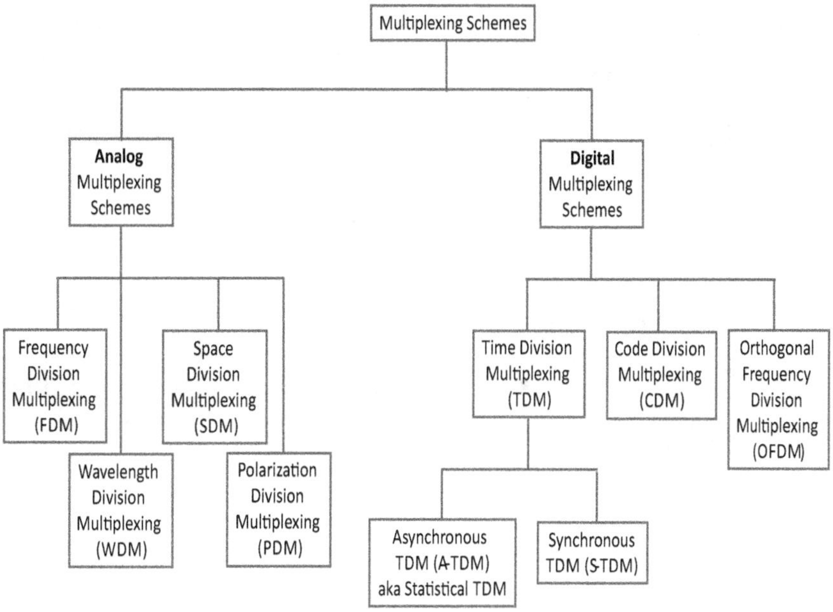

Fig. 8.4 Classification of multiplexing schemes.

division multiple access (PDMA), and multi-user shared access (MUSA). Figure 8.9 illustrates the difference between three multiple access schemes.

An EM wave comprises three orthogonal components: the electric field component, the magnetic field component and the direction of travel of the wave. The direction of the electric field component in an EM wave is called polarisation. Horizontal polarisation (HP) is orthogonal to vertical polarisation (VP), and so is left-hand circular polarisation (LHCP) to right-hand circular polarisation (RHCP). We can multiplex two signals whose polarisations are orthogonal to share the same frequency band at the same time without remarkable interference at the detector. Achieving orthogonality in polarisation is what we refer to as **polarisation division multiplexing** (PDM) in Fig. 8.4.

Wavelength division multiplexing (WDM) and wavelength division multiple access schemes use light pulses instead of electrical signals for information exchange. These multiplexers combine multiple data streams using different wavelengths

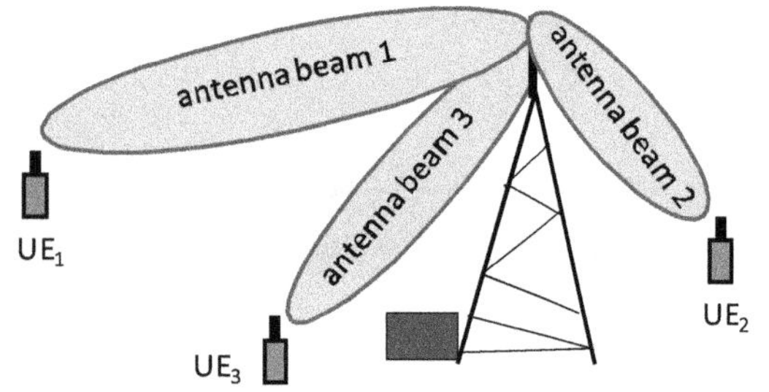

Fig. 8.5 Illustration of SDMA using multi-beam antennas.

into a single beam of light to share a single physical resource. The applications of WDM/WDMA are popular in fiber-optic communications systems. SONET and SDH multiplexers are the two popular fiber-optic based multiplexing hierarchies utilising WDM and dense WDM.

Spatial division multiplexing (SDM): Directional antennas can be used to radiate different geographical directions to create non-overlapping beams (Fig. 8.5). Non-overlapping beams are orthogonal, allowing the beams to carry the information of different users. An example of SDM is shown in Fig. 8.5, in which three orthogonal beams are used to serve three users simultaneously. Space division multiplexing (SDM), or just spatial multiplexing (SM), allows multiple signals to share the same physical radio resource at the same time, but in different geographical locations. There are two principal ways of achieving SDM/SM, namely:

- Use a multi-beam adaptive array antenna (or smart antenna) whose beams are separated spatially so that each beam serves a different user.
- The cellular architecture which reuses the same set of frequencies after each reuse distance D or guard space. The reuse distance is the spatial separation between the co-channel cells. If the reuse distance is large enough then the CCI (co-channel interference), which is here the equivalence of the MAI/MUI, does not interfere with the performance of the system in any remarkable manner.

SDMA is used in cellular communications, for example. In this application, the same set of frequencies or channels are used in multiple cells separated by a minimum geographical distance referred to as reuse distance. Cells using the same set of frequencies are called co-channel cells. We leave the further treatment of this topic to the chapter in this book dedicated to fundamentals of mobile communications. In cellular wireless communications systems SDM/SDMA is an add-on scheme. It usually superimposed on TDM/TDMA, FDM/FDMA or CDMA in the form of frequency reuse after every co-channel reuse distance.

In recap, multiplexing refers to

(1) A number of low bandwidth signals or channels share simultaneously a higher bandwidth transmission medium whose bandwidth is at least equal to the total bandwidth of the signals multiplexed.
(2) For two or more signals to share a transmission link at the same time they should be *orthogonal* in a measurable dimension, i.e., should not overlap in all the domains of time, frequency and code, or a derivative thereof.

8.1.2 *Orthogonality: The Foundation of Multiplexing*

We have mentioned above that traditional multi-access schemes enable multiple users to share communications resources which are orthogonal in frequency, time, spreading code, etc. Thus, signals sharing a communications resource in traditional multi-access schemes are orthogonal (or almost orthogonal) in some measurable dimension. So, *what is orthogonality?* Orthogonality of signals means that the signals do not overlap in all the domains of time, frequency and code, or a derivative thereof. Orthogonality stems from linear algebra. In linear algebra, two vectors $\mathbf{v}_1$ and $\mathbf{v}_2$ are said to be orthogonal if their inner product (aka dot product or scalar product) equals zero. An example of orthogonal vectors is shown in Fig. 8.6(a), while two orthogonal analog waveforms are illustrated in Fig. 8.6(b) and two orthogonal digital waveforms are illustrated in Fig. 8.6(c).

The **inner product** is defined as

$$\mathbf{v}_1 \cdot \mathbf{v}_2 = ||\mathbf{v}_1|| \cdot ||\mathbf{v}_2|| \cos\theta \tag{8.2}$$

where θ is the angle between the two vectors. In Fig. 8.6(a), we have $\mathbf{v}_1 = [0, 2]^T$ and $\mathbf{v}_2 = [2, 0]^T$

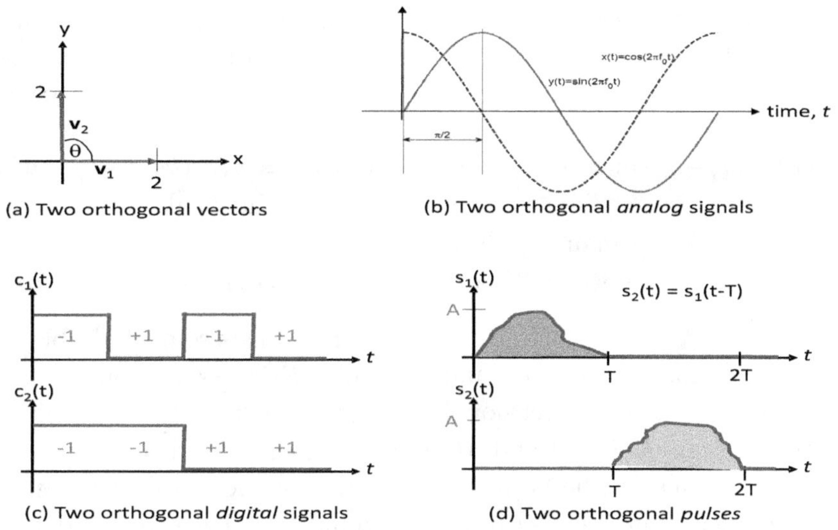

Fig. 8.6 Orthogonal vectors and signals.

and $\theta = \pi/2$. Thus,

$$\begin{aligned}\mathbf{v}_1 \cdot \mathbf{v}_2 &= ||\mathbf{v}_1|| \cdot ||\mathbf{v}_2|| \cos\theta \\ &= 2 \cdot 2\cos(\pi/2) = 4 \cdot 0 = 0. \end{aligned} \quad (8.3)$$

Orthogonal signals are also referred to as *perpendicular signals*. Signals can be treated as vectors. For example, $x(t) = \cos(\omega_0 t)$ and $y(t) = \sin(\omega_0 t)$, where $w_0 = 2\pi f_0$ are orthogonal, because their inner product is zero, i.e.,

$$\begin{aligned} x(t) \cdot y(t) &= \int_0^T x(t)y(t)dt = \int_0^T \cos(\omega_0 t)\sin(\omega_0 t)dt \\ &= \frac{1}{2}\int_0^T \sin(2\omega_0 t)dt = \frac{\cos(2\omega_0 T) - \cos(0)}{4\omega_0} \\ &= \frac{\cos(4\pi) - \cos(0)}{4\omega_0} = \frac{1-1}{4\omega_0} = 0 \end{aligned} \quad (8.4)$$

where $T = \frac{\omega_0}{2\pi}$ is the period of each of the two signals. We can observe in Fig. 8.6(b) that the two signals $x(t)$ and $y(t)$ have a phase difference of $\pi/2$, validating their orthogonality.

Orthogonality measures the similarity between vectors and signals. A signal or vector is perfectly self-similar to itself and therefore its inner product to itself is unity. Orthogonal signals, however, are 100% dissimilar and thus have zero inner product. Orthogonal signals are mutually independent of each other, the feature exploited in multiplexing and multiple access schemes to avoid remarkable mutual interference between signals. The signals sharing a common finite resource can be partially orthogonal so long as their mutual interference do not remarkably increase the detection error of each.

A popular set of orthogonal functions are the harmonic complex exponential functions, which are defined as

$$\begin{aligned}\varphi_m(t) = e^{jm\omega_0 t} &= \cos(m\omega_0 t) \\ &+ j\sin(m\omega_0 t), \ m = 1, 2, \ldots \end{aligned} \quad (8.5)$$

where the fundamental frequency is f_0 and mf_0 are referred to as harmonic frequencies. The inner product of these functions is defined as

$$\begin{aligned}\varphi_m(t) \cdot \varphi_n(t) &= \frac{1}{T}\int_0^T \varphi_m(t)\varphi_n^*(t)dt \\ &= \frac{1}{T}\int_0^T e^{j(m-n)\omega_0 t}dt \\ &= \frac{e^{j(m-n)\omega_0 T} - 1}{j(m-n)\omega_0 T} \\ &= \begin{cases} 1, \ \texttt{if} \ m = n \\ 0, \ \texttt{if} \ m \neq n \end{cases}. \end{aligned} \quad (8.6)$$

This is the reason why the complex exponential functions are called orthonormal basis functions. They are used to express a periodic signal into sums of sinusoids referred to as Fourier series. That is, a real periodic signal $s(t)$, which is periodic in $T = \frac{2\pi}{\omega_0}$, can be expressed as

$$s(t) = \sum_{n=-\infty}^{\infty} c_n \varphi_n(t) = \sum_{n=-\infty}^{\infty} c_n e^{jn\omega_0 t}. \quad (8.7)$$

Equation (8.7) is referred to as *complex Fourier series*. The real version is obtained if we set $c_n = \alpha_n + j\beta_n$ to produce

$$s(t) = c_0 + 2\sum_{n=1}^{\infty} \alpha_n \cos(n\omega_0 t) - \beta_n \sin(n\omega_0 t)\,. \quad (8.8)$$

The orthogonality of signals has many applications in communications systems besides multiplexing. For example, it is used in:

(1) Digital modulation schemes: the in-phase and the quadrature components in two-dimensional modulation schemes, such as M-QAM and M-PSK, are orthogonal signals. The in-phase and the quadrature carriers components can be moderated by different set of data streams.
(2) A signal which is horizontally polarised is orthogonal to another signal which is vertically polarised. Also a left-hand circular polarised signal is orthogonal to a right-hand circular polarised signal. Signals which use orthogonal polarisations can share a transmission medium without interfering with each other.

Let us summarise some properties of basic orthogonal functions.

(1) If the two signals $s_1(t)$ and $s_2(t)$ are orthogonal, then the energy E_x and the power P_x in the composite signal $x(t) = s_1(t) + s_2(t)$ are $E_x = E_1 + E_2$ and $P_x = P_1 + P_2$, respectively, where E_i, $i = 1, 2$ and P_i, $i = 1, 2$ are the energy and power in the original signals.
(2) The two signals $s_1(t) = a$ (i.e., a dc signal) and $s_2(t) = b\cos(k\omega_0 t + \varphi)$ are orthogonal.
(3) The two signals $s_1(t) = a\cos(k\omega_0 t + \varphi_1)$ and $s_2(t) = b\cos(l\omega_0 t + \varphi_2)$, where k and l are integers but $k \neq l$, are orthogonal.
(4) The two signals $s_1(t) = a\sin(k\omega_0 t + \varphi)$ and $s_2(t) = b\cos(k\omega_0 t + \varphi)$, where k is an integer, are orthogonal.

8.1.3 *Medium Access Control Schemes*

Figure 8.7 lists some desirable features of MAC protocols. *Fairness* is about ensuring that each of the users competing for the finite communication resource is not disadvantaged. It does not necessarily mean that every user receives the same service. For example, we have weighted fairness, which can be based on service fee, quality-of-service class of the traffic, etc. *High throughput* means that the effective bit rate experienced by a user should be as high as possible. The *cell edge throughput* is usually used as that is the worst case throughput.

Users and thus their applications desire a *bounded delay*, especially low delay for both channel access and time to deliver their transmitted data. *Low energy consumption* is needed to slow down depletion of the energy of battery-powered devices. This is especially important for devices with non-rechargeable battery and thus their lifetimes are limited by battery depletion time. Prolonging the battery energy depletion time is also important for mobile terminals. We seek MAC schemes which can differentiate traffic of different features and thus requiring different *quality of service* (QoS). The capacity and size of networks are not static but change over time. Thus, protocols which allow the

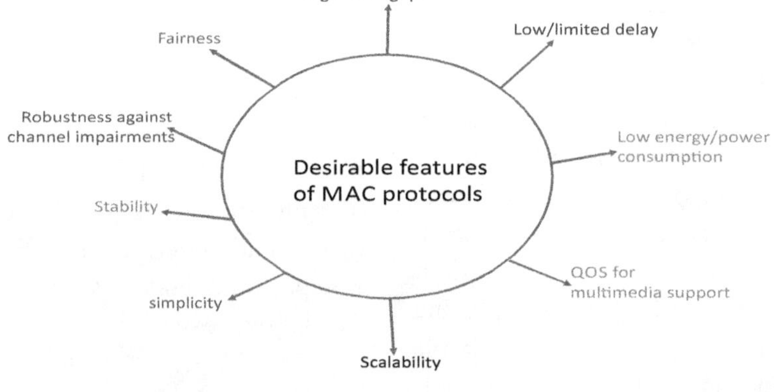

Fig. 8.7 Desirable features of media access control protocols.

network to *scale easily* (i.e., expand or shrink) are desirable.

Simple algorithms are preferred over complicated ones solving the same problem. Simplicity reduces device costs and energy consumption. *Stability* is important in every protocol. Protocols which converge fast and deliver definite results are desirable. Wireless channels impair signals transferred over them via many types of mechanisms. The cumulative effect is called fading. We desire MAC protocols which are *robust*, that are able to operate even amidst deep fading, especially multipath fading.

Although not straightforward, we attempt to classify the basic medium access control (MAC) schemes in this section.

8.1.3.1 *Classification of Multiple Access Control Schemes*

MAC schemes are protocols implemented in either the physical layer (i.e., ISO/OSI Layer 1) and/or the sublayer of the data link control layer (i.e., ISO/OSI Layer 2). MAC schemes arbitrate between multiple users competing for access to a shared communications resource, either a medium or a device. Without an appropriate MAC scheme there would be too much collisions (i.e., multiple access interference or MAI) to make communications possible for systems shared by multiple users. MAC schemes answer two questions: *which* of the multiple nodes is allowed to use the common resource at a given time, and *how* is the resource shared. The art and science of allowing multiple users to share a resource or facility is called *multi-user communications*. In mobile radio communications, for example, multiple mobile stations compete for access to a shared radio spectrum (a medium) or a radio base station (a device).

Although difficult, we attempt to classify MAC schemes in Fig. 8.8. MAC protocols can be synchronous or asynchronous. In a synchronous MAC scheme all nodes are synchronised to a master clock time which is broadcast regularly on a signalling channel often referred to as a beacon. Asynchronous MAC scheme, however, uses a distributed control mechanism to control channel access as nodes are in general not synchronised to any common clock. MAC protocols can also be sender-initiated or receiver-initiated. Furthermore, MAC schemes can be categorised into:

- channelisation schemes (e.g. FDMA, TDMA, CDMA, OFDMA),
- scheduling (e.g. OCASD, PFS, TAOS, WFQ, GPS, FCFS), and
- contention-based (e.g., aloha and variants, CSMA and variants).

Channelisation schemes divide the resource into time intervals (called slots), frequency subbands or

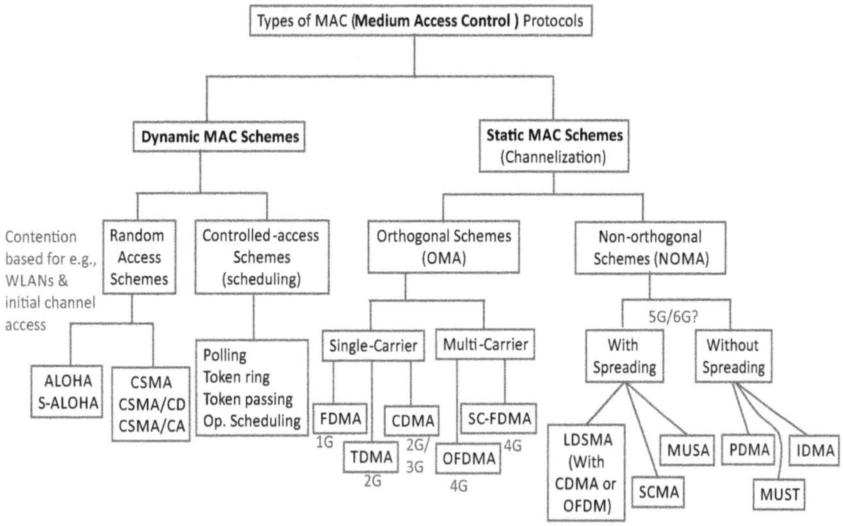

Fig. 8.8 Classification of medium access control schemes with examples.

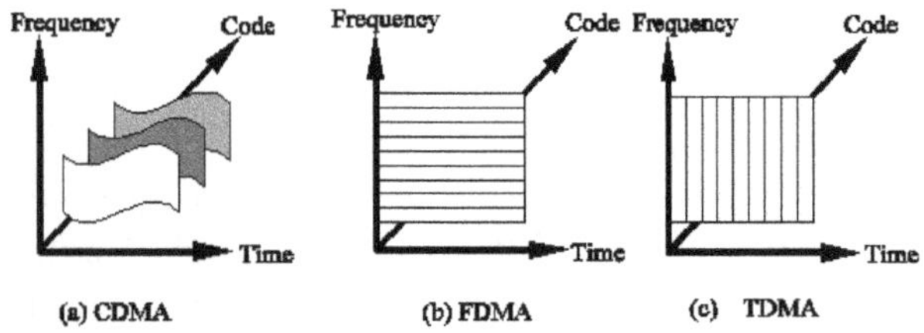

Fig. 8.9 Differentiation between basic channelisation MAC schemes.

use different noise-like sequences (called codes) to allow the sharing of a communications resource by multiple users. Channelisation schemes are usually used in systems, such as cellular networks, which have centralised nodes which can maintain centralised synchronisation across all network users. Figure 8.9 illustrates pictorially the differences between the three basic channelisation MAC schemes.

Controlled-access MAC schemes use an algorithm to allocate resources to different users in an orderly manner. There are no collisions or multiple access interference in systems using controlled-access MAC schemes, such as scheduling. There are different types of scheduling algorithms, such as channel-aware scheduling, link/rate adaptation based scheduling, round robin, polling and token passing. In fact, channelisation MAC schemes are a special class of scheduling, and thus belong to controlled-access schemes.

Random access schemes allow the users seeking access to a common resource to contend for it. Such schemes are usually used in local area networks. They are also used by mobile devices in cellular wireless networks to solicit traffic channels from their serving base stations prior to data transfer. Random access schemes are liable to collisions as more than one user can attempt access to the system about the same time. Therefore, there are various mechanisms used to control collisions in systems using random access schemes.

8.1.3.2 *A Theory on Multiple Access Schemes*

Much of the presentation here is based on the article [Jamalipour *et al.* (2005)], and it is valid for BPSK modulation. Let us define the following:

- $a_k(t)$: multiple access function of user k at time t.
- $b_k(t)$: signal of user k at time t.
- f_k, f_c: carrier frequency.
- T_b, $R_b = \frac{1}{T_b}$: bit period and data rate.
- $b_{i,k}(t) \in \{\pm 1\}$: ith bit in user k data at time t.
- $p_x(t) = 1,\ 0 \le t \le x$ and $p_x(t) = 0$ otherwise: pulse shaping function.
- T_c: chip duration for CDMA spreading codes.

Then, the signal transmitted by user k is

$$s_k(t) = 2\sqrt{P} a_k b_k(t) \cos(2\pi f_c t) \tag{8.9}$$

where

$$b_k(t) = \sum_{i=-\infty}^{\infty} b_{i,k}(t) p_{T_b}(t - iT_b)$$

and

$$a_k(t) = \begin{cases} N_f p_{T_b}(t - kT_b) & : FDMA \\ N p_{T_b}(t - kT_b) & : TDMA \\ \sqrt{2P} \sum_{l=1}^{N_f} a_k^l \cos(2\pi f_l t) & : MC-CDMA \\ \sqrt{2P} \sum_{l=1}^{N_f} a_k^l p_{T_c}(t - lT_b) & : DS-CDMA \end{cases} \tag{8.10}$$

The signature sequence $\mathbf{a}_k = \{a_k^1, a_k^2, \ldots, a_k^N\}$ is periodic with the period $N = T_b/T_c$. The cross-correlation between signals of different users is measured by

$$\int_{T_b} a_k(t) a_l(t) dt < \epsilon, \ k \neq l\,. \tag{8.11}$$

Perfect orthogonality between signals of different users means that $\epsilon = 0$. In order to achieve the same data rate and bandwidth for spreading in time and frequency domains the following condition must be fulfilled $(1+N_f)/T_b = (1+\alpha)/T_c$, yielding $1 + N_f = (1+\alpha)N$, where α is the roll-off factor of the Nyquist filter used for pulse shaping. The achievable generic signal-to-noise-plus-interference ratio (SNIR) is [Jamalipour *et al.* (2005)]

$$SNIR = \frac{1}{\rho\epsilon(K-1) + \frac{N_0}{2E_b}}, \ \texttt{where}$$

$$\begin{cases} \epsilon = 0 & : FDMA \\ \epsilon = 0 & : TDMA \\ \epsilon = N_f/2 & : MC-CDMA \\ \epsilon = (1-\alpha/4) & : DS-CDMA \end{cases} \tag{8.12}$$

where N_0 is background noise power spectral density, E_b is the energy in a bit, and ρ is the average interference cancellation factor which is defined in [Jamalipour *et al.* (2005)] as

$$\rho = \frac{I_{residual}}{I_{total}} \qquad (8.13)$$

where $I_{residual}$ is the average remaining interference power after cancellation, and I_{total} is the total interference power without cancellation.

8.1.4 *Check Your Understanding*

Students are encouraged to check their understanding of the content of this section with the following questions prior to proceeding to the next section.

(1) In groups of two to four students, discuss the difference between multiplexing and multiple access schemes.

(2) Discuss the main importance of multiplexing and multiple access schemes in communications systems.

(3) What is the name of the functionality which performs multiplexing?

(4) What is the opposite process of multiplexing?

(5) At which ISO/OSI protocol layer does signal multiplexing usually occur? What about multiple access schemes?

(6) Discuss the reason(s) why signal demultiplexing is also referred to as deconcentrator.

(7) Discuss the reason(s) why the principle of orthogonality is important in multiplexing and multiple access schemes.

(8) Research to find at least two applications of the orthogonality principle other than in telecommunications networks.

(9) (a) Sketch two analog signals which are orthogonal.
  (b) Repeat Part (a) for digital signals.
  (c) Discuss in which dimension(s) is the orthogonality achieved in both cases.

(10) Assume that the signal $x(t)$ has the complex Fourier coefficients $c_3 = 2+3j$, $c_{-3} = 2-3j$ and $c_n = 0, \forall n \neq \{\pm 3\}$.
  (a) Compute the full expression of $x(t)$.
  (b) Plot $x(t)$ using MATLAB over $-\frac{4\pi}{\omega_0} \leq t \leq \frac{4\pi}{\omega_0}$.
  (c) Compute the fundamental frequency $f_0 = \frac{\omega_0}{2\pi}$ in $x(t)$ if $x(t = t_0) = 0$, $t_0 = 520\,\mu$s.
  (d) List all the harmonic frequencies in $x(t)$.

8.2 FDM and FDMA

Assume that a given communications system uses a total bandwidth of B Hz. This bandwidth is divided into K portions, referred to as channels, each of size W Hz. We require that $B > KW$, and the remaining bandwidth $B - KW$ is used as spaces between the K channels and thus referred to as **guard band**. A rule-of-thumb is to use about 10% of the overall system bandwidth for guard bands. This is the case with the LTE/LTE-A system (see Table 8.2). The size of each guard band is about $W_g = \frac{B-KW}{K+1}$, as there are $K+1$ guard bands around the K channels. Thus, yielding the FDM/FDMA efficiency

$$\eta_f = \frac{KW}{B} = \frac{B - W_g}{B}. \qquad (8.14)$$

Every user in the system is assigned one of the K channels to use all the time for the duration of his/her communication. In the downlink/downstream/forward link we refer to the scheme described above as FDMA (frequency division multiple access). The same principle, if applied to the uplink/upstream/reverse link, is referred to as FDM (frequency division multiplexing). For example, the POTS/PSTN landline system uses $W = 4$ kHz while the GSM cellular mobile system uses $W = 200$ kHz.

Some applications of FDM/FDMA in practical systems are:

- The legacy telephony system PSTN/POTS switching hierarchy.
- Satellite systems.

Table 8.2 Standard LTE network bandwidths and their properties.

Bandwidth, B (MHz)	Number of PRBs	Useful Bandwidth (MHz)	Guard Band (MHz)	Downlink Subcarriers	Uplink Subcarriers
20	100	18	2	1201	1200
15	75	13.5	1.5	901	900
10	50	9	1	601	600
5	25	4.5	0.5	301	300
3	15	2.7	0.3	181	180
1.4	6	1.08	0.32	73	72

- Cable TV. A television channel multiplexes the three signals video, audio and colour using three different frequencies.
- FM radio broadcasting, including Stereo FM.
- Cellular mobile communications systems use FDM to achieve frequency-division duplexing (FDD) transmissions. First generation (1G) cellular mobile communications systems used mainly FDM.

FDM was developed in the early 1900s for the analog telephony system. FDM allows many channels (i.e., information sources) to simultaneously share a single communications channel by dividing and assigning a portion of the channel's bandwidth to each of the channels in the composite, as illustrated in Fig. 8.10. Thus, FDM achieves orthogonality between signals multiplexed in the frequency domain. Consequently, if $S_k(f)$ and $S_l(f)$ are the frequency domain versions of two signals then, in the *ideal* case, FDM requires that

$$\int_{-\infty}^{\infty} S_k(f)S_l(f)df = \begin{cases} 0, & \texttt{if } k \neq l \\ C, & \texttt{if } k = l \end{cases} \tag{8.15}$$

where C is a real-valued finite constant. Although FDM can transport both analog and digital information, it is usually used in analog communications systems. The architecture of an FDM system multiplexing N signals is illustrated in Fig. 8.10. As often each of the N messages come from a different user, they are also referred to as *channels.* At the FDM transmitter each of the N message signals $s_i(t)$ (Fig. 8.11) is modulated by a corresponding sinusoidal carrier whose frequency is generated by a sinusoidal oscillator. These N modulated signals are then summed and transferred onto the transmission channel, which can be wireless or wireline.

The carrier frequencies $f_{ci}, 1 \leq i \leq N$ are selected in such a way that the modulated message signals do not overlap in the FDM spectrum. As indicated in Fig. 8.11, usually spaces called **guard bands** are created between the individual channels to minimise adjacent channel interference (ACI).

As illustrated in Fig. 8.11(e), in the PSTN/POTS system only one sideband of each voice signal is transmitted to conserve bandwidth. Each voice signal occupies about 3.1 kHz bandwidth (i.e., $B_{k\text{L}} \approx 0.3$ kHz and $B_{k\text{H}} \approx 3.4$) kHz. However, 4 kHz is allocated to each voice signal, resulting in ca. 600 Hz for guard band.

At the receiver RF bandpass filters BPF_i whose passbands are centred around the corresponding frequencies f_{ci} are used to **demultiplex** (i.e., separate) the N channels, which are then *demodulated.* For synchronous demodulators, replicas of the carrier frequencies f_{ci} are required, as shown in Fig. 8.10. As the demodulated signals contain unwanted high-frequency signals (including carrier shifted message signal and noise), the demodulator output is fed into a lowpass filter LPF_i. The cut-off frequency of LPF_i is about the maximum frequency content in the message signal $s_i(t)$. Usually all the message signals multiplexed on a single channel in FDM systems conform to the same modulation format, such as FC-AM, DSBSC-AM, VSB-AM, etc. The 4-channel FDM spectrum shown in Fig. 8.11(e) uses single sideband amplitude modulation, and only the lower sidebands (LSBs) are transmitted. Due to its relatively small bandwidth consumption compared to FM and PM, AM is more popular in FDM systems, especially single sideband (SSB) AM.

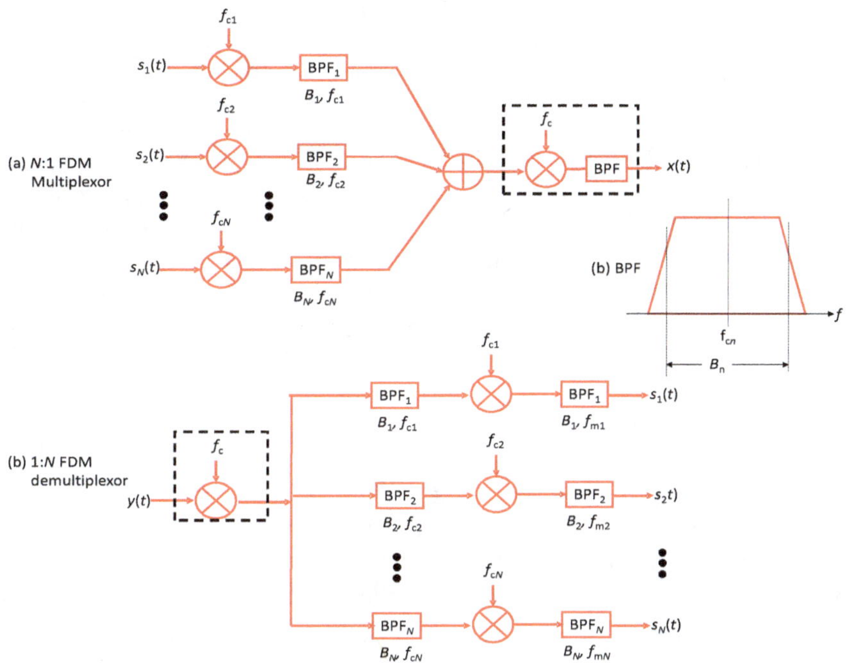

Fig. 8.10 Illustration of the FDM principle.

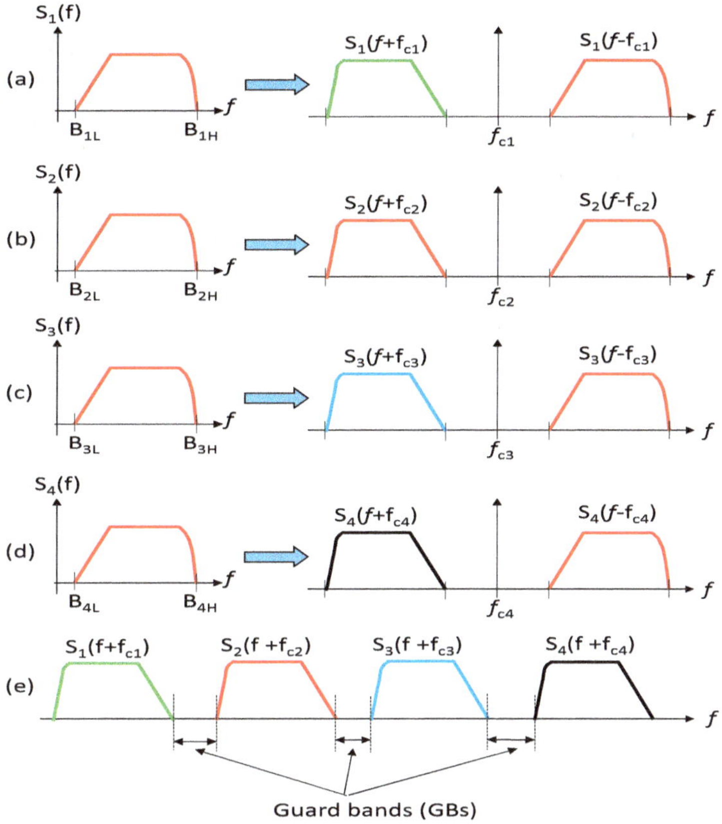

Fig. 8.11 (a)–(d) Four baseband voice signals of the same bandwidth sharing a channel via FDM and their (e) FDM spectrum of lower sidebands: $f_{ck} = f_{ck-1} + 4$ kHz; $B_{k\text{H}} \approx B_{k\text{L}} + 3.1$ kHz.

Table 8.3 North American (AT&T) FDM telephony hierarchy.

Aggregation Level	Number of Voice Channels N	Frequency Range (kHz)	Structure
Group MUX	12	60 - 108	12 voice ch.
Supergroup MUX	60	312 - 552	5 groups
Mastergroup MUX	600	564 - 3 084	10 supergroups
Mastergroup MUX	1 200 - 3 600	312 564 - 17 548	varying
Jumbogroup MUX	3600	564 - 17 548	6 mastergroups
Jumbogroup MUX	10 800	3 124 - 60 566	3 jumbogroups

The main processing performed by the FDM multiplexor and demultiplexor are:

- Analog modulation and demodulation.
- Bandpass filtering using bandpass filters (BPFs) whose center frequencies are at the respective carrier frequencies.

8.2.1 *FDM in PSTN/POTS*

Table 8.3 shows the standard FDM hierarchy used in the North American telephony system. This analog carrier system uses a single-sideband AM modulation to conserve spectrum, and it was designed to use FDM to transfer voiceband signals over high-bandwidth links. In the North American FDM hierarchy, the first-level multiplexor combines $N = 12$ 4-kHz voice signals to form a group. The second-level multiplexor combines $N = 5$ 48-kHz groups to form a supergroup, and so forth. The carrier frequencies for channel i in a

$$\begin{aligned} \texttt{Group is:}\quad & f_{ci} = 58 + 4i,\ i = 1, 2, \ldots, 12 \\ \texttt{Supergroup is:}\quad & f_{ci} = 288 + 48i,\ i = 1, 2, \ldots, 5 \\ \texttt{Mastergroup is:}\quad & f_{ci} = 444 + 240i,\ i = 1, 2, \ldots, 10 \end{aligned} \tag{8.16}$$

The bandwidth of the bandpass filters in the FDM used for POTS/PSTN is $B_n \approx 4$ kHz. This is enough for voice signal bandwidth of about 3.1 kHz (i.e., 0.3–3.4 kHz).

8.2.2 *FDM in 1G Cellular Mobile Systems*

Five major first generation (1G) cellular mobile communications technologies were deployed in the world in the 1980's. These were Nippon Telephone Telegraph (NTT) in Japan, Nordic Mobile Telephone (NMT) in Scandinavia, Total Access Communications Systems (TACS) in UK, American Mobile Phone System (AMPS) in North America, and C450 in Germany (then West Germany). All of the 1G technologies used FDM and frequency division duplexing (FDD). Figure 8.12 summarises the basic features of these technologies for various geographical locations of the world. For example, PLMN operators of AMPS acquired 25 MHz bandwidth and divided it into 416 duplex channels, 21 channels for signalling and 395 channels for voice transmission. The 832 carrier frequencies are located at

$$\begin{aligned} f(k)_{\texttt{uplink}} &= 825 + 0.03k\ \texttt{MHz},\ 1 \le k \le 799 \\ f(k)_{\texttt{uplink}} &= 825 + 0.03(k - 1023)\ \texttt{MHz}, \\ & \qquad 991 \le k \le 1023 \\ f(k)_{\texttt{downlink}} &= f(k)_{\texttt{uplink}} + 45\ \texttt{MHz} \end{aligned} \tag{8.17}$$

It is clear from Fig. 8.12 that the 1G cellular technologies had differing features, rendering them incompatible with each other. Larger frequency deviations were used for audio channels than control channels in all the technologies. This makes sense as user signals require a lager data rate and consequent bandwidth than signalling.

Example

GSM uses the frequency bands shown in Table 8.4. Channel spacing in GSM is fixed at $W = 200$ kHz.

(1) Compute the number of duplex FDM/FDMA channels available in each system.

System Feature	C450	TACS	NTT	NMT	AMPS
Originated country	Germany	England	Japan	Scandinavia	North America
Downlink band (MHz)	461.3-465.74	917-950	870-885	463-467.5	869-894
Uplink band (MHz)	451.3-455.74	872-905	925-940	453-457.5	824-849
DL/UL frequency separation (MHz)	10	45	55	10	45
Number of simplex channels	222	1320 (incl. 21×2 control channels)	600	180	832 (incl. 21×2 control channels)
Channel spacing/ bandwidth (kHz)	20	25	5, 12, 25	25	30
Data rate (kbps)	5.28	8	0.3	1.2	10
Cell radius (km)	5-30	2-20	5-10	1.8-40	2-20
FM frequency deviation audio/control channel (kHz)	±4 / ±2.5	±9.5 / ±6.4	±5 / ±4.5	±5 / ±3.5	±12 / ±8

Fig. 8.12 Basic features of 1G cellular mobile communications technologies.

Table 8.4 GSM frequency bands.

System	Uplink band (MHz)	Downlink band (MHz)	Bandwidth, B (MHz)
GSM 850	824–849	869–894	25
GSM 900	890–915	935–960	25
GSM 1800	1710–1785	1805–1880	75
GSM 1900	1850–1910	1930–1990	60

(2) What type of duplexing does the GSM technology use?
(3) Compute the guard band separating the uplink band from the downlink band of frequencies in each system.
(4) Compute the minimum frequency separation between the uplink and the downlink of a duplex channel in GSM system.

Solution

(1) Each of the system uses duplex channels. The number of full duplex channels are

$$N_{\mathrm{ch}} = \frac{B}{W} = 125, 125, 375, 300$$

for GSM850, GSM900, GSM1800 and GSM1900, respectively. Each of these channels is divided into 8 time slots in GSM.
(2) A different frequency band is used for the uplink and downlink transmissions. Therefore, GSM uses frequency division duplexing (FDD).
(3) The guard band separating the uplink from the downlink in each system is 20 MHz.
(4) The minimum frequency separation between the uplink and the downlink of a duplex channel in each GSM system is 45 MHz, 45 MHz, 95 MHz and 80 MHz for GSM850, GSM900, GSM1800 and GSM1900, respectively.

8.2.3 *Problems in FDM*

FDM is liable to two main weaknesses: intermodulation distortion (IMD) or intermodulation (IM) noise and crosstalk noise. FDM systems usually use guard bands to separate the signals in order to mitigate adjacent channel interference. However, the signals can spread in transmission and overlap in frequency, resulting in crosstalk. More details of crosstalk can be found in the chapter on telecommunications traffic engineering.

IM occurs if any device or component in the transmission medium behaves nonlinearly. Examples of such devices are transformers and high-power amplifiers. A travelling wave tube amplifier (TWTA) used in satellite systems, for example, must be operated in its linear range if FDM is used to avoid IM. The problem in such operation is the low gain, necessitating either a high transmit power or a low effective radiated power. IM arises if multiple signals at different frequencies or a single multi-frequency signal pass through a nonlinear device. Let us illustrate IMD using a simple example.

Assume that the two signals $s_1(t) = \sin(2\pi f_1 t)$ and $s_2(t) = \sin(2\pi f_2 t)$ share a transmission medium using FDM. The composite signal $x(t) = s_1(t) + s_2(t)$ leave the multiplexor and enters the transmission medium. Assume that a power amplifier in the system behaves nonlinearly and squares its input signal. Find the frequencies in the signal $y(t)$ appearing at the input of the FDM demultiplexor.

The signal entering the demultiplexor is

$$\begin{aligned} y(t) &= x^2(t) = [s_1(t) + s_2(t)]^2 \\ &= s_1^2(t) + s_2^2(t) + 2s_1^2(t)s_2^2(t) \end{aligned}$$

which can be simplified to $1 - \frac{1}{2}\cos(4\pi f_1 t) - \frac{1}{2}\cos(4\pi f_2 t) + \cos(2\pi(f_1 - f_2)t) - \cos(2\pi(f_1 + f_2)t)$. Thus, the new frequencies introduced into the multiplexed signals by the transmission medium are: $2f_1$, $2f_2$, $f_1 - f_2$, $f_1 + f_2$ and a dc (i.e., 0 Hz). Interestingly, the original frequencies are missing in the demultiplexor's input waveform. The non-dc IMD produced in this example are called second-order inter-modulation products (IMPs).

8.2.4 *Check Your Understanding*

Students are encouraged to check their understanding of the content of this section with the following questions prior to proceeding to the next section.

(1) Each analog AM radio broadcasting station is allocated a 10-kHz spectrum in the frequency band around 535 kHz to 1605 kHz. The signals of all AM radio stations are based on FDM.

(a) List the possible carrier frequencies for the first three contiguous AM radio stations.

(b) Draw the composite spectrum, indicating guard bands and the spectrum of the individual stations. Research to find the appropriate guard bands used in AM radio.

(c) Sketch a block diagram of the FDM system of three AM radio signals, showing all its components (e.g., local oscillators, the moderator/heterodyne and the filters) with their calibrations.

Hint. Use the LSBs of single-sideband AM modulation.

(2) Use Fig. 8.10 to explain the mathematics involved in the multiplexing and demultiplexing processes.

(3) An FM radio signal is allocated 200 kHz bandwidth, and it comprises the four FDM-multiplexed signals: the 14.96-kHz monaural audio signal, the 19-kHz stereo pilot carrier, the 30-kHz double sideband suppressed-carrier amplitude-modulated stereo audio signal, and a double sideband suppressed-carrier amplitude-modulated signal called radio broadcast data system (RDS) which is centered at 57 kHz. There is a 25-kHz guard band at either side of an AM radio signal to mitigate adjacent channel interference (ACI). An FM radio broadcaster in Australia, called 96three, broadcasts at 96.3 MHz.

(a) Sketch the composite FM stereo radio spectrum of 96three, showing the spectrum of the individual signals and the required guard band frequency locations.

(b) Sketch a block diagram of the 96three FM system indicating the details of the required mixing (or heterodyning), the local oscillators and the filtering.

(4) A radio receiver is tuned to receive the upper sideband of an FM station broadcasting on the carrier $f_c = 96.5$ MHz. The bandwidth of a baseband FM signal is 200 kHz. The frequency of the local oscillator used at the receiver for frequency translation or heterodyning is f_{LO}, where $f_{LO} > f_c$. The demodulation occurs in two stages. First, the signal is converted from the radio frequency (RF) to a lower intermediate

frequency (IF), and then to baseband audio frequency (AF). By using the appropriate block diagrams containing LOs and filters, explain the signal reception process. Indicate the center frequencies and the required bandwidths of the local oscillators and the filters used. Assume that the receiver uses two amplifiers: an RF amplifier and an IF amplifier, which are centred at 96.3 MHz and 5 MHz, respectively.

8.3 TDM and TDMA

Assume that a given communications system has a total bandwidth of B Hz. We decide on a fixed time period of duration $T_\mathtt{f}$ seconds. We refer to $T_\mathtt{f}$ as a **frame duration** or just a frame. We then divide $T_\mathtt{f}$ into K pieces, referred to as **time slots** or timeslots, each of which has the duration $T_\mathtt{s}$, so that $T_\mathtt{f} = KT_\mathtt{s}$. Each user in the system is allowed to use the entire system bandwidth B Hz, but only for one time slot duration, and another user takes turn. Some TDM/TDMA systems (e.g., GPRS and EDGE) allocate multiple time slots in a frame to a single user to increase the data transfer rate. In synchronous TDMA, such as used in GSM, time slots with the same index in multiple consecutive TDMA frames form a TDMA channel. Each active user is allocated a TDMA channel. A user can thus transfer a large file over many time slots in many consecutive TDMA frames. In the downlink/downstream/forward link we refer to the scheme described above as TDMA (time division multiple access). The same principle, if applied to the uplink/upstream/reverse link, is referred to as TDM (time division multiplexing). Each time slot contains a time frame referred to as **guard time** or **guard interval** or **guard period** (GP), which we denote as $T_\mathtt{g}$. No data is transmitted during the GP. Thus, the framing efficiency is

$$\eta_\mathtt{f} = \frac{T_\mathtt{f} - KT_\mathtt{g}}{T_\mathtt{f}} . \tag{8.18}$$

A better formula for the TDMA framing efficiency is

$$\eta_\mathtt{f} = 1 - \frac{(N_\mathtt{t} + N_\mathtt{r})b_\mathtt{g} + N_\mathtt{r}b_\mathtt{r} + N_\mathtt{t}b_\mathtt{p}}{T_\mathtt{f}R_\mathtt{b}} \tag{8.19}$$

where $N_\mathtt{t}$, $N_\mathtt{r}$ are the number of traffic and reference bursts per frame, $b_\mathtt{g}$ is number of bit per guard interval, $R_\mathtt{b}$ is bit rate, $b_\mathtt{r}$ is number of bits in each reference burst, and $b_\mathtt{p}$ is the number of bits in each preamble. A burst is equivalent to one TDMA time slot.

Example

In the GSM system, for example, channel bandwidth is $B = 200$ kHz, frame duration is $T_\mathtt{f} = 4.615$ ms and comprises $K = 8$ time slots. Each time slot comprises 156.25 bits and lasts $T_\mathtt{s} = 0.577$ ms. The GP in each time slot comprises 8.25 bits (Fig. 8.15). Compute the efficiency of a GSM frame.

Solution

A time slot comprises 156.25 bits and lasts $T_\mathtt{s} = 0.577$ ms. Thus, the 8.25 bits in each guard time lasts

$$T_\mathtt{g} = \frac{8.25 \texttt{ bits}}{156.25 \texttt{ bits}} \times 0.577 \texttt{ ms} \approx 30.5\,\mu\mathtt{s} .$$

Therefore, the framing efficiency is

$$\eta_\mathtt{f} = \frac{T_\mathtt{f} - KT_\mathtt{g}}{T_\mathtt{f}} = \frac{4.615 - 8 \times 30.5 \cdot 10^{-3}}{4.615} \approx 94.72\% .$$

The GP is used to avoid, or at least minimise, the overlapping of neighbouring bursts in A TDM/TDMA frame. Two factors contribute to the potential overlapping of bursts. These are echoes or signal reflections and signal propagation delays. These do not affect detection performance so long as the GP is longer than the time taken for the echoes to reach the receiver. This means that the length of the GP should be at least equal to the delay spread or the length of the impulse response of the channel (aka *channel order*) in question. A longer GP reduces a system's transmission efficiency, but can cope with a longer delay spread. The choice of the GP, then, requires a trade-off between bandwidth efficiency and tolerable channel delay spread. For this reason, some communications standards allow a choice between multiple GPs. For example, the IEEE802.11 allows the choice between two GPs, namely 0.4 μs and 0.8 μs, while IEEE802.11 uses only 0.8 μs. The DVB-T COFDM allows a choice between four GPS, namely $\{\frac{1}{32}, \frac{1}{16}, \frac{1}{8}, \frac{1}{4}\} \times T_s$, where T_s is the symbol duration.

Example

The GSM system uses 8.25 guard periods in each time slot. How much can the clocks in the system be out of synchronisation and still be able to detect the data without errors?

Solution

Guard periods are used to maintain synchronisation of the receivers between different slots and frames. TDMA/FDD systems intentionally provide several time slots of delay between the forward and reverse time slots for a particular user. Each bit in GSM's TDMA lasts $577\mu s/156.25 \approx 3.6928\mu s$. This is the maximum tolerable clock drift.

TDM was invented by a French Engineer by name Jean-Maurice-Emile Baudot around 1894 for the communications system at that time, the telegraph. The unit *baud* used to measure symbol rate was named after this inventor as an honor. TDM requires the orthogonality between signals sharing a communication resource in the time domain. Thus, if $s_k(t)$ and $s_l(t)$ are the time domain versions of the two signals then, in the ideal case, TDM requires that

$$\int_{-\infty}^{\infty} s_k(t)s_l(t)dt = \begin{cases} 0, & \texttt{if } k \neq l \\ C, & \texttt{if } k = l \end{cases} \quad (8.20)$$

where C is a real-valued finite constant. TDM can be divided into synchronous TDM and asynchronous TDM.

Besides the reduced transmission efficiency owing to the guard intervals, data sources in TDM/TDMA system require a highly stable and accurate clock in order to maintain a perfect synchronisation between the clocks of the transmitter (e.g., base station) and the receiver (e.g., mobile station). A mismatch of the clock frequencies causes *bit slips.* Bits are deleted in the receive clock slower than the source clock. However, spurious bits are inserted into the received data if the receive clock is faster than the source clock. Clocks used in TDM/TDMA systems can be pendulum, atomic or quartz crystal.

The accuracy of a clock is degraded over time and the causes can be aging, bit stuffing, imperfections in materials used to manufacture clock, temperature changes or external influences. Clock accuracy degradation caused by aging, temperature changes and slaving inaccuracies is referred to as **clock wander**. Clock accuracy degradation caused by phase noise and bit stuffing is referred to as **clock jitter**. All the components in the communications network have clocks which are synchronised to a very reliable master clock. In the GSM system, the master clock is that of the gateway mobile switching center (GMSC). The higher the data rate, the stricter the required clock accuracy, as a little clock drift results in multiple bit errors.

The base stations in a cellular network need not be perfectly synchronised on a time slot basis. However, synchronisation between cells are good for detection and interference management.Each base station periodically broadcast data (including synchronisation burst) to enable mobile devices to synchronise to them.

8.3.1 *Synchronous TDM*

An example of the applications of synchronous TDM (S-TDM) is found in T1 and E1 multiplexing hierarchies. The S-TDM is also called *fixed-assignment TDM/TDMA*. The S-TDM multiplexor allocates the same amount of time to each of the devices it serves in a given time frame. This is the case whether a device has data/signals to send or not. For this reason, S-TDM can waste resources as the time slots allocated to an idle device is wasted. It is worthy of emphasis that the word *synchronous* in S-TDM comes from the fact that time slots in a TDM frame/cycle are pre-allocated to devices/users in a fixed manner. S-TDM is only good if each user has enough data to fill up the time slot(s) allocated to him/her in each TDMA frame.

Synchronous TDM is illustrated in Fig. 8.13(a) and (b). For simplicity of illustration, we are considering only 4 TDM frames, and the TDM-based system is serving 3 users, U, V and W. In TDM frame 3 users U and V are idle. Although they have

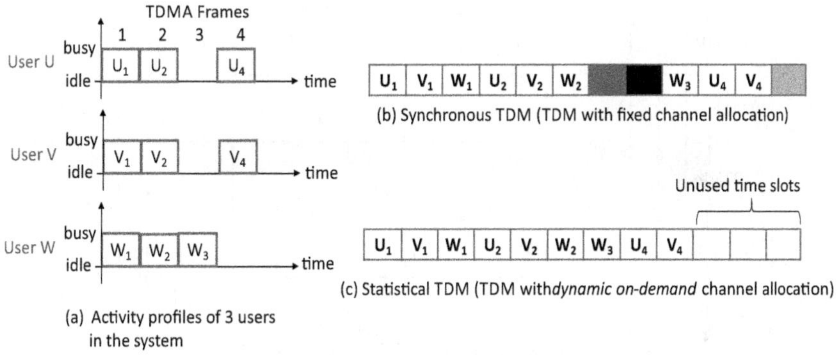

Fig. 8.13 Comparing synchronous TDM with asynchronous TDM.

no data to send, the system has reserved time slots for them, resulting in wastage. A similar trend occurs during frame 4 in which the time slot allocated to user W is wasted,

Timing in Synchronous TDM

Synchronisation TDM has two basic requirements on the multiple data sources sharing a medium. These are:

(1) The data rates of the sources should be rational multiples of each other. For example, if data source i sends data to the multiplexor at the rate R_i bps and the rate of another data source j is R_j bps, then it is expected that $R_i = nR_j$, $i \neq j$ and n is a positive integer. This is not always the case.
(2) The clocks of the data sources are perfectly synchronised. This is not always the case either. The clocks can drift away from each other for many reasons.

A technique used to achieve Point 1 is to use a very reliable master clock and require all data sources to synchronise their clocks to it. A technique used to achieve Point 2 is **pulse stuffing**. Assume that data from n data sources are multiplexed and each source is allocated R_b, $i = 1, 2, \ldots, n$ bps in each TDM frame to obtain the cumulative multiplexed rate of nR_i. During a particular frame if a source is able to deliver only data of rate $R_i < R_b$, then dummy bits must be added to bring the rate to the allocated rate of R_b. This requires the data rate of the stuffed bits to be $R_{ps} = R_b - R_i$ bps. Both the multiplexor and the demultiplexor must know the locations of the stuffed bits inside the TDM frame.

8.3.2 *Asynchronous TDM*

Asynchronous TDM (A-TDM) is also referred to as statistical TDM. A-TDM is used in, for example, Asynchronous Transfer Mode (ATM) networks. A-TDM is more intelligent than S-TDM as the length of the time slot it allocates to each device in a frame duration depends on need. A device which has nothing to send during a given time frame is not allocated time slot in that frame. The price to pay for the resource utilisation efficiency is increased complexity. The nature of A-TDM causes it to support a higher capacity than S-TDM using the same resources.

Asynchronous TDM is illustrated in Fig. 8.13(a) and (c). No time slot is allocated to users U and V and TDM/TDMA frame 3 as they have no data to send. The same treatment is accorded to user W during the TDM/TDMA frame 4 as it is idle. Thus, asynchrounous TDM avoids wastage of communication resource.

8.3.3 *Hybrid FDMA/TDMA Systems*

TDMA can be combined with FDMA to improve the spectral efficiency over that of TDMA and FDMA alone. Also, the hybrid FDMA/TDMA system is also known to offer robustness against tapping and interference better than either FDMA or TDMA. In FDMA/TDMA in system, each user is allowed to use a certain portion of the system

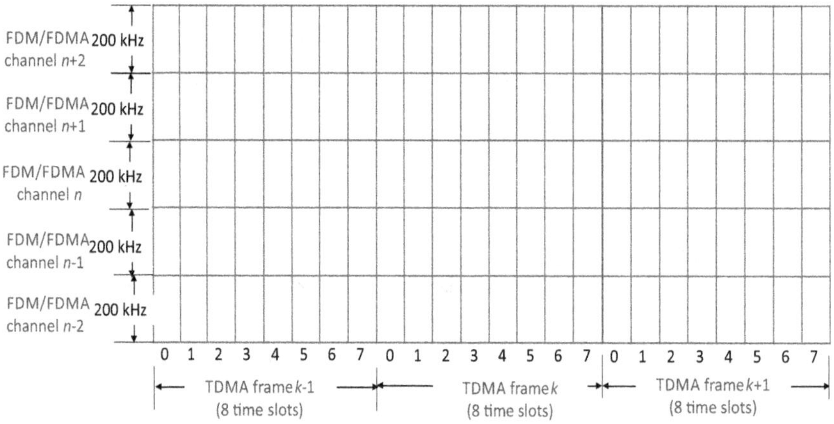

Fig. 8.14 TDMA/FDMA channelisation in GSM (each time slot contains a guard period).

bandwidth for a certain time frame. If B is the total system bandwidth, W is channel bandwidth (or carrier spacing), K is the number of time slots per channel and W_{g} is the total guardband, then we obtain the approximate capacities

$$\texttt{FDMA capacity: } N_{\texttt{FDMA}} = \frac{B - W_{\text{g}}}{W} \texttt{ users}$$

$$\texttt{TDMA capacity: } N_{\texttt{TDMA}} = K \texttt{ users}$$

$$\texttt{FDMA/TDMA capacity: } N_{\texttt{F/TDMA}} = \frac{B - W_{\text{g}}}{W} \times K \texttt{ users}. \tag{8.21}$$

The FDMA/TDMA approach is used in 2G cellular mobile technologies, such as GSM, as illustrated in Fig. 8.14. Each PLMN operator of GSM system acquired a minimum of 25-MHz bandwidth from the its local spectrum regulator and divided it into 124 frequency bands with carrier/centre frequencies at

$$f(k)_{\text{UL}} = 890.2 + 0.2(k-1) \texttt{ MHz}, \; 1 \le k \le 124$$
$$f(k)_{\text{DL}} = 935.2 + 0.2(k-1) \texttt{ MHz}, \; 1 \le k \le 124. \tag{8.22}$$

Thus, for example, channel 1 in GSM 900 uses the pair of carriers at 890.2 MHz (uplink) and 925.2 MHz (downlink) in the band 880–915 MHz (uplink) and 925–960 MHz (downlink). This band is used in LTE as band 8 in EMEA and APAC regions of the world. The bands for the channels are, respectively

$$f(k)_{\text{UL}} - 0.1 \le CH(k)_{\text{UL}} \le f(k)_{\text{UL}} + 0.1, \; 1 \le k \le 124$$
$$f(k)_{\text{DL}} - 0.1 \le CH(k)_{\text{DL}} \le f(k)_{\text{DL}} + 0.1, \; 1 \le k \le 124 \tag{8.23}$$

as there are 100 kHz of guardband at either side of each carrier frequency.

An uplink (UL) and its corresponding downlink (DL) are separated by 45 MHz. Each carrier frequency of 200-kHz width is divided into 8 time slots, each of which is one voice channel (see Fig. 8.15). A sequence of timeslots in contiguous multiple frames make up a channel in TDM/TDMA. A user is assigned the same time slot in each TDMA frame for the duration of the communication. This is illustrated in Fig. 8.15. In this illustration, the data for user 1 is transported in timeslot 5 of each TDMA frame. Each timeslot comprises 156.25 bits, of which 6 bits are tail bits (TBs), 8.25 bits are for the guard period (GP), 26 bits are training sequence (TS) and the remaining $2 \times 57 = 114$ bits are used carry user or signalling data. The 156.25 bits are transferred within the timeslot of 577 μs duration, yielding the raw data rate of 270.79 kbps per time slot. The training bits are used to estimate the near-instantaneous quality or state of the wireless channel.

Example: Data Rate of FDMA versus TDMA

We seek to compare the capacity of FDMA and TDMA over time period of duration T_{f}, which can

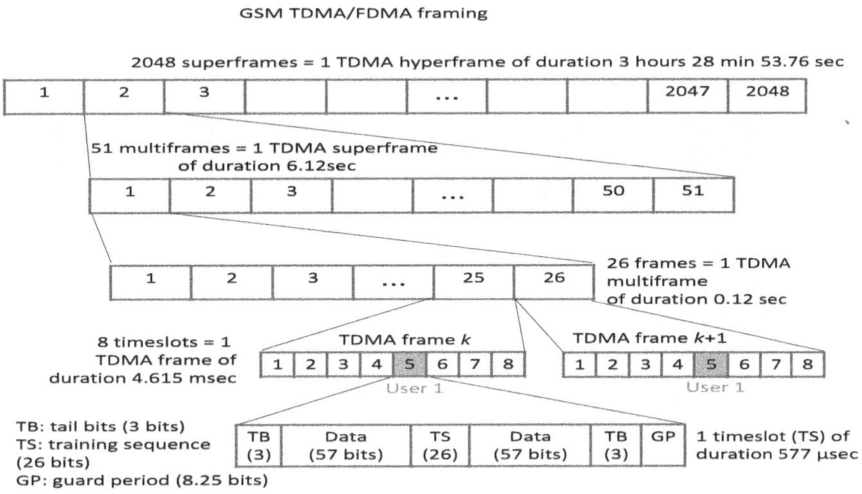

Fig. 8.15 GSM TDMA/FDMA framing.

be referred to as a frame. Assume that a given telecommunications system has the bandwidth B Hz and serves N_u users. The total transmit power in the system at a time is P W. In TDMA each of the N_u users is allocated the entire system bandwidth B Hz but for a fraction $\frac{1}{N_\mathrm{u}}$ of the time $T_\mathtt{f}$, referred to as time slot of duration $\frac{T_\mathtt{f}}{N_\mathrm{u}}$. Once, a TDMA user is given the permission to transmit, it does so with the total system power P W. Thus, the data rate per TDMA user in a time slot (which is equal to the average data rate) is

$$R_{\mathtt{b,TDMA}} = \frac{1}{N_\mathrm{u}} \times B \log_2\left(1 + \frac{P}{BN_0}\right) = \frac{B}{N_\mathrm{u}} \log_2\left(1 + \frac{P}{BN_0}\right). \tag{8.24}$$

In FDMA system, each of the N_u users is allocated the bandwidth $\frac{B}{N_\mathrm{u}}$ to use during the entire frame $T_\mathtt{f}$. The total power in the system is shared among the users so each transmits at $\frac{P}{N_\mathrm{u}}$ W, resulting in the average capacity per user

$$R_{\mathtt{b,FDMA}} = \frac{B}{N_\mathrm{u}} \log_2\left(1 + \frac{P/N_\mathrm{u}}{(B/N_\mathrm{u})N_0}\right) = \frac{B}{N_\mathrm{u}} \log_2\left(1 + \frac{P}{BN_0}\right). \tag{8.25}$$

This means that both TDMA and FDMA achieve the same capacity.

8.3.4 *Check Your Understanding*

Students are encouraged to check their understanding of the content of this section with the following questions prior to proceeding to the next section.

(1) Compute the bandwidth efficiency of one timeslot within a GSM frame.
(2) In GSM system, the uplink and the downlink channels used to serve a given user are separated by three time slots. This means that a mobile station does not transmit and receive at the same time. A mobile station can use this time to measure the quality of the channel. How much maximum time does the mobile station have available for the channel measurements?
(3) In GSM, the mobile station sends its channel measurements to the switch (MSC) every 480 ms or every 960 ms. How long should the coherence time of the channel be for such measurements to be useful?
(4) Discuss whether the TDMA used in GSM is asynchronous or synchronous TDMA.
(5) **In-class group research**: The logical channels used in the GSM system are

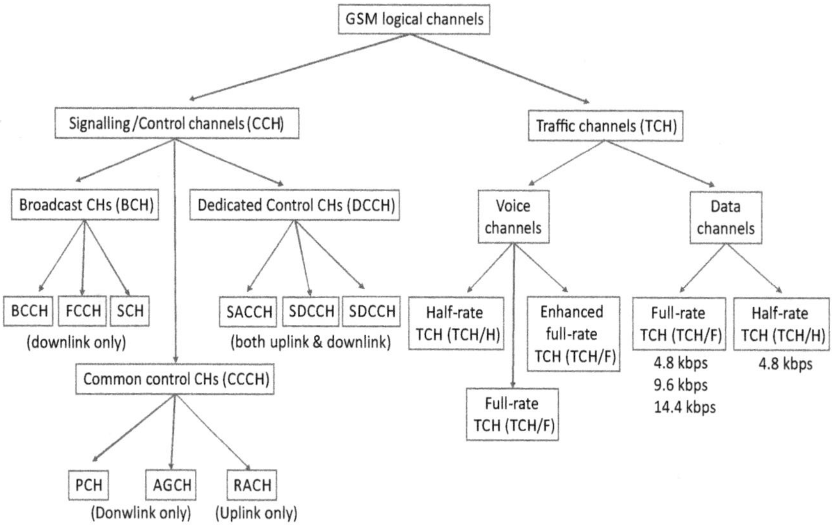

Fig. 8.16 Logical channels used in GSM system.

shown in Fig. 8.16. In groups of 3 to 4 students, research to find the functions of each of the channels. The figure indicates the duplex directions in which each channel is used to initiate the discussions.

(6) (a) Discuss the purposes of the guard period in TDM/TDMA-based systems. (b) Explain the main disadvantage in using guard periods in a communications system. (c) State the needed trade-off in deciding the duration of the guard period in a TDM/TDMA-based communications technology.

(7) (a) Compute the duration of the guard period in each time slot within a TDMA frame of GSM system.
(b) How far away from a mobile user can the farthest signal reflector be located without causing detection problem in a GSM system?

(8) Assume that the effective symbol duration in the 8-MHz DVD-T system is 448 μs. The standard allows its implementers to choose between the guard periods $\{\frac{1}{32}, \frac{1}{16}, \frac{1}{8}, \frac{1}{4}\} \times T_s$, where T_s is the symbol period. With respect to multipath delay spread, find the range of wireless channels tolerable by signals transferred in the DVD-T system.

(9) Synchronous optical network (SONET) is a synchronous TDM used over fiber-optic media. The SONET multiplexer, STS-1/OC-1 with 54.84 Mbps output data rate, reads the input buffers of each of the 672 data sources at 64 kbps. Assume that two of the data sources are slow and can only cope with 40 kbps and 56 kbps. Using appropriate figures and quantitative analysis, illustrate the operation of this multiplexer. Show any required pulse stuffing to achieve synchronisation across the data sources.

8.4 CDM and CDMA

Code division multiplexing (CDM) and code division multiple access (CDMA) are spread spectrum (SS) techniques. All SS schemes expand a signal at the transmitter and compress it to its original bandwidth at the receiver to achieve many advantages discussed later. Figure 8.17 illustrates the principle. The signal $s_1(t)$ occupies the bandwidth $W = B_1 - 0 = B_1$ and has the power spectral density (psd) S_1. Prior to transmission, the signal is expanded to cover the larger bandwidth $B_{ss} = B_2 - 0 = B_2$ but transmitted with the reduced psd S_2. The higher the spreading, the lower the psd of the spread signal becomes. As illustrated

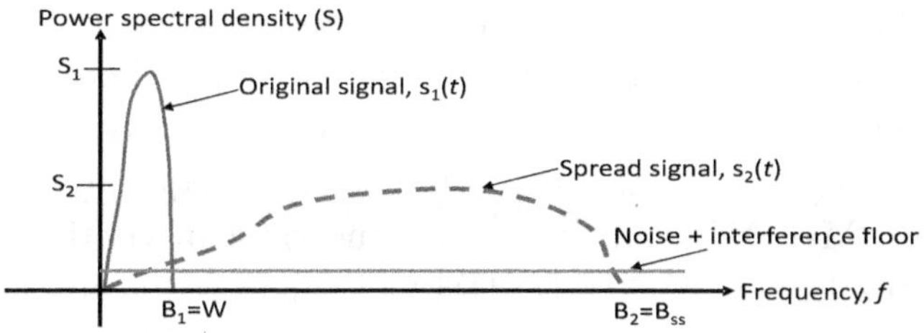

Fig. 8.17 Illustration of the spread spectrum principle.

in Fig. 8.17, the psd of the spread signal is lower than the psd of the cumulative noise and interference in the system. This is exactly one of the desired features in SS techniques.

CDM/CDMA achieve orthogonality between different signals sharing a common resource using *spectrum spreading codes*, referred to as **signature sequences**. The spreading code, made of multiple chips, of a user should be known by both the sender and receiver of the message spread by the code. The codes used by different users may not be perfectly orthogonal in practical systems, but must possess sufficient uncorrelatedness. Thus, each of the multiple users sharing a resource in a CDM system uses the entire system bandwidth all the time. If c_k is the spreading code for user 1's signals in a CDM/CDMA system, then, in the ideal case, we have

$$\langle c_k, c_l \rangle = c_k \cdot c_l = \begin{cases} 0, & \texttt{if } k \neq l \\ C, & \texttt{if } k = l \end{cases} \tag{8.26}$$

where C is a real constant. Orthogonality of pseudo-noise sequences used in spread-spectrum systems become non-orthogonal if they lose their synchronisation in the transmission system.

What about the application of SS techniques in practical systems?

(1) CDMA is applied in most of the 3G cellular mobile technologies, under the umbrella of IMT-2000, and they are:
  (a) IMT Direct Spread (DS-CDMA, aka UTRA-FDD)
  (b) IMT Time Code (TD-CDMA, aka UTRA-TDD)
  (c) IMT Multicarrier (MC-CDMA, aka CDMA 2000)
  (d) IMT Single Carrier (SC-CDMA, aka UWC-136)
  (e) IMT Frequency Time (IMT-FT, aka DECT)
(2) Wireless local area networks (IEEE802.11a, IEEE802.11b, IEE802.11g, IEEE802.11) and Bluetooth.
(3) Satellite communications: all the 24 satellites in the Global Positioning System (GPS) use 10-bit Gold codes to broadcast signals at the same frequency.
(4) Military applications: JTIDS/MIDS family of military radios, Link-16 used by NATO to exchange tactical data. It is believed that the Hollywood actress Hedy Lamarr and pianist George Antheil were the first to describe spread spectrum techniques in 1941 in their U.S. patent 2.292.387. It was later that the USA military adopted it. Two reasons account for this. First, the military held tightly onto the intellectual property. Second, it was economically unfeasible to implement SS techniques in mass-market products until technology developed to a certain level.

An important parameter in SS systems is the **processing gain**, which the ratio

$$G_{\mathrm{p}} = \frac{\texttt{chip rate}}{\texttt{data rate}} = \frac{R_{\mathrm{c}}}{R_{\mathrm{b}}} = \frac{T_{\mathrm{b}}}{T_{\mathrm{c}}} = \frac{B_{\mathrm{SS}}}{W} \tag{8.27}$$

where T_{c} is chip duration, $R_{\mathrm{c}} = \frac{1}{T_{\mathrm{c}}}$ is chip rate and $R_{\mathrm{b}} = \frac{1}{T_{\mathrm{b}}}$ is the bit rate. A *chip* is the smallest continuous waveform in SS techniques. We can increase the noise immunity and all the features of SS techniques by increasing the processing gain. The drawback in using a large processing gain is the increased bandwidth, requiring an appropriate trade off between the two. Another important parameter in SS systems is the **spreading factor** (SF), which is defined as

$$SF = \frac{\texttt{chip rate}}{\texttt{symbol rate}} = \frac{R_{\mathrm{c}}}{R_{\mathrm{s}}} = \frac{T_{\mathrm{s}}}{T_{\mathrm{c}}}. \tag{8.28}$$

The spreading factor equals the length of the spreading code. As the symbol rate cannot be larger than the bit rate, the SF is smaller than or equal to the processing gain. Their relationship is

$$G_{\mathrm{p}} = SF + \texttt{(coding gain)} - \texttt{(upper \& physical layer control bits)} \tag{8.29}$$

The fourth key formula in SS techniques is the relationship between the signal-to-noise ratio (SNR) before and after spreading, and the spreading gain. If N_0 W/Hz is the one-sided psd of the additive noise, then the noise power in the spread system is $N = B_{\mathrm{ss}}N_0$ W. The power in the information signal is $P = E_{\mathrm{b}}R_{\mathrm{b}}$, where E_{b} is the energy in a bit. Let us define the SNR prior to spreading as $SNR_{\mathtt{in}} = \frac{P}{N}$ and the SNR after spreading as $SNR_{\mathrm{out}} = \frac{E_{\mathrm{b}}}{N_0}$. Then, we have

$$SNR_{\mathtt{in}} = \frac{P}{N} \stackrel{!}{=} \frac{E_{\mathrm{b}}R_{\mathrm{b}}}{B_{\mathrm{ss}}N_0} = \frac{E_{\mathrm{b}}}{N_0} \times \frac{R_{\mathrm{b}}}{B_{\mathrm{ss}}} = \frac{SNR_{\mathrm{out}}}{G_{\mathrm{p}}}. \tag{8.30}$$

We can express Eq. (8.30) in decibels as

$$SNR_{\mathrm{out}}\ \texttt{[dB]} = SNR_{\mathtt{in}}\ \texttt{[dB]} + G_{\mathrm{p}}\ \texttt{[dB]}. \tag{8.31}$$

Example

A typical data rate for adaptive multi-rate (AMR) voice, video signals and 384 packet-switched service are 12.2 kbps, 64 kbps and 384 kbps, respectively. WCDMA/UMTS uses a fixed chip rate of 3.84 Mcps. (a) Compare the processing gains of the two signals. (b) Compute the corresponding required signal-to-noise ratio in the transmitted signal if the receiver requires the minimum signal-to-noise ratio of 8 dB.

Solution

Given are: $SNR_{\mathrm{out}} = 8$ dB, $R_{\mathrm{c}} = 3.84$ Mcps and $R_{\mathrm{b}} = \{12.2, 64, 384\}$ kbps. (a) The processing gains are obtained from Eq. (8.27) as $G_{\mathrm{p}} \approx \{25, 17.8, 10\}$ dB, respectively. We obtain the received SNR from Eq. (8.31) as $SNR_{\mathrm{out}} = \{-17, -9.8, -2\}$ dB, respectively.

FDM/FDMA was the first multi-access scheme developed for analogy networks, followed by TDM/TDMA developed for digital networks. The capacity provided by each of these multi-access schemes is fixed for a given network. Secondly, a single channel (which is a single frequency in FDM/FDMA and a set of time slots in contiguous frames in TDM/TDMA) is allocated to each user in a fixed manner for the duration of its communication. These techniques were best suited for constant bit rate applications, such as human voice, but not for bursty data traffic [Jamalipour *et al.* (2005)] which currently dominates communications networks.

Spread spectrum (SS) techniques, and in particular, code division multiplexing (CDM) and code division multiple access (CDMA) were developed to resolve the weaknesses in time and frequency based schemes. CDM/CDMA has a flexible capacity, and its capacity is limited by the multi-user interference (i.e., MUI/MAI) in the system. Cellular mobile systems based on CDM/CDMA has a cluster size of unity as every cell in the network use the same frequency. This idea enables softer handoff of calls between neighbouring cells as well as improving the spectral efficiency over TDM/TDMA and FDM/FDMA. A main problem in CDMA is the **near-far problem**, which is solved using strategies such as transmitter power control.

How does spread spectrum (SS) schemes work? As illustrated in Fig. 8.17, SS techniques spread or stretch a signal, say $s(t)$, with the original bandwidth, say W_1, to occupy a larger bandwidth, say W_2, i.e., $W_2 \gg W1$, at the transmitter. The spreading of a signal does not change its data rate. Thus, from the classical Shannon's capacity formula we have

$$R_{\mathrm{b}1} = W_1 \log_2\left(1 + \frac{P_1}{W_1N_0}\right) \stackrel{!}{=} R_{\mathrm{b}2} = W_2 \log_2\left(1 + \frac{P_2}{W_2N_0}\right). \tag{8.32}$$

If by the nature of the SS technique $B_2 > B_1$, then a consequence from Eq. (8.32) is $P_2 < P_1$. This means that a SS signal has a lower power spectral density (PSD) than its unspread version. The advantage in this is that the SS signal is more robust against both ingress unintentional interference and intentional interference (i.e., jamming) as well as causing a lower egress interference to disturb other users. The low transmit power in SS signals also saves power, which is important especially for mobile handsets. Shannon's formula also indicates that we need a smaller signal-to-noise ratio to achieve the same data rate if the bandwidth is increased.

Other advantages in SS include:

(1) *Multiple access scheme*: this is the focus in this book.
(2) *Covert communications*: the fact that one needs the spreading code to read a SS signal means that keeping the code secret maintains secrecy of the signal. This is important for military applications.
(3) *Jamming resistance* of SS signals is also achieved by the fact that they cannot be decoded unless we know the codes used to spread them at their sources. Attempt to despread a SS signal without the appropriate spreading key does not work.
(4) *Robust against multipath fading*: a SS signal occupies a spectrum much larger than it needs. Channel fading is usually limited to pockets of frequencies. This means that the probability that a remarkable part of the spectrum occupied by a SS signal experiences fading is small, making it robust against fading.
(5) Resistance to eavesdropping as knowledge of the spreading code is necessary to decode SS signals. Also, the lower PSD of SS signals makes their detection difficult.
(6) Inherent clocking and synchronisation: the spreading code used to enable clocking, such as the detection of the beginning and the end of a transmission.
(7) SS systems do not have a fixed capacity. The capacity is limited by only the multi-user interference.

8.4.1 *Pseudo-Random Noise Codes*

Binary sequences generated using *deterministic* algorithms but possessing the features of a sampled random noise are referred to as pseudo-random noise (PRN) sequence, pseudo-random bit sequence (PRBS) or **PN code**. The key properties of random noise expected in a PRBS are:

- Self-orthogonality: a random noise w has unity correlation with itself at zero lag, but zero correlation at a nonzero lag. That is, $R_{ww}(\tau) = \delta(\tau)$. Similarly, a period of a PRBS has a similar feature. The self-orthogonality feature of PRBS is needed to avoid multipath interference.
- Mutual orthogonality: two sequences w and v have the cross-correlation $R_{wv}(\tau \approx 0, \forall\tau)$. This property is needed to mitigate multi-user interference.
- Balance between 1 and 0: we want the number of bit 0 and bit 1 in each period of the PRBS to differ by a maximum of one digit.
- Runs in a period: we refer to a string of continuous values as a *run*. Among the runs of bit 1 and bit 0 in a period of the sequence about 50% of them have length 1, 25% have length 2, 12.5% have 3, 6.25% have length 4, etc. Thus, the percentage of a run with length k bits is $\frac{100\%}{2^k}$. For example, the binary sequence 0010110111100010 has five runs of bit 0 (namely, 00, 0, 0, 000, 0 of run lengths 2, 1, 1, 3 and 1 respectively) and four runs of bit 1 (namely, 1, 11, 1111, 1 of run lengths 1, 2, 4 and 1 respectively).

All SS techniques use PRN codes to achieve the spreading and despreading of signals. There are four basic types of SS techniques. Namely, frequency hopping SS (FHSS), direct sequence SS (DSSS), time hopping spread spectrum (THSS) and hybrid techniques. These SS schemes differ in where in the processing chain that the PN code is used to spread and/or despread the signal. The PRN codes must possess certain futures, such as orthogonality (or at least low cross-correlation), good autocorrelation, bits balancing and period (long period preferable). The more complex the PRN sequences are, the more robust they are. Popular PRN codes include Barker, M-Sequence, Gold and Walsh-Hadamard. A usual, in the following we encode or map the bits as

$$\texttt{bit } 1 \leftrightarrow -1 \texttt{ and bit } 0 \leftrightarrow +1\,. \tag{8.33}$$

We often do not write the plus sign on +1 for brevity.

A periodic binary sequence $\{s[k]\}$ with the period T_p has the normalised discrete autocorrelation sequence $R_s[k]$

$$R_s[k] = \frac{1}{T_\text{p}} \sum_{n=0}^{T_\text{p}-1} s[n]s[n-k]\,. \tag{8.34}$$

The meaning of $s[n-k]$ is a right cyclic shift of the sequence $s[n]$ by k chips. For example, if $s[n] = [-1,+1,-1,+1,-1,+1\ -\ 1,+1]$ then $s[n-3] = [+1,-1,+1,-1,+1,-1,+1,-1]$. Equation (8.34) yields the same results as

$$R_s[k] = \frac{1}{T_\text{p}} \times [(\texttt{number of agreements of sequence}) - (\texttt{number of disagreements in its } k \texttt{ cyclic shifted version})] \tag{8.35}$$

which reduces to

$$R_s[k] = \begin{cases} 1, & k = lT_\text{p},\ l = 0,1,2,\ldots \\ -\frac{1}{T_\text{p}} = -\frac{1}{2^m-1}, & k \neq lT_\text{p} \end{cases} \tag{8.36}$$

where m is the number of flip-flops in the LFSR used to generate the pseudorandom sequence. Figure 8.18 illustrates the autocorrelation sequence of a sequence with period $T_\text{p} = 7$.

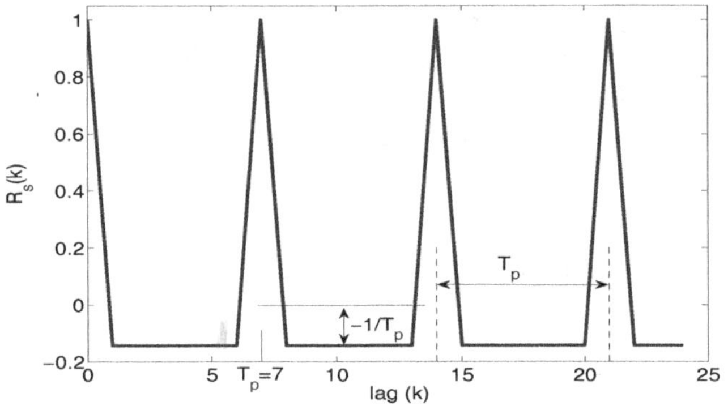

Fig. 8.18 Autocorrelation sequence for a sequence with period $T_\text{p} = 7$ chips.

Walsh-Hadamard Codes

A Walsh-Hadamard code (aka Walsh code, Hadamard code) is a set of $2N$ codewords with binary elements, each of length $2N$ [Walsh (1923)]. They are among the popular codes used in SS for multiple access. They are derived from square matrices referred to as Hadamard matrices. The rows and columns of Hadamard matrices are orthogonal. Thus, perfectly synchronised Walsh codes are perfectly orthogonal. Walsh codes have linear phase, fixed length, are orthogonal with high auto-correlation but low cross-correlation properties. The Hadamard matrix is generated in an iterative manner, which is

$$H_1 = [1],\ H_{2k} = \begin{bmatrix} H_k & H_k \\ H_k & \overline{H}_k \end{bmatrix},\ k = 1,2,3,\ldots \tag{8.37}$$

For example,

$$H_2 = \begin{bmatrix} H_1 & H_1 \\ H_1 & \overline{H}_1 \end{bmatrix} = \begin{bmatrix} 1 & 1 \\ 1 & -1 \end{bmatrix} \begin{matrix} c_1 \\ c_2 \end{matrix},$$

$$H_4 = \begin{bmatrix} H_2 & H_2 \\ H_2 & \overline{H}_2 \end{bmatrix} = \begin{bmatrix} 1 & 1 & 1 & 1 \\ 1 & -1 & 1 & -1 \\ 1 & 1 & -1 & -1 \\ 1 & -1 & -1 & 1 \end{bmatrix} \begin{matrix} c_1 \\ c_2 \\ c_3 \\ c_4 \end{matrix}$$

and

$$H_8 = \begin{bmatrix} H_4 & H_4 \\ H_4 & \overline{H}_4 \end{bmatrix} = \begin{bmatrix} 1 & 1 & 1 & 1 & 1 & 1 & 1 & 1 \\ 1 & -1 & 1 & -1 & 1 & -1 & 1 & -1 \\ 1 & 1 & -1 & -1 & 1 & 1 & -1 & -1 \\ 1 & -1 & -1 & 1 & 1 & -1 & -1 & 1 \\ 1 & 1 & 1 & 1 & -1 & -1 & -1 & -1 \\ 1 & -1 & 1 & -1 & -1 & 1 & -1 & 1 \\ 1 & 1 & -1 & -1 & -1 & -1 & 1 & 1 \\ 1 & -1 & -1 & 1 & -1 & 1 & 1 & -1 \end{bmatrix} \begin{matrix} c_1 \\ c_2 \\ c_3 \\ c_4 \\ c_5 \\ c_6 \\ c_7 \\ c_8 \end{matrix}$$

We can observe that both the row vectors and the column vectors of the Hadamard matrix are mutually orthogonal. Thus, we can obtain N orthogonal PN codes, each of length N, from an $N \times N$ Hadamard matrix. Although we can use either the row or column vectors of the matrix to build the PN codes, our case studies earlier use row vectors. This is common practice.

Applications of Walsh codes include:

- IS-95 (aka cdmaOne) uses 64-bit Walsh codes: code 0 for pilot channel (used for channel quality estimation), code 32 for synchronisation, code 1 for control channel (codes 2 to 7 can also be used for control channel), and the remaining are used for traffic channels.

- cdma2000 uses multiple Walsh codes of different lengths, including 64-bit Walsh codes, as Walsh codes of any length can be found.

Maximal-Length Sequences

Walsh codes are orthogonal, but have high autocorrelation. Spreading sequences with low autocorrelation are needed for wireless channels with large delay spreads caused by multipath signal propagation. A *register* in digital electronics is a special store for data with a quick access. Each register is a flip-flop and so can store only one bit at a time. Multiple registers or flip-flops connected in series and controlled by a common clock to shift the contents of all of them at a clock pulse to the right is called a *shift register*. If the feedback of a shift register is a linear function of a set of its previous states then the shift register is called *linear feedback shift register* (LFSR). The feedback is an exclusive OR (XOR) of certain bits, and XOR is a linear operation.

Pseudorandom binary sequences are usually generated using LFSRs, comprising flip-flops and XOR gates. The initial contents of a LFSR is called a *seed*, while the bits fed back into it are called *taps*. The sequence of the output bits of a LFSR is referred to as its *sequence*. An LFSR containing m registers or stages is called **Rm LFSR**. The sequence generated by a LFSR depends on the three parameters:

- The seed or initial conditions.
- The number of registers or stages in the shift register.
- The feedback loop or taps.

This means that we can use different seeds to obtain different sequences from the same LFSR.

As all operations in a LFSR are linear, we refer to the output of a LFSR as linear recursive sequences (LRS). Binary sequences generated by LFSR are periodic or recursive. A LFSR with m registers (or storage elements) can produce a binary sequence of length up to $2^m - 1$ bits before repeating itself. An LFSR with the period $T_p = 2^m - 1$ is called **maximal length shift register** (MLSR), and the sequence produced by an MLSR circuit is called MLSR sequence (MLSRS) or m-sequence. If the period of an m-stage LFSR is less than $2^m - 1$ then the shift register circuit is non-maximal length. This is because an m-stage LFSR can have up to $2^m - 1$ different combinations of contents, called states, excluding the all-zeros state which produces all-zero sequence or outputs. A different seed just shifts the sequence.

A LFSR can be completely defined by its polynomial in a dummy variable, say D or X. Such a polynomial is referred to as the **characteristic polynomial** or **generator polynomial** of the LFSR and it is written in the form

$$G(D) = g_m D^m + g_{m-1} D^{m-1} + g_{m-2} D^{m-2} + \cdots + g_2 D^2 + g_1 D + g_0 \qquad (8.38)$$

where always $g_m = g_0 = 1$, and $g_k \in \{0,1\}, k \notin \{m, 0\}$. If the characteristic polynomial of an LFSR is a *primitive polynomial*, then the LFSR generates maximal-length sequences.

A LFSR can be implemented in one of two ways: Galois implementation and Fibonacci implementation. Figure 8.19 shows the general architectures of these two implementations of LFSRs. The feedback tap set (aka feedback set or feedback pattern) of the LFSR is expressed as

$$\begin{aligned} \texttt{General representation:}\ & [f_1, f_2, f_3, \ldots, f_{J-1}, f_J] \\ \texttt{Fibonacci implementation:}\ & [f_1, m - f_2, \ldots, m - f_J]_f \\ \texttt{Galois implementation:}\ & [f_1, f_2, f_3, \ldots, f_{J-1}, f_J]_g \end{aligned} \qquad (8.39)$$

where $f_1 = m$ and J is the number of feedback taps, excluding g_0. Note that, by convention, g_0 is not included in the feedback tap set. We note also that $f_1 = m$ is not a feedback per se, but input to the LFSR. The elements in the feedback tap set are ordered from left to right in descending order.

Example of m-sequences

We consider m-sequences generated using a 3-stage LFSR.

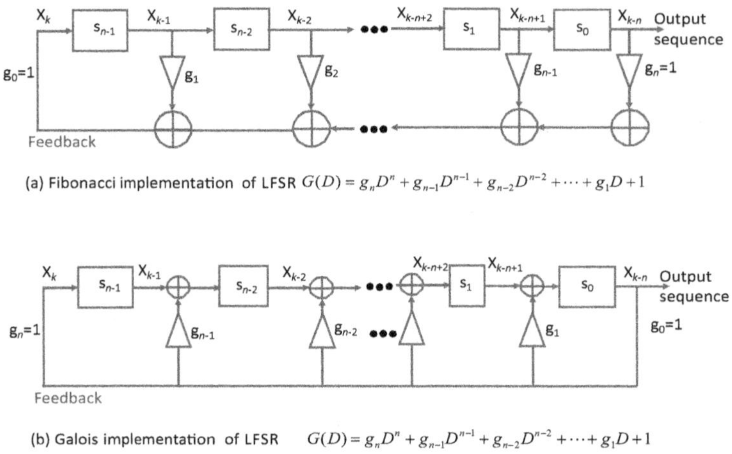

Fig. 8.19 The general architecture for Fibonacci and Galois implementations of LFSRs.

(1) Compute all of the maximal-length feedback tap sets and express them in both Fibonacci and Galois forms.
(2) Draw the block diagram for the circuits used to implement the LFSRs for both Fibonacci and Galois implementations.
(3) Generate the sequences produced by the two characteristic polynomials found using the seed 001. Use the Fibonacci implementation of the LFSR.
(4) Find the PN code corresponding to the sequences generated.
(5) Compute the cross-correlation between the obtained sequences and their cyclic shifts by three chips.

Solution

(1) The length of the m-sequences generated with a LFSR with $m = 3$ registers is $N = 2^3 - 1 = 7$. Therefore, primitive polynomials with degree $m = 3$ which are factors of the polynomial $p(D) = D^7 + 1$ are the **generator polynomials** (aka *connection polynomials*) of the desired m-sequences. They are

$$D^7 + 1 = (D + 1)(D^3 + D + 1)(D^3 + D^2 + 1)\,.$$

The two polynomials are $G_1(D) = D^3 + D + 1$ and $G_2(D) = D^3 + D^2 + 1$. This results in the feedback taps $[3, 1]_f$ and $[3, 2]_f$ for Fibonacci and $[3, 2]_g$ and $[3, 1]_g$ for Galois implementations, respectively.

(2) The block diagram of the circuits used to implement the LFSRs are shown in Fig. 8.20. The defining *linear recurrence* relations of the LFSRs are

$$\begin{aligned} X_k &= X_{k-1} \oplus X_{k-3} \quad \text{for Fig. 8.20(a)} \\ X_k &= X_{k-2} \oplus X_{k-3} \quad \text{for Fig. 8.20(b)} \\ X_k &= X_{k-3} \quad \text{for Fig. 8.20(c)} \\ X_k &= X_{k-3} \quad \text{for Fig. 8.20(d)}\,. \end{aligned} \tag{8.40}$$

(3) The sequence of outputs and states of the two LFSRs are shown in Fig. 8.21. We can observe that one period of the sequence causes the LFSR to pass through all its $2^3 - 1 = 7$ nonzero states. However, the LFSR remains in its all-zero state if the seed is all zeros. For this reason, each of the two LFSRs described by the characteristic polynomials $G_1(D) = D^3 + D + 1$ and $G_2(D) = D^3 + D^2 + 1$ produces maximal-length sequences. The length of the sequence generated is also $2^3 - 1 = 7$. In general, a maximal-length LFSR with m registers generates $2^m - 1$ unique sequences, each of length $2^m - 1$.
(4) The sequences generated are $S_1 = 1001110$ and $S_2 = 1001011$, respectively. As the length of each sequence is $2^m - 1|_{m=3} = 7$, the two LFSRs described by the characteristic polynomials are maximal-length LFSRs. Using the coding `bit 1` $\leftrightarrow -1$ and `bit 0` $\leftrightarrow +1$, we obtain the corresponding PN codes as

$$\begin{aligned} S_1 = 1001110 &\leftrightarrow c_1 \\ &= [-1, +1, +1, -1, -1, -1, +1] \end{aligned}$$

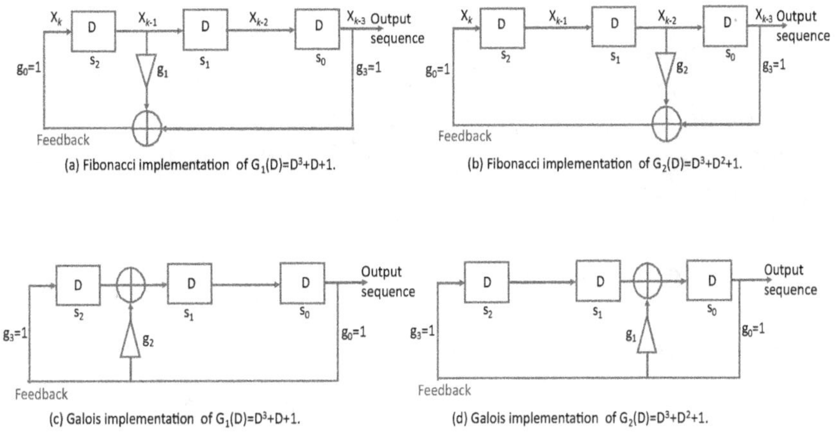

(a) Fibonacci implementation of G1(D)=D3+D+1.
(b) Fibonacci implementation of G2(D)=D3+D2+1.
(c) Galois implementation of G1(D)=D3+D+1.
(d) Galois implementation of G2(D)=D3+D2+1.
Output sequence
Feedback

Fig. 8.20 Fibonacci and Galois implementations of two LFSRs. The modulo-2 additions are performed in hardware using XOR gates.

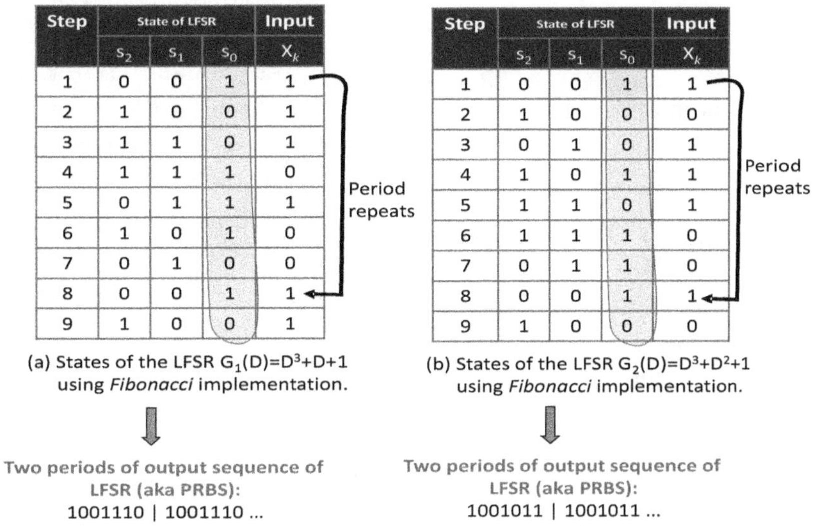

Step	State of LFSR			Input
	S_2	S_1	S_0	X_k
1	0	0	1	1
2	1	0	0	1
3	1	1	0	1
4	1	1	1	0
5	0	1	1	1
6	1	0	1	0
7	0	1	0	0
8	0	0	1	1
9	1	0	0	1

(a) States of the LFSR $G_1(D)=D^3+D+1$ using *Fibonacci* implementation.

Step	State of LFSR			Input
	S_2	S_1	S_0	X_k
1	0	0	1	1
2	1	0	0	0
3	0	1	0	1
4	1	0	1	1
5	1	1	0	1
6	1	1	1	0
7	0	1	1	0
8	0	0	1	1
9	1	0	0	0

(b) States of the LFSR $G_2(D)=D^3+D^2+1$ using *Fibonacci* implementation.

Fig. 8.21 Generating an LFSR sequence using Fibonacci implementation.

and

$$S_2 = 1001011 \leftrightarrow c_2 = [-1,+1,+1,-1,+1,-1,-1]\,.$$

(5) If $c_1[n] = [-1,+1,+1,-1,-1,-1,+1]$ then $c_1[n-3] = [-1,-1,+1,-1,+1,+1,-1]$, and $c_2[n-3] = [+1,-1,-1,-1,+1,+1,-1]$. From Eq. (8.34), we obtain

$$R_{c_1}[0] = R_{c_2}[0] = \frac{1}{7}\sum_{n=0}^{6} c_1[n]c_1[n] = 7/7 = 1\,,$$

$$R_{c_1}[3] = \frac{1}{7}\sum_{n=0}^{6} c_1[n]c_1[n-3] = -1/7\,,$$

and

$$R_{c_2}[3] = \frac{1}{7}\sum_{n=0}^{6} c_2[n]c_2[n-3] = -1/7\,.$$

The achieved results validate Eq. (8.36), which is illustrated in Fig. 8.18.

The number of chips in the m-sequence increases as its period increases. The correlation between the codes/sequences decreases with increasing period. This means that we can generate PRBS which have a very small correlation and thus almost orthogonal by increasing the period of the sequence.

As observable in Fig. 8.20, the Galois implementation of the LFSR uses a fewer number of logic gates in its feedback loop than the Fibonacci, making it faster, and consequently preferable. The Galois implementation is also called modular shift register generator (MSRG) or multiple-return shift register generator (MRSRG), while the Fibonacci implementation is also called simple shift register generator (SSRG).

The m-sequences described earlier are linear as they are generated by LFSRs. Therefore, it is easy to predict them. A popular m-sequence used in CDMA is the Gold codes which were developed by Dr. Robert Gold in 1967/1968. Gold codes use two sets of LFSRs to make it harder to predict the generated sequence. A Gold code uses two LFSRs and adders. Thus, Gold sequences are obtained by combining two pairs of m-sequences. A Gold code is uniquely defined by the two parameters:

- The characteristic polynomial for the two LFSRs.
- The seed for the second LFSR. The seed for the first LFSR is always $\underbrace{000\cdots001}_{\texttt{number of registers in LFSR}}$.

We can use two m-bit LFSRs with m registers to produce $2^{m-1} + 2$ Gold codes. Each of the LFSR uses a characteristic polynomial with the degree m.

Gold sequences are obtained by combining a pair of m-sequences. Another more complicated sequence, called **Kasami sequences**, are obtained by combining a decimated version of one of the two m-sequences that generated the Gold sequence and the Gold sequence itself.

8.4.2 *FHSS and Its Application in Bluetooth*

Bluetooth (BT) is one of the technologies which probably does not need an introduction, because everyone who uses an everyday electronic device most probably uses it. BT is one of the key wireless standards supporting Wireless IoT. The core specifications of BT can be found at its official website https://www.bluetooth.com/specifications/bluetooth-core-specification/. A team of scientists led by Nils Rybbeck at Ericsson Mobile in Lund Sweden started an initiative with two related objectives, namely, to develop:

- A wireless RS232 interface to replace RS-232 cables.
- A wireless headset.

This project led to the development of Bluetooth. A few years later, IBM, Nokia, Toshiba and Intel joined Ericsson, which later led to the formation of the Bluetooth Special Interest Group (SIG) in 1998. The membership of the SIG has grown from five members in 1998 to more than 35,000 in 2019. Members of SIG are into consumer electronics, telecommunications, computing and networking. Indeed, BT has been living up to its name as it has united technology industries. The physical layer and the medium access control (MAC) layer of BT were standardised as IEEE 802.15.1. However, this standard is no more followed. Currently, BT SIG oversees, develops, specify and standardises BT. Every device must satisfy the SIG standards to be marketed as BT device.

How did BT achieve its name? The Intel employee Jim Kardach suggested the name Bluetooth for the then nascent technology. Bluetooth is the translation of the Danish word Blåtand, which is the middle name of King Harald Blåtand Gormsen who ruled over both Norway and Denmark in the 10th century. He is known to have united Danish tribes into one kingdom, just as Bluetooth united communication protocols as well as computing and telecommunications industries.

The developmental history of BT is summarised in Table 8.5. Core specifications are numbered as X.0, where X is a positive integer, except BT 5 which has no decimal point. Revisions of a core specification are indicated X.1, X.2, etc. The first BT standard, Core Specification 1.0, was released in 1998/1999. In 2003, Bluetooth 1.2 was released, which included the adaptive frequency hopping spread spectrum (AFH). BT 3.0 can use a nearby Wi-Fi network to achieve data rate up to 24 Mbps. This is the reason it is referred to as Bluetooth High Speed (HS). Bluetooth Low Energy (BLE) was released in 2010. BLE is also referred to as Bluetooth Smart. Therefore, we have two types of Bluetooth:

Table 8.5 Abbreviated Bluetooth developmental history.

Date	Milestone
1989	Nils Rybbeck (leader), Tord Wingren and Jaap Haartsen at Ericsson Mobile in Lund Sweden started a project to develop wireless headset
1990	Jaap Haartsen wins European inventor award on BT
1997	Örjan Johansson appointed leader of BT development
1997	IBM through Adalio Sanchez joins Ericsson on BT development project
1997	Intel joined BT development and brought Toshiba & Nokia on board
1998	BT SIG formed by Ericsson, IBM, Intel, Toshiba & Nokia
1999	Bluetooth 1.0 core specification (aka Classic BT) with 721 kbps peak rate
1999	mobile headset as first BT device released
2000	BT chipset released
2001	BT-enabled mobile phone, Sony Ericsson T36/T39, released
2001	IBM ThinkPad A30, BT-enabled laptop, was released in October
2003	Bluetooth 1.2 (introduced AFH)
2004	Bluetooth 2.0 core specification, incl. EDR
2009	Bluetooth 3.0 core specification, incl. HS
2010	Bluetooth 4.0 core specification, incl. 1 Mbps LE
2017	Bluetooth 5 core specification, incl. 2 Mbps LE

AFH: adaptive frequency hopping spread spectrum; EDR: enhanced data rate; HS: high speed; LE: low energy

- Classic Bluetooth: BT 1.0 to BT 3.0, including Enhanced Data Rate (EDR).
- Bluetooth Low Energy (BLE): from BT 4.0 and later versions.

Note that most of the modern Bluetooth chips contain both strains of BT radios.

Although BT was meant to be a short-range radio standard to replace RS-232 cables, it is now being used for other purposes, such as short-range wireless ad hoc networking. It is thus a key standard for Wireless Personal Area Networks (WPANs). An ad hoc network is a communications network without pre-planning and fixed infrastructure. Third, a BT device can be configured to serve as an access point to other BT devices.

Although the dominance of BT in short-range data transfer is being challenged since 2015 by the file transfer application called Xender, it still has a wide range of applications. Xender is not a wireless standard. It is an application which uses Wi-Fi air interface to transfer data. For example, in consumer electronics Bluetooth chip is found in cordless mouse, cordless computer keyboards and headsets. Bluetooth is enjoying a growing popularity in medical devices. It is used in blood pressure monitors (e.g. Trisa/Cardio Pro 4.0 and Whithings), pulse oximeters (e.g. iHealth/PO3) and ECG (e.g. Bio-Radio), just to name a few. According to [Baker (2005)], BLE is the most suitable wireless standard for healthcare applications. Another popular application of Bluetooth is in *appcessories*, which are devices (e.g. toys, games and fitness products) controlled by smartphones and tablets over a wireless interface such as Bluetooth. We also have smart clothing, such as Moretek Bluetooth beanie hat which plays music and can attend to calls while keeping the head and ears warm. Other examples of smart clothing are Snapchat's smart glasses, swimsuits with ultraviolet sensors, smart yoga pants and smart fitness socks with advanced textile sensors to track steps. Table 8.6 lists more applications of Bluetooth and various industries. Clearly, Bluetooth is a major wireless standard for IoT. A Bluetooth-enabled keyless padlock can be connected to a home gateway and operated over the Internet.

How does Bluetooth provide security and privacy in communications? Bluetooth devices combine the Personal Identification Number (PIN) and

Table 8.6 Sample Bluetooth applications in various industries.

Industry	Sample Applications
Health	wireless vital signs monitors (sensors for pulse rate, blood glucose and blood pressure), pedometers, pulse oximeters
Appcessories	wearable devices, toys (e.g. iDance Cube), games, fitness activity monitors (e.g. Fitbit with haptic sensors)
Home automation	HVAC & white goods, sensors for security, temperature, smoke detection, carbon monoxide, locking/unlocking of key-less padlocks and doors, smart energy meters and displays
Entertainment	stereo streaming, speakers
Consumer electronics	headsets, mouse, keyboards, mobile phones
Agriculture	animal tagging & tracking, soil moisture sensors
Transportation	automobiles, intelligent transportation systems, proximity monitoring
Smart clothing	Bluetooth Beanies (e.g. Moretek Bluetooth Beanie Hat), smart wrist watches, smart glasses, smart socks, smart swimsuits

a Bluetooth address to authenticate and identify other Bluetooth devices. 128-bit data encryption can be used to further strengthen the security in Bluetooth. The multiple access scheme used by Bluetooth is frequency hopping spread spectrum (FHSS). Bluetooth Classic uses FHSS to change the carrier frequency 1,600 times each second, allowing only synchronised receivers to access the transmitted data. This is an added security and privacy to the data transferred. As mentioned above, BLE (aka Bluetooth Smart) has modified the FHSS to adaptive FHSS (AFH) to increase its resilience against radio frequency interference (RFI). In AFH, BLE does not consider frequencies under interference in the set of frequencies which can be hopped onto. Thus, BLE on the average can use a fewer number of channels than Bluetooth Classic.

Figure 8.22 compares the protocol stacks or architecture of BLE and Classic Bluetooth. We can observe that BT has a complete protocol stack, although traditionally wireless standards are defined by only their two lowest ISO/OSI layers. With BLE, we focus on Bluetooth 5 for two reasons. It is the most current specification, and it is the one most suited for IoT applications.

The Generic Access Profile (GAP) layer in protocol stack takes care of connection functionality. It is responsible for access modes and procedures of the device, such as device discovery, link establishment, link termination, initiation of security features and device configuration. A BT device can be in one of three main states, namely:

- *Standby mode*: This is the initial idle state of a device upon power up.
- *Device discovery mode*: there are three substates in this mode, namely, advertising state, scanning state and initiator state. A device in the advertising state broadcasts its availability for communication.
- *Connection mode*: in this state a device can perform the master role (if it was the initiator and scanner in the discovery state) or slave role (if it was the advertiser in the discovery state).

The Generic Attribute Profile (GATT) is the layer in the BLE protocol stack which enables the application to perform data communication between two devices in the connected state. In reference to the GATT every connected device is in one of these two modes:

- The GATT server: the device containing the database that is being read or written by a GATT client.
- The GATT client: the device that is reading or writing data from or to the GATT server.

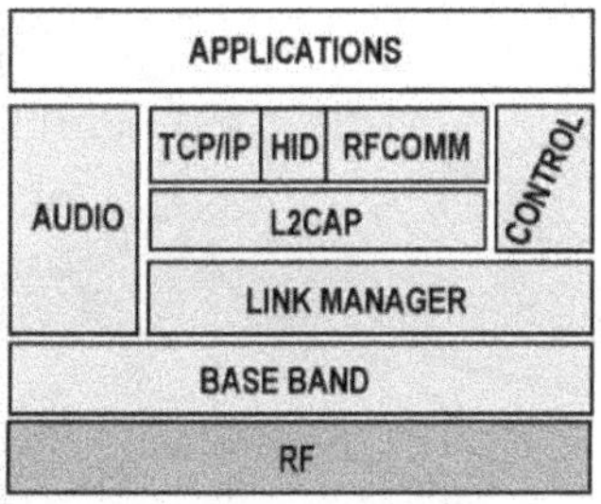

(a) Classic Bluetooth protocol stack

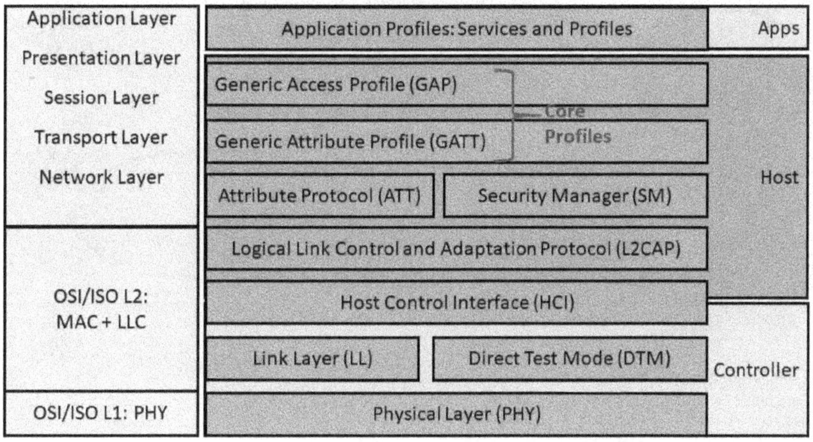

(b) BLE protocol stack vs ISO/OSI 7-layer protocol stack

Fig. 8.22 Bluetooth protocol architectures.

Roles of BLE Devices

As said earlier, our discussions on Bluetooth has focused on BLE as it is the version most suited for IoT. BLE has many admirable features, including:

- **Stateless protocol**: no communications session information is retained by the receiver or server. Therefore, there is no tight dependence between client and server devices.
- **Low-power consumption:** a coin-sized battery can last one year.
- **Very flexible topology.** See Fig. 8.23 for an illustration.
- **Connection versatility:** enables connection to a large variety of devices, e.g. smartphones, tablets, laptop, earpieces.

In terms of functions performed in the GAP layer, devices supporting BLE are divided into the ff four classes:

1. Broadcaster: devices in this mode can only transit signals.
2. Observer: devices in this mode can only receive signals.
3. Peripheral: devices in this role can support slave devices.
4. Central: this is a master device in a piconet and can support multiple connections, as well as initiate connections to peripherals.

These types of device roles are illustrated in Fig. 8.23.

In the broadcaster role, a device can only advertise its capabilities, but cannot connect to other devices. Other devices cannot connect to broadcasters either. Devices in the peripheral role advertise their capabilities and also notify central devices of their readiness to connect to them (i.e. central devices). Observers can only receive signals. Central devices actively scan for peripheral devices which are advertising themselves and then establish connections with them. The main difference between broadcaster and peripheral is that the former cannot be connected to. Also, the main difference between the

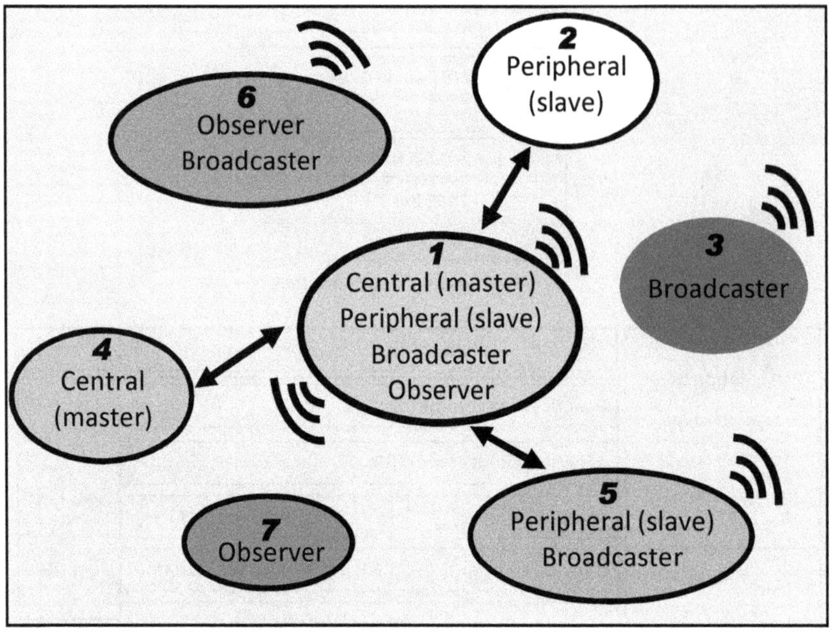

Fig. 8.23 Possible BLE network topology with seven BLE-enabled devices.

observer and the central is that the former cannot initiate connections. Both the central and peripheral can act as either client or server at any given time. A central device can connect to multiple peripherals at the same time. Similarly, a peripheral device can connect to multiple central devices at the same time.

Note that a single device can support multiple roles. This is the reason why BLE supports very flexible network topology. Consider the network topology in Figure 8.23 illustrating the flexibility in BLE topology. There are seven devices, each with a different set of possible roles. For example, device 4 can connect to device 1 as a peripheral, but to devices 2 or 5 as a central or a master.

BLE Addresses

Addresses are used to identify devices, just as humans are identified using names. Physical addresses in Bluetooth are of 48 bits long to conform to the 48-bit universal medium access control (MAC) addresses used in local area networks (LAN). BLE has two types of addresses: random addresses and public addresses. Public addresses are allocated by IEEE. The random addresses have the following basic features:

- They are private addresses but can be resolvable or non-resolvable.
- They are generated by following rules specified by the Bluetooth Special Interest Group (BT SIG).
- They are static address and changed usually only upon power cycling, i.e. resetting or powering down and up again.

8.4.2.1 *Bluetooth Physical Layer*

Bluetooth operates at the 2.4 GHz Industrial Scientific Medicine (ISM) band spanning 2.4–2.4835 GHz. BT uses the band 2.402–2.48 GHz for data transfer, with two guard bands at 2.4–2.402 MHz and 2.48–2.4835 GHz. The ISM band is a license-free band. Therefore, BT shares it (i.e. ISM band) with many other technologies, including garage doors, Wi-Fi (IEEE802.11b, IEEE802.11g and IEEE802.11n), ZigBee and microwave ovens. For example, Wi-Fi uses 2.412–2.495 GHz, while BT uses 2.402–2.48 GHz. However, not all the frequencies are used by every Wi-Fi in every location. Also, BLE uses AFH. These together reduces the likelihood of both Wi-Fi and BT operating at the same frequency, and significantly interfering with each other. Another point is that, with the hopping rate of 1600 per second, BT stays on a given channel for only 0.625 ms for a single-slot transmissions. BT standard uses FHSS scheme to mitigate both inter-technology and intra-technology interference. BLE uses AFH to further mitigate possible radio frequency interference.

The carriers (aka radio frequency or RF channels) in the BT system are defined as

$$\begin{aligned} \texttt{BR/EDR:} \quad & f_k = 2.402 + k/1000 \ \texttt{GHz}, \\ & k = 0, 1, 2, \ldots, 78 \\ \texttt{BLE:} \quad & f_k = 2.402 + k/500 \ \texttt{GHz}, \\ & k = 0, 1, 2, \ldots, 39 \end{aligned} \tag{8.41}$$

Thus, BLE transmissions use 40 carriers spaced 2 MHz apart, resulting in per-channel bandwidth of 2 MHz. The Classic Bluetooth with its basic rate (BR) and enhanced data rate (EDR) transmissions, however, use 79 carriers which are 1 MHz apart. Note that the bandwidth available in the 2.4 GHz band available to Bluetooth differs across countries. For example, the entire 83.5 MHz is available in the USA and many European countries except France which uses only 23 RF channels spaced 1 MHz apart.

In order to minimise radio frequency interference (both susceptibility and emissions), the centre frequencies of the transmitted signal should not deviate from the nominal values by more than these values:

- ±25 kHz for one-slot packet.
- ±40 kHz for two-slot packet.
- ±40 kHz for three-slot packet.

The above are the allowable deviations during packet transfer. Each time slot in Bluetooth lasts 625 μs, and a packet can last one slot, three slots or five slots. Clearly, the accepted deviation depends on the size of the packet being transmitted. The allowable maximum frequency deviation from the centre frequency prior to packet transfer is ±75 kHz. The maximum frequency drift is 400 Hz/μs.

Depending on their maximum output power capabilities, BT devices are divided into three classes, as illustrated in Table 8.7. With the power control ability, a Class 1 device can also operate in the output power ranges of Classes 2 and 3 devices. Note that $P_{\texttt{max}}$ and $P_{\texttt{min}}$ are the effective isotropic radiated powers (EIRPs) at the output of the antenna. Noting that

$$P_{\texttt{min}} \leq EIRP = G + P_{\text{T}} \leq P_{\texttt{max}} \tag{8.42}$$

the actual transmit power P_{T} must be set according to the gain (G) of the antenna used. The chosen transmit power level also depends on the channel quality, which is measured by the received signal strength indicator (RSSI). A lower RSSI requires a higher transmit power. The minimum receiver sensitivity of a Bluetooth device is −70 dBm. This minimum sensitivity is needed to achieve the maximum bit error rate of 0.1%. For a given RSSI and end-to-end signal attenuation, the transmit power must be chosen so that the power in the received signal at the receiver is not less than the minimum receiver sensitivity of −70 dBm.

In order to keep egress RFI to acceptable levels specialised bandpass filters are used in the Bluetooth transmitters to limit the radiation at frequencies outside the necessary bandwidth. The method used to define the allowable maximum power at

Table 8.7 Bluetooth device classes.

Device Class	*P*max	*P*min	Maximum Range	Power Control	Device Examples
Class 1	100 mW	1 mW	100 m	Mandatory	Miccus Home RTX 2.0 Bluetooth adapters, Motorola's headphones
Class 2	2.5 mW	1/4 mW	10 m	Optional	Mobile phones, Headsets, headphones
Class 3	1 mW	N/A	1 m	Optional	headphones

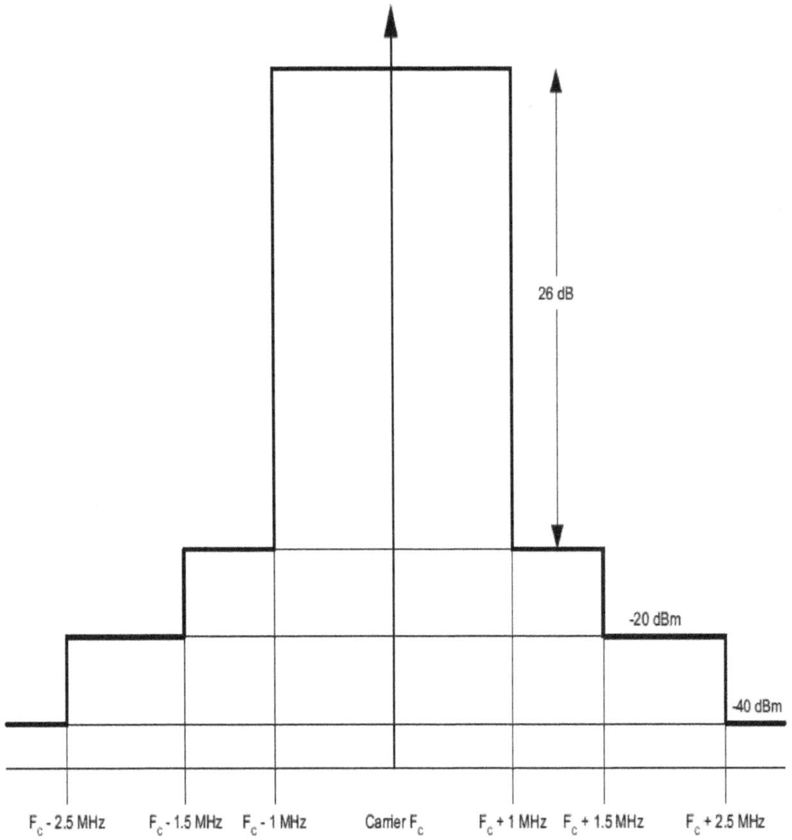

Fig. 8.24 Bluetooth transmitter spectral mask.

which frequency is referred to as spectral mask, which for BT is illustrated in Fig. 8.24. F_C in Fig. 8.24 is the carrier or centre frequency.

BT standard defines the following five physical channels:

(1) Basic piconet physical channel: uses the basic channel hopping sequence. The requirements on the pseudo-random hopping sequence are:
    (a) must have very long period length without repetitive patterns over a short time interval, and
    (b) must distribute the hop frequencies equally over the 79 MHz during a short time interval.
(2) Adapted piconet physical channel: uses the adapted channel hopping sequence. Connected devices which have enabled their AFH use this channel.
(3) Page scan physical channel.
(4) Inquiry scan physical channel.
(5) Synchronisation scan physical channel.

Each of these channels has a defined channel hopping sequence. BT uses pseudo-random RF channel hopping sequence with the maximum hopping rate of 1600 hops/s in the CONNECTION state, the synchronisation train substate, and the synchronisation scan substate and the maximum is 3200 hops/s in the inquiry and page substates. The maximum data transfer rate of serial port and parallel port are 115.2 kbps and 150 kbps, respectively. The speed was about 9.6 kbps at the time Bluetooth was released with the rate of about 721 kbps to 1 Mbps. Thus, Bluetooth was relatively very fast. Figure 8.25 summarises some Bluetooth radio parameters [NI (2016)].

Bluetooth 5 has three variants of physical layer: LE 1M (also in BT 4.0), LE 2M (new) and LE Coded. LE 1M uses GFSK modulation with the minimum frequency deviation of 185 kHz to achieve

Feature	Value
Modulation	**Basic Rate (BR) transmissions:** -Binary GFSK at 1 Msymbol/s -Bandwidth bit period product BT=0.5 -Modulation index: 0.28 to 0.35 (0.32 nominal) **Enhanced Date Rate (EDR) transmissions:** - π/4-DQPSK at 2 Msymbol/s -Bandwidth bit period product BT=0.4 -8DPSK at 3 Msymbol/s -Bandwidth bit period product BT=0.4 **Low Energy (LE) transmissions:** -Binary GFSK with at 1 Msymbol/s -Bandwidth bit period product BT=0.5 -Modulation index: 0.45 to 0.55 (0.5 nominal)
Data rate	BR/EDR: 1-3 Mbps; LE: 1 Mbps
Maximum data throughput	BR: ≤433.9 kps (symmetric/synchronous links), ≤723.2 kbps/57.6kbps (asymmetric/asynchronous links, with 5/1 slot packets in direction) EDR: ≤ 2.1 Mbps LE: 0.27 Mbps
Frequency deviation	±(115 to 175) kHz (BR/EDR) and ±185kHz (LE)
Frequency hopping (depends of state of BT device)	1600 hops/s (standard: CONNECTION, SYN TRAIN & SYN SCAN modes), 3200 hops/s (PAGE mode and INQUIRY mode)
Transmit power	Power class 1: 1 mW to 100 mW for 100 m range Power class 2: 0.25 mW to 2.5 mW for 10 m range Power class 3: 1 mW for 1 m range Lower energy (LE): 0.01 – 10 mW for 50 m range
Receiver Sensitivity (lowest operating power level at specified bit error rate – BER)	Basic rate: - 70 dBm (at 0.1% BER) Enhanced data rate: - 70 dBm (at 0.01% BER) Low energy: - 70 dBm (at 0.1% BER)
Connection time	20 ms (BR/EDR); 2.5 ms (Low energy)
Latency	100 ms (BR/EDR); 3ms (Low energy)
Network topology	Scatternet (BR/EDR); star-bus (LE)

Fig. 8.25 Some Bluetooth radio parameters [NI (2016)].

Table 8.8 Features of the physical layer variants in Bluetooth 5.

	LE 1M	LE 2M	LE Coded (S=2)	LE Coded (S=8)
Requirements	required	optional	optional	optional
Modulation	2-GFSK	2-GFSK	2-GFSK	2-GFSK
Minimum frequency deviation	185 kHz	370 kHz	370 kHz	370 kHz
Symbol rate	1 Mbaud	2 Mbaud	1 Mbaud	1 Mbaud
Data rate	1 Mbps	2 Mbps	0.5 Mbps	0.125 Mbps
Range change factor	1	0.8	2	4
Error detection	CRC	CRC	CRC	CRC
Error correction	no	no	FEC	FEC

1 Ms/s symbol rate. LE 2M uses GFSK modulation with the minimum frequency deviation of 370 kHz to achieve 2 Ms/s symbol rate. Earlier BT specifications use 24-bit cyclic redundancy check (CRC) for only error detection. Forward error correction (FEC) is introduced in Bluetooth 5 via the LE Coded PHY in addition to the CRC. The FEC is introduced to increase the range of Bluetooth 5 over earlier specifications. The FEC a convolutional encoder with the generator polynomials

$$G_0(X)=1+X+X^2+X^3 \text{ and } G_1(X)=1+X^2+X^3. \tag{8.43}$$

The basic features of the three PHY are summarised in Table 8.8. Two versions of FEC encoding are used with the LE Coded PHY: $S = 2$ and $S = 8$. For $S = 2$, the FEC encoder maps each input bit to one output bit, resulting in rate 1 encoder. For $S = 8$, the FEC encoder maps each input bit 0

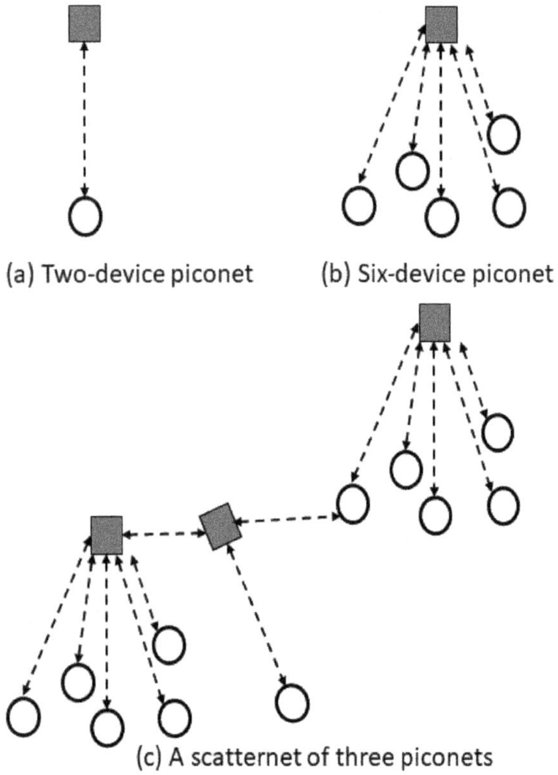

Fig. 8.26 Bluetooth piconets and scatternets.

to the output bit pattern 0011, and input bit 1 into output 1100, resulting in rate 1/4 encoder. As noted in Table 8.8, the rate 1/4 encoded extends the transmission rate by factor 4 compared with Bluetooth 4.0 without FEC.

8.4.2.2 *Bluetooth Piconets and Scatternets*

The network topologies adopted in Classic Bluetooth are piconet and scatternet (see Fig. 8.26). However, latter BT standards (viz BT 4.x and BT 5) use star-bus and mesh topologies but allows multihop links. Bluetooth-enabled devices must first pair up prior to exchanging information. Up to eight devices can pair up to form a network referred to as a *piconet*. As Bluetooth is based on client/server or master/slave networking model, one of the devices in a piconet must act as the master device, while the remaining becomes its slaves. Multiple piconets can interconnect to form a *scatternet*. Thus, one device can belong to multiple piconets. However, a device can serve as a master device in only one piconet at a given moment. This means that a device in a scatternet can serve as a master device in one piconet but a slave device in another piconet. As any device in a piconet can serve as the master, each device can toggle between slave and master status. The master device communicates with only one slave device at a time.

8.4.2.3 *Bluetooth MAC and FHSS/AFH*

The functions of the Bluetooth medium access control (MAC) are provided by the Logical Link Control and Adaptation Protocol (L2CAP) and the link layer. The L2CAP in Bluetooth specifies two types of connections. These are:

- synchronous connection-oriented (SCO) channels, and
- asynchronous connectionless (ACL) channels.

Both SCO and ACL use the client/server and master/slave communications model. The SCO is suited for streaming applications, while the ACL is suited for file transfers. Also, BT devices use ACL to establish the presence of each other prior to negotiating a two-way audio communications over SCO. ACL supports both symmetric and asymmetric data transmissions. Note that ACL can also be used to transfer voice, although not the best. The third popular application of Bluetooth is called *state*, which is the transfer of small amounts of data occasionally. A Bluetooth audio connection transfers data at 64 kbps and supports audio frequencies of up to 4 kHz.

As Table 8.9 shows, the ACL uses two types of frames. The first is the DM frames which use forward error correction (FEC). The second type are the DH frames which do not use FEC. Time is slotted in BT, and each time slot lasts 625 μs because of the hopping rate of 1,600 per second. Each frame can use 1, 3 or 5 slots, as indicated in Table 8.9. BT frames are referred to as data medium-rate (DM) and data high-rate (DH). The number of time slots used by each frame is written against it. For example, DM1 and DM3 use one and three time slots, respectively. Also, in reference to the basic data rate (BDR), the enhanced data rate (EDR) transmits at data rates which are multiples of the BDR rates. These are indicated as prefix. For example, 2-DH3 and 3-DH5 transmit

Table 8.9 Frame types and lengths in Bluetooth asynchronous connectionless channel.

Number of slots	1	3	5
BDR with FEC	DM1 1-18 bytes	DM3 2-123 bytes	DM5 2-226 bytes
BDR without FEC	DH1 1-128 bytes	DH 3 2-185 bytes	DH5 2-341 bytes
2 Mbps EDR	2-DH1 2-56 bytes	2-DH3 2-369 bytes	2-DH5 2-681 bytes
3 Mbps EDR	3-DH1 2-85 bytes	3-DH3 2-554 bytes	3-DH5 2-1023 bytes

DM: data medium-rate; DH: data high-rate; FEC: forward error correction; BDR: basic data rate; EDR: enhanced data rate

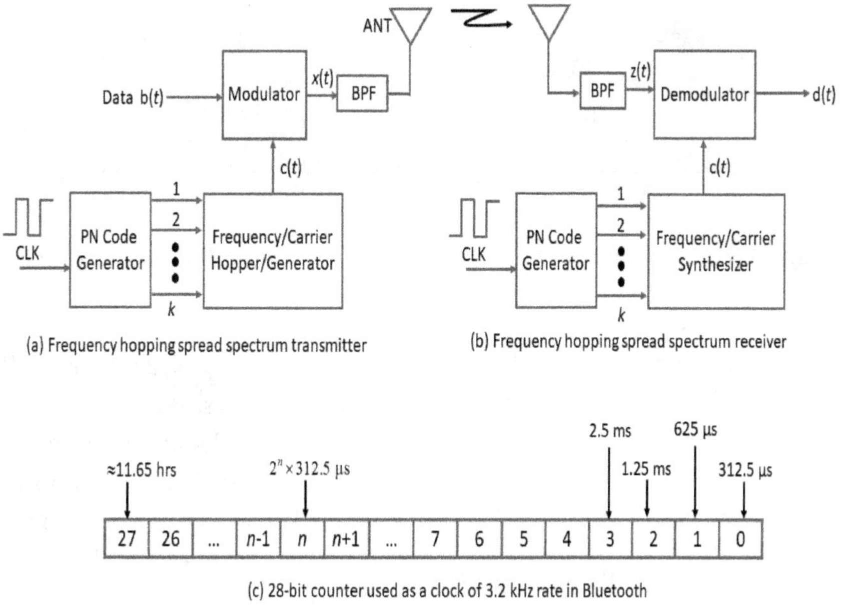

Fig. 8.27 Illustration of FHSS technique.

with 2 Mbps and 3 Mbps using three and five time slots, respectively.

FHSS was invented by Hedy Lamarr and George Antheil in 1941. FHSS is used in GSM, the military systems single channel ground and airborne radio system (SINCGARS) and AN/ARC-164 HAVE-QUICK II. FHSS is also used predominantly in peripheral devices (e.g., mice, keyboards, printers) for short-range data transfer. The general principle of the FHSS technique is illustrated in Fig. 8.27. The signal is spread or encoded at the transmitter and despread or decoded at the receiver.

In FHSS, the available system bandwidth is divided into multiple frequency carriers, and the carrier frequency modulated by the signal of interest is changed during transmission according to a given **hopping pattern**. The hopping pattern is determined by a pseudorandom noise (PN) sequence which is referred to as a code. The PN code used should be known to both the transmitter and the receiver. There are two types of FHSS: slow hopping and fast hopping. Assume that the data rate is R_b and the rate at which the carrier frequency is changed is R_h. The parameter R_h is referred to as *hopping rate*. Let $m > 1$ be a nonzero positive number. Then, we have

$$\begin{aligned} &\texttt{Slow hopping FHSS:} \quad R_\text{b} = mR_\text{h} \Leftrightarrow T_\text{h} = mT_\text{b} \\ &\texttt{Fast hopping FHSS:} \quad R_\text{h} = mR_\text{b} \Leftrightarrow T_\text{b} = mT_\text{h} \end{aligned} \tag{8.44}$$

where T_b is the bit duration and T_h is the chip duration or the time between two consecutive

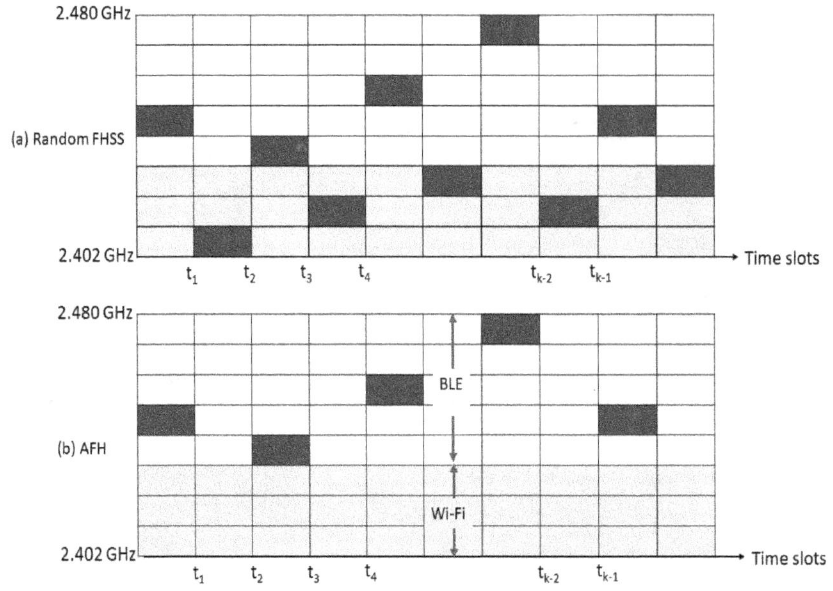

Fig. 8.28 Comparing AFH in BLE with the random FHSS in Classic Bluetooth: $t_k - t_{k-1} = 625\,\mu$s.

frequencies. In slow hopping FHSS several bits are transmitted before the carrier frequency changes. In fast hopping FHSS, however, the carrier frequency changes several times during the transmission of a single bit. Figure 8.25 shows some Bluetooth parameters. We observe from this figure that the standard hopping rate is $R_{\rm h} = 1,600$ hops/s, while the minimum symbol rate is $R_{\rm s} = 1$ Mbaud (for GFSK binary modulation), resulting in $R_{\rm s} = 625R_{\rm h}$. Thus, BT uses slow FHSS. An advantage in slow FHSS is that it is easier to implement coherent detection, allowing the use of phase modulators. Coherent detection is unpopular in fast FHSS based systems, however.

Figure 8.27 illustrates the transceiver using FHSS. The BT transceiver, in the master device of each piconet, uses a hardware called *hop selection kernel* to determine the value of k in order to select a carrier frequency in each hopping period. Note, however, that BLE uses the AFH technique to avoid using some of the 79 carriers if their states are unfavourable for data transmission because of interference.

BT uses a clock with a rate 3.2 kHz, which is a 28-bit free-running digital counter device, to manage the operation of the hop selection kernel. This means that the period of the clock or counting device is $\frac{2^{28}-1}{3.2\ \texttt{kHz}} \approx 23$ **hours**. The BT clock is illustrated in Fig. 8.27(c). The least significant bit of the clock ticks at 312.5 μs. However, it does not affect the operation of the hop selection kernel. The standard hopping period is 625 μs. This results in the hopping rate of 1,600 hops per second, which is the standard hopping frequency.

Figure 8.28 compares AFH in (b) with FHSS in (a). FHSS and other wireless technologies sharing the 2.4 GHz band can be operating at the same time over overlapping bands to interfere with each other. This is illustrated in Fig. 8.28(a). AFH, on the other hand, seeks frequencies being used by other radio technologies sharing the band. It then delete those frequencies for the list of frequencies that it can hop onto. This is illustrated in Fig. 8.28(b).

Up to eight Bluetooth devices can interconnect to form a piconet. One of the devices in a piconet serves as the master (or server) while the rest are slaves (or clients). The role of the master can be assumed by any of the devices in a piconet. Multiple piconets can interconnet to form a larger-range network called **scatternet**. The clock and the 48-bit NIC/MAC/physical address of the master determine the frequency hopping sequence and the timing of all devices in a piconet.

8.4.3 *Check Your Understanding*

Students are encouraged to check their understanding of the content of this section with the following questions prior to proceeding to the next section.

(1) (a) Compute the processing gain and the spreading factor of UMTS. (b) Compare your results with those of IS-95 (aka cdma2000).

(2) Discuss the reason(s) why coherent modulation and detection are easier to implement in slow frequency hopping spread spectrum (FHSS) based systems than in fast FHSS based systems.

(3) A Hadamard matrix has the general dimension $k \times k$.

   (a) What values of k are permissible?
   (b) Use the appropriate Hadamard matrix to construct mutually orthogonal pseudorandom noise codes to serve 10 users at a node of an UMTS/WCDMA system.

(4) Assume that $\mathbf{H}_k$ is a Hadamard matrix of dimension $k \times k$, where k is a positive integer. Show that $\mathbf{H}_k\mathbf{H}_k^T = 2\mathbf{I}_k$, where $\mathbf{I}_k$ is the $k \times k$ identity matrix.

(5) (a) Are the sequences generated from a given linear feedback shift register whose characteristic polynomial is a primitive polynomial perfectly orthogonal? (b) If not, how can we reduce the correction between the sequences?

(6) Discuss how the Fibonacci implementation of linear feedback shift register (LFSR) differs from, and/or is similar to, the Galois implementation.

(7) We have discussed that a maximal-sequence LFSR with m produces $2^m - 1$ unique sequences, each of length $2^m - 1$. Compute the remaining 6 sequences for the case study discussed in Figs. 8.20 and 8.21.

(8) Compute the sequences generated by the two Galois implementations of LFSR shown in Fig. 8.20(c) and Fig. 8.20(d). Prove that the results are the same as those obtained from Fig. 8.20(a) and Fig. 8.20(b).

(9) The 4-stage LFSR described by the generator polynomial $X^4 + X + 1$ is used to generate m-sequences.

   (a) Compute all of the maximal-length feedback tap sets and express them in both Fibonacci and Galois forms.
   (b) Draw the block diagram for the circuits used to implement the LFSRs for both Fibonacci and Galois implementations.
   (c) Generate the sequences produced by the LFSR using the seed 00001. Use the Galois implementation of the LFSR.
   (d) Use the seed 1001 to generate the sequence produced by the LFSR above and compare your results with that obtained in (c).
   (e) Find the PN code corresponding to the sequences generated.
   (f) Compute the cross-correlation between the obtained sequences and their cyclic shifts by 3 chips.

(10) Assume that the average transmit power of each user in a CDMA network is 10 nW, 50 μW and 125 mW, and the system bandwidth is $B_s = 5$ MHz. The thermal noise psd is $N_0 = kT = -174$ dBm/Hz. Compute the average throughput of each user if a NodeB in the network serves 100 users at any given moment. (b) Plot the average throughput of each user as a function of the number of users. (c) Compare the throughput for the three transmit powers given.

(11) We are designing a mobile cellular network using UMTS technology to serve Canberra city in Australia which has area of about $A = 814.2$ $\mathtt{km}^2$ and a population of about

403,468. Assume that 70% of the inhabitants would subscribe to and use the mobile network service, and mobile network will have a 90% coverage of the city. Assume that the maximum number of users, called pole capacity, that each cell in the mobile network can support in the uplink is given by

$$N_u = 1 + \frac{1}{\alpha}\left[\frac{g_s G_p}{(1+\beta)\frac{E_b}{N_0}} - \frac{N}{P}\right].$$

Other parameters of the system are:

- Sectorisation gain, $g_s = 0.8$ per sector.
- System chip rate, $R_c = 3.84$ Mcps.
- System bandwidth, $B_s = 5$ MHz.
- User uplink activity factor, $\alpha = 0.5$.
- Ratio of the power of all interfering cells to the power of cell of interest, β is 0.93 (3-sector cell), 0.67 (omni-directional antenna), 0.4 (microcell).
- Bit energy to one-sided thermal noise power density ratio, $E_b/N_0 = 8$ dB.
- Transmit power of each user equipment at eNodeB, $P = 50$ nW (transmit power control is employed).
- Thermal noise density at eNodeB, $kT \approx -174$ dBm/Hz, where k is the Boltzmann's constant.

We need to compute the following in order to dimension the system described above.

(a) Compute the pole capacity of a cell site using 3 × 120° sectored antennas with the sectorisation efficiency of 80%.

(b) Assume that the efficiency of each sector is 80% and the network can be loaded up to only 85% of its pole capacity. Compute the net capacity of each cell.

(c) Assume that the mobile service provider seeks to maintain a grade of service (GOS) of 2%, how much traffic load in Erlangs can each cell support? Assume that each user is served by one channel.

(d) Assume that, on the average, each user generates Au = 0.4 E traffic into the network. Compute the traffic demand in Canberra in Erlangs.

(e) Compute the number of cells dimensioned above needed to cover the required service area, i.e., 90% of Canberra.

(f) Assuming each cell has a hexagonal shape, how far away from the base station can a mobile user be located and still be able to receive 0.4 E traffic intensity?

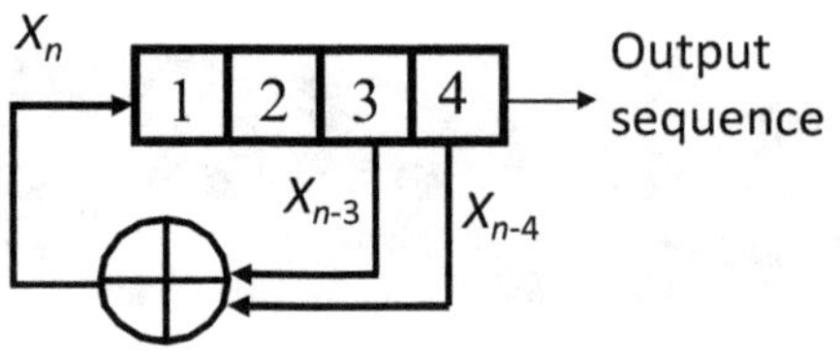

Fig. 8.29 A LFSR circuit.

(12) Study the LFSR circuit in Fig. 8.29.

(a) Write out the characteristic polynomial of the LFSR described by the circuit.

(b) Write down the sequence generated by it.

(13) In wireless communications, what does the ISM band mean?

(14) Discuss what motivated the development of the Bluetooth technology.

(15) Analyse the meaning of the name Bluetooth, including its origin.

(16) Investigate into the mandate of the Bluetooth Special Interest Group (SIG).

(17) Assume that a Bluetooth device transmits a packet of size 12 bytes over a 1-Mbps link to a master device located 120 m away. The operating current over a 125-Ω resistive load is 12 mA.

(a) Compute the one-way latency for the transmission of the packet.

(b) Compute the average output power of the device in both mW and dBm.

(c) How much average energy in mJ did the device dissipate for the entire transaction?

(d) Assume that the Bluetooth device must be powered by a battery. Find the required minimum voltage of the battery.

(18) Give the name for the IEEE standard for Bluetooth.

(19) Bluetooth operates in the 2.4 GHz ISM unlicensed frequency band which is also used by many other radio standards, including microwave ovens, cordless phones, Wi-Fi, garage door openers. Discuss the techniques used by Bluetooth to manage radio frequency interference from these co-existing technologies.

(20) Discuss the various techniques used in Bluetooth radio standard to provide communications security and privacy.

(21) List three radio technologies which share the 2.4 GHz ISM band with Bluetooth.

(22) On the average BLE can use a fewer number of radio frequency channels than Bluetooth Classic. Explain the causes of that.

(23) Broadly categorise the applications of Bluetooth into three with justification.

(24) Assume that a BT device starts to transmit a packet of size 369 bytes over a 2-Mbps link at 2 pm. Fifty microseconds later another BT device begins to transmit a packet of size 56 bytes over a 1-Mbps link. Unfortunately, the two devices are using the same hopping sequence.

(a) How much time do their transmissions collide?

(b) What fraction of the frame being transmitted by the first device will be garbled?

(25) Assume that an antenna with gain 1.5 dBi is connected to a Class 1 Bluetooth device. Neglect all possible losses and compute the maximum power that the radio electronics must feed to the antenna.

(26) A slave device is located 5 m away from its master device in a Bluetooth piconet, in which all devices are Class 2 devices. Data is transmitted from the master to the slave with the receiver sensitivity of −70 dBm. Assume that both devices use perfectly omnidirectional antennas, and the path loss exponent is 2.4. What power must the master feed its antenna to ensure acceptably good communications quality. Neglect any other losses besides propagation losses.

(27) Assume that at a given moment four users are being served in a spread spectrum based wireless network. Assume that users 1 and 3 are communication, while users 2 and 4 are also communicating. Assume that all transmissions reach each user. Table 8.10 shows the spreading codes and data currently being transmitted by each user.

Table 8.10 Spreading codes and backlogged data of users of a spread spectrum system.

Users	**Data to send**	**Spreading code**
User 1 (U1)	1111	00000000
User 2 (U2)	0101	01100110
User 3 (U3)	1100	01101001
User 4 (U4)	1001	01010101

Note that bit 0 is encoded as +1, while bit 1 is encoded as −1.

(a) Using figures, analyse the composite signal that reaches each of the receivers.

(b) Explain in detail how each of the receivers decodes its intended signal correctly.

8.5 Random and Round-Robin Multiple Access Schemes

Random access MAC schemes are used only in the uplink. They are, in general, based on the survival of the fittest principle, which is contention. We survey some of the commonest ones in this section. Random access MAC protocols do not have synchronisation between data sources, as there is no coordination. Each backlogged data source transmits at the full channel access rate R_{b}. Multiple nodes can attempt to use the transmission medium in overlapping time intervals, causing *collision* of packets. Collided packets are lost and must be retransmitted. Thus, the main objectives of random MAC protocols are collision detection and recovery upon collision. Advantages in random access schemes include:

(1) Reduced signalling overhead as no need for communications resource request and allocation.
(2) Low latency, as no time wasted to wait for resource allocation prior to data transfer.
(3) Scalability and flexibility.
(4) Simplicity and hence easy implementation.

Contention-based MAC schemes are thus suited for the exchange of low-rate, low-latency, bursty and small packets. These features make them attractive for massive machine type communications (mMTC), as in the Internet of Things (IoT). However, the collisions caused by the inherent contention reduces resource utilisation efficiency, including wastage in device power.

8.5.1 *Aloha and Slotted Aloha*

Aloha was developed by Norman Abramson and his team in the early 1970s at the University of Hawaii, USA, to network over satellite links the computers of the university. Aloha and its variants are simple schemes. However, they have poor throughput owing to their inherent message collisions. There are three basic types of Aloha: pure Aloha (p-Aloha), slotted Aloha (s-Aloha) and reservation Aloha (r-Aloha). The Aloha protocol works as follows: A node having data sends it at will without regard to other nodes' activities. It then sets a timer, called acknowledgement time (ACK), and expects a response from the receiver within this time. The node retransmits the data if either it receives no ACK before the time expires, or it receives a negative acknowledgement (NAK). The data receiver sends a NAK if the transmission collides with the transmissions of other nodes. Let us summarise the assumptions and parameters used by the Aloha protocol:

- b: number of bits in each frame/packet transmitted by a node. Messages transmitted by all nodes have this same fixed size.
- R_{b}: while transmitting data, each node transmits at this fixed rate, measured in bits per second.
- λ_{s}: the average rate at which packets are transmitted *successfully* by all nodes in the system.
- λ_{f}: the average rate at which packets are transmitted *unsuccessfully* by all nodes in the system.
- $\lambda = \lambda_{\mathrm{f}} + \lambda_{\mathrm{s}}$: average cumulative traffic arrival to the shared medium.
- T_{p}: time in seconds taken to transmit each packet, where

$$T_{\mathrm{p}} = \frac{b}{R_{\mathrm{b}}}\,. \tag{8.45}$$

Let us define the probability of a node transmitting a packet successfully (i.e., no collisions) as

$$P_{\mathrm{s}} = \frac{\lambda_{\mathrm{s}}}{\lambda_{\mathrm{f}} + \lambda_{\mathrm{s}}} = \frac{\lambda_{\mathrm{s}}}{\lambda} \tag{8.46}$$

and the normalised total offered traffic load as

$$G = \frac{b\lambda}{R_{\mathrm{b}}} \overset{(8.45)}{=} \lambda T_{\mathrm{p}}\,. \tag{8.47}$$

We assume that the rate at which packets are supplied to the shared transmission medium is Poisson distributed with rate λ. The probability that N packets are generated during the time frame τ is

$$P_n = Pr[n \texttt{ packets in } \tau] = \frac{(\lambda\tau)^n e^{-\lambda\tau}}{n!}\,. \tag{8.48}$$

A period of time in which collisions can occur is referred to as the **vulnerability period**. The con-

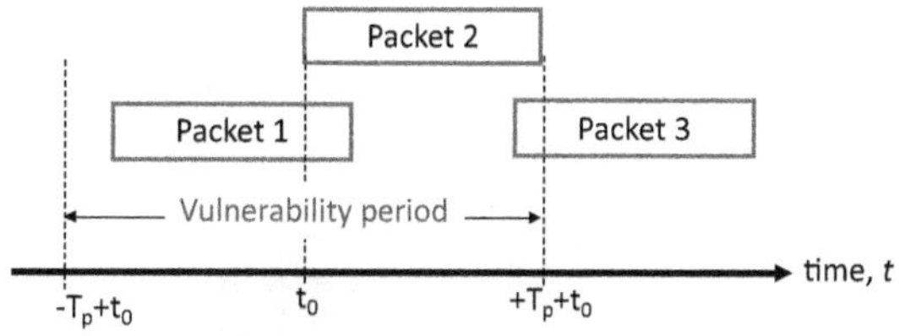

(a) Pure Aloha: packet 3 collides with the tail while packet 1 collides with the beginning of packet 2 within the vulnerability interval.

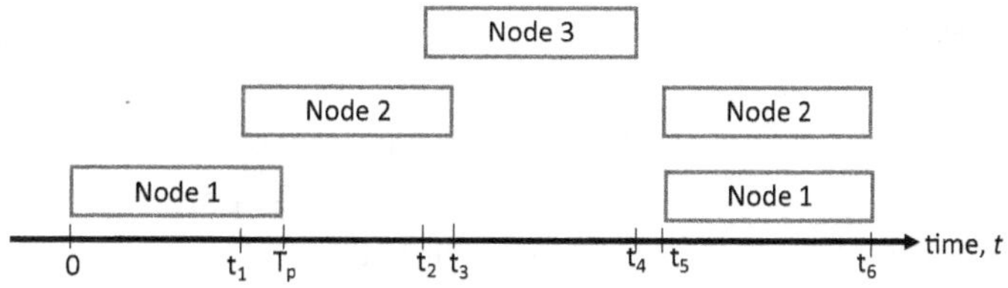

(b) Pure Aloha: transmissions by three computers with both partial and full collisions

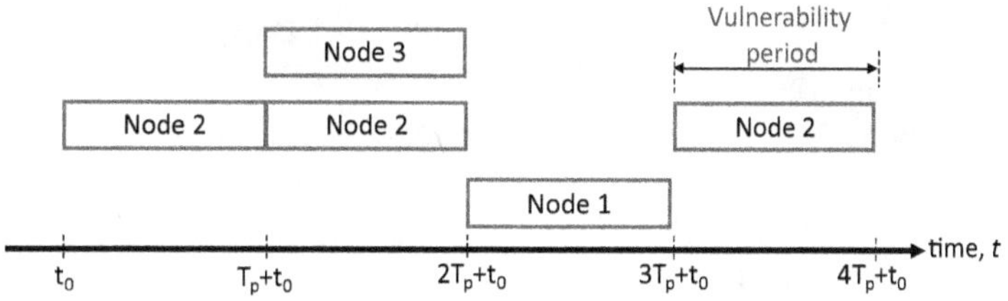

(c) Slotted Aloha: transmissions by three computers with only full collisions

Fig. 8.30 Illustration of the principle of Aloha and slotted Aloha.

dition for the successful transmission probability is that there are no transmissions during the vulnerability period illustrated in Fig. 8.30(a) for pure Aloha and in Fig. 8.30(c) for slotted-Aloha. Aloha system is liable to both partial and full packet collisions (Fig. 8.30(b)). Plugging $n = 0$ and $\tau = 2T_\text{p}$ in (8.48) yields

$$\begin{aligned} P_\text{s} = P_0 &= Pr[0 \texttt{ packets in } 2T_\text{p}] \\ &= \frac{(\lambda\tau)^0 e^{-2\lambda T_\text{p}}}{0!} = e^{-2\lambda T_\text{p}} \,. \end{aligned} \tag{8.49}$$

The *throughput* is defined as

$$\begin{aligned} S_\texttt{aloha} &= \textbf{(offered load)} \times P_\text{s} \\ &= Ge^{-2\lambda T_\text{p}} \overset{(8.47)}{=} Ge^{-2G} \,. \end{aligned} \tag{8.50}$$

As obvious from Eq. (8.50), the throughput of Aloha decreases with the vulnerability period. Reducing the vulnerability period will reduce the collisions and increase the throughput. By allowing nodes to transmit data only at the beginning of a time slot, slotted-Aloha reduces the vulnerability period of pure Aloha by a factor of two as illustrated in Fig. 8.30(c). Also, a slotted-Aloha system is liable to only full collisions, but not partial collisions (Fig. 8.30(c)). The backoff period in slotted-Aloha upon a collision is an integer multiple of the time slot duration. Thus, the throughput of slotted-Aloha results from Eq. (8.50) straightforward as

$$S_\texttt{s-aloha} = Ge^{-G} \,. \tag{8.51}$$

These results are compared with those of carrier-sense multiple access schemes in Fig. 8.31. As seen in the figure, $S_\texttt{s-aloha}$ achieves its maximum value at $G = 1$, while $S_\texttt{aloha}$ achieves its maximum value at $G = 0.5$.

Example: Aloha versus Slotted Aloha

We seek to compare the throughput of pure Aloha (aka unslotted Aloha) and slotted Aloha in a system in which the packet length is fixed at 64 bytes

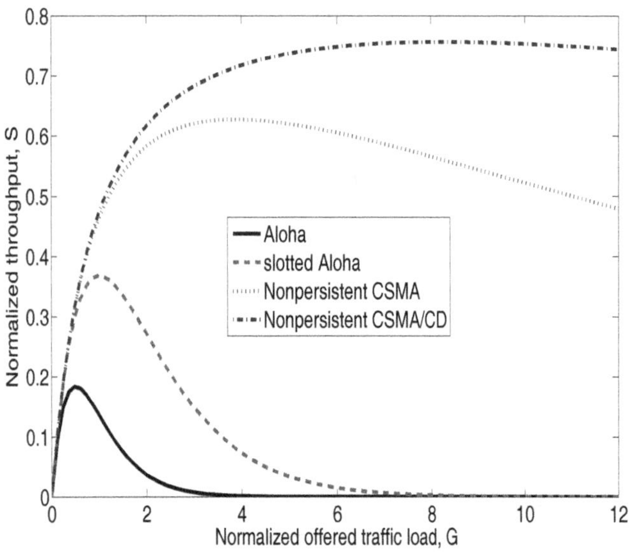

Fig. 8.31 Comparison of random access protocols.

and the data rate is 2 Mbps. In each case assume that the cumulative average packet arrival rate is 1500 packets per second.

Solution

The normalised total traffic is

$$G = \frac{b\lambda}{R_\mathtt{b}} = \frac{(64 \cdot 8) \times 150,000}{2 \cdot 10^6} = 0.384\,.$$

The corresponding throughputs are, respectively

$$S_\mathtt{aloha} = Ge^{-2G} = 0.384 \cdot \exp(-0.768) \approx 0.18$$

and

$$S_\mathtt{s\text{-}aloha} = Ge^{-G} = 0.384 \cdot \exp(-0.384) \approx 0.26\,.$$

8.5.2 *CSMA, CSMA/CD and CSMA/CA*

What does carrier sense multiple access (CSMA) mean? The transmission medium used in a communications system transports or carries data or signals. It is thus referred to as a *carrier* in local area networks (LANs). Also, each node or computer attached to a LAN is called a station (STA). In order to avoid collisions, a station which has data to send to another station first listens or monitors the shared carrier to ensure that no other signals are occupying it. The monitoring is called *sensing*. Multiple access (MA) just means a protocol which arbitrates between multiple stations sharing the carrier.

A LAN can be switched or non-switched. Switched LANs interconnect nodes using switches and/or routers which are ISO/OSI Layer 2 and 3 devices, respectively. Non-switched LANs, however, interconnect nodes using repeaters and/or hubs which are ISO/OSI Layer 1 devices. These are illustrated in Fig. 8.32. Each port of a switch or router is a separate collision domain. Therefore, the nodes in Fig. 8.32(c) can transmit frames at the same time without collisions. Such a network does not require nodes to implement carrier sensing (CS), collision detection (CD) or collision avoidance (CA). Therefore, backoff algorithms are not used in switched LANs. Un-switched LANs, however, are liable to collisions. Therefore, nodes in the LAN of Fig. 8.32(b) implement backoff algorithms and must also sense the carrier to detect or avoid collisions.

Bandwidth-delay product (BDP) is the product of the round-trip time (RTT) of a frame and the data transfer rate, $R_\mathtt{b}$, i.e.,

$$\begin{aligned} BDP &= RTT \times R_\mathtt{b} = 2t_\mathtt{prop} \times R_\mathtt{b} \\ &= 2b \times a \;\; \mathtt{[bits]} \end{aligned} \tag{8.52}$$

where $t_\mathtt{prop}$ is the one-way signal propagation time, b is the number of bits in a frame, and a, referred to as the **normalised collision window**, is defined as

$$a = \frac{\mathtt{signal\ propagation\ time}}{\mathtt{packet\ transmission\ time}} = \frac{t_\mathtt{prop}}{b/R_\mathtt{b}}\,. \tag{8.53}$$

The BDP indicates how much data can be in transit within the network. Alternatively said, the BDP is the maximum amount of data that can be transmitted before receiving an acknowledgement from the data sink. A network with $BDP > 10^5$ bits is called **long fat network** (LFN). If $a < 1$ or $BDP < 2b$ then the first bit in a packet reaches receiver before its transmission completes. Such a situation is desirable as all other devices sense

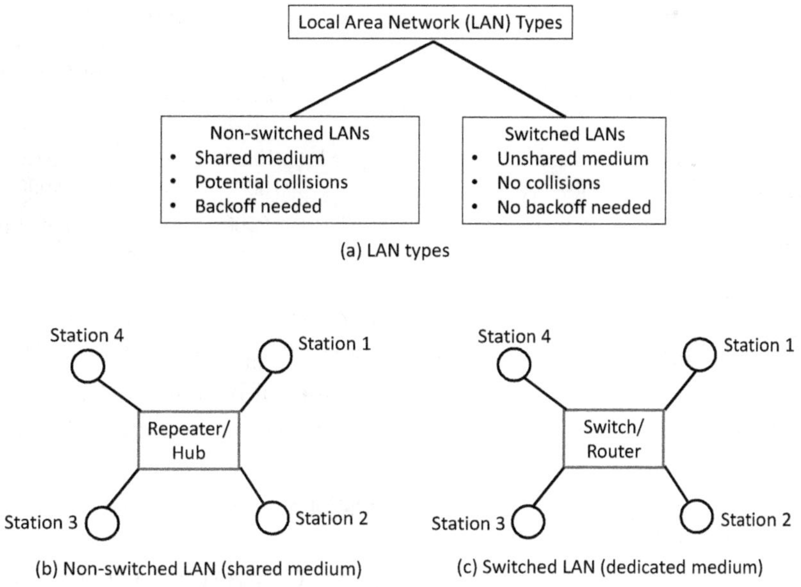

Fig. 8.32 Switched versus non-switched LANs.

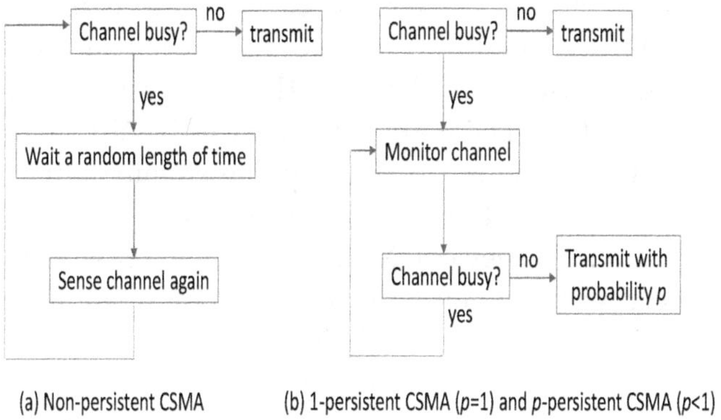

Fig. 8.33 The summarised algorithm for CSMA schemes.

when another is using the medium. This is the basis of all CSMA protocols. The lower the BDP, the better the performance of CSMA and its variants. As Fig. 8.33 illustrates, there are three versions of CSMA schemes.

CSMA

There are three basic types of CSMA: non-persistent CSMA, 1-persistent CSMA and p-persistent CSMA. In *non-persistent CSMA*, a node sends a packet once it senses the medium to be idle. If the medium is busy, then the device backs off (i.e., waits a random length of time) before sensing the channel again. In *1-persistent CSMA*, however, the node continuously monitors a busy channel and transmits its data as soon as it finds it free. The p-persistent CSMA differs from 1-persistent CSMA as it transmits with the probability p when it finds the channel idle.

CSMA/CD

Adding collision detection (CD) to CSMA yields CSMA/CD. The additional features of CSMA/CD not in CSMA are:

- Upon sensing a collision, a node stops transmission altogether and sends a jamming signal through the medium to warn all other devices of a collision. The main problem in pure CSMA is that a node continues to complete the transmission of a packet even upon collision as it does not detect the collision. This causes a wastage of transmission resources.
- The node then waits for a random length of time, referred to as **backoff**, and then re-attempts transmission.
- After every collision the backoff time, that is the amount time waited before re-attempting transmission, is doubled. This procedure is called **binary exponential backoff algo-**

rithm (BEBA). A nodes makes altogether up to 16 attempts (original attempt plus 15 retransmission attempts) and then discards the packet if unsuccessful.

The throughput of the CSMA protocols are

`Non-persistent CSMA:`

$$S = \frac{Ge^{-aG}}{G(1+2a)+e^{-aG}} \tag{8.54}$$

`Non-persistent CSMA/CD:`

$$S = \frac{Ge^{-aG}}{Ge^{-aG}+3aG(1-e^{-aG})+2-e^{-aG}}$$

where a is defined in Eq. (8.52).

Binary Exponential Backoff Algorithm

The Binary Exponential Backoff Algorithm (BEBA) is used to reduce collisions on the shared medium. Time is divided in time periods referred to as time slot or slotTime. The slotTime is the time required to transmit the smallest Ethernet frame. For example, the smallest frame of the 10BaseT Ethernet comprised 512 bits (46 bytes Layer 3 data plus 14 bytes of Layer 2 header plus 4 bytes of layer 2 trailer). Table 8.11 shows the durations of the slotTime for some data rates. After each collision a station waits a random amount of time before re-sensing the carrier or medium. The waiting time is called *backoff*. The backoff after m collisions is

$$Backoff(m) = k \times slotTime\ [\mu s] \tag{8.55}$$

where k is a positive integer picked randomly from the interval $\{0, 1, 2, \ldots, 2^m - 1\}$. The average time of the backoff, measured in time slots, is

$$E[m] = \sum_{n=0}^{2^m-1} n = \frac{2^m-1}{2}. \tag{8.56}$$

Thus, the average backoff is $\frac{2^m-1}{2} \times slotTime$. In practical systems, the maximum number of retransmissions is limited. For example, the CSMA/CD in IEEE 802.3 limit is $m \leq 10$, resulting in the longest backoff of $511.5 \times slotTimes$. In IEEE 802.11 we have $CW_{min} = 31 \leq m \leq 1023 = CW_{max}$. As the number of backoff slots is limited, the BEBA is also referred to as truncated BEBA.

Table 8.11 Time slot durations for different data rates.

Data Rate (Mbps)	SlotTime (bits)	SlotTime (μs)
10	512	51.2
100	512	5.12
1000	4096	4.096
$\geq$ 2500	half-duplex not allowed	

CSMA/CA

Why CSMA/CA? CSMA/CD enhances the operation of pure CSMA for wired systems. However, the carrier sensing used by CSMA/CD to detect or sense frame collisions does not work on wireless channels. Wireless channels are liable to the **hidden terminal problem** (HTP) or **hidden node problem** (HNP). The HTP is illustrated in Fig. 8.34(a). There are three devices X, Y and Z. The transmissions of each device has a limited range beyond which other stations cannot sense or hear. For example, devices X and Z cannot sense each others' transmissions. Device Y, however, can hear the transmissions of both X and Z. Assume that Z is currently transmitting to Y upon sensing the medium idle. As X cannot sense/hear this transmission, it also starts to send frames to Y thinking that the medium is idle. Thus, frames from X and Z collide at Y. This problem is called **hidden terminal problem.**

CSMA/CA is a combination of physical carrier sensing (i.e., CSMA) and *virtual carrier sensing* (i.e., RTS/CTS) mechanism to avoid frame collisions caused by the HNP/HTP. A combination of CSMA/CA and the BEBA is referred to as Distributed Coordination Function (DCF). This is illustrated in Fig. 8.34(c). In CSMA/CA, the frame sender has to sense the medium and wait for it to be idle for a period of time called DCF Interframe Space (DIFS). It then sends a special signalling frame called *ready to send* (RTS) to the intended receiving station. This transmission is heard

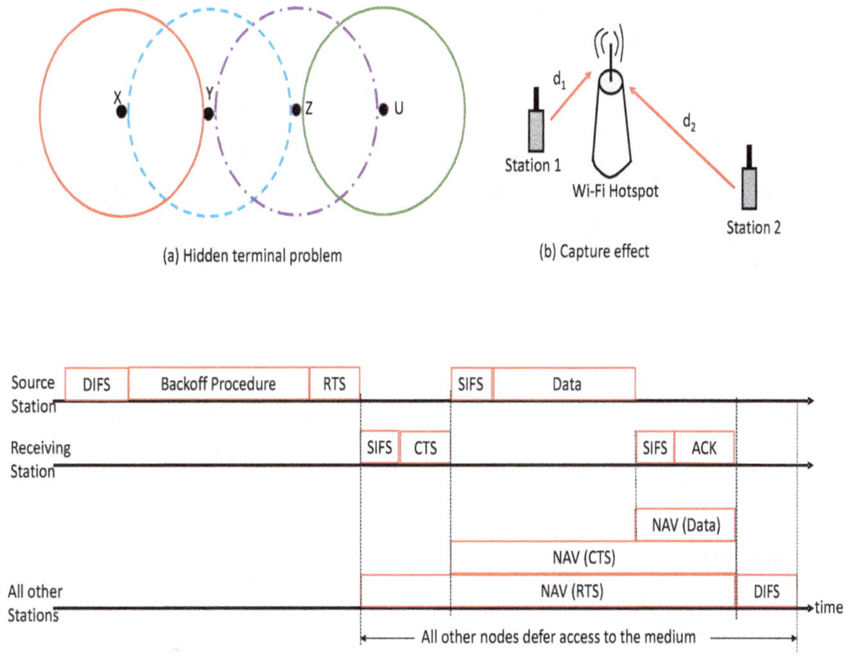

(a) Hidden terminal problem

(b) Capture effect

(c) RTS/CTS virtual sensing access method for 4-way handshake in IEEE802.11

Fig. 8.34 Hidden terminal and exposed terminal problems and capture effect.

by all stations within range of the transmitting station. The receiving station responses with a frame called *cleared to send* (CTS). All stations within the hearing range of the receiving station hear the CTS frame. Both the RTS and the CTS frames contain the time frame in which the medium will remain busy. Each station keeps a counter called *network allocation vector* (NAV), which contains the amount of time required until the medium/carrier becomes idle again. This means that all nodes around the sender and the receiver do not transmit in order to avoid the likelihood of collisions until the value in their NAV decreases to zero.

The station sends its frames upon receiving a CTS. If a station does not receive a CTS upon sending a RTS it does not transmit, but enters the BEB mode. CSMA/CA is used in all IEEE 802.11 protocols (referred to as Wi-Fi), which are the most popular wireless local area networking (WLAN) standards. The short interframe space (SIFS) is the time taken for a frame to be processed by the physical and medium access control layers. It includes the time to receive and process a frame and send its acknowledgement.

Another problem in broadcast wireless systems, such as wireless local area networks (WLANs) is the **exposed terminal problem** (ETP). Let us reconsider Fig. 8.34(a). Devices X and Z can sense the transmissions of device Y. Assume that device U is idle, and device Z wants to send frames to device U. When device Z senses the transmission medium it hears the transmissions of device Y to device X. Therefore, it ceases to transmit to device U, although no collisions could occur if it had progressed with its intentions. Thus, the ETP wastes transmission time.

A third problem in wireless transmission, which is similar to the near-far effect in CDMA systems, is the **capture effect**. Consider Fig. 8.34(b), in which two stations are sharing a Wi-Fi hotspot. Station 1 is much closer to the hotspot than Station 2. Assuming that no transmitter power control is used and both stations are transmitting on the same frequency, then the power in Station 1 transmissions would reach the hotpsot being much higher than that in Station 2 transmissions. This means that the signals from Station 1 can be detected correctly, but not that of Station 2, resulting in unfair sharing of the hotspot. Capture effect is a major problem in FM radio broadcasting.

8.5.3 *Round-Robin Protocols*

In round-robin (RR) protocols, each of the nodes backlogged at any given time is given an access to the channel one after the other by rotation. A node with empty transmit buffers can reject its turn to use the medium. RR schemes can be centralised or distributed. An example of a centralised RR is polling, while token passing is an example of distributed RR.

In polling, the centralised controller broadcasts messages to all nodes. Each message contains the address of the node which is given permission to use the medium. All nodes decode every message, but just discards those not addressed to them. Polling is used in mobile networks. Polling is simple, but wastes resources through the polling messages.

Token passing uses a special bit pattern, called token, to regulate access to the transmission medium. Like a baton, whoever has the token has the permission to use the medium, and then pass it on to the next device. An example of token passing is token ring. Weaknesses in token passing include the possibility of token duplicating, token loss and change of network topology. Also, RR schemes have inherited the inefficiency of synchronous TDMA and work efficiently only when the majority of the nodes have data to send most of the time.

8.5.4 *Check Your Understanding*

Students are encouraged to check their understanding of the content of this section with the following questions prior to proceeding to the next section.

(1) Compare the throughput of pure Aloha and slotted Aloha in a system in which the packet length is fixed at 512 bytes and the data rate is 1 Mbps. In each case assume that the cumulative average packet arrival rate is 500 packets per second.

(2) Explain one practical application of polling and one of token passing round-robin protocols.

(3) A slotted-Aloha system has the throughput 0.4 and packets are injected into the shared medium according to the Poisson process with the parameter λ.

  (a) Find the value of the normalised offer load G.

  (b) Compute the average number of transmission attempts needed to send one packet.

(4) (a) Describe the carrier sense multiple access (CSMA) algorithm.

  (b) Give an example of a communications network which implements CSMA.

(5) CSMA with collision detection (CSMA/CD) was developed to improve upon the performance of CSMA. Describe how CSMA/CD differs from CSMA.

(6) CSMA/CD is implemented in wired communications systems. Wireless communications systems, however, implement CSMA with collision avoidance (CSMA/CD).

  (a) State the differences between CSMA/CA and CSMA/CD.

  (b) Why is it that we cannot use CSMA/CA in wireless systems?

(7) Figure 8.30 illustrates three stations sharing the transmission medium via pure Aloha and slotted-Aloha. Show in which time frames are full and partial frame collisions occurring.

(8) Using Fig. 8.31, compute the maximum throughput of pure Aloha as a function of the maximum throughput of slotted-Aloha.

8.6 OFDM, OFDMA, SC-FDMA and Their Variants

Orthogonal Frequency Division Multiplexing (OFDM) is also referred to as *discrete multi-tone modulation* (DMTM) or *multi-carrier modulation*

(MCM). OFDM and its variants have a wide range of practical applications. OFDMA extends the capability of OFDM from a single user to multiple users. OFDM is counted as one of the most important and popular technologies in the digital communications industry. It is the standard transmission in modern systems, including:

(1) *WLANs:* IEEE 802.11 at 5.8 GHz WLAN and the IEEE 802.11ax wireless local area network (WLAN) standard which is still in research laboratories and whose products are not expected to appear in the market until 2019.
(2) WMAN: Worldwide Interoperability for Microwave Access (WiMAX) or IEEE 802.16.
(3) Television and audio broadcasting: digital video television (DVB) and digital audio broadcasting (DAB) systems in many countries, including Australia, Japan and Europe.
(4) Power-line communications networks
(5) Wired broadband Internet access: asymmetric digital subscriber lines (ADSL) uses Discrete MultiTone (DMT) transmission which has the same principle as OFDM.
(6) Mobile communications: 4G standard LTE and LTE-Advanced, 5G mobile technologies.

Both OFDM and Orthogonal Frequency-Division Multiple Access (OFDMA) have a common basis. They divide a wideband channel with bandwidth, say B_s Hz, into multiple narrowband channels called subcarriers, each with bandwidth, say B_k. The subcarriers in an OFDM/OFDMA signal can have different or the same bandwidth or spacing. In 4G wireless technologies, all subcarriers have the same bandwidth and thus $B_k = W$. 5G wireless technologies are poised to allow flexible subcarrier bandwidth. Although the OFDM subcarriers overlap in frequency, they do not interfere with each other (Fig. 8.43). This is the reason why they are referred to as **orthogonal** subcarriers. Orthogonality among the subcarriers is needed to mitigate inter-carrier interference (ICI), aka inter-channel interference or inter-subcarrier interference.

Why do we subdivide a wideband channel into multiple subchannels with smaller bandwidths?

- The low bandwidth of the subcarriers, compared with the wideband channel, results in low symbol rate (or data rate) which attracts lower inter-symbol interference (ISI) for a better signal-to-noise plus interference ratio (SNIR).
- A low ISI allows the use of simple equalisers to reduce system complexity, which is good for low power consumption and low processing delay.

OFDM is a variety of frequency division multiplexing (FDM). **How does OFDM differ from FDM?** Some of the the basic differences are:

(1) OFDM subcarriers do overlap in frequency but do not disturb each other so there is ideally no ICI, as illustrated in Fig. 8.43. The reciprocal of the subcarrier spacing equals the time needed by the transmitter to perform the inverse fast Fourier transform (IFFT), which equals the symbol duration.
(2) In contrast to FDM, OFDM does not use guard bands in order to increase spectral efficiency.
(3) FDM uses bandpass filters which are not needed in OFDM, reducing the latter's implementation complexity and allowing full implementation by digital signal processors (DSPs).

OFDM is a combination of a digital multi-carrier modulation and multiplexing method. As OFDM uses multiple carrier signals it is a multi-carrier technology. It is superior over single-carrier schemes because it is able to withstand severe channel conditions such as narrowband interference, frequency-selective fading due to multipath without employing complex equalisation filters and attenuation of high frequencies metallic wired transmission media. OFDM simplifies the function of equalisation through a combination of factors:

(1) It divides a wideband channel into a set of narrowband channels so that the OFDM symbol duration (aka symbol time) becomes long relative to the channel's delay spread.
(2) Uses cyclic prefix to combat any residual inter-symbol interference (ISI).

The finite-impulse response (FIR) of a wireless channel has the frequency response in z-transform

Table 8.12 Wireless channel properties.

Channel Impairment	Causes	Important Parameters	Mitigation Techniques
Frequency-selective fading (or ISI in time domain) or flat fading (no ISI)	Multipath signal propagation	Channel delay spread τ_{m}, channel's coherence bandwidth $\left(B_{\mathrm{c}} \propto \frac{1}{\tau_{\mathrm{m}}}\right)$	Equalization, channel coding, diversity
Slow fading or fast fading	Doppler shift due to relative movement between transmitter and receiver	Doppler spread (F_{d}), channel's coherence time $\left(\tau_{\mathrm{c}} \propto \frac{1}{F_{\mathrm{d}}}\right)$	Increase spacing between subcarriers/carriers

as

$$H_c(z) = \sum_{n=0}^{N} h_n z^{-n} \tag{8.57}$$

where z^{-1} represents a unit delay of duration T_s, h_n is the gain of the nth path of the wireless channel and N is called channel's order which indicates its delay spread. This means that the channel spreads to affect N symbols each of duration T_s. Thus, the channel delay spread is NT_s seconds. An equaliser for such a channel has the transfer function

$$H_e(z) = \sum_{m=0}^{M} g_m z^{-m} = \frac{1}{H_c(z)} \tag{8.58}$$

where in general the duration of the equalisation filter's length M is larger than the channel's order N.

8.6.1 *Subcarrier Allocation Methods*

By its nature an OFDM-based system assigns multiple subcarriers to each user. The subcarriers assigned to a given user can be consecutive subcarriers (as done in LTE/LTE-A) or non-consecutive subcarriers (Fig. 8.35). The former case is referred to as localised subcarrier allocation or **localised user multiplexing** (LUM), while the latter case is referred to as **distributed multiple access**, distributed subcarrier mapping, interleaved mapping or **distributed user multiplexing** (DUM). Figure 8.35 illustrates the two subcarrier allocation methods using four mobile users sharing 12 subcarriers equally at an eNodeB.

As discussed earlier, LTE/LTE-A standards allocates a minimum of 12 subcarriers to a single user. We are allocating 3 subcarriers to each user only for simplicity of illustration. DUM is a form of frequency diversity as each user's transmissions are spread over a wider bandwidth. There are two types of symbols in OFDM: **data symbol** or **modulating symbol** and **OFDM symbol**. Data symbols are generated by single-carrier modulation scheme and they modulate the subcarriers. One OFDM symbol (often referred to just as symbol) contains multiple data symbols. DUM spreads the data symbols of a user over the entire system bandwidth and fills the unused subcarriers with nulls (Fig. 8.36). Assume there there are altogether N subcarriers and N_{u} users (or mobile terminals) in the system, and that each user is allocated K subcarriers in each block, where $N = KN_{\mathrm{u}}$, as illustrated in Fig. 8.36. The subcarriers filled with zeros can be assigned to other mobile users. If there is an equal distance between the used or occupied subcarriers, as illustrated in Fig. 8.36, then we refer to the resulting scheme as **interleaved FDMA** (IFDMA). Localised and distributed subcarrier allocation are illustrated in Fig. 8.35(a) and (b), respectively.

Table 8.13 compares LUM and DUM methods with respect to three features: error performance, PAPR and system throughput (or data rate). FDMA with distributed subcarrier allocation has a smaller PAPR than its equivalent with localised subcarrier allocation. The only feature among the three in which LUM outperforms DUM is system throughput. Table 8.12 summarises some wireless channel properties and their mitigating techniques. The potential to achieve a high throughput if combined with channel-state-dependent scheduling (CSD) led LTE/LTE-A standard to adopt LUM mode.

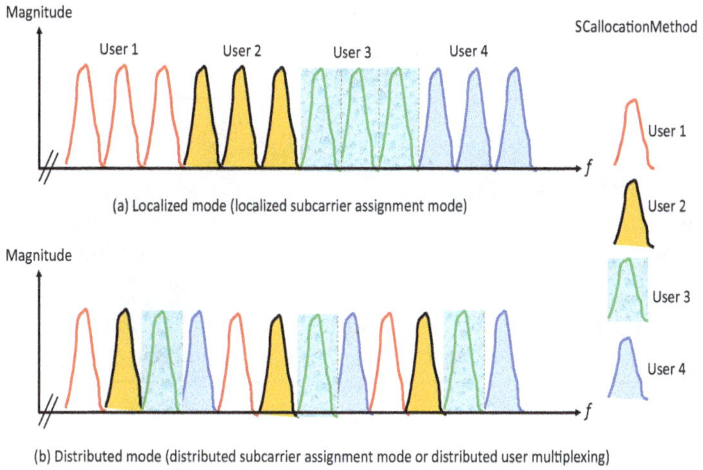

Fig. 8.35 Localised and distributed subcarrier mapping in OFDM.

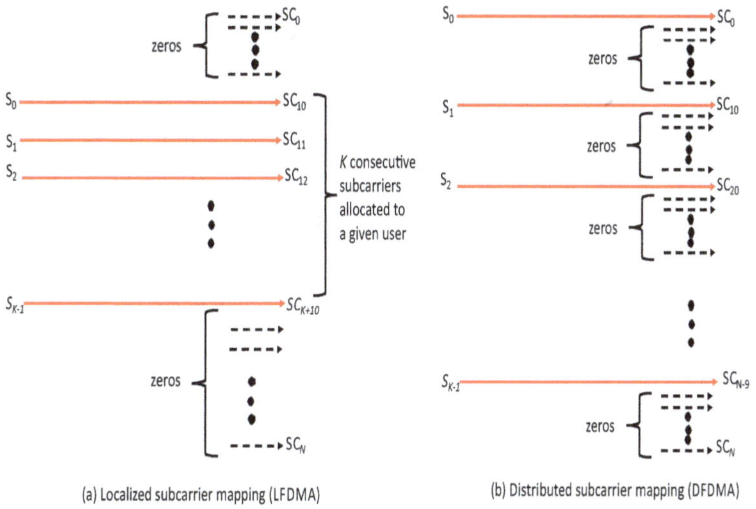

Fig. 8.36 Localised versus distributed subcarrier mapping in OFDM: a user is assigned K of Nsubcarriers in the system.

8.6.2 *OFDM Basics*

As orthogonal frequency division multiplexing (OFDM) is the basis for both OFDMA and SC-FDMA, we discuss its fundamental principles using LTE as a case study. OFDM transmission exploits the features of discrete Fourier transform (DFT), inverse DFT (IDFT) and fast Fourier transform (FFT). See Fig. 8.37. OFDM also exploits linear convolution and cyclic convolution. The *linear convolution* is defined over infinite-length sequences $x[n]$ and $h[n]$ as

$$\begin{aligned} y[n] = x[n] * h[n] &= \sum_{l=-\infty}^{\infty} x[l]h[n-l] \\ &= IDTFT\{X(e^{j\omega})H(e^{j\omega})\} \end{aligned} \quad (8.59)$$

where $X(e^{j\omega})$ is the discrete-time Fourier transform (DTFT) of the sequence $x[n]$. It is defined as

$$\begin{aligned} \texttt{DTFT:}\ X(e^{j\omega}) &= \sum_{l=-\infty}^{\infty} x[l]e^{-j\omega} \\ \texttt{IDTFT:}\ x[n] &= \frac{1}{2\pi}\int_{\alpha_0}^{\alpha_0+2\pi} X(e^{j\omega})d\omega\,. \end{aligned} \quad (8.60)$$

The *cyclic convolution* (aka *circular convolution*) of two finite-length sequences $u[n]$ and $v[n]$, each of length L is defined as

$$\begin{aligned} r[n] = u[n] \otimes v[n] &= \sum_{l=0}^{L-1} u[l]v[n-l] \\ &= IDFT\{U[k]V[k]\} \end{aligned} \quad (8.61)$$

Table 8.13 Localised versus distributed subcarrier allocation methods.

Feature	Localised Subcarrier Allocation	Distributed Subcarrier Allocation
Subcarriers assigned to a user	Contiguous	Distributed
Error performance	Poorer due to lack of frequency diversity	Better due to inherent frequency diversity which mitigates frequency selective fading caused by multipath signal propagation.
PAPR	Larger	Smaller
System throughput	Increases remarkably if combined with channel-state-dependent scheduling (CSD).	CDS has little effect on system throughput due to inherent frequency diversity.

	Input Signal	Output Signal	
		DFT/FFT	IDFT/IFFT
(a)	Time-domain signal	Frequency-domain signal	Frequency-domain signal
	Frequency-domain signal	Time-domain signal	Time-domain signal

$$Y(f), s(t) \xrightarrow{} \boxed{\text{FFT}} \xrightarrow{y(t), S(f)} \boxed{\text{IFFT}} \xrightarrow{} Y(f), s(t)$$

(b)

$$Y(f), s(t) \xrightarrow{} \boxed{\text{IFFT}} \xrightarrow{y(t), S(f)} \boxed{\text{FFT}} \xrightarrow{} Y(f), s(t)$$

(c)

Circular convolution: $z[n]=x[n] \otimes h[n] = \text{IDFT}\{X[k]H[k]\}$

Linear convolution: $z[n]=x[n] * h[n] = \text{IDTFT}\{X(e^{j\omega}) \cdot H(e^{j\omega})\}$

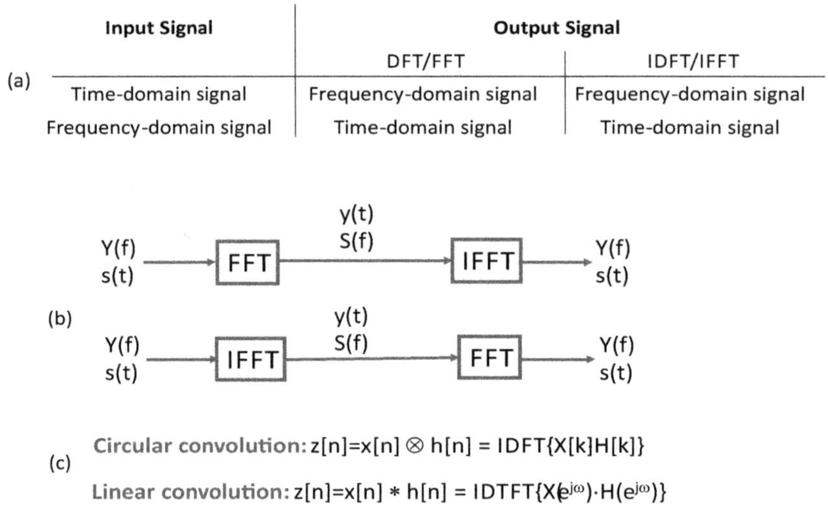

Fig. 8.37 Application of convolution, DTFT, DFT and FFT in OFDM.

where the DFT and its pair are defined as

DFT: $U[k] = \dfrac{1}{L} \displaystyle\sum_{l=0}^{L-1} u[l] e^{-j2\pi kl/L}, \ k=0,1,\ldots,L-1$

IDFT: $u[l] = \dfrac{1}{L} \displaystyle\sum_{k=0}^{L-1} U[k] e^{j2\pi kl/L}, \ l=0,1,\ldots,L-1.$

(8.62)

The transfer of a signal through a transmission medium (or channel) can be modelled as a linear filtering process, in which the signal is processed or filtered by the channel. Similarly, the processing of a digital signal on a digital signal processor (DSP) can be modelled as a filtering process. Filtering process in the time domain is a convolution of the signal $x[n]$ to be filtered and the transfer function $h[n]$ of the signal's processor. The processor here can be the DSP or the transmission medium, as examples. Signal processing in the frequency domain is often much easier to handle than doing so in the time domain. We have observed in Eq. (8.61) that cyclic convolution is equivalent to the inverse DFT or FFT. Therefore, we have to find strategies which allow us to compute the linear convolution through the cyclic convolution in order to utilise DFT or FFT and their inverse. Assume that we seek the linear convolution of the two sequences $x[n]$ of length L_x and $h[n]$ of length L_h. Then, the procedure for doing so through cyclic convolution is as follows:

(1) **Compute the length of the resulting sequence:** The number of elements in the resulting sequence is $N_1 = L_x + L_h - 1$. Therefore, we zero pad each of the two sequences involved to the length $N \geq N_1$. Usually, we select a number which is a power of two in order to speed up the processing as the DFT is usually computed using fast Fourier transform (FFT). Thus,

```
1. clear all;
2. xn=[1 0 -4 2]; hn=[3 1 5];
3. Lx=length(xn); Lh=length(hn);
4. N1=Lx+Lh-1; N=2^ceil(log2(N1));
5. % Zero padding
6. xnp=[xn zeros(1,N-Lx)];
7. hnp=[hn zeros(1,N-Lh)];
8. % Compute the DFTs
9. Xkp=fft (xnp); Hkp=fft (hnp);
10. % Multiple the two DFTs
11. Ykp=Xkp .* Hkp;
12. % Linear convolution via cyclic 13.
convolution
13. display('===========')
14. display('Linear convolution via
cyclic convolution')
15. ynp=ifft (Ykp)
16. % Direct linear convolution
17. display('===========')
18. display('Direct linear convolution')
19. yn=conv(xn,hn)
```

Fig. 8.38 Computing linear convolution through cyclic convolution.

select

$$N = 2^{ceil(\log_2(L_x+L_h-1))}\,. \tag{8.63}$$

(2) **Zero padding**: Append $N - L_x$ zeros to the sequence $x[n]$ and $N - L_h$ zeros to the sequence $h[n]$. Let us refer to the zero-padded sequences as $h_p[n]$ and $x_p[n]$, respectively.

(3) **Compute the DFTs**:

$$Xkp = DFT\{h_p[n]\} \text{ and } Hkp = DFT\{h_p[n]\}\,. \tag{8.64}$$

(4) **Multiply the two DFTs**:

$$Ykp = Xkp \times Hkp\,. \tag{8.65}$$

(5) Compute the linear convolution:

$$yp[n] = IDFT\{Xkp \times Hkp\}\,. \tag{8.66}$$

Figure 8.38 shows MATLAB code that can be used to compute the linear convolution needed for linear filtering through the cyclic convolution. The output of line 15 of the code in Fig. 8.38 is the result from the cyclic convolution, while the results of the direct linear convolution comes from line 19 of the code. We observe that the results are the same. The only difference is the $N - L_x - L_h + 1$ zeros, which is 2 here, appearing at the end of line 15.

8.6.2.1 *Justification for OFDM*

OFDM was developed to mitigate three problems in wireless transmissions: cost and scarcity of wireless bandwidth, inter-channel interference (ICI), and inter-symbol interference (ISI). The design constraints in OFDM are [Roessler (2016)]:

$$\begin{aligned} T_{\text{cp}} &\geq \tau_{\text{d}} \text{ and} \\ T_{\text{u}} &= \frac{1}{\Delta f} \gg \tau_{\text{d}} : \texttt{for ISI avoidance } (a) \\ f_{\text{D}} &\ll \Delta f : \texttt{to reduce ICI } (b) \\ T_{\text{cp}} &\ll \frac{1}{\Delta f} : \texttt{for spectral efficiency } (c) \end{aligned} \tag{8.67}$$

where T_{cp}, T_{u}, Δf, f_{D} are cyclic prefix (CP) duration, useful OFDM symbol time, OFDM subcarrier spacing and maximum Doppler spread, respectively. An OFDM transmission does not need to use a CP. However, the OFDM in LTE/LTE-A uses CP. For this reason, it is referred to as cyclic prefix OFDM (CP-OFDM). Our discussions focus on CP-OFDM as LTE/LTE-A is used as our case study. Wireless bandwidth is expensive and limited. OFDM considers this by achieving a high spectral efficiency through the overlapping of orthogonal subcarriers in frequency rather than creating guard bands between them, and using a cyclic prefix which is just 7.05% to 25.05% of its symbol duration.

The subcarrier spacing of OFDM used in LTE/LTE-A is fixed at $\Delta f = 15$ kHz and relates to the useful symbol duration as

$$\Delta f \times T_{\mathrm{u}} = 1\,. \tag{8.68}$$

This relationship is needed to achieve orthogonality between the subcarriers. We need a large symbol duration to combat inter-symbol interference. However, we also need a large subcarrier spacing to reduce interference between subcarriers arising from Doppler spread at high mobile user travelling speed. A trade-off between these two competing requirements result in $\Delta f = 15$ kHz.

The Doppler spread is caused by relative movement between the two antennas communicating. Doppler spread causes ICI. OFDM reduces ICI by ensuring that the subcarrier spacing is much larger than the Doppler spread in the wireless system. Guard bands are also used to mitigate ICI. Apart from Germany which has, in general, no speed limit on its highways (called Autobahns), the world's speed limit is about 130 km/h in NT Australia and 140 km/h (85 mph) in Texas USA. The highest frequency band in LTE/LTE-A is 5855-5925 MHz, which is the unpaired Band 47 for time division duplexing (TDD) mode of operation. The maximum Doppler spread for these limits is $f_{\mathrm{D}} = \frac{v}{c} f_{\mathrm{C}} \approx 768.06$ Hz, which is less than 5.2% of the LTE/LTE-A subcarrier spacing of $\Delta f = 15$ kHz. Note that the LTE standard allows a subcarrier spacing of 7.5 kHz. However, as at Q3 2018 no PLMN uses this smaller spacing. The LTE standard has been designed to allow a speed of up to about 350 km/h. Even with this speed, the maximum Doppler spread is less than 13% of Δf at 6 GHz, fulfilling condition (8.67b).

The delay spread caused by multipath signal propagation manifests in the time domain as ISI and in the frequency domain as frequency-selective fading. OFDM uses a cyclic prefix and long symbols to mitigate ISI. With these techniques in place, only a simple one-tap equaliser can be used at the receiver to mitigate any residual ISI. The OFDM principle is illustrated in Fig. 8.39. We are using LTE to explain OFDM. Unlike previous generations of cellular mobile technologies, LTE/LTE-A support the six different system bandwidths $B_{\mathrm{s}} \in \{1.4, 3, 5, 10, 15, 20\}$ MHz. A PLMN operator can obtain from a local spectrum regulatory authority any of these bandwidths, depending on availability and the desire of the PLMN operator. In wideband single-carrier system such as WCDMA/UMTS, the bandwidth is not divided, but used to transmit each symbol, as illustrated in Fig. 8.39(a) and (b). We assume $B_{\mathrm{s}} = 20$ MHz in our case study. The symbol duration (aka symbol time) in the single-carrier case is $T_{\mathtt{sc}} = 1/B_{\mathrm{s}} = 0.05\,\mu\mathtt{s}$. The symbol duration is so small that a delay spread as small as 0.04 μs affects a large portion of the symbol, causing a remarkable ISI.

In LTE, the bandwidth $B_{\mathrm{s}} = 20$ MHz is divided into 1,200 subcarriers each of bandwidth $B_k = 15$ kHz, resulting in the fixed symbol time $T_{\mathtt{mc}} = 1/B_k \approx 66.7\,\mu\mathtt{s}$. This symbol duration is much larger than the delay spread of 0.04 μs. Therefore, ISI has no remarkable effect on the symbols, as illustrated in Fig. 8.39(c) and (d). We note that the delay spread or time dispersion in practical cellular mobile systems is much larger than the 0.04 μs. For example, it is about 15 to 20 μs in hilly and rural terrain, and about 5 μs in urban areas. Note that OFDM symbols used in LTE/LTE-A have much larger duration than these typical multipath channel time dispersion. The ratio

$$g_{\mathtt{mdg}} = \frac{\begin{array}{c}\texttt{symbol duration in}\\ \texttt{multi-carrier transmission}\end{array}}{\begin{array}{c}\texttt{symbol duration in}\\ \texttt{single-carrier transmission}\end{array}} = \frac{T_{\mathtt{mc}}}{T_{\mathtt{sc}}} \tag{8.69}$$

is the equivalent of the spreading gain in CDMA systems. We refer to it here as **multi-carrier despreading gain** (MDG). Its value lies in the range 9.3 to 133.3 in LTE/LTE-A.

8.6.2.2 *Cyclic Prefix in OFDM, OFDMA and SC-FDMA*

We need to understand how time is measured in LTE/LTE-Advanced and their framing structure in order to meaningfully understand the cyclic prefix (CP). The time units used in LTE/LTE-A is

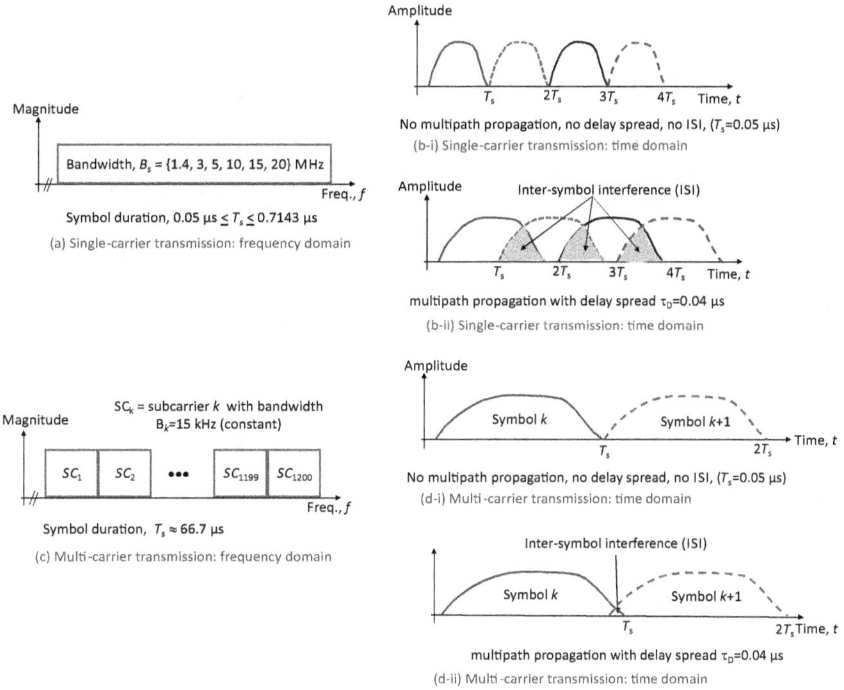

Fig. 8.39 Illustration of the OFDM principle.

Table 8.14 LTE/LTE-A time units.

Time Unit Name	Value/Duration
Basic time unit (T_0)	$\frac{\frac{1}{15\ \text{kHz}\times 2048}=1}{30720}$ ms
Frame ($T_{\mathtt{frame}}$)	$307200T_0 = 10$ ms
Half-frame	$153600T_0 = 5$ ms
Sub-frame (= TTI)	$30720T_0 = 1$ ms
Slot	$T_{\mathtt{slot}} = 15360T_0 = 0.5$ ms
Symbol duration (T_s)	$\frac{0.5\ \text{ms}}{7}$ for short/normal CP $\frac{0.5\ \text{ms}}{6}$ for long/extended CP

TTI (Transmit Time Interval) is the time interval for radio resource scheduling.

compiled in Table 8.14, while the frame structure is illustrated in Fig. 8.40. Time in LTE/LTE-Advanced is measured in units of $T_0 = \frac{1}{15\ \mathtt{kHz}\times 2048} = \frac{1}{30720}$ ms.[4] Each radio frame lasts $T_{\mathtt{frame}} = 307200 \times T_0 = 10$ ms. One frame is divided into 10 sub-frames, each of duration $T_{\mathtt{sf}} = 30720 \times T_0 = 1$ ms. The smallest time that can be allocated to a user for information transfer in both the downlink and uplink directions is the sub-frame time. For this reason, the sub-frame is also referred to as *transmission time interval* (TTI). TTI is the smallest time interval that an eNode can schedule a user's transmissions in both uplink and downlink. It is also the shortest duration of decodable transmitted data. Each radio frame in LTE/LTE-Advanced is divided into 20 time slots (TSs), each lasting $T_{\mathtt{slot}} = 15360T_0 = 0.5$ ms.

Each OFDM symbol contains a block or group of data symbols. Therefore, OFDM is liable to these two types of interference:

- **Intra-symbol interference (IaSI)**: this is the interference between the data symbols or subcarriers within one OFDM symbol. It is also referred to as inter-channel interference (ICI), inter-carrier interference (ICI) or self-interference. This interference is mitigated by using mutually orthogonal subcarriers.
- **Inter-block interference (IBI)**: this is the interference between different OFDM symbols.

[4]We note that standard LTE/LTE-A documents use the symbol T_s instead of T_0. This can be confused with symbol duration, the reason for deviating from the norm.

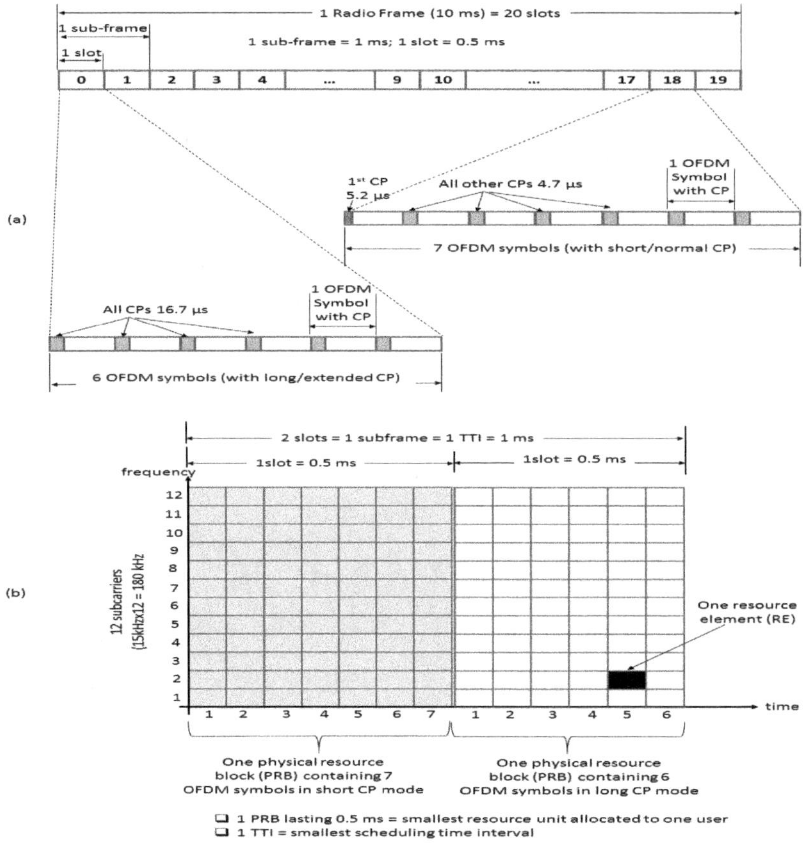

Fig. 8.40 LTE frame structure and resource blocks with the CPs.

IBI is a special form of inter-symbol interference (ISI). It is mitigated using the cyclic prefix, and a OFDM symbol duration which is much longer than average wireless channel delay spread to render the channel non-frequency selective.

This means that each is liable to two types of interference: inter-symbol interference (ISI) and inter-carrier interference (ICI), which is also referred to as intra-symbol interference (IaSI) or self-interference (SI).

ISI is mitigated using symbols with large duration of about $T_{\mathrm{s}} = 66.7\,\mu\mathrm{s}$ and equaliser. How does OFDM, OFDMA and SC-FDMA mitigate IBI? The cyclic prefix (CP) is developed to do the job. The CP is a guard interval pre-appended to each OFDM symbol, as illustrated in Fig. 8.41(a). The content of the CP is the same as the tail (or end part) of the OFDM symbol containing it.

Consider the multipath channel scenario illustrated in Fig. 8.41(c). In this example, three copies (i.e., original and two reflections) of a transmitted symbol or signal reaches the user equipment (UE). Each of these three signals travels a different path between the transmitter and the receiver and thus experiences a different channel gain and propagation delay. The longest path is the reflected path with the approximate length $d_3 = d_{31} + d_{32}$, while the shortest path is the direct path with the length d_1. The difference between these two paths yields the channel's delay spread. Thus, the delay spread in distance is

$$\Delta d = d_3 - d_1 = d_{31} + d_{32} - d_1 \,. \tag{8.70}$$

Assuming that the radio signal travels with the speed of light $c \approx 3 \cdot 10^8$ m/s, the delay spread measured in time is

$$\tau_d = \frac{\Delta d}{c} = \frac{d_{31} + d_{32} - d_1}{3 \cdot 10^8} \; \texttt{[sec]} \,. \tag{8.71}$$

The length of the CP depends on the channel's delay spread, and hence the radio signal propagation environment. The length of the CP in time

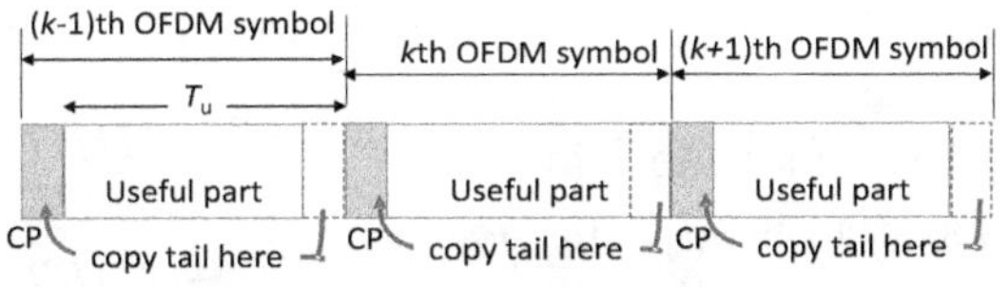

(a) OFDM symbols with CP (CP in each symbol is a copy of the same symbol's end part)

Cyclic Prefix (CP) Type	Application of cyclic prefix (CP)	Length	Samples (at 32.55 ns sampling rate)	β_{cp}	OFDM symbols in a time slot (TS) of 0.5 ms
normal/ short, Δf=15 kHz	Remaining symbols in TS	4.7 µs	144	6.6%	7
	1st symbol in time slot (TS)	5.2 µs	160	7.2%	
extended/ long, Δf=15 kHz	For all symbols in TS	16.7 µs	512	20%	6
Δf=7.5-kHz subcarrier spacing	For all symbols in TS	33.3 µs	1024	33.3%	3

(b) Properties of LTE/LTE-Advanced cyclic prefix (CP)

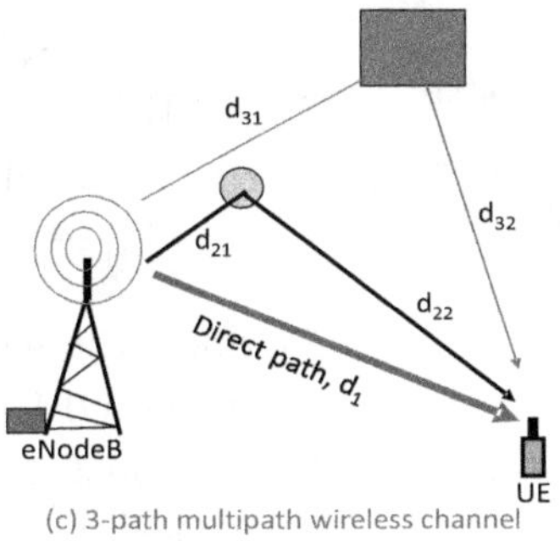

(c) 3-path multipath wireless channel

Fig. 8.41 OFDM symbols' cyclic prefixes (CPs) for ISI mitigation.

should be at least equal to the delay spread of the wireless channel. As compiled in Fig. 8.41(b), LTE has four types of CPs differ by their durations. The normal CP has two values: 4.7μs for the first symbol and 5.2μs for the remaining symbols in a time slot of duration 0.5 ms. The main reason for using a different length for the CP in the first symbol is to make the duration of a time slot, measured in $T_0 = \frac{1}{30720}$ ms units, divisible by 15,360. Plugging these CPs into Eq. (8.71) means that the path lengths can differ by up to 1.4 km for the short CP, and up to 5 km for the long CP. The short CP is meant for urban propagation environments using small cells or high-rate applications. The long CP is for rural areas typically with large cells or low data rate applications and multi-cell broadcasts. The main issue in the CP is the reduction of resource utilisation efficiency as it is an error control mechanism. If the duration of the CP is T_{cp}, where $T_{\text{cp}} \in \{T_{\text{ncp}}, T_{\text{ecp}}\}$ and the duration of the useful part or payload portion of the symbol is T_{u}, then the duration of the OFDM symbol is

$$T_{\text{s}} = T_{\text{u}} + T_{\text{cp}} \,. \tag{8.72}$$

In LTE/LTE-A we have $T_{\text{u}} = \frac{1}{15\ \text{kHz}} = 2048T_0 \approx 66.7\,\mu$s. The fraction of the symbol time used for the CP is expressed as

$$\beta_{\text{cp}} = \frac{T_{\text{cp}}}{T_{\text{s}}} = \frac{T_{\text{cp}}}{T_{\text{u}} + T_{\text{cp}}} \,. \tag{8.73}$$

All commercial LTE/LTE-A systems as at Q3 2018 use the normal CP and subcarrier spacing of 15 kHz, the reason why our case studies are based on them.

The impulse response of a multipath channel with K paths is

$$h(t) = \sum_{k=0}^{K-1} h[k]\delta(t - \tau_k) \tag{8.74}$$

where $h[k]$ and τ_k are the path gain and the signal propagation delay of the k-th path. The length of the impulse response, K, is referred to as **channel order**. The delay spread is the difference between the minimum and the maximum of the path delays.

Example: A 3-tap ISI Channel

Assume that the symbol transmitted at time l in a given wireless communications system is $s[l]$, and the channel perturbs signals with the AWGN $w[l]$ and three-path propagation made of two reflected paths and one direct path.

(1) Assuming that the received signal at time l is $r[l]$, write an equation to describe the input and the output characteristics of the channel.
(2) Analyse how the channel disturbs communications signals.

Solution

(1) The situation is equivalent to the multipath channel described in Eq. (8.74) with $K = 2$. Therefore, the input and output characteristics of the channel is

$$r[l] = h[0]s[l] + h[1]s[l-1] + h[2]s[l-2] + w[l]. \quad (8.75)$$

(2) Each transmitted symbol is disturbed by two most recently transmitted symbols before it. For example, as observed in Eq. (8.75), the symbol transmitted at time l is disturbed by the additive noise and the symbols transmitted at times $l-1$ and $l-2$.

The CP has another important purpose beyond IBI mitigation. It converts the linear discrete-time convolution into a discrete-time cyclic convolution to simplify OFDM implementation. The simplified transceiver stems from the fact that we can now compute linear convolution via cyclic convolution using IDFT/IFFT. If the sequence with the CP $x[n]$ is transmitted over a multipath wireless channel with the impulse response $h[n]$ we obtain the output sequence

$$r[n] = x[n] \otimes h[n] \overset{DFT}{\leftrightarrow} R[k] = H[k]X[k],\ n = 0, 1, \ldots. \quad (8.76)$$

This means that we can retrieve the transmitted signal samples $x[n]$ at the receiver from the DFT of the received sequence through the point-wise IDFT operation

$$x[n] = IDFT\{X[k]\} = IDFT\left\{\frac{R[k]}{H[k]}\right\},\ k = 0, 1, \ldots. \quad (8.77)$$

The division $\frac{R[k]}{H[k]}$ in Eq. (8.77) removes any channel distortion from the received signal in order to retrieve the transmitted signal. It is therefore a frequency-domain equalisation, which is simpler to implement.

Example: Linear to Cyclic Convolution by CP

Assume that we transmit the four QPSK symbols $S[0]$, $S[1]$, $S[2]$, $S[3]$ using OFDM over a two-tap multipath wireless channel with the impulse response $h[n] = h_0\delta[n] + h_1\delta(n-1)$.

(1) Show the sequence of data symbols transmitted in the OFDM symbol, including the CP.
(2) Demonstrate that the addition of the CP has caused the wireless channel to convert linear convolution to cyclic convolution.

Solution

We neglect any additive noise in the channel in this analysis. The IDFT of the given data symbols is

$$s[n] = \frac{1}{M}\sum_{m=0}^{M-1} S[m]e^{j\frac{\pi}{2}mn} = \frac{1}{4}\sum_{m=0}^{3} S[m](j)^{mn},\ n = 0, 1, 2, 3 \quad (8.78)$$

We obtain the samples $s[0]$, $s[1]$, $s[2]$, $s[3]$. We use the last symbol as the cyclic prefix to obtain the symbol block

OFDM data block with CP:

$s[3]$, $s[0]$, $s[1]$, $s[2]$, $s[3]$

The *prefix* in CP originates from the fact that we have pre-appended the last data symbol to the data samples, while *cyclic* comes from the cycling operation. Transmitting the symbol block over a 2-tap multipath channel without additive noise We obtain the receive signal

$$r[l] = h_0 s[l] + h_1 s[l-1]$$
$$= \sum_{k=0}^{K-1} h_k s[l-k] = h[l] \otimes s[l] \qquad (8.79)$$

where here K = 2. This is a cyclic convolution between the input sequence and the channel's response. We can further analyse Eq. (8.79) to make it even clearer. It follows from Eq. (8.79) that

$$l=0:\ r[0] = h_0 s[0] + h_1 \underbrace{s[-1]}_{=s[3]} = h_0 s[0] + h_1 s[3]$$
$$l=1:\ r[1] = h_0 s[1] + h_1 s[0]$$
$$l=2:\ r[2] = h_0 s[2] + h_1 s[1]$$
$$l=3:\ r[3] = h_0 s[3] + h_1 s[2]\,. \qquad (8.80)$$

We can rewrite Eq. (8.80) in matrix-vector form as

$$\mathbf{r} = \mathbf{H}\mathbf{s}^T = \begin{bmatrix} h_0 & 0 & 0 & h_1 \\ h_1 & h_0 & 0 & 0 \\ 0 & h_1 & h_0 & 0 \\ 0 & 0 & h_1 & h_0 \end{bmatrix} \begin{bmatrix} s[0] \\ s[1] \\ s[2] \\ s[3] \end{bmatrix}. \qquad (8.81)$$

We can clearly observe the cyclic shift to the right in the rows of the channel matrix **H**. Thus, the insertion of one symbol as the CP has converted the channel's linear convolution into a circular convolution which is implementable in IDFT. Note that the number of symbols used as CP depends on the channel order. A multipath channel with longer delay spread requires a longer CP to convert the linear to circular convolution.

8.6.2.3 *JOFDM Waveform Fundamental Principles*

OFDM is a combination of multi-carrier digital modulation and multiplexing. LTE allows a choice between QPSK, 16-QAM and 64-QAM modulation, depending on the quality of the wireless link. The better the channel, the higher the data rate and bandwidth efficiency achievable (Fig. 8.42). A lower level modulation is more robust against channel impairments, but has a lower bandwidth efficiency, as illustrated in Fig. 8.42. Link quality in the zone AB is not good enough for signal transfer. QPSK is chosen modulation for the channel quality in zone BC. Link quality in the zones CD and beyond D are suited for 16-QAM and 64-QAM, respectively. The principle of selecting a modulation and coding scheme to suit an instantaneous channel quality is referred to as **link adaptation** or **rate adaptation**. Symbols from any of these three modulation schemes are used to modulate the subcarriers of OFDM in LTE/LTE-A. OFDM multiplexes signals which are orthogonal in frequency, just as FDM. The difference, however, are: the signals multiplexed belong to the same signal and originate from the same source.

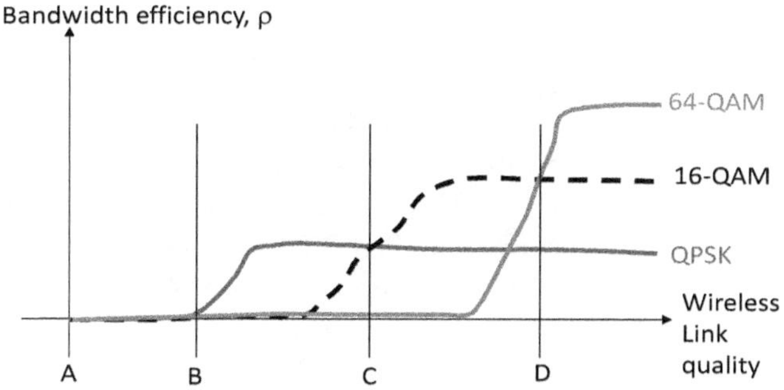

Fig. 8.42 Wireless link adaptation in LTE.

Let Δf be the subcarrier spacing. In LTE/LTE-A, we have $\Delta f = 15$ kHz (most popular) or $\Delta f = 7.5$ for both uplink and downlink. As $\Delta f = 15$ kHz is the popular choice, our discussions are based on this. The sampling frequencies used in the IFFT/FFT of LTE's OFDM for each system bandwidth are given in Table 8.15. The *sampling period* for the 20-MHz bandwidth is $T_{\texttt{sampling}} = 1/f_{\texttt{sampling}} \approx 32.55$ ns. The sampling frequencies and the subcarrier spacing of LTE/LTE-A were chosen to enable one clock circuitry to run multi-mode (e.g., LTE, UMTS, CDMA2000) mobile terminals. This is the reason why the sampling frequencies are multiples of 3.84 MHz, which is the chip rate of WCDMA. Using sampling rates which are multiples of 3.84 MHz also enables the backward compatibility of LTE/LTE-A to UMTS/WCDMA and other 3G radio standards. The number of samples used to compute the IFFT/FFT for each OFDM symbol are also shown in Table 8.15. Each is obtained from

$$N = \lfloor \texttt{(OFDM symbol duration)} \times \texttt{(sampling frequency)} \rfloor = \lfloor 66.7\mu\text{s} \times f_{\text{s}}\ \texttt{[MHz]} \rfloor\,. \qquad (8.82)$$

Table 8.15 Some LTE physical layer parameters.

System bandwidth, B_s (MHz)	1.4	3	5	10	15	20
Sampling frequency, f_s (MHz)	1.92	3.84	7.68	15.36	23.04	30.72
F_s as multiples of 3.84 MHz	0.5	1	2	4	6	8
Number of IFFT/FFT samples, N	144	308	512	1024	1536	2048
Number of used subcarriers, K	72	180	300	600	900	1200
Number of resource blocks, N_{rb}	6	15	25	50	75	100
Transmission bandwidth, B_e (MHz)	1.08	2.7	4.5	9	13.5	18
Guard band, $B_s - B_e$ (MHz)	0.32	0.3	0.5	1	1.5	2
Bandwidth efficiency (%)	77.14	90	90	90	90	90

Twelve subcarriers make one resource block (RB) of 180 kHz. The smallest bandwidth that can be allocated to one user is one RB lasting 0.5 ms. The smallest unit defined in LTE/LTE-A is the resource element (RE). One RE comprises one OFDM subcarrier during one OFDM symbol interval. Thus, there are $12 \times 6 = 72$ REs in one resource block using long/extended CP and $12 \times 7 = 84$ REs in one resource block using short/normal CP.

Some physical layer parameters of LTE/LTE-A are shown in Table 8.15. The percentage of the system bandwidth used by the subcarriers (including the dc subcarrier) is

$$\texttt{occupied bandwidth} = \frac{(\texttt{number of subcarriers}) \times 15\ \texttt{kHz}}{\texttt{system bandwidth}}. \quad (8.83)$$

The occupied bandwidth is equivalent to the bandwidth efficiency shown in row 9 in Table 8.15. Its value is either 77.14% for the 1.4 MHz system bandwidth, and 90% for the remaining. This means that 10% or 22.86% of the LTE/LTE-A system bandwidth is used as guard bands (GBs), as shown in row 8 in Table 8.15. The size and the positions of the guard bands for the 15-MHz and 20-MHz system bandwidths are illustrated in Fig. 8.46. Thus, bandwidth efficiency in LTE/LTE-A are reduced by the guard bands at the ends of the system bandwidth and by the cyclic prefix in each OFDM symbol.

What is the purpose of the guard bands? They are used to reduce the out-of-band (OOB) power that the OFDM signal leaks into adjacent systems as interference. OFDM signal contains sidelobes, which decay asymptotically as $1/f^2$, where f is frequency [Farhang-Boroujeny (2011)]. The 3GGP standard requires the adjacent channel leakage power ratio (ACLR) or adjacent channel interference ratio (ACIR) of not more than 45 dB. This means that the OFDM signal level within an adjacent band should be at least 45 dB less than the signal it disturbs. Unfortunately, an OFDM can generate as high as 20 dB interference to neighbouring systems. So the guard bands are used to reduce the OOB leakage to acceptable levels. Therefore, the OFDM subcarriers at the ends of the system bandwidth (e.g., 100 and 132 subcarriers for the 15-MHz and 20-MHz system bandwidths, respectively) are not used for data transfer, but as guard bands.

One OFDM symbol contains two or more data symbols. Each data symbol modulates one OFDM subcarrier. This means that the number of bits carried over one subcarrier depends on the digital modulation scheme chosen, which also depends on the channel quality. Two, four and six bits per subcarrier for QPSK, 16-QAM and 64-QAM, respectively. The waveform of a sinusoidal signal is periodic but has infinite duration. Thus, each OFDM subcarrier has infinite duration in time. As this is practically non-realisable, each subcarrier waveform is truncated to the symbol duration of $T_s \approx 66.7\,\mu\text{s}$. The truncation is performed by the *windowing function* (aka rectangular pulse shaping filter) $x(t) = b\Pi(\frac{t}{T_s/2})$ illustrated in Fig. 8.43(c). The spectrum of the windowing function is the sinc function $X(f) = bT_s\texttt{sinc}(2fT_s)$. The windowing in time domain makes the signal finite in time domain, but infinite duration in the frequency domain. The

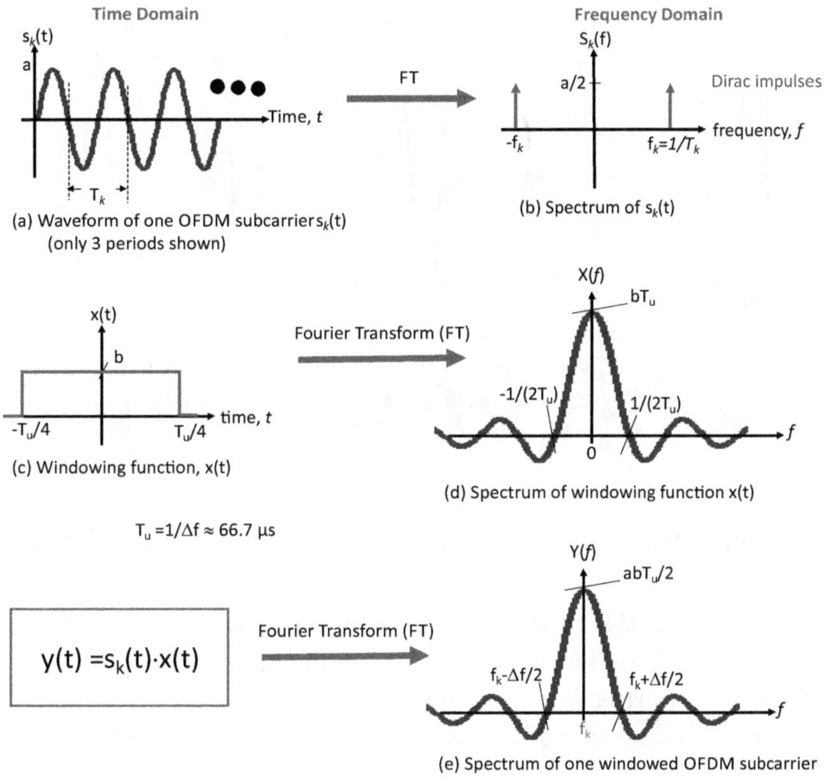

Fig. 8.43 The basic principles of OFDM waveforms.

opposite of this is also true. Assume that a subcarrier with amplitude a is at the frequency f_k, then the waveform representing the k-th subcarrier is

$$s_k(t) = a\sin(2\pi f_k t),\ k = 1, 2, \ldots, K \tag{8.84}$$

where we have assumed a zero phase for each subcarrier. The signal $s_k(t)$ is shown in Fig. 8.43(a), while its spectrum $S_k(f)$, obtained via Fourier transform, is shown in Fig. 8.43(b). The relationship between $S_k(f)$ and $s_k(t)$ is $s_k(t) \overset{\mathcal{FT}}{\leftrightarrow} S_k(f)$. Note that $S_k(f)$ is a Dirac delta function, i.e., $S_k(f) = \frac{a}{2}\delta(f - f_k)$. In the following, we shall use only the positive frequency components.

The windowing operation requires the multiplication of $s_k(t)$ by $x(t)$ to obtain the truncated subcarrier waveform $y_k(t)$, i.e.,

$$\begin{aligned} y_k(t) &= s_k(t) \times x(t) \\ &= a\sin(2\pi f_k t) \times b\Pi\left(\frac{t}{T_s/2}\right). \end{aligned} \tag{8.85}$$

Note that the duration of the windowing function is the same as the OFDM symbol period of $T_s = 66.7\,\mu$s. Applying the multiplication property of Fourier transform on Eq. (8.85) yields the convolution in the frequency domain

$$\begin{aligned} y(t) &= s_k(t) \times x(t) \overset{\mathcal{FT}}{\leftrightarrow} Y_k(f) \\ &= X(f) * S_k(f) = X(f) * \frac{a}{2}\delta(f - f_k) \\ &= \frac{a}{2}X(f - f_k). \end{aligned} \tag{8.86}$$

Therefore, the spectrum of each subcarrier is the same as the spectrum of the windowing function, which is a sinc function, shifted by the frequency of the subcarrier. It is represented analytically as

$$\begin{aligned} Y_k(f) &= \frac{a}{2}X(f - f_k) \\ &= \frac{abT_s}{2}\mathtt{sinc}[2(f - f_k)T_s],\ k = 1, 2, \ldots, K. \end{aligned} \tag{8.87}$$

This is modulation. The spectrum of the windowed subcarrier $Y_k(f)$ is illustrated in Fig. 8.43(e).

Each of the 12 modulated subcarriers in a resource block has the form in Eq. (8.87). The difference between them is their differing carrier

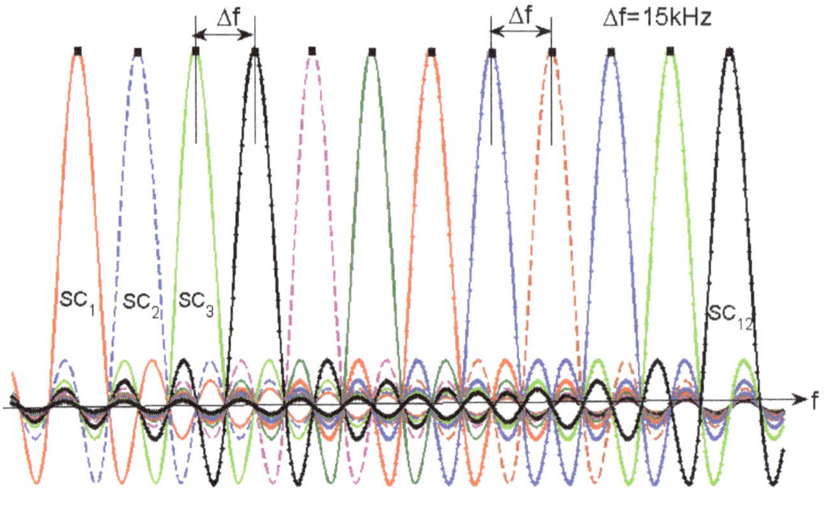

Fig. 8.44 Spectrum of one OFDM symbol comprising 12 subcarriers as in one LTE/LTE-A RB.

frequencies f_k. The waveforms of the 12 subcarriers in LTE/LTE-A resource block are illustrated in Fig. 8.44. The main lobe of subcarrier k is centred at its frequency f_k. We can observe that the peak of each subcarrier carrying a data symbol occurs at a frequency where all the adjacent subcarriers achieve their nulls. Thus, sampling at the peak of one subcarrier to measure its amplitude and phase does not disturb any subcarrier, eliminating inter-channel interference. This is the reason why we say that the subcarriers are mutually orthogonal.

The number of subcarriers in OFDM are also shown in Table 8.15. These subcarriers are used for the four purposes:

- Data transfer over the data subcarriers.
- Pilot subcarrier (used for channel quality estimation).
- Identify the center frequency of the system bandwidth using the dc subcarrier. The dc subcarrier is used on only the downlink, the reason the number of subcarriers in the downlink is always one more than those in the uplink.
- For guard bands using the guard band subcarriers.

8.6.2.4 *JOFDM Signal Generation*

A simplified *analog* implementation of OFDM transmitter is illustrated in Fig. 8.45. Our discussions focus on only the OFDM part of the transmitter. The data symbols from a single-carrier modulator, such as QPSK, 16-QAM and 64-QAM, are fed into a bank of analog modulators. In this example, we assume that each user is allocated k subcarriers. Thus, each OFDM symbol for a user comprises the k data symbols $S_0, S_1, \dots, S_{k-1}$. Each symbol modulates one OFDM subcarrier. The n-th subcarrier has the form $c_n(t) = \sin(2\pi f_n t)$, $n = 0, 1, 2, \dots, k-1$. We use only normalised subcarriers with unity amplitude and zero phase for simplicity of illustration. Be reminded that the subcarriers are mutually orthogonal over the OFDM useful symbol duration, i.e.,

$$\int_{iT_u}^{(i+1)T_u} c_l(t)c_j(t)dt = 0, \ \forall l \neq j\,. \tag{8.88}$$

Let X_n^i be the data symbol modulating the nth subcarrier during the ith OFDM symbol. Therefore, the waveform for the modulated nth subcarrier is $p_n(t) = X_n^i c_n(t) = X_n^i \sin(2\pi f_n t)$, $n = 0, 1, 2, \dots, k-1$, resulting in the baseband OFDM signal without the CP

$$\begin{aligned} s(t) &= \sum_{n=0}^{k-1} X_n^i \sin(2\pi f_n t) \\ &= \alpha \sum_{n=0}^{k-1} \sin(2\pi f_n t + \phi_n), \\ & iT_u \le t \le (i+1)T_u \end{aligned} \tag{8.89}$$

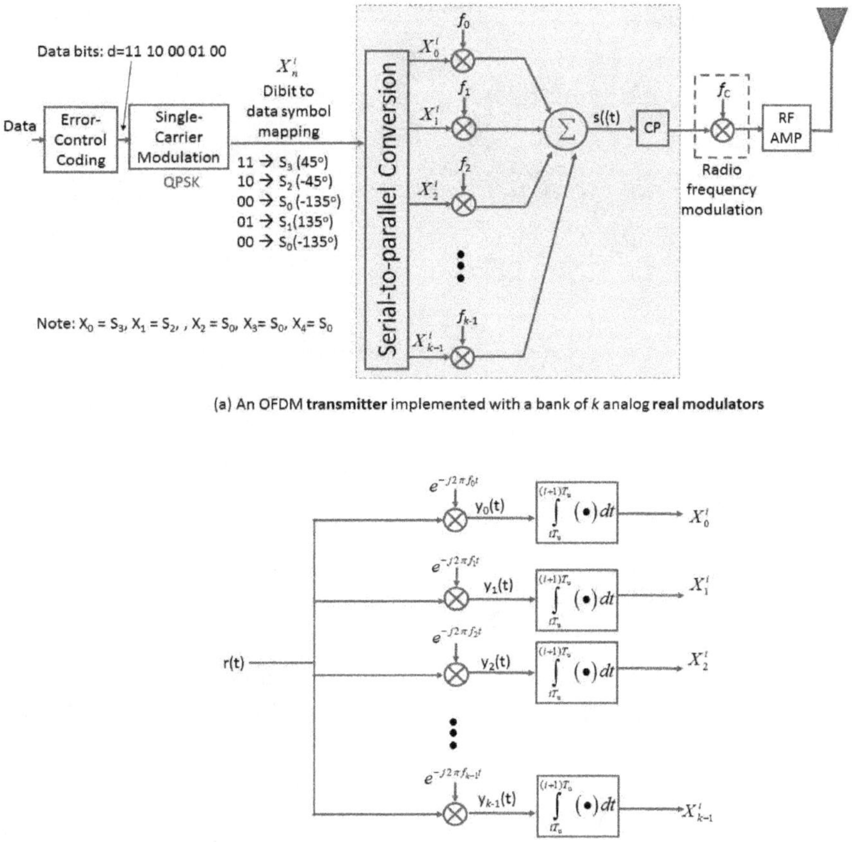

(a) An OFDM **transmitter** implemented with a bank of *k* analog **real modulators**

(b) An OFDM **receiver** implemented with a bank of *k* analog **complex demodulators**

Fig. 8.45 A simplified analog OFDM transmitter.

where $X_n^i \in \{S_0, S_1, S_2, S_3\}$, $\phi_n \in \{\pm 45°, \pm 135°\}$, $f_n = F_L + n\Delta f$, $n = 0, 1, \ldots, k-1$, $\Delta f = 15$ kHz and F_L depends on the bandwidth acquired by the PLMN operator. The carrier frequency, f_c, in Fig. 8.45 is usually in the middle of the band allocated to a PLMN operator. Figure 8.46 illustrates the position of the carrier frequency in relation to the subcarriers and the subcarriers used as guard intervals. As illustrated in Fig. 8.46, $F_L = -9$ MHz in 20-MHz system bandwidth, but $F_L = -6.75$ MHz in 15-MHz system bandwidth. In the band B28 the carrier frequency can be $f_c = 700$ MHz so that $F_L = -6.75$ becomes $F_L = -6.75 + 700 = 693.25$ MHz. Thus, a negative frequency in Fig. 8.46, just indicates that it lies to the left of the carrier frequency.

Let us assume that the symbols S_k in Eq. (8.89) are produced by QPSK. That means that each dibit maps onto one QPSK symbol. Assuming Gray coding then we obtain the mapping of dibits onto QPSK waveforms

$$\begin{aligned} 00 \leftrightarrow S_0(t) &= \frac{1}{\sqrt{T}} \sin(2\pi f t - 135°) \\ &= \frac{1}{\sqrt{T}} \sin(2\pi f t + 225°) \\ 01 \leftrightarrow S_1(t) &= \frac{1}{\sqrt{T}} \sin(2\pi f t + 135°) \\ 10 \leftrightarrow S_2(t) &= \frac{1}{\sqrt{T}} \sin(2\pi f t - 45°) \\ &= \frac{1}{\sqrt{T}} \sin(2\pi f t + 315°) \\ 11 \leftrightarrow S_3(t) &= \frac{1}{\sqrt{T}} \sin(2\pi f t + 45°)\,. \end{aligned} \tag{8.90}$$

In the following, we assume that $T = 1$ and that we are transmitting the error-coded data $d = 1110000100$. Thus, $k = 5$ the QPSK data symbols are transmitted in the order S_3, S_2, S_0, S_1, S_0. The last symbol is used as the cyclic prefix, so

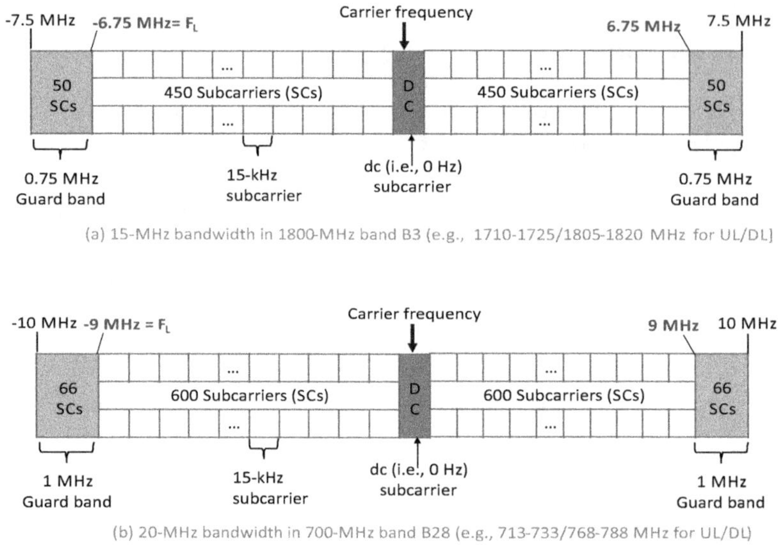

Fig. 8.46 Subcarriers and guard subcarriers in 15-MHz and 20-MHz system bandwidths.

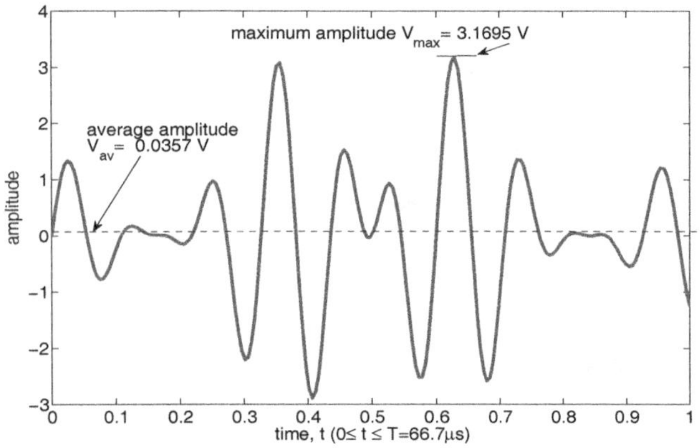

Fig. 8.47 An OFDM waveform with six symbols and six subcarriers.

the data symbol sequence in the OFDM symbol is $S_0, S_3, S_2, S_0, S_1, S_0$. In this case, the OFDM symbol with the CP comprises six data symbols. The OFDM waveform Eq. (8.89) with the above data is shown in Fig. 8.47. We can observe the differing amplitudes within the symbol period of 66.7 μs. This is the property which tags OFDM with the high peak-to-average power ratio (PAPR). The peak voltage is about 3.1695 V, while the average voltage is about 0.0357 V, yielding a high PAPR of about 3,142.5.

Equation (8.89) is a form of inverse discrete Fourier transform (IDFT). Therefore, instead of implementing it using a bank of analog modulators and mixers, we implement it using IDFT. IDFT is simpler than analog modulation and can also be implemented on a digital computer using digital signal processing tools. To further speed up processing and reduce complexity, IFFT algorithms are used. Figure 8.48 shows a simplified OFDM transceiver (transmitter and receiver) implementing Eq. (8.89) using IDFT/IFFT.

In order to make it clearer how Eq. (8.89) lend itself to IDFT/IFFT implementation, let us sample it. The sampling rate is $f_s = N\Delta f$, leading to the sampling period $T_{\text{samp}} = \frac{1}{f_s} = \frac{1}{N\Delta f}$. Thus, the sampling instants are $t = lT_{\text{samp}} = \frac{l}{N\Delta f}$. Plugging this into Eq. (8.89) yields

$$\begin{aligned} s_l = s(lT_s) &= \sum_{n=0}^{k-1} S_n \sin\left(\frac{2\pi nl}{N}\right) \\ &= \sum_{n=0}^{k-1} S_n \phi_n[l],\ l = 0, 1, \ldots, k-1 \end{aligned} \tag{8.91}$$

where we have noted that $f_n/\Delta f = f_n T_s = n$, and $\phi_n[l] = \sin\left(\frac{2\pi nl}{N}\right)$.

We can also use complex modulators in Fig. 8.45 to obtain the alternative form to Eq. (8.89) as

$$\begin{aligned} s(t) &= \sum_{n=0}^{k-1} X_n^i e^{j2\pi f_n t} \\ &= \sum_{n=0}^{k-1} X_n^i e^{j2\pi \Delta f n t},\ iT_u \le t \le (i+1)T_u\,. \end{aligned} \tag{8.92}$$

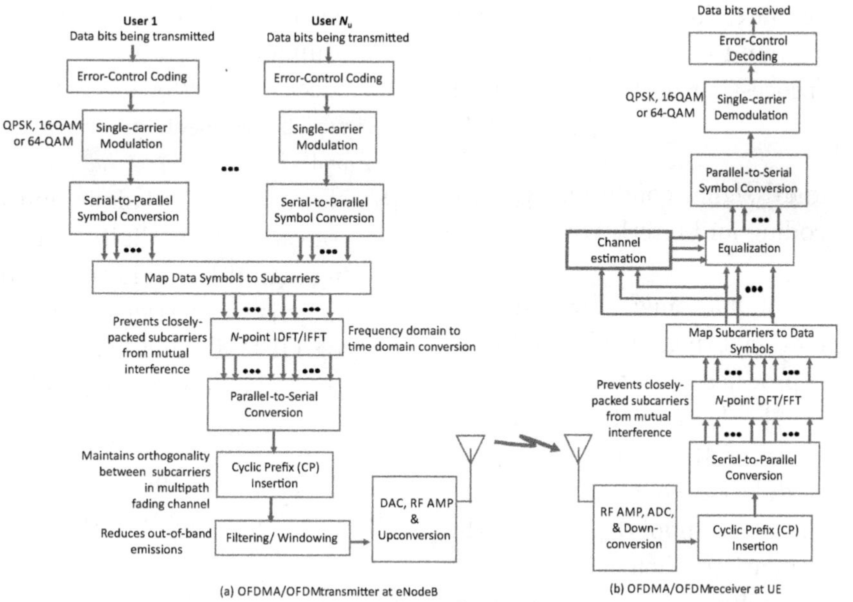

Fig. 8.48 A simplified OFDM and OFDMA transceiver (N is number of subcarriers).

We now need to sample Eq. (8.92) in order to be able to process it on digital systems. Sampling must usually fulfill the Nyquist sampling theorem which requires us to sample at the rate $f_s \geq 2 \times f_{\mathtt{max}}$, where $f_{\mathtt{max}}$ is the maximum frequency in $s(t)$. As observed in Fig. 8.43, the OFDM signal is sinc function which ideally has infinite maximum frequency. So, what is $f_{\mathtt{max}}$? We select a sampling rate which exceeds the so-called **essential frequency** in $s(t)$ which sufficiently fulfills the Nyquist condition. We set

$$f_{\mathtt{max}} \approx (\mathtt{number\ of\ subcarriers\ in\ OFDM\ signal}) \times \Delta f = k\Delta f \tag{8.93}$$

and the sampling rate and the discrete sampling times

$$f_s = N\Delta f \;\mathtt{and}\; t = \frac{l}{f_s} = \frac{l}{N\Delta f} \tag{8.94}$$

where $N > k$. Note that $k\Delta f$ is the minimum or nominal frequency of the OFDM signal containing k subcarriers. Plugging Eq. (8.94) into Eq. (8.92) yields

$$\begin{aligned} s_l &= \sum_{n=0}^{k-1} X_n^i e^{j2\pi\frac{ln}{N}} \\ &= \sum_{n=0}^{N-1} Y_n^i e^{j2\pi\frac{ln}{N}}, \; iT_u \leq t \leq (i+1)T_u \end{aligned} \tag{8.95}$$

where

$$Y_n^i = \begin{cases} X_n^i, & 0 \leq n \leq k-1 \\ 0, & k \leq n \leq N-1 \end{cases}$$

Note that Eq. (8.95), which is the sampled OFDM signal, is an N-point IDFT of the block of k data symbols X_n^i, $n = 0, 1, \ldots, k-1$. The values of N and the sampling rates for different system bandwidths standardised in LTE/LTE-A are compiled in Table 8.15. Clearly, from this table the sampling rate does not need to be an integer.

Comparing Eqs. (8.91) and (8.95) with Eq. (8.61) reveals that they are circular convolution, which can be realized using IDFT. However, the output of the wireless channel is a linear convolution of the input signal and the channel's response. Therefore, we desire linear convolution, but not circular convolution. The linear convolution Eq. (8.59) can be realised using DTFT, which is not realisable. Our desire now is how to compute the linear convolution via the circular convolution so that we can apply IDFT, rather than IDTFT. Interestingly, adding one data symbol as cyclic prefix to compute the OFDM symbol converts the linear convolution to circular convolution to enable us to apply DFT/IDFT. The same results can be achieved by zero padding the finite sequence input into the circular convolution algorithm. The procedure used to compute linear convolution via circular convolution is found in Appendix A.

The sequence of processing at the transmitter are:

(1) Perform all standard signal conditioning, such as error control coding and interleaving, on the user data bits.
(2) Bits to data symbol mapping: Perform single carrier modulation using QPSK, 16-QAM or 64-QAM to produce the data symbols $\{S_0, S_1, S_2, \ldots, S_{k-1}\}$. Each of these symbols are taken from the set of M symbols in the chosen multi-level modulation scheme. The number of bits in each data symbol depends on the chosen digital modulation scheme.
(3) One-to-one data symbol to subcarrier mapping: use each data symbol to modulate only one OFDM subcarrier. In this step the amplitude and the phase of each subcarrier are changed by those of one data symbol in order to shift each data symbol into the appropriate frequency space.
(4) Perform N-point IDFT on the modulated subcarriers to produce the sequence $\{s_0, s_1, s_2, \ldots, s_{k-1}\}$. The IDFT is performed using IFFT algorithms to speed up processing. Note that an IDFT/IFFT is performed on each OFDM symbol separately, which comprises multiple data symbols from a single-carrier modulator.
(5) Perform parallel-to-serial conversion to arrange the k sequence in series.
(6) Add a cyclic prefix (CP) to the serialised sequence to obtain the OFDM symbol, comprising k data symbols and a cyclic prefix. The OFDM symbol is then $\{s_{k-1}, s_0, s_1, s_2, \ldots, s_{k-1}\}$. Here, we have pre-appended a copy of the last data symbol which is s_{k-1} as the CP. This is the reason for the 'prefix' in CP.
(7) The next operation is filtering, windowing or pulse shaping to reduce out-of-band emissions, which is a form of egress electromagnetic interference (EMI). Reducing the out-of-band energy spillage has two advantages: (a) reduction of egress EMI and (b) conservation of mobile terminals' battery power. Nyquist pulses are commonly used for the filtering. The negative effect of the filtering is that it increases PAPR, which we seek to reduce, especially in the uplink.
(8) Now, the OFDM signal must be amplified, then converted from digital into an analog signal using digital-to-analog converter (DAC). The final stage is the carrier modulation to shift the analog OFDM signal to the radio frequency (RF) band allocated to the PLMN operator.

Note that OFDM can be implemented in either analog or digital domain. The opposite processes to the above occur at the receiver.

8.6.2.5 *OFDM Peak-to-Average Power Ratio*

Multi-carrier technologies such as OFDM and OFDMA transmit multiple subcarriers in their symbol. The disturbance of the wireless channel on signals usually depends on the frequency. Thus, the transmit power and data rate over each subcarrier in an OFDM symbol are usually different. Assume that each OFDM symbol uses N subcarriers (or data symbols). Let the power level of each subcarrier be P_k, $k = 1, 2, \ldots, N$. Then the average of these powers is $P_{\mathsf{av}} = \frac{1}{N}\sum_{k=1}^{N} P_k$. Let the power in the transmitted data symbol with the largest amplitude be $P_{\mathsf{peak}} = \max\{P_k\}$. Then the peak-to-average power ratio (PAPR) is defined as

$$PAPR = \frac{P_{\mathsf{peak}}}{P_{\mathsf{av}}} = \frac{P_{\mathsf{peak}}}{\frac{1}{N}\sum_{k=1}^{N} P_k}. \tag{8.96}$$

Assume that a mobile device transmits $P_{\mathsf{av}} = 250$ mW and the PAPR is 8 dB, then the device must transmit the maximum power of $10^{0.8} \times 250$ mW $\approx$ 1.6 W.

An OFDM waveform with only six subcarriers is illustrated in Fig. 8.47. We can observe the wide variations in the amplitude caused by constructive and destructive combination of the subcarrier waves. The PAPR of this waveform is about 3,142.5 or 35 dB. If the OFDM signal is amplified using a linear amplifier then there will be no distortion. However, distortion occurs if the amplifier is nonlinear. A distortion in the time domain results in a distortion in the the power spectrum of the signal.

Distortion in the power spectrum causes signal spreading and power leakage into adjacent frequency bands, resulting in adjacent channel interference (ACI). The basic technique used to cope with PAPR is amplifier linearisation. However, operating power amplifiers in their linear range produces a smaller gain than the gain in the nonlinear operation mode. A lower gain requires a higher transmit power to maintain a desired effective radiated power (ERP). This is manageable by the eNodeBs or base stations, but not the mobile devices. Transmitting at a high power drains the batteries of mobile devices quicker, and it may also pose health hazards.

PAPR increases with the length of the IDFT in the OFDM. To see this let us make the following analysis. The K-point IDFT samples are $s[k] = \frac{1}{K}\sum_{n=0}^{K-1} S[n]e^{j2\pi kn/N}$. We assume that $s[k]$ falls across a resistor of 1-ohm resistance to produce the power $P_k = E\{|s[k]|^2\}$. Let us assume that $S[n] = \alpha, \forall n$. Then the average power in $s[k]$ is

$$\begin{aligned} P_{\text{av}} = E\{|s[k]|^2\} &= \frac{1}{K^2}\sum_{n=0}^{K-1} E\{|S[n]|^2|e^{j2\pi kn/K}|^2\} \\ &= \frac{1}{K^2}\sum_{n=0}^{K-1} E\{|S[n]|^2 = \frac{\alpha^2}{K}\,. \end{aligned} \tag{8.97}$$

Also,

$$\begin{aligned} s[0] &= \frac{1}{K}\sum_{n=0}^{K-1} S[n]e^{j2\pi\cdot 0\cdot n/K} \\ &= \frac{1}{K}\sum_{n=0}^{K-1} S[n] = \frac{1}{K}\times K\alpha = \alpha\,. \end{aligned} \tag{8.98}$$

Thus, the peak power is $P_{\text{peak}} = |s[0]|^2 = \alpha^2$, resulting in

$$PAPR = \frac{P_{\text{peak}}}{P_{\text{av}}} = \frac{\alpha^2}{\alpha^2/K} = K\,. \tag{8.99}$$

Clearly, the PAPR increases with K. Thus, the large PAPR in OFDM is attributed to its IFFT component. Each user in an LTE/LTE-A network is allocated a minimum of 12 subcarriers (i.e., $K \geq 12$), resulting in $PAPR \geq 10.7$ dB.

Example: PAPR

Assume that the single-carrier modulation symbols entering the IDFT/IFFT engine of an OFDM transmitter are

$$S[0] = 2 + j,\ S[1] = 1 - 2j,\ S[2] = 1 + j,\ S[3] = 2\,.$$

Compute the PAPR in the system.

Solution: PAPR

Applying Eq. (8.97), we find the average power in all the symbols as

$$P_{\text{av}} = \frac{1}{4^2}\sum_{k=0}^{3}|S[k]|^2 = \frac{5+5+2+4}{16} = 1\ \texttt{W}\,.$$

Applying Eq. (8.98), we obtain the peak value

$$s[0] = \frac{1}{4}\sum_{k=0}^{3} S[k] = \frac{6}{4} = 1.5\ \texttt{V}\,.$$

Thus, we obtain $PAPR = \frac{P_{\text{peak}}}{P_{\text{av}}} = \frac{1.5^2}{1} = 2.25$.

8.6.3 *OFDMA in LTE/LTE-A*

OFDMA is used in the downlink of LTE/LTE-A and 5G NR. **How does OFDM differ from OFDMA?** First, as discussed in the introduction section, traditionally OFDM would be used only in the downlink while OFDMA would be used in the uplink. Second, unlike OFDM, OFDMA enables the allocation of subcarriers to users in both time and frequency scales. All the subcarriers in OFDM are assigned to a single user at a time unless it synergises with a multiple access scheme such as TDMA (each user uses all subcarriers but for a time slot) or OFDMA (each user uses a subset of the subcarriers). Such a static radio resource sharing would not allow channel-state-dependent scheduling. This is illustrated in Fig. 8.49. In this illustration three users are active at a cell which has 12 subcarriers numbered 1 through 12. Time is also divided into five slots numbered 1 through 5. OFDM allocates all the 12 slots to user 1 in time slot 1, to user 2 in time slots 2, 3 and 4, and to user 3 in time slot 5. OFDM does this allocation, although the channel conditions in the 12 subcarriers can be different at the location of a user.

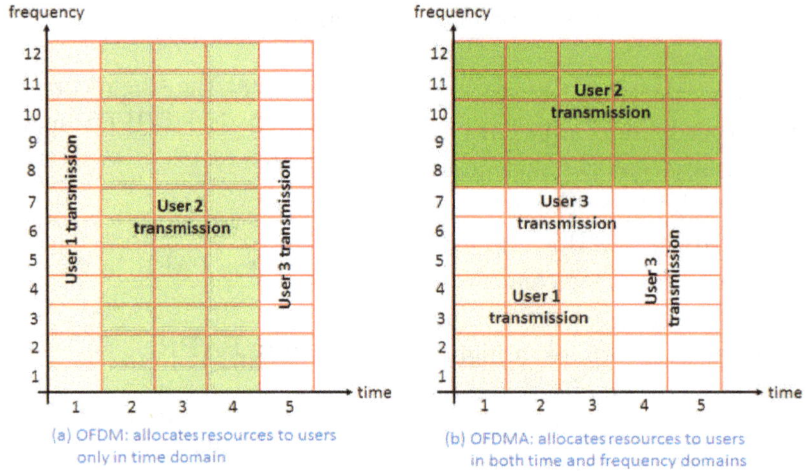

Fig. 8.49 Differences in resource allocation between OFDM and OFDMA.

In OFDMA, a subset of the subcarriers can be allocated to a user at a given time. This means that multiple users can share the subcarriers at a given time, as illustrated in Fig. 8.49(b). For example, all the three users are allocated portion of the subcarriers in time slots 1, 2 and 3. Also, users 2 and 3 share the 12 subcarriers in time slots 4 and 5. Thus, OFDMA allows radio resource allocation along both time and frequency, combining FDMA and TDMA. This enables differentiated transmission rates among multiple users, depending on the quality of each subcarrier's fading state at each user. Each subcarrier experiences different fading situation at even the same user.

Like any thing, OFDMA and OFDM also have some drawbacks. We have discussed its large peak-to-average power ratio above. OFDMA and OFDM are also disturbed by frequency offsets or instability caused by Doppler spread and any instability in the electronic circuitry of the local oscillator. Electronics instability is caused by aging and/or manufacturing defects. Doppler spread results from any relative movement between the two antennas communicating. Any drift in frequency causes the subcarriers to lose their mutual orthogonality and overlap, in a situation referred to as inter-carrier interference (ICI). We can observe in Fig. 8.44 that the OFDM spectrum comprises multiple sync functions, each of which has infinite bandwidth. Therefore, each symbol emits energy outside its allocated frequency band. This out-of-band emissions is referred to as **spectral regrowth**. Also, consecutive OFDM symbols are mutually independent, causing a time-domain discontinuity. This time-domain discontinuity translates to spikes in the frequency domain.

8.6.4 *SC-FDMA in LTE/LTE-Advanced and 5G*

The single-carrier frequency division multiple access (SC-FDMA) is used in the uplink of both 4G and 5G mobile cellular technologies. The downlink of both technologies, however, use OFDMA. Why? There are two main reasons: SC-FDMA has a smaller PAPR but a higher system throughput than OFDMA. The large PAPR in OFDMA would require mobile terminals to transmit high powers if used in the uplink. Transmitting high powers is avoided for several reasons, including: safety of humans, and mobile battery energy conservation. What does the *single carrier* in SC-FDMA originate from? Unlike OFDMA, data symbols do not directly modulate the subcarriers in SC-FDMA. As illustrated in Fig. 8.50, rather, the frequency-domain samples of the data symbols modulate the subcarriers because of the DFT operation. For that reason, one data symbol is spread over multiple subcarriers. The group of subcarriers modulated by each data symbol is viewed as one frequency band transferring symbols in series as in a conventional FDMA system. Also, localised subcarriers allocation method

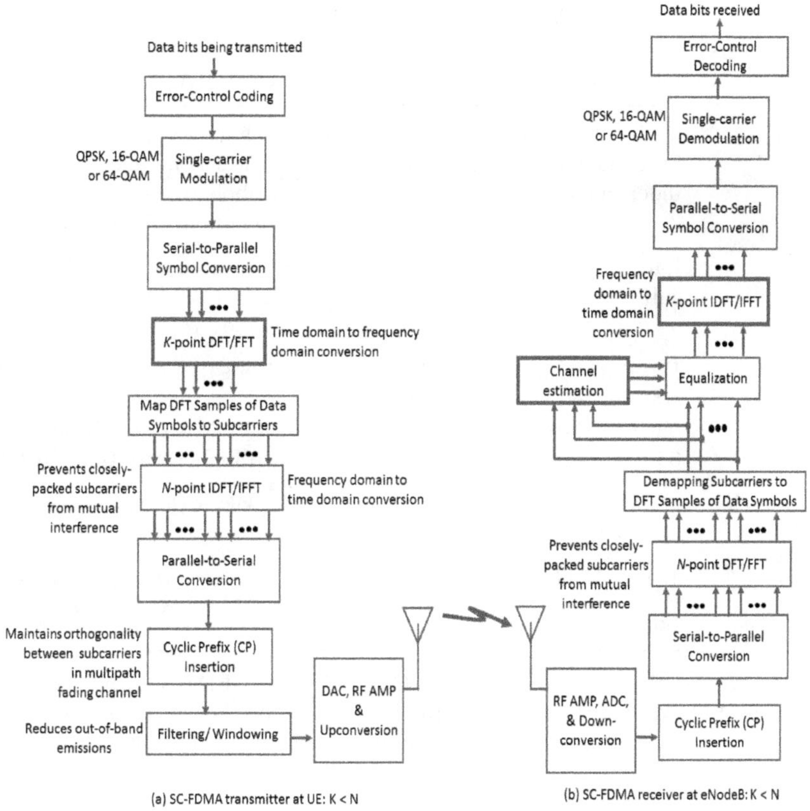

Fig. 8.50 A simplified SC-FDMA transceiver (K is number of subcarriers allocated to one user, N is all subcarriers in the system).

whereby a user is allocated *contiguous* subcarriers without any gaps between them is used. Compared with OFDMA, SC-FDMA has these advantages attributed to its use of one carrier:

- Less affected by frequency offsets caused by Doppler spread.
- Less sensitive to frequency instability caused by electronics aging.
- It allows the usage of nonlinear high-power amplifiers to conserve the power of mobile terminals.

A weakness in SC-FDMA attributed to its use of a single carrier is that an eNodeB needs a block equaliser to mitigate ISI on the uplink. SC-FDMA is compared with OFDMA in Table 8.16.

Maintenance of the orthogonality between subcarriers is important for the operation of SC-FDMA to avoid inter-user interference or MAI. Any misalignment between the timing of users in the uplink must not be larger than the duration of the cyclic prefix used.

A simplified transmitter of SC-FDMA is illustrated in Fig. 8.50. The data is first processed in the baseband, including error control coding. It is then modulated using any of the single-carrier digital modulation schemes QPSK, 16-QAM and 64-QAM. The symbols generated by any of these digital modulation schemes are in the time domain. We refer to each as a **sub-symbol** or **data symbol** of SC-FDMA. As the smallest radio resource that can be allocated to a single user is the resource block made of N subcarriers, each SC-FDMA symbol or signal comprises N *consecutive* sub-symbols of the digital modulation scheme used.

The time-domain data symbols from the chosen digital modulator are usually filtered in time domain before being converted into the frequency domain via discrete Fourier transform (DFT). The DFT is usually computed using a fast Fourier transform (FFT) algorithm. It is the DFT component in SC-FDMA which sets it apart from OFDMA. It is also the component which makes SC-FDMA a single-carrier transmission. SC-FDMA is also

Table 8.16 Comparing OFDMA and SC-FDMA.

Feature	OFDMA	SC-FDMA
Subcarrier modulation	Each subcarrier is modulated by a different data symbol independently	Each subcarrier is modulated by a linear combination of all data symbols transmitted at the same time due to the DFT.
PAPR	Large as it uses multiple parallel subcarriers	Low, same as that of one QPSK/16-QAM/64-QAM data symbol
Transmission of QPSK/16-QAM/ 64-QAM data sub-symbols	Data sub-symbols in each RB are transmitted in parallel within one OFDMA symbol time of 66.7μs	Data sub-symbols in each RB are transmitted in series within one SC-FDMA symbol time of 66.7μs at a higher rate than OFDMA
Bandwidth of each data symbol	15 kHz	15 kHz $\times$ (Number of carriers in RB)
Data rate	$15 \log_2 M$ kbps	$15N_{sc} \log_2 M$ kbps
System throughput	Lower	Higher

PAPR: peak-to-average power ratio. As each OFDMA or SC-FDMA symbol contains multiple QPSK/16-QAM/64-QAM symbols, we refer to the symbols of the latter as sub-symbols or data symbols. M is the number of symbols in the modulator, and N_{sc} is the number of subcarriers in a resource block (RB).

referred to as Discrete Fourier Transform Spread OFDM (DFT-s-OFDM or DFT-S-OFDM) because of the DFT involved. SC-FDMA is also referred to as linear precoded OFDMA (LP-OFDMA). The frequency-domain symbols are mapped onto OFDM subcarriers to the desired frequencies in the overall system bandwidth. The subcarriers modulated by the symbols are converted back into the time domain using an inverse FFT (IFFT) algorithm. The IFFT is needed to prevent the closely-packed subcarrier from mutual interference. We then insert the cyclic prefix (CP) to mitigate intersymbol interference (ISI) and keep the subcarriers mutually orthogonal. ISI occurs in the time domain and it is equivalent to frequency-selective fading in the frequency domain. Some filtering and windowing are performed in order to reduce out-of-band emissions. Radio frequency (RF) amplification and digital-to-analog (DAC) is needed prior to sending the signal over the antenna. The SC-FDMA receiver reverses the process performed in the transmitter.

How does SC-FDMA reduce its PAPR compared with OFDM and OFDMA? The high amplitude or power variations in OFDM signal arises from the fact that the data symbols directly and independently modulate subcarriers. The power variations in SC-FDMA signal is reduced by first mixing the data symbols prior to the standard OFDMA processing. The mixing operation is achieved in the DFT/FFT.

8.6.5 *OFDM Extensions*

Technologies which use more than one carrier to transfer a user's data are referred to as multi-carrier systems or multi-channel systems. OFDM is the simplest of all multi-carrier modulation (MCM) methods proposed for 4G and 5G mobile networks. Owing to its popularity, many OFDM-based transmission technologies have been proposed, which are more complicated than OFDM itself. Each of the beyond-OFDM proposals addresses at least one of the weaknesses in OFDM: the use of cyclic prefix (CP), fixed CP, requirement of time synchronisation of users especially in the uplink, and fixed subcarrier spacing. We briefly sample a few of them to conclude our discussion of OFDM.

Coded OFDM is described in, for example, [Choi (2012)]. Coded OFDM is the modulation method used in terrestrial digital video broadcasting (DVB-T) standard. The name was coined as the scheme combines forward error correction (FEC) and modulation.

Filtered OFDM (f-OFDM): is an asynchronous OFDM which does not require users in the uplink to be synchronised in time. Furthermore, unlike OFDM, f-OFDM allows variable subcarrier spacing, variable-length cyclic prefix, etc. Also, each subcarrier is filtered to reduce inter-carrier interference. The flexibility in f-OFDM allows its parameters to be optimised for individual users based on their traffic type and channel qualities.

OFDM in 5G: OFDM with cyclic prefix (CP-OFDM) is a holy grail in modern communications industry. However, as discussed earlier, there are identified weaknesses in OFDM, including:

(1) *Tight synchronisation*: the clocks of all devices at a site must be near-perfectly synchronised with that of the serving base station, especially in the uplink.
(2) *Inflexible parameters*: fixed symbol duration and fixed subcarrier spacing.
(3) Reduced spectral efficiency: uses cyclic prefix (CP) as guard intervals which reduces spectral efficiency.

There is much effort to address these weakness in OFDM. The CP-OFDM is also adopted in the air interface of 5G New Radio (NR). A remarkable difference between the OFDM used in 5G compared with that in 4G is that 5G allows variable subcarrier spacing (SCS). The available SCSs in 5G are

$$SCS_k = 15 \times 2^k \ \texttt{kHz}, \ k \in \{0, 1, 2, 3, 4\}. \quad (8.100)$$

This means that the symbol duration T_s also varies as

$$T_s = \frac{1}{SCS_k} = \frac{0.5^k}{15} \ \texttt{ms}, \ k \in \{0, 1, 2, 3, 4\}. \quad (8.101)$$

The term **numerology** in cellular communications technology is used to refer to the symbol duration and the SCS. We leave the discussions of these topics to signal processing texts, such as [Luo and Zhang (2016)] and [Choi (2012)].

8.6.6 *LTE Channels and Scheduling*

We briefly review resource allocation in LTE in this section, beginning with a review of LTE protocol architecture and its channels. LTE protocol architecture is illustrated in Fig. 8.51. The protocol architecture is divided into the user plane (U-plane) and control plane (C-plane) above the packet data convergence control protocol (PDCP) sublayer. The C-plane is used to manage signalling and control data, while the U-plane is used for user data transfer. Of particular importance in this chapter is the MAC sublayer. LTE/LTE-A MAC sublayer has many functions, and it is said to be the brain of the technology. Its functions include:

(1) Scheduling. Both uplink and downlink scheduling are performed by the eNB, but not the UEs. The scheduler allocates one or more resource blocks to each users. The UE then allocates scheduled resource blocks to its bearers. *The specific scheduling algorithm is left open in the LTE/LTE-A standard for implementer. Therefore, we will survey some of the proposals below.*
(2) Selection of the size of a transport block.
(3) Management of retransmission using the hybrid ARQ algorithm.
(4) Multiplexing and prioritisation of logical channels.
(5) Dynamic selection of modulation scheme from QPSK, 16-QAM and 64-QAM based on the CQI received from the physical layer. Signalling and control channels always use QPSK as it is the robust against channel impairments.

8.6.6.1 *LTE Channels*

An understanding of LTE channels and protocol architecture is important to properly understand its resource allocation. LTE channels and their mapping are illustrated in Fig. 8.51. A channel can be a control channel (CCH) or traffic channel (TCH). CCHs are used for signalling transfer, while TCHs are used to transfer user data. All channels in LTE/LTE-A are divided into three categories:

- *Logical channels*: indicate the **purpose and type** of information transmitted over the air,

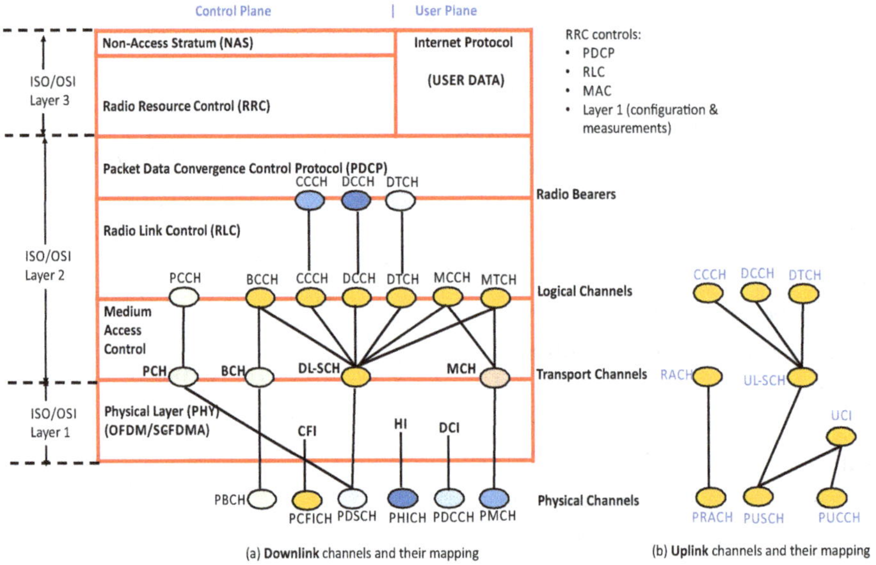

Fig. 8.51 LTE protocol architecture and types of channels.

such as broadcasting, control and signalling, multicasting and traffic/payload. There are seven for downlink and three for uplink. Comprising both control channels and traffic channels, these channels transfer data between the radio link control (RLC) sublayer and the medium access control (MAC) sublayer, both of which belong to ISO/OSI Layer 2.

- *Transport channels*: decide **how** the information is to be transmitted over the air by, for example, selecting modulation method, interleaving type and transport block size used to transfer the data. There are four for downlink and two for uplink. Comprising both control channels and traffic channels, these channels transfer data between the medium access control (MAC) sublayer and the physical layer. One or multiple logical channels are mapped onto one transport channel.
- *Physical channels*: handles all signal processing functions and knows exactly the location of resource elements, and **where** the transmitted data is. There are six for downlink and three for uplink. Comprising both control channels and traffic channels, these channels transfer data over the wireless channel between two LTE devices. One or two transport channels are mapped onto one physical channel.

Here is a brief description of each of the logical channels:

(1) PCCH (paging control channel) is used to alert devices/UEs of incoming calls.
(2) BCCH (broadcast control channel) is used to broadcast Master Information Block (MIB) periodically. The MIB is transmitted periodically by each eNB, regardless of the presence or absence of UEs. The payload part of MIB has 24 bits for the purposes:
    (a) 3 bits for system bandwidth, which is needed by EUs to decode any physical layer channel.
    (b) 3 bits for PHIC information: 2 bits to encode the number of PHIC groups (N_g) used in the first symbol, and 1 bit to indicate if the PHIC is normal or extended.
    (c) 8 bits for system frame number, which UEs use for both initial and periodic synchronisation of their clocks with that of the eNB.
    (d) 10 bits are reserved for future use.
(3) CCCH (common control channel) is used to send signalling and control information to multiple UEs.
(4) DCCH (dedicated control channel) is used to send signalling and control information to a particular UE.

(5) DTCH (dedicated traffic channel) is used to send payload data to a particular UE.
(6) MCCH (multicast control channel) is used to send signalling and control information to a group of UEs.
(7) MTCH (multicast traffic channel) is used to send payload data to a group of UEs.

Note that CCCH, DCCH and DTCH are duplex logical channels.

The transport channels are:

(1) PCH (paging channel) for paging. It carries the data of the logical PCCH channel.
(2) BCH (broadcast channel) carries the data of the logical BCCH channel. Therefore, the MIB are transported by it.
(3) DL-SCH (downlink shared channel) carries the data of six logical channels for both signalling and payload transfer (Fig. 8.51).
(4) MCH (multicast channel) transports the data of two multicast logical channels.
(5) RACH (random access channel) is used by UEs for initial access to the wireless network.
(6) UL-SCH (uplink shared channel) transports the data of all the 3 uplink logical channels. Therefore, it transports both signalling and user payload data.

LTE/LTE-A have five downlink (viz. PBCH, PDSCH, PMCH, PDCCH and PHICJ) and three uplink (viz. PRACH, PUSCH and PUCCH) physical channels. These are:

(1) PBCH (physical paging broadcast channel) transmits the data from the broadcast transport channel, BCH. It is thus the conduit for the MIB.
(2) PDSCH (physical downlink shared channel) transfer data from the two transport channels PCH and DL-SCH. It transfers user data and the system information block (SIB).
(3) PMCH (physical multicast channel) transports the data from the transport multicast channel. It is used mainly for multimedia broadcast multicast service (MBMS).
(4) PDCCH (physical downlink control channel) carries downlink control information or indicator (DCI) at the physical layer. It informs UEs about paging, the scheduling of the PDSCH (information on block needed for uplink data transfer by UEs), hybrid ARQ AK/NAK, uplink power control information, and uplink grant. Every UE must decode this channel successfully to be able to receive or transmit data.

    As illustrated in Table 8.15, an LTE/LTE-A system has between 6 to 100 resource blocks. These resource blocks are allocated in both time and frequency. Therefore, a mechanism is needed for the receiver to find out the appropriate information needed to decode the received signal, such as which of the received resource blocks contains which data and which encoding and modulation schemes were used by the transmitter. These control information are carried in the DCI. The receiver must first decode the DCI in order to find the information needed to decode the actual signal transferred over the PDSCH channel. Information other than resource allocation transferred over the DCI are shown in Fig. 8.52. These include transmitter power control (TPC) command, channel quality indicator (CQI) Report Request or channel state information (CSI) Report Request. The format of the 4-bit CQI is shown in Fig. 8.55. The DCI also conveys to UEs the Transmitted Precoding Matrix Indicator (TPMI) which is the codebook (precoding matrix) index used for transmissions to that UE on the PDSCH. The UE needs the TPMI to decode the data sent to it over the PDSCH. As the DCI carries many types of information, a different DCI format is designed to carry certain information (Fig. 8.52). The DCI is used for both uplink (over PUSCH) and downlink (over PDSCH) scheduling.
(5) PHICH (physical hybrid ARQ indicator channel) transmits the 1-bit hybrid ARQ (HARQ) status to inform UEs of the outcome of data transfer: "0" for ACK indicating success, while "1" for NAK indicates failure. The 'hybrid' in HARQ means a combination of automatic repeat request (ARQ) and forward error correction and it is used only for unicast transmissions

DCI Format	Purpose	Major Contents
Format 0	UL Grant. Resource Allocation for UL Data	RB Assignment,TPC,PUSCHHopping Flag
Format 1	DL Assignment for SISO	RB Assignment,TPC, HARQ
Format 1A	DL Assignment for SISO (compact)	RB Assignment,TPC, HARQ
Format 1B	DL Assignment for MIMO with Rank 1	RB Assignment,TPC, HARQ,TPMI, PMI
Format 1C	DL Assignment for SISO (minimum size)	RB Assignment
Format 1D	DL Assignment for Multi User MIMO	RB Assignment,TPC, HARQ,TPMI,DL Power Offset
Format 2	DL Assignment for Closed Loop MIMO	RB Assignment,TPC, HARQ, Precoding Information
Format 2A	DL Assignment for Open Loop MIMO	RB Assignment,TPC, HARQ, Precoding Information
Format 2B	DL Assignment for TM8 (Dual Layer Beamforming)	RB Assignment,TPC, HARQ, Precoding Information
Format 2C	DL Assignment for TM9	RB Assignment,TPC, HARQ, Precoding Information
Format 3	TPC Commands for PUCCH and PUSCH with 2 bit power adjustment	Power Control Only
Format 3A	TPC Commands for PUCCH and PUSCH with 1 bit power adjustment	Power Control Only
Format 4	UL Assignment for UL MIMO (up to 4 layers)	RB Assignment, TPC, HARQ, Precoding Information

Fig. 8.52 Formats of downlink control indicator (DCI).

over the unicast channels. The ARQ algorithm used is stop-and-wait.

(6) PCFICH (physical control format indicator channel): conveys to UE the format of the signal being received, indicates the number of OFDM symbols used for the PDCCH which lies between 1 and 3. Carrying the control format indicator (CFI), the PCFICH is sent in the first symbol of every sub-frame.
   (a) UEs must be able to decode this channel in order to decode the PDCCH and PHICH, and PDCCH is necessary to decode the PDSCH. It informs UEs of the format of the signal being received.
   (b) Contains number of OFDM symbols in the current subframe used for PDCCH, which can be 1, 2 or 3 symbols (or 4 for systems with less than 3 MHz bandwidth). The payload of PCFICH is the control format indicator (CFI).
(7) PRACH (physical random access channel) transports the RACH in the uplink.
(8) PUSCH (physical uplink shared channel) transports the UL-SCH and UCI (uplink control information). Thus, PUSCH is used to convey both signalling and user data on the uplink.
(9) PUCCH (physical uplink control channel) carries the UCI (uplink control information) if the UE has no data to send in a given TTI. It conveys messages such as scheduling request, CQI (see Fig. 8.55), modulation scheme and HARQ ACK/NAK to eNB.

8.6.6.2 *Scheduling in LTE*

As discussed earlier, the schedulling function at the eNB of LTE/LTE-A allocates resources to each user in multiples of the physical resource block (PRB). One PRB lasts 0.5 ms and contains 12 subcarriers totaling 180 kHz. Also, the smallest schedulling time interval is the 1-ms transmit time interval (TTI). As illustrated in Fig. 8.40(b), each duplex channel requires a minimum of 2 PRBs within each TTI. Therefore, the duplex PRB is referred to as 180 kHz $\times$ 1 ms resource block. Figure 8.40(a) shows that each PRB carries 6 or 7 OFDM symbols. Therefore, as each subcarrier is modulated by one data symbol (aka modulation symbol), there are either 72 or 84 modulation symbols in each PRB. RBs are virtual resources which are transported in the actual resources, the PRBs, at the physical layer. As shown in Table 8.15, the number of PRBs available in TTI for the uplink or downlink is between 6 (for 1.4-MHz system bandwidth) and 100 (for 20-MHz system bandwidth).

The specific algorithm used by the scheduler in LTE is not standardised. IT is left as an implementation choice. Figure 8.53 illustrates the signalling for downlink scheduling, while Fig. 8.54 shows the equivalent for uplink scheduling. The uplink scheduling can be persistent or semi-persistent. In the former method, the UEs can send data at any time. The semi-persistent scheduling, however, needs uplink Grant allocating PRBs from the eNB before the UE can data transfer.

The downlink scheduling process is illustrated in Fig. 8.53. The steps are [Capozzi *et al.* (2013)]:

(1) Signals which exist only within the physical layer and are known to both the transmitter and the receiver are referred to as **reference signals**. Examples of reference signals are Channel State Information Reference Signal (CSI-RS) and cell specific reference signal (CS-RS). Each UE decodes the reference signals in

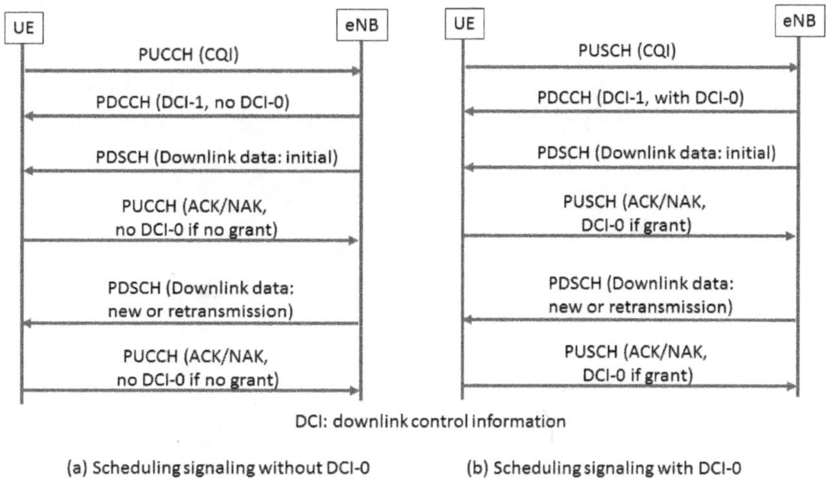

Fig. 8.53 Signaling flow for downlink scheduling in LTE.

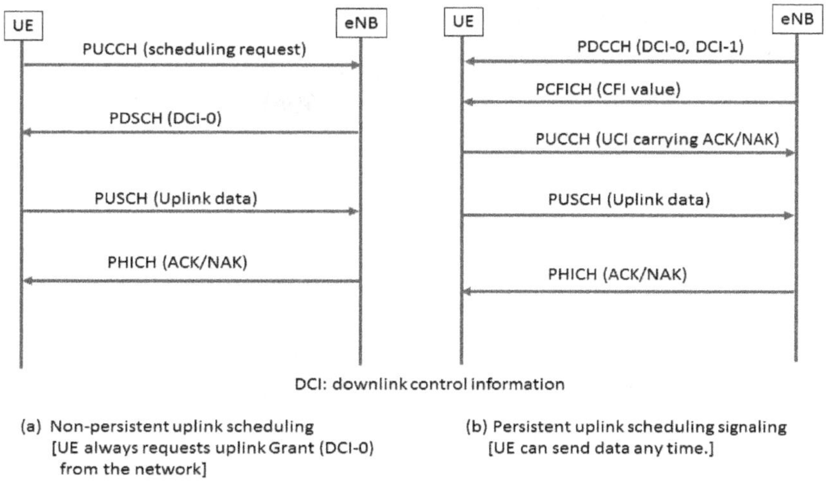

Fig. 8.54 Signalling flow for uplink scheduling in LTE.

order to compute the CQI. It then sends the CQI to the eNB using PUCCH channel.

(2) The eNB uses the received CQI to allocate PRBs in both uplink and downlink to UEs. The CQI is also used by the adaptive modulation and coding (AMC) module to select the best modulation and coding scheme (MCS) matching the CQI. The eNB sends these information to the UEs through the DCI using PDCCH channel.

(3) Data transfer and acknowledgement of transmission starts and continues until completion. Note that an UE may need more than one TTI to complete its data transfer.

[Capozzi *et al.* (2013)] discusses an LTE scheduling scheme which maximises a utility function. Let the utility function for the ith user and PRB k in the lth TTI be

$$U_{i,k}(l) = function(QSI_i, CQI_{i,k}, RAH_i, RBS, QCI_i) \quad (8.102)$$

where for user i

- QSI_i is the queue status information. For example, the longer the queues at user i, the higher its scheduling priority could be to reduce end-to-end delays and the chance of queue overflows.

- $CQI_{i,k}$ is the channel quality information on PRB k at the ith user. For example, we can allocate resource blocks to users based on their channel qualities to maximise throughput.
- RAH_i is the resource allocation history of the ith user. For example, we can prioritise a user which has been starved of service in order to maintain fairness.
- RBS indicates the receiver buffer state, which can be used to prioritise receivers with the largest buffer space to avoid possible buffer overflows.
- QCI_i indicates quality of service (QoS) requirements of the data of user i.

We then allocate the kth PRB to user j in the lth TTI which fulfills [Capozzi *et al.* (2013)]

$$j = \arg\max_{i,k} U_{i,k}(l)\,. \tag{8.103}$$

The serving eNB takes the decision in Eq. (8.103) in each TTI and informs UEs through the DCI/PDCCH the PRBs allocated to them for downlink transmission over PDSCH and uplink transmissions over PUSCH.

We sample some of the utility functions defined in the open literature. We have the **Required Activity Detection with Delay Sensitivity (RAD-DS)** [Monghal *et al.* (2010)]

$$U_{i,k}(t) = U_i^{RR}(t) \cdot RA_i(t) \cdot DS_i(t) \tag{8.104}$$

where $U_i^{RR}(t)$ is the round robin utility metric, $RA_i(t)$ is the required activity factor of the traffic and $DS_i(t)$ is the delay sensitive factor of the traffic involved. The RD-DS is a dynamic scheduler which supports VoIP.

The **Exponential Proportional Fair (EXP/PF)** utility function is defined as [Basukala *et al.* (2009)]

$$U_{i,k}(t) = \exp\left(\frac{\alpha_i D_{HOL,i} - X}{1+\sqrt{X}}\right) \cdot \frac{d_k^i(t)}{\bar{R}^i(t-1)} \tag{8.105}$$

where $\bar{R}^i(t)$ is past average throughput achieved by the user i until time t, $d_k^i(t)$ is the expected data rate for the ith user at time t on the kth resource block,

$$X = \frac{1}{N_{\mathrm{rt}}} \sum_{i=1}^{N_{\mathrm{rt}}} \alpha_i D_{HOL,i} \tag{8.106}$$

$D_{HOL,i}$ is the delay of the head of line packet, τ_i is the delay threshold for the ith user, δ_i is the acceptable probability for the ith user that a packet is dropped due to deadline expiration,

$$\alpha_i = -\frac{\log \delta_i}{\tau_i}\,. \tag{8.107}$$

This rule is an adaptation of the exponential rule to support multimedia applications.

Modified largest weighted delay first (M-LWDF) algorithm is a typical packet scheduling algorithm which supports hybrid real-time services over wireless networks. Its utility function is [Ramli *et al.* (2009)]

$$\begin{aligned} U_{i,k}(t) &= \alpha_i D_{HOL,i} \cdot U_{i,k}^{PF}(t) \\ &= \alpha_i D_{HOL,i} \frac{d_k^i(t)}{\bar{R}^i(t-1)}\,. \end{aligned} \tag{8.108}$$

The utility function for the **Generalised Proportional Fair (GPF) scheduling** is [Wengerter *et al.* (2005)]

$$U_{i,k}(t) = \frac{[d_k^i(t)]^{\xi}}{[\bar{R}^i(t-1)]^{\psi}} \tag{8.109}$$

where the two parameters ξ and ψ modify, respectively, the impact on the allocation policy of the instantaneous data rate and of the past achieved throughput. In the special case of $\xi = \psi = 1$, the GPF reduces to the standard PF scheduling.

The utility function for the **LOG-Rule** is defined in [Sadiq *et al.* (2009)] as

$$U_{i,k}(t) = b_i \log(c + a_i D_{HOL,i}) \cdot \Gamma_k^i \tag{8.110}$$

while that for the **EXP-Rule** is defined in [Sadiq *et al.* (2009)] as

$$U_{i,k}(t) = b_i \exp\left(\frac{a_i D_{HOL,i}}{c + \sqrt{\frac{1}{N_{\mathrm{rt}}}\sum_j D_{HOL,i}}}\right) \cdot \Gamma_k^i \tag{8.111}$$

where Γ_k^i is the spectral efficiency the kth subcarrier (or sub-channel) at user i and a_i, b_i and c are tunable parameters. The optimal set of parameters proposed in [Sadiq *et al.* (2009)] and [Capozzi *et al.* (2013)] is

$$a_i \in \left\{\frac{5}{0.99\tau_i}, \frac{10}{0.99\tau_i}\right\};\quad b_i = 1/E\{\Gamma^i\};\quad c = 1\,. \tag{8.112}$$

8.6.7 *Check Your Understanding*

Students are encouraged to check their understanding of the contents of this section with the following questions prior to proceeding to the next section.

(1) State two features in LTE/LTE-A system which reduces the bandwidth efficiency.
(2) How many data symbols (aka modulation symbols) are in each LTE/LTE-A physical resource block?
(3) Discuss the differences between logical channels, transport channels and physical channels in LTE/LTE-A.
(4) Explain the meaning and purposes of reference signals in wireless networks, such as in LTE/LTE-A.
(5) The 5G New Radio standard has the four reference signals: bi-directional demodulation reference signal (DMRS), uplink sounding reference signal (SRS), downlink channel state information reference signal (CSI-RS) and phase tracking reference signal (PTRS). Research to find the purpose of each of these reference signals.
(6) Discuss the main reasons why OFDM-based transmission was chosen for LTE/LTE-A instead of single-carrier transmissions used in its precursor technologies such as CDMA, TDMA and FDMA.
(7) Describe the meaning of transmission time interval (TTI) in wireless technologies such as LTE/LTE-A, WCDMA and 5G NR.
(8) How does the TTI relate to the sub-frame in LTE/LTE-A?
(9) LTE/LTE-A defines resource element (RE). What is the meaning and purpose of the RE in LTE/LTE-A?
(10) OFDM and its variant OFDMA pride themselves of good spectral efficiency. Quantitatively justify the validity of this claim or otherwise.
(11) Reflect on the two possible implementations of OFDM: using a bank of analog modulators and mixers or IDFT/IFFT. Which choice is used in practice and why?
(12) Using illustration, describe the resource block (RB) in LTE/LTE-A.
(13) Discuss the *main basis* for the selection of the length of the cyclic prefix in OFDM and its variants.
(14) What constraints must be considered in the choice of the subcarrier spacing in OFDM and its variants?
(15) Describe the causes of inter-symbol interference (ISI) in communications systems.
(16) Explain the mechanisms used in OFDM and its variants to mitigate ISI and inter-block interference (IBI).
(17) Subcarriers in multicarrier transmissions such as OFDM, OFDMA and SC-FDMA are orthogonal at the transmitter. Yet, they are liable to inter-channel-interference (ICI). Discuss the cause of ICI in such systems.
(18) Describe the strategies used in OFDM and its variants to combat inter-block interference (IBI).
(19) Discuss the reason why OFDMA is used in the downlink but not in the uplink of LTE/LTE-Advanced.
(20) Discuss the features in SC-FDMA which make it preferable to OFDMA in the uplink of LTE/LTE-A and 5G New Radio.
(21) Describe the differences between FDMA and single-carrier FDMA (SC-FDMA) used in the uplink of LTE/LTE-Advanced mobile technologies.
(22) Describe the meaning of distributed user multiplexing (DUM), also referred to as distributed multiple access, as used in the uplink of LTE/LTE-Advanced. Include any advantages in DUM in your discussion.
(23) Although the localised subcarrier assignment method has a larger PAPR than the distributed subcarrier assignment method, it is the method adopted

in the LTE/LTE-A standard. Explain the reason behind the choice.

(24) Assume that OFDMA were used in the uplink of a given cellular mobile wireless system. The time synchronisation between the signals from all users reaching the base station should not differ by more than the duration of the cyclic prefix (CP). Discuss the reason(s) behind this restriction.

(25) Assume that a wireless channel used for mobile communications has the delay spread of $\tau = 1\,\mu$s. Compute the number of previously transmitted symbols which causes ISI to the most currently transmitted symbol if the data rate is:

(a) $R_b = 500$ bps.
(b) $R_b = 100$ kbps.
(c) $R_b = 1$ Gbps.

Compare the results of QPSK and 16-QAM modulation schemes. Illustrate your answer graphically.

(26) Multiple subcarriers carry one data symbol in SC-FDMA, but only one subcarrier carries each data symbol in OFDM and OFDMA. Explain what causes the difference.

(27) Assume that a mobile user wants to transfer the pre-conditioned data x = 11010000100110110010 using OFDM over the two-tap multipath wireless channel with the impulse response $h[n] = 0.1\delta(n) + 0.5\delta(n-1)$. The single-carrier modulation used is 16-QAM. Compute the OFDM waveform, assuming that each user is allocated five subcarriers in each TTI (or scheduling period).

(28) Assume that we seek to transmit the data 1110001010 using OFDMA over LTE/LTE-A air interface. Assume that the data is already error-control coded and ready for single-carrier modulation.

(a) Using QPSK, generate the data symbols.
(b) Apply IDFT/IFFT and the the results from (a) and add the cyclic prefix to produce the baseband OFDM symbol.

(29) Analyse the difference between ICI, IBI and ISI in OFDM transmission.

(30) Compute the maximum PAPR in an OFDM-based communications system which allocates 24 subcarriers to user as a unit.

(31) Discuss the differences between the OFDM used in 4G and the one used in 5G mobile technologies.

(32) Figure 8.45(b) illustrates an analog OFDM demodulator. If the input, $r(t)$, to this circuit equals the signal $s(t)$ in Fig. 8.45(a), prove that the demodulator works as expected.

(33) Let N be the number of samples used in the IFFT/FFT in OFDM, and K be the number of subcarriers in one OFDM symbol.

(a) State the analytical relationship between N and K.
(b) Justify your decision.

8.7 Channel-Aware Multiple Access Schemes

All modern wireless networks, especially mobile networks, periodically measure the state of the wireless channel at each mobile user. The measured channel state is referred to by various names, including channel quality indicator (CQI) and channel state information (CSI). Clearly, CQI is a form of CSI. The CQI can be obtained from downlink pilot signals or via a process called channel sounding. An inherent feature in a wireless channel is that its condition or quality varies over many degrees of freedom. The variations in a wireless channel depend on many factors, such as time, operating frequency, user mobility, non-stationary clutter and geographical location of mobile users. The variations are caused by shadowing, multipath propagation, path loss, Doppler spread, interference from other systems, etc. For this reason, wireless channels are said to be spatio-temporary varying. Wireless systems which consider users channels' state when allocating resources to them are referred to as channel-aware schemes. There are three basic types of channel-aware schemes (CASs):

- **Constant data rate schemes** adapt the transmit power to suit the instantaneous channel condition in order to transmit data at a constant rate regardless of the channel condition. This is achieved via power control. Devices with poor channel conditions transmit higher powers than those at good channel conditions.
- **Link adaptation schemes** (aka rate adaptation schemes) mitigate the channel variations using, e.g., smart antennas (transmit/receive diversity, multipath combining), interleaving, interference suppression to increase the spectral efficiency of point-to-point links. They allow devices to select a transmission rate depending on their channel conditions.
- **Opportunistic schemes** exploit the channel variations. They sometimes even induce more variations in the channel using, for example, beamforming, in order to increase multi-user diversity [Viswanath *et al.* (2002)].

Opportunistic communications is a system-level design, while link adaptation is a link-level design of wireless communications networks.

Scheduling is a dynamic algorithm of allocating finite radio resources to users competing for the latter. It decides which among the active or backlogged users should receive or transmit at which time. Metrics used in channel-aware scheduling include:

(1) *System throughput* or effective data rate, which is sometimes referred to as system capacity or sum throughput. The higher the data rate, the higher the achievable throughput.
(2) Fairness: there are many types of fairness metrics, such as max-min fairness, proportional fairness, utilitarian fairness, temporal fairness and Jain's fairness index (aka Raj Jain's equation). Let $f_l(t)$, $l = 1, 2, \ldots, L$ be the throughput or fraction of service received by user or queue or traffic class l up to the present time t, then the Jain's fairness index (JFI) is defined as [Jain *et al.* (1984)]

$$JFI = \frac{(\sum_{l=1}^{L} f_l(t))^2}{L \sum_{l=1}^{L} f_l^2(t)} \,. \tag{8.113}$$

(3) *Quality-of-service* (QoS) requirements of users' traffic (e.g., delay bound, jitter, reliability (e.g., bit error rate, packet error rate, maximum packet drop count) and average throughput).
(4) *Stability* of the scheduling algorithm and the queues, which can be proved using, e.g., the Lyapunov drift stability theory.
(5) *Residual time* of a queue: which is the remaining time that traffic in a queue can wait to receive service without violating its delay constraints.
(6) Energy efficiency or power consumption: transmitting at a lower rate requires longer transmission times. Staying in the transmission mode for long increases the energy consumption. Energy-efficient protocols are very crucial for the realisation of the Internet-of-Things (IoT) and massive machine type communications (mMTC).

One of the wireless channel models used to analyse the performance of channel-aware MAC is the auto-regressive (AR) model, which is defined as

$$\gamma_m(t) = a\gamma_m(t-1) + (1-a)w_m(t) \tag{8.114}$$

where $\gamma_m(t)$ is the signal-to-noise ratio (SNR) of user m's link at time t, a, $|a| < 1$ is the wireless channel correlation factor, and w_ms are iid normal distributed random noise with mean μ_m and variance σ_m, respectively. Consequently, the instantanous Shannon capacity of user m's link is

$$R_m(t) = B \log_2(1 + \gamma_m(t)) \,. \tag{8.115}$$

Channel-dependent scheduling can occur in either the time domain or in the frequency domain. If the channel variations are faster in the frequency domain than in the time domain, then frequency domain scheduling can yield a better system throughput than time-domain scheduling. We discuss the basic principles of the two types of channel-dependent scheduling in the time domain using case studies.

By its nature, a channel-aware resource allocator requires CSI of users. This is usually obtained via a feedback channel, rendering it non-instantaneous. An easier way is to use as the input to the scheduling algorithm a statistical user channel states rather than the actual instantaneous CSI. Assume that multiple multipath components reach a maximum-ratio combining receiver, which obtains the total signal-to-noise ratio (SNR) for user m as γ_m. If $\gamma_m \leq \gamma_{0,m}$ then data transfer cannot occur. The parameter $\gamma_{0,m}$ is referred to as *optimal cutoff SNR*, while the **outage probability** is defined as

$$P_{\text{out}} = F_{\gamma_m}(\gamma_{0,m}) = \int_{-\infty}^{\gamma_{0,m}} f_{\gamma_m}(y)dy \qquad (8.116)$$

where $F_{\gamma_m}(y)$ and $f_{\gamma_m}(y)$ are the cumulative distribution function (CDF) and probability distribution function (pdf) of user's m SNR. If the channel bandwidth is B and the data rate is adapted to the channel variations, then the optimal rate adaptation (ORA) with constant transmitter power is [Annamalai *et al.* (2010)]

$$R_m = \frac{B}{\log_e 2}\int_0^{\infty} log_e(1+x)f_{\gamma_m}(x)dx$$

$$= \frac{B}{\log_e 2}\int_0^{\infty} \frac{1-F_{\gamma_m}(x)}{1+x}dx\,. \qquad (8.117)$$

Equation (8.117) is referred to as the *ergodic capacity* of user's m channel. The CDF and pdf can be that of Nakagami-m or Rayleigh fading statistics.

8.7.1 *Link Adaptation Schemes*

The *link* here means wireless channel. Link adaptation schemes (LASs) view channel variations as evil and combat them. LASs allow users to receive or transmit data regardless of their instantaneous channel conditions. They select the data rate resulting from a combination of modulation and error-control coding commensurate with a user's channel condition. For this reason, LASs are also referred to as *rate adaptation*. Assume that two users u_1 and u_2 with allocated transmission rates $R_1(t)$ and $R_2(t)$, respectively, are served at a base station. If the channel condition of u_1 is better than that of u_2, then $R_1(t) > R_2(t)$, and a lower-order modulation and stronger error control is used on transmissions to/from u_2.

There are two stages in rate adaptation:

- Estimate the qualities of users' channels.
- Select the transmission rate commensurate with the channel quality.

The main reason to transmit at low rate when channel state is bad is to reduce the bit error rate. For example, at the signal-to-noise ratio of 10 dB, 16-QAM at 4 Mbps achieves the bit-error rate (BER) of 10^{-7}, while BPSK at just 1 Mbps achieves only 10^{-1}.

8.7.1.1 *Automatic Rate Fallback*

The automatic rate fallback (ARF) is a simple rate adaptation algorithm that was designed for WaveLAN-II WLAN by Lucent Technologies (now part of Alcatel-Lucent) [Kamerman and Monteban (1997)]. ARF was latter adopted in the IEEE 802.11 standard. The physical layer (PHY) of this standard supports multiple data rates through different combinations of channel coding and modulation schemes. The various coding and modulation combinations supported are shown in Table 8.17. The different data rates achieve different error performance over a given wireless channel condition. A mobile device selects from a pool the best data rate commensurate with its channel condition to maximise the throughput.

The IEEE 802.1g physical later is based on OFDM. The cyclic prefix in the OFDM has the length $T_{\text{cp}} = 0.8\,\mu\text{s}$, while the useful symbol duration is $T_{\text{u}} = 3.2\,\mu\text{s}$, resulting in the symbol duration $T_{\text{s}} = 4\,\mu\text{s}$ and the corresponding OFDM symbol rate of $R_{\text{s}} = 250\,\texttt{kbaud}$. The subcarrier spacing of the OFDM is 312.5 kHz and they are allocated to users using the distributed user multiplexing mode. Each of the data rate in Table 8.17 is obtained by multiplying R_{s} by the number of bits per symbol.

How does ARF work? The transmission of a frame or packet can either be success or failure. Let N_{s} and N_{f} denote the count for successful and failed transmissions, respectively.

Table 8.17 Various modulation, coding and data rate of OFDM used in IEEE 802.11g.

Modulation Method	Channel Coding Rate	Number of Data Bits per OFDM Symbol (m)	Data Rate (Mbps)	Code Bits per Subcarrier	Code Bits per OFDM Symbol
BPSK	1/2	24	6	1	48
BPSK	3/4	36	9	1	48
QPSK	1/2	48	12	2	96
QPSK	3/4	72	18	2	96
16-QAM	1/2	96	24	4	192
16-QAM	3/4	144	36	4	192
64-QAM	2/3	192	48	6	288
64-QAM	3/4	216	54	6	288

- Upon each successful transmission we update the counters as: $N_{\mathtt{s}} = N_{\mathtt{s}} + 1$ and $N_{\mathtt{f}} = 0$.
- Upon each failed transmission we update the counters as: $N_{\mathtt{s}} = 0$ and $N_{\mathtt{f}} = N_{\mathtt{f}} + 1$.
- After a given number of consecutive successful transmissions (say, S), transmit at the next higher rate. Stay with the higher rate if the first transmission succeeds, else return to the previous rate.
- After a given number of consecutive failed transmissions (say, F), transmit at the next lower rate.

8.7.2 *LTE Link Adaptation Through AMC*

The LTE/LTE-A standard has three modulation schemes QPSK, 16-QAM and 64-QAM and 15 coding rates ranging between 1/12 and 11/12. A set of modulation and coding commensurate with the instantaneous channel quality to meet a required BLER (block erasure rate) is selected. This is referred to as Adaptive Modulation and Coding (AMC). If the channel condition at a subcarrier of a user is good then a higher modulation scheme (i.e. 16-QAM or 64-QAM) and a more efficient channel coding are used to maximise throughput. On the other hand, if the channel condition is poor then a more robust modulation (i.e., QPSK) and a low-rate channel coding scheme are used to minimise errors and retransmissions. UEs use the channel quality indicator (CQI) signalling to report their channel conditions to the serving eNB. Based on the received CQI, the eNB selects the optimised AMC from the list in Fig. 8.55. The optimised AMC is the most spectral efficient MCS decodable by the relevant UE with less than 10% BLER. A combination of coding rate and modulation scheme enables us to find the appropriate modulation and coding scheme (MCS) and the number of resource blocks (RBs) for each CQI index. Note, however, that the LTE/LTE-A standard does not have a one-to-one mapping between the CQI and the MCS. There are many MCS for each CQI, and it is left for implementers to decide.

Let the number of contiguous resource blocks allocated to user i in a given TTI (i.e., a scheduling period) be $N_{\mathtt{rb}}$, $\gamma_{i,k}$ be the signal-to-noise plus interference ratio (SNIR) of subcarrier k of user i, P_i be the total power allocated to user i and $CSI_{i,k}$ be the channel state information of subcarrier k of user i. Then, we have

$$\gamma_{i,k} = \frac{P_i}{12N_{\mathtt{rb}}} CSI_{i,k}. \tag{8.118}$$

An example of the CSI for the uplink using multiple receive antennas and maximal ratio combining is [Abu-Ali *et al.* (2014)]

$$CSI_{i,k} = \frac{G_{\mathtt{UE}} G_{\mathtt{eNB}} \sum_{l=1}^{L} |H_{i,k}^{l}|^2}{PL_{\mathtt{u}} \cdot \sigma_{\mathtt{n}}^2 F \Delta f} \tag{8.119}$$

where $G_{\mathtt{UE}}$ and $G_{\mathtt{eNB}}$ are the gains of the user equipment and base station, respectively, F is the noise figure of the eNB receiver, $\sigma_{\mathtt{n}}^2$ is the additive noise density per Hertz, $PL_{\mathtt{u}}$ is the path loss and $|H_{i,k}^{l}|^2$

Radio Bearer Index (CQI)	PDSCH Modulation	Modulation Order (Bits/ Symbol)	Channel Coding Rate x 1024	Bearer Efficiency (bits/symbol)
0	Out of range			
1	QPSK	2	78 (≅1/12)	0.1523
2	QPSK	2	120 (≅1/9)	0.2344
3	QPSK	2	193 (≅1/6	0.377
4	QPSK	2	308 (≅1/3)	0.6016
5	QPSK	2	449 (≅ 1/2)	0.877
6	QPSK	2	602 (≅3/5)	1.1758
7	16-QAM	4	378 (≅1/3)	1.4766
8	16-QAM	4	490 (≅1/2)	1.9141
9	16-QAM	4	616 (≅3/5)	2.4063
10	64-QAM	6	466 (≅1/2)	2.7305
11	64-QAM	6	567 (≅ 1/2)	3.3223
12	64-QAM	6	666 (≅3/5)	3.9023
13	64-QAM	6	772 ≅ 3/4)	4.5234
14	64-QAM	6	873 (≅5/6)	5.1152
15	64-QAM	6	948 (≅11/12)	5.5547

Fig. 8.55 4-bit CQI and its choices of modulation and coding in LTE link adaptation [ETSI-3GPP (2016)].

is the normalised gain of the path on subcarrier k from user i to antenna l of the eNB. Assuming minimum mean squared error (MMSE), the effective SNIR over the entire set of PRBs allocated to the ith user in the uplink is [Calabrese (2006)], [Abu-Ali *et al.* (2014)]

$$\gamma_i = \left[\frac{1}{\frac{1}{12N_{\mathtt{rb}}} \sum_{k=1}^{12N_{\mathtt{rb}}} \frac{\gamma_{i,k}}{\gamma_{i,k}+1}} - 1 \right]^{-1}. \qquad (8.120)$$

Using Eq. (8.120), we can compute the available data rate as $R_i(t) = 12N_{\mathtt{rb}} \log_{10}(1+\gamma_i)$. With this rate we can select the appropriate combination of coding and modulation schemes from the standardised list in Fig. 8.55 to achieve the expected block error rate (BLER).

8.7.3 *Opportunistic Scheduling*

Fading is an inherent feature in wireless channels. The wireless channels of multiple users sharing a base station undergo independent fading. This means that at any scheduling time there is a high chance that the channel states are different: some of them have good channel conditions, some have average conditions, while others may be in deep fade. The variations of channel states of users depending on their relative positions to the serving base station is referred to as **multi-user diversity**. The gain achieved in exploiting multi-user diversity is referred to as multi-user diversity gain (MDG). The higher the number of users sharing a base station, the higher the expected MDG. Also, the larger the channel variations, the higher the expected MDG.

In wireless communications, scheduling is the method by which radio resources are allocated to users by the serving base station. Scheduling schemes which prioritise wireless users with relatively good channel state while allocating resources are referred to as opportunistic scheduling (OS). Opportunistic scheduling arise from the fact that it exploits good channel states to maximise system throughput. The origin of OS in recent years is attributed to [Knopp and Humblet (1995)]. The main motivation behind opportunistic scheduling is to maximise the overall system throughput. However, some OS schemes consider other metrics such as fairness and stability. We discuss principles of OS using some algorithms as case studies.

8.7.3.1 *MaxSNR Scheduling*

A simple OS scheme is the maximum signal-to-noise ratio (MaxSNR) scheduler, which is also referred to as Max-C/I scheduler or MaxRate scheduler [Bonald (2005)]. The MaxSNR scheduling picks the user among all active users in the system at the scheduling period (aka TTI in LTE/LTE-A) which has the best instantaneous channel quality or signal-to-noise ratio, or equivalently, the best feasible data rate. In the scheduling period t, Max-C/I schedules the user m^* fulfilling

$$m^* = \arg\max_m R_m(t) \tag{8.121}$$

where $R_m(t)$ is the instantaneous data rate of user m. We review in the next sections two OS schemes which fall between the Max-C/I (i.e., aggressively channel aware) and round-robin (non-channel aware) schemes.

8.7.3.2 *MaxWeight and ExpRule Scheduling*

In the following, let us note the definitions:

- $R_l(t)$: achievable data rate of user l, $l = 1, 2, \ldots, L$ at time t.
- $Q_l(t)$: queue length of user l, $l = 1, 2, \ldots, L$ at time t.
- γ_l, a_l: arbitrary positive constants.
- $l^*(t)$: the user scheduled at scheduling instant t.

The *MaxWeight* computes the product of the instantaneous queue length and data rate as the utility function and uses the scheduling policy [Andrews *et al.* (2004)]

$$\textit{MaxWeight}: \ l^*(t) = \arg\max_l R_l(t)Q_l(t) \tag{8.122}$$

while the ExpRule scheduling policy is [Lee *et al.* (2004)]

$$\textit{ExpRule}: \ l^*(t) = \arg\max_l \gamma_l R_l(t) e^{\frac{a_l Q_l(t)}{\beta + Q(t)^\eta}} \tag{8.123}$$

where β, $\eta \in (0, 1)$ and $Q(t) = \frac{1}{L}\sum_{l=1}^{L} a_l Q_l(t)$.

8.7.3.3 *Proportional Fair Scheduling*

The proportional fair (PF) scheduler assumes that the number of users L is fixed and each user's queue is fully backlogged. Let $I_{k=l} = 1$ if $k = 1$ and zero otherwise be the indicator function, $T_l(t)$ be the throughput of lth user up to time t, then the PF scheduling rule is [Viswanath *et al.* (2002)]

$$l^* = \arg\max_i \frac{R_l(t)}{T_l(t)} \quad \text{(a)}$$

$$T_l(t+1) = \left(1 - \frac{1}{t_c}\right)T_l(t) + \left[\frac{R_l(t)}{t_c}\right]_{I_{l=l^*(t)}} \quad \text{(b)}$$

where t_c is the averaging time window. The PF scheduler is implemented in the downlink of CDMA 1xEV-DO technology.

8.7.3.4 *ROCP and W-SRPT*

Riding on the channel peak (ROCP) and wireless shortest remaining processing time first (W-SRPT) are two opportunistic scheduling schemes which consider the size of backlogged files or traffic. Let us note the definitions [Hu *et al.* (2004)]:

- $R_k(t)$: achievable data rate on channel to user k at time t.
- $E[R_k]$: average data rate of user k.
- X_k: residual backlogged file size of user k at time t.

The **W-SRPT scheduler** picks for service the user which has the least expected transmission time. The W-SRPT scheduling rule is

$$k^* = \arg\min_k \left(\frac{X_k(t)}{E[R_k]}\right). \tag{8.124}$$

The ROCP scheduling rule is

$$k^* = \arg\min_k \left(\frac{R_k(t)}{E[R_k]}\right). \tag{8.125}$$

8.7.3.5 *Traffic Aided Opportunistic Scheduling*

Traffic-aided opportunistic scheduling (TAOS) is a set of scheduling algorithms which combine file size and channel variations to optimise the total time required to transfer a file [Hu *et al.* (2004)]. Let us note the definitions [Hu *et al.* (2004)]:

- $R_k(t)$: achievable data rate on channel to user k at time t.
- $E[R_k]$: average data rate of user k.
- U_k: the average throughput (i.e., data rate) of user k.
- F_k: the initial backlogged file size of user k.

- $X_k(n)$, $X_k(t)$: remaining size of backlogged file of user k during the nth scheduling time slot or time t.
- $M(t)$: the number of remaining users in the system at time t.
- $I_k(t)$: the rank of user k when users are ordered in ascending order based on their ratio $\frac{X_k(t)}{E[R_k]}$ at time t.

Then the respective scheduling rules for TAOS-1 are [Hu *et al.* (2004)]:

$$\texttt{TAOS-1:} \quad k^* = \arg\min_k \left(\frac{F_k U_k(t)}{R_k(t)} \right) \tag{8.126}$$

$$\texttt{(TAOS-1a):} \quad k^* = \arg\min_k \left(\frac{F_k}{R_k(t)} \right) \tag{8.127}$$

$$\texttt{(TAOS-1b):} \quad k^* = \arg\min_k \left(\frac{X_k}{R_k(t)} \right) \tag{8.128}$$

where $U_k(t)$ is updated as

$$U_k(t+1) = \begin{cases} (1-1/t_c)U_k(t)+(1/t_c)R_k(t), & k=k^* \\ (1-1/t_c)U_k(t), & k \neq k^* \end{cases}. \tag{8.129}$$

The parameter t_c is the sliding observation window, which is measured in the number of scheduling time slots.

TAOS-2 algorithm first computes the ratio $\frac{X_k(t)}{E[R_k]}$ for all backlogged users. It then ranks them in the descending order. Using $I_k(t)$ as the rank of user k at time t, it defines the function

$$D_k(t) = I_k(t) - 1 - [M(t) - I_k(t) + 1]\left(\frac{R_k(t)}{E[R_k]} - 1 \right). \tag{8.130}$$

TAOS-2 then uses the scheduling rule at time t to pick a user for service [Hu *et al.* (2004)]

$$k^*(t) = \arg\min_k \left(\frac{F_k}{R_k(t)} \right). \tag{8.131}$$

8.7.3.6 *OCASD: Multi-user Diversity Based OS*

The Opportunistic Channel-Aware Scheduling with Service Differentiation (OCASD)[Gyasi-Agyei (2005)] is a cross-layer, downlink OS algorithm which allocates resources to users based on their link state and traffic quality-of-service (QoS) requirements. Traffic of users are categorised according to their QoS and queued appropriately. OCASD exploits multi-user diversity, that is the location-dependent wireless channel quality to maximise the overall system throughput. The penalty in OCASD is the possibility of delaying resource allocation to users experiencing bad channel conditions. Therefore, OCASD is suited for only data traffic which can tolerate delays.

The OCASD model is illustrated in Fig. 8.56, while its parameters are defined in Table 8.18. In the depicted scenario three active users at a cell over three time slots (ts) is considered. In slots ts_A and ts_C user 1 is experiencing a deep fade, while user 2 and user 3 are in deep fades in ts_B and ts_C, respectively. A non-channel-aware scheduler, such as round robin (RR), in the worst case scenario would schedule packets to the mobiles in the order shown in Fig. 8.56 (a-iii), causing 100% packet error rate (PER) and zero throughput. The round robin scheme allocates radio resources to users in turn (i.e., one after the other) without regard to their channel states. It thus assigns resources to users in the fairest possible how, albeit at the cost of poor system throughput. An opportunistic scheduler, on the other hand, can achieve 0% PER using the scheduling order in Fig. 8.56 (a-ii). By postponing a transmission to an MH until its link hits near its peak, some delay is introduced. However, depending on the latency time scale used, this delay can be offset by the reduced transmission time when data transmission occurs at a higher data rate.

The basic design principles of OCASD are:

(1) Packets belonging to the same traffic class (TC) based on QoS requirements but to be scheduled to different users are queued in the same logical queue at the base station. Packets in each queue are arranged in the order of arrival to the queue. Packet (re)ordering in a queue can also be based on (earliest) delay deadlines.
(2) Only the HOL (head of line, i.e., packet at the beginning of a queue) packet in each queue is considered in each scheduling decision.
(3) w_i and d_i of each $P_{HOL,i}$ and $R_m(t)$ of the user to receive $P_{HOL,i}$ are jointly used in the OCASD scheduling policy Eq. (8.132). w_i and d_i improve the short-term performance of the policy for real-time traffic, compared to a

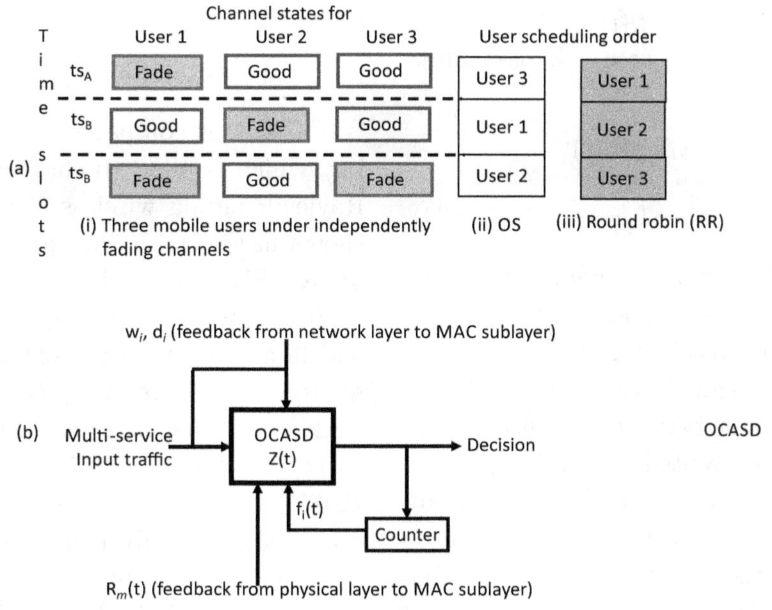

Fig. 8.56 The model for OCASD opportunistic scheduling [Gyasi-Agyei (2005)].

Table 8.18 OCASD scheduling parameters used in time slot t.

Parameter	Value
$P_{\texttt{HOL},i}(t)$	Head-of-line (HOL) packet in queue TC_i at time t.
$L_i(t)$	Size of HOL packet in traffic class or queue TC_i at time t.
w_i	Weight assigned to packets belonging to queue TC_i, $i = 0, 1, \ldots, c-1$.
d_i	Delay bound (deadline) of packets belonging to queue TC_i.
$B_i(t)$	Fraction of service received by queue TC_i up to time t.
$f_i(t) = \frac{B_i(t)}{B(t)}$	Average amount of traffic scheduled from TC_i up to time t.
$R_m(t)$	Predicted downlink data rate of wireless channel of user m, $m = 1, 2, \ldots, M$
S_i^t	Throughput of TC_i at time t.
$s_i^T(Z)$	Average throughput received by TC_i up to time T under policy Z.
t_c	Latency time scale, determines how long a queue can wait for service.
$b_i^T(Z)$	Fraction of time slots received by TC_i up to time T under policy Z.

greedy opportunistic scheme based solely on $R_m(t)$. We expect the IP layer to communicate to the MAC layer (where OCASD operates) the traffic QoS-related parameters w_i and d_i in the IP packet header field.

At the scheduling time t, the OCASD algorithm schedules a packet from i^* bound to user m^* which fulfills the **OCASD scheduling policy**

$$\{i^*, m^*\} = \arg\max_{i,m}\{Z_{i,m}(t)\} \tag{8.132}$$

where the heuristic cost function is defined as

$$Z_{i,m}(t) = [|w_i - f_i(t)| \cdot B(t)]^{\texttt{sgn}(w_i - f_i(t))} \frac{R_m(t)}{d_i} \tag{8.133}$$

where $B(t) = \sum_{i=0}^{c-1} B_i(t)$ and $f_i(t) = B_i(t)/B(t)$. We use the exponentially-weighted low-pass filter to update $B_i(t)$ as

$$B_i(t+1) = (1 - 1/t_c)B_i(t) + \frac{L_i(t)}{t_c}\mathbf{1}_{\{Z^t=i\}}\,. \tag{8.134}$$

Owing to the sign function or signum function $\mathtt{sgn}(w_i - f_i(t)) \in \{-1, 0, +1\}$, traffic class i is penalised at time t if it has received cummulative service in excess of its assigned weight w_i in $[0, t)$. Note that the expression being exponentiated in Eq. (8.133) should be prevented from falling below unity in order to achieve the aim of penalising an aggressive traffic class and ensuring weighted temporal fairness. This is the reason for the factor B(t) in Eq. (8.133).

For simplicity we write Z instead of $Z_{i,m}(t)$. The average throughput of traffic class i up to time T under the policy Eq. (8.133) is then

$$s_i^T(Z) = \frac{1}{T}\sum_{t=1}^{T} S_i^t \mathbf{1}_{\{Z^t=i\}}\,. \tag{8.135}$$

Let $s^T(Z) = \sum_{i=0}^{c-1} s_i^T(Z)$, then the best-case overall system throughput up to time T is

$$s(Z) = \lim\,\sup\{s_i^T(Z)\}\,. \tag{8.136}$$

OCASD is compared with round-robin scheduling, which is blind to channel conditions, in a flat Rayleigh fading wireless channel. The results are shown in Fig. 8.57. We can observe the superiority of OCASD over RR in throughput performance.

It is assumed that each user generates a packet within a time slot with size chosen from the probability mass function $p_l(x) = 0.13\delta(x-96) + 0.12\delta(x-144) + 0.42\delta(x-192) + 0.01\delta(x-720) + 0.01\delta(x-960) + 0.27\delta(x-1104) + 0.01\delta(x-1248) + 0.03\delta(x-1536)\}$ [Gyasi-Agyei (2003)], where $\delta(n-n0) = 1$ only if $n = n0$ and zero otherwise, $p_l(x) = Pr\{l = x \texttt{ bytes}\}$ and $\sum_x p_l(x) = 1$. Each generated packet is then randomly assigned to one of the c queues. The rationale behind this classification is that packets belonging to even the same application may require different QoS levels. We use a user speed of $v = 4.5$ km/h and eight multipaths in the fading model, and set $c = 4$ traffic classes. The fairness of OCASD is compared in Fig. 8.58 using the same data as those used in Fig. 8.57.

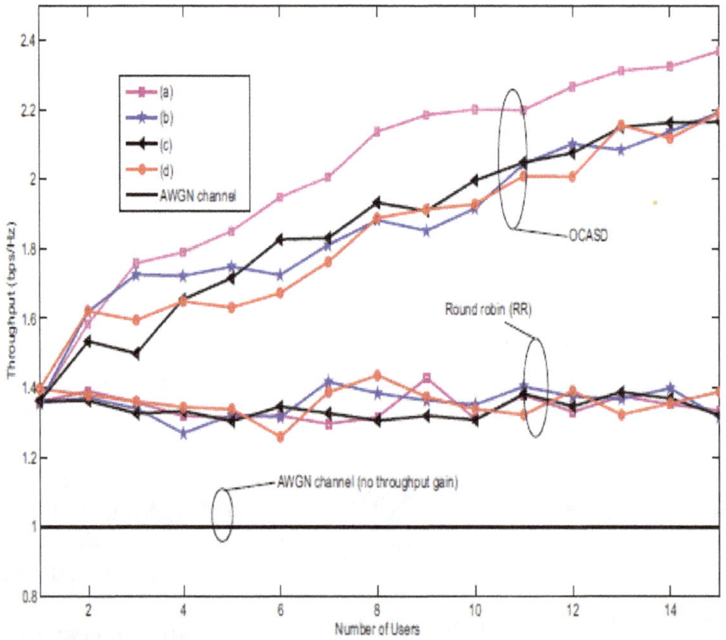

Fig. 8.57 Throughput comparison: (a) $w_i = 1/c, d_i = 1$ **sec**$, \forall i$, (b) $\mathbf{w} = [0.0009, 0.009, 0.09, 0.9001], d_i = 1$ **sec**, (c) $w_i = 1/c, d = [10, 1, 0.1, 0.01]$ s and (d) $\mathbf{w} = [0.0009, 0.009, 0.09, 0.9001], \mathbf{d} = [10, 1, 0.1, 0.01]$**sec**. [Gyasi-Agyei (2005)].

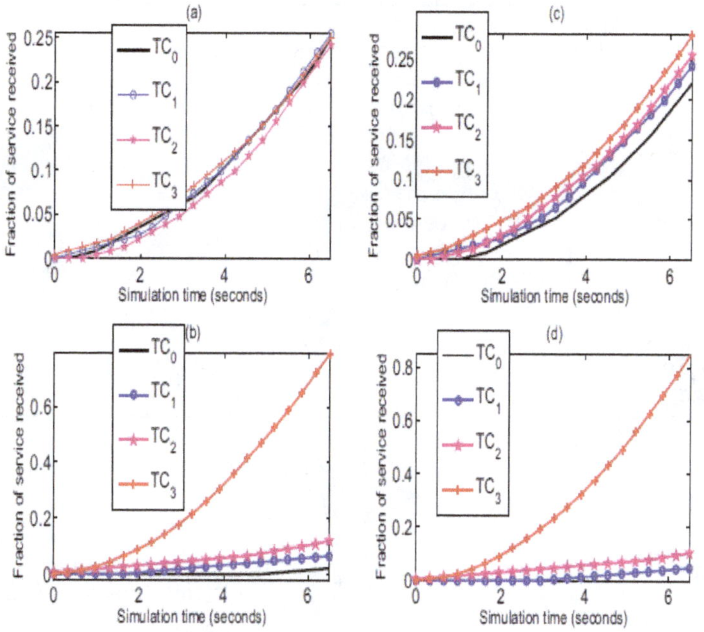

Fig. 8.58 Fairness comparison: (a) $w_i = 1/c, d_i = 1$ sec, $\forall i$, (b) $\mathbf{w} = [0.0009, 0.009, 0.09, 0.9001], d_i = 1$ sec, (c) $w_i = 1/c, d = [10, 1, 0.1, 0.01]$ s and (d) $\mathbf{w} = [0.0009, 0.009, 0.09, 0.9001], \mathbf{d} = [10, 1, 0.1, 0.01]$ sec. [Gyasi-Agyei (2005)].

8.7.3.7 *BLOT: Cross-Layer Opportunistic Scheduling*

Like OCAST, Best Link Lowest Throughput first (BLOT) is a cross-layer OS algorithm. Its model is shown in Fig. 8.59. BLOT defines the utility function

$$f(x,t) = f_1(x,t) \cdot f_2(x,t) = \frac{R_m(t)}{B_i(t) + \epsilon} \quad \text{(a)} \tag{8.137}$$

$$B_i(t+1) = (1 - 1/t_c)B_i(t) + \frac{L_i(t)}{t_c} I_i(t) \quad \text{(b)} \tag{8.138}$$

where ϵ is a small positive number used to prevent division by zero. The *indicator function* is defined as $I_i(t) = 1$ if packet from queue i is scheduled at time t, otherwise $I_i(t) = 0$. The function $f_2(x,t)$ is a measure of equity that is used to guarantee minimum service level to multiple data services active concurrently at a given user. The function $f_1(x,t)$ measures the instantaneous feasible data transfer rate of users. The BLOT scheduling rule is

$$\begin{aligned} x_t^* &= \arg\max_x \{f(x,t)\} \Rightarrow (i^*, m^*)_t \\ &= \arg\max_{i,m} \left\{ \frac{R_m(t)}{B_i(t) + \epsilon} \right\}. \end{aligned} \tag{8.139}$$

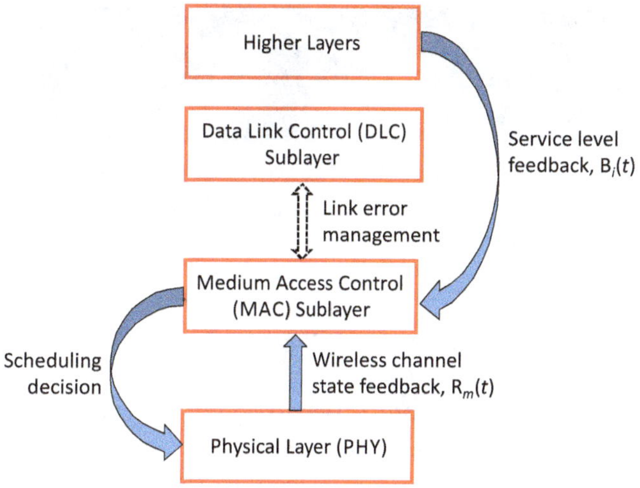

Fig. 8.59 Model for BLOT cross-layer scheduling [Gyasi-Agyei and Kim (2006)].

The average sum throughput for user m up to time t is

$$S_m(t) = \frac{1}{t} \sum_{i=1}^{c} \sum_{l=0}^{t} R_m(l) I_{im}(l) \tag{8.140}$$

where $I_{im}(t) = 1$ if queue i of user m is picked for service in scheduling time slot t; otherwise, $I_{im}(t) = 0$. Similarly, the average sum throughput

```
1. While time slot τ ≤ t_c do
        2. For m ∈ M (M is the set of backlogged users at time t)
                3. For i ∈ B (B is the set of non-empty queues at time t)
                        compute f(x,t) = R_m(t)/[B_i(t) + ε], x = (i,m)
                end
        end
4. Find connection x* ≠ arg max_x {f(x,t)}
5. If max{size{x*} = 1}, pick connection x+ ∈ x*
        B_i+ (t + 1) = (1 – 1/t_c)B_i+(t) + L_i+(t)I_i+(t)/t_c
  else B_i*(t + 1) = (1 – 1/t_c)B_i*(t) + L_i*(t)I_i*(t)/t_c
 end
6. t → t + 1
7. Goto step 1
end
```

Fig. 8.60 Pseudocode for BLOT assuming the availability of the $R_m(t)$s by some channel estimation technique. [Gyasi-Agyei and Kim (2006)].

for queue i up to time t is

$$S_i(t) = \frac{1}{t}\sum_{m=1}^{M}\sum_{l=0}^{t} R_m(l)I_{im}(l)\,. \tag{8.141}$$

The BLOT algorithm is compiled in Fig. 8.60.

8.7.4 *Check Your Understanding*

Students are encouraged to check their understanding of the content of this section with the following questions prior to proceeding to the next section.

(1) Discuss the meaning of the term *multi-user diversity*.
(2) Explain multi-user diversity gain, and how it can be enhanced.
(3) *In-cass discussions:* channel-aware scheduling methods are usually cross-layer in nature. Discuss.
(4) Compare how wireless channel variations are handled in opportunistic resource allocation with rate adaptation.
(5) Discuss the differences between the two resource allocation methods: rate adaptation schemes and opportunistic scheduling.
(6) What is the main motivation behind opportunistic scheduling.
(7) Critically analyse the two OS schemes discussed in the section: BLOT and OCASD. State the differences and similarities between them.
(8) Use MATLAB, R or any software to simulate the BLOT algorithm in Fig. 8.60. Compare BLOT with Max-C/I and round robin scheduling with respect to:
   (a) System throughput.
   (b) Per-user fairness.
   (c) Per-user throughput.
(9) Discuss the advantages and disadvantages in channel-aware resource allocation to users.
(10) Both IEEE 802.11g and LTE/LTE-A use OFDM transmission. Using illustrations, show how the subcarrier allocation to users in these two technologies differ.

8.8 Non-Orthogonal Multiple Access Schemes

In theory all the multiple access (MA) schemes discussed so far use orthogonality (or near orthogonality) among signals of different users to eliminate multi-user interference (MUI) or multiple access interference (MAI). Modern and future networks must support massive connections (as in e.g., Internet of Things), high spectral efficiency, ultra-low transmission latency to achieve virtual reality, low energy consumption, etc. Traditional MA schemes requiring orthogonality among signals sharing the radio resource cannot meet these requirements. For this reason, researchers are aggressively looking into non-orthogonal multiple access schemes (NOMA) [Saito *et al.* (2013)], [Kim *et al.* (2013)], [Choi (2017)]. This section surveys a few of the popular NOMA proposals, especially those which contended for adoption in 5G cellular mobile technology.

OMA schemes, especially channelisation types, are suited for streaming applications such as human voice, but not for bursty data traffic which is dominant in modern communications networks. Another justification for NOMA schemes is that the signals sharing a physical radio resource in an OMA scheme can lose their orthogonality through the transmission medium. For example, it is known that the subcarriers allocated to each user in an OFDMA system can undergo a different carrier frequency offset (CFO). The differing CFOs among the subcarriers endanger the orthogonality among them, causing inter-carrier interference (ICI) and MUI/MAI. This means that differing CFO can destroy the orthogonality that we have spent radio resource to acquire at the transmitter. Also, channel impairments destroy the orthogonality between spreading sequences used in spread-spectrum communications. Common causes of MUI are distortion in the channel and out-of-cell interference. Therefore, MUI and ICI cancellation or mitigation methods are employed in multi-user detection. Popular methods used to mitigate MUI/MAI in NOMA schemes include successive interference cancellation (SIC) and superposition coding.

Advantages in NOMA over OMA include [Aldababsa *et al.* (2018)]:

(1) Equitable service to all users, low latency and massive connectivity: NOMA does not keep some users on hold while serving the rest for any reason. This means guaranteeing fairness to all users. By not delaying service to any user, transmission delays are reduced.
(2) A NOMA scheme has a higher level of granularity in resource allocation than an equivalent OMA scheme, enabling it to support a larger number of connections simultaneously. This feature is important for massive machine type communications as in the Internet of Things (IoT).
(3) *Higher throughput and spectral efficiency:* NOMA assigns the same frequency to multiple users whose channels are most probably in different conditions. At a given time instant, some may be in deep fading while others may be experiencing favourable channel conditions. If the multiuser interference is appropriately handled by, e.g., SIC, then the prospect of achieving throughput and spectral efficiency higher than OMA is increased. Thus, NOMA can achieve the **sum-rate capacity** of a wireless channel.
(4) *Non-binary spreading sequences*: NOMA schemes based on spread spectrum use spread sequences whose elements are not necessarily binary. Therefore, NOMA has a wider spectrum of spreading sequences than OMA schemes.

Like anything, *NOMA has some drawbacks* and the identified ones include [Aldababsa *et al.* (2018)]:

(1) Each user in NOMA-based system has to decode the signals of some of the other users else it cannot decode its signal. This increases processing delay and receiver complexity. Orthogonality of signals sharing a resource using OMA enables good system-level performance even with simple receivers as MUI/MAI is ideally zero. NOMA, however, requires advanced receivers as the intentionally-introduced MUI/MAI must be cancelled prior to signal detection.
(2) NOMA requires each mobile station to communicate its channel state information to the base station. This increases the signaling load on the network.
(3) In order to decode its signal, a receiver must use successive interference cancellation (SIC). Any errors in the SIC at a user increases the error probability in successive decoding. In order to reduce this error we reduce the number of users sharing the communications resource.

(4) NOMA schemes work better if there are remarkable differences between the channel gains of the multiple users sharing a resource. This requires a restriction on the number of users.

Practical Applications of NOMA

In fact, NOMA has already been deployed in practical cellular communications systems. For example, CDMA used in 3G systems are known to use spreading codes with small correlation between them. However, these codes are not perfectly orthogonal. Therefore, CDMA is a NOMA scheme. Layered division multiplexing (LDM) is adopted in ATSC 3.0 (Advanced Television Systems Committee) standard for digital TV [Zhang *et al.* (2016a)]. Multiuser superposition transmission (MUST) [Lee *et al.* (2016)], a NOMA scheme, is included in 3GPP LTE Releases 13 and 14 for downlink transmissions [3GPP (2015b)], [NTT DOCOMO (2015)]. NOMA is currently being investigated for adoption in 5G New Radio within the 3GPP LTE Release 15 [ZTE-CATT-Intel-Samsung (2016)], [Aldababsa *et al.* (2018)]. The possibility of adopting NOMA in future communications systems is on the increase.

Example

Multiple access schemes (MAS) are necessary in all multiuser communications systems.

(1) Briefly state the main difference between the two classes of MAS: orthogonal multiple access (OMA) schemes and non-orthogonal multiple access (NOMA) schemes.
(2) State two examples of OMA schemes and two examples of NOMA schemes.
(3) Which of the two MAS philosophies is recent trend and which is a traditional method?
(4) State two advantages of NOMA over OMA schemes.

Solution

(1) OMA schemes mitigate MUI/MAI by requiring the signals sharing a physical radio resource to be orthogonal or near orthogonal in a measurable dimension. Signals sharing a radio resource in NOMA are not orthogonal. Popular methods used to mitigate MUI/MAI in NOMA schemes is successive interference cancellation (SIC) and superposition coding. OMA schemes, compared with NOMA schemes, are spectrally inefficient owing to their orthogonality requirement and the use of guard bands and guard periods.
(2) Examples of OMA schemes are FDMA, TDMA, CDMA, OFDMA, while examples of NOMA are SCMA (sparse code multiple access), PDMA (pattern division multiple access), MUSA (multiuser shared access), IDMA (interleave division multiple access) and LDSMA (low-density signature multiple access).
(3) Traditional multiple access schemes (including those used in 4G cellular networks) are OMA, while modern trend is NOMA.
(4) Potential advantages of NOMA schemes over OMA schemes include higher spectral efficiency, lower transmission latency (needed for virtual reality) and low energy consumption.

8.8.1 *Basic Principles in NOMA*

The basic principles of NOMA include successive interference cancellation (SIC) of multi-user interference (MUI) and superposition coding. We discuss some basic principles of NOMA in this section. Let us note the following definitions prior to proceeding:

- $CSQ_k(t)$: indicate the channel state quality of user k at time t.
- $g_k = |h_k(t)|^2$: channel gain of user k, where $h_k(t)$ is the corresponding channel coefficient at time t.
- β_k: fraction of the time-frequency communications resource in the system allocated to user k in a scheduling period in OMA. For a 2-user case, we assume that $\beta_1 = \beta$ and thus $\beta_2 = 1 - \beta$.
- $P_{\rm T}$: the total power transmitted by the base station in the downlink.
- P_k: the power allocated to user k at a given time t, so that $P_{\rm T} = \sum_k P_k$.

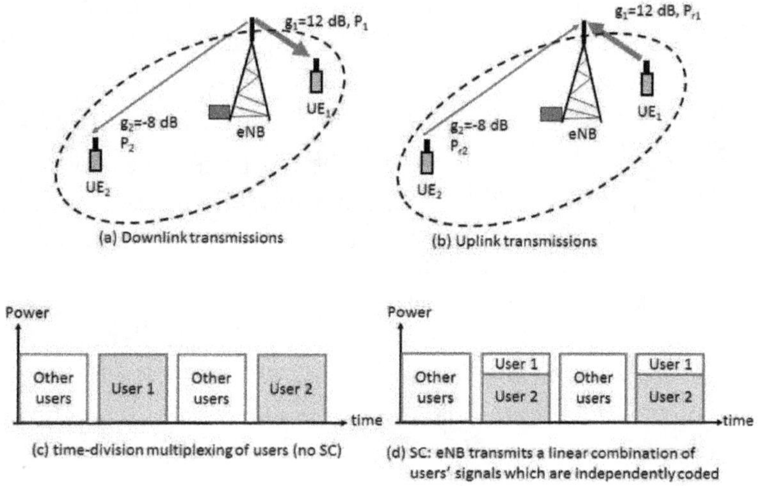

Fig. 8.61 Illustrating the NOMA principle with two users at a cell.

- a_k: fraction of the total power allocated to user k in a scheduling period, where $\sum_k a_k = 1$. For a 2-user case, we assume that $a_1 = a$ and thus $a_2 = 1 - a$. Thus, $P_1 = aP_T$ and $P_2 = (1-a)P_T$.

Let us use Fig. 8.61 to illustrate NOMA assuming, for simplicity, that there are only two active users at a cell served by base station. User 1 has a better channel gain than user 2, and thus it is allocated a smaller power than user 2 by the base station. Let the signal transmitted by the base station to user k be $s_k(t)$. All users receive the superimposed transmitted signal

$$s(t) = \sum_{k}^{K} \sqrt{P_k} s_k(t) \tag{8.142}$$

where $K = 2$ is the number of users at a cell. The baseband signal received by user k is then

$$\begin{aligned} r(t) &= h_k s(t) + w_k = \sum_{k=1}^{K} h_k \sqrt{P_k} s_k(t) + w_k \\ &= \sum_{k=1}^{K} x_k(t) + w_k \end{aligned} \tag{8.143}$$

where $w_k \sim \mathcal{N}(0, \sigma^2)$ is the background noise.

8.8.1.1 *Superposition Coding*

The origin of superposition Coding (SC) is attributed in Cover and Thomas [Cover and Thomas (2006)]. It is a non-orthogonal *downlink* transmission method which exploits channel ordering to achieve the capacity region of a scalar Gaussian broadcast channel. The main aim of SC is to communicate two messages to two different users simultaneously using the same time-frequency communications resource. The two message recipients are referred to as best user or near user (i.e., user closest to the serving base station) and worst user or far user (i.e., user farthest away from the serving base station). The two messages are encoded differently, and thus said to be in two layers. The layering is usually based on the different fractions of the base station power allocated to the superimposed signals. The best user can decode both messages (referred to as individual codewords in the cloud) while the worst user can decode only the message in the coarse layer (referred to as the cloud) [Cover and Thomas (2006)]. SC is a multi-user transmission method which intentionally introduces interference among user signals at the transmitter, using a library of off-the-shelf point-to-point channel codes [Vanka *et al.* (2012)]. The main motivation behind SC is the improved spectral efficiency compared with orthogonal transmission schemes.

Let us consider the cell of a cellular communications network shown in Fig. 8.61. There are K users in the cell, but we use only the two shown in the figure to illustrate the SC principle. User 1 is closer to the service base station that user 2. Therefore, the channel gain of user 1 is higher (i.e., 12 dB) than that of user 2 (i.e., −8 dB). Commensurate with the near-far terminology in cellular networks, we refer user 1 as the near user (NU) and user 2 as the far user (FU). Let $s_k(t)$, $k = 1, 2$ be the signal sent to user k. Signals sent from the base station to FU are also heard by NU. If FU can successfully decode its signal $s_2(t)$ then NU can most probably decode $s_2(t)$ too. However, the FU cannot decode the signal $s_1(t)$ sent to NU. SC exploits this channel ordering as its basic principle. The base station encodes the signals of the two users differently, combine them linearly and transfer them at the same time using possibly different power levels. This is superposition coding. The receiver can use SIC to decode the two signals. The SC method can be generalised to $K > 2$ users.

In Fig. 8.61(c) the signals of users are multiplexed in time so there is no MUI/MAI as in all OMA schemes. In Fig. 8.61(d), however, the eNB first encodes each user's signal independently, combines them linearly and then transmits in the same time-frequency communication resource. The superposition coding arises from the fact that the signals of multiple users are combined or superimposed and transmitted in the same time-frequency communication resource. Usually, the user with poorest link state is allocated the highest power to achieve fairness in throughput. Therefore, in Fig. 8.61(d), user 2 is allocated a higher power than user 1. User 2 signals can be so encoded that they can be decoded even in the presence of user 1 signals. User 1 with a strong signal (or good channel) can detect user 2 signal, deduct it from the superimposed signal it receives, and decode its signal.

For the sake of making our discussions self-contained, we review the mathematical formulation of SC here. More details can be found in [Cover and Thomas (2006)] and [Vanka *et al.* (2012)]. We assume that the wireless channel is a broadcast or downlink channel with the AWGN noise w_k, $k = 1, 2$. Let s be the input signal to the wireless broadcast or downlink channel, r_k, $k = 1, 2$ the channel's output at user k, and h_k channel gain at user k. Then, the outputs of the Gaussian broadcast channel are:

$$\begin{aligned} r_1(i) &= h_1 s(i) + w_1(i) \ \ \texttt{(a)} \\ r_2(i) &= h_2 s(i) + w_2(i) \ \ \texttt{(b)} \end{aligned} \tag{8.144}$$

where $s(i)$ is computed in Eq. (8.146). Let the sequence of channel's inputs and outputs be, respectively,

$$(s(i)),\ (r(i)),\ i = 1, 2, \ldots, n\,.$$

The total transmit power of the base station is P_{T}, of which the fraction a is allocated to user 1 (i.e., the user with strong channel). The steps in SC are then:

(1) Channel coding: map each set of $k_k = \lfloor nR_k \rfloor$, $k = 1, 2$ bits to a codeword with length n, i.e.,

$$\texttt{User 1:} \quad f_1 : \{0,1\}^{\lfloor 2^{nR_1} \rfloor} \longrightarrow \mathcal{C}^n$$

$$\texttt{User 2:} \quad f_2 : \{0,1\}^{\lfloor 2^{nR_2} \rfloor} \longrightarrow \mathcal{C}^n$$

where for $n \longrightarrow \infty$

$$R_1 < B \log_2\left(1 + \frac{a|h_1|^2 P_{\mathrm{T}}}{N_0 B}\right) \quad \texttt{and}$$

$$R_2 < B \log_2\left(1 + \frac{(1-a)|h_2|^2 P_{\mathrm{T}}}{N_0 B}\right). \tag{8.145}$$

Compute the linear combination

$$s(i) = \sqrt{a} s_1(i) + \sqrt{1-a} s_2(i)\,. \tag{8.146}$$

Transmit $s(i)$ over the downlink broadcast channel so that each user receives it.

(2) Perform a single-user decoding to estimate the signal of user 2 (i.e., far user) from the channel output Eq. (8.144(a)) by treating $s_1(i)$ as Gaussian noise, i.e.,

$$g_2 : \mathcal{C}^n \to \{0,1\}^{2^{nR_2}}\,. \tag{8.147}$$

(3) Perform a successive interference cancellation (SIC) decoding

$$g_{2,1} : \mathcal{C}^n \to \{0,1\}^{2^{nR_1}}\,. \tag{8.148}$$

(a) Decode user 2 signal using the single-user decoder g_2.

(b) Use the knowledge of a h_1 and estimate of $s_2(i)$ using the far-encoder f_2 to cancel user 2 signal from channel output r_1, i.e.,

$$\begin{aligned} \bar{r}_1(i) &= r_1(i) - h_1\sqrt{1-a} s_2(i) \\ &= h_1 \sqrt{a} s_1(i) + w_1(i)\,. \end{aligned} \tag{8.149}$$

(4) Decode user 1 signal via the single-user decoding

$$g_1 : \mathcal{C}^n \longrightarrow \{0,1\}^{2^{nR_1}}\,. \tag{8.150}$$

Clearly, successive interference cancellation or successive decoding is needed in SC. We therefore discuss it in the next section.

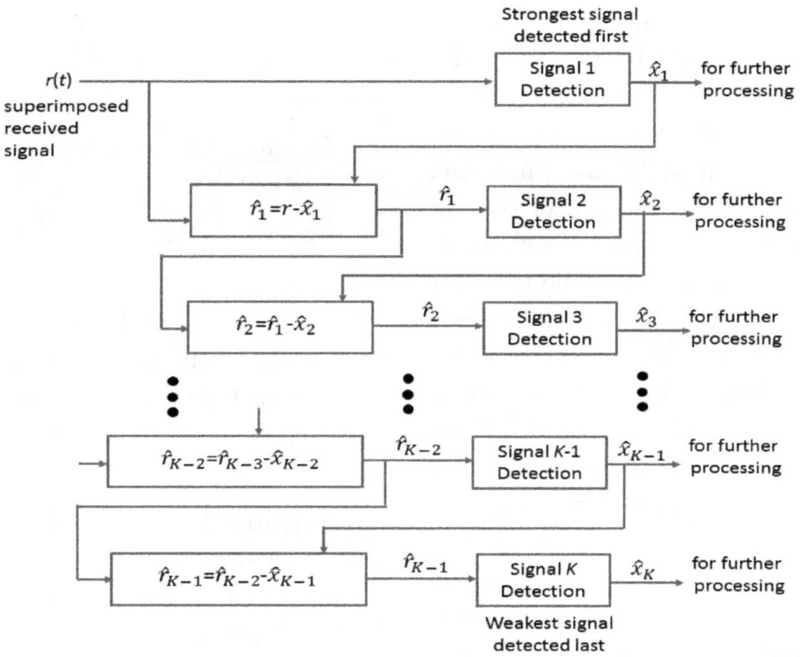

Fig. 8.62 The principle of successive interference cancellation in NOMA: assume that signal i is stronger than signal j if $i < j$.

8.8.1.2 *Successive Interference Cancellation*

Successive interference cancellation (SIC) is a series of single-user decoding, and it has been previously studied in information theory under the name *onion peeling* and *stripping* [Gallager (1994)], [Choi (2017)]. In NOMA, signals of users at various power levels are superimposed and transmitted. This is power division multiplexing (PDM). Also, users have different signal-to-noise plus interference ratios (SNIR) owing to one or both of these factors:

- Near-far effect or distance from user to the service base station.
- Differing allocated powers depending of user's channel state.

Decoding of a signal is possible only if the SINR is higher than the decision threshold. The principle underpinning SIC is to decode a signal with high enough SINR, then use the decoded signal to model the strongest interference and then subtract it from the received superimposed signal. That way the weaker signal with low SINR can be decoded as well.

The SIC principle is illustrated in Fig. 8.62. In this figure, the strength of the signals are ordered in decreasing order. That signal i is stronger than signal j if $i < j$). SIC decoding is performed in the order of decreasing SNIR or decreasing signal strength. The signal of the user with the best channel condition is decoded last, as it is allocated the least power. Power and rates are allocated based on their channel states and priority level. SIC exploits the fact that the most reliable interference cancellation is to cancel the signal of the strongest user.

8.8.1.3 *Sum Capacity Maximisation*

NOMA achieves the capacity region of uplink (aka multi-access channel) and downlink (aka broadcast channel) using a combination of superposition coding at the transmitter and successive interference cancellation (SIC) at the receiver. The users served by a given base station are categorised according to their channel conditions and hence power levels. More power is allocated to user groups in poor wireless channel states. Thus, if $CSQ_i(t) > CSQ_j(t)$ then $P_i(t) < P_j(t)$. Signals of users with high power can be decoded by treating signals of users with low powers as interference. Signals of users with low powers, however, have to first deduct the high-power signals from their received signals prior to decoding. This is the rationale behind successive interference cancellation (SIC) receivers.

As illustrated in Fig. 8.61, we assume here that user 1 has the channel gain of 12 dB, while user 2 is at a poorer channel with the gain of −8 dB. User 1 can be seen as being around the cell center or close to the serving base station while user 2 is at the fringe of the cell. Note that the normalised capacity or capacity per Hertz of bandwidth is the spectral efficiency, ρ. SIC decoding in NOMA is done in order of decreasing channel gains. Thus, user 1 decodes the signal to user 2 and subtracts it from the superimposed signal that it receives, and then decodes its signal. We assume that the SIC decoding is perfect and that user 1 completely cancels the interfering signal from user 2. Therefore, using the above definitions and Shannon's channel capacity, we can formulate the capacity bounds of both users for *OMA schemes* based on frequency division multiplexing as

$$R_{1,FDM} < C_{1,FDM} = \beta B \log_2\left(1 + \frac{a g_1 P_{\mathrm{T}}}{\beta B N_{0,1}}\right)$$
$$R_{2,FDM} < C_{2,FDM} = (1-\beta) B \log_2\left(1 + \frac{(1-a) g_2 P_{\mathrm{T}}}{(1-\beta) B N_{0,2}}\right) \tag{8.151}$$

and OMA based on time division multiplexing as

$$R_{1,TDM} < C_{1,TDM} = \beta B \log_2\left(1 + \frac{a g_1 P_{\mathrm{T}}}{B N_{0,1}}\right)$$
$$R_{2,TDM} < C_{2,TDM} = (1-\beta) B \log_2\left(1 + \frac{(1-a) g_2 P_{\mathrm{T}}}{B N_{0,2}}\right) \tag{8.152}$$

where $N_{0,k} = \kappa T$ is the one-sided power spectral density of the background noise in kth user's receiver, T is the temperature in Kelvin, and κ is the Boltzmann constant. The capacity bounds for both users for *NOMA schemes* are

$$R_{1,NOMA} < C_1 = B \log_2\left(1 + \frac{a g_1 P_{\mathrm{T}}}{B N_{0,1}}\right)$$
$$R_{2,NOMA} < C_2 = B \log_2\left(1 + \frac{(1-a) g_2 P_{\mathrm{T}}}{B N_{0,2} + a g_2 P_{\mathrm{T}}}\right). \tag{8.153}$$

The data rates of the two-user NOMA with superposition coding but without interference cancellation are

$$R_{1,noSIC} < C_1 = B \log_2\left(1 + \frac{g_1 P_1}{g_1 P_2 + B N_{0,1}}\right)$$
$$R_{2,noSIC} < C_2 = B \log_2\left(1 + \frac{g_2 P_2}{g_2 P_1 + B N_{0,2}}\right) \tag{8.154}$$

where $P_2 = (1-a)P_{\mathrm{T}}$ and $P_1 = aP_{\mathrm{T}}$. The base station controls the parameter a in Eqs. (8.151) and (8.153) or P_1/P_2 in order to achieve the throughput required by each user to successfully decode its signal.

Definition: The **capacity region** of the two-user downlink broadcast channel in Fig. 8.61(a) is the closure of the set of achievable data rates (R_1, R_2). This is illustrated in Fig. 8.63.

The normalised sum-rate of NOMA and OMA are compared in Fig. 8.63 for all possible power allocation ratios, i.e., $0 \leq a \leq 1$. We can observe that NOMA always has a higher sum-rate than OMA. For example, NOMA sum-rate at point A is 2.0965 bps/Hz while that of OMA at point B is only 1.4631 bps/Hz when over 80% of the power is allocated to user 1, i.e., $a = 0.8195$. The highest achievable sum-rate for the given parameters are 1.5789 bps/Hz for OMA and 2.311 bps/Hz for NOMA. This occurs if all the power is allocated to user 1.

8.8.1.4 *Factor Graph*

A factor graph is a **bipartite** graphical representation of the relationship between two sets of nodes:

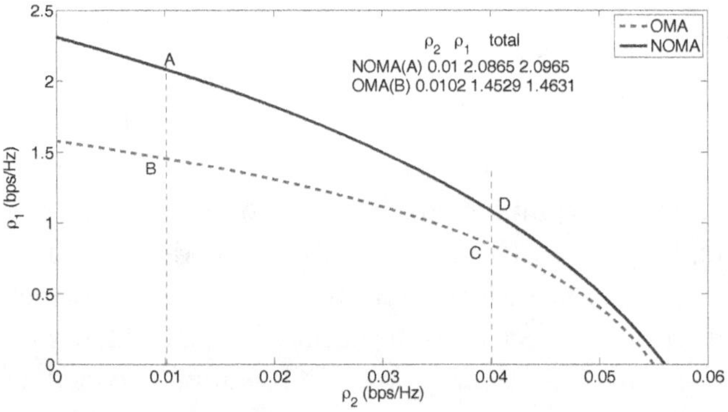

Fig. 8.63 Comparing downlink capacity regions of NOMA with OMA: $P_{\mathrm{T}} = 0.25$ W, $B = 1$ Hz, $\beta = 0.5$, $|h_1|^2/N_{0,1} = 10^{1.2}$, $|h_2|^2/N_{0,2} = 10^{-0.8}$.

(a) Factor graph

$$\mathbf{F}_{4 \times 6} = \begin{pmatrix} 1 & 0 & 0 & 1 & 0 & 1 \\ 1 & 1 & 1 & 0 & 0 & 0 \\ 0 & 1 & 0 & 1 & 1 & 0 \\ 0 & 0 & 1 & 0 & 1 & 1 \end{pmatrix}$$

(b) LDS indicator matrix describing the factor graph

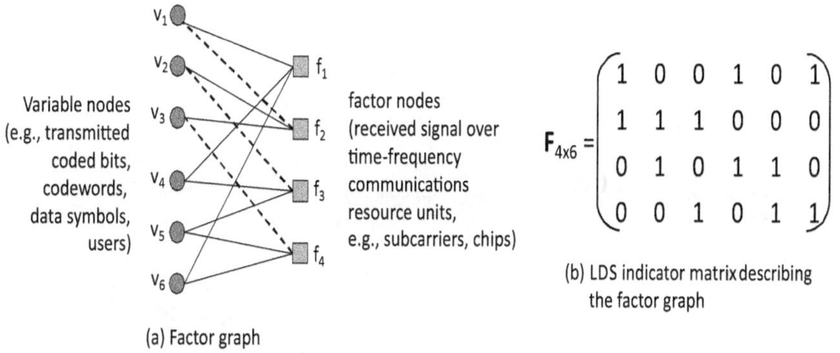

Fig. 8.64 Generalisation of signal spreading using factor graph: $d_v = 2$, $d_c = 3$, $K = 6$ users, $N_c = 4$ communications resource elements.

variable nodes v_i and factors nodes v_l [Kschischang *et al.* (2001)]. A variable node is also referred to as **layer node**, while a factor node is also called **resource node**. Edges connect factor nodes to variable nodes. Being a bipartite graph, a pair of nodes belonging to the same set in a factor graph are not connected, as illustrated in Fig. 8.64(a).

The factor graph has wide application areas. For example, in optimisation problems, the variable nodes are the variables in the problem, while each factor node is a local function of the variable nodes connecting to it via edges. In spread-spectrum based NOMA schemes, the factor graph is a generalisation of the spreading of signals.

As illustrated in Fig. 8.64(a), the variable nodes represent the transmitted coded bits, codewords or data symbols. A factor node represents the received signal over a time-frequency communications resource. For example, in LDS-CDMA, a factor node represents the received signal at each chip. In LDS-OFDM, however, a factor node represents the received signal at each subcarrier. As shown in Fig. 8.64(b), the factor graph can also be represented by a factor matrix, as there is a one-to-one mapping between them.

Factor graphs aid the design of receiver algorithms, such as message passing algorithm (MPA). It also shows pictorially the different types of mapping that can occur between the two sets of nodes. We can observe from the factor graph or factor matrix how many variable nodes share a factor node. Multiple variable nodes can share a factor node in NOMA to improve spectral efficiency. This is the reason why signals sharing a resource in

NOMA are non-orthogonal. For example, we observe in Fig. 8.64 that each factor node is shared by three variable nodes. Thus, $d_c = 3$. For example, users/symbols 1, 4 and 6 share the resource element 1 and thus causes interference to each other. We also observe that each variable node (aka user or symbol) is spread over two factors nodes (i.e., resource elements). Thus, $d_v = 2$. Usually, in LDS we expect

$$d_c \ll K \text{ and } d_v \ll N_c. \tag{8.155}$$

8.8.2 *Low Density Signature Multiple Access*

Low-density signature multiple access (LDSMA) [Hoshyar and Tafazolli (2008)] is a modified CDMA scheme, as it uses low-density spreading (LDS) codes [Hoshyar *et al.* (2010)]. Spreading codes in LDS can even be selected randomly. A spreading code which has a large number of its chips being zero is called *sparse code*. A non-sparse code is called *dense code*. First, conventional multi-carrier CDMA transmission uses dense spreading codes, while LDS-CDMA uses sparse spreading codes. Second, conventional CDMA spreads symbols in the time domain, while LDS-CDMA spread symbols in the frequency domain. Third, chips in a spreading code used in conventional CDMA are binary and real, but they can be non-binary and complex in LDS-CDMA. Each user uses a low-density signature sequence to spread its data symbols in LDSMA. Therefore, like conventional CDMA, multiple users can use the same set of carriers at the same time. The receiver exploits the low-density

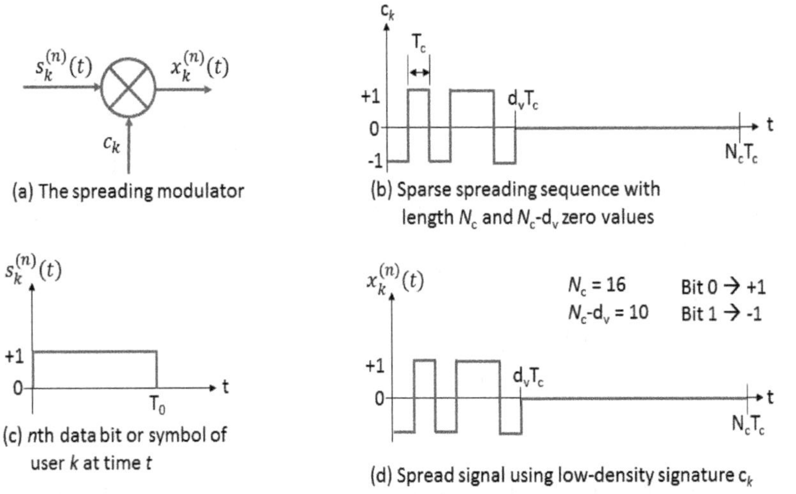

Fig. 8.65 Illustrating the principle of low-density signature (LDS) multiple access.

feature in the spreading codes used to decode the received signal iteratively using soft-in-soft-out multiuser decoding algorithm based on message passing algorithm (MPA).

The principle of LDSMA is illustrated in Fig. 8.65. Assume that K users are sharing a communication facility at a cell. For a given user k, $k = 1, 2, \ldots, K$, let its nth signal transferred at time t be $s_k^{(n)}(t)$. Thus, the vector of data symbols of user k is

$$\mathbf{s}_k = [s_k^{(1)}, s_k^{(2)}, \ldots, s_k^{(M_k)}]^T, \; k = 1, 2, \ldots, K \,. \tag{8.156}$$

Let the pseudorandom code vector used to spread the ith data symbol of user k be

$$\mathbf{c}_k^i = [c_k^{(i,1)}, c_k^{(i,2)}, \ldots, c_k^{(i,N_c)}]^T, \; k = 1, 2, \ldots, K \tag{8.157}$$

where $c_k^{(i,j)} \in \{0, 1\}$. Note that the LDS-OFDM described in [Hoshyar *et al.* (2010)] uses symbols, but $s_k^{(n)}(t)$ can also be a bit. Each symbol contains one or more bits, depending on the data modulation scheme in use. The number of chips in $\mathbf{c}_k^i$ is assumed to be N_c for all users, and that most of the chips in c_k are zero. This is the reason why c_k is referred to as low-density signature (LDS) sequence. This feature is the foundation of LDSMA schemes. In the example illustrated in Fig. 8.65, we have $N_c = 16$ chips, the number of nonzero chips is $d_v = 6$ and the number of zero chips being $N_c - d_v = 10$, fulfilling the requirement $N_c > d_v$.

An example of LDSMA, specifically LDS-OFDM, is illustrated in Fig. 8.66. Here, there are three users and each user transmits two symbols (i.e., $M_k = 2\,\forall k$) in the studied transmission period. Therefore, $K = 3M_k = 6$ symbols are modulating four subcarriers. Each symbol is also spread over two subcarriers. Therefore, each subcarrier is modulated by more than one symbol, resulting in interference. A different power can be allocated to each symbol from each user. Also, symbols modulating the same subcarrier can be allocated different levels of power. The level of mutual interference depends on:

- The channel gain of each subcarrier at each user.
- The amount of power allocated to each symbol modulating each subcarrier.

An important parameter used in LDSMA is the **system overload**, which is quantified by the **system overload factor**

$$\lambda_O = \frac{K}{N_c} \,. \tag{8.158}$$

The overload factor of the example in Fig. 8.66 is

$$\lambda_O = \frac{\texttt{number of symbols superimposed}}{\texttt{number of resource elements}} = \frac{K}{N_c} = \frac{6}{4} = 1.5 \,. \tag{8.159}$$

A conventional CDMA system uses dense spreading sequences and requires orthogonality between

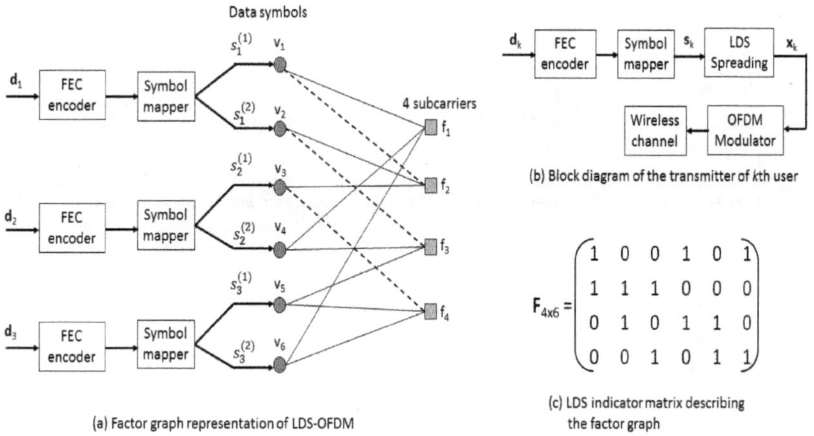

Fig. 8.66 LDS-OFDM model: $M_k = 2\,\forall k$, $K = 3M_k = 6$, $d_v = 2$, $d_c = 3$, $N_c = 4$.

users. Thus, $K \leq N_c$. LDS-CDMA (or LSDMA), however, achieves $K \geq N_c$.

The matrix of the spreading vectors used by user k to spread all its M_k symbols, referred to as **signature matrix** is

$$\mathbf{C}_k = (\mathbf{c}_k^1\, \mathbf{c}_k^2 \cdots \mathbf{c}_k^{M_k}) \in C^{N_c \times M_k},\ k = 1, 2, \ldots, K\,. \tag{8.160}$$

Thus, as illustrated in Fig. 8.66, the data symbol vector of user k after the LDS spreading is

$$\mathbf{x}_k = (x_k^1, x_k^2, \ldots, x_k^{N_c})^T = \mathbf{C}_k \cdot \mathbf{s}_k,\ k = 1, 2, \ldots, K\,. \tag{8.161}$$

The spreading matrix for the entire system with K users is an $(N_c \times \sum_{k=1}^{K} M_k)$-matrix, where, again, M_k is the number of data symbols at user k. As in [Hoshyar *et al.* (2010)], let $\mathcal{J}_{n|k}$ denote the set of user k data symbols sharing the same subcarrier n, then

$$\mathcal{J}_{n|k} = \{(i, l) : c_l^{i,n} \neq 0,\ l = k\}\,. \tag{8.162}$$

Thus, the number of data symbols of all users sharing the same chip or subcarrier n is

$$\mathcal{J}_n = \bigcup_{k=1}^{K} \mathcal{J}_{n|k} = \{(i, k) : c_k^{i,n} \neq 0\}\,. \tag{8.163}$$

Note that the cardinality of the set $\mathcal{J}_n$, which is the number of data symbols which share the same subcarrier, equals d_v, i.e., $|\mathcal{J}_n| = d_v$. The sparse spreading code for symbols sharing the same subcarrier have a nonzero chip in the same position. The user k transmitted symbol on the nth subcarrier is

$$x_k^n = \sum_{(i,l)\in\mathcal{J}_{n|k}} s_l^{(i)} c_l^{(i,n)}\,. \tag{8.164}$$

If the channel gain of subcarrier n at user k is h_k^n, then the received signal over subcarrier n is

$$r_n = \sum_{(i,k)\in\mathcal{J}_n} h_k^n c_k^{(i,n)} s_k^{(i)} + w_n\,. \tag{8.165}$$

The receiver reverses the processes illustrated in Fig. 8.66(b). A multi-user detector such as MAP can be used to detect each user's symbol after the OFDM demodulator.

8.8.3 *Sparse Code Multiple Access*

The Sparse Code Multiple Access (SCMA) is a hybrid scheme combining CDMA and OFDMA schemes. It combines two processes: bit to data symbol mapping and bit spreading [Nikopour and Baligh (2013)] by mapping bits directly onto a multidimensional codeword of an SCMA codebook set. Thus, the spreading is not performed over data symbols, but over bits. Each user is assigned a dedicated codebook. SCMA contended for adoption in 5G cellular mobile technologies. SCMA is a generalisation of low-densiy spreading multiple access, and their differences and similarities are shown in Fig. 8.67(a). SCMA maps symbols into higher-dimensional complex symbols, and then multiplex them over a fewer number of orthogonal subcarriers. The basic features of SCMA are:

(1) Each user is given a unique codebook containing codewords. Each codeword represents a spread transmission layer, and each set of coded data bits from a user is directly mapped onto a sparse codeword in its codebook. A sparse

Feature	LDS	SCMA
Spread code type	sparse	sparse
Multiple access	yes, using signature codes	yes, using codebooks
Coding gain (constellation shaping gain)	no, coded data modulate QAM symbols	yes, coded data carried over complex-valued codewords
Degree of freedom	*K* sparse spreading codes, each of length *n*	*K* codebooks, each with *M* codewords. Length of each codeword is *n*, which equals the dimension of complex constellation, e.g., *n*=2 for QPSK, *n*=4 for 16-QAM
Receiver algorithm	symbol-based MPA	codeword-based MPA
Modulation & symbol spreading	done separately	combined
Main design task	signatures	codebook with multi-dimensional codewords

(a) Comparison between LDS and SCMA

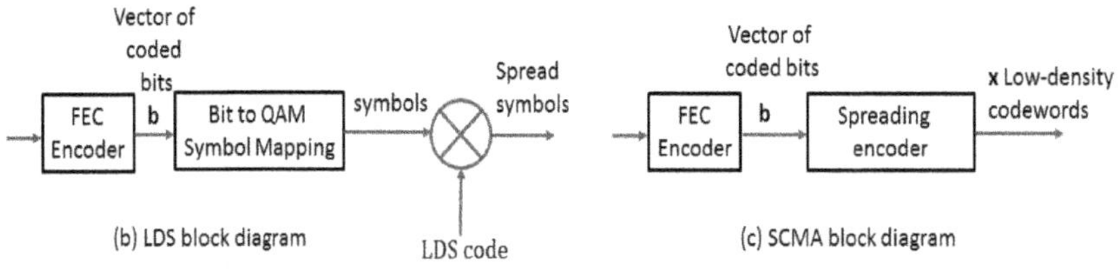

Fig. 8.67 Comparing SCMA with LDS.

codeword has most of its elements being 0 bit. The elements in the codeword can be complex valued.

(2) Multiple data streams belonging to different users can be transmitted over the same OFDM subcarriers. The sparsity of the codewords means that only a few data symbols modulate the same OFDM subcarrier, reducing the mutual interference. Therefore, message passing algorithm (MPA) can be used to decode the signals at the receiver.

(3) If the length of each sparse codeword (equivalently, the number of time-frequency communications resources) is n and the number of overlaid layers or users is K, then the overload factor is $\lambda_O = \frac{K}{n}$.

(4) The complexity of the detection of signals sharing a communications resource using SCMA depends on two factors:

   (a) The number of projection points per dimension in the multi-dimensional constellations. The dimension equals the length of the codewords.

   (b) The sparseness of the codewords used.

SCMA has three stages: encoding, multiplexing and decoding. We review these sequentially.

8.8.3.1 *SCMA Encoding*

Let $\mathbb{B}$ be the set of binary numbers, $\mathbb{B}^m$ be the set of binary vectors of length m and $\mathbb{C}$ be the set of complex numbers. In the encoding step each user maps the vector $\mathbf{b}$ of length $d_v = \log_2 M$ bits onto the sparse codeword $\mathbf{x}$ of length or dimension n, i.e., we compute the function

$$f : \mathbb{B}^{\log_2 M} \to \mathcal{X} \subset \mathbb{C}^n \tag{8.166}$$

which is defined by the equation

$$\mathbf{x} = f(\mathbf{b}), \ \mathbf{x} \in \mathcal{X} \tag{8.167}$$

where the cardinality $|\mathcal{X}| = M$ is the number of codewords in each user's codebook. The n-dimensional codeword $\mathbf{x}$ has $d_v \ll n$ nonzero elements. Let $\mathbf{c} \in \mathcal{Y} \subset \mathbb{C}^{d_v}$ be an d_v-dimensional complex-valued constellation point such that

$$g : \mathbb{B}^{\log_2 M} \to \mathcal{Y}, \ \mathbf{c} = g(\mathbf{b})\,. \tag{8.168}$$

The next step is to map the d_v-dimensional complex-valued constellation point onto an n-dimensional SCMA codeword, i.e.,

$$f :\equiv \mathbf{V}g, \quad \texttt{where } \mathbf{V} \in \mathbb{B}^{n \times d_v}\,. \tag{8.169}$$

All the elements in $n-d_v$ rows of $\mathbf{V}$ are zeros. Thus, each codeword in a codebook contains 0 in the same $n-d_v$ positions, and a maximum of d_v elements in each codeword of length n are nonzero, which are specified in the matrix $\mathbf{V}$. With M codewords in each codebook, and each codeword having n elements, the maximum number of codebooks (also called **layers** or users) is

$$K = \binom{\log_2(M)}{d_v} = \binom{n}{d_v} = \frac{n!}{(n-d_v)!d_v!}\,. \tag{8.170}$$

Note that the length of the codewords equals the dimension of the complex constellation of the modulation scheme used. For example, $n = 2$ for QPSK and $n = 4$ for 16-QAM.

8.8.3.2 *SCMA Multiplexing*

Multiplexing is achieved in SCMA by requiring each user to create and use a unique codebook (aka layer). The K layers (or codebooks) for K users of SCMA encoder is defined as

$$\mathcal{S}_k(\mathbf{V}_k,\, g_k,\, M_k,\, d_{v,k},\, n),\ k = 1, 2, \ldots, K\,. \tag{8.171}$$

As in [Nikopour and Baligh (2013)], we assume that all layers or codebooks have the same constellation size or number of codewords in it, i.e., $M_k = M$, and all codewords have the same number of nonzero elements, i.e., $d_{v,k} = d_v\ \forall k$. Thus, the representation of the SCMA code in Eq. (8.171) reduces to

$$\mathcal{S}([\mathbf{V}_k]_{k=1}^K,\, [g_k]_{k=1}^K;\, K, M,\, d_v,\, n)\,. \tag{8.172}$$

The constellation function g_k produces the constellation set $\mathcal{Y}_k$ with $M_k\,(= M)$ codewords or alphabets, each of length d_v. The mapping matrix $\mathbf{V}_k$ then maps each d_v-dimensional constellation point onto an SCMA codeword of length n to form the codebook (or set of codewords) $\mathcal{X}_k$, $k = 1, 2, \ldots, K$. SCMA codewords share n orthogonal communications resources, such as MIMO spatial layers or OFDM subcarriers.

As observable in Fig. 8.67(c), the transmitted SCMA codeword of user k is $\mathbf{x}_k = (x_{1k}, x_{2k}, \ldots, x_{nk})^T$. Over an AWGN channel, the received signal vector is

$$\mathbf{r} = \sum_{k=1}^K \mathtt{diag}(\mathbf{h}_k)\mathbf{x}_k + \mathbf{w} = \sum_{k=1}^K \mathtt{diag}(\mathbf{h}_k)\mathbf{V}_k g_k(\mathbf{b}_k) + \mathbf{w} \tag{8.173}$$

where $\mathbf{w} \sim \mathcal{CN}(0, N_0\mathbf{I})$ is the AWGN vector, and $\mathbf{h}_k = (h_{1k}, h_{2k}, \ldots, h_{nk})^T$ is the vector of channel gains of layer/user k. Note that $\mathtt{diag}(\mathbf{h}_k)$ is a diagonal matrix whose ith diagonal element equals h_{ik}. The special case in which $\mathbf{h}_k = \mathbf{h}$, $\forall k$, (8.173) reduces to

$$\mathbf{r} = \mathtt{diag}(\mathbf{h}) \sum_{k=1}^K \mathbf{x}_k + \mathbf{w} = \mathtt{diag}(\mathbf{h}) \sum_{k=1}^K \mathbf{V}_k g_k(\mathbf{b}_k) + \mathbf{w}\,. \tag{8.174}$$

We obtain the overloading factor $\lambda_O = K/n$ by multiplexing K layers (or signals of users) over n communications resources.

8.8.3.3 *SCMA Representation as a Factor Graph*

The SCMA algorithm can be represented by a factor graph comprising of factor nodes and variable nodes. An example is shown in Fig. 8.68. The received signal over the ith resource element over an AWGN channel can be modelled as

$$r_i = \sum_{k=1}^K h_{ik}x_{ik} + w_i,\ i = 1, 2, \ldots, n\,. \tag{8.175}$$

By the nature of SCMA the codewords $\mathbf{x}_k$, $k = 1, 2, \ldots, K$ are sparse. Thus, only a few of them

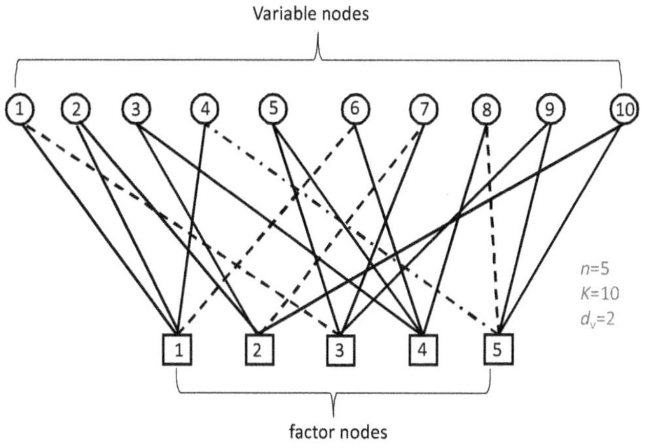

Fig. 8.68 The factor graph of an SCMA scheme.

share the ith communications resource element. Let

$$\mathbf{f}_k = \mathtt{diag}(\mathbf{V}_k\mathbf{V}_k^T),\ k = 1, 2, \ldots, K \quad (8.176)$$

be the binary indicator vector for the kth layer (or user). The set of nonzero elements in $\mathbf{f}_k$ are the communications resource elements used by layer (or user) k. Therefore, the SCMA code $\mathcal{S}$ in Eq. (8.172) can be represented by the **factor graph matrix** (aka **indicator matrix of the factor graph**)

$$\mathbf{F}_{n\times K} = (\mathbf{f}_1, \mathbf{f}_2, \ldots, \mathbf{f}_K) \in \mathbb{B}^{n\times K} \quad (8.177)$$

where n is the number of communications resources or the length of the codewords, and K is the number of layers or users. The total number of layers sharing a resource element is

$$\begin{aligned}\mathbf{d}_f &= (d_{f1}, d_{f2}, \ldots, d_{fn})^T \\ &= \sum_{k=1}^{K}\mathbf{f}_k = \sum_{k=1}^{K}\mathtt{diag}(\mathbf{V}_k\mathbf{V}_k^T). \quad (8.178)\end{aligned}$$

Note that $d_{\mathrm{v}} = d_{fi}$ is the number of layers using the ith resource element. Thus, there are $d_{fi} - 1$ interferers.

Note that the kth layer node is connected to the ith resource element iff the element in the ith row and kth column of $\mathbf{F}$ is non-zero, i.e., $(\mathbf{F})_{ik} = f_{ik} = 1$. For all i, let the set of layer nodes connected to the ith resource element be defined as

$$\mathcal{L}_i = \{k|(\mathbf{F})_{ik} = 1,\ \forall k\}. \quad (8.179)$$

Also, let for all k the set of resource nodes/elements being used by the kth layer node be

$$\mathcal{R}_k = \{i|(\mathbf{F})_{ik} = 1,\ \forall i\}. \quad (8.180)$$

Using the factor graph representation of SCMA, we can rewrite the received signal Eq. (8.175) as

$$r_i = \sum_{k\in\mathcal{L}_i} h_{ik}x_{ik} + w_i,\ i = 1, 2, \ldots, n. \quad (8.181)$$

Now, we have the following four equivalent representations for SCMA:

$$\{\mathbf{F}_{n\times K},\ [\mathbf{V}_k]_{k=1}^K,\ [\mathcal{L}_i]_{i=1}^n,\ [\mathcal{R}_k]_{k=1}^K\}. \quad (8.182)$$

In the following, we use an example to illustrate SCMA algorithm.

SCMA Example

Figure 8.68 illustrates the factor graph describing a SCMA. Study the figure and answer the following questions:

(1) How many communications resource elements, n, are available in the entire system?
(2) Evaluate the maximum number of communications resource elements that each layer (or codeword) is spread over.
(3) Compute the number layers or unique codebooks, K, in the system.
(4) What is the dimension of the indicator matrix, $\mathbf{F}$, that uniquely describes the SCMA represented by the factor graph in Fig. 8.68?
(5) Compute the indicator matrix, $\mathbf{F}$, equivalent to the factor graph in Fig. 8.68.
(6) What is the maximum number of layers/users which share a given resource element in the system?
(7) Compute the overload factor, $\lambda_O = \frac{K}{n}$, of the system. Explain your results.
(8) Compute $\mathcal{L}_i$, $i = 1, 2, \ldots, n = 5$, i.e., the set of layer nodes connected to each resource element.
(9) Compute $\mathcal{R}_k$, $k = 1, 2, \ldots, K = 10$, i.e., the set of resource nodes being used by the kth layer node.
(10) What is the dimension of the constellation points of the SCMA? Explain the SCMA process.
(11) Design an example of the SCMA codebooks.

Solution

(1) There are five factor nodes or resource nodes in Fig. 8.68. Therefore, the number of communications resource elements in the system is $n = 5$.
(2) Each column of $\mathbf{F}_{n\times K}$ has only two nonzero elements. Thus, $d_{\mathrm{v}} = 2$. This means that each layer is spread over two communications resource elements.
(3) The number of layers (or unique codebooks) in the SCMA is $K = \binom{n}{d_{\mathrm{v}}} = \binom{5}{2} = 10$.
(4) The dimension of the factor graph indicator matrix $\mathbf{F}$ is $n \times K = 5 \times 10$.

(5) The indicator matrix equivalent to the factor graph is

$$\mathbf{F}_{n\times K} = (\mathbf{f}_1, \mathbf{f}_2, \ldots, \mathbf{f}_{10}) = \begin{bmatrix} 1 & 1 & 0 & 1 & 0 & 1 & 0 & 0 & 0 & 0 \\ 0 & 1 & 1 & 0 & 0 & 0 & 1 & 0 & 0 & 1 \\ 1 & 0 & 0 & 0 & 1 & 0 & 1 & 0 & 1 & 0 \\ 0 & 0 & 1 & 0 & 1 & 1 & 0 & 1 & 0 & 0 \\ 0 & 0 & 0 & 1 & 0 & 0 & 0 & 1 & 1 & 1 \end{bmatrix}. \quad (8.183)$$

(6) We have $\mathbf{d}_f = \sum_{k=1}^{K} \mathbf{f}_k = (4, 4, 4, 4, 4)^T = d_{\mathbf{f}}$. Therefore, the maximum number of layers (or users) which share a given resource element in the system is 4. Thus, there are $d_{\mathbf{f}} - 1 = 3$ interfering signals on each resource element, which is less than n.

(7) The overload factor of the system is $\lambda_O = \frac{K}{n} = \frac{10}{5} = 2$. That is 10 sparse codewords belonging to 10 users (or layers) are transmitted using 5 orthogonal resource elements, such as OFDM subcarriers. A massive connectivity, as in Internet of Things or massive machine-to-machine (mM2M), requires $\lambda_O \gg 1$. We can obtain the desired overload factor by varying the spreading factor (i.e., length of the codewords, n) and the number of nonzero elements in the codewords.

(8) We can read off from the indicator matrix or factor graph the set of layer nodes connected to each resource element as

$$\{\mathcal{L}_1\}, \{\mathcal{L}_2\}, \{\mathcal{L}_3\}, \{\mathcal{L}_4\}, \{\mathcal{L}_5\} = \{1, 2, 4, 6\}, \{2, 3, 7, 10\}, \{1, 5, 7, 9\}, \{3, 5, 6, 8\}, \{4, 8, 9, 10\}. \quad (8.184)$$

(9) The set of resource nodes being used by the kth layer node is

$$[\mathcal{R}_k]_{k=1}^{K=10} = \{(1, 3), (1, 2), (2, 4), (1, 5), (3, 4), (1, 4), (2, 3), (4, 5), (3, 5), (2, 5)\}. \quad (8.185)$$

(10) The dimension of the complex codebook (or constellation points) of the SCMA equals the length of the codewords, i.e., $n = 5$. As $d_{\mathrm{v}} = 2$, each set of 2 error-coded data bits are mapped onto an 5-dimensional codeword. That is, we map each 2-bit symbol to 5-dimensional symbol. This is spreading. $n - d_{\mathrm{v}} = 3$ elements in each codeword are zeros. There are $K = 10$ codebooks or layers, and each codebook contains $M = 2^{d_{\mathrm{v}}} = 2^2 = 4$ codewords. Therefore, $K = 10$ codewords or layers are spread over $n = 5$ resource elements, as illustrated by the factor graph in Fig. 8.68.

(11) We have $K = 10$ layers or codebooks, $M = 4$ codewords in each codebook, $n = 5$ dimension or size of each codeword, $d_{\mathrm{v}} = 2$ nonzero elements in each codeword. All the 4 codewords in each codebook have $n - d_{\mathrm{v}} = 3$ zeros in the same positions (Fig. 8.69). As shown in Eq. (8.182), we have four different but equivalent ways of constructing the n-dimensional codebooks for all the $K = 10$ layers. We use the indicator matrix in Eq. (8.183) to design the codebooks in two stages.

The first stage is to use the information in the indicator matrix in (8.183) to assign zeros to the appropriate rows of each of the $K = 10$ codebooks. For example, the first column of $\mathbf{F}$ in Eq. (8.183) has zeros in rows 2, 4 and 5. Therefore, we assign 0 to the corresponding rows of codebook for layer/user 1. The second column of $\mathbf{F}$ in Eq. (8.183) has zeros in rows 3, 4 and 5. Therefore, we assign 0 to the corresponding rows of codebook for layer/user 2. This trend is repeated to fill the all-zeros rows in all the 10 codebooks, as illustrated in Fig. 8.69.

Now, each codebook has two entries not yet filled. We can fill each of these two entries with either 0 or 1. Therefore, we have four combinations, which is just enough to uniquely assign a 5-bit code to each 2-bit data stream. The final codebooks are shown in Fig. 8.70. This codebooks are shared between the transmitter and the receiver. The codewords in each codebook are unique. However, the same codeword can be assigned to different 2-bit data stream in different codebooks. Such an occurrence is called **codeword collision**. For example, layer 1 (or user 1) maps 2-bit data stream to 5-dimensional codewords as:

$$\mathbf{b} = \{0, 0\} \longrightarrow \mathbf{x}_{1,1} = 00000;$$
$$\mathbf{b} = \{0, 1\} \longrightarrow \mathbf{x}_{1,2} = 00100;$$
$$\mathbf{b} = \{1, 0\} \longrightarrow \mathbf{x}_{1,3} = 10000;$$
$$\mathbf{b} = \{1, 1\} \longrightarrow \mathbf{x}_{1,4} = 10100.$$

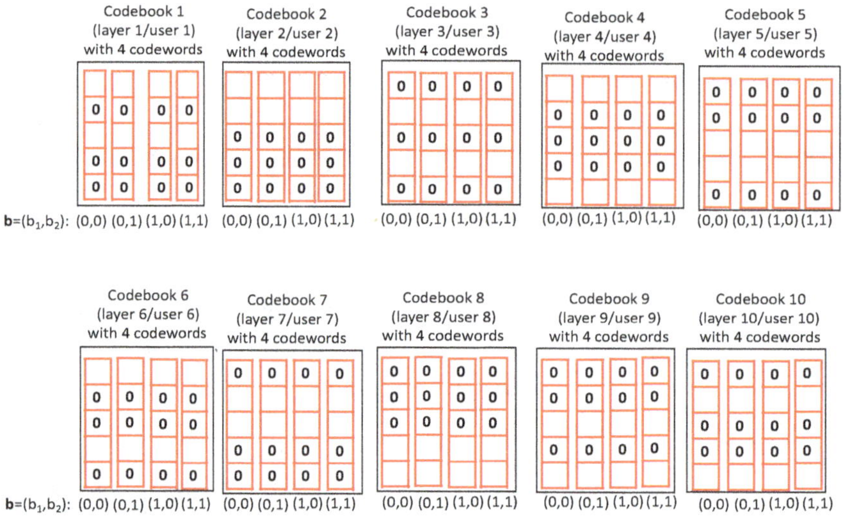

Fig. 8.69 Stage 1: designing the codebook and encoding in SCMA.

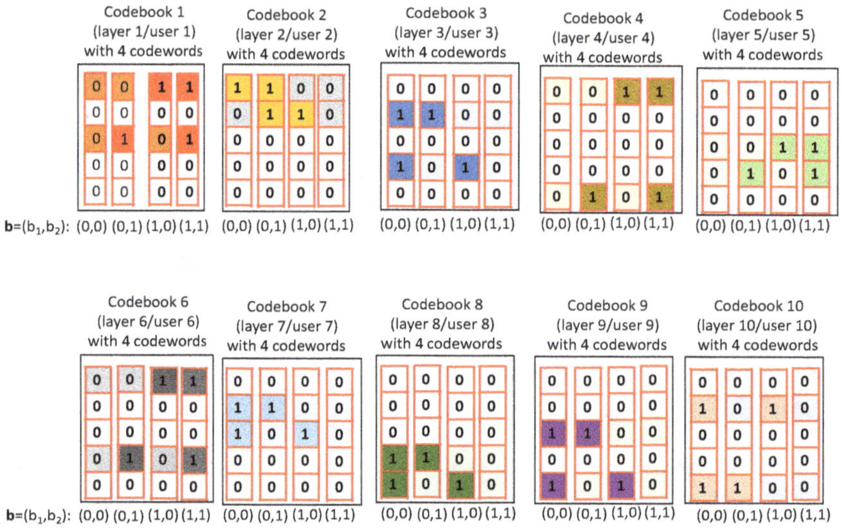

Fig. 8.70 Stage 2: designing the codebook and encoding in SCMA.

Figure 8.71 shows the ten 2-bit symbols and their corresponding codewords multiplexed onto five resource elements in a given time frame. We can observe that a maximum of four symbols interfere with each other.

8.8.4 *Multi-User Shared Access*

Multi-user shared access (MUSA) is based on spread-spectrum technology, and it was a candidate scheme for 5G cellular networks [Yuan *et al.* (2016)]. The symbol of each users transmission is spread using a short, complex-valued spreading code. MUSA is a NOMA scheme as the spreading codes are not orthogonal. MUSA reduces system complexity, reduces signalling related to scheduling request and resource assignment, reduces latency and resource wastage by:

- Allowing users to choose their pseudorandom spreading codes by themselves in order to eliminate code coordination by a centralised device, such as a base station.
- Using grant-free transmissions. In **grant-free transmission**, a mobile device transmits data into the channel without the requirement

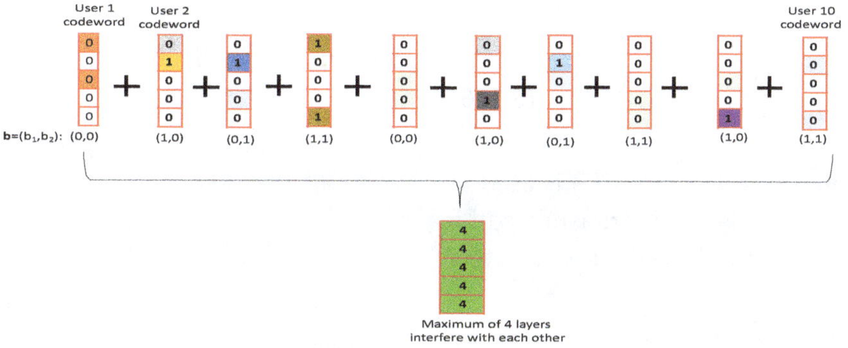

Fig. 8.71 Multiplexing of $K = 10$ codewords over $n = 5$ resource elements.

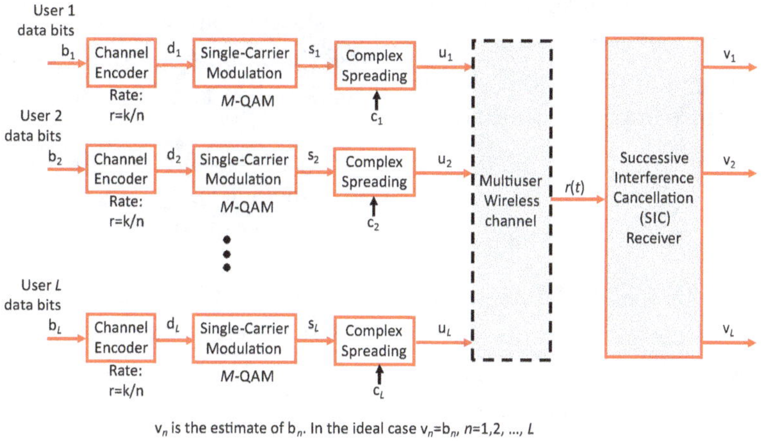

Fig. 8.72 MUSA transceiver structure. [Yuan *et al.* (2016)].

to first request resources and wait for their allocation by a base station. This reduces data transfer delays and eliminates the signalling needed for the request and allocation of resources.

Figure 8.72 illustrates the MUSA model [Yuan *et al.* (2016)]. The MUSA processing proceeds sequentially as follows:

(1) Assume that L users are served by one base station and their stream of data bits are b_l, $l = 1, 2, \ldots, L$. Each user's data bits are error-protected using a rate $r = k/n$ channel encoder. For example, user l feeds its k-bit data block $b_l = (b_{l,1}, b_{l,2}, \ldots, b_{l,k})$ into the encoder to obtain encoded bits $d_l = (d_{l,1}, d_{l,2}, \ldots, d_{l,n})$.

(2) The coded bits modulate the carrier of a single-carrier modulator, such as M-QAM, where $M = 2^m$, to produce the data symbols

$$s_l = (s_{l,1}, s_{l,2}, \ldots, s_{l,\frac{n}{\log_2 M}}),\ l = 1, 2, \ldots, L\,.$$

(3) Each data symbol is spread in the time domain using a complex-valued, short spreading code $c_l = (c_{l,I} + jc_{l,Q})$ of length N_c to obtain the spread signal u_l, i.e.,

$$\begin{aligned} u_l &= c_l \times s_l \\ &= (u_{l,1}, u_{l,2}, \ldots, u_{l,\frac{nN_u}{\log_2 M}}),\ l = 1, 2, \ldots, L\,. \end{aligned}$$

According to the original proposal, the spreading codes c_l, $l = 1, 2, \ldots, L$ can be non-binary but always complex valued and the elements of c_l can be taken from the the set [Yuan *et al.* (2016)]

$$\{0, 1, 1 + j, j, -1 + j, -1, -1 - j, -j, 1 - j\}\,.$$

(4) The signal u_l is transmitted over the wireless channel.

Assume that $h_{l,t}$ is the channel gain/coefficient of lth user at discrete time t and $w_{l,t}$ is the additive channel noise, then the received signal can be expressed as

$$r_t = \sum_{l}^{L} h_{l,t} u_l + w_{l,t} \,. \tag{8.186}$$

Successive interference cancellation (SIC) is used in the receiver to retrieve the data bits transmitted by each user from r_t. In the ideal case without detection error, we have $v_l = b_l$, $\forall l$.

Each user in MUSA chooses its own spreading code independent of other users. An inherent issue in this is that the spreading codes chosen by two or more users can be the same, causing **spreading code collisions**. Compared with traditional CDMA, MUSA uses shorter spreading codes to reduce code collisions, system complexity and consequential power consumption.

8.8.5 *Pattern Division Multiple Access*

MUI and inter-carrier interference (ICI) are eliminated in conventional multiple access schemes via strict orthogonality between the signals/users sharing a radio resource. The multi-user detector used in the original pattern division multiple access (PDMA) proposal is successive interference cancellation (SIC) using the believe propagation (BP) algorithm (BPA). However, multi-user detection based on the message passing algorithm can also be used. PDMA gained support from companies such as Qualcomm, Huawei, ZTE and DTmobile for adoption in 5G cellular mobile systems. PDMA is described in [Dai *et al.* (2018)], [Chen *et al.* (2017)].

Assume that K users or signals share n orthogonal physical radio resource elements, which can be subcarriers in OFDMA, codes in CDMA, timeslots in TDMA, etc. Assume that $n < K$. PDMA defines the **overload factor** as

$$\alpha = \frac{\texttt{number of users}}{\texttt{number of physical resources shared}} = \frac{K}{n} > 1 \,. \tag{8.187}$$

Let us define a vector called *pattern vector* for user k as

$$\mathbf{s}_k = [s_{1k}, s_{2k}, \ldots, s_{nk}]^{\mathrm{T}}, \; k = 1, 2, \ldots, K \,.$$

The pattern vectors for the K users sharing a radio resource make up the *pattern matrix*

$$\mathbf{S} = [\mathbf{s}_1 \mathbf{s}_2 \cdots \mathbf{s}_K] = \begin{bmatrix} s_{11} & s_{12} & \cdots & s_{1K} \\ s_{21} & s_{22} & \cdots & s_{2K} \\ \vdots & \vdots & \ddots & \vdots \\ s_{n1} & s_{n2} & \cdots & s_{nK} \end{bmatrix} \in \mathbb{R}^{n \times K} \,. \tag{8.188}$$

The pattern matrix $\mathbf{S}$ comprises groups of users' patterns, and users with nonzero elements in its ith row share the ith physical resource element. PDMA allocates physical resource elements to users in such a manner that an appropriate level of *disparity* among the data symbols of the K users is achieved in two levels/dimensions:

- Power disparity at the physical resource element level.
- Diversity disparity at the symbol level.

These two disparities between symbols of different users are exploited by the multi-user detector at the receiver to achieve the dual purposes:

(1) Eliminate MUI/MUI, although the users sharing the block of resources are not orthogonal.
(2) Retrieve transmit diversity at the receiver.

User multiplexing in PDMA can be achieved in power domain, code domain or space domain, or a combination thereof. Let us consider PDMA using code domain multiplexing in the uplink of a cellular communications network. In such a case, the pattern matrix Eq. (8.188) is interpreted as spreading matrix. Under these assumptions [Dai *et al.* (2014)], [Wang *et al.* (2015)]:

- The number of the overlapped spreading sequences with the same number of 1's should be as small as possible.
- The number of groups with different number of 1's in their spreading code should be minimised.

Under these two conditions, we obtain the maximum number of users that can be supported with the n orthogonal subcarriers as

$$K = \binom{n}{1} + \binom{n}{2} + \cdots + \binom{n}{n} = 2^n - 1 \,. \tag{8.189}$$

Samples of the pattern matrix are [Dai *et al.* (2014)]

$$n = 2,\ K = 3:\ \ \mathbf{S}_{2\times 3} = \begin{bmatrix} 1 & 1 & 0 \\ 1 & 0 & 1 \end{bmatrix} \tag{8.190}$$

and

$$n = 3,\ K = 7:\ \ \mathbf{S}_{3\times 7} = \begin{bmatrix} 1 & 1 & 0 & 1 & 1 & 0 & 0 \\ 1 & 1 & 1 & 0 & 0 & 1 & 0 \\ 1 & 0 & 1 & 1 & 0 & 0 & 1 \end{bmatrix}. \tag{8.191}$$

There are other popular NOMA schemes that we are unable to discuss in order to keep short a chapter which is already lengthy. We have interleave division multiple access (IDMA) [Ping *et al.* (2006)]. In IDMA, each user first spreads its symbols and then apply unique interleave pattern on them. Conventional DS-CDMA interleaves the symbols prior to spreading them. There is layered division multiplexing (LDM) [Zhang *et al.* (2016b)]. The LDM is adopted in ATSC 3.0 (Advanced Television Systems Committee) standard for digital TV [Zhang *et al.* (2016a)]. There is also bit division multiple access (BDMA), which is based on bit division multiplexing [Jin *et al.* (2013)]. Multi-user superposition transmission (MUST)[Lee *et al.* (2016)] is included in 3GPP LTE Releases 13 and 14 for downlink transmissions [3GPP (2015b)], [3GPP (2015c)], [NTT DOCOMO (2015)].

8.8.6 *Check Your Understanding*

Students are encouraged to check their understanding of the content of this section with the following questions prior to proceeding to the next section.

(1) Discuss the differences between sparse spreading code and dense spreading code. Compare their advantages and disadvantages.
(2) Reflect on the main motivation for NOMA schemes.
(3) In a tabular form, compare NOMA with OMA schemes. Include both their similarities and differences.
(4) Discuss the basic principles of NOMA schemes.
(5) Explain the reason why spreading code collision can occur in multi-user shared access (MUSA) scheme.
(6) Cite examples of practical systems which have adopted NOMA schemes.
(7) Comparing the receivers of NOMA and OMA schemes, which of them is expected to have a higher complexity? Justify your answer.
(8) State three differences between conventional multi-carrier CDMA and LDS-CDMA transmissions.
(9) How many unique layers or codebooks can we design for SCMA if the cardinality of each codebook is $M = 20$? Assume that the signal constellation diagram of the digital modulation scheme inherent in the SCMA has 256 points.
(10) By studying the data in Fig. 8.67, discuss the differences between LDSMA and SCMA.
(11) Figure 8.73 illustrates the factor graph describing a NOMA scheme. Study the figure and answer the following questions:
    (a) How many communications resource elements, n, are available in the entire system?
    (b) Evaluate the maximum number of communications resource elements that each layer (or codeword) is spread over.
    (c) Compute the number layers or unique codebooks, K, in the system.
    (d) What is the dimension of the indicator matrix, $\mathbf{F}$, that uniquely describes the SCMA represented by the factor graph in Fig. 8.73?
    (e) Compute the indicator matrix, $\mathbf{F}$, equivalent to the factor graph in Fig. 8.73.
    (f) What is the maximum number of layers/users which share a given resource element in the system?
    (g) Compute the overload factor, $\lambda_O = \frac{K}{n}$, of the system. Explain your results.

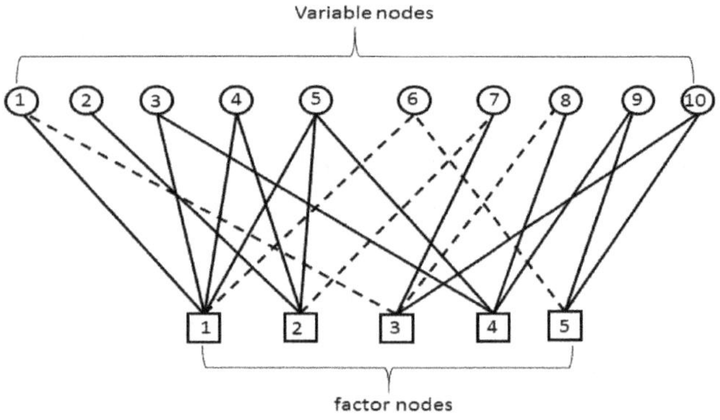

Fig. 8.73 The factor graph for a given non-orthogonal multiple access scheme.

(h) Compute $\mathcal{L}_i$, $i = 1, 2, \ldots, n = 5$, i.e., the set of layer nodes connected to each resource element.
(i) Compute $\mathcal{R}_k$, $k = 1, 2, \ldots, K = 10$, i.e., the set of resource nodes being used by the kth layer node.
(j) What is the dimension of the constellation points of the SCMA? Explain the SCMA process.
(k) Design an appropriate set of codebooks for all the layers in the SCMA.

8.9 Review of Important Terms and Acronyms

Acronyms, abbreviations and important terms		
LTE	LTE-A	resource sharing
factor graph	superposition coding	successive interference cancellation
low-density spreading	sparse code	sparse code multiple access
multi-user superposition transmission (MUST)	pattern-division multiple access (PDMA)	layered-division multiplexing
aloha (s-aloha)	asynchronous TDMA	cross-layer network design
carrier sensing	channel-ware scheduling	link adaptation
multi-carrier transmission	multiple access	multiplexing
multiuser diversity	non-orthogonal MAC	OCASD scheduler
OFDM OFDMA, SC-FDMA	opportunistic communications	opportunistic scheduling
orthogonal MAC	pseudo-random code	radio resource
radio resource management	rate adaptation	single-carrier transmission
spread spectrum	synchronous TDMA	subcarrier

8.10 Supplementary Reading

(1) H. Vincent Poor and Zhiguo Ding, *Multiple Access Techniques for 5G Wireless Networks and Beyond*, Springer, 2019.
(2) Ye Geoffrey Li and Gordon L. Stuber, *Orthogonal Frequency Division Multiplexing for Wireless Communications*, Springer, 2006.
(3) Alex Brand and Hamid Aghvami, *Multiple Access Protocols for Mobile Communications: GPRS, UMTS and Beyond*, Wiley, 2002.
(4) Klaus Grobe and Michael Eiselt, *Wavelength Division Multiplexing: A Practical Engineering Guide*, Wiley, 2013.
(5) Ronald K. Jurgen, *Multiplexing and Networking*, SAE International, 2006.
(6) Kamil Sh. Zigangirov, *Theory of Code Division Multiple Access Communication*, Wiley-IEEE Press, 2004.
(7) Sudip Misra and Barun Kumar Saha and Sujata Pal, *Opportunistic Mobile Networks: Advances and Applications*, Springer, 2016.
(8) David Tse and Pramod Viswanath, *Fundamentals of Wireless communications*, Cambridge University Press, 2005.
(9) Fa-Long Luo and Charlie (Jianzhong) Zhang (eds.), *Signal Processing for 5G: Algorithms and Implementations*, Wiley-IEEE Press, Chichester, 2016.

Chapter 9

Principles of Cellular Mobile Systems Design

Chapter's Learning Outcomes

Mobile communication networks enable users to move from one location to another while actively using the services provided by the network. Why is competence in cellular mobile network design important in IoT? Three standards developed by the 3GPP for mobile cellular communications have been adopted to interconnect devices in the IoT. These are classified among the long-range IoT connectivity standards. They are the Long-Term Evolution for machinetype communications (LTE-M), Narrowband IoT (NB-IoT) and Extended Coverage GSM IoT (EC-GSM-IoT). The move enables incumbent mobile operators to leverage their investment. EC-GSM-IoT is based on the 2G GSM technology, while LTE-M and NB-IoT are based on the LTE/LTE-A 4G technology. A company called Neul in the UK, which was acquired by Huawei Technologies in September 2014, is one of the major organizations promoting the use of cellular technologies as IoT connectivity. Neul focuses on using NB-IoT to provide deep coverage connectivity between ultra-low power IoT devices.

This chapter introduces readers to terrestrial mobile wireless cellular communications principles, systems and design. In particular, we study:[1]

(1) Fundamentals of mobile cellular networks and their design, including channel allocation strategies, frequency reuse distance, cluster size, mobile device and SIM card, cell site selection, acquisition and planning.
(2) Interference and power management strategies, including power control, near-far effect, ACI and CCI management strategies, and discontinuous transmissions.
(3) Coverage and capacity management, including cell splitting, sectoring, mobile base stations, segmentation and dualisation, load balancing using cell breathing.
(4) Mobility management, including paging, roaming, handoff and location updating.
(5) A review of 5G network, including OSS/BSS and architecture.

9.1 Introduction

Wireless communications enable the transfer of information without using wires. There are two types of wireless networks: satellite and terrestrial. Terrestrial wireless networks can also be either fixed wireless or mobile wireless networks. Mobile communication networks enable users to move from one location to another while actively using the services provided by the network. Mobile cellular communications networks are based on two fundamental principles: frequency reuse and frequency agility. Frequency reuse is the technique of reusing the same frequency at different geographical locations within the coverage area which are separated by a minimum distance called reuse distance or guard space. Frequency agility is the ability of a mobile unit to operate on different frequencies within given bands of the electromagnetic spectrum. This enables a mobile unit to switch from one channel to another, a process referred to as handoff.

[1] "Not everything that counts can be counted, and not everything that's counted counts" — Albert Einstein.

Mobile communications devices are now part of the citizens of the Internet-of-Things. The handsets are used to record enormous amounts of data of various forms. We use hand-held devices to take photos which are exchanged through social media, as pedometers to count our daily steps, to record speeches, etc. Further, we use mobile phones to collect bio-data, such as oxygen levels, heart rate and blood pressure. These data can be automated to be transferred over the Internet to another device. These devices also implement the language of the Internet, the Internet Protocol (IP). In fact, the 5G mobile technology being installed currently is poised to accelerate the development of IoT.

Wireless communications went through its humble beginnings of using smoke signals, through telegraph to its present state of using electromagnetic signals to provide cellular mobile services. A wireless network allowing its users to move while in active communication is referred to as a **mobile wireless network**. A mobile network run by one mobile service provider in one country is referred to as a *public land mobile network* (PLMN). The owner of the infrastructure of a PLMN is referred to as **PLMN operator** or **carrier**. A company which leases a PLMN infrastructure owned by another company to provide mobile communications services is referred to as a **mobile virtual network operator** (MVNO), **carriage service provider** (CSP), mobile other licensed operator (MOLO), or virtual network operator (VNO). There is a growing number of MVNO. Currently, the telecommunication infrastructure manufacturers (e.g., Ericsson and Huawei) manufacture, install and manage much of the infrastructure so that the service providers run their service without owning any physical infrastructure. This trend is a form of infrastructure-as-a-service (IaaS).

Most of the topics discussed in other chapters of this book are valid for mobile cellular communications as well. There are a few topics, though, which are peculiar to cellular communications. The most fundamental of them is the cellular concept. The design objectives of a mobile cellular network are: high system capacity, high bandwidth or spectral efficiency, high quality of service and high power efficiency. Assume that a given cellular mobile system has the following features: network efficiency is G, cluster size is $C = 7$, spectral efficiency is η, cell area is $A_{\texttt{cell}}$ m^2, and the system bandwidth is B Hz. The relationship between these parameters is given in [Jamalipour *et al.* (2005)] as

$$\rho = \frac{N_{\mathrm{u}}G}{A_{\texttt{cell}}[\texttt{m}^2]B[\texttt{Hz}]C} \;\; [\texttt{Erlangs/m}^2\texttt{/Hz}]\,. \qquad (9.1)$$

Mobile communications systems have the following two design choices:

(1) Mount high-power radio transmitters on tall towers so each covers a large geographical area.
(2) Mount low-power radio transmitters on short towers (even below rooftop) so each covers a small geographical area.

Option 1 used to be the norm in the early days of mobile communications. In modern times it may be used only in mobile systems servicing sparsely-populated geographical areas, or in point-to-point microwave links. Option 1 achieves a low system capacity and low radio spectrum efficiency, as there would be excessive interference if frequencies were reused unless there is thousands of kilometers separation between radio towers using the same set of frequencies. The second option underpins the **cellular concept**.

Table 9.1 summarizes the developments in mobile cellular communications technologies, by sampling some of their key features. An important feature of a mobile technology is the operating frequency. As at Q3 2018, commercial 5G systems have been released in Qatar, Lesotho, Finland, Estonia and USA (Houston, Indianapolis, LA and Sacramento) by Verizon. We can observe several developments in mobile cellular technologies from this table. First, it takes about 10 years to move from one generation of technology to another with continuing improvement within a given generation of technology. Second, the core network has changed from the PSTN/POTS (public-switched Telephone Network / Plain-Old Telephony System) to the Internet. This has been made possible by packetizing all forms of information and transferring them via packet switching instead of circuit switching. Third,

Table 9.1 Key features of mobile cellular communications technologies across generations.

Feature	1G	2G	2.5G/2.75G	3G/3.5G/3.75G	4G	5G
Radio technologies	NMT, AMPS, TACS, NTT, C450	GSM, cdmaOne (IS-95)	GPRS/EDGE	UMTS/W-CDMA, cdma2000, UMTS-TDD	LTE/LTE-Advanced, Mobile WiMAX, Wi-Fi	NR
Core network	PSTN	PSTN	PSTN	Internet	Internet	Internet
Base station name	BS	BTS	BTS	Node B	eNodeB (or eNB)	gNodeB, MBS, SBS
Release year	1979 (Japan)	1991 (Finland)	2000/2003 (Germany/US)	2001 (Japan)	2009 (Sweden)	2018 (Qatar, Lesotho, Finland, Estonia, USA)
Average handset cost	\$3, 995	\$900	\$400	\$350	\$600	?
Latency (ms)	?	500-1000	300-500	100-500	< 100	1
Spectrum (MHz)	800-900	850, 900, 1800, 1900, 825-849	850-1900	800-2500	many	3-300 GHz
Carrier spacing (kHz)	30	200, 1250	200	5000, 1250	15	$15 \times 2^k, k = 0 - 4$
Channel bandwidth (MHz)	0.03	0.2, 1.25	0.2	5, 1.25, 1.6	1.4, 3, 5, 10, 15, 20	1000 × 4G
Data rate (kbps)	1-2.4	9.6-64	64-384	384-5000	1-50 Mbps	1-10 Gbps
Killer application	Analog voice	Digital voice, text messages	Digital voice, messaging, packet data	Multimedia	All-IP based multimedia	All-IP based multi-service
Transmission	Circuit switching	Circuit switching	Packet switching	Packet switching	Packet switching	Packet switching
Multiple access	FDMA	TDMA, CDMA	TDMA, CDMA	W-CDMA	OFDM, MC-CDMA, LAS-CDMA, Network-LMDS	NOMA, MUSA, SCMA, RSMA, PDMA, GFDM
Modulation schemes	PM, FM	GMSK	GMSK, 8-PSK	QPSK, 16-QAM, 64-QAM	QPSK, 16-QAM, 64-QAM	FBMC, UFMC, f-OFDM, GFDM
Antenna technology	Passive				MIMO	Massive MIMO
Features	No security, low spectral efficiency	Low data rate, poor data services, wasteful switching	Best-effort service	Improved data services, but wanting Internet access	Improved latency, low battery life, costly hardware	Low delay, high data rate, better QOE

G: Generation of technologies; gNodeB: next-generation NodeB; BS: base station; BTS: base transceiver station MBS: macrocell BS; SBS: small-cell BS; FM/PM: frequency/phase modulation; GSM: Global System for Mobile Communications, originally Groupe Special Mobile; UMTS: Universal Mobile Telecommunications System; LTE: Long-Term Evolution; NR: New Radio; WWWW: World Wide Wireless Web; CN: core network; GFDM: Generalized Frequency Division Multiplexing; f-OFDM: Filtered OFDM we observe that the data rate increases over the generations of technologies as transmission delay (aka latency) decreases.

We can also observe that every technology coins its own terminology. From 1G through to 3G, the channel/carrier spacing increased from 30 kHz to 5 MHz to meet the needed data rate. However, from 4G this trend of increasing carrier spacing to support higher data rates has been reversed. This has been made possible by moving away from single-carrier to multi-carrier modulation techniques and carrier aggregation. Carrier aggregation in LTE-Advanced allows up to five channels to be combined to serve a single user. Given the channel bandwidth in Table 9.1, each user can be served with up to 6,15, 25,50, 75 and 100 MHz of bandwidth.

Another interesting feature in the evolution of mobile cellular networks is their architecture. The architecture of 1G and 2G networks had long chain of hierarchies. However, modern cellular technologies are increasingly adopting flatter architectures. This is obvious in Figs. 9.1 and 9.2. Cellular mobile technologies developed after 3G systems are referred to generically as Beyond 3G (B3G), and LTE is the first most popular B3G. LTE integrates the radio controller function into the eNodeB in order to achieve flat architecture. A flat architecture creates a tight interaction between the different protocol layers of the radio access network (RAN) to reduce latency and improve efficiency. Also, as a new development, 4G network allows direct communication between the eNodeBs. This eases the amount of switching and call processing delays. Such an architecture has opened the door for machine-to-machine (M2M) communications.

A mobile network has the four basic components:

(1) The mobile handset.
(2) The base station.
(3) The backhaul or fixed network.
(4) The mobile switch or gateway.

A mobile wireless network has a wireless part and a fixed part. The former is called wireless access network or air interface, while the latter is called core network or backbone network. A mobile user, through the mobile terminal, accesses a mobile network through the base station. This is illustrated in Figs. 9.1 and 9.2. The device used by mobile users to access the mobile service is called mobile handset, mobile phone, mobile host (MH), user equipment (UE), etc. The name for device which connects the wireless and fixed components of a mobile wireless network is given a different name across the generations of technologies. As listed in Table 9.1, it is referred to as radio base station (RBS) in 1G, base transceiver station (BTS) in 2G, Node B (or NB) in 3G, evolved NodeB (eNB) in 4G and next generation NodeB (gNB or gNodeB) in 5G. We will use base station (BS) as a generic name in this text.

The two directions of the wireless link connecting the user via the user terminal and the network via the base station have different names. The direction from the network (or base station) to the user is referred to as **downlink** or **forward link**. The direction from the mobile user to the network (or base station) is referred to as **uplink** or **reverse link**. The field strength in the transmission in the uplink is about the same as that in the downlink. Such a condition is referred to as **reciprocity principle**. However, as the base station and mobile station are liable to different levels of noise and interference, the received signal-to-noise rations are not reciprocal. In time-division duplexing (TDD) mode, the reciprocity assumption of the wireless link is used to estimate the channel quality.

Antennas used in cellular mobile networks are usually mounted on top of hills/mountains and skyscrapers. However, in some geographical locations, such natural and man-made heights for optimal radio sites may not be readily available. Therefore, real estate must be acquired to mount tall towers on the ground. One can observe what I call a *tower farm* in such locations. For this reason, some governments have mandated the sharing of tower sites and also the usage of towers which blend with the environment. Tower or cell-site sharing also saves telecommunications operators money. Figure 9.3 shows such an aesthetically made mobile antenna tower which the author photographed during a visit to the campus of Kwame Nkrumah University of Science and Technology (KNUST) located

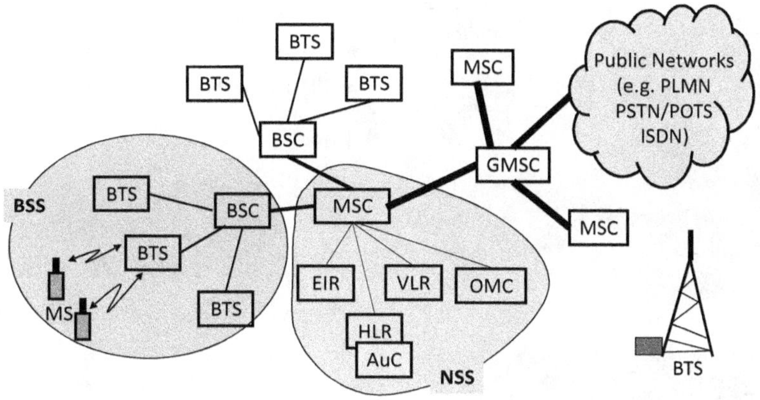

Fig. 9.1 GSM/2G network architecture.

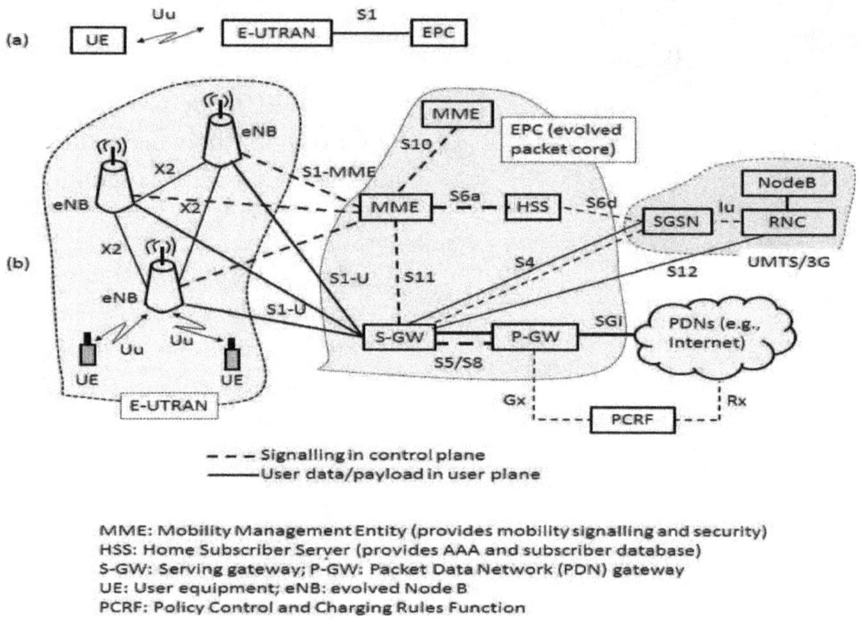

Fig. 9.2 LTE-Advanced/4G network architecture.

in the second largest city in Ghana called Kumasi. You can observe that the tower looks like a natural tree, blending easily with its environment.

An interesting thing to observe from the antenna mast in Fig. 9.3 is that multiple antennas are mounted on the tower, and some of them are at different heights. We shall discuss this again later in this chapter. A reason for mounting antennas at different heights on the same tower relates to the desired coverage. The higher the antenna from the ground, the larger its coverage. This means that the same cell site can be used to implement cells of different sizes. The idea of placing multiple antennas at the same site but at different heights is a form of **space diversity**, aka antenna diversity or spatial diversity. Antenna diversity is used to mitigate multipath-caused frequency-selective fading, by receiving and combining multiple copies of the transmitted signal. However, antenna diversity cannot be done in any arbitrary manner.

In order to ensure that the signals received at the antennas are not remarkably correlated, the multiple antennas must be separated vertically by at least 0.5λ (Fig. 9.4). A rule-of-thumb is 2λ to 8λ vertical separation if antenna polarizations are not orthogonal, and at least 0.25λ vertical separation

Fig. 9.3 An artificial tree used as mobile antenna tower to blend with its environment.

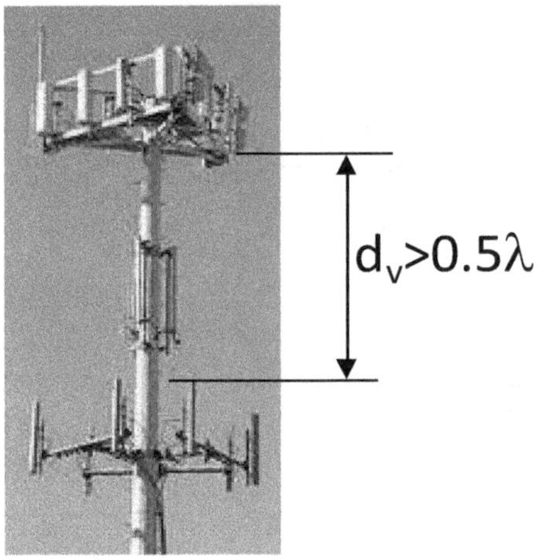

Fig. 9.4 An antenna configuration for spatial diversity.

if the antennas are additionally separated by 90° from each other to avoid polarisation/directionality effects on the radio link. The required minimum vertical separation also depends on the height of antennas above the ground level. Another rule-of-thumb is at least eight wavelengths (8λ) separation for antennas placed 30 m above ground, but eleven wavelengths (11λ) separation for antennas placed 50 m above ground level. Typical base station antenna heights are:

- 40 m to 80 m for country side roads.
- 20 m to 60 m for urban areas.
- 20 m to 30 m for suburban areas.
- 30 m to 80 m for country side areas.

The ITU-R provides the formula for vertical isolation in decibels for *vertical dipole antennas* as [ITU (2011)]

$$I_v\,[\mathtt{dB}] = 28 + 40\log(d_v/\lambda), \; d_v > 10\lambda \qquad (9.2)$$

where d_v is the vertical distance in meters between the antennas operating at the wavelength λ in meters. Given the required antenna isolation, we can find the required vertical separation between the antennas. Antenna isolation should be as large as possible to increase the antenna's efficiency. Mobile antennas sharing a ground plane as used in antenna's diversity can operate at an isolation lower than -10 dB, albeit at lower efficiency.

Of course, there are multiple reasons other than antenna's diversity to mount multiple antennas on the same tower, such as:

(1) *Sectorisation:* Antennas are mounted on the same height but separated by an angle of $360°/S$, where S is the number of sectors.
(2) *Technology co-location:* Antennas can be used for different radio technology, such as some for LTE/LTE-A, some for UMTS/WCDMA
(3) *Infrastructure sharing:* Antennas are used by multiple PLMN operators, as we now have infrastructure sharing in which the infrastructure can be owned by a third-party company.

The larger cells formed by the antennas mounted at higher levels can serve as umbrella cells to reduce the number of handovers for vehicular users. Also, the antennas located at lower level can be decommissioned when less system capacity is required, but made operational when the cell becomes congested. We can also observe that multiple antennas are mounted at the same height on the tower. Such antennas can be used to achieve cell sectoring.

Besides thermal noise, mobile cellular wireless systems are liable for several types of disturbance called interference. These include:

(1) Intersymbol interference (ISI), which also affects wired networks.
(2) Intermodulation distortion (IMD).

(3) Adjacent channel interference (ACI).
(4) Co-channel interference (CCI).
(5) Near-far-effect (NFE).

These types of disturbances are discussed in the appropriate sections of this book.

9.1.1 *Check Your Understanding*

Students are encouraged to check their understanding of the content of this section with the following questions prior to proceeding to the next section.

(1) The device which enables a user to access a 4G/LTE cellular mobile network's service is called
  (a) eNodeB (eNB)
  (b) NodeB (NB)
  (c) Next generation NodeB (gNB)
  (d) Base transceiver station (BTS)
(2) The device which enables a user to access a 5G cellular mobile network's service is called
  (a) eNodeB (eNB)
  (b) NodeB (NB)
  (c) Next generation NodeB (gNB)
  (d) Base transceiver station (BTS)
(3) The air interface standard for 5G cellular mobile networks is called
  (a) UMTS
  (b) Long-Term Evolution (LTE)
  (c) High Speed Downlink Packet Access (HSDPA)
  (d) New Radio (NR)
(4) Explain the meaning of carrier aggregation (CA) in mobile communications. Describe the benefit in using CA in LTE-Advanced, although earlier mobile technologies do not.
(5) What are the two advantages in the sharing of mobile cellular antenna tower sites among multiple operators that are mentioned in this section?
(6) What does the term *antenna tower farm* mean?
(7) **In-class research:** Why is it that 4G mobile networks are less hierarchical than 3G and 2G mobile networks?
(8) Discuss the reasons why multiple antennas are sometimes mounted at the same height on the same antenna tower.
(9) Discuss the reasons why multiple antennas are sometimes mounted at different heights at the same cell site.
(10) What type of disturbance or noise is specific to mobile cellular systems?
(11) Assume that an eNodeB uses two sets of antennas which are mounted one set above the other on the same antenna tower. What should be the minimum vertical separation between the two sets of antennas for proper system functioning?
(12) How do the names used for base station differ across 1G, 2G, 3G and 4G mobile cellular networks?
(13) Research to find at least one mobile cellular network design software with industry support.
(14) Describe the reciprocity principle as applied in mobile communications.

9.2 The Cellular Architecture

The electromagnetic spectrum is inelastic and congested, requiring efficient use of wireless bandwidth. There are many ways of achieving efficient usage of a given wireless bandwidth so that maximum number of people can be served at the same time. These include spread spectrum, single sideband and the cellular concept ([Lee (1993)], p. 181). The basic principle underlying cellular mobile networks is **channel reuse** or **frequency reuse** in which the same set of channels (made of timeslots, spread spectrum pseudo-noise sequence, or radio frequencies) are used more than once in the same mobile network. The entire geographical area to be served by a cellular network is called **coverage area**. In order to achieve frequency reuse, we must partition the coverage area into regions, each of which is referred to as a **cell**. The geographical region over which the transmissions of a base station is strong enough for a mobile station to detect defines the cell. The word *cellular* in the name *cellular communications* originated from the division of a coverage area into cells.

The cellular concept was developed by Bell Labs in the USA in 1971. While the cellular concept increases system capacity and enables lower transmit powers to prolong inter-recharge time of mobile phone's battery, it also has some drawbacks. For example, network vendors need to acquire and mount more radio infrastructure. The mobility of mobile users needs to be managed, and more signalling can be used to manage the transfer of ongoing calls between radio nodes.

Each cell in a cellular network is served by at least one radio node called base station (BS). The BS is given various names. It is called base transceiver station (BTS) in 2G/GSM system, Node B in 3G/UMTS system, evolved Node B (eNodeB or eNB) in 4G/LTE/LTE-Advanced networks and New Radio (NR) in 5G networks which are expected to be commercially available in 2020. Also various names are used for the user terminal, including mobile station (MS), mobile terminal (MT), terminal equipment (TE), mobile handset (MH) and user equipment (UE).

Cells which use the same set of channels are referred to as **co-channel cells**. There should be a sufficient geographical separation between co-channel cells to minimise a type of interference called **co-channel interference** (CCI). Neighbouring cells are thus never co-channel cells. The distance between co-channel cells is called **co-channel reuse distance** or **frequency reuse distance**. As the frequency reuse distance has the sole purpose of reducing CCI, it is also referred to as **guard space**. This compares with the guard interval and guard band used in TDM/TDMA and FDM/FDMA, respectively. This distance is measured from the centers of the co-channel cells. The main reason for adopting the cellular architecture is to achieve as high as possible spectral efficiency and **system capacity**. The smaller the reuse distance, the higher the spectral efficiency and system capacity, albeit at the cost of increased CCI. Using larger reuse distance reduces the CCI, but at the cost of lower spectral efficiency and system capacity. Thus, the reuse distance is a design parameter requiring optimisation. Channel reuse was developed to reduce the problem of limited and expensive bandwidth.

Conceptual system design and analysis traditionally uses hexagonal-shaped cells. Practical coverage of a base station, however, is far from being that shape. A realistic cell coverage shape is circular if the antenna is ideally omnidirectional.

There are many types of channels in a cellular network. One type is control channels, which also has several types. Two of them are forward control channel (FCC) and reverse control channel (RCC). A user equipment, once powered on, hooks onto the strongest control channel provided by its serving base station. It scans periodically for a better control channel to hook onto. Every base station advertises periodically its identity for user equipments in its service area to hook onto it.

How does a call reach a user equipment (UE)? The base station directs the UE to hook onto an available pair of forward traffic channel (FTC) and reverse traffic channel (RTC) so it can initiate a call. The base station also sends an alert message to the UE over the FTC to instruct the UE to ring so its user can know a call has arrived.

9.2.1 *Shape of Cells and Cell Sizes*

The shape of the cells in a cellular network is assumed to be the hexagon in the presentation of cellular systems, which is a geometrical shape with six equal faces or edges. Practical cell shape is usually far from being a hexagon. Assuming the base station uses an omni-directional antenna, a circular cell shape would appear more realistic than the hexagonal shape. Why is it that hexagons are used for illustration of cellular architecture most of the time? There are two main reasons, which are:

(1) Hexagons tessellate, leaving no coverage gaps or overlaps. For good quality of service (QoS) reason it is preferred that wireless cellular systems do not have coverage gaps. For the

reason of optimising the usage of wireless resource (i.e., spectrum) and radio infrastructure, coverage overlaps are also not the favorite of radio system designers. Thus, from a wireless user's point of view, coverage overlaps are preferred to coverage holes/gaps. However, service providers have a dilemma if they have to choose between coverage gaps (i.e., maximum resource efficiency) and coverage overlaps (i.e., maximum service quality and user satisfaction to prevent churn). Note, however, that the overlap of the footprints of multiple cell sites is necessary to increase the probability of call handoff success.

(2) A hexagon is one of the geometrical figures which covers almost the surface area of a circle/sphere. An ideal antenna, which is a theoretical construct, illuminates a spherical area with uniform intensity. An ideal omnidirectional antenna is a uniform isotropic radiator with 0 dB gain in all directions.

(3) The hexagonal shape makes systematic cellular system design and analysis simpler and tractable. The real-life amorphous shape of the coverage of base station antennas cannot be used. Figure 9.5 shows other cell shapes that are considered for the modelling and analysis of cellular communications systems.

Common shapes used for cells in cellular networks are illustrated in Fig. 9.5. The dots in the figure represent base station sites. The base station is positioned near the center of a cell in center-excited cell. We have edge-excited cell in which the three base stations are positioned at three of the size corners of a hexagon. Edge-excited cell sites use directional antennas to sectorise the cells, while omnidirectional antennas are suited for center-excited cells.

Cell sizes in a given cellular network do not have to be and are usually not the same. The sizes of cells serving a geographical area do not have to be the same, and in practice that is the case. This is illustrated in Fig. 9.7. The signal strength contours indicate the actual coverage of a cell. The actual cell's coverage, referred to as *footprint*, depend on many factors, including:

(1) Transmit power and receiver sensitivity (aka receiver threshold or RF sensitivity).
(2) Antenna height, antenna gain and antenna radiation pattern.
(3) Obstacles in the propagation path.
(4) Atmospheric attenuation.
(5) Terrain or topography (i.e., dimensions) of obstructions in the coverage area, such as dessert, hill, mountain, open area, oasis, forest and water.
(6) Morphology of coverage area, e.g., shape, form, structure and distribution of obstructions.

The capacity or number of users to be served in a given geographical area. Smaller cells are used as the demand for capacity increases. The cells of a cellular system can be divided into multiple cells in order to increase system capacity. The name for such a process is **cell splitting**.

Cells are classified based on their sizes into:

- *Macro-cell*: also written as macrocell, is usually used in sparsely-populated areas, remote areas or for point-to-point microwave links. The radius of a macrocell can be in the range 1 to 3 km (for small macrocell) and 3 to 40 km (for large macrocell).
- *Micro-cell*: also written as microcell, is usually used in densely populated areas requiring a high network capacity. Typical radius for microcells is in the range 200 m to 2 km. A base station with a microcell coverage can transmit powers below 20 mW, and can be mounted at rooftop height.
- *Pico-cell*: also written as picocell, is usually used to cover small areas, such as a hotspot or a room in buildings. Typical radius for a picocell is less than 4 m to 200 m. A base station with a microcell coverage can transmit powers below 20 mW, and can be mounted below rooftop.
- *Femto-cell*: was originally coined to describe indoor cellular coverage such as small business offices and homes. For this reason, some people call it personal base station coverage. Femtocells are installed by users themselves, but not by the telecom service provider. This reduces costs for both stakeholders. Thus, femtocells in cellular networks function almost like the PABX/PBX in

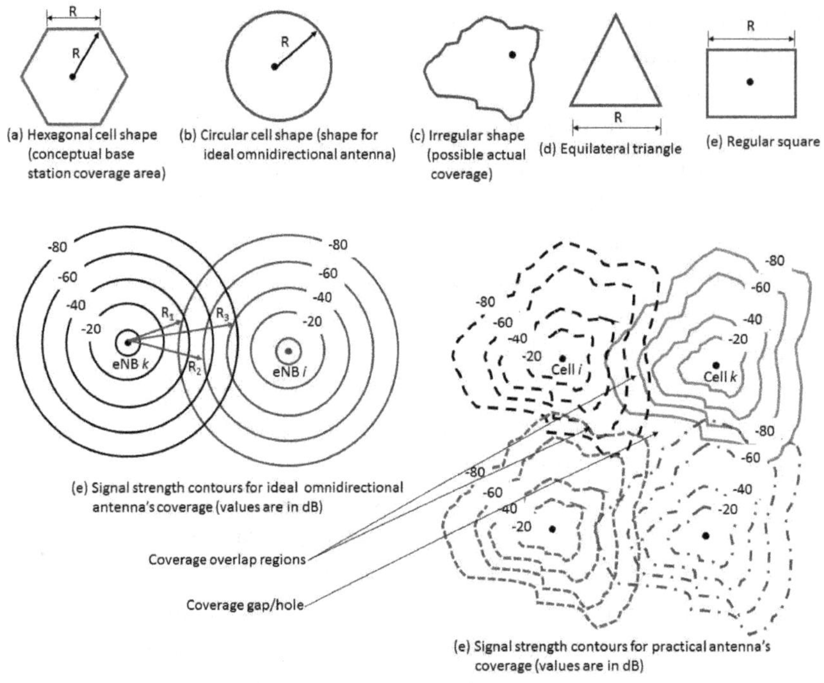

Fig. 9.5 Shapes and coverage patterns of cells in mobile cellular networks.

the wired telephone network, as they can offload traffic from the main cellular network. Typical radius for a femtocell is less than 4 m.

- *Umbrella cell*: usually covers several microcells to support vehicular users using the mobile system while moving at fast speeds. Communicating over microcell zones while moving at high speed causes frequent handoff, something which increases the probability of forced call termination. Using only macrocells may not provide enough capacity. This is the reason why large cells called umbrella cells are overlaid on microcell areas. An umbrella cell uses a set of channels different from those of the microcells within it to minimise interference.

These types of cells are illustrated in Fig. 9.6. Note, however, that the figure is not drawn to scale. Also, some macrocells in 1G mobile networks could have radius as large as 40 km. Mobile stations in 2G cellular mobile system could transmit powers in the range 0.8 W to 20 W. In modern cellular systems, however, the average transmit power of mobile devices is about 250 mW (about 24 dBm), while eNodeB transmit powers are in the range 46 dBm.

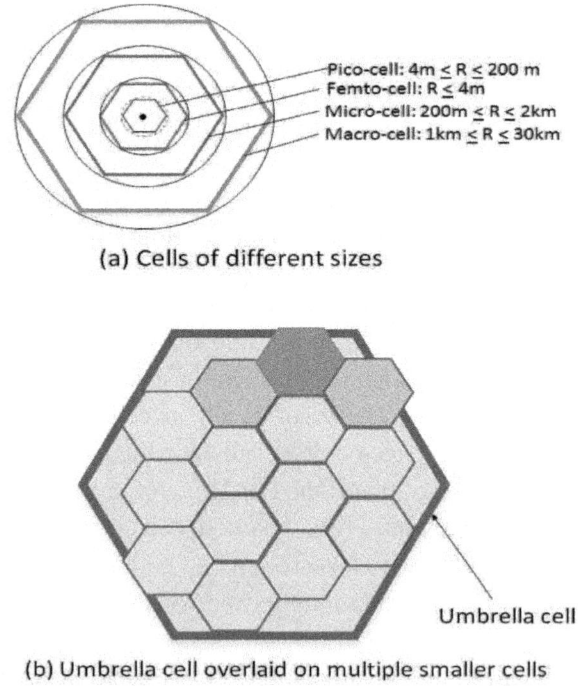

Fig. 9.6 Types of cell sizes used in mobile cellular networks (R is cell radius).

Example: Hexagonal Geometry

Find the dimensions of the six sides and the area of the hexagon enclosed in a circle of radius R.

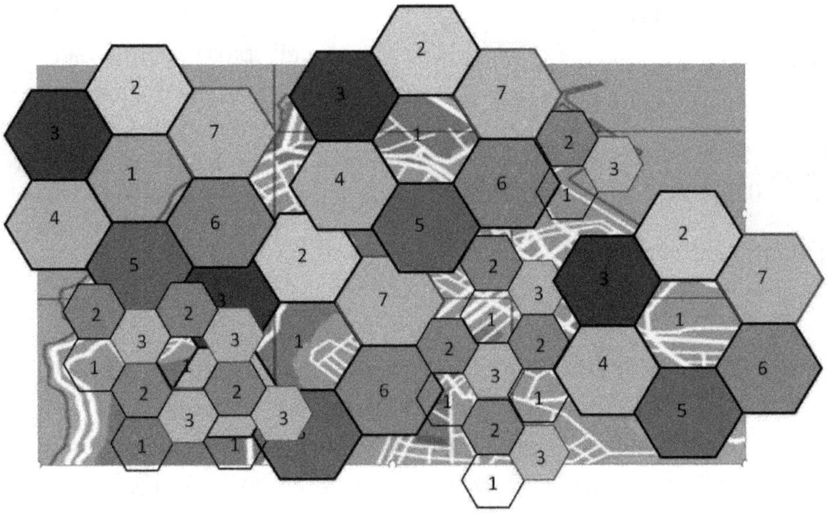

Fig. 9.7 Overlay of frequency reuse plan over the map of a cellular coverage area.

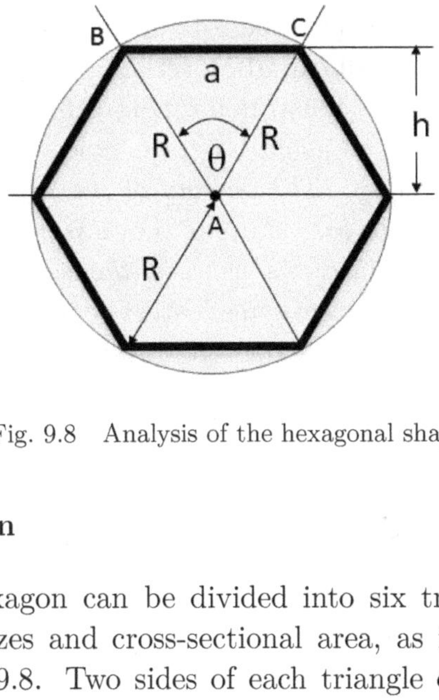

Fig. 9.8 Analysis of the hexagonal shape.

Solution

The hexagon can be divided into six triangles of equal sizes and cross-sectional area, as illustrated in Fig. 9.8. Two sides of each triangle equals the radius R of the circle enclosing the hexagon. The third side can be obtained from the cosine law or sine law. Applying the cosine law on the triangle ABC in Fig. 9.8, we obtain

$$a^2 = R^2 + R^2 - 2R \cdot R\cos\theta = R^2 \Leftrightarrow a = R. \quad (9.3)$$

We have noted in Eq. (9.3) that the side of a triangle cannot be a negative value, and that $\theta = 60°$ and thus $\cos\theta = 0.5$. This means that each of the three sides of triangle ABC is equal to R (i.e., equilateral triangle), which is the same as each side of the hexagon.

Assume that the area of each of the six equilateral triangles making up the hexagon is A_t, and the area of the hexagon is A_h. The height of triangle ABC is $h = R\cos 30° = \frac{\sqrt{3}}{2}R$. The area of triangle ABC is $A_t = \frac{1}{2} \times a \times h = \frac{\sqrt{3}}{4}R^2$. Now, it is easy to observe from simple geometry that the hexagon's area is

$$A_h = 6A_t = 6 \times \frac{\sqrt{3}}{4}R^2 = \frac{3\sqrt{3}}{2}R^2. \quad (9.4)$$

Example: Number of Cell Sites for a Coverage Area

The size of Melbourne city in the Victorian State of Australia is about 9,990 square kilometers. Assume that a telecommunications operator called Telstra has 100% mobile cellular coverage of entire Melbourne, what is the maximum number of cell sites that Telstra needs to acquire if it uses (a) micro-cells, each of radius 500 m? (b) macro-cells, each of radius 20 km? Assume that each cell has a hexagonal shape.

Solution

The surface area of a hexagon was computed in Eq. (9.4) as $A_h = \frac{3\sqrt{3}}{2}R^2$. Thus, the area of a microcell of radius $R = 500$ m is $A_{mi} = \frac{3\sqrt{3}}{2}R^2 = \frac{3\sqrt{3}}{2} \times 0.5^2 \approx 0.6495$ $\texttt{km}^2$. The area of a macrocell of radius $R = 20$ km is $A_{ma} = \frac{3\sqrt{3}}{2}R^2 = \frac{3\sqrt{3}}{2} \times 20^2 \approx 1,039.23$ $\texttt{km}^2$. Therefore, the number of microcell sites needed to cover the entire Melbourne is

$$N_{mi} = \frac{A}{A_{mi}} = \frac{9990}{0.6495} \approx 15,381 \qquad (9.5)$$

while the number of macrocell sites needed for the same coverage area is

$$N_{ma} = \frac{A}{A_{ma}} = \frac{9990}{1,039.23} \approx 10\,. \qquad (9.6)$$

Clearly, we can observe the difference between using more sites for microcells than macrocells. However, if the capacity needed is high, the operator has no choice than to use microcells, or sometimes even picocells. We also realise that an operator can use a few cell sites to cover a huge geographical area so long as the propagation environment allows signals to travel that far.

Mobile networks are classified into **coverage-limited networks** and **capacity-limited networks**. The capacity (i.e., number of simultaneous users) of a coverage-limited network is so small that the size of the cell can be made as large as propagation losses allow. The sizes of cells in a capacity-limited network, however, are made as small as possible to support the needed high capacity. Sometimes cell partitioning, sectoring or mounting of temporary cells are needed to support the capacity. So, in principle, cell sizes in coverage-limited networks are larger than those in capacity-limited networks.

9.2.2 *Cell Planning and Cluster Size*

The procedure used to to divide a given bandwidth into channels, divide a geographical area to be served into cells, and assign a set of channels to each cell to keep the CCI and ACI (adjacent channel interference) within a cluster within acceptable limits is called **frequency planning** or **cell planning**. The set of cells using all the channels available in the cellular network is called cluster size. Coverage gaps or coverage holes are locations within the coverage area where there is no mobile service in the network. Cell can also overlap. Coverage holes and overlaps are illustrated in Fig. 9.5(e). The cell planning aims to avoid coverage holes and reduce coverage overlap. *The overlap of the footprints or coverages of two base stations is necessary for handoff of communications between the base stations.* Table 9.2 summarises some features of coverage hole (aka blackspot) and coverage overlap.

Table 9.2 Coverage hole and coverage overlap and their features.

Feature	Some Causes	Features
Coverage hole	Poor network planning	Poor or no mobile service
	Terrain-related obstructions in region	Call dropouts
	Non-optimal antenna angle and tilt	Noise during communication
	Mobile is too far away from base station	Difficult hearing callee
		Very variable signal quality
		Slow connection during data transfer
		Increased mobile battery drain
		Connection timeouts
Coverage overlap	Crosstalk	Good for handoff
	CCI, inter-cell interference	Increased CCI
	Intentional for effective handoff	Requires more cell sites and thus increases costs to service provider

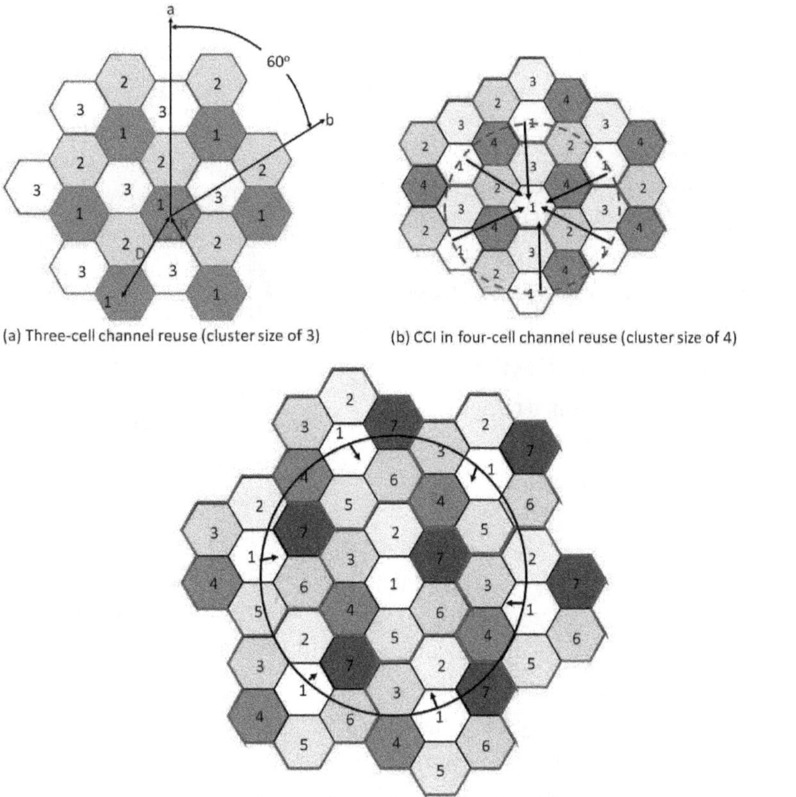

(a) Three-cell channel reuse (cluster size of 3)

(b) CCI in four-cell channel reuse (cluster size of 4)

(c) Seven-cell channel reuse (cluster size of 7)

Fig. 9.9 Frequency plans for cellular architectures in cluster sizes of 3, 4 and 7.

The steps in frequency planning are:

(1) Settle on a cluster size.
(2) Decide on channel bandwidth and frequency allocations.
(3) Assign frequencies to cells in a cluster.

Frequency planning is done under two constraints: minimise CCI between clusters, and minimise ACI within a cluster. The cluster size C is not chosen arbitrarily. It is determined from the relationship

$$C = i^2 + ij + j^2,\ i \geq 0, j \geq 0\,. \tag{9.7}$$

The distance between co-channel cells is referred to as channel **reuse distance**, and denoted D in Fig. 9.9. It can be verified from the hexagonal geometry that

$$D = \sqrt{3C}R = \sqrt{3(i^2 + ij + j^2)}R\,. \tag{9.8}$$

Figure 9.9 shows cellular architectures of a system using cluster sizes of $C = 3$, 4 and 7. The pattern formed by the hexagonal-shaped cells, as shown in Fig. 9.9, is referred to as **honeycomb** in reference to the prismatic wax cells built by honey bees to store their larvae, honey and pollen. Current cellular networks commonly use $C = 3, 4, 7, 12, 21$. We can observe that the reuse distance increases as the cluster size increases. We define the **co-channel reuse ratio** (CRR), which is also referred to as **channel reuse ratio**, as

$$q = \frac{D}{R} = \sqrt{3C} = \sqrt{3(i^2 + ij + j^2)}\,. \tag{9.9}$$

We also define the **frequency reuse factor** (FRF) as

$$FRF = \frac{1}{C} = \frac{1}{i^2 + ij + j^2},\ i \geq 0, j \geq 0\,. \tag{9.10}$$

Note that some people use the term frequency reuse factor as synonymous to the cluster size. The set of co-channel cells nearest to a given cell is called first-tier or tier-one co-channel cells. Those are the co-channel cells with the strongest CCI. For that reason, we give special attention to the tier-one co-channel cells. They can be located by following the simple procedure:

(1) Move i cells along any series of cells.
(2) Turn 60° in anti-clockwise direction.
(3) Then move j cells.

Assume that a cellular mobile service provider has acquired from the local radio frequency regulator a total bandwidth of $B = 20$ MHz and the cellular technology uses channel/carrier spacing of $W = 15$ kHz as with LTE-Advanced. Then disregarding any guard bands, the total number of carriers available to the service provider is $B/W \approx 1333$ simplex channels. A pair of channels (called duplex channel) is needed for two-way communications. The total number of duplex channels in the system is

$$N_{\mathsf{ch}} = \frac{1}{2} \times \frac{B}{W} \approx 666\,. \tag{9.11}$$

Without the cellular architecture the total network capacity (i.e., maximum number of users at a time) would be 666. Assume that the cellular system uses the cluster size of C and that the ith cell in a cluster is given N_i duplex channels. Then we have

$$N_{\mathsf{ch}} = \sum_{i=1}^{C} N_i = 666\,. \tag{9.12}$$

Note that the number of duplex channels N_i allocated to each cell in a cluster does not have to be the same. If we repeat the cluster $N_{\mathtt{cl}}$ times in order to fully cover the coverage area, then the system capacity increases from $N_{\mathsf{ch}} = 666$ to

$$N_u = N_{\mathtt{cl}} N = N_{\mathtt{cl}} \sum_{i=1}^{C} N_i\,. \tag{9.13}$$

If the same number of duplex channels are allocated to the cells in a cluster, then Eq. (9.13) becomes

$$N_u = N_{\mathtt{cl}} C N_i\,. \tag{9.14}$$

Therefore, the capacity of the cellular network is directly proportional to the number of clusters $N_{\mathtt{cl}}$ used to cover the coverage area. Geographical areas requiring low network capacity, such as lowly-populated areas are served with a larger cluster size than densely-populated geographical areas. This is illustrated in Fig. 9.7. A trade-off between total system capacity and interference level must be made in cellular network design. The choices are:

(1) *Small cluster size for higher capacity*: we can use a smaller cluster size C to boost the system capacity while keeping the size of cells intact. This is because reducing the cluster size while keeping cell size fixed requires more clusters to cover the given coverage area. As Eq. (9.14) shows, increasing the number of clusters $N_{\mathtt{cl}}$ used to cover a given area increases the overall system capacity.
(2) *Large cluster size for low interference*: a small co-channel interference results in a better system performance. According to Eqs. (9.9) and (9.8), increasing the cluster size increases the reuse distance to reduce the co-channel interference. However, using a larger cluster size reduces the number of clusters needed to cover a given geographical area, which reduces the overall system capacity.

9.2.3 *Cell Site Selection, Acquisition and Planning*

The base station can be positioned at the center of the cell, leading to center-excited cell, as shown in Fig. 9.10(a). The base station can also be located at the edge of the cell, leading to edge-excited cell, as shown in Fig. 9.10(b). The base station can also be placed at the corner of the cell, leading to corner-excited cell, as shown in Fig. 9.10(c).

The radio propagation environment around antennas usually has a nonuniform terrain. For that reason the radiation pattern of a base station's antenna has an amorphous shape illustrated in Fig. 9.5(c) and (e). The first step in cell design is to determine the expected minimum received signal strength (RSS) at cell fringes. We then determine the size of the cell using the minimum RSS, the transmit power, the local terrain, antenna gain and antenna tower height. We need to select and

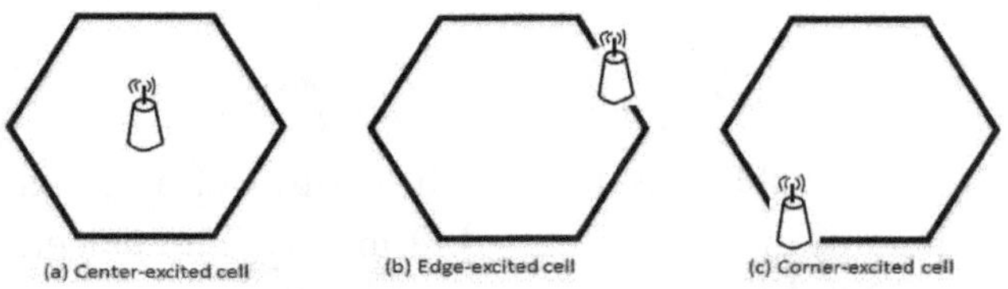

Fig. 9.10 Three common positions of a base station in a cell.

acquire all the cell sites needed for a system. Remember that for various reasons, such as costs, the optimal cell site may not be acquirable.

9.2.4 *Co-Channel Interference and Reuse Distance*

A cellular communications network is liable to a special form of disturbance called co-channel interference (CCI). CCI occurs in only wireless cellular networks owing to frequency reuse. Like all other communications systems, cellular networks are also prone to background noise, man-made noise and atmospheric noise. A network in which the non-interference type of noise dominates interference is referred to as **noise-limited network**. This can be the case, for example, in a point-to-point microwave link which is used by only a pair of devices. Cellular networks are usually **interference-limited networks** as CCI and ACI outweigh background noise. Again, the main cause of CCI is the frequency reuse technique, a mechanism for a higher spectral efficiency.

Depending on the size of the coverage area and cluster size used, a cellular network can use more than two clusters. Such a situation results in more than two sets of co-channel cells. However, the CCI in a given cell is dominated by the CCI from the first tier of co-channel cells. This is illustrated in Fig. 9.11.

In the following assume that all base stations transmit the same power P_T, sizes of cells in a cluster are the same, all co-channel base stations are at the same distance D away from the mobile of interest, and that radio propagation effects is uniform in the coverage area with the constant path loss exponent γ. There are many different types of path loss models depending on the radio propagation environment, as discussed in the chapter on wireless channels and radio propagation. For simplicity, we adopt here the path loss model

$$P_R = P_0\left(\frac{d_0}{d}\right)^{\gamma} \tag{9.15}$$

where P_R is the received power at a distance d away from the transmitter and P_0 is the received power at a close-in reference distance d_0 in the far-field region of the transmitting antenna. Let I_k be the power level of kth interfering base station at the reference mobile station. Therefore, using Eq. (9.15) we obtain the expression

$$I_k = P_0\left(\frac{d_0}{D}\right)^{\gamma}, \quad k = 1, 2, \ldots, K \tag{9.16}$$

where K is the number of tier-one co-channel cells. The least power received by the reference mobile station occurs if it is located at the edge of the cell which is at the distance R away from the center of the cell, resulting in the power level

$$P = P_0\left(\frac{d_0}{R}\right)^{\gamma}. \tag{9.17}$$

We therefore obtain the signal-to-interference power ratio (SIR), which is also referred to as carrier-to-

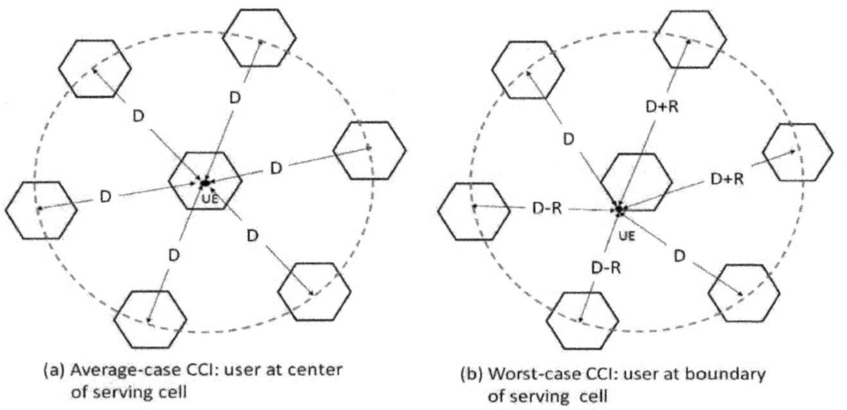

Fig. 9.11 First-tier co-channel cells and CCI.

interference (C/I)

$$SIR = \frac{P}{N_0 + \sum_{k=1}^{K} I_k} \approx \frac{P}{\sum_{k=1}^{K} I_k} = \frac{1}{K} \frac{P_0\left(\frac{d_0}{R}\right)^{\gamma}}{P_0\left(\frac{d_0}{D}\right)^{\gamma}}$$
$$= \frac{1}{K}\left(\frac{D}{R}\right)^{\gamma} = \frac{q^{\gamma}}{K} = \frac{(3C)^{\frac{\gamma}{2}}}{K}. \quad (9.18)$$

We have neglected the background noise term N_0 in Eq. (9.18) as it is usually relatively small compared to the CCI. We could also derive Eq. (9.18) by using instead the following formulas for the signal of interest and those of the interfering signals

$$P = c\frac{P_T}{R^{\gamma}}, \; I_k = c\frac{P_T}{D^{\gamma}}, \; k = 1, 2, \ldots, K \quad (9.19)$$

where c is a radio propagation constant. From Eq. (9.18) we can express the frequency reuse distance in the form

$$D = (K \times SIR)^{\frac{1}{\gamma}} R. \quad (9.20)$$

A cellular network design which meets acceptable CCI level must consider both Eqs. (9.8) and (9.20) in selecting the appropriate frequency reuse distance as

$$D = \max\{\sqrt{3C}R, \, (K \times SIR)^{\frac{1}{\gamma}} R\} \quad (9.21)$$

where $C = i^2 + ij + j^2$ for positive integers i and j.

Example

A cellular system which uses microcells of radius 500 m and the cluster size of 3 requires the minimum signal-to-interference ratio (SIR) of 24 dB.

(1) Compute the average reuse distance D if the path loss exponent is 2 or 4.
(2) Compute the worst-case reuse distance D for path loss exponents of 2 and 4.

Assume in each case that the CCI far outweighs all other types of noise in the system, and that no sectoring is used in the system.

Solution

We are given $SIR = 24$ dB, which is about 251.2 in ratio form. There are $K = 6$ co-channel interfering cells in a cluster size of $C = 3$ system without sectoring, as observed in Fig. 9.11(a). Let SIR_{γ}, $\gamma = 2, 4$ be the SIR for path loss exponents 2 and 4, respectively. Then, from Eq. (9.18), we obtain

$$SIR_{2a} = \frac{1}{6}\left(\frac{D}{R}\right)^2 = \frac{1}{6}q^2 \stackrel{!}{=} 251.2,$$
$$SIR_{4a} = \frac{1}{6}\left(\frac{D}{R}\right)^4 = \frac{1}{6}q^4 \stackrel{!}{=} 251.2 \quad (9.22)$$

for the average reuse distance case, and

$$SIR_{2w} \approx \frac{1}{6}\left(\frac{D-R}{R}\right)^2,$$
$$SIR_{4w} \approx \frac{1}{6}\left(\frac{D-R}{R}\right)^4 \quad (9.23)$$

for the worst-case situation. In the worst-case, the mobile is closer at the edge of its serving cell and thus distance R closer to the interfering co-channel cells, while in the average case it is assumed located at the middle of its serving cell. Plugging in the given values, we obtain, respectively

$$D_{2a} = \sqrt{6 \times SIR_{2a}}R$$
$$= \sqrt{10^{2.4} \times 6}R \approx 19.41 \text{ km } (a)$$
$$D_{4a} = (6 \times SIR_{4a})^{\frac{1}{4}} R$$
$$= (10^{2.4} \times 6)^{\frac{1}{4}} R \approx 3.1 \text{ km } (b) \quad (9.24)$$
$$D_{2w} = D_{2a} + R \approx 19.91 \text{ km } (c)$$
$$D_{4w} = D_{4a} + R \approx 3.6 \text{ km } (d)$$

We observe that the reuse distance is smaller for larger path loss exponent, as the strength of signals reduces faster as the path loss exponent increases. Also, we observe that the reuse distance is larger in the worst-case CCI than in the average-case CCI, as expected.

Example: SINR Computation

Compute the SINR (i.e., signal-to-interference-plus noise ratio) in a cellular system using cluster size of 3. Assume that the path loss exponent is 3.5 for radio propagation in all positions in the coverage area, temperature around the mobile user is 29°C, 5000 m cell radius, 20 MHz channel bandwidth, transmit power of each base station is 30 W and received power 2500 m away from each base station is 240 mW.

Solution

The Boltzmann's constant is $k \approx 1.38 \times 10^{-23}$ J/K. Temperature of $T = 29°C$ is equivalent to 302 K. Thus, the thermal noise power is $N_0 = kTB = 1.38 \times 10^{-23} \times 302 \times 20 \times 10^6 \approx 8.34 \times 10^{-14}$ W or -100.8 dBm. The frequency reuse distance is $D = 5\sqrt{3 \times 3} = 15$ km. The power of the base station of interest at the mobile station is

$$P = P_0\left(\frac{d_0}{R}\right)^{3.5} = 0.24\left(\frac{2500}{5000}\right)^{3.5} = \frac{0.24}{2^{3.5}} \approx 21.2 \texttt{ mW}.$$

The power of each interfering base station at the mobile station is

$$I_k = I = P_0(d_0/D)^{3.5} = 0.24(2500/15000)^{3.5} = \frac{0.24}{6^{3.5}} \texttt{ W}, k = 1, 2, \ldots, 6.$$

Therefore, the SINR is

$$SNIR = \frac{P}{N_0 + \sum_k I_k} = \frac{P}{N_0 + 6I} = \frac{0.24 \times 2^{-3.5}}{8.34 \times 10^{-14} + 6 \times 0.24 \times 6^{-3.5}} \approx 7.8$$

or 8.92 dB. We observe that the power in the background thermal noise N_0 is much smaller (by more than six orders of magnitude) than the co-channel interference. This is the reason why mobile cellular networks are interference-limited systems, and the effect of thermal noise is often neglected.

We note that Eq. (9.18) is valid only for the average case CCI in which the mobile user is located at the center of its serving cell, as illustrated in Fig. 9.11(a). As illustrated in Fig. 9.11(b), the worst-case CCI occurs if the mobile user is located at a fringe of the serving cell. In which case the SIR for a seven-cell clustering is

$$SIR = \frac{R^{-\gamma}}{2(D-R)^{-\gamma} + 2D^{-\gamma} + 2(D+R)^{-\gamma}} = \frac{1}{2(q-1)^{-\gamma} + 2q^{-\gamma} + 2(q+1)^{-\gamma}} \quad (9.25)$$

where we have used the definition $q = D/R$. Common values of SIR used in practical systems lie in the range 13 dB to 18 dB. However, the 2G technology GSM, for example, works fine with 7 dB SIR but uses frequency hopping to reduce interference. Seven decibel SIR can use cluster size as low as three if the path loss exponent is four or worse.

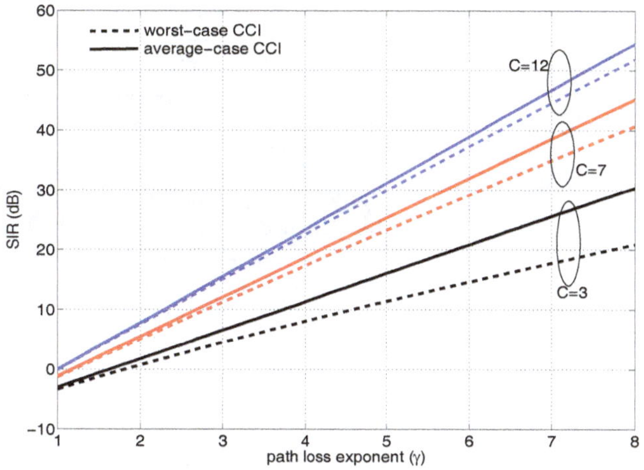

Fig. 9.12 Variation of SIR with path loss exponent γ and cluster size C.

The SIR for both average-case CCI and worst-case CCI according to Eqs. (9.18) and (9.25) have been plotted in Fig. 9.12 as a function of the path loss exponent (γ) and the cluster size C. We observe from the figure that the SIR increases with increasing cluster size. However, as noted above, increasing the cluster size reduces the network capacity, requiring an optimised tradeoff between capacity and maximum acceptable CCI. The SIR also increases with the path loss exponent. This means that we can make the cells smaller if the coverage area is poor for radio signal propagation. Such a situation can lead to a coverage-limited mobile network. We also observe from Fig. 9.12 that, as the path loss exponent and cluster size increase, the difference between the average-case SIR and worst-case SIR increases. The difference is just 1.0409 dB at $\gamma = 2$ (i.e., free-space path loss case) to about 9.5 dB at $\gamma = 8$.

Example

Typical path loss exponents used for mobile cellular system design are:

$$\begin{aligned} &\texttt{indoor line-of-sight} \\ &\texttt{radio propagation}: 1.6 \le \gamma \le 1.8 \\ &\texttt{free-space radio propagation}: \gamma = 2.0 \\ &\texttt{urban areas radio propagation}: 2.7 \le \gamma \le 3.5 \\ &\texttt{suburban areas} \\ &\texttt{radio propagation}: 3.0 \le \gamma \le 5.0 . \end{aligned} \quad (9.26)$$

Table 9.3 SIR versus path loss and cluster size case study.

Environment	γ	Path loss (dB/dec)	SIR_{ac} (dB)	SIR_{wc} (dB)
Indoor LOS	1.6–1.8	16–18	2.8–4.12	2.5–3.76
Free-space	2.0	20	5.44	5.01
Urban areas	2.7–3.5	27–35	10.07–15.36	9.36–14.25
Suburban areas	3.0–5.0	30–50	12.05–25.27	11.21–23.23

LOS: line-of-sight, γ: path loss exponent, dec: decade, SIR_{ac}: average-case CCI, SIR_{wc}: worst-case CCI.

Use Fig. 9.12 to compute the range of SIR achievable in each of the four types of radio propagation environment above for cluster size of 7. Comment on your results.

Solution

The results are compiled in Table 9.3. All the values were read off Fig. 9.12.

As CCI is the most important disturbance in mobile cellular communications systems, various mechanisms have been developed to keep it under acceptably low levels. Strategies used include:

- Increasing the reuse distance, as discussed above.
- Use directional instead of omnidirectional antennas.
- Tilt antenna beams.
- Reduce the height of antenna.
- Chose different sites/locations to mount base stations.

Depending on the operating environment, some of the above strategies may be easier or cheaper to implement than the rest. For example, acquiring a new site to mount a base station can be very challenging in some geographical locations.

Example: Frequency Reuse Distance

A mobile cellular network requires a minimum SIR of 18 dB to meet its minimum performance. A path loss exponent of 3.5 is applicable in the service area using a cluster size of 3. If the cell radius used is 1 km, what should be the frequency reuse distance?

Solution

Given parameters in the question are: $R = 1$ km, $C = 3$, $\gamma = 3.5$, $SIR = 18$ dB $= 10^{1.8} \approx 63.1$.

The formula relating the reuse distance D, the cluster size C and the cell radius R is given in Eq. (9.8) as

$$D_1 = \sqrt{3C}R = \sqrt{3 \times 3} \times 1 \text{ km} = 3 \text{ km}.$$

Another formula relating the reuse distance D, the signal-to-interference ratio (SIR), the signal propagation path loss exponent γ and the cell radius is given in Eq. (9.20) as

$$\begin{aligned} D_2 &= (K \times SIR)^{\frac{1}{\gamma}} R \\ &= (6 \times 10^{1.8})^{\frac{1}{3.5}} \times 1 \text{ km} \approx 0.29 \text{ km}. \end{aligned}$$

The minimum reuse distance which will fulfil the SIR requirement is

$$D = \max\{D_1, D_2\} = 3 \text{ km}.$$

Using $D = 0.29$ km would cause the SIR to fall below 18 dB.

9.2.5 *Adjacent Channel Interference*

Adjacent channel interference (ACI) is the interference between channels using neighbouring frequencies in the frequency spectrum. Sources of ACI include imperfect filters in the radio receivers which allow spurious emissions from adjacent frequencies to leak into the passband of other channels. Another source of ACI is insufficient frequency separation between frequencies being used at close geographical separation. ACI can be reduced by using appropriate filtering at the receiver. Another ICI-reduction method is the use of large enough guard bands between frequencies being used in neighbouring cells. There are two types of ACI: in-band ACI and out-of-band ACI. Logically, attempting to reduce in-band ACI through filtering would also filter out the signal of interest as both use the same frequency band. Out-of-band ACI, however, can be reduced using appropriate filtering, as well as applying the techniques used to reduce CCI.

Example: Frequency Planning

Assume an operator has acquired a 20-MHz band to roll out a mobile cellular network based on LTE-Advanced technology using a cluster size of $C = 4$. The bandwidth of each channel is 1.4 MHz. Assume that one channel is used for signalling. A guard band of 150 kHz is needed to keep the ACI to an acceptable level. Place the traffic channels in the cells such that both the adjacent channel interference (ACI) within a cluster and CCI between clusters are minimised.

Solution

The bandwidth of each channel is given as $W = 1.4$ MHz. The total system bandwidth is $B = 20$ MHz. Let the total number of channels in the network be k. We need $k - 1$ guard bands between the k channels. Thus, we obtain k from the relationship

$$1.4k + (k-1) \times 0.15 = 20 \Leftrightarrow k = 13\,.$$

One of the channels must be reserved for signalling. This means that we have $N_\mathrm{T} = 12$ traffic channels. With the cluster size of $C = 4$, each cell receives $n = N_\mathrm{T}/C = 12/4 = 3$ traffic channels. Let us number the traffic channels as

$$f1, f2, f3, \ldots, f12\,.$$

We divide these 12 channels into the four channel groups $G0$, $G1$, $G2$ and $G3$ as

$$\begin{aligned} G0 &= \{f1,\ f5,\ f9\} \\ G1 &= \{f2,\ f6,\ f10\} \\ G2 &= \{f3,\ f7,\ f11\} \\ G3 &= \{f4,\ f8,\ f12\}\,. \end{aligned} \tag{9.27}$$

Figure 9.13 shows the frequency plan. In order to further minimise ACI, the serving base station using channel group G0, for example, can avoid assigning f1 to a mobile station closer to the cell using the channel group G1 as they have adjacent frequencies. If the transmissions can follow the mask shown in Fig. 9.13(b) then there will be no ACI as the neighbouring channels overlap only within the guard bands. However, the illustration in Fig. 9.13(b) is the ideal case in which the adjacent channels are orthogonal, which is far from the truth in practice. Using a large guard band results in a low ACI, but at the cost of poor bandwidth utilisation.

There are a number of strategies used to reduce ACI, including:

(1) Use band-pass filters with sharp transition at the receiver to filter out out-of-band ACI. Some receivers use at their frontend filters with a high Q factor (i.e., high selectivity) to block ACI.
(2) Avoid using adjacent channels in adjacent cells. This avoids the so-called secondary source of ACI. However, this is only possible if the cluster size is large. We observe in Fig. 9.13(a), for example, that this is not possible as the cluster size is only 3.
(3) Assign adjacent channels to different cells to avoid the so-called primary source of ACI, as done in Fig. 9.13(a). This is referred to as channel interleaving.
(4) Different modulation schemes have different out-of-band radiation patterns. Using those with small energy spillout of their passbands (i.e. out-of-band radiation) reduces ACI. This is the reason why MSK (minimum shift keying) is used in GSM, for example.
(5) We can separate the uplink transmissions from downlink transmissions using TDD (time-division duplexing) or FDD (frequency-division duplexing).
(6) We can control the transmit power of mobile devices via the base station to reduce ACI.

The above strategies can be summarised into three: appropriate filtering, geographical separation and frequency separation.

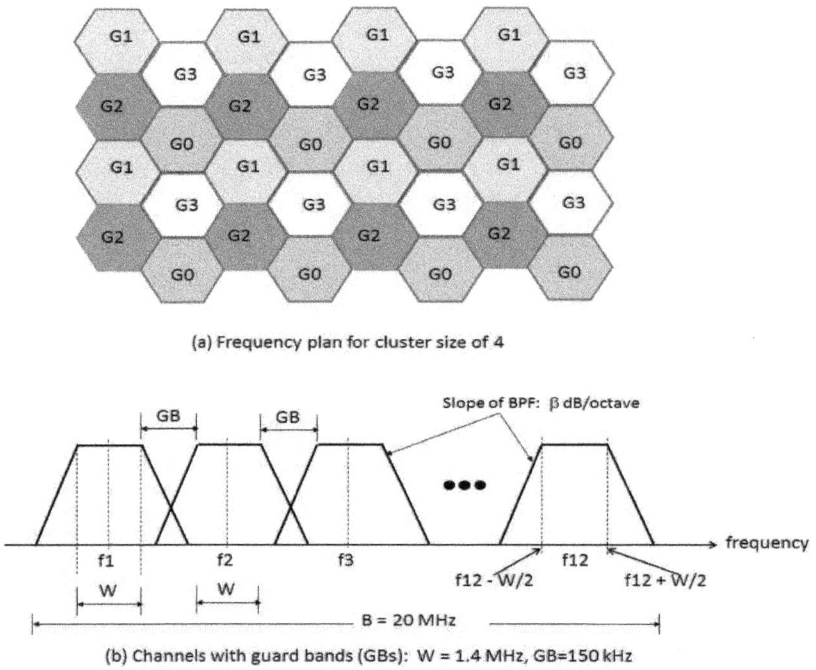

Fig. 9.13 Frequency planning for minimum ACI and CCI (fk, $k = 1, 2, \ldots$ are carrier frequencies).

9.2.5.1 *Reducing ACI by Frequency Separation and Filtering*

Let us reconsider Fig. 9.13(b). Assume that the bandpass filter (BPF) used at the radio node's intermediate frequency has the roll-off of β dB/octave, as shown in Fig. 9.13(b). Assume that the signal of interest is using the carrier frequency (or center frequency) of $f1$. Let us consider another signal transmitted at the carrier frequency $f2$. As the BPF reduces the power in the signal transmitted at $f1$ outside its designated passband by β dB/octave, the loss (or isolation) in power at $f2$ is

$$L_{fs} = \beta \log_2 \frac{f2}{f1} \approx 3.32\beta \log_{10} \frac{f2}{f1}. \qquad (9.28)$$

Example: Channel Separation

Assume that a mobile cellular system uses channels of bandwidth $W = 1.4$ MHz and the roll-off of the bandpass filter used at the radio base station is 12 dB/octave. If the desired power isolation between neighbouring frequencies is 36 dB, compute the required separation between neighbouring channels.

Solution

Channel bandwidth is $W = 1.4$ MHz, required power isolation between neighbouring frequencies is $L_{fs} = 36$ dB, and the BPF's roll-off is $\beta = 12$ dB/octave. Plugging the given values into Eq. (9.28) yields

$$36 = 12 \log_2 \frac{f2}{f1} \Leftrightarrow f2 = 2^3 f1 = 8f1.$$

The frequency separation between neighbouring channels is

$$\Delta f = f2 - f1 = 7f1 = 9.8 \text{ MHz}.$$

9.2.6 *Near-Far Effect*

The near-far effect (NFE), also referred to as near-end-far-end effect, is the situation in which multiple users are being served by the same base station, but the user of interest is farther away from the base station than the other users. This means that the power levels reaching the base station from the undesired users mask that of the desired user. The NFE is illustrated in Fig. 9.14, in which UE_1 is the user interesting to the eNodeB, and UE_2 is the strongest interfering user.

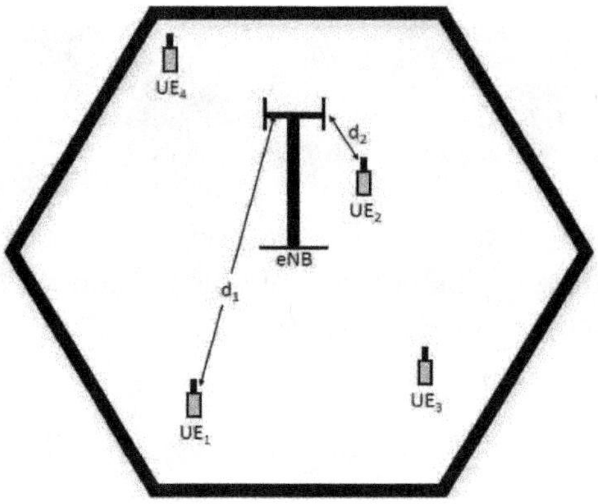

Fig. 9.14 Illustrating the near-far effect.

As shown in Fig. 9.14, the UE_1 and UE_2 are located at distances d_1 and d_2 respectively from the eNB, where $d_2 \ll d_1$. Let the transmit power from UE_i, $i = 1, 2$ be P_t and the power in the signal received at the eNodeB from UE_i, $i = 1, 2$ be $P_{r,i}$. We assume that all mobiles transmit signals with the same power level. Note that the path loss $L(d)$ at the distance d is $L(d) \propto d^{\gamma}$, or $L(d) \propto 10\gamma \log_{10} d$ dB, where γ is the propagation path loss exponent. We define the near-far-effect ratio (NFER) as ([Lee (1993)], p. 152)

$$NFER = \frac{\texttt{path loss of near-end mobile}}{\texttt{path loss of far-end mobile}} = 10\gamma \log_{10} \frac{d_2}{d_1} . \quad (9.29)$$

The NFE is reduced using transmit power control. Another way of reducing the NFE is to require the interfering signals to use different carrier frequencies which are far apart from each other. In order to exploit the second option, we must equate Eqs. (9.28) and (9.29) to find the minimum frequency separation needed to offset the received power difference resulting from the NFE. Thus

$$\frac{1}{NFER} \overset{!}{=} L_{fs} \Leftrightarrow 10\gamma \log_{10} \frac{d_1}{d_2} = 3.32\beta \log_{10} \frac{f2}{f1} \Leftrightarrow$$

$$f2 = \left(\frac{d_1}{d_2}\right)^{3.012\frac{\gamma}{\beta}} \cdot f1 . \quad (9.30)$$

Example

Assume that $\gamma = 3.5$, $\beta = 12$ dB/oct and $d_1/d_2 = 3$, (a) compute the NFER, and (b) find the needed frequency separation to offset the NFER.

Solution

(a) From Eq. (9.29) we obtain $NFER \approx -16.7$ dB.
(b) From Eq. (9.30), we obtain

$$f2 = \left(\frac{d_1}{d_2}\right)^{3.012\gamma/\beta} f1 = 3^{3.012\times 3.5/12} f1 \approx 2.63 f1 .$$

The frequency separation required is thus approximately $f2 - f1 = 2f1$.

9.2.7 *Channel Allocation*

The channels used in mobile cellular networks are of two basic types: control channels (CCHs) and traffic channels (TCHs). The CCHs are used to maintain the system, but are not used to transfer users' payload. Mobile handsets use it to maintain contact with the base station whose signal strength is strongest at their locations. Mobile stations also use the CCHs to acquire TCHs from the network. The actual payload of users are transferred over TCHs. Practically, each cell site needs only one control channel but multiple traffic channels. Also, the frequency reuse distance for control channel is larger than that of traffic channels.

The process of assigning channels to cells in a cellular network is called **channel allocation** or **channel assignment**. This can be done in a static or fixed manner, leading to fixed channel allocation (FCA) schemes. Channels can also be assigned to base stations in a dynamic manner, leading to dynamic channel assignment (DCA) methods. There many proposals for both FCA and DCA methods. While FCA schemes reduce signalling load on the network, they are unsuitable if the capacity needed in the cells of a cluster is not uniform.

In a FCA scheme, each cell in a cluster is assigned a fraction of the channels in the network. In a **strict FCA**, a cell cannot accept a call request if all its assigned channels are engaged, leading to **call blocking**. In a **flexible FCA** scheme, however, a cell with all its channels engaged can borrow channels from neighbouring cells to serve a new call. There are several methods used to achieve channel borrowing, which is an inter-cell procedure. The cell receiving the borrowed channel is called **acceptor cell**, while the cell lending channels is called **donor cell**. A requirement of channel borrowing strategies

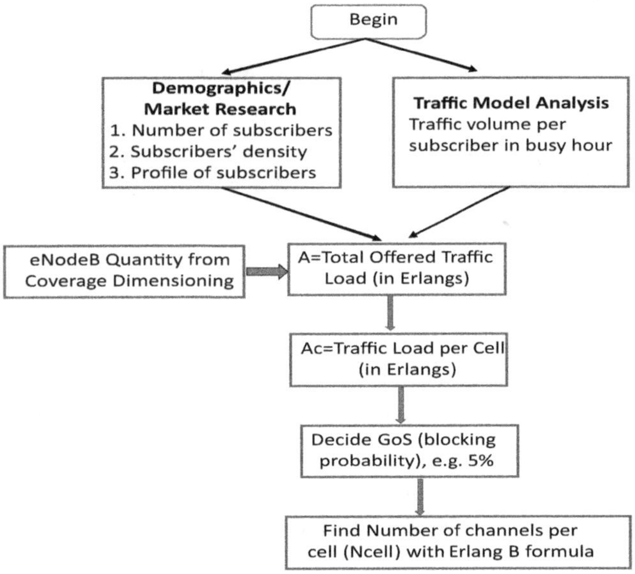

Fig. 9.15 An example of non-iterative channel assignment procedure.

is that it should not affect the service quality that the donor cell is offering to its ongoing calls.

In DCA methods, channels are not allocated to base stations in a fixed manner. Rather, any time a base station has to service a call it has to request channels from its serving switch. Channel allocation is performed systematically using an algorithm considering metrics, such as:

- Minimising CCI and ACI.
- Keep call blocking rate low.
- Keep call dropping rate low.

Dynamic channel allocation schemes provide a better radio resource utilisation and lower blocking rate than static channel allocation. However, DCA algorithms require more signalling and processing at the switching center than FCA methods. The switching center must keep track of signal strengths of channels and traffic load distribution, among other things. Figure 9.15 shows a procedure which can be used to assign a number channels to cells of a cellular network in a fixed manner. This procedure requires us to have already performed *coverage dimensioning* to find the number of cells needed for the coverage area, as well as the radius of each cell. This is discussed in the section on coverage-limited network design case study.

9.2.8 *Check Your Understanding*

Students are encouraged to check their understanding of the content of this section with the following questions prior to proceeding to the next section.

(1) The main reason for using the cellular architecture is

(a) Reduce co-channel interference in cellular system.
(b) Acquire more bandwidth from the local RF regulator.
(c) Reduce adjacent channel interference in cellular system.
(d) Boost spectral efficiency and system capacity through frequency reuse.

(2) Which of the following statements is wrong about mobile cellular systems?

(a) The signal-to-interference ratio (SIR) does not depend on the radius of cells.
(b) The SIR depends on the cluster size used.
(c) The radius of cells depends on the required capacity but limited by handoff frequency.
(d) Mobile cellular systems are not affected by adjacent channel interference.

(3) What is the reuse distance for cluster size of three and cell radius R?

(a) 9R.
(b) 3R.
(c) $\sqrt{3}R$.
(d) 6R.

(4) Increasing the cluster size in a cellular system

(a) Increases the co-channel interference.
(b) Reduces the co-channel interference.
(c) Increases the system capacity.
(d) Reduces the system capacity.

(5) If the radio propagation path loss exponent is γ, which of the following is the reuse factor, q?

(a) R/D.
(b) D/R.
(c) $\sqrt{3N}R$.
(d) $(6 \times SIR)^{\gamma}R$.

(6) Frequency reuse is an example of which multiple access scheme?

(a) TDMA.
(b) OFDMA.
(c) SDMA.
(d) CDMA.

(7) The practical shape of the radiation pattern of a base station's antenna is

(a) Circular.
(b) Amorphous.
(c) Rectangular.
(d) Hexagonal.

(8) Discuss the differences between coverage-limited cellular mobile network and capacity-limited cellular mobile network.

(9) State the three reasons why the six-sided hexagonal shape is used as the radiation pattern of a base station in the design and analysis of mobile cellular networks, although that shape does not reflect the reality.

(10) (a) Compute the area enclosed by a hexagon inscribed in a circle of radius R as a function of R.
(b) What is the length of each side of the hexagon?

(11) A cellular system uses a cluster size of three. Compute the reuse distance D as a function of the radius R of the circle in which a hexagonal cell is inscribed.

(12) Cellular systems using channels with 15 kHz bandwidth seek the minimum signal-to-interference ratio (SIR or $\frac{P}{I}$) of 24 dB. Assume that P is the power in the signal from the base station of interest, N_0 is the power of background noise, and I is the total power in all signals received from interfering base stations. If $\frac{N_0}{P} = -18$ dB, find $\frac{P}{N_0+I}$.

(13) Assume that $\frac{P}{N_0+I} = 80$, $N_0/I = -8$ dB, $P = 3$ W, compute N_0.

(14) If $\frac{P}{N_0+I} = 24$ dB, $I = -95$ dBm, $P = 3$ W, what is N_0?

(15) Compute $SIR = \frac{P}{N_0+I}$ in decibels if $I/P = -9$ dB and $P/N_0 = 48$.

(16) How many interferers are in the first-tier co-channel cells if the cluster size is 3, 4, 7 or 12?

(17) Compute the SNIR (i.e., signal-to-interference-plus noise ratio) in a cellular system using cluster size of 12. Assume that the path loss exponent is 3.4 for radio propagation in all positions in the coverage area, thermal noise power of $N_0 = kTB$, 500 m cell radius, 15 kHz channel bandwidth, transmit power of each bases station is 45 dBm and received power 300 m away from each base station is 200 mW. Assume that the ambient temperature (T) is 300 K. The Boltzmann's constant is $k \approx 1.38064852 \cdot 10^{-23}$ m^{2}kg s^{-2} K^{-1}. Assume that no cell in the network uses sectoring.

(18) The path loss exponents indicating the attenuation of radio signals depends on the propagation environment. Typical values used for mobile cellular system design are given in Eq. (9.26). Use Fig. 9.12 to compute the range of SIR achievable in each of the four types of radio propagation environments given in Eq. (9.26) for cluster sizes of 3 and 12. Comment on your results.

(19) The bandpass filter used at an eNodeB to reject out-of-band radiations has the roll-off of 8 dB/octave. Assume that each channel has the bandwidth of 20 MHz, and the signal-to-interference ratio (SIR) in reference to adjacent channel interference should be at least 14 dB.

(a) What should be the minimum frequency separation between adjacent carriers? *Hint.* Review Fig. 9.13.
(b) If the radio propagation path loss exponent is 2.5, compute the minimum geographical separation between two co-channel users needed to achieve the same SIR of 14 dB.

9.3 Coverage and Capacity Management

Coverage and capacity are related. This is because a network covering a large geographical area has the chance of serving more users than the other with a smaller coverage area, so long as people live in the area. Designers of mobile cellular networks increasingly seek strategies to improve the number of users, referred to as **network capacity** or just capacity, that can be served simultaneously using possibly the same set and amount of resources. Common strategies used to boost the network capacity and coverage are:

- Frequency reuse and acquisition of more frequencies to boost capacity: frequency reuse reuses each set of frequencies in cells separated by the frequency reuse distance. Acquiring more frequencies from the regulator is another option. However, this is usually an expensive choice considering the cost in acquiring bandwidth. Also, the needed spectrum may not be available.
- Cells on wheels to boost coverage.
- CCI reduction: any method which reduces the CCI increases the capacity. Methods include:
  (1) Antenna beam tilting.
  (2) Cell sectoring.
  (3) Microcell zoning, which is sectoring without reducing trunking efficiency and without increasing handoff, call switching and control signalling overhead.
  (4) Power control of transmissions from users by the serving base station.
- Reduction in cluster size to increase capacity.
- Cell splitting to boost capacity.
- A congested cell can borrow channels from neighbouring cells to increase its capacity.
- Repeaters to provide coverage, especially in hard to reach locations.
- Cell on wheels (COW): a temporary cell can be used to increase both coverage and capacity.
- Installation of new base stations to fill coverage holes and greenfield areas.
- Increasing the tower height of a base station increases the coverage area.
- Increasing the effective radiated power (ERP) of a base station increases its coverage.

The mobile cellular operator usually seeks to increase the network's capacity when the blocking probability exceeds the acceptable level.

9.3.1 *Cell on Wheels and Rapid-Deployment Unit*

Cell on wheels or cell tower on wheels (COW or C.O.W.) is a portable base station which is usually used for a temporary cellular mobile network coverage. For this reason, it is also referred to as mobile temporary cell, portable cell tower or mobile cell site. There are four basic applications of COW, all of which are temporary solutions:

- To boost capacity and coverage at, or media coverage of, a big event, such as festivals, sports, trade fairs and festive periods (e.g., Easter and Christmas) which attract large crowds of people.
- To provide wireless network coverage at emergency situations, such as base stations being damaged by natural disasters such as floods, storm, tsunamis, earthquake and bushfires. A COW is used while the destroyed permanent base station is under recovery.
- To provide coverage during the construction of a permanent base station.
- To provide wireless network coverage in locations where a site for permanent placement of a base station is not possible. COWs of this nature can be used over an extended period of time.

Their nature demands that they are mounted on vehicles, especially used by the army at war front. Manufacturers of COW include ICS Industries with headquarters in Melbourne Australia and CellSite Solutions, LLC in USA. Figure 9.16 shows an

Fig. 9.16 Mini-cell on wheels.

example of a mini COW. The bottom COW is powered by solar cells, while the bottom one is power by mains power supply or battery bank. COW has many different sizes, depending on the required coverage and transmit power, one of which is the mini COW. The mini COW is small enough to be towed by a standard four-wheel drive vehicle.

Commissioning a COW is simple and fast as one does not require any difficult-to-acquire right-of-way. Also, if main power (three-phase power) supply is not available in the area where COW is being commissioned then either solar cells or battery bank can be used. Solar-cell-powered COW is called solar COW (or SCOW). As a COW can be commissioned quicker than conventional base station and is portable, it is also called rapid-deployment unit (RDU). The connection of a COW to the core telecommunications network may be the dominating cost in setting up a COW, especially if the COW is set up in a remote venue.

9.3.2 *Cell Splitting*

This is the process of subdividing a given cell into multiple smaller cells. Cell splitting reduces the size

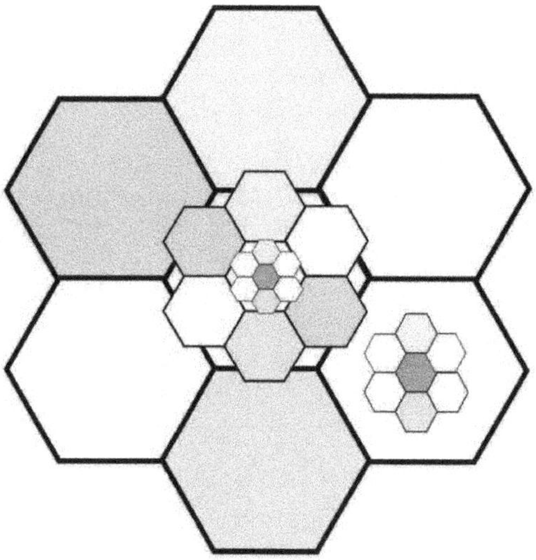

Fig. 9.17 Illustration of cell splitting.

of cells, but does not change the frequency reuse factor $q = \frac{D}{R}$ by maintaining the frequency plan. Usually, the reason for the splitting of a cell into multiple smaller cells is congestion, which means that the cell which is divided is not able to cope with its offered traffic load. Each split cell is served by a base station with smaller antenna gain and smaller transmit power to limit its coverage area and reduce CCI. Figure 9.17 illustrates the process of splitting a cell.

Example

Assume that the transmit power used in a macrocell of radius R_1 in a mobile cellular network is $P_{t,1}$. This cell is congested and thus split into multiple cells, each of smaller radius R_2 so that $R_1 > R_2$. Find the transmit power $P_{t,2}$ in each of the new smaller cells.

Solution

Note that, in order to reduce interference, we want the received power to be the same regardless of the location of the mobile users or the size of a cell. Therefore, the received powers are $P_{r,2} = P_{r,1} = P_r$, where $P_{r,i}$ is the received power in a cell of radius R_i, $i = 1, 2$. The relationship between the transmit and receive powers are $P_{r,i} \propto P_{t,i}R_i^{-\gamma}$, where γ is the radio signal propagation path loss exponent.

Table 9.4 Reduction in transmit power as a function of cell radius and path loss exponent in cell splitting.

γ	$\frac{P_{t,1}}{P_{t,2}}$ (dB) $R_1/R_2 = 2$	$R_1/R_2 = 3$	$R_1/R_2 = 4$
2.0	6.02	9.54	12.04
2.5	7.53	11.93	15.05
3.0	9.03	14.31	18.06
3.5	10.53	16.70	21.07
4.0	12.04	19.09	24.08

Therefore, we have

$$P_{t,2} = P_{t,1}\left(\frac{R_2}{R_1}\right)^{\gamma}. \tag{9.31}$$

Practical values for the path loss exponent lie in the range $2 \leq \gamma \leq 4$. Assume that each cell is divided into two, three or four smaller equal cells. The ratio of the transmit powers for the smaller and larger cells are compiled in Table 9.4. For example, if $\gamma = 2$ the transmit power in two smaller cells is a factor eight less than the transmit power in the larger undivided cell.

Now, how does the network capacity increase through cell splitting? Consider a large cell with radius R_1 which is partitioned into multiple equal-sized cells, each of radius R_2, where $R_1 > R_2$. The coverage area A_i of a cell of radius R_i is $A_i \propto R_i^2$, $i = 1, 2$. Assume that the number of cells needed to cover a mobile network's service area of total size A m^2 is $N_{\texttt{cells},i}$. Thus,

$$\begin{aligned} N_{\texttt{cells},1} &= \frac{A}{A_1} \propto \frac{A}{R_1^2} \\ N_{\texttt{cells},2} &= \frac{A}{A_2} \propto \frac{A}{R_2^2}. \end{aligned} \tag{9.32}$$

Thus,

$$N_{\texttt{cells},2} = N_{\texttt{cells},1}\left(\frac{R_1}{R_2}\right)^2. \tag{9.33}$$

Assume that each cell is assigned $N_{\texttt{ch}}$ duplex channels. Note that cell partitioning does not change the frequency plan nor the frequency reuse pattern and that both the original cell and each of the smaller cells use the same number of channels. Let $N_{\texttt{users},i}$ be the capacity of the network using cells, each of radius R_i, $i = 1, 2, \ldots$. Then, we have

$$\begin{aligned} N_{\texttt{users},1} &= N_{\texttt{cells},1} \times N_{\texttt{ch}} \quad \texttt{(a)} \\ N_{\texttt{users},2} &= N_{\texttt{cells},2} \times N_{\texttt{ch}} = N_{\texttt{users},1}\left(\frac{R_1}{R_2}\right)^2 \quad \texttt{(b)}. \end{aligned} \tag{9.34}$$

We can now observe from Eq. (9.34) that the smaller the cells used, the larger the capacity of the entire network. This is the reason cells are partitioned into smaller cells when more capacity is needed. We have so far explained two advantages in cell splitting: reduced transmit power to prolong time between mobile phone battery recharging and boost of network capacity. An interesting question is: does cell splitting have any drawbacks? The answer is a resounding yes. First, cell splitting produces smaller cells. This results in more handoff and thus more signalling. Second, as each cell is served by a base station, cell splitting increases the number of radio sites and associated costs.

Example

A mobile cellular system uses 70 cells to cover a given geographical area. Each cell is assigned 40 duplex radio channels. The network became increasingly congested such that the call blocking and call dropping rates increased beyond bounds. The operator realised the need to boost the capacity of the network to prevent churn through cell splitting. Therefore, the telecommunications operator has tasked us to compare the capacities and costs involved in partitioning each cell into two, three or four equal parts.

Solution

The capacity of the original cell is obtained from Eq. (9.34)(a) as

$N_{\texttt{users},1} = N_{\texttt{cells},1} \times N_{\texttt{ch}} = 70 \times 40 = 2,800$ users.

Partitioning each cell into two, three or four equal parts yields the ratios $(R_1^2/R_2^2)^2 = \{4, 9, 16\}$. The resulting capacities is obtained from Eq. (9.34)(b) as

$$\begin{aligned} N_{\texttt{users},2} &\overset{(9.33),(9.34)b}{=} N_{\texttt{users},1} \times \{4, 9, 16\} \\ &= \{11200,\ 25200,\ 44800\}. \end{aligned}$$

The higher the capacity, the higher the costs in acquiring radio infrastructure and land to mount the radio base stations. Therefore, we select the lowest capacity that is enough to meet demand. We can also over-dimension the network for future expansion.

9.3.3 *Cell Sectoring*

The technique of using directional antennas to divide the illumination area of a cell into multiple divisions is referred to as **cell sectoring**. Sectoring is cheaper to implement than cell splitting as no additional base stations are required. A base station either uses omnidirectional or sectored antennas. By dividing a cell into multiple sectors, cell sectoring reduces the co-channel interference for the same cluster size as the number of co-channel interferers are reduced. The reduction in CCI is exploited to either reduce the frequency reuse distance to increase system capacity or increase the signal-to-interference ratio.

An omnidirectional antenna ideally radiates all directions with equal intensity. Therefore, its beamwidth is 360°. If we divide a cell of a mobile cellular network into, say n_s, sectors and illuminate each sector using a directional antenna then the beamwidth of each antenna becomes ideally

$$\theta_s = \frac{360^\circ}{n_s}. \qquad (9.35)$$

Figure 9.18(a) and Fig. 9.20(a) show cells divided into three sectors each, requiring directional antennas with 120° beamwidth. Figure 9.18(b) and Fig. 9.20(b) show cells divided into six sectors each, requiring directional antennas with 60° beamwidth. Figure 9.19 shows two examples of sectored cells.

The CCI reduction factor depends on the cluster size and the number of sectors used. For example, sectoring cells in a seven-cell cluster into three sectors, each illuminating 120°, reduces the number of co-channel interferers from 6 to 2. A cell divided into 6 sectors, each illuminating 60°, reduces the number of co-channel interferers from 6 to 1. Figure 9.18 illustrates the sectoring of a cell into two 180°, three 120° and six 60° sectors, respectively. A cell can also be sectored into four 90° sectors.

Cell sectoring reduces the number of (first-tier) co-channel interferers and increases the frequency reuse in order to increase network capacity. It is

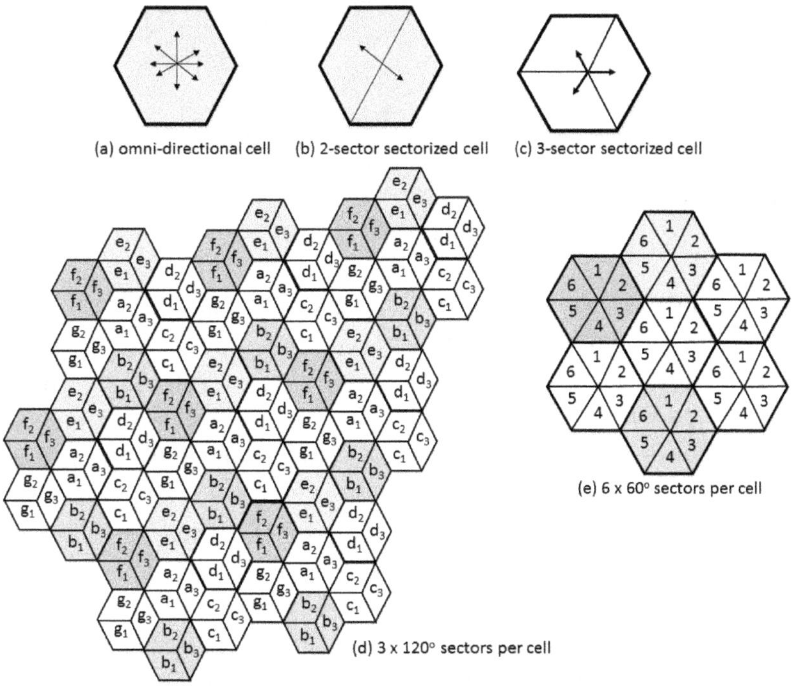

Fig. 9.18 Cell sectoring illustration: co-channel sectors have the same number.

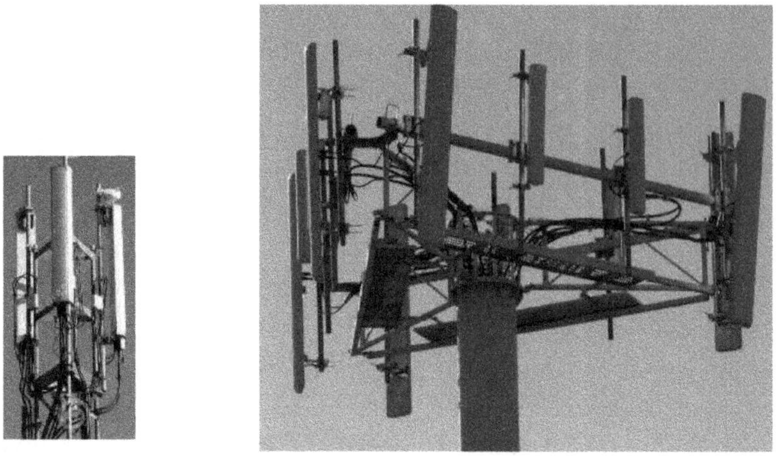

(a) A three-sector cell

(b) Co-located technology site or/and infrastructure sharing site

Fig. 9.19 Cell sites with sectoring and infrastructure sharing.

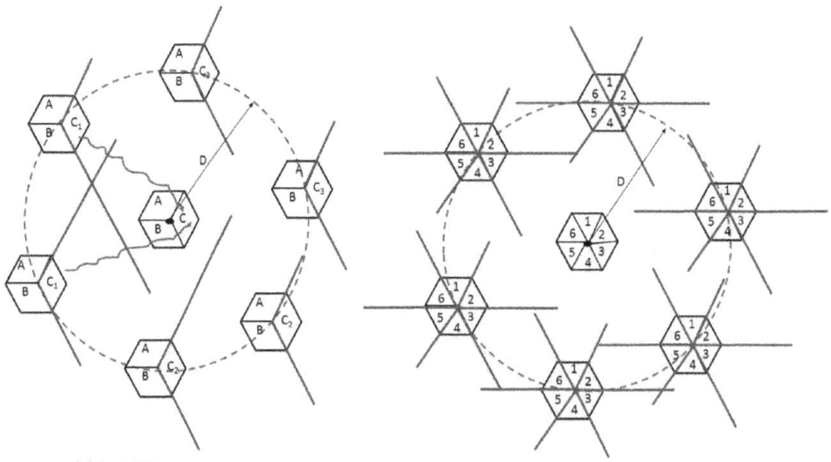

(a) 3 x 120° sectors

(b) 6 x 60° sectors

Fig. 9.20 Interfering sectors in sectored cells: co-channel sectors have the same number (D: first-tier frequency reuse distance).

worthy of remembrance that, just as the footprint of an omnidirectional antenna is irregular, it is impossible to confine a directional antenna to illuminate perfectly a given angle, such as 60° or 120°. Also, practical antennas have side lobes which is another source of interference. This means that our presentation here is using the ideal case for the sole purpose of easy illustration.

The frequencies or channels allocated to the cell which is sectored must be shared among the sectors. If the number of channels in a cell is $N_{\texttt{f,cell}}$ and the cell is divided into n_s equal sectors, then each sector has $N_s = \frac{N_{\texttt{f,cell}}}{n_s}$ channels. The set of frequencies or channels in a cell is shared among the sectors in the cell. For example, if there are k frequencies in a cell and it is sectored into three equal sectors (i.e., $n_s = 3$), each beaming 120°, then each sector receives $k/3$ frequencies. This is illustrated in Fig. 9.18(d) for a cluster size of seven. Note that $a_1 + a_2 + a_3 = k$, for example. The frequencies in the sectors of a given cell do not overlap.

A first-tier co-channel cell in cellular system using the cluster size of seven and three sectors per cell is shown in Fig. 9.20. Each cell has a set of three antennas, each illuminating one of the three sectors. Sectors indicated with the same letter are co-channel sectors. We can observe that, among the first-tier six co-channel cells of the center cell, the illumination angles of the antennas in only two of the sectors, namely those two indicated as C_1, affect the sector at the center using the set of frequencies marked C. This means that the number of co-channel interference has been reduced from six to two to increase the SIR.

We can observe from Fig. 9.20 that the co-channel interference is reduced by two main factors: number of co-channel interferers reduces, and co-channel interferers are located at the back of the signal of interest. Again, sectors with the same number are co-channel sectors. We observe in Fig. 9.20(a) that only the radiation angles of the two sectors marked C_1 affect the sector C located in the cell's center. Also, in Fig. 9.20(b), each of the six sectors in the center cells is affected by the radiations of a single co-channel sector. There are, of course, some disadvantages in sectoring, including:

- The need for more antennas. This also requires a stronger antenna tower to carry the increased weight or load.
- The reduction of the coverage area from a cell to a sector thereof for each subgroup of channels causes more handovers. Also, the need for intra-cell or inter-sector handover arises.
- Decreased trunking efficiency as the number of channels in the pool are subdivided into small-sized pools.
- Increased system complexity.

Example

A mobile cellular system using a cluster size of $C = 7$ and an omnidirectional antenna has been working well with the SIR of about 15 dB. The grade of service of the cellular network has deteriorated beyond the acceptable level owing to congestion. Therefore, the telecommunications service provider needs to boost its network capacity by about a factor 2 to curtail churn at the lowest possible cost using sectoring and the appropriate cluster size. You are tasked to bring proposals for discussion and decision-making. Assume that the radio propagation path loss exponent in the geographical area served by the mobile network is $\gamma = 3.5$.

Solution

We are given the radio propagation path loss exponent as $\gamma = 3.5$. We need to simulate Eq. (9.18) to find the combination of cluster size C and the number of sectors S which yields at least a capacity increase of 100%. We should first note the approximate relationship between the cluster size C and the capacity or number of simultaneous users N_{u}, which is

$$N_{\mathrm{u}} \propto \frac{1}{C} . \tag{9.36}$$

This means that, if a system using the cluster size C_1 achieves the capacity $N_{\mathrm{u},1}$, then its capacity $N_{\mathrm{u},2}$ for cluster size C_2 will be approximately

$$N_{\mathrm{u},2} = \frac{C_1}{C_2} N_{\mathrm{u},1} \tag{9.37}$$

resulting in the capacity boost factor

$$\eta_C = \frac{N_{\mathrm{u},2}}{N_{\mathrm{u},1}} = \frac{C_1}{C_2} . \tag{9.38}$$

These are the values compiled in the last column of Table 9.5 using cluster size 7 as the reference.

The relationship between the path loss exponent γ, the cluster size C, and the number of first-tier co-channel interferers K was obtained in Eq. (9.18) as

$$SIR = \frac{(3C)^{\frac{\gamma}{2}}}{K} .$$

The number of co-channel interferers reduces from six to two for three-sector cell and from six to one for six-sector and four-sector cells, as illustrated in Fig. 9.20. Table 9.5 compares the possible design parameters fulfilling the conditions stated in the problem formulation. It is clear from the table that we need to use the cluster size of three with three sectors per cell to achieve the needed minimum capacity and the minimum SIR.

Given the grade of service (or blocking probability) of 2% and the number of channels per

Table 9.5 Analysing effects of cell sectoring on trunking efficiency, SIR and cell capacity: $\gamma = 3.5$, $GOS = 2\%, N_\text{T} = 105$ duplex channels.

Cluster size, C	# Sectors, n_s	# Co-channel interferers, K	$SIR = 10\log_{10} \cdot [(3C)^{\frac{\gamma}{2}}/K]$ (dB)	# Channels per cell, $N_\texttt{cell} = \frac{N_\text{T}}{C}$	# Channels per sector, $N_\text{s} = \frac{N_\texttt{cell}}{n_s}$	Sector capacity, A_s (Erlangs)	Cell capacity, $A_\text{c} = A_\text{s} \times n_s$ (Erlangs)	η_C
9	1	6	17.267	11	11	5.842	5.842	7/9
9	3	2	22.039	11	3	0.6022	1.8066	7/9
9	6	1	25.049	11	1	0.0204	0.1224	7/9
7	1	6	**15.357**	15	15	9.010	**9.010**	1
7	3	2	20.129	15	5	1.657	4.971	1
7	6	1	23.139	15	2	0.2235	1.341	1
4	1	6	11.104	26	26	18.38	18.38	7/4
4	3	2	15.875	26	8	3.627	10.881	7/4
4	6	1	18.886	26	4	1.092	6.552	7/4
3	1	6	8.918	35	35	26.44	26.44	7/3
3	3	2	**13.689**	35	11	5.842	**17.526**	7/3
3	6	1	16.699	35	5	1.657	9.942	7/3

sector, the sector capacities in Column 7 of Table 9.5 were read off Erlang B table. As Table 9.5 shows, the total cell capacity in Erlangs at $C = 3$, $n_s = 3$, $SIR = 13.689$ dB is 17.526 E, which is about twice the capacity achieved with $C = 7$, $n_s = 1$, $SIR = 15.357$ dB. For a given cluster size, we can observe from Column 4 in Table 9.5 that the SIR increases as the number of sectors in the cell increases owing to decrease in CCI. However, as the last column in the table shows, cell capacity reduces. This is what we mean by decreasing trunking efficiency in cell sectoring.

We can find the total number of users supportable in a cell if we know the traffic intensity generated by each user. For example, assume that each user on the average makes 4 calls per day and that each call lasts 3.5 minutes. This translates to the traffic intensity per user of $A_u = (\texttt{4 calls per day}) \times (3.5\ \texttt{minutes/call})/(\texttt{minutes in a day}) = 0.0097$ E. The cell capacity in terms of the number of users has increased from 9.010 E/0.0097 E = 928 to about 17.526 E/0.0097 E = 1802 users, which is a capacity boost by a factor of 1.94.

Example

An eNodeB in a LTE-Advanced-based mobile cellular network serves a number of users, each of which on the average generates a single call of duration three minutes during the busy hour. Each blocked call is not queued but cleared. The grade of service of the network is 2%. The number of duplex traffic channels assigned to the cell is $N_{\texttt{cell}} = 21$.

(1) How many users can the eNodeB serve simultaneously if no sectoring is used?
(2) How many users can the eNodeB serve simultaneously if the cell is divided into:
   (a) Three sectors each covering 120°?
   (b) Six sectors each covering 60°?

   Assume that channels are shared equally among the sectors of a cell.
(3) Compare the trunking efficiencies of the three cases studied above.

Solution

The average traffic generated by each user is $A_u = 3/60 = 0.05$ E. The GOS is 0.02.

(1) With the number of channels in the pool being $N_{\texttt{cell}} = 21$, and $GOS = 0.02$ we read off the Erlang B table the traffic handling capacity of the cell as $A = 14.0$ E. The number of users that the system can support is $A/A_u = 14/0.05 = 280$.
(2) Sectored cell case:
   (a) Three sectors per cell means that each sector receives $N_{\texttt{s}} = \frac{N_{\texttt{cell}}}{n_{\texttt{s}}} = 21/3 = 7$ traffic channels. With $GOS = 0.02$, $N_{\texttt{s}} = 7$, we read off the Erlang B table the maximum capacity of a sector as 2.94 E. This means that each sector can support up to $A/A_u = 2.94/0.05 = 58.8$. Thus, the entire cell supports up to $3 \times 58.8 = 176.4$ users. As we cannot have a fractional user, the actual maximum capacity of the cell using three sectors is $3 \times 58 = 174$ users.
   (b) With six sectors per cell, each sector receives $21/6 = 3.5 \approx 3$ duplex traffic channels. From the Erlang B table, the maximum supportable capacity is 0.6 E, resulting in $0.6/0.05 = 12$ users. This means that the total cell capacity is $6 \times 12 = 72$ users.
(3) Trunking efficiency in non-sectored case is $\zeta_a = (1-P_b)A/N_{\texttt{cell}} = 0.98(14.0/21) \approx 0.65$, that in the three-sector case is $\zeta_i = (1 - P_b)A/N_{\texttt{cell}} = 0.98(3 \times 2.94/21) \approx 0.41$, while that in the six-sector cell is $\zeta_{ii} = (1 - P_b)A/N_{\texttt{cell}} = 0.98(6 \times 0.6/21) \approx 0.17$.

It is clear from the above example that the trunking efficiency (i.e., the amount of actual carried traffic intensity per channel using the same number of duplex traffic channels in a cell) reduces as the number of sectors in the cell increases. This is one of the penalties in using sectoring.

Sectorisation Gain and Efficiency

We have discussed above that cell sectorisation is used to reduce interference and increase capacity. Unfortunately, the radiation patterns of the sectors of an antenna are not perfectly isolated from each other. There are some overlaps. Therefore, the interference is reduced by a factor less than the *number of sectors* S used. Let β_o and β_s be the interference factors for omni-directional cell and sectorized cell, respectively. The **sectorisation gain** is then defined

$$g_s = \rho_s \times S \tag{9.39}$$

where ρ_s is the **sectorisation efficiency**, which is defined as

$$\rho_s = \frac{1+\beta_o}{1+\beta_s} \leq 1\,. \tag{9.40}$$

Note that $\rho_s \leq 1$ as $\beta_o \leq \beta_s$.

9.3.4 *Microcell Zoning*

As discussed earlier, the drawbacks in sectoring include increased frequency of handover, which in turn overloads the control channels in the system as well as increasing switching load. Sectoring also reduces trunking efficiency. A solution to these weaknesses in sectoring is microcell zoning (aka zone microcell). The microcell zoning concept is used to distribute the coverage area of a cell, as well as extending cell's boundary to coverage holes. Through the appropriate placement of base station antennas, microcell zoning boosts capacity by reducing the CCI.

In microcell zoning a cell is divided into multiple smaller cells referred to as zones [Lee (1991)], [Rappaport (2002)], of say four zones, as illustrated in Fig. 9.21. All the zones in a microcell zone are controlled by a single base station, thus making a cell. This enables a mobile connection to maintain the same transmission channel as the user moves between zones of a cell, preventing the need for handoff through a device (e.g., switch) outside the cell. As illustrated in Fig. 9.21, the zones in a cell are connected to the serving base station via wired media (e.g., coaxial or fiber optic cable) or microwave links. Each zone is served by a pair of transmitter and receiver (TX/RX) circuit, referred to as **zone transmitter**. However, the zone transmitters are low-power devices as each covers a small area compared to the base station. Confining the transmissions of a radio node to a smaller area reduces co-channel interference which is exploited to reduce the cluster size or to increase the signal-to-interference ratio. Remember that capacity increases if we reduce the cluster size.

A mobile user in a microcell zone is served by the same set of radio resources as it moves across zones. The channels allocated to the cell are not divided as it is in cell sectoring, maintaining the trunking efficiency. Rather, the base station switches the channel used by a mobile station as it relocates from one zone to another in the microcell zone. This way, the channel moves with the mobile user without the involvement of a switching device.

As in sectoring, microcell zoning has some drawbacks. These include:

- It needs additional antennas.
- Also, base stations employing the microcell zoning concept must be capable of switching a call between zones within a given cell, something

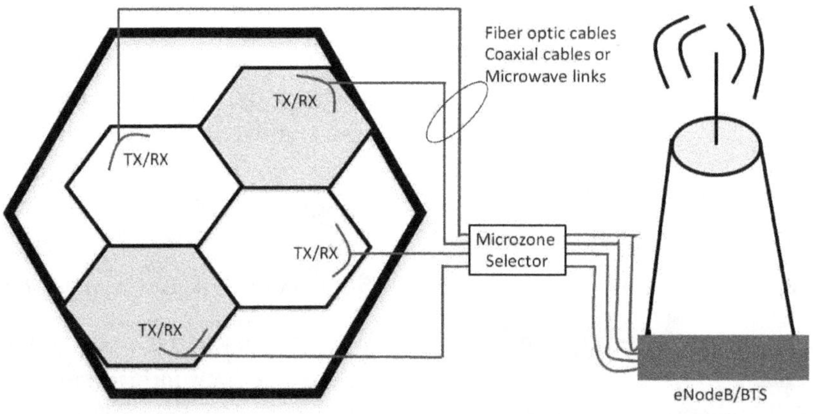

Fig. 9.21 Illustrating the microcell zoning concept.

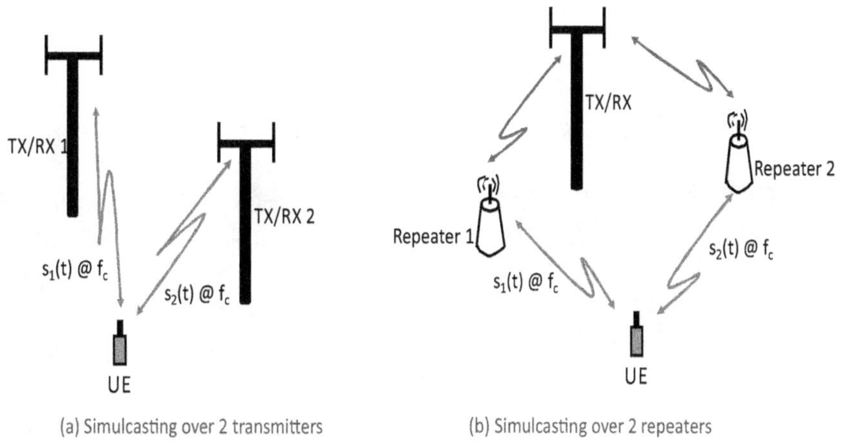

Fig. 9.22 Illustrating simulcast repeater concept.

which translates into increased complexity in base stations.

9.3.5 *Simulcast Repeater*

The word *simulcast* is a concatenation of the two words *simultaneous* and *broadcast*. It is used in wireless communications to refer to the concept of multiple transmitters (e.g. radio base stations or repeaters) transmitting at the same time the same signal at the same frequency using the same modulation method. This concept is illustrated in Fig. 9.22. For the sake of simplicity, we assume that only two signals reach the receiver at a given time. Simulcast is popularly used in wireless paging systems, although it is uncommon in cellular mobile networks.

By its nature, simulcasting produces multipath signals at the receiver as at least two transmitters have overlapping coverage. What then are the advantages in simulcasting? First, it is used when we have insufficient frequencies or channels. Second, it can be used to improve coverage of the radio system by filling up coverage holes. Third, it simplifies mobile-to-mobile communications of dispatching in a coverage area [Lee (1993)], [Rappaport (2002)]. The received signal in a simulcast system can be represented as

$$r(t) = \sum_{k=1}^{n} a_k \cos(2\pi f_{c,k} t + \varphi_k) \qquad (9.41)$$

where for multipath signal k

- a_k: amplitude of the carrier signal.
- $f_{c,k}$: frequency of the carrier signal.
- $\varphi_k = \Delta_k \cos(2\pi f_k t + \theta_k) + \phi_k$: is the phase of the carrier signal, f_k is the amplitude of the modulating signal, Δ_k is the peak deviation of FM/PM modulator, and ϕ_k is the carrier phase delay.

In reference to the situation shown in Fig. 9.22, we have $n = 2$. At the transmitters all the parameters of the carrier signals and the message signals are the same. However, the transmission medium can change any combination of them to result in detection problem. If the carrier frequencies reaching the receiver differ, then beat frequencies and their harmonics are produced. In order to ensure that all transmitters use exactly the same carrier frequency we employ one of the following:

- Customised matched crystal oscillators. The problem in this approach is the need to re-tune the crystal oscillators as ageing causes them to drift.
- Rubidium atomic clocks. The problem in this approach is high cost.
- Mount a GPS receiver at each transmitter site to train the local oscillator. This is currently the most popular approach.

A main problem in simulcasting is time delay interference (TDI) arising out of the multipath signals owing to the inherent coverage overlap of the transmitters. The good news is that TDI becomes a problem if and only if two conditions are fulfilled at the same time, namely:

- the difference between the *time delay* of the multipath signals reaching a receiver is large, and
- the difference between the *amplitudes* of the multipath signals reaching a receiver is large.

The effect of TDI on system performance depends on the modulation and signalling scheme used. The rule-of-thumb is that the TDI must be kept within 31% of the symbol period. Using directional antennas at the transmitters is a solution to TDI. Some practical systems control TDI by using omnidirectional antenna at the central site, and directional antennas at the surrounding sites which are configured not to illuminate in the direction of the central site. With a good timing (such as that of GPS satellite) for synchronisation between the transmitters, however, one of the multiple signals reaching a receiver at any time will be stronger than all the remaining multipath signals and thus be able to capture the receiver. Another drawback in simulcast is the high cost, arising from two factors: (a) the fact that each radio site must have a dedicated repeater or transmitter for each of the frequencies used in the network, and (b) the use of complex control and timing hardware for the synchronisation between the transmitters.

9.3.6 *Transmit Power Control*

In transmit power control the base station instructs the mobile stations to transmit signals with power levels according to their relative distances from it (i.e., the base station). A mobile closer to the base station transmits at a lower power than another mobile far away from it in such a manner that the power levels in their signals reaching the base station are of similar magnitudes.

Power control is important in cellular mobile communications systems, especially those systems using spread-spectrum multiple access (SSMA) techniques. The reason is that the capacity of SSMA-based systems is soft, and limited by interference. Power control is used to reduce interference, boost system capacity, solve the near-far problem and prolong the time between the recharging of the battery of mobile handsets. Discontinuous transmission (DTX) is another mechanism used to prolong battery recharge intervals of mobile handsets.

Example: Transmit Power Control

Consider two active users u_1 and u_2 in a cell of a cellular mobile system operating at $f_c = 2.1$ GHz in a moment in which the propagation path loss exponent is $\gamma = 2.5$. Assume that the distances between the users and the serving base station are respectively $d_1 = 50$ and $d_2 = 400$. Assume that the transmit powers of the users are respectively P_{t1} and P_{t2}. We seek what should be the ratio $r_p = \frac{P_{t2}}{P_{t1}}$ in order for their uplink transmissions to achieve the same SIR (signal-to-interference ratio) at the serving base station.

Solution

The relationship between the transmit power P_t, receive power P_r, propagation distance d and the path loss exponent $\gamma = 2.5$ is $P_r = P_t(\frac{\lambda}{4\pi d})^\gamma$. Note that the transmissions of each mobile user is an interference to the other user at the serving base station. Thus, the condition to achieve the same SIR at the base station is that the powers in their signals are the same on reaching the base station, i.e., $P_{r1} = P_{r2}$. Thus,

$$1 = \frac{P_{r2}}{P_{r1}} = \frac{P_{t2}}{P_{t1}}(d_1/d_2)^\gamma$$

resulting in

$$r_p = \frac{P_{t2}}{P_{t1}} = (d_2/d_1)^\gamma = (400/50)^{2.5} \approx 181$$

or 22.6 dB. This means that the far user must transmit about 22.6 dB more power than the near user to achieve the same SIR at the base station.

9.3.7 *Segmentation and Dualisation*

These are two techniques employed if additional cells are needed within the frequency reuse distance of a cellular mobile network to increase capacity without fully splitting cells. What is the purpose of **segmentation**? It can be used to reduce CCI and fill coverage holes in a cellular network. Segmentation violates the frequency reuse plan by enabling

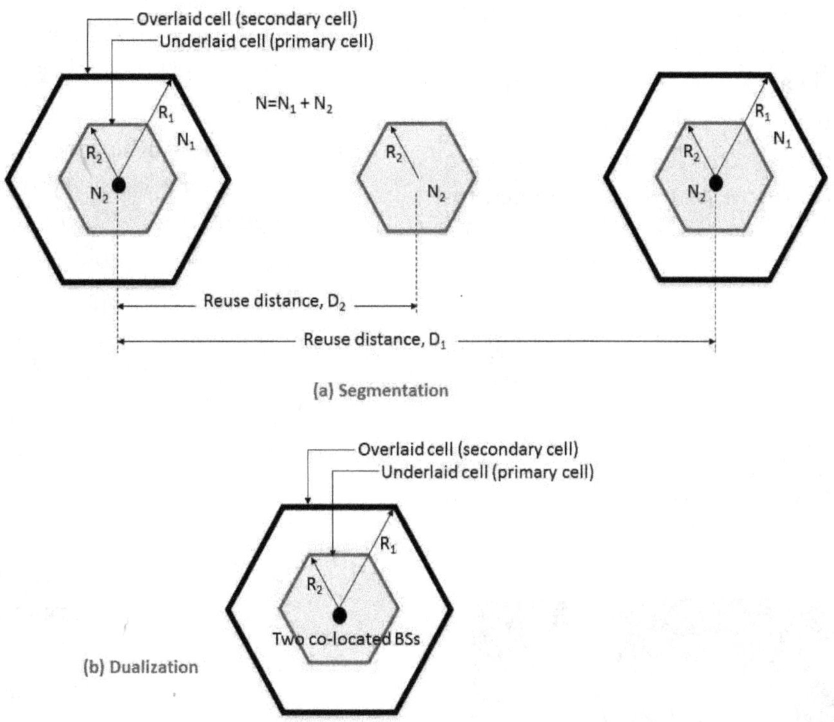

Fig. 9.23 Illustrating cell segmentation and dualisation in cellular mobile networks.

frequency reuse within the original frequency reuse distance, and thus increasing the CCI. The set of frequencies are divided into non-overlapping subsets or segments so that each cell within the reuse distance is assigned one subset in order to reduce the CCI. The division of the available channels into segments reduces trunking efficiency and thus capacity. This is a drawback in segmentation.

As illustrated in Fig. 9.23(a), the number of frequencies available to a cell N is divided into two non-overlapping subsets N_1 and N_2 so that $N = N_1 + N_2$. The set of frequencies N_2 serves the underlaid segment (or smaller cell or primary cell), while the set of frequencies N_1 serves the overlaid segment (or larger cell or secondary cell). Channels are moved from the larger cell into the smaller cell as the latter's traffic demand increases. This process is repeated until no channel remains in the larger cell and then dissolved.

Dualisation (Fig. 9.23(b)) is used instead of a full cell splitting when capacity demand on a cell increases to a level that cell splitting is appropriate. In dualisation two base stations (BS) are installed in a cell, a high-power BS and a low-power BS. The low-power BS illuminates only the small cell (aka primary cell), while the high-power BS illuminates larger cell (aka secondary cell).

9.3.8 *Load Balancing Using Cell Breathing*

Cells in a mobile cellular network may be configured with the same size and resources. Some of the cells can, however, be in demand more than others. For this reason, a cell can be congested while its neighbour cells are not. The coverage area of a cell can be reduced to reduce the number of users served in the time of congestion, while its neighbour cells which are less congested are expanded to absorb the excess load. A cell that shrank because it was overloaded can expand at another time when it is less congested. The technique in which the coverage area of a cell is shrunk and expanded based on traffic load is referred to as **cell breathing**. Thus, cell breathing makes a trade-off between capacity and coverage.

Cell breathing is used mainly in 2G and 3G mobile cellular networks, especially those based on spread-spectrum multiple access scheme. The capacity of networks, such as 3G UMTS, based on code-division multiple access (CDMA) is limited by interference. A minimum signal-to-interference ratio, say $SIR \geq 15$ dB, is required for an optimum operation. The SIR reduces as the interference increases. The interference increases, however, as the number of users in a cell increases. Reducing the coverage area of a cell reduces the number of users served and thus the interference. Cell coverage reduction also increases the strength of signals received by user equipment from the serving Node B in 3G CDMA networks. Networks whose capacity is not limited by (intra-cell) interference, such as LTE, does not use cell breathing.

9.3.9 *Check Your Understanding*

Students are encouraged to check their understanding of the content of this section with the following questions prior to proceeding to the next section.

(1) **In-class discussions:** What is COW in cellular mobile communications? Briefly describe the four basic application areas of COW.

(2) Discuss and differentiate between the strategies used to increase the capacity of mobile cellular networks.

(3) In what ways does cell splitting differ from cell sectoring? Also, discuss in what ways the two processes are similar.

(4) Discuss in groups the pros and cons in cell splitting.

(5) Discuss in groups the pros and cons in cell sectoring.

(6) **In-class group research**: research to find the differences between segmentation and dualisation in cellular mobile networks.

(7) Describe the differences between cell splitting and dualisation as used in cellular mobile networks.

(8) Describe the differences between cell splitting and segmentation as used in cellular mobile networks.

(9) A mobile cellular network operator seeks to boost its capacity over the entire coverage area by a factor of 5 using cell splitting. Discuss how that can be achieved.

(10) Consider a mobile cellular network based on cluster size of 19 operating in an environment with path loss exponent of 2.8. Each cell is assigned 10 traffic channels. If microcells of radius 500 meters are used, compute the following quantities in both ratio form and in decibels:

   (a) The maximum traffic load that the network can support without exceeding the grade of service of 2%.

   (b) The signal-to-interference ratio without sectoring.

   (c) The SIR when 60° sectors are used in each cell.

   (d) The SIR when 120° sectors are used in each cell.

(11) A given mobile cellular system has the following features:

   - Coverage area is 50,000 square kilometers
   - Population of the area is 100,000 people, of which only 80% can afford a mobile service.
   - Operator has acquired 20 MHz bandwidth and mobile network uses 1.4 MHz channel spacing without guard bands. Each cell uses one channel for signalling and the rest for payload traffic.
   - Each mobile user uses the network only 3 minutes each hour, on the average.
   - The farthest point a base station's signal has to reach is 1 km.
   - Cluster size used in 21.

   Assume that the network does not queue calls (i.e., BCC system), and compute the following:

(a) The total number of duplex channels used in the mobile cellular network.
(b) The maximum possible number of traffic channels in each cell. Assume channels are shared equally among the cells in a cluster.
(c) The amount of traffic capacity of (i) each cell, and (ii) the entire network.
(d) The grade of service of the network.
(e) The trunking efficiency of the network, and each cell.

(12) A given mobile cellular system has the following features:

- Coverage area is 50,000 square kilometers
- Population of the area is 100,000 people, of which only 80% can afford a mobile service.
- Operator has acquired 20 MHz bandwidth and mobile network uses 1.4 MHz channel spacing without guard bands. Each cell uses one channel for signalling and the rest for payload traffic.
- Each mobile user uses the network only 3 minutes each hour, on the average.
- The farthest point a base station's signal has to reach is 1 km.
- Cluster size used in 21.

Assume that the network queues calls which arrive at a busy time (i.e., BCD system), and compute the following:

(a) The total number of duplex channels used in the mobile cellular network.
(b) The maximum possible number of traffic channels in each cell. Assume channels are shared equally among the cells in a cluster.
(c) The amount of traffic capacity of (i) each cell, and (ii) the entire network.
(d) The percentage of time that an arriving call has to be queued.
(e) The probability that an arriving call is held in a queue more than 5 seconds.
(f) The trunking efficiency of the network, and each cell.

(13) Reconsider the above two questions using 60° sectored cells throughout the network.

9.4 Mobility Management

Arguably, the most important feature in mobile wireless networks is the ability of the user to change location while using the services provided by the network. In order to provide the needed service to users seamlessly, the mobile network must track the current locations of its users, whether momentarily active or inactive. Owing to the level of movements, network users are classified into *fixed users* (i.e. no movement), *nomadic users* (i.e. moving at walking speed) and *mobile users* (i.e. moving in a car, train or airplane). CelPlanner^{TM} software uses this nomenclature. The process of doing so is referred to as **mobility management**. The functions in mobility management can be subdivided into location management (LM), cell selection (CS), and handoff (HO) management (see Fig. 9.24).

An UE which changes its location by moving from the coverage area of one eNodeB to that of another eNodeB can be in one of two states:

- *Inactive state*: the UE is powered on, but not currently in use to exchange payload.
- *Active state*: the UE is powered on, and currently exchanging payload.

Both states on a UE causes it to change eNodeB. Switching from one eNodeB to another while the UE is in active state is referred to as **handoff** or **handover**. However, switching from one ENodeB to another while the UE is in inactive state is referred to as **cell selection**. Whether in active state or inactive state, the UE receives broadcast signals from usually multiple eNodeBs. The UE selects or chooses one of the cells/eNodeBs based on either proximity or the strength in the signal received. Note that cell selection management functions

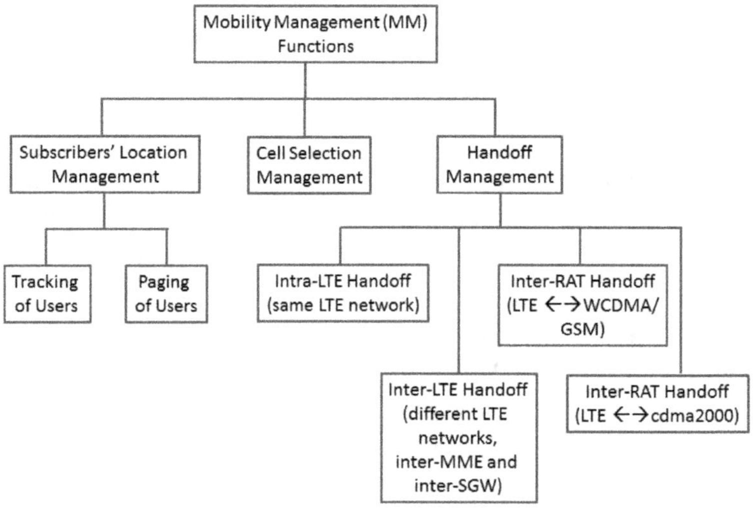

Fig. 9.24 Mobility management functions in LTE/LTE-Advanced.

include cell search, cell identification, cell selection and cell reselection.

As the traffic exchanged in mobility management is not payload, but signalling, an objective in mobility management strategies is to minimise it (i.e., signalling traffic). Signalling traffic sources are:

(1) Handoff or session management signalling (ca. 61.9%).
(2) Paging signalling (ca. 28.7%).
(3) Tracking area updating (TAU) signalling (ca. 4.9%).
(4) IRAT (inter-radio access technologies) signalling (ca. 2.13%).
(5) Active mobility signalling (ca. 1.2%).
(6) Attach/detach signalling (ca. 1.08%).
(7) All other signalling, e.g., cell selection (ca. 0.1%).

The percentage values stated against each signalling type were estimated from LTE MME in US cities.[2]

The LTE architecture is illustrated in Fig. 9.2. We can observe its flatness compared to those of 1G, 2G and 3G mobile systems. For example, the E-UTRA comprises only eNodeB and UE. LTE uses no centralised node. For this reason, the eNodeB is tasked with all the air interface processes. The control plane functions are managed by the MME (mobility management entity), while the S-GW (serving gateway) manages all the user plane functions. The LTE's equivalent of 3G's home location register (HLR) is the HSS (home subscriber server). One of its main functions is the storage of data on subscribers. The functions of the VLR (visitor location register) in 3G/2G are performed by the MME in LTE/LTE-A networks. PLMN operators have implementation choices. For example, if the S9 interface is implemented then the visited PCRF (policy control and charging rules function) uses it to request a roaming user's QoS policies from its home PCRF.

As mobility management is a user plane function, only the MME but not the S-GW is engaged. The main processes involved in mobility management are:

- **Handover** or **handoff**, also referred to as **small-scale mobility** or automatic wireless link transfer, is the process in which an active mobile connection is transferred from a radio resource to another. The two sets of resources involved can be handled by the same base station (resulting in intra-cell handoff) or different base stations (resulting in inter-cell handoff).
- **Roaming**, also referred to as **large-scale mobility**, is a type of mobility which allows a mobile

[2]Unfortunately, the author has missed the source of this data so readers may use it as just an information.

user to move to a geographical location which is outside the coverage area of his subscriber (aka home network) and still be able to use the mobile service in an unnoticeable manner. An example is movement across cities. Roaming is enabled by two processes: registration and location tracking. With registration it is the UE which takes steps to update its current whereabouts to the network. Location tracking, however, is initiated and managed by the network to find the current location of the mobile users.

- **Location updating** is the process used by the mobile device to announce its identity and current location to the base station serving the geographical area its user moves to. This status update is transferred to the home network. The home network is the PLMN owned by a user's service provider where the latter has its subscription.
- **Paging** is the process used by the network to alert the UE of events of many different natures, such as to receive an incoming call and emergency messages.

9.4.1 *Mobile Device, SIM Card and Identification Numbers*

A mobile user accesses a mobile network service using a mobile device, which is referred to variously, including mobile phone, mobile handset, mobile host (MH), mobile station (MS) and user equipment (UE). There are many manufacturers of mobile devices, and each manufacturer provides different kinds of handsets. Devices used to access mobile communications service has been categorised, as found in Table 9.6. The categorisation is based on features such as the maximum data rate at which the device can connect to the network. The connectivity rates depend on the quality of the electronics (or wireless modem) inside the device, as well as the category of radio technology that it supports.

There are many types of identifications needed for the purpose of mobility management, billing and user identification, and the identification of network elements. These identifications are grouped into:

- Identification of mobile subscribers: IMSI, SUPI, SUCI, GUTI, S-TMSI, etc.
- Identification of mobile devices: IMEI, IMEISV, MSISDN, MSIDN (Mobile Station International Data Number), MSRN (Mobile Station Roaming Number) for PSTN/ISDN routing, etc.
- Identification of base stations and network: cell identity (CI), cell global identification (CGI), base station identity code (BSIC), home NodeB (HNB) name, HNB unique identity, identification of HSS (home subscriber servers), etc.
- Identification of location areas: Location area identification (LAI), routing area identification (RAI), etc.

Some of the identifications are peculiar to the upcoming 5G mobile technology. Details of the various types of identifications can be found in the 3GPP TS 23.003 V15.4.0 specification of June 2018. We sample only a few of the identifications below.

An integrated circuit referred to as subscriber identification/identity module (SIM) or SIM card for short, must be obtained from a mobile service provider and inserted into the mobile phone for it

Table 9.6 Some categories of mobile handsets and supportable data rates.

Category	Download (Mbps)	Upload (Mbps)	Device Example
Cat 3	100	50	Apple iPhone 5
Cat 4	150	50	Samsung Galaxy S5, Apple iPhone 6
Cat 6	300	50	Galaxy Alpha, OnePlus 3T, Samsung Galaxy S6, Apple iPhone 6s
Cat 9	450	50	Samsung Galaxy Note 7
Cat 11	600	50	Hotspot devices
Cat 16	1000	50	Hotspot devices

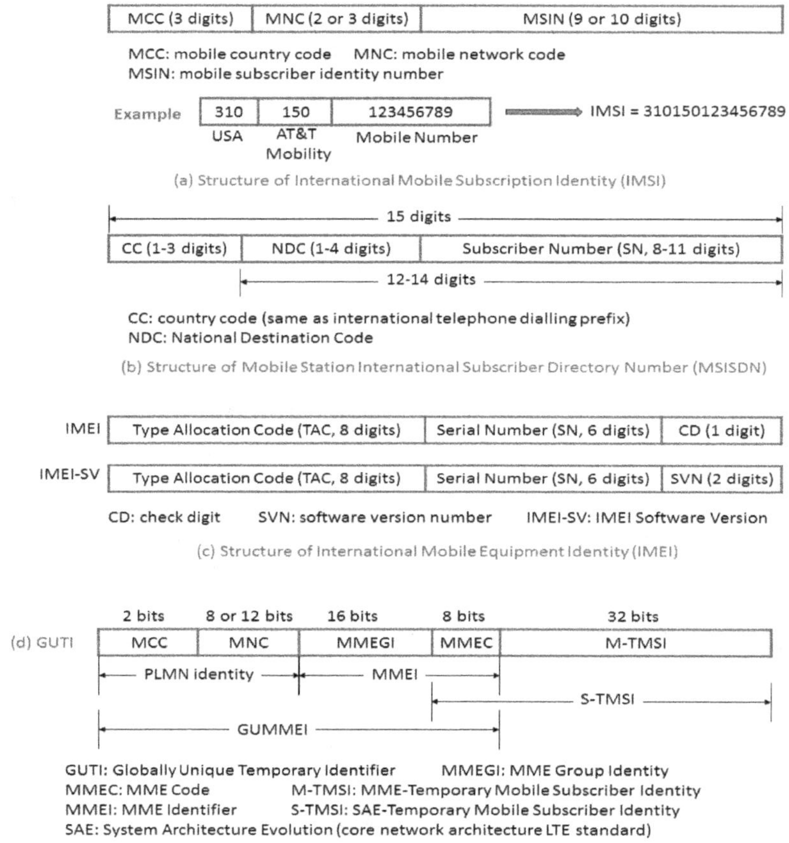

Fig. 9.25 International identification numbers of users and user equipment.

to connect to the mobile service. Data stored on the SIM card include the international mobile subscription identity (IMSI), which globally identifies the mobile via the user equipment. The IMSI is used to route calls to appropriate mobile phones. Other data stored on the SIM card are the service provider's identity and password (or key) needed to encrypt and decrypt data. Of course, a SIM card also has some memory to store user data. The IMSI and its components are illustrated in Fig. 9.25(a). Note that the digits are decimal digits. As 4 bits are needed to represent each decimal digit, the 15 decimal digits in the IMSI are encoded as 60 bits in electronic devices.

The use of SIM cards started with 2G cellular mobile technologies around 1992, and it was released with the first digital mobile phone, Nokia 1011 for GSM networks. For safety reasons, a mobile phone without an active SIM card is allowed to call only emergency numbers. With the increasing desire to reduce the weight and size of mobile phones, the size of SIM cards has undergone continuous reduction from the early full-size SIM through thumbnail-sized standard-SIM, micro-SIM of 2003 to the current nano-SIM which was released in 2012. In fact, the SIM slot in some modern mobile phones is so small that they can use only micro-SIM or nano-SIM.

It is important to note that some mobile phones do not have SIM card slot. Such phones are usually manufactured for a specific mobile service provided. I remember that the phone that I was using in Australia in 2002 did not work when I travelled to South Korea. This is because operators there at that time tied the handset to their service. Some mobile phones have SIM slot but locked to a certain mobile service provider. There are shops which unlock the phones so user can use them to access the service of competing operators. I am unsure whether this practice is legal. The IMSI comprises three parts: the mobile country code (MCC), the mobile number code (MNC) and the mobile sub-

Region	1st digit of MCC	Country	Operator	Some MCCs	Some MNCs
Test networks	0	Test network		001	01
Europe	2	Germany	Vodafone D2	262	09
		Finland	TeliaSonera	244	91
		Sweden	Telia/TeliaSonera	240	01
		UK	Vodafone	234	27
North America & Caribbean	3	Canada	Bell Mobility	302	610
		USA	Verizon	311	13, 270, 271, 276, 281, 287, 481, 486,
Asia & Middle East	4	Bangladesh	Airtel/Warid	470	07
		India	Bharti Airtel Limited	404	10, 45
		UAE/Dubai	Etisalat	431	02
Oceania	5	Australia	Telstra	505	01, 11, 71, 72
			Optus (SingTel)		02
			Vodafone		03, 06, 07
			Dept of Defence		04
			RailCorp		13
			AAPT/TPG		14
			VicTrack		16
		Samoa	Samoatel Mobile	549	27
		Singapore	SingTel	525	7
Africa	6	Ghana	MTN	620	01
		Ghana	Vodafone		02
		Ghana	Airtel/ZAIN		06
		Nigeria	MTN	621	30
		South Africa	MTN	655	10, 12
South & Central America	7	Brazil	Brazil Telcom	724	16
Worldwide	9	For satellite, maritime, aircraft communications and expedition			

Fig. 9.26 A sample of global public land mobile network (PLMN) identifiers.

scriber identification number (MSIN). The MCC uniquely identifies the country in which the mobile subscription is made, while the MNC uniquely identifies the home PLMN providing the subscription.

A sample of MCC and MNC are shown in Fig. 9.26. Different organisations manage the three components of the IMSI:

- MCC: The ITU allocates and manages the MCC.
- MNC: National numbering plan administrations allocate and manage this. For example, in Australia the MNCs are managed by ACMA (Australia Communications and Media Authority), and in Ghana they are managed by the NCA (National Communications Authority).
- MSIN: Allocated to subscribers by each PLMN operator providing the mobile subscription.

Only the MCC and the MNC are needed for international identification of roaming mobile users.

Another number used to uniquely identify mobile users globally is the MSISDN (Mobile Station International Subscriber Directory Number). The MSISDN is very similar to the numbering plan used in PSTN/POTS and both are defined in the ITU-T recommendation E.64. It is made of 1-3-digit country code (CC), the national destination code (NDC) and the subscriber number (SN). The CC is the same as the international telephone prefix numbers for countries, and it is used to identify the country in which the mobile user has the mobile service subscription. For example, the CC for Australia is 61, 233 for Ghana, 44 for UK, 1 for USA and Canada and 32 for Belgium. The MSISDN and its components are illustrated in Fig. 9.25(b).

Note that every mobile subscriber is assigned these two numbers IMSI and MSISDN, each of which has the maximum size of 15 digits. The MSISDN is *assigned to* the SIM card, but the IMSI is *burned into* the SIM card. As these two numbers are linked to the SIM card, a mobile user can acquire multiple of them through multiple subscriptions. The number that we dial while making a call is the MSISDN. The easiest way to remember the difference is that IMSI is the SIM card, while the MSISDN is the mobile number. A subscriber can change one of these two numbers but maintain the other.

Every mobile device (not the user) is also identified during manufacturing by a number called IMEI (international mobile equipment number), which is also called mobile equipment identity (MEI). It is usually written at the back of the device or inside of its battery compartment. IMEI is the serial number of the device, and has nothing to do with user identification. In order to reduce the crimes and theft relating to mobile devices, it is unlawful in many countries to change the IMEI of a device, and/or possess a tool which can be used to change it. We can find the IMEI of a mobile phone by just typing on the screen *#06#. Operators use the IMEI and the TAC to block stolen mobile devices so they cannot be used.

Figure 9.26(c) shows the format of the IMEI. It is made of an 8-digit TAC (type allocation code), a 6-digit serial number (SN) and either a 1-digit check digit (CD) or a 2-digit Software Version Number (SVN). The check digit is computed using the Luhn algorithm or Luhn checksum formula. As illustrated in Fig. 9.26(c), there are two versions of IMEI: the 15-decimal-digit IMEI and the 16-decimal-digit IMEI Software Version (IMEISV). The SN uniquely identifies each mobile device of a given model, while the TAC uniquely identifies a particular model of a given manufacturer. This means that every mobile device manufacturer must acquire a new uniquely allocated 8-decimal-digit TAC for each model of mobile device that it manufactures. The TAC is allocated by the GSMA, headquartered in London UK, through its approved national reporting bodies. The GSMA is made of worldwide mobile operators and mobile device manufacturers. Note, however, that a given model of mobile handset can be allocated multiple TACs, depending on such things as the location of manufacture and model revision.

Figure 9.26(d) shows the format of another mobile identification number. It is called GUTI (Globally Unique Temporary Identifier). The GUTI is a temporary number allocated to a mobile handset by the MME on these two occasions:

- During the TAU (tracking area updating) process.
- When an UE first attaches to the LTE network.

What is the importance of the GUTI? The IMSI is a very important number, and therefore measures are taken to ensure its privacy and security. For that reason, it is not transmitted over the air interface. For example, if a bad person gets access to someone else's IMSI he can use it to access mobile services at the cost of its authorised owner. Therefore, the GUTI is used in lieu of the IMSI. Like the IMSI, the GUTI uniquely identifies mobile users globally. The international part of the identification numbers are allocated and managed by the ITU-T, while the national parts are managed by national telecommunications authorities or regulators. For example, MNCs are managed by ACMA in Australia and by NCA in Ghana. The format and components of the GUTI are shown in Fig. 9.26(d). The MMEI identifies the MME which allocated the GUTI to the UE. The MMEC identifies an MME in MME group or pool. A pool or group of MMEs are identified by the MMEGI. The GUMMEI is used for communication between the old and the new MMEs about the UE it is assigned to.

9.4.1.1 *eUICC and Embedded SIM for M2M Communications*

The embedded Universal Integrated Circuit Card (eUICC), which is also called embedded SIM (eSIM) or soft-SIM, is a new technology in wireless communications. Mobile users are identified in PLMNs by their SIM cards. Traditional SIM is a physical card which is bound to a specific mobile service provider. This requires a user to acquire a new SIM anytime one switches between Mobile Network Operators (MNO). Such a traditional SIM will be too restrictive to use in machine-to-machine communications. Therefore, the eUICC has been developed. An eSIM enables multiple SIM card profiles on a

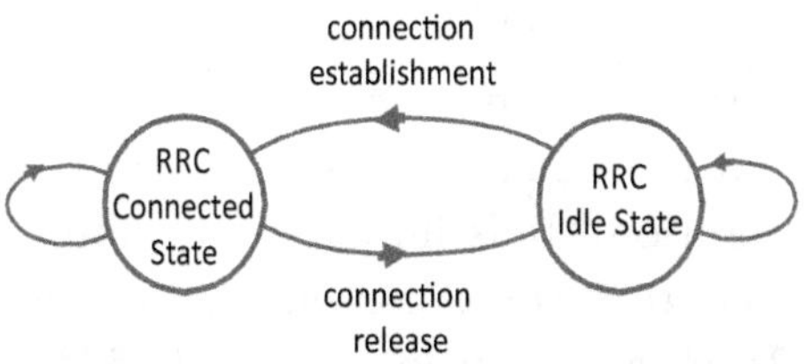

Fig. 9.27 Simplified states and state transitions of an UE which is powered on.

single card. With eSIM both initial provisioning and changing of MNO occur Over-The-Air (OTA) remotely.

9.4.2 *Paging and User Tracking*

Every UE which is powered on can be in one of the two states or modes: idle state or connected state (Fig. 9.27). It is important to note that being in the idle state does not mean that the UE does not perform any activity. An idle UE is connected to the evolved packet core (EPC), but not the eNodeB. Over the air interface, UEs connect to their serving eNodeBs via the radio resource control (RRC) protocol. According to [ETSI-3GPP (2016)], the UE performs the following functions in the idle mode:

(1) Scans and selects the appropriate public land mobile network (PLMN) from the list of those available.
(2) Monitors the paging channel.
(3) Receives broadcast messages.
(4) Reporting and registration of tracking area to the NAS (non-access stratum), which is a functional layer in the protocol stack in the UE/EPC interface.
(5) The measurement and selection of an appropriate cell (or eNodeB) to attach to.

An active (i.e., non-idle) UE is said to be in RRC-connected, EMM-Registered or ECM-Connected state. EMM means EPS mobility management, while ECM means EPS connection management. Evolved packet system (EPS) is used to refer to both the the LTE air interface and the evolved packet core (EPC) which is the core network for LTE. In the active state, the UE performs functions in both the user plane and the control plane. User plane functions include transmission and reception of data, measuring and reporting of channel state information (CQI) to the eNodeB, and the monitoring of control channels. The control plane functions include transmission and reception of data and the measurement of signal qualities of neighbouring cells.

9.4.2.1 *Paging in LTE*

paging in LTE Paging is the process by which the network, either the eNodeB or the MME (mobility management entity), sends alerts of different events to the user equipment. Paging in LTE is used for the following purposes:

(1) To alert all UEs in a given geographical area of an emergency situation, such as tsunami and earthquake. The notification is referred to as Cell Broadcasting in GSM and WCDMA, but Earthquake and Tsunami Warning System (ETWS) in LTE.
(2) To alert an idle UE to initiate a mobile-terminating SMS, packet-switched data connection request or fall-back circuit-switched connection request.
(3) To prompt a UE to re-acquire system information.
(4) To send the Commercial Mobile Alert Service (CMAS) notification to all UEs in a given geographical area. CMAS communicates a highly emergency information. It is sent, for example, to communicate an eminent extreme-level or severe-level threats, a child abduction alert (referred to AMBER[3] alert), or by people of high authority in a geographical area.

Note that both CMAS and ETWS are public warning systems (PWSs) in LTE. Mobile handsets usually sleep in order to conserve their batteries while in the idle state. Every mobile network communicates system information to mobile devices it serves. Part of the system information is the paging cycle, which is the periodic times that mobile devices in

[3]AMBER, meaning America's Missing: Broadcast Emergency Response, was named after Amber Hagerman, a 9-year-old boy who was abducted and murdered in USA in 1996. The term Amber alert is now used globally to broadcast urgent and important information.

idle state should wake up and expect to receive signals from the serving base station. The paging cycle is communicated at the time to mobile attaches to the network. Thus, an idle mobile device wakes up at the beginning of a paging cycle to monitor for any paging message. It re-enters into the sleep mode in the absence of a paging message. The presence of a paging message causes the mobile handset to request a connection to the serving base station.

In 4G LTE network, the paging messages can originate from the eNodeB or the MME. Paging from the eNodeB are for purposes such as to inform UEs of any change in system information, emergency notifications (e.g., warnings against earthquake or tsunami). Paging originating from the MME is used to alert mobile devices of incoming connection calls and text messages, as well as all other services subscribed to by the user.

9.4.2.2 *Mobile User Tracking*

An operator of an LTE-based network, during network roll-out, divides its cells or eNodeBs into *logical* groups referred to as tracking area (TA). The geographical area referred to as TA in LTE is called routing area or location area in 3G and 2G cellular mobile networks. A TA is a **paging area**. At any point in time every UE must be registered with only one TA and the UE/TA biding is recorded in the serving MME. The MME paging message to an UE is broadcast to all cells belonging to the TA currently attached to the UE. Traditionally, TAU was triggered if an UE moves across two different TAs. In order to minimise signalling load, TAU occurs if an UE moves to a TA whose TAI is not included in its TAI list.

The PLMN operator identifies each TA using a unique 16-digit number called tracking area code (TAC). Each mobile operator is identified globally by its 20-bit or 24-bit public land mobile number (PLMN). The PLMN is made of 12-bit Mobile Country Code (MCC) and an 8-bit or 12-bit Mobile Network Code (MNC). Both MNC and MCC are defined by the ITU-T in its Recommendation E.212. All mobile service providers, whether virtual mobile network operator (VMNO) or traditional operator who owns infrastructure is issued one or more MNCs. As observable in Fig. 9.26, the same operator acquires a different set of MNC and MCC, depending on the number of networks and the country of operation. It is important to note that not all PLMNs are actually for public use. For example, VicTrack and RailCorp networks in Australia, based on GSM-R radio technology, are private networks. They are used by the railway companies for the communication between railway regulation control centers and the trains.

From Fig. 9.26 we can construct the PLMNs for Vodafone in Australia as 50503, 50506 and 50507, while the PLMN of Vodafone in Ghana is 62002. As each decimal digit is encoded using four binary digits, we need 20 binary digits to encode each of the 5-digit decimal numbers in each PLMN ID. This is the reason why 20 to 24 bits have been allocated for the PLMN in the TAI (Fig. 9.28).

The PLMN identity (comprising the MNC and the MCC) and the TAC make up the tracking area identifier (TAI), whose format is illustrated in Fig. 9.28. The TAI of a TA is globally unique and all cells or eNodeBs in a TA have the same TAI. The TAC is 16 bits long, but usually written in four hexadecimal digits in the form 0x0001, 0x0002, 0x0003, etc. Computer compilers are informed of a number being written hexadecimal notation by preceding the number with the two characters 0x. We have written each in four digits as each hexadecimal digit needs four bits to be encoded. Each PLMN operator has the freedom to encode the TAC the way it wants it. For example, a PLMN operator in Australia can use the first 4 bits (i.e., one hexadecimal character) in the TAC to encode the nine states and territories in the country, while the remaining 12 bits (i.e., three hexadecimal characters) are used to divide the coverage of each state. This would mean that the network could have $2^{12} = 4096$ tracking areas within each state.

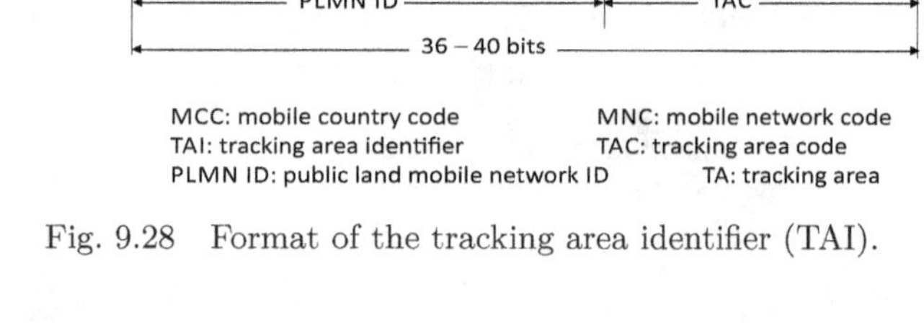

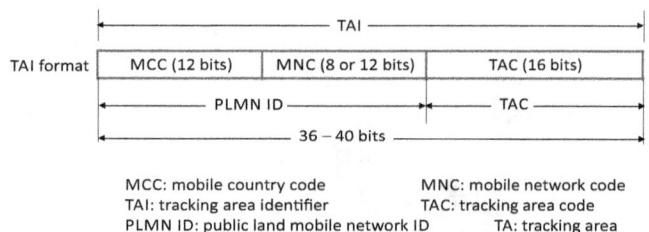

Fig. 9.28 Format of the tracking area identifier (TAI).

State	Total Area (km²)	Population (June 2017)	First TAC Hex digit	Range of TAIs Used in TA	Coverage area of TA (km²)	Range of TA (km)
Western Australia	2,529,875	2,580,400	0x0	0x0000 – 0x0FFF	617.65	15.42
Queensland	1,730,648	4,928,500	0x1	0x1000 – 0x1FFF	422.52	12.8
Northern Territory	1,349,129	246,100	0x2	0x2000 – 0x2FFF	329.38	11.3
South Australia	983,482	1,723,500	0x3	0x3000 – 0x3FFF	240.11	9.6
New South Wales	800,642	7,861,100	0x4	0x4000 – 0x4FFF	195.47	8.7
Victoria	227,416	6,323,600	0x5	0x5000 – 0x5FFF	55.52	4.63
Tasmania	68,401	520,900	0x6	0x6000 – 0x6FFF	16.70	2.54
Australian Capital Territory (ACT)	2,358	410,300	0x7	0x7000 – 0x7FFF	0.576	0.48
Jervis Bay Territory (ACT)	73	401	0x8	0x8000 – 0x8FFF	0.0178	0.09

Fig. 9.29 TAI coding example for a fictitious PLMN operator in Australia.

Figure 9.29 illustrates an example TAI coding. We can observe from the figure that the areas of the TAs range from pico-cells (90-m radius) to macro-cells (15.42-km radius). Therefore, they are theoretically possible. However, in order to minimise tracking area updating (TAU) and signalling load on the network, no PLMN operator would create TAs of pico-cell size. The range of each TAC in Fig. 9.28 is equal to the equivalent radius of a hexagon with the same surface area as the TA. The surface area of each TA equals the area of the state divided by 2^{12}

Each UE in the active state is known at the cell level by its serving eNodeB. However, an UE in the idle state is known at the TA level by the MME. Regardless of whether an UE is in the RRC-connected state or RRC-idle state, it can initiate a tracking area update (TAU). The difference is that the TAU concludes only when the UE enters the RRC-connected state. Many events trigger TAU. According to the LTE standard, TAU occurs at the occurrence of any of the events:

- Each UE stores a pool of TAIs, referred to as TAI list. Movement of an UE between TAs whose TAIs are in its stored TAI list does not require TAU. This is meant to reduce signalling load associated with TAU. However, if a UE moves from one TA into another TA whose TAI is not included in its stored TAI list, then TAU must be made.
- Indicating that the UE has selected a Closed Subscriber Group (CSG) cell whose CSG identity is not included in the UEs allowed CSG list.
- When the timer T3412 expires. Whenever a UE connects to an eNodeB, the latter sets a timer called T3412 and sends to the UE as part of the Attach Accept message during the attach procedure. This timer is kept in the UE so it can use it to periodically notify the network of its availability.
- Registering for EPS (evolved packet system) service after an inter-system change. For example re-selecting from UMTS/CDMA to LTE.
- Re-registering to LTE after circuit-switch-fallback is completed.
- Upon the UE receiving the 'Load Balancing TAU Required' message from the eNodeB for the release of the current RRC. This is used to achieve MME load balancing.
- When UE-specific DRX (discontinuous reception) cycle is updated.
- Registering with the circuit-switched (CS) domain for non-EPS service when the UE is already attached for EPS services. This involves completing an IMSI attach as part of the TAU procedure.

A TAU procedure is illustrated in Fig. 9.30. A summary of the procedure is:

(1) UE notifies the network through the appropriate eNodeB when it moves from one MME pool/group to another using the TAU message over the RRC signalling connection. The MMEs in an LTE network are divided into groups or pools. Each MME pool area has one or more tracking areas (TAs). The content of the signalling message is the TAU containing the two important parameters:
  (a) The TAC (tracking area code).
  (b) UE's temporally but globally unique identity, which is the GUTI.
(2) The eNodeB forwards the TAU request to the MME, as shown in Fig. 9.30.

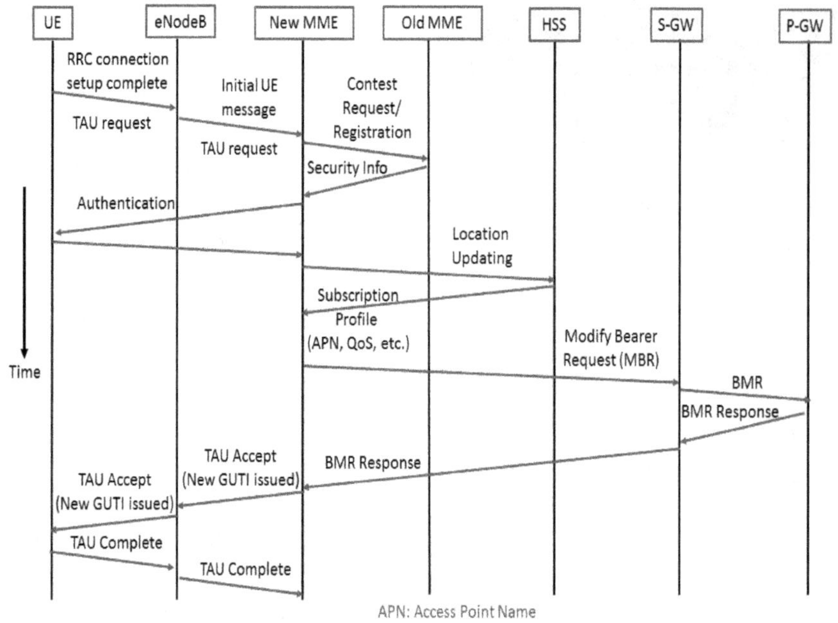

Fig. 9.30 Procedure for a tracking area updating (TAU) in LTE.

9.4.3 *Handoff*

The process of transferring an ongoing mobile communication from the serving resource (e.g., base station, sector) to another resource is referred to as handoff or handover (HO). There are four steps to handoff: HO initiation, resource reservation, HO execution and HO completion. There are several different algorithms for each of these four steps in the HO process. Several factors can contribute to the need for a handoff, but all of them focus on the service quality experienced by the mobile user, and/or for load balancing. Factors causing handoff include:

(1) Relocation of the mobile user owing to movement.
(2) Too low or poor received signal strength (RSS), which can lead to high bit error rate or call dropping.
(3) Better signals from other base stations.
(4) Too large distance between mobile station and the serving base station.
(5) The need for traffic load balancing to relief a base station of congestion.
(6) Too high co-channel interference.

Based on the role of the base station in executing it, handoff schemes can be classified into:

- Mobile-assisted handoff (MAHO): under the instruction of the network, the mobile handset measures and then forwards to the network the strength of signals it receives from surrounding base stations. Based on this information, the network executes handoff if the criteria for handoff are fulfilled. MAHO is used in 4G LTE networks and 2G cellular GSM (Global System for Mobile Communications), for example. In 2G cellular mobile systems (e.g., GSM) the measurement performed by the mobile station is called received signal strength indicator (RSSI). The measurement performed by the mobile station in 3G cellular mobile systems (e.g., WCDMA) is called received signal code power (RSCP). In 4G cellular mobile systems (e.g., LTE/LTE-A) the measurements performed by the user equipment are called reference signal received power (RSRP) and reference signal received quality (RSRQ). The typical range of values of the RSRP are −44 dBm to −130 dBm. Note that RSRQ is measured in dB, but the RSRP, RSSI and RSCP are measured in dBm. In LTE/LTE-A, the RSRP equals the received signal strength (RSS) of a specific cell or eNB, while the RSRQ is the ratio

$$RSRQ = \frac{\texttt{RSS from target cell}}{\texttt{RIP at the UE}} \quad (9.42)$$

where RIP (received interference power) is the sum of the powers received from all interfering cells, which are cells other than the target cell. In MAHO, the mobile device measures the RSSI, RSCP, RSRP and RSRQ, whichever is appropriate for the cellular technology, and forward to the base station for handoff execution.

- Mobile-controlled handoff (MCHO): the mobile handset monitors the transmissions from all base stations within its range continuously. It then initiates handoff to a new base station when the criteria for handoff are fulfilled. MCHO is used in IS-41 (Interim Standard 41) for example.
- Network-controlled handoff (NCHO): multiple base stations monitor continuously mobile users' transmissions. The network then initiates handoff to another base station if the handoff criteria in use are fulfilled. NCHO is used in AMPS and CT-2, for example.

As illustrated in Fig. 9.24, LTE has to handle handoff within a given operator's network, and between LTE networks managed by different operators. These are inter-LTE handoffs. A handoff between two networks using the same air interface standard, such as inter-LTE handoff, is referred to as **horizontal handoff**. LTE also has to handle handoff to and fro two types of earlier radio access technologies (RATs). Namely, those standardised by 3GPP (e.g. GSM and WCDMA) and those standardised by other organisations, such as cdma200. A handoff between two networks using different air interface standards, such as handoff between LTE and WCDMA, is referred to as **vertical handoff**.

Based on how many base stations that a mobile is connected to during handoff, handoff algorithms are divided into:

- **Hard handoff algorithms**. These are also referred to as **break-before-make** as the radio resources of the old base station are released before the connection to the new base station completes. The chance of a call or session ending abruptly is higher in hard HO. An abrupt ending of a session is called **forced termination** or **session/call dropping**. A HO algorithm seeks to avoid or at least minimise *session/call dropping probability*. LTE/LTE-A use only hard HO for many reasons, such as:

  (1) The LTE architecture is flat and so has no centralised node to coordinate soft HO.

  (2) LTE is based on OFDM. Therefore, neighbouring cells use different set of OFDM subcarriers, requiring the UE to resynchronise to a new set of OFDMA subcarriers whenever it switches from one cell to another. The UE would need multiple transceivers in order to synchronise to multiple sets of OFDM subcarriers at the same time. UEs usually have only a single transceiver, making soft HO in LTE/LTE-A impossible. Neighbouring cells in spread-spectrum based cellular technologies, such as WCDMA/UMTS and cdma2000, use the same carrier frequencies but different set of spreading codes. This enables a UE to communicate with two NodeBs to achieve soft HO. LTE/LTE-A does not allow adjacent-cell frequency reuse, which is permissible, and in fact, the norm in 3G.

  (3) Soft HO was used to resolve cell-edge reception challenges in CDMA-based technologies. An eNodeB of LTE/LTE-A uses a different set of subcarriers for uplink and downlink communications, making the transmissions orthogonal. This orthogonality of carrier frequencies eliminates intra-frequency interference.

- **Soft handoff algorithms**. These are also referred to as **make-before-break** as the radio resources of the old base station are not released until the connection to the new base station completes successfully. For this reason, session/call dropping rate is expected to be lower than in hard handoff algorithms. Unlike predecessor technologies, LTE does not support soft handoff. Soft HO occurs between two antenna sectors of two different NodeBs. Soft HO is used in spread-spectrum based cellular technologies, such as WCDMA/UMTS and cdma2000.

- **Softer handoff algorithms**. Softer HO is soft handoff between two sectors belonging to the same NodeB in 3G cellular mobile technologies, such as WCDMA/UMTS and cdma2000. For example, two adjacent sectors belonging to a NodeB of a WCDMA network can receive two versions of a transmission from one mobile station owing to multipath signal propagation. The NodeB then uses its RAKE receiver to combine the signals.

Intra-cell handoff is handoff within a cell, such as moving from one cell sector to another sector of the same base station. **Inter-cell handoff** is handoff from one cell or base station to another.

The distance of a mobile unit from a base station can be estimated by measuring power levels. Base station to mobile unit distance estimates also enable the estimation of the direction in which the mobile unit is moving within the cell. Using the GPS timing in the mobile network, we can determine the rate at which the handoffs are made. We can then estimate the speed at which the mobile unit is moving using the handoff rate and the distance moved.

As said earlier, the chance of an ongoing call being terminated prematurely is referred to as **dropping rate** or **forced termination probability** (P_f). The event in which the mobile network denies access to a user who initiates a mobile call is called blocking, and the rate of that happening is called blocking probability (P_b). The blocking probability is referred to as grade of service (GOS) in teletraffic engineering. These two parameters are some of the most important metrics used to analyse the performance of handoff algorithms. We seek to minimise P_f without increasing P_b. Mobile network designers use strategies to keep P_b and P_f within acceptable limits. An average user accepts call blocking than call dropping. For this reason, reducing P_f at the cost of increasing P_b is preferable by a mobile network operator.

The success of handover depends on many parameters, such as the speed at which a mobile user is moving and cell sizes. The performance metrics which must be considered while designing handoff algorithms include:

- Low forced termination probability (P_f), aka call dropping rate (P_d).
- Low call blocking probability (P_b).
- High call completion probability (P_c).
- Handover probability (P_{ho}), i.e., the probability that a call undergoes handoff prior to completion.
- Handover blocking probability (P_h), i.e., the probability that a handoff process fails.
- Low duration of handover interruption (t_{hi}), i.e., the period of time during handover in which the ongoing call is not connected to any base station.
- Handover rate (n_H), i.e., number of handoffs per unit time. The smaller the cell, the higher the expected value for n_H.
- Low handover delay (t_{ho}), i.e., the time taken to from the point that handover is needed to the point that handover is executed.

An ongoing call is dropped if the new base station to accept the call has no idle resources to serve the call being transferred to it, and the signal strength of the serving base station has fallen below the mobile's receiver sensitivity level. Two of the mechanisms used to keep P_B within limits are handoff queuing and guard channels. **Guard channels** are a set of channel, which is a fraction of the channels of the mobile network, which its reserves absolutely for handoff calls. This means that a handoff call drops if and only if all the guard channels are busy and there are no non-guard channel which is idle. In **handoff queuing** the mobile network keeps the handoff call on hold as it looks for resources to serve it. Of course, a queued handoff call can still drop under certain conditions.

A problem relating to handoff is **cell dragging**. It occurs if a slowly-moving mobile user gets closer to a new base station than its currently serving base station, but the signal strength from the serving base station is still strong enough (above the handover threshold) to keep the call. The level of the received signal strength (RSS) at the mobile user may be due to line-of-sight link. The time duration in which a mobile user stays within a given cell or base station before the call is handed over to another cell is called **dwell time**.

We use Fig. 9.31 to illustrate handoff. In this figure, the mobile user is currently being served by

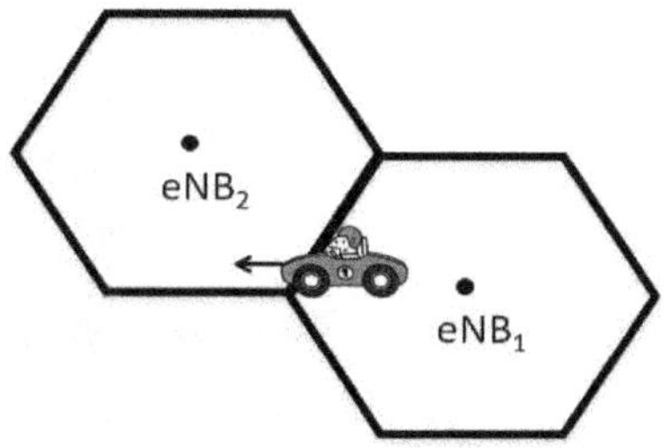

(a) Mobile user moving across cells during a call

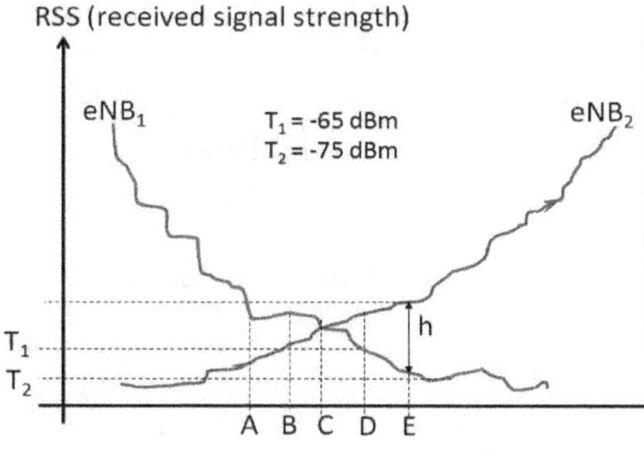

Location	Ms RSRP from eNB1 (dBm)	Mn RSRP from eNB2 (dBm)	SIR at UE (dB)
A	-40	-68	18
B	-47	-60	13
C	-50	-50	0
D	-65	-45	-20
E	-70	-39	-31

RSRP: reference signal received power

(b) RSRP dynamics in

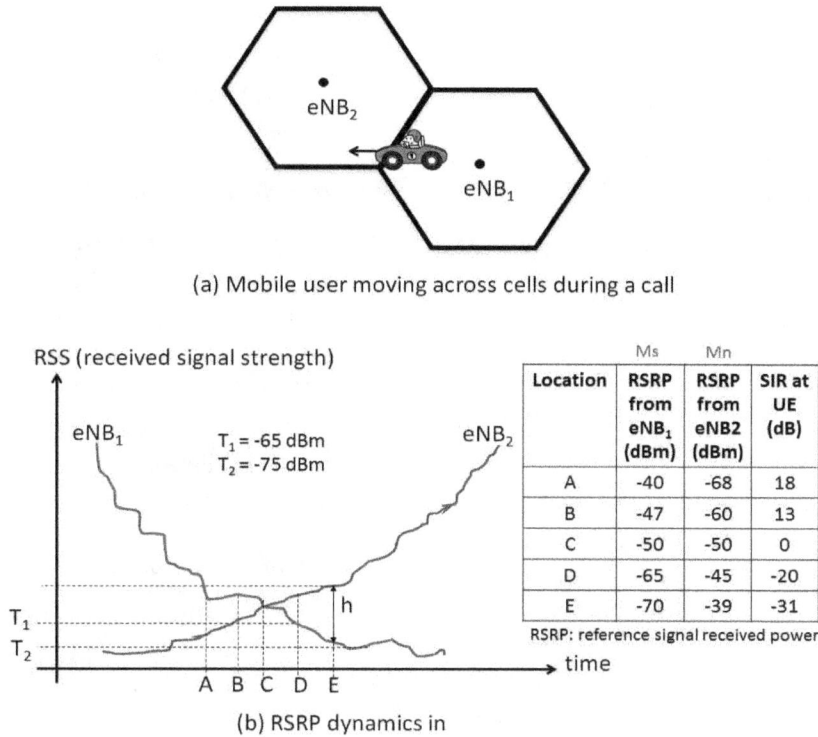

Fig. 9.31 An ongoing mobile call requiring a handoff (h: handoff hysteresis margin).

the eNB_1. However, it is increasingly moving away from the coverage area of eNB_1 to that of eNB_2. Let us observe the dynamics in the received signal strengths from both base stations eNB_1 and eNB_2, as well as the dynamics in the signal-to-interference ration (SIR). The SIR drops from 45 dB at location A which is very close to the serving eNB_1 to -90 dB at location E which is far away to the serving base station.

An important question in handoff algorithms is at which of the instants A, B, C, D and E in Fig. 9.31 is most appropriate to execute the handoff, assuming that radio resources are available in the new cell? A formula used to initiate HO in LTE/LTE-A is ([3GPP (2018)], p. 163)

$$M_s + FO_s + CO_s + Off < M_n + FO_n + CO_n - h \tag{9.43}$$

where

- h: **handover hysteresis margin**. Measured in decibels (dB), this margin is used to prevent or at least minimise the ping pong effect in handovers. This margin is illustrated in Fig. 9.31. The strength of the signal in the new base station should be higher than that of the old base station by this margin on the least.
- M_s, M_n: measurement results by UE of the serving eNB and new/neighbour eNB, respectively, which can be either RSRP (in dBm) or RSRQ and RS-SINR in dB.
- FO_s, FO_n: frequency-specific offset of the serving cell and new/neighbour cell frequency, respectively.
- CO_s, CO_n: cell-specific offset of the serving cell and new/neighbour cell, respectively.
- Off: handoff offset parameter.

Handoff algorithms can be classified based on the criteria or parameters used. The parameters used in HO decisions and algorithms include:

(1) Power/energy based parameters: UE battery power, mean battery power, UE power class.
(2) Received signal strength (RSS) at UE of signals received from eNodeBs.
(3) The minimum required RSS for service continuation.

(4) Received signal received quality (RSRQ)
(5) Handoff hysteresis margin (HHM), h.
(6) Path loss between UE and eNodeB.
(7) UE speed of motion.
(8) Mobility pattern of UE.
(9) Dwell/residence time of UE in the cell.
(10) QoS requirements of traffic being exchanged during the handoff.
(11) Bit error rate (BER).
(12) Load of cell (available resources in cell), cell capacity.
(13) SINR: average SINR target of UE, SINR of serving cell.
(14) Cost function relating two or more of the above parameters.

Reference [Zhang *et al.* (2016a)] discusses a handoff algorithm for the transition between a femtocell and a macrocell based on the cost function

$$f(i) = \frac{\alpha(i)C_{\mathtt{max}}(i)P_0(i)}{\log[e\beta(i) + N_{\mathtt{UE}}(i)]} \qquad (9.44)$$

where $i \in \{s, n\}$ denotes a cell with s denoting the currently serving eNB and n being the neighbour/new eNB the mobile may hand over to, $\alpha(i)$ is a factor used to tune the cost function's value, $\beta(i)$ is a factor adapted selected to match the type of cell, $C_{\mathtt{max}}(i)$ is the capacity of cell i, $N_{\mathtt{UE}}(i)$ is the number of UEs in cell i, $P_0(i) \in \{RSRP(i),\ RSRQ(i)\}$ denotes the state of cell i. The HO procedure continues if and only if the following condition is fulfilled

$$f(n) > f(s) + h(i) \qquad (9.45)$$

where $h(i)$ is the HO hysteresis margin for cell $i \in \{s, n\}$.

As discussed earlier, LTE/LTE-A uses hard handover. The occurrence of an important thing is called event. An example is the change in signal strength. The event in which the signal quality of a neighbour eNB becomes better than that of the serving eNB by an offset is called Event 3. A handoff can be executed in Event 3 if it remains so for at least a time period referred to as TTT (Time to Trigger). A simple hard HO algorithm used in LTE, which is also referred to as **power budget handover algorithm**, is [Aziz and Sigle (2009)], [Lin (2003)]

$$\begin{aligned} &Enter\ Event\ 3: RSRP_{\mathtt{n}} - h > RSRP_{\mathtt{s}} \\ &Leave\ Event\ 3: RSRP_{\mathtt{n}} + h > RSRP_{\mathtt{s}} \end{aligned}. \qquad (9.46)$$

A HO is executed if and only if condition (9.46) is fulfilled for a period of time as long as at least the TTT. Both the TTT and h are design parameters used to minimise or even avoid a ping-pong effect in handoff. The larger they are the less frequent the HO. The possible range of values are

$$\begin{aligned} &0\ \mathtt{dB} \le h \le 30\ \mathtt{dB} \\ &TTT \in \{0, 40, 64, 80, 100, 128, 160, 256, \\ &\qquad 320, 480, 512, 640, 1024, 1280, \\ &\qquad 2560,\ and\ 5120\}\ \mathtt{ms} \end{aligned}$$

Example

An algorithm proposed for a hard handoff in LTE is given in Eq. (9.46) where RSRPn and RSRPs are the reference signal received powers measured from the new eNB and the serving eNB, respectively, by the user equipment, and h is the handover hysteresis margin. The range of values for the optimum hysteresis margin is found to be $4\ \mathtt{dB} \le h \le 4.5\ \mathtt{dB}$. By studying the scenario in Fig. 9.31, concisely describe which region of Fig. 9.31(b) should the handoff between eNB1 and eNB2 occur and why. Assume that the links of the two eNBs have been in the state shown in Fig. 9.31 for a time frame which is at least equal to the time-to-trigger (TTT) the handoff used to avoid a ping-pong effect. Here, TTT is assumed to be 80 ms.

Solution

The reference signal received power (RSRP) of the neighbour eNBs exceeds that of the serving eNB by 20 dB in the range C and D. This is where the handoff should occur as the hysteresis margin is exceeded, and the TTT is also exceeded.

9.4.4 *Check Your Understanding*

Students are encouraged to check their understanding of the content of this section with the following questions prior to proceeding to the next section.

(1) Explain the meaning of cell dragging in cellular mobile networks.
(2) Discuss handover hysteresis margin, and its importance in the handoff process in cellular mobile networks.
(3) Describe the ping-pong effect in cellular mobile networks, and state one way of minimising it.
(4) Discuss the difference between the paging originating from the eNodeB and paging originating from the MME in an LTE network.
(5) Discuss the meaning of a user equipment (UE) being in an idle state. What are the activities performed by a UE in the idle state?
(6) Research to find the two largest public land mobile network (PLMN) providers in your country of residence. Then, find the PLMN identifier for each of them in a tabular form.
(7) Discuss the main difference between registration and location tracking in mobile networks.
(8) What is the maximum number of tracking areas that an LTE network can have?
(9) Compute the maximum number of mobile country codes available to the ITU-T for the unique identification of PLMNs.
(10) How many different PLMNs are uniquely identifiable in each country?
(11) Compute the maximum number of mobile subscribers that a given PLMN operator can support with a single MNC.

9.5 Cellular Mobile Network Design Case Study

We use a case study to discuss the two complementary perspectives of designing a mobile cellular network: coverage dimensioning (for coverage-limited network) and capacity dimensioning (for capacity-limited network). The procedures are illustrated in Fig. 9.32. Assume that the coverage dimensioning procedure estimates the number of cell sites needed to cover a given geographical area as $N_{\texttt{cov}}$, while the capacity dimensioning procedure estimates same as $N_{\texttt{cap}}$, then we need

$$N_{\texttt{cells}} = \max\{N_{\texttt{cov}}, N_{\texttt{cap}}\} \tag{9.47}$$

cell sites to cover the area.

A cellular mobile network at any point in time is either a **thermal noise-limited network** or an **interference-limited network**. Networks based on spread spectrum multiple access schemes, such as cdma2000 and WCDMA, are known to be interference-limited networks. In a *noise-limited network*, the performance (i.e., range or coverage) of the system is constrained by thermal noise as the latter dominates the interference in the system. In such a network, the bit error rate (BER) decreases exponentially as the signal-to-noise ratio (SNR) increases. Therefore, we can increase the transmit power to increase the SNR and the coverage. Thermal noise-limited networks are also referred to as **coverage-limited networks**.

In *interference-limited network*, however, the co-channel interference (CCI) outweighs the thermal noise, and the BER decreases linearly with the SNR. Thus, increasing the transmit power has unremarkable effect on the BER. Interference-limited networks are also referred to as **capacity-limited networks**. Assume that the thermal noise power is $P_{\texttt{th}}$, while the interference power is $P_{\texttt{int}}$. In traditional cellular mobile networks which use carrier frequencies below 3 GHz, the ratio of $P_{\texttt{int}}$ to $P_{\texttt{th}}$, referred to as interference-to-noise ratio (INR), is a function of the frequency reuse distance D and the transmit power spectral density (PSD), i.e.,

$$INR = \frac{P_{\texttt{th}}}{P_{\texttt{th}}} = function(D, PSD)\,. \tag{9.48}$$

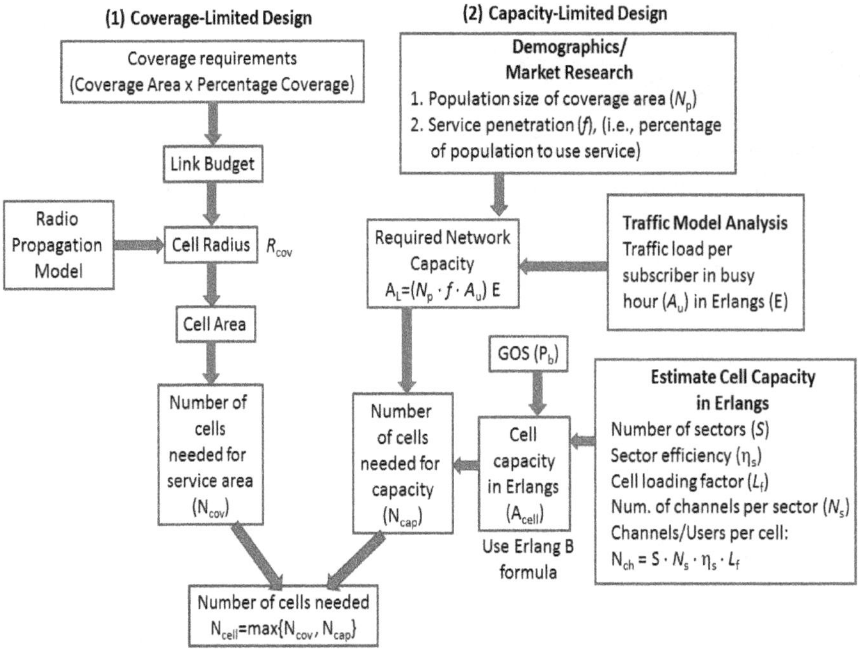

Fig. 9.32 Two complementary methods of cellular mobile network design.

However, this relation has not been proven to work in mmWaves communications using carrier frequencies above 3 GHz. This area is currently an open research topic.

9.5.1 *Topography, Morphology and Their Resolution*

The meaning of the words terrain, topography and morphology appears somewhat confusing in the literature. Some people use terrain and topography as synonymous, while others refuse to do so. Still, some people use morphology in situations that others use terrain or topography. In this book, we use terrain and topography as synonymous and differentiate them from morphology. However, we should remember that all three words describe the same object but from different perspectives. Table 9.7 summarises the differences.

The natural and artificial features on Earth's surface described in topographical maps are grouped into seven categories:

(1) Relief: mountains, hills, slopes, contours (for depressions).
(2) Vegetation: trees, wooded areas.
(3) Hydrography: rivers, lakes, streams, swamps, falls, rapids.
(4) Transportation: bridges, roads, railways, airports, airfields, seaplane anchorages.
(5) Boundaries: geographical, administrative, recreational, territorial, international.
(6) Culture: electric power lines, towers/buildings, pipelines, urban development.
(7) Toponymy: names of places, boundaries, water features and landforms.

Terrain is largely used as synonymous to topography.

Example

The parameters shown in Fig. 9.33(a)–(c) are important for mobile cellular network design. Therefore, software packages, such as CelPlanner®, used to design cellular mobile networks define them.

(1) (a) Briefly explain the difference between 24k image and 100k image in topography.
(2) Concisely explain how 1 sec topography (or morphology) data differs from 3 sec topography (or morphology) data.

Table 9.7 Comparing and contrasting topography and morphology.

Topography	Morphology
Stems from the words 'topos' and 'graphia' in Greek meaning 'place' and 'writing', respectively.	Stems from the words 'morphemes' and 'logos' in Greek meaning 'internal structure' and 'study', respectively.
First use occurred in geography to mean *precise, quantitative* description of natural and man-made features in a place or surface, e.g. surfaces of the Earth, moon, asteroids, planets.	First use occurred in linguistics to describe internal structure of words. *Qualitative* scientific study of shape/form, structure/texture and distribution of objects in a surface, and cause for a shape.
Answers the question 'what are the dimensions of the features on a surface?'	Answers the question 'what does an object look like?'
E.g. finding heights, sizes and roughness of features of a surface.	Qualitative description of 3D shape of a surface.
Found from optical profilometry, atomic force microscopy, stylus profilometry.	Found from scanning electron microscopy, optical microscopy.

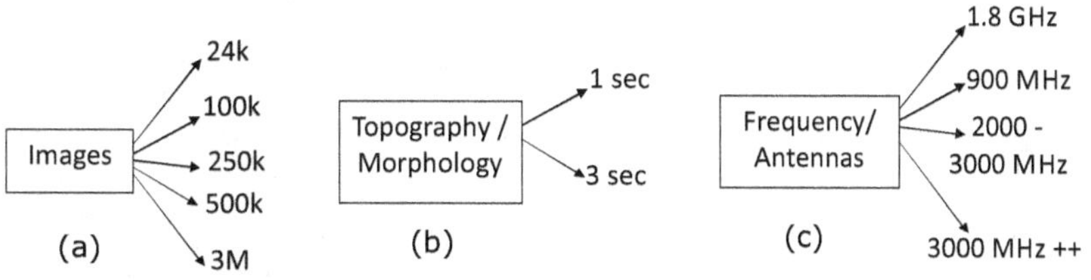

(d)

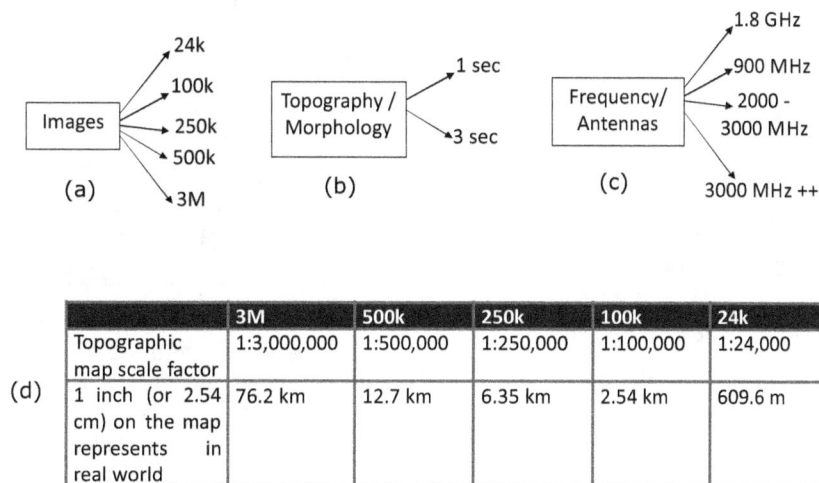

	3M	500k	250k	100k	24k
Topographic map scale factor	1:3,000,000	1:500,000	1:250,000	1:100,000	1:24,000
1 inch (or 2.54 cm) on the map represents in real world	76.2 km	12.7 km	6.35 km	2.54 km	609.6 m

Fig. 9.33 Some interesting parameters in mobile cellular network design.

(3) How does a 900-MHz antenna differ from a 1.8-MHz antenna?

Solution

(1) Xk measures the amount of details in an image. The smaller the scale factor, the more detailed the image is. For example, a 24k image is finer/better than a 100k image. This is quite opposite to the convention used to measure the quality of images in digital photography. The scaling of geographical images can be done in either inches or centimeters (cm). North American data usually uses the former scaling. In such a scaling, 24k means 1 inch on a map equals 24,000 inches in real world. Using the conversion of 1 inch equals about 2.54 cm, we obtain the data in Fig. 9.33(d).

(2) Topographic data provide information about the elevation of the surface of the Earth. The *sec* is an abbreviation for *second* (or more precisely, *arcsecond*). One arcsecond (or second)

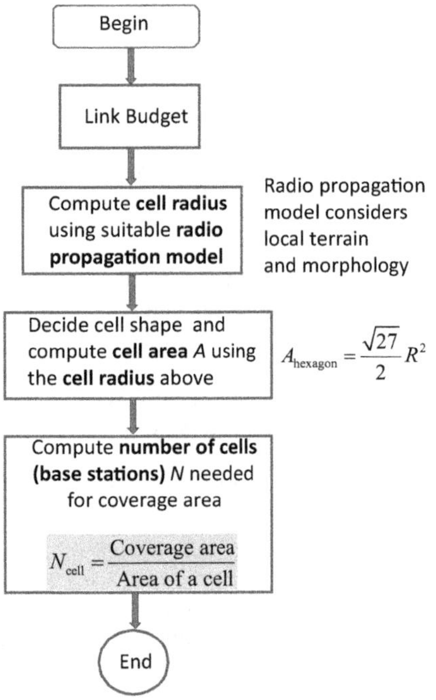

(a) Coverage-limited mobile network design algorithm

Parameter	Value (decibels)
eNodeB/BTS transmit power, P_t = 20 W	+43.0103 dBm
Cable loss, L_c = 2 dB	-2 dB
BTS gain, G_t	+15 dBi
Other gains	+10 dB
Slow fading margin (aka shadow fading margin) for 90% cell edge coverage and slow fading standard deviation of 8 dB, L_{sf} =	-10.2 dB
Interference margin, L_{int}	-2 dB
Path loss, L_p	??
Obstructions penetration loss, L_{pen}	-18 dB
Human body loss, L_{body}	-1 dB
UE antenna gain, G_r	+0 dB
Total power level at receiver, P_{Tot}	34.8103 dBm - L_p

(b) An example of link budget

Fig. 9.34 Mobile cellular network coverage dimensioning.

latitude is about 30.83 m on the Equatorial plane, while 3 sec is about 92.5 m. The seconds specify the resolution of the topographical/morphological data, which shows the amount of details in the data. One second resolution is better than 3 sec resolution.

(3) Every antenna is developed to operate effectively in a range of frequencies (or a band of frequency spectrum). For example, a 900-MHz antenna operates in its efficient mode at frequencies around 900 MHz. Ideally, such an antenna should not pick signals with frequencies outside its dimensioned band.

9.5.2 *Coverage-Limited Network Design*

We are designing a cellular mobile network to cover a geographical area whose four corners are defined by the coordinates (37.3639° S, 144.3983° E), (37.3639° S,145.5279° E), (38.2633° S,144.3983° E) and (38.2633° S, 145.5279° E). The LTE technology is used. The algorithm and the network data in Fig. 9.34 are used to calculate the downlink budget of the LTE network. The remaining system data are:

- Carrier frequency, f_C: 850 MHz (i.e., LTE band 5)
- Bandwidth, B: 20 MHz
- eNB antenna tower height, h_{bs}: 40 m
- UE antenna height, h_{ms}: 1.5 m
- UE (receiver) sensitivity, P_{UES}: -104 dBm

The operator has found that the radio propagation in the entire coverage area (i.e., Melbourne) of the mobile cellular network can be described by the Hata (aka Okumura-Hata) radio propagation model

$$L_p = 69.55 + 26.16 \log_{10} f + (44.9 - 6.55 \log_{10} h_{bs}) \log_{10} d_{TR} - 13.82 \log_{10} h_{bs} - A_c \quad (9.49)$$

with the mobile antenna correction factor for large cities being

$$A_c = \begin{cases} 9.29(\log_{10}(1.54 h_{ms}))^2 \\ \quad -1.1, & 150 \le f \le 200 \text{ MHz} \\ 3.2(\log_{10}(11.7554 h_{ms}))^2 \\ \quad -4.97, & 200 < f \le 1500 \text{ MHz} \end{cases}$$

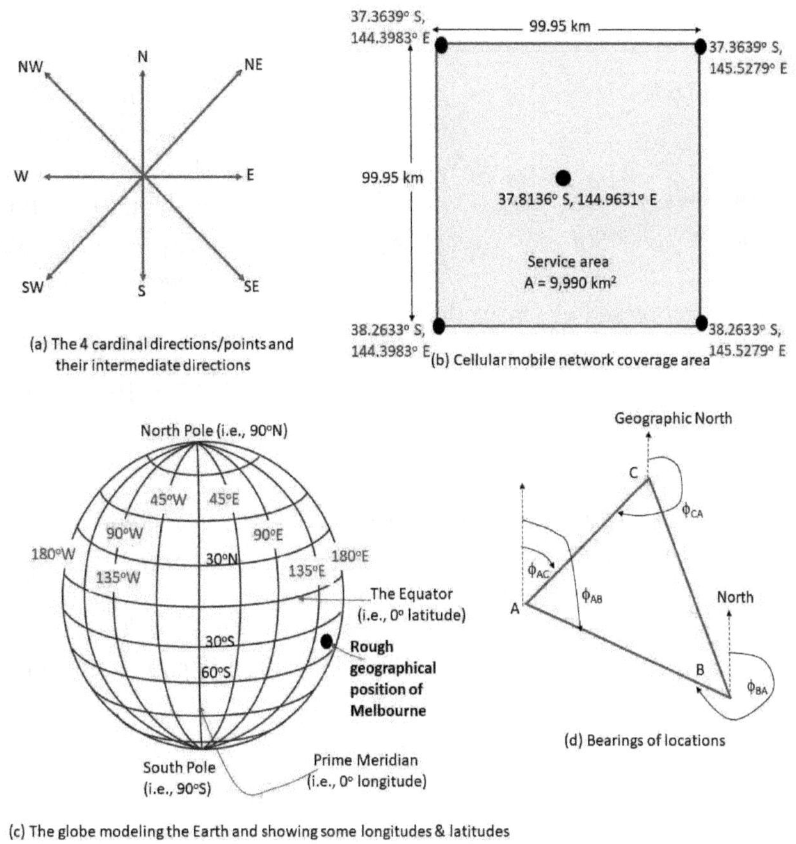

Fig. 9.35 Cardinal directions and the globe showing some latitudes and longitudes.

where L_{p} is the total path loss in decibels (dB), frequency is in MHz, antenna heights are in meters, and $d_{\text{T-R}}$ is the distance between the eNodeB and the user equipment (UE) in km, which is the maximum cell radius. We use the following steps to illustrate the design process.

(1) Make a link budget and use it to compute the maximum path loss L_{p} that the link between the eNodeB and the UE can tolerate for an acceptable link operation in both decibels (dB) and ratio form.

(2) Using the results in Part (1), we compute the maximum radius of each cell, $R_{\mathtt{cell}}$, used in the cellular network to ensure optimum system operation. We need to find a propagation model suitable for the service area. Here, we are using the Hata path loss model given in Eq. (9.49).

(3) We compute the number of cells needed to cover the service area specified above, assuming 100% coverage. We assume that all cells have the same size and each has the hexagonal shape.

We neglect all other losses not mentioned here.

The Design Process

We first need to find the surface area of the geographical area to be covered by the cellular mobile network using the given geographical coordinates. The coverage area is illustrated in Fig. 9.35(b). This requires us to remember some geography lessons in high school. Figure 9.35 shows all that we need, i.e., the longitudes and latitudes. We also need to know how to convert geographical coordinates into distances. Longitudes are imaginary circles which intersect the south and north poles and the equator. Longitude zero degree is called prime meridian. Latitudes are imaginary circles connecting the east to the west and parallel to the equator. The range of values for the longitude and the latitude are (Fig. 9.35):

$$\begin{aligned} 180^\circ \ \mathtt{W}\ (\mathtt{or} - 180^\circ) &\leq \textit{longitude} \\ &\leq 180^\circ \ \mathtt{E}\ (\mathtt{or} + 180^\circ) \\ 90^\circ \ \mathtt{S}\ (\mathtt{or} - 90^\circ) &\leq \textit{latitude} \\ &\leq 90^\circ \ \mathtt{N}\ (\mathtt{or} + 90^\circ) \end{aligned} \quad (9.50)$$

The location of a place is expressed as (latitude, longitude) or $\binom{latitude}{longitude}$. Both latitude and longitude are specified using the sexagesimal notation, comprising arc degrees (denoted, °), arc minutes (denoted, ′) and arc seconds (denoted, ″), and altogether referred to as DMS (namely, degrees, minutes, seconds). The DMS relate as follows:

$$1^\circ = 60' = 3600'' \ \mathtt{and}\ 180^\circ = \pi \ \mathtt{radians}\,. \quad (9.51)$$

A rule-of-thumb used to convert one degree of latitude and one degree of longitude into kilometers are:

$$\begin{aligned} 1^\circ \ \mathtt{of\ lat.} &\approx 110.567\ \mathtt{km\ (Equator)} \\ &\quad - 111.699\ \mathtt{km\ (poles)} \\ 1^\circ \ \mathtt{of\ longitude} &\approx 111.32 \times \cos(\mathtt{latitude} \\ &\quad \mathtt{in\ radians})\,. \end{aligned} \quad (9.52)$$

Using the above relations, we obtain the surface area of the location to be served by the cellular mobile network as $99.95\mathtt{km} \times 99.95\mathtt{km} = 9{,}990\mathtt{km}^2$ (see Fig. 9.35(b)). We have assumed $1^\circ\ \mathtt{of\ latitude}\ \approx 111.133\ \mathtt{km}$. Thus, $99.95\ \mathtt{km} \approx 0.8994^\circ$ latitude. We computed the longitudes using Eq. (9.52). We now use this surface area to proceed with the system design.

(1) The coverage dimensioning objective is to allow maximum possible path loss in order to obtain maximum link range or cell radius. This is achieved when the power in the received signal equals the receiver sensitivity. Thus, we need to equate the received power obtained from the link budget (see Fig. 9.34(b)) to the receiver sensitivity, which is

$$\begin{aligned} P_{\mathtt{UES}} = -104\ \mathtt{dBm} &\overset{!}{\approx} 34.81\ \mathtt{dBm} \\ &- L_{\mathrm{p}} \Leftrightarrow L_{\mathrm{p}} \approx 138.81\ \mathtt{dB}\,. \end{aligned} \quad (9.53)$$

(2) In order to find the cell radius or communications range, we need to compute the path loss resulting from the system's parameters and the propagation model. We then equate the results to the path loss computed from the link budget. We are using the Hata propagation model quoted in Eq. (9.49). With the given system parameters we obtain

$$\begin{aligned} A_{\mathrm{C}} &= 3.2(\log_{10}(11.7554 h_{\mathtt{ms}}))^2 - 4.97|_{h_{\mathtt{ms}}=1.5\mathtt{m}} \\ &\approx 6.725 \times 10^{-4}\ \mathtt{dB} \end{aligned}$$

and

$$\begin{aligned} L_{\mathrm{p}} &= 69.55 + 26.16 \log_{10} f \\ &\quad + (44.9 - 6.55 \log_{10} h_{\mathtt{bs}}) \log_{10} d_{\mathtt{TR}} \\ &\quad - 13.82 \log_{10} h_{\mathtt{bs}} - A_{\mathrm{c}} \\ &\approx 124.0425 + 34.4065 \log_{10} d_{\mathtt{T\text{-}R}}\,. \end{aligned} \quad (9.54)$$

Equating (9.53) and (9.54) yields the needed results as

$$\begin{aligned} &124.0425 + 34.4065 \log_{10} d_{\mathtt{T\text{-}R}} \\ &\overset{!}{=} 138.8103 \Leftrightarrow d_{\mathtt{T\text{-}R}} \approx 2.7\ \mathtt{km}\,. \end{aligned}$$

A cell with radius 2.7 km is large enough to call a macrocell.

(3) A cell with radius 2.7 km in the ideal case covers the geographical area $A_{\mathtt{cell}} = \frac{\sqrt{27}}{2} R_{\mathtt{cell}}^2 \approx 18.94\ \mathtt{km}^2$. The number of such cells, $N_{\mathtt{cell}}$, needed to cover 100% of the service area is

$$\begin{aligned} N_{\mathtt{cell}} &= \frac{\mathtt{size\ of\ service\ area}}{\mathtt{area\ of\ a\ cell}} \\ &= \frac{9{,}990\ \mathtt{km}^2}{18.94\ \mathtt{km}^2} = 528\,. \end{aligned}$$

This completes the coverage-limited system design, in which we use the system parameters and the appropriate path loss model to find the number of eNodeBs needed to cover a geographical area of a given size.

Note that the coverage dimensioning in practical systems is more involved than what has been presented. For example, the size of cells are not of the same size in practical systems. This is because cell sizes depend on many factors, such as:

(1) Link budget, which depends on system parameters (transmit power, antenna gains, etc.) and path loss, which also depends on the terrain (morphology and topology) of the specific geographical area.
(2) Local traffic load which depends on user density, traffic pattern of users, service penetration, etc.
(3) The propagation delay that the PRACH (Physical Random Access Channel) Preamble has been configured to handle, which lies in the range 15 km (for PRACH Preamble Format 0) and 108 km (for PRACH Preamble Format 3). UEs use the PRACH to perform random access procedure which is needed for many purposes, such as seeking an initial access or synchronisation with the cellular mobile network.

9.5.3 *Capacity-Limited Network Design*

The effective planning of the capacity of a network requires subscriber forecasting to know the estimated number of users, traffic forecasting to find out the expected traffic load which depends on the types of applications that the users of the network will generate, and the QoS requirements of each of the applications to run over the network. Network planning also seeks to achieve certain objectives, such as:

(1) Resource optimisation and load balancing so that the resources are provisioned depending on their need.
(2) Satisfactory quality of experience (QoE), which is judged by service users.
(3) Cost optimisation to keep CAPEX within budget. For example, some CAPEX can be postponed.
(4) Scalability so that the network can expand in many dimensions over time based on need. Areas of scalability include coverage of the network and the type of services and use cases of the network. For example, the current mobile networks must support video transfer (e.g., ViLTE), voice (e.g., VoLTE), broadcasting such as the eMBMS (evolved multimedia broadcast multicast services) over LTE, over the top (OTT) and IoT (Internet of Things). Earlier networks did not need to support these applications. Modern networks must have an inherent capability to support future applications that are unthinkable at the time of roll-out.

We can use the procedure shown in Fig. 9.32 and Fig. 9.15 to dimension the capacity. Alternatively, we can dimension the capacity without input from coverage dimensioning. We then compare the number of eNodeBs or cells needed in both coverage dimensioning and capacity dimensioning. If the results from both do not tarry then we need to iterate to adjust one until they match.

It is important to note that users and traffic is usually not uniformly distributed in a network. Therefore, the sizes and capacities of cells in a practical cellular network may not be the same. Roughly about 15% of the cells may carry 50% of the traffic load in a network. This is the reason why some cells can be congested while others are idle. This is the rationale behind load balancing and channel borrowing strategies. Furthermore, traffic load changes over time. The 60-minute period in the day that the network experiences maximum load is traditionally called busy hour (BH). The BH is known to carry about 7% of the total network traffic in a 24-hour period.

Example

Assume that a cell of an LTE network using six sectors (i.e., $S = 6$) has the uplink capacity of $A_{\texttt{cell}} = 60$ Mbps, cell loading factor of $f_{\texttt{cell}} = 80\%$ and peak-to-average traffic load of $\eta = 25\%$. If each user has on the average traffic demand of $A_{\text{u}} = 40$ kbps, how many users can the cell support simultaneously?

Solution

The designed effective capacity of the cell is $A_{\texttt{cell,e}} = A_{\texttt{cell}} \times f_{\texttt{cell}} = 60 \times 0.8 = 48$ Mbps. Thus, the cell capacity is

$$N_{\text{u}} = \frac{A_{\texttt{cell,e}} \times S}{A_{\text{u}}(1+\eta)} = 5760 \texttt{ subscribers.}$$

9.5.4 *Check Your Understanding*

Students are encouraged to check their understanding of the content of this section with the following questions prior to proceeding to the next section.

(1) Consider the five geometrical shapes shown in Fig. 9.5(c) and (e). With justification, describe which of them is most suitable to model the footprint of (i.e., the shape of the area illuminated by) an omni-directional antenna used at a radio base station or eNodeB in practical mobile cellular network.

(2) Assume that an operator has acquired a 20-MHz bandwidth from a local radio frequency spectrum regulator to roll out a PLMN based on LTE-Advanced technology using a cluster size of $C = 4$. In LTE, the smallest unit of frequency resource which can be allocated to one user is called a physical resource block (PRB). One PRB occupies 180 kHz of frequency, lasts 0.5 ms in time, and it is made of either twelve 15-kHz OFDM subcarriers or twenty-four 7.5-kHz OFDM subcarriers. The operator divides the 20 MHz into 100 PRBs occupying 18 MHz. The remaining 2-MHz bandwidth is used as guard bands. Place the traffic channels in the cells such that both the ACI (adjacent channel interference) within a cluster and CCI (co-channel interference) between clusters are minimised. Explain your cellular system design (i.e., cell planning and frequency planning) using appropriate diagrams.

   (a) Compute the total number of channels k that are available in the entire mobile cellular network.

   (b) Compute the number of channels n in each cell if the available channels are shared equally among the cells in a cluster.

   (c) Number the channels or carriers as f1, f2, f3, f4, etc. and group them as G0, G1, etc. and then assign channel groups to cells in a cluster in a manner that the ACI within a cell is minimised.

(3) Consider the illustration in Fig. 9.36.

   (a) (i) State the name of the region marked A. (ii) State one cause of that region. (iii) Describe the problems caused by such a region in the coverage area.

   (b) (i) State the name of the region marked B. (ii) State one cause of that region. (iii) Describe the problems caused by such a region in the coverage area.

(4) Consider the illustration in Fig. 9.37, which was obtained from CelPlanner^{TM}.

   (a) Briefly describe what CelPlanner^{TM} software application is.

   (b) Describe precisely the meaning of the illustration in Fig. 9.37.

(5) Consider Fig. 9.38.

   (a) Describe what a cluster size is. State which cluster size is used in Fig. 9.38.

   (b) Show in Fig. 9.38 the frequency reuse distance D. Describe the importance of the frequency reuse distance in mobile cellular networks.

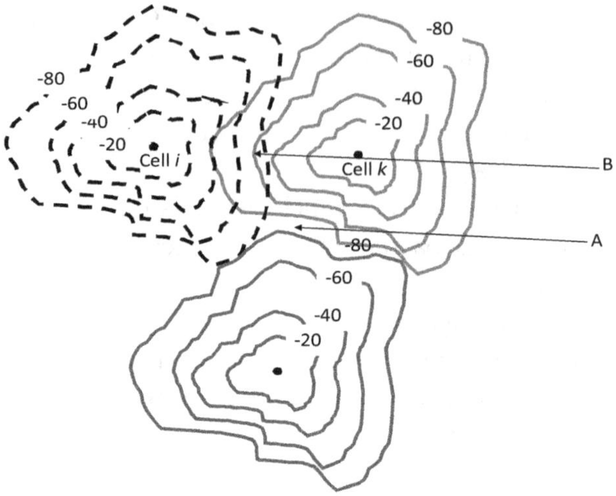

Fig. 9.36 Non-ideal cell planning.

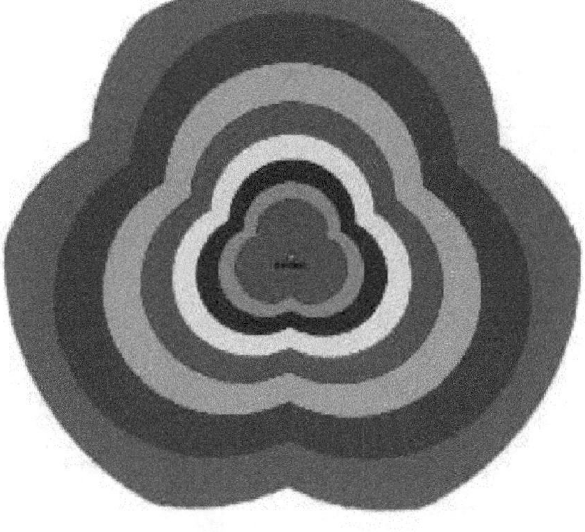

Fig. 9.37 The radiation pattern of a base station.

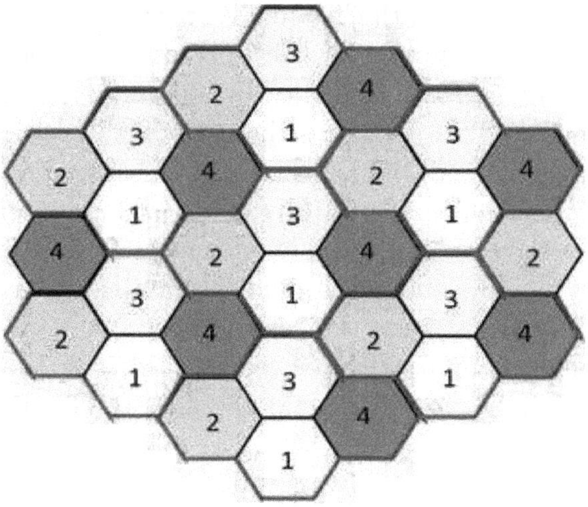

Fig. 9.38 A cell planning example.

(c) Assume that the mobile cellular network using the cell sites' plan shown in Fig. 9.38 uses microcells, each of radius 500 m, and requires the minimum signal-to-interference ratio (SIR) of 18 dB to operate with acceptable performance. Compute the average reuse distance D if the path loss exponent is 3.5.

(6) The size of a given city is about 5000 square kilometers. Assume that a telecommunications operator has 80% mobile cellular coverage of the entire region, estimate the maximum number of cell sites the Telco needs to acquire if it uses:

(a) Micro-cells, each of radius 100 m?
(b) Macro-cells, each of radius 15 km?

Assume that each cell has a hexagonal shape.

(7) Two strategies used in mobile cellular network design to achieve the same purpose are illustrated in Fig. 9.39. We seek to differentiate between these two strategies according to the illustration.

(a) What do we seek to achieve with the two strategies?
(b) State the name of each strategy
(c) Describe the differences between the two strategies.
(d) State two disadvantages in each strategy with justification.

(8) Discuss the reasons why base stations are placed on high heights (e.g., rooftops, masts) in sparsely-populated areas, but are positioned below rooftop in densely-populated areas.

(9) A cell in a cellular mobile network has a radius of 50 km and 20 duplex traffic channels, each of bandwidth 30 kHz. Each of the users make on the average two calls, each lasting three minutes during the busy hour (BH). The network maintains a grade of service of 5%.

(a) Compute the maximum capacity of the cell in terms of traffic intensity in Erlangs.
(b) Compute the maximum capacity of the cell in terms of number of users.
(c) Assuming a circular cell shape, compute the spectrum efficiency of the single-cell cellular network. The spectrum efficiency (SE) is defined as

$$SE = \frac{\texttt{total traffic intensity [Erlangs]}}{cellarea[\texttt{km}^2] \times \texttt{total system bandwidth [MHz]}}. \quad (9.55)$$

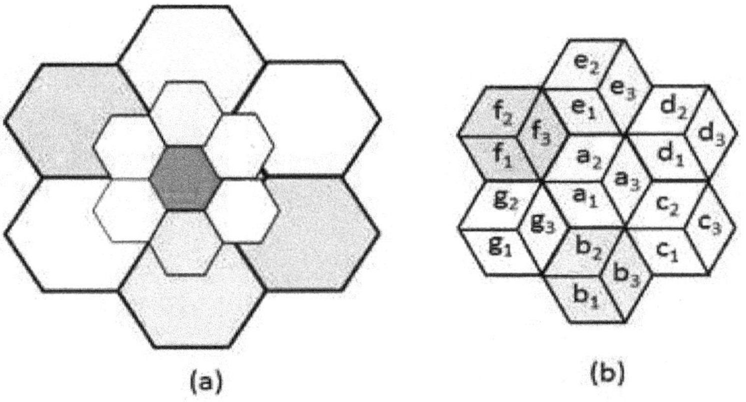

Fig. 9.39 Two strategies used in mobile cellular network design.

9.6 Overview of 5G Networks

The world is currently using the 4th Generation (4G) of mobile communications technology, referred to as Long-Term Evolution (LTE), and preparing a commercial rollout of 5G. There were multiple standards for all generations of mobile network technologies preceding 5G. For example, 3G has WCDMA and CDMA2000, and 4G has WiMAX (IEEE 802.16m) and LTE/LTE-Advanced. 5G includes a new air interface, called 5G New Radio (5G-NR), as well as harmonising existing technologies, such as WiMaX, Wi-Fi and LTE. Thus, 5G is a hybrid multi-technology. 5G improves upon the features of 4G. Figure 9.40 summarises some of the differences between the two technologies. A combination of the following parameters is referred to in 5G terminology as *numerology*:

(1) Frame size/duration.
(2) Subframe size/duration.
(3) Slot size/duration.
(4) Subcarrier spacing.
(5) Symbol duration.
(6) Cyclic prefix duration.

5G comes with many new features, such as millimeter waves (mmWaves), massive multiple-input multiple-output (MIMO), full duplex, beamforming and small cells called femto cells. What are mmWaves? These are wireless signals transmitted over frequencies above 6 GHz. Such signals have wavelengths in the millimeter range. The reason for using mmWaves is that the sub-6GHz bands are crowded and it is hard to find a big enough bandwidth to support the high data rate expected in 5G. Unfortunately, mmWaves cannot penetrate obstacles and are easily absorbed by objects in the wireless link. This mandates the use of small cells (or femto cells). MIMO is discussed in Chapter 7. One of the advantages in MIMO is the increase in data rate. However, using many antennas and transmitting omnidirectionally would increase the interference in the system. This is the reason why beamforming has been introduced in 5G. With beamforming, we use directional antennas to illuminate selected locations.

The 5G air interface is called New Radio (NR). 5G-NR can operate at frequencies up to 300 GHz in theory. However, the frequency bands currently approved by the ITU for 5G-NR, is divided into two:

- Frequency Range 1 or FR1 (sub-6 GHz): Radio signal transmission in this band is the best, but it is so crowded that 5G could not find enough bandwidth in it. The largest channel bandwidth in this band is 100 Mhz. 5G PLMN operating in the non-stand-alone (NSA) mode use this band. Example implementations are done by SK Telecom, KT Corporation and LU U Plus in South Korea. All these networks operate at 3.5 GHz.
- Frequency Range 2 or FR2 ($f > 24$ GHz): Bands approved by ITU as at Q2 2019 are 24.25–27.5 GHz, 37–40.5 GHz, 42.5–43.5 GHz, 45.5–

Feature	LTE/4G	5G Network
Network architecture	inflexible	flexible (due to SDN, NFV, LINP)
Core network	Evolved Packet Core (EPC)	5G Core (5GC)
Handoff	Break-before-make	Make-before-break (eMBB, URLLC)
Data rate	Up to 1.5/3 Gbps (UL/DL)	Up to 10/20 Gbps (UL/DL)
Latency	60 - 98 ms	$\leq$ 1 ms
Capacity/connected devices	10,000 devices per square kilometer	1 million devices per square kilometer
Maximum mobility	350 km/h	500 km/h
Device battery life	~ 1 year	~ 10 years (for sensors)
Base station	eNodeB (eNB)	gNodeB (gNB) divided into Distributed Unit (DU) & Central Unit (CU)
Carrier frequency	0.45 ~ 3.8 GHz, 5 GHz	0.6 ~ 40 GHz
Spectrum efficiency	90% of channel bandwidth	98% of channel bandwidth
Channel bandwidth	1.4 – 20 MHz	Below 6 GHz: 5MHz – 100 MHz Above 6 GHz: 0.050 – 1 GHz
Subcarrier spacing	MBMS + Unicast: 15 kHz MBMS: 7.5, 1.25 kHz	Below 6 GHz: 2^k x 15 kHz, k=0,1,2 Above 6 GHz: 2^k x 15 kHz, k=2,3,4
Carrier aggregation	Up to 32 CC	Up to 16 CC
Frame structure	Slot format: fixed 1 slot duration: 0.5 ms 1 subframe = 2 slots: 1 ms 1 radio frame: 10 ms	Slot format: semi-statically and dynamically configurable 1 slot: 2^k x 0.125 ms, k=0,1,2,3 1 subframe = 1 ms 1 radio frame: 10 ms
Multiplexing/multiple access scheme	Downlink: OFDM Uplink: DFT-S-OFDM (SC-FDMA)	Downlink: OFDM Uplink: OFDM, DFT-S-OFDM
Channel coding	Turbo coding (data) TBCC (signalling/control)	LDPC (data) Polar code (signalling/control)
HARQ	Transport block (TB) based	Transport block (TB) based and code block group based
MIMO	SU-MIMO: 8 antennas ports MU-MIMO: 2 antenna ports	SU-MIMO: 8 antennas ports MU-MIMO: 16 antenna ports
Beamforming	Up to 8 layers (digital, no analog)	Up to 12 layers (digital & analog)

Fig. 9.40 Comparison between 4G/LTE and 5G technologies.

47 GHz, 47.2–50.2 GHz, 50.4–52.6 GHz, 66–76 GHz and 81–86 GHz. These bands support mmWaves. These bands have huge bandwidths except that the mmWaves cannot travel far, mandating small cells (called femto cells). Channel bandwidths vary between 50 MHz and 400 MHz. Verizon is using 28 GHz and AT&T is using 39 GHz.

In order to optimise the good propagation at sub-6 GHz and the high data rates at mmWaves, Huawei has embarked on the so-called multi-layer spectrum approach. In this approach the spectrum is divided into three regions:

- *Super data layer*: transfers signals in the FR2 bands to achieve very high data rates within small cells. One of the problems associated with small cells is the large number of cells sites required and their corresponding real estate acquisition/rentals. PLMN providers seek to mount the base stations on public utility poles. However, some City Councils seek to charge operators about \$30,000 to \$45,000 as an application fee to mount just one micro-base station on a utility pole! It is also noted that, owing to the delay associated with the acquisition of a local permit, it takes 18 to 24 months to mount a single micro-base station ([Sanou (2018)], p. 16).
- *Coverage and capacity layer*: transfers signals with frequencies within 2 to 6 GHz to optimise coverage and capacity/data rate.
- Coverage layer: uses sub-2 GHz to transfer signals over a wide area or deep indoors.

ITU requires 5G networks to support the following three classes of use cases (see also Fig. 9.41):

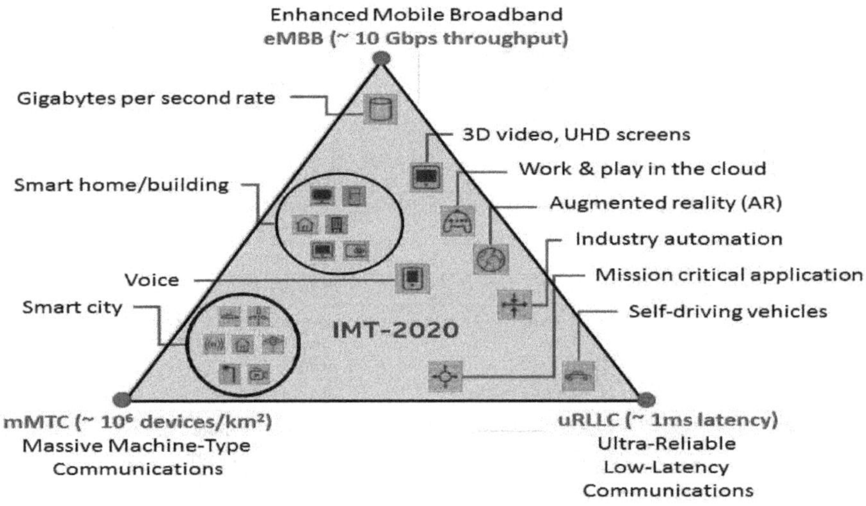

Fig. 9.41 The three basic 5G services defined by ITU-R [Sanou (2018)].

(1) **Enhanced/Evolved Mobile Broadband (eMBB)** to support bandwidth-hungry multimedia applications with data rates in excess of 10 Gbps, such as:

 (a) Immersive 360° experience (viz. virtual reality or VR and augmented reality or AR).
 (b) Enhanced indoor and outdoor broadband connectivity.
 (c) Enterprise collaboration.
 (d) High-definition (HD) videos.
 (e) Sharing of pre-recorded and live videos.
 (f) Real-time gaming.
 (g) Movement of data to and between data centers by Over-the-top (OTT) application providers.[4] OTT services can be accessed via apps on mobile devices or computers.

(2) **Ultra-Reliable Low-Latency Communications (uRLLC)** with less than 1 ms latency to support delay-sensitive and loss-intolerant use cases such as:

 (a) Remote surgery (aka telesurgery): a use case of telepresence in which a doctor uses a robot surgical system to perform surgery on a patient while in physically different locations.
 (b) Immersive experience: virtual reality (VR) and augmented reality (AR).
 (c) Communications in a connected autonomous cars (CAV), i.e. a network of driverless vehicles.
 (d) Tactile Internet.
 (e) Smart grids.
 (f) Remote patient monitoring and telehealth.
 (g) Industrial automation.

(3) **Massive Machine-Type Communications (mMTC)** to support services requiring high density of connections (i.e. more than $10^6/\mathtt{km}^2$ connected devices), such as:

 (a) IoT with massive connected devices of varying nature.
 (b) Smart city and smart home.
 (c) Remote monitoring.
 (d) Asset tracking.
 (e) Smart agriculture.
 (f) Energy monitoring.

These three types of services have different requirements. Therefore, 5G network is being built on a virtualised environment in order to realise the three different service types.

[4]An OTT is a company which supplies streaming media (e.g. radio, audio, TV, webcast, film and Youtube) as standalone products to clients over the Internet to be played in real time rather than storing it for playback. Examples of OTTs are YouTube, Amazon Web Services (AWS), Netflix, Hulu Plus, Apple, Facebook Watch, and DirectTV. Users of OTT service do not need to subscribe to satellite pay-TV service.

9.6.1 *5G System Architecture*

The 5G System (5GS) is a *cloud-native* architecture as it adopts cloud computing philosophy end-to-end. 5GS comprises the access network (AN), which can be wireless or wired, and the core network (5GC). A link interconnecting a pair of elements in the radio access network (RAN), made of base stations and UE in LTE/LTE-A, is called **fronthaul**. The set of all fronthauls is the fronthaul network. A link interconnecting a RAN and the 5GC is called backhaul. The set of such links is called **backhaul network**. Following the current trend in network, the 5GC uses four complementary technologies to achieve network softwarisation, namely:

- **Network function virtualisation (NFV)**: NFV brings an end to the implementation of network functions as dedicated hardware appliances. Rather, functions are implemented as instance on a generic hardware which is referred to as Common Hardware for all Apps (CHAP) or commercial off-the-shelf or commercially-available off-the-shelf (COTS) hardware. NFV enables fast and low-cost network changes, including upgrades.
- **Software-Defined Networking (SDN)**: SDN enabled the transition from hardware-controlled to software-controlled networks for several advantages. For example, network functions/elements can be reconfigured in real time to improve service quality, resilience and overall performance.
- **Network slicing**: This is the partitioning of a physical network infrastructure into multiple logical instances, called slice. A main advantage in slicing is the reduced CAPEX and OPEX in creating network services. Slices can support different radio access networks.
- **Cloud Radio Access Network (C-RAN)**: 5G cells using mmWaves are small. This means that a large number of cells must be mounted. A motivation for C-RAN is to reduce the costs of deploying these 5G small cells. C-RAN also reduces costs on cooling, heating, space, power and test access of the network. C-RAN exploits cloud computing and virtualisation philosophies. The signal processing functions performed at base stations in traditional PLMNs are now replaced by centralised processing units (CPUs) positioned in the cloud.

  Figure 9.42 illustrates the C-RAN concept. Conventionally, each base station contains a router, a baseband unit (BBU) and the Radio Remote Head (RRH). The BBU and the RRH are connected via the Common Public Radio Interface (CPRI). The CPRI is a full transceiver or just an antenna. In C-RAN, we move the BBUs and the routers from the base stations and position them together at a central site. The next step is to virtualise the BBUs and the router using COTS. The CPRI in the distributed RAN is up to about 60 m long, while those in the C-RAN can be up to 20 km long. The CPRI is an example of fronthaul links.

An evidence of how PLMN operators are exploiting SDN and NFV is the investment being made by Telefonica in its UNICA programme in partnership with Ericsson.

The 5G NFV is being standardised by ETSI. According to Pages 19–20 in [ETSI-3GPP (2018)], 5G System (5GS) architecture comprises 18 network functions. Before proceeding, let us compare the functions of 5G core network in Fig. 9.43 with those of 4G/LTE core network in Fig. 9.2. This is done in Table 9.8. Something is remarkable in the 5GS. That is the clear separation between the user plane (UP) and the control plane (CP) functions. The UP is used to transfer payload or user traffic, while the CP handles signalling and control traffic, as usual. This separation enables network agility, scalability, deployability and fast time-to-market of new network services. Also, the monolithic elements have been divided into smaller functions to enable customisation.

Figure 9.43 shows the non-roaming 5G system architecture. The UPF supports the user plane functions, while the control plane functions are supported by the AMF, AUSF, UDM, PCF, NEF and AF.

The purposes of the 5G network functions are discussed in, for example, the ETSI standard document [ETSI-3GPP (2018)]. We quote from this document the purposes of some of the network

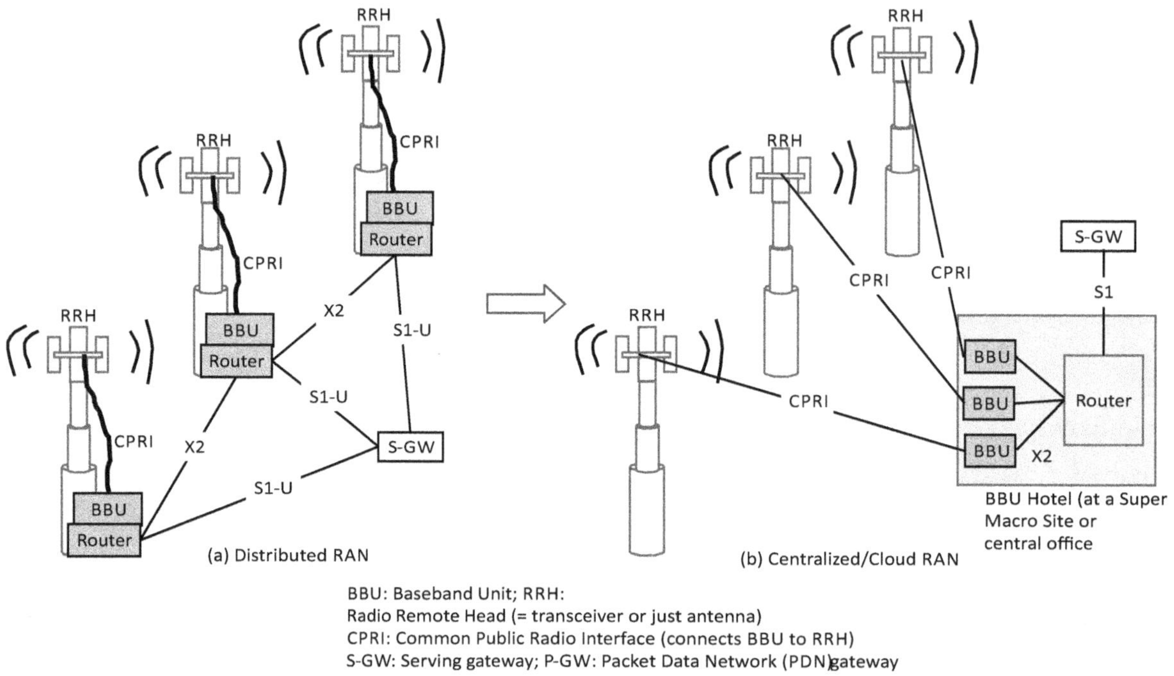

Fig. 9.42 Illustration of the cloud RAN (C-RAN) concept.

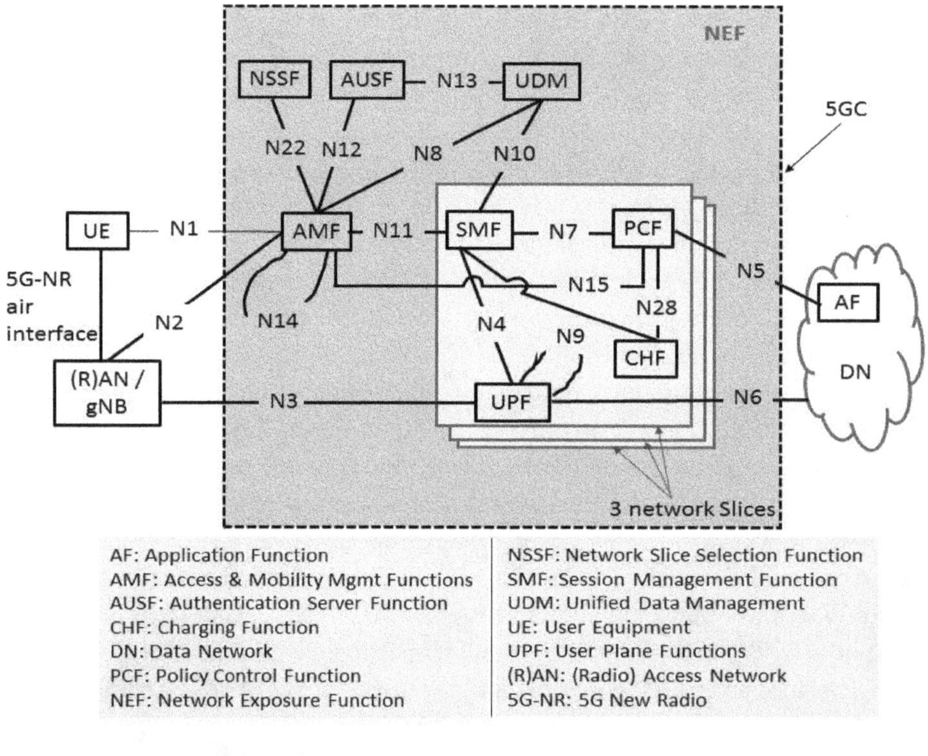

Fig. 9.43 Non-roaming 5G system architecture (cf. Fig. 4.2.3-2 in [ETSI-3GPP (2018)]).

Table 9.8 Mapping between functions in 4G/LTE and 5G core networks.

Functions	4G/LTE	5G
User plane functions	User plane functions in SGW and PGW	UPF
Control plane functions	Control plane functions of SGW, PGW and MME	SMF
Policy and charging control	PCRF	PCF
Application functions (AF)	AF	AF
User data management	HSS	UDM
Network access and mobility management	Authentication and mobility functions in MME	AUSF and AMF

functions. One instance of the AMF can support some or all of these functions:

(1) Termination of radio access network control plane interface (N2).
(2) Termination of Non-Access Stratum or NAS (N1), NAS[5] ciphering and integrity protection.
(3) Registration management.
(4) Connection management.
(5) Reachability management.
(6) Mobility management.
(7) Lawful intercept[6] (LI) for AMF events and interface to LI System.
(8) Provide transport for Session Management (SM) messages between UE and Session Management Function (SMF).
(9) Transparent proxy for routing SM messages.
(10) Access authentication.
(11) Access authorisation.
(12) Provide transport for Short Message Service (SMS) messages between UE and SMS Function (SMSF).
(13) Security Anchor Functionality (SEAF) as specified in TS 33.501 [29].
(14) Location services management for regulatory services.
(15) Provide transport for location services messages between UE and Location Management Function (LMF) as well as between RAN and LMF.
(16) 4G Evolved Packet System (EPS) Bearer ID allocation for interworking with EPS.
(17) UE mobility event notification.

A single instance of **Session Management Function** (SMF) can perform all or some of the following functions [ETSI-3GPP (2018)]:

(1) Session Management: session establishment, modify and release, including maintenance of tunnel between UPF and access network node.
(2) UE IP address allocation and management (including optional authorisation).
(3) DHCPv4 (server and client) and DHCPv6 (server and client) functions.
(4) Address Resolution Protocol (ARP) proxying as specified in IETF RFC 1027 and/or IPv6 Neighbour Solicitation Proxying as specified in IETF RFC 4861 functionality for the Ethernet Protocol Data Units (PDUs). The SMF responds to the ARP and / or the IPv6 Neighbour Solicitation Request by providing the MAC address corresponding to the IP address sent in the request.
(5) Selection and control of UP function, including controlling the UPF to proxy ARP or IPv6 Neighbour Discovery, or to forward all ARP/IPv6 Neighbour Solicitation traffic to the SMF, for Ethernet PDU Sessions.
(6) Configures traffic steering at UPF to route traffic to proper destination.
(7) Termination of interfaces towards policy control functions.

[5]5G NAS (Non-Access Stratum) includes procedures related to 5GMM (5GS Mobility Management) and 5GSM (5GS Session Management) on the 5GS (5G system).

[6]Lawful Interception (LI) is a regulatory requirement that operators must satisfy as a legal obligation toward local Law Enforcement Agencies (LEA) and Government Authorities. LI deals with security control by intercepting information transferred over an operator's network.

(8) Lawful intercept (for SM events and interface to LI System).
(9) Charging data collection and support of charging interfaces.
(10) Control and coordination of charging data collection at UPF.
(11) Termination of SM parts of NAS messages.
(12) Downlink data notification.
(13) Initiator of AN specific SM information, sent via AMF over N2 to AN.
(14) Determine Service and Session Continuity (SSC) mode of a session.
(15) Roaming functionality:
   (a) Handles local enforcement to apply QoS SLAs (VPLMN).
   (b) Charging data collection and charging interface (VPLMN).
   (c) Lawful intercept (in VPLMN for SM events and interface to LI System).
   (d) Support for interaction with external Data Network (DN) for transport of signalling for PDU session authorisation/authentication by external DN.

A single instance of a **User Plane Function** (UPF) can support some or all of the following functions [ETSI-3GPP (2018)]:

(1) Anchor point for Intra-/Inter-RAT mobility (when applicable).
(2) External PDU session point of interconnect to data network.
(3) Packet routing and forwarding (e.g. support of uplink classifier to route traffic flows to an instance of a data network, support of branching point to support multi-homed PDU session).
(4) Packet inspection (e.g. application detection based on service data flow template and the optional Packet Flow Descriptions (PFDs) received from the SMF in addition).
(5) User plane part of policy rule enforcement, e.g. gating, redirection, traffic steering.
(6) Lawful intercept (UP collection).
(7) Traffic usage reporting.
(8) QoS handling for user plane, e.g. UL/DL rate enforcement, reflective QoS marking in DL.
(9) Uplink traffic verification (Service data flow to QoS flow mapping).
(10) Transport level packet marking in the uplink and downlink.
(11) Downlink packet buffering and downlink data notification triggering.
(12) Sending and forwarding of one or more "end marker" to the source NG-RAN node.
(13) ARP proxying as specified in IETF RFC 1027 and/or IPv6 Neighbour Solicitation Proxying as specified in IETF RFC 4861 functionality for the Ethernet PDUs. The UPF responds to the ARP and/or the IPv6 Neighbour Solicitation Request by providing the MAC address corresponding to the IP address sent in the request.

The purposes of the Policy Control Function (PCF) includes: support of the unified policy framework to govern network behaviour; provision of policy rules to Control Plane function(s) to enforce them; and assessment of subscription information relevant for policy decisions in a Unified Data Repository (UDR). The Authentication Server Function (AUSF) supports authentication for 3GPP access and untrusted non-3GPP access as specified in the Technical Specification 33.501.

The **Application Function** (AF) interacts with the 3GPP Core Network in order to provide services to support, for example, the following:

- The influence of application on traffic routing.
- Accessing Network Exposure Function.
- Interacting with the Policy framework for policy control.

Application functions trusted by a 5G network operator are allowed to directly interact with the appropriate network functions. Untrusted AFs, however, interact with the appropriate network functions through the Network Exposure Function (NEF). Thus, the NEF exposes the network functions in the 5Gs to allow their access by external systems. NEF has several purposes, such as: (a) enabling troubleshooting within and between PLMN operators, and (b) enabling analytic tools to extract data. The NEF supports at least these request types:

(1) Policy/charging requests to invoke policy/charging logic.
(2) Requests' monitoring to notify state changes.

The process of creating multiple logical network instances (with possibly different QoS) over the same physical network infrastructure is referred to as *network slicing*. A network function in 5G used to allocate a set of appropriate network slices to users or UE is the Network Slice Selection Function (NSSF). Other functions of the NSSF include:

- Determining the allowed Network Slice Selection Assistance Information (NSSAI) and, if needed, the mapping to the Subscribed Single NSSAI (S-NSSAIs).
- Determining the Configured NSSAI and, if needed, the mapping to the Subscribed S-NSSAIs.
- Determining the AMF Set to be used to serve the UE, or, based on configuration, a list of candidate AMF(s), possibly by querying the NRF.

9.6.1.1 *5G Service Based Architecture*

The control plane of 5G core network has adopted the Service Based Architecture (SBA) to achieve a remarkable improvement compared with 4G/LTE's evolved packet core (EPC). Figure 9.44 shows the non-roaming version of this SBA (cf. Fig. 4.2.3-1 in [ETSI-3GPP (2018)]). In the SBA, all control plane functions interact via a bus architecture, referred to as **Service Based Interface** (SBI). The SBI enables each control plane function to access the functions provided by other control plane functions. Each control plane function makes its functions in the form of services available over an *open* Application Programming Interface (API). This is not possible in LTE owing to its *closed* APIs. The API or SBI for each network function is given a name. This name follows a simple convention. It begins with the letter N and then the acronym for the specific network function written in lower case as subscript. For example, the SBI of the network functions PCF and UDM are respectively $\mathrm{N}_{\texttt{pcf}}$ and $\mathrm{N}_{\texttt{udm}}$.

All functions in the 5GC are stored in a registry referred to as *Network Resource Function* (NRF), whose functionality is akin to Domain Name Server in the computer networks. Each network function registers itself and its services in a data store via the NRF. Also, each network function learns about other network functions by querying the NRF. NRFs belonging to different PLMN operators can interface with each other.

The protocol stack of SBI is shown in Fig. 9.44(b). SBI uses JavaScript Object Notation (JSON) format described in IETF RFC 8259 as the serialisation protocol. Control plane messages are transferred using HTTP/2 protocol instead of DNS, Diameter or GTP used in 4G/LTE. In the absence of an alternative security mechanism, Transport Layer Security (TLS) will be used to protect messages at the SBI transport layer. The SBI protocol stack indicates a stronger convergence between computer and telecommunications networks and technologies.

9.6.2 *5G Network Management and Orchestration: NFV*

The set of activities used to maintain a network in its (optimum) operating state is referred to as network management. An end-to-end network management is an inter-administrative process, requiring collaboration. A communications network comprises many functions or components, some at any time can perform abnormally. The failure of a network function can result in degraded service or even outage. Network management seeks to identify potential failures, locate them, and then fix them in time. Network management, also has to ensure security and privacy. There are alternative standards for network management, one of which is the FCAPS model, which originated from ISO/ITU-T Telecom Management Network framework. FCAPS is the acronym for: Fault, Configuration, Accounting, Performance and Security.

5G network management uses software-defined networking (SDN), network function virtualisation (NFV) and cloud computing for two types of decoupling: decoupling of the control and data planes, and the decoupling of network functions from specialised, dedicated hardware to improve scalability, agility and time-to-market new services.

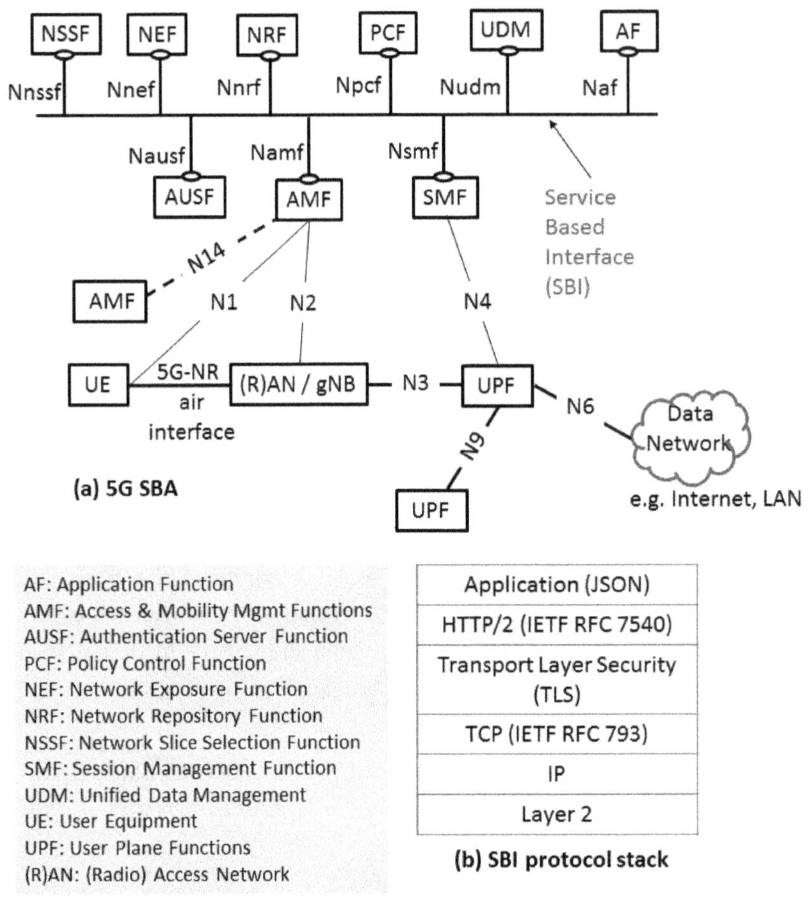

Fig. 9.44 Non-roaming Service Based Architecture for control plane of 5G core network (cf. Fig. 4.2.3-1 in [ETSI-3GPP (2018)]).

The conventional philosophy of implementing a network function in a specialised, dedicated hardware is nicknamed *ossification* [Li and Chen (2015)], [Ma *et al.* (2018)]. The term arose from the fact that any modification in the network function requires a new hardware owing to the rigid philosophy.

Network functions include switching using switches, routing using routers, security provisioning using firewalls, virus scanners, intrusion detection systems and spam protection, traffic analysis using deep packet inspection (DPI), signalling using Internet Protocol Multimedia Subsystem (IMS) and Session Border Controller (SBC), tunneling using IPSec/SSL virtual private network gateways, and charging and billing using the OSS/BSS. NFV borrowed ideas from virtualisation and cloud computing technologies in Information Technology. NFV is thus founded on the four technologies: softwarisation, orchestration, automation and virtualisation. A complementary technology to NFV is software-defined networking (SDN). The focus in NFV is the optimising of network services. SDN, however, focuses on separating the forwarding and control planes in order to obtain a centralised view of the network. NFV can be realised using either virtual machines (VMs) or application containers. Table 9.9 summarises some of their differences.

Why NFV? Traditional network functions are implemented using specialised hardware or appliances. Each of these hardware is built to fulfill a specific purpose or function. Implementing network functions in hardware causes many issues, including:

- The cost of deploying a 5G network can range from \$6.8 million for a small city to \$55.5 million in a large city with high population density. This prohibitive CAPEX in 5G deployment may

Table 9.9 Comparison between a virtual machine and a container.

Virtual Machine	Application Container
Can run an operating system (OS), called *guest OS*, different from its host OS	Must use the same OS as the host machine
Hardware is virtualised to run multiple OS instances	OS is virtualised to enable multiple workloads run on a single OS instance
Each VM has its own image file	Multiple containers can share an host OS kernel and, usually, the libraries and binaries
Heavy weight (e.g. gigabytes in size)	Light weight (e.g. megabytes in size) and thus portable
Takes minutes to start VM	Takes only seconds to start

cause/increase the *digital divide*, as PLMN operators may install 5G in only locations with high potential on return-on-investment (ROI).

- The launch of a new service requires an additional hardware, which in turn requires space, power supply, etc.
- Management of devices life-cycle: The useful lifetimes of network devices are becoming increasingly shorter owing to fast-paced technological innovations. This trend negatively affects return on investment (ROI), making investment in network infrastructure unattractive. In effect, hardware-based network function implementation throttles technology uptake unless communications service providers are pushed by competition for market share.
- Hardware-based network functions cannot be easily moved to other geographical locations. This raises risk concerns as mounted infrastructure which is underutilised cannot be re-located elsewhere easily.
- Non-agile network: lack of scalability.
- Accelerated service innovation and provisioning: it takes a longer time and higher cost to design, deploy, and manage hardware-based than software-based networking services.
- It does not allow a quick response to market demand.

NFV solves all the above issues.

NFV began in October 2012 through a proposal to ETSI by 13 PLMN operators, namely:

(1) Colt Technology Services Group Ltd and British Telecom (BT) from UK.
(2) Orange S.A. from French.
(3) Telecom Italia from Italy.
(4) Deutsche Telekom from Germany.
(5) Telefonica Telecom from Colombia
(6) Verizon, CenturyLink Inc. and AT&T from USA.
(7) Telstra from Australia.
(8) KDDI and NTT from Japan.
(9) China Mobile.

These communications service providers aimed to reduce their costs (both in CAPEX and OPEX) and simplify their operations. In 2012 a special group, called *ETSI Industry Specification Group for Network Functions Virtualisation* (ETSI ISG NFV), was formed within ETSI to develop the architecture and the requirements for the virtualisation of various functions within communications networks. The work of the ETSI ISG NFV has resulted in the ETSI-standardised NFV architecture in Fig. 9.45. Membership of ETSI ISG NFV exceeds 300, and includes major network operators, major telecommunications equipment vendors, service providers, IT vendors and technology providers. Under NFV the hardware-based functions eNB, MME, HSS, S-GW, P-GW, PCRF, Switch and Router in the LTE/4G architecture in Fig. 9.2 have become veNB, vMME, vHSS, vS-GW, vP-GW, vPCRF, vSwitch and vRouter virtual machines, respectively. In 2014, the Open Platform for NFV (OPNFV) was established by the Linux Foundation with the support of several major telecom infrastructure manufacturers and service providers.

The modern trend adopted in 5G, referred to

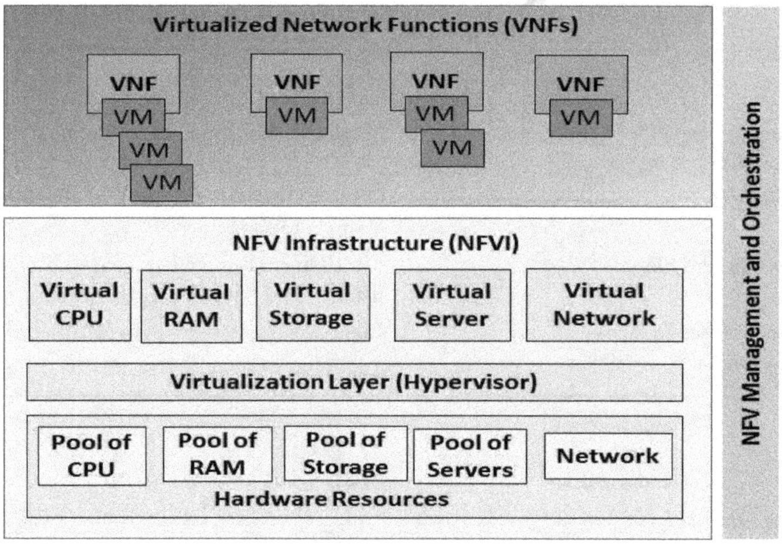

Fig. 9.45 ETSI NFV architecture.

as NFV, is a migration from a special-purpose hardware implementing a single function to a generic hardware which can be programmed and re-programmed to implement on-demand several network functions. This also means that network services or functions are made hardware agnostic. NFV also means that network functions are implemented as applications (Apps) which run on a software. Applications implementing network functions are realised by virtual machines (VMs) hosted on the generic hardware (i.e. CHAP or COTS). Note that an application can run on multiple VMs. A VM is created and run by a middleware called hypervisor, which can be implemented in software, firmware or hardware.

9.6.3 *5G OSS/BSS*

Communications service providers (CSPs) aim to increasingly improve customer experience to raise revenue by reducing/avoiding churn, finding new customer and/or launching new services and products. These objectives are achieved with the aid of the Operational or Operations Support Systems (OSS) and Business Support Systems (BSS). As these two subsystems work hand-in-hand to provide an end-to-end service, they are usually written together as BSS/OSS, OSS/BSS or B/OSS. Communications systems evolve or even revolve over time. Therefore, the OSS/BSS are also expected to adapt to the dynamics. Next generation networks are poised to handle voluminous data at high data rates, exploit automation, and orchestrate services on demand using virtualised infrastructure. These features of modern networks require OSS/BSS to be unified, virtualised and automated in order to meet dynamic customer needs. The requirements on modern OSS/BSS can be summarised into:

(1) Network Operations Center (NOC) changes from reactive maintenance to preventative maintenance. this is enabled by the exploitation of artificial intelligence (AI) techniques.
(2) NOC changes from focusing on maintaining network infrastructure to maintaining services.
(3) OSS layers move from being silos and manual management to unified and automated entity to support *zero-touch provisioning* of services in real time.
(4) Dynamic network slicing and service level agreements in networks require adaptation by the OSS/BSS.
(5) OSS/BSS must exploit softwarisation (network function virtualisation or NFV and software-defined networking or SDN), just as 5G does.

The features of 5G network requires a dramatic change in the OSS and BSS used in previous generation of networks. An OSS and BSS which can be orchestrated in real time is required in order to turn 5G technologies into services and services into revenue.

9.6.3.1 *Operational Support Systems*

The OSS are a set of computing infrastructure used to manage a communications network. Four of the functions of the OSS are:

(1) Network inventory,
(2) Service provisioning, including service activation and de-activation,
(3) Network configuration, and
(4) Fault management.

The OSS is responsible for network planning, network building/construction and commissioning, optimisation, monitoring and troubleshooting. The OSS contains the network operating center (NOC). NOC used to manage the operation of the radio base stations. However, today the NOCs have become service operating centers (SOCs) as they focus on the services traversing the network.

As discussed earlier, a 5G network divides its capacity into network slices. Each slice maps into one of the three services: uRLLC, eMBB and mMTC. Each slice conforms to a specific set of business rules and policies, which are kept in the service catalog, to enable fast onboarding of new services. It is the task of the OSS to orchestrate and manage these slices to reduce service delivery delays.

9.6.3.2 *Business Support Systems*

The BSS manages the business operations in relation to customers. The BSS provides the four main management functions:

(1) Customer relationships management (CRM), including retention, updating of customer profile, and management of fault and problems.
(2) Order management, including acceptance and dispatching of orders.
(3) Revenue management, including charging and billing.
(4) Product management.
(5) Inventory management.

9.6.4 *Check Your Understanding*

Students are encouraged to check their understanding of the content of this section with the following questions prior to proceeding to the next section.

(1) Explain meaning of the phrase *digital divide* as used in ICT. Justify the reason why the 5G technology can cause or increase digital divide.
(2) State at least two drivers of 5G network.
(3) 3G mobile technologies used make-before-break handoff algorithm. 4G uses break-before-make handoff algorithm. The two use cases eMBB and uRLLC in 5G, however, use make-before-break handoff algorithm. Explain why this switch is necessary.
(4) Explain the term *network slicing* as used in 5G.
(5) Discuss the differences between the core networks of 4G and 5G.
(6) Analyse 5G New radio (5G-NR) and show how it differs from LTE eNodeB.
(7) List the names of the three basic use cases of 5G network. For each use case, state the basic features and two examples of killer applications with justification.
(8) Discuss how the OSS differs from BSS in PLMNs.
(9) **In-class group research:** In groups of 2-4 students, research and discuss Logically-Isolated Network Partitions (LINP) and its application in 5G networks.
(10) 5G core networks exploit two types of decoupling. Explain the difference(s) between the two.
(11) Explain the meaning of the term *ossification* as used in PLMNs.
(12) 5G core network separates CP functions from UP functions. Discuss three benefits in such a network architecture.

9.7 Review of Important Terms and Acronyms

Acronyms, abbreviations and important terms		
C-RAN	SBA	SBI
softwarization	SDN	NFV
ACI interference	base station	capacity management
capacity-limited network	cell	cell breathing
cell dragging	cell on wheels	cell planning
cell sectoring	cell splitting	center-excited cell
channel allocation	channel reuse	cluster size
co-channel interference	co-channel reuse factor	co-channel reuse ratio
corner-excited cell	coverage management	coverage/service area
coverage-limited network	downlink/forward link	dualisation
dwell time	edge-excited cell	eNodeB (eNB)
femto-cell	forced termination	frequency reuse
frequency reuse distance	frequency reuse factor	frequency reuse plan/pattern
handoff/handover	handoff hysteresis	cell selection
horizontal handoff	IMSI/IMEI/MSISDN	interference-limited network
load balancing	LTE/LTE-A	macro-cell
MCHO/NCHO/MAHO	micro-cell	microcell zoning
mobile station	mobility management	morphology
nano-cell	near-far effect	near-far effect
noise-limited network	paging	picocell
PLMN	power control	radio access technology
reuse distance	roaming	segmentation
session dropping	SIM card	simulcast repeater
SNIR/NIR	space diversity	terrain
topography	tracking area (TA)	tracking area code
tracking area updating	umbrella cell	uplink/reverse link
user equipment	user tracking	vertical handoff

9.8 Supplementary Reading

(1) Leonhard Korowajczuk, *LTE-A, WiMAX 2.2 and WLAN (4G/5G): Network Design, Optimisation and Performance Analysis*, Wiley, 2019.

(2) Afif Osseiran and Jose F. Monserrat, *5G Mobile and Wireless Communications Technology*, Cambridge University Press, 2016.

(3) Nishith Tripathi and Jeffrey H. Reed, *Cellular Communications: A Comprehensive and Practical Guide*, Wiley-IEEE Press, 2014.

(4) T. S. Rappaport, *Wireless Communications: Principles and Practice*, 2nd ed. Singapore: Pearson Education, Inc., 2002.

(5) K. Feher, *Wireless Digital Communications: Modulation and Spread Spectrum Applications*, Upper Saddle River, NJ: Prentice Hall, 1995.

(6) S. Haykin and M. Moher, *Modern Wireless Communications*, Singapore: Pearson Education, Inc., 2002.

(7) Yong Li and Pan Hui, *Opportunistic Networking: Models, Protocols, and Applications*, CRC Press, 2019.

Chapter 10

Point-to-Point Microwave System Design

Chapter's Learning Outcomes

We discuss in this chapter the design of point-to-point terrestrial microwave radio links. Microwave communications can occur close to the Earth's surface and be called terrestrial microwave communications or in the sky and be called satellite microwave communications. Microwave communications can also be achieved using guided or unguided transmission medium. Microwave communications occur in line-of-sight mode. Owing to the small wavelengths, microwaves communication over long distances require high power transmitters, low-loss transmission lines, high gain antennas and sensitive receivers. This chapter introduces readers to the fundamentals of terrestrial point-to-point microwave radio link design, and in particular, we study:[1]

(1) General introduction to terrestrial microwave radio communications and radio link engineering.
(2) Basic components of a terrestrial microwave radio communications system, noting that the end-to-end link is a system in itself.
(3) Types of transmission lines used in terrestrial microwave radio communications, namely coaxial cables and waveguides.
(4) The **link power budget** to determine transmit power and antenna gains.
(5) The **link height budget** to determine the height of antennas.
(6) The site-specific environmental factors (weather, geography, interference) affecting terrestrial microwave radio systems.

10.1 Introduction

Unlike in satellite communications, terrestrial systems transfer signals over the earth's surface between a transmitter and a receiver on land. Conventionally, communications systems and devices operating in the ultra high frequency (UHF) band 0.3–3 GHz, equivalent to the wavelengths $10 \leq \lambda \leq 100$ cm, are called **microwave** systems and devices. However, some systems operating in the very high frequency (VHF) band 30–300 MHz, equivalent to the wavelengths $1 \leq \lambda \leq 10$ m, are also referred to as microwave systems. Most of the frequency bands are further divided into sub-bands. For example, the VHF band is subdivided into VHF low band and VHF high band. The VHF high band is 148–174 MHz, or $2.027 \geq \lambda \geq 1.724$ m, while the VHF low band is 30–40 MHz. The wavelengths are calculated using the free-space propagation relation

$$\begin{aligned} \text{velocity} &= \text{frequency} \times \text{wavelength or } c \\ &= f \times \lambda \end{aligned} \tag{10.1}$$

where c is the speed of light in free space.

Electromagnetic (EM) waves with frequencies in the range 0.3 GHz to 300 GHz (equivalently 1 m to 1 mm) are classified as microwaves, but the range currently used for practical communications is 2–40 GHz. The transmission of information using EM waves at microwave frequencies is referred to as microwave communications. An experimental 64-km microwave link was built in 1931 over the English channel (to interconnect southern England and northern France.

[1]The reverence of nature is the beginning of knowledge, but the unwise despises wisdom and instruction.

Long-distant information transfer can be achieved over these three ways: optical glass fiber links, coaxial copper cables, or wirelessly using microwave links. There are two types of microwave communications links. These are terrestrial microwave links and satellite microwave links. Microwave communications require the transmit and receive antennas to be in line of sight. Western Union constructed its first microwave link and made its first transmission between Pennsylvania and New York in 1945. The transfer of telephony traffic between cities over microwave links began in the 1950s. Around 1960s satellite microwave links gained commercial entry. Currently, terrestrial microwave communications is facing fierce competition with satellite wireless and cellular communications. Microwave path design requires two main budgets:

- **Power budget** to determine the required transmit power and antenna gains.
- **Height budget** to determine the required effective height of the antenna masts.

The space between the transmitting and the receiving antennas is referred to as link or path. In the design of a microwave link we usually know certain parameters, such as:

(1) The length of the link d_{T-R}, which is established knowing the two sites to be interconnected.
(2) The operating radio frequency f or the wavelength λ, which is acquired from a local radio regulatory body, usually an arm of the government. The available bandwidth is higher the higher the frequency. However, the free-space attenuation and other disturbances increase with the frequency. The operating frequency depends on many factors, such as its availability (unless it is in the license-exempt industrial, medical scientific (ISM) band), expected link range, desired bandwidth and antenna to be used.
(3) The required link data rate or capacity R_b, which depends on the services to be supported by the link, the available system bandwidth B and the signal-to-noise plus interference ratio (SNIR). According to the Shannon-Hartley's theorem, the limit of R_b is

$$R_b < B\log_2\left(1 + \frac{P_s}{N+I}\right) \qquad (10.2)$$

where I is the cumulative power of all interfering signals, P_s is the power in the desired signal, and N is the thermal noise power. There are several ways of increasing R_b. We can do so by decreasing I through the use of antenna with smaller side lobes and back lobes, or use frequency with less disturbance. We can also increase R_b by using antenna with higher gain. This results in larger antenna which requires stronger tower and cost. Another way of increasing R_b is to increase the transmit power. Such a choice can cause interference to other systems and also increase operational cost owing to increased energy usage. We can now observe any strategy taken needs a compromise, the reason why telecommunication experts are friends of optimization techniques.
(4) The required link availability or reliability, which is usually about 99.999% of the time (i.e. five-nines). This means that less than 5.3 minutes per year of link outage can be tolerated.
(5) Link latency (τ), which is the time taken to send data end-to-end. This is equal to half the link's round-trip time. Best link performance requires $\tau < 2 - 3$ ms.
(6) Local environmental factors, including the geography (i.e. topography and morphology), weather (e.g. rain fall rate, snow, hail, temperature), and potential sources of radio frequency interference.
(7) Bit error rate (BER), which is ratio of bits received in error to total number of bits transmitted over a given time frame. The BER should not be greater than 10^{-8} to 10^{-6}.

These are input variables into the design of the microwave link. Based on these input parameters we need to find the appropriate dimensions of the output variables, which include:

- Antenna parameters, e.g. gain or directivity, beamwidth, size (diameter), mount height, polar-

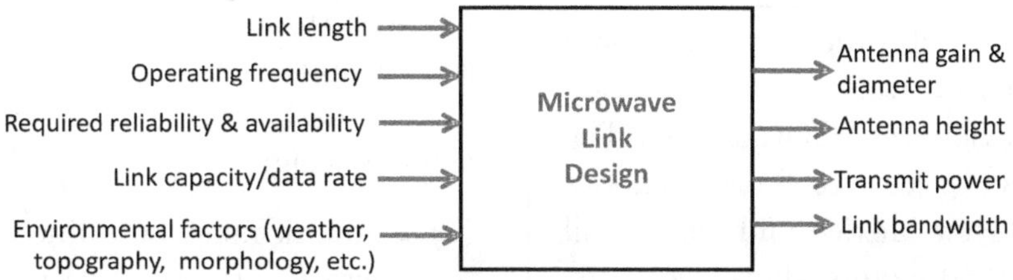

Fig. 10.1 Microwave link design model.

isation, cross-polarisation discrimination (XPD), tilt, reflection coefficient and voltage standing wave ratio (VSWR).

- Transmit power, which partly determines the radio electronics to be used.
- The link bandwidth, which depends on the operating frequency. Higher frequencies usually have higher bandwidth.

The relationship between the input and the output parameters of the link design is illustrated in Fig. 10.1.

As illustrated in Fig. 10.9, every microwave link has the basic components:

- Transmitter electronics.
- Receiver electronics.
- Antennas.
- Transmission line (aka feedline or feed line) interconnecting the antennas to the electronics.

Case Study: The World's Longest All-IP Microwave Link

The national video and audio broadcasting operator in Equatorial Guinea, called RTV EG, found microwave link to be the most cost-effective option (compared with satellite and fiber-optic links) to interconnect two pints. The two points are Pico Basile on the island of Bioko (the highest summit with an altitude of 3,011 m) and Monte Douba. The distance between the two locations separated by the sea is 241 km. Long wireless links like this one are liable to two main challenges:

- **ducting**, which is caused by the differing humidities and temperatures in the atmospheric layers, and
- **multipath fading**, which is caused by the reflection of radio waves by the surface of the water. Being a good reflector of radio waves, waves reflected off the surface of the water reach the receiver with different amplitudes and phases compared to the line-of-sight radio wave and cause multipath fading.

The solution of the above challenges faced by long over-water radio links requires robust extreme weather-proof long-haul radios and good engineering. **Ceragon Networks Ltd**, headquartered in Tel Aviv in Israel, used an in-house manufactured long-haul radios with space diversity combining to provide a robust solution to RTV EG. The link is designed with a large link margin in order to cope with the high dynamic range of the multipath fading. According to Ceragon Networks, this 241-km microwave link is the longest all-IP microwave over-water path in the world.

Another long-haul microwave link over water is the ExploreAir LR all-outdoor microwave radio system developed by **Exalt Wireless** headquartered in California USA. It is used to construct 235-km microwave link over water between an island in the Mediterranean and the coast of Lebanon. This is a 500-Mbps full duplex link. The link operates at 7 GHz (i.e. 6–43 GHz band), and uses field replaceable diplexers.

Eclipse link: As a third case study, **Aviat Networks** and its agent, Telecomunicaciones y Sistemas S.A., has commissioned 193-

km microwave radio link over water and land in Honduras for Central American Corporation for Air Navigation Services (COCESNA). The link, connecting Las Cruces (1600 meters above mean sea level) in New Mexico and Dixon Hill on Roatan Island (250 meters above mean sea level) in Caribbean, is used for air traffic control over the Central American territory. Reliability in air traffic control systems is very important thus requiring 99.9995% link reliability, and residual bit error rate of 10^{-13}. The link operates at 6 GHz and uses QPSK modulation to deliver 40-Mbps data rate. One of the challenges in the installation of this link was how to overcome very strong winds in the locality which can cause misalignment of the antennas and cause poor link quality.

The basic components used in microwave wireless systems are also used in other radio systems, but with some differences. The technologies differ depending on the operating frequencies or wavelengths. Components and techniques which function well at the high microwave frequencies (i.e., low wavelengths) do not work at the lower frequencies (longer wavelengths) used in other radio systems, such as mobile cellular communications networks. Examples of features differentiating high-frequency from low-frequency systems are the types of antennas and transmission media. Table 10.1 provides some comparison.

The design of a microwave link must consider the following:

(1) The link must be a line-of-sight (LOS) link to operate. Strategies used to increase the probability of maintaining the needed LOS include:
   (a) Avoiding all obstacles or/and
   (b) Using antenna masts that are taller than all obstacles in the path with clearance for earth curvature bulge and growth of foliage.
(2) Flat terrain can create undesirable signal reflections.
(3) Precipitation can absorb or scatter some of the energy in a microwave signal.
(4) The growth of foliage in the microwave path can weaken marginally strong signal and degrade the link's reliability and quality.

There are many other issues which need consideration in the design and operation of a microwave link to prevent unnecessary failures. Major causes of microwave link failures include:

- Rain outage (predictable and therefore acceptable) in access links above about 10 GHz.
- Equipment failure within the MTBF (Mean Time Between Failure) period.
- Maintenance error or manual intervention (e.g., failure of a locked-on module or path).

Table 10.1 Some features differentiating microwave systems from low-frequency systems.

Feature	Microwave systems	Low-frequency radio systems
Transmission lines used	Ordinary wires would degrade the signal adversely so special wires and cables: waveguides (i.e., hollow pipes) or coaxial cables are used.	Ordinary wires and cables, such as twisted pairs, coaxial cables, fiber-optic are used
Antennas used	Parabolic (aka dish) antennas are used. These are highly directional antennas required in order to tightly focus the transmitted energy, and also to receive signals mainly from one specific direction. Thus, a small power can be used for long-distance transmissions.	Parabolic antennas would be too large so they are not used. Antennas of mobile handsets are generally omnidirectional.

- Infrastructure failure (e.g., antenna, batteries, towers, power system).
- Low fade margin in non-diversity links.
- Power fade (long-term loss of fade margin) in paths above about 6 GHz.
- Obstructions, e.g. buildings, electric power lines, mountains, hills, trees and high-voltage infrastructure.

10.1.1 *Check Your Understanding*

Students are encouraged to check their understanding of the content of this section with the following questions prior to proceeding to the next section.

(1) What is the difference between the two phrases *line-of-sight* and *free-space* in wireless communications?
(2) Two main types of budgets must be performed in the design of microwave links. What are they and how does one differ from the other?
(3) What is an electromagnetic (EM) wave?
(4) What are microwaves with respect to signal or message transmission?
(5) Why is it that the distance travelled by an electromagnetic wave decreases as its frequency increases?
(6) **In-class research:** In groups of three to five students research to find the differences and similarities between the three long-haul microwave links discussed in this section.

10.2 Microwave Link Availability and Reliability

We expect the radio link to be available when we need to use it. As discussed earlier, the industry benchmark is the 5-nines availability. The reliability of radio links are dimensioned using approximate models. We discuss some models proposed by the literature. In the following let us define the outage probability as P_w. This is the probability (or fraction of time) that the link's quality falls below a given fade margin F in dB. Let A be the availability of the link, i.e. the fraction of time that the link works as expected.

A link availability formula has been provided by Lenkurt (see p. 60 of [White (1970)]) as

$$\begin{aligned} A &= (1 - P_w) \times 100\% \\ &\approx (1 - 10.42ab \times 10^{-6} f d_{T-R}^3 10^{-F/10}) \\ &\quad \times 100\% \end{aligned} \tag{10.3}$$

where a is the terrain factor, b is the climate factor, f is the operating frequency in GHz, d_{T-R} is the link distance in km, and F is the given fade margin in decibels. Some estimated values for the climate factor b are $\frac{1}{2}$ for Gulf coast or similar hot and humid areas, $\frac{1}{4}$ for normal interior temperate or northern areas and $\frac{1}{8}$ for very dry areas. Some values given for the terrain factor are $a = 4$ for very smooth terrain, including overwater, $a = 1$ for average terrain with some roughness and $a = \frac{1}{4}$ for mountainous, very rough, or very dry terrain. The formula in Eq. (10.3) assumes that no antenna diversity is used.

Figure 10.2 uses the reliability model in Eq. (10.3) to compare the required fade margin to achieve a given link reliability for 50 km and 80 km links operating at 3 GHz or 9 GHz, respectively. Clearly, the higher the distance or operating frequency, the higher the required fade margin to achieve a given radio link reliability.

Antenna diversity or space diversity can be used to improve the reliability of the link. If the receiver uses two antennas which are separated vertically by the distance h_v meters then the space diversity improvement factor (SDIF) over the no-diversity case is

$$SDIF \approx 4 \times 10^{-6} f h_v^2 10^{\frac{F_l}{10}} d_{T-R}^{-1} \tag{10.4}$$

where F_l is the smallest fade margin of the two antennas in dB, d_{T-R} is the path length in km, and f is frequency in GHz. The link availability with antenna diversity then increases to

$$\begin{aligned} A &= \left(1 - \frac{P_w}{SDIF}\right) \times 100\% \\ &\approx \left(1 - 2.605abd_{T-R}^4 h_v^{-2} 10^{-\frac{F+F_l}{10}}\right) \\ &\quad \times 100\% \,. \end{aligned} \tag{10.5}$$

The ITU-R has also proposed reliability models for various situations in its report [ITU (2015)].

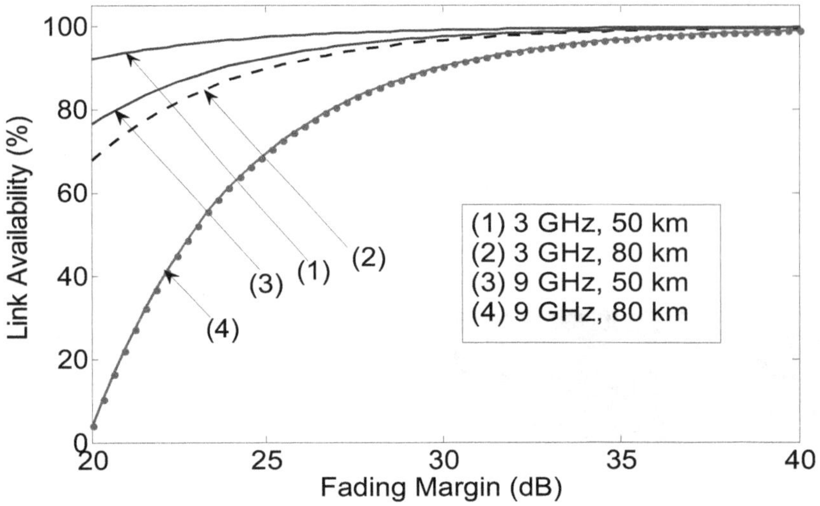

Fig. 10.2 Microwave link reliability versus fade margin ($a = 4, b = 1/2$).

The model applicable for detailed link design is (see pp. 7 and 8 of [ITU (2015)])

$$P_{w-itu} = Kd_{T-R}^{3.4}(1+|\epsilon_p|)^{-1.03}f^{0.8} \times 10^{-0.00076h_L - \frac{F}{10}}100\% \qquad (10.6)$$

where

- f is frequency in GHz.
- $h_L = \min\{h_R, h_T\}$, i.e. minimum of the two antenna heights.
- $|\epsilon_p| = \frac{|h_R - h_T|}{d_{T-R}}$ is the magnitude of the path inclination measured in millirad or 0.001 radian.
- P_w is the percentage of time that the fade depth F in dB is exceeded in the average worst month.
- $K = 10^{-4.4-0.0027dN_1}(10+s_a)^{-0.46}$ is the geoclimatic factor accounting for the average worst fading. It is obtained from the fading data of the geographic area where the radio link operates. s_a is the roughness of the local terrain and dN_1 accounts for point refractivity index.

The availability of the radio link is then

$$A_{itu} = 100 - P_{w-itu}. \qquad (10.7)$$

The ITU-R's model is valid for the following situations $6 \leq s_a \leq 850$ m and uses $s_a = 1$ if the surface roughness is less than 1, $0.45 \leq f \leq 37$ GHz but usable for up to 45 GHz, $7.5 \leq d_{T-R} \leq 185$ km, $|\epsilon_p| \leq 37$ mrad, $-860 \leq dN_1 \leq -150$ and $17\ \mathtt{m} \leq h_L \leq 2.3$ km.

10.2.1 *Check Your Understanding*

Students are encouraged to check their understanding of the content of this section with the following questions prior to proceeding to the next section.

(1) Consider a 5-km microwave link operating at 4 GHz. This link requires the availability of 99.999%. The Lenkurt reliability model in Eq. (10.3) classifies the climate into hot humid, normal temperate and mountainous, and the terrain into very smooth, average and very round.

  (a) Using the model (10.3), find the required fade margin for each of the nine combinations of terrain and climate that enables the radio link to achieve the desired reliability. Present your results in a table form.

  (b) By how much do we improve the above reliabilities if the receiver side of the link uses two antennas separated vertically by $\lambda/2$?

(2) Repeat the above question using the ITU-R reliability model.

10.3 Determination of Antenna Height

We determine the appropriate height of a microwave antenna through **height budget**, which accounts for anything in the link which can block the EM wave. The factors that account for the physical height of the antenna used in microwave systems are:

(1) The height and geographical location of the most tallest obstruction.
(2) Fresnel zones and their clearance.
(3) The bulge of the Earth's curvature.
(4) The bending effect of the EM wave beyond the optical horizon.

10.3.1 *Fresnel Zones and Their Clearance*

EM wave fronts spread out around the line-of-sight path as they move from the transmitting to the receiving antenna. The spreading covers a region which has the appearance and shape of an ellipse. This region of spread waves around the LOS path is called **Fresnel zone** or **F-zone**. Fresnel (pronounced with a silent 's') was named after Monsieur Augustin-Jean Fresnel, a French civil engineer and physicist who was renowned for his research contributions into the wave theory of light. The Fresnel region is divided into uncountable number of zones, so we have the first Fresnel zone (FFZ), the second Fresnel zone (SFZ) and so forth. However, only the first few zones are of practical importance in wireless communications. The FFZ is also referred to as the primary Fresnel zone as that is the region containing the largest portion of the power in the EM wave.

Let us summarise some basic features of the Fresnel zones:

- Waves from transmitter (TX) to receiver (RX) spread out. Some are refracted by obstacles and reach the RX but takes longer than the direct path.
- The refracted waves whose paths are longer than the direct path by odd multiples of the wavelength λ travel longer and enhance the direct path, while those longer by even multiples of λ cancel out the direct path.
- The multi-half-wavelength signals form ellipses (called Fresnel zones) around the direct path.
- Each Fresnel zone has an ellipsoidal shape. In FFZ or zone 1 the refracted signal is 0 to 90 degrees, in SFZ it is 90 to 270 degrees, in TFZ it is 270 to 450 degrees, and so on, out of phase with the direct signal.

Why do we desire that the Fresnel zones, especially the FFZ, be cleared of obstructions? First, microwaves have small wavelengths and thus can be absorbed by even rain droplets, such as found in foliage or trees. Second, if the spread EM waves touch obstacles they can be reflected and reach the receiving antenna with the direct LOS path. The reflected path can be out-of-phase with the direct LOS path and cause destructive interference by reducing the received signal's power. This occurs in even-numbered Fresnel zones. If they are in-phase then a desirable effect results, which is constructive interference in which the strength in the received EM wave increases. *Constructive interference* occurs in odd-numbered Fresnel zones. We prevent obstructions-caused destructive interference by providing a clearance between the dominant obstacle and the FFZ.

Fresnel zone n is the ellipse made of all points where the length of a reflected or diffracted path differs from the direct EM wave path by the amount $n\lambda/2$. The existence of Fresnel zones require that a point-to-point link must have clearance of FZs in addition to LOS. It has been established that, in order to ensure a reliable LOS microwave link performance, at least 60% of the FFZ (from the line of sight path to the outer boundary of the FFZ) must be cleared of obstructions. This is the reason why Fresnel zone is also sometimes referred to as *RF line-of-sight*. The link's reliability is not remarkably affected if obstacles cover up to 20% of the FFZ. However, the impact of obstructions become remarkable as they obstruct more than 40% of the FFZ. Why is clearance of the FFZ more important than the higher-order Fresnel zones? The FFZ contains much of the energy in the EM wave. Therefore, the intensity of any reflections within the FFZ is high and causes the highest disturbance at the receiver. Consider the illustration in Fig. 10.3.

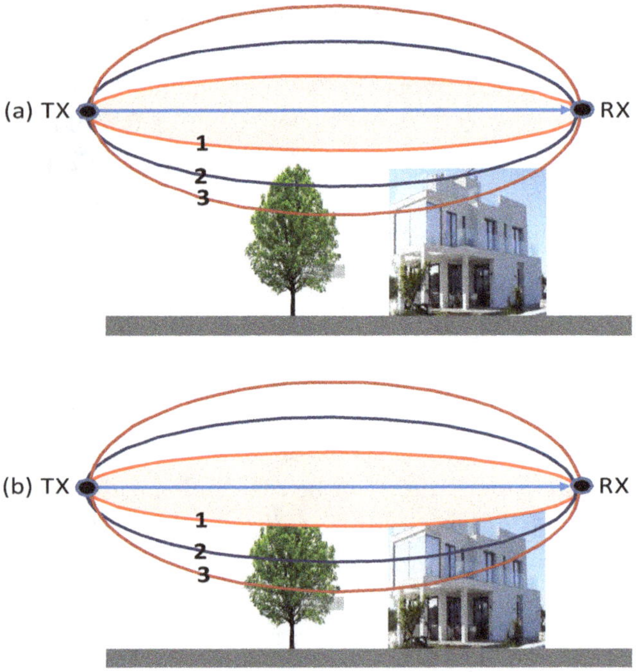

Fig. 10.3 FFZ clearance.

The situation in (a) is alright, but not that in (b) as the FFZ is significantly disturbed.

In order to provide the appropriate clearance of the dominant obstacle, we need its height as well as the radii of the Fresnel zones. We find the Fresnel zones' radii through knife-edge geometry and note that:

- The Fresnel zones are the successive regions where the secondary waves caused by diffraction of obstructions have total path lengths greater than the total path length of the direct LOS path by $n\lambda/2$.
- The surface of the point along which the length of the diffracted path is longer than the direct LOS path by a multiple of half-wavelength determines the radii of the Fresnel zones.

Figure 10.4 illustrates the knife-edge diffraction showing the first three Fresnel zones. In the figure h is the vertical distance between the direct LOS path and the diffracted path. Applying Pythagoras theorem in linear geometry to the left triangle APY we can obtain the length of the left part of the diffracted path as $L_f = \sqrt{d_1^2 + h^2}$. Doing similarly to the right-hand side triangle PYZ gives the length of right part of the diffracted path as $L_r = \sqrt{d_2^2 + h^2}$. Thus, the total length of the diffracted path is $L_d = \sqrt{d_1^2 + h^2} + \sqrt{d_2^2 + h^2}$. With the total length of the LOS path being $d = d_1 + d_2$, we obtain the difference of the lengths of the two paths as

$$\begin{aligned} \Delta &= \sqrt{d_1^2 + h^2} + \sqrt{d_2^2 + h^2} - d_1 - d_2 \\ &= d_1\sqrt{1 + \left(\frac{h}{d_1}\right)^2} + d_2\sqrt{1 + \left(\frac{h}{d_2}\right)^2} - d_1 - d_2 \,. \end{aligned} \tag{10.8}$$

We are going to find an approximate for Eq. (10.8) by using the trick $\sqrt{1+y} \approx 1 + \frac{y}{2}$ if $y << 1$. Thus,

$$\begin{aligned} \Delta &\approx d_1\left[1 + \frac{1}{2}\left(\frac{h}{d_1}\right)^2\right] + d_2\left[1 + \frac{1}{2}\left(\frac{h}{d_2}\right)^2\right] - d_1 - d_2 \\ &= h^2\frac{d_1 + d_2}{2d_1 d_2} \,. \end{aligned} \tag{10.9}$$

We have understood above that the surface of the point along which the length of the diffracted path is longer than the direct LOS path by a multiple of half-wavelength determines the radii of the Fresnel zones, i.e.

$$\begin{aligned} \texttt{If } h = F_n &:\Rightarrow \Delta \approx h^2\frac{d_1 + d_2}{2d_1 d_2} \\ &\stackrel{!}{=} \frac{n\lambda}{2} \Leftrightarrow F_n \approx \sqrt{\frac{n\lambda d_1 d_2}{d_1 + d_2}} \end{aligned} \tag{10.10}$$

where F_n is the radius of the n-th Fresnel zone and n is a positive integer. We have illustrated the Fresnel zones for $n = 1, 2, 3$ in Fig. 10.4. The successive regions in which the path lengths of the diffracted EM waves (aka secondary waves) differ from the length of the direct LOS EM wave relative to the receiver by integer multiples of $\lambda/2$ define the Fresnel zones.

As illustrated in Fig. 10.4(b), the maximum radius for each Fresnel zone occurs if $d_1 = d_2 = d$. In which case the radius obtained in Eq. (10.10) becomes

$$F_{\texttt{max},n} = \sqrt{\frac{n\lambda d}{4}} \approx 8.66\sqrt{\frac{d_{T-R}\ \texttt{[km]}}{f\ \texttt{[GHz]}}}\ \texttt{[meters]}\,. \tag{10.11}$$

Considering Fig. 10.4(b) and assuming that the transmit and receive antennas are of the same

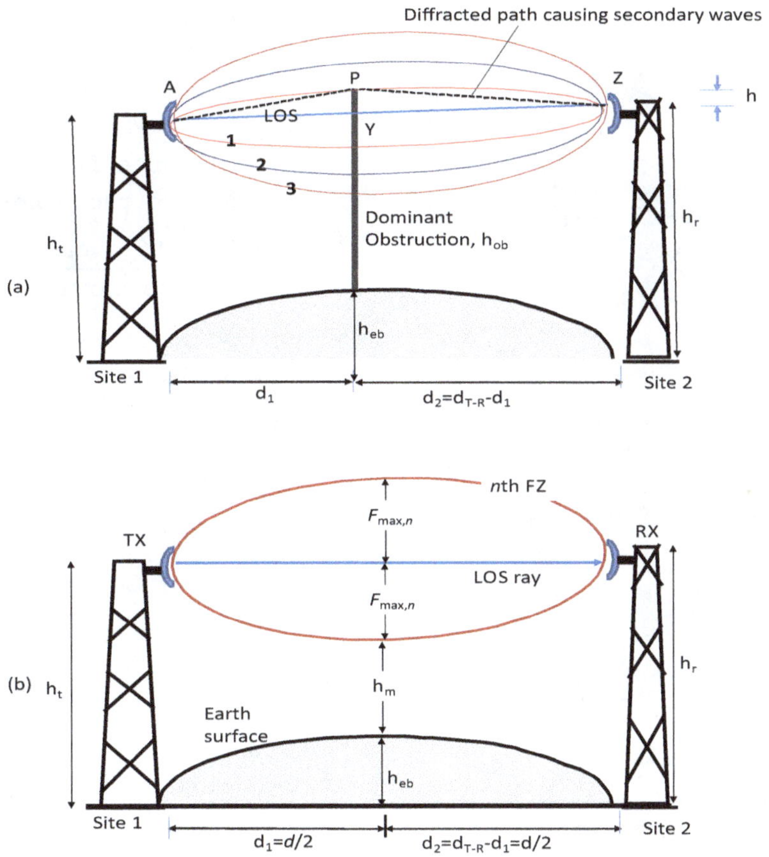

Fig. 10.4 Knife-edge diffraction: (a) showing the first three Fresnel zones, and (b) condition for maximum Fresnel zone radius.

height (i.e., $h_t = h_r = h_a$), the nth Fresnel zone does not touch the ground if and only if

$$F_{\mathtt{max},n} + h_{eb} < h_a \tag{10.12}$$

where, again h_{eb} is the Earth's bulge measuring the curvature of the Earth for wireless communications purposes. Let us for now assume that there is no Earth bulge, i.e., $h_{eb} \approx 0$. This is called flat-earth case.

We seek to use to study the minimum antenna heights required for the clearance of the FFZ at different frequencies and different distance between receiver and transmitter in Fig. 10.5.

We can observe that a taller antenna is needed as the distance between the receiver and transmitter increases, if all else is kept constant. The same pattern is true as the operating frequency increases. For example, a 4-m antenna height is alright at 1 km receiver/transmitter distance at 5.8 GHz, but at least 6-m antenna height is needed at 2.4 GHz. As the range increases to, say 2 km, we need at least 5-m antenna towers at 5.8 GHz but over 8-m towers at 2.4 GHz. This means that operating a LOS microwave link at a higher frequency is better as it saves us from erecting tall towers.

10.3.2 *The Earth's Bulge*

In telecommunications the Earth's bulge (h_{eb}) is used to refer to the circular segment or curvature of earth profile which blocks long-distant communications (See Fig. 10.6). As observed in Fig. 10.5, the radius of the Fresnel zone increases with the transmission distance. Therefore, the longer the distance between the transmitter and receiver of a radio link, the higher the effect of the Earth's curvature. The Earth's bulge can cut into the FFZ if the heights of the antenna are not appropriately dimensioned. The Earth's bulge correction factor in meters is determined from the equation

$$h_{eb} = 0.078 d_1 d_2 \ \mathtt{[m]} \tag{10.13}$$

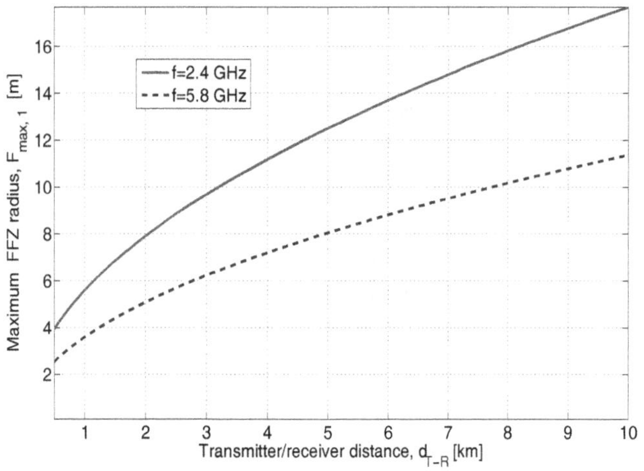

Fig. 10.5 Antenna heights required for FFZ clearance.

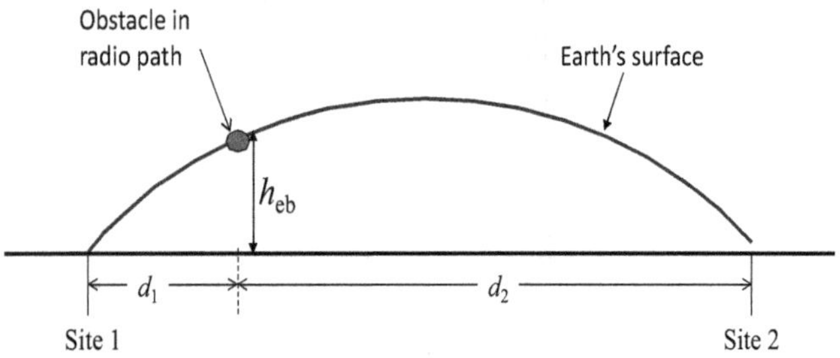

Fig. 10.6 Illustration of the Earth's bulge.

where d_1 and d_2 should be in kilometers, but h_{eb} is in meters. It may be noted that the maximum earth's bulge is $h_{eb} \approx 0.02 d_{T-R}^2$, and it occurs if $d_1 = d_2 = \frac{d_{T-R}}{2}$.

10.3.3 *Ray Bending and k-Factor*

Figure 10.7 contrasts two horizons: radio horizon and the optical horizon. The optical horizon is determined by how far our human eye can see. The optical horizon, however, is determined by how far the antennas can see. Radio line-of-sight means that both the transmitting and receiving antennas see each other. For this reason, the radio horizon is also referred to as the RF line-of-sight, and it is the most important in radio communications. Note further that optical or visual LOS does not necessarily mean radio LOS. The bending of the EM wave beyond the optical horizon is accounted for in the link design by the k-factor, which is defined as

$$k = \frac{\begin{array}{c}\texttt{distance from TX antenna}\\ \texttt{to optical horizon}\end{array}}{\begin{array}{c}\texttt{distance from TX antenna}\\ \texttt{to radio horizon}\end{array}} = \frac{d_o}{d_r} = \frac{\texttt{effective earth radius}}{\texttt{true earth radius}}. \quad (10.14)$$

The k-factor is not constant as the radio horizon distance changes with the atmospheric refractive index. The atmospheric refractive index n depends on the atmospheric pressure P [hPa], the absolute temperature T [K], and the density of water vapour ρ [g/m^3]. The relationship is

$$n = \frac{\texttt{speed of EM waves in vacuum}}{\texttt{speed of EM waves in the medium}}. = 1 + 10^{-6} \times \frac{77.6}{T}\left(P + \frac{4810}{216.7}\rho\right), \; f \leq 100 \texttt{ GHz}. \quad (10.15)$$

Note that P, T and ρ all depend on height. If the optical horizon is farther away than the radio horizon then $k > 1$, otherwise $k < 1$. In particular, if $k = \frac{4}{3}$ then we have from Eq. (10.14) the situation in which the effective earth radius equals $\frac{4}{3}$ of the true earth radius. This is referred to as four-third earth radius. Table 10.2 provides some values of the k-factor and their meanings.

The substandard refraction is caused by the increasing air dielectric constant with height, resulting in the upward bending of EM waves. The super-refraction, on the other hand, can be caused by meteorological conditions such as a remarkable decrease in air humidity or temperature rise with increasing height [Navy (1972)]. We have the *standard atmosphere* at the temperature of 273.15 degrees Kelvin or zero degrees Celsius, absolute pressure of 100 kPa or 1 bar and air density of 1.2922 kg/m^3. The rule-of-thumb used at standard atmosphere for the optical and radio horizon distances are

$$d_o = 0.46\sqrt{h} \texttt{ [km] and } d_r = 0.49\sqrt{h} \texttt{ [km]} \quad (10.16)$$

where h is the height of the antenna tower in meters. This results in $k = \frac{D_o}{D_r} \approx 0.9388$.

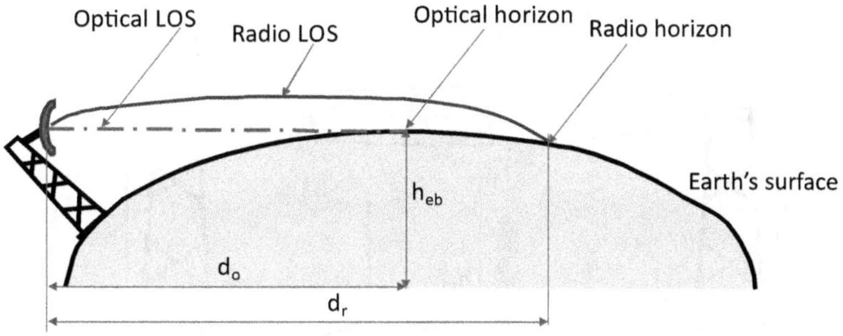

Fig. 10.7 Radio horizon versus optical horizon.

Table 10.2 Some values of the k-factor accounting for ray bending.

k-factor value	Type of atmosphere/meaning
$\frac{1}{2}$	Substandard atmosphere, i.e. the Earth bulges
1	Homogeneous atmosphere, i.e. no bending of waves
$\frac{4}{3}$	Standard atmosphere, i.e. EM wave bends slightly
∞	Superstandard atmosphere with super-refraction
Negative	We presume the Earth is plane, but not curved
> 1	(Optical horizon) $>$ (radio horizon), ray bends toward the earth, radio horizon is larger than the optical horizon, commonly $k = 4/3$ used
< 1	(Optical horizon) $<$ (radio horizon), radio horizon is less than optical horizon, commonly $k = 2/3$ is used

Example

An organisation operates a microwave link at 4 GHz to interconnect its two sites separated by 10 km. During the operational life of the system someone erected a storey building of effective height 260m 2 km away from one of the sites, resulting in unreliable operation. What should be the antenna height to restore a reliable operation of the system. For simplicity, assume that the antennas at both ends of the link have the same height.

Solution

We illustrated the problem using Fig. 10.8. The wavelength at 4 GHz is $\lambda = \frac{c}{f} = 3/40$ m. The k-factor accounting for the ray bending can be evaluated using the rule-of-thumb as $k = \frac{d_o}{d_r} \approx 0.94$. We have $d_{T-R} = 10$ km, $d_1 = 2$ km and thus $d_2 = d_{T-R} - d_1 = 8$ km. The correction factors are then

`Earth's bulge and ray bending:`

$$h_{ebk} = \frac{0.078 d_1 d_2}{k} \approx 1.33 \text{ m} \tag{10.17}$$

$$\texttt{FFZ radius:} \quad F_1 \approx \sqrt{\frac{\lambda d_1 d_2}{d_1 + d_2}} =\approx 0.35 \text{ m} \tag{10.18}$$

The height h of the antenna tower should be

$$\begin{aligned} h = &\ (\texttt{height of obstruction}) + (\texttt{FFZ radius}) \\ &+ (\texttt{Earth's bulge and ray bending}) \\ &+ \texttt{margin} \\ = &\ 260 + 0.35 + 1.33 + 12 \text{ m} = 273.7 \text{ m}. \end{aligned} \tag{10.19}$$

We need to add a margin h_{cl} to cater for any foliage that may grow on top of the specular obstruction. In the example earlier, we have set $h_{cl} = 12$ m. The height budget is summarised in Table 10.3.

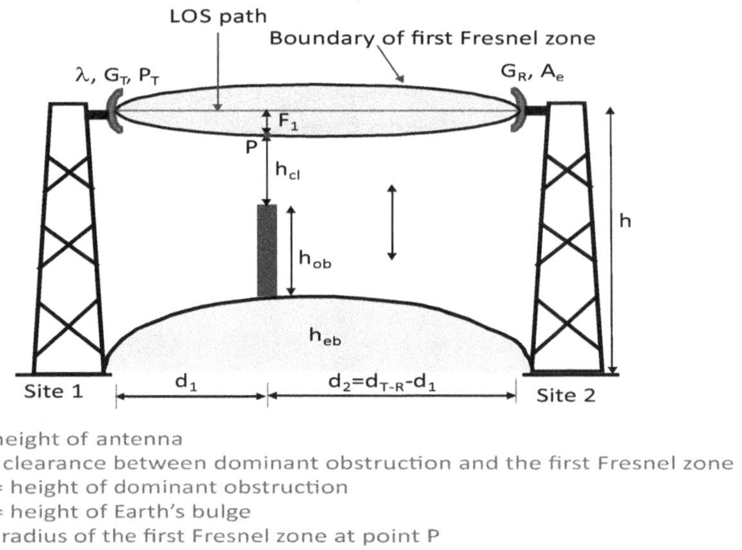

Fig. 10.8 Determination of antenna height.

Table 10.3 Link height budget.

Parameter	**Value**
Earth bulge with ray bending	1.33 m
Dominant obstacle height	260 m
Radius of FFZ	0.35 m
Clearance margin	12 m
Total height of antenna	273.7 m

10.3.4 *Check Your Understanding*

Students are encouraged to check their understanding of the content of this section with the following questions prior to proceeding to the next section.

(1) What is Earth's bulge and how does it affect terrestrial microwave link design?
(2) What causes Fresnel zones?
(3) What are their shapes?
(4) Which Fresnel zone has the maximum effect on radio system design?
(5) What does Fresnel zone clearance mean?
(6) What factors must be considered in determining the height of an antenna used in LOS microwave link?
(7) What is k-factor in terrestrial microwave link design?

10.4 LOS Microwave Link Power Budget

The predominant loss in microwave power budget is the free-space path loss. However, several losses other than the free-space loss must be considered in the calculation of a link's power budget. Failure to do so makes the system unreliable. Considering these losses makes the system conservative in situations when these additional losses do not apply.

10.4.1 *Free-Space Path Loss*

As illustrated in Fig. 10.9, a microwave radio system has as basic components transmitting and receiving antenna often mounted on masts of appropriate heights, radio electronics and transmission line which connects the electronics to the antenna. The transmission line commonly used is coaxial cable or waveguide, both of which are usually made of copper metal. Waveguides are preferable at microwave frequencies in order to reduce power losses.

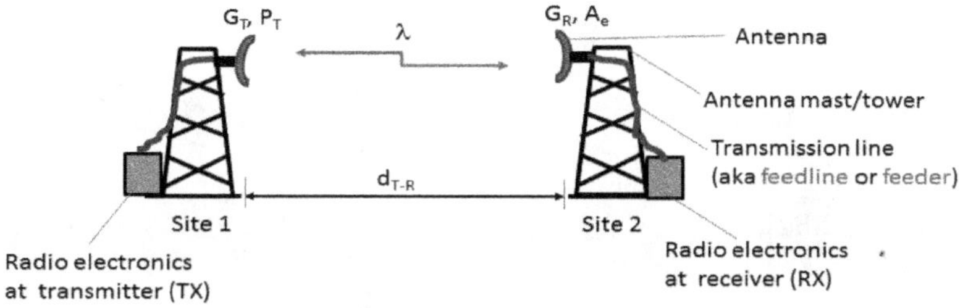

Fig. 10.9 LOS point-to-point microwave path (ideal case).

We start our analysis of the LOS point-to-point (P2P) microwave link or path by deriving the free-space path loss (FSL). Consider Fig. 10.9, in which two sites (namely, Site 1 and Site 2) are interconnected via a P2P LOS microwave link. The two sites can be, for example, two local area networks (LANs) at two premises of an organisation. The two sites can also be a radio base station and the core network of a cellular network. In the latter case the microwave link serves as a backhaul link. For simplicity of analysis, we make the following assumptions:

(1) The antenna masts are of equal heights (h).
(2) The link's length is so small that the waves see the Earth as being flat.
(3) The electromagnetic (EM) wave propagates in free space, i.e. air without any obstructions. Note that the atmosphere is not a vacuum because it contains some matter.
(4) There is no other signal loss besides that caused by EM wave spreading in free space.
(5) The antennas at both the receiver and transmitter have no directivity or gain, i.e. $G_R = G_T = 1 = 0\ dB$.

The parameters in the figure are the gain of the transmit antenna G_T, the gain of the receive antenna G_R, the operating wavelength in meters λ, the distance between the transmit and the receive antennas in kilometres d_{T-R}, the power in the EM waves leaving the transmitting antenna, and the effective area of the antenna. As illustrated in the figure, we are using an aperture antenna.

The power in the signal leaving the transmitter at Site 1 towards the receiver creates an electric field whose intensity or strength at a distance d_{T-R} from the transmitter is

$$E_d = \frac{P_T G_T}{4\pi d_{T-R}^2}\ \texttt{Wm}^{-2}. \tag{10.20}$$

The power incident on the receive antenna with the effective aperture $A_e = \eta A_a$ is

$$P_R = E_d A_e = E_d A_a \eta = \frac{P_T G_T A_a \eta}{4\pi d_{T-R}^2}\ \texttt{W}. \tag{10.21}$$

The relationship between the gain and the effective area of an aperture antenna is

$$A_a = \frac{\lambda^2 G_R}{4\pi}\ \texttt{m}^2. \tag{10.22}$$

Plugging Eq. (10.22) into Eq. (10.21) results in

$$P_R = \frac{\lambda^2 P_T G_T G_R \eta}{(4\pi d_{T-R})^2}\ \texttt{W}. \tag{10.23}$$

The path loss is defined as

$$\begin{aligned} FSL &= \frac{\texttt{transmitted power}}{\texttt{received power}} = \frac{P_T}{P_R} \\ &= \frac{(4\pi d_{T-R})^2}{\lambda^2 G_T G_R \eta}. \end{aligned} \tag{10.24}$$

Equation (10.24) is usually written in decibels (dB). We do so by taking the logarithm to base ten and multiplying by ten, i.e.

$$\begin{aligned} FSL(dB) &= 10\log_{10}(4\pi d_{T-R})^2 \\ &\quad - 10\log_{10}(\lambda^2 G_T G_R \eta)|_{G_T=G_R=\eta=1} \\ &= 20\log_{10}(4\pi) + 20\log_{10} 1000 \\ &\quad + 20\log_{10} d_{T-R}[km] - 20\log_{10}\lambda[m] \\ &\approx 81.98 + 20\log_{10} d_{T-R}[km] \\ &\quad - 20\log_{10}\lambda[m]. \end{aligned} \tag{10.25}$$

The FSL is often also written in terms of operating frequency instead of the operating wavelength. By noting that $c = f\lambda$, where the speed of light in vacuum is $c \approx 3 \cdot 10^8$ m/s $= 0.3$[km MHz], we can rewrite the path loss as

$$FSL = \left(\frac{4\pi d_{T-R}}{\lambda}\right)^2 = \left(\frac{4\pi d_{T-R} f}{c}\right)^2 = \left(\frac{4\pi d_{T-R}[\mathtt{km}] f[\mathtt{MHz}]}{0.3}\right)^2 \quad (10.26)$$

which in decibels is

$$\begin{aligned} FSL(dB) &= 20\log_{10}\left(\frac{4\pi}{0.3}\right) + 20\log_{10} d_{T-R}[\mathtt{km}] \\ &\quad + 20\log_{10} f[\mathtt{MHz}] \\ &\approx 32.44 + 20\log_{10} d_{T-R}[\mathtt{km}] \\ &\quad + 20\log_{10} f[\mathtt{MHz}]\,. \end{aligned} \quad (10.27)$$

Thus, the path loss relates to the transmission distance and frequency as:

$$FSL \propto (d_{T-R} \cdot f)^2\,. \quad (10.28)$$

This is the reason why signals at microwave frequencies (i.e. high frequencies and low wavelengths) undergo more path loss than low-frequency signals. If the climate and the environment had no effect on the EM wave, and there were no losses in the transmission media then the total loss in the link would be as given in Eqs. (10.26) and (10.27). Unfortunately, such an ideal condition does not exist. Thus, microwave LOS path design must consider other factors, which are the topics in the ensuing sections of this chapter. The FSL for four different microwave frequencies are shown in Fig. 10.10.

Example

A radio transmitter which is rated $P_T = 10$ W is used to send data to another point farther away from it. The system operates at $f = 8$ GHz.

(1) If the system has been designed to work reliably only if the FSL does not exceed 150 dB, how far could the transmitting and receiving antennas be apart?
(2) Assuming that the only losses in the system is the FSL, what should be the gains of the receiving and transmitting antennas if the receiver sensitivity[2] is $P_r = -101$ dBm. Assume that the two antennas are identical and that the fade margin used is 15 dB.

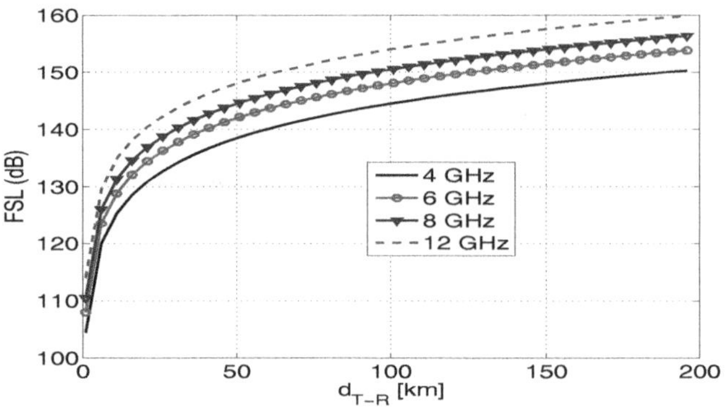

Fig. 10.10 Free-space path loss at different frequencies and distances.

Solution

(1) From Fig. 10.10 if $f = 8$ GHz and $FSL \leq 150$ dB, then $d_{T-R} \leq 100$ km. We can obtain the same results from Eq. (10.27).
(2) We assume transmitting at maximum power which is $P_T = 10$ W or 40 dBm. We are told in the question that $G_T = G_R = G$. The power budget for the microwave link is

$$\begin{aligned} P_r &= P_T + G_T + G_R - L - (\mathtt{fade\ margin}) \Leftrightarrow \\ G &= \frac{1}{2}[P_R - P_T + FSL + (\mathtt{fade\ margin})] \\ &= 12dB\,. \end{aligned} \quad (10.29)$$

10.4.2 *Case Study: Cambium Point-to-Point 50650 Radio*

The Cambium Point-to-Point (PTP) 50650 radio can be used to build a point-to-point microwave link. The link can be either near LOS (line-of-sight) or non-LOS. Some of the parameters of the PTP50650 radio are (see Fig. 10.11 for more parameters):

- Maximum transmit power for BPSK modulation: $P_t = 27$ dBm.
- Transmit and receive antenna gains: $G_t = 23$ dBi, $G_r = 23$.
- Operating frequency band: 5.15–5.25 GHz, with the carrier being at the centre of the band.
- Receiver sensitivity: $S_r = -98$ dBm with channel bandwidth of 5 MHz.

[2]The receiver sensitivity, also referred to as the receiver threshold or RF sensitivity, is the minimum power that should be in an arriving signal for the receiver to be able to detect at a given bit error rate.

System Parameter/Feature	Value
Radio system model	PTP50650
Frequency bands (GHz)	4.9 – 6.05 (viz., 4.94-4.99, 5.15-5.25, 5.25-5.35, 5.47-5.725, 5.725-5.85, 5.825-6.05)
Channel bandwidths (MHz)	5, 10, 15, 20, 30, 40 & 45
Spectral efficiency (bps/Hz)	≤ 10
Maximum transmit power	27 dBm (BPSK), 23dBm (256-QAM)
Antenna gain (flat panel)	Up to 23 dBi
Receiver sensitivity	-98 dBm
Modulation	Adaptive: BPSK to 256QAM
Duplexing	TDD
Transmission range	200 km
Latency (one way)	1-3 ms
Layer 2 frame size	Up to 9600 bytes
Line rate	Up to 850 packets per second
User data throughput (variable)	Up to 450 Mbps (2x2, 256-QAM, 45-MHz channel)
Wind speed survival	322 kph (or km/h)
Operating temperature	-40 °C - +60 °C (incl. solar radiation)
Feeder cable length	100 m (PoGE), up to 300 m (fiber)
Dust-water intrusion protection	International Protection (IP) IP66 & IP67
Source of data: www.cambiumnetworks.com	

Fig. 10.11 Some parameters of Cambium Network's PTP50650 radio.

Example

Assume that $L_t = 1.0$ dB is the feeder/transmission line loss at transmitter, and $L_r = 0.2$ dB is the feeder/transmission line loss at receiver. Assume further that the sum of all other loses not mentioned in this case study plus the fade margin is 18 dB.

(1) Using the appropriate link budget of the system, compute the maximum path loss (PL) that the microwave link can withstand and still operate as expected.
(2) Using your results in Part (1), find the maximum range $d_{\text{T-R}}$ that the microwave link can operate and still achieve its dimensioned reliability of 99.999%.

Solution

(1) We find the maximum acceptable path loss from the link budget. We do it in tabular form for easiness (see Table 10.4). From the link budget in Table 10.4, the power in the received signal is about $P_r = 53.8 - PL$ dBm. This power must be at least equal to the receiver sensitivity, S_r. Thus,

$$\begin{aligned} P_r &= 53.8 - PL \geq S_r \\ &= -98 \texttt{ dBm} \Leftrightarrow PL = 151.8 \texttt{ dB}. \end{aligned} \tag{10.30}$$

(2) The carrier frequency is usually at the middle of the operating frequency band, i.e., $fc = 5.15 + ((5.25 - 5.15)/2)$ GHz $= 5.2$ GHz. This results in the wavelength $\lambda = 3/52$ m. All other losses and fade margin have been already considered in the above analysis. Thus, the only loss left is the free-space path loss, which is defined as

$$\begin{aligned} PL = FSL &= \left(\frac{4\pi d_{\text{T-R}}}{\lambda}\right)^2 \Leftrightarrow d_{\text{T-R}} \\ &= \frac{\lambda\sqrt{PL}}{4\pi} \approx 178.61 \texttt{ km}. \end{aligned} \tag{10.31}$$

Comment: The Cambium Network's point-to-point microwave radio system, PTP50650, can cover the maximum range of 200 km (see Fig. 10.11).

10.4.3 *EM Wave Attenuation by Atmospheric Gases*

The atmosphere is not a vacuum but full of gases. These gases can absorb part or all the energy in EM wave, depending on the weather, operating wavelength and propagation distance. Gases with remarkable effects on EM waves are water vapour and oxygen. These have the respective chemical symbols H_2O and O_2. The atmospheric attenuation for terrestrial paths (or paths which are slightly inclined but close to the ground) is given by [ITU (2016a)]

$$\begin{aligned} A_{atm}\,\texttt{[dB]} &= \gamma d_{T-R} = (\gamma_o + \gamma_w) d_{T-R} \\ &= 0.1820[N''_{O_2}(f) + N''_{HO_2}(f)] d_{T-R} \\ &= 0.1820 \sum_k [(S_k F_k)_{O_2} + (S_k F_k)_{HO_2}] d_{T-R} \\ &\quad + N''_D(f) d_{T-R} \end{aligned} \tag{10.32}$$

where

- γ_o: attenuation due to dry air (oxygen, pressure-induced nitrogen and non-resonant Debye attenuation) in dB/km
- γ_w: attenuation due to water vapour in dB/km
- d_{T-R}: path length in km

Table 10.4 Link budget case study.

Parameter	Positive/gain	Negative/loss
Transmit power, P_t	+27 dBm	
Transmit antenna feeder loss, L_t		−1 dB
Transmit antenna gain, G_t	+23 dBi	
Path loss (PL)		−PL
Receive antenna gain, G_r	+23 dBi	
Receive antenna feeder loss, L_r		−0.2 dB
Other losses (Lo) and fade margin		−18 dB
Sub-total	$S_p = +73$ dBm	$S_m = -(19.2 + \text{PL})$ dB
Resultant received power, P_r		$S_p + S_m = 53.8 - \text{PL}$ dBm

- $N''_D(f)$: measured in dB/km, is the dry air continuum caused by two factors: pressure-induced nitrogen attenuation above 100 GHz, and non-resonant Debye spectrum of oxygen below 10 GHz
- f: frequency in GHz
- p: dry air pressure in hPa
- ρ: density of water vapour in g/m^3
- T: temperature in Kelvins (K)
- S_k: the strength of the k-th oxygen (O_2) or water vapour line (HO_2)
- F_k: the k-th spectral line shape factor for oxygen or water vapour

The dry air continuum is given by

$$N''_D(f) = 9 \times 10^4 \frac{fp}{T^2}\left[\frac{6.14 \times 10^{-5}}{d[1 + (f/d)^2]} + \frac{1.4 \times 10^{-12} p(300/T)^{1.5}}{1 + 1.9 \times 10^{-5} f^{1.5}}\right] \quad (10.33)$$

where d is the width parameter of the Debye spectrum, which is

$$d = 5.6 \times 10^{-4}\left(p + \frac{\rho T}{216.7}\right)(300/T)^{0.8}. \quad (10.34)$$

The strength of the oxygen and water vapour spectral lines are respectively

$$S_k = \begin{cases} a_1 \times 10^{-7} p\left(\frac{300}{T}\right)^3 e^{a_2(1-300/T)} & \texttt{for } O_2 \\ \frac{b_1 \rho T}{2167}\left(\frac{300}{T}\right)^{3.5} e^{b_2(1-300/T)} & \texttt{for } HO_2 \end{cases}. \quad (10.35)$$

The formula for the spectral line shape factors are

$$F_k = \frac{f}{f_k}\left[\frac{\Delta f - (f_k - f)\delta}{(\Delta f)^2 + (f_k - f)^2} - \frac{\Delta f - (f_k + f)\delta}{(\Delta f)^2 + (f_k + f)^2}\right] \quad (10.36)$$

where f_k is the frequency of the k-th spectral line for both oxygen and water vapour, δ is the correction factor for interference effects, which is given by

$$\delta = \begin{cases} \left(a_5 + \frac{300}{T} a_6\right) \times 10^{-4}\left(p + \frac{\rho T}{216.7}\right)^{0.8} & \texttt{for } O_2 \\ 0 & \texttt{for } HO_2 \end{cases} \quad (10.37)$$

and Δf is the width of the spectral line, which is given by

$$\Delta f = \begin{cases} a_3 \times 10^{-4}\left[p\left(\frac{300}{T}\right)^{0.8-a_4} + \frac{330\rho}{216.7}\right] & \texttt{for } O_2 \\ b_3 \times 10^{-4}\left[p\left(\frac{300}{T}\right)^{b_4} + \frac{b_5 \rho T}{216.7}\left(\frac{300}{T}\right)^{b_6}\right] & \texttt{for } HO_2 \end{cases}. \quad (10.38)$$

For simplicity of illustration, we have not considered in Δf above Zeeman splitting of oxygen and Doppler spreading of water vapour lines. The factors a_k and b_k are found in pages 10–12 of the ITU-R document [ITU (2016a)]. Figure 10.12 shows the atmospheric attenuation per kilometer over frequency. The figure was obtained by writing MATLAB® code to implement the above formulas. The values for a_k's, b_k's and f_k's are found in the ITU-R report [ITU (2016a)]. We can observe the dependence of the attenuation on several parameter, including the temperature and operating frequency. The MATLAB® code and function used to plot Fig. 10.12 are shown in Figs. 10.13 and 10.14, respectively.

10.4.4 *Knife-Edge Diffraction Losses*

One of the electromagnetic wave propagation mechanisms is diffraction, in which areas which are not in

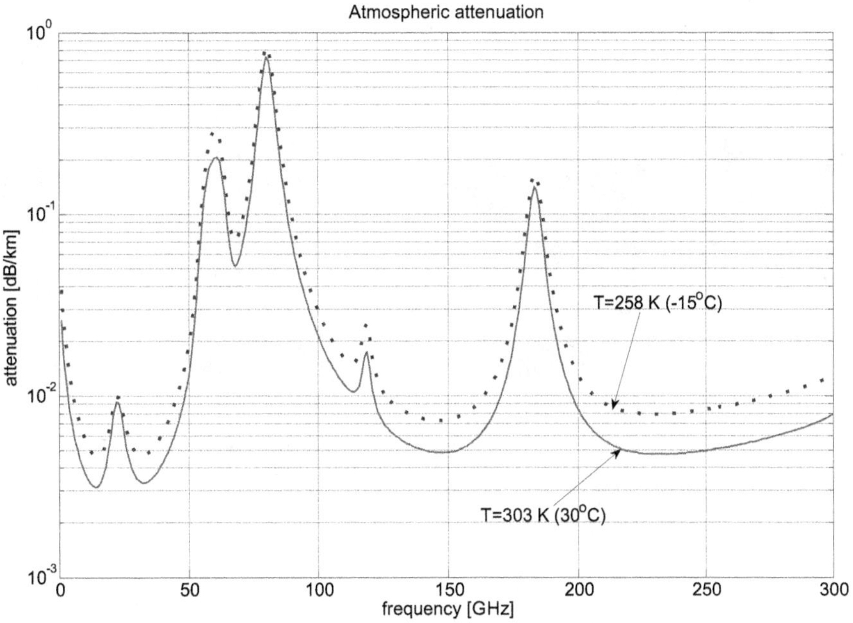

Fig. 10.12 Atmospheric attenuation for $p = 1013.25$ hPa and $g = 7.5\ g/m^3$.

```
% Computation of attenuation due to the atmosphere using
% the model in ITU-R P.676 Report
clc; clear all;
f=0:1:300;  % frequency in GHz
T=288;  % temperature in Kelvin = 15 oC
p=1013.25;  % dry air pressure in hPa
g=7.5;  % density of water vapour in g/m3
% Import comma-separated value (CSV) formatted H2O file into
MATLAB
H2Ofile= 'C:\Users\agyasiagyei\Documents\H2OAttSpectroscopicData.cs
v';
H2Odata = csvread(H2Ofile);
[rowH2Odata, colH2Odata]=size(H2Odata);
fkH2O=H2Odata(:,1); % picks data in first column
BH2O=H2Odata(:,2:colH2Odata); % leaves out data in first column
% Import comma-separated value (CSV) formatted O2 file into MATLAB
O2file= 'C:\Users\agyasiagyei\Documents\O2AttSpectroscopicData.csv'
;
O2data = csvread(O2file);
[rowO2data, colO2data]=size(O2data);
fkO2=O2data(:,1); % picks data in first column
AO2=O2data(:,2:colO2data); % leaves out data in first column
%
At=zeros(1,length(f)); % attenuation vector
for n=1:length(f)
    [Nd,Gsum] = atmattenuation(AO2,BH2O,T,p,g,fkH2O, fkO2,f(n));
    At(n)=0.1820*Gsum + Nd;
end
Y1=semilogy(f, At,'-b');
set(Y1,'LineWidth',2); set(gca,'fontsize',18);
xlabel('frequency [GHz]'); ylabel('attenuation [dB/km]')
title('Atmospheric attenuation')
grid
```

Fig. 10.13 MATLAB® code used to plot Fig. 10.12.

```
function [Nd,Gt] = atmattenuation(AO2,BH2O,T,p,g,fkH2O, fkO2,f)
%function "atmattenuation" computes atmospheric loss in dB/km
% USAGE: [Nd,Gt] = atmattenuation(AO2,BH2O,T,p,g,fkH2O, fkO2,f)
% AO2: matrix of spectroscopic constant factors for O2
% BH2O: matrix of spectroscopic constant factors for H2O
% p: dry air pressure in hPa
% T: temperature in Kelvins K
% g: density of water vapour
% f: frequency in GHz
% fkH2O: vector of frequencies of the k-th spectral line for H2O
% fkO2: vector of frequencies of the k-th spectral line for O2
% Nd: dry air continuum
% Gt: normalized attenuation per km
%=======================
% compute width of Debye spectrum
d=5.6*10^(-4)*(p + g*T/216.7)*(300/T)^0.8;
% compute dry air continuum
Nd=9*10^4*f*p/T^2*(6.14*10^(-5)/d/(1+(f/d)^2) + ...
    1.4*10^(-12)*p*(300/T)^1.5/(1+1.9*10^(-5)*f^1.5));
GO2=0; GH2O=0; % initializations
for i=1:size(AO2,1)
    a=AO2(i,:);
    SkO2=a(1)*10^(-7)*p*(300/T)^3*exp(a(2)*(1-300/T));
    deltaO2=(a(5)+300*a(6)/T)*10^(-4)*(p+g*T/216.7)^0.8;
    DfO2=a(3)*10^(-4)*(p*(300/T)^(0.8-a(4)) + 330*g/216.7);
    FkO2=f/fkO2(i)*((DfO2-(fkO2(i)-f)*deltaO2)/(DfO2^2+(fkO2(i)-
f)^2)-(DfO2-(fkO2(i)+f)*deltaO2)/(DfO2^2+(fkO2(i)+f)^2));
    GO2=GO2 + SkO2*FkO2;
end
for j=1:size(BH2O,1)
    b=BH2O(j,:);
    SkH2O=b(1)*g*T/2167*(300/T)^3.5*exp(b(2)*(1-300/T));
    deltaH2O=0; % water vapour interference correction factor
    DfH2O=b(3)*10^(-4)*(p*(300/T)^b(4) +
b(5)*g*T/216.7*(300/T)^b(6));
    FkH2O=f/fkH2O(j)*((DfH2O-(fkH2O(j)-
f)*deltaH2O)/(DfH2O^2+(fkH2O(j)-f)^2) - ...
        (DfH2O-(fkH2O(j)+f)*deltaH2O)/(DfH2O^2+(fkH2O(j)+f)^2));
    GH2O=GH2O + SkH2O*FkH2O;
end
Gt=GO2 + GH2O;
end
```

Fig. 10.14 MATLAB® function used in the code to plot Fig. 10.12.

line-of-sight with the transmitting antenna can still receive radio signals. The basis of diffraction is the *Huygen's principle*, which states that all points on a wavefront behave as sources of secondary wavelets. A special case of diffraction, referred to as knife-edge diffraction, presumes that the obstruction in the LOS path causing the diffraction is sharp or narrow like the edge of a sharp knife. The knife-edge diffraction enables the estimation of the loss due to diffraction effects in a conservative manner.

The diffraction loss (DFL) is also referred to as diffraction fading or obstruction loss. This is caused by any obstructions in the LOS microwave path. The DFL depends on terrain and vegetation, the locations of the transmitting and receiving antennas relative to the dominating obstruction, and the nature of the obstruction. For a given path clearance, the DFL varies from a minimum value for a single knife-edge obstruction to a maximum value assuming that the Earth were smooth and spherical. The DFL over an average terrain with losses larger than 15 dB is approximately (p. 9 of [ITU-R P.530-16]) or p. 32 of [Freeman (2007)]

$$A_d = 10 - 20\frac{h}{F_1}, \; f > 10 \; \texttt{GHz} \tag{10.39}$$

where h in meters is the height difference between the tallest obstruction and the LOS ray, and F_1 is the radius of the first Fresnel zone (FFZ), which is obtained from Eq. (10.10) as

$$F_1 \approx \sqrt{\frac{\lambda d_1 d_2}{d_1 + d_2}} \approx 17.3\sqrt{\frac{d_1 d_2}{(d_1 + d_2)f}} \; \texttt{[m]}. \tag{10.40}$$

Here, f is the operating frequency in GHz and d_1 and d_2 are distances in km from the TX/RX to the obstruction, respectively.

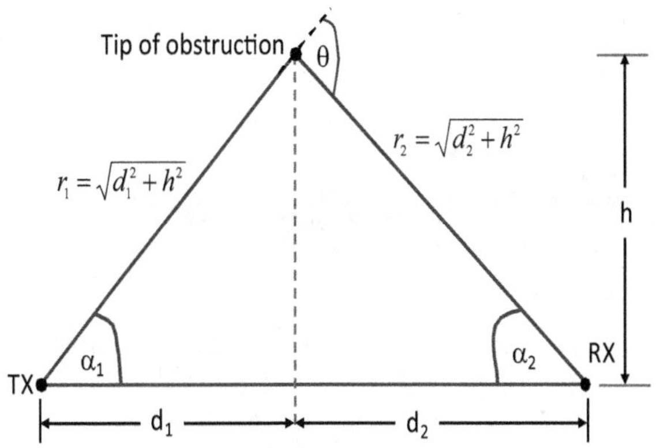

Fig. 10.15 Approximate knife-edge diffraction geometry.

We use the simplified geometry shown in Fig. 10.15 to derive the path difference between the LOS path and a single diffracted path. The Pythagoras theorem has been applied to find the lengths shown in the figure. As we did in the Fresnel zone radius calculation, let Δ be the path difference. Then from the geometry in Fig. 10.15, we obtain

$$\begin{aligned}\Delta &\approx r_1 + r_2 - (d_1 + d_2)\\ &= \sqrt{d_1^2 + h^2} + \sqrt{d_2^2 + h^2} - (d_1 + d_2)\\ &\approx \frac{(d_1 + d_2)h^2}{2d_1 d_2}\end{aligned} \tag{10.41}$$

where we have applied the approximation $\sqrt{1+x} \approx 1 + x/2$ if $x \ll 1$. For simplicity we have assumed that the heights of the transmitting and receiving antennas are the same in Fig. 10.15. This allows us to focus on the angles involved in the diffraction. Note that the angle of diffraction is $\theta = \alpha_1 + \alpha_2$. For a small ϕ, we have the approximation $\tan(\phi) \approx \phi$. We also note that usually the height of the obstruction relative to the LOS path is very small compared with the distances d_1 and d_2. Thus, we use the following approximations:

$$\begin{aligned}\alpha_1 &= \tan^{-1}\left(\frac{h}{d_1}\right) \approx \frac{h}{d_1}\\ \alpha_2 &= \tan^{-1}\left(\frac{h}{d_2}\right) \approx \frac{h}{d_2}\\ \theta = \alpha_1 + \alpha_2 &\approx \frac{h}{d_1} + \frac{h}{d_2} = \frac{(d_1 + d_2)h}{d_1 d_2}\end{aligned} \tag{10.42}$$

We define the *Fresnel-Kirchhoff diffraction coefficient* (aka *Fresnel-Kirchhoff diffraction parameter* or just *Fresnel-Kirchhoff coefficient*) as

$$\nu = h\sqrt{\frac{2(d_1 + d_2)}{\lambda d_1 d_2}}\,. \tag{10.43}$$

Note that h and thus ν is positive if the tip of the obstruction causing the knife-edge diffraction is taller than the altitude of the direct line-of-sight path, and it is negative otherwise. Noting the approximations in Eq. (10.42), we can also express Δ as

$$\Delta = \frac{h}{2}(\alpha_1 + \alpha_2) = \frac{h}{2}\theta\,. \tag{10.44}$$

This enables us to define ν as a function of the parameters in the knife-edge diffraction geometry as

$$\nu = h\sqrt{\frac{2(d_1 + d_2)}{\lambda d_1 d_2}} = \sqrt{\frac{2h\theta}{\lambda}} = \sqrt{\frac{2d}{\lambda}\alpha_1\alpha_2}\,. \tag{10.45}$$

Let E_0 be the intensity of the electric field at the receiver produced by the LOS path without any ground reflection nor knife-edge diffraction. Also, let $E'_{\mathsf{d},n}$ be the intensity of the electric field at the receiver produced by a knife-edge diffracted wave. Assume that the magnitudes of these two fields are equal so that we focus on their phase difference. The ratio between the intensities of these two fields is

$$\frac{E'_{\mathsf{d},n}}{E_0} = e^{-j\kappa\Delta} = e^{-j\frac{\pi}{2}\nu^2}\,. \tag{10.46}$$

Equation (10.46) is for a single ray. However, from Huygens principle, multiple rays diffract from the obstacle towards the receiver. The sum of these rays results in the *complex Fresnel integral*

$$\begin{aligned}F(\nu) &= \frac{E_d}{E_0} = \frac{1+j}{2}\int_v^{\infty} e^{-j\frac{\pi}{2}t^2}\,dt\\ &= \frac{1+j}{2}\left[\int_v^{\infty}\cos\left(\frac{\pi}{2}t^2\right)dt - j\int_v^{\infty}\sin\left(\frac{\pi}{2}t^2\right)dt\right]\end{aligned} \tag{10.47}$$

where $E_d = \sum_n E'_{\mathsf{d},n}$. Note that $\lim_{\nu\to-\infty}\frac{E_d}{E_0} = 1$, meaning that the path has no obstructions. The diffraction loss is given by

$$J(\nu) = -20\log_{10}|F(\nu)|\,. \tag{10.48}$$

Note again that $|F(\nu)|$ is the magnitude of the normalised electric field produced at the receiver relative to the LOS path. The diffraction loss $J(\nu)$ is referred to as *Fresnel-Kirchhoff loss.* As a closed-form solution of Eq. (10.47) has not yet

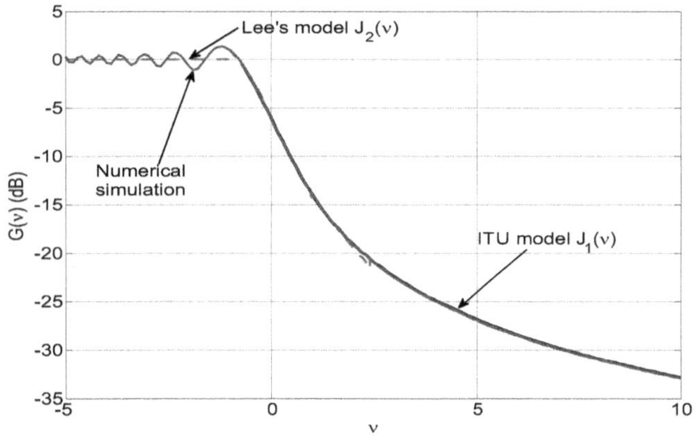

Fig. 10.16 Approximate knife-edge diffraction gains $-J(\nu)$.

been found, the ITU-R has provided an approximate model for this knife-edge diffraction loss in dB for $\nu > -0.5\sqrt{2} \approx -0.7$ as [ITU (2016b)]

$$J_1(\nu) = 6.9 + 20\log_{10}[\sqrt{(\nu-0.1)^2+1} + \nu - 0.1], \ \nu > -0.5\sqrt{2}. \quad (10.49)$$

W.C.Y. Lee has also provided an approximation of $J(\nu)$ in decibels as [Lee (1985)]

$$J_2(\nu) = \begin{cases} 0, & \nu \le -1 \\ -20\log_{10}(0.5-0.62\nu), & -1 \le \nu \le 0 \\ -20\log_{10}(0.5e^{-0.95\nu}), & 0 \le \nu \le 1 \\ -20\log_{10}[0.4-\sqrt{0.1184-(0.38-0.1\nu)^2}], & 1 \le \nu \le 2.4 \\ -20\log_{10}\left(\frac{0.225}{\nu}\right), & \nu > 2.4 \end{cases}. \quad (10.50)$$

The two approximations $J_1(\nu)$ and $J_2(\nu)$ of the Fresnel-Kirchhoff loss are compared with results from numerical simulations in Fig. 10.16. Figure 10.16 was obtained using the MATLAB code shown in Figure 10.17. In fact, Fig. 10.16 shows the diffraction gain, which is $G(\nu) = -J(\nu)$. Note that a negative gain equals a loss, and vice versa. As the values in the graphs are negative, they reflect losses. We can observe that the two approximations are very close to each other, as well as with the simulation results, in the regions that they are defined. The diffraction loss is smaller if the obstruction is shorter than the line of sight EM wave. This is the region in Fig. 10.16 where $\nu < 0$. The diffraction loss increases as the height of the obstruction relative to the LOS path increases in the positive direction.

```
% Plotting of knife-edge diffraction losses
clear all; clc; % clears up
v0=linspace(-5,-1); J0=zeros(1,length(v0));
% ITU Model J_1(\nu)
v=linspace(-0.5*sqrt(2)+eps, 10);
J1=6.9 + 20*log10(sqrt((v-0.1).^2+1) +v -0.1);
figure(1)
Y1=plot(v,-J1,'-b'); set(Y1,'LineWidth',2.5);
set(gca,'fontsize',18); xlabel('\nu')
ylabel('G(\nu) (dB)'); grid
hold on
% Lee's Model J_2(\nu)
v1=linspace(-1.0+eps,0-eps);L1=20*log10(0.5-0.62*v1);
v2=linspace(0,1-eps); L2=20*log10(0.5*exp(-0.95*v2));
v3=linspace(1,2.4-eps); L3=20*log10(0.4 - ...
    sqrt(0.1184-(0.38-0.1*v3).^2));
v4=linspace(2.4,10); L4=20*log10(0.225./v4);
v=[v0 v1 v2 v3 v4]; J2=[J0 L1 L2 L3 L4];
Y2=plot(v,J2,'--r'); set(Y2,'LineWidth',2.5);
% Numerical integration using MATLAB functions
% fresnelc and fresnels
 v=-5:0.1:10; Lv=length(v); J=zeros(1,Lv);
for k=1:Lv
    J(k) = ((fresnelc(inf)-fresnelc(v(k))) ...
        -j*(fresnels(inf)- fresnels(v(k))))*(1+j)/2;
end
Y3=plot(v,20*log10(abs(J)));
set(Y3,'LineWidth',2.5);
hold off
legend('ITU model','Lee model', 'Numerical
Simulation')
```

Fig. 10.17 MATLAB code used to obtain Fig. 10.16.

The phase difference between the LOS direct path and the dominant reflected path is

$$\alpha = \kappa\Delta = \frac{2\pi\Delta}{\lambda} = \frac{\pi}{2}\nu^2 \quad (10.51)$$

where $\kappa = \frac{2\pi}{\lambda}$ is the wave number. The phase difference is also referred to as the *electrical length* of the path difference.

The phase difference causes either of two types of interferences: constructive interference and destructive interference. A perfect constructive interference occurs if the phase difference is integer multiples of 2π, i.e. $2\pi n$. Thus, $\alpha = \frac{\pi}{2}\nu^2 = 2\pi n$, or $\nu^2 = 4n$. This means that all signals reaching the receiver have the same phase, resulting in zero phase difference. Destructive interference occurs if the direct LOS path is out of phase with the diffracted paths. A perfect destructive interference results if the phase difference is π (i.e. 180°) or odd multiples of it. That is $\alpha\frac{\pi}{2}\nu^2 = (2n-1)\pi$, or $\nu^2 = 2(2n-1)$. This gives the following situations between the direct path and the dominant reflected path

$$\nu^2 = \begin{cases} 2(2n-1), \ n = 1, 2, \ldots & \texttt{destructive interference} \\ 4n, \ n = 1, 2, \ldots & \texttt{constructive interference} \end{cases}. \quad (10.52)$$

We remark that n must be a positive integer in the above equations. Thus, constructive interference occurs if $\nu = 4, 8, 12, 16, \ldots$, while destructive interference occurs if $\nu = 2, 6, 10, 14, \ldots$.

An additional attenuation is caused by the curvature of the obstacle causing the knife-edge diffraction. ITU-R models this attenuation as [ITU (2016b)]

$$T(R, h, \lambda) = \left[8.2 + \frac{12h}{R}\left(\frac{\pi R}{\lambda}\right)^{\frac{2}{3}}\right] m^b \qquad (10.53)$$

where R is the radius of the obstacle causing the knife-edge diffraction,

$$m = R^{\frac{2}{3}}\left(\frac{\lambda}{\pi}\right)^{\frac{1}{3}} \frac{d_1 + d_2}{d_1 d_2}$$

and

$$b = 0.73 + 0.27\left[1 - \exp\left(-\frac{1.43h}{R}\left(\frac{\pi R}{\lambda}\right)^{\frac{2}{3}}\right)\right].$$

Note that $\lim_{R \to 0} T(R, h, \lambda) = 0$.

Example: A LOS microwave link connecting two campuses of an organisation separated by 5 km operates at 18 GHz. The two antennas are mounted on masts of equal height of 40 m above the ground level. During the operation of the link, a building of height 44 m was erected in the LOS path 2 km away from one of the campuses.

(1) What is the height of the obstruction above the LOS path?
(2) Calculate the Fresnel-Kirchhoff diffraction parameter ν caused by the obstruction of the LOS link.
(3) Find the Fresnel-Kirchhoff loss caused by the obstruction.
(4) What is the phase difference between the LOS path and the dominant diffracted path?
(5) Does the obstruction cause constructive or destructive interference?

Solution:

(1) The height of the obstruction above the LOS path is $h = (44 - 40)\ \mathtt{m} = 4\ \mathtt{m}$.
(2) Given in the problem are $f = 18$ GHz, $d_1 = 2$ km and $d_2 = 5 - 2 = 3$ km. The Fresnel-Kirchhoff diffraction parameter is $\nu = h\sqrt{\frac{2(d_1+d_2)}{\lambda d_1 d_2}} \approx 1.2649$.
(3) Given $\nu \approx 1.3$, we can obtain the Fresnel-Kirchhoff loss from Fig. 10.16 as 15.5278 dB, while Eq. (10.50) yields 15.5282 dB. This validates the equivalence of the two formulas $J_1(\nu)$ and $J_2(\nu)$ in Eqs. (10.49) and (10.50), respectively.
(4) The phase difference between the LOS path and the dominant diffracted path is $\alpha = \frac{\pi}{2}\nu^2 \approx 0.8\pi$ or 144°.
(5) We have $\nu^2 \approx 1.6$. From Eq. (10.52), we observe that 1.6 is closest to 2 (i.e. the smallest possible value of ν^2 resulting from $n = 1$). This means that the obstruction causes destructive interference. Generally, if the dominant obstruction is closest to the transmitter then it potentially causes destructive interference.

Figure 10.18 illustrates the diffraction effects of obstructions with the same height, but differing positions d_1 on the LOS path away from the transmitter. We can observe that $\nu^2 < 2$ if the operating frequency is $f = 2$ GHz. This means that the obstruction causes destructive interference at that frequency regardless of its position along the LOS path. At 18 GHz the link experiences destructive interference if the obstruction is located at $500 \leq d_1 \leq 4,500$ m away from the transmitter. However, constructive interference can occur at distances $d_1 \leq 250$ m and $d_1 \geq 4,750$ m. Thus, the effects of the obstruction on the link depends on its position and the operating frequency.

10.4.5 *Fade Margin*

Wireless channels cannot usually be expected to behave the same over time. For this reason, a fade margin is always considered in the link budget to make the design conservative. Roger Freeman has proposed in p. 75 of his book [Freeman (2007)] the formulas compiled in Table 10.5. The following definitions apply in the table

- A: fading depth in decibels.
- f: carrier frequency in gigahertz.
- d: length of the radio link in kilometers.
- P_f: the probability of a fading depth exceeding the given fade margin A during one-year period.

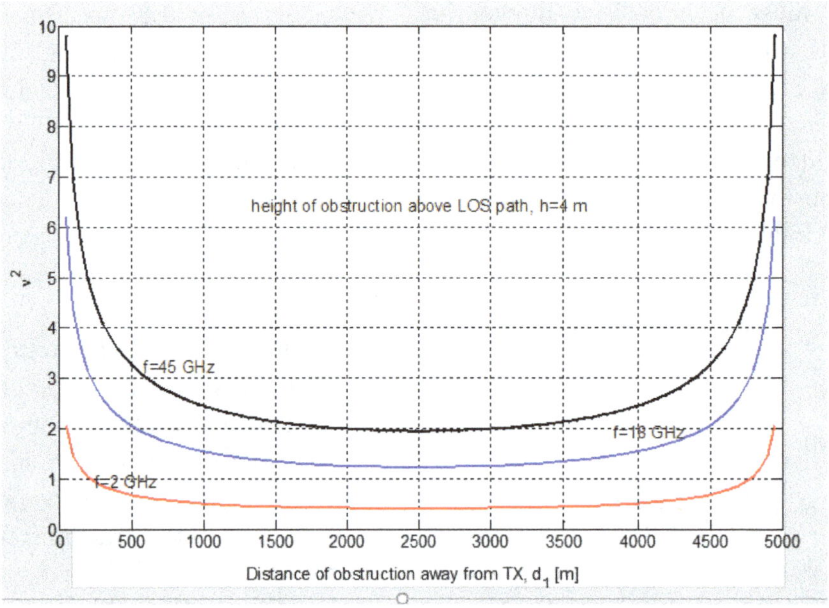

Fig. 10.18 Fresnel-Kirchhoff coefficient as a function of frequency and distance.

Table 10.5 Approximate fade margin formulas [Freeman (2007)].

Link Type	Approximate Fade Margin Formula	Radio Link Characteristics
A	$P_{fA} = 16 \times 10^{-7} fd^2 10^{-A/10}$	Favorable for EM wave propagation
B	$P_{fB} = 8 \times 10^{-7} fd^{2.5} 10^{-A/10}$	Causes average fading
C	$P_{fC} = 2 \times 10^{-7} fd^3 10^{-A/10}$	Causes severe fading

Example: Find the required fading margin for a microwave link spanning a distance of 2 km and operating at 4 GHz which requires 99.999% availability. Assume that the operating environment experiences an average fading per year.

Solution: 99.999% availability means outage probability of $P_f = 0.001\% = 10^{-5}$. We need to rearrange $P_f = 8 \times 10^{-7} f d^{2.5} \times 10^{-A/10}$ to yield

$$A = -10 \log_{10}\left[\frac{P_f \times 10^7}{8 f d^{2.5}}\right].$$

Plugging in the given values results in $A \approx 17.4$ dB.

There are other effects that must be considered in the engineering of a point-to-point microwave link. These include rain attenuation, climate factor, and diversity (e.g. frequency and space diversities). The losses caused by these effects are considered in the selection of the fade margin.

10.4.6 *Check Your Understanding*

Students are encouraged to check their understanding of the content of this section with the following questions prior to proceeding to the next section.

(1) What does link budget mean in microwave system design?
(2) What is free-space path loss? What factors determine it?
(3) See Fig. 10.10. What is the FSL experienced on a microwave link operating at 8 GHz and interconnecting two two radio base stations 5 km apart?
(4) Assume that a microwave link operates at 84 GHz in a location where the dry air pressure is 1013.25 hPa, temperature is 30°C and the density of water vapour is 7.5 g/m^3. From Fig. 10.12 what attenuation due to the atmosphere must be in the link's power budget?
(5) What is the maximum radius of the second Fresnel zone?
(6) What is diffraction loss?

(7) A microwave link operating at 4 GHz spans a distance of 2 km. What is the maximum diffraction loss experienced by a microwave signal transmitted over the link if it (i.e. the signal) hits an obstruction 200 m away from the transmitting antenna? Assume that the obstruction is 50 cm taller than the LOS or direct path.

(8) Assuming that $h/d_1 \ll 1$ and $h/d_2 \ll 1$, prove that the path difference (aka excess path length) is given by Eq. (10.41).

10.5 Transmission Lines

What is a transmission line? It is an electrical conductor which is used to transfer electrical signals between two points, called generator (i.e. signal source) and load (i.e. signal sink). Examples of transmission lines are microstrip line, two wires in parallel, optical fiber, coaxial cable and waveguide. We discuss only the latter two in this chapter. In radio communications, as discussed here, a transmission line connects a radio transmitter (or receiver) to the transmitting (or receiving) antenna. We review in this section the basics of the two types of transmission lines commonly used to interconnect the antenna and the radio electronics. These are coaxial cable and waveguide.

The signal is transmitted in a form of electrical energy along transmission lines as they are made of conducting metals, usually copper material. Part of the energy in the signal gets lost in transit for three main reasons:

- **Losses caused by skin effect**: Every metallic material is non-ideal and thus possesses some resistance which absorbs part of the energy passing through them into heat. A magnetic field is generated whenever a current passes through a metal. As the operating frequency increases the magnetic field pushes the current from inside towards the outer edges or circumference or "skin" of the inner conductor, reducing the metal's layer carrying the current. As the current-carrying area of the wire decreases its resistance increases according to the relation

$$R = \frac{\rho l}{A_e} \ [\Omega] \qquad (10.54)$$

where A_e is the effective area, i.e. current-carrying area, l is the length of the wire, and ρ $[\Omega/\text{m}]$ is the electrical resistivity of the conductor material. This phenomenon is referred to as *skin-effect*. Skin-effect causes electric power to dissipate in the transmission line as heat, resulting in what is referred to as *skin-effect loss.* This is the reason behind the frequency-dependent resistance of a cable. The frequency-dependent resistance caused by skin effect degrades the transition of voltages on long-distant transmission lines. The depth of the skin-effect is given by

$$\delta = \sqrt{\frac{\rho}{\pi f \mu}\left[\sqrt{1 + (2\pi f \rho \epsilon)^2} + 2\pi f \rho \epsilon\right]} \qquad (10.55)$$

where f is the frequency in Hz, μ in H/m is the permeability of the magnetic property of the conductor, ϵ is permittivity of the conductor material and ρ in Ω/m is resistivity of the conductor's material, which is the reciprocal of the conductor's conductivity σ S/m. Note that δ is measured from the surface or "skin" of the conductor towards inside of it. Thus, a larger δ means smaller skin effect. For a coaxial cable with two conductors, δ is measured from the surface of the inner conductor. The intensity of the electric field at a distance z from the conductor's surface is proportional to $e^{-z/\delta}$.

A waveguide has more walls than coax cable. For example, a rectangular waveguide has four walls, and thus a large surface area to carry the current through it, reducing the effect on it by skin-effect. Waveguides are often plated or coated with metals such as silver or gold to further reduce the resistance caused by the walls in order to further reduce its losses.

- **Dielectric loss:** Parallel lines like coax use a dielectric material as insulator to separate the two conductors. The dielectric material draws part of the current passing through the line, which we call *leakage current* and the cause of *dielectric loss.* Waveguides use only one conductor filled with air. Although air is also a dielectric material, its losses are negligible compared with the losses caused by the materials used to separate two conductors. This is the reason why dielectric losses in waveguides are negligible.

- **RF radiation loss:** An electric current passing down a wire creates a magnetic filed, which in turn creates an electric field. Part of the energy in these fields radiates out of the cable, especially if it is not shielded, causing EM radiation which can disturb other signals. In this way, the cable works inadvertently as an antenna, and this behaviour becomes more pronounced at the higher frequencies in which microwave communications occur or at high voltages. A waveguide constrains the current flowing through it inside its inner walls, preventing radiation losses.

In recap, a waveguide has no radiation losses, very little skin-effect losses, and a negligibly low dielectric losses, compared with two-wire lines such as coax and parallel lines. This is the reason why it is the preferred choice in high-frequency applications. Unfortunately, waveguides are expensive and thus not used as much as coaxial cables.

10.5.1 *Coaxial Cable*

A coaxial cable (aka coax) is a transmission line with two wires: an inner wire and an outer wire which are separated by a dielectric material with the dielectric constant ϵ, as illustrated in Fig. 10.19. Like any transmission line, a coax connects a source to a load. At the transmitting side the transmit antenna is the load while the signal source is the source. At the receiving side, however, the receiving antenna acts as the source. The outer conductor of a coax is earthed to prevent radio frequency interference (RFI), both ingress RF to disturb signal of interest or egress RFI which can disturb other signals. The earthing of the outer conductor also serves as the reference voltage to the inner conductor. As the outer conductor is earthed, coax cable is an *unbalanced* transmission line.

Popular coax cables are those with the characteristic impedance of $Z_0 = 50\ \Omega$ (e.g. RG 8A/U for high-frequency applications) and $Z_0 = 75\ \Omega$ (e.g. RG 306A/U for low-frequency applications). However, there are also 93-ohm, 95-ohm, 125-ohm

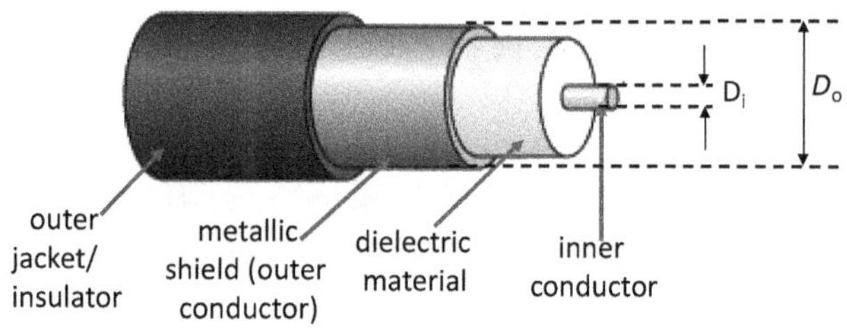

Fig. 10.19 Coaxial cable: D_i and D_o are the diameters of the inner and the out concentric conductors.

and 185-ohm coax cables. The characteristic impedance of a coax cable with the dielectric constant (aka relative permittivity) ϵ_r is

$$Z_0 = \frac{138}{\epsilon_r}\log_{10}(D_o/D_i)\ [\Omega]$$
$$= \frac{60}{\epsilon_r}\log_e(D_o/D_i)\ [\Omega]. \qquad (10.56)$$

The resistance of a copper wire of diameter D_i in meters and length l in meters at microwave frequency f in Hertz (Hz) is given by

$$R_{cu} = 82\times 10^{-9}\sqrt{f}\frac{l}{D_i}\ [\Omega]. \qquad (10.57)$$

The attenuation experienced by a signal moving along a coax cable at frequency f MHz is given by

$$\gamma_{coax} = a_1 + a_2 f^{0.5} + a_3 f\ \texttt{[dB/km]} \qquad (10.58)$$

where $\{a_k\}$ depends on the physical parameters of the cable.

Transmission lines have other effects on signals besides attenuation. For example, the signal propagation velocity along the line can depend on the frequency of the signal. We know from Fourier theory that practical signals are composed of multiple sinusoids at different frequencies and amplitudes. A distortion called *dispersion* occurs if the different frequencies in a given signal travel at different speeds along the line.

10.5.2 *Steady-State AC Analysis of Transmission Lines*

A transmission line such as coax is analysed using two-port network theory. We divide the entire transmission line of length L into small segments of uniform length Δl, and then model each segment as an alternating current (AC) circuit in steady state with the distributed parameters or elements:

- L': combined inductance of both conductors per unit length, measured in Henry per meter or H/m.
- C': capacitance of the two conductors per unit length, measured in farad per meter or F/m.
- G': conductance per unit length of the dielectric material separating the two conductors, measured in Siemens per meter or S/m.
- R': combined resistance per unit length of both conductors, measured in ohms per meter or Ω/m.
- $\sigma_c, \mu_c, \epsilon_c$: the conductivity, permeability and permittivity of the conductors.
- σ, μ, ϵ: the conductivity, permittivity and permeability of the insulating material separating the two wires.

For air we have $\mu_0 = \mu_{air} = 4\pi \times 10^{-7}$ H/m and $\epsilon_0 = \epsilon_{air} = 8.85$ pF/m. The parameters for other materials are expressed as $\epsilon = \epsilon_r \epsilon_0$ and $\mu = \mu_r \mu_0$. These parameters depend on the physical and geometric properties of the cable. They are

$$\begin{aligned} R' &= \sqrt{\frac{f\mu_c}{\pi\sigma_c}\frac{D_i + D_o}{D_i D_o}} \\ L' &= \frac{\mu}{2\pi}\log_e(D_o/D_i) \\ C' &= \frac{2\pi\epsilon}{\log_e(D_o/D_i)} \\ G' &= \frac{2\pi\sigma}{\log_e(D_o/D_i)} \end{aligned} \tag{10.59}$$

We can observe that $L'C' = \mu\epsilon$, which gives the velocity at which EM waves propagate the transmission line as

$$v_p = f\lambda = \frac{1}{\sqrt{L'C'}} = \frac{1}{\sqrt{\mu\epsilon}} = \frac{2\pi f}{\kappa} \tag{10.60}$$

where $\kappa = 2\pi/\lambda$ is the wave number.

A transmission line connecting a signal source to a signal sink is illustrated in Fig. 10.20(a). It is a two-port network with port a-b at the generator and port c-d at the sink. The sink is usually referred to as load. Here the load is that of an antenna. Thus, $Z_{ant} = Z_L$, and Z_i is the input impedance which is discussed further. Figure 10.20(b) shows a model that is used to analyse a transmission line. Let us consider one segment of the coax cable with the infinitesimally small length Δl, as shown in Fig. 10.21. The capacitance, resistance, inductance and conductance are in reality mixed up and distributed along the length of the transmission line. For the sake of analysis, however, we view them as separated and each lumped together, as shown in Figs. 10.20 and 10.21. The lumped elements approach distributed parameters as Δl tends to zero. Prior to further analysis, let us refresh our minds with the equations relating the voltage $u(t)$ and current $i(t)$ across a resistor of resistivity R [Ω], capacitor of capacitance C [F], inductor of inductance L [H] and a conductor of conductance G [S] through the generalised Ohm's law, namely,

$$\begin{aligned} &\texttt{Resistor:}\ u_R(t) = i_R(t)R \\ &\texttt{Capacitor:}\ i_c(t) = C\frac{du_C(t)}{dt} \\ &\texttt{Inductor:}\ u_L(t) = L\frac{di_L(t)}{dt} \\ &\texttt{Conductor:}\ u_G(t) = i_G(t)/G\,. \end{aligned} \tag{10.61}$$

We need to apply the Kirchhoffs' voltage law (KVL) and Kirchhoffs' current law (KCL) noting the above voltage-current relationships to analyse the circuit in Fig. 10.21. The KVL states that the sum of all voltages inside a loop equals zero. Applying the KVL to loop A in Fig. 10.21 yields the equation

$$u(x,t) = R'\Delta l i(x,t) + L'\Delta l\frac{\partial i(x,t)}{\partial t} + u(x+\Delta l, t)$$

which can be rearranged into

$$\frac{u(x,t) - u(x+\Delta l,t)}{\Delta l} = R' i(x,t) + L'\frac{\partial i(x,t)}{\partial t}\,.$$

By noting the limit $\lim_{\Delta l} \frac{u(x+\Delta l,t)-u(x,t)}{\Delta l} = \frac{\partial u(x,t)}{\partial x}$, we can arrive at

$$-\frac{\partial u(x,t)}{\partial x} = R' i(x,t) + L'\frac{\partial i(x,t)}{\partial t}\,. \tag{10.62}$$

The KCL states that the sum of all currents entering and those leaving a point in an electrical circuit equals zero. Applying this law to point Q in Fig. 10.21 results in

$$i(x,t) = G'\Delta l u(x+\Delta l,t) + C'\Delta l\frac{\partial u(x,t)}{\partial t} + i(x+\Delta, t)$$

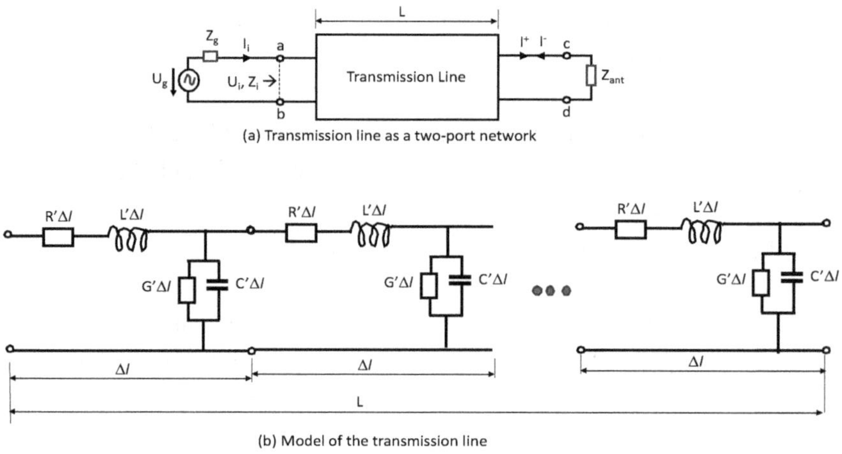

Fig. 10.20 Modelling of a transmission line at high frequency.

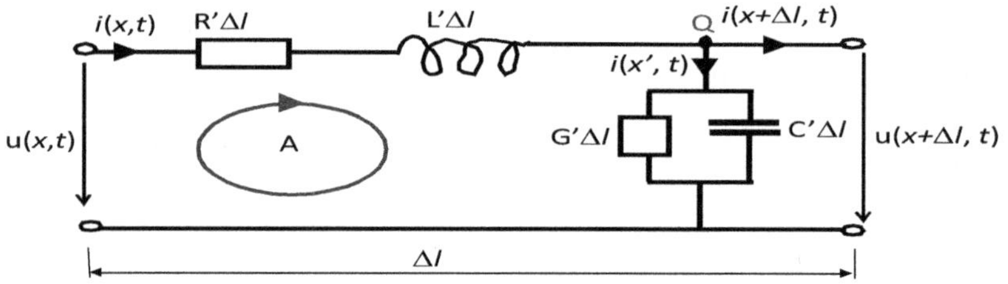

Fig. 10.21 Transmission line section.

which can be rearranged and simplified to

$$-\frac{\partial i(x,t)}{\partial x} = G'u(x,t) + C'\frac{\partial u(x,t)}{\partial t}. \quad (10.63)$$

Note that as Δl tends to zero $u(x+\Delta l, t)$ and $i(x+\Delta l, t)$ turn to $u(x,t)$ and $i(x,t)$, respectively. We note that Eqs. (10.62) and (10.63) are referred to as **coupled transmission line equations** which are generally valid for all transmission lines, including coax cables. We can combine the two equations into second-order differential equations as follows. Take the partial differentiation of Eq. (10.63) with respect to t and Eq. (10.62) with respect to x and combine them to obtain the **decoupled transmission line equation**

$$\frac{\partial^2 u(x,t)}{\partial x^2} = R'G'u(x,t) + (R'C' + L'G')\frac{\partial u(x,t)}{\partial t} + L'C'\frac{\partial^2 u(x,t)}{\partial t^2}. \quad (10.64)$$

We can find a similar decoupled equation in terms of the current $i(x,t)$. Let us assume that the voltage signal $u(x,t)$ is sinusoidal, i.e.

$$u(x,t) = ae^{-\alpha x}\cos(\omega t - \kappa x) = Re\{ae^{j(\omega t - \kappa x) - \alpha x}\} \quad (10.65)$$

and define the phasor of wave propagating in positive x direction as

$$U(x,t) = ae^{j(\omega t - \kappa x) - \alpha x} = ae^{j\omega t - \gamma x}. \quad (10.66)$$

Here, a is the signal amplitude at $x = 0, t = 0$, α is the attenuation constant, and κ is called the wave number, phase constant or phase change, and $\gamma = \alpha + j\beta$ is the wave's propagation constant. γ determines the phase and amplitude of each wave component as it propagates along the transmission line. κ measures the magnitude of the EM wave vector. It is sometimes written as β and thus

$$\kappa = \beta = \frac{2\pi}{\lambda}. \quad (10.67)$$

In the following analysis we shall use $U(x,t)$ instead of $u(x,t)$ for simplicity reasons, and note that the actual voltage is the real part of $U(x,t)$, i.e. $u(x,t) = Re\{U(x,t)\}$. The needed partial differentials of $U(x,t)$ are

$$\begin{aligned}
\frac{\partial U(x,t)}{\partial x} &= -(\alpha + j\kappa)U(x,t) \ (a)\\
\frac{\partial^2 U(x,t)}{\partial x^2} &= [(\alpha^2 - \kappa^2) + j2\alpha\kappa]U(x,t) \ (b)\\
\frac{\partial U(x,t)}{\partial t} &= j\omega U(x,t) \ (c)\\
\frac{\partial^2 U(x,t)}{\partial t^2} &= -\omega^2 U(x,t) \ (c)\,.
\end{aligned} \tag{10.68}$$

Substituting Eqs. (10.68) and (10.66) into Eq. (10.64) we obtain

$$\begin{aligned}
&[(\alpha^2 - \kappa^2) + j2\alpha\kappa]U(x,t)\\
&\quad = [(R'G' - L'C'\omega^2) + j(R'C' + L'G')\omega]U(x,t)\,.
\end{aligned} \tag{10.69}$$

We need to compare the imaginary and real parts of both sides of this complex-valued equation to obtain two real equations with the two unknown variables κ and α, i.e.

$$\begin{aligned}
\texttt{Real terms: } \alpha^2 - \kappa^2 &= R'G' - L'C'\omega^2 \ (a)\\
\texttt{Imaginary terms: } 2\alpha\kappa &= (R'C' + L'G')\omega \Leftrightarrow \alpha\\
&= \frac{(R'C' + L'G')\omega}{2\kappa} \ (b)
\end{aligned}$$

Substituting (b) into (a) yields

$$4\kappa^4 + 4(R'G' - L'C'\omega^2)\kappa^2 - (R'C' + L'G')^2\omega^2 = 0 \tag{10.70}$$

which has the solution

$$\kappa^2 = \frac{L'C'\omega^2 - R'G'}{2}\left[1 + \sqrt{1 + \frac{(R'C' + L'C')^2\omega^2}{(L'C'\omega^2 - R'G')^2}}\right]. \tag{10.71}$$

By noting that at microwave frequencies $\lim_{f\to\infty} \frac{1}{\omega} \to 0$, when $R' \to 0$, we have $\kappa \approx \omega\sqrt{L'C'}$. In this special case of $R' = 0$ the attenuation constant of the line becomes $\alpha = \frac{R'C' + L'G'}{2\sqrt{L'C'}}$. We can now evaluate the generalised attenuation constant of the transmission line as

$$\begin{aligned}
\alpha &= \frac{(R'C' + L'G')\omega}{2\kappa}\\
&= \frac{(R'C' + L'G')\omega}{2\sqrt{\frac{L'C'\omega^2 - R'G'}{2}\left[1 + \sqrt{1 + \frac{(R'C' + L'C')^2\omega^2}{(L'C'\omega^2 - R'G')^2}}\right]}}\,.
\end{aligned} \tag{10.72}$$

Thus, the attenuation of a signal which propagates the line length l m is $20\log_{10} e^{-\alpha l}$ dB. Given the dimensions of a cable, we can compute R', G', L', C' using the formulas in Eq. (10.59) and then use the values to find Z_0, γ, κ, α, and the wave propagation velocity v.

10.5.3 *Propagation Constant and Characteristic Impedance*

We can write the current passing through the transmission line just as we did for the voltage signal as

$$i(x,t) = Re\{ae^{j(\omega t - \kappa x) - \alpha}\} = Re\{I(x,t)\}\,. \tag{10.73}$$

Thus we can rewrite the coupled transmission line Eqs. (10.62) and (10.63) in the form

$$\begin{aligned}
&(10.66)\ \texttt{and}\ (10.73)\ \texttt{in}\ (10.63):\\
&-\frac{\partial I(x,t)}{\partial x} = [G' + j\omega C']U(x,t) \ (a)\\
&-\frac{\partial^2 I(x,t)}{\partial x^2} = [G' + j\omega C']\frac{\partial U(x,t)}{\partial x} \ (b)\\
&(10.66)\ \texttt{and}\ (10.73)\ \texttt{in}\ (10.62):\\
&-\frac{\partial U(x,t)}{\partial x} = [R' + j\omega L']I(x,t) \ (c)\\
&-\frac{\partial^2 U(x,t)}{\partial x^2} = [R' + j\omega L']\frac{\partial I(x,t)}{\partial x} \ (d)
\end{aligned} \tag{10.74}$$

These set of equations can be decoupled to obtain the two second-order differential equations

$$\begin{aligned}
&\texttt{Insert (10.74c) into (10.74b):}\\
&-\frac{\partial^2 I(x,t)}{\partial x^2} = \gamma^2 I(x,t) \ (a)\\
&\texttt{Insert (10.74a) into (10.74d):}\\
&-\frac{\partial^2 U(x,t)}{\partial x^2} = \gamma^2 U(x,t) \ (b)
\end{aligned} \tag{10.75}$$

where we have used the definition

$$\gamma = \alpha + j\beta = \sqrt{(R' + j\omega L')(G' + j\omega C')} \tag{10.76}$$

as the complex wave propagation constant, β is the phase constant in rad/m and α is the attenuation in Neper/m or Np/m of the wave. These wave equations are solved by assuming that two waves propagate along the transmission line but in opposite directions. The wave propagating along the transmission line is a sum of two waves: one from left to right (i.e. the incident wave), and the one from right to left (i.e. the reflected wave). Thus, the phasor representations of the current and the voltage waves are respectively

$$I(x) = I_0^+ e^{-\gamma x} + I_0^- e^{\gamma x} = I^+(x) + I^-(x) \quad (10.77)$$

$$U(x) = U_0^+ e^{-\gamma x} + U_0^- e^{\gamma x} = U^+(x) + U^-(x)\,. \quad (10.78)$$

We note that I_0^+ and U_0^+ are the amplitudes at $x = 0$ of the incident wave which travels in the forward direction, i.e. from generator to load. Also, I_0^- and U_0^- are the amplitudes at $x = 0$ of the reflected wave which travels in the backward direction, i.e. from load to generator. Plugging Eq. (10.78) into Eq. (10.74b) yields

$$I(x) = \frac{\gamma}{R' + j\omega L'}[U_0^+ e^{-\gamma x} - U_0^- e^{\gamma x}]$$

which can be rearranged into

$$\begin{aligned} Z_0 &= \frac{U_0^+ e^{-\gamma x} - U_0^- e^{\gamma x}}{I(x)} = \frac{R' + j\omega L'}{\gamma} \\ &= \sqrt{\frac{R' + j\omega L'}{G' + j\omega C'}} \\ &= \sqrt{\frac{\mu}{\epsilon}}\,. \end{aligned} \quad (10.79)$$

The parameter Z_0, measured in ohms, is called the **characteristic impedance** of the transmission line.

10.5.4 *Reflection Coefficient* Γ

Every transmission line has a characteristic impedance Z_0, which is also called **surge impedance**. Popular values of Z_0 for coax cables are 50 ohm and 75 ohm. Usually, transmission lines are used to transfer a signal in the form of voltage or current from a source to a device. The device is called load. Each device has an impedance that we refer to as load impedance and denoted Z_L. If $Z_0 = Z_L$ then the load absorbs all the voltage falling across it or current passing through it. If $Z_0 \neq Z_L$ then part of the energy in the incident signal is reflected back to the source or generator. The voltage on the load is the sum of the incident and the reflected voltages, whereas the current passing through the load is the difference between the incident and the reflected currents. We define the parameter called the reflection coefficient of a line as

$$\Gamma = \frac{\texttt{voltage reflected off the load}}{\texttt{voltage incident on the load}} = \frac{U^-}{U^+}\,. \quad (10.80)$$

The incident and the reflected waves written in phasor form in Eqs. (10.77) and (10.78) are illustrated in Fig. 10.22. Let us note that

$$U_L = I_L Z_L, I_0^+ = \frac{U_0^+}{Z_0}, I_0^- = -\frac{U_0^-}{Z_0}\,.$$

The voltage and current at the load are obtained from Eqs. (10.77) and (10.78) by setting $x = 0$ as illustrated in Fig. 10.22, i.e.

$$U_L = I_L Z_L = U(x=0) = U_0^+ + U_0^-$$

$$I_L = I(0) = I_0^+ + I_0^- = \frac{U_0^+}{Z_0} - \frac{U_0^-}{Z_0}\,.$$

From these two equations we obtain

$$U_0^+ = \frac{1}{2} I_L (Z_L + Z_0) \ \texttt{and} \ U_0^+ = \frac{1}{2} I_L (Z_L - Z_0)\,. \quad (10.81)$$

Using the definition in Eq. (10.80) the reflection coefficient at the load is

$$\begin{aligned} \Gamma_L = \Gamma(x=0) &= \frac{U_0^-}{U_0^+} = \frac{Z_L - Z_0}{Z_L + Z_0} \\ &= \frac{z_L - 1}{z_L + 1} = |\Gamma| e^{j\theta_L} \end{aligned} \quad (10.82)$$

where $z_L = Z_L/Z_0 = x + jy$ is the normalised load impedance. The reflection coefficient is also referred to as **s11**. The expression

$$-20 \log 10(|\Gamma_L|) \ \texttt{[dB]} \quad (10.83)$$

is referred to as the **return loss**. The reflection coefficient or the return loss indicates how much of the power incident on a load is reflected back. Ideally, we want $\gamma = 0$ so that no power is reflected off the load into the transmission medium. In order to achieve a low reflection coefficient the antenna's impedance must match the impedance of the transmission line. At the radio receiver, the combined impedance of the wireless medium and the transmission line must match the impedance of the antenna to achieve small power reflection.

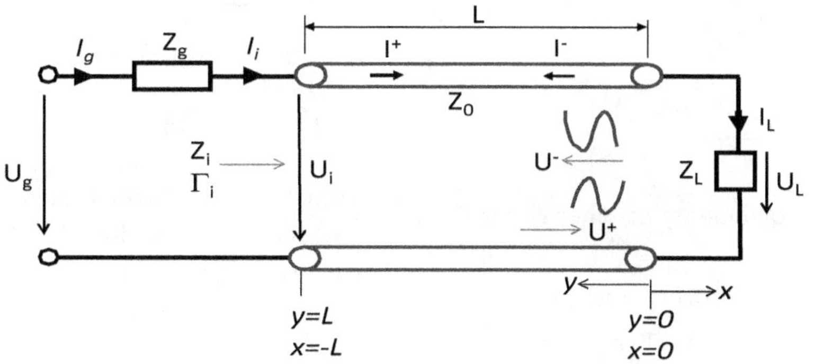

Fig. 10.22 Derivation of reflection coefficient and VSWR.

Special cases of the load with impedance Z_L terminating a transmission line with the characteristic impedance Z_0 are:

- $Z_L = 0 \Rightarrow \Gamma = -1$: short circuit. Total signal reflection occurs, but the voltage and currents assume opposite polarities, resulting in net zero voltage at the load. The current into the load, however, doubles.
- $Z_L = \infty \Rightarrow \Gamma = 1$: open circuit. We have total signal reflection with the same polarity as the incident wave. This results in doubling the voltage at the load, but zero current into the load.
- $Z_L = Z_0$: terminating impedance equals the characteristic impedance of line. Thus, in the ideal case no part of the voltage incident on the load is reflected, i.e. $\Gamma_L = 0$. This situation is called *matching line.*
- $Z_L = j\omega L$ (entirely inductive load) or $Z_L = \frac{1}{j\omega C}$ (entirely capacitive load): the entire energy in the signal incident on the load returns back to the source in the ideal, resulting in reflection coefficient of unity, i.e. $\Gamma = 1$.

10.5.5 *Generalised Voltage, Current and Impedance on Line*

From Fig. 10.22 we can observe the following

$$U_0^- = \Gamma_L U_0^+,\ I_0^+ = \frac{U_0^+}{Z_0},\ I_0^- = -\frac{U_0^-}{Z_0} = -\Gamma_L \frac{U_0^+}{Z_0}\ (a)$$
$$U(y=L) = U_i,\ I(y=L) = I_i\ (b) \tag{10.84}$$

Plugging Eq. (10.84a) into Eqs. (10.77) and (10.78) yields

$$I(x) = \frac{U_0^+}{Z_0}(e^{-\gamma x} - \Gamma_L e^{\gamma x})\ (a)$$
$$U(x) = U_0^+(e^{-\gamma x} + \Gamma_L e^{\gamma x})\ (b)\,. \tag{10.85}$$

From these two set of equations we can compute the impedance at any point on the transmission line as

$$Z(x) = \frac{U(x)}{I(x)} = \frac{e^{-\gamma x} + \Gamma_L e^{\gamma x}}{e^{-\gamma x} - \Gamma_L e^{\gamma x}}$$
$$= Z_0 \frac{1 + \Gamma_L e^{2\gamma x}}{1 - \Gamma_L e^{2\gamma x}}\,. \tag{10.86}$$

A parameter related to the impedance is the admittance of a line, which is defined as

$$Y(x) = \frac{1}{Z(x)} = Y_r(x) - jY_i(x)\,. \tag{10.87}$$

By substituting $\Gamma_L = \frac{Z_L - Z_0}{Z_L + Z_0}$ and noting that $2\cosh x = e^x + e^{-x}$, $2\sinh x = e^x - e^{-x}$, $\tanh x = \frac{\sinh x}{\cosh x}$, we can rewrite Eq. (10.86) as

$$Z(x) = \frac{U(x)}{I(x)} = Z_0 \frac{Z_L + Z_0 \tanh(\gamma x)}{Z_0 + Z_L \tanh(\gamma x)} \tag{10.88}$$

where, again, $\gamma = \alpha + j\beta$ is the complex-valued wave propagation constant. The corresponding reflection coefficient is

$$\Gamma(x) = \frac{Z(x) - Z_0}{Z(x) + Z_0}\,. \tag{10.89}$$

It is an easy task to prove that $Z(x) = Z_0 \frac{1+\Gamma(x)}{1-\Gamma(x)}$. Applying Kirchhoff's voltage law to Fig. 10.22 also provides

$$U_g = I_i Z_g + U_i\,. \tag{10.90}$$

Plugging Eqs. (10.90) and (10.84) into Eq. (10.99) and simplifying we obtain

$$U_0^+ = \frac{U_g Z_0 e^{-j\gamma L}}{(Z_g + Z_0)(1 - \Gamma_g \Gamma_L e^{-j2\gamma L})} \tag{10.91}$$

where the reflection coefficient at the generator or voltage source is $\Gamma_g = \frac{Z_g - Z_0}{Z_g + Z_0}$. Substituting Eq. (10.91) back into Eqs. (10.77) and (10.78) and remembering Eq. (10.84) yields the generalised formula for the voltage and current along the line as

$$U(x) = \frac{U_g Z_0 e^{-j\gamma L}}{(Z_g + Z_0)(1 - \Gamma_g \Gamma_L e^{-j2\kappa L})} \times e^{j\gamma x}(1 + \Gamma_L e^{-j2\gamma x}) \tag{10.92}$$

$$I(x) = \frac{U_g e^{-j\gamma L}}{(Z_g + Z_0)(1 - \Gamma_g \Gamma_L e^{-j2\gamma L})} \times e^{j\gamma x}(1 - \Gamma_L e^{-j2\gamma x}). \tag{10.93}$$

The two equations in Eq. (10.92) are the standing voltage and standing currents on the transmission line.

10.5.6 *Properties of Lossless Transmission Line and VSWR*

For a lossless line $\alpha = 0$ and thus $\gamma = j\beta$. Also, $\tanh(\gamma x) = \tanh(j\beta x) = j\tan(\beta x)$. This reduces Eq. (10.88)

$$Z(x) = \frac{U(x)}{I(x)} = Z_0 \frac{Z_L + jZ_0 \tan(\beta x)}{Z_0 + jZ_L \tan(\beta x)}. \tag{10.94}$$

For a short-circuited line transmission line in which $Z_L = 0$ the impedance reduces to

$$Z(x) = jZ_0 \tan(\beta x) = jZ_0 \tan\left(2\pi \frac{x}{\lambda}\right). \tag{10.95}$$

The characteristics of such a lossless line which is short circuited is depicted in Fig. 10.23. We can observe the changing nature of the impedance. The behaviour of the line depends on its length L as follows:

- **inductive** if $\frac{2n}{4}\lambda < L < \frac{2n+1}{4}\lambda$, $n = 0, \pm 1, \pm 2, \ldots$
- **capacitive** if, $\frac{2n-1}{4}\lambda < L < \frac{2n}{4}\lambda$, $n = 0, \pm 1, \pm 2, \ldots$
- **short circuit** if $L = \frac{2n}{4}\lambda$, $n = 0, \pm 1, \pm 2, \ldots$
- **open circuit** if $L = \frac{2n+1}{4}\lambda$, $n = 0, \pm 1, \pm 2, \ldots$

We obtained the complex propagation constant of the EM wave as

$$\gamma = \alpha + j\beta = \sqrt{(R' + j\omega L')(G' + j\omega C')}. \tag{10.96}$$

Ideally, a lossless line has $R = G = 0$. However, we assume a transmission line is lossless if $R' \ll \omega L'$ and $G' \ll \omega C'$, which yields

$$\gamma = \alpha + j\beta \approx j\omega\sqrt{L'C'} \text{ and }$$
$$Z_0 = \sqrt{\frac{L'}{C'}} = \frac{U(0,t)}{I(0,t)}. \tag{10.97}$$

That is, the source with voltage $U(0, t)$ sees the transmission line as a resistor of resistance $Z_0 =$

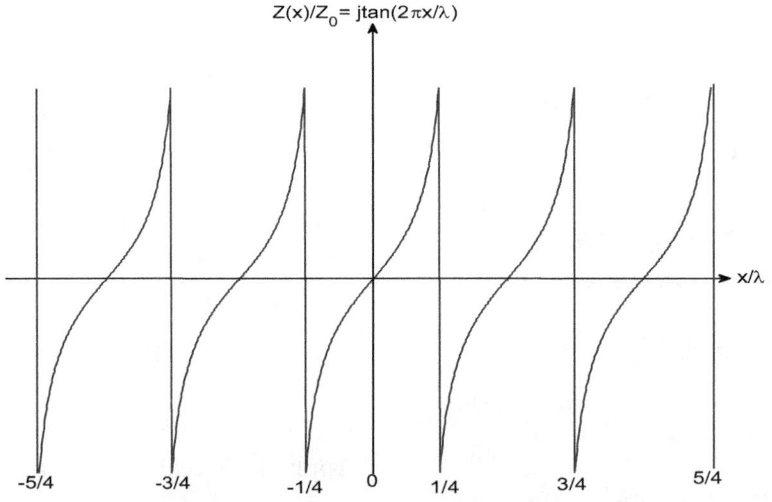

Fig. 10.23 Impedance along a lossless short-circuited transmission line.

$\sqrt{\frac{L'}{C'}}$. Thus, $\alpha = 0$ and $\beta = \omega\sqrt{L'C'} = \omega\sqrt{\mu\epsilon}$. By noting the relationship between the wavelength and β, i.e. $\lambda = \frac{2\pi}{\beta}$, we find the speed of wave propagation along the transmission line as

$$v_p = f\lambda = \frac{2\pi f\lambda}{2\pi} = \frac{\omega\lambda}{2\pi} = \frac{1}{\sqrt{L'C'}} = \frac{1}{\sqrt{\mu\epsilon}}. \tag{10.98}$$

This means that the wave propagation speed does not depend on radio frequency so long as L' and C' do not change with frequency. Thus, all frequency components in a signal which propagates along the transmission line do so with the same speed. This is a desirable feature, as the line causes no distortion to signals, i.e. $U(x,t) = U(0, t - \tau)$, where the delay is $\tau = x/v_p = x\sqrt{L'C'}$. Such a transmission medium is called non-dispersive line.

Example

Assume in Fig. 10.22 that $U_g = 120\cos(2\pi \times 60t)$ [V]. Assume that the length of the transmission line is $L = 1$ km. Assume velocity factor of 0.66 for the cable. Then the time that it takes the wave to reach the load (i.e. propagation delay) is $\tau = \frac{L}{0.66\times c} \approx 5.1$ μ sec, where $c \approx 3 \times 10^8$ m/s. Thus, the voltage reaching the load at time $t = 0$ is $U_L(t = 0) = 120\cos(2\pi \times 5.1\ \mu\texttt{sec}) \approx 119.9998$ V.

For a lossless transmission line we have $\gamma = j\beta$ as $\alpha = 0$ and $U_0^- = \Gamma U_0^+$. Thus, the two phasors become

$$U(x) = U_0^+(e^{-j\beta x} + \Gamma e^{j\beta x}) \text{ and}$$
$$I(x) = \frac{U_0^+}{Z_0}(e^{-j\beta x} - \Gamma e^{j\beta x}). \tag{10.99}$$

The magnitude of $U(x)$ is

$$|U(x)| = |U_0^+|\sqrt{1 + 2|\Gamma|\cos(2\beta x + \theta_\Gamma) + |\Gamma|^2} \tag{10.100}$$

where we have used $\Gamma = |\Gamma|e^{j\theta_\Gamma}$ and the trigonometric identity $\cos(u+v) = \cos(u)\cos(v) - \sin(u)\sin(v)$. The maximum of $|U(x)|$ occurs when the cosine wave achieves its maximum, which is 1. This requires that $2\beta x + \theta_\Gamma = 2i\pi$, while i is an integer. The minimum occurs when the cosine wave achieves its minimum value of -1. The condition for this is $2\beta x + \theta_\Gamma = (2i + 1)\pi$. We thus have

$$\begin{aligned}
|U(x)|_{min} &= |U_0^+|\sqrt{1 - 2|\Gamma| + |\Gamma|^2} \\
&= |U_0^+|(1 - |\Gamma|),\ x = \frac{\lambda}{4}\left(\frac{\theta_\Gamma}{\pi} + 2i + 1\right) \\
|U(x)|_{max} &= |U_0^+|\sqrt{1 + 2|\Gamma| + |\Gamma|^2} \\
&= |U_0^+|(1 + |\Gamma|),\ x = \frac{\lambda}{2}\left(\frac{\theta_\Gamma}{2\pi} + i\right).
\end{aligned} \tag{10.101}$$

The ratio of the maximum to minimum amplitude of the standing wave is referred to as voltage standing wave ratio (VSWR), i.e.

$$\begin{aligned}
S = VSWR &= \frac{|U(x)|_{max}}{|U(x)|_{min}} \\
&= \frac{|U_0^+|(1 + |\Gamma|)}{|U_0^+|(1 - |\Gamma|)} = \frac{1 + |\Gamma|}{1 - |\Gamma|}.
\end{aligned} \tag{10.102}$$

It is worthy of reminder that $|\Gamma| \leq 1$ and $S \geq 1$. Also, we note that Γ can be any positive or negative complex number, while the VSWR is always a positive real number. The VSWR is also referred to as standing wave ratio (SWR). A VSWR of unity is the minimum and the ideal case of zero reflection. This requires that the antenna or the load is perfectly matched to the transmission line. The VSWR is infinite for reflection coefficient of unity. A VSWR of about two or less is considered a good enough matched line. A higher VSWR causes more losses (i.e. reflected power) as power absorbed by the load is reduced. Also a higher SWR means that more power is reflected back to the receiving electronic circuitry, which can be damaged thereby.

Example: A lossless coax cable with $Z_0 = 50\ \Omega$ is connected to the load $Z_L = R_L - \frac{j}{\omega C_L}$ at a distance 0.03 m, where $R_L = 75\ \Omega$ and $C_L = 5$ pF. The operating wavelength is 0.3 m.

(a) What percentage of the energy incident on the load is reflected back to the source?
(b) What is the impedance at the input end of the transmission line?
(c) Compute the VSWR on the line. Assume a relative velocity of 0.66 for the cable.

Solution: Given in the question are $Z_0 = 50\ \Omega$, $\lambda = 0.3$ m, $R_L = 75\ \Omega$ and $C_L = 5$ pF. Thus $f = 0.66 \times 3 \times 10^8/0.03 = 6.6$ GHz. Thus, $Z_L = 75 - j/(2\pi \times 6.6 \times 10^9 \times 5 \times 10^{-12}) \approx 75 - j4.8$. The reflection coefficient gives the fraction of reflected energy. It is

$$\Gamma_L = \frac{0.5 - j0.1}{2.5 - j0.1} \approx 0.1981 + j0.0479$$
$$= 0.2038\angle 13.6^\circ .$$

We note that $L = 0.03\ m = 0.1\lambda$, and so $\beta L = \frac{2\pi}{\lambda} \times 0.1\lambda = 0.2\pi$. The input impedance is

$$Z_i = Z(x = 0.03\ \mathtt{m}) = Z_0 \frac{Z_L + jZ_0 \tan(\beta x)}{Z_0 + jZ_L \tan(\beta x)}$$
$$= Z_0 \frac{75 - j4.8 + j50 \tan(0.2\pi)}{50 + j(75 - j4.8)\tan(0.2\pi)}$$
$$\approx 49.1368 - j20.5870 = 53.2752\angle - 22.7^\circ .$$

The generator or signal source sees the impedance of the 50-Ω line terminated by the load of Z_L as $Z_i \approx 49.1368 - j20.5870 = 53.2752\angle -22.7^\circ$. Therefore, we can observe that the termination of a line by a load changes the characteristic impedance of the line. The VSWR on the line is

$$VSWR = S = \frac{1 + |\Gamma|}{1 - |\Gamma|} = \frac{1 + 0.2038}{1 - 0.2038} \approx 1.5 .$$

Transmission line stubs: we have observed that signal reflection back to the source occurs if there is a mismatch in the impedances of the line and the load. The reflected voltage adds up to the forward voltage. As a voltage change results in impedance change, we can exploit reflection to modify the characteristics of a transmission line. Thus, we can vary the reflection coefficient by varying the load impedance Z_L and polarity or phase of the reflected wave by the choice of the length of the transmission line used. This idea is exploited by using an appropriate piece of transmission line to realise the functions of capacitors, inductors and resistors in **matching networks**. Table 10.6 gives examples of how transmission line stubs are used to realise the functions of a capacitor, inductor, short circuit and open circuit.

Table 10.6 Realising LRC functions using stubs.

Stub length	Z for $Z_L = \infty$	Z for $Z_L = 0$
$\frac{\lambda}{4}$	$0\ \Omega$	∞
$\frac{\lambda}{8}$	$-jZ_0\ \Omega$	jZ_0

Example: The voltage $u_g(t) = 60\cos(3.96\pi \times 10^9 t - \frac{\pi}{4})$ is supplied by a generator with the internal resistance $Z_g = 2\ \Omega$ to an antenna over a 100-Ohm coax cable of length 250 mm. Assume the relative wave velocity along the cable is 0.66, and the speed of waves in vacuum as 3×10^8 m/s. The antenna has the impedance $Z_L = 150 + j75$.

(1) Calculate the wave reflection coefficient at the generator Γ_g.
(2) Calculate the voltage reflection coefficient at the antenna Γ_L.
(3) Find the formula for the instantaneous voltage along the cable $u(x,t)$.
(4) Find the formula for the instantaneous current along the cable $i(x,t)$.
(5) What is the power in the voltage signal from the generator?
(6) How much power is in the signal that the generator delivers to the antenna?

Solution: The given question is illustrated in Fig. 10.24. We are given in the question: $Z_0 = 100\ \Omega$, $Z_g = 2\ \Omega \to z_g = Z_g/Z_0 = 0.02$, $Z_L = 150 + j75 \to z_L = Z_L/Z_0 = 0.15 + j0.75$, $v_r = 0.66$ and $f = 1.98$ GHz. Thus, $\lambda = v_r * c/f = 0.1$ m, $\kappa = 2\pi/\lambda = 0.2\pi$, length of cable $L = 0.25\ \mathtt{m} = 2.5\lambda$ and $\kappa L = 5\pi$. We need to convert the given generator voltage into a phasor, i.e. $u_g(t) = 60\cos(3.96\pi \times 10^9 t - \frac{\pi}{4}) = Re\{60e^{-j\frac{\pi}{4}}e^{j\omega t}\}$, resulting in the phasor $U_g = 60e^{-j\frac{\pi}{4}}$, Thus,

$$U_0^+ = \frac{U_g Z_0 e^{-j\kappa L}}{(Z_g + Z_0)(1 - \Gamma_g \Gamma_L e^{-j2\kappa L})}$$
$$= \frac{60e^{-j\pi/4} \times 100e^{-j5\pi}}{(2 + 100)(1 + 0.96 \times 1.5e^{j(\pi/3 - 10\pi)})}$$
$$\approx -4.56 + j27.65 = 28.03e^{j0.55\pi}. \qquad (10.103)$$

(1) We have $\Gamma_g = \frac{Z_g - Z_0}{Z_g + Z_0} = -98/102 \approx -0.96$.
(2) Load reflection coefficient is $\Gamma_L = \frac{Z_L - Z_0}{Z_L + Z_0} = \frac{-0.85 + j0.75}{0.15 + j0.75} = 0.7436 + j1.2821 \approx 1.5\angle 60^\circ$. This

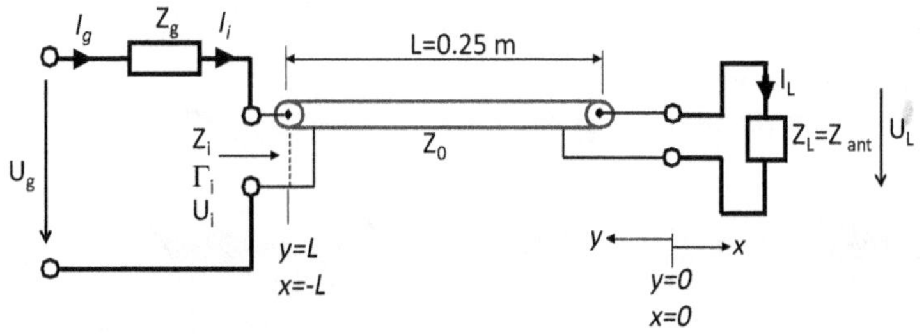

Fig. 10.24 Transmission line analysis example.

is impractical as a load cannot reflect more power than what reaches it!

(3) $U(y) = 28.03e^{j0.55\pi}e^{j0.2\pi}(1+1.5e^{j\pi/3}e^{-j0.4\pi}) \approx -43.07 + j54.8 = 69.7e^{j0.712\pi}$. Thus, $u(y,t) = Re\{U(y)e^{j\omega t}\} = 69.7\cos(\omega t + 0.712\pi)$.

(4) $I(y) = 0.2803e^{j0.55\pi}e^{j0.2\pi}(1-1.5e^{j\pi/3}e^{-j0.4\pi}) \approx 0.0308 - j0.1544 \approx 0.16e^{-j78.7^\circ}$. Thus, $i(y,t) = Re\{I(y)e^{j\omega t}\} = 0.16\cos(\omega t + 78.7^\circ)$.

(5) Power in the generator signal is $P_g = |u_g(t)|^2/Z_g = 1.8$ kW.

(6) Power at the input of the transmission line is $P_i = \frac{1}{2}Re\{I_i^* U_i\} = \frac{1}{2}|I_i|^2 Re\{Z_i\}$. Note that $I_i = \frac{U_g}{Z_g+Z_i}$ and $Z_i = Z_0\frac{Z_L+jZ_0\tan(\kappa L)}{Z_0+jZ_L\tan(\kappa L)}$. For a lossless line, the power delivered to the load equals the power entering the line.

10.5.7 *Smith Chart*

We cannot conclude discussions on transmission lines without mentioning a special chart called **Smith chart**. Prior to the proliferation of computers and software, Smith chart developed by Phillip Smith in 1939 at Bell Labs, was used to analyse alternating current circuits, antennas and transmission lines. This chart has many uses. For example, it can be used to compute the SWR, find Z_L given the reflection coefficient and vice versa, study impedances or admittances, scattering parameters, find the input impedance given the load impedance and vice versa, used in impedance matching of lines, and the design of components like oscillators and amplifiers. The Smith chart has two advantages over manual calculations: its graphical feature enables us to visualise the frequency-dependent impedance locus, and it is less error prone.

In the following we develop the analytical theory underpinning the Smith chart. In general, the impedance and the reflection coefficient of a transmission line are complex variables. Let us break the normalised impedance into its real $z_r(x)$ and imaginary $z_i(x)$ parts as

$$z(x) = \frac{Z(x)}{Z_0} = z_r(x) + jz_i(x) \tag{10.104}$$

and do similarly for the reflection coefficient as

$$\Gamma(x) = \Gamma_r(x) + j\Gamma_i(x)\,. \tag{10.105}$$

We can express the voltage reflection coefficient found in Eq. (10.80) as

$$\Gamma(x) = \frac{Z(x) - Z_0}{Z(x) + Z_0} = \frac{z(x) - 1}{z(x) + 1}\,. \tag{10.106}$$

We can rearrange Eq. (10.106) as

$$z(x) = \frac{1+\Gamma(x)}{1-\Gamma(x)} \overset{(10.105)}{=} \frac{1+\Gamma_r(x)+j\Gamma_i(x)}{1-\Gamma_r(x)-j\Gamma_i(x)}\,. \tag{10.107}$$

We multiply through both the numerator and denominator of the right-hand side of Eq. (10.107) by $1-\Gamma_r+j\Gamma_i$ to obtain

$$\begin{aligned} z(x) &= z_r(x) + jz_i(x) \\ &= \frac{1-\Gamma_r^2(x)-\Gamma_i^2(x)+j2\Gamma_i(x)}{[1-\Gamma_r(x)]^2+\Gamma_i^2(x)}\,. \end{aligned} \tag{10.108}$$

Now, we need to equate the real and imaginary terms in both sides of Eq. (10.108) and then rearrange each into the form of the equation of a circle. A circle of radius c centered at (a,b) has the equation $(x-a)^2+(y-b)^2=c^2$ in the two-dimensional (x,y) coordinate system. Equating the real parts of both sides of Eq. (10.108) yields

$$z_r = \frac{1-\Gamma_r^2-\Gamma_i^2}{(1-\Gamma_r)^2+\Gamma_i^2}$$

which, upon simplification, becomes

$$\left(\Gamma_r - \frac{z_r}{1+z_r}\right)^2 + \Gamma_i^2 = \left(\frac{1}{1+z_r}\right)^2. \tag{10.109}$$

Equation (10.109) is a set of circles with radii $\frac{1}{1+z_r}$ and centered at $(\frac{z_r}{1+z_r}, 0)$ in the (Γ_r, Γ_i) coordinate system. Equating the imaginary parts of both sides of Eq. (10.108) yields

$$z_i(x) = \frac{2\Gamma_i(x)}{\Gamma_i^2(x) + (\Gamma_r(x) - 1)^2}$$

which, upon simplification, becomes

$$(\Gamma_r - 1)^2 + (\Gamma_i - 1/z_i)^2 = (1/z_i)^2 . \quad (10.110)$$

Equation (10.110) are a set of circles with radii $1/z_i$ and centered at $(1, 1/z_i)$ in the (Γ_r, Γ_i) coordinate system. The superposition of the two sets of circles in Eqs. (10.109) and (10.110) results in the **Smith chart**.

From the above relations we can summarise some of the basic features of the Smith chart as:

- At the origin or center of the Smith chart we have $\Gamma(y) = 0$, i.e. the load is matched to the transmission line.
- The numbers on the outermost scale or circle of the Smith chart are wavelengths. A distance of $\lambda/2$ along the transmission line is equivalent to one revolution around the Smith chart.
- Horizontal circles represent constant resistances.
- The centre of the Smith chart is located at $(\Gamma_r, \Gamma_i) = (1, 0)$, i.e. on the real axis.
- All points on a circle with its centre at the origin have the same reflection coefficient and VSWR.
- Impedances plotted in the Smith chart are normalised by the characteristic impedance Z_0, while admittances are normalised by $Y_0 = 1/Z_0$.
- Vertical circles represent constant reactances.
- Anti-clockwise movement along the Smith chart moves us from the generator towards the load, while clockwise movement along the chart moves us from the load towards the generator.
- All points on the same circle in the Smith chart have the same reflection coefficient. The radius of the outermost circle has the radius of unity, which is the maximum possible reflection coefficient.
- Points above the horizontal line indicate inductive impedance, while points below the horizontal line indicate capacitive impedance.

Figure 10.25 is a Smith chart showing the five impedances $Z_1 = Z_0$, $Z_2 = 150 + j150$, $Z_3 = 150 - j75$, $Z_4 = 7.5 = 0.1Z_0$ and $Z_5 = 10Z_0$ which are normalised by the characteristic impedance $Z_0 = 75\ \Omega$.

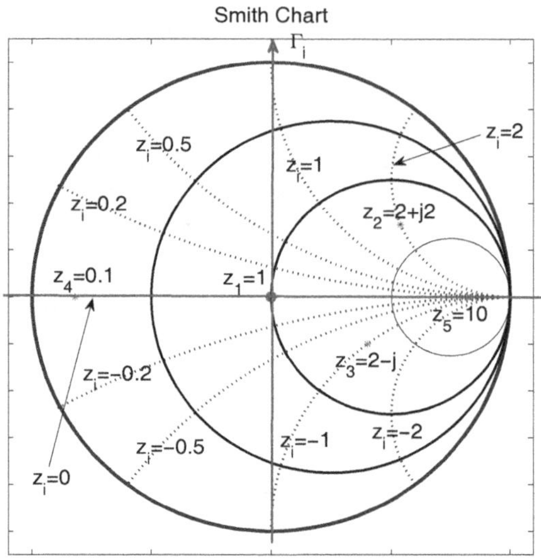

Fig. 10.25 Smith chart showing impedances normalised by $Z_0 = 75\ \Omega$.

Example: A lossless 75-ohm transmission line connects an antenna with the input impedance of $Z_L = 75 + j150$ to a generator as shown in Fig. 10.26. Thus, the load is the antenna.
Use both manual calculations and the Smith chart to find the

(a) normalised input impedance $z_{in} = Z_{in}/Z_0$ at the distance $L = 0.4\lambda$ away from the load,
(b) reflection coefficient at the antenna Γ_L,
(c) admittance of the antenna $Y_L = 1/Z_L$,
(d) VSWR, and
(e) locations on the transmission line, measured from the load, where the standing wave achieves its minimum value (y_{min}) and the maximum value (y_{max}).

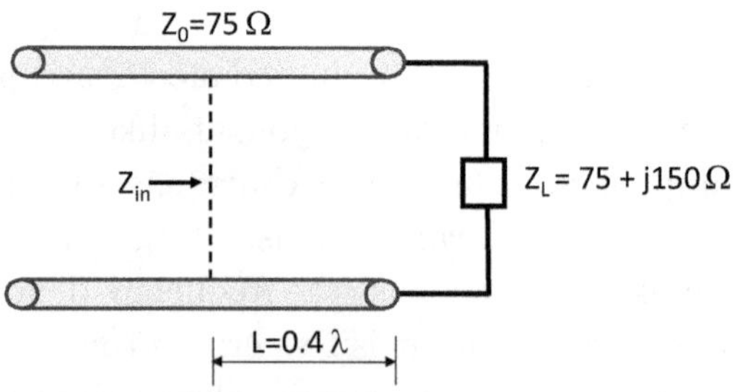

Fig. 10.26 Smith chart illustrative example.

Solution: (a) The normalised load is $z_L = Z_L/Z_0 = 1 + j2$ and $\beta y = \frac{2\pi}{\lambda} \times 0.4\lambda = 0.8\pi$, and $\tan(0.8\pi) \approx -0.73$. This gives the normalised input impedance at the distance $L = 0.4\lambda$ away from the load as

$$z_{in} = \frac{z_L + j\tan(\beta y)}{1 + jz_L \tan(\beta y)}$$

$$= \frac{1 + j2 + j\tan(0.8\pi)}{1 + j(1 + j2)\tan(0.8\pi)} \approx 0.23 + j0.59\,.$$

We need to plot the given load on the Smith chart. This is the intersection between the circle with the constant normalised impedance of $j2$ Ω and that with the constant normalised impedance of 1 Ω as shown in Fig. 10.27. To find z_{in} at distance $L = 0.4\lambda$ away from the load, we need to move the normalised load z_L clockwise from its location along the constant $|\Gamma(y = 0.4\lambda)|$ circle by the distance 0.4λ. The constant $|\Gamma(y = 0.4\lambda)|$ circle is the circle with its center at the origin and radius $|\Gamma(y = 0.4\lambda)|$, i.e. distance from the center (or origin) to the load. The impedance at the point after moving the load by $L = 0.4\lambda$ is the normalised input impedance z_{in}, which is roughly $z_{in} \approx 0.24 + j0.58$. We obtain Z_{in} as $Z_{in} = z_{in}Z_0$.

(b) The reflection coefficient is $\Gamma_L = \frac{z_L - 1}{z_L + 1} = \frac{j}{1+j} \approx 0.707\angle 45°$. The magnitude of the reflection coefficient, i.e. $|\Gamma_L|$ is the distance from the centre to the load, which can be read off the Smith chart as approximately 0.71. Note that one revolution around the Smith chart equals $\frac{\lambda}{2}$ movement along the transmission line. Thus, $\theta_\Gamma = \frac{(0.25-0.1875)\lambda}{0.5\lambda} \times 2\pi = 0.25\pi$. Note that we can also read this angle directly from the outer circle of the Smith chart.

(c) The normalised admittance of the load is $y_L = \frac{1}{z_L} = \frac{1}{1+j2} = 0.2 - 0.4i$, resulting in the admittance $Y_L = y_L Y_0 = \frac{y_L}{Z_L} = 0.0027 - j0.0053$. On the Smith chart y_L is located at the distance $\frac{\lambda}{4}$ away from the load in a clockwise direction. As a full revolution along the chart equals $\frac{\lambda}{2}$, y_L is located 180° from z_L or opposite it using the centre as the reflection point. As shown in Fig. 10.27, this results in the point of intersection of the lines where $z_i = 0.4$ and $z_r = 0.2$. Thus, $y_L = 0.2 - 0.4i$.

(d) We have VSWR $= \frac{1+|\Gamma_L|}{1-|\Gamma_L|} = \frac{1+\frac{1}{\sqrt{2}}}{1-\frac{1}{\sqrt{2}}} \approx 5.82$. The steps to find the VSWR from the Smith chart are:

- Plot the normalised load on the Smith chart.
- Draw the constant VSWR circle. The centre of this circle is the origin or the centre of the Smith chart. The radius of the VSWR circle is the distance from the centre to the load. All impedances on this circle have the same VSWR.
- Read the VSWR from the Smith chart. This is the point on the circle lying on the real line which is the horizontal line passing through the centre of the chart. Note that the VSWR circle cuts the horizontal line on two locations. The point on the VSWR circle at the right of the origin is the VSWR.

In Fig. 10.27 we can observe that the VSWR falls between 5 and 6 and quite close to 6, confirming our manual result.

(e) From Eq. (10.101) we found that the maximum value and the minimum value of the voltage standing wave occur at the respective distances y_{max} and y_{min} measured from the load towards the generator, where

$$y_{max} = \frac{\lambda}{2}\left(\frac{\theta_\Gamma}{2\pi} + k\right) \text{ and}$$

$$y_{min} = \frac{\lambda}{4}\left(\frac{\theta_\Gamma}{\pi} + 1 + 2k\right), \quad k = 0, 1, 2, \ldots$$

where k is a positive integer. Note that θ_Γ is the phase of the reflection coefficient in radians, which in the given question is $45° = \frac{\pi}{4}$ radians. The first minimum and maximum standing wave occurs if $k = 0$, resulting in $y_{max} = \frac{\lambda}{16} = 0.0625\lambda$ and $y_{min} = \frac{5\lambda}{16} = 0.3125\lambda$, respectively. On the Smith chart y_{max} is found by moving from the load to the positive real axis in a clockwise direction. This results in $y_{max} = (0.25 - 0.1875)\lambda = 0.0625\lambda$. The value of y_{min} is just the point where the VSWR circle cuts the negative real axis which is $\lambda/4$ away. Thus, $y_{min} = \lambda/4 + y_{max} = \frac{5\lambda}{16}$. We have observed that it is relatively easier faster and less error prone to calculate a transmission line's parameters using the Smith chart than doing so manually.

10.5.8 *Waveguides*

A waveguide is a hollow tube made of an electrically-conductive material used as a transmission line at microwave frequencies. Wavelengths of

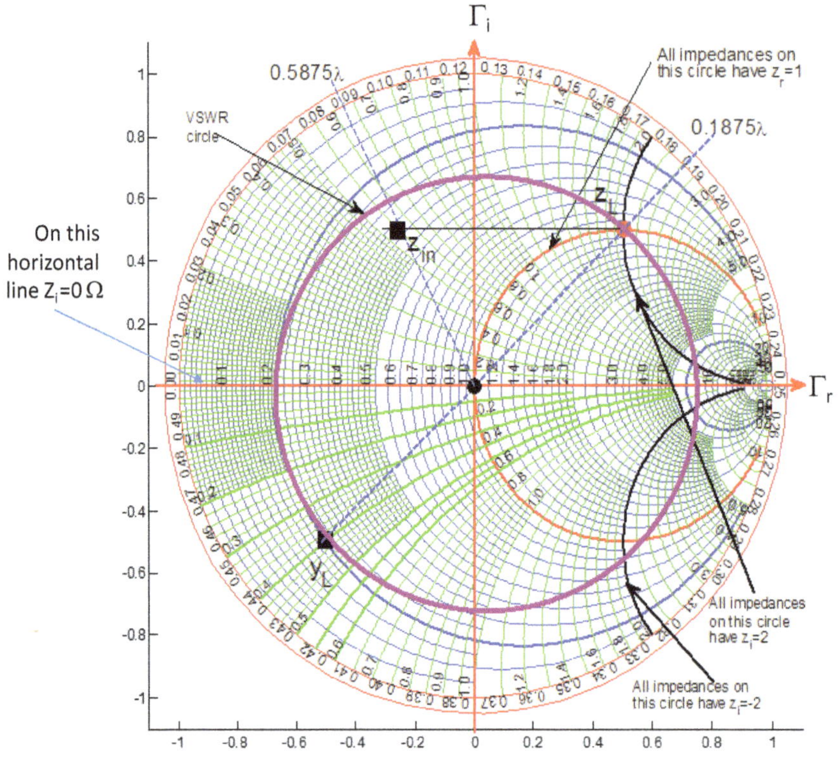

Fig. 10.27 Smith chart illustrative example for $Z_0 = 75\ \Omega$.

signals conveyed over waveguides must be small compared to the dimensions of the cross-section of the waveguide. Thus, waveguides are impractical to use at low frequencies. A waveguide has just a single conductor filled with space. The hollow space in a waveguide is air which is a dielectric material and thus behaves as a capacitance distributed over the length of the waveguide. The metallic conductor functions as inductance distributed over its length.

An electromagnetic (EM) wave comprises three vectors: the vector for the electric field (E-field), the vector for the magnetic field (H-field) and the vector for the propagation direction of the wave (V-field). Both the E-field and H-field propagate in the direction of the wave movement. The mode of an EM wave is its shape or form. EM waves propagate along transmission lines having two conductors (e.g. coax cable) in TEM mode. TEM means transverse electromagnetic or transverse electric and magnetic. It means that the three vectors of the EM wave are always at right angles (i.e. transverse) with each other as they move. Two conditions must be fulfilled for a wave to move in TEM mode:

- Wavelength of the wave must be larger than the size of the cross-section dimensions of the transmission line. This is the case at low frequencies.
- The transmission line must have two conductors as it is in coax cable.

As a waveguide has only one conductor, it cannot serve as the conduit for TEM mode propagation of EM waves according to Maxwell's equation. Either the E-field vector or the H-field vector is transversal to the wave direction vector when a wave propagates in a waveguide. This is referred to as TM (i.e. transverse magnetic) or TE (i.e. transverse electric) mode. In the TM mode only the magnetic field vector is transversal to the wave's direction, while in the TE mode the E-field is the one which is perpendicular to the wave's direction. The TE wave is also called H-waves, while the TM wave are also referred to as E-waves.

TE waves are denoted by TE_{mn} where $0 \leq m, n \leq \infty$ are positive whole numbers which indicate the mode of waves propagating along the waveguide. However, in practice only a limited combination of m and n is possible depending on the dimensions and shape of the waveguide. EM wave propagation along transmission lines are studied by finding solutions to the Maxwell's equation, which we skip here. These equations have multiple solutions, referred to as modes, eigenfunctions of the Maxwell's equations.

Every waveguide functions as a high-pass filter with a certain cut-off frequency f_C. Frequencies below the f_C cannot pass through it while all frequencies above f_C pass through unattenuated, in the ideal case. The cut-off frequency of a waveguide depends on the size of its cross-section and its shape. Multiple modes can propagate along a line, and the one with the smallest cut-off frequency is called **dominant mode**. Usually, only the dominant mode is used for signal transfer. We can choose the dimensions of a waveguide so as to allow only the dominant mode to exist. Waveguides can be identified as circular, square, rectangular and elliptical waveguides. The cut-off frequency decreases as the size of the waveguide increases. The cut-off frequencies for some waveguides are:

`rectangular in` TE_{10} `mode:`

$$f_C = \frac{1}{2x_0\sqrt{\mu\epsilon}} = \frac{c}{2x_0} \quad (a)$$

`circular with internal radius` r_0`:`

$$f_C = \frac{1.8412}{2\pi r_0\sqrt{\mu\epsilon}} = \frac{1.8412c}{2\pi r_0} \quad (b) \tag{10.111}$$

where μ and ϵ are the permeability and permittivity of the material in the hollow space, x_0 is the dimension of lager side of the rectangular cross-section and c is the speed of the EM wave in the hollow space.

We can realise a definite attenuation by operating a waveguide below its cut-off frequency which results in a real wave propagation constant. We need to replace $\gamma = \alpha + j\beta$ that is used in the analysis of two-wire transmission lines with $\gamma_{mn} = \alpha_{mn} + j\beta_{mn}$ when analysing waveguides.

The equivalent of a two-wire transmission line's characteristic impedance in waveguide is the wave impedance. About 500 Ω is assumed for a rectangular waveguide. Approximate formulas used are

$$Z_{TE} = \mu f \lambda \quad (a)$$
$$Z_{TM} = \frac{1}{\epsilon\lambda f} \quad (b) \tag{10.112}$$

10.5.9 *Check Your Understanding*

Students are encouraged to check their understanding of the content of this section with the following questions prior to proceeding to the next section.

(1) What is a transmission line?
(2) Briefly discuss and differentiate between the types of transmission lines discussed in this section.
(3) What are the three main types of losses that a signal potentially experience while propagating along a transmission line? What are the causes of the losses?
(4) Discuss the meaning of reflection coefficient.
(5) What are the potential effects of non-zero VSWR on a transmission line?
(6) What are the typical characteristic impedances of coaxial cables?
(7) What is the reason why a cable used in local area networking is usually terminated with a resistor having the resistance equal to the characteristic impedance of the cable?
(8) What is transmission line *stub* and what is its importance?
(9) How does a waveguide differ from coaxial cable? Discuss all their similarities and difference.
(10) What is Smith Chart? Discuss the applications of it.
(11) We want only the wave in the TE_{10} mode to propagate along a rectangular waveguide. What should be the dimensions of the waveguide? What is the required cut-off frequency of the waveguide?

10.6 Microwave Design Software and Vendors

There are a plethora of software packages that are available for the design of microwave systems. We sample a few of them in this section. **LineCalc** is a tool in the ADS software package which enables the calculation of impedances and dimensions of various transmission lines, including microstrip lines and waveguides. ADS (Advanced Design System), formerly called Microwave Design System, is a product of Keysight Technologies. ADS is an automation software for electronic design. Keysight Technologies issues free licenses to universities for education purposes. Another microwave link design software is LINKPlanner which is developed and maintained by Cambium Networks. LINKPlanner runs on both Microsoft Windows and Mac computing platforms. Another microwave link design software is iQ.link® whose licensor is Comsearch, a ComScope's company.

Several companies manufacture microwave radio components while other companies focus on design and installations. For example, in Australia we have Murray Wireless, RF Tool Co, Wave1, Tomco Technologies, Vector Telecom and ATI Australia. Global manufacturers of microwave infrastructure include Ericsson, Huawei, Alcatel, Bi-Directional Microwave Systems, and Global Microwave Systems, Inc.

10.7 Review of Important Terms and Acronyms

antenna	aperture of antenna
free space	Fresnel zones
line-of-sight	Earth's bulge
diffraction loss	k-factor
voltage standing wave ratio (VSWR)	link power budget
link height budget	refraction coefficient (Γ)
path loss	free-space path loss
transmission line	feeder line
lossless line	stub
coax cable	waveguide
wave propagation constant	feeder line
characteristic impedance Z_0	surge impedance
wavelength	wave number
smith chart	internal resistance of generator
line attenuation constant α	phase constant β
skin-effect	skin-effect loss
leakage current	dielectric loss
radiation loss	radome

10.8 Review Questions and Problems

(1) What do we mean by line-of-sight (LOS) signal propagation in wireless communications?

(2) Why is it that LOS signal propagation is required for the proper functioning of a microwave radio communication system?

(3) What do we mean by the phrase *free space* in wireless communications?

(4) Analyse the expression for κ in Eq. (10.67) and note that the wave velocity is $v = \frac{\omega}{\kappa}$. Which components of a multi-frequency signal travel faster along a coax cable?

(5) Calculate the resistance in ohms per meter (Ω/m) experienced by a signal at 2 GHz which propagates a coax cable. Assume that the center conductor of the coax cable has the diameter of 2 mm.

(6) Consider the voltage signal $u(x,t) = 120\sin(4\pi t \times 10^6 - \frac{\pi}{3}x)$ propagating along a 50-Ω coaxial cable.

(a) What is the frequency of the signal?
(b) Calculate the wave number.
(c) At what velocity is the signal propagating along the cable?
(d) What is the velocity factor of the coax cable?
(e) How long does the voltage signal take to move from the source to a load attached to the line 1 km away?

(7) Explain the meaning of skin-effect in transmission lines. What is the cause of it? What is the meaning of skin depth? A coaxial cable used as a transmission line transfers signals of wavelength 0.3 m. The permeability of copper is circa $\mu_{cu} \approx 1.256629$ μH/m, while its resistivity is $\rho_{cu} = 1.72 \times 10^{-8}$ Ωm. How far away from the surface of the inner conductor does the current flow?

(8) An antenna is usually rated to operate in a given frequency range with specified VSWR (or SWR). Low VSWR is desired. Assume that a given wideband antenna operating in 800 MHz to 2.4 GHz is rated to achieve VSWR of less than 3. What range of reflection coefficient at the antenna is permitted in order for the antenna to fulfill its VSWR promise?

(9) **Design project:** A company has two sites which are separated geographically by a distance d km. Each site has a local area network (LAN). The company has tasked you to design a point-to-point microwave link to interconnect the two LANs of its two campuses. The microwave link is expected to be highly reliable and of very high-quality (i.e. lower than expected latency, higher throughputs, high link stability). Your design must consider the basic features of a microwave link design, such as:

- Link availability and reliability (target: 99.999%), and how to improve these.
- Link bandwidth and data transmission rate.
- Link signal-to-noise ratio (target: at least 20 dB).
- Link latency (how long does it take to transfer data over the link, target 2-3 msec).
- Bit error rate (target: less than 10^{-6}).
- Operating frequency of the link and the reasons why you picked that frequency.
- Analysis of the profile of the path or link.
- The distance between the two sites being interconnected by the microwave link.
- Link budget, including path loss calculations, fading, interference, Fresnel zones.
- Any requirement of line-of-sight or otherwise.
- Economic aspects of the link (e.g. CAPEX and OPEX).
- Equipment acquisition and commissioning issues.
- Required various testing before mass commercial application of the link.
- Discuss any software package used to design microwave link.
- State companies which manufacture equipment for microwave links.
- State all necessary operating conditions under which your system operates as expected.

(10) A microwave link operating at 18 GHz interconnects two sites separated by a distance of 15 km. The transmitting side (TX) has the parameters: transmit power $P_t = 400$ mW, dish antenna gain $G_t = 20$ dBi and transmission line loss $L_t = 1.2$ dB. The parameters at the receiver (RX) are: transmission line loss $L_r = 0.5$ dB and antenna gain $G_r = 23$ dBi. Data is sent from transmitter (TX) to the receiver (RX).

(a) Do a link budget to find the power (P_r) in the signal received by the receiver (RX).
(b) From the link budget, what should be the minimum sensitivity S_r of the receiver in order for the link to work as expected?
(c) Assume that someone erects a tall building in the path of the microwave link discussed above after it had been launched so that the received signal has power level less than the receiver sensitivity, resulting in high bit error rates. Explain how you would solve the problem in the most techno-economical way.

(11) A LOS microwave link is used to interconnect two campuses of a company which are separated by 10 km. Assume that the system parameters are $P_t = 320$ mW transmit power, antenna gain is $G_t = 15$ dB, receive antenna gain is $G_r = 23$ dBi, transmission line losses $L_t = 2.5$ dB (at transmitter) and $L_r = 0.28$ dB (at receiver). Both antennas are aperture antennas. The effective area A_e of a parabolic dish antenna relates to its gain G as $A_e = \frac{\lambda^2 G}{4\pi}$, where $\lambda = 0.03$ m is the wavelength of the radio signal. What is the intensity of the electric field at the receiving antenna?

(12) The impedance of a lossless transmission line which is open, i.e. not terminated or very long, is given by $Z(x) = -jZ_0 \cot(2\pi x/\lambda)$, where Z_0 is the characteristic impedance of the line.

(a) Write MATLAB® program to plot this impedance. Calibrate the horizontal axis as x/λ and the vertical axis as $Z(x)/Z_0$.
(b) Under what circumstance does the line behave as a capacitor?
(c) What should be the length of the wire used if we expect it to behave as an inductor?
(d) Under what circumstance does the line behave as a short circuit?
(e) Under what circumstance does the line behave as an open circuit?

(13) **Student project:** Design a waveguide which functions as antenna. *Hint.* Use simple tin and copper wires.

(14) What is the characteristic impedance Z_0 of free space given its permeability μ_0 and permittivity ϵ_0?

(15) Calculate the range of values for the half-power beamwidth of a 2-meter parabolic antenna used in a microwave link operating at 4 GHz.

(16) Calculate the gain of a 2-meter parabolic antenna used in a microwave link operating at 24 GHz. Assume an aperture efficiency of 0.6.

(17) While setting up a local area network an IT Technician mistakenly terminated a 75-ohm transmission line with a 100-ohm resistor. A generator feeds a 220-MHz signal down the cable towards the terminator which is 37.5 mm away. The signal propagates with a speed of 0.66 relative to waves' travelling speed in vacuum.

(a) Compute the fraction of the incident power that the terminator absorbs.
(b) Find the VSWR along the cable.
(c) Calculate the impedance that the generator thinks the transmission line has. (Note: this is the input impedance.)

(18) A microwave link uses the Jirous 38-cm parabolic antenna JRE-28 EXTREM UPB at both ends. The antennas operates in the frequency band 10.3–11.7 GHz and each has a gain of 28.3 dBi. Assuming that all losses are negligibly small, how far can the antennas be separated if the maximum transmit power is 50 kW? Other parameters of the antenna are: SWR not exceeding 1.3, 6° 3dB beamwidth, and $2x45°$ or horizontal and vertical.

(19) A 50-ohm coax cable connects an electronic equipment to an antenna with the input impedance $Z_L = 150+j75$ ohms. Assume that the output impedance of the electronic equipment matches that of the coax cable perfectly. Use the Smith chart to find the following and then validate your answers using manual calculations.

(a) Is the load matched to the transmission line or not? Explain your answer.
(b) Do we expect any standing wave to occur on the line? Explain your answer.
(c) The input impedance Z_{in} at a distance $\lambda/8$ away from the load.
(d) Compute the reflection coefficient at the load, Γ_L.
(e) Compute the VSWR.
(f) Find the admittance at the load, Y_L.
(g) The distance from the load towards the generator where the voltage standing wave achieves its maximum value.
(h) The distance from the load towards the generator where the voltage standing wave achieves its minimum value.

(20) **Design project:** Find two points separated geographically not more than 50 km and design a point-to-point microwave link to in-

terconnect them. The link should fulfill the following specifications:

- $SNR > 20$ dB
- Availability of at least 99.999%
- Bit-error rate of not exceeding 10^{-8}
- Latency of not more than 2-3 ms
- Data rate of at least 10 Mbps

Design and state all other link parameters. Do a site survey and quantify the various potential losses and obstructions and use them to make the link budget: both power budget and height budget. Specify the data on the equipment used at both ends of the link. State clearly the sources of losses, such as atmospheric loss, diffraction losses, transmission line losses and fade margin.

(21) A 4-km line-of-sight link operating at 4 GHz uses transmit and receive antennas mounted on masts of heights 30 m and 45 m, respectively. A building has been erected at distance 500 m away from the transmitter.

(a) What should be the maximum height of the building with respect to the line-of-sight path so that the diffraction loss does not exceed 10 dB?

(b) Assume that a wireless signal is intercepted by a knife-edge object. Analyse the effects of the obstruction on the signal. Is the interference constructive or destructive?

(22) The gain of the transmitter and receiver of a 5-km microwave link are respectively 15 dB and 20 dB. If the link operates at 2.1 GHz and the maximum transmit power is 2 W, how much power is in the received signal?

(23) State the two main reasons why aperture antennas mounted on towers are usually covered with a radome.

(24) Explain the main reason why line-of-sight, point-to-point microwave links usually use only aperture antennas, e.g. horn antennas, cassegrain antennas or simple parabolic antenna.

10.9 Supplementary Reading

(1) ITU-R P.676-11 Report, Attenuation by atmospheric gases, Sept 2016.

(2) David Pozar, *Microwave Engineering*, John Wiley & Sons Inc., 2013.

(3) Ahmad Shahid Khan, *Microwave Engineering: Concepts and Fundamentals*, CRC Press, 2014.

(4) *Microwave Communication Basics.* Available at www.commscope.com.

(5) Frank Gustrau, *RF and Microwave Engineering: Fundamentals of Wireless Communications*, ohn Wiley & Sons Inc., 2012.

(6) Luigi Moreno, *Point-to-point radio link engineering*, Technical Report, 2013. Available at www.heraldpro.com.

Chapter 11

Radar Systems Fundamentals

Chapter's Learning Outcomes

Radar is an electronic device which uses microwave radio waves to detect and measure relevant parameters of both stationary and moving objects, referred to as **targets**. Targets tracked with a radar include precipitation, terrain, airplane, missile, ship, spacecraft, car and human being. The parameters of targets that are estimated using the radar include range or distance, radial velocity (for moving targets), altitude, and the direction of movement (for moving targets). A more recent use case of radar is radar as devices of the Internet-of-Things (IoT). For example, data collected by a radar mounted on one ship can be shared with other ships over the Internet. This chapter introduces readers to the fundamentals of radar system design, its operation and its applications, and in particular, we study:

(1) Radars as devices for the Internet-of-Things (IoT).
(2) General information about radars, e.g. radar types, major manufacturers and their application areas.
(3) Fundamental principles underpinning a radar's operation.
(4) Fundamental parameters of radars, e.g. radar cross section, radar's range, range resolution, blind range, range ambiguity, Nyquist frequency, Nyquist velocity and velocity aliasing.
(5) Kalman filtering, one of the popular algorithms used in radar signal processing to estimate interesting parameters of targets amidst noise.

11.1 Introduction

A radar is an electronic measuring equipment used to detect objects and their parameters using radio signals. Radar is an acronym coined from RAdio Detection And Ranging by Lt Commander S. M. Tucker of the United States Navy around 1940 during the WW2. The history around the development of radar is not very clear as there are differing accounts. No single person can be claimed to have invented the radar. Some people prepared the way for its invention, while others developed further the ideas of their predecessors. A Scottish scientist called James Clerk Maxwell developed the theory underpinning the behaviour and propagation of radio waves around the 1860s. Part of his discovery is the fact that a radio wave has both magnetic and electric fields, the reason why it is referred to as electromagnetic (EM) wave. For this reason, the basic equations underpinning EM wave propagation are referred to as *Maxwell's equations*. Heinrich Hertz, a German physicist, used Maxwell's theory to practically produce and receive EM waves in 1886. In this, he also discovered that part of the energy in EM waves reflect off metallic surfaces. In his honour the unit of radio frequency is named after Hertz. Interestingly, it is said that Heinrich Hertz could not perceive at that time any practical use or relevance of his discovery.

The Italian Gugliemo Marconi successfully transmitted EM waves over a long distance in 1897. German Christian Hulsmeyer made the first radar test in 1904. In 1917, the American Nikola Tesla proposed the radar concept. In 1921, the American Albert Wallace invented the vacuum tube called

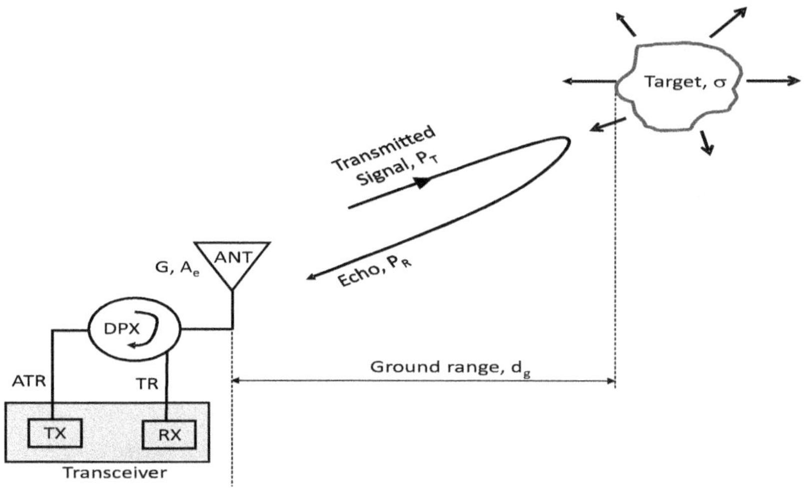

Fig. 11.1 Basic operation of monostatic radar systems.

cavity magnetron for high-power EM wave transmissions. Great Britain released the first radar system, called Chain Home, in 1930 to detect bombs from the German army during World War 2 (1939–1945). This ground-based radar system used long radio waves requiring a huge antenna of size 110 meters. Airborne radar requires much more transmit power than ground-based ones. For this reason, their invention could not occur until Hetcalf and Hahn invented the klysteron in 1936 in America.

The operation of a monostatic radar system is illustrated in Fig. 11.1. It comprises a transmitter (TX) and receiver (RX) which are co-located as the transceiver, the object under surveillance, which is referred to as the **target**, and a duplexer (DPX). The ATR (anti-transmit/receive) switch decouples the antenna from the transmitter during echo reception periods when its use is not needed. As the transmit and receive directions use frequencies very close to each other, narrowband filters are used.

Radar's operation is based on the reflection of EM waves. The radar sends signals or EM waves towards the target. The signal hits the target which reflects part of the energy. Part of the reflected signal, referred to variously as **echo**, **backscatter** and back re-radiation, returns to the radar receiver. The radar processes the echo to extract relevant information about the target, such as:

- Slant range of targets, d_R: this is the distance between the radar and the target at the time that it scatters the incident signal. It is computed using a pulse's round-trip time.
- Direction, bearing or angular position of targets: from the angle at which the echo reaches the radar, called the angle of arrival (AOA). Both the azimuth (i.e. horizontal direction) and elevation (i.e. vertical direction) of the target can be measured.
- Radial velocity of targets: relative to the radar's antenna from the Doppler shift in the frequency of the echo. The Doppler shift is the change in the operating frequency caused by the target's movement relative to the radar.
- Shape and type of the target: for example, a weather radar system is able to determine the type of precipitation such as rain, snow or fog. For this reason, such a radar system is referred to as *precipitation radar*.
- Material composition of targets.
- Size or radar cross-section of a target.

The target's size and shape are factors of the strength of the echoes. The target's location are found from its slant range and bearings relative to the radar. Any undesirable echoes is referred to as *clutter* [Aerospace and Electronic Systems Society (2008)]. Clutter is produced from various sources such as birds, precipitation, ground, the sea, meteors and aurora. It is bad if clutter is able to generate an echo which is stronger than the echo from the target under surveillance.

Although a given radar system may be able to measure more than one parameter about a target, in general, a single radar system does and cannot measure all the parameters listed earlier, at least not with precision. Clearly, the radar operation does not work if the target refuses to scatter any of the energy in the incident signal towards the radar. Alternatively, the radar operation fails if the energy in the backscatter P_R is so small that the radar's receiver cannot detect or sense it. This is the reason why a stealth aircraft (with poor signal reflection) is used by combatant forces. Unfortunately, the stealth technology depends on the operating frequency and does not work at VHF frequencies. Such targets are covered by materials with near zero EM wave reflection coefficient to inhibit their detection by radars or any other radio positioning system. Objects made of wood or any nonconducting materials (such as fishing boats) produce much weaker echoes than objects made of conducting materials such as steel. Radar systems operating at optical frequencies, i.e. terahertz range, are referred to as *ladar* (from LAser raDAR).

Every radar system has some parameters which are constant. These include the carrier frequency, pulse width and the pulse repetition period. We discuss these in greater detail subsequently. The frequency bands used by radar systems are shown in Table 11.1. Frequencies below 10 GHz are used for surveillance. Frequencies between 10 GHz and 22 GHz are used by fighter airplanes, while 22–60 GHz are used for helicopter attacks. Anti-tank missiles use the upper frequency band 60–180 GHz. Of course, as shown in Fig. 10.12, atmospheric attenuation prevents the usage of parts of the radio frequency spectrum. As observed from Table 11.1, radar systems operate at high (aka microwave) frequencies because:

- The higher the operating frequency, the smaller the wavelengths. Also, the smaller the wavelength, the smaller the size of targets detectable by the radar. Thus, higher operating frequencies provide higher resolution. A radar system can be designed to be *frequency agile* by varying the radio frequency used to exchange signals so that an enemy cannot easily interfere with the radar's operation.
- The smaller the wavelength, the smaller the size of the antenna at the same gain or directivity.

Table 11.1 Radar frequency bands.

Frequency bands used in radar systems	
Band name	**Frequency range**
High Frequency (HF)	3-30 MHz
Very High Frequency (VHF)	30-300 MHz
Ultra High Frequency (UHF)	300-3000 MHz
L (L for "long" wave)	1-2 GHz
S (S for "short" wave)	2-4 GHz
C (C for "compromise" between S and X bands)	4-8 GHz
X (X for cross)	8-12 GHz
Ku (German: "kurz unter" meaning "below short")	12-18 GHz
K (German: "kurz "meaning "short")	18-27 GHz
Ka (kurz above)	27-40 GHz
Q	33-50 GHz
U	40-60 GHz
V (very high)	50-75 GHz
E	60-90 GHz
W (next alphabetic character to "V")	75-110 GHz
D	110-170 GHz
G	140-220 GHz
Y	220-325 GHz

Table 11.2 Classification of radar systems.

Classification of radar systems	
Purpose of radar	Track, search, search and track
Waveforms used	Continuous wave (CW)[unmodulated, frequency-modulated CW (FMCW)] and pulsed (without compression, pulse compression)
Antenna architecture used	Monostatic radar (uses a single antenna for TX and RX), bistatic radar such as passive radar (uses separate antennas for RX and TX), multi-static radars (e.g. MIMO radar) uses more than 2 antennas
Application area	Ground-based, marine-based (e.g. Naval-based), airborne, missile, space-based
Usage	Cruise control (e.g. in cars), navigation, surveillance, missile control, air policing and defense, weapon control, weather monitoring, traffic control (road/speed, airport), material testing, ecology (e.g. animal and insect migration detection), remote sensing , reconnaissance and battlefield , climate monitoring, health monitoring

The operation of radar systems is not affected by environmental factors such as precipitation, dust and dirt.

11.1.1 *Radar Types and Their Applications*

Radar systems can be classified in various ways, some of which are shown in Table 11.2. A radar system can be primary radar or secondary radar. A primary surveillance radar (PSR) extracts information on a target using echo from the target. A secondary surveillance radar (SSR) does not rely on the echo from a target. Rather, the target cooperates with the radar by responding to interrogations from the radar. For this reason, the SSR radar is able to acquire more information from the target than what the PSR can. For example, the SSR can acquire the altitude and identity of targets, which cannot be done using a PSR. In order to function as expected, the SSR requires the cooperation of the target. Thus, SSR works only in friendly environments such as controlling and monitoring the movements of airplanes and ships. The military uses SSR to identify foes or friends. A friendly target is expected to respond positively to an interrogation from the SSR. A target which refuses to cooperate is identified as a foe. Clearly, a mis-identification can occur. A secondary radar system used to identify friends and foes is referred to as Identification of Friends and Foes (IFF).

Based on the antenna architecture used, radar systems can be classified into monostatic or bistatic. A monostatic radar uses a single antenna for both signal transmission and the reception of echoes, as illustrated in Fig. 11.1. A bistatic radar, however, uses separate antennas for each purpose, as illustrated in Fig. 11.9. Figure 11.2 shows the generalised block diagram of a monostatic radar system. It has receiving and transmitting units, duplexer (or circulator) to separate the transmitting and receiving electronics, a waveform generator at the transmitter and the antenna. The transmitter is usually a solid-state device or a vacuum tube (e.g. magnetron or klystron, which are high-power devices). There are also signal processing chips, data extraction and processing units and a display. The transmission line interconnecting the antenna, the transceiver and the duplexer is usually a waveguide or coaxial cable. We shall discuss in more details these components in appropriate sections of this book. In the following discussions we assume that the EM waves travel through the ether with the speed of light in vacuum, i.e. $v = c \approx 3 \cdot 10^8$ m/s. In general, $v \approx \frac{c}{\epsilon_r}$, where ϵ_r is the relative permittivity of the transmission medium.

A radar system can just track targets, only detect targets, only search targets or perform any combinations of these three functions. Search radars use antennas with wide beamwidth to scan 360° looking for targets. Track radars, however, use narrow beamwidth antennas to scan small angles. Radar systems which measure bearing (i.e. position) and range of targets are called two-dimensional (2D) radar, while those which measure in addition targets' altitude are called 3D radar. Radars used by the military can find the range (up to circa

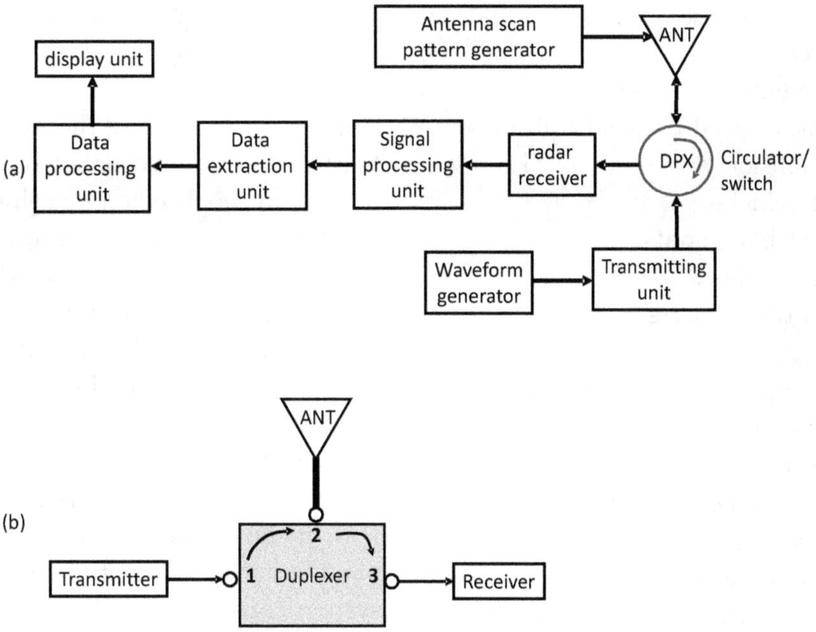

Fig. 11.2 Generalised block diagram of a monostatic radar system.

480 km), direction of motion, speed and position of moving targets.

Radar systems have a wide range of applications. A common application area is for air traffic control (ATC) at airports. In modern times most radio sensing is done using radars, especially on airplanes and satellites. Radar applications range from monitoring open-pit underground mining through marine monitoring, and land vehicles to space surveillance. Forecast indicates that the market for radar applications in only the military will reach $13 billion by the year 2020. The military uses radar systems for guidance and surveillance of weapons and for foe/friend identification.

Pico-radar systems or nano-radar systems used to monitor living beings are of increasing interest and importance. Wearable radar systems are becoming popular in recent times. They are used to guide the movements of vision-impaired people, for example. In such applications, radar systems are supplanting guide dogs. Wearable radar systems are also used to monitor the health of human organs. Figure 11.3 illustrates the block diagram of a wearable radar system used to monitor human heart's condition. In this application, the target is the human heart which is sounded to check its

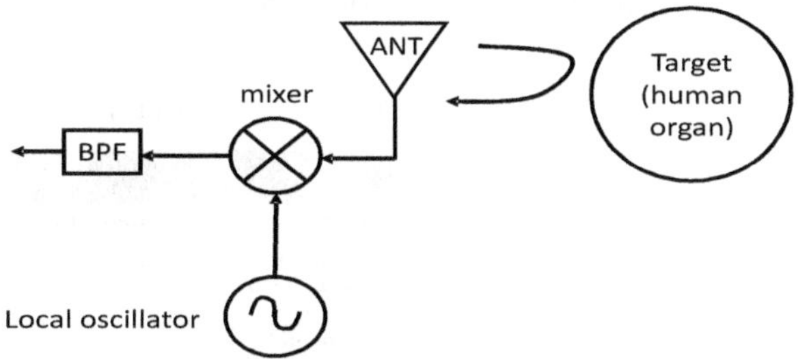

Fig. 11.3 Block diagram of a wearable radar for health monitoring [Fletcher and Kulkarni (2010)].

condition. The main components of the radar system are the bandpass filter (BPF), the mixer, the antenna (ANT) and the local oscillator. These elements are discussed in appropriate sections of this book.

Radars are now being used in increasing manner in land-based vehicles. Automotive radars operating in the W-Band are used to detect and classify objects with a high level of precision to aid in decision making to avoid danger and accidents. An automotive radar is a major component in Advanced Driver Assistance Systems (ADAS). Automotive radars operate in the W-band and the ISM (industrial scientific and medical) bands. We can use an automotive radar system for many purposes, including:

- Detect vulnerable road users, such as children and vision-impaired people.
- Detect and warn against collision.
- Keep track of vehicle speed to avoid collisions on the road. The collision avoidance is also used during parking of vehicles.
- Monitor and detect blind spots.
- Assist while changing lane.
- Provide rear cross-traffic alerts.

A more current application of radars is **radar IoT** in which radars are devices in the Internet-of-Things (IoT). An example system in this radar use case is the microwave Doppler radar Miros SM-050 manufactured by Miros Group. This radar is mounted on ships at sea to measure sea wave and water surface current. This data is used for two purposes:

- Find the true ship's speed through water (StW) for optimisation of fuel consumption.
- Monitor hull stress to ensure the safety of the ship and its cargo.

The sensed data using the radar can be transferred to a remote station over the Internet. Data collected by a radar mounted on one ship can be shared with other ships over the Internet. Smart IoT parking garages which track and record unoccupied parking spots employ radar systems.

A sample of major manufacturers of radar systems is compiled in Table 11.3. In Table 11.3 AESA means Active Electronically Scanned Array, while CW means continuous wave. Manufacturers usually focus on radars for specific applications. For example, Weibel in Denmark focuses on the manufacturing of continuous wave Doppler radar systems operating at X-band frequencies. NASA used radar systems from Weibel to detect debris coming off space shuttles during launching time. An application of radar in education has been gaining momentum lately under a class of radar

Table 11.3 Major vendors of radar systems.

Company	HQ/Origin	Founding Year	Sample Products
Rheinmetall	Germany	1889	Meteo-Measurement-Radar Systems
Honeywell International	USA	1906	Weather **radar** systems for any aircraft, e.g. Primus 440/660/880 family of radars
Raytheon Company	USA	1922	Military products incl. marine, airborne and space radars, e.g. APG-73, APG-82(V)1 AESA
Rockwell Collins	USA	1933	Weather radar systems: TWR-850/WXR-840, RTA-4100
Saab Group	Sweden	1937	Military defense & civil security systems (monopulse Fire Control Radar, FCR, dual polarised Doppler radar for weather forecasting)
Leonardo S.p.A (formerly Finmeccanica)	Italy	1948	Vixen 500E AESAradar, Leonardo's Vixen 500E AESAradar
General Dynamics Corporation	USA	1952	X-band Space Time Adaptive Radar (XSTAR)
Israel Aerospace Industries (IAI)	Israel	1953	Ground Surveillance Radar Family (e.g. ELM-2105, ELM-2105)
Airbus Group	France	1970	Civil and military aeronautical products, incl. COBRA radar for weapon location, ASR-NG radar for airport surveillance
Weibel Scientific	Denmark	1977	X-band CW Doppler radar
Reutech Radar Systems (Subsidiary of Reunert)	South Africa	1987	RSR 904 radar for anti-poaching applications
Northrop Grumman	USA	1994	military aircraft, autonomous and space systems, radar (e.g. AN/APG-81 AESARadar, APG-83 SABR)
Lockheed Martin Corporation	USA	1995 (merger)	ground-based air surveillance radars mostly for military use, e.g. AN/TPS-59, AN/FPS-117, AN/TPS-77, TPY-X
BAE Systems	UK	1999	8-DS CW Doppler tracking radar
Thales Group	France	2000	air defense and surveillance radar (e.g. GROUND Master 400)

systems called **didactic radars** which operate in the license-exempt ISM frequency bands. An example is the Didactic Primary Radar (DPR) which is the basic component of the Sky-Radar teaching system.

11.1.2 *Check Your Understanding*

Students are encouraged to check their understanding of the content of this section with the following questions prior to proceeding to the next section.

(1) What does the acronym *radar* mean?
(2) Who invented the radar technology and around which year?
(3) In radar terminology, explain the meaning of target and backscatter.
(4) What parameters of a target can we use a radar system to estimate?
(5) Why do we operate radar systems at high frequencies (usually in the gigahertz range)?
(6) Discuss how a didactic radar system can be of benefit to your studies into radar.
(7) Using a sketch showing its main components, describe the operation of a radar system.
(8) What are the ranges of wavelengths of signals used in S-band and C-band radar systems?
(9) What is frequency-agile radar system? What advantages do we get from the frequency agilility?
(10) How does a monostatic radar differ from a bistatic radar?
(11) **In-class research:** In collaboration with other students, search for at least five applications of radar technology.
(12) **In-class research:** Which company is the largest manufacturer of radar technologies? When was it established? Where is the company headquartered? Which radar technologies does the company manufacture? Which country or organisation is the main buyer of the company's products?

11.2 Pulsed Radar System and Range Limits

Most radar systems are pulsed, and the first radar system was pulsed. A pulse is a signal with a short duration in time, but has a very high amplitude. An example of natural pulses is lightening which has a very short duration but high power or amplitude. The operation of a pulsed radar system is illustrated in Fig. 11.4. A pulsed radar sends a pulse of peak power P_T for the time period τ. The time taken to transmit a pulse τ is referred to as *pulse duration* or *pulse width*. The radar then rests for the period $t_{rest} = T_p - \tau$, as illustrated in Fig. 11.4. It is clear that a pulsed radar system operates with amplitude-modulated waveforms, and specifically, on-off keying. The pulse is repeated every T_p seconds. Thus, T_p is referred to as pulse repetition period (PRP) or radar's cycle time. The reciprocal is called pulse repetition frequency or rate (PRF/PRR), which is the number of pulses emitted by the radar antenna per second, i.e.

$$R_p = \frac{1}{T_p} \,. \tag{11.1}$$

11.2.1 *Radar Range and Angular Resolution*

We use the geometry shown in Fig. 11.5 to compute the radar's range. This is the slant distance d_R between the radar's antenna and the target. The radar computes the range by measuring the time in seconds taken by the signal to leave the radar, hit the target and return to the radar. This time is referred to as round-trip time (RTT). Assuming that the EM waves travel with the speed of light which is approximately 3×10^8 m/s, the radar computes the range as

$$d_R = \frac{c \cdot RTT}{2} \ \texttt{[m]} \,. \tag{11.2}$$

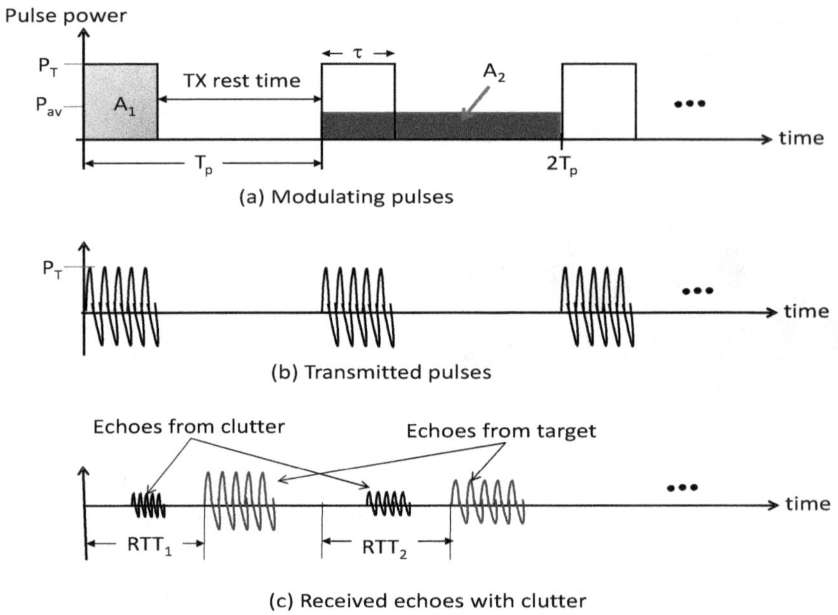

Fig. 11.4 The operation of a pulsed radar system.

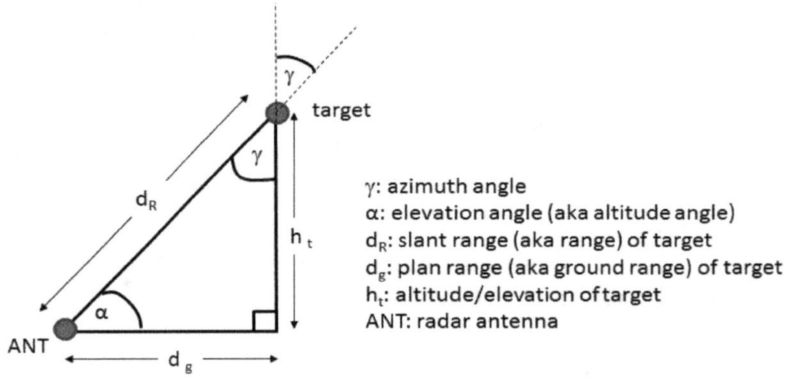

Fig. 11.5 Geometry for a radar system.

This distance can also be measured in radians or degrees as

$$d_R = \frac{c \cdot RTT}{2} \times \frac{2\pi}{\lambda}$$
$$= \pi f \cdot RTT \text{ [radians] or } f \cdot RTT \cdot 180^\circ \quad (11.3)$$

by noting that an EM wave covers 360° after covering a distance equal to its wavelength, i.e. $1\lambda \equiv 2\pi \text{ rad} \equiv 360^\circ$. Note from Fig. 11.5 the difference between the two ranges: ground range d_g and the slant range d_R. It is the latter that is measured by the radar. The slant range is the length of the line-of-sight path between the radar antenna and the target. The elevation angle is also referred to as *radar angle*.

The direction of the target is measured by exploiting the directivity of the radar antenna. The directivity measures how the antenna is able to concentrate the power in the EM wave that it transmits in a given direction. The higher the directivity, the higher the gain of the antenna. The higher the directivity of the antenna, the narrower its beamwidth. The size (or aperture) of the antenna increases as we decrease its directivity. This is the reason why radar antennas are huge.

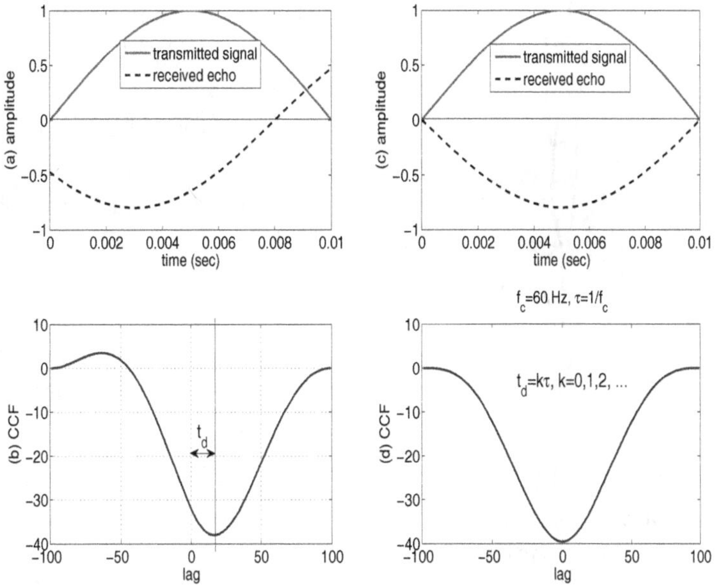

Fig. 11.6 Cross-correlation between transmitted signal and received echo [$\tau = T_c = 20$ ms].

How does the radar determine the RTT that is needed to compute the range of targets? There are several available methods. Let us assume that the transmitted signal is $x(t)$. The received echo corresponding to $x(t)$ takes the RTT to return to the radar. That means we can represent the received echo signal by $y(t - RTT)$. A simple method used to find the RTT is to compute the cross-correlation function (CCF) between $x(t)$ and $y(t - RTT)$. The time $z = t_0$, measured from zero, at which the maximum of the CCF occurs is the RTT. The CCF between the two signals is

$$R_{xy}(z) = \int_t x(t)y^*(t - z)dt \qquad (11.4)$$

where y^* is the complex conjugate of y. Figure 11.6 illustrates the principle where the pulse $x(t) = \sin(2\pi t \times 50\ \texttt{Hz})$, $0 \leq t \leq \tau$ is transmitted and the received echo due to only distance-related delay is $y(t) = \alpha \sin[2\pi(t - t_d) \times 50\ \texttt{Hz}]$, $0 \leq t \leq \tau$. The delay in Fig. 11.6(b) is 16 lags or $\frac{4}{5}\tau = 8$ msec. The echo RRT or delay in Fig. 11.6(d) is either zero or integer multiples of the pulse width. Figure 11.6(b) is the CCF of the two signals in Fig. 11.6(a), while Fig. 11.6(d) is the CCF for the signals in Fig. 11.6(c). The range and velocity of a moving target can also be estimated by performing the series of operations: pass the received echo and transmitted signal through a mixer, low-pass filter the output of the mixer and then perform Fast Fourier Transform (FFT).

We determine both the angle of elevation α and the azimuth angle γ from the radar to the target by finding the pointing direction of the highly directive antenna at the time of reception of the echo. The measurement accuracy increases with the radar antenna's directivity as the strength in the echo is increased. The accuracy also increases with the operating frequency. As illustrated in Fig. 11.5, the accurate determination of the target's direction of travel requires the true north direction. Modern radar systems use GPS (i.e. Global Positioning System) satellites to determine the true north direction.

The radar sometimes has to be able to see two targets as distinct which have the same slant range and radar cross-section (RCS) and are separated by a small angle. The smallest angle that the two targets should be separated for the radar to be able to see them as distinct is referred to as its *angular resolution*. The smaller the beamwidth of the antenna, the better the angular resolution. Antennas with large beamwidth cannot resolve targets which are very close to each other. This situation is illustrated in Fig. 11.7. The antenna (a) with large beamwidth

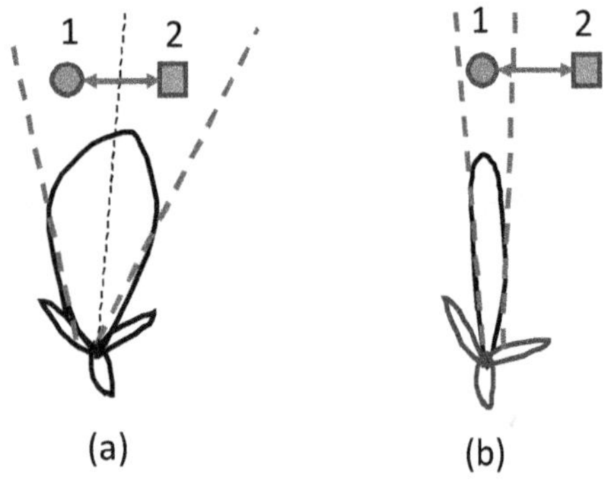

Fig. 11.7 Angular resolution of close targets.

cannot resolve the two objects 1 and 2 as distinct, unlike the highly-directive antenna (b) with smaller beamwidth. Given the 3-dB beamwidth θ_B and range d_R, the angular resolution is

$$\Delta\theta = 2d_R \sin(\theta_B/2). \tag{11.5}$$

If h_0 is the height of the radar's antenna above the Earth's surface, R is the radius of the Earth at the latitude of the range r, $R_{eff} = \frac{4}{3}R_e$ is the effective Earth radius accounting for atmospheric refraction, then the altitude of the radar's beam is

$$h_r = h_0 - R_{eff}^2 + \sqrt{r^2 + R_{eff}^2 + 2rR_{eff}\sin(\alpha)}\,. \tag{11.6}$$

Example: An airplane maneuvering into an airport is at the altitude 64 m and slant range 3 km away from the monostatic radar used for air-traffic control (ATC) at the airport.

(1) Calculate the time taken for a pulse sent from the radar to hit the airplane.
(2) Find the elevation angle of the airplane in degrees.
(3) Compute the azimuth angle of the airplane in degrees.
(4) How far is the airplane's footprint away from the airport's radar system?
(5) If the radar operates at 2.4 GHz, how much distance in radians does the airplane need to cover to land at the airport?

Solution: Below are the solutions using the formulas discussed above.

(1) Assume that the EM wave propagates with the speed of light $c = 3 \cdot 10^8$ m/s. Then $t = d_R/c = 10$ μs.
(2) From Fig. 11.5 we obtain $\sin(\alpha) = \frac{h_t}{d_R} = \frac{64\ \mathtt{m}}{3\text{km}} \approx 0.0213 \Leftrightarrow \sin^{-1}(0.0.0213) \approx 1.2°$.
(3) From Fig. 11.5 we obtain $\gamma = 90° - \alpha \approx 90° - 1.2° = 88.8°$.
(4) The distance from the airplane's footprint to the airport's radar system is the ground range shown in Fig. 11.5, which can be obtained from the Pythagoras theorem in geometry as $d_g = \sqrt{d_R^2 - h_t^2} \approx 2.9993$ km.
(5) From Eq. (11.3) we obtain $\pi f \cdot \frac{RTT}{2} = \pi \cdot 2.4 \times 10^9\ \mathtt{Hz} \times 10\ \mu\mathtt{s} = 24000\pi$.

11.2.2 *Maximum Unambiguous Range*

The maximum unambiguous range (MUR) of a radar is the maximum range of a target which ensures no interference occurs between a transmitted pulse and the echo of a previous pulse. The MUR is determined by the pulse repetition period T_p. Thus, the radar range is not just the value obtained from the radar equation derived in Section 11.3. The limiting maximum range is the the maximum unambiguous range, which is less than or equal to the range found from the radar equation. It also depends on two parameters of the echo: the time that the echo of the target farthest away from the radar reaches the radar (i.e. the round-trip-time or RTT), and the time taken to receive the echo which is the pulse duration τ. The MUR is also referred to as first range ambiguity (FRA). Also, radar systems (i.e. monostatic radars) which use a single antenna to both transmit the high-power pulses and receive the weak echos cannot transmit and receive at the same time. Thus, the pulse repetition period (PRP) must be dimensioned to ensure that it is not shorter than the time taken by the echo from the farthest target to reach the radar's receiver, plus the time to process the echo, i.e.

$$\begin{aligned} T_p &\geq RTT_{max} + pulse\ duration \\ &= \frac{2d_R}{c} + \tau \Leftrightarrow d_R \leq \frac{(T_p - \tau)c}{2}\,. \end{aligned} \tag{11.7}$$

We emphasise that $d_{R_{max}} = \frac{(T_p - \tau)c}{2}$ defines the farthest distance that a pulse can travel before the

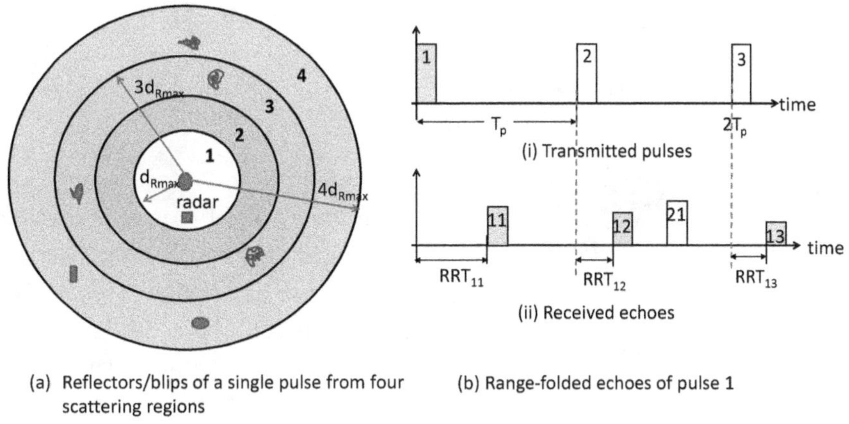

Fig. 11.8 Illustration of range folding in pulsed radar.

radar sends its succeeding pulse. Modern radar systems exploit phased array antenna system to avoid ambiguous range detection. This is achieved by tracking the desired target and avoiding signal reflections off clutter.

The locations of objects (whether target or clutter) which can reflect power are called *scattering regions*. Echoes which arrive at the radar after transmitting the next pulse are located in the scattering region $d_{R_{max}} + d$ away from the radar and are referred to as second-trip echoes. Those which return after the third pulse in the third cycle has been transmitted are located in the scattering region $2d_{R_{max}} + d$ away from the radar and are referred to as third-trip echoes, fourth-trip echoes are located in the scattering region $3d_{R_{max}} + d$ away from the radar, and so forth. These scattering regions are illustrated in Fig. 11.8. The figure also shows **blips** which are dots shown in the radar display to represent targets. Each blip represents a single target. There are various techniques to isolate the multi-trip echoes from degrading the performance of the radar system. If the PRF is fixed then the radar cannot differentiate between ranges $d + kd_{R_{max}}$, where k is a positive integer and $d_{R_{max}} = \frac{(T_p-\tau)c}{2}$. Echoes from these scattering regions are referred to as *range-folded echoes* and they are sources of range ambiguities in the pulsed radar's range estimation. This is because the radar mistakenly thinks that echoes from the scattering regions 2, 3 and 4, for example, are generated by 2nd, 3rd and 4th transmitted pulses, respectively, rather than from the 1st pulse. These are illustrated in Fig. 11.8(b). In Fig. 11.8(b)-ii, pulses 11, 12 and 13 are echoes arising from the same transmitted pulse 1. Their respective actual round-trip times (RRT) are RRT_1, RRT_2+T_p and RRT_3+2T_p. However, the radar thinks that their RTTs are respectively RRT_1, RRT_2 and RRT_3 and confuse echo 12 with the echoes of pulse 2 while echo 13 can be confused with the echoes of pulse 3.

Example: A pulsed radar system uses pulses occupying 25 kHz bandwidth and the pulse repetition frequency of 50 kHz. The peak transmit power is 200 W. Find the maximum unambiguous range of the radar system.

Solution: Given in the problem are the signal bandwidth $B = 25$ kHz and the pulse repetition frequency of $f_p = 50$ kHz. Thus, the pulse repetition period is $T_p = 1/f_p = 200$ μs. For optimum performance the pulse duration τ should be $\tau \approx \frac{1}{B} = 40$ μs. From the right-hand side of Eq. (11.7) we have $d_{max} = \max\{d_R\} = \frac{c(T_p-\tau)}{2} = 24$ km. Thus, the radar can detect targets that are not farther away from it than 24 km. □

M. L. Skolnik found that the pulse width for the optimum radar performance is [Skolnik (1980)], [Chang (2000)]

$$\tau \approx \frac{1}{B} \tag{11.8}$$

where B is the bandwidth in Hertz of the radar signal. This means that the smaller the pulse duration, the larger the bandwidth and hence the data transfer rate. However, a small pulse width results in a smaller average transmitted power [see (Eq. 11.12)] and thus smaller range. Chirp pulses are used to resolve this dilemma.

11.2.3 *Minimum Unambiguous Range*

The minimum unambiguous range is also referred to as *range resolution*, and denoted by Δd. The range resolution has two related interpretations:

- A target which is closer to the radar than the distance Δd cannot be detected by the radar. For this reason, the distance $0 < d < \Delta d$ is referred to as the *blind range.*
- Two targets with the same radar cross-section σ which are separated by a distance less than Δd cannot be distinguished by the radar as distinct objects.

The width of the pulse τ determines the minimum distance that a target should be for the radar to be able to detect it. A monostatic radar uses a duplexer (see Section 2.3.7) to switch its single antenna between the transmit and receive circuitry. Let us assume that it takes t_{on} to do so. We must consider this recovery time when computing the minimum range of the radar. Thus, the minimum range detectable by the radar is obtained from

$$\begin{aligned} &\textit{pulse duration} + \textit{receiver recovery time} \\ &\quad \le \textit{round} - \textit{trip time} \end{aligned}$$

or

$$\begin{aligned} \tau + t_{on} \le RTT_{min} = \frac{2d_{min}}{c} \Leftrightarrow d_{min} \\ \ge \frac{(\tau + t_{on})c}{2}. \end{aligned} \tag{11.9}$$

We have assumed in Eq. (11.9) that the time needed to process an echo is approximately equal to the pulse duration. Thus, the effective range d_R of the radar is

$$\texttt{Radar's range:} \frac{(\tau + t_{on})c}{2} \le d_R \le \frac{(T_p - \tau)c}{2} \tag{11.10}$$

where $c \approx 3 \times 10^8$ m/s is the speed that EM waves travel in space. We note that $t_{on} = 0$ for a bistatic radar system which uses a separate antenna for transmission and echo detection. The radar's range is akin to the human eye which cannot see objects too close or too far away. Thus, a shorter pulse width results in lower detectable minimum range, but the average power transmitted by the radar is reduced. Also, a smaller pulse width results in a better range detection accuracy or resolution. This means that radar systems which use large pulse widths are disadvantaged in the minimum target distance that they can detect the target. From $\tau + t_{on} \le RTT_{min} = \frac{2d_{min}}{c}$, we obtain

$$\delta_R = \Delta d = d_{min} = (\tau + t_{on})c/2 \,. \tag{11.11}$$

It is worthy of note that the minimum range d_{min} defined in Eqs. (11.9) and (11.11) is also referred to as the **range resolution** of the radar system. The relationship between T_p, τ, the peak transmit power P_T and the average transmit power is

$$\begin{aligned} &\frac{\texttt{pulse repetition period}, T_p}{\texttt{pulse width}, \tau} \\ &\quad = \frac{\texttt{peak power}, P_T}{\texttt{average power}, P_{av}} \,. \end{aligned}$$

This means that

$$P_T \tau = P_{av} T_p \,. \tag{11.12}$$

Thus, the larger the pulse width τ, the larger the average power radiated by the radar to increase the range. However, from Eq. (11.11) increasing the pulse width also increases the d_{min} so that the radar cannot detect closed-range targets. Thus, we have the situation

	minimum unambiguous range	maximum unambiguous range	maximum range
pulse width (τ) increases	increases	decreases	increases
PRF (f_p) increases	unchanged	decreases	unchanged

Equation (11.12) results from the two areas A_1 and A_2 shaded in Fig. 11.4 being equal. The duty cycle of the pulse is

$$duty\ cycle = \frac{on - period}{cycle\ period} = \frac{\tau}{T_p} = \frac{P_{av}}{P_T}. \tag{11.13}$$

A common range for the duty cycle is $0.1 \leq \frac{\tau}{T_p} \leq 0.4$ and $T_p \geq 10$ μs. We can observe that the longer the pulse width, the higher the average transmitted power and thus the longer the range. Unfortunately, longer pulse repetition time yields lower average transmitted power.

Example: A pulsed radar system uses pulses of width 5 μs and the pulse repetition period of 20 μs. The peak transmit power is 200 W. Assume that the radar uses separate antennas for signal transmission and the reception of backscatters.

(1) Calculate the range resolution of the radar system.
(2) Find the duty cycle of the radar's pulses.
(3) How much power is transmitted by the radar on the average.

Solution: Given are $\tau = 5$ μs, $T_p = 20$ μs and $P_T = 200$ W. Note that for a bistatic radar using separate antennas for transmission and reception there is no recovery time to switch between transmission and reception and thus $t_{on} = 0$. Therefore,

(1) Range resolution: $d_{min} = \frac{c\tau}{2} = 3 \cdot 10^8$ m/s $\times$ $5 \cdot 10^{-6}$ s/2 $= 75$ m. Thus, any pair of targets which are not separated by at least 750 m cannot be distinguished by the radar system as different.
(2) The duty cycle is $\tau_p = 5/20 = 0.25$ or 25%.
(3) The average transmitted power is $P_{av} =$ (duty cycle) $\times$ peak power $= 0.25 \cdot 200$ W or 50 W.

11.2.4 *Pulse Integration*

Some radar systems, such as the search and tracking radars, use scanning beams. A pulsed radar used for such a purpose usually sends more than one pulse to the target during each of its scanning cycle. The amount of time, measured in time units, that the radar antenna's beam illuminate a target during each scanning period is referred to as dwell time (t_D). The rate at which the antenna scans r_s is called ration rate or scanning rate and it is measured in degrees per time unit (e.g. degrees per second) or revolutions per time unit (e.g. revolutions per minute or RPM or rpm). Let us define m as the number of echoes received at the radar receiver per scanning cycle as the result of multiple pulsing of the target. Further, let θ_B be the 3-dB horizontal beamwidth, measured in degrees, of the radar's antenna. Then we have the relationships

$$t_D = \frac{\theta_B}{r_s} \tag{11.14}$$

and

$$m = t_D \times PRF = \frac{\theta_B}{r_s} \times PRF = \frac{\theta_B}{r_s T_p}. \tag{11.15}$$

Example: A pulse radar system has the pulse repetition period of $T_P = 5$ ms, an antenna with a 3-dB beamwidth of 2°, and an antenna scanning rate of 2 rpm. How many echoes return to the radar's receiver for integration during a scan cycle?

Solution: Given in the problem are $T_P = 5$ ms, $\theta_B = 2°$, and $r_s = 2$ rpm, which is equal to $2 \times 360° \times s^{-1}/60 = 12°/\text{s}$. Thus, $m = \frac{\theta_B}{r_s T_p} = 83.5$. The antennas illuminate the target with 83 pulses in each of its scanning cycle.

The illumination of a target with multiple pulses by the radar has three main applications or relevance:

(1) It enables the reception and integration or summing of multiple echoes to enhance target detection.
(2) It is exploited to track moving targets.
(3) It is used to achieve the desired search pattern.

11.2.5 *Check Your Understanding*

Students are encouraged to check their understanding of the content of this section with the following questions prior to proceeding to the next section.

(1) In radar terminology, what is clutter?
(2) Discuss the meaning of range ambiguity in radars' operation.
(3) What does 'first range ambiguity' mean?
(4) Explain a radar's blind range with examples.
(5) What are range-folded echoes? What problems do they cause?
(6) Discuss the meaning of scattering regions in a radar's operation.
(7) A pulse radar system generates 5-μs pulses at the rate of 40 kHz. Its antenna has the 3-dB beamwidth of 1°, and an antenna scanning rate of 5 rpm. How often does the radar's beam illuminate a target during a scan cycle?
(8) A monostatic pulsed radar system uses pulses of width 8 μs and the pulse repetition period of 32 μs. The peak transmit power is 300 W.
    (a) Calculate the range resolution of the radar system.
    (b) Find the duty cycle of the radar's pulses.
    (c) Compute the average power output of the radar assuming unity gain.
    (d) Calculate the maximum unambiguous range of the radar system.

11.3 Radar Range Equation

We consider the special case of a bistatic radar system which uses a separate antenna system for transmission and reception, as illustrated in Fig. 11.9. We assume that the distance travelled by the EM wave from radar's transmit antenna to the target is d_1, while that between the target and the receive antenna is d_2. The transmitting side has the antenna gain G_T and the peak transmit power P_T. The receive antenna gain is G_R, the effective area A_e and antenna radiation efficiency $\eta < 1$. In the following discussion, we assume that:

(1) The polarisation of the antennas match perfectly the polarisation of the EM waves. Thus, there is no polarisation losses.
(2) There is no reflection losses at both antennas and at the target.
(3) Transmit antenna has 100% radiation efficiency.
(4) The transmit antenna has maximum directional radiation and the receive antenna has maximum directional reception.

If the transmit antenna sends EM waves with peak power P_T and antenna gain G_T then the intensity of the electric field vector reaching the target d_1 away is

$$E_1 = \frac{P_T G_T}{4\pi d_1^2}. \tag{11.16}$$

This results in the power which illuminates the target with the effective scattering radar cross-section (RCS) or echo area σ m^2 as

$$P_1 = \sigma E_1 = \frac{\sigma P_T G_T}{4\pi d_1^2}. \tag{11.17}$$

The target scatters the power P_1 to create electric field in a spherical region. The intensity of the electric field vector reaching the receive antenna of the radar system d_2 away is

$$E_2 = \frac{P_1}{4\pi d_2^2} = \frac{\sigma P_T G_T}{(4\pi)^2 d_1^2 d_2^2}. \tag{11.18}$$

The power that this electric field creates in the receive antenna is

$$P_R = A_e E_2 = \frac{\lambda^2 G_R \eta}{4\pi} E_2 = \frac{\lambda^2 \sigma P_T G_T G_R \eta}{(4\pi)^3 d_1^2 d_2^2}. \tag{11.19}$$

Equation (11.19) is referred to as the **radar range equation**. The radar can detect or *sense* an incoming echo only if its power level is above a certain threshold, say S_{min} watts. The power level S_{min} is referred to as the **receiver sensitivity**. Thus, a proper functioning of the radar requires $P_R \geq S_{min}$. The lower the value of S_{min}, the higher the receiver's sensitivity. The higher the receiver sensitivity, the larger the detectable range. However, as

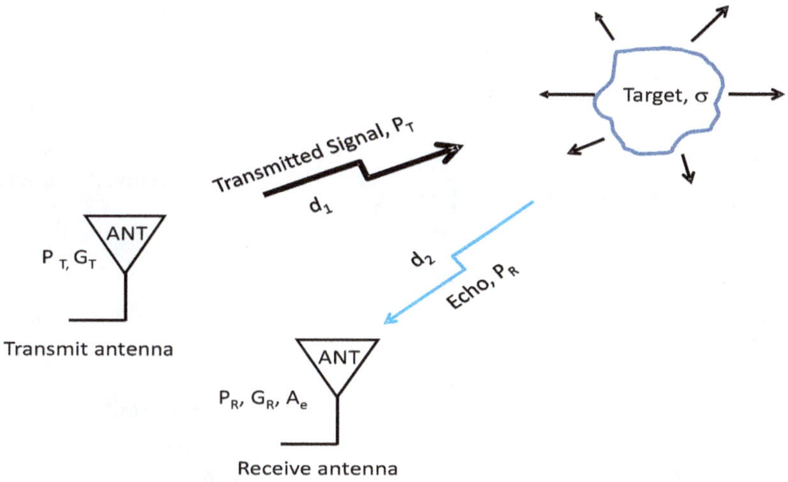

Fig. 11.9 The operation of a bistatic radar system.

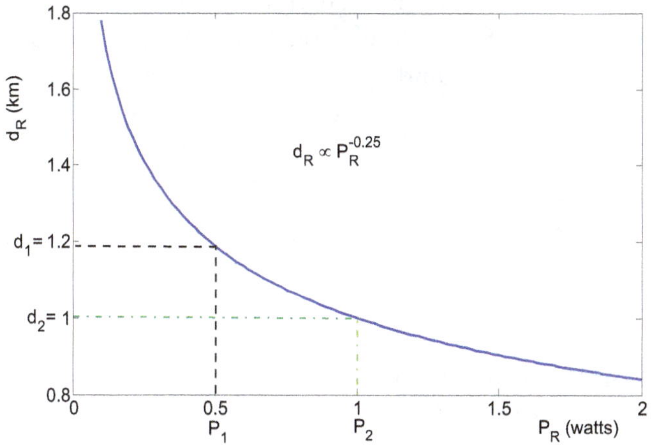

Fig. 11.10 Dependence of range on receiver sensitivity.

the detectable range increases the dynamic range (i.e. minimum to maximum power levels of echoes) of the radar receiver also increases. As $d_R \propto P_R^{-\frac{1}{4}}$, a reduction of S_{min} by 3 dB increases the range by 20%. This is illustrated in Fig. 11.10.

Now, let us assume the special case in which $d_1 \approx d_2 = d_R$. This assumption is especially valid for a monostatic radar system. Substituting this into Eq. (11.19) enables us to find the radar's slant range as

$$d_R = \left[\frac{\lambda^2 \sigma P_T G_T G_R \eta}{(4\pi)^3 P_R}\right]^{\frac{1}{4}}. \tag{11.20}$$

The maximum range d_{max} is achieved if $P_R = S_{min}$. There is also another limit on the range of a radar system, which is the horizon. Targets below the horizon (locations below the straight line from radar and tangent to the Earth's surface) cannot be detected by the radar. If $h_{\texttt{target}}$ is the altitude of the target and $h_{\texttt{radar}}$ is the altitude of the radar, then a rule-of-thumb on radar's horizon range are

$$d_{\texttt{rh}}\ \texttt{[km]} \approx 1.2576\sqrt{h_{\texttt{radar}}\ \texttt{[m]}} \tag{11.21}$$

for the range to horizon and

$$d_{\texttt{rbh}}\ \texttt{[km]} \approx 1.2576\left(\sqrt{h_{\texttt{radar}}\ \texttt{[m]}} + \sqrt{h_{\texttt{target}}\ \texttt{[m]}}\right) \tag{11.22}$$

for the range beyond horizon over the Earth's curvature. The relations Eqs. (11.21) and (11.22) are plotted in Fig. 11.11.

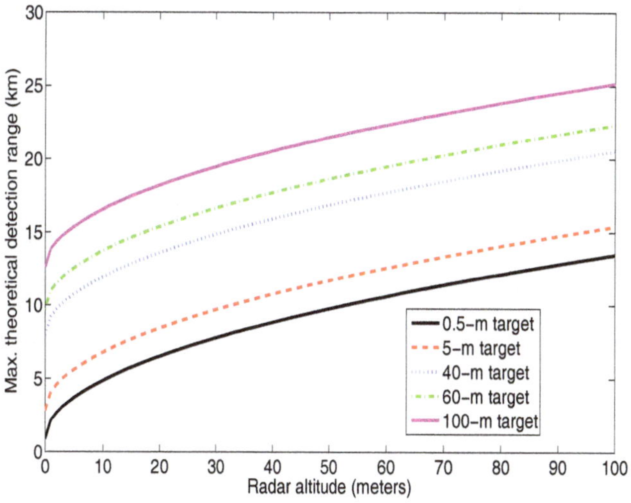

Fig. 11.11 Radar maximum range as a function of target and radar heights.

Example: Consider the radar system illustrated in Fig. 11.9 operating at 2 GHz. Assume that the distances between the target and the radar's transmitter and receiver are $d_1 = 20$ km and $d_2 = 20.02$ km, respectively. The peak transmit power is 150 W, and transmit antenna gain of $G_T = 30$ dBi. The receive antenna is a 4-m dish with the aperture efficiency $\eta = 0.55$. The target is a commercial airliner with the radar cross-section $\sigma = 80$ m^2.

(1) What is the intensity E_1 of the electric field created by the radar around the target?
(2) How much of the transmitted power is intercepted by the airliner P_1?
(3) What is the intensity of the electric field created by the backscatter at the radar E_2?
(4) What is the power P_2 in the echo reaching the radar antenna?
(5) Now, assuming that $d_1 = d_2 = d_R$ and the radar receiver's sensitivity is $S_1 = 3.3$ nW, calculate the maximum range d_1 of the radar.
(6) What is the maximum range d_2 if the receiver's sensitivity is $S_2 = 6.6$ nW.

Solution: The wavelength at 2 GHz is $\lambda = c/f = 0.15$ m.

(1) $E_1 = \frac{G_T P_T}{4\pi d_1^2} \approx 29.8\ \mu\mathtt{Wm}^{-2}$.
(2) $P_1 = \sigma E_1 = 80\ \mathtt{m}^2 \times 29.8\ \mu\mathtt{Wm}^{-2} \approx 2.4$ mW.
(3) $E_2 = \frac{P_1}{4\pi d_2^2} \approx 476.5pWm^{-2}$.
(4) $P_2 = E_2 A_e = \frac{\eta\pi D^2}{4} E_2 \approx 3.3$ nW.
(5) Given are: $S_1 = 3.3$ nW, antenna with $D = 4$ m diameter and $\eta = 0.55$ radiation efficiency, and $\lambda = 0.15$ m. Thus, antenna's gain is $G = \frac{4\pi A_e \eta}{\lambda^2} = (\frac{\pi D}{\lambda^2})^2 \eta \approx 3860$. From Eq. (11.20) we obtain $d_1 = \left[\frac{\lambda^2 \sigma P_T G^2}{(4\pi)^3 S_1}\right]^{0.25} \approx 5$ km.
(6) The receiver's sensitivity has increased by a factor 2 or 3 dB. Thus, the new range has decreased by a factor of 20%, i.e. $d_2 = d_1/2^{0.25} \approx 4.2$ km.

We can observe how low the power in the receive echo is compared to the transmit power. This is the reason why radar systems need sensitive receivers and high-gain antennas. In order for the above system to detect the range of 5 km, the power in the received echo must not be less than 3.3 nW.

$$F = \frac{\text{Receiver's input SNR}}{\text{Receiver's output SNR}} = \frac{S_i/N_i}{S_o/N_o}$$

Fig. 11.12 Effects on noise from radar's receiving circuitry.

11.3.1 *Radar Equation with Receiver's Noise*

Every electronic receiver generates thermal noise which defines its output signal-to-noise ratio, as illustrated in Fig. 11.12. The thermal noise power at the input of the receiver is $N_i = kTB$, where B is the signal bandwidth in Hertx, T is the equivalent noise temperature in Kelvin, and $k \approx 1.38 \times 10^{-23}$ J/K is the Boltzmann's constant. Noting that P_R in Eqs. (11.19) and (11.20) is the power at the input of the radar's receiver and thus $P_R = S_i$, we can express it as

$$F = \frac{\textit{input SNR}}{\texttt{output SNR}} = \frac{S_i/N_i}{S_o/N_o} \Leftrightarrow S_i = P_R$$

$$= FN_i S_o/N_o = FkTB\frac{S_o}{N_o} \qquad (11.23)$$

where F is referred to as noise figure. The minimum of the expression

$$S_i = FkTB\frac{S_o}{N_o} \qquad (11.24)$$

which occurs when S_o/N_o achieves its minimum value is referred to as the *receiver's sensitivity*. It is the minimum power in the input signal that is required for a proper operation of the radar. Another related term is the *minimum operational sensitivity* (MOS), which is $S_i/G_R = FkTB\frac{S_o}{N_o}/G_R$, where G_R is the gain of the receiver's antenna. A receiver which is able to detect low-level signals is more sensitive than the other which cannot. Unfortunately, the more sensitive the receiver is, the more susceptible it is to potential jamming. Substituting Eq. (11.23) into Eq. (11.20) yields the radar's maximum slant range considering the receiver noise as

$$d_{max,1} = \left[\frac{\lambda^2 \sigma P_T G_T G_R \eta}{(4\pi)^3 FkTB\frac{S_o}{N_o}}\right]^{\frac{1}{4}}. \qquad (11.25)$$

Example: Compute the maximum detectable range for a radar system with the following parameters: antenna gains $G = G_R = G_T = 30$ dBi, noise figure $F = 1.6$ dB, signal-to-noise ratio at the input of the radar's detector $SNR_o = 15$ dB, $P_T = 200$ W, $\sigma = 80$ m^2, $f = 2.45$ GHz, equivalent noise temperature $T = 288.15$ K (i.e. 15°C), Boltzmann's constant $k = 1.38 \times 10^{-23}$J/K, system bandwidth $B = 600$ kHz and antenna aperture efficiency $\eta = 0.55$.

Solution: Noting that $\lambda = c/f =$ and plugging the given parameters into Eq. (11.25), we obtain $d_{max,1} = 27.94$ km.

11.3.2 *Effects of Antenna Scanning Rate, Losses, Noise and Multiple Echoes on Range*

As discussed under pulsed radar systems, usually more than one echo is processed during a scanning cycle. If the number of pulses integrated is m then the radar's range equation becomes

$$d_{max,2} = \left[\frac{m\lambda^2\sigma P_T G_T G_R \eta}{(4\pi)^3 FkTB\frac{S_o}{N_o}}\right]^{\frac{1}{4}}. \qquad (11.26)$$

We still have to consider losses in the system in the final range equation. There are losses due to transmission line (e.g. skin-effect loss, radiation loss, dielectric loss), impedance mismatch losses, etc. Let us cover all these losses with the parameter L_{tot}. This yields the final maximum detectable slant range

$$d_{max,3} = \left[\frac{m\lambda^2\sigma P_T G_T G_R \eta}{(4\pi)^3 FkTB\frac{S_o}{N_o}L_{tot}}\right]^{\frac{1}{4}}. \qquad (11.27)$$

Equation (11.27) indicates that the range increases as the number of echoes m processed increases. However, m increases as the scanning rate of the radar's antenna decreases. This is the reason why the slower the antenna's beam rotate, the longer the detectable range. Let us use an example to clarify this important fact.

Example: A pulsed radar system operates with the following parameters: 500-μs pulse repetition period (PRP), 40% duty cycle, antenna's 3-dB beam width $\theta_B = 1.6°$, antenna gains $G = G_R = G_T =$ 30 dBi, antenna aperture efficiency $\eta = 0.55$, noise figure $F = 1.6$ dB, signal-to-noise ratio at the input of the radar's detector $SNR_o = 15$ dB, $P_T =$ 200 W, $\sigma = 80$ m^2, $f = 2.45$ GHz, equivalent noise temperature $T = 288.15$ K (i.e. 15°C), cummulative system losses $L_{tot} = 0$ dB, Boltzmann's constant $k = 1.38 \times 10^{-23}$J/K, and system bandwidth $B = 600$ kHz.

(1) How many pulses does the radar transmit to hit the target if its antenna scans targets at
   (a) 4 rpm?
   (b) 20 rpm?
(2) Compute the maximum unambiguous range of the radar installation.
(3) Compute the maximum detectable range at antenna's scanning rate of 4 rpm.
(4) Compute the maximum detectable range at antenna's scanning rate of 20 rpm.

Solution:

(1) From Eq. (11.15) we have $m = \frac{\theta_B}{r_s T_p}$. We need to convert rpm to degrees per second as 4 rpm $= 4 * 360°/60$ s $= 24°$/s, and 20 rpm $= 20 * 360°/60$ s $= 120°$/s. Thus,
   (a) $m = \frac{1.6°}{24° \text{s}^{-1} \times 0.5 \times 10^{-3} \text{ s}} \approx 130$ pulses.
   (b) $m = \frac{1.6°}{120° \text{s}^{-1} \times 0.5 \times 10^{-3} \text{ s}} \approx 20$ pulses.
(2) The maximum unambiguous range is $d_{ur} = \frac{(T_p - \tau)c}{2} = \frac{(1-\tau/T_p)cT_p}{2} = 45$ km.
(3) At the antenna's scanning rate of 4 rpm we integrate $m = 130$ pulses. Plugging the given parameters into Eq. (11.27) yields the maximum detectable range as $d_{max,3} = \left[\frac{m\lambda^2\sigma P_T G_T G_R \eta}{(4\pi)^3 FkTB\frac{S_o}{N_o}L_{tot}}\right]^{\frac{1}{4}} \approx 943.5$ km, which is $130^{\frac{1}{4}} d_{max,1}$.
(4) At the antenna's scanning rate of 20 rpm we integrate about $m = 20$ pulses. Plugging the given parameters into Eq. (11.27) yields the maximum detectable range as $d_{max,3} = \left[\frac{m\lambda^2\sigma P_T G_T G_R \eta}{(4\pi)^3 FkTB\frac{S_o}{N_o}L_{tot}}\right]^{\frac{1}{4}} \approx 59.1$ km, which is $20^{\frac{1}{4}} d_{max,1}$.

We can observe the following from our example above. The maximum range is only 27.9 km if we use a single pulse to track a target. This range increases to circa 59.1 km with the integration of 20 pulses and 943.5 km with the integration of 130 pulses. Unfortunately, the maximum unambiguous range of the radar is just 45 km, rendering the integration of pulses not useful.

11.3.3 *Check Your Understanding*

Students are encouraged to check their understanding of the content of this section with the following questions prior to proceeding to the next section.

(1) Consider a pulsed radar system operating at 6 GHz using 4-m dish antenna with the radiation efficiency $\eta = 0.6$ and the following parameters: pulse repetition frequency $PRF = 80$ kHz, pulse width $\tau = 1.25$ μs, $T = 15°$ C, bandwidth $B = 250$ kHz, transmit power $P_T = 200$ W, antenna scanning rate of 4 rpm, noise ratio $f = 4$, and total system losses $L_{tot} = 1.8$ dB. The target is a jet airliner with the RCS $\sigma = 80$ m^2. The Boltzmann constant is about $\kappa \approx 1.38 \times 10^{-23}$ Watts/K/Hz. Assume that the radar pulses the target only once during each scanning cycle. The radar receiver has the sensitivity of about 10^{-13} Watts.

(a) Calculate the duty cycle of the system.

(b) What is the average power output of the radar system?

(c) What is the minimum unambiguous range of the radar?

(d) Calculate the maximum unambiguous range of the radar.

(e) Find the maximum range of the radar from the radar range equation.

(f) Ascertain which of the ranges found above is the limiting maximum range.

(g) If the antenna's 3-dB beam width is $\theta_B = 70\frac{\lambda}{D}$, where D is the antenna's diameter, how many pulses does the radar send towards the target in each scanning cycle?

(h) Based on the range found in Part (f) compute the actual power in the signal at the input to the radar receiver.

(i) What is the SNR at the input of the radar receiver? *Hint.* Use the range found in Part (f) to compute the actual power in the signal at the input to the radar receiver.

(j) What is the signal-to-noise ratio (SNR) at the output of the radar receiver?

11.4 Radar Cross-Section

The radar cross section (RCS) is an indication of the ability of an object to scatter EM waves incident on it towards the radar receiver. The RCS of an object depends on its electrical properties (e.g. reflection coefficient) and physical properties (e.g. shape, size and roughness). Let E_i be the power density of the electromagnetic wave incident on a scattering object. Upon incident on an object, part of the power density in the incident signal is scattered. Let E_s be the power density of the part of the scattered EM wave which propagates in the direction of the radar receiver. Then, the radar cross section (RCS), usually represented by the symbol σ, is defined as

$$\begin{aligned}\texttt{RCS:}\ \sigma &= (\texttt{Projected cross section}) \\ &\quad \times Reflectivity \times Directivity \\ &= \frac{4\pi d^2 \cdot E_s}{E_i}\ [\texttt{m}^2] \end{aligned} \tag{11.28}$$

where d is the distance from the backscatterer to the radar. The *reflectivity* is a measure of the fraction of the power incident on the target that is reradiated (scattered) by the target. The *directivity* indicates how the scattered power is focused in the direction of the radar compared with the ideal isotropic radiator. Thus, the higher the intensity of the scattered signal reaching the radar, the larger the RCS of the scattering object. The RCS of a target classified as electrically large with perfectly reflecting surface area A_e which is directly illuminated by the radar is

$$\sigma = \frac{4\pi A_e^2}{\lambda^2}. \tag{11.29}$$

The RCS depends on several parameters, including the:

- Physical geometry of the target, including the physical dimensions and shape.
- Polarisation of the wave incident on the target.
- Electrical properties of the surface of the target.
- Angle of incidence of the EM wave.
- Operating wavelength or frequency.
- Orientation of the target at the time of EM wave incidence.

The dimension of the RCS is usually squared meter or m^2. However, normalised RCS can be given in the units of decibels per squared meter (dBm^2 or dBsm), where σ `dBsm` $= 10\log_{10}(\sigma$ `m`$^2)$. It can also be given in RCS per squared wavelength in decibels or RCSλ^{-2} in dB. The RCS can be varied through two methods: materials and shape. Using materials with low reflection coefficient makes the target difficult to track. Shapes which are not flat nor concave but rounded are good for minimising the reflection coefficient.

In some situations a radar can monitor multiple targets in a given area. In such a case differential scattering coefficient σ^0 is used instead of the RCS relating to an individual target. Also referred to as normalised radar cross-section or backscatter coefficient, σ^0 is the average radar cross-section of a set of targets per unit area. The analytical definition of σ^0 is [Moore (2008)]

$$\sigma^0 \approx \frac{P_{av}(4\pi)^3 R^3 \sin(\theta)}{P_T \lambda^2 G_0 \phi_0 r_R} \tag{11.30}$$

where R is the range, P_T is the transmitted power, P_{av} is average power returned, ϕ_0 is antenna beamwith, r_R is short-range resolution. The assumptions underlying Eq. (11.30) include the echo from the ground which contributed by a large number of scattering elements whose phases are independent, σ^0 is essentially constant and the difference in range across a resolution element is negligible.

The RCS of a sphere of radius r meters and that of a flat surface with dimensions w and h at sufficiently high frequencies where $\lambda \ll$ *Range* are respectively

$$\texttt{Sphere:}\quad \sigma = \pi r^2,\ \lambda \ll r \tag{11.31}$$

$$\texttt{Flat surface:}\quad \sigma = 4\pi\left(\frac{wh}{\lambda}\right)^2. \tag{11.32}$$

The maximum RCS of a cylinder of radius r and height h is

$$\sigma_{max} = \frac{2\pi r h^2}{\lambda^2}. \tag{11.33}$$

Thus, the RCS of a sphere is independent of frequency at such special conditions. Table 11.4 shows the RCS of some more targets.

Table 11.4 RCS of sampled targets at microwave frequencies.

Target	RCS/σ (m^2)
Small aircraft	1
Medium jet airplane	40
Jumbo jet	100
Helicopter	3
Large ship (5x10^6 kg)	10,000
Human being	1
Flying insect	0.00001 – 0.0001
Bicycle	2
Automobile (small truck)	100 - 200
Winged missile	0.1
Large fighter airplane	6

There are strategies used to reduce the RCS of an object. These include covering the target's surface with materials which absorb EM waves. A target can also generate EM waves to destructively interfere with radar's transmission to inhibit its detection. A third strategy is to use shapes that are poor reflectors of waves. These include round surfaces and tilt. Flat surfaces should be avoided if the RCS is to be reduced.

11.4.1 *Check Your Understanding*

Students are encouraged to check their understanding of the content of this section with the following questions prior to proceeding to the next section.

(1) What is the radar cross-section (RCS)?
(2) Discuss the factors which determine the RCS of an object.
(3) What are the techniques used in stealth objects?
(4) Calculate the RCS of a 4-meter square target made of perfect reflecting material at 12 GHz frequency.

11.5 Doppler Radar

The basis of the Doppler radar system is the change in operating frequency caused by relative movement between the radar and the target. This relative movement causes a change in the carrier frequency of the echo or backscatter relative to the transmitted signal. This frequency shift or spectral change is referred to as *Doppler shift* and the underlying principle *Doppler effect*. You might have observed that you hear a louder siren when an active ambulance approaches you than when it passes by you immediately. The physical phenomenon behind it is the Doppler effect. The carrier frequency of the sound signal reduces as soon as the ambulance passes by you. A Doppler radar system can use either of the waveforms: pulses or continuous wave. In the following assume:

- f: frequency of radar signal in Hertz
- d_R: slant range, i.e. distance between radar and target in meters
- λ: wavelength of radar signal in meters
- v: velocity of target relative to radar
- f_D: Doppler frequency or Doppler shift
- $\omega_D = 2\pi f_D$: Doppler angular frequency
- $v_r = \frac{d}{dt}d_R(t)$: target's radial velocity, which is the velocity of the target relative to the line passing through the middle of the main beam of the radar's radiation pattern, where d_R is the radial range
- α: angular direction of movement of the target relative to radar antenna's main beam

The number of wavelengths covered in the round-trip distance of $2d_R$ is $n_\lambda = \frac{2d_R}{\lambda}$. The radar signal covers an angle of 360° or 2π after moving a distance equal to one wavelength. Thus, the phase change in the signal after covering the round-trip distance is

$$\theta_r = 2\pi \times \frac{2d_R}{\lambda} = \frac{4\pi d_R}{\lambda}. \tag{11.34}$$

Noting that $2\pi f_D = \frac{\partial d_R}{\partial t}$, we obtain the Doppler shift equivalent to one round-trip distance as

$$2\pi f_D = \frac{\partial d_R}{\partial t} = \frac{4\pi}{\lambda}\frac{\partial d_R}{\partial t} = \frac{4\pi v_r}{\lambda}. \tag{11.35}$$

Thus,

$$f_D = \frac{2v_r}{\lambda} = \frac{2\ v\cos(\alpha)}{\lambda} = 2\frac{f}{c}v\cos(\alpha) \tag{11.36}$$

or alternatively

$$v_r = \frac{\lambda f_D}{2} = \frac{cf_D}{2f} \quad \texttt{for } v_r \ll c. \tag{11.37}$$

Example: An application of Doppler radar is shown in Fig. 11.13 for the tracking of the speeds of motorists by the police. The radar operates at $f_0 = 2.5$ GHz.

(1) Express the Doppler frequency f_D caused by the movement of the car relative to the radar as a function of the parameters given in the question, i.e. α, f_0, v, v_r and the speed of EM waves c.
(2) If $\alpha = \pi/4$ and $v = 90$ km/hr, what is the resulting Doppler shift?

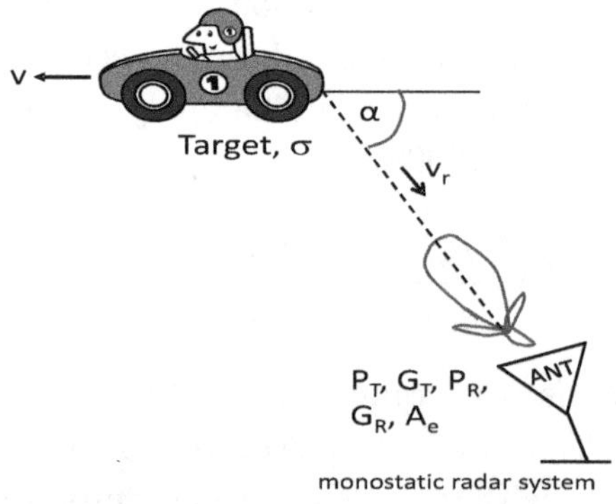

Fig. 11.13 S-band CW radar application example.

(3) What is the frequency in the received signal which reaches the radar's receive antenna?

Solution:

(1) The vehicle is moving away from the radar. Thus, the Doppler frequency must have a negative sign, i.e. $f_D = -\frac{2f_0 v_r}{c} = -\frac{2f_0 v\cos(\alpha)}{c}$.
(2) Given are $\alpha = \pi/4$ and $v = 90$ km/hr or 25 m/s. Thus, the Doppler shift is $f_D = -2\times 2.5\times 10^9 \times 25 \times \cos(\pi/4)/(3\cdot 10^8)$ Hz ≈ -416.6 Hz. This is the frequency of the signal at the output of the radar's IF mixer.
(3) The frequency of the signal reaching the input of the radar's receiver is $f = f_0 - f_D \approx$ 2.08 GHz. The target is moving away from the radar system so the frequency of the backscatter is expected to be smaller than that of the incident signal sent from the radar system. The opposite occurs if the target were moving towards the radar. This explains the reason for the change in the siren immediately after an operational ambulance passes us by.

It is important to emphasise that radars can measure only the radial velocities of targets. The radial velocity is the velocity along the line passing through the boresight of the antennas main beam. This means that an object which moves but in the direction which is perpendicular to, or at right angles with, the radar's beam is seen by the radar is stationary as $v_r = 0$. For this reason, some authors say velocity instead of radial velocity of targets. Just as with the range ambiguity there is velocity ambiguity, in that the radar can measure up to a certain speed without ambiguity. The maximum radial velocity that the radar can measure accurately but without ambiguity is referred to as *Nyquist velocity*. An equivalence of the Nyquist velocity is the *Nyquist frequency*, which is the maximum measurable Doppler shift without ambiguity. The velocity ambiguity or frequency ambiguity is referred to as *velocity aliasing* as the velocity is folded over the Nyquist velocity. The Nyquist velocity is the speed which produces the phase shift of π radians or 180° in the backscatter in reference with the transmitted signal, i.e.

$$V_N = \frac{\lambda f_{D,max}}{2} \Leftrightarrow f_{D,max} = 2V_N/\lambda \qquad (11.38)$$

where $f_{D,max}$ is the maximum Doppler shift equivalent to the Nyquist velocity V_N. Given a signal with the maximum frequency component of f_m, the Nyquist sampling theorem requires that we sample the signal using the sampling frequency $F_s \geq 2f_m$. In the Nyquist sampling theorem $f_N = 2f_m$ is referred to as the *Nyquist frequency*. In the radar's operation the pulse repetition frequency (PRF) equals the Nyquist frequency and the maximum Doppler shift corresponds to the maximum frequency in the signal being processed, i.e. f_m. Thus, $PRF = 2f_{D,max} = 2\times\frac{V_N}{\lambda}$, resulting in the Nyquist velocity

$$V_N = \pm PRF \times \frac{\lambda}{4} \qquad (11.39)$$

where PRF is the pulse repetition frequency (f_p). The interval $\pm V_N = \pm f_p \times \frac{\lambda}{4}$ is referred to as *Nyquist co-interval*. Radial velocities of targets which exceed or equal to V_N cause estimation ambiguity. Now we have the following two formulas relating the PRF of a Doppler radar system

`Unambiguous velocity:`

$$v_r \leq v_{max} = V_N = \frac{\lambda f_p}{4} \qquad (11.40)$$

`Unambiguous range:`

$$d_R \leq d_{max} \overset{(11.7)}{=} \frac{(T_p - \tau)c}{2} = \frac{(1-\tau f_p)c}{2f_p} \qquad (11.41)$$

We observe in Eq. (11.40) that as the PRF increases the unambiguous velocity increases, but the unambiguous range decreases. Also, we can decrease the aliasing by increasing the wavelength or increasing the frequency. However, such a measure decreases the unambiguous range. This effect is referred to as **Doppler dilemma**, which is a problem with only radar systems based on the Doppler shift. All radar systems, however, suffer from the range ambiguity issue. Figure 11.14 illustrates the Doppler dilemma graphically, which is obtained from the relation

$$v_{max} = V_N = \frac{c\lambda(1-\tau f_p)}{8d_{max}}. \qquad (11.42)$$

Example: Consider a Doppler radar operating at 78 GHz with the maximum blind range of 250 m. The pulses used by the radar has the duty cycle of 0.1.

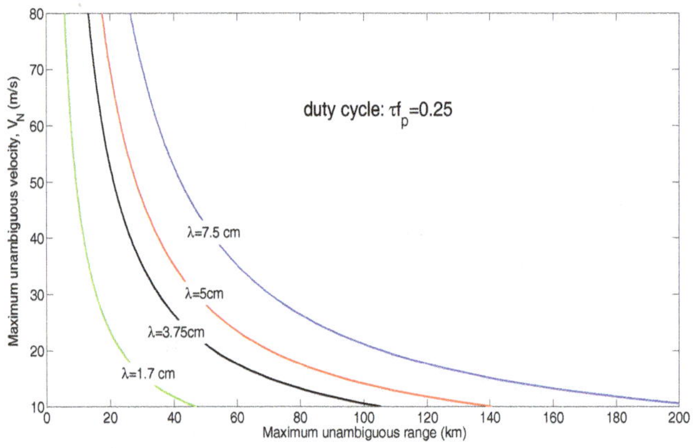

Fig. 11.14 Graphical illustration of Doppler dilemma.

(1) What is the width of pulses that this radar system uses?
(2) Calculate the pulse repetition frequency f_p of the radar system?
(3) Compute the maximum Doppler shift that the radar can measure without ambiguity.
(4) Find the Nyquist velocity of the radar system.

Solution: We have $\lambda = \frac{c}{f} = 1/260$ m.

(1) We need to find τ given $d_{min} = 250$ m and $c = 3 \cdot 10^8$ m/s. We know that $d_{min} = c\tau/2 \Leftrightarrow \tau = 2d_{min}/c \approx 1.67$ μs. We have assumed that the radar system is bistatic and so $t_{on} = 0$ s.
(2) The duty cycle equals $\frac{\tau}{T_p} = \tau f_p$, where T_p is the pulse repetition period and f_p is the pulse repetition frequency. Thus, $f_p = \frac{0.1}{\tau} \approx 53.3$ kHz.
(3) The maximum unambiguous Doppler shift is the Nyquist frequency, which is $f_N = f_p/2 \approx 26.7$ kHz.
(4) The Nyquist velocity is the maximum target velocity that the radar can measure with acceptable accuracy, which is $V_N = \pm\frac{f_p\lambda}{4} \approx \pm 51.25$ m/s. Thus, the radar under discussion can measure target velocities up to 51.25 m/s without ambiguity. However, target's radial velocities of 100 m/s and 90 m/s will be aliased as -2.5 m/s and -12.5 m/s, respectively. Velocities -100 m/s and -90 m/s will be aliased as 2.5 m/s and 12.5 m/s, respectively. These foldover velocities are illustrated in Fig. 11.17.

Note that the wind speed of 408 km/h was recorded of Tropical Cyclone Olivia on 10 April 1996. Thus, radars with high Nyquist frequencies are preferred especially for locations which experience high wind speeds. We can increase the V_N by using lower frequencies with longer wavelengths, increasing the pulse repetition frequency (PRF) or using specialised scanning methods such as dual PRF. Unfortunately, as discussed earlier, increasing the PRF increases range-folding caused by multiple scattering regions with delayed echoes. Smaller PRF also potentially caused Doppler velocity aliasing. Thus, an optimum trade-off between range folding and Doppler velocity aliasing is required. The needed trade-off between the maximum target range and the maximum Nyquist frequency is referred to as **Doppler dilemma**. Algorithms have been developed to detect and correct aliased Doppler velocities.

A popular application area of the Doppler radar is in weather nowcasting and forecasting. In this application, the radar is used to estimate radial wind velocity relative to the radar by processing backscatters from hydrometeor or precipitation (i.e. rain, hail, snow, drizzle and graupel) in order to estimate of the chance of rain, cyclones, thunderstorms, etc. A clear-air backscatter generated by illuminating insects, for example, is used to improve the accuracy in radial wind velocity estimation. Doppler radar's velocity estimation accuracy is hindered by echoes from clutter, such as the sea and ground.

For example, the Australian BOM uses a network of about 60 Doppler radars operating in the S-band and C-band to monitor the weather (including flash flooding and thunderstorms) in the country. The antennas of these radars use plan position indicator (PPI) scanning method to scan 360° at a fixed elevation angle relative to the ground [Rennie (2012)]. The antennas have beamwidths of 1° for 250-m resolution and 2° for 500-m resolution, respectively. The detectable slant ranges between 150 and 300 km. Table 11.5 is a compilation of some data on some of the Doppler radars used for weather prediction by the Australian BOM [Rennie (2012)].

The parameters of BAE Systems, Inc. Doppler radar system is shown in Table 11.6 as a case study.

Table 11.5 Data on some Doppler radars for weather forecasting.

Count	Height (m)	Band S: 2-4 GHz C: 4-8 GHz	θ_{3dB}	Range Resolution ($c\tau/2$)	Nyquist Velocity V_N (m/s)	Range (km)
1	30	S	1°	250 m	39.9	150
2	14	S	1°	250 m	52.2	150
3	146	C	1°	250 m	26.6	150
4	449	S	1° (2°)	500 m	31.2	300
5	64	C	1°	250 m	26.6	150
6	195	S	1°	250 m	26.1	300
7	699	S	1° (2°)	500 m	39.0	200
8	175	S	1°	250 m	52.1	150
9	211	S	1° (2°)	500 m	39.0	200
10	375	S	1°	500 m	19.5	200
11	62	C	1°	250 m	26.5	150
12	511	C	1°	250 m	26.7	150

Table 11.6 Radar case study: BAE Systems' 8-DS CW Doppler Radar.

Data on BAE Systems, Inc.'s 8-DS Continuous-Wave Doppler Radar System	
Parameter	**Value**
Transmitter type	Solid state
Antenna type	Microstrip array
Antenna polarization	Linear
Transmitter power	≤ 320 watts
Maximum antenna gain	41 dBi
Antenna beamwidth	4.4 x 4.4 to 1.1 x 1.1 degrees
Antenna dimensions	Variable, depending on power and beamwidth
Transmitter frequency	10.250 - 10.450 GHz (i.e. X-band radar)
Transmitter source	Phase-locked DRO (PLDRO)
Transmitter frequency stability	±5 PPM (better stability optional)
Receiver configuration	Coherent I/Q
Noise figure	< 3 dB
Receiver bandwidth	0.7 – 210 kHz (10 – 3000 m/s)
Receiver option	Multi-frequency target ranging
Designed for continuous operation	100% duty at elevated temperatures
Motion angular velocity	30 Deg/s
Elevation angular velocity	30 Degrees/sec
Azimuth angular acceleration	30 Degrees/sec^2
Elevation angular acceleration	30 Degrees/sec^2
Elevation range	-5° – 90°
Timing and synchronization	IRIG– B/GPS
Azimuth range	± 240 Deg - Continuous Azimuth Movement Available

These parameters shall be studied subsequently. The acronym IRIG-B in Table 11.6 refers to a time code, developed by an arm of the US army called Inter-Range Instrumentation Group (IRIG), which has become standard formats for the transfer of precision timing information. The pulsing rate in IRIG-B is 100 PPS (pulses per second) with 10-ms index count interval.

Moving Target Indicator

A pulsed Doppler radar system can also detect a Doppler shift in the received echo, in addition to a target's range, and exploit it in two main ways:

- Use the Doppler shift to discriminate between a small moving target amidst a large clutter.

- To estimate the velocity of a moving target relative to the radar's antenna.

A pulsed radar system used for the above two purposes in addition to range determining is referred to as *Moving Target Indicator* (MTI). Applications of MTI include determining an airplane amidst the clutter of clouds, a ship amidst a large clutter of sea waves, and a little car amidst the clutter of bigger cars and environmental structures.

11.5.1 *Check Your Understanding*

Students are encouraged to check their understanding of the content of this section with the following questions prior to proceeding to the next section.

(1) Discuss the meaning of Doppler ambiguity.
(2) Which type of radar systems are liable to Doppler ambiguity?
(3) Let us study Fig. 11.14. Which is the best frequency range to use for a radar system which seeks to detect high ranges and high velocities? Justify your answer.
(4) Discuss the meaning of continuous-wave radar and how it operates.
(5) Study Fig. 11.16. Discuss how the Doppler frequency depends on target's velocity and the operating frequency of the radar system.
(6) Study the curves in Fig. 11.15. If the PRF is 50 kHz and the radar uses pulses of 2 microseconds duration, what is the maximum achievable unambiguous range?
(7) **In-class research:** Search for the types of radar systems used in your city of residence to monitor the local weather. Discuss the basic parameters of the radar system, including its: Doppler velocity, Doppler frequency, minimum unambiguous range, maximum unambiguous range, transmit power, antenna gain and antenna beam width.

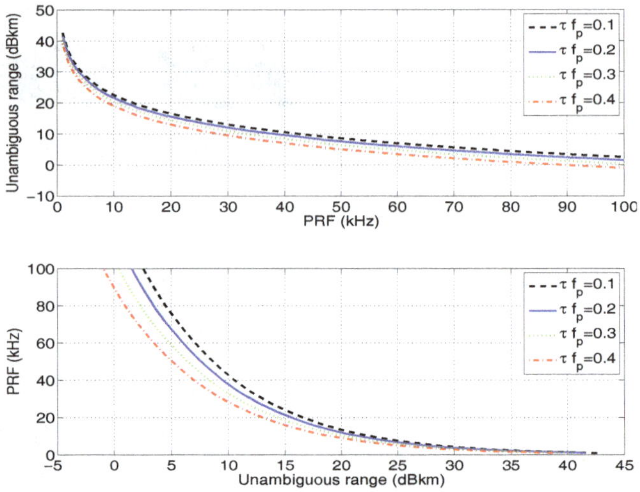

Fig. 11.15 PRF versus unambiguous range and pulse duty cycle.

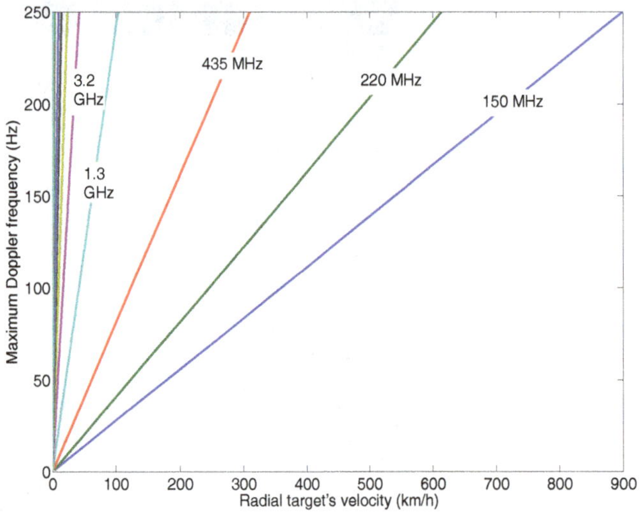

Fig. 11.16 Maximum Doppler shift versus Nyquist velocity.

(8) Study the curves in Fig. 11.14.
   (a) At what radio frequency should we operate a radar system requiring the maximum unambiguous range not less than 70 km and the Nyquist velocity not less than 52.2 km/s?
   (b) Calculate the pulse repetition frequency of the system found in (a).
   (c) Compute the suitable pulse repetition period and the pulse width of the radar system.
   (d) How does the MTI radar differ from pulsed radar?

PRF	Doppler ambiguity	Range ambiguity
low	yes	no
medium	yes	yes
high	no	yes

(a) Doppler and range ambiguities.

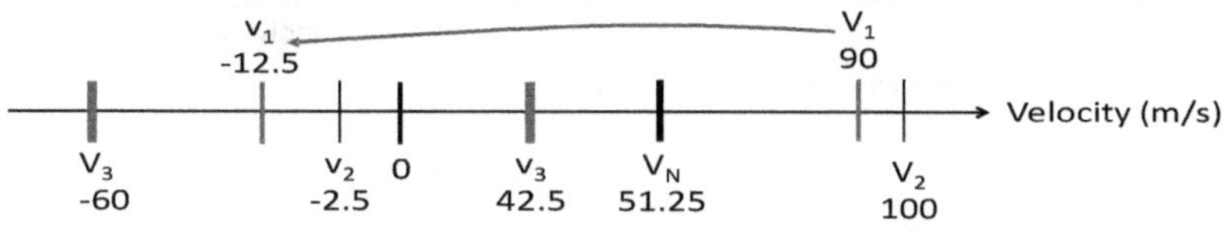

(b) Three velocities V_k and their aliases v_k.

Fig. 11.17 Velocities V_k and their aliased counterparts v_k.

11.6 Continuous-Wave Radar and Pulse Compression

The continuous-wave (CW) radar system transmits a high-frequency signal continuously, the reason for its name. The continuous transmission required in CW radar makes it necessary to use separate antennas for transmission and reception. The unmodulated CW radar transmits signals with fixed amplitude and fixed frequency. Thus, its transmitter does not shut down during the reception of echoes, in contrast to pulsed radar systems. Such an operation causes two issues:

(1) The transmitted signal with very high power can feedback into the sensitive receiver to destroy it. This is prevented by creating a spatial separation between the transmit and the receive antennas. For this reason, CW radar systems are usually bistatic radar systems using separate transmit and receive antennas. The Doppler frequency shift in the received echoes can also be used to create frequency separation between the transmit and receive antennas by exploiting the principle of frequency-division duplexing.
(2) How to assign the multiple received echoes to a time system for run-time estimation of target's speed.

There are two variants of CW radar systems: unmodulated CW radar and frequency-modulated CW radar.

11.6.1 *Unmodulated CW Radar*

The unmodulated CW radar system is simpler than pulsed radar and it can measure only the radial velocity of targets but not their range using the Doppler shift. Another weakness in the unmodulated CW radar system is that it cannot differentiate between echoes from multiple objects. The unmodulated CW radar uses patch antennas for small bandwidths or horn antennas for large bandwidths. The UCM radar can use the strength in the received echo to only guess the size and range of a target.

A popular application of the unmodulated continuous wave (UCW) radar is for the detection of the speed of cars by the police. Radars used for such purposes are referred to as radar speed gun, speed gun or radar gun. Of course, a speed gun is a Doppler radar system.

Two alternative transceivers for the unmodulated CW radar are illustrated in Fig. 11.18, namely the direct conversion receiver (aka zero-intermediate-frequency receiver) in (a), and the superheterodyne receiver in (b). The word heterodyne means mixing two signals to produce two signals called sidebands, namely, lower sideband and upper sideband. In both architectures the part of the transmitted signal is used as the local oscillator. The transceiver for the unmodulated CW radar system can demodulate the received echo to convert it from radio frequency (RF) directly to audio frequency (AF) without an intermediate demodulation step. The Doppler shift being measured is usually

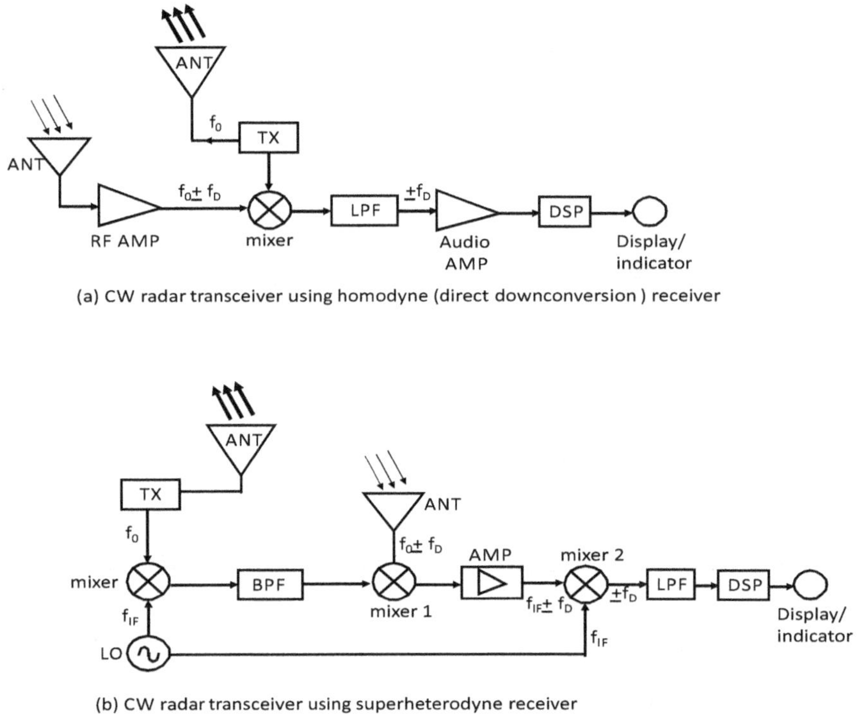

Fig. 11.18 Transceivers for bistatic continuous-wave radar.

in the AF range. Every electronic circuit (including the mixer) generates a random noise called flicker noise whose power density falls with frequency. The power density of the flicker noise P_{fn} relates to frequency f as $P_{fn} \propto 1/f$. Thus, flicker noise is a low-frequency noise. For this reason, the flicker noise at the output of the mixer can overshadow the demodulated echo containing the Doppler shift (which is also a low frequency) in a direct conversion receiver. For this reason a superheterodyne receiver was invented. The superhet first down converts the received echo at RF frequency $f_0 \pm f_D$ into echo with the intermediate frequency (IF) $f_{IF} \pm f_D$. The flicker noise arising from the RF mixer (i.e. mixer 1) is blocked by the IF amplifier, which is a bandpass filter. The IF amplifier boosts the echo signal by about 30 to 40 dB. For this reason, the strength of the flicker noise arising from the IF mixer (i.e. mixer 2) is too low to disturb the echo signal.

We use the trigonometric identities $2\cos A \cos B = \cos(A - B) + \cos(A + B)$ and $2\sin A \sin B = \cos(A - B) - \cos(A + B)$ to analyze the operation of mixers to study the transceivers shown in Fig. 11.18. Signals transmitted by the radar towards a target operates at the frequency f_0, while the corresponding echo's frequency is $f_0 \pm f_D$. The positive Doppler frequency is for targets moving towards the radar, while the negative Doppler applies if the target moves otherwise. Thus, in the ideal case, the input signal to the mixer in Fig. 11.18(a) is

$$\begin{aligned} s_1(t) &= 2\cos(2\pi f_0 t)\cos[2\pi(f_0 \pm f_D)t] \\ &= \cos(2\pi f_D t) + \cos[2\pi(2f_0 \pm f_D)t]\,. \end{aligned}$$

The low-pass filter (LPF) is tuned to block the frequency component $2f_0 \pm f_D$ and produce the output $s_2(t) = \cos(2\pi f_D t)$. We have neglected the flicker noise at the output of the mixer. The transceiver in Fig. 11.18 works similarly. The output of the mixer and the LPF $s_2(t) = \cos(2\pi f_D t)$ is referred to as *beat signal* because its frequency is the difference of the frequencies of the two input signals. For this reason, the Doppler shift f_D is also referred to as *beat frequency*. In general, we refer to the difference signal $s_3(t) = s_1(t) - s_2(t)$ as beat signal if $s_1(t)$ and $s_2(t)$ have comparable amplitudes and slightly different frequencies. The frequency of $s_3(t)$ then becomes the beat frequency.

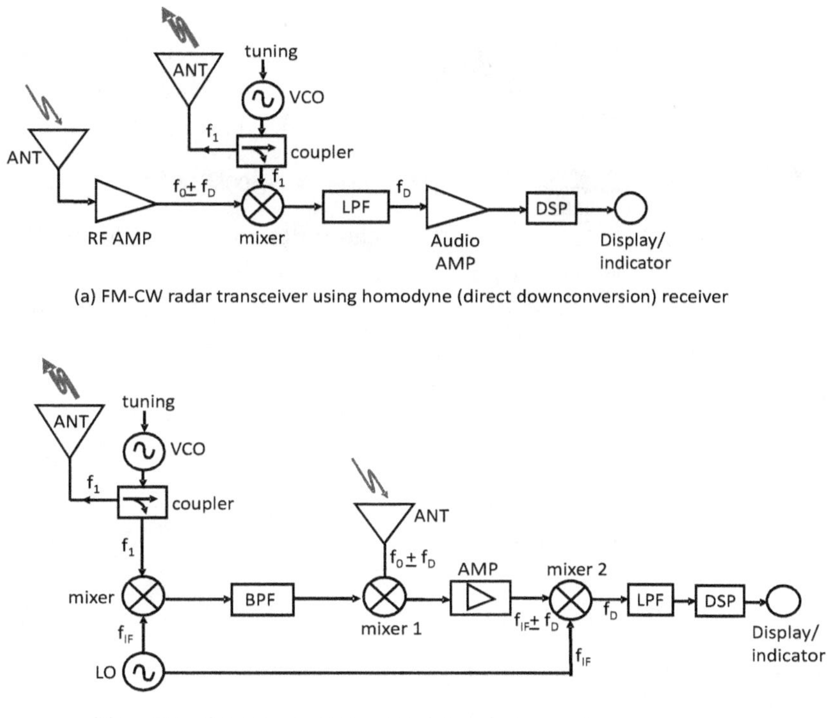

(a) FM-CW radar transceiver using homodyne (direct downconversion) receiver

(b) FM-CW radar transceiver using superheterodyne receiver

Fig. 11.19 Block diagrams of FMCW radar transceivers.

11.6.2 *Frequency-Modulated CW Radar*

As discussed earlier, the unmodulated CW radar system can measure only the target velocity radial to the radar antenna's beam, but not the target's range. The reason is that the timing information needed to measure range is missing in unmodulated CW signal. The CW radar transmits continuously so there is no way that it can measure the round-trip time of a given transmission without ambiguity. The frequency-modulated CW radar changes or modulates the frequency of the carrier signal, as illustrated in Fig. 11.19. The frequency change provides the timing information needed to measure the range of targets, while the Doppler shift is still exploited to measure their radial velocities. A pulsed radar using pulses with 100% duty cycle can be used to produce amplitude-modulated CW radar. The pulsed radar system modulates the amplitude, while the FMCW radar uses frequency modulation. We can also modulate the phase of the carrier for the needed timing information.

Although it is simple, the FMCW radar has several applications, including monitoring debris on airport runway, measuring the altitude of airplanes when landing at airports, and monitoring of motorists' speeds by the Police. FMCW radar is also used for automobile driver safety and assistance. FMCW radars are also used as weather radar profiler (aka wind profiler). The problem with it is its weak interference management ability. Consider a situation in which many automobiles mounted with FMCW radars are in a dense traffic situation. Mutual interference can reach unacceptable levels. For this reason, other modulation schemes, such as deterministic OFDM and pseudo-random modulators are investigated for automobile radar applications, which is an area of current increasing interest owing to the advent of driverless automobiles. A sample of FMCW radar systems is given in Table 11.7.

Some of the desirable features of FMCW radar are:

(1) As it transmits continuously even during echo reception, it has to be bistatic, i.e. uses separate antennas for transmission and echo reception. The advantage in this is that it avoid the existence of blind range.

Table 11.7 Sample FMCW radar systems.

Minimum Range	Maximum Range	Transmit Power	Bandwidth	Example
2.5 m	1.2 km	100 mW	65 MHz	Broadband radar
3 m	500 m	4 mW	50 MHz	DPR886 S-band radar
150 m	75 km	1.4-4 kW	1 MHz	Magnetron-based radar for marine use
100 m-1.5 km	15-250 km	30 W	50–500 kHz	WERA(over the horizon (OTH) Remote Ocean Sensing Radar)
4 km	120 km	1.4 kW	400 kHz	76N6 Clam Shell Low Altitude Acquisition Radar for low RCS targets

(2) The peak transmitted power equals its average power because of the continuous transmission. This means that less power is used compared with pulsed radar.
(3) It is simple, especially the linear FMCW (LFMCW) type. This results in low-cost implementation. Its signal processing is also fast as it uses the fast Fourier transform (FFT).

Just as in the unmodulated CW, the main drawback in FMCW radar lies in its main principle of transmitting and receiving at the same time. As noted above, the transmit power is very high, while the receive power is very low. The receiver can be destroyed if part of the transmit power find its way to it. For this reason very strict isolation between the two parts of the transceiver is very essential. FMCW radar uses a **chirp signal** [Aerospace and Electronic Systems Society (2008)], which is a signal whose frequency changes with time during its transmission. We have *up-chirp signal* if its frequency increases with time, and *down-chirp signal* if its frequency decreases with time. Fast-chirp FMCW is probably the most popular modulation scheme used in automobile radar systems as of mid-2018.

The complex envelope of a frequency-modulated (FM) carrier wave can be represented as

$$\tilde{s}(t) = A_c e^{j\phi(t)} \tag{11.43}$$

where $\phi(t)$ is the phase and A_c is the amplitude. The corresponding real envelope is $R(t) = |\tilde{s}(t)| = A_c = constant$. As mentioned earlier, the frequency of the carrier wave is modified or modulated in FM using another signal referred to as the *modulating signal*. The phase of the FM carrier wave using the modulating signal $m(t)$ is

$$\phi(t) = k_f \int_{-\infty}^{t} m(x)dx \tag{11.44}$$

where k_f is the frequency deviation constant (aka modulation sensitivity) measured in radians per volt per second if the modulating signal is a voltage waveform. The FM signal has the form

$$s(t) = A_c \cos[2\pi f_c t + \phi(t)] = A_c \cos[\theta(t)] \tag{11.45}$$

where $\theta(t) = 2\pi f_c t + \phi(t)$. The instantaneous frequency of the FM signal is

$$f(t) = \frac{1}{2\pi}\frac{d\theta(t)}{dt} = f_c + \frac{1}{2\pi}\frac{d\phi(t)}{dt} = f_c + \frac{k_f}{2\pi}m(t)\,. \tag{11.46}$$

The frequency deviation is

$$f_d(t) = f(t) - f_c = \frac{1}{2\pi}\frac{d\phi(t)}{dt} = \frac{k_f}{2\pi}m(t) \tag{11.47}$$

resulting in the maximum or peak frequency deviation

$$\Delta f = \max\{f_d(t)\} = \frac{1}{2\pi}\max\left\{\frac{d\phi(t)}{dt}\right\} = \frac{k_f}{2\pi}\max\{m(t)\}\,. \tag{11.48}$$

Thus, the transmitted FM signal is

$$s(t) = A_c \cos[2\pi f(t)t] = A_c \cos[2\pi f_c t + k_f t m(t)]\,. \tag{11.49}$$

Sample values for the Δf used in FM broadcasting are 75 kHz for 200 kHz bandwidth, 5 kHz for 25 kHz bandwidth and 2.5 kHz for 12.5 kHz bandwidth. The modulating signal can have several forms, and the popular forms are linear, sinusoidal, triangular, square, staircase and sawtooth waveforms.

Frequency modulation (FM) is often also referred to as *sweep* or RF sweep. The rate at which the operating frequency of the carrier wave is changed between a maximum value $f_{max} = f_0 + \Delta f$ and a minimum value $f_{min} = f_0 - \Delta f$ is called *sweep rate*, and it is measured in Hertz/second. The sweep rate is the gradient of the modulating signal $m(t)$. The *sweep bandwidth* is defined as $f_{sb} = f_{max} - f_{min}$.

11.6.2.1 *Linear Modulating Signal with Stationary Target*

We consider the case of using a linear modulating signal $m(t) = \frac{f_b}{t_d}t$ to obtain the instantaneous frequencies illustrated in Fig. 11.20. We assume that the target is stationary and thus there is no Doppler shift. The signal transmitted at time $t_0 = 0$ is $s(t)$ and its corresponding echo y(t) is received

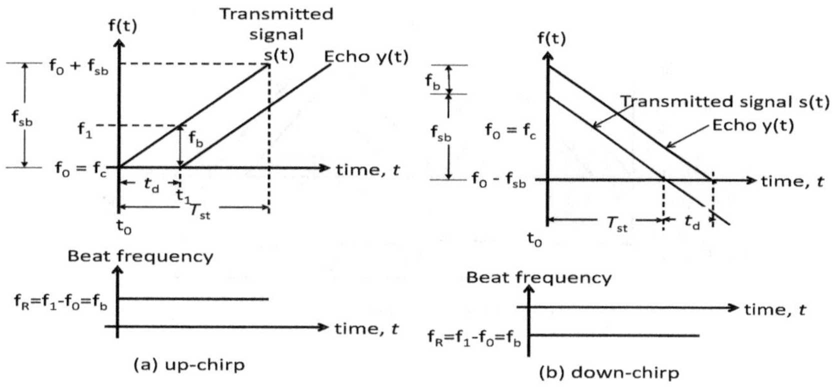

Fig. 11.20 Frequency waveforms for FMCW radar with linear modulating signal.

$t_d = t_1 - t_0 = t_1$ later. At the time t_1 that the echo $y(t)$ reached the radar the transmit frequency has been changed from $f_0 = f_c$ to $f(t = t_1) = f_1 = f_0 + f_b$. Thus, we have

$$s(t) = A_c \cos\left[2\pi f_1 t + \frac{k_f f_b}{t_d} t^2\right] \tag{11.50}$$

and

$$y(t) = B_c \cos\left[2\pi f_0 (t - t_d) + \frac{k_f f_b}{t_d}(t - t_d)^2\right]. \tag{11.51}$$

The two signals $s(t)$ and $y(t)$ are passed through the mixer and then the low-pass filter to produce the beat frequency f_b which is processed to obtain the range d_R. From Fig. 11.20 we obtain the gradient

$$\frac{f_b}{t_d} \stackrel{!}{=} \frac{f_{sb}}{T_{st}} \Leftrightarrow t_d = \frac{f_b T_{st}}{f_{sb}}. \tag{11.52}$$

Plugging $t_d = \frac{2d_R}{c}$ into Eq. (11.52) we obtain the range

$$d_R = \frac{f_b T_{st} c}{2 f_{sb}} = \frac{(f_1 - f_0) T_{st} c}{2 f_{sb}}. \tag{11.53}$$

We know the sweep time T_{st}, the sweep bandwidth f_{sb} and the speed of light $c = 3 \cdot 10^8$ m/s so we can compute the target range by finding the beat frequency f_b at the output of the mixer at time t_1. The beat frequency is the absolute difference between the frequencies of the transmitted signal and the received echo, i.e. $|f_1 - f_0|$.

11.6.2.2 *Triangular Modulating Signal with Moving Target*

Let us consider a target which is moving towards the radar system. The modulating signal has a triangular waveform, as illustrated in Fig. 11.21. The voltage-controlled oscillator (VCO) generates an up-chirp for half of the modulation or sweep period and down-chirp for the other half of the sweep period within the sweep bandwidth $2\Delta f$. The echo returns to the radar after the round-trip propagation time delay t_d. In case the target is moving relative to the radar's main beam the frequency in the echo is changed by both the Doppler shift and the propagation delay. Thus, there are two beat frequencies at the output of the mixer and LPF, namely

$$\texttt{Increasing frequency:}\quad f_{bd} = f_b + f_d \tag{11.54}$$

$$\texttt{Decreasing frequency:}\quad f_{bu} = f_b - f_d. \tag{11.55}$$

The target velocity is embedded in the Doppler shift f_d, while the target's range is found from the beat frequency f_b. We obtain these two frequencies from Eq. (11.54) as

$$f_d = \frac{1}{2}(f_{bd} - f_{bu}) \tag{11.56}$$

$$f_b = \frac{1}{2}(f_{bd} + f_{bu}). \tag{11.57}$$

Thus, we can obtain the radial velocity of the target as

$$v_r = \frac{f_d \lambda}{2} = \frac{(f_{bd} - f_{bu})\lambda}{4} = \frac{c(f_{bd} - f_{bu})}{4 f_0} \tag{11.58}$$

and the target range as

$$d_R = \frac{c f_b T_{st}}{2 f_{sb}} = \frac{c(f_{bd} + f_{bu}) T_{st}}{2 f_{sb}}. \tag{11.59}$$

We obtain a similar results for a target moving away from the radar by using

$$f_{bu} = f_b + f_d \;\texttt{and}\; f_{bd} = f_b - f_d \tag{11.60}$$

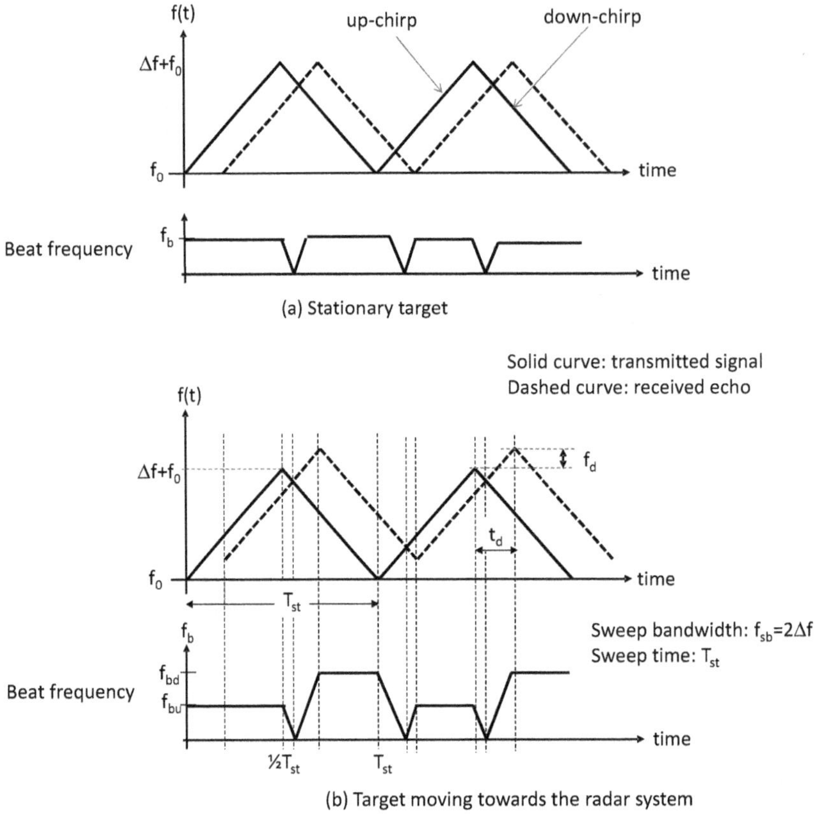

Fig. 11.21 Frequency waveforms for FMCW radar with triangular modulating signal.

as the Doppler shift has a negative sign. The range resolution of the FMCW radar is

$$\delta_r = \frac{c}{2\Delta f}\,. \tag{11.61}$$

Example: Consider the frequency waveforms shown in Fig. 11.22(a) which is used for by W-band FMCW radar to track a target in motion.

(1) Sketch the beat frequency below the given curves.
(2) Compute the Doppler shift in the received echo.
(3) Estimate the round-trip time (RTT) of the echo signal t_d.
(4) Is the target moving away or towards the radar system?
(5) At what radial velocity has the target been travelling?
(6) How far away is the target from the radar?

Solution:

(1) The beat frequency is shown in Fig. 11.22(b).
(2) Each vertical division in the figure is 20 MHz. The Doppler shift is one division as the level of the received frequency is one division higher than the transmitted signal. Thus, $f_d = 1\ \texttt{div} \times 20\ \texttt{MHz/div} = 20$ MHz.
(3) The RTT is two divisions. As each horizontal division equals 15.625 microseconds, we have $t_d = 2\ \texttt{div} \times 15.625\ \mu\texttt{s/div} = 31.25\ \mu$s.
(4) The target is moving towards the radar system as the received frequency is higher than the transmitted and thus the Doppler shift is positive.
(5) We can read off Fig. 11.22(b) the following: $T_{st} = 8\ \texttt{div} \times 15.625\ \mu\texttt{s/div} = 125\ \mu$s, $f_{sb} = 8\texttt{div} \times 20\ \texttt{MHz/div} = 160$ MHz, $f_{bu} = 2\Delta f - f_d = 140$ MHz and $f_{bd} = 2\Delta f + f_d = 180$ MHz. Thus, $f_{bd} + f_{bu} = 300$ MHz and $f_{bd} - f_{bu} = 40$ MHz. Therefore the radial velocity is

$$v_r = \frac{c(f_{bd} - f_{bu})}{4f_0}$$

$$= \frac{3 \cdot 10^8\ \texttt{m/s} \times 40 \times 10^6\ \texttt{Hz}}{4 \cdot 77 \times 10^9\texttt{Hz}}$$

$$\approx 38.96\ \texttt{km/s}\,.$$

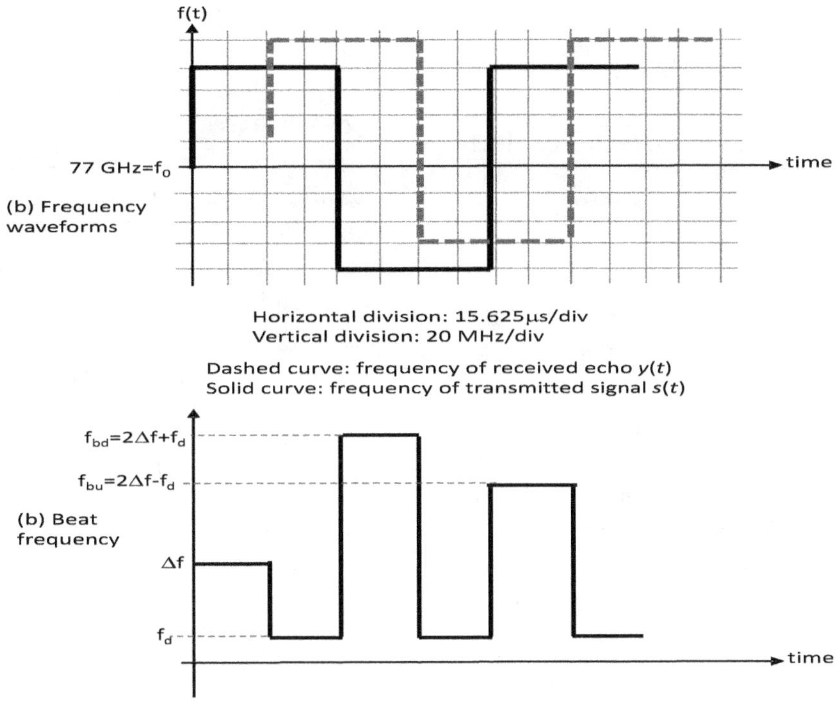

Fig. 11.22 Frequency waveforms for FMCW radar with square modulating signal.

(6) The range is

$$d_r = \frac{c(f_{bd} + f_{bu})T_{st}}{2f_{sb}} \approx 35.2 \text{ km}.$$

11.6.3 *Pulse Compression Radar*

The idea of pulse compression is to use pulses of different widths for transmission and reception in order to increase the range resolution while maintaining an acceptable signal-to-noise ratio at the radar's receiver. Pulses sent from the radar are longer than pulses received as echoes. The main reason is that a short arriving pulse enables us to limit the number of range bins that received echoes fall in. In other words, a short pulse allows the separation of targets close to each other, as well as targets close to the radar.

Let us use a simple example to illustrate pulse compression by neglecting any Doppler shift in the received echo for now. Assume that a pulsed radar transmits the pulse

$$m(t) = \begin{cases} B\cos(2\pi f_c t), & k\tau \le t \le (k+1)\tau \\ 0, & \texttt{otherwise} \end{cases} \tag{11.62}$$

The corresponding echo arrives at the radar after the round-trip time t_d and an amplitude attenuation of A and thus has the form

$$y(t) = \begin{cases} AB\cos[2\pi f_c(t - t_d)] & k(\tau - t_d) \le t \\ \quad +n(t), & \le (k+1)(\tau - t_d) \\ 0, & \texttt{otherwise} \end{cases} \tag{11.63}$$

where $n(t)$ is the additive noise signal. The cross-correlation function between $m(t)$ and $y(t)$ with $A = B = 1$ is plotted in Fig. 11.6 to obtain the round-trip delay t_d. Let the instantaneous respective power and energy in $m(t)$ be $P_m(t)$ and E_m, and those of $y(t)$ be $P_y(t)$ and E_y. Then we have

$$P_m(t) = |m(t)|^2 = B^2, \qquad E_m = \int_{k\tau}^{(k+1)\tau} P_m(t)dt = \tau B^2 \tag{11.64}$$

$$P_y(t) = |y(t)|^2 = (AB)^2, \qquad E_y = \int_{k\tau}^{(k+1)\tau} P_y(t)dt = \tau(AB)^2. \tag{11.65}$$

Let σ_n^2 be the energy in the additive noise. This yields the signal-to-noise ratio (SNR) at the radar's receiver as

$$SNR = \frac{\texttt{signal energy}}{\texttt{noise energy}} = \frac{\tau(AB)^2}{\sigma_n^2}. \quad (11.66)$$

We obtained the resolution of radars in Eq. (11.11) as

$$\Delta d = d_{min} = (\tau + t_{on})c/2. \quad (11.67)$$

We observe in Eqs. (11.66) and (11.11) that increasing the pulse duration τ increases the SNR but decreases the range resolution. The resolution is better with a short pulse, but a longer pulse is needed for good SNR and large range. A pulse with a very narrow duration but a very high amplitude can achieve the purpose. However, the generation of such a pulse is difficult because of limiting practical and physical constraints. The solution to this dilemma is *pulse compression*, in which we transmit a pulse with the width that is long enough to achieve a good SNR, but also results in acceptable range resolution. Strategies used to achieve the pulse compression include:

- Chirp: change the radio frequency of the pulse from low to high usually linearly during its transmission. This is linear frequency modulation, discussed earlier.
- Matched-filtering: digital radar systems use matched filters to achieve the purpose.
- Phase modulation, either binary or poly-phase coding.

These topics are more of radar signal processing topics so we leave them for treatment in books dedicated for such topics.

11.6.4 *Pseudo-Noise Modulated Continuous-Wave Radar*

We have studied two waveforms used in radars: pulse and continuous wave. Both have certain undesirable features. A pulse of a short duration can avoid the range ambiguity problems. However, its low average transmit power and velocity ambiguity are its drawbacks. A continuous wave has a good average transmitted power and velocity measurements, but unsatisfactory range measurements. Thus, pseudo-random noise sequence, such as shown in Fig. 11.24(a), were developed as the pattern used to modulate an aspect of the carrier wave. A popular class of pseudo-random noise sequence is the maximum-length sequence (aka MLS or m-sequences). An m-sequence is usually generated using linear feedback shift registers (LFSR) with m registers. With m-registers $2^m - 1$ unique polar binary sequences, each of length m, is produced as the all-zeros vector is excluded. The example shown in Fig. 11.24(a) uses $m = 8$ and thus each period T_p of the sequence has $m = 8$ chips, each of duration $\tau = \frac{1}{B}$, as illustrated in Fig. 11.24(a).

Pseudo-random noise sequences have the following features compared with pulse and continuous wave waveforms [Chandler (1964)]:

- large average- to peak-power ratio
- unamabiguous measurement of range to large ranges
- unambiguous measurement of velocity (Doppler frequency) to high velocities
- fine range resolution
- fine velocity resolution
- resistance to both sophisticated and power jamming
- difficult for an enemy to detect because of its peak-free, noise-like spectrum.

As can be observed in Fig. 11.24(b), the null bandwidth of the pseudo-noise waveform is f_c. This results in the range resolution

$$\Delta R = \frac{c}{2B} = \frac{c}{2f_c} \quad (11.68)$$

and the maximum unambiguous range

$$d_{R,max} = \frac{cT_p}{2} = \frac{cm}{2f_c}. \quad (11.69)$$

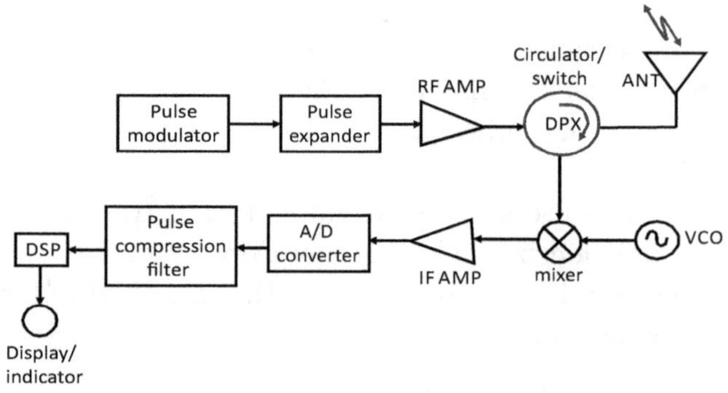

Fig. 11.23 Pulse compression radar system.

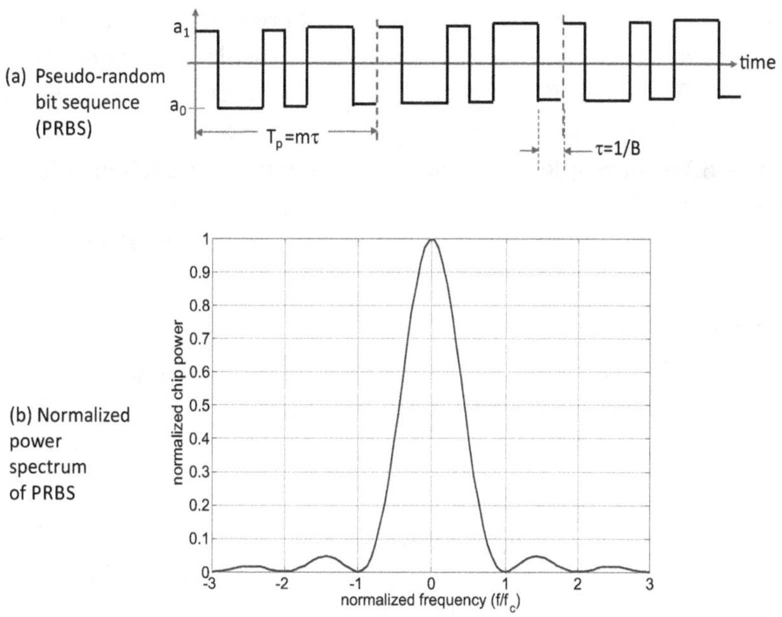

Fig. 11.24 Pseudo-random bit sequence $[B = f_c = \frac{1}{\tau}]$.

Thus, decreasing the duration of each chip increases f_c to improve both the range resolution and the maximum unambiguous range.

11.6.5 *Check Your Understanding*

Students are encouraged to check their understanding of the content of this section with the following questions prior to proceeding to the next section.

(1) What is beat frequency?
(2) Explain the meaning of beat signal.
(3) What is a chirp signal?
(4) Explain sweep bandwidth as used in FMCW radar systems.
(5) What is pulse compression in radar systems?
(6) Discuss the differences and similarities between the strategies used to achieve pulse compression.
(7) Figure 11.23 shows a block diagram for a radar system using compressed pulses. Analyse the input and output of each of the subsystems in the radar's transceiver.

11.7 Synthetic Aperture Radar

Originally known as Doppler beam sharpening, Synthetic Aperture Radar (SAR) is a technique used in radars to achieve a high azimuthal spatial resolution, which is desirable in remote sensing. Conventional radar systems use scanning antennas and are referred to as Real Aperture Radars (RAR) or Side-Looking Airborne Radars (SLAR). The minimum azimuth distance that a pair of objects must be separated for the radar to be able to resolve them as distinct is referred to as the *azimuth resolution*, and denoted δ_a. The lower the value of δ_a, the higher and the better the resolution as objects can be closer to each other and still be observed by the radar as distinct.

If the half-power azimuth beamwidth of the radar's antenna with the diameter D_a is θ_B, the operating wavelength is λ, then the azimuth resolution of SLAR for the slant range d_R is [Skolnik (1980)]

$$\delta_{SLAR} = \theta_B d_R = \frac{\lambda d_R}{D_a}. \tag{11.70}$$

SLAR does not use the SAR principle, and thus possesses azimuth resolution that is lower than that of SAR. The azimuth resolution of SLAR also depends on the range and decreases as the range increases. The poor resolution of SLAR led to the invention of the coherent radar and Doppler beam scanning principle by Carl Wiley in 1951 [Wiley (1954)], [Wiley (1985)]. Carl Wiley's invention was further developed to the SAR principle. The azimuth resolution of SAR with the synthesised antenna beamwidth θ_{SAR} is

$$\delta_{SAR} = \theta_{SAR} d_R = \frac{D_a}{2} \tag{11.71}$$

which is independent on the range.

Synthetic Aperture Radar (SAR) differs from traditional radars in three main ways: it is used mainly to take images of objects, it is mounted on moving objects such as aircraft, space shuttle and satellite, and it achieves a better spatial resolution by exploiting the movements of the radar's antenna over the target's location. As the mounting platform moves, the consecutive transmit/receive times become different positions. By coherently combining the multiple received echoes from a target, we can construct or synthesise a virtual aperture, called **synthetic aperture**, which is much larger/longer than the actual dimensions of the physical antenna being used. This is the basis of the name SAR.

SAR has been a major invention for the development of spaceborne and airborne radar systems for imaging of objects. During the 1950s to 1960s, the application of SAR principle was confined within the military. The civillian use of SAR started in the 1970s and 1980s and led to the launch of the first civilian SAR in 1978 called Seasat. SAR has supplanted SLAR since the 1950s as the modern imaging radar. The SAR principle has wide applications, including the shuttle imaging radar with the C-band, X-band and L-band radars (SIR-C/X-SAR) since 1994, the shuttle radar topography mission (SRTM) operating with X-band and C-band radars since 2000, and the C-band Radatsat Constellation-1/2/3. There are also bi-static radars exploiting the SAR principle, such as TerraSAR-X and TandDEM-X.

SAR techniques include:

- Interferometry which is used to derive the topography of surfaces.
- Differential interferometry, which is used to measure displacements of the Earth's surface.
- Polarimetry, for better retrieval of a target's parameters.
- Holographic tomography, used to generate 360° image of volumetric scatterers.
- Polarimetric SAR interferometry (Pol-InSAR).

The SAR radar principle usually uses chirp pulses. That is, a waveform which has a short duration τ, high power, constant amplitude, but its instantaneous frequency $f_i(t)$ changes linearly over time t. If the chirp rate is r_c Hz/sec and f_0 is the center frequency, then the instantaneous frequency is

$$f_i(t) = f_0 + r_c t, \quad tk\tau \leq (k+1)\tau, \; k = 0, 1, 2, \ldots \tag{11.72}$$

resulting in the chirp bandwidth $B_c = r_c \tau$. The pulse repetition frequency (PRF) used by SAR is a few hundreds of hertz for airborne applications, but a few kilohertz for space-borne applications. The swath width is about a few km to 20 km in airborne applications, but 30 to 500 km in space-borne applications.

The platform on which the radar is mounted and thus the radar antenna travels the distance $d = vT_p$ between the transmission of a pair of chirp pulses, as illustrated in Fig. 11.25. Four positions of the radar's antenna at the times that it transmits a pulse are shown in the figure as 1, 2, 3 and 4. These positions are separated from each other by the constant distance $d = vT_p$. All the positions at which the radar sends pulses to its target make up a virtual antenna array of length $L_a = \sum_k vT_p$, which is the synthetic aperture. The SAR length is larger than the length of the physical antenna being used by SAR.

Example: An X-band (i.e. $f = 8.45$ GHz) radar system uses 4-m antenna to track targets with maximum of 5 km. Compute the azimuthal resolution if (a) conventional radar operation is used and (b) SAR principle is used. (c) Which of the two types of radar operations is better and why?

Solution: (a) From Eq. (11.70) we have $\delta_{SLAR} = \theta_B d_R = \frac{\lambda d_R}{D_a} \approx 153$ m. (b) From Eq. (11.71) we

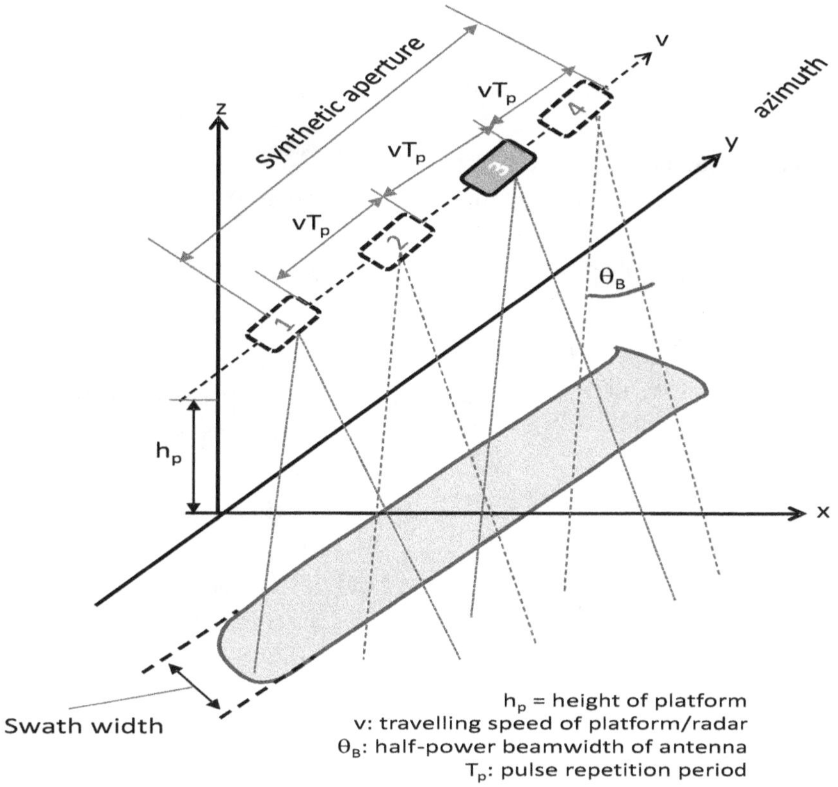

Fig. 11.25 Formation of the synthetic aperture of SAR.

have $\delta_{SAR} = \theta_{SAR} d_R = \frac{D_a}{2} = 2$ m. (c) The SAR is better as it can detect targets which are separated spatially by just 2 m. The SLAR requires a minimum separation as much as 153 m for a 5-km range.

11.7.1 *Check Your Understanding*

Students are encouraged to check their understanding of the content of this section with the following questions prior to proceeding to the next section.

(1) What does the word 'synthetic' in SAR mean?

(2) Discuss the advantages of SAR over conventional radar systems.

(3) Who invented the SAR principle?

(4) Explain how the SAR principle works.

(5) An L-band (i.e. $f = 1.45$ GHz) radar system uses 4-m antenna to track targets with the maximum range of 120 km. Compute the azimuthal resolution if (a) conventional radar operation is used and (b) SAR principle is used. (c) Which of the two types of radar operations is better and why?

(6) What does Doppler beam sharpening mean?

11.8 Probabilities of Detection and False Alarm

A radar's signal can be jammed by an enemy in an attempt to reduce the ability of the radar's receiver to detect a target from the received echo. The radar system can use a combination of a plethora of techniques to avoid a jamming signal. These techniques include *frequency agility* in which the radar changes its operating frequency according to an algorithm or pattern that is uneasy to figure out. Another technique is spread spectrum, in which the radar spreads its transmission over a wide bandwidth so that a jamming signal occupying a smaller bandwidth has no effect on it. In fact, frequency agility is a form of spread spectrum, called frequency hopping spread spectrum. The burn-through range is the distance at which a specific radar can discern targets through the external interference being received.

A radar system can mistake a clutter for a target. It can also mistake a target as a clutter. These are the two types of errors in a radar's operation. The first type is referred to as false alarm, while the latter is called mis-detection. Each of these error types can cause a big problem. For example, if the radar installation used by an army at war mis-detects a friendly force as a foe, then it can result in the killing of its own soldiers. This is catastrophe caused by a false alarm. Also, if an army at war fails to detect an enemy force then it can be defeated. This is catastrophe caused by a mis-detection.

We use the radar receiver circuit illustrated in Fig. 11.26 for the analysis in this section. In this figure, $s(t)$ and $m(t)$ are information-bearing voltage signals, while $n(t)$ and $w(t)$ are voltages of the additive noise. The receiver shown uses an envelope detector, which does not require the phase of the signal. This means that the radar system is doing non-coherent detection of echoes. The drawback in the envelope detector is that it converts any quadrature noise at its input into phase modulation. For this reason, the non-coherent detector requires a higher signal-to-noise ratio to achieve the same detection performance as its equivalent coherent detector.

Radar systems are usually expected to detect targets in the presence of noise or clutter. The noise has two sources: internal noise generated inside the receiver or external noise which together with the useful signal enter the receiver through the antenna. Internal noise originate from thermal mobility of the electrons in the ohmic sections of the receiver, and it is referred to as Johnson noise or thermal noise. The power of the thermal noise at the input of the envelope detector is

$$P_w = kTB_{IF}F = \sigma_w^2 \texttt{ watts} \tag{11.73}$$

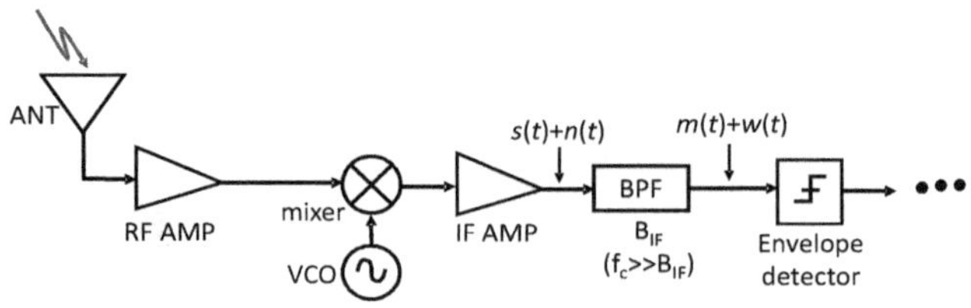

Fig. 11.26 Block diagram of a radar's receiver.

where F is the noise figure of the receiver, and B_{IF} is the noise bandwidth.

The radar's detector uses a threshold logic. Assume that a given voltage threshold is say V_{th}, and the instantaneous voltage is $V(t)$. To decide whether there is a target in the scanned area or not the radar compares $V(t)$ with V_{th}. If the strength $V(t)$ in the received echo exceeds the voltage threshold V_{th}, the radar claims the presence of a target, otherwise not. Unfortunately, noise and clutter can generate echoes with strength exceeding the threshold. Such a situation is referred to as *false alarm*. This is illustrated in Fig. 11.27. In this figure, τ_k is the $k-$th dwell time of the noise voltage $w(t)$ at a level above the threshold voltage V_{th} in the absence of a target, and t_k is the time intervals between a pair of noise envelope voltages which cross the threshold voltage.

The probability density functions (pdf) of the noise envelope voltages shown in Fig. 11.26 are respectively

$$f_n(n) = \frac{1}{\sqrt{2\pi\sigma_n^2}} e^{-\frac{n^2}{2\sigma_n^2}} \tag{11.74}$$

and

$$f_w(w) = \frac{w}{\sigma_n^2} e^{-\frac{w^2}{2\sigma_n^2}} \tag{11.75}$$

where σ_n^2 is the variance of the noise voltage, and w is the amplitude of the envelope of the noise voltage at the output of the audio filter. The radar's decisions are classified into four, as illustrated in Table 11.8.

The false alarm probability or rate is defined as

$$P_{fa} = \frac{\begin{array}{c}\texttt{no. of false targets per}\\ \texttt{pulse repetition time}\end{array}}{\texttt{number of range cells}}$$

$$= Pr\{V_{th} < w < \infty\} = e^{-\frac{V_{th}^2}{2\sigma_n^2}} = e^{-\frac{snr}{2}} \tag{11.76}$$

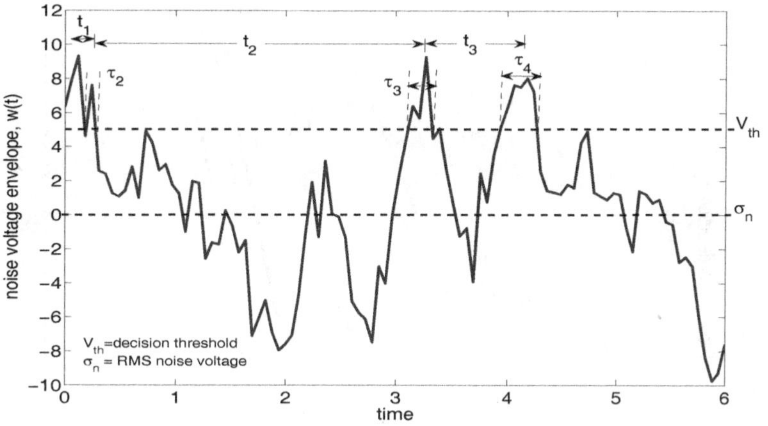

Fig. 11.27 Noise voltage at a radar detector's input.

Table 11.8 Radar decision types.

	Target detected	No target detected
Presence of target	Correct detection (P_d) or true positive	Type II error: missed detection (P_{md}) or false negative
Absence of target	Type I error: false alarm (P_{fa}) or false positive	True negative

where we have defined the signal-to-noise ratio as

$$snr = \frac{V_{th}^2}{\sigma_n^2}. \tag{11.77}$$

We can achieve the false alarms of 10^{-6}, 10^{-8}, 10^{-10} and 10^{-12} at the respective signal-to-noise ratios of about 14.4 dB, 15.7 dB, 16.6 dB and 17.4 dB. The average of the inter-threshold crossing times is referred to as false-alarm time T_{fa}, which is defined as

$$\tau_{fa} = \lim_{M\to\infty} \sum_{m=1}^{M} t_k = \frac{1}{B_{IF}} e^{-\frac{V_{th}^2}{2\sigma_n^2}} = \frac{1}{B_{IF}} P_{fa}\,. \tag{11.78}$$

The false alarm probability is also referred to as *Type I error*. We can increase the threshold V_{th} in order to reduce P_{fa}, but that increases the probability of missed detection P_{md}, which is claiming there is no target when there is one.

We assume that the useful signal $s(t)$ at the input of the IF bandpass filter is a sinusoid with the amplitude β and frequency f_{IF} which is the same as the centre frequency of the filter, i.e. $s(t) = \beta\cos(2\pi f_{IF} t)$. Then, the pdf of the signal $m(t)$ at the output of the envelope detector is a Rician distribution and it is given by

$$f_m(x) = \frac{x}{\sigma_n^2} \exp\left(-\frac{x^2+\beta^2}{2\sigma_n^2}\right) I_0\left(\frac{x\beta}{\sigma_n^2}\right) \tag{11.79}$$

where $I_0(x) = \sum_{k=0}^{\infty} \frac{1}{k!\Gamma(k+1)}(x/2)^{2k}$ is the zeroth-order modified Bessel function (aka hyperbolic Bessel function) of the first kind with the argument x, and $\Gamma(x)$ is the gamma function. The probability of correct detection of a target by the radar is

$$P_d = \frac{\texttt{number of correctly-detected targets}}{\texttt{all possible blips}}$$

$$= Pr\{V_{th} < x < \infty\} = \int_{V_{th}}^{\infty} f_m(x)dx \tag{11.80}$$

which has no closed-form solution. We shall use numerical simulations in MATLAB to plot graphs of P_d. The desirable values lie in the range $P_{fa} < 10^{-6}$ and $P_d > 90\%$.

Assume that $w(t) = w_r(t) + jw_i(t)$ is a complex-valued white random noise with Gaussian distribution, i.e. $w(t) \sim \mathcal{CN}(0, \sigma_w^2)$, and $w_i(t), w_r(t) \sim \mathcal{N}(0, \sigma_w^2/2)$. If the transmitted signal is a pulse and

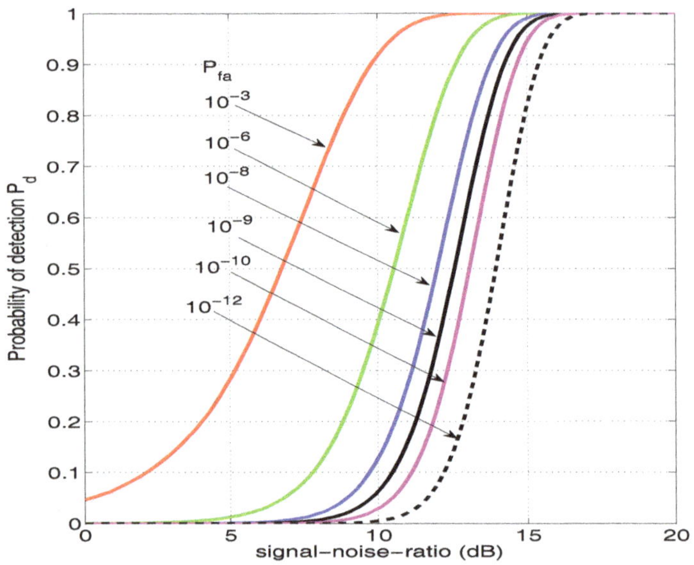

Fig. 11.28 P_d versus P_{fa} and SNR.

the echoes are detected coherently, then the detection probability becomes

$$P_d = \frac{1}{2}[erfc^{-1}(2P_{fa}) - \sqrt{snr}]. \quad (11.81)$$

We can clearly observe the dependence of the detection probability on the false alarm rate. Figure 11.28 compares the P_d at different P_{fa} over the signal-to-noise ratio. We clearly observe that a higher P_d requires a higher P_{fa}, and vice versa. Thus, an optimum trade-off is necessary.

11.8.1 *Check Your Understanding*

Students are encouraged to check their understanding of the content of this section with the following questions prior to proceeding to the next section.

(1) A radar system is liable to the two types of detection errors as it confuses between a clutter and a target: false alarm and misdetection. Explain the difference between the two error types. Also, explain the dependence of one error type on the other.
(2) What is burn-through range of a radar system?
(3) How do we decrease the false alarm rate? What penalties occur if we do so?
(4) Discuss the meaning of non-coherent detection of signals? What advantages and disadvantages are in such a signal detection technique?
(5) Study Fig. 11.28. What SNR does a coherent detector need to achieve a maximum false alarm rate of 10^{-6} and a minimum detection probability of 0.9?
(6) Assume that the noise at the input of a radar's detector has the pdf $f(v) = \frac{1}{v}e^{\frac{-v}{\overline{v}}}$. In this model $\overline{v}$ is the average noise voltage. The threshold voltage above which the radar claims the presence of a target is $V_{th} = 25$ dB. Compute the probability of false alarm P_{fa} if $\overline{v} = 18$ dB.

11.9 Radar Displays and Indicators

A radar system presents the information that it extracts from target on displays or indicators, referred to as scope. There are many types of scopes which are numbered from A-scope through R-scope. In addition there are Range Height Indicator (RHI) and Plan Position Indicator (PPI). The PPI is the commonest display, and it shows the azimuth angle

Table 11.9 Classification of radar displays.

Display Type	Information Displayed	Feature
PPI	Slant range using concentric circles and 0-360° bearing	Circular shape, for weather surveillance
A-scope, M-scope, O-scope & R-scope	Target's range horizontally and echo amplitude vertically.	Instrumentation
B-scope	range on vertical axis and bearing on horizontal axis	Rectangular shape, for tracking of missiles
C-scope & D-scope	azimuth angle on horizontal axis and elevation angle on vertical axis	
E-scope & RHI	Range on horizontal axis and elevation angle or altitude on vertical axis	
F-scope & G-scope	Elevation angle on vertical axis and azimuth on horizontal axis	rectangular
H-scope	Modified B-scope, has range on vertical axis and azimuth on horizontal axis	slope is proportional to elevation angle.
I-scope	target appear as a circle whose radius is the range when the antenna looks at it.	
K-scope & N-scope	Modified A-scope where target appears as pair of vertical deflections.	
P-scope	Measures azimuth angle	

or bearings from 0° to 360° as it monitors the entire space around it. It displays range of targets around its circular shape, and the range is measured from its center outwards. Table 11.9 compares the basic features of radar scopes.

Figure 11.29 illustrates the PPI display. In this figure α is the bearing of the target illustrated. A target is represented as a dot in the display. Such a representation of a target is referred to as *blip*. Actually, targets with strong echoes are shown as bright spots, while targets with weak echoes (smooth and/flat surfaces) are represented by dark areas. The range of a target is measured from the centre to the blip representing the target. The bearing is measured from the true north position in a clockwise direction to the blip.

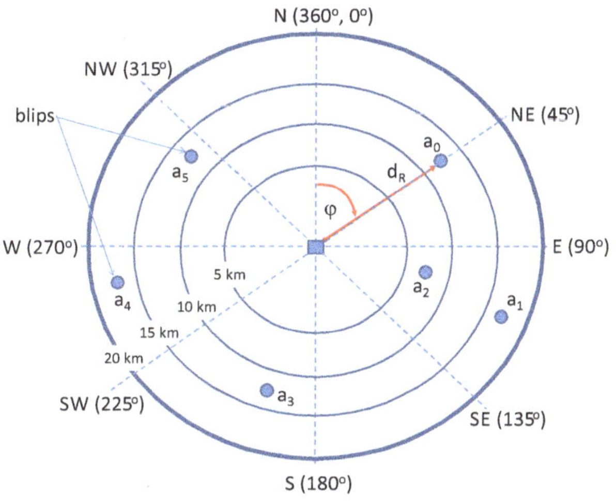

Fig. 11.29 Illustration of plan position indicator display.

11.9.1 *Check Your Understanding*

Students are encouraged to check their understanding of the content of this section with the following questions prior to proceeding to the next section.

(1) With respect to radar displays and indicators, what is a blip?

(2) Estimate the range and bearing of each of the targets shown in Fig. 11.29 as blips a_1 and a_5.

(3) **In-class research:** Research to find how each of the radar display types listed in Table 11.9 operates. Point out their differences and which applications prefer which type of display.

11.10 Radar Signal Processing: Kalman Filtering

Kalman filter, named in honour of Rudolf Kalman, is also referred to as recursive least squares (RLS) filter or information filter. It is a linear, optimal estimator of parameters of interest from indirect, noisy (i.e., inaccurate and uncertain) observations. It is *recursive* so that new measurements can be processed as they arrive, contrasting it from batch processing algorithms requiring the availability of all needed prior to processing. Kalman filter is an *optimal* predictor of parameters because it minimises the mean-squared error of the estimated parameters in Gaussian-distributed noise environment. Actually, the filter's parameters are estimated from the minimisation of the mean-squared error of the parameters of interest. It is also called *filter* as it extracts the target's parameters of interest from clutter. In non-Gaussian-distributed noise environment, the Kalman filter is the best linear estimator. Kalman filtering is popularly used for data prediction and one of its applications is in the estimation of the parameters of a target being tracked by a radar system amidst noise.

Let us explain some terminology used in Kalman filtering. These include recursive, optimal and covariance. Assume that we seek to measure the distance of a target moving at the constant speed of v m/s towards a stationary point. Let us assume that at time t the distance between the stationary point and the moving target is s_t meters, then at time T seconds earlier the distance was s_{t-T} meters. Therefore, the relationship between the current distance and the previous distance is

$$s_t = s_{t-T} - v \cdot T = s_{t-T}[1 - vT/s_{t-T}] = \alpha s_{t-T} \tag{11.82}$$

where $\alpha = 1 - \frac{vT}{s_{t-T}} < 1$. Note that s_t is the current state and s_{t-T} is the previous state of the moving target. In the following, we view time as a sequence of discrete steps, each of duration T, and refer to time with n instead of the continuous time variable t. We can thus rewrite Eq. (11.82) as

$$s_n = \alpha s_{n-1}\,. \tag{11.83}$$

Equation in the form (11.82), in which a current value depends on the previous value, is referred to as *recursive equation* or *recursion*. Recursion is the verb form of the word *recur*, meaning *come back* or *return* to previous value. The formula (11.82) can be used, for example, to measure the distance of an arriving plane from the airport. Kalman filter gives the best estimate of the data of interest in the mean-squared error sense by minimising the mean squared error between the estimated and the actual data. This is what is referred to as *optimal*.

Let us interpret s_n as the parameter of a target under surveillance which we seek to estimate using a radar system. Let w_n be an additive noise sample with mean m_n and standard deviation σ_n. The noise w_n models any inaccuracy in the measuring process caused by both measuring error and systematic error. Then, the parameter estimation problem can be written as

$$x_n = \alpha_n s_n + w_n,\ n = 0, 1, 2, \ldots \tag{11.84}$$

where α_n is a scaling factor. The parameter n is discretised time. Kalman filtering has a wide range of applications, something which is attributed to the following features:

(1) Kalman's method of finding the optimal mean-squared error (MSE) filter exploits state space techniques, an approach which enables the filter to be used for the three purposes:
   (a) A predictor. This is the feature used in radar signal processing to estimate the parameters of targets.
   (b) A filter.
   (c) A smoother.
(2) The use of state space techniques to define the Kalman filter results in a simple implementation of the filter in the discrete domain.
(3) Kalman's method of finding the optimal MSE filter does not need to compute the impulse response of the filter.

11.10.1 *Heuristic Mean-Squared Error*

Let the estimate of s_n in Eq. (11.84) be $\hat{s}_n$. The objective in Kalman filtering is to find $\hat{s}_n$ with the minimum possible error. We define the error term as $e_n = s_n - \hat{s}_n$. We seek to express the error term as a positive and monotonically increasing function.

The square of the error term fulfills both condition. Thus,

$$g(e_n) = (s_n - \hat{s}_n)^2 \, . \tag{11.85}$$

A generalised solution is obtained by operating on the average value of $g(e_n)$, which leads to the mean-squared error (MSE) equation

$$E[g(e_n)] = E[e_n^2] = E[(s_n - \hat{s}_n)^2] \, . \tag{11.86}$$

Minimising Eq. (11.86) leads to the minimum mean-squared error (MMSE) estimation of Kalman filter.

11.10.2 *Maximum Likelihood Mean-Squared Error*

An alternative way of deriving the Kalman filter parameters is via the maximum likelihood estimator, which is the estimate $\hat{s}_n$ of s_n which maximises the likelihood or the probability s_n, namely $\max\{Pr[x_n|\hat{s}_n)]\}$. If $w_n \sim \mathcal{N}(0, \sigma_n^2)$, then we have

$$\max\{Pr[x_n|\hat{s}_n)]\} = \beta_n \exp\left[-\frac{(x_n - \alpha_n \hat{s}_n)^2}{2\sigma_w^2}\right]$$

which leads to the **maximum likelihood function**

$$\begin{aligned} Pr[x|\hat{s}] &= \prod_n \max\{Pr[x_n|\hat{s}_n]\} \\ &= \prod_n \beta_n \exp\left[-\frac{(x_n - \alpha_n \hat{s}_n)^2}{2\sigma_n^2}\right] \end{aligned} \tag{11.87}$$

where β_n is a normalisation constant. Taking the logarithm of Eq. (11.87) we obtain the log-likelihood function

$$\log Pr[x|\hat{s}] = -\frac{1}{2}\sum_n \frac{(x_n - \alpha_n \hat{s}_n)^2}{\sigma_n^2} + \log(\beta_n) \, . \tag{11.88}$$

The filter which minimises the mean-squared error term in Eq. (11.88) is the optimal filter being searched for.

11.10.3 *State Space Derivation of Kalman Filter*

Let us define the following parameters that we will use in the ensuing Kalman filtering algorithm:

- $\mathbf{s}_n \in \mathcal{R}^{N\times 1}$: state vector of the process at time n.
- $\hat{\mathbf{s}}_n \in \mathcal{R}^{N\times 1}$: estimate of the state vector $\mathbf{s}_n$ at time n.
- $\hat{\mathbf{s}}'_n \in \mathcal{R}^{N\times 1}$: previous estimate of the state vector $\mathbf{s}_n$.
- $\mathbf{A} \in \mathcal{R}^{N\times N}$: state transition matrix of the process from the state at time n to the another state at time $n+1$.
- $\mathbf{w}_n \in \mathcal{R}^{N\times 1}$: sample at time n of unknown white noise process whose covariance matrix $Q = E[\mathbf{w}_n\mathbf{w}_n^T]$ is known.
- $\mathbf{v}_n \in \mathcal{R}^{M\times 1}$: random vector of measurement error at time n with the covariance matrix $\mathbf{R} = E[\mathbf{v}_n\mathbf{v}_n^T]$.
- $\mathbf{z}_n$: measured data at time n.
- $\mathbf{B} \in \mathcal{R}^{M\times N}$: models the noise-free relationship between the state vector $\mathbf{s}_n$ and the measured data $\mathbf{z}_n$. It is assumed to be stationary over time.
- $\mathbf{P}_n \in \mathcal{R}^{N\times N}$: error covariance matrix at time n.

The covariance matrices $\mathbf{Q}$ and $\mathbf{R}$ are assumed to be stationary over time. We define the state vector of a random process as

$$\mathbf{s}_{n+1} = \mathbf{s}_n + \mathbf{w}_n, \; n = 1, 2, \ldots \tag{11.89}$$

We seek to estimate the state vector $\mathbf{s}_n$ from the measured data

$$\mathbf{z}_n = \mathbf{B}\mathbf{s}_n + \mathbf{v}_n, \; n = 1, 2, \ldots \tag{11.90}$$

where $\mathbf{v}_n$ is the measurement noise. The error covariance matrix at time n is defined as

$$\mathbf{P}_n = E[\mathbf{e}_n\mathbf{e}_n^T] = E[(s_n - \hat{s}_n)(s_n - \hat{s}_n)^T] \, . \tag{11.91}$$

If $\hat{\mathbf{s}}'_n \in \mathcal{R}^{N\times 1}$ is the earlier estimate of the state vector $\mathbf{s}_n$, then we can formulate the updating equation

$$\hat{\mathbf{s}}_n = \hat{\mathbf{s}}'_n + K_n(\mathbf{z}_n - \mathbf{B}\hat{\mathbf{s}}'_n) = \hat{\mathbf{s}}'_n + K_n\mathbf{i}_n, \; n = 1, 2, \ldots \tag{11.92}$$

where K_n is called **Kalman gain** and

$$\mathbf{i}_n = \mathbf{z}_n - \mathbf{B}\hat{\mathbf{s}}'_n \tag{11.93}$$

is referred to as the measurement residual or the innovation. Substituting $\mathbf{z}_n$ in Eq. (11.90) into Eq. (11.92) yields

$$\hat{\mathbf{s}}_n = \hat{\mathbf{s}}'_n + K_n(\mathbf{z}_n - \mathbf{B}\hat{\mathbf{s}}'_n) = \hat{\mathbf{s}}'_n + K_n(\mathbf{B}\mathbf{s}_n + \mathbf{v}_n - \mathbf{B}\hat{\mathbf{s}}'_n) \, . \tag{11.94}$$

Substituting Eq. (11.94) into Eq. (11.91) gives

$$\begin{aligned} \mathbf{P}_n = E[\mathbf{e}_n\mathbf{e}_n^T] = E[\{(\mathbf{s}_n - \hat{\mathbf{s}}'_n)(\mathbf{I} - K_k\mathbf{B}) - K_n\mathbf{v}_n\} \\ \times\{(\mathbf{s}_n - \hat{\mathbf{s}}'_n)(\mathbf{I} - K_n\mathbf{B}) - K_n\mathbf{v}_n\}^T] \, . \end{aligned} \tag{11.95}$$

Noting that the error term $\mathbf{s}_n - \hat{\mathbf{s}}'_n$ is uncorrelated with the measurement noise $\mathbf{v}_n$, Eq. (11.95) becomes

$$\mathbf{P}_n = E[\mathbf{e}_n\mathbf{e}_n^T] = (\mathbf{I}-K_n\mathbf{B})\mathbf{P}'_n(\mathbf{I}-K_n\mathbf{B})^T+K_n\mathbf{R}K_n^T \tag{11.96}$$

where $\mathbf{R} = E[\mathbf{v}_n\mathbf{v}_n^T]$ is the covariance matrix of the measurement error at time n $\mathbf{v}_n$, and

$$\mathbf{P}'_n = E[(\mathbf{s}_n - \hat{\mathbf{s}}'_n)(\mathbf{s}_n - \hat{\mathbf{s}}'_n)^T] \tag{11.97}$$

is the earlier estimate of $\mathbf{P}_n$.

Let us note the following features of $\mathbf{P}_n$:

(1) It is the updating equation for the error covariance matrix.
(2) Its diagonal elements σ_{kk}^2 are the mean square errors $\sigma_{kk}^2 = E[e_k e_k^T]$.
(3) The trace of a matrix is the sum of the elements in its main diagonal. The trace of $\mathbf{P}_n$, which is written $tr[\mathbf{P}_n]$, is the sum of the mean-squared errors. This means that minimising the trace minimises the MSE. This is what we are exploiting in the following.

We need to find the trace of $\mathbf{P}_n$, differentiate it with respect to the Kalman gain K_k and then set the results to zero to find the optimum Kalman gain. Expanding Eq. (11.96) yields

$$\begin{aligned}\mathbf{P}_n = \mathbf{P}'_n - K_n\mathbf{B}\mathbf{P}'_n - \mathbf{P}'_n\mathbf{B}^T K_n^T \\ + K_n(\mathbf{B}\mathbf{P}'_n\mathbf{B}^T + \mathbf{R})K_n^T .\end{aligned} \tag{11.98}$$

The trace of any matrix $\mathbf{U}$ is equivalent to the trace of its transpose, i.e., $tr[\mathbf{U}] = tr[\mathbf{U}^T]$. We exploit this identity now. Thus,

$$\frac{\partial tr[\mathbf{P}_n]}{\partial K_n} = -2(\mathbf{B}\mathbf{P}'_n)^T + 2K_n(\mathbf{B}\mathbf{P}'_n\mathbf{B}^T + \mathbf{R}) \stackrel{!}{=} 0 \tag{11.99}$$

resulting in the **optimum Kalman filter gain**

$$K_k^* = (\mathbf{B}\mathbf{P}'_n)^T(\mathbf{B}\mathbf{P}'_n\mathbf{B}^T + \mathbf{R})^{-1} = (\mathbf{B}\mathbf{P}'_n)^T\mathbf{S}^{-1} \tag{11.100}$$

where $\mathbf{S} = \mathbf{B}\mathbf{P}'_n\mathbf{B}^T + \mathbf{R}$ is the measurement prediction covariance matrix associated with the measurement residual (aka innovation). Now, we need to substitute the optimum gain Eq. (11.100) into Eq. (11.96) or Eq. (11.98) to obtain

$$\mathbf{P}_n = (\mathbf{I} - K_n^*\mathbf{B})\mathbf{P}'_n . \tag{11.101}$$

Equation (11.101) is the updating equation for the error covariance matrix with the optimal Kalman filter gain. We compute the estimate $\hat{\mathbf{s}}_n$ of the state variable $\mathbf{s}_n$ using the three equations (11.101), (11.100) and (11.92) after observing $\mathbf{z}_n$.

Lastly, we need to find equation for the state projection which enables us to compute the current state from past states, or estimate the future state from the current state. Let

$$\hat{\mathbf{s}}'_{n+1} = \mathbf{A}\hat{\mathbf{s}}_n . \tag{11.102}$$

We use the projection of the error term into the future in order to find the projection of the error covariance matrix into the future. That is

$$\begin{aligned}\mathbf{e}'_{n+1} &= \mathbf{s}_{n+1} - \hat{\mathbf{e}}'_{n+1} \\ &= (\mathbf{A}\mathbf{s}_n + \mathbf{w}_n) - \mathbf{A}\hat{\mathbf{s}}_n \\ &= \mathbf{A}\mathbf{e}_n + \mathbf{w}_n .\end{aligned} \tag{11.103}$$

With Eq. (11.103), we can define the updating

$$\begin{aligned}\mathbf{P}'_{n+1} &= E[\mathbf{e}'_{n+1}\mathbf{e}'^{\,T}_{n+1}] \\ &= E[(\mathbf{A}\mathbf{e}_n + \mathbf{w}_n)(\mathbf{A}\mathbf{e}_n + \mathbf{w}_n)^T] \\ &= \mathbf{A}\mathbf{P}_n\mathbf{A}^T + \mathbf{Q}\end{aligned} \tag{11.104}$$

where we have noted in the derivation of Eq. (11.104) the noise $\mathbf{w}_n$ occurs between the transition from time n to time $n+1$ while the error term $\mathbf{e}_n$ is accumulated up to time n. Thus, $\mathbf{w}_n$ and $\mathbf{e}_n$ are orthogonal in time and therefore not correlated.

Summary of Kalman Filter

Given the observed data $\mathbf{z}_n = \mathbf{B}\mathbf{s}_n + \mathbf{v}_n$, $\mathbf{B}$ and covariance matrix of the noise $\mathbf{v}_n$, Kalman filter estimates the state $\mathbf{s}_n$ using the steps:

(1) Compute the Kalman gain:
$$K_n = \mathbf{P}'_n\mathbf{B}^T(\mathbf{B}\mathbf{P}'_n\mathbf{B}^T + \mathbf{R})^{-1}.$$
(2) Update the estimate:
$$\hat{\mathbf{s}}_n = \hat{\mathbf{s}}'_n + K_n(\mathbf{z}_n - \mathbf{B}\hat{\mathbf{s}}'_n).$$
(3) Update the covariance matrix:
$$\mathbf{P}_n = (\mathbf{I} - K_n\mathbf{B})\mathbf{P}'_n.$$
(4) Projection into the future state:
$$\hat{\mathbf{s}}'_{n+1} = \mathbf{A}\hat{\mathbf{s}}_n \text{ and } \mathbf{P}_{n+1} = \mathbf{A}\mathbf{P}_n\mathbf{A}^T + \mathbf{Q}$$

where $\mathbf{Q} = E[\mathbf{w}_n\mathbf{w}_n^T]$ and $\mathbf{R} = E[\mathbf{v}_n\mathbf{v}_n^T]$. Note that the initial value $\mathbf{P}'_0$ cannot be the zero matrix. It can be set to the identity matrix.

11.10.4 *Check Your Understanding*

Students are encouraged to check their understanding of the content of this section with the following questions prior to proceeding to the next section.

(1) Discuss the differences and similarities between the MMSE estimator (11.86) and the maximum likelihood estimator (11.87).
(2) Discuss the reasons why Kalman filter is popular in estimation and target tracking.
(3) Prove the validity of Eq. (11.95).
(4) Derive Eq. (11.96) to prove its validity.
(5) Prove the validity of Eq. (11.101).
(6) Derive Eq. (11.104).

11.11 Review of Important Terms and Acronyms

radar IoT	pulsed radar	continuous-wave radar
ladar	radar cross-section	pulse repetition rate
pulse width	pulse duration	pulse repetition time
duty cycle	pulse integration	Doppler radar
echo	backscatter	re-radiation
monostatic radar	bistatic radar	radial velocity
range	range ambiguity	range resolution
range bins	scattering region	blip
Kalman filter	angular resolution	antenna beamwidth
moving target indicator	chirp pulse	Doppler shift
pulse compression	false alarm	synthetic aperture radar
Nyquist frequency	Nyquist velocity	superheterodyne receiver

11.12 Review Questions and Problems

(1) What is the meaning of the acronym radar? That is, write radar in full.
(2) Who coined the acronym *radar* and at what time?
(3) Assume that the radar cross section (RCS) of a target being monitored is $\sigma = 50\text{m}^2$. We want to know how far this object can be away from the radar installation and still be detectable by the radar. Compute the maximum range of the radar described above when: peak transmit power $P_t = 320$ W, transmit frequency $f = 10.25$ GHz, antenna gain $G = 41$ dBi, total noise figure $F = 3$ dB, $T = 290$ K operating temperature, Boltzmann's constant $k = 1.38 \cdot 10^{-23}$ J/K, radar receiver's bandwidth $B = 200$ kHz, and the minimum signal-to-noise ratio at the output of the radar's detector is 20 dB. Assume no losses in the system.
(4) A pulse radar uses modulating pulses of 20-microsecond duration and pulse repetition frequency of 2 kHz.
   (a) What should be the optimum bandwidth B to be used by the system?
   (b) What is the cycle time of the pulsed radar?
   (c) Calculate the duty cycle of the pulse.
   (d) What should be the appropriate maximum slant range (R) of the radar system in order not to confuse echoes from a target with pulses generated by the radar?
   (e) Assuming that we are using a bistatic radar system, what should be the minimum range that a target can be and still be detectable by the radar?
   (f) The peak radiated power from the radar is $P_t = 25$ kW. What is the average transmitted power?
(5) A pulsed radar system operating at an airport is expected to detect the range and the position of aircrafts within 96 km radius by the microwaves reflected from the aircraft's surface back to the antenna of the radar system. Assume that T_p is the pulse repetition period, τ is the width of the modulating pulse, R is the slant range of the radar (i.e. maximum distance of a detectable target from the radar), and t_R is the pulse's round-trip time, i.e. the time taken by pulse to hit the target and return to the radar as echo.
   (a) What should be the appropriate pulse repetition period (Tp) of the radar system in

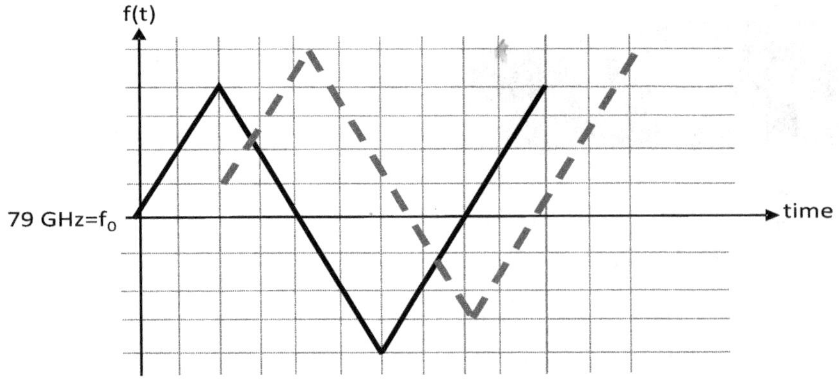

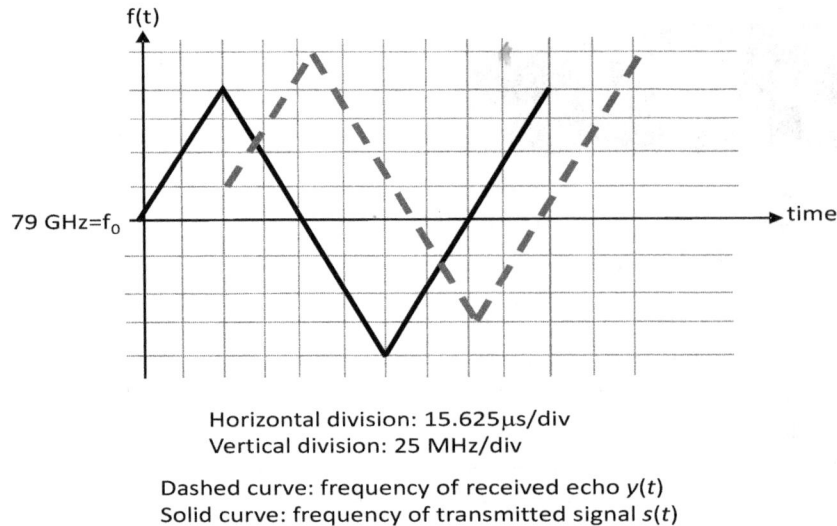

Fig. 11.30 Frequency waveforms for FMCW problem.

order not to confuse echoes from the target with pulses generated by the radar? Assume $\tau = 10^{-6}$ s.

(b) The peak radiated power from the radar is $P_t = 25$ kW. What should be the appropriate width of the modulating pulse in order to obtain an average power of 2.1 kW at the output of the radar's antenna?

(c) Calculate the duty cycle of pulses used by the pulsed radar?

(6) What is range folding in pulsed radar operation, and how can it be mitigated?

(7) Discuss the nature and causes of Doppler velocity aliasing in continuous-wave radar operation, and how it can be avoided.

(8) What does the term Doppler dilemma mean?

(9) A pulsed monostatic pulsed radar uses pulses of duration 5 μs. The duplexer takes 0.4 μs to switch between transmit and receive antennas. This time is referred to as recovery time. What is the minimum distance that an airplane must be away from the radar for it to be detected unambiguously?

(10) Consider the frequency waveforms shown in Fig. 11.30 which is used for by W-band FMCW radar to provide safety in land-based automobile.

(a) Sketch the beat frequency below the given curves.

(b) Compute the Doppler shift in the received echo.

(c) Estimate the round-trip time (RTT) of the echo signal t_d.

(d) Is the target moving away or towards the radar system?

(e) At what radial velocity has the target been travelling?

(f) How far away is the target from the radar?

(11) Figure 11.31 shows two planes, one arriving (Fig. 11.31 (a)) and another departing (Fig. 11.31 (b)) an airport. Assume that they are monitored by the primary airport radar system operating at 2.8 GHz (i.e., S-band), 25 kW peak and 2.1 kW average transmit powers.

(a) Assume that the radar system used at the airport is a pulsed radar. Compute the duty cycle of the pulse used by the radar system.

(b) If a Doppler frequency of 432.0988 kHz is measured by the radar, at what speed is each of the two airplanes travelling?

(c) Compute the carrier frequency of each of the signals which reflect off the airplanes and return to the radar.

(d) Assume that the radar cross section of each of the airplanes is 100 m^3, radar antenna gain is 20 dB, equivalent noise temperature is 23 °C, 200 MHz radar system

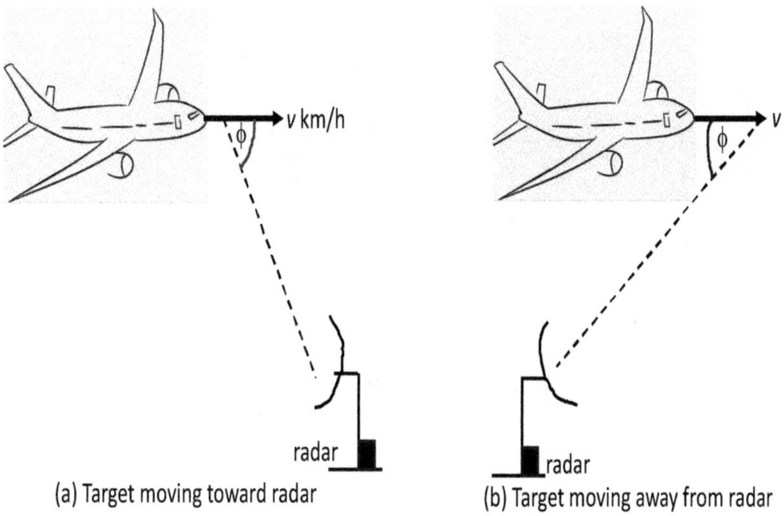

Fig. 11.31 Tracking of aeroplanes at the airport ($\phi = pi/6$).

bandwidth, 8 dB radar receiver noise figure and eight pulses are integrated. If the range of the airport surveillance radar system is 96 km, compute the minimum signal-to-noise ratio required for the radar system to operate as expected.

(12) The S-band pulsed radar system DEFENDER S850/S1000, manufactured by Enterprise Electronics Corporation (EEC), operates at 2700–3000 MHz. The radar uses pulses of width 0.2–2.0 *mu*sec and pulse repetition frequency 200–2400 Hz. Using high-power coaxial magnetron, the radar achieves a peak transmit power of 850 kW/1000 kW. The antenna with 8.5-mm diameter has the half-power beamwidth of 0.95°, minimum gain of 45 dB and maximum scanning speed of 10 rpm.

(a) Compute the range of the radars pulse repetition period.
(b) Compute the maximum number of echoes that the radar integrates in each scan cycle.
(c) Compute the (i) minimum and the (ii) maximum range of the radar system.

11.13 Supplementary Reading

(1) Merrill I Skolnik, *Radar Handbook*, McGraw-Hill Education, 2008.
(2) Bassem R. Mahafza, *Radar Systems Analysis and Design Using MATLAB*, Chapman and Hall/CRC, 2013.
(3) Merrill I Skolnik, *Introduction to Radar Systems*, McGraw-Hill Education, 2002.
(4) Mark A. Richards (ed.), James A. Scheer (ed.) and William A. Holm (ed.), *Principles of Modern Radar: Basic Principles*, SciTech Publishing, 2010.
(5) William L. Melvin (ed.), James A. Scheer (ed.), *Principles of Modern Radar: Advanced Techniques*, SciTech Publishing, 2012.

Chapter 12

Satellite Communications

Chapter's Learning Outcomes

Wireless communications enables us to transfer information between two entities and/or locations over the ether. There are two types of wireless communications: terrestrial wireless communications with all communications entities located on the Earth, and satellite communications (satcom) with a segment hanging in the sky. Although satcom requires line-of-sight between space segment and Earth stations, it has many advantages over terrestrial wireless communications. For example, its wide coverage or footprint makes it the largest wireless wide area network. This feature positions satcom to be the backbone for Internet of Things (IoT). Also, the cost of transferring data over satellites is independent on distance. The functions of a satellite depends on the equipment it carries. For example, a TV satellite carries a TV station. This chapter presents the fundamental principles in satcom. Several case studies of modern operational satellite systems for mobile IoT, television broadcasting, navigation and direction finding, surveillance, environmental monitoring, etc. are also studied. The presentation in this chapter is specific to satcom, making it a good introduction into the subject matter. In particular, topics studied include:[1]

(1) Basic principles in satellite communications, including types of orbits and orbital laws, such as Kepler's laws of planetary motion and Isaac Newton's laws of universal gravity.
(2) Satellite applications and use cases, such as satellite IoT and mobile satellite.
(3) Noises affecting satellite link and the satellite link budget.
(4) Major satellite manufacturers, launching companies, launching methods and vehicles, and best launching sites.
(5) Satellite Internet of Things (IoT) and how artificial satellites and drones are enabling the realisation of a full-scale IoT.
(6) Case studies of popular mobile satellites enabling mobile communications and IoT.

12.1 Introduction

What is a satellite? National Aeronautics and Space Administration (NASA) of USA government defines the orbit and the satellite as

> An **orbit** is a regular, repeating path that one object in space takes around another one. An object in an orbit is called a **satellite** or space vehicle (SV).

We have natural satellites and artificial satellites (aka man-made satellites). Examples of natural satellites are the planets (Earth, Mars, Neptune, Jupiter, Venus, Uranus, Saturn, Mercury), comets and asteroids in the solar system which revolve around the Sun. The moon is a natural satellite which orbits the Earth. Natural satellites are studied in Astronomy. Information transfer, which is the objective of this book, however, studies artificial satellites.

The principle of satellite communications (satcom) is illustrated in Fig. 12.1. An artificial *satellite*

[1] A prudent person doesn't act without sufficient knowledge and reflection.

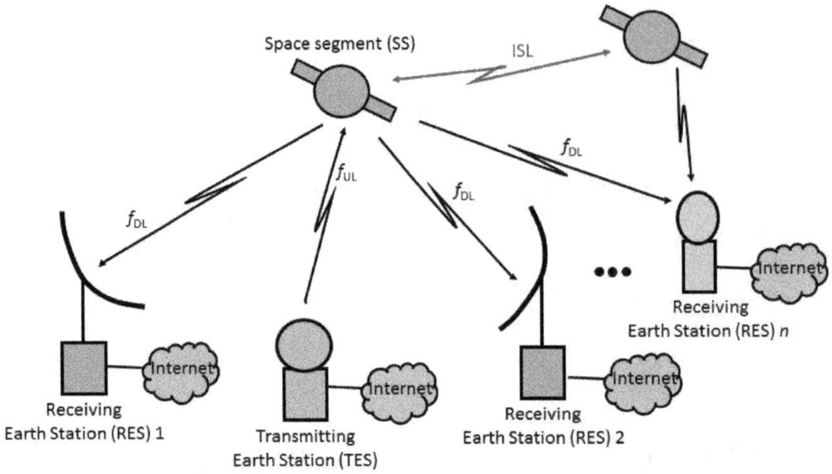

Fig. 12.1 The principle of satellite communications (ISL: inter-satellite link).

system comprises a space segment (SS) and two or more earth stations. We refer to the space segment as **satellite**. How does the ground station(s) communicate with the space segment? A third component called *Tracking, Telemetry and Control* (TTC), is used for that purpose. Thus, a satellite system has actually three main components. The earth stations are radio transceivers located on the surface of the Earth, while the space segment hangs in the sky. Usually, one of the earth stations serves as the transmitting station, where signals are sent to the space segments. The space segment reflects signals so that other earth stations can receive them. Today, many automobiles are delivered with factory-fitted GPS satellite receivers, which receive directions from GPS satellites. There are also GPS receivers sold in isolation.

The entire geographical locations which can detect the transmissions of a single satellite in space is called the **satellite's footprint**. Figure 12.2 illustrates a satellite's footprint. The higher the altitude of the space segment, the larger the footprint of the satellite. This is the reason why ideally only three geostationary satellites can cover the entire Earth.

Satcom has a wide range of applications. The satellite can carry different types of equipment. In fact, the applications of a satellite depends on the type of equipment it carriers. Satellite are currently used for Internet access, remote sensing, weather forecasting, TV broadcasting, navigation (e.g., GPS), environmental monitoring, etc. Satellite communications has both advantages and disadvantages. Its disadvantages include:

(1) Long signal propagation delay, referred to as latency, owing to the long path. Low-earth orbit (LEO) has the shortest latency, while geostationary earth bit (GEO) and geosynchronous orbit (GSO) have the longest latency.
(2) Large path loss owing to the long path.
(3) High chance of unsuccessful launch into orbit.
(4) Challenges in repairing a faulty satellite in orbit.
(5) Disposal of expired satellites.

Fig. 12.2 The footprint of a satellite.

The main advantages in satcom is the large footprint of one satellite. The higher the altitude, the larger the footprint. For this reason, a satellite system can cover every spot on the Earth. Second, the cost of transmitting signals over a satellite link does not depend on the length of the link. This is not so for wired communications links. For example, the data rate of an xDSL link depends on the distance from the service provider's closest switch. Third, satellite systems can operate in locations where, for any reason, terrestrial communications systems cannot.

12.1.1 *Satellite Applications and Satellite IoT*

Satcom has a wide range of applications. The functions of a satellite depends on the equipment (referred to as its *payload*) that it carries. For example, a TV satellite carries a TV station. Also, a weather forecasting satellite carries equipment which can measure large-scale storms, tornadoes, clouds, hurricanes, etc. and broadcasts their photos to the Earth. Of course, a satellite can carry multiple equipment to serve multiple purposes. This is often the case. A satellite can be used where there are no roads and electricity, such as in remote areas, mountain tops, deep valleys and during disasters.

The applications of artificial satellites can be categorised into four, as illustrated in Fig. 12.3. They are signal reflection, security, Earth observation and scientific studies. Signal reflection is used for communication, broadcasting, positioning, navigation and timing (PNT). National security comes under monitoring, something which is now very important so we have separated it from other services. Earth observation is also a form of monitoring. ECHELON is operated by the alliance UKUSA which is a security agreement between United Kingdom, USA, Australia, Canada and New Zealand. This agreement is also referred to as **Five Eyes**. Figure 12.3 also gives examples of satellite systems used to support each use case.

According to the September 2018 report written by BRYCE Space and Technology for the Satellite Industry Association (SIA) [Bryce (2018)] there were 1,738 artificial on-orbit operational satellites, and that, on the average, 272 satellites are launched into orbits each per year since 2013. The distribution of the satellites according to their use cases is shown Fig. 12.4, while their distribution according to their size and weights is shown in Fig. 12.5,

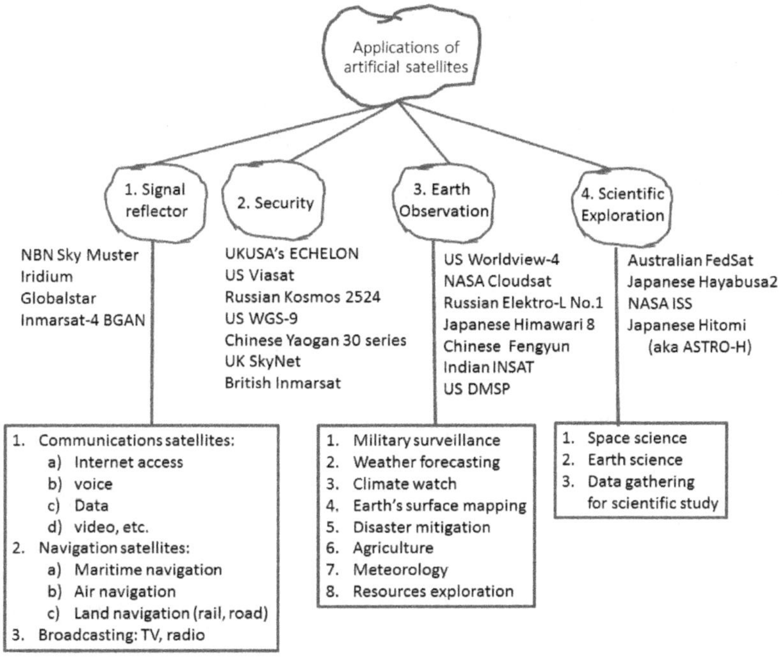

Fig. 12.3 Satellite communications applications (BGAN: Broadband Global Area Network; FedSat: Australia's Federation Satellite).

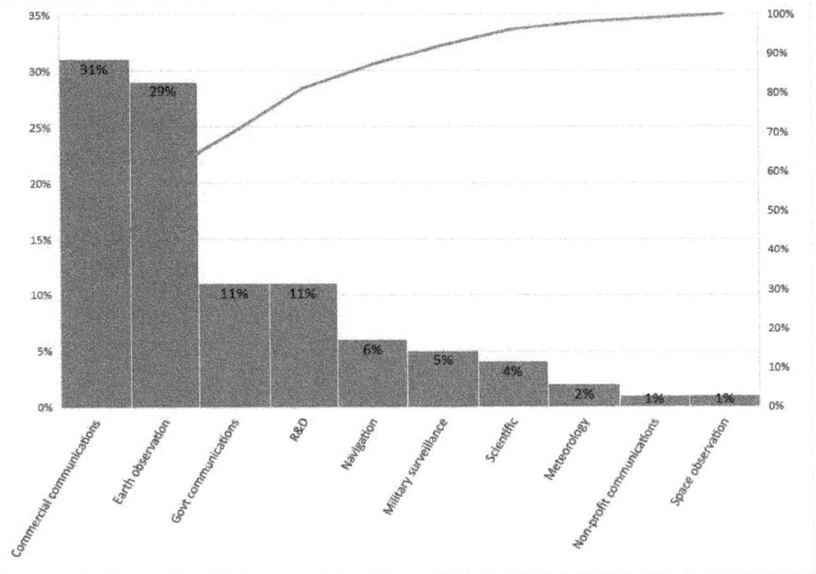

Fig. 12.4 Classification of on-orbit satellites by their use cases.

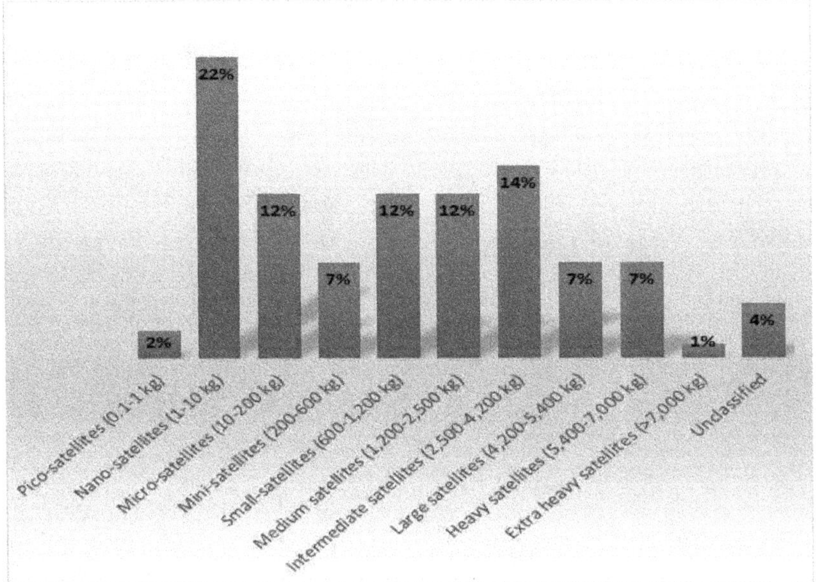

Fig. 12.5 Classification of on-orbit satellites by their weights or sizes.

respectively. It is certainly impossible to describe each of the many active satellites. Therefore, we discuss general principles of satellite communications and study a few of them as case studies in the course of our discussions.

According to the report [Bryce (2018)], the satellites weighing less than 1.2 tonnes are mostly placed on LEO orbits. There are 531 GEO satellites among the 1,738 on-orbit satellites, and the USA alone operates 803 of the satellites. The same report found that 10% of on-orbit satellite anomalies occur within two months of being launched. For this reason, several organisations have launched business cases to develop satellite mission extension vehicles (MEV) and Space Drones which can be used to refuel, repair and even assemble satellites in space. These organisations include Northrop Grumman, Effective Space, DLR Deutsche Orbitale

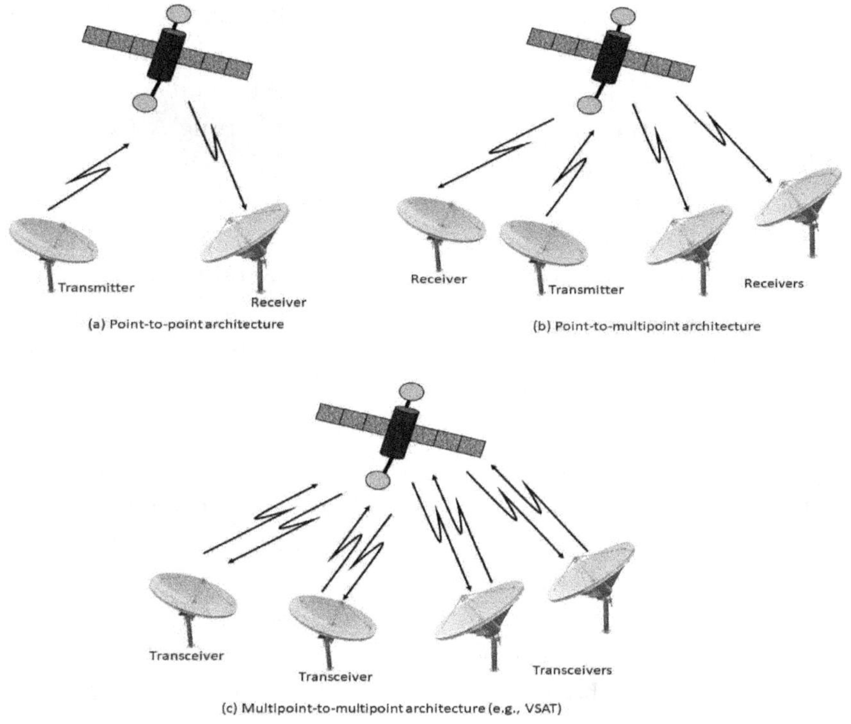

Fig. 12.6 The three types of satellite configurations.

Servicing Mission (DEOS), Restore-L, and NASA GSFC Robotic Refueling Mission on International Space Station.

There are three types of satellite services:

- Point-to-point (P2P): only one earth station feeds the satellite with signals. The satellite, on the other hand, relays signals to only one earth station. This configuration is illustrated in Fig. 12.6(a).
- Point-to-multipoint (P2M): only one earth station feeds the satellite with signals. The satellite, however, relays signals to two or more earth stations. This configuration is illustrated in Fig. 12.6(b).
- Multipoint-to-multipoint (M2M): multiple earth stations feed the satellite with signals. The satellite, also, relays signals to multiple earth stations. This configuration is illustrated in Fig. 12.6(c). A VSAT architecture is M2M as all earth stations must be able to transmit and receive signals.

Satellite services can be broadly categorised into fixed satellite service (FSS), broadcast satellite service (BSS), mobile satellite service (MSS) and satellite-enabled Internet of Things (referred to as satellite IoT). Among these the killer application as at 2018 is satellite IoT. IoT is a disruptive technology. For this reason, we separate this application of satellite and discuss it in the next section.

12.1.2 *Satellite IoT and Mobile Satellite Case Studies*

Gartner's report claims that there will be more than 500 IoT-enabled smart devices by 2022. These smart devices include cars, kitchen appliances, electricity meters in smart grid, land, sea and air vehicles and sensors in a sensor network. The currently anticipated killer application in satellite communications is satellite IoT. Traditional satellite operators are upgrading their systems to support IoT. There are also new entrants starting off with IoT as their main business case. Notable satellite IoT initiatives include:

(1) **Google Project Loon** providing Internet access in remote and rural locations, and disaster zones. This is a network of solar-powered balloons filled with Helium and air mixture

and floating in the stratosphere by winds with speeds between 5 and 20 mph. Solar energy recharges the batteries on board which are deployed in the night. The footprint of each balloon from the altitude of about 10 to 60 km is about 40 km in diameter. The lifespan of each balloon is about two years. Loon operates in the 2.4 and 5.8 GHz ISM bands.

(2) **OneWeb Satellites**: this is a partnership project between Airbus and OneWeb to launch 900 LEO satellites to provide world-wide high-speed Internet access. For more details see https://onewebsatellites.com/about-us/

(3) **Facebook Aquila drones for Internet access**: this project intended to use solar-powered drones to deliver high-speed Internet services in remote areas. A drone is an unmanned aerial vehicle (UAV) or airplane, which is controlled by a ground station. Facebook terminated this project in June 2018 to join Airbus HAPS project using the so-called atmospheric satellites (aka pseudo-satellites).

(4) **Mustang project**: this is a hybrid IoT as it aims to combine LEO satellites and Sigfox terrestrial communications technologies to support worldwide communications. It is being developed as a partnership between the three France-based organisations Sigfox, Sysmeca and CEA-Leti research institute, and Airbus Defense and Space. Mustang operates in the 868 MHz and 915 MHz ISM bands. The French government supports Mustang.

(5) **Thuraya satellite**: owned by Yahsat (with Mubadala as its parent company) in United Arab Emirates and Hughes, Thuraya currently uses five satellites to serve over 160 countries. The first, called Al Yah 1, was launched in April 2011 on a geosynchronous **orbit *with*** 35,802 km apogee altitude, located at 52.5°E, 11.6 kW payload and operates in C-band, Ku-band and Ka-band. The second is Al Yah 2, launched in 2012, located at 47.5°E, 9.7 kW payload and operates in Ka-band. The third satellite in operation is Al Yah 3, launched in 2018, located at 20°W, 7.5 kW payload and operates in Ka-band. The first two satellites were manufactured by EADS Astrium & Thales Alenia Space with the third manufactured by Orbital ATK but launched by French's Arianespace SA using Ariane 5 ECA rocket.

(6) **Starlink**: with the SpaceX as its parent company and owned by Elon Musk, Starlink is a LEO satellite network under development to provide broadband, reliable wireless Internet access and Earth observation worldwide. One spectacular feature of Starlink is that its satellite will return to the Earth after their lifetime rather than becoming a space junk, a traditional way of dumping expired satellites. The satellites are launched using SpaceX's Falcon 9 rocket launch vehicle. The orbital altitude is 550 km, but can range between 330 and 580 km. Each satellite weighs about 260 kg. SpaceX intends to launch about 12,000 satellites by mid-2020s, with 122 of them already deployed in November 2019. The total investment is about US$10 billion.

Owing to its wide coverage, reliability, integration with land networks, costs comparable with terrestrial wireless and growing data rates, satellites are needed to realise the large-scale IoT. Of particular importance for satellites are areas where terrestrial system are unavailable for any reason.

Inmarsat has now partnered with some public land mobile network (PLMN) operators, such as Vodafone, to provide mobile cellular and satellite IoT services globally. The Inmarsat BGAN M2M terminal is used to realise machine-to-machine management of fixed and mobile assets. Globecomm satellite operator partnered with AT&T in 2016 to deliver satellite IoT services called industrial IoT (IIoT) for asset management. Assets being managed by IIoT include oil pipelines, tractors on farms or a generator used in a desert.

12.1.2.1 *Iridium Internet of Things*

Iridium is the name of the chemical element with atomic number 77. This satellite system was so named as it originally intended to run 77 satellites. It now has 66 active satellites in six orbital planes covering almost every spot in the world, making

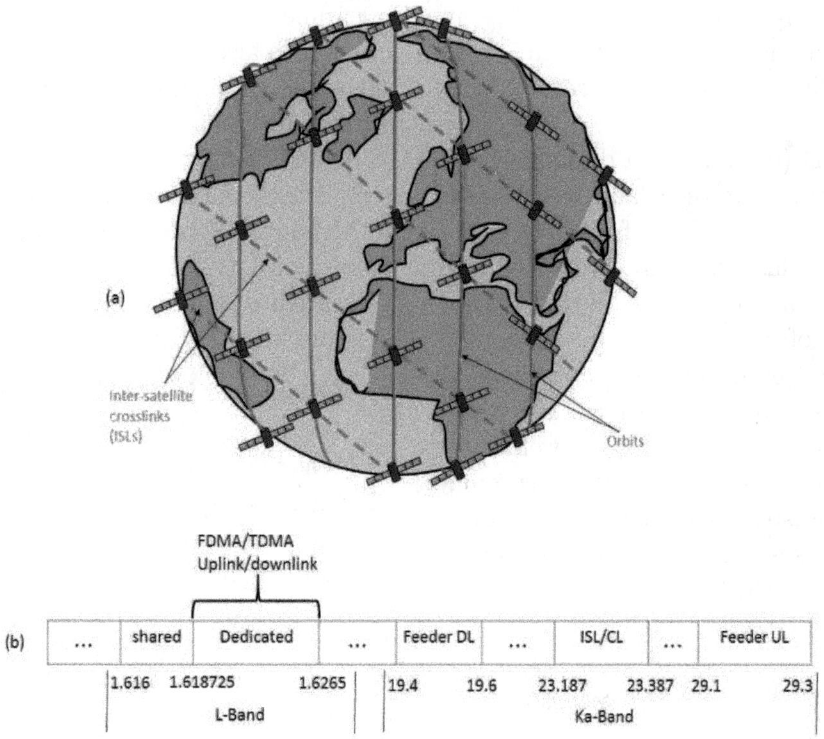

Fig. 12.7 The Iridium satellite system: (a) configuration, (b) frequency bands.

it the largest commercial satellite network in the world. The second generation fleet of Iridium satellite, called Iridium NEXT, were launched using the SpaceX Falcon 9 rocket. With the Iridium NEXT fleet, Iridium serves as the space-borne IoT backbone. Iridium can then be used to track ships and airplanes, track assets, manage fleet, remote monitor things, and command and control 'things' everywhere in the world in the air, land and sea. Iridium NEXT has machine-to-machine (M2M) communication capability. Iridium has developed Iridium Edge operating in the band 1.616–1.6265 GHz, an IoT solution, to support these use cases.

Iridium has an interesting history. The parent company started in mid-1980s and became formerly registered in 1991, but became operational commercially not until 1 November 1998. After about 18 years it is easy to understand that the technologies involved could already be dated. Iridium struggled for enough subscriptions. This was partly due to the cost of subscription. A minute phone call was billed US$6 to US$30, and the handset was sold at about US$3000. Iridium eventually filed Chapter 11 bankruptcy in Aug 1999. Luckily, the government bailed it out in December 2000. This bailout allowed investors to buy it at about only US$20 million, although its assets were worth about US$6 billion.

The Iridium constellation is illustrated in Fig. 12.7. Iridium uses the cellular architecture with the frequency reuse factor of 12. Each satellite has 48 spot beams, and each beam serves one cell, resulting in a cluster size of 48. Its total system capacity is 172,000 users. This system combines both FDMA (frequency-division multiple access) and TDMA (time-division multiple access) multiple access schemes. Iridium has four inter-satellite links (ISL), called cross links. Communication is handed off between satellites over these ISLs in the frequency band 23.187 to 23.387 GHz at the data rate of 25 mbps, as shown in Fig. 12.7(b). There are 66 active satellites in six orbital planes. There are also 15 spare satellites: six already in-orbit and nine on the ground. Each in-orbit satellite is connected to four other satellites: two ISLs in the same orbital plane and one in each adjacent orbital plane. Thus, Iridium is a space-borne partial mesh network of satellites, as illustrated in Fig. 12.7(a). The ISLs enable traffic switching between the satellites in space. Therefore, an Iridium connection can work without the availability of a terrestrial communications service.

Iridium uses TDD (time-division duplexing) for a direct communication between land-based users and the space segment over the L frequency band 1,616–1,626.5 mHz. This 10.5 mHz is divided into 240 channels which are separated by 41.67 kHz, each of bandwidth of 31.5 kHz. The remaining bandwidth is used as guard bands to accommodate Doppler shifts caused by the relative motion between a satellite and a ground station, as well as for inter-modulation products. There is also a communication between the space segment and the gateway to the terrestrial system using FDD (frequency-division duplexing) in which 29.1–29.3 GHz is the uplink and 19.1–19.6 GHz is the downlink.

12.1.2.2 *INMARSAT S EAN Satellite Systems*

Headquartered in London, UK, International Maritime Satellite Organisation (INMARSAT) was launched in 1979 under the directive of the International Maritime Organization (IMO). The IMO is the United Nations (UN) body in charge of maritime industry. INMARSAT has extended its mission to also cover aeronautical communications, such as communications between airplanes and ground stations.

INMARSAT S EAN is an S-band satellite system launched for the European Aviation Network (EAN) in 2017 to support communications on airplanes with terrestrial networks. The footprint of the satellite covers Europe and Middle East. Service provided by INMARSAT S EAN are:

(1) Mobile and fixed broadband services which support email, Web browsing, videoconferencing, telemedicine and live broadcasting.
(2) Satellite telephone with voice quality comparable with that of PSTN/POTS.
(3) Uses a high-gain antenna (HGA) to deliver symmetric IP data communications of up to 432kbps per background channel. Four channels per aircraft can be aggregated to achieve 2.8 mbps data rate.
(4) Always-on IP-based streaming service with $2^n \times$ 8 kbps data rate, where $n = 0, 1, 2, 3, 4$.
(5) Satellite system connected to on-board Wi-Fi network and terrestrial network based on Long Term Evolution (LTE) technology.

Table 12.1 Basic features of the Optus A-series and B-series satellites.

Feature	A-Series System	B-Series System
Physical structure	Cylindrical (Hughes HS 376)	Cubic body with solar wings (Hughes HS 601)
Dimensions	6.4m by 2.2m in diameter	21 m across extended solar panels
Dry weight	540 kg	1,270 kg
Stabilisation method	Spin stabilised	3 axis body stabilised
Solar power supply	1,000 W (at end of life)	3,000 W (at end of life)
Battery capacity	Full operation during eclipse	Full operation during eclipse
Geostationary life	10 years (A3), 7 years (A2)	13 years
Inclined life extension	5 years (nominal)	5 years (nominal)
Number of transponders	15 Ku-band	15 linearised Ku-band, 1 L-band
Transponder bandwidth	45 mHz	54 mHz
Transponder power	12 W (11 transponders), 30 W (4 transponders)	50 W (Ku-band transponder), 150 W (L-band transponder)
Communications payloads	Ku-band communications	Ku-band and L-band communications, UPC beacon (Ku-band), Ka-band beacon

Table 12.2 Basic features of the Optus-10 satellite.

Feature	Value
Manufacturer	Space Systems Loral
Geostationary lifespan	15 years
Inclined life extension	5 years (nominal)
Dry mass[2]	1.677 tones
Dimensions	24.8 m across extended solar panels
Orbit	GEO
On-orbit stabilisation method	3 axis momentum stabilised
Payload	Ku band repeaters (i.e., 24 linearised transponders each with 8 switchable beams) with switchable FSS and BSS
Ground control stations	Belrose in Sydney (main) and Lockridge in Perth
Location of satellite	152° E to 164° E with ±0.07° tolerance
Earth station antenna diameter	Up to 7 m
Satellite antenna	2× 2.4 m offset Gregorian-shaped reflectors
Number of transponders	24 in Ku band
Bandwidth per transponder	36 mHz
FSS band (divided into 24 40-MHz channels)	14.0-14.5 GHz (uplink), 12.25-12.75 GHz (downlink)
BSS band (divided into 24 40-MHz channels)	17.3-17.8 GHz (uplink), 11.7-12.2 GHz (downlink)
Communications payload	Ku-band communications, Ku-band UPC beacon
Transponder power	Up to 133 W
Battery	Full operation during eclipse

FSS: Fixed Satellite Service; BSS: Broadcast Satellite Service

The INMARSAT S EAN was built by Thales Alenia Space in France. It was launched into orbit in 2017 by Arianespace Space SA, headquartered in France, using rocket launcher.

12.1.2.3 *OPTUS Satellite System*

The National Broadband Network (NBN) is the network being currently deployed in Australia to replace the legacy PSTN/POTS to deliver high-speed multimedia Internet access. The NBN has both terrestrial wireless and satellite wireless components. In 2014, the NBN Co. subcontracted SingTel Optus, a telecom operator in Australia, to supply the satellite component.

The OPTUS satellite system is a domestic satellite system serving the entire Australia and New Zealand. The system can also cover Norfolk Island, Cocos Island, Papua New Guinea,Lord Howe Island and Christmas Island. The OPTUS satellite system comprises the six satellites: B-series (i.e., B3), C-series (i.e., C1), D-series (i.e., D1, D2 and D3) and Optus-10. The A-series and B-series satellites were manufactured by Hughes Aircraft Company in the United States of America. The C-series and Optus-10 satellites were manufactured by SSL (Space Systems Loral), a subsidiary of Maxar Technologies. The D-series satellites were manufactured by a USA company called Orbital Sciences Corporation, which merged with Alliant Techsystems in 2014 to form Orbital ATK, Inc. Some features of the A-series and B-series satellites are summarised in Table 12.1, while Table 12.2 summarises some data of the Optus-10 satellite.

[2] *Dry mass* or dry weight of a vehicle is the mass of its hardware without any consumable such as fuel (propellant) and oxidizer. The consumables altogether is referred to as *wet mass*. Note that the wet mass of a satellite reduces as as soon as it is releases from the launching vehicle and turns to zero ideally at the end of its lifespan. The dry mass, however, remains the same in the ideal case.

Example

Using the data in Table 12.2 of Optus-10 satellite, compute:

(1) Beamwidth, θ, of the space segment antennas.
(2) Gain, G, of the space segment antennas. Assume that the antenna aperture efficiency is $\eta \approx 0.7$.

Solution

The middle uplink frequencies are $f_{\mathtt{fss}} = 14.25$ GHz for the FSS and $f_{\mathtt{dss}} = 17.55$ GHz for DSS services, respectively. The equivalent wavelengths are $\lambda_{\mathtt{fss}} \approx 0.0211$ and $\lambda_{\mathtt{dss}} \approx 0.0171$, respectively.

(1) The antennas mounted on the satellite are two Gregorian-shaped reflectors. Both the Gregorian and the Cassegrain are types of parabolic antennas. The beamwidth θ in degrees, the wavelength λ in meters and the diameter D in meters of a parabolic antenna are related by

$$\theta \approx \frac{70\lambda}{D} = \frac{70c}{Df}. \tag{12.1}$$

From Table 12.2 we read $D = 2.4$ m. Plugging these values into Eq. (12.1), we obtain

$$\theta_{\mathtt{fss}} \approx 0.61^\circ \ \texttt{and} \ \theta_{\mathtt{dss}} \approx 0.5^\circ .$$

(2) The parabolic antenna gain is given by

$$G = \eta\left(\frac{\pi D}{\lambda}\right)^2 \approx 0.7\left(\frac{\pi D}{\lambda}\right)^2. \tag{12.2}$$

Plugging the appropriate values into Eq. (12.2), we obtain

$$G_{\mathtt{fss}} \approx 49.5 \ \texttt{dB and} \ G_{\mathtt{dss}} \approx 51.3 \ \texttt{dB}.$$

We observe that the beamwidths are small in order to obtain a high gain to compensate for the large path loss endured by the long link.

12.1.3 *Satellite Multiplexing and Multiple Access Schemes*

Multiplexing and multiple access schemes are used for several purposes as discussed in the chapter dedicated to the topic. Satcom uses the following schemes:

- Permanently assigned multiple access (PAMA) scheme.
- Demand assigned multiple access (DAMA).
- Single channel per carrier (SCPC).
- Multiple channel per carrier (MCPC).
- FDMA (e.g., in GLONASS).
- CDMA (e.g., in Globalstar).

For example, Globalstar uses CDMA with 16.5 MHz channel bandwidth. Odyssey combines CDMA and FDMA. It divides a 16.5 mHz bandwidth into three FDMA channels, each with bandwidth 4.83 mHz. CDMA is applied on each of the three FDMA channels. Iridium combines TDMA with FDMA. The TDMA uses 90-ms frame comprising four times slots. PAMA, DAMA, SCPC and mCPC are not discussed in Chapter 7. Therefore, we review them here.

12.1.3.1 *PAMA and DAMA*

In DAMA the satellite uses a pool of transmission resources (i.e., times slots, frequencies or codes) to serve a pool of terminals. Resources are allocated to terminals in a dynamic manner following a request by the terminals. Terminals use slotted-aloha (s-aloha) method to request resources in the uplink. DAMA is simple and it is employed in very small aperture terminals (VSAT).

In PAMA the satellite network assigns transmission resources to terminals on a permanent basis like a leased line. This means that the resources allocated to terminals are wasted when they are idle, reducing resource utilisation efficiency. PAMA, also referred to as fixed assigned multiple access or FAMA), is therefore more suited for satellite systems serving a few terminals (about 10), which have predictable traffic (e.g., broadcast satellite communication services) and stable Earth station to satellite link conditions.

While a contention scheme (e.g., slotted-aloha) is used to request channels in DAMA, PAMA is contention-free. The contention-free feature enables the control channel access and transmission delays. Channels in either FAMA or PAMA can be based on frequency, time slots, codes, etc. They are allocated to gateway terminal, satellite cells (for cellular satellite systems) and satellites.

12.1.3.2 *SCPC and MCPC*

In SCPC each earth terminal is allocated a dedicated carrier frequency with a fixed bandwidth. SCPC is usually used for point-to-point links in mesh topology. It is good for high-capacity, always-on traffic such as TV and radio broadcasting. However, link capacity is wasted if the terminal cannot use the entire bandwidth available on the channel. SCPC is actually a leased line in satellite communications.

The multiple channel per carrier (MCPC) builds upon SCPC by allowing multiple sources or users to share a single frequency using time division multiplexing (TDM). MCPC thus improves SCPC in resource utilisation efficiency. Multiple frequency (MF) TDMA (MF-TDMA) applies TDMA over multiple frequencies. All terminals share a pool of frequencies or channels. In this scheme no resources are allocated to a terminal on a fixed basis. Thus, no resource is wasted unlike in SCPC. However, MF-TDMA uses guardbands which reduce resource utilisation efficiency.

How can we exploit the high efficiency in SCPC and the flexibility in MF-TDMA. The answer is with Mx-DMA (cross-dimensional multiple access). This scheme changes the coding, modulation, transmit power, carrier frequency and symbol rate in real time in response to channel conditions, traffic demand, and the network QoS management. Mx-DMA uses SCPC principle to allocate resources to terminals.

12.1.4 *Satellite Modulation Techniques*

Satellite communications systems use simple modulation schemes such as BPSK, QPSK, 8-PSK and 16-QAM. For example, Odyssey, Iridium and Globalstar use QPS. These schemes are discussed in the chapter on modulation and demodulation methods.

12.1.5 *Satellite Frequency Bands*

The signal transmission direction from the earth station to the space segment is called *uplink* or *reverse link*, while the opposite direction is called *downlink* or *forward link*. The same terminology is used in terrestrial wireless communications.

Satellite communications signals must be able to penetrate the ionospheric layer of the atmosphere. This has effect on the operating frequency. Signals transmitted at too low frequencies (or too large wavelengths) are reflected back to Earth by the ions in the ionosphere, and thus cannot reach the space segment. The atmospheric attenuation of radio signals increase with frequency. This means that the signals will be too weak to reach the satellite if we use too high frequencies. An optimum trade-off between these two extremes is needed.

Unlike in terrestrial wireless communications, the uplink frequency is always higher than the downlink frequency in satellite communications. This can be observed in Table 12.3. Why? In order to keep the weight of the space segment as low as possible, high-power transmitters are not used. Also, antenna size reduces with increasing frequency. So, a smaller antenna can be used to receive signals at high frequencies. This means that we need to use lower frequencies to achieve a desirable coverage. This is the reason why the transmission from the space segment uses a lower frequency.

12.1.6 *Check Your Understanding*

Students are encouraged to check their understanding of the content of this section with the following questions prior to proceeding to the next section.

(1) **In-class group discussions**: In groups of 3 to 4 students research to find two use cases of satellite IoT which are not mentioned in this book. Preferably, propose new use cases of satellite IoT.

(2) Discuss the difference between wet mass and dry mass of an artificial satellite for communications.

(3) **In-class group research**: In groups of 2 to 4 students research the meaning and examples of the following expressions.

   (a) Disruptive technology.

   (b) Killer application.

Table 12.3 Frequencies for satellite communications.

Band	Uplink Frequencies (GHz)	Downlink Frequencies (GHz)	Sample Applications
L (1-2 GHz)	1.6265-1.6605 (Inmarsat), 1.62135-1.6265 (Iridium)	1.525-1.559 (Inmarsat), 1.62135-1.6265 (Iridium)	GPS, Iridium, INMARSAT, Thuraya, Globalstar (UL), Odyssey (UL)
S (2-4 GHz)	2.048 and 5.005	2.225	Galileo global navigation satellite system (GNSS), Globalstar (DL), Odyssey (DL)
C (4-8 GHz)	5.925-6.675	3.7-4.2	Intelsat-20
X (8-12 GHz)	7.9-8.395	7.25-7.745	Military use, radar applications for air traffic control, vehicle speed detection, maritime vessel traffic control, weather monitoring and defense tracking
Ku (12-18 GHz)	FSS: 14.0-14.8, DSB: 17.3-18.1 (Europe); FSS: 14.0-14.5, DBS: 17.3-17.8 (America)	FSS: 10.7-11.7, DBS: 11.7-12.5 (Europe); FSS: 11.7-12.2, DBS: 12.2-12.7 (America)	DBS and FSS
Ka (17-40 GHz)	27.5-31.0	17.7-21.2	MSS, BSS, radars on military aircrafts

FSS: fixed satellite service using GEO; DBS: direct broadcast service; BSS: Broadcasting-Satellite Service; MSS: Mobile-Satellite Service; EAN: European Aviation Network; UL/DL: Up-/down-link

Relate your discussions to artificial satellites.

(4) In your own words define:
   (a) An orbit.
   (b) Satellite.
(5) State the mathematical formula relating the frequency and wavelength of a wireless signal. State the unit for each parameter in the equation.
(6) List the three broad services provided by artificial satellites.
(7) Discuss the reason why the uplink frequency is higher than the downlink frequency in satellite communications, which is the opposite in terrestrial mobile wireless communications.
(8) Explain the reason why satellite communications requires a line of sight (LOS) between the ground stations and the satellites. Give an example of situations in which you encountered a problem to receive satellite signals because of the lack of LOS.
(9) Compare the need for the ground station antenna to track the satellite in LEO, MEO and GEO systems.
(10) List the two conditions limiting the frequencies used for satellite communications.
(11) **In-class research**: in groups of 2 to 4 students research to find the:
   (a) Number of satellites currently deployed in space.
   (b) Number of spare satellites in orbits, but not deployed.
   (c) Applications of artificial satellites in your country of residence.
   (d) Companies and regulatory body in satellite communications in your country of residence.
(12) **In-class group research**: In groups of 2 to 4 students research the meaning and examples of the following expressions:
   (a) Fixed satellite service (FSS).
   (b) Mobile satellite service (MSS).
   (c) Broadcast satellite service (BSS).

12.2 Satellite Orbits and Orbital Laws

A communications satellite in space is not static but moves around the Earth. The path along which

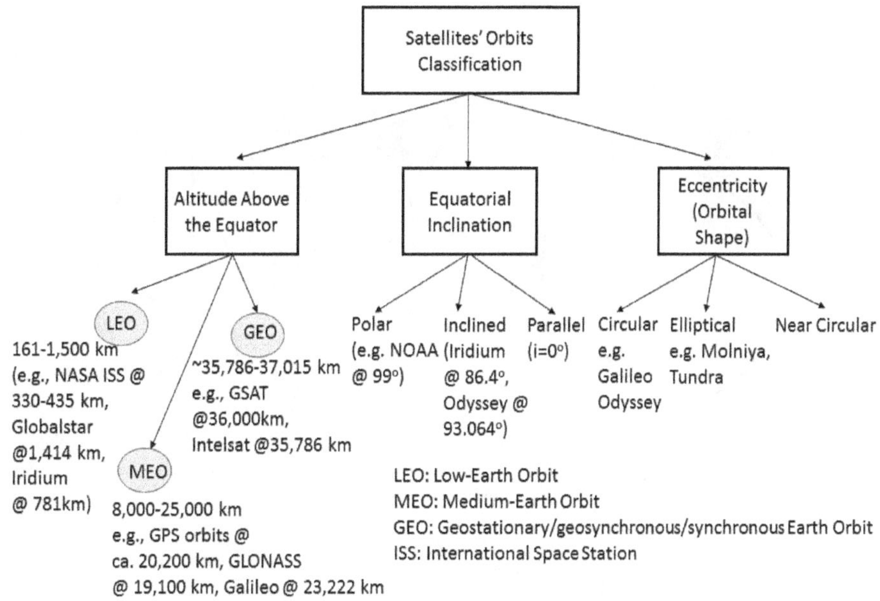

Fig. 12.8 Classification of satellite orbits.

a satellite moves is called **orbit**. Figure 12.8 attempts to classify satellite orbits along the three lines: altitude above the Earth surface, inclination to the equator, and its shape. With respect to its vertical distance above the earth surface, satellite orbits can be classified into geosynchronous orbit (GSO), medium-earth orbit (MEO) and low-earth orbit (LEO). With respect to its shape, satellite orbits can be classified into circular orbits and elliptical orbits. Satellite systems with circular orbit include Globalstar with 52° inclination, Odyssey with 55° inclination and Iridium with 86.4° inclination. Those with elliptical orbits include Molniya and Tundra.

12.2.1 *Measurement of Time*

Time measurements in satellite communications requires a proper understanding of bearings. If we use the position of the sun in the sky relative to the Earth to measure time then we have **solar time**. The **solar day** is the time taken by the sun to move from a particular location and return to it. It lasts 24 hours. The **sidereal day** is the time taken by the Earth to complete one revolution about its axis relative to fixed distant stars. The sidereal day is about four minutes shorter than the solar day. It lasts 86,164 s or about 23.93444 h. The **sidereal period** of a satellite is the time it takes to return to the same position relative to fixed distant stars. The universal time (UT) is the mean solar time measured in reference to longitude 0°.

12.2.2 *Key Parameters of Satellites Orbit*

In the following, we assume that the centre of the orbit coincides with the centre of the Earth, which is point C in Fig. 12.9. There are nine key parameters which describe the orbit of a satellite. These are called *classical orbital elements*. They are:

(1) Semi-major axis.
(2) Semi-minor axis.
(3) Perigee.
(4) Apogee.
(5) Anomaly.
(6) Ascending and descending nodes.
(7) Argument of perigee.
(8) Inclination.
(9) Eccentricity.

Most of these features are basic features of the geometrical figure of ellipse. Figure 12.9 shows an elliptical orbit. In the two-dimensional Cartesian

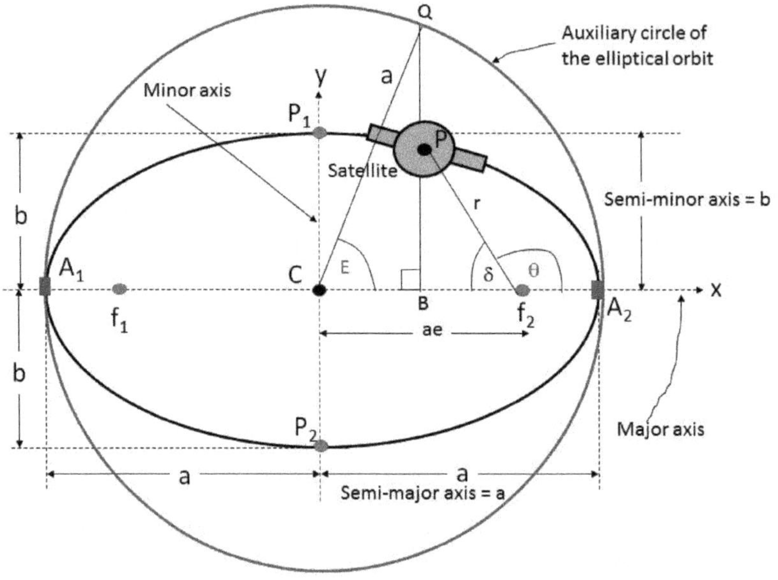

Fig. 12.9 Parameters of a satellite orbit.

coordinate system the ellipse can be described by the equation

$$1 = \frac{x^2}{a^2} + \frac{y^2}{b^2}, \quad a \geq b > 0 \,. \tag{12.3}$$

The linear **eccentricity** of an ellipse is defined as the distance from its centre to one of its two foci divided by the distance from that focus to a vertex. Using the parameters of the ellipse in Fig. 12.9 the eccentricity is defined analytically as

$$e = \frac{|Cf_2|}{|Pf_2|} = \frac{\sqrt{a^2 - b^2}}{a} \,. \tag{12.4}$$

If the eccentricity of an orbit is zero then it is a circular orbit as $a = b$ in Fig. 12.9. The area A_e and the circumference C_e of an ellipse with the eccentricity e are respectively

$$A_e = \pi ab \text{ and } C_e = 4a \int_0^{\pi/2} \sqrt{1 - e^2 \sin^2(\phi)} \,. \tag{12.5}$$

Note that $C_e = 2\pi a$ if $e = 0$ which is the circumference of a circle with radius a.

The semi-major axis, semi-minor axis, perigee and apogee are illustrated in Fig. 12.9. The **perigee** is the point in the orbit where the satellite is closest to the Earth (i.e., points P_1 and P_2 in Fig. 12.9). The **apogee** is the point in the orbit where the satellite is farthest away from the Earth (i.e., points A_1 and A_2 in Fig. 12.9). Thus, the lowest and highest altitudes of a non-circular orbit are the perigee and apogee, respectively. A satellite in orbit passes through the equatorial plane at two positions. The passage from the northern to the southern hemisphere is called **descending node**. The passage from the southern to the northern hemisphere, however, is called **ascending node**, Ω, as illustrated in Fig. 12.10(c). The angle between the perigee and the ascending node is referred to as the **argument of perigee** (see Fig. 12.10(d)).

Anomalies are angles used to calculate the position of an object moving along an elliptical orbit. There are four types of anomalies: true anomaly θ, eccentric anomaly E, elliptic anomaly (aka intermediate anomaly) and mean anomaly M. An orbit for which all the orbital elements are constant except the mean anomaly is called **Keplerian orbit**. As can be observed in Fig. 12.9, we need either the true anomaly or the eccentric anomaly to find the position P of a satellite on its orbit. Note that

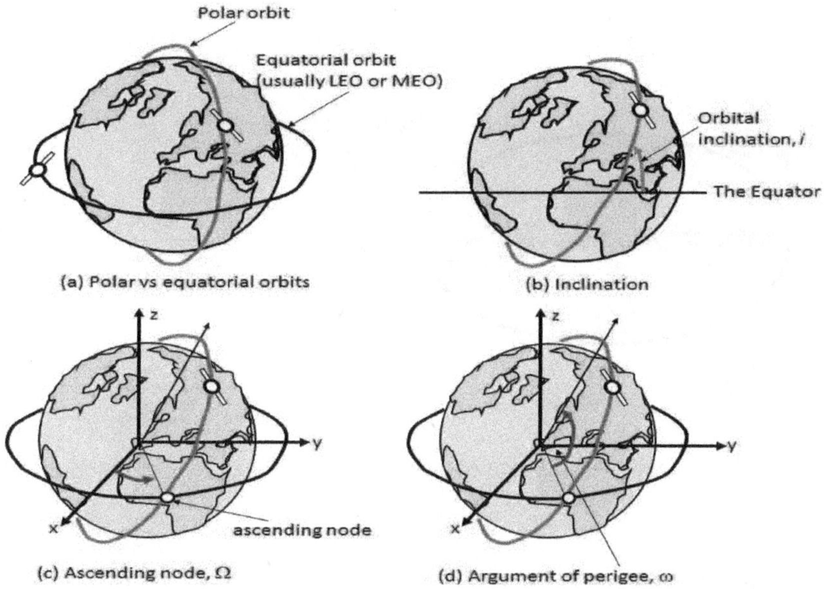

Fig. 12.10 Some types of orbits and orbital elements.

the length of the arc C-Q in Fig. 12.9 equals the semi-major axis a. Also, the length of the arc $P-f_s$ equals the orbital radius r. In the following we use some geometrical tricks to compute the eccentric anomaly E and the true anomaly θ.

From Fig. 12.9 we obtain

$$|Bf_2| = |Cf_2| - |CB| = ae - a\cos(E)\,. \tag{12.6}$$

Also,

$$\begin{aligned}|Bf_2| &= r\cdot\cos(\delta) = r\cdot\cos(\pi-\theta)\\ &= -r\cdot\cos(\theta)\,.\end{aligned} \tag{12.7}$$

Equating (12.6) and (12.7) yields

$$\cos\theta = \frac{a(\cos E - e)}{r}\,. \tag{12.8}$$

The elliptical equation in polar form is

$$r = \frac{a(1-e^2)}{1+e\cos\theta} = \frac{a(1-e)(1-e)}{1+e\cos\theta}\,. \tag{12.9}$$

At the apogee $\theta=\pi$, resulting in $r=a(1+e)$. At the perigee $\theta=0$, resulting in $r=a(1-e)$. Substituting Eq. (12.8) into Eq. (12.9) and simplifying results in

$$E = \cos^{-1}\left(\frac{a-r}{ae}\right)\,. \tag{12.10}$$

The equation relating the mean anomaly (M) and the eccentric anomaly (E) is

$$M = E - e\sin(E) = n(t-t_{\mathrm{p}}) \tag{12.11}$$

where t is called *epoch time* (i.e., the time at which the orbital elements are observed), t_{p} is the time instant that the satellite passes the perigee, and

$$n = \sqrt{\frac{GM}{a^3}} = \frac{2\pi}{T} \tag{12.12}$$

is called the **mean motion constant**. Equation (12.11) is called **Kepler's equation**.

The angle made by the orbital plane of a satellite with the Earth's equatorial plane is called **inclination** and denoted i, as illustrated in Fig. 12.10(b). With respect to the inclination, satellite orbits are classified into four: polar orbit with 90° inclination, **equatorial orbit** or parallel orbit with 0° inclination, **retrograde orbit** with inclination in the range 90° to 180°, and **prograde orbit** with inclination in the range 0° to 90°. Polar orbits are usually LEO orbits. A polar orbit passes over or near both the north and south poles. Satellites on polar orbits are used for weather monitoring (e.g., USA NOAA series, EU Metop series and Chinese F1-series), communications (e.g., Iridium), etc. They have an altitude of about 1,000 km and a period of 12 hours. Highly-inclined orbits are used to cover locations with high latitudes.

An orbit is called **Sun-synchronous orbit** if it fulfills the two special features:

- almost a polar orbit, and

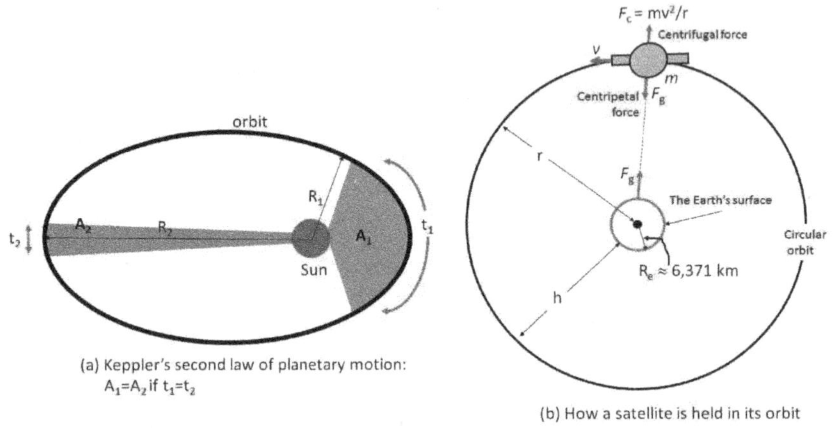

Fig. 12.11 Orbital mechanics: r is orbit radius, h is orbit altitude, R_e is Earth radius, v is satellite velocity, m is satellite mass, F_c is centrifugal force, F_g is gravitational force.

- a satellite orbiting it passes over a given location on the Earth's surface at the same local solar mean time.

12.2.3 *Orbital Equations, Kepler's Laws and Newton's Laws*

We review the forces and laws affecting satellites in this section. In the following, let us note the definitions:

- r: radius of orbit in km.
- h: orbital altitude in km.
- T: orbital period in seconds/minutes/hours.
- $R_e \approx 6,371$ km, 6,378 km at the Equator: Earth radius.
- v: satellite orbital velocity in m/s, which is tangential to the orbit.
- m: satellite mass in kg.
- $G \approx 6.673 \times 10^{-11}$ $\mathtt{Nm^2/kg^2}$: Earth's gravitational constant.
- $M \approx 5.9722 \times 10^{24}$ kg: mass of the Earth.
- F: force, measured in Newtons (N).
- P: momentum of a moving object, measured in kgm/s.

12.2.3.1 *Centripetal and Centrifugal Forces*

An object with mass m in curvilinear motion with the tangential velocity v is pulled towards the center of the path by a force. This force is called **centripetal force**. It depends on the mass m, velocity v and the distance r of the object from the centre of the path. It is defined as

$$F_c = \frac{mv^2}{r} \tag{12.13}$$

where $\frac{v^2}{r}$ is called **centripetal acceleration**. The centripetal force is illustrated in Fig. 12.11(b). We say that the centripetal force is an action force. Its reaction force or apparent force or fictitious force is called **centrifugal force**. As shown in Fig. 12.11(b), the two forces are equal in magnitude, but opposite in direction. The direction of the centripetal force is always towards the instantaneous centre of the orbit or curvilinear path.

12.2.3.2 *Kepler's Laws of Planetary Motion*

A German astronomer called Johannes Kepler published three laws which describe planetary motions in the solar system. Kepler published the first two laws in 1609 and the third one in 1619. In the application of these laws in satellite communications, we substitute the sun for the Earth and the planet for the satellite. These laws are called Kepler's laws of planetary motion, and they are:

(1) **First Kepler's law**: All planets move about the Sun in elliptical orbits, having the Sun as one of the foci. This law establishes the shape of a spacecraft's (e.g., satellite) orbit.
(2) **Second Kepler's law or the Law of areas**: A radius vector joining any planet to the Sun sweeps out equal areas in equal lengths of time. This law is illustrated in Fig. 12.11(a), where

$R_2 \gg R_1$, $t_1 = t_2$ and so $A_1 = A_2$. This second law establishes the velocity at which spacecrafts move along their orbits. Specifically, a satellite's orbital speed varies with its distance from the Earth. This second law can be used to compute the location of a staellite in its orbit.

(3) **Third Kepler's law**: The squares of the sidereal periods (of revolution) of the planets are directly proportional to the cubes of their mean distances from the Sun. If T is the period of a planet revolving around the sun and r is the mean distance between the planet and the sun, then

$$T^2 \propto r^3 \Leftrightarrow \frac{T^2}{r^3} = \texttt{constant}\,. \tag{12.14}$$

We can derive Eq. (12.14) as follows. The gravitational acceleration on the satellite in orbit is

$$a_{\rm g} = \frac{G(M+m)}{r^2} \tag{12.15}$$

while the centripetal acceleration is

$$a_{\rm c} = \frac{v^2}{r} = \frac{(r\omega)^2}{r} = \frac{(r \times 2\pi f)^2}{r} = \frac{4\pi^2 r}{T^2}\,. \tag{12.16}$$

These two accelerations must be the same for the satellite to remain in orbit. Thus,

$$\frac{G(M+m)}{r^2} \overset{!}{=} \frac{4\pi^2 r}{T^2} \Leftrightarrow \frac{T^2}{r^3} = \frac{4\pi^2}{G(M+m)} = \textit{constant}\,. \tag{12.17}$$

Thus, the constant in the Kepler's third law (12.14) is $\frac{4\pi^2}{G(M+m)}$, as all parameters in it are constants for a given satellite.

12.2.3.3 *Isaac Newton's Three Laws of Motion*

In 1686, Sir Isaac Newton, an English philosopher, studied Kepler's laws and published the findings of his research into gravity and the motion of planets. Among the findings are the three laws of motion, which we quote below.

First law of motion: "A body at rest will remain at rest, and a body in motion will remain in motion unless it is acted upon by an external force." This law is also referred to as the *law of inertia*. A satellite would move in a straight line and leave its orbit if no external force were exerted on it.

Second law of motion: "The force acting on an object is equal to the mass of that object times its acceleration." If m in kilograms is the mass of an object moving with the velocity v m/s, then its momentum, measured in kgm/s, is

$$P = mv\,. \tag{12.18}$$

This second law is also framed as: "the rate of change of the momentum of a moving object is proportional to the force exerted on it, and it follows the same direction as the force." Thus, mathematically, we have

$$F = \frac{dmv}{dt} = d\frac{dv}{dt} = ma\,. \tag{12.19}$$

Third law of motion: "For every action (or force), there is an equal and opposite reaction." The third law is popularly known as Newton's *universal law of gravity*. This law postulates that two bodies attract each other with a force of equal magnitude but opposite in direction. This force is called *force of gravity*. The force of gravity is proportional to the product of the masses of the two bodies, but inversely proportional to the distance between them. Analytically, the force of gravity experienced by the satellite in orbit is

$$F_{\rm g} = \frac{GMm}{r^2}\,. \tag{12.20}$$

This force results from the pulling effect of the Earth on the satellite. The sun and the moon also pull the satellite towards them. This cumulative force is referred to as **centripetal force**. The gravitational force arising from the universal law of gravity must be equal to the gravitational force for the satellite to remain in its orbit, i.e.,

$$F_{\rm c} = F_{\rm g}\,. \tag{12.21}$$

In the following, we assume that the orbit is circular. Equating (12.20) to (12.13), we obtain the satellite's **orbital velocity**

$$v = \sqrt{\frac{GM}{r}} = \sqrt{\frac{GM}{R_{\rm e}+h}} \approx \sqrt{\frac{6.673 \times 10^{-11}\,\texttt{m}^3\texttt{s}^{-2}\texttt{kg}^{-1} \times 5.9722 \times 10^{24}\ \texttt{kg}}{r}} \approx \frac{19,963,088.6}{\sqrt{(R_{\rm e}+h)\ [\texttt{m}]}}\ \texttt{m/s}\,. \tag{12.22}$$

As $v \propto 1/r$, the second Kepler's law is confirmed. Noting that M and G are natural constants, the speed depends on only the altitude of the orbit. A circular orbit has a constant radius and so is the orbital speed. An elliptical orbit, however, has a varying radius. From Eq. (12.22) we observe that the satellite moves fastest at the perigee where it is closest to the Earth.

A more general formula for the orbital velocity is

$$v = \sqrt{\frac{2a - r}{ar} \cdot GM} \tag{12.23}$$

where a is the semi-major axis or the elliptical orbit. Note that, for a circular orbit $a = r$ so that Eq. (12.23) reduces to Eq. (12.22).

The acceleration of the satellite is

$$a = \frac{v^2}{r} = GMr^2\,. \tag{12.24}$$

Note that the acceleration of the satellite along its orbit is equal to the acceleration of gravity of the satellite, which is the rightmost expression in Eq. (12.24).

We can express the orbital velocity in terms of the orbital period as

$$v = \omega r = 2\pi f r = \frac{2\pi r}{T}\,. \tag{12.25}$$

Plugging Eq. (12.25) into (12.22), we obtain the **orbital period** for a circular orbit

$$T = \sqrt{\frac{4\pi r^3}{GM}} = \sqrt{\frac{4\pi (R_e + h)^3}{GM}}\,. \tag{12.26}$$

Example: Orbital Velocity and Orbital Period

The altitudes of the orbits of:

- NASA International Space Station (ISS) satellites' is 330-435 km,
- GPS satellites is about 20,200 km, and
- Intelsat-20 satellites is 35,786 km.

(1) Assuming a circular orbit, compute the speed at which each satellite in these three systems should revolve around the Earth in order to remain in orbit.
(2) Assuming a circular orbit, compute the equivalent orbital period for each of the satellite systems.

Solution

Note that the given values are the orbital altitudes, h. We need to add the Earth's radius to each of them to find the corresponding orbital radius.

(1) We just need to convert the altitudes into meters and plug them into Eq. (12.22) to obtain the required orbital velocities. The results are:
   (a) For NASA International Space Station (ISS) satellites' at 330 km, we have
   $$v = \frac{19,958,600.7}{\sqrt{330,000 + 6371000\ \texttt{[m]}}} \approx 7,711.8\ \texttt{m/s} \approx 27,762.5\ \texttt{km/h}\,.$$
   (b) For GPS satellites, we have
   $$v = \frac{19,958,600.7}{\sqrt{20,200,000 + 6371000\ \texttt{[m]}}} \approx 3,872.8\ \texttt{m/s} \approx 13,942\ \texttt{km/h}\,.$$
   (c) For Intelsat-20 satellites, we have
   $$v = \frac{19,958,600.7}{\sqrt{35,786,000 + 6371000\ \texttt{[m]}}} \approx 3,074.6\ \texttt{m/s} \approx 11,068.6\ \texttt{km/h}\,.$$

(2) We need to plug M, G and r into Eq. (12.26) to obtain the orbital periods. that is:
   (a) NASA ISS satellites' at $r = 330 - 435$ km: $90.99\ \texttt{min} \leq T \leq 93.14\ \texttt{min}$.
   (b) GPS satellites at $r = 20,200$ km: $T \approx 11.98$ hours.
   (c) Intelsat-20 satellites at $r = 35,786$ km: $T \approx 23.93$ hours.

12.2.4 *Satellite Round-Trip Delay*

One of the drawbacks in satellite communications is its long propagation delay owing to its long path between satellite and earth stations. An electromagnetic wave travels approximately with the speed of light, i.e., $v \approx c \approx 3 \cdot 10^8$ m/s. With the orbital altitude h, the two-hop delay (from Earth station to satellite and back to Earth station), aka end-to-end latency or round-trip time (RTT), is

$$t_{\texttt{RTT}} = \frac{2h}{c}\,. \tag{12.27}$$

12.2.5 *Choice of Orbital Altitudes: Van Allen Belts*

Can we launch a satellite onto an orbit of any altitude? Theoretically yes, but practically no. The lowest possible orbital altitude is limited by many forces, including:

(1) The Earth's atmospheric drag/friction. The lower the orbit, the higher the drag.
(2) The gravity of the moon.
(3) Solar wind.
(4) Magnetic fields.
(5) Irregularity in the Earth's gravity.

The largest of these forces is the atmospheric drag (or resistance), which would be too high to counteract for sub-LEO altitudes below 160 km. The atmosphere contains gas molecules, which collide with the satellite and slow it down. The lowest altitude for practical orbits is the perigee of highly elliptical orbits. For example, the perigee of the Russian Sputnik 1 (4 October 1957 to 4 January 1958 LEO satellite) was at the altitude of 215 km. A large quantity of thrust would be needed to counteract the atmospheric drag and maintain an orbital velocity which is high enough to keep a satellite below 160 km altitude. The problem is that the air/atmospheric resistance absorbs part of the energy of the spacecraft, the reason why it slows it down. The absorbed energy heats up the area around the spacecraft and encase it in a superheated plasma.

The second issue to consider with the choice of altitude is the **Van Allen Belts**. These are two concentric rings (referred to as inner belt and outer belt) of energetic charged particles trapped by the Earth's magnetic field and surrounding the planet. They were discovered by Explorer 1 spacecraft in 1958. They were named in honour of James Alfred Van Allen, the scientist who led the team which processed the radiation data collected by Explorer 1. The outer belt consists of mainly highly energetic electrons, while the inner belt comprises mainly energetic protons. The radiations in these two belts are so high that satellites and astronauts must not stay in them for long, and avoid it if possible.

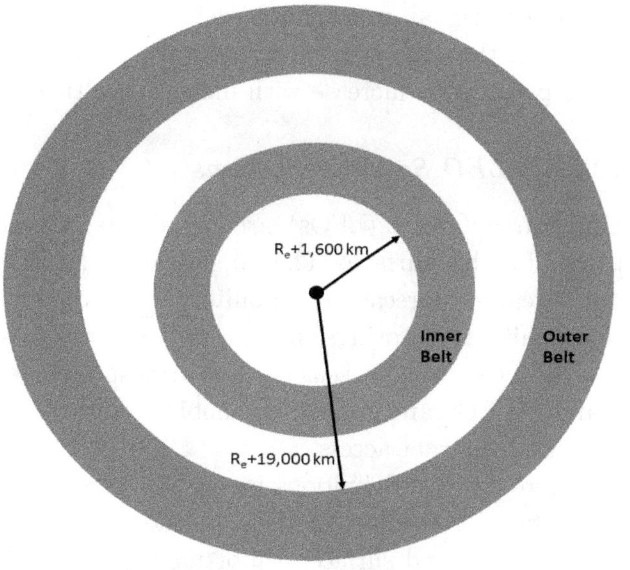

Region	Altitude
LEO	161 – 1,500 km
Inner Van Allen Belt	1,600 – 13,000 km
MEO	8,000 – 25,000 km
Outer Van Allen Belt	19,000 – 40,000 km
GEO	35,786 – 36,000 km

Fig. 12.12 A rough illustration of Van Allen radiation belts.

The damages caused by the charge particles in the Van Allen Belts varies between lethal and none at all, depending on (a) the duration of being exposed to them, (b) the density of the charged particles which varies across each belt, and the energy deposited by the charged particles. Astronauts find strategies to traverse them. They also protect themselves against radiations, extreme temperatures, etc. using spacesuit. NASA uses a pair of spacecraft called Radiation Belt Storm Probes (RBSPs) to study these radiation belts. Figure 12.12 is a rough illustration of the Van Allen radiation belts in relation to satellite orbits. Some sources use the distance from the centre of the Earth $(1.4 - 1.8)R_e$ and $(2.8 - 5.0)R_e$ as the regions for the inner belt and outer belt, respectively. We can observe that, while the LEO for practical satellites lie below the inner belt, parts of MEO and GEO lie in the Van Allen Belts. Those regions have less radiation intensity.

The third issue affecting the choice of orbital altitude relates to the link design issues, which include: (a) the required satellite footprint, (b) the acceptable signal propagation delay and (c) the path loss that the system can sustain. All these three parameters increase with increasing altitude.

12.2.6 *LEO Satellite Systems*

Low Earth Orbits (LEOs) are the orbits closest to the Earth's surface. The applications of LEOs are: weather forecasting, monitoring (land, e.g., road traffic monitoring, air and sea), communication services, remote sensing and atmospheric research. LEOs are popular in mobile communications and Internet access.

There are two definitions for LEOs. First, they are defined according to their altitude or height above the ground surface. Unfortunately, the altitude of a satellite orbit is not a constant, especially for elliptical orbits. It is known to vary as much as 30 km for even circular orbits. A common definition is that the altitude of a LEO should be such that the orbital period is $T = 128$ minutes. This leads to the second definition of LEOs. It is defined as an orbit whose: (a) centre is the Earth, (b) has an eccentricity less than 0.25 and (c) revolves around the Earth at least 11.25 times each day of 24-hour duration. The average lifespan of a LEO satellite is about five years.

Three types of LEO systems are identifiable: little LEOs, big LEOs and broadband LEOs. Figure 12.13 shows examples of each type and their use cases. As shown in Fig. 12.8, LEOs of NASA's International Space Station (ISS) are at 330 km to 435 km altitude. Globalstar is a LEO with 1,414 km altitude and 5,850 km footprint diameter. The altitude of Iridium 159 satellite, launched on 25 July 2018, is about 783.3 km (perigee) and 786.4 km (apogee). The footprint of an Iridium satellite is about 4,700 km in diameter. The period of this satellite is 100.4 minutes with 7,155 km semi-major axis. Telesat, headquartered in Ottawa, Canada, runs LEOs at about 1,000 km altitude. Figure 12.13 also shows the differences between the services provided by each of the three types of LEOs. Of course, we can use a broadband LEO to provide paging service. One of the best features in LEOs is their low latency, making them suitable for broadband Internet access and mobile satellite communications. Table 12.4 compares some basic operators of popular LEO satellite systems.

12.2.7 *GEO Satellites Systems*

A geosynchronous orbit (GSO) is the orbit with the highest altitude that is used for communications.

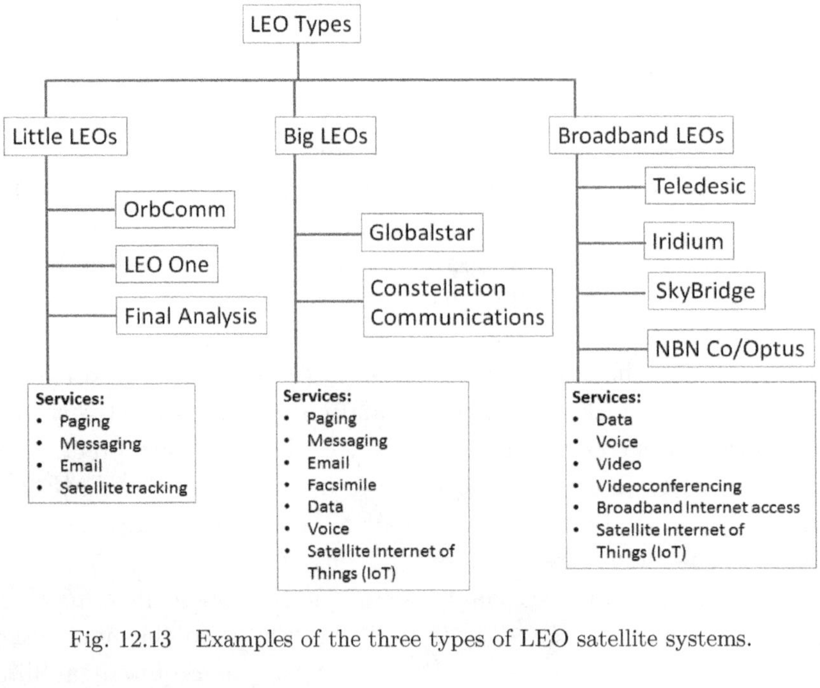

Fig. 12.13 Examples of the three types of LEO satellite systems.

Table 12.4 Comparison of some LEO satellite systems.

Satellite system	Teledesic	SkyBridge	Iridium	Globalstar	Constellation Comms.	Orbcomm	LEO One	Final Analysis
Ownership	Motorola, McCaw, Bill Gates, Boeing, Marconi	Alcatel, Loral	Motorola	Loral, Qualcomm, Loral	Bell Atlantic, Raytheon, Orbital Sciences	Orbital Sciences, Orbcomm	LEO One	General Dynamics Info Systems, Final Analysis
Launch year	2004	2001	1998	2000	2001	1996	2002	2001
Diameter of footprint (km)	1,412	3,000	4,700	5,850				
Data rate (kbps)	20,00/64,000	2,000/20,000	2.4	2.4, 4.8, 9.6 (voice), 7.2 (data)	2.4 (voice), 28.8 (data)	2.4/4.8	2.4-9.6/24	–
Number of satellites	288	80	66	48	46	48	48	38
Number of Earth stations	–	200	15-20	100-210	–	10	20	7
Antenna beams per satellite	64	50	48	16	24 or 32			
Multiple access			FDMA/TDMA	CDMA				
Channel bandwidth				16.5 MHz				
Modulation			QPSK	QPSK				
Orbit altitude (km)	1,375	1,469	781	1,414	2,000	825	950	1,000
Number of orbital planes	12	2	6	8	8	3	8	7
ISL	Yes (8)	No	Yes (4)	No	No	No	No	No
Feeder frequency (GHz)	28.6-29.1/ 18.8-19.3	Ku band	29.1-29.3/ 19.4-19.6	5.091-5.25/ 6.875-7.055	5.091-5.25/ 6.924-7.075	0.148-0.15/ 0.137-0.138 & 0.4	0.148-0.1505/ 0.40015-0.401	VHF/UHF
Mobile frequency (GHz)	–	–	1.616-1.6265/ 1.616-1.6265	1.61-1.6265/ 2.4835-2.5	2.4835-2.5/ 1.61-1.6265	0.148-0.15/ 0.137-0.138 & 0.4	0.148-0.15/ 0.137-0.138	VHF/UHF/ 0.137-0.138

Data rates are given as uplink rate/downlink rate

As shown in Fig. 12.8, the altitude of GSO is about 35,786 km to 37,015 km. A GSO which is circular and has 0° inclination is called *geostationary Earth orbit* (GEO). As such an orbit lies directly above the equator, it is also called *geosynchronous equatorial orbit*. A GSO has a non-zero degree inclination. GEO is also referred to as **Clarke orbit**, in honor of Arthur C. Clarke, who postulated the possibility of using GEO for information exchange in 1945.

The time taken by a satellite to revolve around the Earth once is called `period`, T. The period of a GEO satellite is the same as the rotational period of the Earth about the sun, which is about 24 hours, according to science. For this reason, a GEO satellite appears stationary relative to the Earth. This is the reason why satellite television (TV) antennas mounted in homes do not need to track the satellites. The satellites for TV broadcasting usually use GEOs. Examples of GEO satellite systems are global satellite (GSAT) and Intelsat. The average lifespan of a GEO satellite is about 12 to 15 years.

Scientific Dilemma

Science appears to collide with religion on this. Science postulates that the Earth revolves around the sun with the velocity at the equatorial plane being

$$v_0 = \omega_e R_e = \frac{2\pi}{T_{\mathtt{sid}}} R_e \approx 1,674.32 \ \mathtt{km/h} \tag{12.28}$$

where $R_e \approx 6,378$ km is the Earth's radius at the equator, and $T_{\mathtt{sid}}$ is the sidereal day, defined as

$$T_{\mathtt{sid}} = \frac{365.25}{365.25 + 1} \times 24 \ \mathtt{h} = 23.9345 \ \mathtt{h}. \tag{12.29}$$

Religion, however, says the opposite. This has been a longstanding debate. The religious-biased view has strong evidences as recorded in, for example, Joshua 10:13, Isaiah 38:8, Psalm 50:1, Job 26:7 and Malachi 4:2. Although much has already been done, more research is needed to validate the scientific viewpoint.

Figure 12.14 categorises GEO satellite systems into four. It also provides countries where each is headquartered. Note that by the nature of satcom each of these systems covers more than the country hosting its headquarters. The frequency bands used by each type is also indicated against it. In particular, the MUOS used by the USA DoD uses the frequency band 300 MHz to 320 MHz for its uplink and 360MHz to 380 MHz for its downlink. We can learn from this system and even from Table 12.3 that it is not always that the uplink frequency is higher than the downlink frequency.

12.2.8 *MEO Satellites Systems*

GEO has a large footprint, requiring only three satellites to cover the Earth. Tracking is also not as important as it is in LEO satellites which move very fast. Propagation delay is long in GEO and shortest in LEO. Medium Earth Orbit (MEO) provides a good balance between LEO and GEO. Some MEOs are highly elliptical orbits (HEO), while some are nearly circular. The HEOs provide good coverage of high-latitude locations, such as northern Canada, Arctic Ocean, Greenland, Russia and northern Siberia. GEOs provide good coverage only in regions near the equator, i.e., low-latitude regions.

Examples of MEOs are GPS, Galileo, GLONASS, ICO, Odyssey at 10,354 km altitude and Ellipso satellite systems. A MEO satellite has a lifespan of at least 10 years, but less than that of GEOs. The altitude of MEOs lies roughly in the region 2,000 to 35,786 km. Owing to their altitude, MEO satellites travel faster than those of GEOs but slower than LEOs. This means that antenna tracking is needed in MEOs but not in GEOs.

Example

In this example, we compare LEO (i.e., Globalstar and Iridium), MEO (i.e., Galileo, GPS and GLONASS) and GEO (i.e., Intelsat-20 and GSAT-11) satellites with respect to their:

(1) Round-trip propagation delay or latency $t_{\mathtt{prop}}$.
(2) Antenna size/diameter $D_{\mathtt{a}}$.

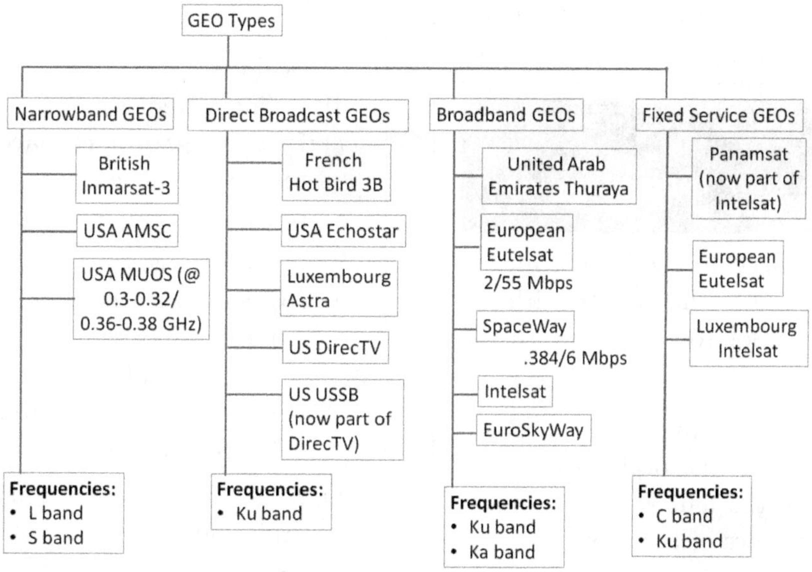

Fig. 12.14 Examples of the four types of GEO satellite systems (AMSC: American mobile satellite system, MUOS: Mobile User Objective System, USSB: United States Satellite Broadcasting).

Table 12.5 Comparing some features of LEOs, MEOs and GEOs.

Satellite	Altitude (km)	Uplink Frequency (GHz)	RTT (msec)	Antenna Diameter (m)
Globalstar	1,414	5.1705	9.5	8.123
Iridium	781	28.75	5.2	1.4609
Galileo	23,222	2.048 or 5.005	154.8	20.5078 or 8.3916
GSAT-11	36,000	12-18	240	3
Intelsat-20	35,786	6.3	238.6	6.6667

RTT: round-trip time

GSAT-11, weighing 5,854 kg, was launched by India on 5 December 2018, 18:16 UTC using Ariane 5 VA-246 rocket vehicle.

Solution

We compute the RTT using Eq. (12.27) and the antenna diameter D_a using the relation

$$D_a \approx \frac{70\lambda}{\theta} = \frac{70c}{\theta f} \propto \frac{1}{f} \tag{12.30}$$

where θ is the antenna beamwidth in degrees, f is frequency and c is the speed of wave propagation in vacuum. The antenna beamwidth of Galileo (or GNSS) is about 0.5°. The solution to the questions are compiled in Table 12.5.

In summary then, we have observed that LEO orbits have low latency and low path loss. The low path loss allows the use of low-gain and low-directivity antennas. These antennas are small (the lower the gain or directivity, the smaller the antenna), reducing device cost and size. The problem with the low directivity lies with ingress and egress electromagnetic interference (EMI). LEO terminals also require steerable antennas. This can be done either remotely or manually. Knowing the high volumes of earth stations which must be steered, some of which are remotely located, it would be too impractical to use manually-steered antennas at terminals to track the fast-moving LEO satellites.

Path loss and path latency are high with GEOs. These features require highly directive antennas with high gains to achieve the needed effective radiated power. The good news is that the high directivity causes low egress and also susceptible to a low ingress EMI. GEO satellites also require theoretically no tracking.

12.2.9 *Check Your Understanding*

Students are encouraged to check their understanding of the content of this section with the following questions prior to proceeding to the next section.

(1) The altitude (i.e., height above the Earth surface) of a non-circular orbit varies along the orbit. State the names for the lowest and highest altitudes of the orbit.
(2) State the two conditions that a geosynchronous Earth orbit must fulfill to become a geostationary Earth orbit.
(3) The Swedish satellite called Odin, launched on 20 February 2001, is used for aeronomy and astrophysics. Some of its parameters at the epoch time of 11 December 2018 01:56:36 are: Sun-synchronous polar orbit with altitude 535 km at perigee and 551 km at apogee, 97.5836° inclination, eccentricity 0.0011155, 0.5128° right ascension of ascending node, and 250 kg dry mass. For this satellite system, calculate the:
    (a) Orbital velocity, v.
    (b) Orbital period, T.
    (c) Number of revolutions per day, $N_{\mathbf{rev}}$.
    (d) Gravitational acceleration on the satellite, $a_{\mathbf{g}}$.
    (e) Centripetal force acting on the satellite, $F_{\mathbf{c}}$.
    (f) Momentum of a satellite, P.
    (g) Semi-major axis of the orbit, a.
    (h) Eccentric anomaly of the satellite, E.
    (i) Mean anomaly of the satellite, M.
    (j) If the epoch time is 11 December 2018 01:56:36, what time does the satellite pass the perigee?
(4) Discuss the three parameters in Kepler's equation with respect to their importance in computing the location of a satellite in orbit.
(5) Why is it that a satellite orbiting an elliptical orbit moves fastest at the perigee?
(6) Define a Keplerian orbit.
(7) **In-class group research**: in groups of 2 to 4 students research to find at least one practical example of:
    (a) Polar orbiting satellites.
    (b) Sun-synchronous orbiting satellites.
    (c) Elliptic orbits.
    (d) Equatorial orbits (aka tropical orbits).
    (e) Geosynchronous orbits.
    (f) Geostationary orbits.
(8) State the two conditions which a satellite orbit must fulfill to be classified as Sun-synchronous orbit.

12.3 Satellite Manufacturers, Payload and Launching

We discuss a sample of major satellite manufacturers, launching methods and vehicles, and payload in this section. In 1957, Russia (then USSR) launched the world's first artificial satellite called Sputnik 1. This satellite orbited a LEO with elliptical shape and lasted about three months (4 October 1957 to 4 January 1958) and fell onto the atmosphere. On 3 November 1957 Russia launched Sputnik 2. USA launched Explorer 1 (aka Early Bird) on 1 February 1958 and Courier 1B on 4 October 1960. Courier 1B was the world's first solar-powered satellite. Sputnik 1 and 2 were powered by silver-zinc batteries. Italy launched San Marco 1 (aka San Marco A) on 15 December 1964. COMSAT launched its MARISAT 1 on 19 February 1976 to provide communications services to ships. Manufactured by Hughes Communications, MARISAT is the world's first mobile satellite system. It is GEO with altitude 35,867 km (at perigee) and 35,703 km (at apogee). These were the humble beginnings of the artificial satellite industry.

12.3.1 *Major Satellite Manufacturing Companies*

The first artificial satellite, Sputnik 1, was manufactured by Energia Corporation in Russia. Energia Corporation also manufactures ballistic missiles, manned spacecrafts and space station components. Jet Propulsion Laboratory (JPL) USA manufactured Explorer 1. The satellite industry is growing with 1,459 operational satellites globally by 31 December 2016. According to the report [Technavio (2018)] by Technavio, between 2018 and 2022 satellite manufacturing and launch market will be worth about $25 billion by 2022 globally. A new use case of satellite is Internet-of-Things (IoT). An example is OneWeb satellites being built by Airbus Defense and Space for OneWeb. Some of the world's largest satellite manufacturers are:

(1) Boeing: founded in 1916 and headquartered in Chicago, USA.
(2) Airbus: founded in 1970 and headquartered in Toulouse, France.
(3) Lockheed Martin: founded in 1995 and headquartered in Bethesda, Maryland, USA.
(4) Thales Alenia Space: founded in 2002 and headquartered in Cannes, France.
(5) SpaceX: founded in 2002 and headquartered in Hawthorne, CA, United States.
(6) Rockwell Automation: founded in 1903 and headquartered in Milwaukee, Wisconsin, USA.
(7) SSL (formerly Space Systems/Loral): founded 1957 and headquartered in Palo Alto, USA.
(8) Northrop Grumman Innovation Systems (acquired Orbital ATK in 2018): founded 2015 and headquartered in Dulles, VA, USA.
(9) Mitsubishi Electric Corporation: founded in 1960 and headquartered in Chiyoda, Japan.
(10) JSC Information Satellite Systems: founded in 1959 and headquartered in Zheleznogorsk, Russia.

12.3.2 *Satellite Launch Companies and Launch Vehicles*

The dynamics of the satellite industry have affected the manufacturing of launch vehicles and launching services. Prior to 2010 manufacturers and launchers were public companies mostly owned by governments. Currently, the space industry is open to competition with the increase of privately-owned manufacturers and launchers. Globally major launch companies include Arianespace of France, SpaceX of USA, Sea Launch of Switzerland using the Russian-Ukrainian Zenit rocket, International Launch Services of United States using the Russian Proton launch vehicle, Mitsubishi Heavy Industries of Japan, United Launch Alliance of USA.

There are two types of launch vehicles:

- Space shuttles, also called Reusable Launch Vehicles (RLV).
- Rockets, also called Expendable Launch Vehicles (ELV).

The cost of launching a satellite into orbit is about 33% of the total cost of a satellite. Satellite launch has a high chance of failure. A rocket is any vehicle whose motion is governed by Isaac Newton's third law of motion. Specifically, a rocket is propelled or boosted by a rocket engine. A rocket engine is a reaction engine, because it is pushed to the opposite direction as it expels combusted fuel in one direction. The main difference between a rocket and a space shuttle is that the former can be used only once, while the latter is reusable. Note that a space shuttle is also lifted off ground into space using rockets, such as Space Shuttle Solid Rocket Boosters (SRBs). A space shuttle is designed to return to the earth like an airplane.

Until 1973, NASA used Saturn V rocket launch vehicle. China has a family of rocket launch vehicles called Long March. There are Russian Soyuz-U, Indian Geosynchronous Satellite Launch Vehicle (GSLV) and US Delta-II (now retired) launch vehicles. Arianespace manufactures a series of rocket launch vehicles, including Ariane5, which is used to launch big satellites onto GEOs. One launch using Ariane5 costs $165 to $220 million. Falcon 9 is a rocket launch vehicle manufactured by SpaceX. With a payload of about 22,800 kg, it costs $50 million to $62 million to launch one satellite using Falcon 9. Atlas V is a rocket launch vehicle manufactured by United Launch Alliance, USA. Table 12.6 compares some launch vehicles used to launch very small satellites [Satellite Industry Association (2017)]. It can cost between $50 million $400 million to launch a satellite into its orbit.

Other launch companies include:

Table 12.6 Very small vehicles used to launch satellites [Satellite Industry Association (2017)].

Launch Vehicle Name	Electron	LauncherOne	VectorH	VectorR
Manufacturer	Rocket Lab, USA	Virgin Orbit, USA (formerly Virgin Galactic)	Vector Space Systems, USA	Vector Space Systems, USA
LEO Capacity	150 kg	400 kg	100 kg	60 kg
First flight	2016	2017	2019	2018
Price	$32,667	$25,000	$30,000	$25,000

- International Launch Services (ILS), headquartered in Reston, Virginia, USA.
- Blue Origin, headquartered in Kent, Washington, USA.
- Boeing Launch Services (BLS).
- China Great Wall Industry Corporation (CGWIC).
- Lockheed Martin Commercial Launch Services (LMCLS).
- Mitsubishi Heavy Industries (MHI) manufactures launch vehicles as well as launching satellites.

Airborne vehicles need either a jet engine or a rocket engine for propulsion. Both engines move vehicles according to Newton's third law of motion by emitting gases in a direction opposite to the desired direction of vehicle movement. Either of them needs an oxygen to burn its fuel in order to move. Jet engines do not carry oxygen, as they get the needed oxygen from the air. Unfortunately for jet engines, there is no oxygen at altitudes above about 20 km. Consequently, jet engines cannot be used for satellites. Rockets have to carry an oxider just as they carry the fuel. There are three types of fuels for rocket engines:

- Solid propellants: used by solid-propellant rockets.
- Hybrid propellants: used by hybrid-propellant rockets.
- Liquid propellants: used by liquid-propellant rockets.

Examples of each type of fuel is shown in Fig. 12.15. The rockets' industry began with solid rockets. Currently, the popularity of liquid and hybrid rockets is higher than solid rockets and it is growing owing to their efficiency and controllable nature.

The type of fuel used by a rocket depends on its type and the altitude from which it is launched. For example, Falcon 1 and Falcon 9 rockets use kerosene and liquid oxygen. This fuel is used by rockets which lift off from the ground and thus liable to the atmospheric pressure. The Space Shuttle orbiter uses liquid oxygen and liquid hydrogen. Deep space rockets need liquids which can withstand temperatures and pressures over a long time period. This is the reason why they use hydrazine and nitrogen tetroxide.

12.3.3 *Satellite Launching Methods*

The process of moving a satellite from ground to its orbit is called launching. How are satellites placed on their orbits? Two main methods are used to launch satellites into orbits:

- Space Shuttle launch: requires astronauts to travel space.
- Rocket launch: no space flight by humans needed.

The satellite in each launch method is a payload or cargo to the launch vehicle. The launch vehicle in the beginning has to move vertically in order to penetrate the atmosphere with smallest possible energy. The vehicle is oriented horizontally after covering an altitude of about 193 km by firing small rockets in it. Thereafter, the satellite is separated from the launch vehicle by firing more rockets. The satellite is then controlled remotely to get it into orbit.

After being built, a satellite must pass a robustness test prior to being launched onto its orbit. This pre-launch test is referred to as **shake-and-bake test**. Literally, this test ensures that the electronics and everything in the satellite can withstand two

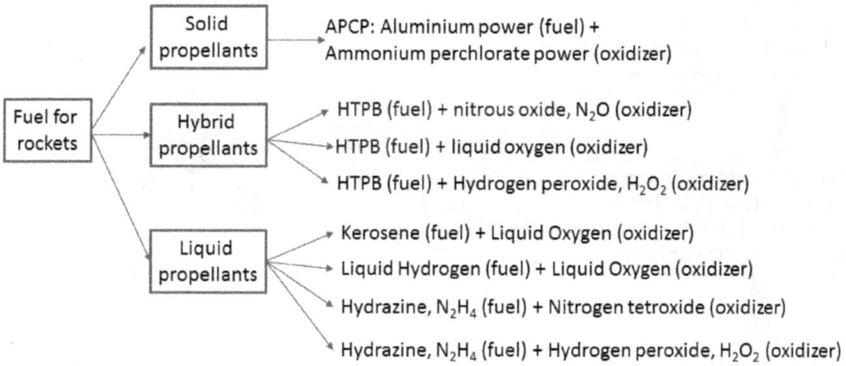

Fig. 12.15 Types of rocket fuel and examples (HTPB: Hydroxyl-terminated polybutadiene; APCP: Ammonium perchlorate composite propellant).

things: extreme temperatures, and extreme pressures or shaking.

Why? Temperatures in the satellite orbit can be very hot or very cold. The average temperatures of LEOs lie in the range −170 °C to 123 °C, while MEOs and GEOs experience even higher temperatures of about −250 °C to 300 °C. Heat is transferred by three ways: radiation, conduction and convection. In space or satellite orbits only radiation by the sun is possible. An object heats up only if it absorbs energy in the form of photons, but cools down if it emits photons. This principle is used for the thermal design of material for satellite manufacturing. Every artificial satellite needs a thermal control system (TCS) which is appropriate for its orbit. There are three types of TCS: passive (has no power supply and no moving components), partially passive and active (requires power supply and has moving components) TCSs. During the launch of a satellite it undergoes enormous amount of shakes and vibrations. The satellite and all its subsystems must be able to endure this.

The most energy-consuming part of satellite launch is moving it from ground to an altitude of about 32 km. Launching a satellite onto a LEO can be done in one stage. However, launching a satellite onto a MEO or GEO occurs in stages. Why? Multiple fuel tanks instead of one are used. A fuel tank is released from the rocket and thrown away to reduce the weight being carried in order to reduce fuel consumption. Thus, the lunch vehicle's cargo reduces as it moves up. You may remember the sad fate of one of NASA's space shuttle called Challenger which occurred on 28 January 1986. It broke apart to kill all the seven astronauts on-board when the solid rocket burned a hole in the tank and ignited the hydrogen and oxygen. GEO satellite launch occurs in four stages. This is illustrated roughly in Fig. 12.16.

12.3.4 *Satellite Launch Locations*

The weight that must be moved comprises the weight of the rocket, atmospheric drag, payload and fuel. The smaller the weight, the smaller the fuel used to move the launcher. Spending much of the fuel on-board the satellite in launching reduces the amount left for station-keeping which reduces the lifespan of the satellite. The launching location also depends on the orbit to place the satellite. Certain locations are best for satellite launch for two reasons.

First, the inclination of a satellite depends on the latitude of the launch location. A higher latitude yields a larger inclination. For example, NASA launches satellites onto equatorial orbits using Ariane rocket from Guiana Space Centre in French Guyana. This location has the latitude of about 3.9339°N, which is close to the equator. As discussed earlier, the inclination of GEO is zero degrees. This means that part of the fuel on the satellite must be used to bring the inclination to zero if it is not launched at a location along the equator.

Second, after being detached from the launch vehicle the force or boost needed to propel the

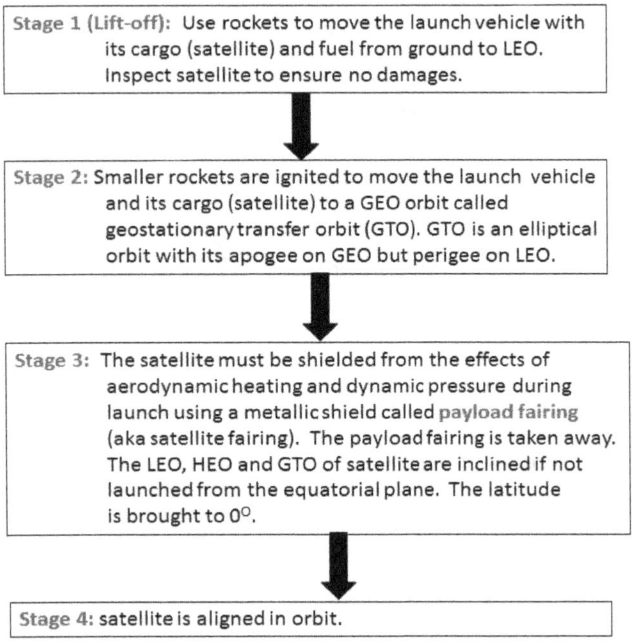
Stage 1 (Lift-off): Use rockets to move the launch vehicle with its cargo (satellite) and fuel from ground to LEO. Inspect satellite to ensure no damages.
Stage 2: Smaller rockets are ignited to move the launch vehicle and its cargo (satellite) to a GEO orbit called geostationary transfer orbit (GTO). GTO is an elliptical orbit with its apogee on GEO but perigee on LEO.
Stage 3: The satellite must be shielded from the effects of aerodynamic heating and dynamic pressure during launch using a metallic shield called payload fairing (aka satellite fairing). The payload fairing is taken away. The LEO, HEO and GTO of satellite are inclined if not launched from the equatorial plane. The latitude is brought to 0°.
Stage 4: satellite is aligned in orbit.

Fig. 12.16 Rough illustration of the stages in launching a satellite onto a GEO.

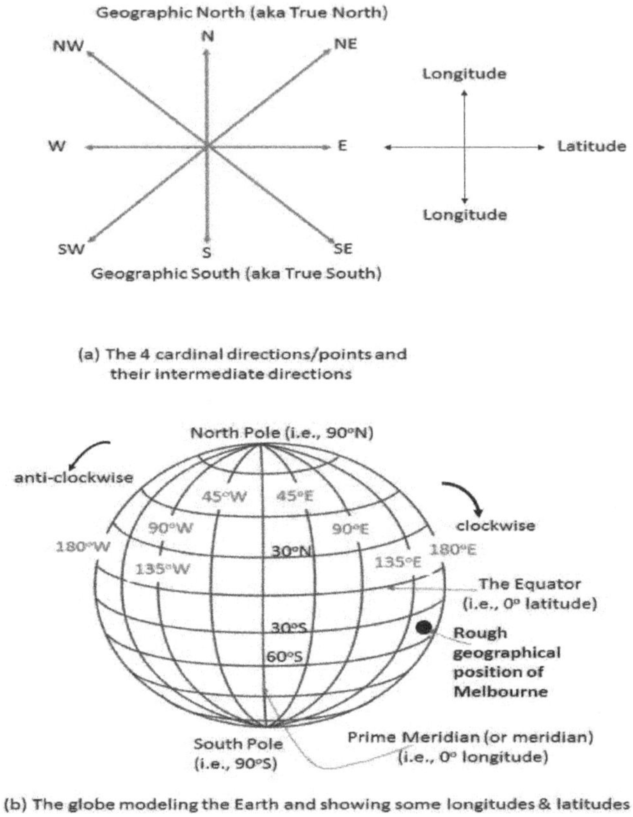
Geographic North (aka True North)
N
NW
NE
W
E
SW
S
SE
Geographic South (aka True South)
Longitude
Latitude
Longitude
(a) The 4 cardinal directions/points and their intermediate directions
North Pole (i.e., 90°N)
anti-clockwise
clockwise
45°W
45°E
90°W
90°E
180°W
180°E
135°W
135°E
30°N
30°S
60°S
The Equator (i.e., 0° latitude)
Rough geographical position of Melbourne
South Pole (i.e., 90°S)
Prime Meridian (or meridian) (i.e., 0° longitude)
(b) The globe modeling the Earth and showing some longitudes & latitudes

Fig. 12.17 Longitudes, latitudes and true north.

rocket forward depends on the Earth's rotational velocity, which depends on the geographical location. Equatorial regions have the maximum boost owing to the fastest Earth rotation there.

Geographic locations (e.g. satellite launch locations) are usually identified using bearings, longitude and latitude. These are summarized in Fig. 12.17.

Some of the popular launch locations are:

(1) Algeria with 31.09951° N, 2.83581° W: France launched 230 rockets with weight up to 18 tonnes from 1947 to 1967.
(2) Vikram Sarabhai Space Centre, Thiruvananthapuram (thumba), Kerala, India with 8.5314° N, 76.869° E: India has launched more than 2,000 rockets since 1971.
(3) Akita Rocket Range, Japan 39.57148° N, 140.05785° E: launched 81 rockets between 1956 and 1990.
(4) Baikonur Cosmodrome, Tyuratam, Kazakhstan: Russia has launched more than 1,000 rockets with weights up to 2,400 tones since 1957.
(5) Churchill Rocket Research Range, Manitoba: Canadian army launched more than 3,500 rockets in 1954–1998.

12.3.5 *Check Your Understanding*

Students are encouraged to check their understanding of the content of this section with the following questions prior to proceeding to the next section.

(1) How was the first artificial satellite sent into space powered?
(2) What is the name of the world's first solar-powered artificial satellite?
(3) What is the term used to refer to the robustness test performed on a satellite before it is launched into orbit? Justify the reason for the term.

(4) Discuss what happens to the rocket used to launch a satellite after launching it.
(5) Explain the reason(s) why satellites are launched using rocket engines instead of jet engines.
(6) Launching of a satellite onto orbit, especially GEO, is done in at least two stages. Explain the technical reason(s) behind it.
(7) In-class research: in groups of 2 to 4 students, research on the Internet or books to find:
    (a) Currently, the world's largest satellite launching company.
    (b) Currently, the world's largest manufacturer of satellites.
    (c) Globally, the most popular satellite launching locations. State the reasons why those locations are used.
    (d) Currently, the world's largest manufacturer of launch vehicles for satellites.
    (e) The percentage of satellites launched using space shuttle compared with those launched using rockets.
(8) Describe the main difference between a rocket and a space shuttle used to launch satellites onto orbits.
(9) The propellant for a rocket comprises two substances. What are they?

12.4 Satellite Space Segment Subsystems

A satellite communications system comprises two main subsystems: the Earth segment (aka ground segment) comprising one or more Earth stations and the space segment, called satellite. The space segment, on the other hand, comprises two subsystems: the payload (aka platform) and the bus. A payload is designed to support a specific mission, while a bus can support different types of payloads.

12.4.1 *Satellite Payload*

The payload (specifically, the communications payload) is the equipment which provides the actual functions of the satellite, such as the transponders. There are many types of satellite payloads, such as communications, navigation, imaging and hosted payload. The hosted payload is also called hitchhiking or piggybacking, communication payload, imaging payload and navigation payload. Regardless of the type of payload, each must be able to receive, transfer and maybe process signals. Therefore, each payload contains the two main components:

- Repeater: comprises microwave receivers, filters, power amplifiers, baseband processing, switching networks (e.g., in Iridium satellite system) and RF multiplexers. Contained within the repeater are the transponders. A *transponder* is a device which intercepts a wireless signal and transmits a different signal.
- Antennas: comprises reflectors, support structure, pointing mechanisms, feeds and feed networks. The antennas, usually with multiple spot beams, converts the signals from/to the repeater between electrical signal and electromagnetic waves.

Figure 12.18 is the block diagram of OPTUS-10 satellite communications' payload [OPTUS (2018)]. That of A series and B series satellites is shown in Fig. 12.19. The payload has another meaning. During the launch of a satellite by a rocket, the latter keeps the former. Therefore, the satellite is a payload to the rocket.

12.4.2 *Satellite Bus*

Also called platform, the bus is the support system of the payload. It contains the support vehicle and control subsystems which enable the payload to fulfill its objectives as a microwave repeater in space. Each bus is built to fulfill three requirements: power budget, mass budget and heat disposal. The mass budget indicates the total weight of the payload supportable by the bus. The total power or energy that is available in the bus to support the payload is the power budget. The components in the bus include:

(1) Telemetry, Tracking, Command and Monitoring (TTCM) subsystem.
(2) Power and Antenna subsystem (PAS).

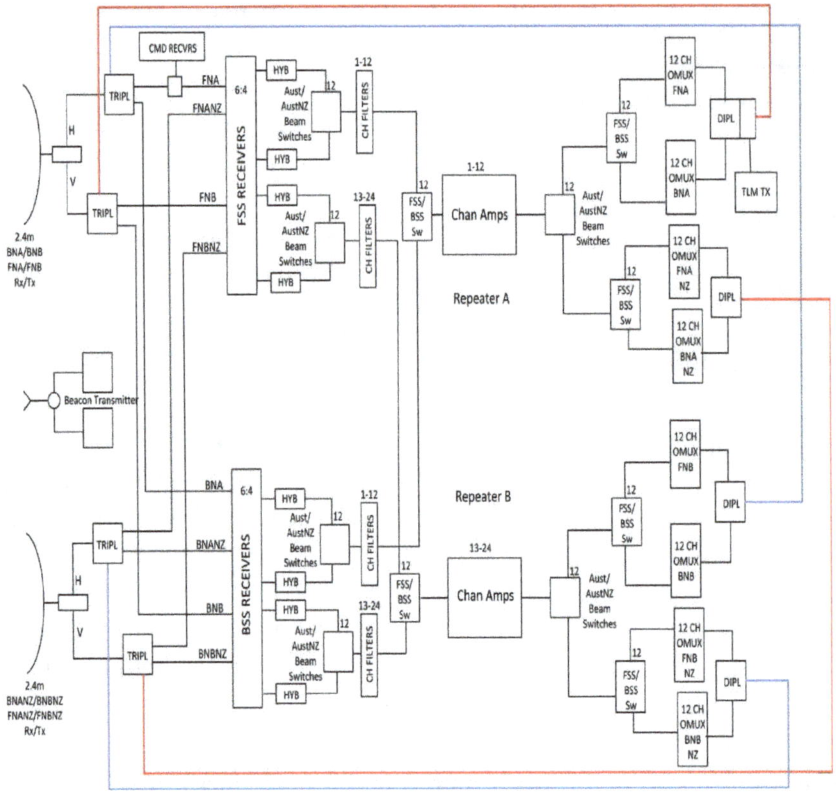

Fig. 12.18 Block diagram of OPTUS-10 satellite communications' payload [OPTUS (2018)].

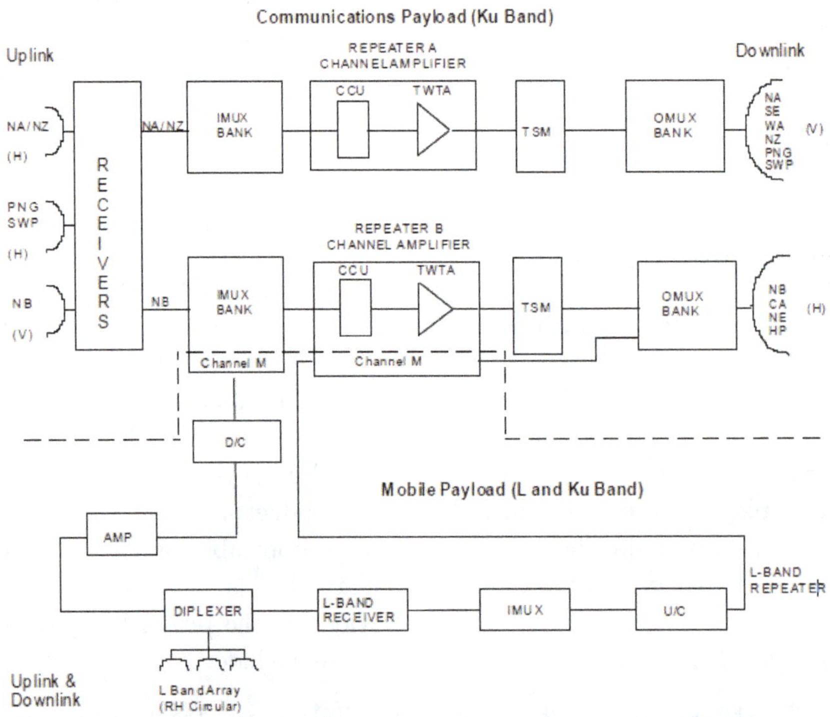

Fig. 12.19 Communications payload of OPTUS satellite series A and B [OPTUS (2018)]. IMUX/OMUC: input/output multiplexer, CCU: channel control unit, TSM: transmit beam switch matrices, D/C: downconverter, U/C: upconverter.

(3) Attitude and Orbit Control (AOC) subsystem, comprising altitude control subsystem and orbit control subsystem.
(4) Thermal control to maintain the electronics and other components within a safe temperature range over the life of the satellite.
(5) Frame structure to keep all subsystems together and protect them during launch and after deployment on orbit.
(6) Wiring harnesses, radiation shielding.

12.4.2.1 *The TTCM Subsystem*

This subsystem is described by other names, such as tracking, telemetry, command and ranging (TTC&R). The TTCM is used to control the satellite in orbit remotely. This subsystem has four functions: telemetry, tracking of the satellite in orbit, commanding and monitoring of the satellite. Telemetry is concatenation of two Greek words *tele* (meaning remote) and *metron* (meaning measure). The telemetry subsystem on the satellite is used to collect data and sent over distance to the earth monitoring stations. The satellite control center on the ground uses the tracking subsystem to monitor the position and orbit of a satellite. The command subsystem can be used to move a satellite from one orbit onto another orbit, for example. It is also used to control the subsystems of the satellite, such as switching them off or on.

12.4.2.2 *The Power and Antenna Subsystem*

The power and antenna subsystem has two functions: supplying the payload of the needed power to run the electronics on it, and antennas for the transmission and reception of electromagnetic signals. Power is supplied to a satellite in orbit mainly by solar energy. Therefore, the satellite has solar panels. The satellite also needs rechargeable batteries or accumulators as secondary power source. The batteries are used for two purposes: for launching of the satellite and during eclipses of the sun (i.e., satellite is hidden from the sun by the earth or moon) when solar energy is not present. The solar panels are not commissioned until after the satellite has been launched onto its orbit. The solar energy is used to recharge the on-board batteries. Two types of solar arrays are found in satellites: solar sail (aka rectangular solar arrays) and cylindrical solar arrays. The former has a better exposure to sunlight than the latter.

Some of the components on-board a satellite can be shut down when there is a prolonged period of time in which the sun is unavailable to illuminate the solar panels. The Earth casts its shadow on LEO satellites on the average about 45% of its period. GEO, on the other hand, experiences the Earth's shadow effect only during the equinox of the sun, and it lasts less than 70 minutes, i.e., about only 4.9% of its orbital period.

Antennas are designed for specific systems and specific frequencies of operation. Antennas deployed on satellites are of four types: array antenna, wire antenna (either monopole or dipole), reflector antenna (e.g., parabolic reflector) and aperture antenna (e.g., horn antenna). Only the wire antenna is omnidirectional, the remaining three are directional antennas. We leave the detailed treatment of these to the chapter on antennas and MIMO.

12.4.2.3 *The AOC Subsystem: Orbital Station-Keeping*

Attitude and Orbit Control (AOC) subsystem uses rocket motors to maintain the satellite in its orbit, and to maintain on-board antennas pointing towards the right earth station. The main components of the AOC are stabilisation, attitude sensors, and steering actuators. The three forces — gravitational forces from Sun/Moon, deviations of the gravitational force of the Earth from that of a homogeneous sphere, and solar radiation pressure and air-drag — are referred to as **non-Keplerian forces**. These forces attempt to violate Eq. (12.22) and keep a satellite out of its orbit. Station-keeping is used to counteract their effects. One of two awful situations can happen. These are:

$$\texttt{Satellite is too slow:}\quad F_g > F_c \Leftrightarrow v < \sqrt{\frac{GM}{r}}$$

$$\texttt{Satellite is too fast:}\quad F_g < F_c \Leftrightarrow v > \sqrt{\frac{GM}{r}} \tag{12.31}$$

First, if the satellite slowed down the Earth would pull it towards itself. In such a situation the satellite would fall down to the Earth. Second, if the satellite is too fast then the sun would pull it away from the orbit to itself. Such a situation would cause the satellite to move in a straight line into outer space. The process through which these two unpleasant situations is avoided is referred to as **orbital station-keeping** or **satellite station-keeping**. The velocity needed by an object to leave the Earth and go into outer space is called **escape velocity**. It is defined as

$$v_e = \sqrt{\frac{2GM}{R_e}} \approx 11.186 \text{ km/s}. \qquad (12.32)$$

Two methods are used to intentionally slow down the orbital velocity of a satellite. This is used, for example, to move the satellite from one orbit onto a lower orbit. The first is **retroburning**, which is the process of firing rockets in the opposite direction of the satellite's movement. The second method is **aerobraking**, which is achieved using the dragging effect of gas molecules in the atmosphere or planet.

A satellite (or the space segment of a satellite system) contains rocket motors with different forces. The rocket motors with low force is called **thruster**. Thrusters are used to change the pointing direction of a satellite or move it slightly. Thrusters are used to keep a satellite in its orbit using the principle of Newton's third law of motion: *for every action, there is an equal and opposite reaction.* The most powerful thrusters are called *bi-propellant mixtures* with an example being the mixture of dinitrogen tetroxide (N_2O_4) as the oxidant and monomethylhydrazine ($CH_3N_2H_3$) as the fuel. These two chemicals combust when they come in contact with each other.

The TTCM on the earth station monitors the satellite's position in space. If any anomaly is detected then the TTCM commands the orbital control subsystem on-board to make the appropriate correction. One of two methods are used to control the altitude of a satellite and stabilise it. These are the **three-axes method** and **spinning the satellite method**. As its name says, the three-axes method controls the three axes in which a space vehicle (e.g., satellite, airplane) is free to rotate. These are:

- **Yaw axis** (aka vertical axis): nose of satellite rotates left or right about a vertical axis towards the centre of the Earth.
- **Pitch axis** (aka lateral or transversal axis): satellite rotates about an axis which is at right angles to the orbital plane.
- **Roll axis** (aka longitudinal axis): rotation about an axis passing through the tail and nose of the satellite towards the direction of satellite's movement in the orbital plane.

These axes are illustrated in Fig. 12.20.

12.4.3 *Satellite Transponder Subsystem*

The most important component of the space segment is the **transponder**, which is a portmanteau word from **transmitter-responder**. The transponder has a triple function: works as a signal *repeater*, amplifier and a *frequency translator*. It is a repeater because it regenerates signals from their received noisy versions. It is an amplifier because it boosts the power in the regenerated signal. It is a frequency translator because it receives signals at one frequency in the uplink, say f_{UL}, but sends the regenerated signal over the downlink at a different frequency, say f_{DL}.

There are two types of transponders:

- bent pipe transponders and
- regenerative transponders.

These transponders are contrasted in Fig. 12.21. For example, Globalstar and Odyssey use bent

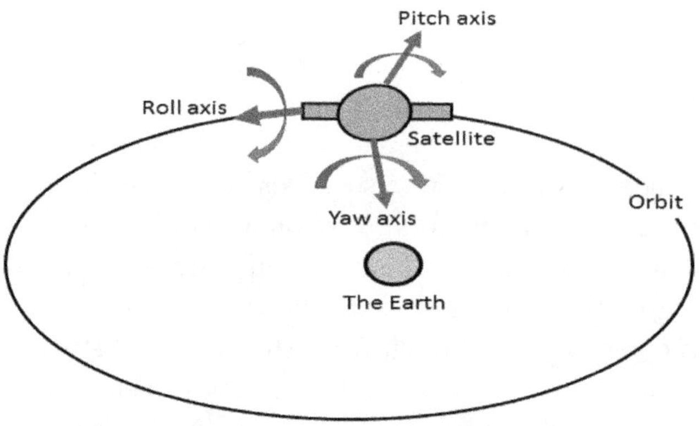

Fig. 12.20 The three rotational axes of a spacecraft.

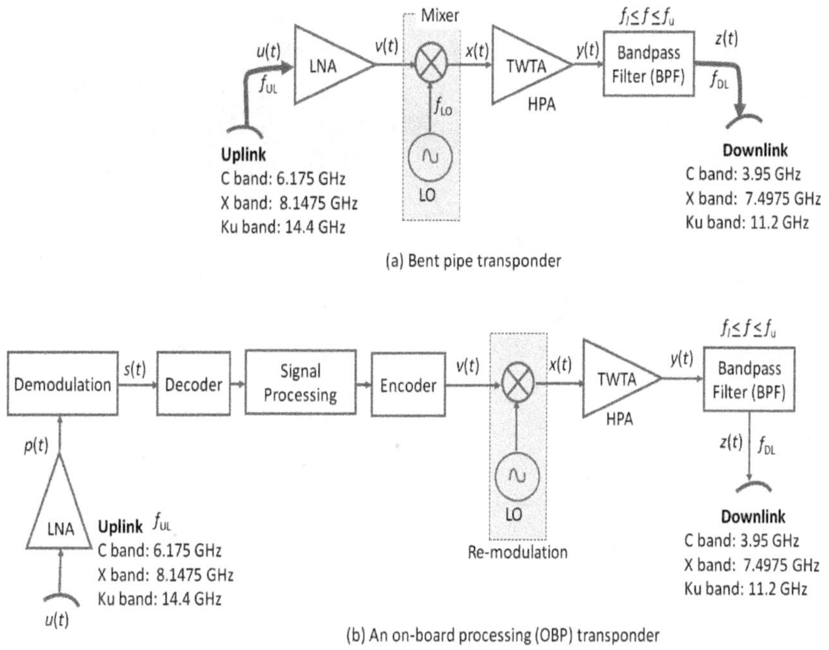

Fig. 12.21 Block diagrams of the two basic types of satellite transponders (HPA: high-power amplifier, TWTA: travelling wave tube amplifier).

pipe transponders, while Iridium uses regenerative transponders.

12.4.3.1 *Bent Pipe Transponder*

Bent pipe transponders are traditional transponders which are very simple. The only functions performed are amplification and frequency translation. The only advantages of the bent pipe over the regenerative transponder is its (a) simplicity and (b) the ability to process both analog and digital signals. A bent pipe transponder is illustrated in Fig. 12.21(a). It comprises:

- A low noise amplifier (LNA).
- A mixer for the frequency translation. It comprises a local oscillator to generate the needed frequency, and a multiplier.
- Travelling wave tube amplifier (TWTA).
- A bandpass filter (BPF).

The uplink and downlink use difference frequencies to minimise interference. In order to reduce intermodulation distortion caused by crosstalk, the amplifiers can be operated in their linear range. However, the gain in the linear range is lower than in the nonlinear operation.

How do we dimension the amplifiers and the filters in the bent pipe for it to deliver the required performance? Let us assume that both the LNA and TWTA in the bent pipe transponder in Fig. 12.21(a) are linear and produce no intermodulation distortion. In which case the signals $u(t)$ and $v(t)$ are both single-tone signals with the frequency f_{UL}. Also, the signals $x(t)$ and $y(t)$ are both single-tone signals with the frequency f_{DL}. Without loss of generality, let us define the single-tone signals

$$\begin{aligned} v(t) &= a_{\mathrm{v}} \cos(2\pi f_{\mathrm{UL}} t) \ \texttt{and} \\ g(t) &= a_{\mathrm{g}} \cos(2\pi f_{\mathrm{LO}} t)\,. \end{aligned} \tag{12.33}$$

Then, the output of the mixer is

$$\begin{aligned} x(t) = v(t) \cdot g(t) &= a_{\mathrm{v}} a_{\mathrm{g}} \cos(2\pi f_{\mathrm{UL}} t) \cos(2\pi f_{\mathrm{LO}} t) \\ &= \frac{1}{2} a_{\mathrm{v}} a_{\mathrm{g}} [\cos(2\pi (f_{\mathrm{UL}} - f_{\mathrm{LO}}) t) \\ &\quad + \cos(2\pi (f_{\mathrm{UL}} + f_{\mathrm{LO}}) t)]\,. \end{aligned} \tag{12.34}$$

We have invoked the product-to-sum identity of trigonometry in Eq. (12.34). The transponder has to convert the uplink frequency f_{UL} to the downlink frequency f_{DL}. Therefore, we can establish the general relation

$$f_{\mathrm{DL}} = f_{\mathrm{UL}} - f_{\mathrm{LO}} \Leftrightarrow f_{\mathrm{LO}} = f_{\mathrm{UL}} - f_{\mathrm{DL}}\,. \tag{12.35}$$

The BPF is so designed to block the sum frequency term in Eq. (12.34) so that the signal $z(t)$ contains only the difference frequency term.

Example

Assume that a Ku-band satellite uses the uplink frequency $f_{\text{UL}} = 14.4$ GHz and the downlink frequency $f_{\text{DL}} = 11.2$ GHz. In order to achieve the desired operation of the bent pipe transponder, compute the:

(1) Local oscillator frequency.
(2) Pass band of the bandpass filter (BPF).

Solution

(1) Using Eq. (12.35), the local oscillator must generate the frequency

$$f_{\text{LO}} = f_{\text{UL}} - f_{\text{DL}} = (14.4 - 11.2)\ \text{GHz} = 3.2\ \text{GHz}\,.$$

(2) The BPF must pass the difference frequency term, but gets rid of the sum frequency

$$f_{\text{sum}} = f_{\text{UL}} + f_{\text{DL}} = (14.4+11.2)\ \text{GHz} = 25.6\ \text{GHz}\,.$$

A BPF with centre frequency

$$f_c = f_{\text{UL}} - f_{\text{LO}} = 11.2\ \text{GHz}$$

and bandwidth less than

$$B = 2(f_{\text{sum}} - f_c) = 28.8\ \text{GHz}$$

does the job. Usually, much less bandwidth is required and used, as we just need to pass a single-frequency signal.

12.4.3.2 *On-Board Processing Transponders*

More sophisticated transponders are the On-Board Processing (OBP) transponders, aka **regenerative transponders**. These perform all the functions of the bent pipe and additional ones. Their disadvantages compared to the bent pipe is their complexity and cost. The block diagram of a regenerative transponder is shown in Fig. 12.21(b). The regenerative transponders perform on-board processing (OBP) besides amplification and frequency translation. They perform demodulation and remodulation, multiplexing and demultiplexing, decoding and re-encoding, beam switching and routing of signals between satellites. All satellites used for in-orbit satellite networks to enable intersatellite signal handoff, called intersatellite link (ISL), must have OBP functions. Consequently, such satellites need regenerative transponders. Owing to their OBP functions, regenerative transponders are also referred to as **processing transponders**. There are three basic types of OBP, which are:

(1) Support processing: performs functions such as switching of beams of adaptive antennas, demodulation, and re-modulation.
(2) Intermediate frequency (IF) or radio frequency (RF) switching: switching can be controlled by ground stations.
(3) Baseband processing and switching/routing, which can be controlled by the Earth station or autonomous. Functions performed include demodulation, demultiplexing, error detection and correction, signal routing, antenna pointing/re-pointing, removal of header information, data buffering, inter-satellite handoff.

12.4.4 *Check Your Understanding*

Students are encouraged to check their understanding of the content of this section with the questions below prior to proceeding to the next section.

(1) In satellite communications, describe what a transponder is.
(2) Describe how an on-board processing (OBP) transponder differs from a bent pipe transponder.
(3) Describe your understanding of the term orbital station-keeping or **satellite station-keeping**.
(4) Discuss the meanings and purposes of retroburning and aerobraking in satellite communications.

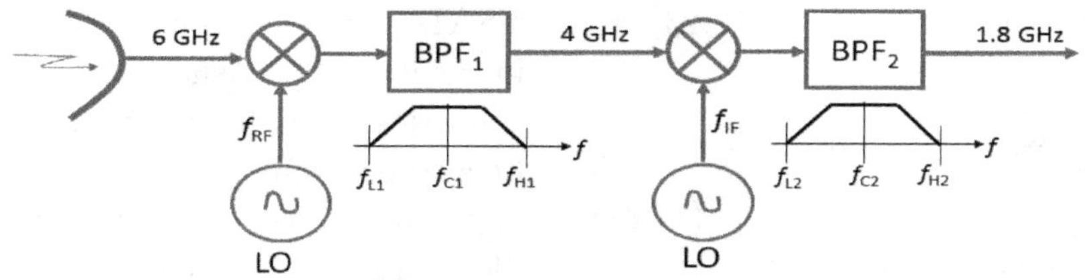

Fig. 12.22 Block diagram of a simplified wireless receiver.

(5) Using examples, describe the meaning of *payload* in satellite communications.
(6) Define the term *payload fairing* as used in satellite communications.
(7) Study the communications payload of OPTUS-10 satellite shown in Fig. 12.18. Write a four-page summary of all the components and functions in the payload.
(8) Consider the wireless receiving circuit in Fig. 12.22. Assume that there is no spurious or intermodulation distortion, and dimension the electronic elements.
 (a) Mixer design: find f_{RF} of the RF mixer.
 (b) BPF design: find f_{L1}, f_{H1} and f_{C1} of the RF BPF.
 (c) BPF design: find f_{L2}, f_{H2} and f_{C2} of the IF BPF.
 (d) Mixer design: find f_{IF} of the IF mixer.

12.5 Earth Stations

The earth stations can be divided into consumer equipment and network infrastructure. The former includes antennas/dishes, satellite radio equipment and mobile satellite terminals. The network infrastructure includes very small aperture terminals (VSAT), gateways and control stations.

12.5.1 *Earth Station Antenna Look Angles*

Two angles are needed to locate the position of a satellite and point an Earth station antenna to to it for optimum antenna gain. These angles are together called antenna **look angles**, and they are defined as:

- Elevation angle, β: angle measured in the plane perpendicular to the horizon (from the horizontal plane to the satellite orbital plane). Horizontal is 0^0, vertically downwards is $-90°$ and vertically upwards is $90°$.
- Azimuth angle (AZ), α: measured clockwise from the north in the plane parallel to the horizon. Thus, north is 0^0 and south is $180°$.

Finding the right look angles is very important for non-GEO satellites requiring the Earth station antennas to track them. We need to know the sub-satellite point (SSP), which is the point where a straight line joining the satellite and the centre of the Earth first cuts the Earth. The SSP is also referred to as *nadir*. We also need to know latitudes, longitudes, law of cosines, law of tangents and law of sines. Applying the law of cosines on Fig 12.23, we obtain

$$d = \sqrt{R_{\text{e}}^2 + r^2 - 2R_{\text{e}}r\cos(\phi)}\,. \qquad (12.36)$$

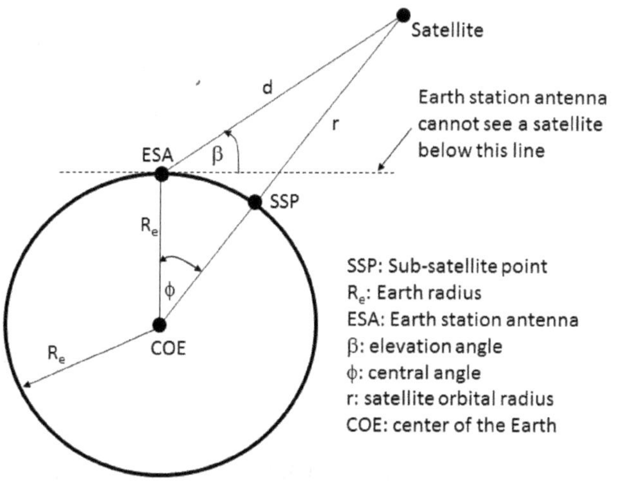

Fig. 12.23 Deriving the elevation angle of a satellite earth station antenna.

Also, applying the law of sines on Fig 12.23, we obtain

$$\frac{r}{\sin(\beta+90^\circ)} = \frac{d}{\sin(\phi)} \Leftrightarrow \sin(\beta+90^\circ) = \frac{r}{d}\sin(\phi)\,. \tag{12.37}$$

By noting that $\sin(\beta+90^\circ) = \cos(\beta)$ and plugging Eq. (12.36) into Eq. (12.37), we obtain the elevation angle

$$\begin{aligned}\cos(\beta) &= \frac{r}{d}\sin(\phi)\\ &\overset{(12.36)}{=} \frac{\sin(\phi)}{\sqrt{1+\left(\frac{R_e}{r}\right)^2 - 2\frac{R_e}{r}\cos(\phi)}}\,.\end{aligned} \tag{12.38}$$

Let L_{es} and l_{es} be the latitude and the longitude of the earth station antenna, and L_{ssp} and l_{ssp} be the latitude and the longitude of the sub-satellite point. Then,

$$\begin{aligned}\cos(\phi) &= \cos(L_{\mathsf{es}})\cos(L_{\mathsf{ssp}})\cos(l_{\mathsf{ssp}} - l_{\mathsf{es}})\\ &\quad + \sin(L_{\mathsf{es}})\sin(L_{\mathsf{ssp}})\,.\end{aligned} \tag{12.39}$$

With

$$\varphi = \mathtt{atan}\left(\frac{\tan(l_{\mathsf{ssp}} - l_{\mathsf{es}})}{\sin(L_{\mathsf{es}})}\right) \tag{12.40}$$

we obtain the azimuth for a GEO if the earth station is located in the Northern Hemisphere as

$$\begin{aligned}&\texttt{Satellite lies SE of the earth station: } \alpha = 180^\circ - \varphi\\ &\texttt{Satellite lies SW of the earth station: } \alpha = 180^\circ + \varphi\end{aligned} \tag{12.41}$$

and if the earth station is located in the Southern Hemisphere

$$\begin{aligned}&\texttt{Satellite lies NE of the earth station: } \alpha = \varphi\\ &\texttt{Satellite lies NW of the earth station: } \alpha = 360^\circ - \varphi\,.\end{aligned} \tag{12.42}$$

12.5.2 *VSAT*

Very small aperture terminal (VSAT) is a full-duplex (or transceiver) satellite earth station which uses parabolic antennas with a small diameter (about 0.75 m to 3.8 m). A VSAT using a parabolic antenna with diameter less than 0.6 m is called a micro-VSAT. The name VSAT originated from the size of the antenna. However, it is currently used to refer to the entire satellite ground station system. VSATs used in Europe and North America operate in the Ku band. Asian, Latin American and African countries use C band VSAT. A VSAT network can be a receive-only (popularly used for TV reception) or for both receive and transmit. A VSAT linking distributed corporate LAN must be able to transmit and receive. A major issue in VSAT is finding the optimum location to position the VSAT antenna. A system similar to VSAT is Television Receive Only terminals (TVRO). However, TVRO is unidirectional which can receive signals from satellites but cannot transmit to them.

As illustrated in Fig. 12.24, a VSAT connects to an Earth station called central hub over a satellite link. The link interconnecting a satellite earth station to a terrestrial (switched) network, such as the Internet, is called **backhaul**. VSAT provides Internet access in remote locations. A company with branches located in different places can use VSAT technology to interconnect its local area networks (LANs) to form a wide-area network (WAN) using the configuration illustrated in Fig. 12.6(c). A third popular application of VSAT is point-of-sale transactions using credit cards.

VSAT vendors include Direcway, Israeli Gilat technology, ViaSat and Spacenet. The basic components of a VSAT are:

- Antenna: needed to convert signals between electrical and electro.
- Low-noise block (LNB): comprises a low noise amplifier (LNA) and frequency down converter.
- Block upconverter (BUC): comprises a power amplifier (PA) and RF upconverter using a local oscillator (LO).
- Feed: comprises a tuned radio frequency receiver (TRF) or transmit-reject filter, feed horn and orthomode transducer (OMT), waveguide and a circular tube.
- Satellite MODEM for signal conversion between audio frequency (AF) and intermediate frequency (IF).
- Interfacility links (IFLs): the cables interconnecting the outdoor unit (ODU) with the in-door unit (IDU). The IDU is usually the satellite MODEM. The ODU comprises everything else, such as the antenna dish, the feed, BUC and LNB.

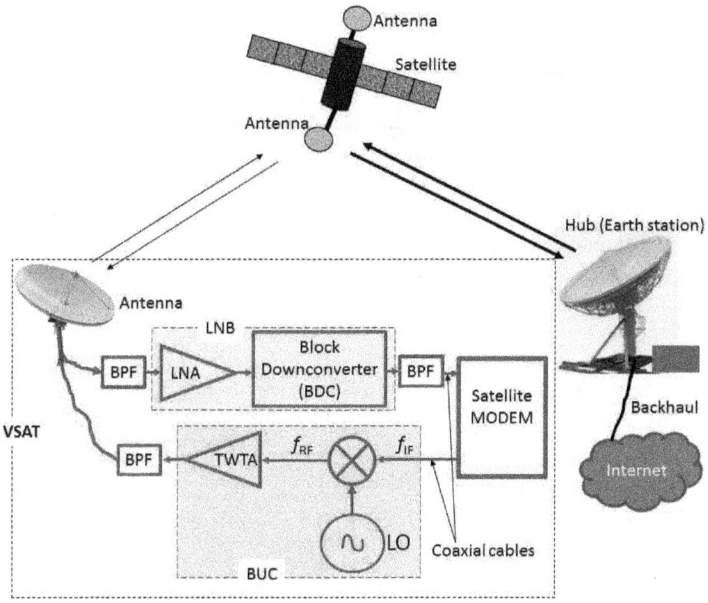

Fig. 12.24 A VSAT architecture.

The TRF comprises one or multiple tuned radio frequency (RF) amplifier stages followed by a demodulator (or detector).

Figure 12.24 illustrates a VSAT architecture. In the uplink direction, the satellite MODEM modulates the baseband signal at audio frequency (AF) and convert it into intermediate frequency (IF) signal. The block upconverter (BUC) then converts the IF signal into RF signals and then amplify it using the power amplify(PA), which is usually a travelling wave tube amplify (TWTA). Typical power ratings of BUCs are 3–6 Watts.

In the downlink direction, the LNB amplifies the RF signal using the LNA before downconverting it into IF signal over its block downconverter. The satellite MODEM then converts the IF signal into AF signal to enable baseband processing. The satellite MODEM is also the interface between the satellite system and users' consumer devices, such as laptop, desktop and facsimile. The MODEM is centrally managed from the service provider's Network Operations Center (NOC). The 'block' in BUC and block downconverter (BDC) indicates that the device converts an entire block or band of frequencies into another band of frequencies.

Orthomode transducer (OMT) is a type of duplexer, and it is also referred to as polarisation duplexer. Both the transmitter and the receiver parts of the VSAT system is served by a single feed horn. A passive microwave device used to interface these two components (i.e. the transmitter and receiver sides) to enable them to share the same single feed horn is called OMT. Polarisation is the direction of the electric field component of the radio signal. Common polarisations for VSAT feeds are linear co-polarisation, circular cross-polarisation and linear cross-polarisation. The OMT can separate or combine two signal paths with orthogonal polarisations. The basic parameters of a VSAT antennas is summarised in Table 12.7.

12.5.3 *Gateways*

The gateway is also referred to as hub or teleport. It is a ground station which interfaces the communication between subscribers, as well as between the satellite subscribers and terrestrial communications networks. Figure 12.25 is a block diagram of a satellite gateway. We leave the description of the components of the gateway as a research exercise to the reader. Every satellite network usually has one control center, but multiple gateways. For example, Iridium has 12 gateways scattered around the

Table 12.7 VSAT antenna case study: Intellian v60G Ku-band VSAT Antenna.

Antenna Parameter	Value
VSAT antenna type	3 axis: azimuth, elevation, cross-level
Size	0.845 m × 0.78 m
Weight	59.5 kg
Reflector diameter	0.6 m
Input voltage	100-240 VAC
Transmit gain	38.1 dBi (mid-band)
Receive gain	35.8 dBi (mid-band)
System G/T	> 13.64 dB/K (clear sky with 30° elevation)
BUC output power	4 W, 8 W, 16 W
Polarisation	Cross-polarisation and co-polarisation or cross-polarisation only
Elevation range	−10° to 100°
Satellite modem interface	Ethernet port, RS-232C, I/O ports

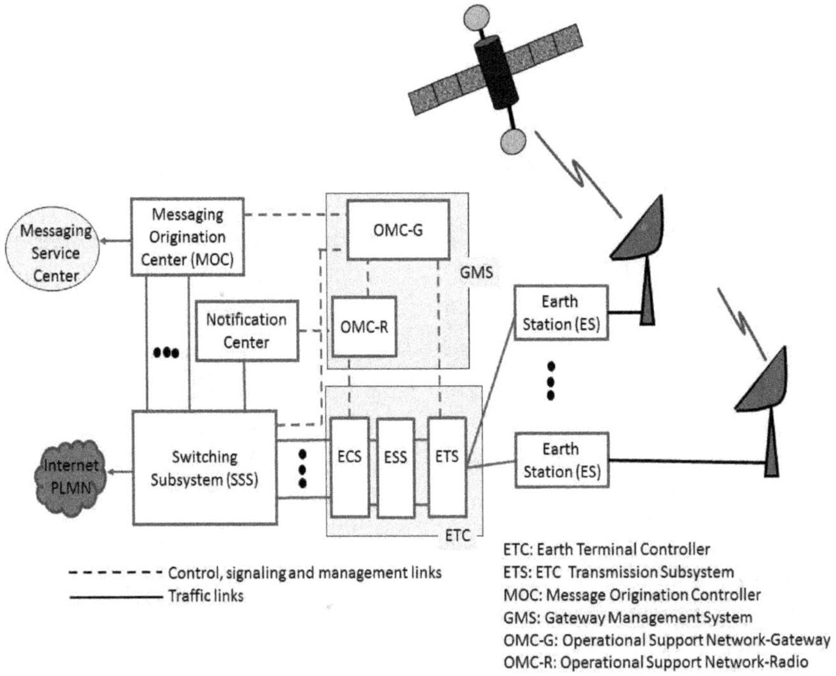

Fig. 12.25 The basic architecture of a satellite gateway.

world, but only one control center located in Chandler, Arizona, USA. Globalstar has 24 gateways, seven of which are located in North America.

12.5.4 *System Control Segment*

The system control segment (SCS) provides the central management function for the satellite network or system. It provides global operation, support, and control services for the space-borne segment constellation, delivers satellite tracking data to gateways, and performs the termination control function of messaging using the Message Termination Center (MTC). The SCS of Iridium comprises three main components: telemetry, tracking and control (TTAC), Control Facilities (CF) and Operational Support Network (OSN).

12.5.5 *Check Your Understanding*

Students are encouraged to check their understanding of the content of this section with the following questions prior to proceeding to the next section.

(1) State the differences between anomaly and look angles in satellite communications.
(2) List three popular applications of the VSAT system.
(3) Search to find three vendors of VSAT services in your city of residence.
(4) Clearly state the differences between BUC and LNB as used in VSAT.
(5) What is the main reason for the word *block* in BUC?
(6) Assume that a VSAT system uses a Gregorian-shaped antenna with the parameters listed in Table 12.7. Compute:
  (a) The uplink mid-frequency corresponding to the antenna's gain.
  (b) The downlink mid-frequency corresponding to the antenna's gain.
  (c) Compute the beamwidth of the antenna in both the uplink and downlink directions.
  (d) Compute the total system temperature in both the uplink and downlink directions.
(7) Find four major vendors of VSAT infrastructure globally.
(8) Can we replace the BPFs in the VSAT architecture of Fig. 12.24 with low pass filters (LPFs)? Justify your answer.
(9) Discuss the purpose in the satellite MODEM in the VSAT architecture of Fig. 12.24.
(10) **In-class group study**: Study the gateway whose block diagram is shown in Fig. 12.25. In groups of 2 to 4 students, research to find the functions of each of the components of the gateway.
(11) **In-class group research**: In groups of 2 to 4 students compare and contrast the low noise amplifier (LNA) and the travelling wave tube amplifier (TWTA).
(12) What are look angles? Discuss the importance of look angles for a satellite earth station.
(13) Derive the condition(s) under which the elevation angle β of an earth station equals the central angle ϕ.
(14) Search to find the number of gateways and their locations for the OPTUS satellite system used in Australia and New Zealand.

12.6 Satellite Link Design: Link Budget

A satellite link or connection comprises the uplink (i.e., link from the Earth station to satellite) and the downlink (i.e., link from the space segment to the Earth station). A main aspect of the link design is the **link budget**. We shall use the architecture in Fig. 12.26 for our discussions on satellite link design. In link budget we subtract all losses from all gains. Knowing the satellite orbital altitude and the target bit error rate (aka link performance), the link budget informs us about the parameters of the satellite transceiver (antenna size and gain, noise factors, transmit power, receiver sensitivity, etc.) and the link margin required to offset all the losses (some of which must be estimated) in the link to meet the required link availability or outage.

The satellite link is affected by many variables: some are natural while others are man-made. Figure 12.27 summarises the factors constituting the losses to be considered in a satellite's link budget. Factors affecting the link design and thus the link budget include:

- The path loss: caused by signal spreading in space, but not attenuation.
- Rain attenuation.
- Atmospheric absorption.
- The terrain.
- Transceiver parameters: gain and efficiency of antenna gain, transmit power, feeder cable losses.

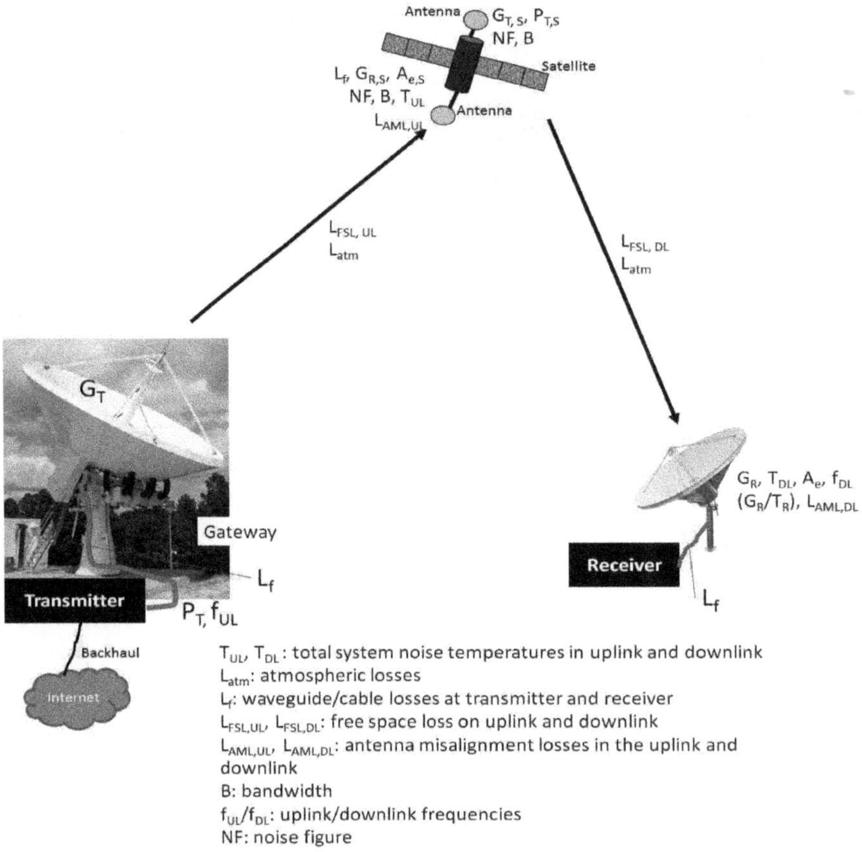

Fig. 12.26 An architecture for satellite link budget.

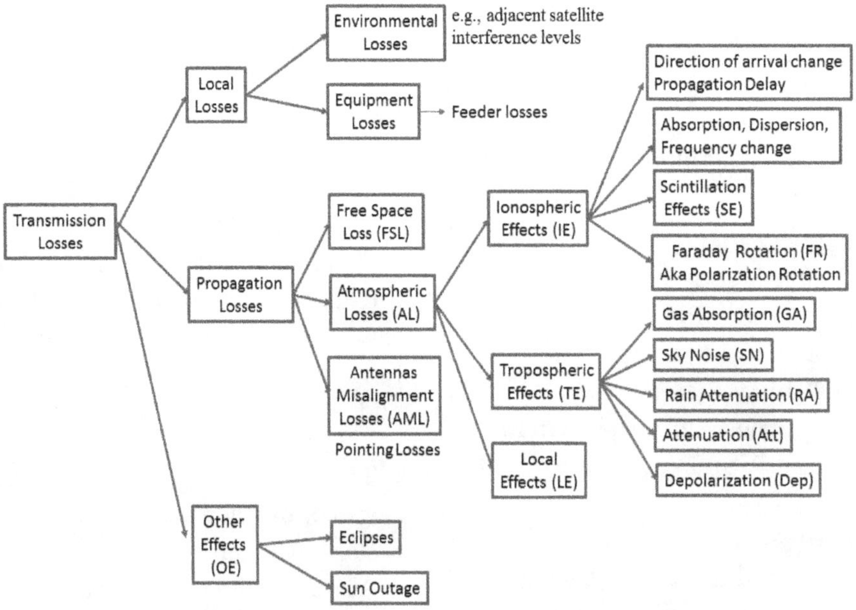

Fig. 12.27 Losses affecting a satellite link design.

- The longitude and latitude of the uplink and downlink of the earth stations.
- Downlink and uplink frequencies.
- Required data rate.
- Modulation scheme (e.g., BPSK or QPSK).
- Error correction type. Satellites usually use FEC with the rate 1/2 or 3/4.
- Expected bit error rate (BER), which dictates the minimum required E_b/N_0.

Most of these link factors were discussed in Chapter 10 on point-to-point microwave communications, and in Chapter 8 on terrestrial mobile communications. The reason is that most of these factors affect every wireless communications links. The factors that are peculiar to satcom are the ionospheric effects which are not remarkable in terrestrial wireless links. There are three figures of merit in satellite link design. These are:

- Effective isotropic radiated power (EIRP).
- Receiver gain to system temperature ratio, G/T.
- Carrier-to-noise ratio, C/N or C/N_0.

The gain-to-noise-temperature ratio (G/T) is an important figure of merit used to characterise the performance of antennas. Here, G is the gain of the receive antenna at the receive frequency while T is the equivalent noise temperature in kelvins (K) of the receiving system. It is defined as

$$G/T\ \texttt{[dB/K]} = \texttt{antenna gain [dB]} - 10\log_{10}(T)\ \texttt{[dB/K]}\,. \quad (12.43)$$

Components of the system noise temperature include:

(1) Antenna noise temperature.
(2) Noise temperatures of all components from the antenna terminal to the receiver output, including waveguides, cables, LNA and block downconverter.

The EIRP is defined as

$$EIRP = P_T G_T \quad (12.44)$$

which in decibels is equivalent to

$$EIRP\ \texttt{[dBW]} = P_T\texttt{[dBW]} + G_T\ \texttt{[dB]}\,. \quad (12.45)$$

The carrier-to-noise ratio is defined as

$$\begin{aligned} C/N_0 &= \frac{P_R}{k \cdot T} = \frac{P_T G_T G_R / L_{tot}}{k \cdot T} \\ &= EIRP \cdot \frac{G_R}{T} \cdot \frac{1}{k \cdot L_{tot}} \end{aligned} \quad (12.46)$$

which in decibels become

$$C/N_0 = EIRP\ \texttt{[dBW]} + G/T\ \texttt{[dB/K]} - L_{tot}\ \texttt{[dB]} - k\ \texttt{[dBW/K/Hz]}\,. \quad (12.47)$$

Equation (12.47) is valid for both the uplink and the downlink. In this equation L_{tot} includes all the losses mentioned in Fig. 12.27.

The free space path loss (FSL) was derived in Chapters 8 and 10 as

$$L_{FSL} = \left(\frac{4\pi h}{\lambda}\right)^2 \quad (12.48)$$

where h is the link distance which equals the altitude of the orbit on which the satellite is located. Satellite links are usually symmetrical so h holds for both uplink and downlink. The FSL is caused by the spreading of the signal between the transmitter and the receiver, but not by attenuation. A situation called **sun outage** occurs if the sun's radiation on a satellite is too strong that the signal-to-noise ratio is too low to detect satellite signals by the earth stations. The sun outage occurs if the satellite reaches a point in the middle between the earth and the sun. There are a plethora of software packages available to do the link budget. Figure 12.28 shows an Excel spreadsheet that can also be used to that effect.

12.6.1 *Check Your Understanding*

Students are encouraged to check their understanding of the content of this section with the following questions prior to proceeding to the next section.

(1) List the components in a satellite link.
(2) Describe what a link budget is.
(3) What is the importance of the G/T ratio in satellite link design?

Parameter	Units	Value
Transmit power (P_T)	dBm	57.92
Transmit antenna gain (G_T)	dB	
EIRP (P_TG_T)	dBm	
Speed of light in vacuum ($c=3\cdot10^8$m/s)		
Frequency (f=1.625 GHz)	dBHz	92.1085
Channel bandwidth (B=31.5 kHz)	Hz	44.9831
System noise temperature (T=15°C)	dBK	24.5939
Boltzman's constant (k)	dBm/K/Hz	-198.6
Thermal noise power (N_0=kT)	dBm/Hz	-174.0061
Thermal noise power (N=kTB)	dBm	-129.023
Antenna feeder loss	dB	-1
Distance /altitude (h)	m	781000
Free space loss	dB	
Rain attenuation	dB	-0.12
Atmospheric absorption/losses	dB	-1
Polarization loss	dB	-3
Antenna misalignment loss	dB	-1
Receive antenna gain (G_R)	dB	
Receiver G/T	dB/K	
Carrier-to-noise ration (C/N = EIRP+(G/T)-losses-k-B)	dB	
Bit rate (R_b)	bps	
E_b/N_0 @ BER of 10^{-3}	dB	
uplink margin	dB	16

Fig. 12.28 An example of spreadsheets for the satellite's uplink budget.

(4) Research to find examples of the environmental losses listed in Fig. 12.27 which affects a satellite's link.

(5) Research to find some of the factors contributing to equipment losses.

(6) The performance of a satellite link can be adversely affected by the locality of the earth stations. Reflect and state some of these local effects.

(7) **In-class research**: In groups of 2 to 4 students, research and write a report of *Faraday rotation* and how it affects a satellite's link quality, especially at frequencies below 10 GHz.

(8) **In-class research**:

(a) In groups of 2 to 4 students, research to find the missing values in Fig. 12.28 for the *Iridium* satellite system.

(b) Perform the uplink budget.

In-class research:

(a) In groups of 2 to 4 students, research to find the appropriate values in Fig. 12.28 for the OPTUS-10 satellite system.

(b) Perform the uplink budget.

12.7 Current Research Areas in Satellite Communications

Humans will continually possess the desire to see beyond what is now known in order to discover new things or improve upon the currently known. This is also true for satellite communications. For example, the following are some of the ongoing research topics in satcom:

(1) *Higher data rates*: the higher the data rate, the better the quality of communications, especially for the transfer of images and videoconferencing. Higher data rates require higher frequencies. However, the higher the frequency, the higher the signal propagation path losses.

(2) *Light-weight satellites*: lifting a weight into space is difficult and cost much fuel. Therefore, research into finding *weightless* spacecrafts remains unabated.

(3) *Reusable launch vehicles*: we seek launch vehicles which can be used to launch multiple vehicles, especially unmanned types. Bringing a vehicle back to Earth after launching a satellite is very difficult even for manned launched vehicles. Rockets used to launch satellites so far can be used only once. Some rockets return to the Earth but are unusable, while some remain in space as junks.

(4) **Electric spacecrafts**: we prefer electrically-powered spacecrafts to chemically-propelled types. Spacecrafts powered by electricity use much less propellant than chemically-powered rockets.

12.8 Review of Important Terms and Acronyms

Acronyms, abbreviations and important terms		
aerobraking	anomaly	apogee
astronaut	azimuth angle	backhaul
bent pipe transponder	big LEO	BPF
broadband LEO	BSS	BUC
bus	centrifugal force	centripetal force
Clarke orbit	DAMA	downlink
earth station	elevation angle	escape velocity
figure-of merit (G/T)	free space loss (FSL)	FSS
gateway	GEO	Globalstar
gravitational force	ground station	inclination
IoT	Iridium	ISL
Keppler's laws	launch vehicle	LEO
link budget	link margin	little LEO
LNA	LNB	local oscillator (LO)
look angles	mass budget	MEO
MSS	Newton's laws	OPTUS
orbit	orbital elements	orbital mechanics
orbital period	orbital velcocity	PAMA
payload	perigee	pitch axis
power budget	processing transponder	propellant
rain attenuation	regenerative transponder	repeater
retroburning	rocket	roll axis
round-trip delay	satellite	satellite footprint
solar panels	space shuttle	space vehicle
SPC	spinning the satellite method	three-axes method
transponder	TWTA	uplink
Van Allen Belts	VSAT	yaw axis

12.9 Supplementary Reading

(1) Shree Krishna Sharma and Symeon Chatzinotas, *Satellite Communications in the 5G Era*, The Institution of Engineering and Technology, 2018.

(2) Sastri L. Kota and Kaveh Pahlavan, *Broadband Satellite Communications for Internet Access*, Springer, 2003.

(3) Roger Cochetti, *Mobile Satellite Communications Handbook*, John Wiley, 2014.

(4) Louis J. Ippolito Jr., *Satellite Communications Systems Engineering: Atmospheric Effects, Satellite Link Design and System Performance*, John Wiley, 2017.

(5) Gerard Maral and Michel Bousquet, *Satellite Communications Systems: Systems, Techniques and Technology*, John Wiley, 2010.

(6) Timothy Pratt and Charles W. Bostian, *Satellite Communications*, John Wiley, 2002.

Chapter 13

Wireless Positioning Fundamentals

Chapter's Learning Outcomes

Have you ever wondered how to get to a desired geographical location from your current location? That is what positioning, localisation or navigational systems do. Wireless positioning systems exploit inherent features of wireless signals emitted by radio beacons of known locations to estimate the direction, position (latitude and longitude) and altitude, and velocity of an object on the Earth's surface. Wireless positioning systems have a wide range of use cases. For example, a wearable real-time locationing system enables the tracking of in-patients to avoid spreading of infectious diseases, as well as finding the whereabouts of nurses and doctors at a hospital. The wireless connectivity used for positioning can be terrestrial or satellite wireless. Satellites used for positioning include the Global Positioning System (GPS), GLONASS and Galileo. This chapter has two main parts: first it introduces readers to the fundamental principles in wireless positioning. We then review, as case studies, some of the popular positioning systems. In particular, we study:[1]

(1) Abbreviated history of positioning systems.
(2) Basic principles underpinning wireless positioning systems, i.e., TOA, TDOA, DOA, RSSI, triangulation and trilateration.
(3) Basic types and case studies of positioning systems, e.g., VOR, DME, LORAN, GPS and their applications.
(4) Direction finding networks: e.g., LoJack, Wireless E911 and US Rescue 21.
(5) Review of ILS and MLS used for air traffic management at airports.

13.1 Introduction

As we know, many English words have multiple meanings which depend on context. The words positioning, navigation and localisation are used as synonymous terms by some people, while others see the subtle differences between them. Let us examine some of their definitions.

- *Navigation* is the process or activity of accurately ascertaining one's position and planning and following a route — *Oxford English Dictionary.*
- *Positioning* is the process of finding the geographical coordinates (or latitude and longitude) of a point in the universe with respect to a (fixed) reference system or coordinate system.
- *Direction finding* is the process of using bearings to determine the direction towards the physical location of a target (or a source of radio frequency energy). The target may be cooperative or non-cooperative, moving or stationary [Denisowski (2019)]. See Fig. 13.1 for examples of targets.
- *Radiolocation* is the process of finding the physical location of a wireless transmitter.
- *Localisation* is the positioning plus information about the environment where the point is located.

Wireless navigational systems use the properties of radio signals to estimate the direction, position (latitude and longitude) and altitude, and velocity

[1]Three things are unaffordable to miss: wisdom and understanding, counsel and might, and knowledge and the reverence of nature.

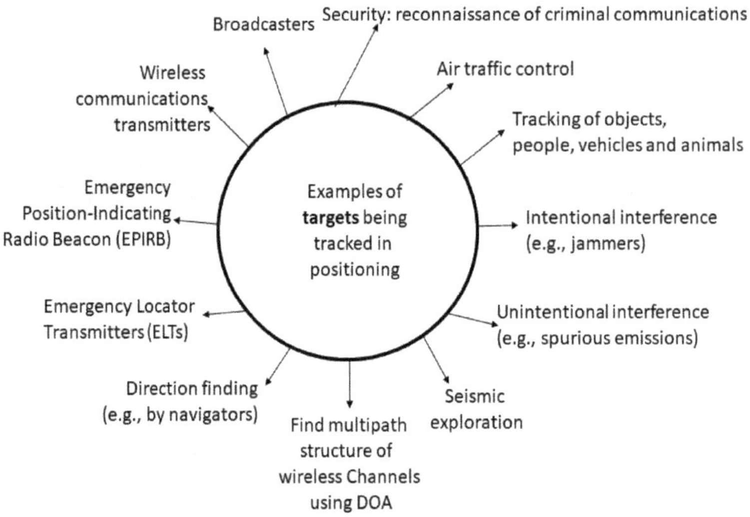

Fig. 13.1 Application examples of wireless positioning systems.

of an object on the Earth's surface. For these reasons, they are also referred to as radio positioning systems. Techniques used to find the position of an object using radio-frequency signals is also referred to as radiolocation. Radio navigational systems, such as the Global Positioning System (GPS), has applications beyond finding the position of an object on the Earth. For example, GPS is being used for the calibration of other navigational systems, to synchronise computer networks and synchronise motion picture equipment. GPS is currently the positioning system with the highest accuracy of about 3.3 m. The next is LORAN C (about 180 m), DECCA (about 200 m), TRANSIT and TACAN (about 400 m), VOR and DME (about 900 m) and OMEGA (about 2.2 km). Measurements received from transmitters with well-known geographical locations are used to estimate the location of an entity. The reference transmitters are referred to as **anchor nodes** or **access nodes**, while the the node which is being located is referred to as the **target node**.

13.1.1 *Categorisation of Positioning Technologies*

We attempt to categorise wireless positioning systems in Fig. 13.2. There are two basic classes of radio navigational systems:

- Terrestrial radio navigational systems: use transmitters mounted on land. Examples are LORAN (LOng-Range Aid to Navigation), RDF (radio directional finding, ILS (instrument landing system) and MLS (microwave landing system).
- Satellite radio navigational systems: use transmitters mounted in space. Examples are GPS (Global Positioning System) which is owned and managed by the US DoD/Army, Galileo Global Navigation Satellite System (GNSS) which is owned and managed by the European Union, and GLONASS (GLObal Navigation Satellite Systems) owned by the Russian Federation.

An earth-based positioning system has a limited coverage, while a space-based position system can have a global coverage. Second, an earth-based positioning system usually uses ground waves which cannot be used to measure altitude, rendering it two-dimensional. It measures only the latitude and the longitude of locations. A space-based position system measures in addition the altitude, making it three-dimensional. Therefore, earth-based positioning systems are not the best for aviation. Third, land-based systems use fixed reference points to compute locations through the *intersection of circles*. Space-based systems, however, use moving satellites as reference points to compute locations through the *intersection of spheres*. The accuracy

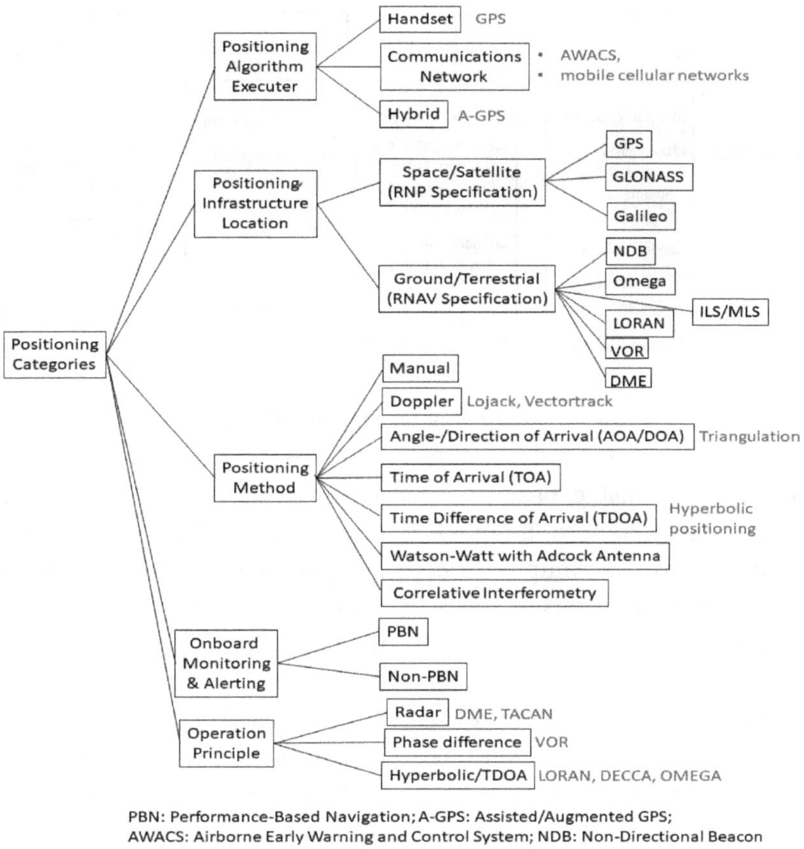

Fig. 13.2 Classification of wireless positioning systems.

of measured data from space-based positioning systems depends on how accurate the positions of the satellites can be estimated.

Positioning systems can also be categorised into handset-based, network-based and hybrid systems. In *handset-based positioning*, the target node is usually a portable device such as a GPS receiver which can be integrated into the device. The target node extracts its position from signals it receives from anchor nodes which have known locations. A popular example of handset-based positioning technologies is GPS. In *network-based positioning*, multiple anchor nodes forward their measurements from signals received from the target node to a central node to estimate the target's location. Examples of network-based positioning technologies are mobile cellular networks and Airborne Early Warning and Control System (AWACS). There are *hybrid positioning technologies* like Augmented/Assisted GPS (A-GPS) which store satellite orbital information in databases for retrieval over terrestrial networks to expedite localisation.

Positioning systems can be differentiated based on their operation principle. Some are based on radar principles, such as DME and TACAN. Systems like VOR exploit the phase difference between the received signal and a reference signal. We have LORAN, DECCA and Omega which are based on hyperbolic principles arising from time difference of signal arrivals.

13.1.2 *Evolution of Wireless Positioning Technologies*

Navigational systems have come of age. They have evolved from primitive stone age navigational instruments through celestial and LORAN to the modern sophisticated satellite-based navigation. Figure 13.3 attempts to summarise these developments. In the stone age people relied on dead reckoning as the method of finding one's bearings.

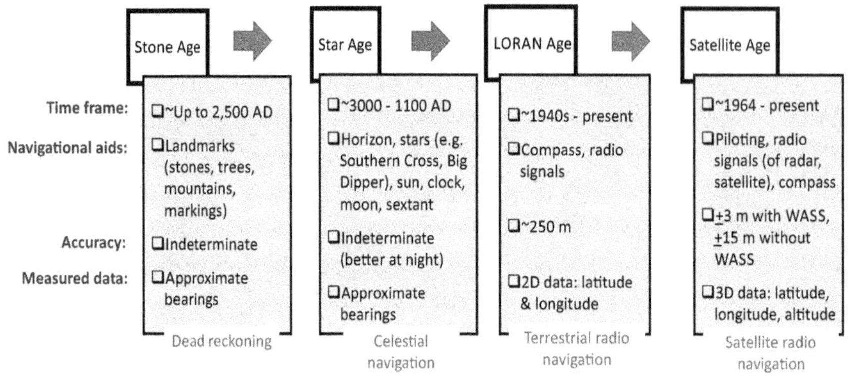

Fig. 13.3 Evolution of navigational technologies.

Dead reckoning is the process of using past position (e.g., based on landmarks) to determine one's current position. The inputs to the computation of the current position were past positions as reference points, time, speed and direction of travel. One can clearly see the weakness in this process, which is error propagation. The estimation of the previous locations in error affects the current estimation. Therefore, one could be very much off course after a while. During the stone age even date of births of people were remembered using reference points. I remember my grand mum saying something like "this person was born three days after that president came to power."

Celestial navigation relied on the horizon, specular star constellations, the moon and the sun as its navigational aids. Clearly, navigators then needed to know their geography quite well. For example, the Southern Hemisphere were identified through its Southern Cross star constellation, while the Northern Hemisphere was identified through the *Big Dipper* star constellation. Navigators estimated their location by computing the angle between the horizon and sky-borne object (i.e., stars, moon and sun). The tools used to achieve this included sextant. A sextant is used to measure the angle between the horizon and a celestial body (e.g., planets, stars, moon and sun). The estimated angle is then used to read off a chart the corresponding geographical position. A main weakness in celestial navigation is that it does not work if it is cloudy and during the day.

Satellite-based navigation is the most current and most globally useful navigational system. It began with the Transit satellite (aka NAVSAT meaning navigational satellite or NNSS meaning national navigational satellite system) which was put into operation in 1964 by the US Navy. The most popular satellite-based positioning system is currently the Global Positioning System (GPS) with the Wide Area Augmentation System (WAAS) which proves the accuracy of GPS.

Some of the navigational methods used many centuries ago are still in use today. For example, navigators still rely on fixed visual references for navigation, a navigational method referred to as `piloting`. The problem in this method is the requirement of good visibility. We have observed from Fig. 13.3 that all navigational systems have some things in common. For example, all of them need some reliable reference points whose locations are known.

Wireless positioning technologies are used in a myriad of industries and a myriad of applications within a given industry. For example, the military use it to locate electronic order of battle (EOB) and hostile forces. The law enforcers use positioning system to track people and vehicles, for spectrum deconfliction or to find the location of secret transmitters.

13.1.3 *Basic Principles of Wireless Positioning*

The basis of wireless positioning technologies is the ability to estimate the location of a geographical point if we know distances or angles from that point to known locations or reference points. We exploit the variations in amplitude, phase or frequency of the radio signal received from the source whose location is being sought. In satellite-based positioning technologies, the reference points are the moving satellites in space. Terrestrial positioning technologies, however, use fixed radio towers or beacons as reference points. We use the time a signal from a reference point takes to reach a radio detector to estimate the distance between the two points.

(1) A radio device receives signals from a minimum of two to four transmitters with known locations. The receiver then extracts information from the received signals to estimate its location. Alternatively, anchor stations can receive signals from a target and forward the information to a central station to compute the target's location.
(2) The radio receiver and the transmitters are usually expected to be in line-of-sight (LOS), i.e., linked by a direct path, for the estimated data to be accurate or even be valid. This is because the receiver has no way of knowing any detours experienced by signals between it and the transmitters. All wireless positioning techniques are affected by the radio wave properties studied in Chapter 5. These include signal reflection, refraction, absorption, background noise (aka ambient noise) and especially, multipath signal propagation.

13.1.4 *Latitude, Longitude and Bearings*

A position on the earth is usually stated in terms of its latitude, longitude and altitude. Latitudes and longitudes are measured in arc degrees (°), arc minutes (′ or arcmin) and arc seconds (″ or arcsec). Therefore, a proper understanding of them is important in navigation. Latitudes and longitudes were briefly discussed in Section 9.5.2 (Coverage-Limited Network Design) in Chapter 9 on Cellular Mobile Systems Design Principles.

The *bearing*, also known as *azimuth*, of a point is the horizontal angle measured clockwise from the geographical north direction to the point. We use bearings in positioning systems to find the direction towards a signal. Let us review Fig. 9.35(d). Let ϕ_{ij} be the bearing of location j with respect to location i. Then in reference to the points A, B and C in Fig. 9.35(d) we can write the bearings

$$\begin{aligned} 45^\circ &< \phi_{AC} < 90^\circ, \\ 90^\circ &< \phi_{AB} < 180^\circ \\ 180^\circ &< \phi_{CA} < 270^\circ, \\ 270^\circ &< \phi_{BA} < 300^\circ\,. \end{aligned} \tag{13.1}$$

Let us pause to discuss some applications of wireless positioning in aviation.

13.1.5 *Navigation Specifications: RNAV and RNP*

Two sets of regulations govern all aspects of aircraft operations in civil aviation. These are Instrument Flight Rules (IFR) and Visual Flight Rules (VFR). IFR-based navigation relies on instruments for guidance, while VFR-based navigation relies on natural visual guidance. Of course, VFR is safe only in good weather conditions.

Area Navigation (RNAV), previously called Random Navigation, is a set of IFR used for navigation. RNAV systems include Omega, VOR, LORAN and DME. As positioning systems geared more towards satellite-based systems, International Civil Aviation Organisation (ICAO) is transitioning from RNAV to Required Navigation Performance (RNP) or RNP Authorisation Required (RNP AR). RNP is equivalent to GNSS-based RNAV. Therefore, RNP is a subset of RNAV. Let us note the following definitions by ICAO:

- Performance-based navigation (PBN) is an area navigation based on performance requirements for aircraft operating along an ATS route, on an instrument approach procedure or in designated airspace.
- *Navigation specification*: a set of aircraft and aircrew requirements needed to support performance-based navigation operations within a defined airspace. There are two kinds of navigation specification:
  - *Area navigation (RNAV) specification*: a navigation specification based on area navigation that does not include the requirement for performance monitoring and alerting.

– *Required navigation performance (RNP) specification*: A navigation specification based on area navigation that includes the requirement for performance monitoring and alerting.

Therefore,

$$RNP = RNAV + \texttt{on-board performance monitoring} + \texttt{alerting} \quad (13.2)$$

The performance being monitored onboard of NAVAID based on RNP are accuracy and integrity of measurements, continuity and system availability.

The Barometric-Vertical Navigation (Baro-VNAV) is a type of RNAV system which computes the vertical guidance for the pilot using the barometric altitude information from the aircraft's altimeter.

13.1.6 *Check Your Understanding*

Students are encouraged to check their understanding of the content of this section with the following questions prior to proceeding to the next section.

(1) State the reason why space-borne navigational systems are superior to ground-based systems in their application in aviation.
(2) Discuss two differences and two similarities between land-based and space-borne radio navigational systems.
(3) List examples of the navigational aids used during the stone age.
(4) List examples of the navigational instruments used during the star age.
(5) **In-class discussion:** the navigational aids used prior to the radio age were primitive and required more skills to operate compared with modern navigational instruments. Discuss.
(6) Discuss the purpose of Big Dipper and Southern Cross star constellations in navigation.
(7) Explain the meaning of dead reckoning. What is the main weakness in dead reckoning?
(8) Explain the meanings of VFR and IFR and differentiate between them.
(9) Discuss the differences between RNAV and RNP.
(10) Research to find which electronic approach and landing system(s) are used at the airport closest to your suburb of residence.

13.2 Positioning Methodologies

We survey some of the most popular methods used to find the position of, and directions to, a location. We start with the most primitive method, which is manual direction finding. All the positioning systems apply one or a combination of the following parameters of wireless signals: time of arrival (TOA), time difference of arrival (TDOA), angle of arrival (AOA) or direction of arrival (DOA), and received signal strength indicator (RSSI). Kalman filtering can be used to fuse multiple positioning methods to increase accuracy and reliability. Note, however, that RSSI method works in both line-of-sight (LOS) and non-LOS (NLOS) situations, while AOA/DOA, TOA and TDOA require LOS.

In positioning, we classify nodes into target node and reference nodes or anchor nodes. The locations or coordinates of the reference nodes are assumed known, which are used to find the location of the target node, in a process referred to as positioning or localisation. The positioning of a target can be done either by the target or by the reference nodes through a master node.

13.2.1 *NDB, ADF and Manual Direction Finding*

A device which has been intentionally designed to attract attention to a specific location is called a

beacon. A wireless transmitter with a known fixed location which broadcasts signals periodically or continuously at a specific frequency is referred to as a **radio beacon**. The information broadcast by the radio beacon is often its location and/or identity. A radio beacon which transmits signals with equal power in all directions or 360° is referred to as **non-directional beacon** (NDB). NDBs operate in the frequency band 190–535 kHz. NDBs usually broadcast their identity using 1 to 3-character Morse code.

Direction finding systems can be manual or automatic (aka automatic direction finder or ADF). An ADF receiver is an instrument which continuously and automatically displays the relative bearing from a location to a suitable known radio station. ADFs (such as those onboard aircrafts) normally tune to NDBs and use their bearings for navigation purposes.

A manual direction finder (DF) is a radio instrument equipped with a directional antenna, such as log-periodic or Yagi-Uda, which moves or rotates the antenna until the detected signal reaches its peak amplitude. The bearing at the peak signal amplitude is the direction of the signal being detected.

Figure 13.4 illustrates the components of two direction finding architectures. The basic components are antennas (one each for very high frequency or VHF, ultra-high frequency or UHF and tremendously high frequency or THF), a receiver with USB ports, a processor and a microphone. It has interfaces for connection to a standard computer.

In manual DF, we determine the bearings to detected radio signals from a radio transmitter. We superimpose all the bearings. The intersection of the bearing lines is the estimated location of the radio transmitter. This method is called triangulation, which is discussed later in this chapter.

A NDB is a low frequency, ground-based omnidirectional radio transmitter used as an instrument approach at airports and offshore platforms or drill ships for petroleum exploration. ADF is the instrument used to detect signals transmitted by the NDB onboard aircrafts. International Civil Aviation Organisation (ICAO) requires the ADF signal to be transmitted uninterrupted 24/7 basis in the frequency band 190 to 1800 kHz worldwide. Using the ADF, a pilot determines the direction to the NDB relative to the aircraft.

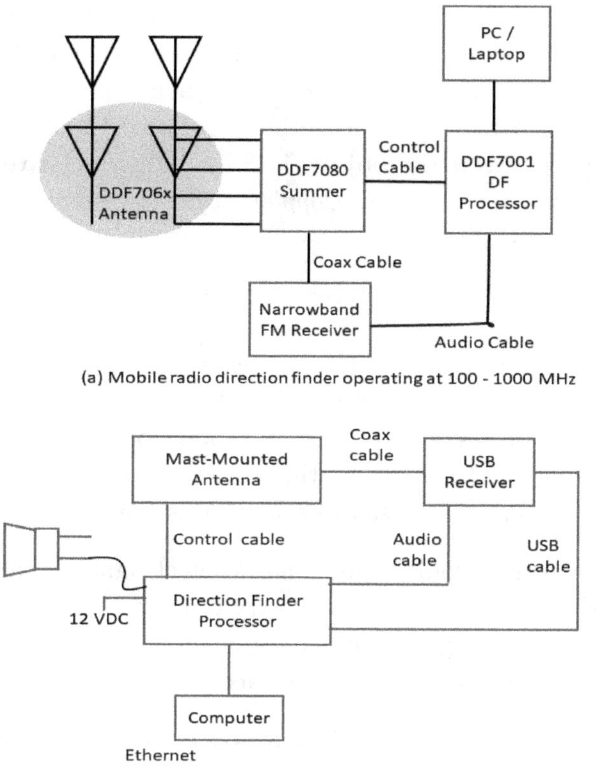

(a) Mobile radio direction finder operating at 100 - 1000 MHz

(b) Fixed site radio direction finder operating at 115 - 1000 MHz

Fig. 13.4 Block diagram of direction finding instruments [Doppler Systems (2019)].

NDBs which transmit powers in the range 25 to 125 watts can cover a range up to circa 185.2 km. However, some NDBs transmit 500–1000 watts power to cover a longer range. As it is reliable and cheap, NDB is the most widely used navigational aids at airports. Besides being used for airplane approaches, the NDB is used as Locator Outer Markers (LOM) for Instrument Landing Systems (ILS).

13.2.2 *Doppler Shift-Based Direction Finding*

Named in honour of the Austrian physicist and mathematician Christian Andreas Doppler (1803–1953), Doppler shift is a form of frequency modulation which is caused by the relative motion between two objects. Doppler-based positioning systems were first used in 1941. The relative motion between the two objects changes the frequency in the signal transmitted from one object to the other. An increase in frequency indicates moving towards each other, while a decrease in frequency indicates moving away from each other. Assume that the angle between the objects is θ, and their relative velocity is v, then the *Doppler shift* between them is

$$f_{\mathrm{d}} = \Delta f = \frac{v}{c} f_0 \cos\theta = \frac{v}{\lambda}\cos\theta \tag{13.3}$$

where f_0 is the operating frequency, $c \approx 3 \cdot 10^8$ m/s and λ is the operating wavelength.

In case there is no relative motion between the two objects we must synthesise a motion. Two of the popular ways used to achieve this are:

- By using a rotating antenna to detect signals from the other relatively stationary object.
- By switching sequentially between a set (e.g., 4) of antennas at a very fast pace.

Each of the two methods produces a set of Doppler pauses which are used to synthesise the so-called Doppler sine wave [Denisowski (2019)].

> Doppler effect is exploited to find the direction of a signal source. For example, if the frequency in the detected signal is increased then we know that we are moving towards the signal source. Lojack and Vectortrac are Doppler based tracking systems used to track stolen vehicles by the police. A vehicle equipped with a Lojack transceiver (called vehicle locator unit or VLU), for example, can be tracked by the police. Upon reporting a loss, the Lojack network activates the VTU in the vehicle to enable its tracking by the police. Of course, the police should have access to the Lojack system.

As we shall study in the chapter on radar systems, a popular application of Doppler shift is the determination of an object's travelling velocity. Assume that a target-tracking antenna has beamwidth of $\Delta\alpha$ and transmits signals at the frequency f_0. Assume that there are many different types of objects (referred to as backscatters) in the region illuminated by the antenna's beam. There is always a

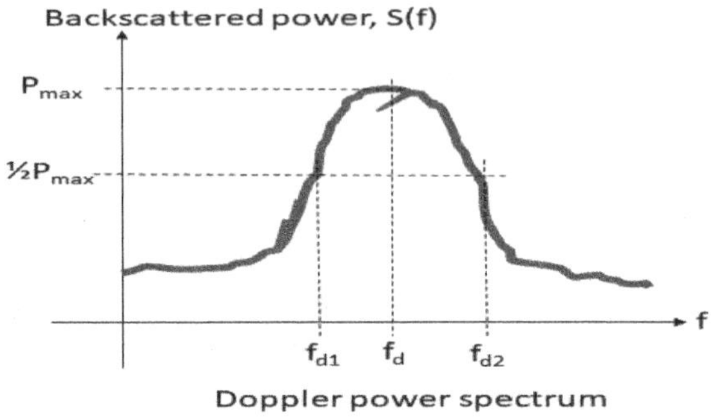

Fig. 13.5 Doppler power spectrum.

background noise (aka ambient noise) in the system which is Gaussian or normal distributed with mean zero and variance σ. Each backscatter produces a different phase shift or Doppler shift, and the phases have random distributions. Therefore, the cumulative return signal appears like a narrowband signal superimposed on the Gaussian-distributed background noise. The resultant backscattered power and Doppler shift, referred to as Doppler power spectrum, has a Gaussian shape (Fig. 13.5). The amplitude of the distribution is a function of the antenna/radar parameters, and the mean frequency is the Doppler shift f_{d} .

Under the assumption of uniform backscattering in the region illuminated by the antenna, and the returns from the nearest and farthest illuminated points are the same, the Doppler power spectrum peaks at the frequency

$$f_{\mathrm{d}} = \frac{2v\cos\theta}{\lambda}. \tag{13.4}$$

As illustrated in Fig. 13.5, the half-power bandwidth of the Doppler power spectrum is the difference between the upper (f_{d1}) and lower (f_{d2}) half power points of the spectrum, i.e.,

$$\begin{aligned}\Delta f_{\mathrm{d}} &= f_{\mathrm{d1}} - f_{\mathrm{d2}} \\ &= \frac{2v}{\lambda}[\cos(\theta - \Delta\theta/2) - \cos(\theta + \Delta\theta/2)] \\ &= \frac{2v}{\lambda}\sin(\theta)\cdot\sin\left(\frac{\Delta\theta}{2}\right) \approx \frac{v\Delta\theta}{\lambda}\sin(\theta).\end{aligned} \tag{13.5}$$

Let us define the relative width of the Doppler power spectrum as

$$\gamma_w(\theta) = \frac{\Delta f_{\mathrm{d}}}{f_{\mathrm{d}}} \overset{(13.4),(13.5)}{=} \Delta\theta \cdot \mathtt{tan}]\,\theta. \tag{13.6}$$

If the backscatter power is $S(f)$, then the total average power is

$$\bar{P}_r = \int_{-\infty}^{\infty} S(f)df\,. \qquad (13.7)$$

Example

Assume that the beamwidth of the antenna of a tracking radar system operating at 3 GHz is 4°. The relative width of the Doppler power spectrum is measured as 20%. Compute:

(1) The mean angle at which the strongest backscatter returns to the radar antenna.
(2) Moving velocity of the target if the Doppler power spectrum peaks at 10 Hz.

Solution

We are given $\gamma_w(\theta) = \frac{\Delta f_\mathrm{d}}{f_\mathrm{d}} = 0.2$, $f_0 = 3$ GHz, $f_\mathrm{d} = 500$ Hz and $\Delta\theta = 4°$.

(1) From Eq. (13.6) the mean angle at which the strongest backscatter returns to the radar antenna is

$$\theta = \mathtt{tan}^{-1}\left[\frac{\gamma_w(\theta)}{\Delta\theta}\right] = \mathtt{tan}^{-1}(0.05) \approx 0.05\,.$$

(2) The wavelength is $\lambda = c/f = 0.1$ m. Thus,

$$v = \frac{f_\mathrm{d}\lambda}{\cos\theta} = 100\cos 0.05 \approx 100.1\ \mathtt{m/s}\,.$$

13.2.3 *Power-Based Positioning: RSSI*

A common metric used in received signal power-based positioning is the received signal strength (RSS) or the RSS indicator (RSSI). This method uses triangulation to locate a target. It exploits the fact that the attenuation experienced by a radio signal depends on the distance that it travels. Therefore, knowing the wireless channel and the transmitted signal power enables us to estimate the distance travelled by a signal. This travel distance is the distance between the transmitter and the receiver. Knowing the distances from the target to at least three reference nodes enable us to find the position of the target through triangulation.

Positioning algorithms based on RSS or RSSI require knowledge of the channel or at least its statistics. Let us define the following:

- P_T: power in the transmitted signal.
- $P_{\mathrm{R},k}$: power in signal received by node k at the distance d_k away from the transmitter.
- γ: path loss exponent, which indicates how fast the power of RSS decreases with distance from the transmitter.
- d_0: known reference distance away from the transmitter.
- $P_0(d_0)$: known power in the transmitted signal at the reference distance d_0.
- $X_\sigma \sim \mathcal{CN}(0, \sigma^2)$: Gaussian r.v. with mean 0 and variance σ^2 modelling shadow fading.
- G_T, $G_{\mathrm{R},k}$: gain of the transmitter and receiver, respectively.

The standard formula in decibels (dBm) used for the received power is the Friis free-space propagation formula

$$P_{\mathrm{R},k} = P_0(d_0) + 10\gamma\log_{10}\frac{d_0}{d_k} + X_\sigma\ \mathtt{[dBm]},\ k = 1, 2, \ldots \qquad (13.8)$$

which is equivalent to

$$d_k = d_0\left[\frac{P_0(d_0)X_\sigma}{P_{\mathrm{R},k}}\right]^{\frac{1}{\gamma}},\ k = 1, 2, \ldots, K\,. \qquad (13.9)$$

An alternative form to Eq. (13.8) for free-space signal propagation is

$$P_{\mathrm{R},k} = \left(\frac{\lambda}{4\pi d_k}\right)^2 P_\mathrm{T}G_\mathrm{T}G_{\mathrm{R},k},\ k = 1, 2, \ldots \Leftrightarrow d_k = \sqrt{\frac{\lambda^2 P_\mathrm{T}G_\mathrm{T}G_{\mathrm{R},k}}{16\pi^2 P_\mathrm{R}(d_k)}}\,. \qquad (13.10)$$

Multiple anchor points or reference points with known coordinates can receive a signal transmitted by a target node. Each anchor node can estimate its distance d_k, $k = 1, 2, \ldots$ from the target if we know channel's parameters. We then use these measured distances and triangulation to compute the unknown location.

Another model for the received signal strength is [Song (1994)]

$$P_{\mathrm{R},k}(d_k) = C_k P_\mathrm{T} d_k^{-\gamma} = C_k P_\mathrm{T}\|\mathbf{x} - \mathbf{x}_k\|_2^{-\gamma},\ k = 1, 2, \ldots, K \qquad (13.11)$$

where C_k accounts for antenna gain, antenna height and other factors affecting signal propagation, $\mathbf{x}$ is the coordinates of the target and $\mathbf{x}_k$ is the coordinates of the reference point k.

Let $\mathbf{z} = (x, y)^T$ be the coordinates of the target being sought after, and $\mathbf{z}_k = (x_k, y_k)^T$, $k = 1, 2, \ldots, K$ be the coordinates of the known kth reference point. The coordinates of the reference points and the distances from the target to the reference points as in Eq. (13.9) or Eq. (13.10) are known, then we can locate the target using the set of equations

$$(x_k - x)^2 + (y_k - y)^2 = d_k^2, \; k = 1, 2, \ldots, K\,. \tag{13.12}$$

Let us build the difference equations

$$\begin{aligned} d_l^2 - d_K^2 &= (x_l - x)^2 + (y_l - y)^2 \\ &\quad - [(x_K - x)^2 + (y_K - y)^2], \\ &\quad l = 1, 2, \ldots, K - 1 \end{aligned} \tag{13.13}$$

which after simplification becomes

$$\begin{aligned} &(x_l - x_K)x + (y_l - y_K)y \\ &= \frac{d_K^2 - d_l^2 + x_l^2 - x_K^2 + y_l^2 - y_K^2}{2}, \\ &\quad l = 1, 2, \ldots, K - 1\,. \end{aligned} \tag{13.14}$$

The set of $K - 1$ equations in (13.14) can be formulated in vector-matrix form as

$$\mathbf{A}\mathbf{z} = \mathbf{b} \tag{13.15}$$

where $\mathbf{z} = (x, y)^T$,

$$\mathbf{b} = \frac{1}{2}\begin{bmatrix} d_K^2 - d_1^2 + x_1^2 - x_K^2 + y_1^2 - y_K^2 \\ d_K^2 - d_2^2 + x_2^2 - x_K^2 + y_2^2 - y_K^2 \\ \vdots \\ d_K^2 - d_{K-1}^2 + x_{K-1}^2 - x_K^2 + y_{K-1}^2 - y_K^2 \end{bmatrix}$$

and

$$\mathbf{A} = \begin{bmatrix} x_1 - x_K & y_1 - y_K \\ x_2 - x_K & y_2 - y_K \\ \vdots & \\ x_{K-1} - x_K & y_{K-1} - y_K \end{bmatrix}.$$

With these parameters the least mean square estimation of the target location is

$$\hat{\mathbf{z}} = (\hat{x}, \hat{y})^T = (\mathbf{A}^T\mathbf{A})^{-1}\mathbf{A}^T\mathbf{b}\,. \tag{13.16}$$

Like any method, positioning based on RSS/RSSI has known weaknesses and strengths. These are summarised in Table 13.1.

Estimating d_k Using Polynomial Regression

In polynomial regression of order N, we estimate the distance d_k between the target and reference point k as [Buchanan and Turner (1992)], [Zekava and Buehrer (2012)]

$$\hat{d}_k = \sum_{n=0}^{N} c_n \times RSS_k^n, \; k = 1, 2, \ldots, K. \tag{13.17}$$

If d_k is the true distance then we obtain the estimation error

$$e_k = d_k - \hat{d}_k = d_k - \sum_{n=0}^{N} c_n \times RSS_k^n\,. \tag{13.18}$$

A method of least squares computes the coefficients c_n which jointly minimise the sum square of the estimation error, i.e.,

$$E(c_0, c_1, \ldots, c_K) = \sum_{k=1}^{K} |e_k|^2 = \sum_{k=1}^{K} |d_k - \hat{d}_k|^2\,. \tag{13.19}$$

Table 13.1 Advantages and disadvantages in positioning based on RSS/RSSI.

Advantages in RSS/RSSI	Disadvantages
No time synchronisation between nodes needed	Any power control in the system makes positioning inaccurate
No need for additional hardware to devices	Multipath signal propagation affects the received signal rendering inaccurate distance measurements
Works better than TOA and TDOA in non-line-of-sight situations	Three reference nodes with known locations needed

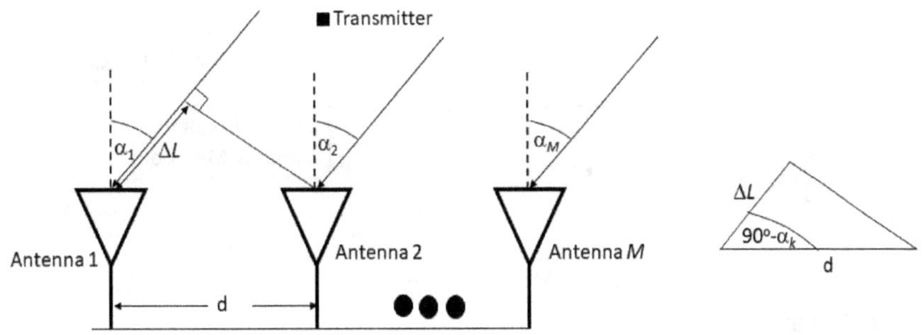

Fig. 13.6 Determining the angle of arrival using an antenna array.

We find the coefficients c_n by solving the set of K simultaneous equations

$$\frac{\partial E(c_0, c_1, \ldots, c_K)}{\partial c_m} = -2\sum_{k=1}^{K} RSS_k^m \left(d_k - \sum_{n=0} c_n \times RSS_k^n\right) \stackrel{!}{=} 0\,. \quad (13.20)$$

13.2.4 *Space-Based Positioning: AOA and DOA*

The two popular algorithms used in space-based positioning are the angle of arrival (AOA) and direction of arrival (DOA). Both methods employ antenna arrays or sectorised antennas. Let us use the array antenna with M elements in Fig. 13.6 to explain them. Note that many algorithms used to estimate both DOA and AOA rely on received signal strength indicator (RSSI) measurements.

13.2.4.1 *Angle of Arrival*

Loosely speaking, the angle of arrival (AOA) is just the bearing of a signal relative to a radio receiver. Strictly speaking, AOA is the angle at which a radio signal arrives at an array antenna. How do we measure the AOA of a radio signal? We need to determine the power in the signal impinging on one or more elements of an array antenna. This is the reason why AOA is also referred to as *power of arrival* (POA) [Denisowski (2019)]. Thus, AOA is a method of using received powers in a signal reaching an antenna (or multiple antennas) to determine the bearing of a radio signal and thus its source.

Consider Fig. 13.6 in which the same signal from the same transmitter impinges on the two elements of the array. The signals travel different paths with different distances. The path difference is shown in Fig. 13.6 as ΔL. In this section, we assume that $\alpha_i = \alpha_j = \alpha, \forall i \neq j$. Using geometric analysis we can obtain $\Delta L = d\cos(90° - \alpha) = d\sin\alpha$. Therefore, the corresponding phase difference is

$$\Delta\alpha = \frac{2\pi\Delta L}{\lambda} = \frac{2\pi d\sin\alpha}{\lambda}\,. \quad (13.21)$$

Let the phase of the signal reaching the array element k, $k = 1, 2$ be θ_k, $k = 1, 2$. Then,

$$\Delta\theta = \theta_1 - \theta_2 = \frac{2\pi d\sin\alpha}{\lambda} \Leftrightarrow \alpha = \sin^{-1}\left[\frac{(\theta_1 - \theta_2)\lambda}{2\pi d}\right]. \quad (13.22)$$

Therefore, knowing just the phase difference $\Delta\theta = \theta_1 - \theta_2$ between signals impinging the elements of the array antenna, we can use Eq. (13.22) to compute the angle of arrival (aka incidence angle) α of the EM wave. Note that d and λ are system parameters, and thus are known. The only limitation in this process is that coherent detection of the signals reaching the antenna elements is required. The consequence is that the elements in the array antenna must be perfectly synchronised in both phase and clock rate.

The type of antennas used to measure the AOA are of two basic types:

- A highly directional antenna (e.g., aperture antenna) which is rotated slowly.
- An array antenna with at least four elements.

The larger the number of elements in the antenna array, the more accurate is the measured AOA.

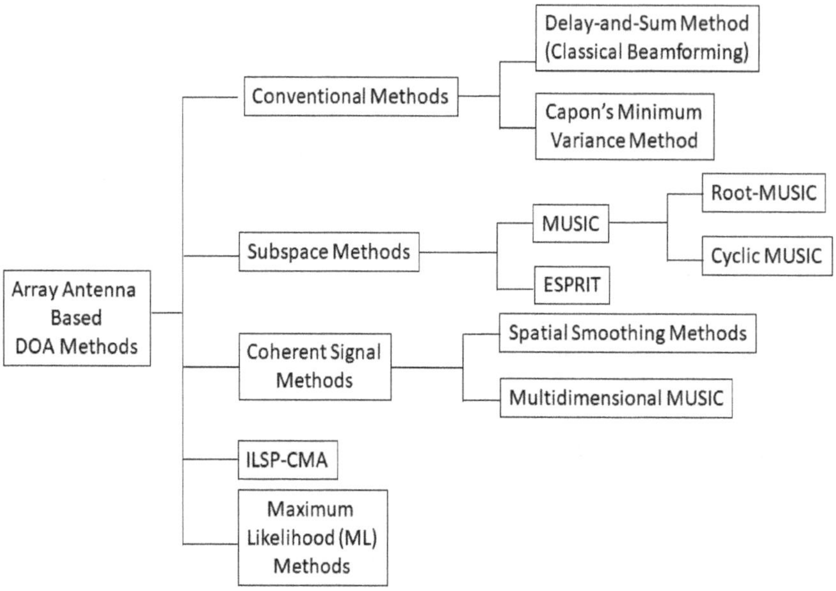

Fig. 13.7 Some algorithms used to estimate the direction of arrival (DOA).

Also, the slower the directional antenna is rotated, the better the accuracy of the AOA if a single antenna is used. The main disturbance on AOA measurement is multipath signal propagation.

13.2.4.2 *DOA and AOA: MUSIC*

The direction of arrival (DOA) is usually estimated using the angle of arrival (AOA) of signals. The AOA is the angle at which a signal incidents at an array antenna. Several algorithms have been proposed for the estimation of the DOA [Muhamed (1996)]. Examples are listed in Fig. 13.7. These include MUltiple Signal Classification (MUSIC) [Schmidt (1979)], [Schmidt (1986)], Estimation of Signal Parameters via Rotational Invariance (ESPRIT) and Iterative Least Squares Projection-based Constant Modulus Algorithm (lLSP-CMA) [Parra (1995)]. The maximum likelihood algorithms are described in, for example, [Ziskind and Wax (1988)]. Some spatial smoothing methods are discussed in, e.g., [Shan *et al.* (1985)]. The Capon's minimum variance method is described in [Capon (1969)]. For the purpose of brevity, we discuss only MUSIC.

MUSIC is a set of theoretical and experimental methods used to estimate multiple parameters of multiple wavefronts impinging on an antenna array with say M elements [Schmidt (1979)], [Schmidt (1986)]. The estimation is done using measurements of signals or wavefronts received at the array elements. MUSIC is used for the asymptotic estimation of these multiple parameters:

(1) Number of wavefronts or signals impinging the array antenna, say D.
(2) Directions of arrival (DOA), $\alpha_1, \alpha_2, \ldots, \alpha_D$, of the wavefronts or signals at the array antenna.
(3) Polarisations of the wavefronts.
(4) Strength of noise or interfering signals.
(5) Strengths and cross-correlations amongst the directional waveforms.

Special forms of MUSIC are the conventional interferometry, multiple frequency estimation and monopulse direction finding using co-located antennas.

MUSIC achieves a high resolution by exploiting the eigen structure of the input covariance matrix. An assumption underlying the MUSIC algorithm is that the signals arriving at the elements of the antenna array are not coherent, meaning that they are not fully correlated. With this assumption, the input covariance matrix of the measured data has a full rank. Let us further define the following parameters prior to proceeding:

- M: number of elements of the array antenna.
- D: number of wavefronts or signals arriving at the array antenna.
- $\mathbf{w} = [w_1, w_2, \ldots, w_M]^T \sim \mathcal{CN}(0, \sigma_w^2 \mathbf{I})$: additive complex white Gaussian noise vector.
- $\mathbf{x} = [x_1, x_2, \ldots, x_M]^T$: measured input data from the M elements of the array antenna.
- $\mathbf{f} = [f_1, f_2, \ldots, f_D]^T$: vector of D wavefronts impinging the antenna array.
- α_i, $i = 1, 2, \ldots, D$: DOA of the ith wavefront at the antenna array. These are the variables being sought in DOA estimation.
- $\mathbf{a}(\alpha_j) = [a_{1j}, a_{2j}, \ldots, a_{Mj}]^T$, $i = 1, 2, \ldots, D$: *mode vectors* or *steering vectors* of the antenna array. Measuring $\mathbf{a}(\alpha_j)$ is referred to as calibrating the antenna array.

The steering vector for a uniformly-spaced linear antenna array, as illustrated in Fig. 13.6, whose elements are isotropic antennas is given by ([Muhamed (1996)], p. 11)

$$\begin{aligned} &\mathbf{a}^T(\alpha_i) \\ &= \begin{bmatrix} 1 & e^{j\kappa d \cos\alpha_i} & e^{j2\kappa d \cos\alpha_i} & \cdots & e^{j(M-1)\kappa d \cos\alpha_i} \end{bmatrix}, \\ &\qquad i = 1, 2, \ldots, D \end{aligned} \tag{13.23}$$

where $\kappa = \frac{2\pi}{\lambda}$ is the wave number for a given wavelength λ.

If $\mathbf{a}(\alpha_i) \neq \mathbf{a}(\alpha_j)$ for $\alpha_i \neq \alpha_j$, then we can find the DOAs α_j) knowing $\mathbf{a}(\alpha_j)$. However, if $\mathbf{a}(\alpha_i) = \mathbf{a}(\alpha_j)$ for $\alpha_i \neq \alpha_j$, then Type 1 ambiguity results. Using the definitions above, the MUSIC algorithms begin with the characterising equations of the measured data $\mathbf{x}$, i.e.,

$$\mathbf{x} = \mathbf{A}\mathbf{f} + \mathbf{w} \tag{13.24}$$

where

$$\mathbf{A} = \begin{bmatrix} \mathbf{a}(\alpha_1) & \mathbf{a}(\alpha_2) & \cdots & \mathbf{a}(\alpha_D) \end{bmatrix} = \begin{bmatrix} a_{11} & a_{12} & \cdots & a_{1D} \\ a_{21} & a_{22} & \cdots & a_{2D} \\ \vdots & \vdots & \ddots & \vdots \\ a_{M1} & a_{M2} & \cdots & a_{MD} \end{bmatrix}$$

is the matrix of array steering vectors.

From Eq. (13.24), we can obtain the covariance matrix of the received or measured data as

$$\begin{aligned} \mathbf{R}_{xx} &= \mathbb{E}\{\mathbf{x}\mathbf{x}^H\} = \mathbf{A}\mathbb{E}\{\mathbf{f}\mathbf{f}^H\}\mathbf{A}^H + \mathbb{E}\{\mathbf{w}\mathbf{w}^H\} \\ &= \mathbf{A}\mathbf{R}_{ff}\mathbf{A}^H + \sigma_w^2 \mathbf{I}\,. \end{aligned} \tag{13.25}$$

We have assumed in the derivation of Eq. (13.25) that the noise vector $\mathbf{w}$ and wavefronts vector $\mathbf{f}$ are uncorrelated and thus $\mathbb{E}\{\mathbf{A}\mathbf{f}\mathbf{n}^H\} = \mathbb{E}\{\mathbf{w}\mathbf{f}^H\mathbf{A}^H\} = 0$. In general, the matrix $\mathbf{R}_{ff}$ is positive definite. If the D incident wavefronts are uncorrelated then $\mathbf{R}_{ff}$ is a diagonal matrix.

If $D < M$, $\mathbf{A}$ has a full column rank, and $\mathbf{R}_{ff}$ is non-singular, then $\mathbf{A}\mathbf{R}_{ff}\mathbf{A}^H$ is a positive definite matrix with the rank D. The consequence of this is that $M - D$ of the eigenvalues of $\mathbf{A}\mathbf{R}_{ff}\mathbf{A}^H$ are zero. Let $\lambda_1, \lambda_2, \ldots, \lambda_M$ and $\mathbf{v}_1, \mathbf{v}_2, \ldots, \mathbf{v}_M$ be the eigenvalues and the eigenvectors of $\mathbf{R}_{xx}$. Then, it follows from Eq. (13.25) that $M - D$ smallest eigenvalues of of $\mathbf{R}_{xx}$ are equal to σ_w^2. If the eigenvalues of $\mathbf{R}_{xx}$ are ordered in decreasing order of their magnitudes, then

$$\lambda_{min} = \lambda_{D+1} = \lambda_{D+2} = \cdots = \lambda_M = \sigma_w^2\,. \tag{13.26}$$

If $\widehat{N}$ is the number of times that λ_{min} repeats, then the estimate of the number of wavefronts impinging on the antenna array is

$$\widehat{D} = M - \widehat{N}\,. \tag{13.27}$$

If $\mathbf{v}_k$ and $\lambda_k = \sigma_w^2$ are the eigenvectors and the eigenvalues of $\mathbf{R}_{xx}$, then it follows from the eigenvalue problem

$$\begin{aligned} &\mathbf{R}_{xx}\mathbf{v}_k = \sigma_w^2 \mathbf{v}_k, \\ &k = D+1, D+2, \ldots, M\,. \end{aligned} \tag{13.28}$$

From Eq. (13.25), we obtain

$$\begin{aligned} &\mathbf{A}\mathbf{R}_{ff}\mathbf{A}^H\mathbf{v}_k = 0, \\ &k = D+1, D+2, \ldots, M\,. \end{aligned} \tag{13.29}$$

As $\mathbf{R}_{ff}$ is non-singular and $\mathbf{A}$ has a full column rank, it follows from Eq. (13.29) that

$$\mathbf{A}^H\mathbf{v}_k = 0,\ k = D+1, D+2, \ldots, M\,. \tag{13.30}$$

This means that the columns of the antenna array steering matrix $\mathbf{A}$ are perpendicular to the eigenvectors $\mathbf{v}_k$ of of $\mathbf{R}_{xx}$, i.e.,

$$\{\mathbf{v}_{D+1}, \mathbf{v}_{D+2} \ldots, \mathbf{v}_M\} \perp \{\mathbf{a}(\alpha_1), \mathbf{a}(\alpha_2), \ldots, \mathbf{a}(\alpha_D)\}\,. \tag{13.31}$$

We have learned from the above that M eigenvectors of the convariance matrix $\mathbf{R}_{xx}$ are divided into the two orthogonal subsets or orthogonal subspaces:

- The set $\mathbf{v}_1, \mathbf{v}_2, \ldots, \mathbf{v}_D$ spans the *signal subspace* (referred to as the *principal eigen-subspace*). These D eigenvectors are parallel to the antenna array steering vectors corresponding to the directions or arrival of the wavefronts.
- The second set $\mathbf{v}_{D+1}, \mathbf{v}_{D+2}, \ldots, \mathbf{v}_M$ spans the *noise subspace* (aka *non-principal eigen-subspace*). These $M - D$ eigenvectors are orthogonal to the antenna array steering vectors as shown in Eq. (13.31).

Therefore, we can estimate the DOAs of the D wavefronts by searching through all the possible antenna array steering vectors to find those which are perpendicular to the noise subspace in accordance with Eq. (13.31).

Let us define the $M \times N$, $N = M - D$ matrix $\mathbf{E}_N$ whose columns are the $N = M - D$ noise eigenvectors $\mathbf{v}_{D+1}, \mathbf{v}_{D+2}, \ldots, \mathbf{v}_M$ (i.e., vectors spanning the noise subspace)

$$\mathbf{E}_N = \left[\mathbf{v}_{D+1} \;\; \mathbf{v}_{D+2} \;\; \cdots \;\; \mathbf{v}_M\right]. \tag{13.32}$$

The square of the Euclidean distance from the vector $\mathbf{u}$ to the signal subspace is $d^2 = \mathbf{u}^H \mathbf{E}_N \mathbf{E}_N^H \mathbf{u}$. We then define the MUSIC spatial spectrum [Schmidt (1986)]

$$P_{MU}(\alpha) = \frac{1}{d^2} = \frac{1}{\mathbf{a}^H(\alpha)\mathbf{E}_N\mathbf{E}_N^H\mathbf{a}(\alpha)} \tag{13.33}$$

or

$$P_{MU}(\alpha) = \frac{\mathbf{a}^H\mathbf{a}}{\mathbf{a}^H(\alpha)\mathbf{E}_N\mathbf{E}_N^H\mathbf{a}(\alpha)}\,. \tag{13.34}$$

The largest $\hat{D}$ peaks in the plot of Eq. (13.33) or (13.34) as a function of α yields the DOAs of the wavefronts impinging on the antenna array. Using the estimated DOAs, we can estimate the matrix

$$\mathbf{R}_{ff} = (\mathbf{A}^H\mathbf{A})^{-1}\mathbf{A}^H(\mathbf{R}_{xx} - \lambda_{min}\mathbf{I})\mathbf{A}(\mathbf{A}^H\mathbf{A})^{-1}\,. \tag{13.35}$$

In recap, the MUSIC algorithm can be summarised into these steps [Schmidt (1986)]

Step 1: Collect the data $\mathbf{x} = [x_1, x_2, \ldots, x_D]$
Step 2: Estimate the input covariance matrix

$$\widehat{\mathbf{R}}_{xx} = \frac{1}{M}\sum_{m=1}^{M} \mathbf{x}\mathbf{x}^H\,. \tag{13.36}$$

Step 3: Execute the eigen decomposition of $\mathbf{R}_{xx}$, namely

$$\mathbf{V}\mathbf{D} = \widehat{\mathbf{R}}_{xx}\mathbf{V} \tag{13.37}$$

where $\mathbf{D} = \mathtt{diag}(\lambda_1, \lambda_2, \ldots, \lambda_M)$, $\lambda_1 \geq \lambda_2 \geq \cdots \geq \lambda_M$ are the eigenvalues of $\widehat{\mathbf{R}}_{xx}$ with the corresponding eigenvectors' matrix $\mathbf{V} = [\mathbf{v}_1\, \mathbf{v}_2 \, \cdots \, \mathbf{v}_M]$.
Step 4: Compute the multiplicity $\hat{N}$ of the smallest eigenvalue value of $\widehat{\mathbf{R}}_{xx}$ and then estimate the number of wavefronts impinging on the array antenna

$$\hat{D} = M - \hat{N}\,. \tag{13.38}$$

Step 5: Formulate the MUSIC spectrum

$$\widehat{P}_{MU}(\alpha) = \frac{\mathbf{a}^H\mathbf{a}}{\mathbf{a}^H(\alpha)\mathbf{E}_N\mathbf{E}_N^H\mathbf{a}(\alpha)}\,. \tag{13.39}$$

Step 6: Search for the $\hat{D}$ largest peaks in the plot of Eq. (13.39) as a function of α.

13.2.5 *Triangulation, Trilateration and Multilateration*

Geometric principles are versatile in positioning technologies. Positioning algorithms use geometric principles, such as the intersection of lines, circles, spheres and hyperbolas, to estimate the location of an object. We review the popular geometric principles in positioning, namely triangulation, trilateration, multilateration and the intersection of hyperbolas. Triangulation determines an unknown position using measured angles, while trilateration and multilateration use measured distances to known locations. Each artificial satellite in orbit broadcasts its location and time. This information is used by satellite receivers to estimate their location, as done, for example, in GPS.

13.2.5.1 *Triangulation*

Triangulation exploits the geometric properties of triangles to estimate the position of an object. It is believed that Thales of Miletus (624–546 BC) was the first to use the triangulation method. He used it to estimate the radius of the Earth's orbit about the sun. *Triangulation* finds an unknown location using angle measurements to two known locations (aka reference points or anchor points). We need directions or angles from the receiver to two radio transmitters with known locations. This method requires the exact determination of the angles of arrivals (AOAs) of the signals from the transmitters to the receiver. We can use phased array antenna architecture to find the AOAs, for example.

Consider Fig. 13.8 in which a receiver at an unknown location P (referred to as target location) receives signals from two radio transmitters with known locations at P_1 with coordinates (x_1, y_1) and P_2 with coordinates (x_2, y_2) (referred to as reference points), respectively. The unknown location, $P(x, y)$, is at the intersection of the lines drawn from the two transmitters to the unknown location. As the locations of the two transmitters are known, the parameters x, y, d and L in Fig. 13.8 can be estimated. Therefore, the triangle with the vertices P_1, P_2 and P is defined, enabling us to estimate the unknown parameters x, y, d and L and hence the location P.

Applying geometric principles on the triangle in Fig. 13.8 we obtain

$$d = d_1 \sin\varphi = d_2 \sin\beta \tag{13.40}$$

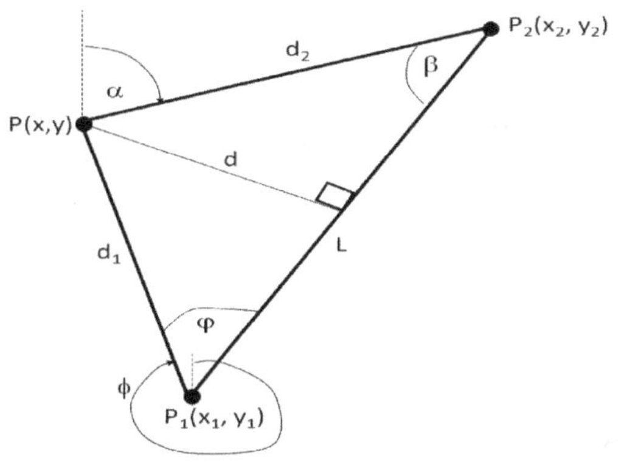

Fig. 13.8 Using triangulation principle to estimate a position.

and

$$L = d_1 \cos\varphi + d_2 \cos\beta \overset{(13.40)}{=} d\frac{\sin(\varphi+\beta)}{\sin\varphi \cdot \sin\beta}. \tag{13.41}$$

Hence, rearranging terms in Eq. (13.41) yields

$$d = \frac{L \sin\varphi \sin\beta}{\sin(\varphi+\beta)} \tag{13.42}$$

where L, referred to as *baseline*, is the distance between the two reference points, and d is the length of the line drawn from the unknown point P to intercept at right angles the line joining the two reference points. We can therefore estimate d knowing L, φ and β. Also, we can formulate the relations

$$\mathtt{tan}\varphi = \frac{d}{d_1 \cos\varphi} = \frac{y-y_1}{x-x_1} \Leftrightarrow \varphi = \mathtt{tan}^{-1}\left(\frac{y-y_1}{x-x_1}\right) \tag{13.43}$$

and

$$\mathtt{tan}\beta = \frac{d}{d_2 \cos\beta} = \frac{y-y_2}{x-x_2} \Leftrightarrow \beta = \mathtt{tan}^{-1}\left(\frac{y-y_2}{x-x_2}\right). \tag{13.44}$$

Note that Eq. (13.43) and (13.44) can also be expressed in the form

$$\mathtt{tan}\varphi = \frac{d}{d_1 \cos\varphi} = \frac{y-y_1}{x-x_1} \Leftrightarrow y = (x-x_1)\mathtt{tan}\varphi + y_1 \tag{13.45}$$

and

$$\mathtt{tan}\beta = \frac{d}{d_2 \cos\beta} = \frac{y-y_2}{x-x_2} \Leftrightarrow y = (x-x_2)\mathtt{tan}\beta + y_2. \tag{13.46}$$

Equating (13.45) to (13.46) and simplifying yields

$$x = \frac{y_2 - y_1 + x_1\mathtt{tan}\varphi - x_2\mathtt{tan}\beta}{\mathtt{tan}\varphi - \mathtt{tan}\beta}. \tag{13.47}$$

Substituting x in Eq. (13.47) into (13.46) yields y. Applying the Pythagoras theorem on the right-angled triangles in Fig. 13.8, we obtain the distances

$$d_1 = \sqrt{(x-x_1)^2 + (y-y_1)^2} \text{ and}$$
$$d_2 = \sqrt{(x-x_2)^2 + (y-y_2)^2}. \tag{13.48}$$

13.2.5.2 *Trilateration*

Multilateration is a technique used to estimate the coordinates of a location by measuring the distance to the target from multiple locations. A special case of multilateration is trilateration. Trilateration is illustrated in Fig. 13.9. The receiver needs at least

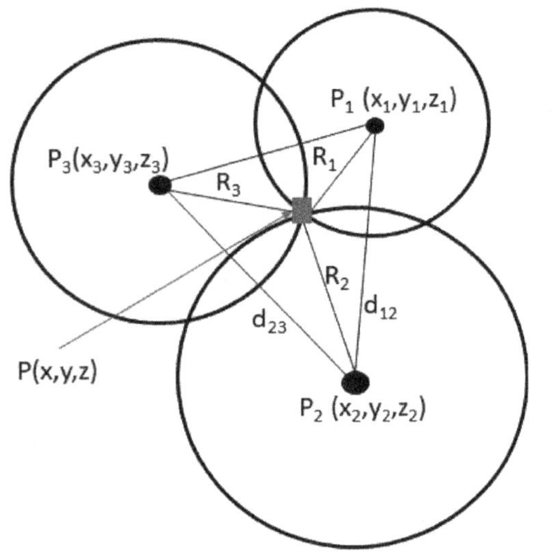

Fig. 13.9 Illustrating trilateration for position determination.

the distances from it to three transmitters of known locations to determine its location. There are alternative methods for estimating the distances, such as received signal strength indicator (RSSI). We can also use time of arrival (TOA) or time of flight[2] (TOF) in situations where we can guarantee synchronisation between the clocks of the receiver and the transmitters.

Let $\mathbf{x}_k = [x_k\, y_k]^T$, $k = 1, 2, 3$ be the known coordinates of the locations of the anchor points (aka reference points). In satellite-based positioning systems, these are locations of satellites. Let t_{0k} be the time that a signal was sent from the anchor point P_k, and t_k be the time that the signal reached the target at the unknown location P. Then, the measured ranges are

$$\hat{R}_k = (t_k - t_{0k}) \cdot c, \; k = 1, 2, \ldots \tag{13.49}$$

where $c \approx 3 \cdot 10^8$ m/s is the propagation speed of wireless signals in the air. If the transmitters are synchronised then $t_{0i} = t_{0j} = t_0$.

Let $\mathbf{x} = [x\, y]^T$ be the true position of the target, and $\hat{\mathbf{x}} = [\hat{x}\, \hat{y}]^T$ be its estimate. The true distance from the target to a known position k is

$$R_k = \|\mathbf{x} - \mathbf{x}_k\| = \sqrt{(x - x_k)^2 + (y - y_k)^2} \tag{13.50}$$

and

$$\hat{R}_k = \|\mathbf{x} - \hat{\mathbf{x}}_k\| = \sqrt{(x - \hat{x}_k)^2 + (y - \hat{y}_k)^2} \tag{13.51}$$

is its estimated/measured range. We can estimate the ranges using the TOF or TOA. The equations of the circles with centres at the reference points and passing through the target location are

$$R_k^2 = (x - x_k)^2 + (y - y_k)^2, \; k = 1, 2, 3\,. \tag{13.52}$$

With these three nonlinear simultaneous equations, we need to solve for x from

$$\sqrt{R_1^2 - (x - x_1)^2} + y_1 = \sqrt{R_2^2 - (x - x_2)^2} + y_2 \tag{13.53}$$

and then substitute for x in

$$R_3^2 = (x - x_3)^2 + (y - y_3)^2 \tag{13.54}$$

to find y.

The problem now is: we know $\mathbf{x}_k = [x_k\, y_k]^T$, $k = 1, 2, 3$ and $\hat{R}_k$, $k = 1, 2, 3$, and we seek $\mathbf{x} = [x\, y]^T$. Solving for the coordinates (x, y) from the above set of nonlinear simultaneous equations is involving. A simpler solution is to estimate the position using an iteration until the estimation error is within an acceptable range.

Let us use a trick to solve for the position of the target. The trick is to compute the difference equations

$$\begin{aligned} \hat{R}_2^2 - \hat{R}_1^2 &= x_2^2 - x_1^2 + y_2^2 - y_1^2 \\ &\quad - 2(x_2 - x_1) - 2(y_2 - y_1) \\ \hat{R}_3^2 - \hat{R}_1^2 &= x_3^2 - x_1^2 + y_3^2 - y_1^2 \\ &\quad - 2(x_3 - x_1) - 2(y_3 - y_1)\,. \end{aligned} \tag{13.55}$$

The vector-matrix form of the pair of difference equations (13.55) is

$$\underbrace{\begin{bmatrix} x_2 - x_1 & y_2 - y_1 \\ x_3 - x_1 & y_3 - y_1 \end{bmatrix}}_{=\mathbf{A}} \underbrace{\begin{bmatrix} x \\ y \end{bmatrix}}_{=\mathbf{z}} = \underbrace{\frac{1}{2} \begin{bmatrix} \hat{R}_1^2 - \hat{R}_2^2 + x_2^2 - x_1^2 + y_2^2 - y_1^2 \\ \hat{R}_1^2 - \hat{R}_3^2 + x_3^2 - x_1^2 + y_3^2 - y_1^2 \end{bmatrix}}_{=\mathbf{b}}\,. \tag{13.56}$$

Therefore, we can estimate the position of the target from

$$\hat{\mathbf{z}} = (\hat{x}, \hat{y})^T = \mathbf{A}^{-1}\mathbf{b} \tag{13.57}$$

[2]The time of flight (TOF) is the time taken by a signal or an object to travel a given distance.

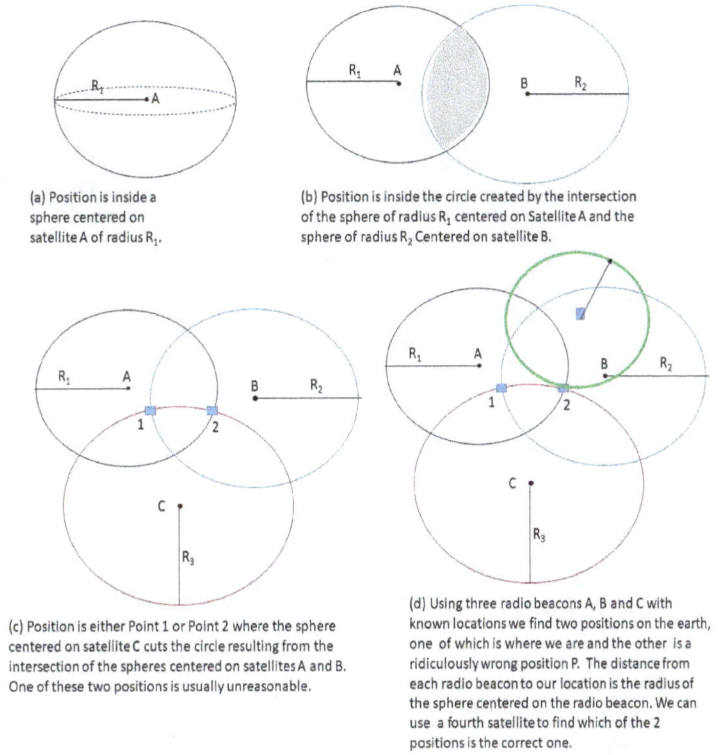

(a) Position is inside a sphere centered on satellite A of radius R_1.

(b) Position is inside the circle created by the intersection of the sphere of radius R_1 centered on Satellite A and the sphere of radius R_2 Centered on satellite B.

(c) Position is either Point 1 or Point 2 where the sphere centered on satellite C cuts the circle resulting from the intersection of the spheres centered on satellites A and B. One of these two positions is usually unreasonable.

(d) Using three radio beacons A, B and C with known locations we find two positions on the earth, one of which is where we are and the other is a ridiculously wrong position P. The distance from each radio beacon to our location is the radius of the sphere centered on the radio beacon. We can use a fourth satellite to find which of the 2 positions is the correct one.

Fig. 13.10 Illustrating multilateration for position determination.

if the matrix $\mathbf{A}$ is invertible otherwise from

$$\hat{\mathbf{z}} = (\hat{x}, \hat{y})^T = (\mathbf{A}^T\mathbf{A})^{-1}\mathbf{A}^T\mathbf{b}\,. \tag{13.58}$$

13.2.5.3 *Multilateration*

Trilateration requires synchronisation between the receiver and the transmitters. This is a limitation so a method called *multilateration*, as used in, for example LORAN, was developed to avoid the need for synchronisation. Like trilateration, multilateration needs signals from four radio transmitters to determine the location. Trilateration finds a location through the intersection of circles and it is seen as a special case of multilateration, which finds a location through the intersection of spheres. Multilateration is illustrated in Fig. 13.10.

In Fig. 13.10(a) the receiver detects signals from only one transmitter so its location lies on a sphere centered on the satellite of radius R_1. In Fig. 13.10(b), it detects signals from two transmitters to limit its location inside a circle created by the intersection of two spheres. Receiving a signal from a third transmitter further limits the position of the receiver to the two points shown in Fig. 13.10(c). We need a fourth transmitter's signal to pick the correct location from these two points. This is shown in Fig. 13.10(d). This explains the reason why a GPS receiver needs a minimum of four satellites to determine its location.

13.2.6 *Kalman Filtering*

Named in honour of its co-inventor, the Hungarian-born American Rudolf Emil Kálmán, Kalman filtering (KF) is also referred to as *linear quadratic estimation (LQE) algorithm.* KF has a wide range of applications. KF is a recursive algorithm which estimates unknown variables in real time from noisy, inaccurate observed time series. The estimate is more accurate than any of the measured data. KF achieves its accuracy by estimating a joint probability distribution over the unknown variables for each time frame. For example, a GPS receiver feeds estimated pseudoranges from satellites' signals into Kalman filters to estimate its position and clock bias. Another application example of KF is in the tracking of objects by radar systems.

A Kalman filter achieves an optimal performance only when the characteristics of the system noise have known statistical properties, i.e., zero-mean, Gaussian, and spectrally white. In practice, the noise characteristics are unknown. For this reason, a KF usually works in two stages. The first stage estimates the static states or initial conditions of the KF. The second stage is the prediction step which estimates the dynamic state. The inputs to the first stage are the raw noisy measured data. The inputs to the second stage are the estimation results from the first stage. For this reason, the KF is referred to as the Extended Kalman Filter (EKF) in the second stage. The output of the second stage are the dynamic states of the variables being estimated.

13.2.7 *Time of Arrival*

The two most popular methods in time-based positioning are the time of arrival (TOA) and the time difference of arrival (TDOA). TOA is the time that a signal transmitted from one node reaches another node. TOA uses the arrival times of signals transmitted from nodes with known locations to estimate distance. Refer to Fig. 13.9 used to illustrate triangulation for how TOA is used to compute distances and hence locations.

Let t_{kl} be the time that the signal transmitted at time t_{0k} by node k (e.g., reference node) reaches another node l (e.g. target node). Then the estimated distance between the two nodes is

$$\hat{R}_{kl} = (t_{kl} - t_{0k})c = \propto t_{kl}, \; l, k = 1, 2 \ldots \qquad (13.59)$$

where c is the speed of wireless signal propagation. Usually, l is fixed so that Eq. (13.59) simplifies to

$$\hat{R}_k = (t_k - t_0)c = \propto t_k, \; k = 1, 2 \ldots, K\,. \qquad (13.60)$$

It is clear from Eq. (13.59) and (13.60) that TOA algorithms work under the three assumptions:

- The positions of the reference nodes should be exact and known. There are only two choices: either GPS-enabled dynamic reference nodes (as with in-orbit satellites) or static nodes (as in mobile cellular base stations).
- Each transmitted signal contains its time stamp t_{0k} indicating the time it left the transmitter.
- Both nodes involved in the estimation of the TOA must be perfectly synchronised. Even a clock drift as small as $1\,\mu\mathtt{s}$ causes about 300 m measurement error.

In a planer or 2D scenario in which the target and the reference nodes lie in the same plane we need three TOA measurements from three reference nodes to estimate the position of a target. However, four TOA measurements are needed in a non-planer or 3D situation.

If $\mathbf{z} = (x, y)^T$ are the coordinates of the target and $\mathbf{z}_k = (x_k, y_k)^T$, $k = 1, 2, \ldots)$ are the coordinates of the anchor points then we have

$$\begin{aligned}
\hat{R}_1^2 &= (x_1 - x)^2 + (y_1 - y)^2 \;\; \texttt{(a)}\\
\hat{R}_2^2 &= (x_2 - x)^2 + (y_2 - y)^2 \;\; \texttt{(b)}\\
\hat{R}_3^2 &= (x_3 - x)^2 + (y_3 - y)^2 \;\; \texttt{(c)} \qquad (13.61)\\
&\;\vdots\\
\hat{R}_K^2 &= (x_K - x)^2 + (y_K - y)^2 \;\; \texttt{(d)}\,.
\end{aligned}$$

We presented a trick while discussing trilateration to solve equations of the form (13.61).

13.2.8 *Hyperbolic Positioning: TDOA*

The second time-based locationing technique is the time difference of arrival (TDOA). In the absence of noise the location of a target in TDOA is found at the intersection of two or more hyperbolas. The foci of the hyperbolas are the locations of the reference radio transmitters. For this reason, TDOA-based position methods are also referred to as hyperbolic positioning methods. The location of the target lies on a hyperbola formed from one TDOA measurement. Therefore, an intersection of hyperbolas formed from multiple TDOAs leads to the target's location. In geometry, a conic section with two fixed points $P_1(x_1, y_1)$ and $P_2(x_2, y_2)$ called foci, such that a point $P(x, y)$ lying on the hyperbola fulfills

$$\begin{aligned}
&||P - P_1||_2 - ||P - P_2||_2\\
&\quad = [(x - x_1)^2 + (y - y_1)^2] - [(x - x_2)^2 + (y - y_2)^2]\\
&\quad = constant \qquad (13.62)
\end{aligned}$$

is called a hyperbola. The standard equation of a hyperbola with center at (a_x, a_y) is

$$\frac{(x-a_x)^2}{a^2} - \frac{(y-a_y)^2}{b^2} = 1 . \tag{13.63}$$

In TDOA, a target node whose location is under study transmits a signal which is received by multiple reference nodes. The path from the target node to the reference nodes are different. Therefore, the signal takes different times to reach the reference nodes. TDOA is the difference between the times taken for the transmitted signal to reach a pair of reference nodes.

Assume that the signal transmitted by the target node at time t_{r} reaches the kth reference node at time t_k. Let d_k be the distance between the kth reference node and the target, $\mathbf{z} = (x, y)$ and $\mathbf{z}_k = (x_k, y_k)$ be the coordinates of the target and kth reference node, respectively. Then, for a *noise-free* system, we have

$$t_k = t_{\mathrm{r}} + \frac{d_k}{c} = t_{\mathrm{r}} + \frac{\sqrt{(x-x_k)^2 + (y-y_k)^2}}{c} \tag{13.64}$$

where c is the speed of signal propagation. Note that Eq. (13.64) is the TOA. Therefore, for $t_l > t_k$ if $l > k$, the TDOA is

$$\Delta t_{kl} = t_l - t_k \overset{(13.64)}{=} \frac{d_l - d_k}{c} = \frac{\sqrt{(x-x_l)^2+(y-y_l)^2} - \sqrt{(x-x_k)^2+(y-y_k)^2}}{c} . \tag{13.65}$$

Comparing Eq. (13.65) with (13.63), we observe that Eq. (13.65) describes a hyperbola. We need another hyperbolic equation from another TDOA to intersect Eq. (13.65) to find the target location $\mathbf{z} = (x, y)$.

The reference node which receives the signal first is usually made the master reference node, so that all TDOAs are measured in reference to it. Therefore, if k is the master reference node, then $t_k \le t_l, \forall l \ne k$. In a coplanar or 2D scenario in which the target and the reference nodes lie in the same plane we need two TDOA measurements from three reference nodes to estimate the position of a target. However, three TDOA measurements from four reference nodes are needed in a non-coplanar or 3D situation.

A careful observation of Eq. (13.65) reveals that TDOA is an improvement over TOA, because:

- Synchronisation between the clock of the target and the reference nodes is not needed in TDOA. Only the reference nodes must be synchronised.
- Signals transmitted do not need to include any time stamp.

13.2.9 *Check Your Understanding*

Students are encouraged to check their understanding of the content of this section with the following questions prior to proceeding to the next section.

(1) Discuss the differences and the relationship between NDB and ADF as used in navigation.
(2) Differentiate between triangulation, trilateration and multilateration methods of finding an unknown location.
(3) Define the angle of arrival (AOA) of a signal.
(4) List the types of antennas that are used to measure the angle of arrival or the bearing of a radio signal and its source. Justify your answer.
(5) Discuss the effect of the number of elements in the antenna array used to measure the angle of arrival on the latter.
(6) Assume that the geographical coordinates of three base stations in a mobile cellular network are $P_0(0,0)$, $P_1(0,200)$ and $P_2(100,0)$. A signal transmitted from a target node located at $P(x,y)$ reached these stations at times $t_0 = 1:00$ pm, $t_1 = t_0 + 20\,\mathrm{ns}$ and $t_2 = t_1 + 10\,\mathrm{ns}$, respectively. Using hyperbolic principles, estimate the location of the target node.
(7) Assume that the three dimensional coordinates of six satellites observed by a GPS receiver at a given time instant are $\mathbf{x}_1 = (0.9390, -1.6265,\ 1.8781) \times 10^7$ m, $\mathbf{x}_2 = (1.7648,\ -0.6423,\ 1.8781) \times 10^7$ m,

$\mathbf{x}_3 = (1.7648,\ 0.6423,\ 1.8781) \times 10^7$ m, $\mathbf{x}_4 = (0.939,\ 1.6265,\ 1.8781) \times 10^7$ m, $\mathbf{x}_5 = (0.939,\ -1.6265,\ -1.8781) \times 10^7$ m and $\mathbf{x}_6 = (0.9390,\ 1.6265,\ -1.8781) \times 10^7$ m [Sarunic (2016)]. Estimate the position of the GPS receiver using:

(a) triangulation/multilateration method.
(b) Kalman filtering with update rate of 0.5 seconds.

13.3 Terrestrial Positioning Technologies

We review in this section some of the positioning systems whose components are located on the ground or mounted on crafts (e.g., airplanes, ships and vehicles).

13.3.1 *Hyperbolic Positioning Systems*

Navigational instruments which find the location of a target through the measurements of time difference of arrivals (TDOAs) and thus the intersection of hyperbolas include LORAN, Omega, GEE and Decca. The Long-Range Navigation (LORAN) was developed by the US Navy. Omega and Decca developed in 1944 by the British Royal Navy. GEE was developed by the British Royal Air Force during the WW2 in 1942.

Decca, operated at 70-130 kHz, was first deployed on 6 June 1944 during the invasion of the Normandy region in northern France. Wireless nodes in a Decca system are grouped into chains. One chain comprises one master station and two or more slave stations. Slave stations are positioned about 80–110 km away from the master station. The accuracy of Decca depends on seasons and time. For example, it achieves 50-m accuracy during the day, but only 200 m at night.

Being launched around 1971, the **Omega navigational system** was the first global navigational aid in the world. GPS caused the shutdown of Omega in 1997. Airplanes and ships could estimate their bearings from signals transmitted from a network of ground-based radio beacons over the frequency band 10 to 14 kHz.

LORAN, operating in the band 90-110 kHz with centre frequency at 100 kHz, was developed by the US DoD during the WW2. There are several versions of LORAN, such as LORAN-A, LORAN-B to LORAN-F. The most current version is the enhanced LORAN (eLORAN). eLORAN is somewhat competitive to satellite-based navigation systems. LORAN-A can achieve a range of 1,120–2,300 km range. The main weakness in LORAN-A was its accuracy as it lacked phase reference. Therefore, LORAN-B was developed to improve upon the accuracy of LORAN-A. Also, LORAN-A transmitters operate in pairs, while LORAN-B transmitters operate in chains. Basically, a chain operation comprises multiple ground-based wireless stations: one serves as a master station or primary station while the remaining serves as slave stations or secondary stations. Owing to the low frequency used, LORAN signals travel as ground waves to cover large distances.

13.3.2 *Inertial Navigation Systems*

The definition of Inertial Navigation Systems (INS) is quite blurry and broad. For example, INS can mean a family of navigational aids or a specific type of navigational instrument. INS use inertial sensors to estimate the acceleration, velocity, position and orientation of a moving object using inertial principles. The most popular inertial sensors or measuring devices are gyroscopes (aka gyros) and accelerometers, which are used to find the location of objects on the Earth's surface.

Attaching a weight to a spring and measuring its deflection and gravitational force realises a simple accelerometer. A gyro estimates the angular velocity of an object relative to the inertial space using measurement methods, such as spinning wheel, Coriolis-effect and Sagnac-effect. Usually gyros and accelerometers are integrated to realise an inertial measurement unit (IMU). Note, however, that some IMUs contain magnetometers and pressure sensors. A magnetometer (or magnetic sensor) is a device used to measure magnetism. A pressure sensor is an instrument that is used to measure the pressure of liquids or gases.

An IMU measures the acceleration of an object in the inertial space and integrate it to find the velocity. Further integration of the velocity yields the position. The attitude of the moving object is found by integrating its angular velocity. We refer to as *navigational equations* the set of equations integrating the measurements from the IMU into velocity, position and orientation/attitude (viz. pitch, yaw, roll). An INS is the combination of an IMU and a computer running the set of inertial equations.

13.3.3 *VOR*

The Very High-Frequency (VHF) Omnidirectional Ranging (VOR) is a line-of-sight instrument which has been used in aviation to estimate the bearing of airplanes to a ground-based VOR facility since the 1960s. There are three types of VOR:

- Low-frequency (LF)/High-frequency (HF) four-course radio ranging which operates in 200–400 kHz band. This system is now obsolete.
- Conventional VOR (C-VOR) operating in 112–118 MHz in the VHF band.
- Doppler VOR (D-VOR), which is an improvement over C-VOR.

Any VOR system comprises an onboard aircraft VOR receiver and a ground component. There are usually duplicates of the ground component, some are mounted at the airport while others are stationed off the airport. Such an architecture is needed to provide pilots guidance for both arriving and departing aircrafts. The onboard equipment comprises three components shown in Fig. 13.11. The cockpit instrument can be one or a combination of the three options shown in Fig. 13.11. As an aircraft can intercept signals from multiple VOR equipment, each VOR broadcasts its unique identifier using Morse code.

With respect to ranging, VORs can be divided into three categories:

- Terminal VOR (T-VOR): has up to circa 3.7 km altitude and 46.3 km range.
- Low-altitude VOR (L-VOR): has up to circa 5.5 km altitude and 74 km range.
- High-altitude VOR (H-VOR): has up to circa 18.3 km altitude and 240.7 km range.

How does a pilot estimate the aircraft's bearing using VOR? The ground station of the VOR, which is aligned with the magnetic north, broadcasts two 30-Hz signals. Both signals have the same phase in the magnetic north direction. However, one of the signals, called reference signal, has a constant phase. The phase of the other signal, referred to as variable signal, changes by 1° for each degree deviation of the azimuthal angle. The VOR receiver onboard the aircraft measures the phase difference between two signals transmitted by the ground equipment of the VOR, and then displays the equivalent radial distance of the aircraft on the OBI, HSI, or RMI.

VOR has some identified weaknesses. First, the VOR equipment cannot be used during mainte-

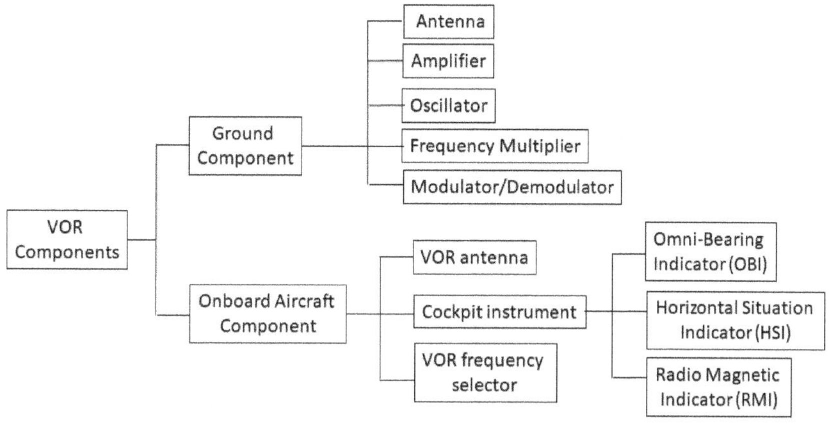

Fig. 13.11 VOR components.

nance. The VOR instrument onboard the aircraft displays erroneous readings when flying near the VOR. This erroneous reading is referred to as *cone of confusion* or *cone of silence*. Third, as a line-of-sight instrument, VOR cannot easily be identified by aircrafts flying in a non-flat terrain. VOR is still in use as at 2019. However, it is just a matter of time and it will be replaced by automatic dependent surveillance-broadcast systems (ADS-B) and/or GPS/WAAS. VOR is often used in conjunction with the Distance Measuring Equipment (DME), which is a type of secondary radar system.

13.3.4 *Distance Measuring Equipment*

The Distance Measuring Equipment (DME) is a navigational instrument used to measure the slant range between an aircraft and a ground-based station at an airport. Some DMEs also display time to reach the airport and the ground speed of the aircraft. The first DME was developed in 1946. The DME is usually co-located with a VOR equipment in an integrated equipment referred to as Rho-Theta. The VOR measures the bearing or the azimuthal angle θ while the DME measures the slant range ρ, as illustrated in Fig. 13.12. At an altitude of about 1.5 km over a flat terrain, DME can measure a slant range of about 96 km. The DME is also used in ILS as beacon marker. One DME can serve about 100 to 200 airplanes at the same time.

The DME functions as a secondary radar system, but in the opposite sense. Both the ground-based equipment (called transponder) and the onboard equipment (called interrogator) must co-operate to achieve the intended results. The interrogator onboard the airplane sends a signal to the transponder which responds by sending a return signal to the interrogator. The interrogator measures the time delay between the transmitted and the received signals, say τ. The interrogator then measures the slant range, ρ, between the two transceivers as

$$\rho = 3 \cdot 10^5 \frac{\tau - 50 \times 10^{-6}}{2} \ \texttt{km}. \tag{13.66}$$

The 50 μs is subtracted because the ground-based transponder waits that amount of time before responding to an interrogation. The division by 2 caters for the fact that the measured time is a round-trip time.

The frequency band 960–1215 MHz in the UHF band has been earmarked for the DME. The interrogator sends its signal with centre frequency in the range 1.025–1.15 GHz, which are divided into 126 channels, each of 1MHz bandwidth. The ground-based DME transponder replies on a frequency that is either 63 MHz higher or 63 MHz lower than the interrogator's frequency. Thus, the DME transponder's signals have the centre frequency in the band 962–1,213 MHz. DMEs are normally paired up with a VOR equipment. Therefore, pilots tune in to DMEs through the VOR. Each DME sends its Morse-coded unique identification at 1.35 kHz between successive VOR identifications at 1.020 kHz.

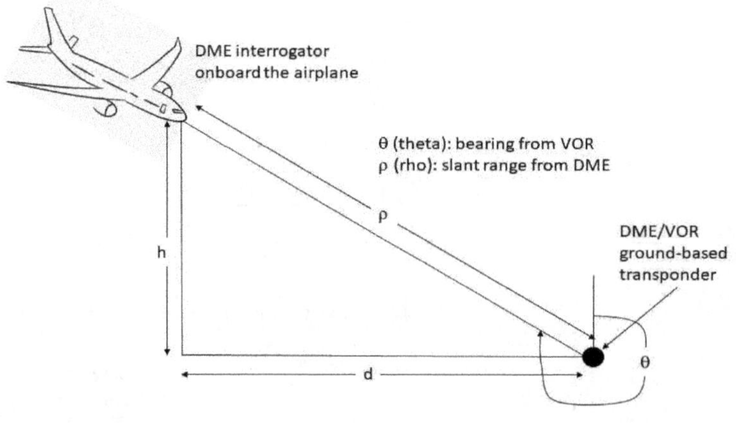

Fig. 13.12 Rho-theta: an integrated VOR-DME equipment.

The DME uses two types of interrogation modes:

- Search mode: the DME interrogator uses this mode to lock onto the ground-based transponder by sending signals at the rate of 150 signals per second. Upon receiving responses to 65% of its interrogations, the DME interrogator switches to the track mode.
- Track mode: the rate of interrogation in this mode is about 30 signals per second. This is to allow the DME to serve more airplanes.

13.3.5 *ILS and MLS*

We have already studied avionics positioning technologies VOR and DME. We add ILS and MLS. A pilot can approach and land an airplane by following either of two sets of rules: Visual Flight of Rules (VFR) or Instrument Flight Rules (IFR). A pilot follows the VFR if the weather is clear enough to ensure good visibility (i.e., over 5-km horizontal and 300-m visibility). In this option the airplane approaches a runway at an airport and lands by visual observation of ground, features and landing lights. How does an airplane approach and land when the pilot cannot see the ground and its environment? During a cloudy and/or misty weather with poor visibility (usually less than 3 miles or about 5 km) a pilot has no choice but to rely on instruments for guidance according to VFR. The popular equipment used for IFR are the:

- Instrument Landing System (ILS).
- Microwave Landing System (MLS).

> I remember my awful experience around 2014. We flew to an airport without a radio landing system (no ILS nor MLS), requiring pilots to rely on physical markings to land. Unfortunately, the visibility was so poor that our aircraft had to hover around the sky for over hour. After failing to land for the 3rd attempt, the pilot announced that he was venturing a fourth attempt, failure of which would require us to return to another airport located circa 250 km away. Guess what? We had a military landing!

13.3.5.1 *ILS*

The Instrument Landing System (ILS) is a ground-based radio navigation system which provides horizontal and vertical distance information to pilots as they approach an airport. The ILS has both ground-based component and a component onboard the airplane (referred to as ILS receiver). In order to use ILS, the pilot needs an ILS receiver and a country-specific ILS chart (aka Approach Plate). The ILS chart contains the data:

- ILS frequency and identification code (2- or 3-letter identifier transmitted on 1.02 kHz for each runway using international Morse code.
- ILS inbound course and glideslope angle and final approach point.
- Minimum descent altitude and ILS categories.
- Go-around or missed-approach procedure.

The main components of the ground-based component of the ILS are:

(1) Directional transmitters called localiser and glideslope for guidance.
(2) Two ways for providing (slant) range to the airport: by marker beacons (inner marker, middle marker, outer marker) at significant points along the approach path, or continuously by a Distance Measuring Equipment (DME). The outer markers indicate the starting area of the flight path or ILS approach to follow for a standard terminal arrival (STAR) or flight procedure.
(3) Lights (approach lights, touchdown lights, centerline lights, and runway lights) to provide visual information).

The *localiser* (aka LOC or LLZ) provides only lateral or horizontal guidance information towards the runway on one of the 40 ILS channels in the frequency band 108.1–111.95 MHz in 0.05 MHz steps. The localiser's antenna is usually mounted beyond the end of the runway, and comprises 13 to 41 directional antennas (e.g., log-periodic or quad antennas), as illustrated in Fig. 13.13(a). The localiser also transmits the ILS facility identification code.

The *glideslope* broadcasts from the side of the runway only the vertical guidance of the aircraft

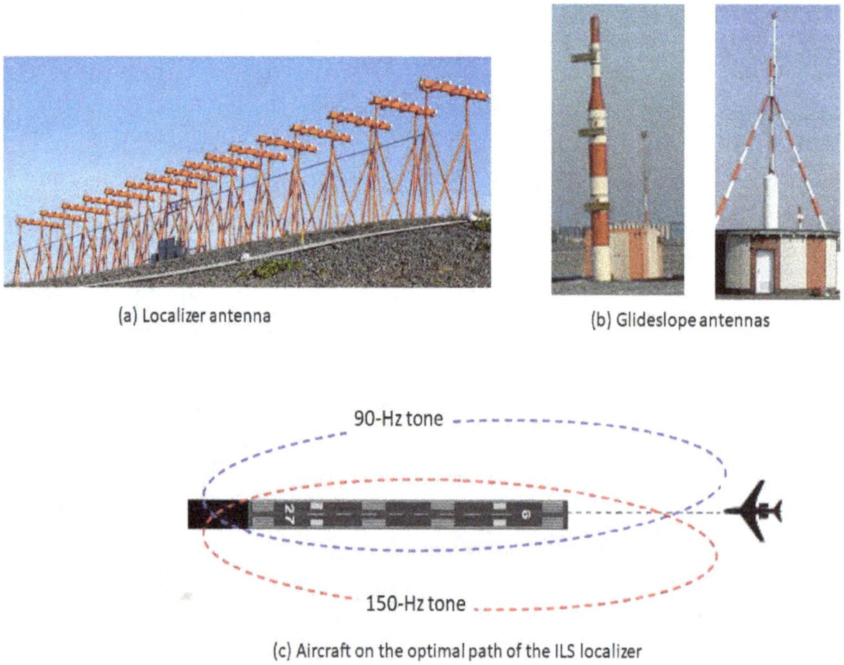

Fig. 13.13 ILS antennas mounted at airports.

relative to the ideal approach. The glideslope operates on one of the 40 ILS channels in the frequency band 329.15–335.0 MHz, and its angle to the runway is usually 3°. The glideslope antenna is illustrated in Fig. 13.13(b). The frequency of the glideslope (also glideslope) is usually paired up with that of the localiser so that a pilot needs to enter only the localiser frequency in the aircraft's instruments to tune to an ILS. The glide scope antenna, with the beamwidth of 1.4°, is located between 230 m and 380 m from the approach end of the runway and offset between 75 m and 198 m from the runway centre line.

The ILS operates using two tones at 150 Hz and 90 Hz which modulate the amplitudes of carriers at ILS frequencies. The difference between the amplitudes of the two tones can determine the relative position of the aircraft from the theoretical descent axis. The ILS receiver on-board the aircraft must demodulate these two signals to retrieve their amplitudes to guide the approach and landing of the aircraft. Figure 13.13(c) illustrates how the two tones are used by the localiser.

ILS has supported the aviation industry since 1929 with the first fully automatic landing using ILS occurring in 1964 at Bedford Airport in the UK. However, the *ILS has these notable weaknesses*:

(1) Terrain can hinder the proper positioning of ILS components. It therefore requires a minimum of 366 m (or 1,200 feet) flat clear terrain in front of the antenna to form a precise signal. Thus, making the use of ILS in hilly terrain and small airports impractical.
(2) The ILS has only 40 channels, causing congestion at busy airports.
(3) Signals of glideslope and localiser are adversely affected by interference caused by reflections off objects near the airport or airport structures (e.g. hanger, snow and tide). The limited number of channels also causes mutual interference between ILS systems operating in a geographical area.

The above listed disadvantages in ILS motivated the development of the MLS.

13.3.5.2 *MLS*

As discussed above, the ILS has some noticeable disadvantages. The Microwave Landing System

(MLS) was therefore developed to replace or at least enhance the ILS. The MLS has the following advantages compared with the ILS:

(1) It allows a variety of approaches from many directions and a variety of glideslopes, including curved-path approaches. This feature increases the traffic supportable by an airport, and allows the deployment of MLS at every site.
(2) MLS signals are not as susceptible to interference caused by signal reflections off objects as ILS.
(3) MLS allows a faster air traffic flow than ILS.
(4) MLS provides a more accurate and reliable guidance than ILS.

The MLS operates on the frequency band 5.031–5.0907 GHz, the reason why it has *microwave* in its name. MLS has 200 channels worldwide, each with 300-kHz bandwidth. However, only one channel (or frequency) is used to transmit all signals, except precision DME (DME/P) which selects frequencies automatically. The ground equipment of MLS has the three components:

- The *distance measuring equipment* (DME) to measure slant range.
- The *azimuthal transmitter* (equivalent to the localiser in ILS) which provides a fan-shaped horizontal approach 40° either side of the runway center line and reach out to 41.67 km (or 22.5 nm).
- The elevation transmitter (equivalent to the glideslope in ILS) of MLS provides a fan-shaped vertical approach from 0.9° to 20° to at least 37 km (or 20 nm).

Although MLS became operational in the 1990s, it could not meet its forecast market acceptance. The failure of MLS is largely due to the development of GPS and WAAS or Differential GPS, which are discussed later in this chapter. MLS had a future in places where GPS was not very much developed until European countries started to use satellite-based landing systems (i.e., GNSS Landing System or GLS) since 9 February 2012.

13.3.6 *Check Your Understanding*

Students are encouraged to check their understanding of the content of this section with the following questions prior to proceeding to the next section.

(1) Describe the components of an inertial measuring unit (IMU).
(2) Describe the operation of an IMU.
(3) Describe the difference between Visual Flight of Rules (VFR) and Instrument Flight Rules (IFR).
(4) What motivated the development of the MLS?
(5) State a reason why the MLS navigational system never achieved its forecast market acceptance.
(6) Describe the *cone of confusion* in VOR.
(7) Describe the two other weaknesses in VOR besides the cone of confusion discussed in the section.
(8) What is the maximum range and altitude that an airplane could be away from a ground-based VOR and still be able to detect its beacons?
(9) Describe the difference between DME and VOR.
(10) Differentiate between the track mode and the search mode of a DME instrument.
(11) Assume that a DME interrogator onboard an airport transmits an interrogating signal at 1.030 GHz. Compute the possible centre frequencies for the response signal from the ground-based DME transponder.
(12) Discuss the purpose of the DME in an instrument landing system (ILS).

13.4 Satellite-Based Positioning Systems: GPS

The Global Positioning System (GPS) is the most popular Global Navigation Satellite Systems (GNSS) in the world. GPS is a world-wide, all-weather satellite-based wireless navigation system. A ground-based GPS receiver estimates its instantaneous position or location and current time from signals it receives from at least four in-view in-orbit GPS satellites. Satellite-based positioning systems are becoming increasingly popular, partly owing to their wide area coverage. For example, the Ground Based Augmentation System (GBAS), referred to as Honeywell SmartPath, is a satellite-based precision landing system which is competitive to the ILS. Other satellite-based positioning systems include:

(1) Global Positioning System (GPS): also known as NAVSTAR, a global system owned and operated by the US Department of Defense (DoD).
(2) Galileo: a global navigation satellite system (GNSS) owned and operated by the European Union (EU). It uses orbits at 23,222 km altitude.
(3) GLONASS: a global system owned and operated by the Russian Federation.
(4) IRNSS: India's Regional Navigation Satellite System.
(5) Quasi-Zenith Satellite System (QZSS): Japanese regional navigation system which started operation in September 2010. It serves Asia and Oceania.
(6) BeiDou-1 (aka regional Compass Satellite Navigation Experimental System) and BeiDou-2 (aka global Compass Navigation Satellite System): a Chinese satellite navigation systems which are still under development.

For brevity of treatment, we shall study only GPS as a case study of GNSS. Data on GPS can be found at its official website *www.gps.gov*. Much of the data in this section were obtained from this website. GPS is used for positioning, navigation and timing (PNT). GPS services are divided into two types: precision positioning service (PPS) for authorised users and standard precision service (SPS) for everyone at no service fee.

Listed in their launching order, GPS satellites are of five generations: Block IIA (2nd generation Advanced), Block IIR (R for Replenishment), Block IIR-M (Modernised), Block IIF (Follow-on) and GPS III/IIIF satellites. The latest satellite, GPS III/IIIF, was first launched into orbit in 2018. There is a wide range of application areas of GPS, including timing, agriculture, aviation, environmental monitoring, marine, rail, public safety and disaster relief, recreation, space, and roads and highways. Refer to Fig. 13.14.

The GPS architecture comprises three components: the space segment, the control segment and the user segment (made of all GPS receivers). The first two segments are developed, maintained and operated by the US Air Force. The user segment, however, has many manufacturers. As GPS is a one-way (aka receive-only or RCVO) system, the user segment comprises only receiving devices, called GPS receivers.

13.4.1 *GPS Space Segment*

The GPS project began in 1973 and became fully operational in 1994/5 with 24 satellites on a MEO at the altitude of circa 20,180 km. The number of in-orbit satellites increased to 27 in June 2011. The orbital period is 11 hrs 58 minutes (i.e., half sidereal day), and the orbits are nearly circular with about 55° inclination to the equatorial plane. GPS uses six equally-spaced orbital planes, each hosting four satellites. Therefore, 24 satellites are in orbit with three in-orbit spares satellites. The US Govt seeks to maintain 24 satellites in orbit 95% of the time. The constellation of the GPS satellites has been made in such a way that a receiver can detect at least four satellites most of the time regardless of geographical position. The only requirement is line of sight.

13.4.2 *GPS Control Segment*

The control segment is a global network of terrestrial facilities which track the GPS in-orbit satellites, monitor their transmissions, perform analysis, and send commands and data to the constellation. As at Q1 2019, the Operational Control Segment (OCS) comprised:

(1) The Master Control Station located in Schriever AFB (Air Force Base), Colorado.
(2) The Alternate Master Control Station located in Vandenberg AFB California. This is a fully

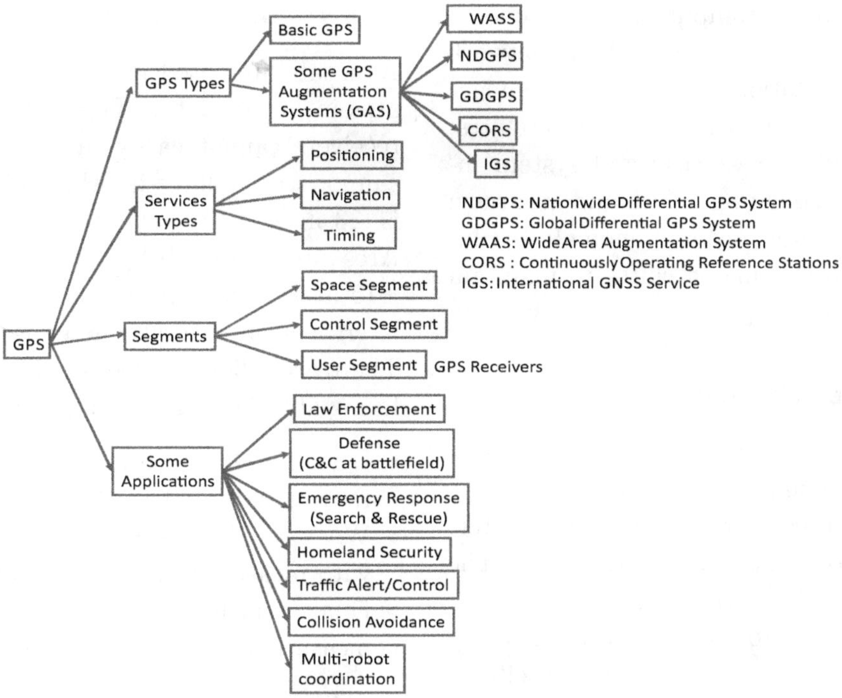

Fig. 13.14 Some features of GPS and its extensions.

operational back-up for the master control station.

(3) 16 monitoring stations, which are divided into:

(a) 10 National Geospatial-Intelligence Agency (NGA) monitoring stations located in Alaska, Australia, Bahrain, Ecuador, NZ, South Africa, South Korea, UK, Uruguay and USNO Washington DC.

(b) Six Air Force Monitoring stations located in Ascension, Cape Canaveral, Colorado, Diego Garcia, Hawaii and Kwajalein.

(4) 11 command and control and ground antennas stations, which are divided into:

(a) Four GPS ground antenna stations located in Ascension, Cape Canaveral, Diego Garcia and Kwajalein

(b) Seven Air Force Satellite Control Network (AFSCN) remote tracking stations located in Diego Garcia, Greenland, Guam, Hawaii, New Hampshire, UK, and Vandenberg AFB California.

The functions of the *ground antenna stations* are: send commands, navigation data uploads and processor program loads to the satellites; collect telemetry; and communicate via S-band and perform S-band ranging to provide anomaly resolution and early orbit support.

Each of the *monitoring stations* is equipped with several sophisticated GPS receivers and atomic clocks to provide the following functions: provide global coverage; tracking of satellites as they pass overhead; collection of navigation signals, range/carrier measurements and atmospheric data; and forwarding of collected data to the master control station. Each monitoring station time stamps its measured data and forwards it to the master control station. The latter uses the data received to estimate parameters of each satellite, including orbital parameters, atmospheric data and clock deviations.

The brain of the GPS system is the *master control station* (MCS). The MCS controls both in-orbit operational and non-operational satellites, but using two different systems. The MCS uses Architecture Evolution Plan (AEP) to control operational satellites but Launch/Early Orbit, Anomaly Resolution and Disposal Operations (LADO) for non-operational satellites. The other main functions of the MCS are: provision of command and control of the GPS constellation; compute the precise locations of the satellites using data from the network of global monitoring stations; generation of navigation messages for upload to the in-orbit satellites; monitoring of satellite broadcasts and system integrity to ensure constellation health and accuracy; perform satellite maintenance and anomaly resolution, including repositioning satellites to maintain optimal constellation

13.4.3 *GPS User Segment*

The user segment is the set of all GPS receivers. Each GPS receiver comprises an antenna and signal processors which estimate parameters of interest from signals broadcast from the satellites. The GPS receiver uses *trilateration* or more precisely, multilateration method, to estimate positions using signals from a minimum of four visible satellites. With a minimum of four visible satellites, a GPS receiver estimates:

- Latitude.
- Longitude.
- Altitude.
- Clock deviation/drift.

Refer to the sections on trilateration and multilateration to find out how the GPS receivers operate. GPS receivers differ in their accuracy and hence pricing.

13.4.4 *GPS Measurement Accuracy*

As a satellite-based positioning system, GPS operates in line-of-sight. This is the reason why GPS receivers cannot easily find their position if they cannot see satellites in orbit. The accuracy of GPS is ± 3 m or ± 10 ns clock accuracy if WASS (Wide Area Augmentation System) is used with GPS and ± 15 m without WASS. The GPS precision is liable to errors from several sources as listed in Table 13.2. There are different types of GPS receivers with differing abilities. They can be categorised into:

- Low-cost GPS receivers, with measurement accuracy of ± 10 m about 95% of the time.
- WAAS-enabled single-frequency GPS receivers, with measurement accuracy of about $\pm 1 - 3$ m.
- Dual-frequency, survey-grade GPS receiver, with measurement accuracy of less than ± 1 cm.

Although the clocks onboard GPS satellites are Cesium atomic clocks with high precision, each clock drifts about 10^{-9} seconds every 3 hours, resulting in measurement error. GPS monitoring stations periodically compute the correct clocks and broadcast them with other GPS signals in order to minimise clock-related errors. GPS receivers use quartz crystal clocks which are less reliable than atomic clocks. They drift about 1 μs every second, requiring receivers to reset their clocks every second. Cesium atomic clocks are too heavy (one can weigh about 20 kgs) and too expensive (e.g., a

Table 13.2 Sources of error in GPS measurements.

Error Source	Error Magnitude
Satellite clock errors	1.5-3.6 m
Orbital errors	< 1 m
Ionospheric effects	5-7 m
Tropospheric effects	0.5-0.7 m
Receiver noise	0.3-1.5 m
Multipath signal propagation	0.6-1.2 m
Selective Availability	Discontinued since 2000
Dilution of precision	GDOP & PDOP < 10 m/m, HDOP & VDOP < 2 m/m
User error	≈ 1 km

DOP: dilution of precision; GDOP: geometric DOP; HDOP: horizontal DOP; PDOP: position DOP; VDOP: vertical DOP.

miniature one costs over $4,000) to be implemented in GPS receivers. The receiver clock errors can be corrected by comparing arrival times of signals from two satellites.

The dilution of precision (DOP) is caused by the improper arrangement of satellites in the sky. The desired positions of four satellites relative to the GPS receiver is that one satellite is directly above it while the remaining three are equally spaced nearer the horizon (but above the mask angle). GPS receivers record several types of DOP, including GDOP, HDOP, VDOP and PDOP which is a combination of VDOP and HDOP. The measurement error due to DOP is minimised when PDOP is minimum.

What is *Selective Availability* (S/A)? The US DoD intentionally degraded the public GPS signals to reduce positioning accuracy to some GPS users. This was a security measure to prevent enemies from using the GPS to attack the US and her allies. During his tenure, President Bill Clinton discontinued the S/A in May 2000 in order to provide accurate positioning for civilian and commercial use globally.

The orbital parameters (e.g., shape of orbit or eccentricity and orbital velocity) of satellite change with time. However, GPS receivers estimate their coordinates relative to these non-static orbital parameters. One of the tasks of the GPS Control Segment is to monitor each satellite and record its position in a document referred to as *ephemeris*. As illustrated in Fig. 13.16, each satellite broadcasts as part of its navigation messages its ephemeris for GPS receivers to use to correct their orbital errors. Multiple ephemeris of multiple satellites is called *ephemerides*.

Two layers in the atmosphere affect the propagation of the satellite signals. These are the ionosphere and the troposphere. The ionosphere contains ions, which are charged particles, which accelerate the carrier signal but slows down the code pattern. The troposphere contains clouds and other weather which also effect the signals. Refraction of EM waves occurs within the ionospheric layers of the atmosphere, which change the orientation and, in particular, the speed of propagation of the waves. This effect then affects the accuracy of time and range measurements. The ionosphere slows down the EM waves and prolong the propagation time (and extend the pseudo range) while advancing the phase of the EM waves (the carrier range), by the same amount. GPS measurements are also affected by multipath signal propagation.

Owing to the multiple sources of error in GPS measurements, such as the satellite and receiver clocks offset (aka bias or mis-synchronisation), ionospheric and tropospheric effects and multipath signal propagation, the measured range is referred to as *pseudo-range* or pseudorange in contrast to the true range, referred to as *geometric range*. The ionospheric portion of the atmosphere contains ions which impede the movement of EM waves and prolong their time to cover a given distance. A rule-of-thumb is to allow about 16 ns tolerance for ionospheric delay. Another layer of the atmosphere, called troposphere, contains weather effects such as clouds, rain and lightning. These slow down the EM waves for about 1.5 ns.

A simple GPS accuracy experiment

This experiment is adapted from [PSU (2019)]. We can determine how accurate our GPS receiver measures its position with this simple example.

(1) Put the GPS receiver in a fixed geographical location for a period of time, say 12 hours.
(2) Record the coordinates of the position as measured by the receiver every 10 minutes, yielding a set of 72 coordinates.
(3) Plot the 72 coordinates.
(4) Draw a circle to enclose as many of the points as possible, at least 95% of them. Measure the radius of the circle.

The measured radius is the positioning error of the GPS receiver used. Of course, the more the data collected, the better the precision of the results.

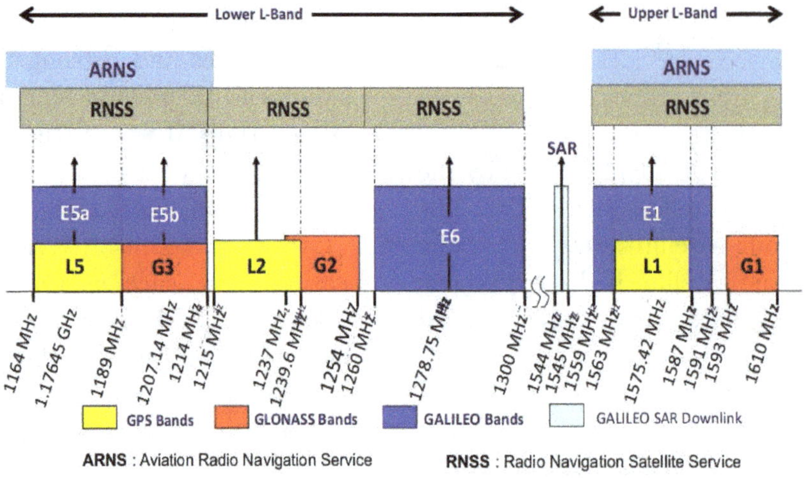

Fig. 13.15 GNSS frequency bands [Subirana *et al.* (2011)].

13.4.5 *GNSS Frequency Bands*

The radio frequency spectrum is divided into bands and each band is usually allocated for specific wireless services. The ITU, through its World Radio Communication Conferences (WRC) held in 2000 and 2003, agreed on the frequency allocations shown in Fig. 13.15 for Global Navigation Satellite Systems (GNSS). This figure was obtained from [Subirana *et al.* (2011)]. We can observe that some of the bands are shared by multiple wireless services, rendering each service prone to mutual radio frequency interference. One of the services provided by the European GNSS, called Galileo, is the Search and Rescue (SAR) service. The Galileo SAR is part of the international Medium Earth Orbit distress alert detection and information distribution satellite network, called COSPAS-SARSAT,[3] which is used to detect and locate emergency beacons activated by adventurers, hikers, airplanes and ships.

13.4.6 *GPS Signals and Messages*

Each GPS satellite broadcasts signals periodically to GPS receivers. Figure 13.16 shows the various signals, while Table 13.3 shows the carrier frequencies used to broadcast each type of signal. What information do the GPS signals communicate to GPS receivers? There are several of them. First, a GPS receiver needs to compute the time taken for a signal to travel from the satellite to it. This requires a *time stamp*, indicating the time on the satellite that the signal was sent. There are multiple oscillators (or clocks) in both the space-born satellite and the earth-based GPS receivers. Reliable time measurements require some corrections from atmospheric effects which delay the signal. So, satellite signals must contain estimates of the wireless channel through which the signal passes to enable some *atmospheric corrective* measures. Another information carried by satellite signals is the *instantaneous position of the satellite*. Further, a GPS receiver must be able to identify which of the 24 to 27 GPS satellites sent a received signal. Therefore, a satellite must send its *identification code*.

GPS signals are divided into code pattern and carrier pattern. These two patterns have different cycles or periods, and are used for positioning, navigation and timing (PNT). Using the standard formula between frequency and wavelength $\lambda = \frac{c}{f}$, the wavelengths of the L1 and L2 carriers are about 19 cm and 24 cm, respectively. As shown in Fig. 13.16, the chipping rates of the P(Y) and

[3]COSPAS means Space System for the Search of Distressed Vessels while SARSAT means Search and Rescue Satellite Aided Tracking System. COSPAS-SARSAT is headquartered in Montreal, Quebec, Canada, and it is a humanitarian cooperative of 45 nations and agencies. A COSPAS-SARSAT beacon is a radio transmitter which can be activated in a life-threatening emergency to summon assistance from government authorities internationally.

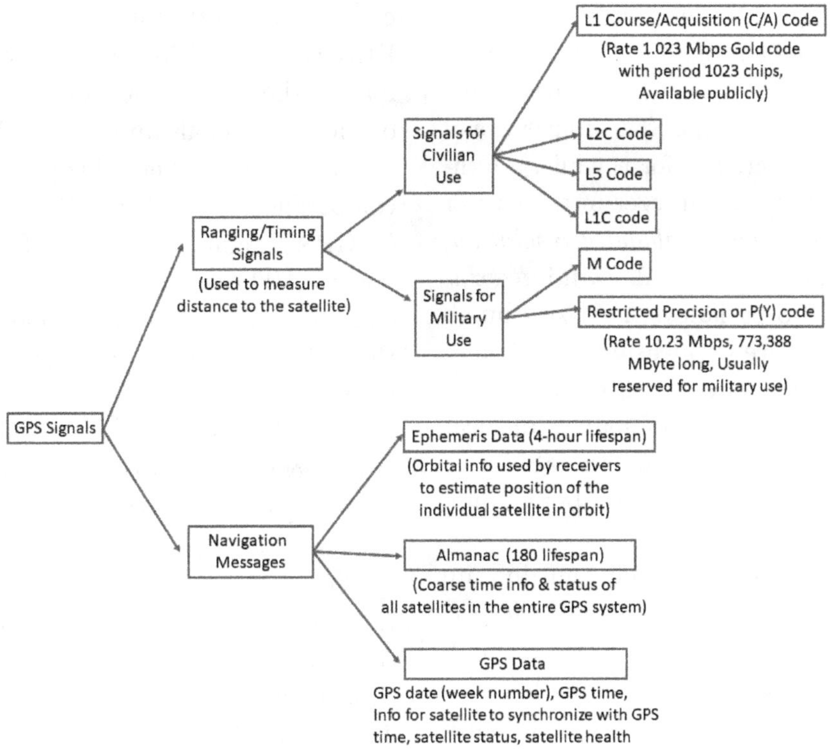

Fig. 13.16 GPS signals and their purposes.

Table 13.3 GPS carrier frequencies and their purposes.

Frequency (GHz)	Carrier Designation	Purpose
0.01023	F_o	Fundamental clock rate (i.e., standard oscillator frequency)
$1.57542 = 154F_o$	L1	Mixture of navigation message, coarse-acquisition (C/A) code for civil use, encrypted precision P(Y) code for military, L1C signal on Block III satellites
$1.22760 = 120F_o$	L2	P(Y) signal for military use, L2C code on Block IIR-M satellites for civil use, navigation message
$1.38105 = 135F_o$	L3	Nuclear Detonation (NUDET) Detection System Payload (NDS), S-band telemetry
$1.379913 = 10.23F_o$	L4	Ionospheric correction
$1.17645 = 115F_o$	L5	Civilian safety-of-life (SoL) signals

C/A PRN codes are 10.23 Mbps and 1.023 Mbps, respectively. Therefore, the corresponding chip durations are $0.1\,\mu$s and $1\,\mu$s. These are equivalent to distances of 0.3 m and 3 m, respectively, assuming the speed of the signals is equal that of light in vacuum. GPS PRN codes are Gold codes which are generated using two 10-bit shift registers.

The distance measured with the carrier pattern is referred to as *carrier phase*. The carrier phase is short. Also, the cycles of the carrier signal are the same, making it difficult to differentiate one from the rest. For these reasons, the number of full cycles that the carrier pattern oscillates between the airborne satellite and the ground-based

GPS receiver is ambiguous. This is referred to as an *integer ambiguity* or an *unknown integer*. How do we resolve the unknown integer? This is currently an ongoing research topic. The conventional method is to track the satellite for a while. Methods used to find the unknown integer in the carrier phase is called *carrier phase ambiguity resolution*, while the mis-calculation of the unknown integer is called *cycle slip* in GPS parlance. A rule-of-thumb for an observable distance or measurement granularity is about 1% of the wavelength of the carrier used. Therefore, in the absence of sources of measurement errors, GPS receivers with the ability to measure the carrier phase can deliver about 2 mm accuracy.

The *code pattern* has effectively an infinite length. The distance measured using the *code pattern* is referred to as *code phase*. A rule-of-thumb for an observable distance is about 1% of the chipping rate of the code used. Hence, in the absence of sources of measurement errors, the code phase could deliver theoretically about 3 mm and 30 mm GPS measurement accuracies over the P(Y) and C/A codes, respectively. Note that GPS receivers achieve much less accuracy. The C/A achieves about ±100 m if selective availability (S/A) is turned on or ±30 m if S/A is turned off. While the code phase is measured by every GPS receiver, only high-precision GPS receivers can measure the carrier phase. The code generated at the receiver and the one received from the satellite could be out of phase. However, the PRN codes are so designed that the receivers can differentiate between their cycles and detect which cycle marks the edge of the timing pulse to eliminate any integer ambiguity.

13.4.6.1 *C/A and P(Y) Codes*

Signals broadcast by GPS satellites are carriers, referred to as Lk, $k = 1, 2, \ldots, 5$, which are modulated by binary signals. The binary signals are unique pseudo-random noise (PRN) codes. For this reason, GPS signals are better referred to as codes. The modulation scheme used is the bi-phase shift keying modulation (BPSK). All the signals are transmitted using left-hand circular polarisation (LHCP), and at low powers. For example, the C/A code is transmitted using 28.8 dBW intensity. The PRN codes used in GPS are so chosen that information theoretic principles enable the amplification of the GPS signals and enable GPS receivers to use small antennas. The added advantages in this are compactness and affordability of the receivers.

GPS satellites broadcast four signals for civilian use. These are the Coarse/Acquisition (C/A) code or signal, the L2C signal, the L5 signal and the L1C signal, in their order of introduction. The C/A signal, which is broadcast over the L1 carrier, is referred to as a *legacy civil signal* or a *legacy civil code* because it was the original signal, while L2C, L5 and L1C signals are called *modernised civil codes* because they were introduced to improve the GPS system. L1C uses multiplexed binary offset carrier (MBOC) modulation scheme. It was introduced in cooperation with EU Galileo to enable international cooperation with other GNSS (European Galileo, Japanese QZSS and Chinese BeiDou), while protecting USA national security.

The L2C signal which is transmitted on the L2 carrier was introduced in 2005 for civilian use. It enables ionospheric correction so that civilian use of GPS achieves an accuracy not worse than military use. Further, it enables faster GPS signal acquisition, larger operating range and enhanced reliability. The L2C signal is transmitted at a power higher than C/A code, allowing a larger range. The US Air Force broadcasts civil navigation (CNAV) messages over the L2C signal.

The satellites broadcast two forms of clock information, C/A code and the restricted precision code or P(Y) code. The C/A code is freely available to the general public, and it is used to modulate only the L1 carrier. Each C/A code, comprising $L = 1023$ binary digits, repeats every 1 ms, resulting in the chip rate of 1.023 Mcp/sec. Therefore, the bit duration is about $T_b = \frac{1}{1023}$ ms $\approx 1\,\mu$s. This bit duration is equivalent to the distance of 293.3 m (i.e., $c \times T_b$). The precision of the GPS measurement is generally about one percent of this bit period, which is about 2.9 m.

The precise code or P(Y) code, however, is reserved for only military use, and it is used to modulate both the L1 and L2 carriers. Each P(Y) code comprises $L = 23,547 \times 10^{11}$ bits which repeats every $T = 266$ days, resulting in the chip rate of 10.25 Mcp/sec (10.23 Mcp/sec is usually advertised). Each bit in a P(Y) code thus lasts circa $0.0976\,\mu\text{s}$, which is equivalent to about 29.3 m in distance. The rule-of-thumb for the precision of the GPS measurement is about one percent of the bit duration, which is about 0.29 m. This makes it clear the reason why the P(Y) code delivers a better precision than using only the C/A code to measure distances.

The GPS system assigns a unique one week's segment of the P(Y) code to each GPS satellite. Thus, each satellite broadcasts the same unique P(Y) code for one week (i.e., mid-night of Sunday to mid-night of Saturday) and then the segment is reset to zero. The P(Y) code is the P-code which is encrypted using W-code sequence to prevent spoofing. The encrypted P(Y) code is referred to as Y-code. Only military GPS receivers can decrypt the Y-code. There is also the M code on Block III-M satellites for military use to enhanced jamming resistance. The M code is transmitted on L2 carrier.

The P(Y) and C/A codes have the following dual purposes:

- Provide time delay measurements to enable a GPS receiver to estimate its distance from the satellite which transmitted the signal. The time differences are used to estimate distances to construct hyperbolae.
- To differentiate received signals from different satellites.

How does the GPS receiver estimate its location? The procedure is: estimate the time taken for a signal to travel from a GPS satellite (aka control point), say k, to the GPS receiver. This time is referred to as the signal propagation delay. Let this time be t_k. This time is extracted by correlating the PRN code received from the satellite with the replica of the PRN code generated locally by the receiver. The location of the maximum of the correlation between these two PRN codes determines t_k. We refer to distances as ranges in GPS terminology. Convert t_k to range (or correctly pseudorange) by multiplying it by the speed of light. A GPS receiver has to measure at least four distances, and plug them into an algorithm to find its location (i.e., X, Y and Z coordinates) and the bias between its clock and that of the space-born GPS satellite. Popular algorithms used to convert the pseudoranges into location include Kalman filtering and trilateration/multilateration.

As shown in Fig. 13.18(b), one of the information carried in the CNAV message is the almanac. A GPS receiver uses the almanac data in the NAV message it receives from the first satellite that it acquires to learn about the satellites that it expects to see. It then loads up C/A codes to use to estimate its distance to each of them.

13.4.6.2 *Navigation Messages*

The content of these messages is continuously updated by the GPS control segment and broadcast to the user via the GPS satellites. As shown in Fig. 13.16, they comprise the following:

(1) Almanac data.
(2) Satellite's orbital data, referred to as ephemeris. This data contains orbital information which GPS receivers use to estimate the position of the individual GPS satellites. The ephemeris signal contains the estimated position of the satellite.
(3) Information about the health of satellite and constellation status.
(4) Offset between Universal Coordinated Time (UTC) and GPS time.
(5) Satellite clock correction parameters.
(6) Ionospheric model parameters so that single-frequency GPS receivers can correct ionospheric time delay.

13.4.6.3 *GPS Civil Navigation Message*

The Civil Navigation (CNAV) messages are the most modern civil messages transmitted by the GPS. Altogether a GPS satellite broadcasts 15 CNAV message types over the two carriers L2C at 1.2276 GHz and L5 at 1.17645 GHz. Only the GPS satellites IIR-M and IIF implement CNAV.

The L2C PRN code has the chip rate of 1.023 Mbps and it comprises the two parts:

- L2CL (Civilian Long) of 1.5-second length and 511.5 kbps chip rate.
- L2CM (Civil Moderate) of 20-millisecond length and 511.5 kbps chip rate.

The L2C signal is generated in a three-step process. First, the CNAV data is added to the L2CM PRN ranging code using modulo 2. Second, the result is then chip by chip time multiplexed with the L2CL PRN ranging code to obtain the L2C code. Third, L2C code is then used to modulate the L2 carrier using BPSK modulation. Figure 13.17 illustrates, in general, how a GPS signal is generated. The L2C signal is broadcast by both IIR-M and IIF GPS satellites.

The L5 code comprises the two PRN ranging codes:

- The in-phase component L5I5.
- The quadrature component L5Q5.

As shown in Fig. 13.17, each of these sub-codes has the chipping rate of 10.23 MHz, the same as the fundamental clock rate of GPS. The L5 signal is broadcast by only GPS IIF satellites. The format and message types of the CNAV are shown in Fig. 13.18 [GPS Directorate (2011)]. Each message has three parts: a 38-bit message (Msg) header, a 238-bit Msg payload and a 24-bit cyclic redundancy check (CRC). The entire CNAV message is error protected using a 24-bit CRC. Both L2C and L5 use the same message format of length 300 bits. However, the 300-bit L2C CNAV message is transmitted in 12 seconds, while the 300-bit L5 CNAV message is transmitted in 6 seconds at a rate of 50 bps. In the figure TOW means time-of-week while SV means space vehicle. Six bits are used to encode the message type, resulting in 64 unique messages. However, only 38 of them are currently defined.

13.4.7 *Satellite-Based Augmentation Systems*

A Satellite-Based Augmentation System (SBAS) combines terrestrial and satellite infrastructure to enhance the reliability and accuracy of positioning information from basic GNSS, such as Galileo

(a)

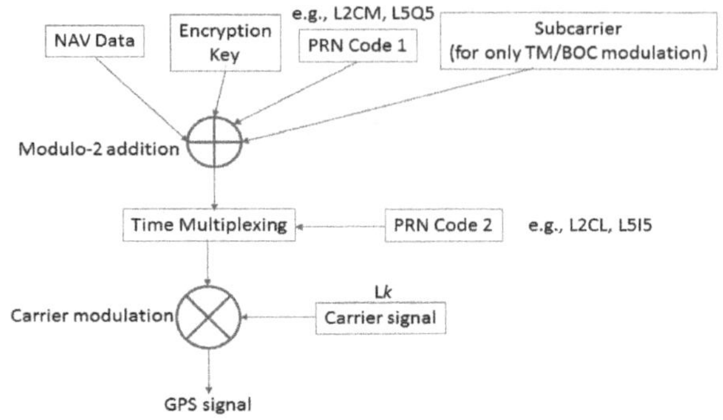

(b)

PRN code	C/A	P(Y)	L1C	L2CM, L2CL	M	L5I, L5Q
Carrier	L1	L1, L2	L1	L2	L1, L2	L5
Center frequency [GHz]	1.57542	1.57542, 1.2276	1.57542	1.2276	1.57542, 1.2276	1.17645
Data rate [bps]	50	50	50	50	-	50
Code length [bits]	1023	619×10^{17}	10230	10230, 767250	-	10230
Chip rate [MHz]	1.023	10.23	1.023	0.5115, 0.5115	5.115	10.23
Multiple access	CDMA	CDMA	CDMA	CDMA	CDMA	CDMA
Modulation	BPSK	BPSK	TMBOC	BPSK	BOC	QPSK

Fig. 13.17 Illustration of GPS signal generation and signal properties (BOC: Binary Offset Carrier, TMBOC: Time-Multiplexed Binary Offset Carrier).

and GPS. Also, SBAS transmit extremely accurate universal time signal, and provides crucial integrity message w.r.t. availability and reliability of GNSS signals. Examples of SBAS are Wide Area Augmentation System (WAAS) improving upon GPS GNSS in North America and Mexico, European Geostationary Navigation Overlay Service (EGNOS) improving upon Galileo GNSS in Europe, the Japanese Multi-functional Satellite Augmentation System (MSAS), and the Indian Geo Augmented Navigation System (GAGAN). GPS enhanced by a SBAS is referred to as GPS augmentation or augmented GPS (A-GPS).

The process by which an aircraft approaching an airport and thus preparing to land estimates its altitude and clearance from obstacles on the ground is called **continuous vertical guidance** (CVG). Controlled flight into terrain (CFIT) is a major hazard in civil aviation. Traditionally, CVG is achieved using technologies such as instrument landing system (ILS), Ground Based Augmentation System (GBAS), Required Navigation Per-

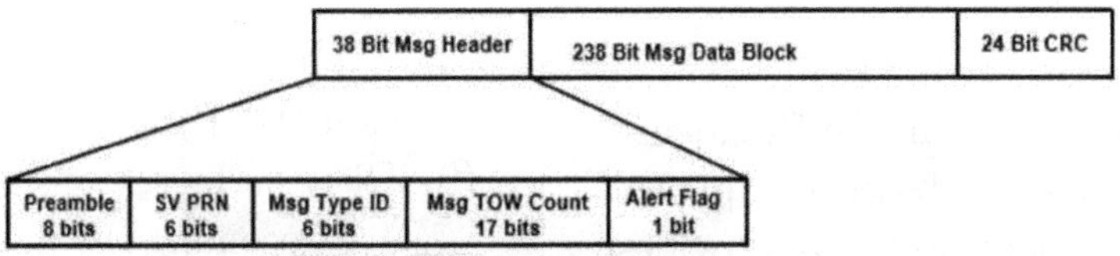

(a) 300-bit CNAV message format

Msg Type	CNAV Message Title	Function/Purpose
0	Default	Default message (transmitted when no msg data is available)
10	Ephemeris 1	SV position parameters for the transmitting SV
11	Ephemeris 2	SV position parameters for the transmitting SV
12	Reduced Almanac	Reduced almanac data packets for 7 SVs
13	Clock Differential Correction	SV Clock differential correction parameters
14	Ephemeris Differential Correction	SV Ephemeris differential correction parameters
15	Text	Text, 29 eight-bit ASCII characters
30	Clock, IONO & Group Delay	SV Clock Correction Parameters, Ionospheric and Group Delay correction parameters (Inter-Signal Correction parameters)
31	Clock & Reduced Almanac	SV Clock Correction Parameters, Reduced almanac data packets for 4 SVs
32	Clock & EOP	SV Clock Correction Parameters, earth orientation parameters; ECEF-to-ECI coordinate transformation
33	Clock & UTC	SV Clock Correction Parameters, Coordinated Universal Time (UTC) Parameters
34	Clock & Differential Correction	SV Clock Correction Parameters, SV clock and Ephemeris differential correction parameters
35	Clock & GGTO	SV Clock Correction Parameters, GPS to GNSS Time Offset parameters.
36	Clock & Text	SV Clock Correction Parameters, Text, 18 eight-bit ASCII characters
37	Clock & Midi Almanac	SV Clock Correction Parameters, Midi Almanac parameters

(b) CNAV message types

Fig. 13.18 Civil navigation (CNAV) message structure and types [GPS Directorate (2011)].

formance Authorisation Required (RNP AR), and Barometric-Vertical Navigation (Baro-VNAV). Unfortunately, only large aircrafts usually install these systems. Thus, a technology is needed to serve the smaller aircraft. This is where Satellite BAS reigns supreme.

How does SBAS operate? First, SBAS monitors signals from GNSS satellites, calculate corrections, and then upload them to a satellite in geostationary orbit which broadcasts the corrections back to the Earth.

13.4.7.1 *WASS, DGPS and LAAS*

The Wide Area Augmentation System (WAAS) is geostationary satellite system developed to improve GPS in three dimensions: integrity, accuracy and availability. WAAS combines a network of ground-based reference stations and GPS to increase measurement accuracy. WAAS broadcasts correction signals from Differential GPS (DGPS) through GEO satellites. DGPS uses a fixed, known position to adjust real-time GPS signals to eliminate pseudo-range errors in position determination. DGPS is the difference between the actual distance measured and the theoretical distance. There are three basic types of DGPS:

- SBAS: Correction messages are sent from geostationary satellites.
- Radio Technical Commission for Maritime Services v.2 (RTCMv2): Correction messages are sent from a static base station to provide 40-80 cm accuracy.
- Real Time Kinetic (RTK): Correction messages are sent from a static base station signal to provide < 2 cm accuracy on RTK-enabled units.

WAAS-enhanced GPS is poised to support all phases of aviation. This means that eventually GPS will supplant instrument landing system (ILS) and microwave landing system (MLS) used at airports. The positioning accuracy of WAAS-enabled GPS can be further improved with Local Area Augmentation System (LAAS).

13.4.8 *ADS-B*

The Automated Dependent Surveillance-Broadcast equipment (ADS-B) is a satellite-based equipment used for air traffic control at airports. It provides extremely accurate information about the locations of airplanes and flight path. The advantages in such a feature include improved aviation safety, reduced geographical separation between aircrafts, increased number of direct flight routes, and reduced operational expenses.

13.4.9 *Check Your Understanding*

Students are encouraged to check their understanding of the content of this section with the following questions prior to proceeding to the next section.

(1) Two of the data carried in the navigation message of GPS satellites are the ephemeris and the almanac. Clearly discuss the differences and the relationship between the two.
(2) State the two reasons why GPS receivers use less reliable quartz clocks, but not atomic clocks.
(3) Explain the meaning of carrier phase and code phase in GPS terminology, and differentiate between the two.
(4) How do we call the signal used in GPS to communicate the estimated positions of the satellites in space?
(5) Discuss the difference between geometric range and pseudo-range in GPS measurements.
(6) Discuss the general nature of signals broadcast by GPS satellites.
(7) Describe the meaning and purpose of *ephemeris* in satellite navigation.
(8) Explain the meaning of dilution of precision (DOP) as used in GPS.
(9) Discuss the reasons why GPS with WAAS are becoming the most preferred positioning systems.
(10) Describe the purpose of Satellite Based Augmentation System (SBAS).
(11) Compare Kalman filtering with trilateration/multilateration algorithms which are used to convert pseudoranges in GPS into location estimation.
(12) Describe and differentiate between the two pseudo-random codes or patterns used to modulate the two carriers (i.e. L1 at 1575.42 MHz and L2 at 1227.6 MHz) of GPS satellites.
(13) Integer ambiguity is a problem in the usage of signals from satellites to estimate a position by radio positioning receivers. Describe the cause(s), effect(s) and solution(s) to *integer ambiguity.*
(14) In the terminology of Global Positioning System (GPS), describe the meaning of *cycle slip.*
(15) Discuss the differences between the legacy civil signals/codes and modernised civil signals/codes in GPS system.
(16) Study the data in Table 13.2. Using the appropriate data from this table, estimate the total delay that a GPS signal undergoes from the satellite to the GPS receiver because of the:

    (a) Ionospheric effects.
    (b) Tropospheric effects.

(17) A ground-based GPS receiver seeks to estimate its pseudorange from a satellite in orbit. It therefore correlated a shifted version of a PRN code with the version received from the satellite. The autocor-

relation attained its maximum upon shifting the locally-generated PRN code by 20 chips. Compute the pseudorange between the receiver and the satellite if the code used is:

(a) C/A PRN code carried by L1 carrier.

(b) L5I PRN code carried by L5 carrier.

Hint. Use the data in Table 13.17.

(18) **Students' project:** Analytics of GPS data/signals using Kalman filtering. **Background:** The Global Positioning system (GPS) is the primary source of information for a broad range of positioning, navigation and timing systems. It is an all-weather, satellite-based radio-navigation system which provides world-wide coverage. A distance measured in GPS is called pseudo-range as it may deviate from the true geometric range or distance because of measurement errors. A GPS receiver estimates receiver-to-satellite pseudo-ranges from signals it receives from GPS satellites in space. It then uses the pseudo-ranges to estimate its instantaneous position. This project seeks efficient algorithms which can be used in GPS receivers to estimate their positions from the estimated satellite- to-receiver pseudo-ranges with a high accuracy. This project explores Kalman filtering as the method to convert pseudo-ranges into positioning estimate. Kalman filtering algorithms for three scenarios will be studied: stationary receiver, receiver with small mobility and receiver with high mobility.

(19) **Students' project:** GPS carrier phase ambiguity resolution techniques. **Background:** The Global Positioning System (GPS) is the primary source of information for a broad range of positioning, navigation and timing systems. It is an all-weather, satellite-based radio-navigation system which provides world-wide coverage. GPS receivers estimate their location by measuring their distances to at least three satellites and feeding the distances into trilateration/multilateration algorithm. Two types of pseudorandom noise (PRN) sequences, referred to as carrier pattern and code pattern, are used for the distance measurements. The distance measured with the carrier pattern is referred to as carrier phase, while distance through the code pattern is called code phase. The carrier phase gives a better positioning accuracy. Unfortunately, the carrier phase is short. Also, the cycles of the carrier signal are the same, making it difficult to differentiate one from the rest. For these reasons, the number of full cycles that the carrier pattern oscillates between the airborne satellite and the ground-based GPS receiver is ambiguous. This is referred to as an integer ambiguity or an unknown integer. Methods used to find the unknown integer in the carrier phase is called carrier phase ambiguity resolution, while the miscalculation of the unknown integer is called cycle slip in GPS parlance. Better carrier phase ambiguity resolution is still being searched for in order to improve GPS measurement accuracy.

13.5 Review of Important Terms and Acronyms

Acronyms, abbreviations and important terms			
access node	ADF	A-GPS	anchor node
AOA	augmented GPS	AWACS	C/A code
celestial navigation	dead reckoning	DECCA	DGPS
dilution of precision	direction finding	DME	DOA
Doppler shift	Galileo GNSS	geometric range	glideslope
GLONASS	GPS	GPS augmentation	hyperbolic DF
ICAO	ILS	INS	IRF
localisation	localiser	LORAN	MLS
multilateration	MUSIC DOA	NAVAID	navigation
NDB	Omega	P(Y) code	PBN
piloting	polynomial regression	positioning	pseudo-range
radio beacon	radiolocation	reference node	RNAV
RNP	RSSI/RSS	SBAS	selective availability
target node	TDOA	TOA	triangulation
trilateration	VOR	VRF	WASS

13.6 Supplementary Reading

(1) Alan Bensky, *Wireless Positioning Technologies and Applications*, Artech House, 2016.

(2) Kegen Yu, *Positioning and Navigation in Complex Environments*, Information Science Reference, 2018.

(3) David Bartlett, *Essentials of Positioning and Location Technology*, Cambridge University Press, 2013.

(4) Stephen Sand and Armin Dammann, *Positioning in Wireless Communications Systems*, John Wiley, 2014.

(5) Kegen Yu, Ian Sharp and Y. Jay Guo, *Ground-Based Wireless Positioning*, Wiley-IEEE Press, 2009.

(6) Simone Frattasi and Francescantonio Della Rosa, *Mobile Positioning and Tracking: From Conventional to Cooperative Techniques*, Wiley-IEEE, 2017.

(7) K. K. Sharma, *Fundamentals of Television and Radar Engineering*, S. K. Kataria & Sons, New Delhi, India, 2009.

(8) Azadeh Kushki and Konstantinos N. Plataniotis, *WLAN Positioning Systems: Principles and Applications in Location-Based Services*, Cambridge University Press, 2012.

Bibliography

3GPP (2009). Evolved Universal Terrestrial Radio Access (E-UTRA): Multiplexing and Channel Coding, Tech. Rep. TS 36.212.

3GPP (2010). E-UTRA: Radio Frequency (RF) Systems Scenarios, Tech. Rep. TR 36.942, 3GPP.

3GPP (2015a). Study on channel models for LTE, Tech. Rep. 3GPP 36.873 (V12.2.0), 3GPP.

3GPP (2015b). Study on downlink multiuser superposition transmission (MUST) for LTE (Release 13), Tech. Rep. TR 36.859, 3GPP.

3GPP (2015c). TP for classification of MUST schemes, Tech. Rep. R1-154984, 3GPP.

3GPP (2018). Evolved Universal Terrestrial Radio Access (EUTRA); Radio Resource Control (RRC); Tech. Rep. 3GPP 36.331 (V15.2.2), 3GPP.

Abu-Ali, N., Taha, A.-E. M., Salah, M. and Hassanein, H. (2014). Uplink scheduling in LTE and LTE-advanced: Tutorial, survey and evaluation framework, *IEEE Commun. Surveys* **16**, 3, pp. 1236–1265.

Aerospace and Electronic Systems Society, I. (2008). IEEE standard for radar definitions, Tech. Rep. IEEE Std. 686-2008, The IEEE.

Agilent Technologies (2018). Digital modulation in communications systems — An introduction, Tech. Rep. Application Note 1298, Agilent Technologies.

Alamouti, S. M. (1998). A simple transmit diversity technique for wireless communications, *IEEE JSAC*, **16**, pp. 1451–1458.

Aldababsa, M., Toka, M., Gokceli, S., Kurt, G. K. and Kucur, O. (2018). A tutorial on nonorthogonal multiple access for 5G and beyond, *Hindawi Wireless Communications and Mobile Computing*, **2018**, pp. 1–24.

Andrews, M., Kumaran, K., Ramanan, K., Stolyar, A., Vijayakumar, R. and Whiting, P. (2004). Scheduling in a queuing system with asynchronously varying service rates, *Probability in the Engineering and Information Sciences* **18**, pp. 191–217.

Annamalai, A., Palat, R. C. and Matyjas, J. (2010). Estimating Ergodic Capacity of Cooperative Analog Relaying under Different Adaptive Source Transmission Techniques, *Proc. IEEE Sarnoff Symposium*, pp. 1–5.

Armstrong, E. H. (1918). Frequency modulation, online, `http://users.erols.com/oldradio`.

Aziz, D. and Sigle, R. (2009). Improvement of LTE Handover Performance through Interference Coordination, in *IEEE 69th Vehicular Technology Conference Spring (VTC Spring 2009)*, pp. 1–5.

Baker, N. (2005). ZigBee and Bluetooth: Strengths and weaknesses for industrial applications, *Computing and Control Engineering*, **16**(2), pp. 20–25.

Basukala, R., Ramli, H. M. and Sandrasegaran, K. (2009). Performance analysis of EXP/PF and M-LWDF in downlink 3GPP LTE system, *Proc. Asian Himalayas International Conf. on Internet* (Kathmundu, Nepal), pp. 1–5.

Bonald, T. (2005). Flow-level performance analysis of some opportunistic scheduling algorithms, *Euro. Trans. Telecommun.* **16**, 1, pp. 65–75.

Bounfour, A. (2015). *Digital Futures, Digital Transformation: From Lean Production to Acceluction* (Springer).

Brandenburg, L. H. and Wyner, A. D. (1974). Capacity of the Gaussian vhannel with memory: The multivariate case, *Bell Syst. Tech. J.* **53**, 5, pp. 745–748.

Brennan, D. (1959). Linear diversity combining techniques, *Proceedings of the IRE* **47**, pp. 1075–1102.

Bryce (2018). State of the satellite industry report, `https://www.sia.org/wp-content/uploads/2017/07/SIA-SSIR-2017.pdf`, accessed: 12th December 2018.

Buchanan, J. L. and Turner, P. R. (1992). *Numerical Methods and Analysis* (McGraw-Hill, New York).

Calabrese, A. (2006). *Adaptive Radio Resource Management for Uplink Wireless Networks*, Ph.D. thesis, Polytechnic University, USA.

Capon, J. (1969). High resolution frequency-wavenumber spectral analysis, *Proceedings of the IEEE* **57**, 8, pp. 1408–1418.

Capozzi, F., Piro, G., Grieco, L. A., Boggia, G. and Camarda, P. (2013). Downlink packet scheduling in LTE cellular networks: Key design issues and a survey, *IEEE Commun. Surveys and Tutorials* **15**, 2, pp. 678–699.

Celplan (2017). 4G network design: Planning and designing 4G networks in 50 steps Tech. Rep., CelPlan Technologies, Inc, Training and Documentation Dept. 1920 Association Dr, 4th Floor, Reston VA 20191, USA.

Chandler, J. P. (1964). An introduction to pseudo-noise modulation, Tech. Rep. TM-64-4, United States Army, Harry Diamond Lab, Washington D.C.

Chang, K. (2000). *RF and Microwave Wireless Systems* (John Wiley, New York).

Chen, S., Ren, B., Gao, Q., Kang, S., Sun, S. and Niu, K. (2017). Pattern division multiple access — A novel nonorthogonal multiple access for fifth-generation radio networks, *IEEE Trans. Veh. Techn.* **66**, 4, pp. 3185–3196.

Choi, J. (2012). *Adaptive and Iterative Signal Processing in Communications* (Cambridge University Press).

Choi, J. (2017). NOMA: Principles and recent results, in *Proc. Int. Symp. on Wir. Commun. Systems (ISWCS)* (Bologna, Italy).

Christensen, C. M. (1997). *The Innovator's Dilemma* (Harvard Business Review Press).

Clark, J. (2016). What is the internet of things? `https://www.ibm.com/blogs/internet-of-things/what-is-the-iot/`.

Clarke, R. H. (1968). A statistical theory of mobile-radio reception, *Bell Systems Technical Journal* **47**, 957–1000.

Cochoy, F. *et al.* (2017) *Digitalizing Consumption: How Devices Shape Consumer Culture* (Routledge).

Cover, T. M. and Thomas, J. A. (2006). *Elements of Information Theory* (John Wiley & Sons, Inc., New York).

Curran, P. (2015). Better materials for better antennas, online, `https://www.analogictips.com/better-materials-for-better-antennas/`.

Dai, X., Chen, S., Sun, S., Kang, S., Wang, Y., Shen, Z. and Xu, J. (2014). Successive interference cancellation amenable multiple access (SAMA) for future wireless communications, *Proc. IEEE Int. on Commun. Systems* (Macau, People's Republic of China), pp. 222–226.

Dai, X., Zhang, Z., Bai, B., Chen, S. and Sun, S. (2018). Pattern division multiple access: A new multiple access technology for 5G, *IEEE Wir. Commun.* **25**, 2, pp. 54–60.

Denisowski, P. (2019). An introduction to radio direction finding methodologies, online, https://docplayer.net/25944983-An-introduction-to-radio-direction-finding-methodologies-paul-denisowski-applications-engineer.html.

Dent, P., Bottomley, G. E. and Croft, T. (1993). Jakes fading model revisited, *Electronic Letters* **39**, 3, 1162–1163.

Deschamp, G. A. (1953). Microstrip microwave antennas, in *3rd USAF Symp. on Antenna Research and Development Program* (University of Illinois, Monticello, Illinois), pp. 18–22.

Doppler Systems (2019). Radio Direction Finders and Direction Finding Systems, online, http://www.dopsys.com/products/components.html.

Dynastream (2014). ANT message protocol and usage, Tech. Rep. Rev. 5.1, Dynastream Innovations Inc., `https://www.thisisant.com/developer/ant/ant-basics#106_tab`,

Erceg, V., Greenstein, L. J., Tjandra, S. Y., Parkoff, S. R., Gupta, A., Kulic, B., Julius, A. A. and Bianchi, R. (1999). An empirically based path loss model for wireless channels in suburban environments, *IEEE Trans. Information Theory* **17**, 7, pp. 1205–1211.

ETSI-3GPP (2016). Physical layer procedures, Tech. Rep. TS 36.213 version 13.0.0 Release 13, ETSI 3GPP, Sophia Antipolis Cedex-France.

ETSI-3GPP (2016). LTE: E-UTRA user equipment procedures in the idle mode, Tech. Rep. TS 36.304 V13.0.0, ETSI 3GPP, Sophia Antipolis Cedex-France.

ETSI-3GPP (2017). 5G: Study on channel model for frequencies from 0.5 to 100 GHz, Tech. Rep. 138 901 V14.0.0, ETSI 3GPP, Sophia Antipolis Cedex-France.

ETSI-3GPP (2018).

Etten, W. V. (1976). Maximum likelihood receiver for multiple channel transmission systems, *IEEE Trans. on Commun.*, **24**, 2, pp. 276–283.

Farhang-Boroujeny, B. (2011). OFDM versus filter bank multicarrier, *IEEE Signal Process. Mag.*, **28**, 3, pp. 92–112.

Fletcher, R. R. and Kulkarni, S. (2010). Wearable doppler radar with integrated antenna for patient vital sign monitoring, in *Proc. IEEE Radio and Wireless Symposium* (Sydney, Australia), pp. 1–5.

Foschini, G. J. (1996). Layered space-time architecture for wireless communication in a fading environment when using multiple-element antennas, *Bell Labs Technical J.* **1**. 2, pp. 41–59.

Foschini, G. J. and Gans, J. (1998). On limits of wireless communications in fading environment when using multiple antennas, *Wir. Pers. Commun.* **6**, 3, pp. 311–335.

Freeman, R. L. (2007). *Radio System Design for Telecommunications* (John Wiley, New York).

Friedlander, B. (2005). Using MIMO to increase the range of wireless systems, in *Proc. IEEE Thirty-Ninth Asilomar Conf. on Signals, Systems and Computers*, pp. 1404–1408.

Gallager, R. (1994). *Communications and Cryptography: Two Sides and One Tapestry*, eds. R. E. Blahut, D. J. Costello, U. Maurer and T. Mittelholzer, An inequality on the capacity region of multiple access multipath channels (Kluwer), pp. 129–139.

Gold, J. (2017). What is IoT? How the internet of things works, https://www.networkworld.com/article/3207535/what-is-iot-how-the-internet-of-things-works.html.

GPS Directorate (2011). Navstar GPS space segment/navigation user segment interfaces, Tech. Rep. IS-GPS-200F, GPS, `http://www.everyspec.com`.

Gronemeyer, S. A. and McBride, A. L. (1976). MSK and Offset QPSK Modulation, *IEEE Trans. Commun.* **COM-24**, pp. 808–820

Gyasi-Agyei, A. (2003). *QoS Enabled IP Based Wireless Networking: Design, Modeling & Performance Analysis*, Ph.D. thesis, Adelaide University, Australia, `http://thesis.library.adelaide.edu.au/public/adt-SUA20060908.111557`.

Gyasi-Agyei, A. (2005). Multiuser diversity based opportunistic scheduling for wireless data networks, *IEEE Communications Lett.*, **9**, 7, pp. 670–672.

Gyasi-Agyei, A. (2019a). *Telecommunications Engineering: Principles and Practice* (World Scientific, Singapore).

Gyasi-Agyei, A. (2019b). *Telecommunications Engineering: Principles and Practice* (World Scientific, Singapore).

Gyasi-Agyei, A. and Kim, S.-L. (2006). Crosslayer multiservice opportunistic scheduling for wireless networks, *IEEE Communications Magazine*, **9**, 7, pp. 50–57.

Haneda, K., Tian, L., Zheng, Y., Asplund, H., Li, J., Wang, Y., Steer, D., Li, C., Balercia, T., Lee, S., Kim, Y., Ghosh, A., Thomas, T., Nakamura, T., Kakishima, Y., Imai, T., Papadopoulas, H., Rappaport, T. S., Jr., G. R. M., Samimi, M. K., Sun, S., Hur, O. K. S., Park, J., Zhang, C., Mellios, E., Molisch, A. F., Ghassamzadah, S. S. and Ghosh, A. (2016). 5G 3GPP-like channel models for outdoor urban microcellular and macrocellular environments, in *Proc. IEEE 83rd Vehicular Technology Conference* (Nanjing, China).

Hasanujjaman, Mondal, H., Biswas, A. and Bhattacharjee, A. K. (2013). Design of single band rectangular patch antenna for satellite application and analysis using different optimizer, *Int. J. of Engineering and Management Research* **3**, 6, pp. 106–109.

Hollenbeck, S. (2016). TAs WHOIS transitions to RDAP, How do we avoid the same mistakes? http://www.circleid.com/posts/20151123_as_whois_transitions_to_rdap_how_do_we_avoid_the_same_mistakes/.

Hoshyar, R., Razavi, R. and AL-Imari, M. (2010). LDS-OFDM an efficient multiple access technique, in *Proc. IEEE 71st VTC Conf. (VTC-Spring)* (Taipei, Taiwan).

Hoshyar, R. and Tafazolli, F. P. W. R. (2008). Novel low-density signature for synchronous CDMA systems over AWGN channel, *IEEE Transactions on Signal Proc.* **56**, 4, pp. 1616–1626.

Howell, J. Q. (1975). Microstrip antennas, *IEEE Trans. on Antennas and Propagation* **23**, pp. 90–93.

Hu, M., Zhang, J. and Sadowsky, J. (2004). Opportunistic beamforming using dumb antennas, *Elsevier Computer Networks* **46**, pp. 505–518,

ITU (2011). Isolation between antennas of IMT base stations in the land mobile service, Tech. Rep. ITU-R M.2244, ITU-R, International Telecommunication Union.

ITU (2015). Propagation data and prediction methods required for the design of terrestrial line-of-sight systems, Tech. Rep. Recommendation ITU-R P.530-16, ITU-R, International Telecommunication Union.

ITU (2016a). Attenuation by atmospheric gases, Tech. Rep. ITU-R P.676-11, ITU-R, International Telecommunication Union.

ITU (2016b). Propagation by Diffraction, Tech. Rep. Recommendation ITU-R P.526-5, ITU-R, International Telecommunication Union, https://www.itu.int/dms_pubrec/itu-r/rec/p/R-REC-P.526-5-199708-S!!PDF-E.pdf.

ITU-R (2005). Specific attenuation model for rain for use in prediction methods, Tech Rep. ITU-R P.838-3, ITU-R, International Telecommunication Union.

ITU-R (2015). Guidelines for evaluation of radio interface technologies for IMT-Advanced, Tech. Rep. ITU-R M.2135-1, ITU.

ITU-R (2017). Probability distributions relevant to radiowave propagation modelling, Tech. Rep. ITU-R P.1057-5, ITU-R, International Telecommunication Union.

Jain, R., Chiu, D. M. and Hawe, W. (1984). A Quantitative Measure of Fairness and Discrimination for Resource Allocation in Shared Systems. Digital Equipment Corporation, DEC-TR-301.

Jakes, W. C. (1994). *Microwave Mobile Communications* (IEEE Press, Piscataway, NJ).

Jamalipour, A., Wada, T. and Yamazato, T. (2005). A tutorial on multiple access technologies beyond 3G mobile networks, *IEEE Commun. Mag.*, 2, pp. 110–117.

Jin, H., Peng, K. and Song, J. (2013). Bit division multiplexing for broadcasting, *IEEE Trans. Broadcasting* **59**, 3, pp. 539–547.

Kamerman, A. and Monteban, L. (1997). WaveLAN-II: A high-performance wireless LAN for the unlicensed band, *Bell Labs Technical Journal* **2**, pp. 118–133.

Kaye, A. R. and George, D. A. (1970). Transmission of multiplexed PAM signals over multiple channel and diversity systems, *IEEE Trans. on Commun. Techn.* **18**, 5, pp. 520–526.

Keysight Technologies (2017). Techniques for advanced cable testing using fieldfox handheld analyzers: Application note, Tech. Rep. 5992-0604EN, Keysight Technologies, www.keysight.com.

Khalighi, M.-A. and Raoof, K. (2002). Capacity of Wireless Communication Systems Employing Antenna Arrays, a Tutorial Study, *Wireless Personal Communications* **23**, 3, pp. 321–352.

Kim, B., Lim, S., Kim, H., Suh, S., Kwun, J., Choi, S., Lee, C., Lee, S. and Hong, D. (2013). Non-orthogonal multiple access in a downlink multiuser beamforming system, in *Proc. IEEE Mil. Commun. Conf. (MILCOM-2013)* (San Diago, USA), pp. 1278–1283.

Knopp, R. and Humblet, P. A. (1995). Information capacity and power control in single-cell multiuser communications, in *Proc. IEEE ICC* (Seattle, WA, USA), pp. 331–335.

Korn, I. (1985). *Digital Communications* (Van Nostrand Reinhold Company, New York).

Kschischang, F. R., Frey, B. J. and Loeliger, H.-A. (2001). Factor graphs and the sum-product algorithm, *IEEE Trans. Info. Theory* **47**, 2, pp. 498–519.

Kuchi, K. and Prabhu, V. K. (1999). Power spectral density of GMSK modulation using matrix methods, in *Proc. IEEE Military Commun.* (Atlantic City, NJ, USA), pp. 1–6.

Kyösti, P., Meinilä, J., Hentilä, L., Zhao, X., Jämsä, T., Schneid, C., Alatossava, M., Bultitude, R., de Jong, Y. and Rautiainen, T. (2008). WINNER II channel models, Tech. Rep. IST-4-027756 WINNER II D1.1.2 V1.2, WINNER.

Lee, H., Kim, S. and Lim, J.-H. (2016). Multiuser superposition transmission (MUST) for LTE-A systems, in *Proc. IEEE ICC Conf.* (Kuala Lumpur, Malaysia), pp. 1–6.

Lee, J.-W., Mazumdar, R. and Shroff, N. (2004). Opportunistic power scheduling for multi-server wireless systems with minimum performance constraints, in *Proc. IEEE INFOCOM* (Hong Kong, China), pp. 1067–1077.

Lee, W. C. Y. (1985). *Mobile Communications Engineering* (McGraw-Hill, New York).

Lee, W. C. Y. (1991). Smaller cells for greater performance, *IEEE Commun. Mag.*, pp. 19–23.

Lee, W. C. Y. (1993). *Mobile Communications Design Fundamentals* (John Wiley & Sons. Inc., New York).

Li, Y. and Chen, M. (2015). Software-Defined Network Function Virtualization: A Survey, *IEEE Access*, Vol. 3, pp. 2542–2553,

Lin, C.-C. (2003). *Handover Mechanisms in 3GPP Long Term Evolution (LTE)*, Ph.D. thesis, University of Technology, Sydney, Australia.

Liu, J., Pare, T., Cheong, M., Zhang, H., Wilhelmsson, L. R., Lee, W., Porat, R., Choudhury, S. and Au, E. (2013). Discussions on penetration loss, Tech. Rep., IEEE 802.11-13/1376r2.

Luo, F.-K. and Zhang, C. (eds.) (2016). *Signal Processing for 5G: Algorithms and Implementations* (John Wiley & Sons, Ltd, Chichester).

Ma, L., Wen, X., Wang, L., Lu, Z. and Knopp, R. (2018). An SDN/NFV Based Framework for Management and Deployment of Service Based 5G Core Network, *IEEE China Communications*, pp. 86–98.

MacCartney, G. R., Rappaport, T. S., Sun, S. and Deng, S. (2015). Indoor office wideband millimeter-wave propagation measurements and channel models at 28 and 73 GHz for ultra-dense 5G wireless networks, *IEEE Access* **3**, pp. 2388–2424.

Monghal, G., Laselva, D., Michaelsen, P. and Wigard, J. (2010). Dynamic packet swcheduling for traffic mixes of best effort and VoIP users in E-UTRAN downlink, in *Proc. IEEE VTC-Spring* (Marina Bay, Singapore), pp. 1–5.

Moore, R. K. (2008). *Radar Handbook*, ed. Skolnik, M. L., chap 12 (McGraw-Hill Education), pp. 12.1–12.35.

Muhamed, R. (1996). *Direction of Arrival Estimation Using Antenna Arrays*, Master's thesis, Virgina Polytechnic Institute and State University, Blacksburg, Virginia.

Mullins, R. and Barros, M. T. (2017). Coginitive network management for 5G, (5G-PPP).

Munson, R. E. (1972). Microstrip phased array antenna, in *22nd Annual USAF Antenna Symposium.*

Munson, R. E. (1974). Conformal microstrip antennas and microstrip phased arrays, *IEEE Trans. on Antennas and Propagation* **22**, 1, pp. 74–78.

Nakagami, M. (1958). The m-distribution, a general formula of intensity of rapid fading, in W. C. Hoffman (ed.), *Statistical Methods in Radio Wave Propagation* (Pergamon Press, Proceedings of a Symposium), pp. 3–36.

Navy, U. (1972). LOS propagation path, Tech. Rep. NAVELEX 0101,112, US Navy, United States Navy Radiocommunications.

NI (2016). Introduction to bluetooth device testing: From theory to transmitter and receiver measurements, Tech. Rep., National Instruments.

Nikopour, H. and Baligh, H. (2013). Sparse code multiple access, in *Proc. IEEE 24th PIMRC Conf.* (London, UK).

Ningdalli, A. and Sujatha, B. K. (2015). FPGA Implementation of $\pi/2$-BPSK and $\pi/4$-QPSK for IEEE 802.15.6 WBAN Standard, *Int. Research J. of Engineering and Technology* **2**, 3, 803–806.

Nordrum, A. (1917). A language for the internet of underwater things, *IEEE Spectrum*, pp. 9–10.

NTT DOCOMO (2015). Deployment scenarios for downlink multiuser superposition transmissions, Tech. Rep. 3GPP R1-152062, NTT DOCOMO, Belgrade, Serbia.

Okumura, T., Ohmori, E. and Fukuda, K. (1968). Field strength and its variability in VHF and UHF land mobile service, *Review of Electrical Communication Laboratory (Tokyo)* **16**, 9-10, pp. 825–873.

OPTUS (2018). OPTUS 10 Satellite Payload Information, unpublished.

Parra, I. P. (1995). *Improved Constant Modulus Algorithm for Smart Antenna Systems*, Master's thesis, The University of Texas at Austin, Texas at Austin.

Petenaude, F. and Moher, M. L. (1992). A new symbol timing tracking algorithm for $\pi/2$-BPSK and $\pi/4$-QPSK modulations, in *Proc. IEEE SUPERCOMM/ICC* (Chicago, IL, USA), pp. 1588–1592.

Ping, L., Liu, L., Wu, K. and Leung, W. K. (2006). Interleave division multiple-access, *IEEE Trans. Wir. Commun.* **5**, 4, pp. 938–947.

Poulos, T. (2015). The internet of things (IoT) in layman's terms, `https://disruptiveviews.com/the-internet-of-things-iot-in-laymans-terms/`.

Pretz K. (2017). Gerhard Fettweis: Developing 5G and beyond, *The Institute* **41**, 1, p. 14.

PSU (2019). GPS error sources, online, `https://www.e-education.psu.edu/geog160/node/1924`.

Raleigh, G. G. and Cioffi, J. M. (1998). Spatio-temporal coding for wireless communications, *IEEE Trans. Wir. Commun.* **46**, 3, pp. 357–366.

Ramli, H., Basukala, R., Sandrasegaran, K. and Patachaianand, R. (2009). Performance of well known packet scheduling algorithms in the downlink 3GPP LTE system, in *Proc. IEE Malaysia International Conf. on Commun.* (Kuala Lumpur, Malaysia), pp. 815–820.

Rappaport, T. S. (2002). *Wireless Communications: Principles and Practice* (Prentice Hall, Ne2w Jersey).

Rappaport, T. S., MacCartney, G. R. Jr., Samimi, M. K. and Sun, S. (2015). Wideband millimeter-wave propagation measurements and channel models for future wireless communication systems design, *IEEE Trans. on Commun.* **63**, 9, pp. 3029–3056.

Rennie, S. J. (2012). Doppler weather radar in Australia, Tech. Rep. CAWCR Tecchnical Report No. 055, Bureau of Meteorology, Australia.

Roessler, A. (2016). 5G waveform candidates: Application note, Tech. Rep. 1MA271-0e, Rohde & Schwarz.

Sadiq, B., Madan, R. and Sampath, A. (2009). Downlink scheduling for multiclass traffic in LTE, *EURASIP J. Wirel. Commun. Netw.* **2009**, 1, pp. 1–18.

Saito, Y., Kishiyama, Y., Benjebbour, A., Nakamura, T., Li, A. and Higuchi, K. (2013). Non-orthogonal multiple access (NOMA) for cellular future radio access, in *Proc. IEEE 77th VTC Conf. (VTC-Spring)* (Dresden, Germany).

Salesforce (2017). SaaS: Software as a service, `https://www.salesforce.com/au/saas/what-is-saas/`.

Samimi, M. K., Rappaport, T. S. and MacCartney, G. R.

(2015). Probabilitistic omnidirectional path loss models for millimeter-wave outdoor communications, *IEEE Wir. Commun. Letters*, **4**, 4, pp. 357–360.

Sandhu, S., Nabar, R., Gore, D, and Paulraj, A. (2019). Introduction to Space-Time Codes, https://web.stanford.edu/group/sarg/sandhu062503.pdf.

Sanou, B. (2018). Setting the Scene for 5G: Opportunities and Challenges, ITU, https://www.itu.int/en/ITU-D/Documents/ITU_5G_REPORT-2018.pdf, ISBN 978-92-61-27591-4.

Sarunic, P. W. (2016). Development of GPS receiver kalman filter algorithms for stationary, low-dynamics, and high-dynamics applications, Tech. Rep., Cyber and Electronic Warfare Division, Defence Science and Technology Group, Australia, `https://www.dst.defence.gov.au/sites/default/files/publications/documents/DST-Group-TR-3260.pdf`.

Satellite Industry Association (2017). State of the satellite industry report, `https://www.sia.org/wp-content/uploads/2017/07/SIA-SSIR-2017.pdf`, accessed: 7th December 2018.

Schmidt, R. (1986). Multiple emitter location and signal parameter estimation, *IEEE Trans. on Antennas and Propagation* **34**, 3, pp. 276–280.

Schmidt, R. O. (1979). Multiple emitter location and signal parameter estimation, in *Proc. of RADC Spectrum Estimation Workshop* (Griffiss AFB, New York), pp. 243–258.

Segata *et al.* (2013). Shadowing or multi-path fading: Which dominates in inter-vehicle communication, Tech. Rep. Technical Report CCS-2013-03, University of Innsbruck, Institute of Computer Science.

Shan, T. J., Wax, M. and Kailath, T. (1985). On spatial smoothing for estimation of coherent signals, *IEEE Trans. on Acoustics, Speech and Sig. Proc.* **33**, pp. 802–811.

Shiu, D. S., Foschini, J., Gans, J. and Kahn, J. M. (2000). Fading correlation and its effect on the capacity of multi-element antenna system, *IEEE Trans. on Commun.*, **48**, 3, pp. 502–513.

Skolnik, M. L. (1980). *Introduction to Radar Systems* (McGraw-Hill, New York).

Song, H. L. (1994). Automatic vehicle location in cellular communications systems, *IEEE Trans. Veh. Techn.* **43**, pp. 902–908.

Statista (2018). Number of worldwide Internet hosts in the domain name system (DNS) from 1993 to 2018 (in millions), https://www.statista.com/statistics/264473/number-of-internet-hosts-in-the-domain-name-system/, accessed: 30th November 2018.

Statista (2019). Internet of things (IoT) connected devices installed base worldwide from 2015 to 2025 (in billions), Online at https://www.statista.com/statistics/471264/iot-number-of-connected-devices-worldwide.

Subirana, J. S., Zornoza, J. J. and Hernandez-Pajares, M. (2011). GNSS signal, online, `https://gssc.esa.int/navipedia/index.php/GNSS_signal`.

Tarokh, V., Jafarkhani, H. and Calderbank, A. R. (1999). Space-time block codes for orthogonal designs, *IEEE Trans. on Info. Theory* **45**, 5, pp. 1456–1467.

Tarokh, V., Seshadri, N. and Calderbank, A. R. (1998). Space-time codes for high data rate wireless communication: Performance criterion and code construction, *IEEE Trans. on Info. Theory* **44**, 2, pp. 744–765.

Technavio (2018). Global satellite manufacturing and launch market 2018–2022, https://www.technavio.com/report/global-satellite-manufacturing-and-launch-market-analysis-share-2018, accessed: 7th December 2018.

Telatar, I. E. (1999). Capacity of multi-antenna Gaussian channels, *Eur. Trans. in Telecom.* **10**, 6, pp. 585–595.

Ungerboeck, G. (1982). Channel coding with multilevel/phase signals, *IEEE Trans. Inform. Theory* **IT-28**, pp. 55–67.

Ungerboeck, G. (1987a). Trellis coded modulation with redundant signal sets Part I: Introduction, *IEEE Communications Magazine* **25**, 2, pp. 5–12.

Ungerboeck, G. (1987b). Trellis coded modulation with redundant signal sets Part II, *IEEE Communications Magazine* **25**, 2, pp. 12–22.

Vanka, S., Srinivasa, S., Gong, Z., Vizi, P., Stamatiou, K. and Haenggi, M. (2012). Superposition coding strategies: Design and experimental evaluation, *IEEE Transactions on Wir. Commun.* **11**, 7, pp. 2628–2639.

Viswanath, P., Tse, D. N. C. and Laroia, R. (2002). Opportunistic beamforming using dumb antennas, *IEEE Trans. Inform. Theory* **48**, 6, pp. 1277–1294.

Walfisch, J. and Bertoni, H. L. (1988). A theoretical model of UHF propagation in urban environments, *IEEE Trans. on Antennas and Propagation* **36**, 12, pp. 1788–1796.

Walsh, J. L. (1923). A closed set of normal orthogonal functions, *American Journal of Mathematics* **45**, 1, pp. 5–24.

Wang, B., Wang, K., Lu, Z., Xie, T. and Quan, J. (2015). Comparison study of non-orthogonal multiple access schemes for 5G, *Proc. IEEE Int. Symp. on Broadband Multimedia Systems and Broadcasting* (Ghent, Belgium), pp. 1–5.

Wengerter, C., Ohlhorst, J. and von Elbwart, A. (2005). Fairness and throughput analysis for generalized proportional fair frequency scheduling in OFDMA, in *Proc. IEE Veh. Tech. Conf., VTC-Spring* (Stockholm, Sweden), pp. 1903–1907.

White, R. F. (1970). Engineering considerations in microwave communications systems, Tech. Rep., Lenkurt Electric Co, Inc., https://hsmm.wikispaces.com/file/view/Engineering+Considerations+for+Microwave+Communications+Systems.pdf.

Wiley, C. A. (1954). Pulsed doppler radar methods and apparatus, U.S. Patent no. 3 196 436.

Wiley, C. A. (1985). Synthetic aperture radars: A paradigm for technology evolution, *IEEE Trans. Aerosp. Electron. Syst.*, pp. 440–443.

Winters, J. H. (1987). On the capacity of radio communication systems with diversity in a rayleigh fading environment, *IEEE JSAC* **5**, 5, pp. 871–878.

Wu, M., Lu, T.-J., Ling, F.-Y., Sun, J. and Du, H.-Y. (2010).

Research on the architecture of internetof things, in *3rd International Conference on Advanced Computer Theory and Engineering (ICACTE)* (Chengdu, Sichuang, China).

Yuan, Z., Yu, G., Li, W., Yuan, Y., Wang, X. and Xu, J. (2016). Multi-User Shared Access for Internet of Things, in *Proc. IEEE 83rd Veh. Tech. Conf.* (Nanjing, China).

Zander, J. and Kim, S.-L. (2001). *Radio Resource Management for Wireless Networks*, (Artech House Publishers).

Zekava, S. A. and Buehrer, R. M. (2012). *Handbook of Position Location: Theory, Practice and Advances* (John Wiley & Sons, New York).

Zhang, H., Wen, X., Wang, B., Zheng, W. and Sun, Y. (2010). A Novel Handover Mechanism Between Femtocell and Macrocell for LTE Based Networks, *Proc. IEEE 2nd Internat. Conf. on Comm. Softw. and Nets*, pp. 28–231.

Zhang, L., Li, W., Wu, Y., Wang, X., Park, S.-I., Kim, H. M., Lee, J.-Y., Angueira, P. and Montalban, J. (2016a). Layered-division-multiplexing: Theory and practice, *IEEE Transactions on Broadcasting* **62**, 1, pp. 216–232.

Zhang, L., Li, W., Wu, Y., Wang, X., Park, S.-I., Kim, H. M., Lee, J.-Y., Angueira, P. and Montalban, J. (2016b). Layered-division-multiplexing: Theory and practice, *IEEE Trans. Wir. Commun.* **62**, 1, pp. 216–232.

Ziskind, I. and Wax, M. (1988). Maximum likelihood localization of multiple sources by alternating projection, *IEEE Trans. on Acoustics, Speech and Sig. Proc.* **36**, 10, pp. 1553–1560.

ZTE-CATT-Intel-Samsung (2016). New study item proposal: Study on non-orthogonal multiple access for NR, Tech. Rep. 3GPP RP-170829, ZTE-CATT-Intel-Samsung, Dubrovnik, Croatia.

Appendix

Cellular Mobile Network Design Using CelPlanner^{TM}
Laboratory 1: Configuring Databases

Learning outcomes

1. How to open an existing project or create a new project.
2. Configure CelPlanner^{TM} directories in order to load background data, such as GIS databases, needed to design a mobile cellular network.

Procedure

How to Create a New Project?

1. Open CelPlanner^{TM} package
2. Select menu: Project → New
   a. Project: Give a name to the project, e.g. Project01, as shown in Fig. A1(a).
   b. System: Select one of the available system of your choice. In the example below we have selected "LTE".
   c. Click OK of to obtain Fig. A1(b)

We can observe from Fig. A1(b) that no database is found. The reason being that none has been loaded yet.

3. Now, click OK again to obtain Fig. A2. We can observe a single sample cell or base station in the figure. By default a newly-created project is automatically allocated one cell.
4. Double click on the cell or base station shown in Fig. A2 to obtain Fig. A3. The parameters are configured with some values by default. We shall learn how to configure the cell or base station in future laboratories. For now, just explore the various configurable parameters and press "OK" to close it so we focus on the learning outcomes of this laboratory.
5. There are two ways of accessing CelPlanner^{TM} directories. First, via the **Project Directories** menu button shown in Fig. A4. This toolbar is yellowish in color and sits between *Vector Elements* and *User Data Files* directories in the toolbar menu. The second way of accessing the project directories is to follow the series of menu tabs:

   Project → Configure → Database Directories

   Again, the directories shown in Fig. A4 are default directories, which must be changed. Changing these default values is referred to as *configuration*, or reconfiguration. Correctly configuring the directories is necessary for the running of projects, else CelPlanner^{TM} will display only erroneous messages at runtime.

6. Now, we need to configure the database directories. Those needed to be configured include:
   a. Antenna
   b. Frequency Table
   c. GIS Databases: Images, Morphology, Regions, Topography and Vectors.
   d. Network
   e. Performance
   f. Predictions
   g. Projects
   h. Subscribers
   i. Traffic
7. Create a folder using *Windows Explorer* for the project and give it a reflective, such *Project01.*

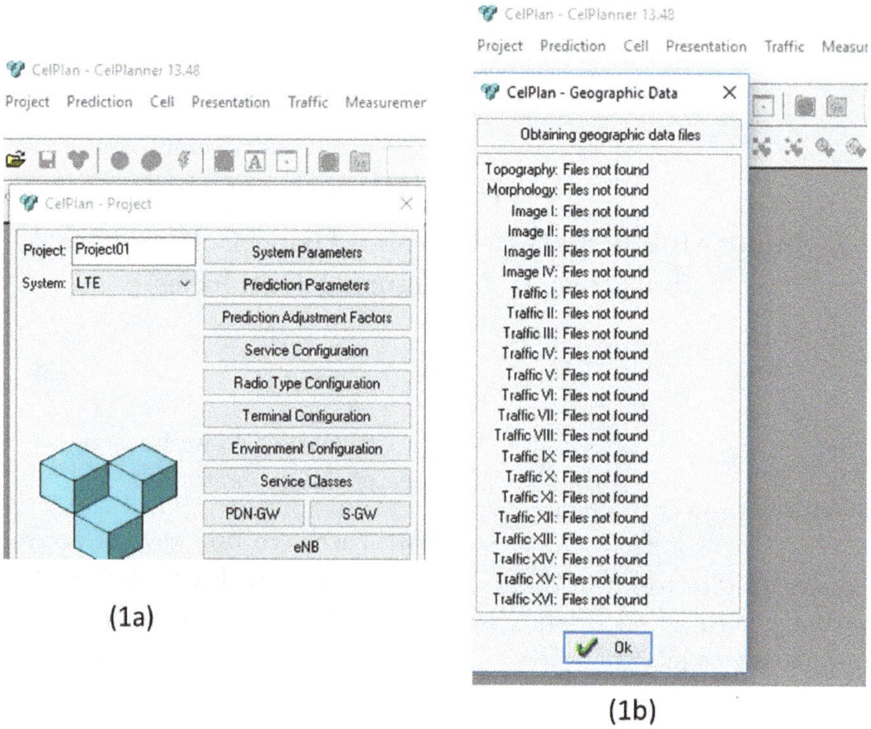

Fig. A1 Creating a new project using CelPlanner^{TM}.

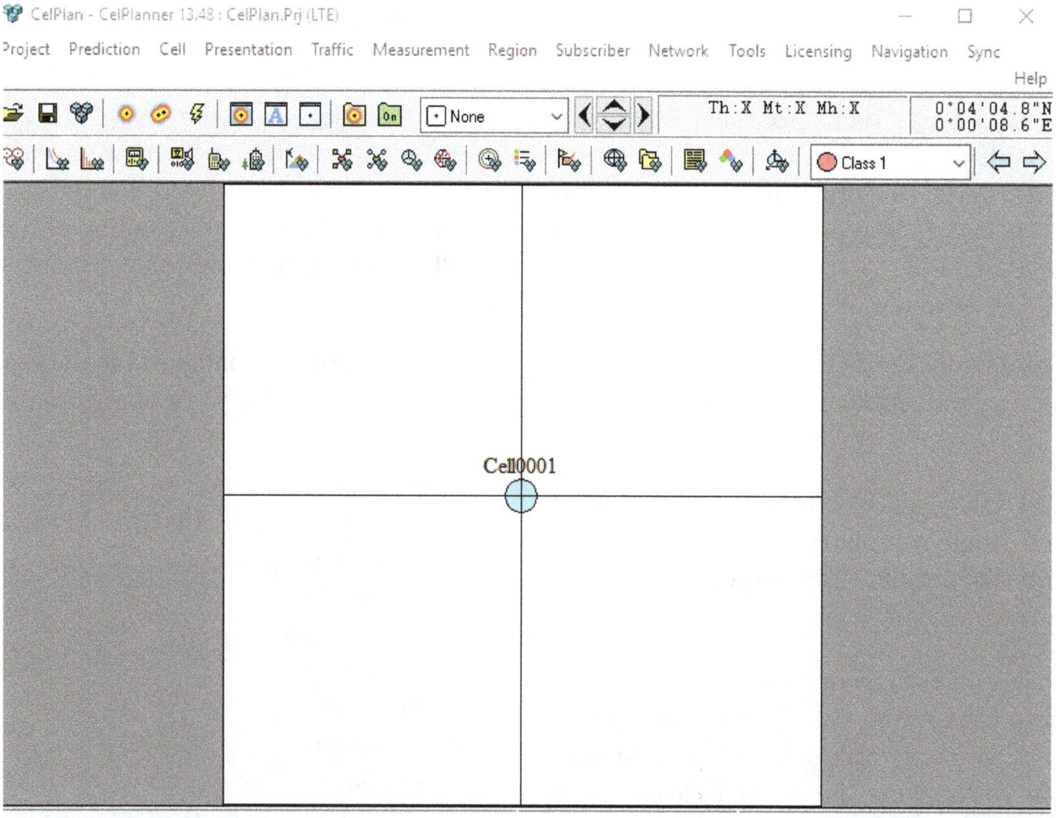

Fig. A2 CelPlanner^{TM} workspace with an automatically-created cell.

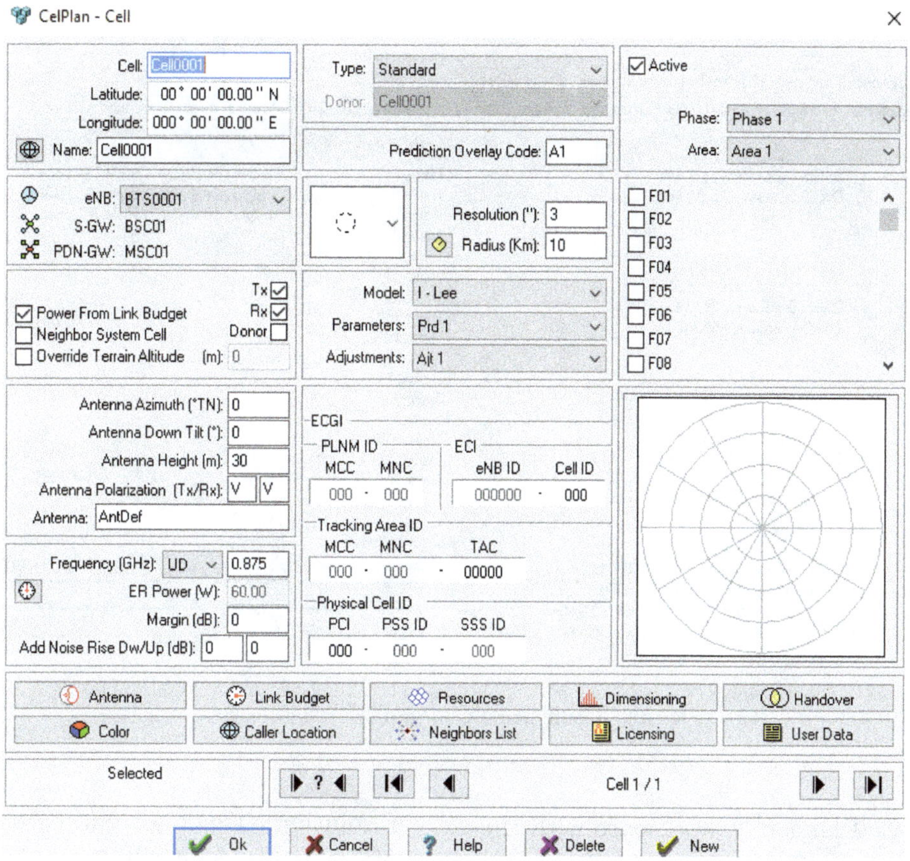

Fig. A3 Parameters to be configured in a cell or base station in CelPlanner^{TM}.

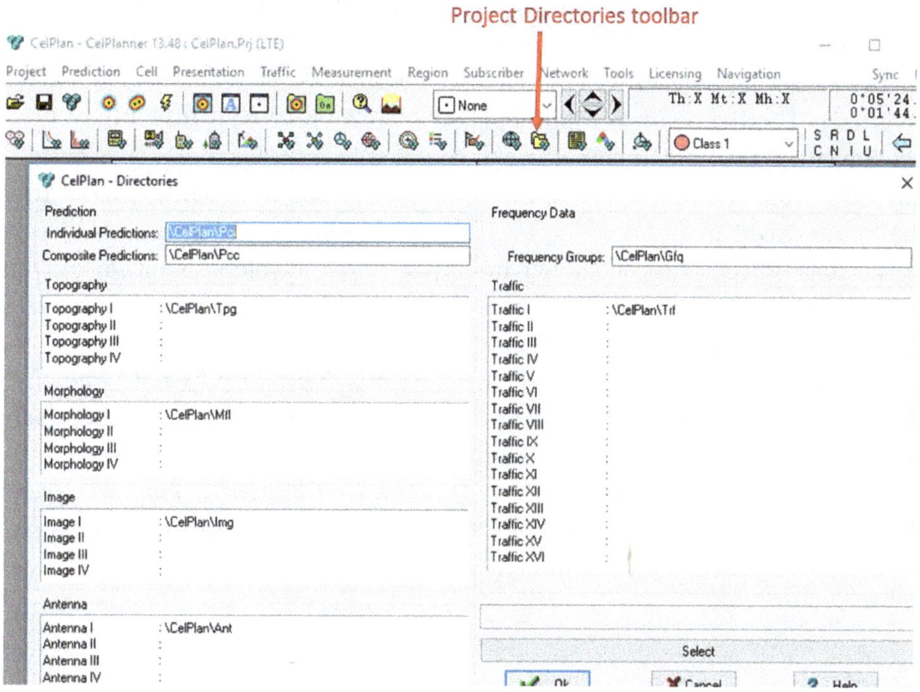

Fig. A4 Default configuration of directories in CelPlanner^{TM}.

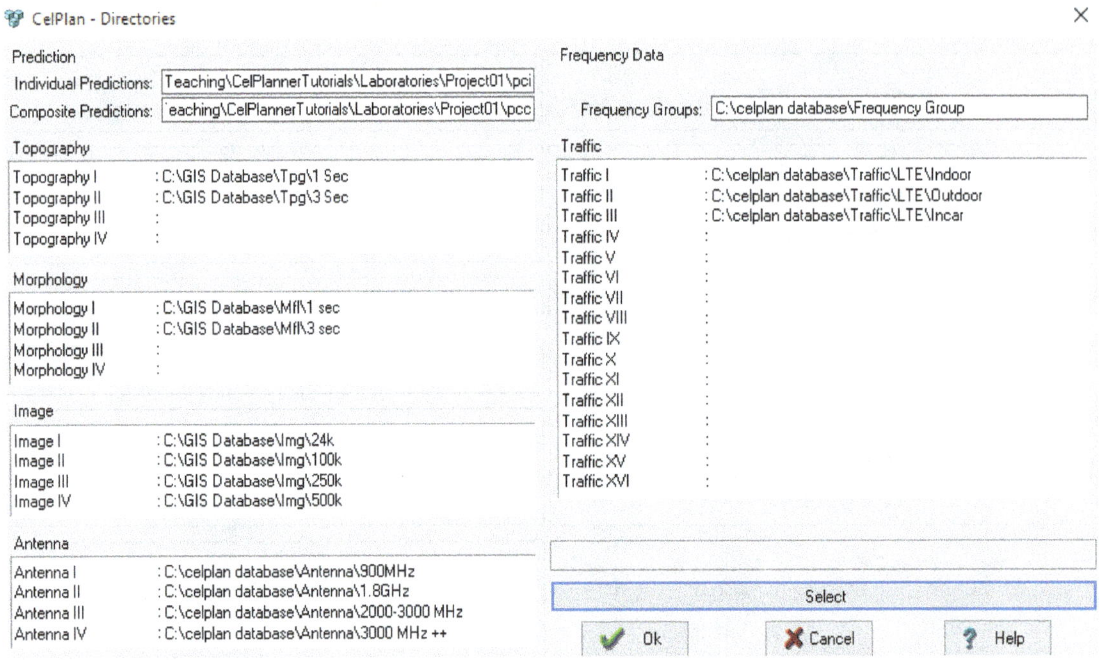

Fig. A5 Configuring CelPlanner™ databases.

a. Save your CelPlanner™ workspace with the chosen name Project01.Prj. CelPlanner™ project files use the file extension .Prj.
b. Create two subfolders inside the folder created previously (e.g. Project01) and call them pcc (meaning prediction configuration composite) and pci (meaning prediction configuration individual). Predictions from individual cells are saved in the folder named pci, while predictions from a group of cells are saved in the folder called pcc.

8. Configure the database directories as shown in Fig. A5. Note that the databases and their location depend on how CelPlanner™ has been installed. To select the subfolder named pcc that you have created, point the mouse button inside the Individual Predictions and click on the button "select" at the bottom of CelPlanner™ workspace. Note that each of the directories contains multiple databases, each with a different resolution. The 1 sec database has a resolution of 30 meters, while the 3 sec database uses 90m resolution. Position the databases in the order of their resolution, as shown in Fig. A5. The 24k image uses a scaling factor of 1:24,000, while the 100k images uses 1:100,000 scale factor. The lower the scale factor, the better the resolution.
9. Click ok to return to Fig. A2.

We have now learned how to create a project and configure the database directories. Only the pcc and pci folders contain output data from the design. The remaining folders contain input data into the design.

It is important to note that the configuration of databases is the first and necessary step for every project done using CelPlanner™. For this reason, some CelPlanner™ may not specifically ask you to configure the databases. It is assumed that you know that.

Cellular Mobile Network Design Using CelPlanner^{TM}
Laboratory 2: Cell Sectoring

Learning outcomes

1) Finding the latitude and the longitude of a location on the Earth using Google Maps for cell/tower installation.
2) Configuration of a cell or base station.
3) Cell-sectoring using CelPlanner^{TM} software package.
4) Generation of 3-sector and 6-sector cells.
5) Discovering functions in CelPlanner^{TM}.

Task

We want to install one 4G/LTE-A radio base station (called eNodeB) in a city in Melbourne Australia called Mernda, whose postcode is VIC 3754. We then divide the cell into three or six sectors of equal sizes.

Assumptions

1) The directories have already been configured (See Laboratory 1).
2) The radio interface used for the wireless network under design is LTE.

Preparatory Theory

Cell sectoring is a method used to reduce co-channel interference (CCI) in mobile cellular systems. The result of reducing the CCI is the increase in overall system capacity. In practical mobile cellular system design, we mount only one antenna mast at each cell, and then sector the cell into n sectors by mounting n directional antennas on the mast. Sectoring in CelPlanner^{TM} software, however, uses a separate base station with a directional antenna mounted on a tower to realise each sector. Therefore, in order to obtain a cell with n sectors, we mount n base stations on n masts at the same geographical location and use a directional antenna at each of the n base stations. The beamwidth of each antenna should be

$$\theta_b = \frac{360^\circ}{n}, \quad n \text{ is the number of sectors}$$

The easiest way to achieve cell sectoring in CelPlanner^{TM} is to mount one base station at the required geographical location, configure it, copy and paste it $n-1$ times. The **antenna azimuth** of the nth sector is

$$\theta_a = \frac{360^\circ}{n}k, \quad k = 0, 1, \ldots, n-1,$$
$$n \text{ is the number of sectors}$$

For example, a three-sector cell uses three directional antennas with 0°, 120° and 240° azimuths, respectively. Thus, only the value for the antenna azimuths are different in base stations used to realise a sectored cell.

Procedure Part A: Opening a project and configuring databases

1. Open CelPlanner^{TM} software by double-clicking it on the desktop.
2. Open the Project menu via Project → New.
3. Give project name as *CellSectoring*.
4. Select *LTE* as the system, which is the radio technology used.
5. Configure the databases as done in Laboratory 1 (see Fig. A6).

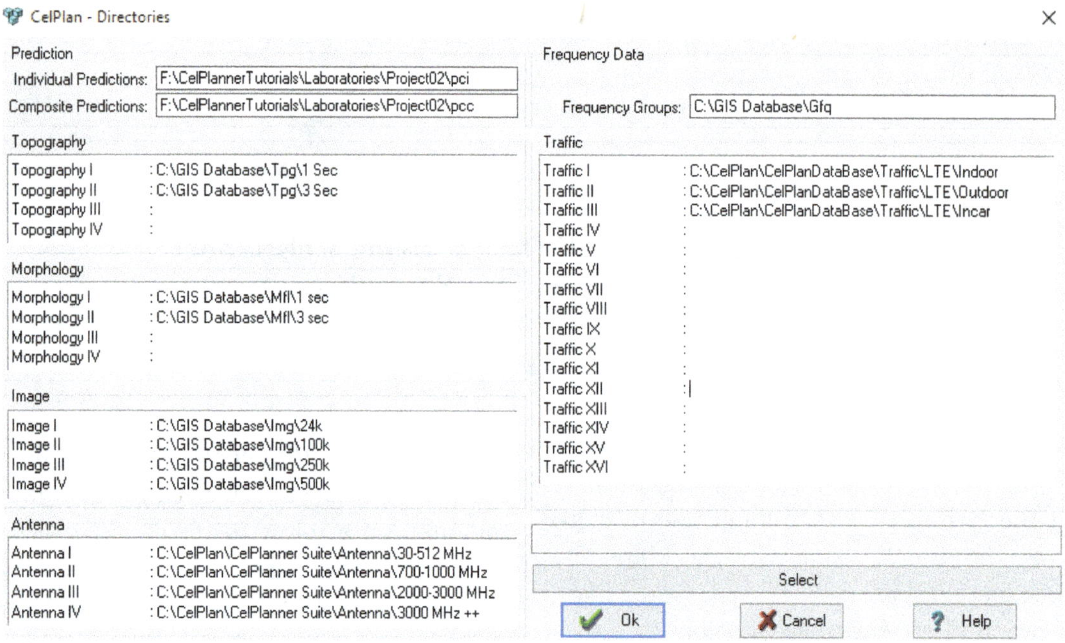

Fig. A6 Configured databases.

You must select the appropriate directories for the two folders: pci and pcc.

6. Click on "Ok" "Ok" now to obtain Fig. A7. As studied in Laboratory 1, CelPlanner™ automatically positions one cell in the workspace. We now need to change the default parameters of the cell. This process is called **cell configuration**.

Procedure Part B: Configuration of a cell/base station

1) In order to configure the parameters of a cell, we need access to the interface containing its parameters, as shown in Fig. A8. There are at least three ways to obtain open this configuration parameters of a cell.
   a. **Project** → Configure → Cell
   b. **Project** → Configure → eNB → Cell
   c. Double click on the cell in the CelPlanner™ workspace.

Executing any of the above series of command opens the window in Fig. A8.

2) **Cell parameters to be configured include:**
   a. Cell name, give a name which is reflective of the location of the cell for easy identification for troubleshooting, etc. For example, I use *MelMernda* for a cell located in a suburb of Melbourne city called Mernda.
   b. Cell location: this parameter specifies the geographical location of a cell (the default parameters are 00°00′ 00.00″ S, 000°00′ 00.00″ E (for Latitude/Longitude). Wikipedia and Google Map provide the latitude and longitude of Mernda as about 37.609° S, 145.093° E, which equals 37°36′ 32.4″ S, 145°5′ 34.8″ E. We need to change the default values to the real position of the cell as part of the configuration process.
   c. Resolution: 3″.
   d. Radius: this is the radius of a cell indicating the transmission range of the base station. For this lab set it to 4 km. Note that Mernda is just a suburb in the Whittlesea City whose total size is about 489.9 km2. Therefore, assuming a hexagonal cell structure, the radius is about 13.74 km.
   e. Model: indicates radio channel propagation model which means how the wireless channel affects signals passing through it. *Choose Lee model.*
   f. Antenna height: 30 m.

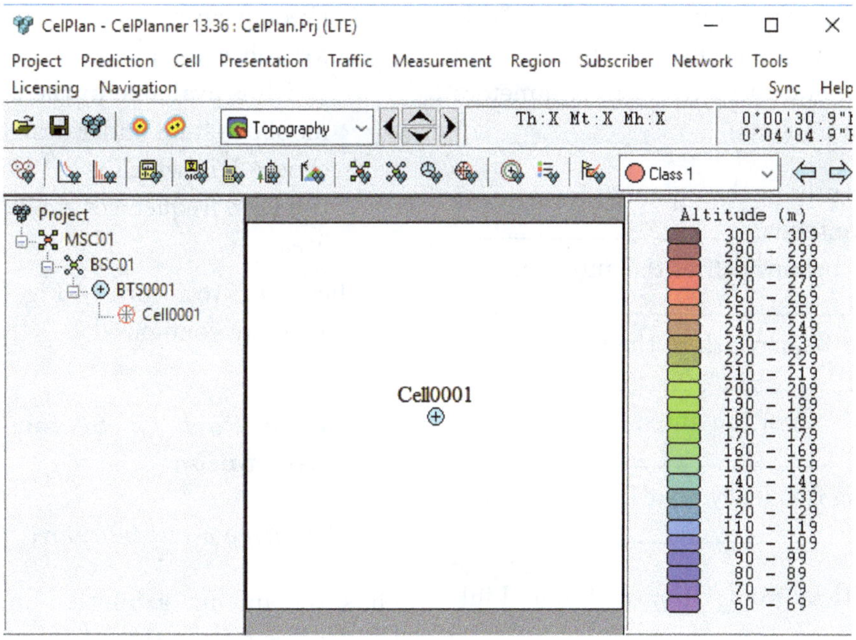

Fig. A7 Initial CelPlanner™ workspace.

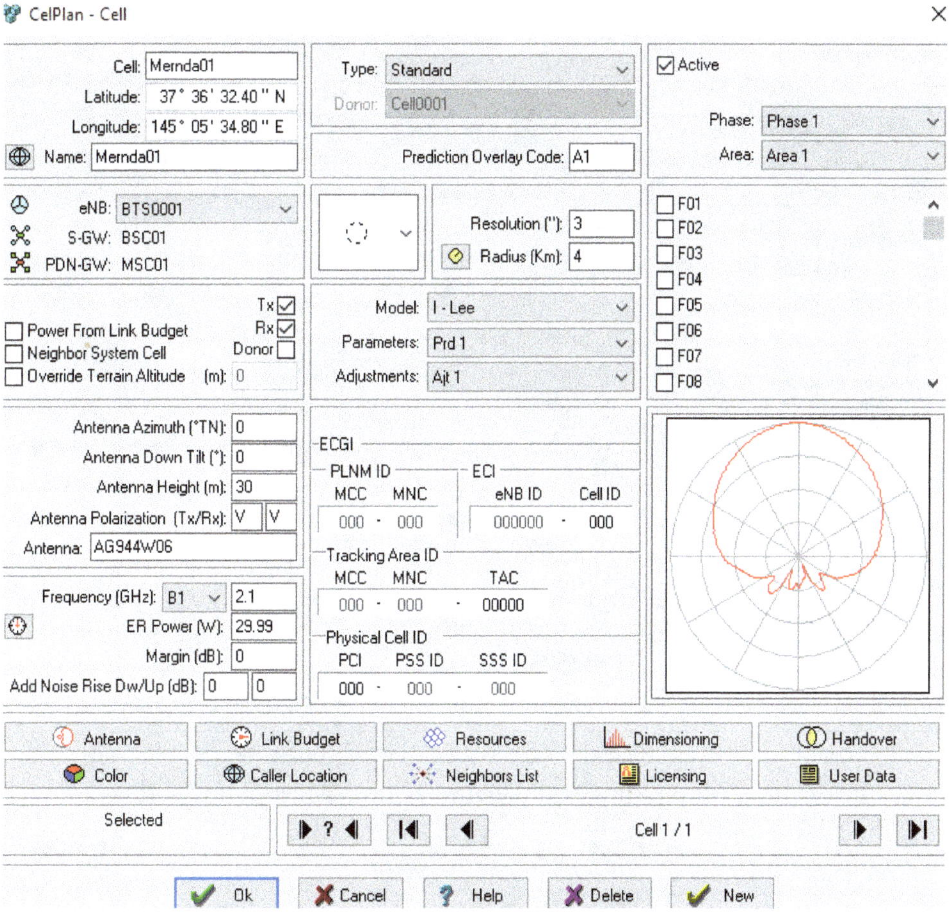

Fig. A8 Cell configuration interface.

g. Antenna: Click on *Antenna* and select Bcd-4506. This is the antenna used by the base station. Note down the following parameters of the chosen antenna:

   i. Manufacturer of the antenna: ________
   ii. Type of antenna: ________________
   iii. Azimuth beamwidth of the antenna: ________________
   iv, Elevation beamwidth of the antenna: ________________
   v. Nominal gain of the antenna: ________________
   vi. Operating frequency band of the antenna: ________________

h. Clear the tick from "Power From Link Budget".
i. ER Power (W): means effective radiated power by the base station. Set it to 30 Watts (Tx power output).
j. Frequency: select one of the frequencies used for the chosen radio technology (here LTE) which is available in the coverage area of the mobile system under design. For example, we choose B1 (which is 2.1 GHz). This is one of the LTE frequencies available in the coverage area.

3) Click "Ok" to return to Fig. A6, but with some parameters configured.

Procedure Part C: Sectorization of a cell/base station

C.1 Sectoring a cell into three sectors

Thus, the antenna azimuthal angles are 0, 120 and 240 degrees for the three base stations.

1. Right-click on the base station to copy it, then paste it using either Ctrl + V or right-clicking

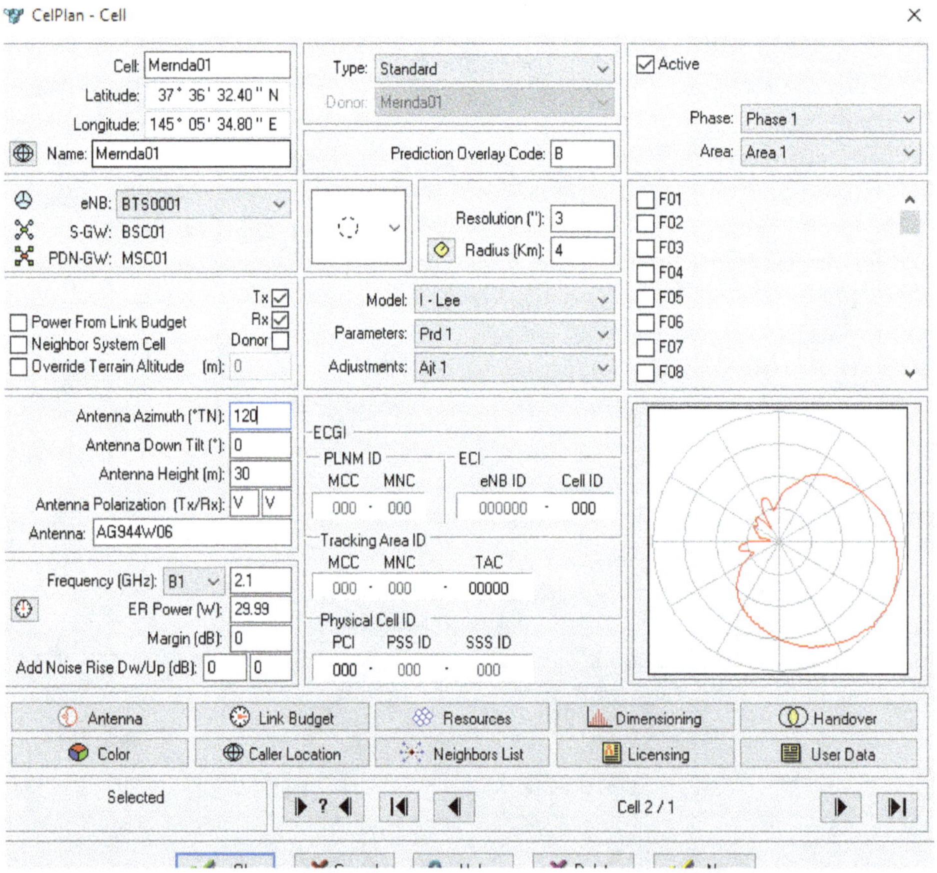

Fig. A9 Creating the second base station as the second sector of a cell.

again and selecting the option "paste" in the pop-up window. The window in Fig. A9 opens.

2. Now, change the Antenna Azimuth to 120 degrees as shown in Fig. A9.
3. Now, open one of the two base stations. Select the button called "New" at the very bottom of the window. This opens a new base station. Give it the name Mernda1_3 and change the Antenna Azimuth to 240 degrees. Leave everything else as there are. Click on "Ok".
4. Save your workspace. It is important to do this periodically.

Note: The same process described above can be followed to sector a cell into any number of desired sectors.

Index

www.ingramcontent.com/pod-product-compliance
Lightning Source LLC
Chambersburg PA
CBHW081209220326
41598CB00037B/6723